IFAC

International Federation of Automatic Control

SYSTEM APPROACH FOR DEVELOPMENT

Other Titles in the IFAC Proceedings Series

ATHERTON: Multivariable Technological Systems
BANKS & PRITCHARD: Control of Distributed Parameter Systems
CICHOCKI & STRASZAK: Systems Analysis Applications to Complex Programs
CRONHJORT: Real Time Programming 1978
CUENOD: Computer Aided Design of Control Systems
DE GIORGIO & ROVEDA: Criteria for Selecting Appropriate Technologies Under Different Cultural, Technical and Social Conditions
DUBUISSON: Information and Systems
ELLIS: Control Problems and Devices in Manufacturing Technology 1980
GHONAIMY: Systems Approach for Development 1977
HAASE: Real Time Programming 1980
HAIMES & KINDLER: Water and Related Land Resource Systems
HARRISON: Distributed Computer Control Systems
HASEGAWA & INOUE: Urban, Regional and National Planning (UNRENAP)
HERBST: Automatic Control in Power Generation Distribution and Protection
ISERMANN: Identification and System Parameter Estimation
ISERMANN & KALTENECKER: Digital Computer Applications to Process Control
JANSSEN, PAU & STRASZAK: Dynamic Modelling and Control of National Economies
LAUBER: Safety of Computer Control Systems
LEONHARD: Control in Power Electronics and Electrical Drives
LESKIEWICZ & ZAREMBA: Pneumatic and Hydraulic Components and Instruments in Automatic Control
MUNDAY: Automatic Control in Space
NIEMI: A Link Between Science and Applications of Automatic Control
NOVAK: Software for Computer Control
O'SHEA & POLIS: Automation in Mining, Mineral and Metal Processing
OSHIMA: Information-control Problems in Manufacturing Technology 1977
RAUCH: Control Applications of Nonlinear Programming
REMBOLD: Information Control Problems in Manufacturing Technology 1979
RIJNSDORP: Case Studies in Automation Related to Humanization of Work
SAWARAGI & AKASHI: Environmental Systems Planning, Design and Control
SINGH & TITLI: Control and Management of Integrated Industrial Complexes
SMEDEMA: Real Time Programming 1977
SUBRAMANYAM: Computer Applications in Large Scale Power Systems
VAN CAUWENBERGHE: Instrumentation and Automation in the Paper, Rubber, Plastics and Polymerisation Industries

SYSTEM APPROACH FOR DEVELOPMENT

Third IFAC/IFIP/IFORS Conference
Rabat, Morocco, 24-27 November 1980

Editors

N. NAJIM and Y. M. ABDEL-FATTAH

Faculté des Sciences, Rabat, Morocco

Published for the

INTERNATIONAL FEDERATION OF AUTOMATIC CONTROL

by

PERGAMON PRESS

OXFORD · NEW YORK · TORONTO · SYDNEY · PARIS · FRANKFURT

U.K.	Pergamon Press Ltd., Headington Hill Hall, Oxford OX3 0BW, England
U.S.A.	Pergamon Press Inc., Maxwell House, Fairview Park, Elmsford, New York 10523, U.S.A.
CANADA	Pergamon Press Canada Ltd., Suite 104, 150 Consumers Rd., Willowdale, Ontario M2J 1P9, Canada
AUSTRALIA	Pergamon Press (Aust.) Pty. Ltd., P.O. Box 544, Potts Point, N.S.W. 2011, Australia
FRANCE	Pergamon Press SARL, 24 rue des Ecoles, 75240 Paris, Cedex 05, France
FEDERAL REPUBLIC OF GERMANY	Pergamon Press GmbH, 6242 Kronberg-Taunus, Hammerweg 6, Federal Republic of Germany

First edition 1981

British Library Cataloguing in Publication Data

IFAC/IFIP/IFORS Conference on System Approach for Development, *3rd, Rabat, 1980*
System approach for development.
1. Underdeveloped areas - Economic conditions - Mathematical models - Congresses
2. System theory - Congresses
I. Title II. Najim, M III. Abdel-Fattah, Y M
IV. International Federation of Automatic Control
V. International Federation for Information Processing VI. International Federation of Operational Research Societies
330.9'172'4 HC59.7

ISBN 0-08-025670-8

Library of Congress Catalog Card no.: 80-41530

These proceedings were reproduced by means of the photo-offset process using the manuscripts supplied by the authors of the different papers. The manuscripts have been typed using different typewriters and typefaces. The lay-out, figures and tables of some papers did not agree completely with the standard requirements; consequently the reproduction does not display complete uniformity. To ensure rapid publication this discrepancy could not be changed; nor could the English be checked completely. Therefore, the readers are asked to excuse any deficiencies of this publication which may be due to the above mentioned reasons.

The Editors

Printed in Great Britain by A. Wheaton & Co. Ltd., Exeter

ERRATUM

The Publisher regrets that Professor Najim's (co-editor) name appears with an incorrect initial, this should read M. NAJIM.

NAJIM & ABDEL-FATTAH: Systems Approach for Development

YSTEM

Technical Committee on Systems Engineering
Technical Committee on Economic and Management Systems
Technical Committee on Social Effects of Automation
Technical Committee on Education

Co-sponsored by

The International Federation of Information Processing (IFIP)
The International Federation of Operational Research Societies (IFORS)

Organized by

Association Marocaine pour le Développement de l'Electronique, l'Informatique et de l'Automatique (A.MA.D.E.I.A.) and Faculté des Sciences, Rabat, Morocco

International Program Committee

S. Abdel-Kader (Egypt)
H. Apter (Switzerland)
A. Van Cauwenberghe (Belgium)
A. A. Concheiro (Mexico)
M. Cuénod (Switzerland)
Y. M. El Fattah (Morocco)
G. Ferraté (Spain)
M. A. R. Ghonaïmy (Egypt)
J. Hebenstreit (France)
J. M. L. Janssen (Netherlands)
S. Kahne (U.S.A.)
D. G. Lainiotis (U.S.A.)
J. A. Lafayette (France)
P. M. Larsen (Denmark)
U. Luoto (Finland)
A. Mansour (Switzerland)
M. Najim (Morocco) (Chairman)
E. Nisenfeld (U.S.A.)
T. Ouazzani (Morocco)
U. Pellegrini (Italy)
C. Roveda (Italy)
I. Sharkass (Egypt)
Y. Sevely (France)
A. Titli (France)
T. Vamos (Hungary)
A. A. Verrijn Stuart (Netherlands)
J. Vlietstra (Netherlands)

National Organizing Committee

J. Abderrazik
F. Bargach
M. Benjilali
H. Benlahsen
R. Benmokhtar
A. El Alj
A. El Jai
A. Hanane
A. Idrissi
I. Khalil (Chairman)
M. Najim
T. Ouazzani
A. Touzani

CONTENTS

Foreword xi

0. TUTORIALS

1. Systems Analysis and Development Possibilities and Limits
M. A. Cuenod xiii

2. Power System Control and Automation
J. P. Waha and M. Ribbens-Pavella xxv

3. Artificial Intelligence, Automatic Control and Development
T. Vámos xxxvii

4. Adaptive Control Systems
Y. M. Abdel-Fattah xlv

5. A Survey on Computer-Aided Planning in Complicate Production Systems
L. Pun, G. Doumeingts, J. Grislain and D. Breuil lv

6. Medical Information Systems
S. E. Aidarous and M. A. R. Ghonaimy lxxix

I. METHODOLOGY

1. On the Drawbacks of a Linear Model in the Decentralization of a Decision Process
P. Mahey 1

2. Vulnerability Analysis of Multiterminal Distribution Systems and Networks
M. Darwish and J. Bernussou 9

3. Quantitative Analysis of an On Line Measurement Equipment for Static State Estimation Purpose
P. Albertos, C. Alvarez and A. Reig 13

4. An Efficient Algorithm for Computer Process Control
P. Albertos, J. A. de la Puente and J. Tornero 21

5. A Comparative Study of Self-Tuning Regulators
G. Favier and P. Guillermin 27

6. Inertia and Multicriteriality as Important Features of Developing Systems
V. C. Matejić 47

II. MODELING

1. Development of System Models for Health Care Efficiency
T. Prasad, M. A. Ismail and V. H. Quintana 53

2. Multivariable Modelling for the Climatic Design Cooling Loads in Built Environments
M. Barkatullah 61

3. Analytical Modelling in Support of Design and Monitoring of Integrated Surveys for Development
W. H. de Man 69

4. An Automatic System for Geological Prospecting Drilling
A. G. Yarusov, G. A. Butkin, E. N. Efimov, V. A. Mazur and N. N. Novik 77

5. Aquifer Dynamics Identification in Natural Gas Storages
G. L. Chierici, G. A. Gottardi and R. P. Guidorzi 87

III. INDUSTRIAL APPLICATIONS I

1. Modeling of a Tunnel Furnace in the Ceramic Industry
D. Barreteau, C. Laguerie, J. P. Babary and M. Kassinopulos 93

2. Modelling and Simulation of a Rotary Cement Kiln
E. A. Dekkiche, P. Prévôt, A. En-Naamani and J. P. Barbier 101

3. On Line Estimation of Clinker Quality in a Rotary Cement Kiln
A. En-Naamani, P. Prévôt, E. A. Dekkiche and J. P. Barbier 111

4. On Line Identification of a Fluidized Bed Heat Exchanger System
T. Jeevanandam and S. S. Lamba 119

5. Identification of an Ethane-Ethylene Distillation Column
A. Andreoni, S. Beghelli, M. Di Martino, R. Guidorzi, L. Pacinotti and R. Rossi 127

6. Development of a New Heat Power System
H. Chou 133

7. Dynamic Modelling and Simulation of Paper Machine Multicylinder Drying Section
J. L. Videau, A. Lemaitre and C. Foulard 143

IV. INDUSTRIAL APPLICATIONS II

1. Stabilization and Control of Absorber Tower Chemical Process
M. Darwish and J. Fantin 153

2. Design of Sensitivity-Reducing Output Feedback Controller for a Distillation Column
N. N. Sorial 159

3. Optimization of Discrete Additions of Substrate During the Batch Biosynthesis of Antibiotics
J. Černetič, S. Strmčnik, M. Pokorny and R. Modic 163

4. Use of a Micro-Computer for the Control of a Phosphate Drying Furnace
M. Najim, K. Najim, D. El Jouhari and L. Radouane 171

5. Optimization of Industrial Textile Dryers
A. Touzani, M. Staroswiecki and C. Melin 181

6. Self-Tuning Regulator Applied to an Industrial Phsophate Drying Furnace
K. Najim and H. Youlal 187

7. Optimization Problems Associated with the Thermal Behaviour of an Epitaxial Reactor
J. P. Babary, A. Benhamou and B. Dahhou 193

8. Multi-Level Optimal Control of a Thermal Process
G. Krzakala, B. Lang and M. Aubrun 203

9. Discrete Optimisation Model for Siting Super Thermal Power Stations
S. R. Gatta, T. Devnath and G. S. Adhar 217

10. Optimal Control of Cutting Processes
A. S. Progrebnyakov, V. A. Gerasimov, G. A. Dorrer, V. Ya. Glizer and M. G. Dmitriev 225

11. Application of a Minicomputer for Direct Numerical Control of Multiple Machine Tools
B. Parhami and K. Alvandi 231

12. Implementation Problems in Multivariable Adaptive Control Structures
G. Bartolino, G. Casalino, F. Davoli and R. Minciardi 237

13. Multiplicity of the Steady-State in Fluidized Bed Reactors-VII. The Effect of Transport and Kinetic Parameters on the Stability and Controlability of a Complex Reaction
S. S. E. H. Elelnashaie, S. S. El Shishini and M. A. El Rifai 245

14. An Application of the Filtering Method in the Identification and Control Stages of a Heat Exchange Process
G. Ratefiarison, C. Humbert and J. Ragot 251

V. MANAGEMENT AND PLANNING

1. Scientific Manpower Planning: A Methodology for Egypt
S. Mahmoud and M. I. Younis 265

2. Planning Process for Introducing Energy Systems into Third World Countries
J. M. Garrison 273

3. Optimal Technology Selection in Rural Roads Construction
F. Lara-Rosano 281

4. Computer-Aided Design of the Implementation of Industrial Facilities
J. Pablo Antun 287

VI. ENERGY

1. Application of Recursive Estimation to Solar Energy
H. Z. Tadros and M. M. A. Salama 293

2. Optimization of the Power of an Array of Photovoltaic Solar Cells
M. Annabi, N. Ben Slama, M. Ksouri and J. Perard 297

3. Optimizing of the Available Solar Cells Power by Suntracking II
A. A. Montasser, A. O. Attia and I. M. El-Dokany 303

4. Microcomputer Systems Allow a Real-Time Parameter Estimation for the Control of a Solar Plant
L. Fortuna, A. Gallo, C. Commarata and M. La Cava 309

VII. WATER RESOURCES AND AGRICULTURE

1. Optimisation of the Use of Water Reserves in Agriculture from the Modelling Water-Soil-Crop System
G. Bel, C. Maertens and J. Puech 317

2. Remote Sensing Systems for Water Resource Management and Agriculture Development
V. Cappellini 323

3. Stochastic Approach to Reservoir Design and Management
M. Nouh 329

4. A Simulation Model for Multiple Regularization of Dams
M. Sbihi, F. Al Youssoufi, R. Alami, A. Hanafi and A. Mezzour 333

5. An Optimal Management Model of an Hydro-Thermal Power System under Irrigational Constraints: The Morocco Case
R. Missaoui, J.-C. Bernard and A. Haurie 343

6. Identification of a Potabilization Process in the Florence City Water Treatment Plant
C. Manfredi and E. Mosca 353

7. Case Study for Automatic Station for Resources Development in Iraq
S. Hussain and S. A. Haddad 361

8. A New Approach to the Simplification of Water-Supply Network Models
S. Strmčnik, D. Matko, F. Bremšak and J. Cernetič 373

9. A New Developed Stream Water Quality Model: A Case Study on the River Nile
M. F. Hassan, M. I. Younis and K. H. Mancy 385

10. The Desalination by the Multi Stage Flash Process and its Control
H. El Hares and M. El Hares 395

VIII. EDUCATION

1. Education and Its Aims
J. Elizondo 403

2. "Purposeful" Physics Curricula: A Systems Model
M. V. Ananthakrishnan 411

3. Training on Electrical Energy System Control: Microsystem Approach
E. H. T. El-Shirbeeny 421

4. Inexpensive Microprocessor Systems, Tutorials and the Future of Computer-Assisted Instruction
R. E. Burns, W. I. Davisson and E. A. Kline 431

5. Automatic Control in Mexico
A. Alonso-Concheiro 441

6. Computer Control Integrated System
J. E. Morales and J. E. Venencia 449

7. Minicomputer Evaluated Microcomputer Programming
D. Calikoğlu 457

IX. HUMAN FACTORS

1. An Approach to the Assessment of Automation Technology
H. K. Jacobsen and P. H. Jesperson 461

2. Automated and Automators: Human and Social Costs of Technological Change
S. L. Hacker 471

3. System Approach for the Transfer of Technology
M. Levi-Jakšić and M. Jakšić 481

4. Optimal Implementation of Process Control in Developing Countries
H. El Hares and F. Tayel 489

X. SOFTWARE, HARDWARE AND MAN-MACHINE COMMUNICATION

1. An Interactive Graphic Software for Complex Systems Modeling
M. Timoulali 493

2. The Operation of Microcomputer Control Systems Performing Hardware Emulation
T. J. Kobylarz 501

3. Language-Dependent Considerations for Computer Applications in Farsi and Arabic Speaking Countries
B. Parhami 507

4. Statistical Distribution of Arabic Letters Aids to the Design of a New Keyboard
S. H. El-Ramly and M. A. El Hamalawy 515

5. A Modeling Tool for the Specifications of a Logical Automatic System: The GRAFCET
M. Blanchard, P. Brard and J. P. Frachet 521

6. LSPEST: Subroutine Package for Identification and Parameter Estimation of Linear Dynamical Models
S. B. Maleshkov, S. I. Markov, I. G. Kalaikov and N. E. Madzarov 531

XI. SIGNAL PROCESSING

1. A More Generalized Wiener Filtering Technique
Heng-Li Kuo 539

2. Characterization and Identification of Noise Covariances in a Stochastic Linear System by Use of Equivalent ARMA Model
G. Favier and P. Guillermin 543

3. On the Frequency Response of M-D FIR Digital Filters
N. C. Hu 553

4. Structure Analysis of Complex Functions: Applications to a Zinc Production Unit
M. Deliris, E. Chacon and A. Titli 561

XII. ROUND TABLE DISCUSSIONS

Round Table 1. Automatic Control Education in Developing Countries
Introduction
P. M. Larsen 571
Experiences from Mexico
A. Alonso-Concheiro 572
Thoughts from Libya
H. el Hares 573
Considerations from Egypt
M. A. R. Ghonaimy 574
The Venezuelan Case
R. A. Padilla 575
Discussions and Conclusions 576

Round Table 2. Control and Instrumentation in Cement and Chemical Industries
M. Cuenod and M. Zelazny 579

Round Table 3. Control in Oil Industry
R. Perret 581

Round Table 4. Economic Modelling
J. M. L. Janssen 583

XIII. TUTORIALS 585

XIV. TECHNICAL SESSIONS 589

FOREWORD

In the last three decades, closely paralleling the modern period in control science, there has been a substantial growth of interest in the development of a "Systems Approach" that can be used for solving industrial, economic, social and environmental problems in developing countries. To summarize the state of the art from the practical as well as the theoretical point of view, the 3rd IFAC/IFIP/IFORS International Conference on "Systems Approach for Development" will be held at Rabat, from November 24 to 27, 1980.

The Conference comes third in a significant and fruitful series. By bringing together many researchers from different countries and with different experiences, along with managers and policy-making people in Governments, the past Conferences revealed that Systems Approach can be considered as an important tool for development.

It is a pleasure to acknowledge the contributions of all the Members of the International Program Committee and their active co-operation.

This Volume contains the papers submitted to the Conference which have been accepted; a few Tutorials, dealing with certain topics in Systems Theory and Applications, are also included. Round Table discussions on a number of themes relevant to the Conference will be included in the Proceedings volume.

According to the topics covered in the Conference, all the papers have been classified into twelve sections:

Tutorials

Methodology

Modeling

Industrial Applications I

Industrial Applications II

Management and Planning

Energy

Water Resources and Agriculture

Education

Human Factors

Software, Hardware and Man-Machine Communication

Signal Processing

SYSTEMS ANALYSIS AND DEVELOPMENT POSSIBILITIES AND LIMITS

M. A. Cuenod

Societe Générale pour l'Industrie-IC, Cointrin, Geneva, Switzerland

Abstract. Any development is implemented through the realisation of projects. The purpose of this survey report is to present the possibilities and the limits of Systems Analysis at the different stages of this realisation, namely:
- Aids to forecast the needs to be fulfilled by a project, and the evolution of these needs.
- Aids to creation and stimulation of imagination and search for new ideas.
- Aids to design for the determination of the optimal data to be given to a project taking into account the aim to be fulfilled and the contraints to be considered, and for the optimal planning of this project in time and space.
- Aids to decision as a support for the numerous decisions involved with the realisation of a project.
- Aids for the execution as help for the management of the execution of a project, namely for the control of the budget and the time schedule of its implementation.
In conclusion some considerations are devoted to the opportunity of developping some new approaches giving the possibility to take into account some qualitative factors which are not easy to be measured but which are significant for the definition of a development strategy.

Keywords. Decision theory, optimal planning, optimal design, management systems, aids to creation.

INTRODUCTION

The development of a country is the result of the implementation of a number of projects. The success of this development is subjected to the selection of optimal projects, of their optimal data and optimal localisation in time and space.
This optimisation depends on many factors and parameters which can be partly only measured; it is for instance difficult to give numerical values to sociological and political criteria linked for example to the creation of new jobs in underdevelopped regions, or the protection of the environment, or the requirements of national independance.
Some people consider that Systems Analysis is not suited to take into account such factors and thus leads to wrong decisions neglecting the actual needs and aspirations of a relevant country.
These criticisms are serious because due to the limited means of developping countries, any error in the definition of their strategy of development can deeply affect their future. They are partly due to the fact that the systems analysts have developped their theory without enough contact with the statesmen responsible for the definition of the politic of development of a country. On the other hand, these statesmen do not trust system analysis to solve the problems they have to face.
The question may in fact be raised whether some recent developments of this theory do meet their true needs, as these were defined 2000 years ago in this formulation attributed to Pericles:
" States should not merely be evaluated in terms of their ability to function efficiently as states, but far more in terms of the types of personality and character they produce among their citizens and of the opportunities they offer to all citizens for individual development." (1)

it is very appropriate to open a dialogue between these two categories of people on the occasion of this symposium.

Due to the consequences of wrong decisions related to the definition of a strategy of development, all means available to prevent such decisions are to be looked for, and the ways opened by the system analysis must be neither over nor under-estimated.

The purpose of this report is to present the possibilities and limits of system analysis applied to the definition of development strategy and the implementation of projects, and to point out that instead of ignoring intuition and creativity or unponderable factors, the modern approaches of systems analysis enable their combination with quantitative economical and technical factors.

Within its prescribed limits, this report intends to describe the aids of systems analysis at the different stages of the implementation of a project as defined by fig. 1, namely:
- aids to forecast
- aids to creation
- aids to design
- aids to decision
- aids to execution

AIDS TO FORECAST

As unpredictable as any forecast can be, still any development strategy is based on the forecast of the needs to be met.
The curves of the evolution of the consumption of specific goods in different countries show that this evolution is subjected to the 4 following types of variations:
- random fluctuations due to the fact that the global evolution of a consumption is the superposition of a high number of individual decisions.
- cyclic variations due to the alternance of the economic conjoncture.
- occasional variations due to important events like wars or a change in the political regime of a country.
- long trend evolution characterising the growing process of the economy.

The observation of the evolution of many socio-economic processes in many countries over a long period, from 5o to 1oo years, enables to make the 3 following remarks:
- important events, such as the two last world wars, do not affect deeply the general trend of the development of a country.
- the juxtaposition of the evolution of consumption of many countries versus increase of their GNP/cap. shows a definite similarity and leads to the formulation of some general rules of growth.
- system analysis gives the possibility to define the growth process with a few number of mathematical models, and also to identify the numerical parameters introduced in these models. (2)

This method has been applied to identify the evolution of steel and cement consumptions in many countries. (3) By recording their consumption per capita versus the GNP/cap., each country can be characterized by a specific curve. The set of these curves covers two S shaped areas given by <u>fig. 2</u>, with the five following periods:

1) <u>period preceding the taking off of the economy</u>
$0 < \text{GNP/cap} < 3oo$ \$/cap.
The main need to be met during this period is food; the population is too poor to use cement or steel. The increase of the consumption of these goods is less than proportional to the increase of the GNP/cap.

2) <u>taking off period of the development of the economy</u>
2oo \$ < GNP/cap < 12oo \$/cap
The main need to be met then is covered by the building industry. The amount of cement used in a country with a 1ooo \$/cap. reaches very high values of up to 4oo, 5oo kg/1ooo \$/GNP, while the growth of steel consumption remains slow.

3) <u>Period of industrialisation</u>
1ooo \$ < GNP < 24oo \$/cap.
The growth of steel consumption becomes higher than the growth of the GNP per capita, while the rate of growth of the cement consumption decreases, which corresponds to the inflection point of the above mentionned S curves

4) <u>Period of saturation</u>
22oo \$ < GNP < 5ooo \$ /cap.
The specific cement consumption reaches its ceiling value, which varies according to the type of countries:
- for mediterranean and mountainous countries: between 55o and 75o kg/cap.
- for northern and flatland countries: between 35o and 5oo kg/cap.
The rate of steel consumption growth decreases, which corresponds to the inflection point of its S curve.

5) <u>Post-industrialisation period</u>
This period begins now for some countries such as Sweden, Switzerland, etc and is characterized by the fact that the specific cement consumption begins to decrease, while the specific steel consumption reaches its ceiling value, which varies for different countries:

- for steel producing countries: 65o to 75o kg/cap.
- for non-producing steel countries: 4oo to 5oo kg/cap.

This system analysis of the economic growth through the comparison of the specific consumption of significant goods, applied to many countries and over a long period, such as it has been sketched above, leads to the following main conclusions:
- a saturation phenomena can be foreseen long before it actually takes place.
- the gravity center of the need for production goods is changing and increases the interdependancy between developping countries and the countries where these production goods are manufactured.

AIDS TO CREATION

The ability to create and to find new ideas is one of the main conditions of the success of any project. New ideas do not emerge by chance or by accident. It is recognised that new ideas result from a collective work, with an appropriate methodology.

The systems approach provides:
- a means of communication, giving a common language and concepts to the different specialists involved in the realisation of a project.

- a clear formulation of the problem to be solved in terms of performances to be reached and of constraints to be taken into account.

- a methodoly described as "inventic" (4), using " invention teams" of 8 to lo people belonging to different fields. This methodology implies:
a) a logic phase for the clear definition of the problem
b) an intuitive phase for the process of creation
c) a critic phase for the evaluation of the results obtained

The system analysis provides also computer programs stimulating the creativity. The principle of these programs is to propose to the "invention team" a set of pairs of concepts selected at random. The team is invited to evaluate the "novelty" of the pair according to certain criteria.
The new combinations proposed by the computers suggest to the team new concepts or ideas. The computers point out the consequences of these new choices. Of itself the computer does not find new ideas, but it gives a support to the creation team and saves time defining a "search strategy", as it gives step by step a survey of the new combinations of ne new ideas already obtained.
Such computer programs are powerful tools to be used in the first stage of the definition of a project.

AIDS TO DESIGN

Defined as "Computer Aided Design"or CAD, numerous methods have been developped and presented in numerous conferences, and specially, among others, by IFAC. (5)
It is not possible, within the limits of this report, to present these methods. We shall only consider the prospect opened by the numerical simulation of a system.

The principle of this simulation is to define a mathematical model, and to proceed with this model to the experiments which are impossible to achieve in reality: it is not possible to realize a system to observe how it works, to undo it and do it again until better results are obtained.

The definition of a mathematical model is one of the main approaches of systems analysis, and has the following advantages:

- it requires a thorough analysis of the structure and the operation of the projected system

- it also requires to express the aim of this system in an "objective function" summarizing the goal to be attained.

- it enables to examine the consequences of the different options and gives a support for the better selection of these options.

- far from ignoring the "intuition", the use of such a model gives a support to make full use of imagination for the design of a system.

The principle of the optimisation through numerical simulation consists in giving different numerical values to the parameters involved by the objective functions and in determining the optimal values of these parameters through a "try and see" procedure.
This method permits, for instance, to give answers to the following questions:

- where are the production centers of a given good to be located in a country?

- In which sequence must they be implemented?

- which characteristics must they be given?

- Is it appropriate to have numerous production centers, in view of:
  . adjusting the production to the evolution of consumption
  . reducing the transportation costs by locating the production near the concumption center
  . increasing the availability and the consequences of the loss of a production center

or is it more appropriate to have few production centers of high capacity to take advantage of the reduction of their specific investments through an increase of size of their installations?

This method has been used for instance for the optimal planning of the cement industry in a country. (6 & 7) The " objective function " in that case is the total cost of the cement consumed in the different consommation centers actualised at the end of the reference period.

This method implies the manipulation of a high number of numerical values. For exeample, a system with lo parameters which can have ten values corresponding to lo^{lo} possible different solutions and this necessitates the use of a computer to be handled.

Many methods like dynamic programming and branch and bound procedure have been developped to find the optimal solution. One way to do it is to apply the "sensibility analysis", that is to vary each parameter one after the other and to determine the influence of these variations on the objective function.

If an optimal solution has been determined on the basis of pure technical and economical considerations, and if for political or social reasons it appears necessary to deviate from the first determined optimal solution, numerical simulation enables to calculate the additional investment costs resulting from this deviation, and to balance them against these political or social advantages.

Econometric models have been developped to simulate in a matrix form the economy of a country. The inputs are all the means necessary for the operation of this economy: manpower, raw material, energy, transportation, communications etc. The outputs are the different kinds of goods and services produced by this economy. Many of these inputs are also outputs: the economic system of a country operates as a closed loop system. In fact, a complete econometric model of a country is a tool too heavy for the practical determination of a development strategy. As indicated by Fig. 3 it is appropriate to make use of reduced models adapted to the economic branch taken into consideration for planning, such as energy, communications, heavy industry etc. This reduced model enables to define a development strategy of this branch. The control variables resulting from this strategy must then be introduced in the general model, which gives the possibility to forecast the impact of the different strategies on the general evolution of the economy of the considered country.

AIDS TO DECISION

Any politic of development implies a sequence of decisions, and its success is highly dependant on the way these decisions have been taken.
Systems theory gives the basis for "concerted decisions" defined as follows:
The decision procedure must be associated to:

  . the preceding steps of design, creation, consultation and elaboration of the decision
  . the following steps of implementation and execution

This procedure must be "adaptative" and must be reviewed and readjusted, taking into account the different impacts of a decision.
The decision is the result of a consensus focussing the efforts of all the people concerned with this decision.
The system approach gives the means to collect, to process and to disseminate data to all concerned.

As indicated by Fig. 4, econometric models can be used as a support for political decisions, using records transmitted through displays and used either in a closed loop, including a panel of politicians and a team of specialists, or in an open loop as an aid for informing the public on the elaboration of political decisions through the television network. (8)

AIDS TO THE EXECUTION

When the decision to implement a project has been taken, the purpose of the "aids to the execution" is to give to those in charge of the management of this project all the requested informations in such a way that the human and material means available for this implementation will be used as efficiently as possible. It implies to determine the optimal use of these means with regards to the three following criteria:
- cost
- time delay

- performance
and for the planning evaluation, regulation and control of these means.

Planning means the definition of the succession of short term goals to be reached and the definition of the elementary tasks to be performed.

Evaluation means the definition of the cost and duration of these elementary tasks , and of what is requested to perform them.

Control means the feedback informations necessary to supervise and eventually to rectify the execution. This feedback necessitates computer programs to check:
- the costs of the different tasks and their impact on the budget of the project.
- the time schedule of their execution
- the transfer of information to all bodies concerned.

As indicated by Fig. 5, computer assisted planning has been developped by use of models evaluating the foreseen performance of an implemented project so that the actual results are made to meet the expected ones.(9)

CONCLUSION

The development of any given country is the result of the implementation of different projects which can be defined as a "set of activities limited in time and space and converging on a given goal".

The definition of a development strategy and the implementation of projects is one domain where system analysis is a proper approach for the optimal use of means generally limited with regards to the needs to be fulfilled. The main characteristic of this approach is to underline the following aspects which are necessary for the success of any project:
- Interdependancy between the elementary tasks implied by the implementation of a project and the goal to be reached.
- Dynamics of the implementation defining the optimal sequence of these elementary tasks.
- Multiplicity of quantitative and qualitative constraints and objectives to be taken into consideration.

System analysis must not be considered as a set of recipes giving automatically a solution, but as an approach which helps to point out the essential from the minor points in the implementation stage of a project, namely:
- forecast of the needs to be fulfilled and of the constraints to be taken into consideration.
- search for new ideas and means
- optimal design of the project to be implemented
- optimal decisions implied by this implementation
- optimal use of the means requested for the realisation of the project.

As mentionned by Karsky (10),
" System dynamic is not a technique nor an esoteric method but a set of approaches which can be used for analysing, clarifying global problems in various fields. There is no"system analyst", but a team belonging to different specialities, having in common some concepts and ways to solve a problem".

This approach is not easy but it is fascinating and stimulating; and this explains the slow but constant development of system analysis over the last decade.

The main purpose of the system approach is to establish a link between engineering professions and social and behavioural sciences to bring forth between them a fruitful dialogue.
As stated recently by Aron Wiener (11)
" Only through their common efforts can the necessary perspective be achieved to inspire a more comprehensive vision of the development of natural resource systems and of their coupling with the socio-political systems and their environment. Such a vision has now become a necessity if we wish to maintain a balance growth of man's material and spiritual welfare and to promote the rise of a stable society".

REFERENCES

1. Karl Deutsch (1966), The nerves of government, Free Press, New York, 191

2. Cuénod (1977) Quelques méthodes d'analyse de la croissance de phénomènes socio-économiques. Revue NT,Nr 5 237-243

3. P.Coïdan and M.Cuénod, (1977) Forecasting cement and steel need in developping countries.
Proceedings de la Conférence de l'IFAC au Caire : Systems approach for development. 247-252

4. A.Kaufmann, M.Fustier, A.Drevet (1973) Paris, Entreprise moderne de l'Edition.

5. Proceedings of the IFAC Symposium on Computer Aided System. (1979) Zürich, Switzerland.

6. M.Cuénod and co-workers. (1973) Approach to the optimal localisation of cement plants and cement distribu-

tion centers in developping countries. Proceedings of the IFAC-IFORS Conference, Algiers, on the Application of systems engineering methods to developping countries.

7. M.Cuénod, A.Duval, M.Kummermann, (1974). Optimisation de la localisation dans le temps et dans l'espace d'un système de production et de distribution.
Proceedings du Congrès de l'AFCET, Paris, 33-42

8. J.G.Balchen.(1979). The need for Computer-aided Design in Modeling and Control of Nontechnical Systems. Proceedings of the IFAC Symposium on Computer Aided Design of Control Systems, Zürich, Switzerland

9. L.Pun, (1979). Computer assisted dynamical plannings for production activities.
Proceedings of the IFAC Symposium on Computer Aided Design of Control Systems, Zürich, Switzerland.

lo. M.Karsky. (1978). La dynamique des systèmes au sein de la SNEA. Bulletin de liaison de la recherche en informatique et automatique de l'IRIA No 41,7-13

11. A.Wiener. (198o). Beyond Resource Systems. Survey report presented at the IFAC Symposium on Water and related Land Resources Systems, Cleveland.

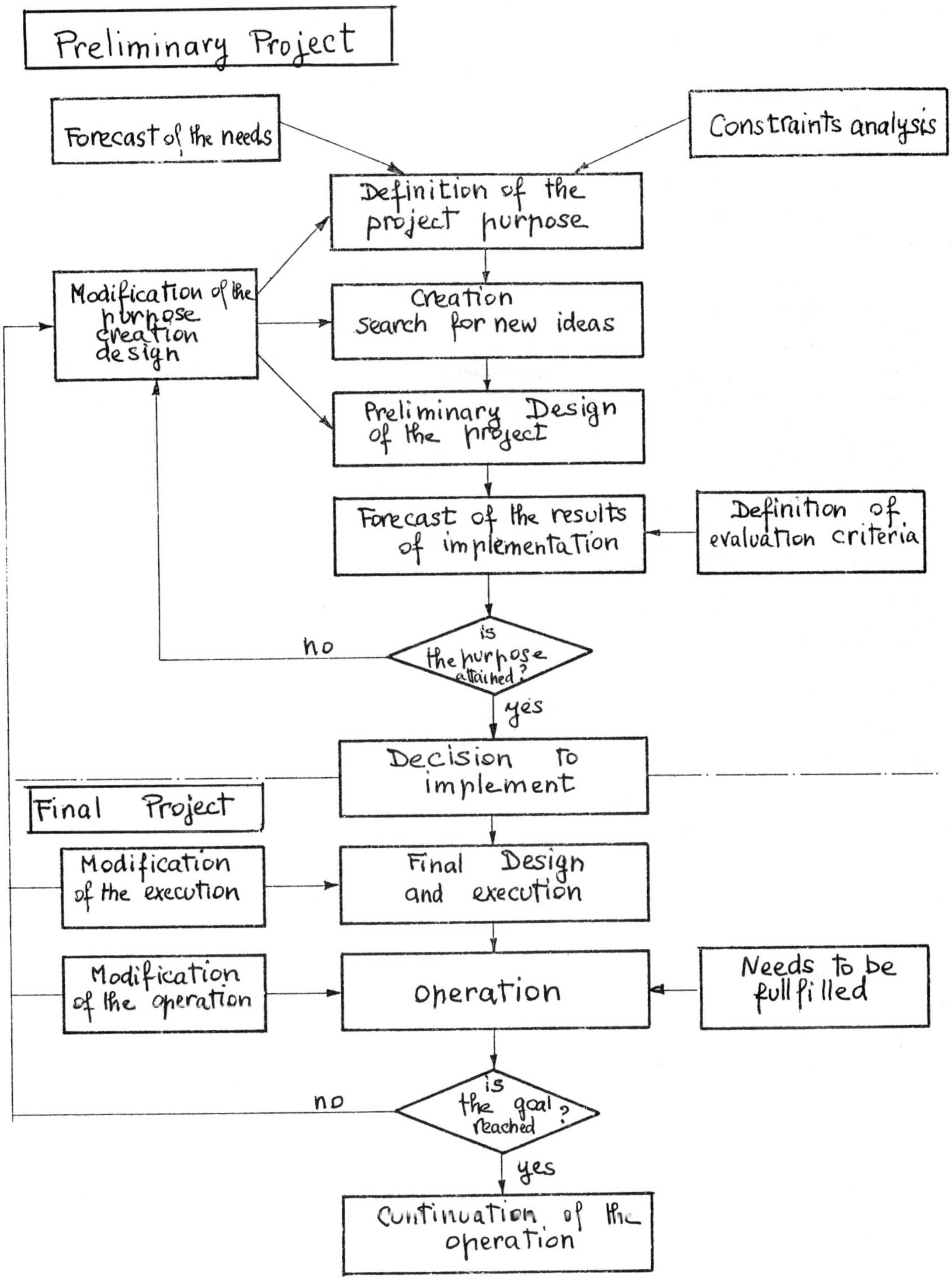

Fig. 1 Scheme of the procedure for the realisation of a project

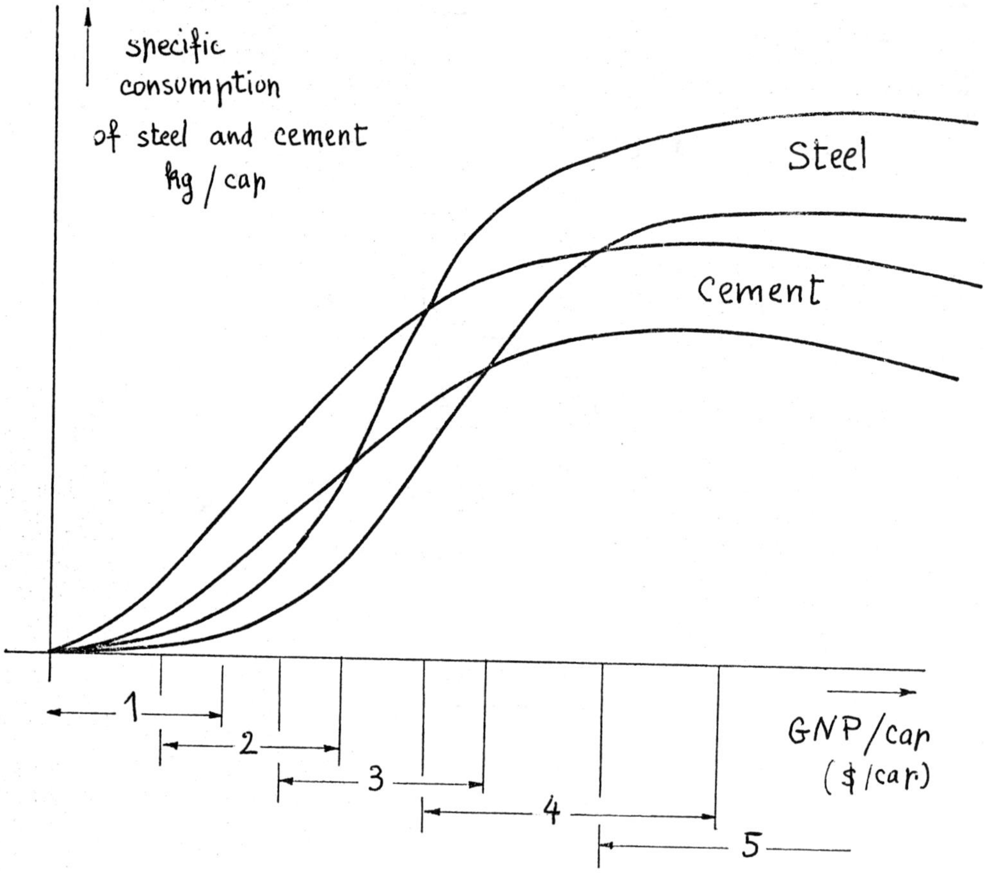

1. Period preceding the taking-off of the economy
2. Taking off period of the development of the economy
3. Period of industrialisation
4. Period of saturation
5. Post-industrialisation period

Fig. 2 Evolution of the specific consumption of steel and cement versus GNP/cap.

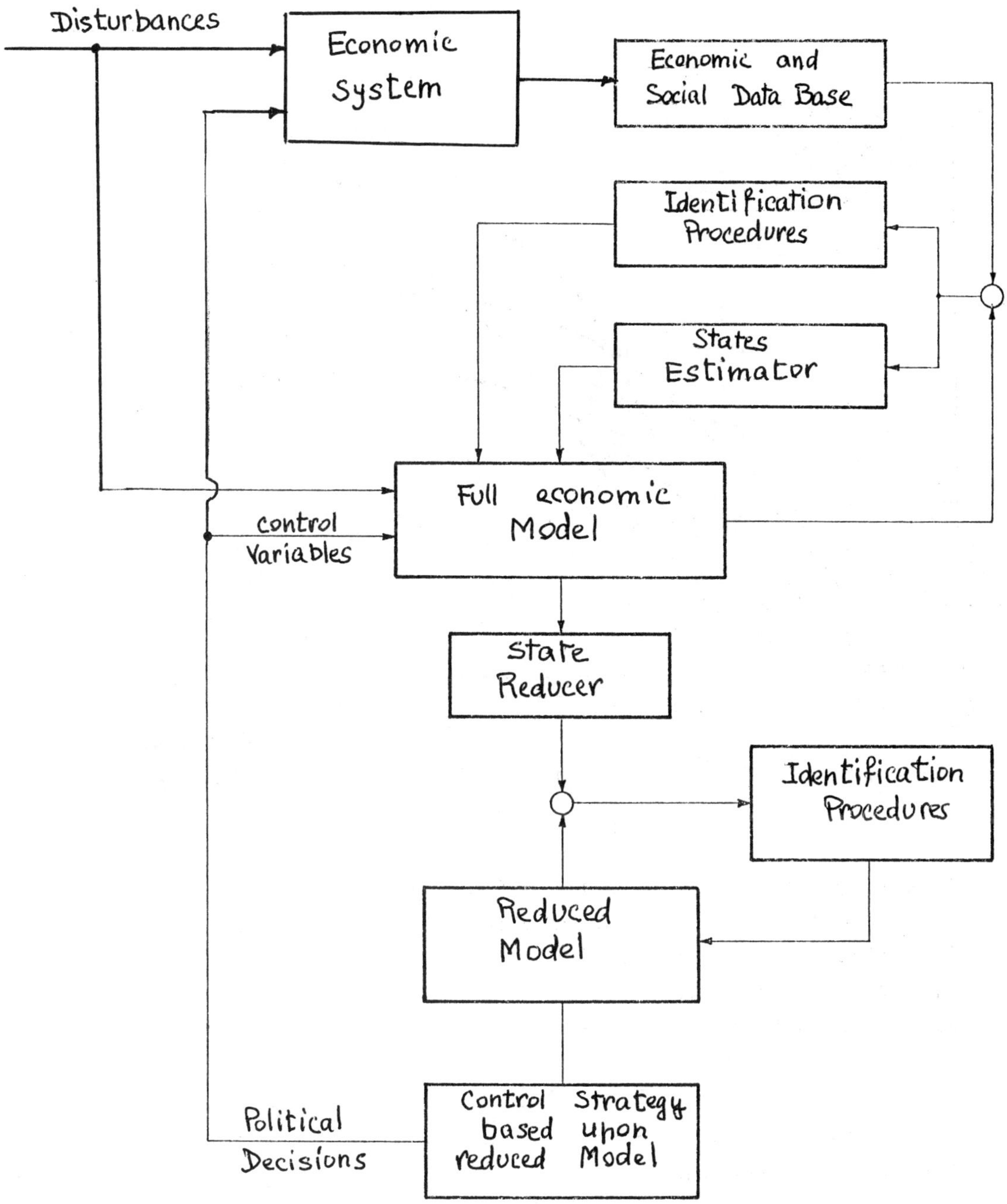

Fig. 3 Principle of the definition of a strategy of development based on economic models.

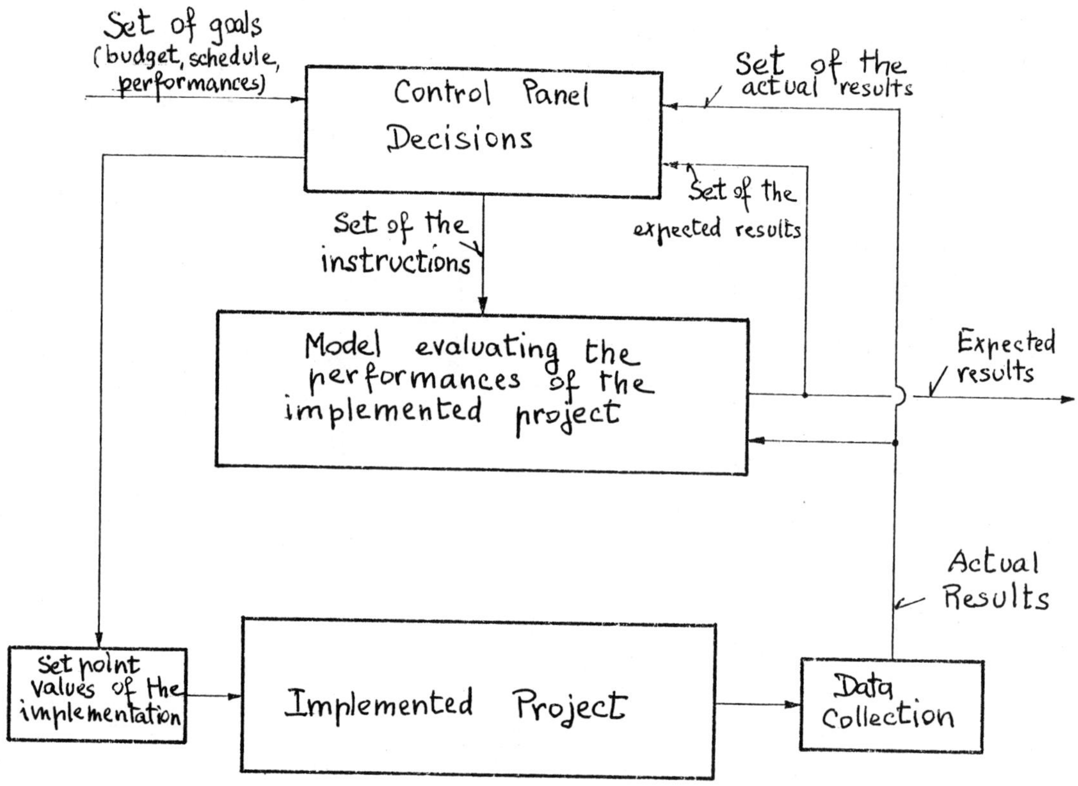

Fig. 5 Principle of the computer assisted planning of project activities

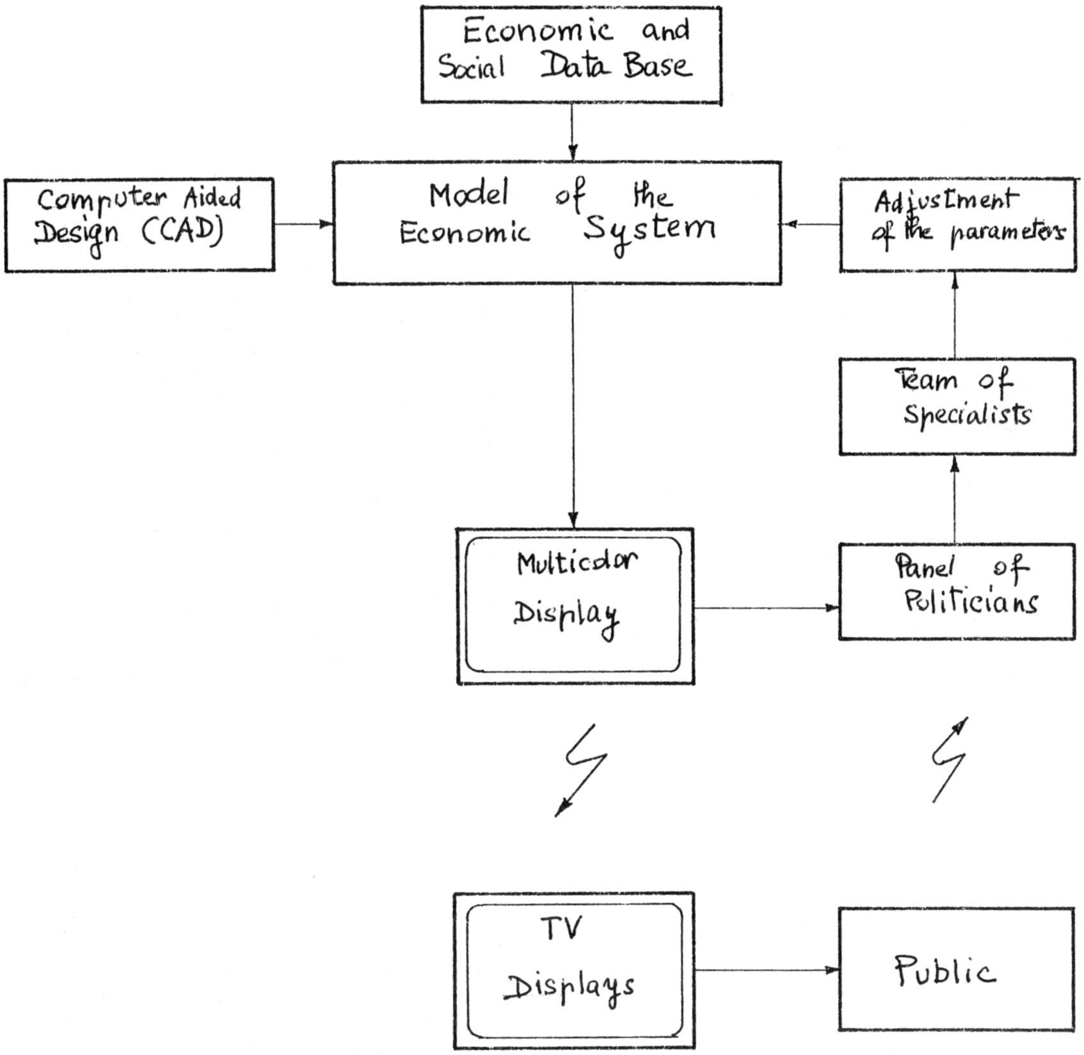

Fig. 4 Principle of the on-line simulation of economic systems as an aid for political decisions.

POWER SYSTEM CONTROL AND AUTOMATION

J. P. Waha* and M. Ribbens-Pavella**

**S. A. Intercom, B-1000 Brussels, Belgium*
***Department of Electrical Engineering, University of Liège, B-4000 Liège, Belgium*

Abstract. In this paper we describe the interplay between problems arising in large-scale electric power systems and their solution by results and methodologies of control systems. Rather than to be exhaustive, we will attempt to impart a flavor of some of these large-scale control questions in power systems. Among them, particular attention will be focused on two specific problems for which we describe recently proposed solutions; these are the construction of static equivalents including reactive power diffusion, and the use of a hierarchical concept for static state estimation of large power systems.

Keywords. Power system control; power transmission; power management; centralized-decentralized control; hierarchical systems; state estimation; static equivalents; reactive power diffusion.

I. INTRODUCTION.

I.1. Generalities

Control and automation of power systems imply modelling, abstraction of the real and economical systems, goals to be reached during the operation in terms of customers'satisfaction (economicity, quality, availability), constraints (technical, economical, environmental or political, social) and means to be used (Quazza, 1977).

The system itself described by state variables some of which are constrained, is operated through control variables, which are also constrained (fuel availability, technical limits), and is under the influence of external facts, normally of random nature. Such "perturbations" are for instance :

- loads fluctuating according to customers' needs, climatical and economical patterns, sociological habits
- outages of different types (plants, network) defined by their amplitude, frequency and duration, sometimes by their "on-off" nature.

A typical power system under a unified responsibility (state owned, private owned) includes generating plants (hydro, nuclear, fossil, gasturbines, etc) in number from tens to several hundreds units, transmission (of bulk power) and subtransmission networks (36kV... 150kV...500kV), interconnections with other systems, sometimes DC links, distribution networks (< 30kV), low voltage networks (110 to 380 V). Loads (or consumers) are fed at high or medium voltage (big industrial loads), private consumers being fed at low voltage (see fig.1). Such a system might have a peak load of 1000 to...50 000 MW, a yearly consumption of 5 to 300 TWh, hundreds of HV lines, cables, a total lenght reaching several thousands km, linked at hundreds to thousands of electrical nodes (substations), network covering areas from 10 000 to...20 000 km^2 or more.

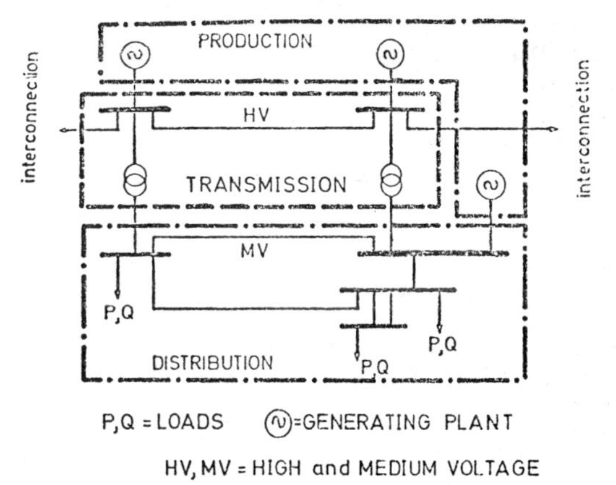

Fig.1. Functional structure of the power industry (From Waha, 1971)

Medium and low voltage networks have sometimes a total length more than 100 000 km, 10^3 to 10^4 nodes, and 10^5 to 10^6 individual customers. Such a large-scale system is multivariable by essence and has to be treated by dichotomical approaches (Glavitsch, 1977) simultaneously in space (geographical approach, voltage level description, functional spliting : transmission, distribution) and time (fast transients at nearly speed of light to long term operation planning including equipment maintenance, hydrological and seasonal effects and fuel contracts. Hierarchical operation and control, centralization and decentralization, coordination bodies are fundamental in such a problem.

Operation of an existing power system, its real-time aspects in direct control, its planning needs (maintenance, fuel, see above) are to be considered in the overall process of expansion and refitting (new equipments, lines, plants) due to normal growth and repair. Control design procedures have to cope with these permanent changes, additions and modifications during the lifetime of the process.

I.2. Statics and dynamics in system operation.

Quazza (1977) and Waha (1971) describe with some details the operation of power systems. Let us stress upon the fact that electrical power at the disposal of customers cannot be stored; primary energy (fossil fuels, nuclear fuel, water, compressed air...) is the only stored energy available. The electrical energy (of AC type) is therefore stored under mechanical form (steam pressure, water levels, kinetic energy of rotating machines) and has to be transformed, transmitted and distributed when needed. Fig. 2 gives, as example, the sketch of a thermal generating plant, boiler, turbine, alternator with its classical feedback loops. For a power system, such a plant is considered as a "production" of real (P) and reactive (Q) power at an electrical node of the network. This power has a given availability linked to planned and forced outages of the plant, a given quality (in terms of frequency and voltage) and economicity (total or marginal price of energy unit-kWh). A first dichotomy consists to treat the power plant alone, describing its internal behaviour; several aspects can be studied as :

- general feedback control of external variables (frequency, voltage, reactive and real power produced) using detailed models, generally linearized, or of internal variables (steam pressure, temperature, furnace under pressure,..) the time constants involved are in these cases ranging from few seconds to several minutes;

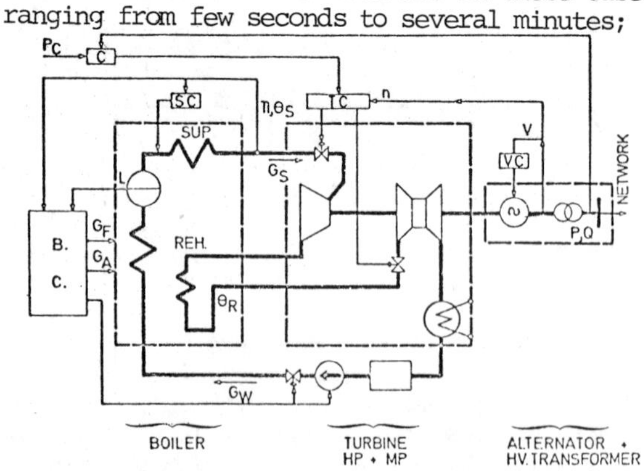

C=General controller *P_C=load set point*
BC=boiler controller *S_C=superheater controller*
TC=turbine controller *R_C=reheater controller (not shown)*
VC=voltage controller
θ_S, θ_R=steam temperature *η=steam pressure* *P,Q=active and reactive produced power*
G_W, G_A, G_S, G_F=water, air, steam, fuel flow

Fig.2. (From Waha, 1971)

- detailed non-linear models for studies of large transients (start-ups, shut-downs, trips, outages); these models can include long term constants linked to thermal effects in metallic envelopes and stresses.

Very few state variables (5 to 10) are involved in certain cases, while numerous ones are needed in others especially when dealing with thermal diffusion processes (modelling by partial differential equations).

Interaction and coordination between generating plants and network needs simplified models of boilers, hydroplants, alternators, when dealing with fast transients (stability problems), or with the so-called slow dynamics (few seconds to few minutes) involving power and voltage oscillations on the system. These phenomena are induced by primary frequency and voltage regulators of turboalternators and plant dynamics.

If we look at the transmission network itself and if we neglect propagation effects due to topology changes or lightning strokes, we may consider that power transmission is done without time lag, the network being purely static (see fig.3). Load and generation patterns will induce flows in the network by Ohm and Kirchhoff laws as a succession of static states.

Outages of elements of the network will be limited by opening breakers under control of protection devices; these actions, purely decentralized, are needed at the first moment to protect people and equipment, and afterwards to guarantee the security of the system, avoiding general black-outs Protection act fast (msec to tenths of second); no centralized control in real time is done.

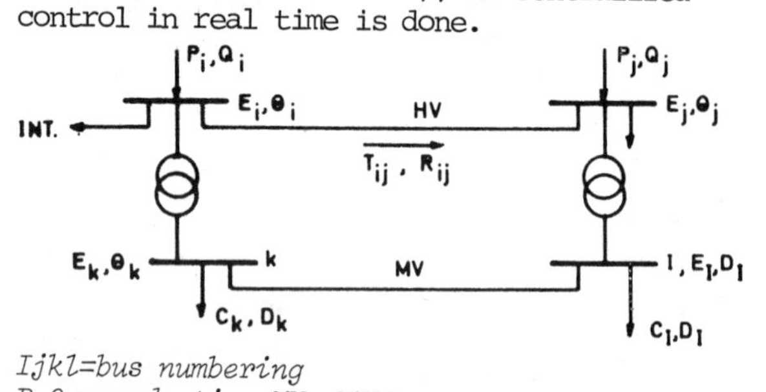

Ijkl=bus numbering
P,Q =production MW, MVAR
C,K =consumption MW,MVAR
E,θ =argument and phase of voltage (kV, rad.) (state variables)
T,R =transits MW,MVAR

Fig.3. Static model of network (From Waha, 1971)

Overloads and thermal problems of network elements (lines, cables, transformers) will contribute to some slow dynamics in terms of minutes or hours, normally controlled by manual procedures at centralized level.

Fuels, water and other scheduling problems, as maintenance, are done on daily, weekly or seasonal basis; centralized techniques are therefore used to reach the economical goals involved.

I.3. General philosophy of power system control and automation.

Table I gives a summary of the dynamics of a power system and a short survey of typical control and automatic actions in system operation. Centralized and decentralized features are described. Fig. 4 gives a sketch according to table I.

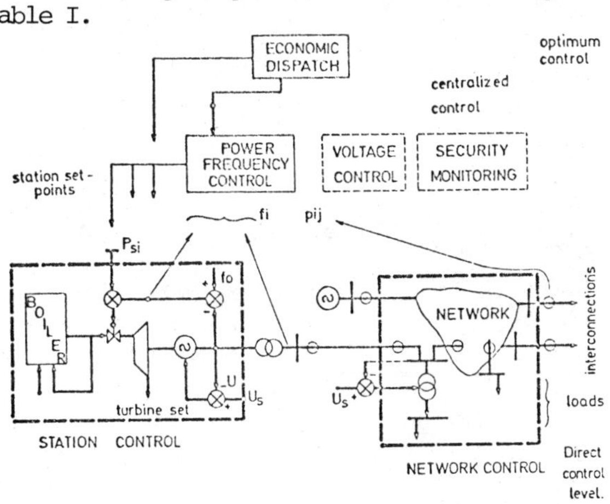

Fig.4. Multilevel aspects of power system operation. (From Waha, 1971)

Fig.5 gives an example of dichotomy and its applications in system control. Generally speaking the overall problem of reaching the operation goals is obtained by direct control, or so-called real-time control with "on-line data", by centralized or decentralized procedures; on the other hand predictive actions for short term (minutes, hours) aim at security, and medium to long studies (hours, days, weeks) deal with planning and scheduling problems

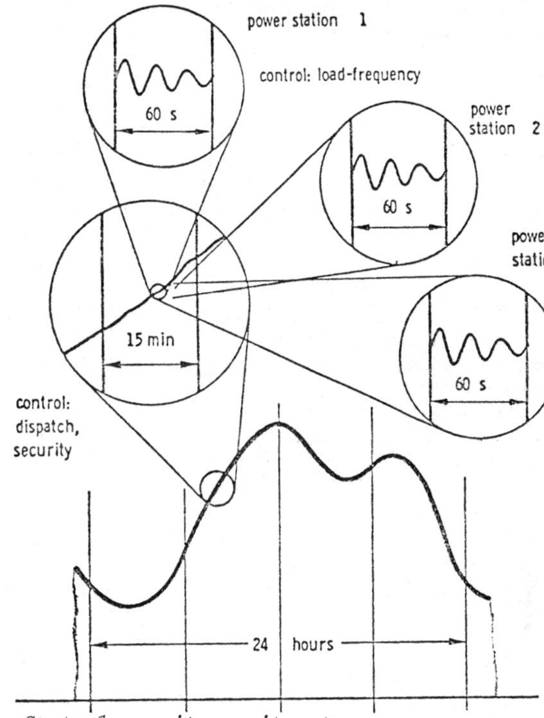

Control : unit commitment
Various time spans in power system control

Fig.5. (From Glavitsch, 1977)

TABLE I

DYNAMICS OF POWER SYSTEMS (TYPICAL VALUES)
SELECTION OF SOME CONTROL AND AUTOMATIC FUNCTIONS

DYNAMICS		CONTROL FUNCTIONS AUTOMATION
Power station		Power Station Control
• Generation	Time constants, periodicity, delays	• Decentralized at power station * Internal state variables (fluid pressure, temperature, levels, oxygen contents, metal temperature ...), controlled by fuel, water, air flows, injection flows, ... * External state variables (frequency, voltage) controlled by steam flow or turbine valving, exciter current)
* Thermal (fossil, nuclear)	sec, min, hours	
* Hydro	sec, few min.	
• Energy management	days, weeks, years	
• Maintenance management	weeks, annual cycles	• Centralized at control center * Set point of real, reactive power (AGC, LFC, secondary voltage control) * Start-up, shut-down, unit commitment
Network		Network control
Transmission	no lag	• Decentralized * protection, restoration * voltage control
Steady-state stability	... sec	
Inertia of turbosets, big loads	sec ...	
Thermal overloads	min - hours	• Centralized * supervision, security, optimal control (economic dispatch) * scheduling, operation planning
Network maintenance	days - weeks	

These predictive studies might use optimal control strategies in some cases; the direct control functions might use optimal or sometimes heuristic techniques according to the needs in time responses or in other words functions of the state of the system.

The overall control process will be updated following the changes in the goals (e.g. the influence of environmental constraints on operation), or following the operational experience (e.g. black-out), or due to economical reasons (prices of fuel). This updating is typically an adaptive procedure which can be in some cases automated (learning procedures). It is clear that this adaptive control phase is very long (it can take years to change the control philosophy of a network), and is imbedded in the overall expansion planning process. The decision to build a high voltage transmission network (supergrid) above existing voltage levels as the 400 kV network in Europe, since 25 years or more, and the decision to install several big nuclear units on some sites, have a definitive impact on system operation and control.

Expansion planning methods used to decide future plant and network element siting, try to simulate future operation of the power system, including their control loops, productive devices and optimal criteria to be met. These methods are therefore very similar to control procedures, including both real-time and extended real-time aspects as for scheduling of maintenance and generation. The main difference lies in the fact that control procedures definition and implementation are based on analysis of several strategies about an existing system with high dimensionality and that expansion planning methods have to deal with the same system besides several other new elements, whose combination of cases can contribute to difficult dimensionality problems.

II. HIERARCHICAL CONTROL OF POWER SYSTEMS.

II.1. The size of power systems, measured in terms of number of elements, state variables, geographical areas, types of generating equipments, and structure of networks leads to the need of a hierarchical concept in its operation, control, and automation. When speaking of hierarchy, it is important to stress upon the number of functional levels, their responsibility, the splitting of several types of actions and the coordination problem which will arise. Hierarchical control imposes some centralizations of responsibilities and some decentralizations at substation and plant levels of action. To limit the size of the problem, appropriate decomposition techniques will have to be used.

II.1.1. Functional description

A first approach of this concept lies upon functional and often geographical splitting :

* generation implantations
* transmission, interconnection network (150-500 kV - meshed networks)
* subtransmission (60, 120 ... kV-meshed, sometimes organized in subnetworks)
* distribution, generally radial (< 30 kV)
* loads and customers.

Let us stress on the fact that the normal growth of a power system along the time implies that generation exists at all levels :

- small and old plants at the lower levels, sometimes at distribution levels,
- big base load unit at the highest voltage level
- peaking units at all possible levels.

The same holds for the functions of the voltage levels (see Kowal and others, 1980) which, from local networks, first interconnected, were then covered by higher voltage, implying a shift of transmission to distribution function, following the state of development of the system.

II.1.2. Time aspects.

A second approach on which the philosophy of centralization and decentralization will be built is the time frame of the phenomena as described in table I. Practically speaking fast transients (propagation effects, short circuit, dynamic stability problems) will be covered by local, generally automatic devices at substation and plant sites. These preprogrammed devices will act from their local information; some teletransmission is sometimes used to link in real-time functions of different sites, but in little quantities (remote shedding, interpretting, teleprotection). Coordination of these devices exists from higher level (by control center through remote control or by maintenance squads), by changing settings (relays), set points, shedding levels..

II.1.3. Operational aspects.

A third approach of this problem is to consider the actions and phenomena involved in operation, actions including the will of the operators but also the external constraints. As external actions we may list :

* load changes
* outages of plants and network elements
* maintenance of elements and construction work for system expansion.

Control actions coping with these external constraints will include :

* loading power plants, starting units
* switching the network
* controlling the system both at the operation planning level (predictive control, feasibility of action, security assessment), and at the action level (direct control) (see fig. 6).

II.1.4. System state.

A fourth description lies in the definition of state of the system (see Dy Liacco, 1975, and fig. 7) :

* normal or preventive
* alert (secure or insecure)
* emergency
* restorative

An important point is to define the type of contingencies considered when analysing the normal, preventive and emergency states, i.e. to list the credible outages and load pattern

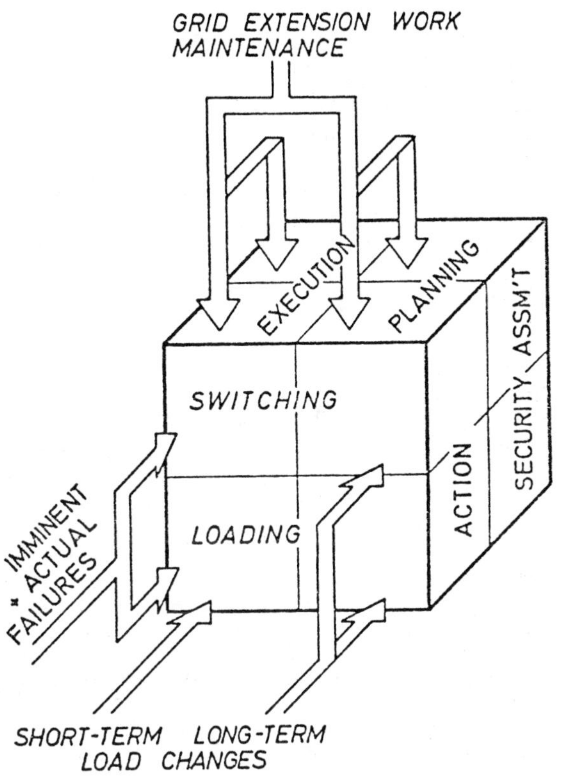

Fig.6. Operational functions of a power system N.B. Time aspects and system states (normal, emergency,...) are not represented explicitely on this sketch. (From Kowal and others, 1980)

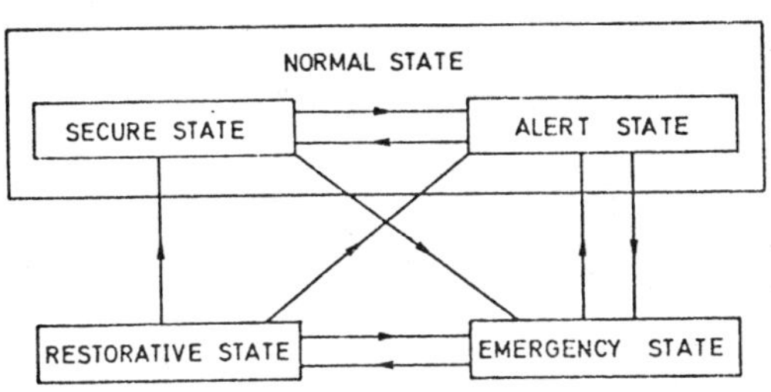

Fig.7. (From Merlin, 1980)

that the power system should withstand. Secondly, when building concepts to control the power system in terms of economy, security and availability it is important to consider the overall system, control loops and control centers included, in a way to analyse at the same time the transitions from one state to the other. This way is conceptually the only one which will manage to design future power systems taking into account automatic features (including control centers) in the overall optimising scheme involved.

II.2. Control center functions; mathematical modelling (see Kowal and others, 1980)

II.2.1. Control center functions depend on type of power systems (hydro, thermal), type of functions (transmission, distribution), type of management and utility organisation (maintenance, manning of substations, availability of manpower ...). Typical functions already exist in every control center or dispatching and may be listed as :

* acquisition (SCADA)[1]
* man-machine dialogue and data presentation
* direct control (real-time) including LFC[2] AGC[3], remote control, secondary voltage control
* extended real-time functions as security assessment, economic dispatch with or whithout security constraints
* scheduling of maintenance
* off-line studies.

Some of these functions are of direct control type, using real-time data. Others are of optimal control type based on real-time and predictive data and statistics; they include sometimes security aspects and prophylactic actions. Finally, long term actions (scheduling) are basically operation planning techniques.

II.2.2. Some of these functions are essentially of dynamic nature (LFC), all the others are mainly static, or at least considered as such. Fig.8 gives as example a flow chart of such a static approach of security procedures; this diagramm could be true for dynamical phenomena too, but to the authors knowledge no realisation exists in the field today.

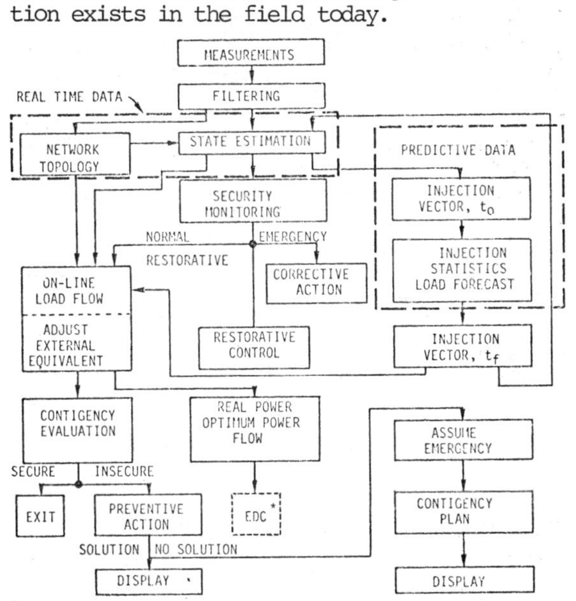

* INCLUDING VOLTAGE AND REACTIVE POWER OPTIMAZATION

Fig.8. (From Dy Liacco, 1979)

Basically the control algorithms which are used reflect the nature of the models available and manageable at dispatching level. A generally accepted description (see Dy Liacco, 1980) is the following :

x : set of state variables
k : set of manupilated or control variables
d : set of disturbances (considered as poorly predictable)
c : system topology or configuration.

(1) $g(x,k,d,c,t)=0$, set of non-linear algebraic equations
(2) $\dot{x} = f(x,k,d,c,t)$, set of non-linear differential equations.

---

[1] Supervision Control and Data Acquisition
[2] Load Frequency Control
[3] Automatic Generation Control

At control center levels, only dynamics ranging from few seconds to hours, days,... are considered due to teletransmission possibilities and computing technology available. Equations (1) and (2) describe then the network as static, but includes slow system oscillations due to governor and automatic voltage regulator interaction among themselves, with loads, and with interconnected areas. Some performance indices (cost, security if quantifiable ...) should be maximized (or minimized) during operation; this leads to the definition of a functional F on a given period :
(3) $F = F(x_f, k_f, q_f, c_f, t_f)$, with initial conditions :
(4) $x(t_o)$
(5) $c(t_o)$
and final conditions at t_f.
The manipulated control vector $k(t)$ should be chosen to (min) maximise (3), taking into account operating constraints.
(6) $h(x,k,d,c,t) \geq 0$.
This description covers fairly well basic system control functions as :

* load frequency control (LFC)
* automatic generation control (AGC)
* economic dispatch
* security analysis
* generation scheduling.

Practically speaking the dimension of the system of equations (1) to (6) does not allow to find a practical solution at one control centre level, for all these problems together. Several practical approaches do exist (R. Mäntynen, 1980 and Delforge, 1978). If we consider the normal state of the system, with some loads and interconnection transits small variations, we may consider that equations (1) to (6) are verified and that (2) describes the dynamical phenomena due to small perturbations of random nature affecting the system. In this case (3) should be a purely economic index, economic dispatch being the normal function to be considered. Long term problems (hydro scheduling, fuel managements) should be considered as an extrapolation of these techniques with the simplified assumption of (2) $\dot{x} = 0$, the system being considered as running in steady state. In this case the supervisory control functions of the dispatching could include security aspects too, which mathematically should be described by violations of the inequalities (6) (overloads,...). The control strategy should therefore avoid to reach this emergency state. Load flow calculations and state estimation from static state variables should then be considered (see fig.8). Economic dispatch including security constraints is also a way to cope with this problem.

Preventive control may next be defined as the set of all prophylactic functions used to keep the system in normal state with regard to a given set of credible contingencies. This problem is more complex than the previous one because it implies not only the analysis of the system at time t, with given topology c, generation and load pattern, but also analysis of possible configurations (line, transformer, plant outages). Simulation tools are normally used in practice for these purposes.

The next level of difficulties is reached when we face some dynamical emergency state (network oscillations, dynamic behaviour of interconnected systems) in which case normally no centralized techniques is really helpfull today. The security of the system is then only assured when decentralized procedures (protections, reclosures,...) are functioning properly, or in other words when the outage circumstances correspond well to the protection scheme design.

Finally, if a black-out occurs, we enter the restorative state. Control centers might in this case be helpfull through their SCADA, remote control functions and sometimes simulation tools. Automatic procedures in emergency and restorative states at centralized level is still open to further theoretical and practical development.

II.3. Centralization versus decentralization.

This problem is linked to utilities organisation and general management (Kowal and others, 1980), ergonomical problems, (quantity of information to be treated by a dispatcher in a control center), short reaction times needed in case of failure (limited speed of available and reliable communication) and technological developments (Delcominette, 1980 and Gubina, 1980). These problems will be described in more detail below. Several enquiries (Kowal and others, 1980) have shown that these problems are of utmost importance on quality and cost of delivered power, but that no general consensus exists. Non technical factors have great influence on the design of a control center hierarchy, number of control center levels, type of functions to be assigned at each level, automatic actions at substations or plant level; let us list among others :

* existing organization
* experience of operating personnel
* financial resources
* operational difficulties (stress during emergency)
* increase of interactions (power exchange)
* influence of security on economicity.

The influence of control centers organization and decentralized automatic devices can be assessed when dealing with security of equipment and delivered power; Kaiser and Glavitsch (1980) give an approach to the way of quantifying such influences.

Paragraph II.2. has shown which kind of functions could and should be centralized. Typically, long term functions (economic dispatch, scheduling) or functions needing the overall system aspect (topology, security assessment) should be centralized. Dynamic emergency situations and restorative functions will be handled by decentralized devices in substations and plants. Nevertheless, the trend is to have coordinated actions on decentralized procedures, changing emergency or restorative routines from centralized extended real-time simulation and result analysis. However, during emergency periods action remain decentralized, due to speed requirements (spinning reserve, load shedding ...).

III. EXAMPLE OF CONTROL OF INTERCONNECTIONS AND SUBNETWORK

III.1. Defining power (sub-) system equivalents

III.1.1. Use and types of equivalents
A power system under responsibility of one organization (state company, power pool,...) is normally operated in parallel with others, through interconnections. These interconnections are of utmost importance in the expansion of the system by their possibilities to limit investments (increase of security, sharing reserves among several utilities). At the same time interconnections allow some better control of frequency and hopefully of voltages. Let us pinpoint the fact that these interconnections might have in some cases dangerous effects by transmitting perturbations (oscillations, risk of general black-outs). It is therefore clear that increased burden is put upon control, supervision and automation (protection ...). A special feature of the overall control organization is that in this case there is no top coordinating body (super-pool or super-national control center), and that control of interconnected networks is done by dispatching at the same level (see Calvaer and others, 1978). Load frequency control is, nowadays,one of the standard ways to increase quality (or frequency) and maintain general security of the interconnections (see Quazza, 1977). Interesting work in this field has been done by using multivariable control theory. Another interesting point is the analysis of security of the interconnected systems using state estimation, on-line load-flows and load prediction. In this case the need of having a "good" description of the external world (equivalent network, equivalent injections) to analyse its own security is important. Nowadays, control centers in operation are using "fixed" sets of equivalent networks, updated on request by manual interventions. Horizontal data transmission between control centers are used to send some information on "external" networks, e.g. information of status of breakers. Other information as the influence on neighbouring systems of "internal outages"(trip of units...) can be transmitted too. Figure 9 describes schematically the decomposition of the overall European network into several national subnetworks denoted by A,B,C, ...(In this fig. only the telecommunication system among control centers has been represented; actually, many other connections established for various pruposes may also, and do, exist).

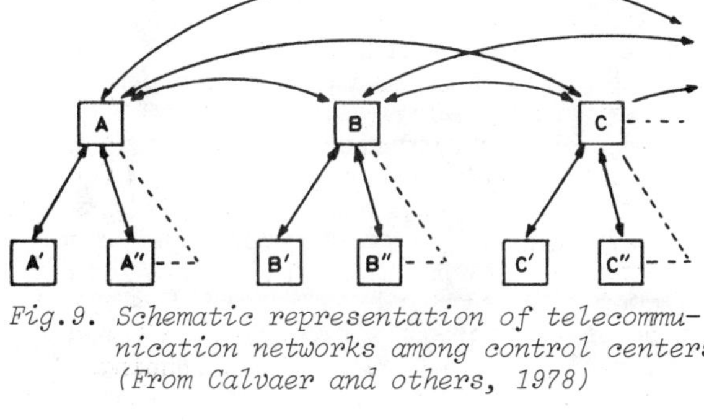

Fig.9. Schematic representation of telecommunication networks among control centers (From Calvaer and others, 1978)

Within the above context, three types of situations may be considered :
(i) evaluating the influence of networks B, C,D,... on the behaviour of network A when a disturbance appears in A
(ii) evaluating the constraints imposed in B, C,D,... when a disturbance occurs in A
(iii) evaluating the consequences in A of disturbance occuring in one of the neighbouring networks B,C,D ...

Following extensive research many approaches are available for obtaining network equivalents in both an off-line and on-line context. Roughly speaking, two quite different types of approaches may be distinguished :
The one is based on idenfication algorithms that do not require data of the external system : they are based on on-line measurements of fluctuations appearing at the boundaries of a control area towards the "unobservable external network". These methods

- can only provide information on past or present situations and are therefore unable to furnish assessment on future new situations;
- don't allow evaluating the influence of disturbances located at the external networks (influence of B,C ... on A) and vice-versa (influence of A on B,C ...)
- presuppose a weak cooperation among centers.

The other approach, already mentioned, advocates the exchange among control centers of their respective equivalents, each established following the hierarchical order. For instance A will construct its own equivalent and will transmit it to B,C,..., from which it will receive their own equivalents.

Whatever the method may be, two kinds of equivalents may essentially be defined : static and dynamic ones. In what follows we shall deal with static equivalents only, i.e. those used when studying small variations of the system motion. Among static equivalents, two different types may be distinguished, depending on whether they aim at the real (active) or at the reactive power fluctuations. If the active power equivalents have been used for quite a long time, defining reactive power equivalents is relatively recent. In what follows we describe the solution proposed and tested by Calvaer (1977, 1978), and by Calvaer and Boulanger (1980).

III.1.2. Static equivalents accounting for the reactive power diffusion.
The problem examined here is the construction of reactive power equivalents to enable evaluating on-line the final state reached by a system subject to a disturbance; let us recall that the standard procedure for solving this latter problem consists in integrating numerically (step-by-step) the non-linear differential equations describing the system's motion during the disturbed period (long-term stability program, including regulators' and governors' action); of course, this latter is an on-line procedure. To construct such an equivalent, the overall system is first decomposed

into two parts (see fig. 10) :

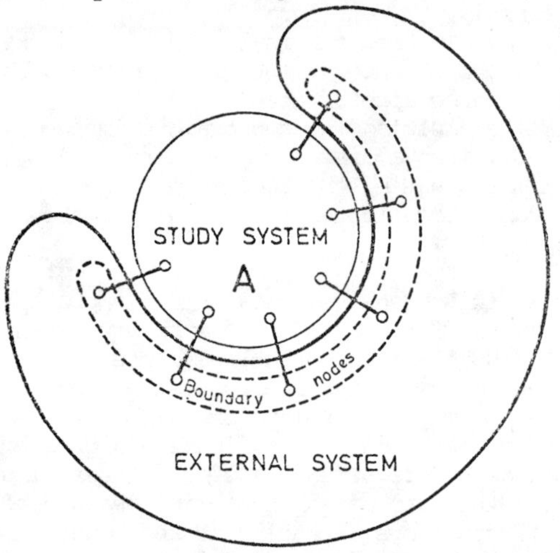

Fig.10.(From Calvaer and Boulanger, 1980)

• the study system (denoted by A in fig.10), in which the disturbances are supposed to take place, and whose behaviour is to be examined in detail

• the external system, represented by its boundary nodes linked to the study system's nodes by tielines (physically materialized by lines, transformers or cables).

When a disturbance (load shedding, generation shedding line tripping or reclosing) occurs in A or on the tielines, the external system contributes to assist A by

* delivering active power by the effect of the machines primary control

* supporting voltages on the tielines by a proper reactive power injection.

The problem lies in establishing equivalents able to represent the

• generators belonging to A

• external system as viewed from the boundary nodes of A in their post-fault configuration.

The solution to this problem as proposed by Calvaer (1977, 1978) consists in replacing :

• each generator belonging to A by two generators in parallel : one of the P,Q type accounting for the active power changes following a disturbance and maintening the same reactive power as before, the other by a P,V type accounting for the reactive power changes

• the external network by a Norton's equivalent, i.e. by a passified network represented by its "sensitivity matrix $\underline{S}$" and by equivalent current injectors.

Fig.11 shows the generator's representation; note that accounting for the reactive power diffusion is based on the notion and computation of equivalent reactances X_{mQ}, X_{mP}.

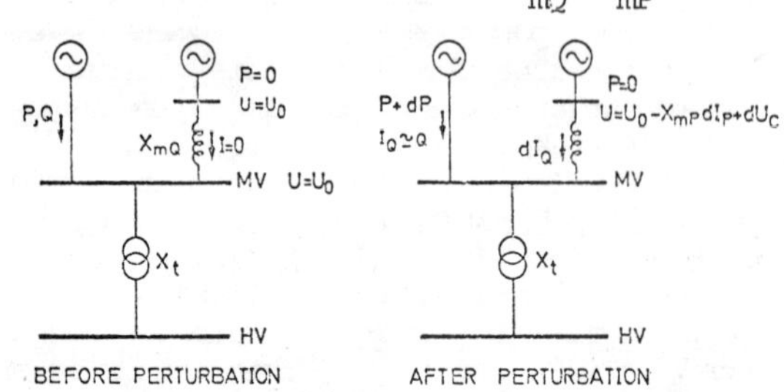

Fig.11.(From Calvaer, 1977)

(For the analytical expression of these latter see Calvaer 1977). Fig.12 represents this external equivalent. This figure calls for some important remarks (Calvaer and Boulanger, 1980) among them let us quote the following :

(i) The computation of the equivalent injectors necessitates that of the DC external equivalent; taking them into account is therefore a trade-off between accuracy and computing simplification.

(ii) Matrix $\underline{S}_{11}$ is poorly affected by those modifications in the external system which are not close to the boundary nodes. This in turn leads to an interesting simplification in the case of low angular differences of voltages between machines buses and boundary nodes, and of high ratios X/R; indeed, it can be shown that in this case $\underline{S}_{11}$ reduces to the susceptance matrix reduced to the external nodes.

Whatever the simplifications of the external representation may be, usage of both internal generators' and external network's equivalents allow obtaining by a simple load flow calculation the final system's state following a disturbance. The advantages with respect to the standard step-by-step procedure are quite obvious.

The above equivalents have been extensively tested on a 15-machine system by means of quite severe disturbances (such as the loss of its biggest machine (500 MVA)), quite heavily loaded and located quite close to a boundary node, (Calvaer and Boulanger, 1980). The results thus obtained have been compared with those provided by the overall system's representation. They have been found in quite good agreement in all cases, especially when equivalent injectors are used; the agreement is almost perfect when using the equivalents after LFC actuation. These tests show with evidence the validity and the interest that the proposed equivalents present for on-line security assessment.

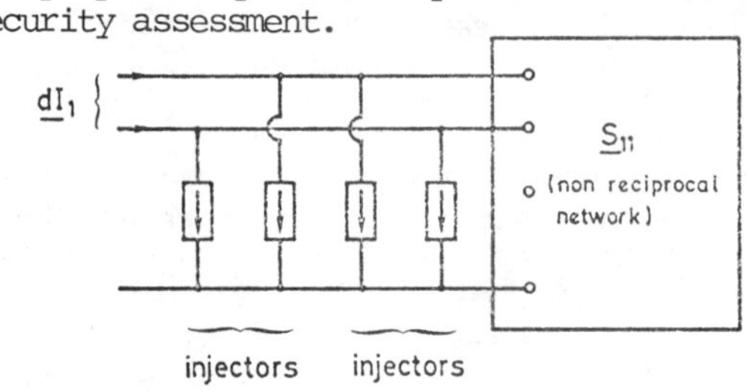

Fig.12. (From Calvaer and Boulanger, 1980)

III.2. Hierarchical state estimation.

Another type of dichotomy might exist inside a utility due to geographical size and functional differences (transmission, subtransmission). For obvious technical, economical and ergonomical reasons all scheduling, supervision and direct control functions may not be centralized at one level (Delfore, 1980 and Kowal, 1980); therefore there exists normally a hierarchy of control centers, each level having different responsibilities, and each lower level having only supervision of a limited area or sub-network. Control centers of the same level have then to be coordinated

together due to common objectives, as scheduling of plants, but also for general network security. In this case, these control centers have the same problems as described above, but coordination is of a different nature, due to higher level control center actions. A basic operation in such a hierarchy of control center, related to degree of centralization (see § II.3), is the mapping of function at each level and the required mathematical tools needed for such a kind of operation.
As an example of this hierarchical approach, let us describe here how this concept is applicable to state estimation (El-Fattah and Ribbens-Pavella, 1976, Van Cutsem and others, 1980a, 1980b).

III.2.1. Introduction.
State estimation is necessary to furnish a reliable data base for the control and monitoring of electric power systems. Estimating the state $\underline{x}$ of a system from redundant measurements $\underline{z}$ affected by errors $\underline{v}$ is a large-scale problem in on-line computer control and information processing. If the state estimator can yield a consistent, reliable and sufficiently accurate on-line interpretation of network state within a reasonably short delay, security monitoring and economic dispatch can be consedirably improved.

The essential qualities required of a state estimation algorithm appear therefore to be reliability, speed, and low computer memory capacity. Besides, detection of bad data, due to faulty meter or telecommunication equipment should be part of the tasks of a state estimation (SE) to insure estimate reliability and to aid diagnostics. Likewise, the estimator should include in some cases the identification of network configuration and neighbouring system "equivalents".

Let n be the number of state variables and m the number of measurements of the observed variables. Using the following notation :

- $\underline{x}$ the n-dimensional state vector; if N is the total number of buses, then n=2N-1 (N voltage magnitudes, N-1 phase angles)
- $\underline{z}$ the m-dimensional measurement vector comprising power flows (active and/or reactive), voltage magnitudes and, in some particular cases, current flows
- $\underline{h}(\underline{x})$ the m-dimensional non linear vector function which expresses the relationship between the state variables and the model values of the measurements
- $\underline{v}$ the m-dimensional error measurement vector due to several partial inaccuracies arising in the course of data acquisition and transmission.

The process is described by :

$$\underline{z} = \underline{h}(\underline{x}) + \underline{v}$$

The estimation problem consists in finding the values of $\underline{x}$ which best fit measurements $\underline{z}$, i.e. yield smallest reziduals $[\underline{z} - \underline{h}(\underline{x})]$; in other words, the best estimate $\hat{\underline{x}}$ of $\underline{x}$ may be defined as the value of $\underline{x}$ minimizing weighted square residuals

$$J = [\underline{z}-\underline{h}(\underline{x})]^T \underline{R}^{-1} [\underline{z}-\underline{h}(\underline{x})]$$

where $\underline{R}$ is the error covariance matrix.

$$\underline{R} = \underline{E}(\underline{v}\underline{v}^T)$$

and where $\underline{E}$ stands for "expectation". When v is Gaussian, $\hat{\underline{x}}$ is the maximum likelihoad estimate.
Setting the derivatives of J with respect to $\underline{x}$ at zero yields

$$-\underline{H}^T(\hat{\underline{x}})\underline{R}^{-1}[\underline{z}-\underline{h}(\hat{\underline{x}})]=0$$

where

$$\underline{H} \triangleq \left(\frac{\partial h_i}{\partial x_j}\right) = \frac{\partial \underline{h}}{\partial \underline{x}}$$

is the (m x n)-dimensional Jacobian matrix.

The above considerations constitute the fundamentals of the well-known weighted least squares (WLS) method.

Many iterative procedures may be used in order to evaluate an estimate $\hat{\underline{x}}$. On the other hand, to meet the essential requirements of a SE many refinements have been proposed. Besides SE techniques themselves they also concern numerical methods, programming techniques and hardware developments. Particular attention has also been given to the data acquisition system.
All these improvements have contributed to the implementation of SE algorithms for medium-sized systems (see Dy Liacco, 1979 and Merlin, 1980). However, large-scale system SE calls for a more appropriate procedure than the "integrated" concept. Thus a means of decomposing a large-scale system into subsystems and of using a hierarchical concept has been sought. In what follows, we outline the fundamentals of this hierarchical SE (HSE) approach as was proposed at first by El-Fattah and Ribbens-Pavella (1976) and then developed and improved by Van Cutsem and others (1980a, 1980b).

III.2.2. Network decompostion.
Let the overall network be decomposed into k areas $A_1, A_2, \ldots, A_k$. In principle, these areas are non-overlapping; hence, each bus of the network belongs to one and only one area; moreover, the areas are interconnected by tielines (physically materialized by lines or transformers). The set of all these tielines constitutes an additional, (k+1) (pseudo)area, the interconnection area. Fig.13 illustrates this decomposition in the particular case k=3.

Let in the i-th area define two kinds of nodes :

- internal nodes : they all belong to A_i and are connected by tielines to other nodes belonging to A_i exclusively
- boundary nodes : every boundary node of A_i is connected, at least by a link, with a node of A_j (j≠i).

Related to the above two kinds of nodes, the following vectors will be used in the sequel :

$\underline{x}_r^i$ the state vector of the internal nodes of A_i

$\underline{x}_c^i$ the state vector (voltage magnitudes and phase angles) of the boundary nodes of A_i

$\underline{x}_i=(\underline{x}_r^{iT},\underline{x}_c^{iT})^T$ the overall state vector of A_i

$\underline{x}_c=(\underline{x}_c^{1T},\underline{x}_c^{2T},\ldots,\underline{x}_c^{kT})^T$ the state vector of the (k+1)-th area.

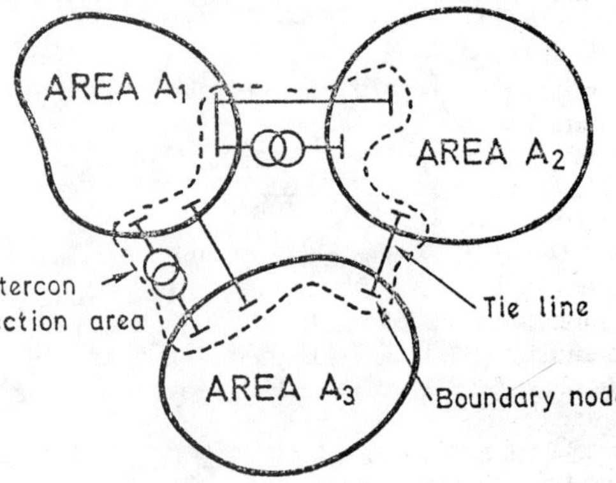

Fig.13. Principle of system decomposition (From Van Cutsem and others, 1980)

It is to be noted that the above decomposition rule is by no means restrictive. Fig. 14 illustrates two quite different typical situations. The first one represents the decompostion of the overall system into k-1 areas of relatively low-voltage level whereas the k-th area consists of the high-voltage level interconnection network. In this case, the main part of the links between areas are transformers; it is a functional decomposition. The second situation concerns the decomposition of a system into subsystems, each comprising various voltage levels. This is the case of utilities pools and constitutes a geographical decomposition. Note also that the basic assumption imposing to a node to belong to only one area is not restrictive either : an adequate usage of isthmus could replace a node belonging to two distinct areas by two nodes, each belonging to one area.

From the foregoing, it can be easily seen that the decomposition of a network is quite arbitrary, and may therefore obey various criteria such as geographical, political, functional, described in the previous paragraph.

III.2.3. Principle of the HSE procedure.
On the basis of the above decomposition scheme a two-level estimation procedure is carried out : at the lower (first) hierarchical level a standard SE is at first performed, simultaneously for all the k areas and independently from each other; the choice of the SE algorithm is free for each area and may differ from one another; it may be dictated by various requirements and depend on several factors, such as the topology, the measurement devices and configuration, etc. The measurement vector $\underline{z}_i$ of A_i ($\forall$ i=1,2,...,k) may therefore be chosen accordingly. The result of these local SE's is the estimated states

$$\hat{\underline{x}}_i = (\hat{\underline{x}}_r^{iT},\ \hat{\underline{x}}_c^{iT}) \quad i=1,2,\ldots,k$$

Part of this result, namely the subvector $\hat{\underline{x}}_c^i$ is then sent to the second (upper) hierarchical level for coordination purposes. Indeed each local SE has been performed with its own reference phase; at this state of the procedure, there is no information concerning the angular differences between the slack buses of the different areas. The role of the second level is precisely the coordination of the $\underline{u}$ vector,

$$\underline{u} = [u_2 \ldots u_k]^T$$

whose components represent the phase angular difference between the reference node of each area and the absolute reference node (arbitrarily taken here as that of area 1, see fig. 15).

To realize this coordination, vector $\underline{u}$ is to be estimated, based on the measurement vector $\underline{z}_c$, which consists of the tieline power flows. As a matter of fact, due to the decomposition and thus to the definition of a slack bus for each area, the tieline power flows cannot be computed on the basis of the $\underline{x}_c^i$ vectors only. Fig.16 illustrates the overall procedure. As can be seen, the second level algorithm is in fact a two-step estimation : a minimization procedure is at first carried out with respect to the $\underline{u}$ variables, while the $\underline{x}_c$ variables are set up to the estimated values deduced from the local estimation $\hat{\underline{x}}_c^i$ ($\forall$i=1,2,..k);

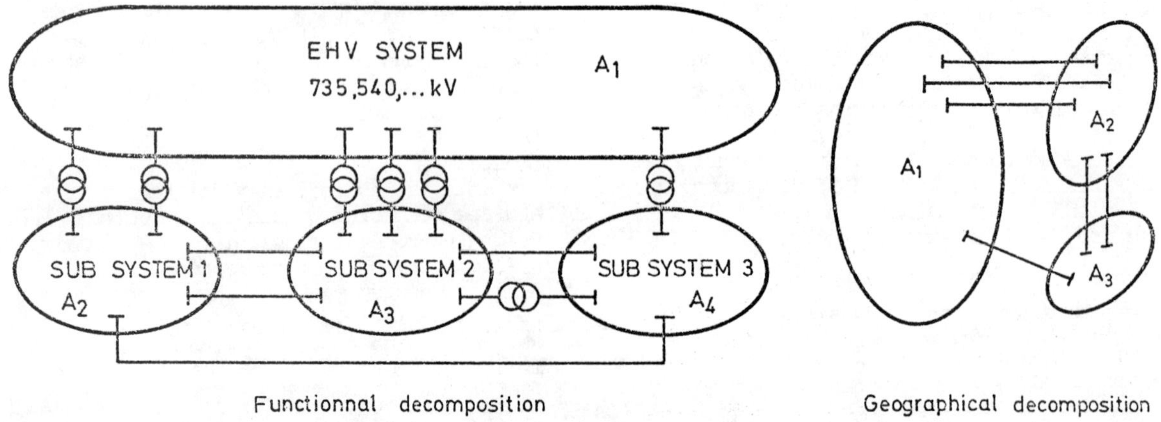

Fig.14. Two decomposition-aggregation systems (From Van Cutsem and others, 1980)

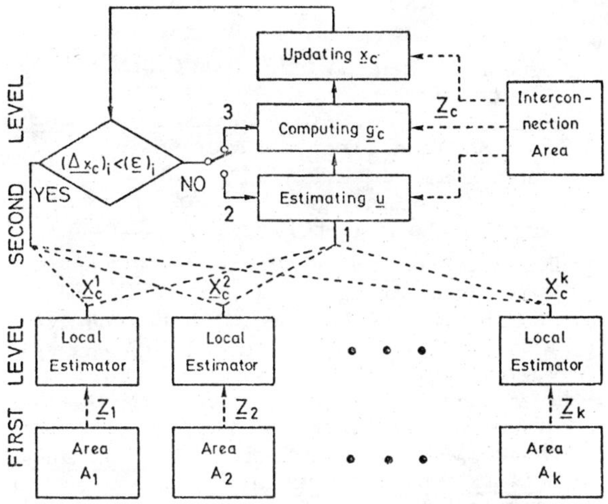

Fig.15. Scheme of two-level static state estimator.
(From Van Cutsem and others, 1980)

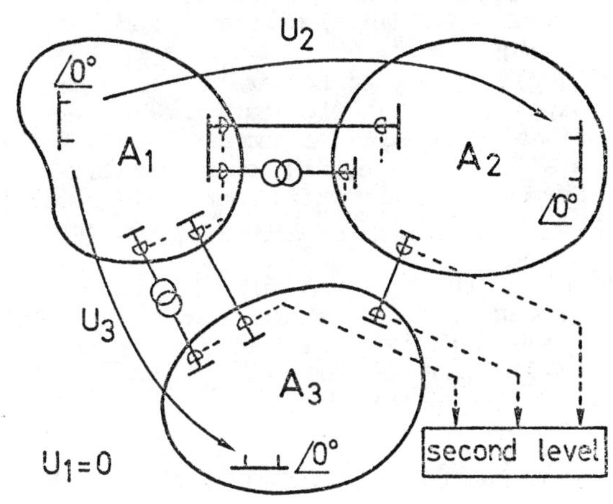

Fig.16. Representation of u and measurements of tielines; -Ð- tieline power flow measurements (P and/or Q)
(From Van Cutsem and others, 1980)

then a reevaluation of $\underline{x}_c$ is performed. The details of this procedure and the way to performing it may be found in Van Cutsem and others (1980b). Let us only note here that the second level process is described by

$$\underline{z}_c = \underline{h}_c(\underline{x}_c,\underline{u}) + \underline{v}_c$$

The quadratic criterion is

$$J_c = [\underline{z}_c-\underline{h}_c(\underline{x}_c,\underline{u})]^T\underline{R}^{-1}[\underline{z}_c-\underline{h}_c(\underline{x}_c,\underline{u})]$$

III.2.4. Comments.

A priori, numerous advantages can be expected from the above outlined approach. Some of them are general, inherent in a hierarchical procedure used in control of large-scale systems; others are specifically related to this approach. The following are the most important.

(i) By means of an appropriate decomposition the data processing cost can be much lower than in the case of the conventional integrated SE (ISE); also high measurement noise can be reduced or avoided.

(ii) At the local level, the smallness of the subsystems' size compared with that of the overall system allows reduction of local computer memory and computing time.

(iii) Because of (ii) and of the fact local SEs can be carried out simultaneously, the first-level estimation is quite fast; it has been shown, that the second-level estimation is also quite fast. Therefore, the HSE procedure allows an appreciable gain in computing time with respect to the ISE procedure.

(iv) The reliability of the HSE is higher than that of the ISE; generally the failure of some measuring devices will not corrupt the accuracy of the state estimation of the whole system but only that of the subsystem to which the devices belong.

(v) When developing a new subarea in the same power system or new interconnections with neighbours, the estimation algorithm is easily embedded without restructuring the whole estimation procedure as in the case of integrated estimation. Similarly, the interruption of the estimation process of a subsystem (which may arise for some reason) does not necessarily involve that of the remaining system.

(vi) From (v), the HSE procedure allows the subsystems to remain quite independent of each other. This in turn leads to another interesting (subsystem) level than at the upper level.

Some of the above advantages have been illustrated through the numerical example treated by Van Cutsem and others (1980) where a 34-bus system has been decomposed into four areas, even if the size of this system was quite small and therefore inappropriate to illustrate fully the advantages of the HSE; among others, an appreciable gain in computing time has surprisingly been observed, despite the small size of the system. The same 34-bus system has also been used by Van Cutsem and others (1980b) to illustrate the suitability and advantages of this HSE for anomalous data analysis. The treatment of a larger system (the Belgian 380-220-150 kV system) is at present under study in order to test the established algorithms and to introduce appropriate improvements.

IV. CONCLUSIONS.

By way of conclusions we would like to point out some peculiarities linked on one hand to the very nature of electric power sysems, on the other hand to economical context of our century.

"An electric power system is the largest physically interconnected system man has invented" (Schweppe, 1978). The only way to control such a system is to decompose it into subsystems and to use hierarchical control strategies.

Research on behaviour of power plants, loads etc, will make it possible to construct mathematical models that approximate actual behaviour; future control systems will use these models in real-time operation.

Computing facilities, including mini- and microcomputer systems are decreasing in cost; therefore, they will be more and more extensively used for power system monitoring and management.

Fuel cost will continue rising; this will justify installing costumer (solar, winding..) generation at the lowest voltage levels, and at the same time nuclear generation at the highest voltage levels. As a consequence, generation plants at the intermediary levels will tend to disappear; this in turn will tend to modify the overall configuration and features of electrical power system transmission-distribution.

Despite the above-predicted changes in the power system picture, the general philosophy described in this paper will still remain valid, provided the overall control scheme is adapted to account for these modifications.

V. REFERENCES.

Calvaer, A.J.(1977). Diffusion of a Reactive Power Perturbation and Some Related Problems. IFAC Symposium on Automatic Control and Protection of Electric Power Systems. Melbourne, Feb. 21-25

Calvaer, A.J.(1978). Echanges de Schémas Equivalent en Temps Réel entre Centres de Contrôle des Réseaux Interconnectés. Cigre 32-04. Aug.30 - Sept.7.

Calvaer, A.J. and P.L. Boulanger (1980). Application of External Equivalents in the Case of Machine Outages in the Study Systems. IFAC Symposium on Automatic Control in Power Generation Distribution and Protection. Sept. 15-19, Pretoria

Delcominette, Delince, Somers and Van Miergroet.(1980). Conduite à distance, conduite locale et automatisme dans les postes : évolution et nouvelles perspectives en Belgique. Cigre 32-09.

Delforge, J.M.,F. Denis, J. Ernould and J.P. Waha.(1978). Organisation de la Conduite des Réseaux en Belgique. Cigre 32-05. Aug.30 - Sept.7.

Dy Liacco, T.E.(1980). Mathematical Problems in Electric Power System Operation. International Conference on Electric Power Problems : The Mathematical Challenge. Seattle, Washington. March 18-20.

Dy Liacco, T.E.(1979). The State of the Art in Control Center Design. Tutorial paper. IFAC Symposium on Computer Applications in Large-Scale Power Systems.New Delhi. Aug.

El-Fattah,Y.M. and M. Ribbens-Pavella.(1976). Multi-Level Approach to State Estimation in Electric Power Systems. Part I. Identification and System Parameter Estimation. IV IFAC Symposium. Tbilisi, USSR. Sept. 21-27, pp. 166-180.

Glavitsch,H.(1977). Power System Control and Protection. The Interaction of New and Existing Concepts in Providing Economic and Reliable Supply. IFAC Symposium on Automatic Control in Power Generation Distribution and Protection.Melbourne. Feb. 21-25.

Gubina,F., A. Ogorelec, M. Plaper, J. Hrovatin.(1980). A concept of Electrical Power Emergency Control by means of Multipurpose Automation. Cigre 32-08

Kaiser, W. and H. Glavitsch.(1980). Effectiveness Evaluation of System Control in Electric Power System by Discrete Modeling and Simulation. Third international symposium. Interlaken, Switzerland. June 25-27

Kowal,J., E.Mariani, L. Petterson, D. Rumpel and J.P. Waha.(1980). Control Center Design. A report of Study Committee 32. Working Group 02, pp. 1-28.

Mäntynen,R. and J. Välsänen.(1980). Development of Power System Control in Imatran Voyma. Cigre 32-06

Merlin,A. (1980). State of the Art on the Use of Computer Aids for Decision making in the Unipède Control Center. Unipède large network committee; experts on load control centers.5/80, 18p.

Quazza, G.(1977). Large Scale Control Problem in Electric Power System. Automatica. Vol.13, pp. 579-593.

Van Cutsem,Th., J.L. Horward, M. Ribbens-Pavella.(1980a).Electrical Power and Energy Systems. Vol. 2,n°2, April. pp.70-8(

Van Cutsem,Th., J.L. Horward and M. Ribbens-Pavella.(1980b). Application of a Two-Level State Estimation for Anomalous Data Analysis in Power Electric Power Systems. IFAC Symposium on Automatic Control in Power Generation, Distribution and Protection. Pretoria. Sept. 15-19.

Waha,J.P. (1971). Survey of Application in Power Plants and Power Distribution Systems. Multivariable Technical Control Systems. Proceedings of the 2nd IFAC Symposium. Düsseldorf, Oct. 11-13, pp. 165-202.

ARTIFICIAL INTELLIGENCE, AUTOMATIC CONTROL AND DEVELOPMENT

T. Vámos

Computer and Automation Institute, Hungarian Academy of Sciences, 1111 Budapest, XI., Kende u. 13/17, Hungary

Abstract. A survey is given on new impacts of artificial intelligence on automatic control. Artificial intelligence offers its tools to automatic control at two decision points: input and control. Visual, voice and information pattern as inputs to control systems are treated, especially for quality control, medical applications. A glimpse on expert systems illuminates the possibilities of automated dispatcher and other decision making. New algorithms solving very complex problems offer several possibilities. A short report on the intelligent robot program of the author's group is given and as a conclusion some considerations on artificial intelligence, control and development.

Keywords. Artificial intelligence; assembling; automatic testing; biomedical; control engineering computer applications; human factors; man-machine systems; pattern recognition; robots.

INTRODUCTION

Views on AI are very different and sometimes extremely biased. In my paper I will present just another personal standpoint, as I did nearly two and a half years ago at our last World Congress in Helsinki /1978/. I think it would be a mistake if only a somehow updated version of that paper would be delivered here. Concerning the basic ideas of that earlier report, I refer only to the Proceedings of the Congress; my task here comprises the following objectives:

- to give an overview of those directions which affected our field in the last few years /an addition to the Helsinki report/;
- to present the results of my own group;
- to attempt a look at the AI-AC problem from the aspect of development /which is the main concern of this conference/.

Starting with the first subject I have to deal a little with the problem of relations between AI and AC. AC is not a closed subject itself, it comprises several areas including classical dynamic theory /esp. linear dynamics/, problems of system synthesis and analysis from the point of view of information-communication and optimal control governed by diversified criteria. In this respect, AI offers its tools to AC at two decisive points: input - as understood in AC language which is problem understanding in terms of AI - and control which is again a real AC terminology, but in computer science this is the processing phase and in AI problem solving. All these look somehow as a play of words - but it intends to give an impression of the parallelity and divergence of approaches.

VISUAL INPUT FOR AUTOMATIC CONTROL

The input problem will be treated in three categories: visual, voice and information-pattern.

Greatest efforts and most significant results can be reported in the visual understanding field. Progress is simplificant in control of robot-manipulator mechanisms, in the inspection quality control of manufacturing processes, the mass-screening for health services. Though many other applications could be enumerated, emphasis is placed upon these three. Concerning visually controlled robots, I shall give you some details later, because this is the topic of my group's activity.

Quality control for manufacturing has taken a main step ahead towards real and important applications: the automated visual control of continuously flowing products, e.g. textile materials /Conners, 1979; Kasdan and Mead, 1978/, metal sheets /Brook and co-workers, 1977; Porter and co-workers, 1979; Saridis and Brandin, 1979/ is realized mostly by special scanning devices, using laser beam, to achieve a relatively noisless, uniform, focused and intensive illumination. The reflected beam - sometimes correlated with the original one - carries the information about the texture of the flowing surface. Many efficient and fast texture analysis algorithms have been developed with different objectives:

- the appropriate texture is a regularly alternating pattern and the fault is some irregularity within that;
- the texture may contain irregular patterns /or this irregularity is even the purpose/ but the pattern elements should be under a certain dimension-limit or the lack of patterns should be eliminated;
- the surface should be uniform /clear, matt, opaque, transparent, gloomy etc./ and irregularities like dirt, crack, inclusion, hole should be avoided.

Other well accomplished methods are available for the inspection of shape and dimensions of mass products, e.g. relays /Kuni and co-workers, 1977/, printed circuits /Sterling, 1979; Jarvis, 1980/, other electrical and precision-engineering components /Hsieh and Fu, 1979; Klein and Breeding, 1979; Batchelor, 1978/, pills etc. These methods use less sophistication than any other task; transparent tables, precise fixtures, preset focused light beams help the mostly simple go/no go test decisions.

The most promising human-oriented application is the automation of mass-screening. This includes X-ray pictures of the lung, other X-ray picture-analysis methods /not mentioning computer-tomography/, blood-cell checking, evaluation of various medical laboratory tests. The understanding of two-dimensional photographic images specified by noisy contours, was one of the first targets in visual pattern recognition, mostly applied for military and related purposes /satellite-photos etc./ - these results could be relatively easily transferred from the most antihuman to the most human applications. The basic problems here are also varied, but mostly not only by the goal itself, but the preference and antecedents of the authors: cells of various forms are identified by brute-force geometric characteristics and structural-geometrical methods, X-ray pictures by regions growing around certain well observable contrast-spots or by thinning the blobs as long as some clear structure can be identified /veins, bones etc./.

Although many papers are published on general problems of visual understanding, vision-psychology which sometimes claims to have found general features of vision, all realistic works relate to certain, very specified applications; these are the successful projects. The reason is clear: human vision is one of the most complex brain activities, for a long time to come only those engineering undertakings are economically and technically feasible - and are proven feasible - which restrict their validity to a well defined purpose. This idea comes up once more later as a more general lesson, when we tackle the expert systems problem.

VOICE INPUT

Considering voice recognition, not too much can be added to the situation as it looked three years ago. The practical limits of about 100 or not too much more words have not changed much. One project tried to shake the word, the Hearsay which claimed to understand about 1000 words but after a period the announcements died out - as we know now, this system tried to find the most separated words instead of separation in a realistic vocabulary. Telephone companies are working hard in voice reply systems. The challenge lies, on one side, in the wonderfully solved voice-synthesis. One of the recent publications of Bell Lab reports on an inquiry system where the names should be spelled separately /Levinson and Shipley, 1980/. A very professional project /De Mori and co-workers, 1979/ leads to the trivial conclusion: a speech understanding which strives to recognize any complex spoken information must apply contextual knowledge /as people do/, knowledge about the syntax and semantics. Comparing with the rapid progress of the reliable and controllable visual devices /e.g. viewdata services/, my earlier conclusion is unchanged, even more justified: a really powerful voice recognition is still far ahead /e.g. an automatic type-writer/, only very special applications seem feasible within a reasonable time: those where a few fixed voice-commands are used and are not practical to replace by any other means /Carnegie Univ., 1976/.

EXPERT SYSTEMS

The most significant progress of the past years in AI has been achieved in the development of expert systems. The idea is the result of a clear, intellectual process, from the very early attempts of general problem solving /Newell and co-workers, 1959/ through the recognition of the problem's complexity to the beauties and compromises of solving realistic tasks and finding practical solution methods. According to our introductory classification, these methods belong more to the control section of the process, although as a whole ideology they cover a significant part of input-processing as well. As mentioned just before, the first gaze at the much dreamed of possibilities of machine problem solving gave an illusion of a very general and powerful tool: the application of logic, combinatorics, graph theory, mathematical linguistics seemed to promise a master-key for all well formulated problems. Unfortunately, most real-life problems are not well formulated, the branchings within the problem solution controlling the intermediate search and decision making steps are even less defined, the complexity is so high that it goes far beyond the limits of the most powerful computing devices. The seemingly perfectly formulated chess-problem was an excellent paradigm. After about 20 years' efforts, the best programs reached the expert-grade, much below the master's level! Another early task, translation of a text from one language to another encountered even more stubborn walls. I gave here a short report on the visual and audio problem too. The next stage was a logical consequence: problems should be limited to such an extent which can be treated and within this limitation the solution should be supported by all available useful knowledge which can be gathered before. This is a rather popularized and cruel formulation of Minsky's frame-concept /1975/, although I hope this gets to the heart of the matter. A frame covers a certain closed world, as it was demonstrated by the celebrated block's world of Winograd /1971/, i.e. a closed set of objects, attributes of these objects, relations between them and actions exercised in this world. All specific frames of activity contain such closed sets. This idea has given an impetus to many experiments on special problem-oriented problem solving: chemical engineering synthesis methods /Buchanan and co-workers, 1969/, pharmaceutical effects /Shortliffe and co-workers, 1973; Warren and Pereira, 1977/, diagnosis of complex diseases /Shortliffe,1976/ /e.g. glaucoma /Weiss and co-workers, 1978/, research methods in mineralogy are only samples of this family.
The extension of the frame-concept is of course rich on further problem solving ideas and methods. An expert-system which is an expert in a certain frame should apply the special reasoning methods usual among the specialists of the field. This reasoning method is mostly not formulized, even not always conscious, the expert of the expert system has to evaluate all these by analysing the special field's paradigm. Not the direct rules, but the usual heuristics are extracted this way - a good example is again chess, where not the simple game-rules but the strategies of a good player should be learnt.
Many earlier methods have been refined in connection with these works: highly sophisticated question-answering languages, new versions of computer-logic realized in problem solving languages based on predicate calculi and other means of logics, structural ideas for the realization of relations as conceptual networks, learning by examples, truth maintenance systems /Doyle, 1978/ etc. Some of these results look, at the first glance, like a very powerful machine problem solving breakthrough, at the second as a small variation of old ideas which can only solve the simplest grammar-school exercises - but the whole progress is steadily exploring new territories and opening new horizons for the human intellectual activity.

HARDWARE BACKGROUND

The hardware background is similarly growing, even faster. Most AI projects are based on some variations of the LISP language as a basic tool. List processing languages handle lists of information by very flexible relation-establishment /pointers, concatenation, partitioning etc./ and by virtue of their recursive nature they are ideal for logical processes. Although LISP is more than 15 years old, it could not find its place in practical applications because of the troublesome implementation on conventional machines. The new LISP machines /Bawden and co-workers, 1978/ - especially if they can be developed as multiconsole devices in the future - are similarly promising tools as several other attempts of building specialized non-conventional, parallel operating computing devices based on the new possibilities of VLSI techniques. Array-processors, special pattern recognition schemes are the first arrivals of this family /Reddy and Hon, 1979/.

COMPUTATIONAL COMPLEXITY

Maybe, this is the heart of the matter. Most problem solving tasks are not trivial because of their dimensions. Complexity adds a qualitative difference to the nature of the problems: a fairly big one cannot be treated as a simple multiplication or enlargement of a similar smaller one. During the last years much has been done in construction of new, fast algorithms, which extend our computational range by orders of magnitude. Real-time calculation of problem analysis tasks is not only a result of improved hardware but also of mathematical excellence! The Fast Fourier Transform is the best known member of this family but several algorithms could be mentioned, one of the last beats being a promising new approach to linear programming. Hardware /e.g. array processors/ and computational methods push several problems which were only theoretically tolerable earlier within the reach of desktop devices. The divergence of methods and applications demonstrates that AI is now a real and effective tool in many tasks of automation. Even office-automation is one growing application field as it marches ahead toward more and more responsible spheres of activity, incorporating management duties. This divergence is accompanied by a convergence too: distributed systems and data bases, high level abstraction devices, man-machine interfaces, automated or semi-automated reasoning are the basic features of a new human environment constructed by humans and, perhaps, for humans.

INTELLIGENT ROBOTS

Now let us have a short look at intelligent robots/Dodd and Rossol, eds., 1979/. The advent of robotics is not so much connected with wages or the general availability of labor force as it is generally thought. Most robots are applied at those working places where conditions for human being are not permissible: high temperature, radiation, poisonous atmosphere, or where the human service does not allow an economic exploitation of an expensive equipment, e.g. forges, presses etc. The lack of adaptivity requires a rigidly fixed application or expensive supplemental tooling. Pattern recognition can add human-like adaptivity to the robot performance. The activity of our group /Báthor and co-workers, 1979/ /as one of the numerous efforts all over the world/ is concentrated on this task: recognition of technical objects learned by the robot before executing the job. The recognition system is therefore a man-machine one, the operator can teach and supervise several equipments. The dialog is helped by a graphic display which is supported by a special real-time graphic computation system visualizing all intermediate steps of recognition, planning the operation of the robot. This graphics masters the model-building of the objects, rotates them and views them from different perspectives to provide all possible images as they can occur on a conveyor belt or assembly table. A fast hidden line algorithm produces not only these views but helps also in planning the robots' movement in avoiding collisions. A full animation procedure can be attained this way.

Recognition is based on matching these learned models and situations with the picture obtained by a TV camera and digitizer picture-processor. At this point - very similar to the human brain's solution - several methods are available and combined to reach the fastest way of recognition.

First the contours of the objects should be obtained,but this is also a variable process; it can be set extremely fast point-by-point, by choosing that level of the digitizing dissimination where the differences are the sharpest - this is somehow a similar procedure to setting the contrast of our TV at an extreme and unrealistic level, where the picture is not convincing enough but the contours are hardest. The usual tool of picture processing: grey-level histogram automates this step. If better noise filtering is needed, another device, a local operator named after Mérő and Vassy can be applied which selects contour elements by the joining of the most likely contour points in short linear sections. The entire contour is found by heuristic algorithms which first assume the branching points of the contours and then find an optimal and feasible path between them. We can follow here a method of AI:i.e. composing a relatively simple and fast approximating, heuristic algorithm in edifice which chooses the most likely resulting elements by checking now the compose, fits /or seems to fit/ to the ultimate goal. Several simple objects - especially if they stand alone and can lay or hang only in some discrete stable situation - can be identified now by very primitive methods: estimation of dimensions, shape characteristics, interior holes etc. Even in those cases, where these brute-force guesses do not lead to identification, they yield some information about the possible variants and this can shorten the later procedure.

A powerful device of AI is applied to this project too: that is the linguistic way of description. These picture grammars describe the objects somehow similarly to the natural language: they formulate sentences in which the nouns are the basic geometric components of the figure and adjectives are attributed expressing the relations of the picture primitives. Such a description is highly independent of dimensions and not very sensitive to the angle of view. Using this method which is an important /and sometimes final/ part of our system, too, the sentences describing the models and those obtained by the processing of the pictures, should be compared. If we have a larger number of possibilities, this comparison process /which is similar to computer language parsing/ can be accelerated by any previous knowledge obtained e.g. by the brute force guess. If a more complicated situation is given, i. e. the objects are complex, some parts hidden to the viewer and the noise due to reflections, shadows, lens- and camera imperfections provide several uncertain interpretations of the contours, we have to apply new additional efforts to obtain a reliable recognition within a reasonable time. In our new experiments we adopt illumination effects of diversely situated and controlled light sources. These illumination routines are controlled by the recognition program and provide a very clear distinction of the different surfaces and sufficient information on their orientation and their relative positions.

Another important feature is the texture of the surfaces. We have spoken about this in detail when discussing the inspection problem.

As it can be understood from this short and rough description, a bunch of methods is collected by the system and a suitable choice and combination is adopted to the particular goal. This is a really good paradigm of AI as a new, powerful possibility for AC. The 80's will call forth these intelligent robots from the various labs and, backed by the latest results of semiconductor and computer technology, they will become a technical reality whether we wish it or not.

CONCLUSION TO DEVELOPMENT

Last comes the most difficult part of my task set by myself: what can be said about all these from the point of view of development, regarding countries where labor force is abundant and the major goal is to create more working places where wages are relatively low if we try to calculate the economic justification of high level automation?

First of all I have to express here once again my personal feeling that it is abominably improper to go anywhere in the world and express opinions, give address from the outside and from some detached platform without a deep knowledge of the language, culture, history, way of life of the different layers of society, reasons of behaviour and especially without having lived long with and within that nation and without taking its destiny upon oneself with full responsibility. This is the most primitive behaviour of arrogant conquerers - antiintellectual, antihuman and disgusting.

It is not a common language in its deep sense what we would like to approach in IFAC - because it is impossible - but a common and correct interface between our people and between specialists of developing countries and of those having started earlier - not in history but only in automatic control.

Approaching from this humble and careful point of view I would like to mention only a few ideas for your consideration:

- several working conditions are really dangerous for human beings and this concerns not only physical but also mental health;
- if a country attempts to exist on the world market, highest level technology is inavoidable in most cases, this is required by the quality demands;
- the gap between requirements of the modern technical world and the educational level, working habits of a people not having many centuries of industrial experience, can be sometimes tightened by new devices of man-machine interfaces: a school of new working discipline;
- the utilization of expensive investments in two or three shifts is sometimes more a problem of getting labor force which is not used to work in shifts, than of acquiring labor force in general;
- in societies which are not highly developed from the technical point of view, a symbiosis of extremely different technical cultures is a general reality, necessity and even

a positive driving force;

- by application of AI methods we can get a more reliable information on individual human performance, behaviour towards new devices, technologies and in this way more individual and locally applicable working conditions can be shaped.

As I emphasized, these are only irresponsible ideas of an alien but we are here to discuss and exchange ideas - to provide a really human interface for mutual understanding and friendship.

REFERENCES

Batchelor, B. G. /1978/. A preliminary note on the automatic inspection of male screw threads. In Proc. of the 3rd Int. Conf. on Automated Inspection and Product Control. Cotswold Press, Oxford. pp. 139-156.

Bawden, A., R. Greenblatt, I. Holloway, T. Knight, D. Moon and D. Weinreb /1978/. The LISP machine. In P. H. Winston and R. H. Brown /Eds./, Progress in AI. MIT, Cambridge, Mass. Vol. II. pp. 345-374

Báthor, M. L. Mérő and T. Vámos /1979/. A knowledge-based interactive robot-vision system. In Proc. of the 6th Int. Joint Conf. on AI. Information Processing Society of Japan, Tokyo. Vol. 2, pp. 920-922.

Brook, R. A., J. F. Claridge and C. Watts /1977/. Automatic inspection of flat strip in the steel industry. In Proc. of the 7th IMEKO Congress. Akadémiai Kiadó, Budapest, Hungary. Vol. III. pp. 379-387.

Buchanan, B. G., G. Sutherland and E. Feigenbaum /1969/. A program for generating explanatory hypotheses in organic chemistry. In B.Meltzer and D. Michie /Eds./, Machine Intelligence, Vol. 4. American Elsevier, N.Y. pp. 209-254.

Carnegie-Mellon Univ. /1976/. Speech Understanding Systems: Summary of Results of The Five-Year Research Effort. Dept. of Computer Science, Pittsburgh, Penna.

Conners, R. W. /1979/. Towards a set of statistical features which measure visually perceivable qualities of textures. In Proc. of IEEE Conf. on PR And Image Processing. IEEE, Chicago. pp. 382-390.

De Mori, R., L. Lesmo and M. Poncini /1979/. The structure of a lexicon for a speech understanding system. In Record of IEEE Int. Conf. on Acoustics, Speech And Signal Processing. IEEE, Washington. pp. 252-255.

Dodd, G. G. and L. Rossol /Eds./ /1979/. Computer Vision and Sensor-Based Robots. Plenum Press, New York-London, 353 pp.

Doyle, J. /1978/. Truth maintenance systems for problem solving. Technical Report AI TR-419, MIT, Cambridge, Mass.

Hsieh, Y. Y. and K. S. Fu /1979/. A method for automatic IC chip alignment and wire bonding. In Proc. of IEEE Conf. on PR And Image Processing. IEEE, Chicago. pp. 101-108.

Jarvis, J. F. /1980/. A method for automating the visual inspection of printed wiring boards. IEEE Trans. on Pattern Analysis And Machine Intelligence, PAMI-2, No. 1.,77-82

Kasdan, H. L. and D. C. Mead /1978/. Automatic visual inspection by optical computing. In Proc. of the 3rd Int. Conf. on Automated Inspection And Product Control. Cotswold, Oxford. pp. 73-80

Klein, C. A. and K. J. Breeding /1979/. Automatic optical identification of faults in bubble memory overlay patterns. In Proc. of IEEE Conf. on PR And Image Processing. IEEE, Chicago. pp. 87-92.

Kuni, A., Y. Oshima, N. Akiyama and K. Isoda /1977/. Study on automatic inspection of defects on contact parts. In Prepr. of the Conf. on Information Control Problems in Manufacturing Technology. Pergamon, Oxford. pp. 95-101.

Levinson, S. E. and K. L. Shipley /1980/. A conversational-mode airline information and reservation system using speech input and output. The Bell System Technical Journal, 59, No.1 , 119-137.

Minsky, M. /1975/. A framework for representing knowledge. In P. H. Winston /Ed./, The Psychology of Computer Vision. McGraw-Hill, N.Y. pp. 211-227.

Newell, A., J. C. Shaw and H. A.Simon /1959/. Report on a general problem-solving program for a computer. In Proc. of the Int. Conf. on Information Processing. UNESCO, Paris. pp. 256-264.

Porter, G. B., T. M. Cipolla and J.L. Mundy /1979/. Automatic visual inspection of blind holes in metal surfaces. In Proc. of IEEE Conf. on PR And Image Processing. IEEE, Chicago. pp. 83-86.

Reddy, D. R. and W. Hon /1979/. Computer architectures for vision.

In G. G. Dodd and L. Rossol /Eds./ /1979/. Computer Vision And Sensor-Based Robots. Plenum Press, New York-London. pp. 169-186.

Saridis, G. N. and D. M. Brandin /1979/. An automatic surface inspection system for flat rolled steel. Automatica, 15, 505-520.

Shortliffe, E. H., S. G. Axline, B. G. Buchanan, T. C. Merigan and S. N. Cohen /1973/. An artificial intelligence program to advise physicians regarding antimicrobial therapy. Computers And Biomedical Research, 6, 544-560.

Shortliffe, E. H. /1976/. Computer-based medical consultations: MYCIN. Elsevier, N.Y.

Sterling, W. M. /1979/. Automatic non-reference inspection of printed wiring boards. In Proc. of IEEE Conf. on PR And Image Processing. IEEE, Chicago. pp. 93-100.

Vámos, T. /1978/. Automatic control and artificial intelligence - invited survey. In Prepr. of the 7th IFAC World Congress. Pergamon, Helsinki. Vol. 4, pp. 2355-2369.

Weiss, S., C. A. Kulikowski and A. Safir /1978/. Glaucoma consultation by computer. Comput. Biol. And Med., 8, 25-40.

Warren, D. H. and L. M. Pereira /1977/. PROLOG: the language and its implementation compared with LISP. SIGPLAN Notices, 12, No. 8., 109-115.

Winograd, T. /1971/. Procedures as representation for data in a computer program for understanding natural language. Technical Report AI TR-17. MIT, Cambridge, Mass.

ADAPTIVE CONTROL SYSTEMS

Y. M. Abdel-Fattah

L.E.E.S.A., Faculté des Sciences de Rabat, B.P. 1014, Rabat, Morocco

Abstract : The paper attempts to summarize some aspects of the recent developments in adaptive systems theory. New stochastic approach for dealing with nonlinear models is presented. The point made is that "randomization" of control policy may help avoid stringent identifiability conditions common in identification-passive adaptation schemes like the self-tuning.

Keywords. Adaptive control, self-tuning, Model-reference, stochastic approximation, parameter-estimation, Markov chains.

Introduction

Stated in fairly general terms the adaptive control problem amounts to controlling an unknown (partially known) system (plant, process, or environment) such as to asymptotically achieve a specific goal. The goal may be considered as the optimization of a certain performance criterion. The principle of operation of adaptive control system is based on the variation of the parameters or the structure as a result of observation and processing the current information in such a way that the adaptive system improves its functioning in the course of time, attaining an optimal state as the final result. The lack of a priori information is compensated by the purposeful handling of the current information.

I. Linear models.

The adaptive controller is characterized by combining a recursive parameter estimation algorithm for a difference equation process and noise model and a suitable control algorithm. If the parameters are those of the plant model, the adaptive sheme is called explicit. Otherwise, i.e. where the estimated parameters coincide with the controller parameters, the scheme is called implicit. The alternatives: indirect and direct, are also used for explicit and implicit, respectively.

Model Reference Adaptive Control (MRAC) and Self Tuning Regulators (STR) are the two principal approaches to the adaptive control problem. In MRAC the objective is to make the output of the unknown plant asymptotically approach that of a given reference model. In STR a design procedure for known plant parameters is first chosen and this is applied to the unknown plant using recursively estimated values of these parameters. Both explicit and implicit schemes for MRAC and STR have been proposed in the literature. Recently there has been a trend to show that all these schemes are basically equivalent, cf. Narendra and Valvani (1979), Egardt (1979), Johnson (1980).
In the following we summarize the results of Egardt (1979) which demonstrate some of the basic features of the adaptive control problem.
The plant is assumed to satisfy the difference equation

$$A(q^{-1})y(t)=q^{-(k+1)}b_oB(q^{-1})u(t)+w(t) \quad (1)$$

where q^{-1} is the backward shift operator, k is a nonnegative integer, and $A(q^{-1})$ and $B(q^{-1})$ are polynomials defined by

$$A(q^{-1})=1+a_1q^{-1}+..+a_nq^{-n}$$
$$B(q^{-1})=1+b_1q^{-1}+..+b_mq^{-m}$$

Furthermore, w(t) is a nonmeasurable disturbance.
The objective of the controller design is to make the difference between the plant output and the reference model output $y^M(t)$ as small as possible, see Fig.1.

The reference output y^M is related to the command input u^M by the reference model, given by

$$y^M(t)=q^{-(k+1)}\frac{B^M(q^{-1})}{A^M(q^{-1})}=q^{-(k+1)}$$

$$\frac{b_o^M+\ldots+b_m^M q^{-m}}{1+a_1^M q^{-1}+\ldots+a_n^M q^{-n}} \qquad (2)$$

where $A^M(q^{-1})$ is asymptotically stable.
The controller configuration is as shown in Fig. 2.

Deterministic design :

In the case when $w(t)=0$, it is possible to have the plant output equal to the reference model output $y^M(t)$. This can be attained as follows :

1) Choose the observer polynomial $T(q^{-1})$ defined by

$$T(q^{-1})=1+t_1q^{-1}+\ldots+t_{nT}q^{-nT}$$

2) Solve the polynomial equation

$$T(q^{-1})A^M(q^{-1})=A(q^{-1})R(q^{-1})+q^{-(k+1)}S(q^{-1})$$

for the unique solutions $R(q^{-1})$ and $S(q^{-1})$ defined by

$$R(q^{-1})=A+r_1q^{-1}+\ldots+r_kq^{-k}$$
$$S(q^{-1})=s_o+s_1q^{-1}+\ldots+s_{nS}q^{-nS},$$

$n_S=\max(n+n_T-k-1,\ n-1)$.

Stochastic design

The same as before with the only exception that the choice of T is not arbitrary. In the case where $w(t)$ is a moving average

$$w(t)=C(q^{-1})v(t)=(1+c_1q^{-1}+\ldots+c_nq^{-n})v(t) \qquad (3)$$

where $v(t)$, $t=0,1,2,\ldots$ are independent, zero mean random variables, it is well known that the optimal choice of the observer polynomial is

$$T(q^{-1})=C(q^{-1})$$

in the sense of minimum variance .

Asumptions

A.1. The number of plant poles n and zeros m are known

A.2. The time delay k is known and the sign of b_o is known. Without loss os generality b_o is assumed positive.

A.3. The plant is minimum phase, i.e. the numerator polynomial $B(q^{-1})$ in (1) has its zeros outside the unit circle.

Class of adaptive controllers

Define the error

$$e(t)=y(t)-y^M(t)$$

To obtain some flexibility in the model structure, a filtered version of the error will be considered. Let Q and P be asymptotically stable polynomials, defined by

$$Q(q^{-1})=1+q_1q^{-1}+\ldots+q_{nQ}q^{-nQ}$$
$$P(q^{-1})=P_1(q^{-1})P_2(q^{-1})=1+p_1q^{-1}+\ldots+p_{nP}q^{-nP}$$

where P_1 and P_2 are factors of P of degress n_{P1} and n_{P2}, respectively.
It is assumed that $P_1(0)=P_2(0)=1$.
Define the filtered error by

$$e_f(t)=\frac{Q(q^{-1})}{P(q^{-1})}e(t)$$

Note that $e_f(t)$ is a known quantity, because $y(t)$ is measured and $y^M(t)$, Q and P are known. It can be shown that $e_f(t)$ satisfies the equation,

$$e_f(t)=\frac{Q}{TA^M}q^{-(k+1)}\Big[b_o\frac{u(t)}{P_1}+b_o(BR-P_2)\frac{u(t)}{P}+S\frac{y(t)}{P}-\frac{TB^M}{P}u^M(t)\Big]+\frac{QR}{TA^MP}w(t) \qquad (4)$$

Deterministic design

Let θ be a vector, containing the unknown parameters of the polynomials $BR-P_2$ and S/b_o and the constant $1/b_o$ as the last element. Note that θ contains the parameters of the controller, so that the adaptive controller is direct (implicit identification).
Define the vector $\phi(t)$ from

$$\phi^T(t)=\Big(\frac{u(t-1)}{P},\ldots\frac{u(t-n_u)}{P};\frac{y(t)}{P},\ldots,\frac{y(t-n_y+1)}{P};\ -\frac{TB^M}{P}u^M(t)\Big) \qquad (5)$$

where

$$n_u=\max(m+k,n_{P2}),\ n_y=\max(n+n_T-k,n)$$

and the number of u- and y- terms are compatible with the definition of θ, Note that the elements of ϕ are known signals.
Using the definitions of θ and ϕ, it is possible to write (4) as

$$e_f(t)=\frac{Q}{TA^M}q^{-(k+1)}\Big(b_o\frac{u(t)}{P}+b_o\,\theta^T\phi(t)\Big)+\frac{QR}{TA^MP}w(t) \qquad (6)$$

Stochastic design

Since C is unknown, it might be desirable to estimate it. Some minor changes are then needed. Concatenate the θ vector with a vector whose elements are the unknow parameters of C/b_o. Also, redefine the ϕ vector as

$$\phi^T(t)=(\frac{u(t-1)}{P},\ldots,\frac{u(t-n_u)}{P};\frac{y(t)}{P},\ldots,\frac{y(t-n_y+1)}{P};-\frac{B^M}{P}u^M(t),\ldots,-\frac{B^M}{P}u^M(t-n)) \qquad (7)$$

The filtered error can then be written as :

$$e_f(t)=\frac{Q}{CA^M}q^{-(k+1)}(b_o\frac{u(t)}{P_1}+b_o\theta^T\phi(t))+\frac{QR}{A^M P}v(t) \qquad (8)$$

Basic Control Scheme

- Estimate the unknown parameters b_o and θ (or some combination of these)
- Use these estimates to determine the control signal.

Error Models.

Deterministic design

Let $\hat{b}_o(t-1)$ and $\hat{\theta}(t-1)$ denote estimates at time t-1 of b_o and θ. Using the model (6), a one step ahead predictor of $e_f(t)$ is defined as :

$$e_f(t/t-1)=\frac{Q}{TA^M}(\hat{b}_o(t-1)\frac{u(t-k-1)}{P_1}+\hat{b}_o(t-1)\hat{\theta}^T(t-1)\phi(t-k-1)) \qquad (9)$$

so that

$$e_f(t)=e_f(t/t-1)+\varepsilon(t)$$

where

$$\varepsilon(t)=\frac{Q}{TA^M}(\tilde{b}_o(t-1)(\frac{u(t-k-1)}{P_1}+\hat{\theta}^T(t-1)\phi(t-k-1))+b_o\tilde{\theta}(t-1)\phi(t-k-1))+\frac{QR}{TA^M P}w(t) \qquad (11)$$

The above equation, relating the plant output error to the controller parameters errors,

$$\tilde{b}_o(t-1)=b_o-\hat{b}_o(t-1),\ \tilde{\theta}(t-1)=\theta-\hat{\theta}(t-1)$$

is called the error model of the adaptive scheme.

Stochastic design

Instead of (9) and (11) we obtain,

$$e_f(t/t-1)=\frac{Q}{CA^M}(b_o(t-1)\frac{u(t-1)}{P_1}+b_o(t-1)\theta^T(t-1)\phi(t-k-1)) \qquad (12)$$

and the error model,

$$\varepsilon(t)=\frac{Q}{CA^M}(\tilde{b}_o(t-1)(\frac{u(t-k-1)}{P_1}+\theta^T(t-1)\phi(t-k-1))+b_o\tilde{\theta}(t-1)^T\phi(t-k-1))+\frac{QR}{A^M P}v(t) \qquad (13)$$

Special cases of the adaptive control scheme :

Ionescu/monopoli(1977) : (Deterministic design)

$P_1=T$ of degree k
P_2 of degree n-1
$Q=P=P_1P_2$ of degree n+k-1.

Self-tuning pole placement- Astrom et al.(1978):(deterministic design).

$Q = TA^M$
$P=P_1=P_2=1$

Astrom and Wittenmark STR (1973): (Stochastic design)

$A^M=1,\ u^M=y^M=0$
$Q=P=1$

Control law :

It would be desirable to choose a control law such that $e_f(t/t-1)$ is equal to zero, because convergence of $e_f(t)$ to zero would then follow from the convergence of $\varepsilon(t)$ to zero (which is the objective of the parameter estimator). This is accomplished by the control law

$$u(t)=-P_1(q^{-1})(\theta^T(t+k)\ \phi(t))$$

as seen from (9). This control law is however noncausal. It is therefore natural to modify it in the following way :

$$u(t)=-P_1(q^{-1})(\theta^T(t)\ \phi(t)) \qquad (14)$$

Difficulties with convergence analysis

There mainly are two problems :

i. to show that the parameter estimator is such that $\varepsilon(t)\to 0$,
ii. to show that $e_f(t)\to 0$. Note that with the causal control law (14), $e_f(t/t-1)$ is not equal to zero in the case $k\neq 0$. This in turn means that it is not easy to conclude that $e_f(t)\to 0$ even if $\varepsilon(t)\to 0$.

The positive-real condition.

Assume for simplicity k=0, $b_o=1$ and that the plant is noise free. If the control law (14) is used, then

$$\varepsilon(t)=H(q^{-1})\ \tilde{\theta}^T(t-1)\ \phi(t-1) \qquad (15)$$

where

$$H(q^{-1})=\frac{Q(q^{-1})}{T(q^{-1})A^M(q^{-1})} \qquad (16)$$

Consider the parameter updating procedure

$$\hat{\theta}(t)=\hat{\theta}(t-1)+\frac{\phi(t-1)}{|\phi(t-1)|^2}\,\varepsilon(t)$$

cf. Ionescu/Monopoli(1977). Then we can write,

$$|\tilde{\theta}(t)|^2=|\tilde{\theta}(t-1)|^2-2\,\frac{\varepsilon(t)(\frac{1}{H}-\frac{1}{2})\varepsilon(t)}{|\phi(t-1)|^2} \quad (17)$$

If $\frac{1}{H}-\frac{1}{2}$ is positive real (PR) the parameter error will eventually decrease. Morover, $\frac{\varepsilon(t)}{|\phi(t-1)|}\to 0$, if $\frac{1}{H}-\frac{1}{2}$ is strictly positive real (SPR). But, it remains to show that the closed-loop signals constituting the vector $\phi(t)$ are uniformly bounded in order to conclude that $\varepsilon(t)\to 0$.

It is straightforward to show that the PR condition can be avoided, cf. Egardt(1979). Thus, let $\bar{x}$ denote the signal obtained by filtering x by $\frac{Q}{TA^M}$ and rewrite (6) as

$$e_f(t)=q^{-1}\left(\frac{\bar{u}(t)}{P_1(q^{-1})}+\theta^T\,\bar{\phi}(t)\right) \quad (18)$$

where the same assumptions as above are used. Now consider (18) as being the model instead of (15). The error model would be,

$$\varepsilon(t)=\theta^T(t-1)\ \bar{\phi}(t-1)$$

If the parameter updating is replaced by

$$\hat{\theta}(t)=\hat{\theta}(t-1)+\frac{\bar{\phi}(t-1)}{|\bar{\phi}(t-1)|^2}\,\varepsilon(t) \quad (19)$$

it follows that

$$|\theta(t)|^2=|\theta(t-1)|^2-\frac{\varepsilon^2(t)}{|\bar{\phi}(t-1)|^2}$$

It thus follows that

$\frac{\varepsilon(t)}{\bar{\phi}(t-1)}\to 0$ without any PR condition.

Of course the boundedness of the closed loop signals is still a problem. The conclusion is that it is possible to eliminate any PR condition (in the noise-free case) if a modified estimation scheme is used.

From the above it is clear that there are two basic problems relative to the design of adaptive control systems . Stability (in the sense of giving rise to bounded control signal and system output), and Convergence (in the sense that the desired performance is achieved $e_f(t)\to 0$.

Parameter Estimation

Deterministic design

In order to get around the PR condition we prefilter the input and output signals by $\frac{Q}{TA^M}$ .So we can rewrite (9) and (11) as ,

$$e_f(t)=q^{-(k+1)}\left(b_o\,\frac{\bar{u}(t)}{P_1}+b_o\theta^T\,\bar{\phi}(t)\right)+\frac{QR}{TA^MP}w(t) \quad (20)$$

$$\hat{e}_f(t/t-1)=\hat{b}_o(t-1)\,\frac{\bar{u}(t-k-1)}{P_1}+\hat{b}_o(t-1)\hat{\theta}^T(t-1)\,\bar{\phi}(t-k-1) \quad (21)$$

$$\frac{\bar{u}(t)}{P_1}=-\,\hat{\theta}^T(t)\bar{\phi}(t) \quad (22)$$

The aim of estimation is to minimize the variance of the error $\varepsilon(t)$,

$$E\,\varepsilon^2(t)=E\,(e_f(t)-e_f(t/t-1))^2$$

which is a function of the parameter estimation errors $\tilde{b}_o(t-1)$, $\tilde{\theta}(t-1)$ Stated as such, we obtain the following stochastic approximation (SA) algorithm as follows,

$$\begin{pmatrix} b_o(t)\\ \theta(t)\end{pmatrix}=\begin{pmatrix} b_o(t-1)\\ \theta(t-1)\end{pmatrix}+\Gamma(t)\begin{pmatrix}\frac{\partial\varepsilon(t)}{\partial b_o}\\ \frac{\partial\varepsilon(t)}{\partial\theta}\end{pmatrix} \quad (23)$$

where $\Gamma(t)$ is an appropriate positive matrix .

$$\frac{\partial\varepsilon(t)}{\partial b_o}=\frac{\bar{u}(t-k-1)}{P_1}+\hat{\theta}^T(t-1)\ \bar{\phi}(t-k-1) \quad (24a)$$

$$\frac{\partial\varepsilon(t)}{\partial\theta}=\beta_o\,\bar{\phi}\,(t-k-1) \quad (24b)$$

where ' $^-$ ' denotes the prefiltering by $\frac{Q}{TA^M}$ as before.

β_o is an a priori estimate of b_o. The importance of the choice of β_o for the stability properties are examined by Egardt (1979).

The matrix $\Gamma(t)$ is taken as ,

$$\Gamma(t)=\frac{1}{r(t)}\,I$$

where

$$r(t)=\lambda r(t-1)+\left(\frac{\partial\varepsilon(t)}{\partial b_o}\right)^2+\left|\frac{\partial\varepsilon(t)}{\partial\theta}\right|^2+\alpha$$

$$0<\lambda<1 \quad ,\ \alpha>0 \quad (25)$$

The above choice of $\Gamma(t)$ basically corresponds to the scheme of optimal learning, cf. Tsypkin(1973),except for the introducing of the 'forgetting factor' λ ; its effect is generally to improve the convergence rate.The constant α is included for numerical reasons. The above estimation algorithm has been used by several authors,e.g. Ionescu/Monopoli (1977).

Stochastic design.

Here we write the filtered error as

$$e_f(t)=q^{-(k+1)}\ \frac{1}{C}\left(b_o\,\frac{\bar{u}(t)}{P_1}+b_o\theta^T\bar{\phi}(t)\right)+\frac{QR}{CA^MP}\,w(t) \quad (26)$$

where ' $^-$ ' now denote filtering by $\frac{Q}{A^M}$.
Note that it is not possible to

filter by $\frac{1}{C}$ since C is unknown.The prediction is calculated as,

$$\hat{e}_f(t/t-1)=\hat{b}_o(t-1)(\frac{u(t-k-1)}{P_1}+\hat{\theta}^T(t-1)\bar{\phi}(t-k-1))$$

The definition of the SA algorithm is still valid with appropriate interpretation of Θ, φ, and ' – ' . The equations are identical and therefore not repeated.

The control in both designs (deterministic , stochastic) is chosen as ;

$$\frac{\bar{u}(t)}{P_1}=-\hat{\theta}^T\bar{\phi}(t)$$

with the appropriate interpretation of Θ; φ and ' – '

Definition

The closed loop system is L^∞-stable if uniformy bounded disturbances (w) and command (u^M) signals give uniformy bounded input (u) and output (y) signals.

Theorem 1 (Egardt (1979)

Consider the plant (1) controlled by the SA algorithm with deterministic or stochastic design. Assume that assumptions A1-3 are satisfied and $b_o<2\beta_o$. If the parameter estimates are uniformly bounded then the closed loop system is L stable.

Theorem 2. (Egardt (1979)

Consider the plant (1) with no noise, i.e. w(t)=0, controlled by the SA algorithm with deterministic design Assume that A1-3 are fulfilled and that $b_o< 2\beta_o$.Then the closed loop system is L^∞- stable.

Theorem 3. (Egard 1979)

Consider the plant (1) with no noise i.e. w(t)=0. Let the plant be controlled by the SA algorithm for deterministic design. Assume that A1-3 hold, $b_o< 2\beta_o$, and α>0. and that the command input $u^M(t)$ is uniformly bounded. Then the output error convergens to zero, i.e.

$$|y(t)-y^M(t)| \to 0, \quad t\to\infty.$$

In the disturbance-free case Theorems 2,3 give a fairly complete picture of how the algorithm behave. Convergence of the output to zero is proved without any a priori assymption of boundedness of closed-loop signals. In the case with disturbances it has been demonstrated (Egardt 1979) that some additional assumptions are needed to guarantee global stability. This means it is necessary to put more conditions on the noise and/or command signal. It does not seem unreasonable that some kind of persistently exciting condition (see e.g. Narendra and Peterson(1980) Morgan/Narandra (1977), Weiss/Mitra (1979) might be sufficient to ensure the boundedness of the parameter estimates.

Discussion.

The error model presented contains the unknown parameters b_o and Θ as a product. This is common for all MRAC. This is in contrast to most model structures used for identification and STR For example, the self-tuning pole placement algorithm in Astrom et al.(1978) uses a model which is linear in the unknown parameter, taken as b_o and $b_o\theta$ In the general case a model which is linear in the unknown parameters can be obtained in the following way .Let Θ denote b_o time the previous Θ vector and write (20) as

$$e_f(t)=q^{-(k+1)}(b_o\frac{\bar{u}(t)}{P_1}+\theta^T\bar{\phi}(t))+\frac{QR}{TA^MP}w(t) \quad (27)$$

This model suggests an alternative to the SA algorithm; Egardt(1979)

Estimation:

$$\theta(t)=\theta(t-1)+\bar{\phi}(t-1)\frac{\varepsilon(t)}{r(t)}$$

$$r(t)=\lambda r(t-1)+|\bar{\phi}(t-k-1)|^2+\alpha \quad (28)$$

$$\varepsilon(t)=e_f(t)-\hat{e}_f(t/t-1)$$

$$\tilde{e}_f(t/t-1)=\beta_o\frac{\bar{u}(t-k-1)}{P_1}+\hat{\theta}^T(t-1)\bar{\phi}(t-k-1)$$

Control

$$\frac{u(t)}{P_1}=-\frac{1}{\beta_o}\hat{\theta}^T(t)\bar{\phi}(t) \quad (29)$$

Note that an a priori estimate β_o of b_o is still used. Also notice that the last element of Θ is known to be unity, but it is necessary to estimate it because β_o might be chosen differently from b_o. The stability of the above algorithm has not been proven for k≠o.

By means of 'error augmentation' and replacing ε(t) by augmented error in the parameter estimation scheme, it is possible in variety of cases to render the adptive scheme globally stable, cf. Narendra and Peterson(1980) Recent works have shown algorithmic equivalence between different discrete adaptive model following schemes, cf. Johnson (1980).

Non-minimum phase Systems

A characteristic feature of the schemes considered are that the zeros are cancelled. This implies that only minimum phase systems can be treated. The underlying design method with cancellation of zeros was chosen because it gives a simple adaptive scheme. A convenient implicit scheme for non-minimum phase systems seem, however, more difficult to give. One possibility to solve the problem is to consider explicit schemes instead, cf.Astrom et al. (1978).A possibility to control non-minimum phase plants with an implicit scheme without cancellation of zero is discussed in Clarke and Gawthrop (1975).

Critique

Assumptions A.1-2 are required by all the known stable adaptive schemes. It is naturally of interest to relax the condition that the orders of the plant model are not underestimated . It would be valuable to have results concerning the control of higher order , or distributed parameter plants. It is naturally of interest to investigate the properties of algorithms which are capable of controlling non-minimum phase systems.

The corresponding assumptions for multivariable systems are likely to be even more restrictive regarding a priori information about the plant that may be needed. Narendra and Peterson (1980) indicated that the adaptive control problem should be reformulated so that less is demanded of the controller allowing less restrictive assumptions to be made . For example instead of requiring perfect model following so that the error goes to zero as $t \to \infty$ emphasis in the future should be on globally stable schemes in which the output error is merely bounded.

Identification-Active Adaptive Control

Since the classic work of Fel'dbaum (1960) (1961) a nonlinear stochastic control theory has been developed . As a consequence of that theory it is clear that even for linear systems with unknown parameters the operations of identification and optimization cannot, in general, be separated without regard for the special structure of the control problem. Indeed, it is difficult to specify without a detailed analysis what properties of an identification procedure can be regarded as optimal in the control context, cf. Sworder (1966),Aoki(1967) El-Fattah (1974).Probing control signal may be needed to enhance system identification and thus approximates optimal dual control performance, cf.El-Fattah and Foulard (1975).Unfortunately,dual adaptive optimal or suboptimal control schemes are generally quite complex. Tsypkin (1971) proposed a variety of ideas for suboptimal adaptive control based on stochastic approximation.

II. Non-linear models

Learning automata can be used as models for adaptive systems. They may solve the adaptive control problem for non-linear models, and in the presence of constraints. We will consider a certain dynamic plant which can be described by the recursion relation.

$$y_{t+1}=F(y_t,u_t,w_t) \ ; \quad t=0,1,2,.. \qquad (30)$$

where y : state of the plant,
u : control,
w : random noise.

We carry out the discretization of the domain of states $Y \subset R^n$, which we will assume to be closed and bounded, into some set of sub-domains $\{Y_i\}$such that

$$Y_i \neq \emptyset, Y_i \cap Y_j = \emptyset \ (i \neq j)(i,j \neq 1,..,r) \bigcup_{i=1}^{r} Y_i = Y$$

The collection of sets Y_i forms a complete partition of the domain Y. The control is assumed to take a finite number of discrete values . The noise w is assumed to be such that a controlled Markov chain model can be assumed. This is the case, for instance, if for each u,

i. F(y,u,w) is a Borel measurable function of y,w;
ii. $y_0,w_0 \cdots, w_t$ are jointly independent random variables for all $t > 0$.

A controlled markov chain model can be stated as follows. Let x_t denote the state of the chain at t:

$$x_t = i \Leftrightarrow y_t \in Y_i; \ i \in C = \{1,..,r\}$$

The transition probability at time t depends upon the control action u_t taken at t and upon a parameter θ. For arbitrary $i \in C$, whenever the chain is in state i, the probability distribution of the next state is

$$(P(i,1,u,\theta),P(i,2,u,\theta),..P(i,r,u,\theta)), \quad u \in U(i)$$

where u is a control action ranging in a finite set U(i) with elements labelled $1,..m_i$.

At each time t, X_t is observed and based upon its value u_t is selected from $U(x_t)$. The parameter θ has the constant value θ^* which is not known in advance, it is known however, that θ belongs to a compact $\Theta \subset R^s$.
It is assumed that a controller selects the control action following a randomized policy D specified by the probabilities,

$$d^1,..,d^r; \ d^i=(d^i_1,..,d^i_{m_i}), i=1,...r$$

where d_k^i denotes the probability of selecting the k-th element of U(i) whenever $x_t=i$. The subclass of policies for which d_k^i is 0 except for exactly one $k=\mu(i)$ is the class of <u>deterministic policies</u>.

Let us define the reward function $c(i,j;k)$ $i,j\in C$, $k\in U(i)$. The performance of the system can be measured by the expected average reward,

$$\phi = \lim_{\to\infty} \frac{1}{T} \sum_{t=0}^{T-1} c(x_t,x_{t+1},u_t) \qquad (31)$$

Under appropriate ergodicity assumptions on the transition probability, ϕ exists and is a function of the policy D and the parameter θ.
The gradient of the expected average reward with respect to policy is defined by the set of vectors.

$$\frac{\partial\phi}{\partial \underline{d}^1}, \ldots, \frac{\partial\phi}{\partial \underline{d}^r}$$

$$\frac{\partial\phi}{\partial d^i} = \left(\frac{\partial\phi}{\partial d_1^i}, \ldots, \frac{\partial\phi}{\partial d_{m_i}^i}\right)^T \quad i\in C$$

The derivatives $\frac{\partial\phi}{\partial d_k^i}$ are defined in El-Fattah and Foulard (1978).

<u>Theorem 4. (El-Fattah 1980)</u>

Let there exist a deterministic policy

$$d_\beta^{\alpha *} \begin{cases} =1, & \text{if } \beta=\mu(\alpha) \\ =0, & \text{if } \beta\neq\mu(\alpha) \end{cases}$$

such that

$$\max\left\{\frac{\partial\phi}{\partial d_\beta^\alpha}(D^*)\right\} = \frac{\partial\phi}{\partial d_{\mu(\alpha)}^\alpha}(D^*) > 0$$

Then D^* is optimal.

<u>Identification-Passive Schemes-Certainty Equivalence</u>

cf. Riordon (1969), Mandl (1971)
The following adaptive schem is intuitively appealing. Based on the observations upto t estimate the parameter θ^* as θ^t, then feed the estimate into the 'deterministic' control law $\mu(x_t;\theta^t)$ which maximizes for the chain described by θ^t, the expected average reward. Like STR an identifiability condition has to be satisfied in order to guarantee convergence of the adaptive control towards the optimum. The following condition is due to Mandl (1971).

<u>Identifiability condition</u>

If θ, $\theta'\in\Theta$, $\theta\neq\theta'$ then there exists an i C, for which

$$(p(i,1,u,\theta),\ldots,P(i,r,u,\theta))\neq(P(i,1,u,\theta'),\ldots P(i,r,u,\theta')) \text{ all } u\in U(i)$$

Under the above condition Mandl shows that the maximum likelihood estimator is strongly consistent for any control law, and consequently the expected average reward converges a.s. to the optimal value.
Mandl's condition might be too restrictive in some practical situations, as indicated by Borkar and Varaiya (1980)

<u>Identification Active Schemes-Learning Control</u>

The following adaptation scheme is proposed in El-Fattah (1980), see fig.3.
At each time t, based on the updated es estimate of parameter θ^t evaluate the gradient

$$\left(\frac{\partial\phi}{\partial d_k^i}(t)\right) \quad i\in C, \; k\in U(i)$$

For each state i determine the component μ_i corresponding to maximum gradient;

$$\max_k \frac{\partial\phi}{\partial d_k^i} = \frac{\partial\phi}{\partial d_{\mu_i}^i}$$

If several control indices satisfy the condition of maximality of gradient, we consider only one of them as μ_i.
At the next time, t+1 adopt the following policy,

$$\underline{d}(t+1)=\underline{d}^i(t)+\gamma(t)(\underline{e}^i(t)-\underline{d}^i(t)), \text{ all } i \; C \quad t=0,1.. (32)$$

where $\underline{e}^i(t)$ is the deterministic policy corresponding to the action μ_i,

$$e_k^i \begin{cases} =1 & \text{if } k=\mu_i \\ =0 & \text{if } k\neq\mu_i \end{cases}$$

$\gamma(t)$ has to satisfy the customary stochastic approximation conditions,

$$1>\gamma(t)>0, \; \Sigma\,\gamma(t)=\infty \; \Sigma\gamma(t)^2<\infty$$

<u>Remarks</u>

- Notice that the 'forcing term' in the learning algorithm (32) converges to zero when, $\theta^t\to\theta^*$, $\underline{d}^i(t)\to\underline{d}^{i*}$ all i, i.e. the controller does not tranquilize until achieving the conditions of optimality.

- The above policy 'reinforcement scheme' ensures, after θ^t becomes in the neighborhood of θ and $\gamma(t)$ becomes sufficiently small, the monotic improvement of the expected average reward (i.e. absolute expediency) :

$$\phi(t+1)\geq\phi(t)$$

The scheme (32) is a generalization of the SA (stochastic approximation) automata, cf. Flerov (1972).

- El-Fattah (1980) presents recursive parameter estimation schemes of controlled Markov chains. Strong consistency and asymptotic behavior of the estimator are also studied.

- In the adaptive scheme (32) it is not necessary to impose identifiability hypothesis like Mandl's. It is only necessary that the parameter estimator be consistent.

REFERENCES

Egardt, B.(1979), Stability of Adaptive Controllers. Berlin: Springer Verlag Lecture Notes in Control and Information Sciences vol. 20.

Narendra, K.S. and L.S. Valvani(1979) Direct and Indirect Model Reference Adaptive Control "Automatica,vol.15, pp. 653-664.

Johnson , C.R.,Jr.(1980) Input Matching, Error Augmentation, Self-tuning and Output Error Identification:Algorithmic Similarities in Discrete Adaptive Model Following to appear in IEEE Trans.Auto. Cont.

Ionescu, T. and R.V.Monopoli(1977) Discrete Model Reference Adaptive Control with an Augmented Error Signal Automatica vol.13,pp. 507-517.

Tsypkin,Ya.Z. (1971) Adaptation and Learning in Automatic Systems. New York:Academic.

Tsypkin Ya.Z. (1973) Foundation of the Theory of Learning Systems. New York:Academic.

Narendra K.S. and B.B. Peterson(1980) Recent Developments in adaptive Control International Symposium on Adaptive Systems Ruhr University, Bochum,Germany

Morgan, A.P. and K.S. Narendra(1977) On the Uniform Asumptotic Stability of Certain Linear Nonautonomous Differential Equations, SIAM J. Control and Optimization, vol.15,pp. 5-24.

Weiss, A. and D. Mitra(1979) Digital Adaptive Filters:Conditions for Convergence Effects of Noise, and Errors arising from Implementation IEEE Trans. on Inf. Theory,vol.25, pp. 637-652.

Astrom, K.J. B. Westerberg and B. Wettenmark (1978) Self-Tuning Controllers Based on Pole placement Design" CODEN:LUTFD2/(TFRT-3148)/1-52/,Dept.of Automatic Control, Lund Institute of Technology, Lund Sweden.

Clarke D.W. and P.J.Gawthrop(1975) Self-tuning controller Proc. IEE vol.122, pp; 929-934.

Fel'dbaum A.A. (1960) Dual Control Theory I-II, Automation and Remote control, Vol.21 pp.1240-1249 1453-1464

Fel'dbaum A.A.(1961) Dual Control Theory, ibid.vol 22, pp. 3-16, 129-142.

Sworder, D.D.(1966) Between Identification and Optimization in Adaptive Control Problems J.Franklin Inst. Vol. 281,pp. 198-213.

Aoki, M. (1967) Commande Optimale de Procede Stochastiques et Adaptàtifs, Revues R.A.I.R.O n° J.3, pp. 5-32.

El-Fattah, Y.M. and C. Foulard(1978) Learning systems : Decision, Simulation and Control.Berlin:Springer Verlag Lecture Notes in Control in Information Sciences vol 9.

El-Fattah (1980) Recursive Estimation and Control in Markov Chains , 13th European Meeting of Statisticians, Bright, England.

Yu.A. Flerov (1972) Some classes of multi-input Automata, J.of Cybernetics vol. 2pp,112-122

Riordon J.S. (1969) An adaptive Automaton Controller for Discrete time Markov Processes, Automatica, Vol. 5,pp.721-730

Mandl, P. (1971) Estimation and Control in Markov Chains, Adv. Appl. Prob., vol.6 pp. 40-60

Borkar V, and P. Varaiga (1980) Adaptive Control of Markov Chains, IEEE Trans. Aut. Cont. vol.AC-24 pp. 953-957.

El-Fattah (1974) Commande Optimale de Procede Stochastiques et Adaptatifs, Revue RAIRO. n°J-2. pp. 5-20

El-Fattah and Foulard (1975) Commande Adaptàtive Sous-Optimale de Systemes Dynamiques Stochastiques Recurrents Lineaires avec des Paramètres Variables Inconnus Revue RAIRO, n° J-3, pp. 5-23.

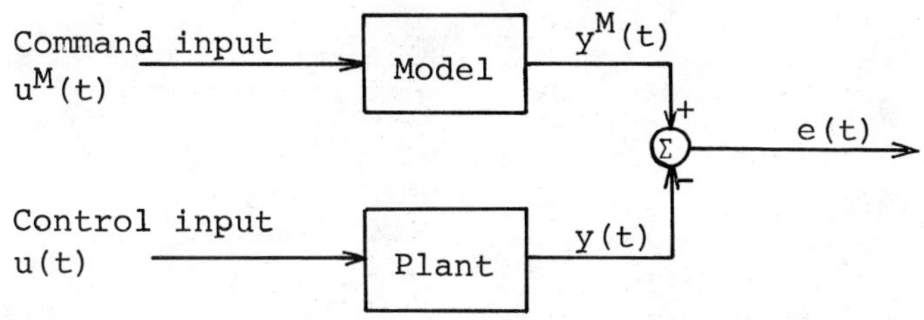

Fig.1.

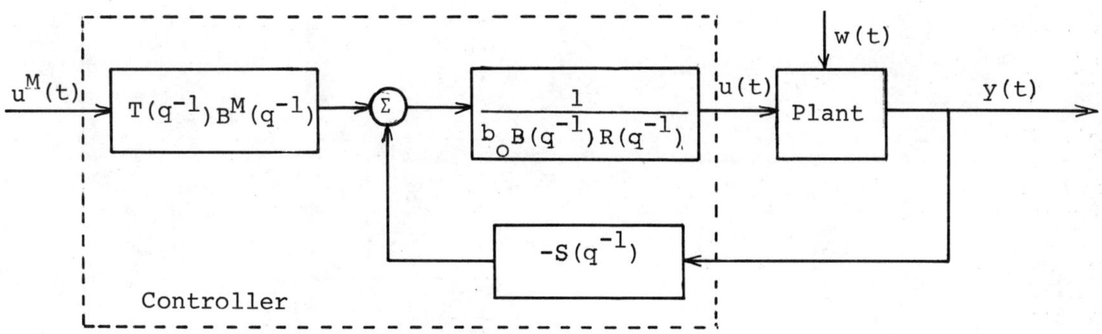

Fig.2.

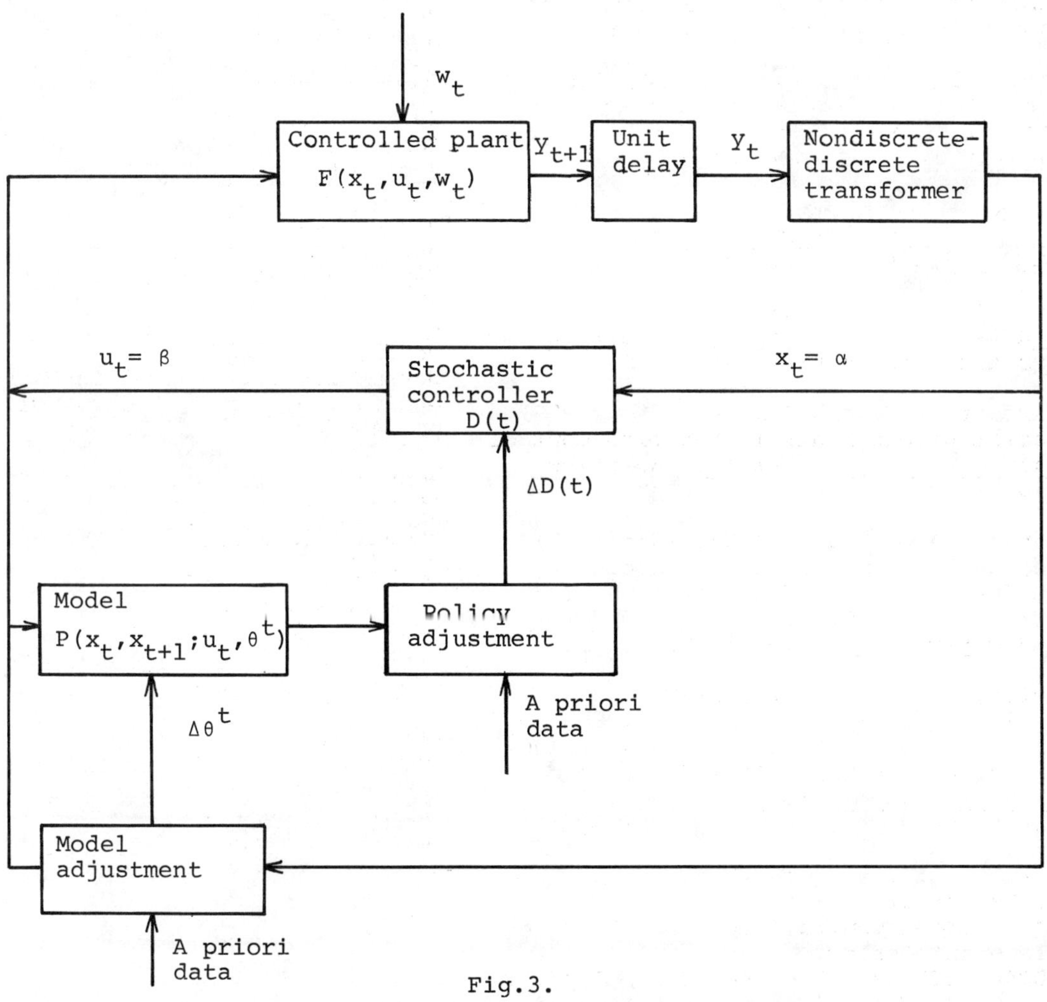

Fig.3.

A SURVEY ON COMPUTER-AIDED PLANNING IN COMPLICATE PRODUCTION SYSTEMS

L. Pun, G. Doumeingts, J. Grislain and D. Breuil

Laboratoire d'Automatique GRAI, Université de Bordeaux 1, 351 cours de la Libération, 33405 Talence Cedex, France

Abstract. A survey based on a methodology for treating systematically computer assisted static and dynamical plannings. Four aspects are discussed. Firstly the comprehension aspect emphasizing the need for a mutual understanding between the users and the analyst. Secondly, the specification aspect emphasizing notably the multiple criteria aspect. Thirdly, the resolution aspect covering both conventional methods and new methods based on the first-order predicates. Fourthly, the implementation aspect based on the various types of interactive controls.

Keywords. Computer-aided planning, production management, multiple criteria, first-order predicates, interactive controls.

1. GENERAL INTRODUCTION

1.2. Past works on computer-aided plannings

In this section, we shall treat the following points : what are the computer-aided plannings, how they are recognized as a special class of control problems, namely convivial controls of discrete organized process ; what has been done for solving these problems ; and what must be done in the future.

In a production process, there are two types of plannings : (1)

(a) Static plannings : establish programs of procurement and manufacturing activities on the basis of given products known (or assumed) sales, known manufacturing techniques and given manufacturing means (workers, machine-tools).

(b) Dynamical plannings : upon perturbations, adjust the old programs to get new adequate programs by modifying the utilization of the resources (materials, machine-tools) and / or the activity sequences.

According to these definitions, Computer-Aided plannings (CAP) are consisting of establishing reference procurement and manufacturing programs, and adjust these programs when there are perturbations, with the help of computers. Such processes involve three systems :

(a) The production system (PS) : for procurement and manufacturing activities

(b) The management system (MS) : for programming

(c) The computer or rather the information system (IS), for informations collecting, processing and presentation.

For control engineers and system analysts, it seems more appropriate to recognize these CAP processes as Convivial Controls of Discrete Organized Activities, i.e. : Harmonious Joint controls exerted by the MS and the IS to plan, to manage the PS activities.

The need of such a recognition is to permit us to get a better insight of the CAP problems which are now control problems. These problems are :

(a) Conventional control problems : modeling, identification, observability, reachability, minimum realization, optimality.

(b) Specific ones due to the involved PS, MS and IS : hierarchical decomposition, multi-criteria decision synchronization, coordination, structure coherence between PS, MS and IS.

The importance of these problems depends on the nature of the PS. There are three main types of PS :

(a) Mass-production

(b) multi-product small-series production (commonly called discrete manufacturing)

(c) unitary production.

The CAP of the type a are relatively simple, because the manufacturing techniques are well defined and organized. The CAP of the type c are too complicate, because the number of the activities and the number of possible perturbations are very high. Therefore, henceafter, when we talk about complicate PS, we refer only to the PS of the type b. In most of the industrialized countries, the importance of this type is more than 50 %.

In the past there have been two types of contributions intending to solve these problems : (a) theoretical contributions (based on mathematical models) ; (b) computer contributions (based on discursive models).

Theoretical contributions. During the last 20 years, there have been at least more than 1000 publications in this field. Dumas P. (2) (1972) has made a survey on heuristic scheduling problems. Rinnoy Kan (3) (1973) and Pun (4) (1974) have made a survey on general scheduling methods. Gelders (5) has made a survey on the use of Branch & Bound principle to machine scheduling problems. Conventional methods are presented in the books of Conway et al (6), and of Baker (7). Aggarwal (8)(1974) and Pun (9) (1974) have made a survey on inventory problems. Minimum-risk type prediction problems and other side problems are treated in most of the books on production control. See for instance Chase & Aquilaine (1973) (100), Ignizio & Gupta (1973) (103), Howard (1975)(99) Ramalingam (1976) (140), Birman et al (1977) (104), Colley et al (1977 (106). In the meantime, system approach also appears. See Gordon (1969) (16), Churchman (1968) (11), Kalman et al (1969) (12), Couger (1974) (13), Lemoigne (1974) (14) and Lemoigne (1977) (15).

The types of problems treated are :

- minimum risk forecasting
- inventory control, determination of economical series
- workshop scheduling (flow-shop, job-shop)
- queueing
- resource allocation
- unitary project management (PERT, CPM)

It is generally agreed (except PERT & CPM) that :

- there is no good coincidence between the theoretical hypothesis and the reality,
- the mathematical formulation is not understood by the users,
- the treated aspect is too partial
- practically there have not been any significant applications.

Computer contributions. These are packages developped either by Computer Companies, or by Service Companies, intending to solve the production management problems such as :

- client records
- requirement determination
- stock book-keeping
- load calculations
- workshop-operation follow-up
- supplier records.

The introduction of these packages in the industries is slow because of the high cost and the high technicity. Anyhow, these packages supply only elements for solving the CAP problems, without giving the methods.

We can notice clearly, looking at the past, a disjunction between the planning theories and the practical computer-aid.

To solve the defined class of convial controls we need therefore a general methodology covering according to the system approach :

(a) understand the user's problem
(b) formulate the problem
(c) model the PS, the MS, the IS and the CAP process
(d) validate the model by identification
(e) formalize the problem mathematically and solve the problem theoretically
(f) develop the decision-aid equipment
(g) implement the equipment and assist the user.

In this survey, we shall concentrate these steps into four large steps :

(a) Structure comprehension
(b) Specification of the problem-elements
(c) Formalization and resolution
(d) Implementation and exploitation.

For each step, we shall analyze the raised questions, the works which have been done on this request, and the things which must be done.

1.2. Computer-Aided Planning in the framework of Computer-Aided Production

Planning process is only one sub-process of the general Production process. The earliest application of the computer in the production process is the Numerical Control of machine-tools. From that, computer successively comes to aid the other sub-process such as : design, drawing, technical operation, manufacturing, robotics, testing, maintenance, etc. In a near future, it is already envisaged to create unmanned machine shops where the computer will aid and control all the sub-processes of the production process.

The planning sub-process is evidently related to the other production sub-processes. Because the common use of the computer, it is important to understand when we search for the solution of the CAP problems, the relation between the CAP and the other Computer-aided processes.

For this, we shall represent (Fig. 1) :

- the product path, i.e. : the various states of the product in a production process
- the activity path, i.e. : the various activities which transform one product-state into another
- the role of the computer for accomplishing these activities.

The various states of the products are :

(a) chosen names of the products after marketing
(b) product prototypes after research and development (R & D)
(c) products in industrial forms suitable for

the markets
(d) product constitution (nomenclature) and
th the associate manufacturing process (often called manufacturing methods or technical processing procedure)
(e) product production programs (manufacturing and procurement programs)
(f) purchased product raw materials and components
(g) final products.

At the present time, the various developped computer aids are :

(a) computer-aided design (electronic circuits, mechanical structures)
(b) computer-aided drafting (drawings of mechanical parts, drawings for printed circuits)
(c) computing-aided manufacturing
  - technical processing procedure
  - machining tool paths
(d) computer-aided machining
  - machine-tool numerical controls
  - industrial robots for manipulations supervising, testings...
(f) Computer-aided testing (for product parts and for final products)
(g) Computer-aided maintenance.

Actually, the Computer-aided plannings (CAP) cover a well-defined field of problems, i.e. : establish and adjust production (procurement and manufacturing programs). However, since there are needs for program adjustments, the tasks imply also the supervising and the modification of the program execution. This is not only a planning, but also a control problem. The expression "Convivial Control" seems therefore more appropriate. From the diagram of Figure 1, we see that the CAP problems are also indirectly related to CA Manufacturing, CA Machinery, CA Testing, CA Maintenance. The whole set would be non treatable. Therefore, in the sequel, we shall assume the disconnexion between the CAP problems and the other computer aids.

2. COMPREHESION

Comprehension is a global state implying the mutual understandings between the users, the system analysts and the problems to be solved, which are actually the CAP problems.

A good comprehension involves :

- a good problem statability from the users i.e. : the problems are stated with a reasonnable completeness, and enough simple and realistic so that their solution has a practical meaning

- a good problem-formulability from the analysts i.e. : understandable by the users and clear enough so to permit formal problem-specification and problem-resolution.

Obviously, a good comprehension state could be eventually reached only after a labourious trial-and-error interactive process between the users and the analyst.

The lack of comprehension is the major reason explaining why during the last 20 years we have failed to apply Operations Research techniques to solve production management problems. In this section, we shall treat the following points : what is the comprehension of the CAP problems, what is the past contribution and some suggestion to establish methodologically the comprehension state.

2.1. Descartes-no and Leibnitz-yes

Whenever complicate systems are concerned, the first reaction of an analyst is to use the Descartes' decomposition approach. This is very misleading. As pointed out by Barthélemy (17), the Descartes' approach is perhaps necessary, but certainly not sufficient. Leibnitz is among the first ones who have warned that one plus one are not always equal to two, i.e. : the total results for solving the problems of the two components do not always solve the problems of the whole.

During the last years, system analysts begin to be aware of the coupling effects. In practice, these effects are not easy to locate. In fact, the decomposition (deduction) process is made in one dimension, the coupling effects are induced in another dimension. New deduction becomes required in this new direction, and so on. Up to now, there does not exist any clear-cut for operating such a deduction-induction analysis.

In the CAP problems, the requirements of the comprehension analysis may be presented at two levels, depending on the size of the system taken into consideration.

(a) PS-level

If we consider only the production system :
Question 1 : how to define the decomposition dimensions with respect to planning problems ?
- product constitution
- product technical manufacturing process
- materials and components
- machines and workshops
- sales cycle
- human operating structure
- planning horizons
- static and dynamical nature of the plannings, ..., etc.

Question 2 : how to define the coupling effects between these dimensions
- between manufacturing and procurement
- between sales and manufacturing
- between human and machine structures
- between information availability and the manufacturing, etc...

(b) CAP-level

If we consider the global CAP process comprising the Production System PS, the Management System MS and the Information System IS,
Question 1 : how to define the dimensions and the coupling effects in each individual systems

PS, MS and IS.
Question 2 : how to define the dimensions of the coupling effects
- between PS and MS
- between PS and IS
- between MS and IS
- and between PS, MS and IS all together.

2.2. Past contribution

(a) On general concepts and methods

Since 1959, Forrester J.W. (18,19) and his theory of "System Dynamics" start to treat non technical complicate systems, non technical as opposed to technical system considered in the Control Theory. The basic ideas are consisting of representing
- variables by nodes
- and relations (mathematical, or discursive) by a connecting line with or without indication on direction.

The system is isolated from the background by uni-directional arrows. Inside, all the inter-relations between the variables are represented. The resulting graph is a model of the reality and can be used for simulation purposes. The method has been applied to study industrial, economical, financial, urban and world systems. The method is of no use to CAP problems, since it does not contain any decomposition criteria or methodology.In 1977, in this respect, Dufour (20) suggests 2 methods for partitioning complex systems, one is based on the theory of information and the other one on the theory of fuzzy systems.

Since the late 1960's, Mesarovic and his team (21,29) study complicate systems and decomposition procedure. The language used is the control language : input & output variables, criteria, constriants, etc. Two concepts are interesting to CAP problems.:

- static decomposition, the physical system is devided into sub-systems, linked together by criteria, control signals or feedback information.
- dynamical decomposition, introducing various control levels : regulating, optimization, parameter adaptation and learning.

Since 1972, Titli and Sigh (23,24,25) have developped hierarchical control and decomposition methods to technical systems. Their interests are on the methods for establishing various types of general criteria on the basis of the Lagrangian functions. These hiearchical calculus, lienar and non linear, are concerned with continuous-time systems, they are of no interest to the discrete manufacturing CAP problems.

(b) On production systems

There have been very few works for establishing a general theory on the hierarchical structure of complex production systems. This is because there are so many varieties of production systems. As pointed out by Brightam (26), problems are ill-structured, in the sense they are novel, specific, one-of-a kind, unfocused and complex.

In some cases, the hierarchy concept is used to treat partial production problems. Applebaum and Shamash (27) have decomposed some electromagnetic devices and their production lines into 2 levels by optimizing a combined economic function. Hoadley & Heyman (28) have established a two-echelon inventory model to minimize the stocking costs. Ruffli (29) and Freeland (30) have worked on 2-level and 3-level production management so that the resulting goal-programming problem is treatable.

In other cases, concepts and methods are suggested to effectuate efficient decomposition. Zenkov and Sorkin (31) suggest aggregation models for representing systems with non-single-valued relations between inputs and outputs. Range of values are considered instead of the values. In this respect, the approach is similar to the fuzzy approach of Dufour (20). Waller (32) suggests a multi-dimensional hierarchy, the describing-language hierarchy.

c) On CAP systems

Most of the workers were only implicitly aware of the relations between the decisional problems and the decomposition problems. They consider some hierarchical structure and then consider formulate an optimization problem, forgetting who decides finally, and how he decides (information problem). Gilmoue & Lawrance (33) in a glass-industry uses the simulation of decision-making behaviour to aid the planning of the production process. Krajewski and Ritzmann (34) emphasizes the importance of information disaggregation (up-down analysis) in the decision-making in manufacturing and service organizations. Erschler et al (35), Doumeingts & Roubellat (36), Pun (37), from 1977, start to consider the production system and the decision system as a whole.

2.3. A methodology for comprehension

a) Basis

Axiom 1 - In CAP, the joining factor is the decisional process, characterized by decision elements DE and decisional sub-processes DSP.
Axiom 2 - The plannings (static and dynamical) in the production system PS are decomposed as multi-level and multi-branch activities. C
Criteria : management problems and associate decisional problems can be specified, relations between the activities must be established so that the decisional processes can be solved.
Axiom 3 - The management system MS is decomposed as structures of managerial activities with respect to the decisional activities.
Axiom 4 - The information system IS is decomposed as structures of information activities with respect to the decisional activities.
Axiom 5 - For each managerial-decisional

problem, in the PS, there must correspond one responsibility in the MS, one information in the IS.

Axiom 6 - The CAP are global MS-IS-PS activities. A CAP problem is solved if there exists at least one trilogical coherence between PS, MS and IS.

Axiom 7 - The trilogical coherence is optimal if there are maximum MS-efficiency and minimum-IS cost for all CAP problems in PS.

b) Tools

Graphes are common tools used as comprehension models of complicate systems. Two categories of graphs exist : (1) block-diagram and signal-flow diagrams found in Control fields, for describing continuous-time systems ; (2) in logical systems, and OR fields, we have PERT, PETRI-nets, DAN (deterministic activity networks), GAN (Generalized activity networks), GERT (Graphical evaluation and review techniques), GRAFCET. See Moder & Phillips (38), Pritsker (39), Newmann & Steinshardt (40), Elmaghraby (41), Blanchard (42), Daclin & Blanchard (43), Peterson (44). All these graphical methods suffer one weak point : there is no symbol representing the supports of the activities, A new graphical method has been developped, GRAI-nets (Graphs with Results and Activities Interrelated) (45, 46, 47).

c) Results

- GRAI-nets basic symbols : a cercle = state or results, an arrow = activity, an arrowed square = support.
- Representation of production systems (Fig. 2). Three branches :
  (1) Manufacturing : g_1 = preparation, g_2 = machining, g_3 = assembling, g_4 = final mounting ;
  Supports : materials, components, technical informations.
  (2) Procurement : h_1 = define suppliers, h_2 = send orders, h_3 = receive materials.
  (3) Information generation : K_1 = define part list, K_2 = establish nomenclature, K_3 = determine technical processing.
- Representation of static planning system (Fig. 3)
  Three levels :
  (1) Long term level, g_1 = determine long-term program according to investment
  (2) Medium term level, (g_{21}, g_{22}, g_{23}) = determine medium-term program for manufacturing, procurement and information generation, according to sales
  (3) Short term level (g_{31}, g_{32}, g_{33}) = determine short-term executing programs according to resources.
- Representation of dynamical planning system (or program adjustment) (Fig. 4), original program, execution, and perturbations, adjustments, modified executing programs.
- Representation of the relations between the production system PS, the production management system PMS, the decision system for management DG and the Decision-Aid system AID.

d) Understanding of the management problems requiring decisions

- Static planning problems LT, MT and ST
- 1 branch regulation problems
- 2 branch regulation and synchronizing problems
- 2-level reuglation and coordination problems
- multi-branch multi-level regulating, synchronizing and coordination problems.

3 - SPECIFICATION

Each planning problem implies a decisional process. Before looking for the solutions, we must be able to specify the decisional sub-processes and the decision elements.

3.1. Decisional sub-processes and decision-elements

A decisional process DP may be looked as a sequence of activities. A complete DP is consisting of 4 decisional sub-processes DSP having each three activities (Fig. 5)

(a) Program establishment (D_0) :
- data acquisition D_{01})
- data structuring D_{02}
- data presentation D_{03}

(b) Situation analysis (D_1)
- supervising D_{11}
- state comparison D_{12}
- fault locating D_{13}

(c) Prospection of possible actions (D_2)
- prospection D_{21}
- determining criteria and constraints D_{22}
- solution evaluation D_{23}

(d) Decision making D_3
- determining strategy and tactics D_{31}
- feasibility study D_{32}
- decision organizing D_{33}.

For control people, a decisional process is obviously similar to an automatic control process. The following table (Tab. 1) summarizes the important properties to be studied in both processes.

Decision elements may be specified by defining a general formal decision model containing explicitly all the elements of all the decison-sub-processes :

$$D = (PG(I), PT, (G, R, X, XF), (J, C, E, Z), (K, \Gamma 1, \Gamma 2, \Gamma 3, W, ST), (CHA, \Gamma 4), PG(I+1))$$

Definition : a decisional process D is an adjusting process which, from an initial program of activities PG(I) upon perturbations PT defines a new program of activities PG(I+1) on using a system model DO = (G,R,X,XF) an evaluation model (J,C,E,Z), a strategy model (K, Γ1, Γ2, Γ3,W,ST) and an implementation model (CHA,Γ4).

TABLE 1 Comparison of Interesting Properties in Control Process and Decisional Process

Property	Control Process	Decisional process
Observability	How to measure system states via the output	How to locate fault via the activity results
Reachability	Is there a control signal capable to bring the output to a value desired	Is there an action capable to bring the results to desired situation
Stability	How to reach the desired value without oscillations	How to reach the desired results without too much adjustment
Optimality	How to design the control signal to get stability and the optimal value of some criterai	How to design the adjusting actions to reach stability and the optimal value of some criteria

The elements of the sub-models are :

System model DO :

X : set of states or results $\underline{x} \subset X$
R : set of supports : $\underline{r} \subset R$
G : set of activities : $\underline{g} \subset G$
$\underline{g} : (\underline{x}_i, r_i) \rightarrow \underline{x}_{i+1}$
$G : X \times R \rightarrow X$
X_f : expected final results, objective of the program PG(I)

Evaluation model D1 :

J : criteria defined by the decision-maker
C : constraints limiting the supports
E : evaluation activities
Z : evaluation results
$E : (J,C, (X,R)) \rightarrow Z$

Strategy model :

K : planning instructions, KO for PG(I), K1 for PG(I+1). K acts on G, R, J, C.
ST : strategy and tactics for defining K STO for PG(I), ST1 for PG (I+1)
Γ1 : decision-maker activities juging the value of his old KO and STO
W : results of the self-judgment Γ1
$\Gamma 1 : (Z,KO,STO) \rightarrow W$
Γ2 : decision-maker activities defining new ST from old experience
$\Gamma 2 : (ST,W) \rightarrow ST$
$\gamma_{2i} : (st_i, w_i) \rightarrow st_{i+1}$
Γ3 : decision-maker activities defining new planning instructions K2 from the new strategy ST2
$\Gamma 3 : (ST,K) \rightarrow K$
$\gamma_{3i} : (st_i, k_i) \rightarrow k_{i+1}$

Planning model :

CHA : field of possible actions, notably on G,R, and eventually on XF
Γ4 : decision-maker activities (or an automatic iteration process) defining the new program from the planning instructions
$\Gamma 4 : (K,PG,CHA) \rightarrow PG$
$\gamma 4 : (K_i, PG(I), CHA) \rightarrow PG(I+1)$

Definition - A convivial control CC is a combined man-machine decisional process :
$CC = (H,SA)$
where we have (Fig. 6) :
- the decision maker :
$H = (K,(CHA,\Gamma 4), (\Gamma 1,\Gamma 2,\Gamma 3,W,ST),PG)$
- the semi-automata :
$SA = (PG,PT,(G,R,X,XF),(J,C,E,Z))$

A convivial Control problem is well defined if all the elements of the CC-process are specified.

We shall treat this specification problem in two parts : the H-part, and the SA-part, In the H-part, since the human decision-maker is involved, the vagueness can only be circumscribed in the decision-aid framework. This will be done in Section 5. The SA-part is conceptually a conventional OR or control dynamical optimization problem. Except two main difficulties : multi-criteria and non-conventional mathematics.

3.2. Past contribution on non-conventional mathematics and multi-criteria

Mathematically, or formally, there are three categories of problems :

(a) Representation of the variables : results (states), supports (parameters), constraints criteria (indices)
(b) Representation of the relations : system activiities, evaluation activities
(c) Representation of the structures : PG, D, CC so that the main properties can be studied : observability, reachability, stability, optimality.

Specification is more concerned with points (a) and (b), resolution is more concerned with (b) and (c). The global aspects of the total set of (a) (b) (c) will be treated in Section 4, here we shall treat the most difficult specification problem : multi-criteria specification. In this respect, our interests in analysing the past contribution are : (1) How the multi-criteria decision-making MCDM processes are stated in general ; (2) How can they be applied in the CAP ; (3) How can the criteria be specified.

Since 1970, the litterature in MCDM is very abundant (48). We shall consider the most important ones :

(a) Books on general decision processes applied to management : Ansoff 1965 (49), Mélèse 1968 (50), Heller 1971 (51), Morris 1972 (52), White 1975 (53), Klingen 1975 (54), Ebert & Mitchell 1975 (55), Eden & Harris 1975 (56), Bridge & Dodds 1975 (57), Hickling (58), Moore & Thomas 1976 (59), Warfield 1976 (60), M&lèse 1979 (61).
(b) Books explicitly on MCDM : Fandel 1972 (62), Wilhelm 1975 (63), Chacko 1976 (64) Keeney & Raiffa 1976 (65), Guigou 1977 (66), Montgolfier & Bertier 1978(67), Hwang & Masud 1979 (68)

(c) Proceedings of symposia on MCDM : Leitmann

1972 (69),Cochrane & Zeleny 1973 (70), AFCET 1974 (71), White 1975 (72), Zadeh et al 1975 (73), Zeleny 1976 (74), Zionts & Thiriez 1977 (75), Gupta et al 1977 (76), Zionts 1978 (77), Fandel & Gal 1980 (78).

(d) Survey articles on some MCDM aspects : Tocher (79), Bernard & Besson (80),Roy (81) Lassibille et Parron (82), Fandel & Wilhelm (83), Roy (84), Steuer (85), Tell (86), Vincke (87), Leeney (88), Zionts & Wallenius (89), Fishburn (90), Rivett (91) Wehrung (92), Farquhar (93), Roy-Vincke-Brans (94), Sweeny et al (95), Mitroff (96) Hirsh et al (97), Tell & Wallenius (98).

The proceedings (c) contain around 300 papers. The surveys (d) refer to around 600 papers (many of them are multi-referenced). The International Abstracts in Operations Research ((IFORS) refers yearly around 100 papers under the item "Decision" The books of the category (a) are generally discursive. The treated subjects are : decision elements and phases, models, required informations, attitudes of the decision-makers. They considered generally one decision-criterion. Many of the authors use the system approach, emphasizing the role of the models. All the books in the category (b) are mathematical : Fandel, Wilhelm, Guigou, Montgolfier & Bertier, Hwang & Masud treat the preferences, utility functions and the goal programming problems, no specific applications are mentionned. Keeney & Raiffa extend their theories to corporate problems. Chacko, in his two volumes, covers corporate, municipal, governmental, societal, military... every situation. Among the proceedings of the category (c), Leitmann is more interested in Differential Games ; Zadeh et al, Gupta et al are more interested in Fuzzy theory. The others are rather related to the OR fields. The surveys of the category (d) are mathematical and methodological ; they treat one of the MCDM aspects : aggregation, preferences, utility function, interactive aspects, programming aspects. As regarding our interests, the past contributions can be summarized as the following.

(a) Clarification in concepts and in terminology

The great variety of the application fields and the associate specialists gives rise to a great confusion in the meaning of the terms. For instance, are quivocal :

- criterion, utility, objective, performance, attribute, value-function, etc.
- decision, action, choice, project, program, alternate, order, etc.

Here, we shall attempt a clarification on the basis of a diagram (Fig. 7) and an example : decision-making to buy a car by a family.

D2 : alternatives (various types of cars)
A : attributes (car characteristics : color, speed, confort, cost...)
PS : scaling possibility
SD2 : scaled alternatives
PAG : aggregating possibility
UF : utility function
BM : board member (family member)
PE : personal evaluation
OP : opinion
PG : grouping possibility
ST : unique conclusion or strategy (final family decision).

At this point, let us emphasize that, in the most general multi-alternative multi-attribute multi-criteria multi-person decisional problem, there are three great problems : attribute sclaing, multi-dimension aggregating and multi-opinion grouping.

b) Clarification in types of decisional process

Decisional processes are different one from the others by their input, finality, criteria deicison makers and output decision. At the present situation, we may enumerate 6 different types :

a) Classification MCDM
b) Choice MCDM (individual or group)
c) Group activity MCDM
d) Individual static activity MCDM
(e) Individual dynamical activity adjustment MCDM
f) Game MCDM.

In Table 2, their characteristics are summarized.

Type	Input	Finality	Criteria	Decision makers	Output decision
Classification MCDM (pattern recognition typology, taxonomy)	Gathered data	Data classification	Dimensions, Topics Types, Characteristics	Analyst	One ways of classifying
Individual or group Choice MCDM	Alternates, symptoms, projects, propositions	Choose	Attributes, qualities. Types of diseases (dimensions), opinions, implications	1 person or 1 group	1 alternate or 1 structure
a) Diagnosis	Symptoms	Diagnostic	Diseases	1 physician	1 name of disease or 1 combination
b) Site selection c) Resource allocation	Sites Projects programs	Select Allocate	Qualities, opinions, Attributes, qualities, Advantages, Opinions, Implications (social, economical)	1 group 1 group of advisors	1 site name 1 allocating structure
Group activity MCDM	Programs of activities	Select	Opinions, Informations, Performances, Constraints	Board members	1 program, 1 orientation, 1 policy
Individual static activity MCDM	General policy, Internal & external informations	Define an activity program	Performances, Constraints	1 director	1 program of activities
Individual dynamical activity adjustment MCDM	Perturbations, Static programs, Executing informations	Define adjustments	Performances, Constraints, dynamical global synchronization	1 director	1 adjustment, 1 sequence of adjustments
Game MCDM	Game rules Game play situations	Win	Maximum gain, Personal strategy	1 group of players	Moves

Table 2 : Various existing MCDM processes (conceptual descriptions)

Note 1 - Choice MCDM has three sub-classes : diagnosis, site selection, resource allocation.
Note 2 - Group activity MCDM is very similar to group choice MCDM. However, they are different in the engagement of the decision makers.
Examples :

- Group activity MCDM, the board of an enterprise chooses one specific product for development, the board is engaged, involved in the later activity.
- Group Choice MCDM : the municipality council chooses one school-site among many suggestions, the council is not engaged in the later construction of the school.

Note 3 - Types a, b, c, are static. Type e is dynamical with one decision maker. Type f is dynamical with 2 or more decision-makers (players).

c) Clarification in MCDM application to Production management

These various types of MCDM can be applied to the various stages of the product circuit (see Fig. 1). In Table 4, we summarize four types of MCDM :

- the applicability in the production management problems
- the criteria
- finality of the decision.

TABLE 3 - Criteria and Finality of Production-Management Problems

Type of MCDM	*Production management problems*	*Criteria*	*Finality of the decisions*
Group Choice	- Initial product - Diversification - Merging - International market	Characteristics (attributes) of the suggestion	Choose one (or several) suggestion
Group activity	- R & D project - Long-term production program - LT procurement program - LT resource program - New collection of products	Both the characteristics of the suggested activities and the activity expected performances	An overall strategy
Individual static activity	- Medium-term and short term planning inventory, scheduling - joint program comprising several branches - coordinated program comprising several levels	Expected performances	Final results of the production system
Individual dynamical activity	Adjustment of the preceding program	Expected performances	Final results of the production system

In our CAP problems, we are particularly interested by the two last types of MCDM. In these MCDM, the main problem is how to aggregate the criteria.

3.3. A methodology for specification

a) Basis

Axiom 1 - In CAP of complicate production systems, the first elements to be specified for enabling later decision-aid are related to the system model and the evaluation model

Axiom 2 - From the comprehension methodology results the general structural understanding of these models, both inside the models and between the models.

Axiom 3 - The role of the specification is to establish, from this general understanding, the model elements in each special case.

Axiom 4 - The specification procedure must be a combination of the up-down decomposition process (Descartes) and the decomposition leads to simple treatable problems. The aggregation introduces the coupling effects.

Axiom 5 - In principle, the CAP are base already on decomposed levels, and branches (Fig. 2 and Fig. 3), the coupling effects come from :

- the resource limitations (constraints) and the programs defined from upper to lower levels.
- the executing performances (criteria) reported from lower to upper levels for synchronization and coordination.

b) Tools

We need formal tools and conceptual tools. As formal tools, we use the automata language and logical algebraic language (See the general forms of the models in Section 3.1).

As conceptual tools, we use various notions developped in the MCDM litterature :

- ordered sets to scale the variables
- binary relations to represent the activities (system and evaluation activities)
- preference, dominance, and other concepts to establish coupling effects between criteria and constraints.

c) Results

Two processes for the specification methodology. (1) Establish basic 1-branch and 2-branch models taking into account coupling effects. (2) Decompose the complete system into these basic structures.

Definition 1. The basic convivial control is a procedure $K \rightarrow Z \rightarrow K$ using a manager H and an equipment A for solving a production management problem specified by the desired results Y and the production planning (X_o, X, G, R) on the basis of a decision process

specified by (C,J,E,Z,W,ST,K).
The same model is valid for all the three types of decisionpprocesses : static, dynamical, and dynamical evolutive. The nuances are the floowing :

TABLE 4 Various Interpretations of the Basic c.c. Model

Decision process	Model nuances
a) Static (plannings)	Z : past informations (ST,K,R,G,E,J,C) well defined, X unique
b) Dynamical (adjustments)	(R,G,J,C,E) well defined ; (Z,X,ST,K) evolue with time
c) Dynamical evolutive (adjustments in varying conditions)	(Z,X,ST,K) evolue with time ; (R,G,E,J,C) are adjusted with successive sets of (Z,K).

The analysis and the synthesis of convivial controls are difficult when complex systems are concerned, which is precisely the case of production management systems. Some help can be obtained by effectuating hierarchical decomposition. For this, we need definitions of typical cases and decomposition method.

Convivial controls of one branch of activities

The branch of activities may be : manufacturing procurements or information generation. By extension, this may be one short, medium or long-term planning. The mathematical statement of the static and dynamical convivial controls will be made on an illustrative example : a fabricating process. Let :

x_{o_j} : the initial program

x_{i_k} : activity results at the step i of the component j

g_i : $(g_i^1, g_i^2, \ldots g_i^k)$ k technical fabricating operations at the step i

r_i^k : k machine-tools and associate supports at the step i

x_f : final results required by the program : quantities, qualities, due dates

x^j : set of results = $(x_1^j, x_2^j, \ldots x_j^f)$

G^k = $(g_1^k, g_2^k, \ldots g_f^k)$ set of activities

C^k : capacity constraints of the machine

J_k : criterion, for instance, stock value

R^k = $(r_1^k, r_2^k, \ldots r_j^k)$ set of supports

γ = $(\gamma_1, \gamma_2, \ldots \gamma_m)$ a sequence of asynchronous disturbances (uncorrelated)

Definition 2. Dynamical convivial control of one branch of activities, upon one disturbance, is a static planning problem (of the generalized) scheduling type), within the same or new horizon, constraints and criteria.

Definition 3. Dynamical convivial control of one branch of activities upon a sequence of disturbances, is either a set of static planning problems, or a set of planning and identification problems.

Convivial control of 2-branch activities connected by a common objective

Consider two branches of activities A and B :

$A = (X_A, G_A, R_A, C_A, J_A)$

$B = (X_B, G_B, R_B, C_B, J_B)$

$H_{AB} : X_A \to R_B$ connecting mappings from A-branch results to B-branch supports

$H_{BA} : X_B \to R_A$ connecting mappings from B-branch results to A-branch activities

The type of convivial control will depned on the relations between the elements of A, B and H_{AB}, H_{BA}.
We distinguish four independant types (Table 5)

TABLE 5 Two-branch Structure Convivial Controls

Structure type (example)	Relations between elements	Convivial control
Type 1 Parallel similar activities A:workshop 1 B:workshop 2	$(X_A, G_A, R_A, C_A, J_A)$ similar to $(X_B, G_B, R_B, C_B, J_B)$	One-branch c.c. with $X = (X_A, X_B), \ldots$ $J = (J_A, J_B)$
Type 2 Parallel different activities A : fabricating B : procurement	$A \cap B = \emptyset$ H_{AB}, H_{BA} introduce time constants	Multi-synchronizing c.c.
Type 3 Series independant activities A : fabricating B : assembling	$A \cap B = \emptyset$ $x_{Af} = x_{BO}$ $M = M(C_A, C_B, J_A, J_B)$ = constant	Multi-scheduling c.c. (with one middle-point synchronization
Type 4 [illegible] inclusive activities A : medium term B : short term	$B \subset A$ $O = O(R_A, R_B, C_A, C_B, J_A, J_B)$ = constant	Two-level coordinating c.c.

Definition 4. There are two kinds of dynamical multi-synchronizing convivial controls : DMS-CC.

a) A slave DMS-CC, upont a sequence of disturbances, is a set of static plannings consisting of taking a sequence of actions $K_{s1}, K_{s2}, \ldots$ modifying only the elements of the slave branch

$K_s = (\delta x_{BO}, \delta G_B, \delta R_B, \delta C_B, \delta J_B)$ so that concerned main supports r_{A12}, r_{A22},... are synchronized to the corresponding A-activities, and so that $x_{AF} = Y$ is satisfied.

b) A concurrent DMS-CC, upon a sequence of disturbances, is a set of static plannings consisting of taking a sequence of actions K_{C1}, K_{C2}..., modifying the elements of the slave branch A, $KC = (\delta x_{BO}, \delta G_B, \delta R_B, \delta C_B, \delta G_A, \delta C_A, \delta J_A)$ so that $x_{Af} = Y$ is satisfied.

Definition 5. There are two kinds of dynamical multi-scheduling convivial controls DMS-CC.

a) A dominance DMS-CC, upon a sequence of disturbances, is a set of static plannings consisting of taking a sequence of actions $K_d = (K_{d1}, K_{d2}...)$ modifyng firstly the free elements R_A, G_A of the dominant branch A so to satisfy the constraints C_A, x_A, Y_a and then the free elements R_B, G_B. The criteria J_A can be eventually relaxed.

b) A coorperative DMS-CC, upon a sequence of distrubances, is a set of static plannings consisting of taking a sequence of actions $K_p = (K_{p1}, K_{p2}...)$ modifying firstly the free elements R, G (either in branch A or in branch B) related to the cheapest C and J, and then succesively the other R, G related to the more expensive C and J.

The 2-level branche situation is very much different than the preceding ones. The structure contains in fact one long-term branch L, but several short-term branches S_i, i = 1, 2, ... m.

$$L = (X_O, X, G, R, C, J, Y)_L$$

$$S_i = (X_O, X, G, R, C, J, Y)_{si}, \quad i = 1,2...m$$

and the distrubances γ may occur at any S_i.

Since S_i is planned within L, it is obvious that in evaluating K_{si}, we must take into consideration

- what has been done, i.e. : the decisions K_{si} to $K_{s(i-1)}$, or the evaluations of the $(C,J)_{s1}$ to $(C,J)_{s(i-1)}$ (past informations)
- what is going to be done, i.e. : the future decisions $K_{s(i+1)}$ to K_{sm} and the future evalutations $(C,J)_{s(i+1)}$ to $(C,J)_{sm}$ (by simulation).

In other words, the decision K_{si} must be consistent with respect to

$$(C,J)_L = \bigcup_{i=1}^{m} (C,J)_{si}$$

Definition 6. A dynamical 2-level 2-inclusive branch coordinating convivial control, upon sequence of distrubances occured at the short term branches, is a set of static plannings consisting of taking some sequence of actions K_{s1}, K_{s2}..., modifying the free elements of the short-term branches.

$$K_{si} = (\delta x_o, \delta G, \delta R, \delta C, \delta J)_{si}$$

so that the set $(C,J)_L = \bigcup_i (C,L)_{si}$, i = 1, 2, ... m, is always logically consistent.

Convivial controls of multi-level multi-branch activities connected by a common objective

This is the most difficult case of controlling complex systems. At the present time, there does not exist any systematic methods of analysis systems (27-36). The synthesis or problem-solving aspect will be discussed in the next section. As for the analysis, the best we can do is :

- decompose the complex system into subsystems as decoupled as possible
- study the conditions making the decomposition possible.

The following decomposition procedure is suggested.

Step 1. Decompose the global complex-system into sub complex-systems (still of a multi-level multi-branch structure) by adopting sub-objectives (per department, per project, per period, etc.).

Step 2. Decompose each sub complex-system into the 4 types of 2-branch structure by considering :

- the similarity of the activities (leading to type 1)
- the master-slave relation depending on the adjustment freedom (leading to type 2a)
- the concurrent relations depending on the manoeuvrability of the free parameters (leading to type 2b)
- the dominance relation depending on the importance given to the branches (leading to type 3a)
- the cooperative relation depending on the marginal economical values of the constraints and criteria (leading to type 3b)
- the inclusive relation depending on the length of the horizons (leading to type 4).

Step 3. Specify the particular conditions of each 2-branch type problem.

Step 4. Specify the residual couplings.

Note . A perfect convivial control requires :

a) the information completeness in time and in space, which requires in turn high investment in information equipments.
b) the logical consistency of the production structure, which requires in turn the exact knowledge of the behaviour of all the human operators.
This is neither possible nor wishable. Convivial controls are therefore tools for successive improvements and behaviour-identifications.

4 - RESOLUTION

4.1. Elements for resolution

The resolution of the CAP problems demands a suitable foramlism and an efficent method. The specificities of these problems are :

- system activities are of a discret nature
- variables are analogical, logical (binary and multi-valued) some ones are defined by range, some others can not be ordered.
- relations are algebraic, or logical.

a) Requirements on the formalism

- understandable by the users (for interactive purposes)
- suitable for building algorithms of resolution
- translatable into computer programs.

b) Types of problems

b1 - Static plannings

- scheduling for one branch of activities
- multi-scheduling for structures containing 2 or more branches.

In favorable conditions, i.e., variables and relations well defined, criteria aggregatable, the problems can be stated as simple programming problems, or VMP (vector-maximization programming) problems.

b2- Dynamical plannings or adjustments

- dynamical reuglating and optimization problems for one branch of acivities
- dynamical multi-branch regulating, optimization and synchronization problems for structure containing 2 or more branches.
- dynamical multi-level regulating, optimization and coordination problems for hierarchical multi-level structures.

The last two types of problems are relatively new in the control field.

4.2. Past contribution

a) Books

- OR approach (scheduling, inventory control, minimum risk deicison, PERT, programming) Howard 1973 (99), Chase & Aquilane 1973 (100) Lambert 1974 (101), Lee & Moore 1975 (102) Ignizio & Gupta 1975 (103), Bierman et al 1975 (104), Morris 1976 (105), Colley Jr et al 1977 (106). These books are interesting for static one-branch plannings.

- Control approach (regulating, dynamical optimization). Banerji & Mesarovic 1970 (107) Albany 1971 (108), Intriligator 1971 (109), Helmer 1972 (110), Bensoussan 1974 (111). In all these works, the basic assumption is the continuous-time systems, which is not our case.

b) Articles

- General works. Golemanov 1972 (112), Porter et al 1976 (113), Bradshaw et al 1976 (114), Gaalman 1976 (115), Mak et al 1976 (116), Axsäter (117), Sakawa 1978 (118), Bamberger 1978 (119). All these works assumes continuous time systems.

- General French works. Campagne 1974 (120), Petit 1975 (121), Erschler 1976 (122), Soenen 1977 (123), Staroswicki 1978 (124).

- GRAI (the author's laboratory) works. Latecoère 1974 (125), Lavergne 1975 (126), Almar 1975 (127), Pun & Boebion (128), Latécoère et al 1975 (129), Mondain 1976 (130), Doumeingts et al 1977 (131), Pun et al 1977 (132), Breuil et al 1978 (133), Breuil 1978 (134), Doumeingts et al 1979 (135). The two last series of works show

the following tendencies :

- attempts to solve discrete manufacturing dynamical problems by using control concepts
- improvements of programming methods for multi-branch or dynamical problems
- efforts towards practical computer packages
- introduction of the decision-aid concepts.

4.3. Suggestion of methods

Up to now, there is no general method capable of solving all the CAP static and dynamical problems, as stated in Section 4.1. From the past contribution (112 to 134) we can extract however three examples each illustrating a typical approach to solving the CAP problems.

Example 1 - Adaptative dynamical scheduling of a workshop (125, 129) (This example illustrates how to solve one-branch dynamical regulating problems).

System :

The factories St Joseph, Bordeaux (France) manufacture boutique-type high quality dress for men and women. Two seasons per year, several hundreds of models per season. The cycle-time is therefore 6-months. The used jersey fabrics are manufactured by the factories themselves, and six months in advance. Delicate coordination problem of fabrics dress models and client-demands. The dress-making is done in two-steps :

- the cutting-step, two lines in parallel with independant activities, one for the fabrics, the other for the lining
- the finishing-step (sleeves, buttons, collars, pockets), several identical lines in parallel.

The factories work on orders. The major part of these orders goes through a normal planning system. About 20 % of the orders are called "special", and they short-cut the normal planning. These special orders require generally a due-date of 15 days.

Problem :

In normal situations, the passing-time of a client order is less than one month, because of these 20 % of special orders (disturbances) this passing time approaches two months. The stock situation of the frabrics is not well known, because there is no continuous recording of workshop movements. How can we improve the system efficiency ?

Analysis and representation :

Adopting conventional control point of view, we find immediat difficulties :

- for defining regulation-reference or optimization criteria
- and, as a consequence, the transfer function or the dynamics.

The finally adopted strategy is a "regulation of waiting-stock levels by adaptive simulation" assuming that, if the waiting stock, measured in operating times, in front of each work station (lady workers) have the same level, then the whole line operates like a synchronized automatic chain (Fig. 3a) finishing workshop ; x_O = input lot ; $(g_1, g_2, \ldots g)$ = workstation activities, support materials ; x_f = finished dresses). The activities model is the operating-time matric (type of work works vs operating times) theoretically known from the records, but readjusted every day. The convivial control activities are then (Fig. 3b) :

- period Θ_1 (every morning), d_1 = determination of the optimal sequence $\underline{os}_1$ with input $(x_O, p)_1$ = normal orders x_O plus the distrubances $\underline{p}$, model $\underline{M}_1$ and equal-level criteria J.
- period Θ_2 (the day), G = $(g_1, g_2 \ldots g)$ ding to finished products $\underline{fp}$ and operat date $\underline{od}$
- period Θ_3 (every morning), ad = adjustment activity, up-dating the model M_1 to M_2

In automata language, the convivial control is specified as the following (Fig. 3c)

ST : model-adaptive equal-level regulation

$K = \delta M = (G_i - M_i)$; $\underline{X}_O = (\underline{x}_O, p)$;

$R = (w_1, w_2 \ldots w_n)$; $(m_1, m_2 \ldots m_n)$;

$X = (\underline{x}_1, \underline{x}_2, \ldots x_n)$; $G = (g_1, g_2 \ldots g_n)$;

$X_f = \underline{fp}$; $Y = \underline{od}$

Resolution :
Two algorithms are implemented as mini-computer packages. The first one, level-equal algorithm, uses a dynamic-programming procedure. The second one, model-up dating algorithm, uses a coincidence-search procedure and a test production program (Fig. 3)

Comments :
In the present convivial control, the (the manager) plays a passive role, collects $\underline{od}$, $\underline{x}_O$, and introduces the

Example 2. Minimum-time synchronization

(127,128) (This example illustraties how to solve 2-branch multi-criteria static planning problems).

System :
The cutting workshop of the Factories St Joseph have two lines : one for the lining and the other for the main cloth. Each line has several work stations. Everyday, there are several hundreds of client orders. All the fabrics go to the main-cloth line, the linings go to the other. At the end of these lines, a synchronizing station assembles the fabric and the lining of the same dress, forms lots and sends them to the finishing workshops.

Problem :
The over-all cutting speed depends on the sequencing of the dresses, because :
- the fabrics are in rolls, at each modification, one must go and fetch the new roll from the stock (transfer time)
- at each size or model-modification, one must change the patterns (accommodation time).

Furthermore, the operations of the 2 lines are different. The pieces arrive therefore at the synchronizing-station at different moments, leading thus to global delays. How to sequence for getting minimum passing times and best global synchronization ?

Analysis and specification :
The structure of the cutting activities is shown in Fig. 10a. The sequences K' and K" are to be defined. G' = $(g'_1, g'_b, \ldots)$ and G" = $(g''_1, g''_2, \ldots)$ represent the cutting activities. $\underline{f}_1$ represents the final lot. The strategy ST is :

a) Since fabrics are very expensive, adopt the following relation :
   fabric line : master
   lining line : slave.

b) Split the problem into three sub-problems Sbp1, Sbp2 and Sbp3.

Sbp1 : from the set of n initial orders, extrat a sub-set n' so that :

$$\underset{n'}{\text{Min}}\ J_1 = \underline{p}^T (\underline{n} - \underline{n}')$$
$$\text{card}\ (n') < k$$

$\underline{p}$ = priority vector ; k = a number characterizing the workshop situation.

Sbp2 : from $\underline{n}$'form a sequence K' for the fabrics so that :

$$\underset{K'}{\text{Min}}\ J_2 = K'^T O$$
$$\underline{b} - \varepsilon < A^T K' < b + \varepsilon$$

O = $f(\alpha_{ij})$, α_{ij} = costs for modifications (fabrics, sizes or models)

A = operating-time matrix

B = optimal waiting level defined by the manager.

Sbp3 : the sequence K' is applied to the fabric line. Choose K" for the lining line so that

$$\min_{K''} J_3 = \Theta(K') - \Theta(K'')$$
$$K = (K',K'')$$

Θ (x) is the passing time when the sequence x is applied.

Resolution :
Sbp1 is solved by enumeration. Sbp2 is solved by truncated integer linear programming. Sbp3 is solved by nonlinear programming with penalization. The decisional process is consisting of :

$x_o = n$
$x_f = K = (K',K'')$
$y = \Delta\Theta = \Theta(K') - \Theta(K'')$ non-synchronization degree
$J = (J_1, J_2, J_3)$
$G = (G', G'', O)$
$C = (p,b)$

Example 3 - Dynamical synchronization and coordination (133,134,135)
This example illustrates how to solve the global CAP problems in a decision-aid approach.

System :
The problem concerns one department of an important French enterprise. This department manfuactures small electromechanical equipments. The simplified struction of the production activities comprises (Fig. 11) :

- one branch of procurement : foundry pieces, electric and mechanical components
- one branch of manufacturing : preparation, assembling, wiring, integrating, control, stocking and expediting.
- one special procurement branch
- one repairing branch.

Problem :

1. Planning and activity scheduling according to a given production program
2. Determining required supports and particularly human and machine resources
3. Fault locating and prospection of possible actions
4. Synchronization between procurement, manufacturing and repairing.
5. Coordination between LT and MT activities.

Specifications :
The adopted convivial control is of the decison-aid type, interactive and conversational where :

- the manager H plays an active role, i.e. : he introduces data, consults, modifies the parameters, asks for evaluations, and decides
- the automata A effectuates the links between the production and the decisional activities.

Resolution :
For communicating with the mini-computer, the langage of the first-order predicate is used. For solving the problem, the procedure of decision-trees is used. Two types of predicates are used :

- predicates realted to the GRAI-nets representing the production structure. They permit to test the results of the activities and to locate the origin of the perturbations.
- predicates related to the decisional process. they permit to modify the state of the production system, and to minimize the effects of the perturbations).

(The details are omitted here due to lack of place. See the references 133-135).

5 - IMPLEMENTATION

5.1. Problems

Stated in a narrow sense, the implementation problem, once the computer system (hardware, software) defined and chosen, is how to put this computer system into use practically.

Stated in a broad sense, the implementation comprises a number of problems :

a) the objectives and their coordination
b) who are the users
c) what are the interactive utilization method
d) centralized computer system or distributed computer systems
e) cost-effectiveness studies
f) development
g) exploitation.

These problems are stated in a logical order. Looking at the present situation, we often find the illogical reality. Some computer is installed somewhere in a company (for instance, for the accounting purposes), then one seeks how, why, for whom this can be extende elsewhere. Therefore, more insights seem appropriate.

a) Objectives

The formulation of the objectives of the CAP is complicated by the following considerations :

- possible relations between the CAP and administrative problems : accounting, personnel management, sales management, budget control... etc.

- possible relations between the CAP and other technical production problems : design, technical processing, testing, maintenance, ... etc (See Fig. 1).

- completeness of the aid within the CAP problems : inventory, procurement, requirement calculations, load calculationqn vairous follow-ups, ...etc. The partial aid on one of these aspects, or on the whole ?

b) Users

- who collects and validates the data ?
- who processes the data ?
- who utilizes the data ?
- one person for each task, or for several ones ?

c) Interactive methods

- passive
- active conversational
- learning and adaptive

d) Computer system configuration

- one centralized large computer system
- completely decentralized small computer systems
- various intermediate combined systems.

e) Cost-effectiveness studies

- definition of the effectiveness criterion
- how to evaluate the costs
- methods for the analysis
- how to carry the experimentations.

f) Development

- how to define the projects
- how to form the project teams (production personnel plus computer analysts)
- how to develop the packages
- how to experiment.

g) Exploitation

- how to train teh personnel
- how to coordinate the users of various level.

5.2. Past contribution

a) Books on applied information systems

Yvon & Semin 1970 (136), Morton 1970 (137), Jacq & Jehamin 1974 (138), Simon Editor 1976 (139), Kochhar 1979 (141), Tardieu & Pascot 1979 (142).

b) Interactive methods

Steuer 1975 (85), Ziouts & Wallenius 1976 (89), Wehrung 1978 (92), Steiner 1975 (144), Pollack 1976 (145), Vincke 1976 (147), Sandwall 1978 (151).

c) Computer system evaluation

Scheer 1978 (145), Sarin 1977 (148), King & Schrems 1978 (149), Luhmann 1978 (150), Berlin 1978 (153), Ein-dor & Segev 1978 (152, 154).

Three aspects : specification and choice of the computer system, package development and exploitation of the computer system, are well developped in the books of the category (a). All the authors stress out the importance of careful design, mutual understanding between user and analyst, and important development effort.

The works on interactive methods are generally related to situation where several decision-makers are implied. The concerned problems are more of a societal or pure economical nature. In production systems, this could be the case of high-level management problems, such as investment, product choice, etc. At the planning level, it must be assumed that there is an unique decision-maker who gives the final judgement. The interactive problem is therefore rather a methodological problem than a problem of criteria.

The cost-benefit evaluation remains an unsolved problem. Even in cases where the computer system (therefore the cost) is well defined, the benefit remains difficult to evaluate. The adoption of computer system in our enterprise for plannings is still a problem of necessity, faith or gambling.

The so-called "integrated production management by computor" has started from 1960. Its practical introduction has been very slow because :

- the computer system is costly (still),
- the really efficient and operational packages are longthy to develop to-day,
- the computer specialists are rare in the industries,
- the problems themselves are difficult to formulate precisely.

Technological progress solves slowly the cost problem. The increasing number of computer specialists solve slowly the second and the third problems. The main problem is the formulation.

5.3 - Suggestion for a methodology

a) Understanding of the aid

Understand the aid means : understand the type of aid required in the CAP, the roles which must be played by the man (managers) and by the machine (computer). For completeness, we must effectuate a global understanding (the over-all decision process) and a detailed understanding (various decisional sub-processes). For this, we indicate in Table 5 various types of global decision-aid and the associate roles of the man and of the machine. In Table 6, the same elements for the detailed understanding.

TABLE 6-Various types of global decision-aids

Type	Role of the computer	Activities of the manager
1. Minimal aid (supervising type)	Simulate the production system model	Evaluate the results. Elaborate strategy and tactics. Find actions
2. Automatised optimization aid (suggestion type)	Simulate system and evaluation models. Propose actions eventually.	Elaborate strategy and tactics. Find final actions.
3. Aid to manager's reflexion (conversational type)	Simulate system evaluation and strategy models	On-line interaction with computer. Find final actions

4. Aid to the improvements of the models (learning type)	Case 3 with adjustable parameters in the models.	On-line interaction. Improve the models. Find final actions.
5. Unmanned machine-shops.	Simulate system, evaluation, strategy and programming models. Find final actions.	Nothing

TABLE 6 - Various types of aids of decisional sub processes

Type of the aid	Information, models and algorithms required for the computer	Manager's intervention
1. To the planning	Past information. Predictions System models	Production program. Support parameters
2. To the supervising	Program to be executed. Follow-up info.	Creation of a checking board Modification of its elements.
3. To the fault locating	Given programs Dynamical data System model Inversion algorithm.	Choice of the suspect support branches
4. To the prospection of the field of actions	System model Support info.	Variations of the activities Variations on the parameters
5. To the evaluating of the actions	System model Evaluation model Resolution algorithms.	Variations of the activities, models, criteria and constraints. Actions during the interactions
6. To the learning	System model Evaluation model Strategy-model	Variations of the activities, models, criteria and constraints. Variations of the strategy

b) Understanding of production and management processes

Once defined the types of aid, we must know the required informations related to the production process, and how is consisted the management process. This can be done by an ascending and a descending analysis.

Descending analysis. Aiming to determine the activity programs (LT, MT, ST), the required supports (resources and informations) for their execution :

- number of the adapted horizons : 3, 4 ?,
- conditions for determining LT programs : covered term period, sales hypothesis, machine and labour investment possibility,
- conditions for determining MT programs : covered term period, resource allocation, optimization criteria, strategies for resource coordination and activity synchronization,
- between the workshops,
- between the manufacturing and the procurement,
- between the stocks and the workshops,
- conditions for determining ST program : manufacturing type (flow shop, job shop),
- requirement for detailed scheduling,
- criteria to be adopted : minimum stock, minimum waiting, maximum productivity, minimum passing time,
- constraints in executing : machine, labour?
- sources of technical, quality control, budget information.

Ascending analysis. Determine from ST to LT the adjusting conditions :

- nature of the possible perturbations,
- Short-Term ST level :
  - . missing materials,
  - . machine cut-down,
  - . labour absence,
  - . program modifications,
  - . error in technical informations,
  - . etc...
- Medium-Term MT level :
  - . ST perturbations not absorbed in the limits of the allocated resources,
  - . modification of sales conditions,
  - . etc...
- Long-Term LT level :
  - . new techniques,
  - . new products,
  - . new market situation,
  - . etc...
- possible actions at the considered horizon level,
- possible actions at the level of the upper horizon :
  - . required informations on the perturbations and on the executing states,
  - . information aggregation methods,
  - . possible fields of actions.

c) Trilogical coherence (155)

The CAP imply three systems : production, management and information systems. The resolution of the CAP problems implies three parallel series of activities : adjustment of the production activities, managing activities, and the information processing activities.

Obviously, the resolution will be correctly done if there is a mutual coherence between these three series of activities.

A categorical approach is suggested in (155) for studying this trilogical coherence. The basic ideas of this approach are the following.

1. Define :
   ADJ : activities for adjusting production programs,
   ROR : activities for reorganizing the management circuits,
   RCF : activities for reconfiguring the information activities.

2. For each type of ADJ, determine the corresponding ROR and RCF.

3. There is a trilogical coherence if for one ADJ-path, there is at least one ROR-path and one RCF-path.

4. There is an optimal trilogical coherence, if the ROR-path and the RCF-path are of minimal distance.

6 - CONCLUSION

In this paper, a system approach has been used for surveying the works on the field of Computer-Aided Plannings. For this, four categories of models have been considered : comprehension, specification, resolution and implementation. From the past contribution, useful results have been extracted. For each category, and for the global problem, a methodology has been suggested for the analysis and the synthesis.

Fig. 1 - Product-path, production system and computer-aids

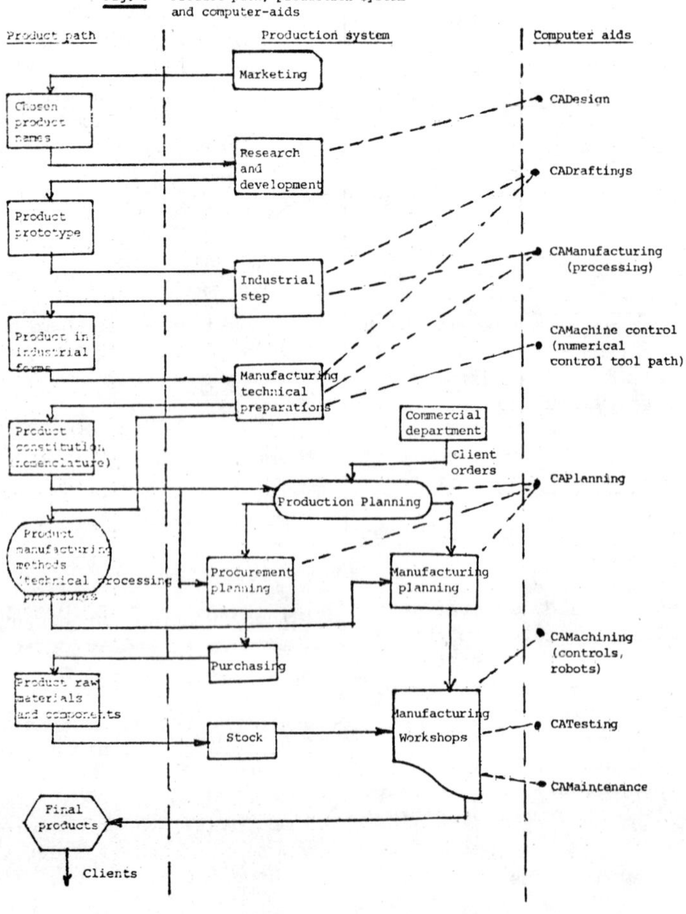

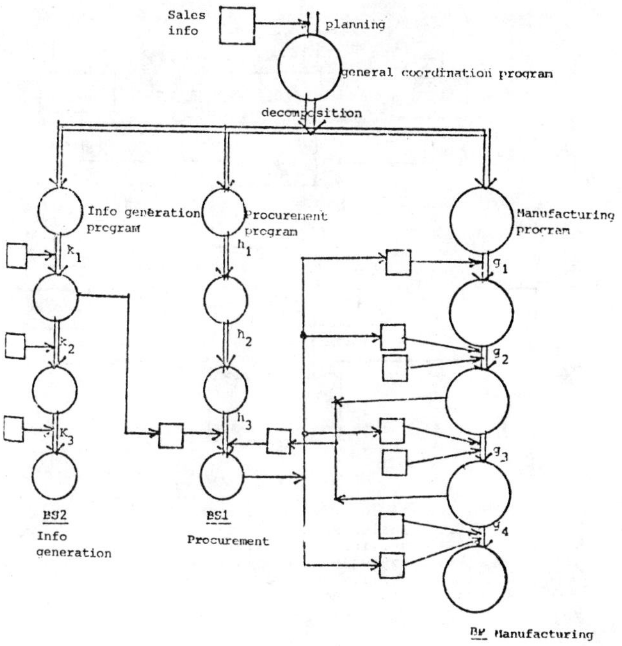

Fig. 2

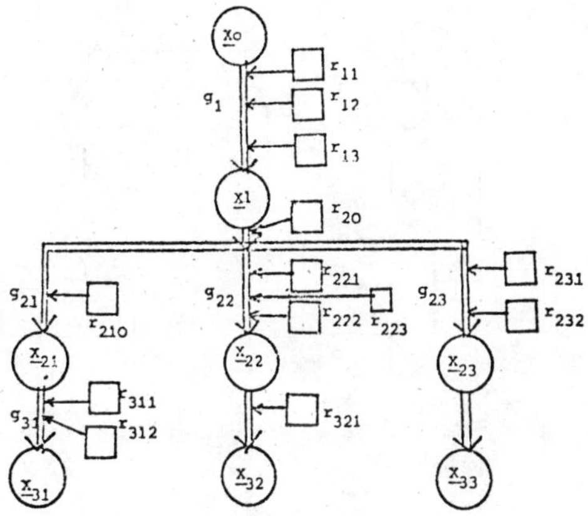

Fig. 3 - Structure of the Static Plannings

Fig. 4 - Structure of the Dynamical plannings

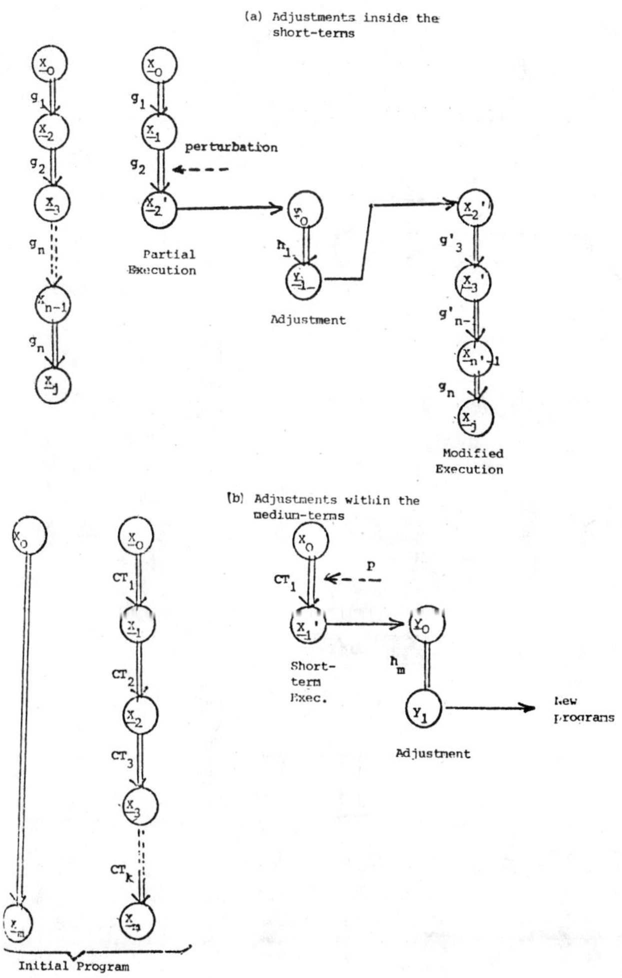

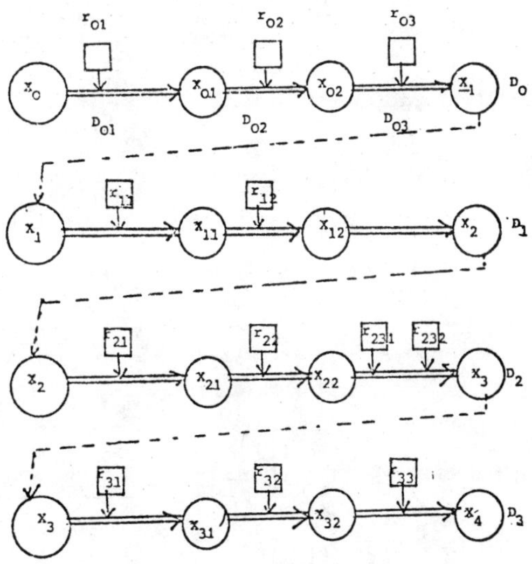

Fig. 5 - Structure of the Decisional Process

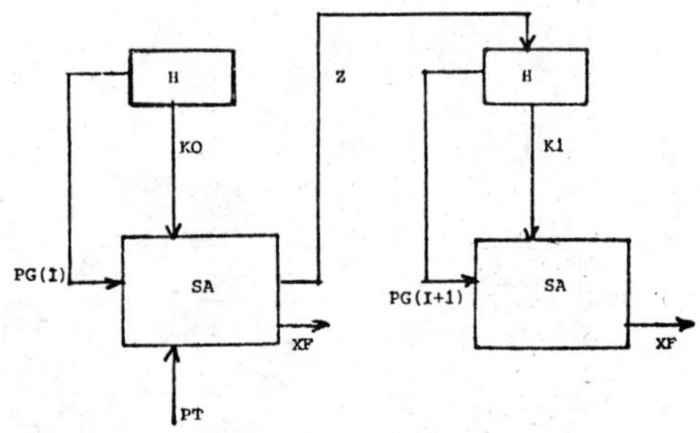

Fig. 6 - Convivial Control Model

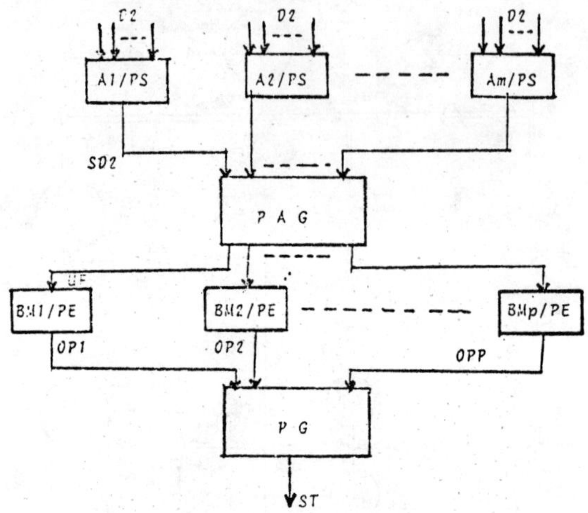

D2 : alternatives - A : attributes
SD2 : scaled alternatives
PS : scaling possibility
PAG : aggregating possibility
UF : utility function
BM : board members
PE : personal evaluations
PG : grouping possibility
OP : opinion
ST : unique conclusion on Strategy

Fig. 7 - Alternative-scaling, attribute aggregation and opinion-grouping process

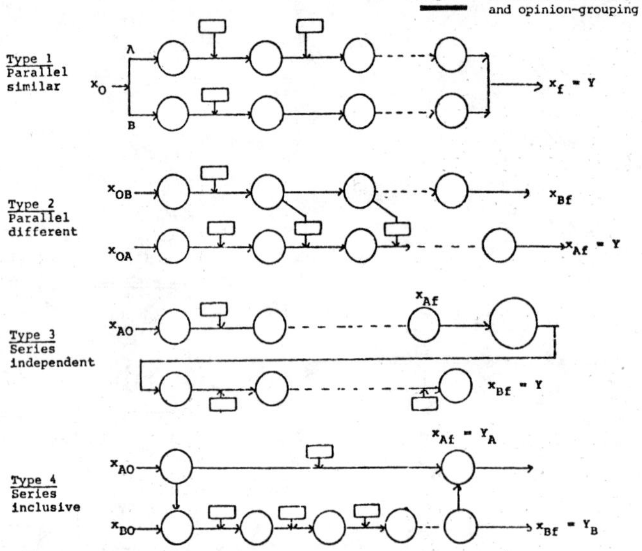

Fig. 8 - Structure-types of 2-branch activities having the same objective

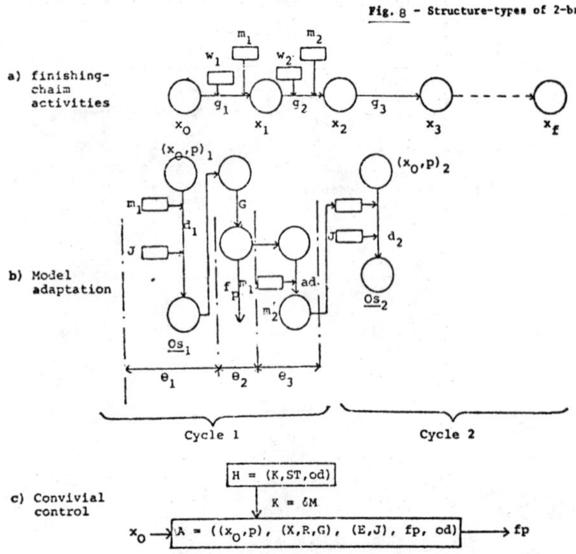

Fig. 9 Dress-making controls

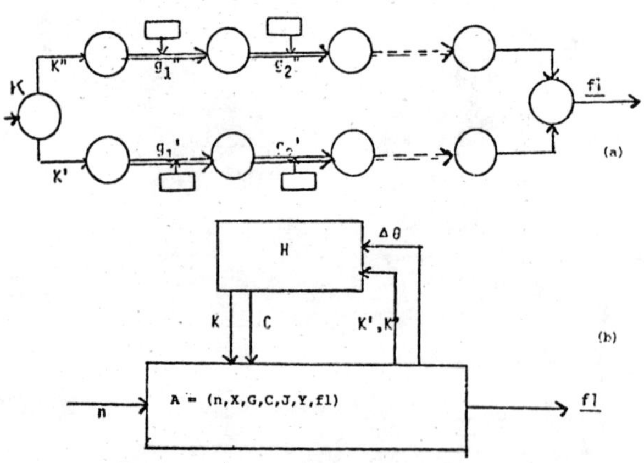

Fig. 10 - Decisional process in the cutting lines
(a) cutting lines
(b) specification of the decisional process

REFERENCES

(1) Pun L.. Computer Assisted Static and Dynamical Plannings for Production Activities. IFAC/CAD Symposium, August 29-31, 1979, Zurich.

(2) Dumas P. (1972). Méthodes d'ordonnancement paramétrique d'un atelier. Rapport 007 ENSTA, Paris.

(3) Rinnoy Kan A.H.G. (1973). Machine Scheduling Problems. Report R/73/4. Interfaculty for Graduate Studies in Management, Rotterdam.

(4) Pun L. (1974). Couplage entre les problèmes et les méthodes de gestion automatisée. Journée de l'AFCET, June 4, Paris.

(5) Gelders F.F. (1974). Application of the Branch and Bound Principle to Machine Scheduling Problems. Note Technique Fac. Sc. Ec. et Soc., Namur.

(6) Conway et all (1967). Theory of Scheduling. Addison Wesley.

(7) Baker K.R. (1974). Introduction to Sequencing and Scheduling. John Wiley.

(8) Aggarwal S.C. (1974). A review of current inventory theory and its applications. Int. J. of Prod. Res., V12, n°4, p. 443.

(9) Pun L. (1974). Problèmes et méthodes mathématiques dans la gestion des matériels. Seminar on Production Management, Seville.

(10) Pun L. (1976). Theories and Facts in Production Mangement Problems. Invited paper, Internet IV, sept. 13-17, Birmingham.

(11) Churchman C.W. (1968). The System Approach. Dell Publ. Co.
French translation : Qu'est-ce que l'analyse par les systèmes. Dunod (1974).

(12) Kalman R.E., P.L. Falb & M.A. Arbib (1969). Topics in Mathematical Theory. Mc Graw Hill.

(13) Couger J.D. & R.W. Knapp Editors (1974). Systems Analysis Techniques. John Wiley.

(14) Lemoigne J.L. (1974). Les Systèmes de décision dans les organisations. PUF, Paris.

(15) Lemoigne J.L. (1977). Les systèmes d'information des les organisations. PUF, Paris.

(16) Gordon G. (1969). System simulation. Prentice Hall.

(17) Klir G.J. Ed. (1972). Trends in General System Theory. Wiley Interscience.

(18) Forrestier J.W.
a) Industrial Dynamics. MIT Press (1961),
b) Principles of Systems. Wright-Allen Press (1968),
c) Urban Dynamics. MIT Press (1969),
d) World Dynamics. Wright-Allen Press (1971),
e) Collected Papers. Wright-Allen Press (1975),
f) The life Cycle of Economic Development, Wright-Allen Press (1973).

(19) Aracil J. (1978). Introduccion a la dinamica de sistemas. Alianza Universidad, Madrid, 302 p.

(20) Dufour J. (1977). Analyse structurale et partition des systèmes complexes. Application aux systèmes macroéconomiques. RAIRO Automatique/ Systems Analysis & Control, V11, n°4, 335-354.

(21) Mesarovic M.D., D. Macko & Y. Takahara (1970). Theory of Hierarchical Multilevel Systems. Academic Press, New York.

(22) Mesarovic M.D. & Y. Takahara (1974). Foundation for the Mathematical Theory of General Systems. Academic Press, New York.

(23) Titli A. (1975). Commande Hiérarchisée et Optimisation des systèmes complexes. Dunod, Paris.

(24) Singh M.G. & A. Titli (1978). System decomposition, optimization and control. Pergamon.

(25) Singh M.G. (1977). Dynamical Hierarchical Control. North Holland.

(26) Brightman H.J. (1978). Difference in ill-structured problem-solving along organizational hierarchy. Decision Sciences, V9, n°1, January, 1-18.

(27) Applebaum J. & Y. Shamash (1977). A method for the optimization of a Production line of electromagnetic devices. Tr IEEE, V. SMC-7, n° 6, June, 443-452.

(28) Hoadley B. & D.P. Heyman (1977). A Turo-Echelon Inventory Model with Purchases, Dispositions, Shipments, Returns and Transshipments. NRLQ (V.S.) 24, march, pp 1-20.

(29) Ruffli T. (1971). A Generalized Goal Decomposition Model. Man. Sc., V17, n° 8, april.

(30) Freeland J.R. & N. R. Baker (1975). Goal Partitioning in a Hierarchical Organization. OMEGA, The Int. J. of Mgmt Sc., V3, n° 6, 673-687.

(31) Zenkov V.V. & L.R. Sorkin (1977). Range Models and Models with variable Coefficients in Hierarchical Control Systems. Automatika ; Telemekhanika, n° 10, pp 139-148, octobre.

(32) Waller R.J. (1976). The Synthesis of Hierarchical Structures : Technique and Applications. Decision Sciences, V7, 659-674.

(33) Gilmour P. & J.M. Lawrance (1977). Simulation of Decision Making behaviour in a process production plan. Int. J. of Prod. Res., V15, n° 4, 351-364.

(34) Krajewski L.J. & L.P. Ritzman (1977). Disaggregation in Manufacturing and Service Organizations : Survey of Problems and Research. Decision Sc., V8, 1-18.

(35) Erschler J., G. Fontan & F. Roubellat (1976). Multi-level Control Structure and Decision-Aid in Production Management Problem. IJPR, V14, n° 2, march, pp 275-284.

(36) Doumeingts G. & F. Roubellat (1979). Système de conduite de production et d'aide à la décision. RAIRO Automatique,

V13, n° 1, pp 77-92.
(37) Pun L. (1977). Approche méthodologique de modélisation en vue de la maîtrise assistée de la production. Congrès AFCET, nov., Versailles.
(38) Moder J.J. &C.R. Phillips (1970). Project Management with PERT & CPM, Van Nostrand.
(39) Pritsker A. Alan B. (1977). Modeling and Analysis using Q-Gert Networks. John Wiley, 420 pp.
(40) Newmann Klaus & Ulrich Steinhardt (1978). GERT Networks and the time-oriented evaluation of projects. John Wiley.
(41) Elmaghraby Salah E. (1977). Activity Networks - Project Planning and Control by Network Models. John Wiley, 443 pp.
(42) Blanchard M. (1979). Comprendre, maîtriser et appliquer le GRAFCET. CEPADUES, Toulouse, 176 pp.
(43) Daclin E. & M. Blanchard (1976). Synthèse des systèmes logiques. CEPADUES, Toulouse.
(44) Peterson J.L. (1977). Petri Nets. Computing Surveys, V9, n° 3, sept.
(45) Pun L. (1977). Vers une méthodologie de l'approche systémique dans l'aide à la gestion de production. Invited paper, Colloque SOGETI, 10 janv.,Liège.
(46) Pun L. (1977). Joint use of predicate and fuzzy logics for solving production management problems. First World Congress on Mathematics at the Service of Man, Barcelone, July 11-15.
(47) Pun L. (1978). Basic formal elements for an efficient management-fuzzy-learning in organizational structure. Third ICCS Symposium, Tokyo-Kyoto, nov. 3-7.
(48) Pun L. (1979). Multi-criteria decision aid-making in production management problems. Third MCDM Symposium, Aug. 20-24, Konigswinter.
(49) Ansoff H.I. (1965). Corporate Strategy. Mac Graw Hill ; Penguin(1968)
(50) Melese J. (1968). La gestion par les systèmes, essai de prascéologie. Ed. Hommes et Techniques, 235 p.
(51) Heller F.A. (1971). Management Decision Making. Van Garcum, 140p.
(52) Morris W.T. (1972). Management for action - Psychological decision making. Reston Publ. Co., 223p.
(53) White D.J. (1975). Decision Methods. John Wiley, 274 p.
(54) Klingen J.S. (1975). Company strategy-A managerial approach for strategic control. Lexington Books, 205 p.
(55) Ebert R.J. & T.R. Mitchell (1975). Organizational decision process - Concepts and Analysis. Crane, Russack & Co, 331 p.
(56) Eden C. & J. Harris (1975). Management decision and decision analysis. Mac Millan, 257 p.
(57) Bridge J. & J.C. Dodds (1975). Managerial decision making. Croom Helm, London, 309 p.
(58) Hickling A (1974). Managing Decisions-The strategic choice approach. Mantec Publ., RUGBY, 58 p.
(59) Moore P.G. & H. Thomas (1976). The Anatomy of Decisions. Penguin Management Texts, 242 p.
(60) Warfield J.N. (1976). Societal systems : Planning, Policy and Complexity. John Wiley, 490 p.
(61) Melese J. (1979). Approches systémiques des organisations. Vers l'entreprise à complexité humaine. Ed. Hommes et Techniques, 158 p.
(62) Fandel G. (1972). Optimale Entscheidung bei mehrfacher Zielsetzung. Springer Verlag, LN on Ec.MS, n° 76, 121 p.
(63) Wilhelm J. (1975). Objectives and multi-objective decision-making under uncertainty. Springer Verlag, LN in Ec. & MS, n° 112, 111 p.
(64) Chacko G.K. (1976). Applied Operations Research / Systems Analysis in Hierarchical Decision-Making (2 volumes). North Holland, 480 p and 555 p.
(65) Keeney R.L. & H. Raiffa (1976). Decision with multiple objectives : preferences and value tradeoffs. John Wiley, 569 p.
(66) Guigou J.L. (1977). Méthodes multidimensionnelles. Dunod, 286 p.
(67) Montgolfier J. et P. Bertier (1978). Approche multicritère des problèmes de décision. Ed. Hommes et Techniques, 248 p.
(68) Huang C.L. & A.S.M. Masud (1979). Multiple-objective decision making - Méthodes & Applications - A state of the art Survey. Springer Verlag, LN on EC-MS, n° 164, 351 p.
(69) Leitmann G. Editor (1972). Multicriteria decision-making and differential games. Plenum Press, 461 p.
(70) Cochrane J.L. & M. Zeleny Editors (1972). Multiple criteria decision-making. MCDM Seminar, oct.26-27, Univ. South California.
(71) AFCET Editors (1974). Aide à la décision. Congrès AFCET, 15-17 mai.
(72) White D.J. Editor (1973. The role and effectiveness of theories of decisions in practice. NATO Conference on Decision, Aug., Luxembourg. Hodder and Stoughton, London, 49 p, (1975).
(73) Zadeh L. et al Editors (1974). Fuzzy sets and their applications to cognitive and decision processes. Proceeding Symposium USA-Japon on Fuzzy sets, Berkeley, June 1-4. Academic Press, 496 p. (1975).
(74) Zeleny D.M. Editor (1975). Multicriteria decision making. Kyoto. Springer Verlag, LN on Ec-MS, n° 123, 345 p. (1976).
(75) Zionts S. & H. Thiriez Editors (1975). Multiple Criteria Decision Making. Second MCDM Symposium, mai 22-23, Jouy-en-Josas. Springer Verlag, LN on Ec-MS, n° 130 (1977).
(76) Gupta M.M. et al Editors (1975). Fuzzy Automata and Decision Processes. Vth IFAC World Congress, Boston. North Holland, 496 p. (1977).

(77) Zionts S. Editor (1977). Multiple Criteria and Problem Solving. Proceedings, Albany. Springer Verlag, LN on EC-MS,n°155, 567 p. (1978).

(78) Fandel G. & T. Gal Editors (1979). Multiple Criteria Decision Making - Theory and Application. Proceedings Hagen/Königswinter, August 20-24. Springer Verlag, LN on EC-MS, n° 177, (1980).

(79) Tocher K.D. (1961). The Role of Models in Operations Research. J.R.S.S. Series A, 124, 2.

(80) Bernard G. et M.L. Besson (1971). Douze méthodes d'analyse multicritère. RAIRO, N V-3, oct.

(81) Roy B. (1971). Problems and methods with multiple criteria functions. Mathematical Programming, V1.

(82) Lassibille G. & C. Parron (1975). Analyse multicritère dans un contexte imprécis. Cahiers IME n° 12, Inst. Math. Economique, Dijon.

(83) Fandel G. & J. Wilhem (1975). On the Theory of Multiple Criteria Decision Making.In (75).

(84) Roy B. (1975). Vers une méthodologie générale d'aide à la décision. Rapport de Synthèse n° 87, SEMA.

(85) Steuer R.Z. An Interactive multiple objective programming procedure employing an algorithm for the vector maximum problem. In (75).

(86) Tell B. Experimental study for 4 multi criteria evaluation methods. In (75).

(87) Vincke Ph. Modélisation des préférences et théorie de l'utilité. In (75).

(88) Keeney R.L. A Group Preference Axiomatization with Cardinal Utility. Man. Sc., V23,n°2, oct. (1976).

(89) Zionts S. & J. Wallenius (1976). An interactive programming method for solving the multiple criteria problem. Man. sc., V22, n°6.

(90) Fishburn P.C. (1977). Multicriteria Choice Function Based on Binary Relations. Op. Res., V25, n°6.

(91) Rivett P. (1977). Multi-dimensional scaling for multi-objective policies. OMEGA, The Int. j. of Man. Sc., V5, n°4.

(92) Wehrung D.A. (1978). Interactive identification and optimization using a binary preference model. Op. Res., V26, n° 2.

(93) Farquhar P.H. (1977). A Survey of Multiattribute Utility Theory and Applications. TIMS 6, 59-83.

(94) Roy B., Ph. Vincke & J.P. Brans (1978). Aide à la décision multicritère. Ricerca Operative, nv serie, n°5.

(95) Sweeny L. et all (1979). Composition vs Decomposition - Two Approaches to Modeling Organizational Decision Processes. Man. Sc., V24, n°14,1491-1499.

(96) Mitroff I. (1977). On being consistent-The Management of Inquiry as a Mulicriteria Decision Problem. TIMS 6, 291-300.

(97) Hirsch G., E. Jacquet-Lagrèze, J. Moscarola (1977). Definition et illustration de quelques concepts utiles à la description et la maîtrise d'un processus de décision. Congrès AFCET, nov., Versailles.

(98) Tell B. & J. Wallenius (1979). A Survey of Multi-criteria Decision Methods and Applications : Utility Theory and Mathematical Programming. The Finish J. of Business Economics, V1, n° 1.

(99) Howard R. (1973). Quantitative Analysis for planning decision. Mac Donald & Evan, 282 p.

(100) Chase R.B. & N.J. Aquilano (1973). Production and operations management. Richard D. Irwing, 691 p.

(101) Lambert P. (1974). La fonction Ordonnancement. Les Editions d'Organisation, Paris, 468 p.

(102) Lee S.M. & L.L. Morre (1975). Introduction to Decision Science. Petrocelli /Charter, 589 p.

(103) Ignizio J. & J.N.D. Gupta (1975). Operations Research in Decision Making. Crane, Russak & Co, New York, 343 p.

(104) Bierman, Bonini & Hausman (1975). Quantitative analysis for business decision. Irwin, Fifth Ed.

(105) Morris W.T. (1976). Engineering economical analysis. Reston Pub., 352 p.

(106) Colley Jr.J.L., R.D. Landel & R. Fair (1977). Production operations-planning & control. Holden-Day Inc. 849 p.

(107) Banerji R.B. & M.D. Mesarovic Editors (1970). Theoretical approaches to non-numerical problem solving. Springer Verlag,LN on EC-MS, n° 28, 466 p.

(108) Albouy M. (1971). La régulation économique dans l'entreprise. Tomes 1 et 2, Dunod.

(109) Intriligator M.D. (1971). Mathematical Optimization and Economic Theory. Prentice-Hall, 508 p.

(110) Helmer J.Y. (1972). La commande optimale en économie. Dunod, Paris, 235 p.

(111) Bensoussan A. (1974). Management Application of Modern Control Theory. North Holland.

(112) Golemanov L.A. (1972). System theoretical approach in the projecting and control of industrial production systems. Thesis Dr of Technology, Helsinky Univ. of Technology.

(113) Porter B., A. Bradshaw & D. Daintith (1976). Synthesis of control for manufacturing systems using the control canonic form. Int. J. of Systems sc. V7, n° 12, 1417-1434.

(114) Bradshaw A., D. Daintith & B. Porter (1976). Synthesis of control policies for production inventory tracking systems. Int. J. of Systems Sc., V7, n° 5, 513-521.

(115) Gaalman G.J. (1976). Optimal linear control of linear production-inventory problems. Int. J. of Systems Sc., V7, n° 9, 1005-1015.

(116) Mar R.L., A. Bradshaw & B. Porter (1976). Stability of production-inventory systems in corporating un-filled - Order backlogs with retarded control policies. Int. J; of Systems Sc., V7, n° 6, 709-721.

(117) Axater S. (1977). Aggregate control using parameters of the production-inventory control systems. Fourth ICPR, Aug., Tokyo.
(118) Sakawa M. (1978). Solution of multi-criteria control problems in certain types of linear distributed-parameter systems by a multicriteria simplex method. J. of Math. Analysis & App. 64, 181-188.
(119) Bamberger W. & R. Isermann (1978). Adaptative on-line steady-state optimization of slow dynamic process. Automatica, V14, pp 223-230.
(120) Campagne J.P. (1974). Analyse et conception d'un système d'information pour la gestion de production. Thèse Dr-Ing., Univ. Claude Bernard, Lyon.
(121) Petit J.F. (1975). Conception et réalisation d'un système informatique pour le contrôle de la production. Thèse Dr-Ing., Univ. Claude Bernard, Lyon.
(122) Erschler J. (1976). Analyse sous contrainte et aide à la décision pour certains problèmes d'ordonnancement. Thèse d'Etat, Univ. Paul Sabatier, Toulouse.
(123) Soenen R. (1977). Etude des systèmes de conduite en temps réel en vue de la commande d'un unité de fabrication. Thèse d'Etat, Univ. Valenciennes et Lille.
(124) Staroswicki M. (1978). Solution of multi-criteria decision problems by means of a particular cooperative equilibrium. State-Dr Thesis, Univ. Lille.
(125) Latecoère L. (1974). Contribution à l'étude d'un système d'ordonnancement. Thèse Dr 3e cycle, Univ. Bordeaux 1.
(126) Lavergne Ch. (1975). Système d'ordonnancement à moyen terme avec optimisation assistée sur mini-ordinateur. Thèse Dr 3e cycle, Univ. Bordeaux 1.
(127) Almar A. (1975). Régulation et optimisation multi-critère du fonctionnement d'un atelier de coupe dans l'industrie de vêtement. Thèse 3e Cycle, Univ. Bordeaux 1.
(128) Pun L. & J. Boebion (1975). A series parallel multi-criteria model for a scheduling problem in dress-making industry. MCDM Symposium, may, Jouy-en-Josas.
(129) Latecoère L., J. Boebion & L. Pun (1975). Model Adaptation Scheduling. Third ICPR, Aug., Amherst, USA.
(130) Mondain M. (1976). Système de calcul rapide des plans de charge à long terme. Thèse Dr 3e Cycle, Univ. Bordeaux 1.
(131) Doumeingts G., L. Pun, M. Mondain & D. Breuil (1977). Decision making systems for the production control planning and scheduling. Fourth ICPR, Aug., Tokyo.
(132) Pun L., J. Boebion & A. Almar (1977). Scheduling method synchronizing two-lines of activities. Fourth ICPR, Aug., Tokyo.
(133) Breuil D., J. Grislain, L. Pun & G. Doumeingts (1978). Trees and Predicates in Production Control Discrete Problems. IFAC Workshop on Control & Management, Sept., Toulouse.
(134) Breuil D. (1978). Contribution à l'étude de l'aide à la décision dans la gestion des systèmes complexes avec application à la gestion de production. Thèse Dr-Ing., Univ. Bordeaux 1.
(135) Doumeingts G., D. Breuil, J. Grislain & L. Pun (1979). Decision-aid and Predicates in Production Control. Computers in Industry, V1, n° 2, Dec., 99-109.
(136) Yvon P.J. & C. Semin (1970). Comment concevoir un système intégré de gestion. Entreprise Moderne d'Edition, 260 p.
(137) Morton M.S.S. (1970). Management Decision Systems-Computer based support for Decision-Making. Division of Research, Havard Business School, Massachussetts, 216 p.
(138) Jacq J.F. & L. Jehanin (1974). La rentabilité des systèmes informatiques dans l'entreprise. PUF Paris, 234 p.
(139) Simon J.C. Editors (1976). Computer Oriented Learning Processes. Noordhoff Leyden, 595 p.
(140) Ramalingam P. (1976). System Analysis for Managerial Decisions - A computer Approach. John Wiley.
(141) Kochhar A.K. (1979). Developement of computer-based production system. Edward Arnold, 274 p.
(142) Tardieu H. & D. Pascot (1979). Conception d'un système d'information. Les Editions d'Organisation, Paris, 192 p.
(143) Grindlay A. (1974). Towards Sophisticated Management Information Systems. INFOR, V12, n°3, Oct., 312-317.
(144) Steiner R.T. An interactive multiple objective programming procedure employing an algorithm for the vector maximum-problem. In (75).
(145) Scheer A.W. (1978). Ein Optimierungsmodell zur Hardwareanswahl. Angnandte Informatik, 104-111.
(146) Pollack M. (1976). Interactive Models in Operations Research - An introduction and some future research directions. Compt & OR, V3, pp 305-312.
(147) Vincke Ph. (1976). Une méthode interactive en programmation linéaire à plusieurs fonctions économqiues. RAIRO, RO, V10,n°6, juin.
(148) Sarin R.K. (1977). Experimental results of an approach for evaluating multi-attribute alternatives. Decision Sc. V8, n° 4, 722-734.
(149) King J.L. & Ed. L. Schrems (1978). Cost-Benefit Analysis in Information System Development and Operation. Computing Surveys, V10, n° 1, March, 19-33.
(150) Luhmann S. (1978). Ziele und Strategien für die Zuteilung von Bearbeitungsplätzen. Angewandte Informatik, 474-481.
(151) Sandewall E. (1978). Programming in an Interactive Environment - The LISP Experience. Computing Surveys, V10, n° 1, March.

(152) Ein-Dor Th. & E. Segev (1978). Organizational context and the success of management information systems. Man. Sc., V24, n° 10, June, 1064-1077.

(153) Berlin V.N. (1978). Administrative experimentation - A methodology for more rigourous "mudding through". Man. Sc. V24, n° 8, April, 789-799.

(154) Ein-Dor Th. & E. Segev (1978). Strategic planning for management information systems. Man. Sc., V24, n°15, Nov., 1631-1641.

(155) Pun L. (1980). Trilogical Coherence in Multicriteria Decision-Aid Making - A Methodological approach to Computer Aided Management. 4th MCDM Symposium, Aug.10-15, Delaware.

(156) Bon L. (1979). La formulation des besoins en information dans les entreprises - Etude linguistique et proposition pour une méthode de classement. Thèse 3e Cycle, Univ. René Descartes, Paris.

MEDICAL INFORMATION SYSTEMS

S. E. Aidarous and M. A. R. Ghonaimy

Electronics and Computers Department, Faculty of Engineering, Ain Shams University, Cairo, Egypt

Abstract

Medical information systems represent a break through in the use of computers in medical care. Areas of application include in-patient and outpatient care, laboratory automation, adminstrative and management as well as planning. Data structures and mode of use of medical information systems by physicians, nurses and adminstrative personnel are considered. Procedures for the selection and implementation of such systems in developing countries are discussed. A case study is presented for the application of such system in a hospital.

INTRODUCTION

Medical knowledge has been expanding rapidly since the begining of this century and new technological hardware has made possible the application of this knowledge to the treatment of patients. This can be manifested in the advent of automation and computers which introduced a new era of medical practice.

At the present there are two basic trends in the use of electronic computers in the medical field. The first one is the analysis and management of biological systems (modelling of biological systems, automation of clinical and laboratory tests, automated diagnosis, prolonged follow-up of patients and prognosis of the course of the disease). The second is the management of various elements of the health care system as a social system (processing of statistical data and primary carriers of scientific and medical information, hospital management, economic analysis, and health prognosis).

Development of data processing in medicine has proceeded along the design and implementation of applications in support of medical care processes. In addition, attention has been given to the business aspect of medicine: accounting, payroll, and billing as well as other functions in support of health care delivery.

In the 1970's attempts have been started to combine the attributes of management science, medicine and informatics to improve the quality and efficiency of health care delivery.

Developing countries suffer from the inavailability and inaccessibility of data rather than its lack. This is reflected on the nations efforts of adequate planning for the country health program. A promising solution for such problem is the creation of a health (medical) information system which coordinates routine data collection, organization and analysis. This will give the decision maker a powerfull tool to plan and evaluate their policies and to suggest possible alternatives.

On the micro level, possible use of information systems in different parts of the health care system will contribute to the performance of the whole system.

In this paper, different areas for the application of medical information systems are presented. The mode of use of such systems by noncomputer personnel is discussed. Impact of the developing countries environment on the procedures for selecting and implementation of medical information systems is considered. A case study is presented.

AREAS OF APPLICATION

The impact of informatics on medical care is most clearly displayed in the enormous increase in data acquisition, recording and analysis in the course of clinical diagnosis. The most fundamental problem is the pateint care in and outside the hospital. This can be achieved through the accurate reporting of patient records, proper identification of tests and services needed, and ordering services according to priorities and availability. In such case, efficient loading of the available services is secured and better patient care will be provided through the fast and proper allocation and performance of facilities.

Automation and loading of different services (laboratory, theatre, X-ray, etc..) represent another important area of application. Scheduling of patients in such services will save both physician and patient times. Day by day schedule can be provided for each service.

The third area of application is the adminstration and management of the system. Automatic billing procedures for patients, pharmacy stock control, payroll and automated service accounts may minimize the cost of health care and facilitate the proper management of the whole system.

In addition to these applications, the availability of such data base will facilitate future planning and development in the system.

DATA STRUCTURE

An example of data structure of a medical information system is given in Fig.1 In this structure three master records are used for patient, physician and date. Different variable records (e.g., tests, medical exams., analysis, etc..) are linked to such records. Addition of further records or links between different records depends on the specific application required. In the given case, a phsician/patient link is established and additional room record is linked to the patient.

The shown data structure serves mainly functions related to patient care. When considering other functions,(adminstrative and management) different structures may be used, or additional records and links will be needed.

CONSIDERATIONS IN THE DESIGN OF HOSPITAL INFORMATION SYSTEMS

Data Entry and Screen Formats

In a hospital information system, it is essential to provide an easy and reliable way of interaction between the data base and the medical, paramedical, and adminstrative personnel. Therefore, it is necessary to avoid, as much as possible, letting the above personnel code the medical and other relevant information forming the different records in the data base. It would be much better to

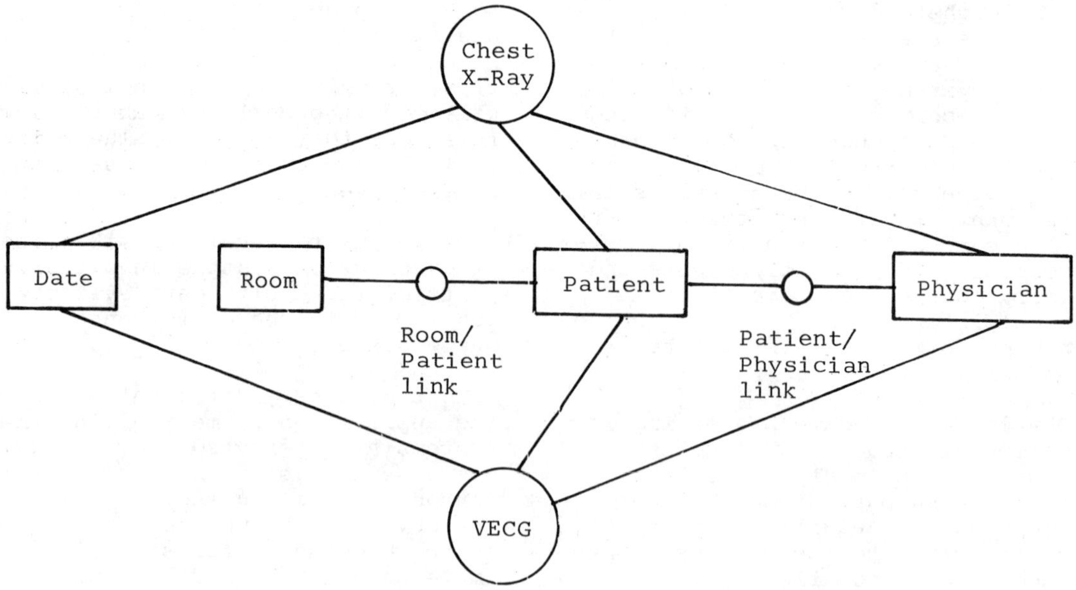

Fig.1 Data Structure of a Medical Information System in a Hospital.

use screen formats appropriate to each record type using extensive validation schemes to insure the correctness of transfering data from the original document to the data base.

Queries

To allow easy interaction with the medical information system, a powerful query system should be provided with extensive explanations whenever possible. When the user is in doubt of any part of his dialogue, he can simply enter a special character (which may be a question mark "?"). The system should display an explanatory quide.

Protection Schemes

Since critical patient records are usually involved in this system, it is necessary to provide stringent accessing schemes to avoid ,

* placement of patient medical records in the wrong place.
* unauthorized manipulation of information regarding some patient medical information.

Most data base management systems available now allow the data base adminstrator to provide different protection levels for different users.

Statistical Analysis

One of the major benefits of having a medical information system is to use the vast amount of data collected to generate regular reports that will be useful for both the medical and adminstrative staff. Also this data base could be used to develop the appropriate models for the health care system.

SELECTION & IMPLEMENTATION

Environment in Developing Countries

In a developing country, it is essential to insure that such system will proceede in an evoluationary manner, so that, its roots will penetrate gradually into the infrastructure.

The following guidelines could be helpful in selecting and then implementing the medical subsystem to be included in the nucleus medical information system:

1. Select a small section of a hospital that has a nucleus for a well-developed manual information system.
2. Develop the structure of the medical data base system taking into consideration the interactive nature of its operation.
3. Develop a suitable query language, preferably performing some experiments using dummy data, on an interactive terminal.
4. Develop a data entry system taking into consideration the critical nature of the data, minimizing as much as possible any manual manipulation or coding of the original data. Take the appropriate measures to validate and check the correctness of the data.

Constrainta and Limitations

The pitfalls of computer usage in medical information systems can be classified into three main types:

1. Machine pitfalls: These are due to the lack of understanding of the role of Electronic Data Processing (EDP) in a hospital. The combination of this type of unawareness coupled with attempts by the staff to acquire status by using computers, form a potent mixture for failure of the EDP and its works. In addition, some personnel are afraid of the results of EDP as regards their security of employment or prospects of advancement, hence a stubborn refusal to cooperate and render the new system a success.

   In many cases, there is undue concentration on the computer hardware without adequate consideration being given to the software aspects design of system, servicing of the installation, periodic review of costs and proper staffing. It is essential that computer staff have access to the data and procedures used throughout the hospital and they are in contact with management staff.

2. Organization pitfalls: These are many, e.g. lack of leadership and knowledge among those charged with responsibility for systems and computers, inadequate definitions of duties and rights of systems and computer units, unsatisfactory correlation between the hospital organization and the EDP system, weak staffing of the computer installation and its ancillaries, over emphasis on vital functions (patient care, clinic scheduling, lab. loading, etc.), and competition between the technical engineering and medical staffs.

   These drawbacks cause damage to various activities in the hospital and may prove quite costly. The lack of support for vital functions and duplication of facilities may bring about substantial financial losses.

3. Staffing pitfalls: These are due to lack of consideration for the human element in EDP, particularly in view of the imbalance between supply and demand of competent computer personnel and the frequent lack of cooperation between management and those involved in manning the installation. What is usually forgotten is that EDP is a continuing effort and computers themselves are vehicles of change.

Skills and Training

As the technological advances are made, new types of health related occupations are being created. Each type of health professional works with the physician and makes his own specific contribution to patient care.

Unfortunatelly, at the present time no-one is quite sure exactly what the functions and responsibilities of allied health workers are within each of the categories. Functions tend to change as new hardware is developed and as medical research progresses.

CASE STUDY

A summary of a case study concerning the nucleus of a hospital information center for one department, namely the cardiology department is now considered.

The data base management system used is TOTAL which is a network structured data base system. TOTAL provides a Data Base Definition Language (DBDL) which allows the user to have an access to a large data base with sophosticated data structures. The user uses a Data Management Language (DML), provided by TOTAL, to access and manipulate the data base. DML commands must be issued via CALL statements in the user programs using any host language which supports subroutine CALL statements. The functions of DML commands include opening and closing of data sets, the serial processing of data sets, and special functions such as data record logging.

Information is stored within data records. The data base is formed by a meaningful association of a number of master data sets and variable data sets. Such association is spread out horizontally and vertically to have a network structure.

There are one single-entry data set, represents the patient record, and five variable entry data sets, represent: ECG test, Chest X-ray Report, Cardiac Cathterization Report, Investigation Sheet and Echocardiogram.

The configuration of these data sets is as follows: (number indicates bytes reserved for each field)

Patient Record (Single-entry)

Patient Number	8
" Name	30
" Address	30
Date of Birth	8
Sex (M or F)	1
Hospital Name	8
Admission Date	8
Discharge Date	8
Room & Bed Number	4
Preliminary Diagnosis	1
5 Tests Performed (Y or N)	5
Haemoglobin	3
Pressure	1
Other Measurements	16

ECG Record (Variable-entry)

Patient Number	8
Date	8
Physician Name	30
Drugs	6
Rate	2
Rythm Regular?	1
Axis (R,L,ODD)	12
Intervals	20
Abnormal Q Waves	6
Ventricular Hypertrophy	6
Infraction	5
Arrythmia	29

Chest X-Ray (Variable-entry)

Patient Number	8
Date	8
Physician Name	30
Posteronetrior View	23
Lateral View	14

Cardiac Cathterization (Variable-entry)

Patient Number	8
Date	8
Physician Name	30
Diagnosis	2
Premedication	4
Interventions	6
Pressure, Oxygen Content & Saturation	248
Oxygen Consumption	5
Systemic Flow	5
Pulmonary Flow	5
Total Pulmonary Resistance	5
Valve Gradients	5
Valve Area	5

Investigation Sheet (Variable-entry)

Patient Number	8
Date	8
Physician Name	30
Haematology	80
Electrolytes & Acid-Base	35
Biochemistry	55
Enzymes	25

Echocardiogram (Variable-entry)	
Patient Number	8
Date	8
Physician Name	30
Clinical Information	30
Aortic Root & Valve	4
Left Atrium	8
Mitral Valve	8
Tricuspid Valve	3
Pulmonary Valve	4
Left Ventricle	23
Ventricular Septum	7
Right Ventricle	9
Pericardium	10
Conclusions	30

The data-base schematic is shown in Fig.2, where Single-entry data sets are represented by squares and Variable-entry data sets are represented by circles.

The flowchart of the main application program is shown in Fig.3

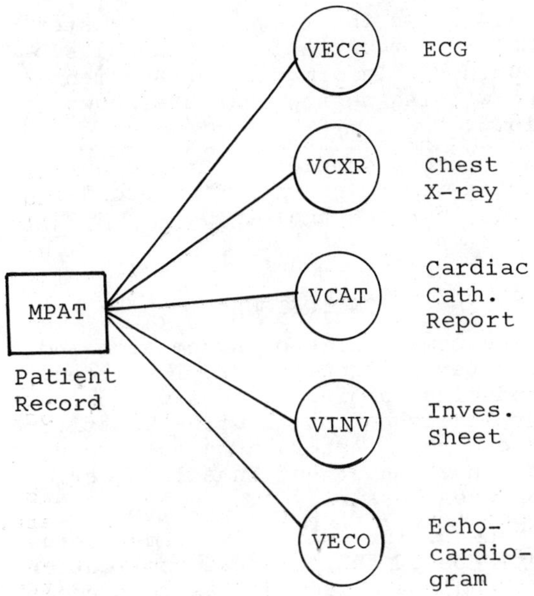

Fig.2 Data-base Schematic

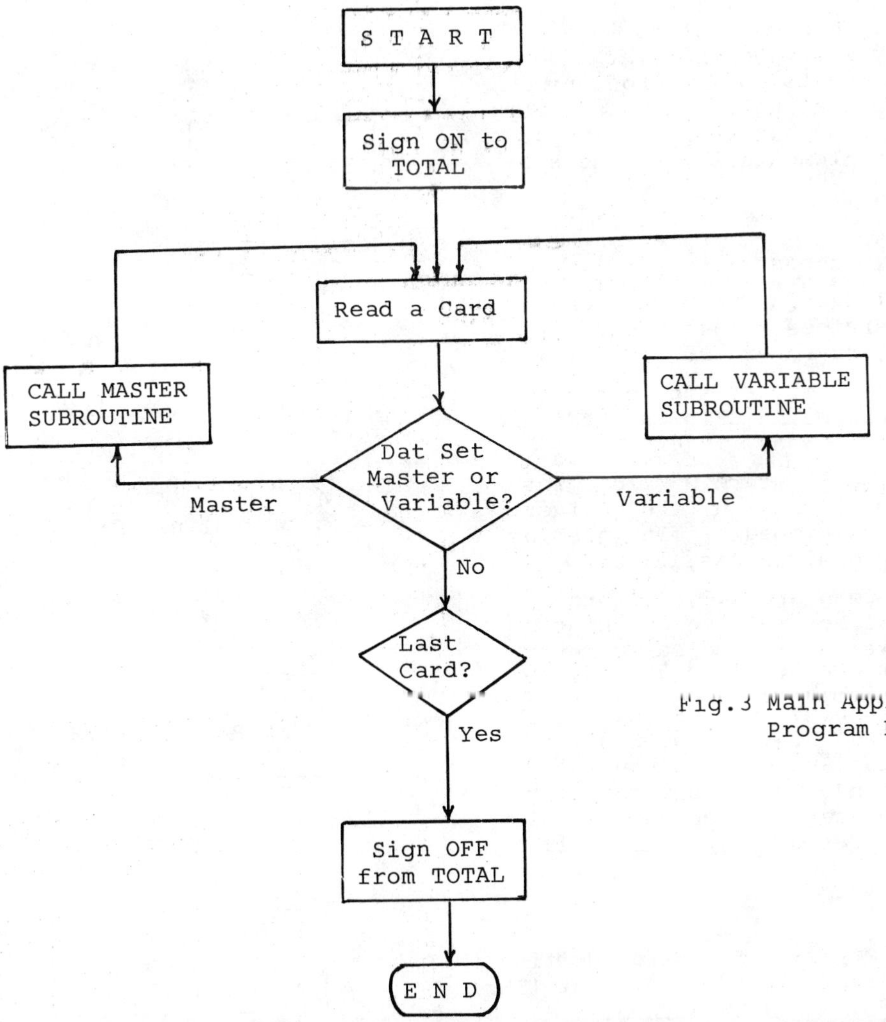

Fig.3 Main Application Program Flowchart.

Fig.4 shows the flowchart of master (MASTER) or variable (VARIABLE) subroutines. Information arrangement and writing subroutines are shown in Fig.5.

A printout sample of the patient master record and its ECG record are given in Fig.6 and Fig.7.

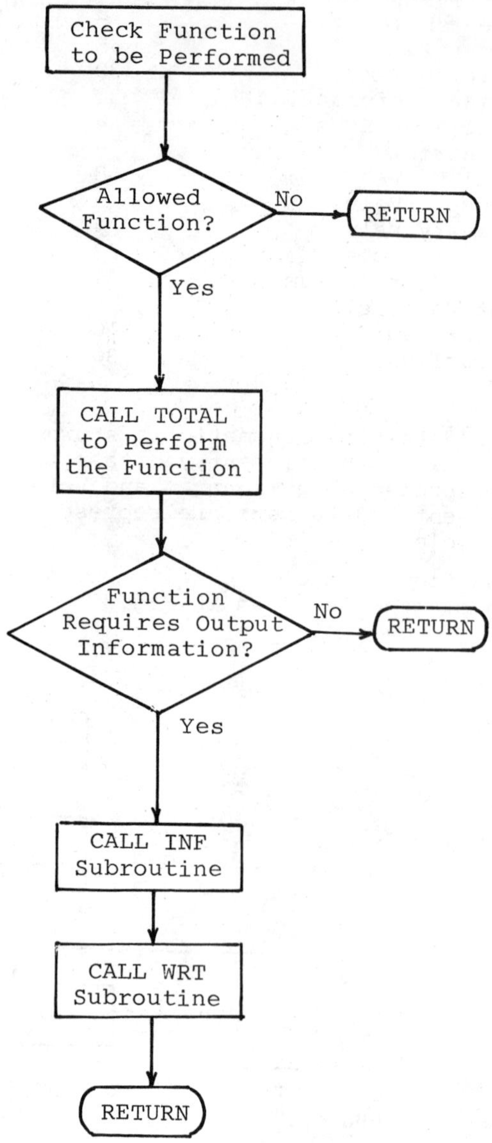

Fig.4 Master and Variable Subroutines Flowchart.

CONCLUSIONS

The combination of automation and computers in health care is already producing complex systems which can tirelessly perform routine repetitive procedures, faster, more accrately, and in a manner not humanly practical nor even possible.

Medical Information Systems represent one of the powerful applications to handle and process the vast amount of data generated in a hospital. The development in such systems facilitates its use by medical and adminstrative personnel while insuring both maintenance and security of the information.

When applying such systems in a developing country, constraints and limitations due to the environment and the personnel have to be carefully considered. They will secure proper design and implementation of the system.

References

1. Special Issue on Technology and Health Services, Proceedings of the IEEE, November 1969.
2. H. Ludwig, Computer Applications and Techniques in Clinical Medicine John Wiley & Sons, 1974.
3. R.F. Rushmer, Future Horizons for Technology in Health Care Delivery, Advances in Biomedical Engineering Vol.6, Academic Press, 1976.
4. J.H. Grossman and J.F. Rockart, A Managerial Perspective on Information Systems in Medical Care Organizations, Ibid.
5. M.A.R. Ghonaimy, H.I. Shahein, S.E. Aidarous and N.A. Nouno, Hospital Information System, The Fourth International Congress for Statistics, Computer Science and Social Research, Cairo, Egypt, Mrach 1979.

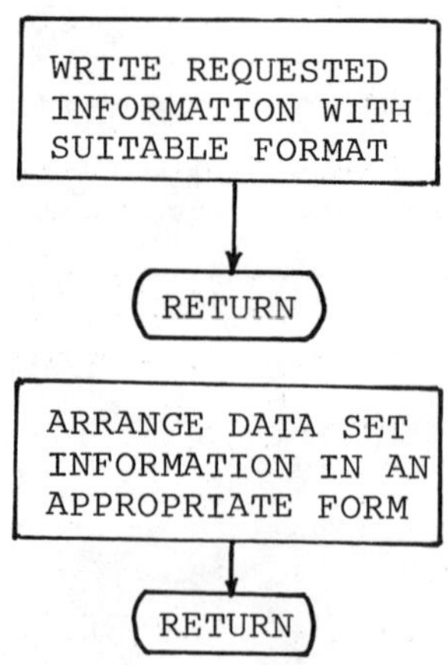

Fig.5 Information Writing (WRT) and Arrangement (INF) Subroutines.

```
***********************************
*                                 *
*   PATIENT IDENTIFICATION REPORT *
*                                 *
***********************************

PATIENT NO.            00000125
NAME                   AHMED EL SAYED
ADDRESS                13 EL HOREA ST.,SHOBRA
AGE                    049
SEX                    M
ARRIVAL DATE           09/05/79
DEPARTURE DATE         12/05/79
ECG TEST               Y
C. CATH. REPORT        Y
CHEST X RAY            Y
INVESTIGATION SHEET    Y
PHYSICIAN NAME         DR. TAREK ADLY
WARD NO.               0023
FLOOR NO.              0004
BED NO.                0002
ROOM STATUS            M
```

Fig.6 Patient Master Record Printout.

```
       ********************                    * ARRYTHMIA
       *  ECG TEST REPORT  *                   * ---------
       ********************                    * .(N) NORMAL SINUS RHYTHM
PATIENT NO.  00000002   DATE  9/03/1980   Y    * .(N) SINUS TACHYCARDIA
PHYS. NAME   DR. RIAD ADLY                1    * .(N) SINUS BRACHYCARDIA
CLINICAL DIAGNOSIS:                            * .(N) LOW ATRIAL RHYTHM
-------- -- -----                              * .(N) JUNCTIONAL RHYTHM
DRUGS: .(N) DIGITALIS    .(Y) QUINIDINE        * .(N) JUNCTIONAL TACHYCARDIA
--- -- .(Y) PROCAINAMIDE.(N) POTASSIUM SALTS* .(N) SUPRAVENT. TACHYCARDIA
       .(Y) DIURETICS    .(N) OTHER ANTIARRH.* .(N) 2 DEG. A-V BLOCK-
RATE:  .(N) BELOW 60   RYTHM:  REGULAR .(-)    * .(N) 2 DEG. A-V BLOCK+WENK. PHEN.
--- -  .(Y) OVER 90    -------                 * .(N) A-V DISSOCIATION
AXIS: (-10 TO +110) /R 002+/L 089+/ODD 025 /* .(N) 3 DEG. A-V BLOCK
-- --                                          * .(N) WONDERING PACEMAKER
INTERVALIS  /QRS  0.08/P-R  0.22/Q-T  0.12/ * .(N) ATRIAL FLUTTER
------ - -  /P     00.17/QTC   00.37           * .(N) ATRIAL FIBRILLATION
ABNORMAL  Q WAVES IN LEADS    000***0000       * .(N) UNDETERMINED RHYTHM
-- --- - -- - ---------                        * .(N) WITH SA ARRYTHMIA
VENTRICULAR HYPERTROPHY RIGHT-GRADE  0000      * .(N) WITH SA BLOCK
----- -- -- --------- LEFT -.(O)VOLTAGE        * .(N) WITH SINUS ARREST
                          VOL.+STRAIN.(N)* .(N) WIH ATRIAL PREMATURE BEATS
INFARCTION     .(N)  ANTERIOR WALL             * .(N) WITH PVCS
--- -- -- -    .(N)  INFERIOR WALL             * .(N) WITH FUSION BEATS
               .(N)  POSTERIOR WALL            * .(N) WITH NODAL ESCAPE
               .(N)  SEPTUM   (N)  LATERAL     * .(N) WITH SYNDROME
```

Fig.7 Patient ECG Record Printout.

ON THE DRAWBACKS OF A LINEAR MODEL IN THE DECENTRALIZATION OF A DECISION PROCESS

P. Mahey

Department of Electrical Engineering, Pontifícia Universidade Católica, Rio de Janeiro, Brazil

Abstract: In the domain of multilevel analysis, the idea of decentralization has a clearly different meaning from that of reducing the size of an optimization problem by decomposition techniques. It is well known that prices only cannot usually be utilized to coordinate a linear economic system. This paper is an attempt to conceptualize the justifications of decomposition techniques against the powerful linear programming tools available on the market. Starting from the classical Dantzig-Wolfe's algorithm, the drawbacks of the "bang-bang" behaviour of linear models are analyzed and quantified, relatively to the desired autonomy of the subsystems and to the information volume treated at each cycle. The problem of justifying expensive and sophisticated algorithms yielding a coherent decentralization is then posed in consideration to their integration in a human decision-making environment.

Keywords. Decision theory; linear programming; multilevel hierarchical systems.

I. Introduction

We are going to discuss along this paper several aspects of the decomposition of optimization models for decentralized organizations. In particular we want to focus on the well-known "bang-bang" behaviour of the completely linear case when a price directive coordination is designed for both decomposition and decentralization purposes. One might be surprised to find here the subject of an extensively discussed polemic since the publication of Dantzig-Wolfe's algorithm in 1961. Very soon, economists have perceived an analogy between decomposition algorithms and multi-level organizational models and a lot of energy has been wasted in order to build algorithms as faithful reproductions of real situations. The auxiliary problems, termed subprograms, may correspond to individual decision-making units, each with its own objectives and resources. The master problem may be interpreted either as a central planning agency as in a socialist economy or as a competitive market as in a capitalist environment. Each iteration of the algorithm represents an information exchange between master and subprograms. Typically the master program would send down marginal prices of the corporate resources and the subproblems would use these prices to compute a local proposal sent back to the master program which may eventually reconsider its former decision in front of the new demand for corporate resources. Distinctions may be made between two interpretations of the iterative process (Kornai, 1973):

- a preparation of the optimal plan.
- a simulation of the price-adjustment process as the Walras's "tatonnement" process of the everyday control of an economic system does.

Besides all confusions which could have been made between these two interpretations, three principal critics have been adressed to the second:

- The first is that "traditional design models, based upon decomposition are highly abstract representations of reality and ignore vital aspect of real organizational considerations, such as the effects of uncertainty, risk sharing, time and informal communication" (Moore, 1979).
- Secondly, in a real-world market, the excess supply of t th period is added to initial stock of the (t+1) th period. The excess demand of the t th period may not be added fully to the demand of the (t+1) th period but can increase it. In most decomposition algorithms, this problem disappears entirely (Kornai, 1973).
- Thirst, the proposals suggesting decomposition algorithms as normative theories of planning are somewhat unrealistic, since no planning process would be organized along the lines of a mathematical program which could easily reach thousands of iterations. Some essays to simulate only few iterations as a suboptimal process have led to very poor results (Christensen and Obel, 1978).

The main argument which justifies the research of decomposition algorithms to solve socio-economic planning problems might be the increasing complexity of such problems and the supposed computational limitations of centralized optimization. Unfortunately, the interest of the big data processing societies in developing more and more sophisticated packages, principally in the field of linear programming, has led to direct methods which are actually quicker and simpler than any of the decomposition methods.

In presenting some disadvantages of a price--directive coordination of linear subprograms, we want to show how experiments and theoretic research on decomposition algorithms can shed light on the mechanisms of socio-economic processes and we hope that this type of system approach will contribute to the development of efficient decomposition programs. The objective of decentralizing the subsystems, as being one of the most relevant feature of modern organizations, will be the principal criterion for comparing and evaluating the performances of some decomposition algorithms.

The purpose of decentralization is to let the subsystems or divisions of an economy calculate their own optimal decisions such that the set of all the optimal decisions is a feasible optimal solution for the global problem. Arrow and Hurwicz (1960) introduced that property as the result of a perfect competition between the divisions (as in the Pareto optimality), i.e. individual profit maximization leads to the maximization of profits for the whole sector. We shall give some conditions of existence of a decentralized decomposition in the general framework of cooperative equilibriums and then go back to the linear case to study proposed solution of this problem found in the literature, beginning with the famous Dantzig-Wolfe's algorithm.

II. Existence of price-directive coordination.

The problem is that of a competitive market of corporate resources in a general decomposable economy.

Let us consider n centers of decision and denote by x_i, $A_i(x_i)$ and $J_i(x_i)$ respectively the variables of decision, the corporate resource consumption and the cost criterion associated with the ith center. We have supposed that the cost criterion is a function separable of the decision variables, so that the coupling between the n centers may be represented by the corporate resources constraints only:

$$\sum_{i=1}^{n} A_i(x_i) \leq 0$$

We are now able to define the so-called corporative equilibrium problem (Pareto optimality) as a mathematical program, or totally centralized problem (P):

$$(P)\qquad \begin{aligned} &\text{Min} \sum_{i=1}^{n} J_i(x_i) \\ &\sum_{i=1}^{n} A_i(x_i) \leq 0 \\ &x_i \in X_i \end{aligned}$$

where X_i is a technological constraints set associated with the ith center. ($X_i \subset E_i$, definition space of the functional J_i)
A_i is an application from E_i to ψ, normed vector space.

The most natural method to solve this problem by decomposition,is to consider it as a model of a manufacturing company consisting of headquarters and n division, the headquarters fixing iteratively a marginal price for the corporate resources consumption of each division. In mathematical programming, that price is generally the same for all the divisions and could be interpreted as a Lagrange multiplier parametrizing the cost functional associated which each division. Let $u \geq 0$ be such a price. Strictly speaking, $u \in \psi^*$, dual space of ψ.
We define now the ith subproblem, $SP_i(u)$, as:

$$SP_i(u)\qquad \begin{aligned} &\text{Min } J_i(x_i) + < u, A_i(x_i) > \\ &x_i \in X_i \end{aligned}$$

where $<\cdot,\cdot>$ denotes duality product between ψ and ψ^* (See Bensoussan et al., 1972).

Let $R_i(u)$ be the set of optimal solution to the ith subproblem. The price-adjustment process or coordination task is effectuated at a superior level (master-program) which modified iteratively the value of u.

Definition 1: The price-adjustment process is called "well-defined" when the following properties are met (Malinvaud, 1967, p. 177).

i) There always exist solutions to the operations according to which the subproblems proposals, the prices and the plan can be determined.

ii) The plan generated by the coordination level is feasible.

Definition 2: The price-adjustment process is called "decentralized" when the following properties are met.

i) The process is well defined

ii) The process converges in a finite number of iterations.

iii) $\exists\, u^* \geq 0$ such that:

$$\forall\, x_i \in R_i(u^*),\ < u^*, \sum_{i=1}^{n} A_i(x_i^*) > = 0$$

If these properties are met, it is clear that for $u = u^*$ the set of solutions of the subproblems $SP_i(u^*)$ is optimal for (P) and reversely if $(x_1^*,\ldots x_i^*,\ldots x_n^*)$ is optimal for (P), then $x_i^* \in R_i(u^*)$ $\forall i = 1,\ldots,n$.

In fact we want to characterize conditions of existence of such a decentralization property:

Assumptions: 1) $X = X_1 \times \ldots \times X_n$ is a closed, bounded subset

2) J_i and A_i are continuous on X_i.

Theorem 1: A necessary and sufficient condition for the price-coordination to be decentralized is that the functional $L(x_1,\ldots x_n, u) = \Sigma J_i(x_i) + \langle u, \Sigma A_i(x_i) \rangle$ has a saddle-point on $\{u \geq 0;\ x_i \in X_i,\ i=1,\ldots n\}$ at $(u^*, x_1^*,\ldots,x_n^*)$ and that the functional $L(x_1,x\ldots,x_n,u^*)$ has a unique minimum on X.

Proof: For the elements of the proof of this theorem, we refer to the results obtained by Potier (1972); p. 255). The addition of the condition of unicity of the minimum of $L(x,u^*)$ ensures the process to be well-defined.
We now define the dual function h(u) as:

$$h(u) = \operatorname*{Min}_{x \in X} L(x,u)$$

Lemma: h(u) is a concave function of u (for the proof see, e.g., Lasdon, 1970). The condition of unicity means that h is differentiable at u^* with partial derivatives

$$\left.\frac{\partial h}{\partial u}\right|_{u=u^*} = \sum_{i=1}^{n} A_i(x_i(u^*))$$

or, equivalently, that $\sum_i A_i(x_i)$ is constant over $R_i(u^*)$, $i=1,\ldots,n$.
A number of sufficient conditions may be given to guarantee unique minima for L(x,u). The simplest is strict convexity. We can see, also, that strict convexity implies unicity of the solution of the $SP_i(u)$, $\forall u$, $\forall i=1,\ldots,n$.

These results join the general and well-known observation that the complete decentralization will not generally be attained in the linear case (see e.g., Baumol and Fabian, 1964; Charnes, Clower and Kortanek, 1967; Lasdon, 1970, p. 162).
This difficulty has led many authors to introduce new modes of coordination based or not on a price-directive iterative process.

III. Coordination procedures for linear decomposition algorithms.

3.1 - The Dantzig-Wolfe's algorithm.

The purpose is not here to develop the decomposition principal of Dantzig-Wolf (See e.g., Dantzig Wolfe, 1961) but to yield a basic framework for our analysis.
The model is that of classical activity analysis, J_i, A_i and X_i being defined linearly in function of the activity levels x_i. We arrive then at the following block-angular structure (fig. 1).

$$\operatorname{Min} \sum_{i=1}^{n} c_i' x_i$$

Subject to:

$$\sum_{i=1}^{n} A_i x_i \leq b_o,$$

$$B_i x_i \leq b_i,\ i=1,\ldots,n$$

$$x_i \geq 0,\ i=1,\ldots,n$$

x_i is an n_i-vector
A_i is a (m_o, n_i) matrix
b_o is an m_o-vector of corporate resources availabilities.
B_i is an (m_i, n_i) matrix
b_i is an m_i-vector

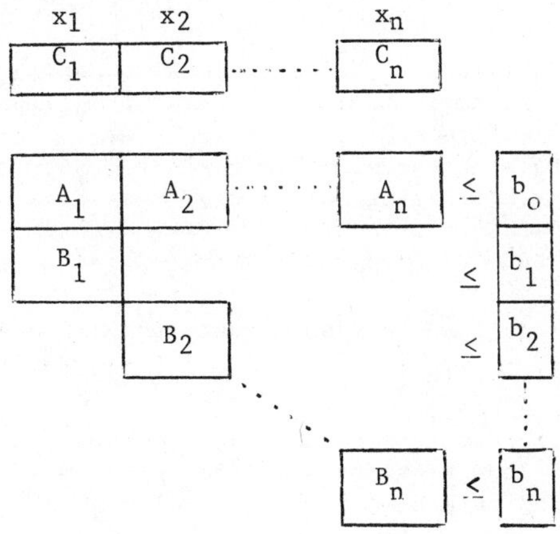

Fig.1-A block-angular decomposable system

Let π^k be the dual m_o-vector associated with the corporate constraints at the k th iteration. The subproblems are defined as

$$\operatorname{Min}\ (c_i' - \pi^{k'} A_i) x_i$$

Subject to: $\quad B_i x_i \leq b_i$

$$x_i \geq 0$$

As dual variables may generally be identified as marginal prices, we see that the decomposition suggests here a price-directive planning procedure. However, if the subproblems are equivalent to that obtained in a classical tatonnement scheme, the superior level has a quite different task of coordination: instead of making a simple updating of prices as guided by the last observation on excess demands, the Dantzig-Wolfe's method accumulates all the information previously generated and so combines efficient production responses in order to find a possible better structure of prices. Thus the Dantzig-Wolfe's process requires much more work at the coordination level, but it is much more robust in the sense of assuming convergence to optimality.

Then, only the coordination level is able to find a feasible solution at each iteration combining eventually various solutions of the same subproblem. As indicated very soon by Baumol and Fabian (1964) the reason of it is that the solution of the subproblems must be an extreme point of the convex polyhedra $X_i = \{x_i \geq 0;\ B_i x_i \leq b_i\}$, and that the global solution will generally not be the union of such points.
Let us detail that last observation in the light of the theorem of section II.

Let us define the dual function $h(u)$ as: $u \geq 0$

$$h(u) = \min_{x_i \in X_i} \{\sum_i c_i' x_i + u'(\sum_i A_i x_i - b_o)\}$$

Suppose now that $\bar{u}$ is such that the n subproblems have a unique solution $x_i(\bar{u})$, extreme point of X_i. The dual function value is then:

$$h(\bar{u}) = \sum_i c_i' x_i(\bar{u}) + \bar{u}' (\sum_i A_i x_i(\bar{u}) - b_o)$$

for all such $\bar{u}$, which is the equation of a hyperplane in the (h.u) space. When u varies in one direction s, some relative cost factors of the nonbasic variables of each subsystem diminish linearly:

$$u = \bar{u} + \rho s, \quad \rho \geq 0$$

$\forall s \in R^{m_o}$, $\exists \rho_1 > 0$ such that: for all $\rho < \rho_1$, $x_i(u) = x_i(\bar{u})$; for $\rho = \rho_1$, one marginal cost of one nonbasic variable of one subsystem (SP_i) reaches zero and the linear subproblem has a nonunique solution; for $\rho > \rho_1$, a pivot operation has modified the solution of the ith subproblem, hence $x_i(u) \neq x_i(\bar{u})$ and $x_j(u) = x_j(\bar{u})$, for $j \neq i$.

Roughly speaking h(u) is a concave piece-wise-linear function. Each hyperface of this concave polyhedra is bounded by edges which represent a pivot operation in the current basis of one of the subproblems. For $u=\bar{u}+\rho_1 s$, the solution of one subproblem is not unique and then h(u) is not differentiable.
*Observe that a good choice for s might be:

$$s_j = \min\{0, \sum_i A_{ij} x_i(\bar{u}) - b_{oj}\}, j=1,\dots,m_o.$$

The solution of the saddle-point problem corresponds to $\max_{u \geq o} h(u)$, hence to a vertex of the concave polyhedra. But an extreme--point of this polyhedra is the intersection of m_o+1 hyperfaces, hence the price-adjustment process will reach the optimal solution with at most m_o simultaneous pivot operations on the basis of the subproblems. Observe that we need not have exactly m_o changes of basis since some of the hyperfaces supported by the hyperplanes $u_j = 0$ could contain the optimal solution.

Then a maximum number of m_o subproblems cannot be decentralized by the price-coordination method. Each time a marginal optimal price is zero, then this number diminishes of one unity (A zero price means a loose corporate constraint, hence a weaker coupling between the subsystems).

Let us go back to the Dantzig-Wolfe master--program with n convexity constraints (Lasdon, 1970, p. 153). A master basis has (m_o+n) rows, hence no more than m_o+n subproblems must be included in the solution, this means that m_o extreme points are to be distributed among the n subproblems. Therefore, a maximum number of m_o subproblems will have more than one extreme point in the global solution.
These two parallel results lead to the following assumption: The dual vector π^k generated by the Dantzig-Wolfe's master program corresponds to an extreme-point of the concave polyhedra of the global dual function h(u), at the optimal solution only.

To show the usefulness of the h(u) function, let us consider the following case. Suppose that there exists i such that $\eta_i \leq \eta$.
Then the linear system:

$$A_i' u + c_i = o$$

may have a nonnegative solution. This means that, if the procedure generates such a price-vector, the ith divisional decision is indifferent to the price allocation and the optimal solution will be interior to the constraint set X_i. Charnes et al. have observed this phenomenon (1967). Let us illustrate this result by a simple example:

Let us consider the following two-blocks decomposable structure:

$$\text{Min} \quad -2x_1 - x_2 - x_3 - y_1 - \frac{1}{2} y_2$$

$$\begin{aligned}
\text{Subject to:} \quad & x_1 + x_2 + 2x_3 + y_1 && \leq \alpha \\
& 3x_1 - x_3 + 2y_1 + 3y_2 && \leq 9 \\
& x_1 + x_2 + x_3 && \leq 3 \\
& 2x_1 - x_2 + x_3 && \leq 2 \\
& y_1 + y_2 && \leq 2 \\
& x_1, x_2, x_3, y_1, y_2 \geq 0
\end{aligned}$$

The polyhedra X_1 and X_2 associated with the x-block and the y-block are shown on Fig.2.a. and fig. 2.b. respectively.

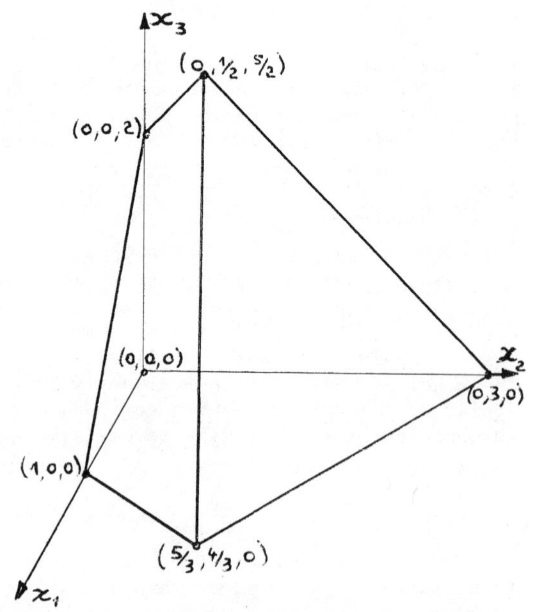

Fig. 2.a

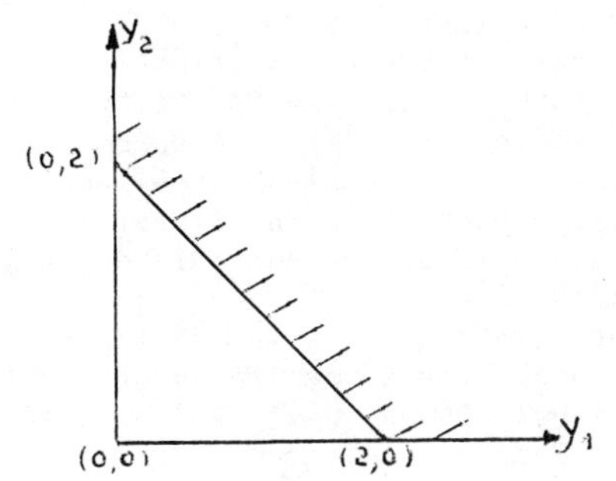

Fig. 2.b

For this simple example, we have:

$n = 2$
$n_0 = 2$
$n_1 = 3$
$n_2 = 2$

Following the notations of section II, the dual function h(u) is:

$$h(u) = \underset{x \in X_1}{\text{Min}} \{(u_1+3u_2-2)x_1+(u_1-1)x_2+(2u_1-u_2-1)x_3\}+$$
$$+\underset{y \in X_2}{\text{Min}} \{(u_1+2u_2-1)y_1+(3u_2-\tfrac{1}{2})y_2\}-2\lambda_1-\alpha\lambda_2$$

In the plane (u_1,u_2), the frontiers of the regions for which the solutions of the subproblems remain unchanged are shown of fig.3. These solutions define for each region a function h(u):

Region	x_1	x_2	x_3	y_1	y_2	h(u)
1	5/3	4/3	0	2	0	$(5-\alpha)u_1$
2	0	1/2	5/2	2	0	$(\frac{15}{2}-\alpha)u_1 - \frac{15}{2}u_2$
3	5/3	4/3	0	0	0	$(3-\alpha)u_1 - 4u_2$
4	5/3	4/3	0	0	2	$(3-\alpha)u_1 + 2u_2$
5	0	3	0	0	0	$(3-\alpha)u_1 - 9u_2$
6	1	0	0	0	0	$(1-\alpha)u_1 - 6u_2$
7	1	0	0	0	2	$(1-)u_1$
8	0	1/2	5/2	0	0	$(\frac{11}{2}-\alpha)u_1 - \frac{23}{2}u_2$
9	0	0	0	0	0	$-\alpha u_1 - 9u_2$
10	0	0	2	0	0	$(4-\alpha)u_1 - 11u_2$

Now, if we choose $\alpha=2$, it is easy to observe that the maximization of h(u) in the regions 3,4,6 and 7 leads to the point A. As h(u) is concave, A is then the saddle-point solution of this problem.
At this corner-point, the solution of the first subproblem lies on the edge which links the extreme points $(\frac{5}{3}, \frac{4}{3}, 0)$ and (1,0,0), and the solution of the second subproblem lies on the edge (0,0)-(0,2). Hence both cannot be decentralized at the optimal solution.

If now we set $\alpha=4$, the saddle-point is in B, which is exactly the intersection of the three dotted lines. The optimal solution is then such that the first subproblem is well decentralized with $x^* = (\frac{5}{3}, \frac{4}{3}, 0)$, but the cost function of the second subproblem is zero and the solution may be interior to X_2. This occurs because $n_2 = n_0$.

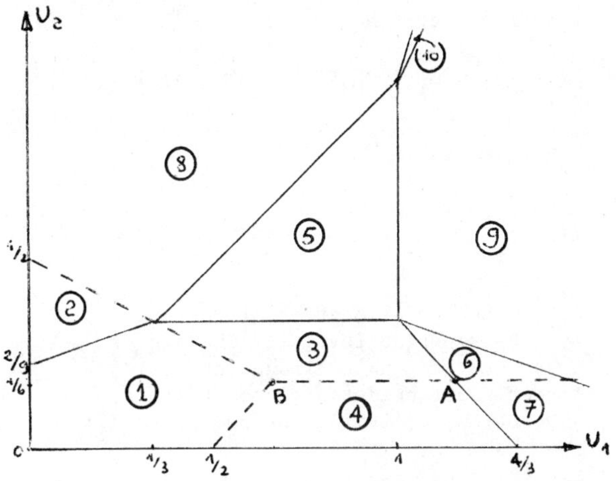

Fig. 3: The darkened lines limit the regions with x(u) constant; the dotted lines limit the regions with y(u) constant.

3.2 - Modifications of the price-adjustment process.

Baumol and Fabian (1964) concluded their analysis about Dantzig Wolfe's algorithm by suggesting the use of suitable nonlinearities in order to avoid the "bang-bang" behaviour of linear models. The price-schedules algorithm of Jennergren (1973) provides us a good example of the choise of these nonlinearities. Jennergren's idea was to associate to each corporate resource a linearly increasing price schedule rather than a constant price. The prices effectively considered by the subdivision manager will depend on the quantity of each resource purchased or delivered by his division. The subproblem can there be formulated as:

$$\text{Min } (C_i-((\pi^k)' + r\, x_i' A_i')Ai)x_i \qquad (3.1)$$

subject to: $x_i \in X_i$

This problem has now a quadratic objective function and Jennergren proved that this procedure is well-defined and decentralized. The headquarters compute the new π^{k+1} as:

$$\pi_j^{k+1} = \text{Max}\{0;\pi_j^k + \alpha[\sum_i A_{ij}\, x_i(\pi^k)-b_{0j}] \qquad (3.2)$$

The process converges to an optimal solution in an infinite number of iterations and precious information on the lower bound of the cost function may be used to stop the procedure for ε-optimality.

The addition of this schedule to the normally constant prices may be interpreted as the addition of preemptive goals in the divisional decision task. Charnes et al. (1967) have introduced this notion in a fundamental paper about coherent decentralization. These goals could be added in the objective function or in the subproblem constraints. The former case leads to subproblems of the form:

$$\text{Min } (C_i - \pi^{k'} A i) x_i + M ||\alpha_i^k - A_i' \, x_i|| \qquad (3.3)$$

Subject to: $x_i \in X_i$

where α_i^k, i= ,...,n, is a partition of the resource vector b_o such that:

i) $x_i \in X_i$ such that $\alpha_i^k = A_i \, x_i$, for all i.

$$\text{ii)} \sum_{i=1}^{n} \alpha_i^k \leq b_o \qquad (3.4)$$

$$\text{iii)} \sum_{i=1}^{n} \pi^{k'} \alpha_i^k = \pi^{k'} b_o \qquad (3.5)$$

α_i^* is then an allocation of resource forced by the superior level. As Charnes et al. have observed, this allocation could be directly affected to the constraints of the subproblem, modifying the polyhedra X_i:

$$\min \, (C_i - \pi^{k'} A_i) \, x_i$$

subject to:

$$B_i \, x_i \leq b_i$$
$$A_i \, x_i = \alpha_i^k \qquad (3.6)$$
$$x_i \geq 0$$

The task of the master-program will be now, not only the determination of the prices π^k, but too the computation of the allocation vector α_i^k. We may observe that, if nonunicity affects the solution of the subproblems, the constraint (3.6) ensures that the process is well-defined. Then, with the choice of "coherent" partions as defined by Charnes (p. 312), the process will be decentralized.

Various attempts have been made to yield coherent allocations combined with well-defined prices and maybe one of the most powerful is the algorithm built by Sengupta and Gruver (1974) which was truly applied to some educational planning problems. In this algorithm the subproblems have the form:

$$\text{Min } (C_i - \pi^{k'} A_i) \, x_i$$

Subject to:
$$B_i \, x_i \leq b_i \qquad (3.7)$$
$$A_i \, x_i \leq \alpha_i^k$$
$$x_i \geq 0$$

This approach is original in the sense that the master-program will be able to identify the subproblems which have a high potential for changing the convex combination of former proposals in a profitable way. Thus, if the greater positive value associated with the jth element of (3.7) (the jth corporate resource) occurs for subproblem i, then an arbitrary large quantity will be allocated to the jth element of α_i^k, allowing the ith subproblem to form a new proposal which uses a greater amount of the jth corporate resource.

We have seen then two general modes of modifying the coordination by prices only. But, as far as decentralized optimization is concerned, both "nonlinearization" of the cost functions and combined allocation of resource have a very similar meaning = considering the imposibility inherent to the linear structure to reach a completely decentralized optimum, the solution generally adopted was to give to the decision centers some additional information about the consequences of their own decisions. It is well known that resource allocation in the constraints is a dual method of transfer pricing. Geoffrion (1970) speaks of dual price-directive and primal resource-directive approaches. In the hierarchical control theory, the terms of unfeasible goal coordination and feasible model coordination are employed (Titli,1975). This is not our objective to look for the good choice between price and resource-allocation since no definite answer to that question could be assumed. Some resource-directive algorithms have seemed to work well in some special economic cases such as capital budgeting allocation method of Maier and Vanderweide (1976), but Moore (1979) has shown under some empirical experiments that the hypothese that budgeting would outperform transfer pricing is not confirmed. Finally, the study of all types of algorithms, primal or dual, which have yielded truly decentralized decision would not be complete without mentioning the min-max algorithms, which we may classify as primal-dual algorithms. These algorithms simply investigate the solution of the saddle-point problem defined in section II. Potier (1972) has developed an adaptation of Uzawa's min-max algorithm to the linear decomposed problem. It seems that the direction-finding procedure $u=\bar{u}+\rho s$ would accelerate the Dantzig-Wolfe's iterative process in the early iterations. A modified steepest-ascent algorithm has been confronted with the models of Jennergren, Sengupta-Gruver and Dantzig-Wolfe in a large-scale production-inventory system. Unfortunately, the diversity and the heterogenous character of these results difficult their publication inside this paper.

IV. Concluding remarks

The laboratory experiments made by Moore(1979) are very useful in the sense that they show us how careful we must be in the search of algorithms which would lead to perfect decentralization in an on-line simulation. Since unmodellable factors such as organizational design and time pressure constraints strongly influence the decision making behaviour, it has little sense to search for an algorithm whose iterations are all better than the sequential decisions of the tatonnement process. Undoubtly, the decomposed algorithms work as an aid to the decision-makers and not as their substitutes. If the mathematical linear decomposed programs cannot lead to complete decentralization, it is not a failure inherent to their mechanisms, but in the contrary, they may give to the decision-makers much information besides the simple task of decentralizing. The study of the dual function is an example of the powerful tools issued from mathematical programming which economists or manager could use to know better the structure of the systems in which they are integrated.

REFERENCES

Arrow, K. and L. Hurwicz (1960). Decentralization and computational in resource allocation. In R. Phouts (Ed.), Essays in Economics and Econometrics, University of North Carolina Press, Chapel Hill, pp. 34 - 104.

Baumol, W. J. and T.Fabian (1964). Decomposition, pricing for decentralization and external economics. Management Sci., 11, 1 - 32.

Bensoussan, A., J.L.Lions and R. Teman(1972). Sur les méthodes de décomposition, de décentralisation et de coordination et applications. Cahiers de l'IRIA, 11, nº 2, 5 - 189.

Charnes, A., R.W. Clower and K. O. Kortanek (1967). Effective control through coherent decentralization with preemptive goals. Econometrica, 35, 294 - 320.

Christensen, J. and B. Obel (1978).Simulation of decentralized planning in two danish organizations using mathematical programming decomposition. Management Sci., 24, 1658 - 1667.

Dantzig, G.B. and P. Wolfe (1961). The decomposition algorithm for linear programs. Econometrica, 29, 767 - 778.

Geoffrion, A.M. (1970). Primal resource-directive approaches for optimizing nonlinear decomposable systems. Oper.Res., 18, 375 - 403.

Jennergren, L.P. (1973). A price schedules decomposition algorithm for linear programming problems. Econometrica, 41, 965 - 980.

Kornai, J. (1973). Thoughts on multi-level planning systems. In L.M. Goreux and A. S. Manne (Eds.), Multi-level planning: Case Studies in México, North Holland, Amsterdam, pp. 521 - 548.

Lasdon, L.S. (1970). Optimization Theory for Large Systems. Mac Millan, New York.

Maier, S.F. and J.H. Vander Weide (1976) Capital budgeting in the decentralized firm. Management Sci., 23, 433 - 443.

Malinvaud, E. (1967). Decentralized procedures for planning. In E. Malinvaud and M.O.L. Bacharech (Eds.), Activity Analysis in the Theory of Growth and Planning, Mac Millan.

Moore, J.H. (1979). Effects of alternate information structures in a decomposed organization: a laboratory experiment. Management Sci., 25, 485 - 497.

Potier, D. (1972). Algorithmes de coordination.Application à la gestion d'unités de production interdépendentes. Cahier de l'IRIA, 11, nº 2, 241 - 375.

Sengupta, J.K. and G.W.Gruver (1974). On the two-level planning procedure under a Dantzig-Wolfe decomposition. Int. J. Syst. Sci., 5, 857 - 875.

Titli, A. (1975). Commande Hierarchisée et Optimisation des Systèmes Complexes. Dunod Automatique, Paris.

VULNERABILITY ANALYSIS OF MULTITERMINAL DISTRIBUTION SYSTEMS AND NETWORKS

M. Darwish* and J. Bernussou**

**Department of Electronics and Communication Engineering, Cairo University, Giza, Egypt*
***Laboratoire d'Automatique et d'Analyse des Systèmes du C.N.R.S., 7, avenue du Colonel Roche, 31400 Toulouse, France*

Abstract. Vulnerability analysis of multiterminal distribution systems and networks when subjected to failure and/or structural perturbations is considered in this paper. A methodology is proposed to allocate and preserve shortages among the various terminals of the system insuring the reliability level. The method is illustrated through application to production distribution system.

Keywords.

INTRODUCTION

Planning multiterminal distribution systems and networks like power systems and other similar supply systems involves the evaluation of expansion alternatives. One of the important aspects in this evaluation is the assessment of the risks of not being able to supply the demands in the future. These risks are influenced by many factors. Some of them are related to the uncertain nature of the demands, others have to deal with the structure of the supply system itself and with the way in which the system might actually be operated in the future (Kennington 1978, Darwish 1979, Duran 1978).

Supply systems are modeled as networks. They consist in general of :

(1) Sources of the commodity, such as electric power generating plants, water resources...etc.

(2) Sinks or demand points where the commodity is consumed, and

(3) Connection links that carry the commodity from sources to the links.

To satisfy the demands, a flow of commodity is sent through the network. This flow has to satisfy the conservation principle, in addition the flow through a link is limited to a maximum amount called the capacity of the link, similarly the sources are limited by their generation capacities (Duran 1978, Darwish 1980, Allan 1977).

Such systems are often subjected to failure and/or structural perturbations, the introduction of redundancy in the system will increase its survivability, but at the same time, increase the cost of construction and maintenance (Allan 1977, Boesch 1970, Dufour 1978). Hence studies related to vulnerability analysis become very important aspect, the objective is a decision aiding methodology to system planning and operation (Darwish 1979, Siljak 1976-77, Warfield 1974).

The uncertainty of the demand need some operating policies to allocate shortages among the various terminals of the system, hence when the system is subjected to failure it is desired to preserve these shortage allocations through proper planning methodologies. In this paper using notions of vulnerability, a methodology is developed to satisfy system demand within certain reliability level.

PROBLEM FORMULATION AND SYSTEM DESCRIPTION

The system under study is assumed to consist of a set of stations connected by interconnection links, and can be represented by weighted directed graph G(V,E), where V is the set of nodes and represent system stations and E is the set of arcs and represent the interconnection links between the stations. It is supposed that each arc E_l ($E_l \in E$ connects the ordered pair (V_i, V_j), V_i, $V_j \in V$) is associated with positive number C_l

called arc capacity and represents the capacity of interconnection link, each node $V_i \in V$ is associated with a requirement value V_i representing station requirements (V_i may be positive (generation or source node), or negative (sink or demand node), or zero (internal node)).

Assuming the graph G(V,E) consisting of :

(a) n nodes giving : (i)n_s source nodes, each denoted by g_i and have generation S_i, hence the subset source nodes is $g = \{g_1, g_2, ..., g_{n_s}\}$ and the generation set :

$$S = \{S_1, S_2, ..., S_{n_s}\}$$

(ii)n_d demand nodes, each denoted by t_i and have consumption d_i, hence the subset demand nodes is :

$$T = \{t_1, t_2, ..., t_{n_d}\} \text{ and}$$

the demand set :

$$d = \{d_1, d_2, ..., d_{n_d}\}.$$

(iii) n_t internal nodes, each denoted by Z_i with zero requirement, hence the subset internal nodes :

$$I = \{Z_1, Z_2, ..., Z_{n_t}\} \text{ and}$$

the set of nodes :

$$V = gUTUI$$

(b) m arcs, each $E_l \in E$ connects two ordered pair (V_i, V_j^l), V_i, $V_j \in V$, and the arc capacity vector is :

$$C = (C_1, C_2, ..., C_m)^T.$$

The problem can always be formulated as : given the generation set S it is required to satisfy the demand set d subject to the arc capacity vector C.

SHORTAGES ALLOCATION POLICIES

The demand of supply system is always subjected to variations, and sometimes system structure will not be able to satisfy all the required demands. Hence, it is not possible in general to avoid the possibility of finding the system short of supply at one or some of its terminals, and it will be important to define policies for allocating shortages to system terminals in a way to maximize the total amount of demand supplied and to keep the expected shortages at each node of the network smaller than a reliability requirement for that node.

Assuming :

X : expected shortage vector,

$$X = (X_1, X_2, ..., X_{n_d})^T$$

Where X_i is the expected shortage at the ith node.

α: reliability requirement vector,

$$\alpha = [\alpha_1, \alpha_2, \cdots \alpha_{nd}]^T$$

Where α_i is the reliability requirement of the ith node and for which it is required always to have shortage within this requirements otherwise the node will be unreliable.

Associated with the shortages X, a shortage cost function f(X).

The shortage allocation policy can be formulated as follows :

Find the vector X minimizing f(X) and subjected to the reliability constraints $X \leq \alpha$ (the relation $X \leq \alpha$ means $X_i \leq \alpha_i$ for all nodes).

This formulation allows to assigne shortages to each demand node which insure the reliability level. However when the system is subjected to failure, it required to maintain and preserve this shortages allocations, this will be the subject of the next section.

VULNERABILITY ANALYSIS TO PRESERVE SHORTAGES ALLOCATIONS

When the system under study is subjected to failure, the shortage allocation policies defined in the previous section may be destroyed , hence it will be important to maintain and preserve these policies. This can be obtained through vulnerability analysis. Let us first state the following definitions :

Definition 1 : a disconnecting set $C(g:t_i)$ is a set $C \subset E$ whose removal from G, separate g from $t_i \in T$.

Definition 2 : a minimum disconnecting set $C^m(g:t_i)$ is a set $C(g:t_i)$; its capacity is minimal. The capacity of $C(g:t_i)$ is denoted by :

$$\mathcal{C}(c) = \sum_{E_\ell \in C(g:t_i)} C(E_\ell)$$

Similarly $C^m(g:t_1, t_2, ..., t_{n_d})$ can be defined as the minimum disconnecting set separating g from $t_1, t_2, ...,$ and t_{n_d}.

Suppose that the demand set d is time varying according to certain statistical distribution. It becomes very important to know whether the system structure can satisfy this demand under failure conditions. Assuming the peak value of the demand of the ith node is $\hat{d}_i$, hence it is required to satisfy the demand ($\hat{d}_i - X_i$ where X_i is the shortage allocated to the ith node according to the policy described in the previous section.

The following theorem gives the necessary and sufficient conditions to satisfy the demand set

$$\bar{d} = \{ \bar{d}_1, \bar{d}_2, \dots, \bar{d}_{n_d} \}$$

(where $\bar{d}_i = \hat{d}_i - X_i$), proof of the theorem and illustrations are contained in (Darwish, a-b).

Theorem 1 : the necessary and sufficient conditions to satisfy the demand set $\bar{d}$ is the condition (I) given by :

$$\forall_{t_i \in T} \mathcal{C}(C^m(g:t_i)) \geq \bar{d}_i$$

Condition (I)

$$\forall_{t_\ell, t_k \in T} \mathcal{C}(C^m(g:t_\ell, t_k)) \geq \bar{d}_\ell + \bar{d}_k$$

$$\mathcal{C}(C^m(g:t_1, t_2, \dots, t_{n_d})) \geq \sum_{i=1}^{nd} \bar{d}_i$$

If $C^m(g:t_\ell)$ and $C^m(g:t_k)$ are disjointed for all t_ℓ, $t_k \in T$, then :

$$\forall_{t_i \in T} \mathcal{C}(C^m(g:t_i)) \geq \bar{d}_i$$

is the necessary and sufficient conditions to satisfy the demand set $\bar{d}$.

Now, if some parts of the system are subjected to failure, then it is very important to know whether the system demand is still satisfied or destroyed under this failure. This is a problem of vulnerability analysis, first we define the following.

Definition 3 : the critical set of order 1 is the set of interconnection links S_1^c, $S_1^c = \{ E_1 \in E$: eliminating any one of them, the resulting system does not satisfy the conditions of theorem 1$\}$.

Similarly, the critical set of order is defined as : $S^c = \{ E_1 \in E$: eliminating any γ of them, the resulting system does not satisfy the conditions of theorem 1$\}$.

From practical point of view, it is seldom to have $\gamma > 1$ simultaneous failures, so we consider the case of single failure at a time and denote the critical set in this case by S^c.

Theorem 2 : the shortage allocation policies are preserved under failure iff (if and only if) the set S^c is an empty set and in this case, the system is said to be invulnerable.

The proof follows directly from the argument that, if the set $S^c = \{\phi\}$, this means that removal of any interconnection link in the system does not violate the conditions of theorem 1, thus system requirements are satisfied, hence shortage allocation policies are preserved under failure, and vice versa if $S^c \neq \{\phi\}$, that is there exist failures for which shortage allocations are violated, and this completes the proof. Using theorems 1 and 2, the following studies can be performed :

1) Determination of the critical set of interconnection (set of interconnections removing any of them disturbe the shortage allocation policies).

1) Determination of system structure which can preserve shortage allocation policies.

APPLICATION

To illustrate the previous aspects, consider the production distribution flow network shown in Fig. 1.

The system datas are the following :

The set of arcs :

$$E = \{ E_1, E_2, \dots, E_{15} \}$$

$$g = \{ V_1, V_5 \} \quad T = \{ V_2, V_6, V_8 \}$$

$$I = \{ V_3, V_4, V_7 \} \text{ and } S = \{ 30, 20 \}$$

$$C = (10,20,20,25,15,5,10,20,20,20,30,10,10,20,20)^T$$

assuming the peak values for the demand are : $\hat{d}_1 = 22$, $\hat{d}_5 = 55$, $\hat{d}_8 = 23$, hence $\hat{d} = \{22,55,23\}$.

The shortage allocation policy allows to determine X_1, X_5, X_8 which represent shortages assigned to nodes V_2, V_6, V_8, these values are given by :

$X_1 = 3$ - $X_5 = 5$ and $X_8 = 2$, hence the set $\bar{d} = \{19,50,21\}$.

Then, we can get the following one node minimum disconnecting sets :

$$\mathcal{C}(C^m(g:V_2)) = C(E_1) + C(E_5) + C(E_a) = 45$$

$$\mathcal{C}(C^m(g:V_6)) = C(E_6) + C(E_4) + C(E_{11}) + C(E_3) = 80$$

$$\mathcal{C}(C^m(g:V_8)) = C(E_{13}) C(E_3) + C(E_{12}) = 40$$

Where C (E_1) is the capacity of the arc E_1, but :

$$C^m(g:V_2) \cap C^m(g:V_6) = \{\phi\}$$

$$C^m(g:V_2) \cap C^m(g:V_8) = \{\phi\}$$

while : $C^m(g:V_6) \cap C^m(g:V_8) \neq \{\phi\}$

hence we need to calculate :

$$\mathcal{C}(C^m(g:V_6,V_8)) = C(E_4)+C(E_6)+C(E_3)+C(E_{11}) + C(E_{12}) = 90$$

$$\mathcal{C}(C^m(g:V_2,V_6,V_8)) = C(E_1)+C(E_3)+C(E_5) + C(E_6 + C(E_9) + C(E_{11}) + C(E_{12}) = 110$$

Hence to satisfy the demand $\bar{d}$ it is required that :

$$\mathcal{C}(C^m(g:V_2)) \geq 19$$

$$\mathcal{C}(C^m(g:V_6)) \geq 50$$

$$\mathcal{C}(C^m(g:V_8)) \geq 21$$

$$\mathcal{C}(C^m(g:V_6,V_8)) \geq 71$$

$$\mathcal{C}(C^m(g:V_2,V_6,V_8)) \geq 90$$

which is true, then this system structure satisfy the demand set $\bar{d}$. Then using definition 3 and theorem 2, we obtain the following critical set :

$$S^c = \{E_4, E_{11}\}$$

CONCLUSION

Vulnerability analysis of multiterminal systems and networks when subjected to failure has been considered in this paper. The study allows to allocate shortages to system terminals, besides strategies for planning system structure preserving the desired shortage allocation policies are developed. The present study helps the system designer and decision maker to properly choose and characterize system structures as well as extension planning policies.

REFERENCES

Allan, R.N. and De Oliveria, M.F. (1977). Reliability modelling and evaluation of transmission and distribution systems. Proc. IEE, Vol. 124, n° 6.

Boesch, F.T. an Thomas, R.E. (1970). On graph of invulnerable communication nets. IEEE. Trans. on circuit theory, Vol. CT-17, n° 2, pp 183-197.

Darwish, M. and Bernussou, J.(1979-80)
-a- Vulnerability as decision aiding to system planning and operation, 5th EHCAC conference on automatic control and systems engineering, Cairo.
-b- Vulnerability and structural perturbations of network flow systems. IFAC symposium on LSSTA TOULOUSE - FRANCE.

Dufour, J. and Gilles, G. (1978). A Methodology for structural analysis and partition of dynamic complex systems-IFAC 7th triennal world congress, Helsinki, pp 1385-1392.

Duran, H. (1978). Reliability in multiterminal distribution systems and networks - Ph.D Standord university.

Kennington, J.L. (1978). A survey of linear cost multicommodity networks flows. Operation research, Vol. 26, n° 2, pp 209-236.

Siljak, D.D. and Vukcevic, M.B.(1976-1977).
-a- Large scale systems : stability, complexity, reliability. Journal of Franklin Institute, Vol. 31, n° 2, pp 49-69.
-b- Vulnerability of dynamic systems. IFAC workshop, TOULOUSE.

Warfield, R.N. (1974). Toward interpretation of complex structural models, IEEE Trans. Syst. Man and Cyber, Vol. 4, pp 405-417.

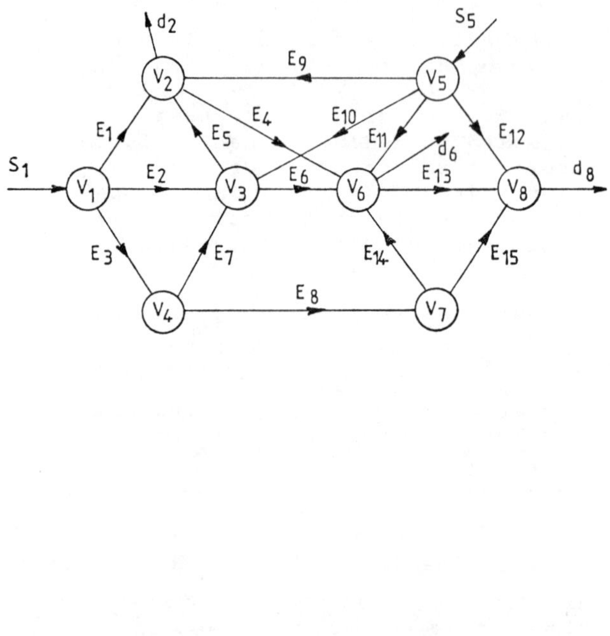

QUANTITATIVE ANALYSIS OF AN ON LINE MEASUREMENT EQUIPMENT FOR STATIC STATE ESTIMATION PURPOSE

P. Albertos*, C. Alvarez and A. Reig****

**Automatic Control Department, E.T.S.I.I. de la Universidad Politécnica de Valencia*
***Electric Power Engineering Department, E.T.S.I.I. de la Universidad Politécnica de Valencia*

Abstract. In a real physical system, the aplication of real-time functions usually requires the on-line state estimation in order to smooth the measures and obtain the state in an accurate way.

From a practical point of view, a previous problem to be solved is the measurement system definition, that is, the determination of number, location, type and precission of the measurement devices needed to compute the state within a given precission.

In this communication, the measurement equipment quality is analysed through the trace of the estimated state covariance matrix. Two new parameters attached to any measurement signal are defined. They make possible the qualitative and quantitative study of the precission measurement system when any change in the system configuration happens.

Then an algorithmic procedure for measurement system design is developed.

Finally, the results are applied to the power systems case.

INTRODUCTION

The problem of data processing, from the control point of view may be characterized by the following features:

1) The state variables (that is, those which in minimal number describe the system, knowing the inputs) are not, usually, directly measurables.

2) The measurements are corrupted by a random observation noise.

The measured magnitudes must be processed by a state estimator in order to obtain the state of the system. The estimator, usually optimal with respect to a performance index, supply the estimated state with a bounded error, given a measurement set. Then, the measurement equipment definition, (i.e. type, location and precission of the measurements) in order to obtain the state within an acceptable precission, is a fundamental problem to be solved.

The solution to the above raised problem demands:

- Knowledge of the influence of each measurement in the state estimation accuracy.

- Determination of which measurements are fundamental (those whose deleting make impossible the state computation) to plan their substitution by some approximative values. (pseudomeasurements).

- Set up a systematic way for the

measurements system design, taking a redundant and accurate measurement set as the initial one, and modifying it in order to reduce the cost, the state estimation errors being within given limits.

PROBLEM FORMULATION

Basic Concepts. We will define our system in the framework of modern control theory. Let us denote by $x(k)$ the n-dimensional state vector, by $y(k)$ the p-dimensional output vector and by $u(k)$ the m-dimensional input vector at the time k. The system dynamic behaviour at this time k is defined by:

$$x(k+1)=f[x(k),u(k)]+\varepsilon(k) \quad (a)$$
$$y(k)=h[x(k)]+v(k) \quad (b) \qquad (1)$$

where f and h are, usually, non linear functions.

The state is afected by random disturbances $\varepsilon(k)$ and moreover, it is not directly measurable. The measurements give information enough to the state computation. These measurements are afected by random observation noise v which covariance matrix pxp dimensional, R, is asumed known [1].

From the point of view of the real-time system operation there are two problems to be solved:

1) State computation from the measurement set.

2) System control, generating the inputs in order to drive the state in the desired way.

We will focusse our attention to the first problem. It will be stated in a more general approach as shown in fig. 1 . The problem to be treated here is the determination of which measurement system, with a reasonable cost supply the best estimated state.

The estimation algorithm can be defined following diferents approaches [2,3]. The estimator structure is usually as depicted in fig. 2.

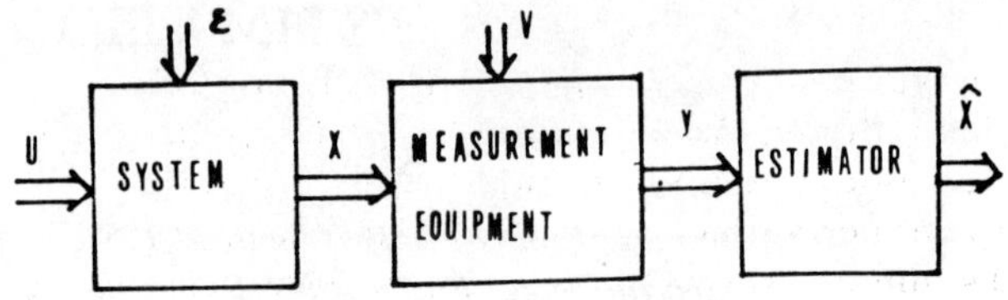

fig. 1. State Computation Approach

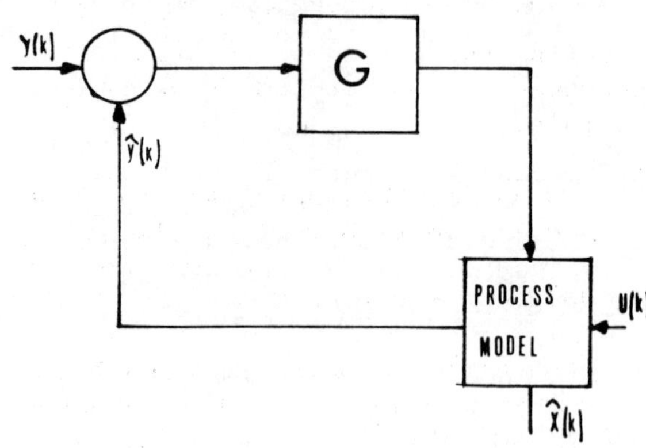

fig. 2

where the gain matrix G may be computed in different ways. In these estimators, the state is updated when a new measurement set is received, according to it and all the past information.

Obviously, the state estimation accuracy strongly depends on the quality, number and location of metering devices.

Stationary linear systems.

The dynamic behaviour of industrial processes under normal operating conditions is much more slower than that of metering and signal processing. Then they can be modelled by stationary linear systems.

The state and output equations can be reduced to

$$x(k+1)=x(k)+\varepsilon(k) \quad (a)$$
$$y(k)=H\,x(k)+v(k) \quad (b) \qquad (2)$$

where H is a constant matrix.

In such a case, the state vector is considered as a subset of the dynamic one either because some information needed to the dynamic analysis is unnecesary in the stationary case, or because groups of dynamic state variables can be substituted by a unic static variable.

The problem of the state estimation in these system is treated in two different ways:

1) Considering the system as a simplified dynamic case, with system matrix I (identity matrix), then applying the general estimation algorithms implemented in real-time mode.

2) Evaluating the state only through the information received in the present time. The system model reduced to equation 2 b.

In such a case, the condition $p \geq n$ is required, and the problem becomes the minimization of a quality index.

The second way is usually referenced as static state estimation.

The state estimation problem concerned with digitally controlled stationary processes may be focussed in a different way. Obviously, the non linear optimization problem is solved by iterations, starting with an initial state and proceeding untill convergence is achieved. Then, each measurement sampling allows the state computing, the last computed state taken as initial one. The procedure, tracking algorithm, usually needs one iteration a sampling period. Fig. 3 describes the problem. Some static estimators can be easily adapted to this new approach.

Let us built the cost index (performance index PI) according to the weighted least squares theory. Letting $\hat{x}$ be the estimated state, the relations:

$$\mathrm{cov}(\hat{x}^t\hat{x}) = (H^t R^{-1} H)^{-1} = \Sigma_x \qquad (3)$$

and

$$\hat{x} = \Sigma_x H^t R^{-1} y \qquad (4)$$

holds.

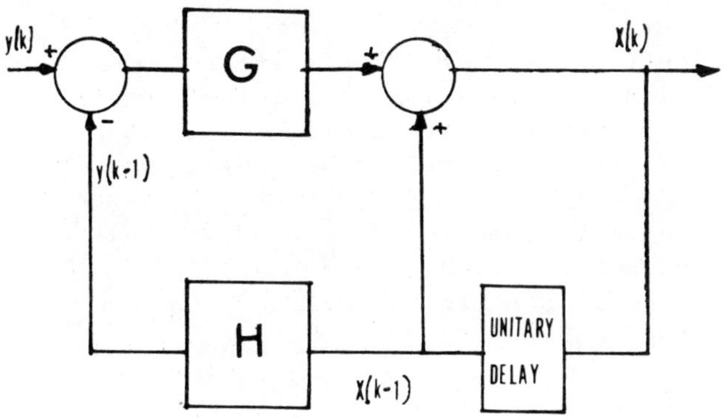

fig. 3.

In this type of systems, the basic observability condition is equivalent to the existence of the H pseudoinverse. That is

$$\mathrm{Det}\ (H^t H) \neq 0 \qquad (5)$$

In the following we will consider linear stationary systems.

<u>Qualitative analysis of the measurement system</u>:

The measurement equipment is defined by the type, location and precision of each measurement. These questions may be related to some parameters in order to a proper measurement set definition.

Conventionally, has been used as a quality parameter the redundancy defined as:

$$\eta = p/n \qquad (6)$$

It presents a serious inconvenient, that is the lack of information about the performances of each measurement device

The matrix covariance (3) is generally adopted to point out the measurement set quality.

In this way, the parameter trace Σ_x [4] will be considered as an escalar quality index.

A measurement will be noted as <u>fundamental</u> if the system becomes unobservable when removed

QUANTITATIVE ANALYSIS OF THE MEASUREMENT SYSTEM

Let us consider the state estimation of an stationary physical system with an initial measurement system which makes it observable.

We will suppose that the noise covariance matrix R is diagonal. That is a good asumption in many of the real applications.

So that, the inverse of the estimated state covariance matrix Σ_x^{-1} can be expressed by

$$\Sigma_x^{-1} = \sum_{i=1}^{p} \alpha_i \, f_i \, f_1^t \tag{8}$$

α_i being the weight coeficient of measurement i (i.e. the inverse of the R diagonal element i) and f_i the i-row of H matrix.

As stated early, the influence of any measurement weight modification in the global error estimation (tr Σ_x) is a principal feature of the corresponding device in order to decide its accuracy improvement or cancellation.

<u>Sensitivity parameter</u>:

This parameter, γ_i, is defined as the sensitivity of tr Σ_x to a change in the i measurement precission.

That is

$$\gamma_i = \frac{d\,tr\,\Sigma_x}{d\,\alpha_i} \tag{9}$$

According to the matrix operation laws, the following expression

$$\frac{d\,tr\,\Sigma_x}{d\alpha_i} = tr\,\frac{d\,\Sigma_x}{d\alpha_i} = tr\left(-\Sigma_x \frac{d\,\Sigma_x^{-1}}{d\alpha_i}\Sigma_x\right) \tag{10}$$

holds.

From expression (8) and by substitution in (10)

$$\gamma_i = -\,f_i^T\,\Sigma_x\,\Sigma_x\,f_i \tag{11}$$

results.

From this equation, the sensitivity of the state estimation error with respect to a measurement precission change can be evaluated easily.

The parameter γ_i, attached to the precission of every measurement device, is related with the measurement equipment configuration. However, it does not depend either on the amount of the precission increment (or decrement) or on the functionament point.

<u>Observability parameter</u>:

Let us suppose a step change $\Delta\alpha_i$ in the i measurement precission, the resultant state covariances matrix being Σ'_x.

According to expression (8) it can be written as

$$\Sigma'_x = (\Sigma_x^{-1} + f_i\,f_i^t\,\Delta\alpha_i)^{-1} \tag{12}$$

Applying to (12) the matrix inversion lemme.

$$\Sigma'_x = \Sigma_x - (\Sigma_x f_i f_i^T \Sigma_x)\,\frac{\Delta\alpha_i}{1 + f_i^T \Sigma_x f_i \Delta\alpha_i} \tag{13}$$

Let us define a new scalar parameter k_i as

$$k_i = f_i^t\,\Sigma_x\,f_i \tag{14}$$

The increment of the covariance matrix can be written

$$\Delta\Sigma_x = -\,(\Sigma_x f_i f_i^T \Sigma_x)\,\frac{\Delta\alpha_i}{1 + k_i\,\Delta\alpha_i} \tag{15}$$

and then, by (11)

$$\Delta\,tr\,\Sigma_x = -\,\gamma_i\,\frac{\Delta\alpha_i}{1 + k_i\,\Delta\alpha_i} \tag{16}$$

that is, the global error increment can be easily evaluated when ever a change $\Delta\alpha_i$ in the precission of the i measurement occurs.

In fig. 4 the relationship $-\Delta tr \Sigma_x / \gamma_i$ as a function of $\Delta\alpha_i$ is shown.

If the increment $\Delta\alpha_i$ in the modified measurement i could be

$$\Delta\alpha_i = -1/K_i \qquad (17)$$

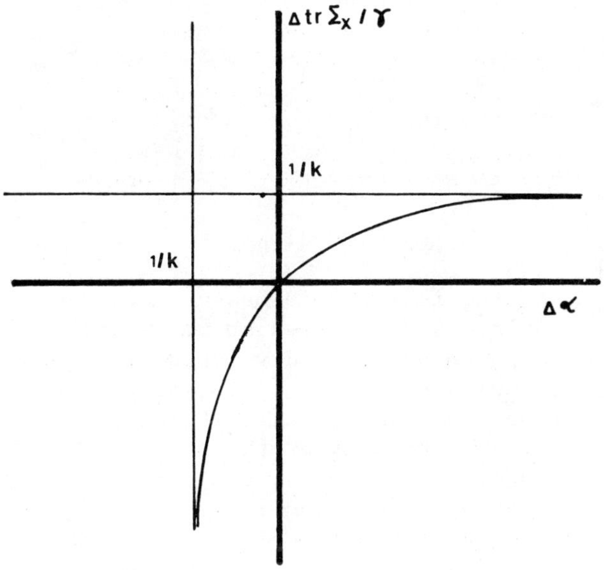

fig 4

that is, if

$$\alpha_i + \frac{1}{k_i} \geqslant 0 \qquad (18)$$

the system could become unobservable for such a critical increment (i.e. the parameter tr Σ_x becomes ∞).

Let as define the observability parameter β_i:

$$\beta_i = 1 - k_i \alpha_i \qquad (19)$$

When the β-parameter corresponding to measurement i, is equal to cero, this measurement is fundamental: That is, the critical increment corresponds to the measurement elimination.

As another interesting conclusion from fig. 4, the largest improvement in the state precision we can get modifying the measurement i corresponds to

$$|\Delta tr \Sigma_x|_{max} = \gamma_i / k_i \qquad (20)$$

MEASUREMENT EQUIPMENT DESIGN

The above results, allow us the formulation of an algorithm for the measurement equipment design, which can be summarized:

1. Adopt as initial measurement equipment that constituted by all the possible measurement with the best usual preccision. Set bounds to tr Σ_x.

2. Compute de β-parameter for each measurement and establish which are fundamental if any.

3. Compute the relation $\frac{\alpha_i \gamma_i}{\beta_i}$, i.e. the increment in tr Σ_x if the measurement i was deleted.

4. According to the above result and economical criterious remove the measurement with largest cost/precission relationship, being tr Σ_x within limits.

5. If any measurement deleting gives a not allovable global error increment stop. Otherwise go to step 2.

Each time that step 2 is realized, establish the necesary redundances and emergencies for fundamental variables in order to fullfill security performances.

ELECTRIC POWER SYSTEM APPLICATION

The above results have been obtained under the assumption of system model linearity. Nevertheless, these results could be applied to non linear systems if linearization around the functionament point is allowed, that is, expansioning h(x) by Taylor series.

$$h(x)=h(x_o)+H(x-x_o)+ \text{ higher order terms} \qquad (21)$$

the higher order terms are negligeable.

Let us consider the electric power systems case in which the application of the results from pharagraphs 3 and 4 are especially well suited.

As ilustration, the simple system shown in fig. 5 is considered. A generation node and a load one are interconected through a line. The real and imaginary components of node voltages are adopted as state variables.

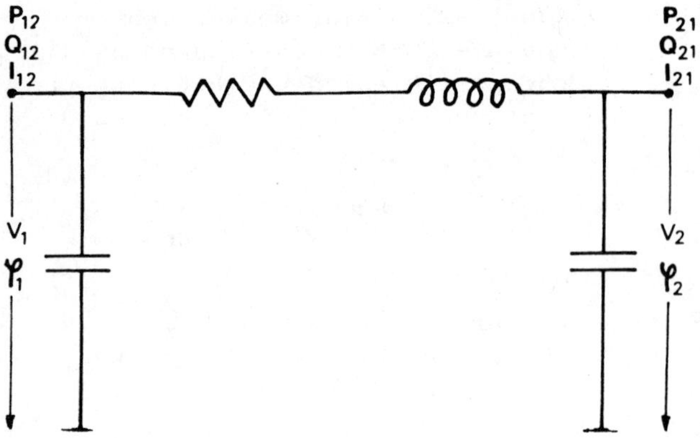

fig 5

The initial measurement set consists in the node voltage module, inyected power, current and power factor at each node.

Table 1 shows the results of the measurement design algorithm application, in a functionament point of 20% full load, where the less important measurement is deleted each iteration. All the measurement devices are taken as class 1.

As can be seen, the power factor and current measurement are "bad" with respect to power and voltage nodes.

The results in other functionament points are similar in the way of the measurement deleting importance.

CONCLUSION

An algorithm to the measurement equipment design for state estimation purposes has been developed. Although it has been developed in the case of linear systems models, it is well suitable to many of the physical systems in which the state estimation implementation is needed. In this way, the electric power system is considered.

Iteration number	MEASUREMENTS									
	V_1	V_2	p_{1-2}	p_{2-1}	q_{1-2}	q_{2-1}	i_{1-2}	i_{2-1}	φ_{1-2}	φ_{2-1}
0	- .2466482 .6731522 - .3897954	- .3783049 .5053901 - .7963196	- .2229294 .5048588 - .0205763	- .2934163 .5437840 - .0251436	-1.1458517 .2366429 - .2256344	-2.4832587 .5373744 - .2153353	- .0003184 .9993872 - .0003185	- .0002921 .9994179 - .0002922	- .0000004 .9999971 - .0000002	- .0000012 .9999954 - .0000006
1	- .2466482 .6731522 - .3897955	- .3783051 .5053899 - .7963204	- .2229300 .5048584 - .0205764	- .2934176 .5437836 - .0251438	-1.1458564 .2366417 - .2256364	-2.4832548 .5373739 - .2153352	- .0003184 .9993872 - .0003185	- .0002921 .9994179 - .0002922		- .0000012 .9999954 - .0000006
2	- .2466486 .6731519 - .3897963	- .3783055 .5053897 - .7963214	- .2229305 .5048573 - .0205765	- .2934164 .5437826 - .0251437	-1.1458522 .2366412 - .2256361	-2.4832692 .5373723 - .2153371	- .0003184 .9993872 - .0003186	- .0002921 .9994179 - .0002922		
3	- .2467999 .6730661 - .3900851	- .3785804 .5052058 - .7971900	- .2228872 .5048009 - .0205748	- .2931693 .5437487 - .0251241	-1.1481612 .2364301 - .2262927	-2.4840534 .5373617 - .2154093	- .0003187 .9993868 - .0003189			
4	- .2469667 .6729705 - .3904043	- .3788789 .5050073 - .7981325	- .2228411 .5047396 - .0205730	- .2928997 .5437115 - .0251027	-1.1505342 .2362252 - .2269570	-2.4850534 .5373461 - .2155023				
5	- .2480945 .6708021 - .3934548	- .3802353 .5029402 - .8042819		- .9221857 .0998353 - .4304318	-1.1395273 .2182326 - .2433187	-2.4743991 .5081897 - .2268894				
6	- .4231463 .5565395 - .8088476	- .5518036 .4234991 -1.3361306		-1.1029760 .0004168 -123.322987	- .3889218 .0195447 - .7272655					
7		-1.7161552 .0000000 **********		-1.1660294 - .0000000 **********	- .7621776 .0000000 **********					

Table 1. the shown parameters are

γ_i

β_i

$\Delta tr\, \Sigma_x$ if the measurement was removed

REFERENCES

1. Schweppe, F.C., Wildes, J. "Power System static State Estimation". Part I, II, III, IEEE Trans on PAS 1970, vol. 89 pp. 120-135.

2. Haudschin, E. "Real time Control of Electric Power Systems". Elsevier Scientific Publishing Co. Amsterdam 1972.

3. Alvarez, C., Albertos, P. "A new algorithm for real-time Estimation of the State of a Power System". Acepted in the IFAC Symp. "Power Generation, distribution and protection". Pretoria 1980.

4. Phna, K., Dillon, T.S. "Optimal choice of measurements for state stimation". 1977. Power Industry Computer Applications Conference.

AN EFFICIENT ALGORITM FOR COMPUTER PROCESS CONTROL

P. Albertos, J. A. de la Puente and J. Tornero

Cátedra de Automática, E.T.S.I.I. Universidad Politécnica de Valencia

Abstract. In this paper a new easy to compute algorithm is introduced. It is suitable for process identification, optimal control and disturbances modelling. The estructural properties of the algorithm allows its use for computer process control.

The identification problem is stated as real time estimation of the weighting sequence parameters.

The disturbances are modelled by a state space representation which is updated by the same algorithm at every sample time.

The control action is computed in open-loop, and a feedback of the prediction error is added to the set point.

The algorithm has been sucessful applied to some simulated and experimental processes.

INTRODUCTION

The improvement in computer reliability as well as the development of special purpose minicomputer systems have determined important growth in computer process control applications.

The largely developed modern control theory gives sucessfull results in control system design, suitable to aerospace and servomechanism problems. It is not the same for industrial processes control.

As pointed out by Foss [1] the industrial processes dynamics and control objetives are quite different from those stated in modern control theory. Then a gap exists between the control problems in the industrial framework and those usually studied by researchers.

Clearly, the problem must be formulated again from the begening, bearing is mind the features of computers, i.e. fast computation and large and fast storage capability, then, the following questions arise:

- Which kind of models best represent the process' dynamics.
- How measurement and structural disturbances can be considered.
- Which are the most suitable performance indexes to measure the control system quality.
- The control algorithm parameters are so intuitive as PID parameters are?.

In this paper we present a general purpose algorithm suitable for:

- process identification
- optimal control
- disturbances modelling

In the next chapter we state the features demanded for the new algorithm. Afterwards we introduce the algorithm and how it can be applied to fullfilled the control problems above mentioned.

The computational properties and the implementation on a microcomputer based system are then considered. Finally some applications on simulated and laboratory systems are shown.

INDUSTRIAL PROCESS MODELLINE AND CONTROL

The process model.

State space representation is not useful in process control applications. Non linearities, pure time delays, loop interactions and uncertainty in models advocates for a more flexible system representation.

Some research groups [2] [3] have coming again to the impulse response representation. The classical drawbacks such us high parameter number, are minimized by the use of computers. On the other hand there are some advantages to deal with industrial processes. The discretized output y (k) of a system can be expressed in a non-recursive form, as a weighted sum of the N_m past inputs, u (k),

$$y(k) = \sum_{i=0}^{N_m-1} h(i).\, u(k-i) \qquad (1)$$

Then the model is linear both in the inputs and parameters. Of course, the number of parameters must be large enough to represent a period of time greater them the system time response. Null values of h (i) for lowest valves of i denotes the presence of a pure time delay.

The generalization to multivariable system is straightforward, as can be seen writting (1) in a vectorial form

$$y = h' u \qquad (2)$$

where means matrix transposition.

Disturbances and uncertainity.

The process nonalinearitia make model parameters vary when a set point change occurs. On the other hand process and model outputs are not the same for any applied input owing to the presence of disturbances as well as uncertainities in the model.

The assumption of a N_d-order system to model the process disturbances is a common practice. Then, the expected difference between process and model output, p (m), can be expressed by the autonomous system

$$p(k+1) = \sum_{i=0}^{Nd-1} g(i).p(k-i) \qquad (3)$$

with disturbed and unknown initial values. The weighting coefficients g (i) arranged as elements of a vector, the predicted error may be written as

$$p(k+1) = g' \cdot q \qquad (4)$$

in a similar way to (2).

Control strategy.

The process control usually does not requiere to reach a given target or to optimize strictly a performance index.

The uncertainities in the process above considered demand for closed loop control. Moreover control parameters must be adjusted on plant in order to either avoid control saturations or modify system stiffness.

Then, the control algorithm to be applied in (2) must be computed in an easy-to-understand way.

Set points are constant most of the time, and a compromise must be reached between regulation errors and control efforts.

GENERAL PURPOSE ALGORITHM

Some previous work [3] [4] and the practical efficiency shown on a significant number of simulated processes and laboratory experiments allow us to propose a general purpose algorithm. It is suitable for system identification, disturbances modelling and open loop control.

The algorithm can be written in the general form

$$\Delta x = v \frac{\alpha - y' Z}{\beta + y' y} \qquad (5)$$

where α, β are scalars; x, v r-dimensional vectors and y, z p-dimensional vectors.

a) Identification.

Let us consider a system given by

$$S = h'_o \; u \qquad (6)$$

and the model to be determined by (2).

The algorithm (5) updates recursively the model parameters, h, making optimal the performance index

$$J_i = \beta_i^2 \; \Delta h' \cdot \Delta h + (S - u'(h + \Delta h))^2 \qquad (7)$$

The index represents a metric distance on the space of output and model parameters . The meaning of any parameters and variable on (5) is summarized in table 1.

b) Open loop control.

Given a process modelled by (2), let us consider the following optimization index [4] :

$$J = \beta_c^2 \Delta u' \cdot \Delta u + \sum_{i=1}^{N_o} \left[C(i) - h'(u(i) + \Delta u) \right]^2 \qquad (8)$$

where N_o = optimization period of time,

C = set point

u = control previously computed

This index compute an optimal value Δ u that minimise the diference between the references and the output in a period of time.

We assume Δu as an No-dimensional vector where all the components are equal.

H' . U (i) = T (i) is the forcasted output in k+i with Δu = 0.

$$w(i) = \sum_{j=1}^{i} h(j)$$

c) Disturbances.

Let us now consider the real system represented by

$$g = h' \; u + p$$

The disturbance, p, being modelled by (4). According to the past errors the disturbance model parameter can be updated by the algorithm (5) minimizing the index

$$I = \beta_d^2 \cdot \Delta g' \cdot \Delta g + (p - q'(g + \Delta g))^2$$

similar to that considered in the identification case.

The algorithm parameters are summarized in table 1.

d) Closed-loop control.

In order to take into account the actual system output, the predicted disturbance value (4), can be added to the known reference value. Then, the control signal computed now by the open loop control algorithm will adequately command the process. If the optimization period considered in that algorithm is, No, greater than one, the disturbance prediction computation (4) must be evaluated No times.

Algorithm (5)	α	β	x	v	r	y	z	p
Identification	S	β_i^2	h	u	N_m	u	h	N_m
Disturbance	p(k)	β_d^2	g	q	N_d	q	g	N_d
Control	w'c	β_c^2	u	1	1	w	T	N_o

table 1.

COMPUTATIONAL ASPECTS

As previously stated, the algorithm can be written on the general form (5). In order to implement the full closed-loop control scheme, the algorithm must be computed twice, together with a convolution product, once for computing the control action and again for updating the disturbance model. The convolution product is performed in order to compu- ter the process model output and to predict the next value of the disturbance variable.

A dimensional flowcharting [5] of the computation process can be seen in figure 1.

The total number of operations is approximately $3\,N_o + (N_o + 1)\,N_i + 3\,N_d$ additions and products, with N_o equal to the optimatizacion period, N_i the length of the impulse response an N_d the length of the disturbance model. For a typical implementation with $N_i = 10$, $N_d = 5$ and $N_o = 5$ a total of 90 products and additions are performed. The computation time on a PDP-11/34 minicomputer (no floating-point instructions) is less than 100 ms, the algorithm being programmed in FORTRAN IV.

From these data a comparison can be made with other methods. PID algorithms are of course much simpler and need much less computing time as the number of operations involved is fixed (6 products and 6 additions). More complex algorithm are neverthless comparable. For example, the adaptive regulation described by Unbehaner end others [6] requieres about 140 products and additions for a third order system.

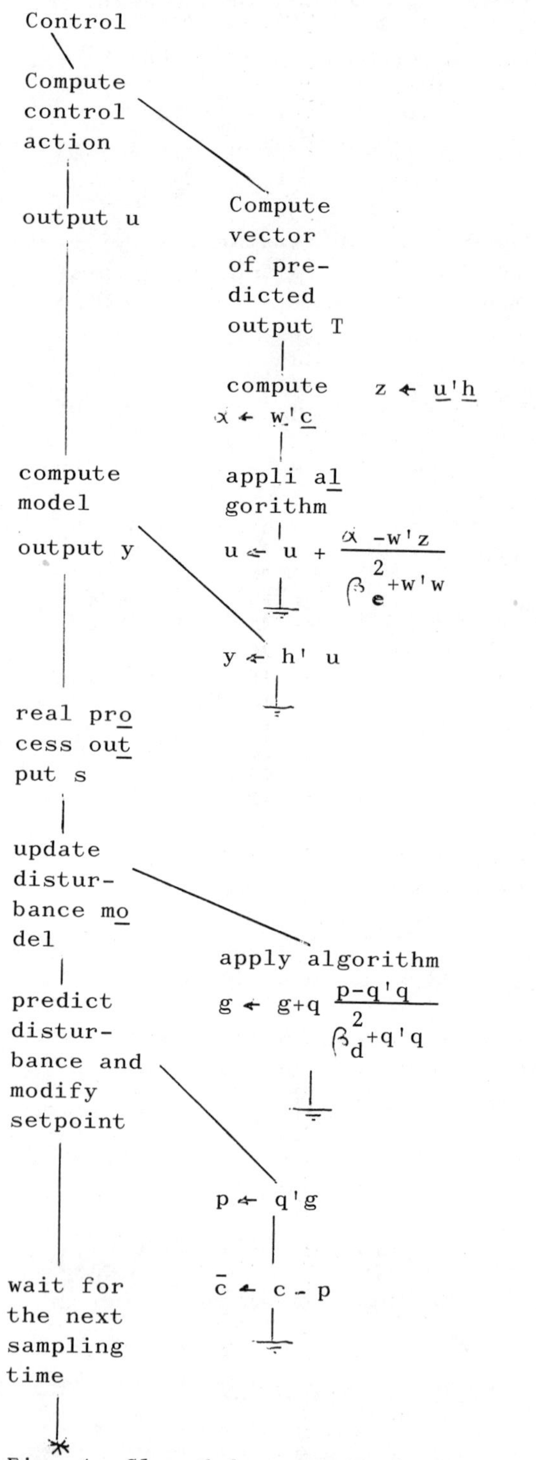

Fig. 1. Closed-loop control scheme. Dimensional flowcharting.

Identificacion algorithm.

a. Real system. The identification algorithm has been tested on several low noise sustems, being observed in the whole experience the same convergence and the same convergence characteristic was observed in all the experiments. The similar behaviour is consecuence of the linear parameter structure.

In all these experiments the structural parameter β_i^2 has an optimal value which is close to zero in the sense of the highest convergence speed.

For the particular case of a first order system consisting of an RC passive network with a 0.3 seconds time constant, taking 10 samples of the impulsional response, the identification algorithm output obtained is:

0.28,0,20,0.14,0.10,0.07,0.05,0.035 0.025,0.018,0.012

as could be expected fromt the theoretical model.

Another interesting result is that the convergence speed of the impulsie response is almost inversely related to the number of points. The number of iterations in order to get less than 1% difference between the system and model parameters, with respect to the sample number is shown in fig. 2.

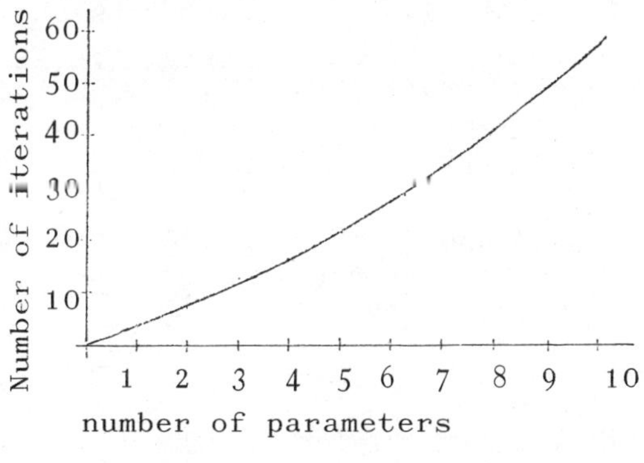

Fig. 2. Iteration/number of parameters

b. Simulated system with noise. Some experiments have been developed on systems of different orders obtaining similar results.

It can be concluded that the convergence speed is function of the applied noise amplitude and the structural parameter as depicted in fig. 3.

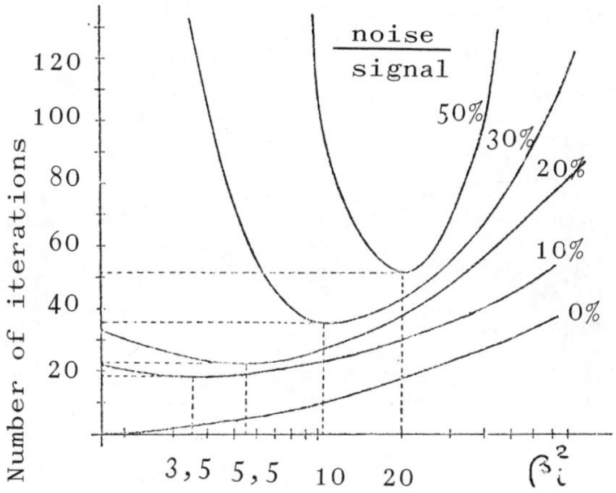

Fig. 3. Convergence speed

It can be observed in this graphic that the structural parameter has a value which minimizes the number of iterations needed to achieve the convergence for each noise amplitude.

Moreover for small structural parameter values, the system parameters swing around the final value and convergency could not be reached. On the other hand a highest structural parameter value makes the algorithm slow but never unstable.

The weight of the structural parameter is also conditioned to the input amplitude level.

Control algorithm.

The control algorithm has the same structure that the identification one, the difference being the involved variable set. The above structural parameter is now the control parameter, β_c^2, which limits the highest variation in the system inputs.

a. Open loop.

The open loop control problem is focused on the assumption that all the system parameter are known. Then the control problem can be treated separately.

Several experiences have been developed on different order simulated systems obtaining a good trayectory tracking for all of them. In the particular case of a third order system with a low noise level, the relationship between the highest

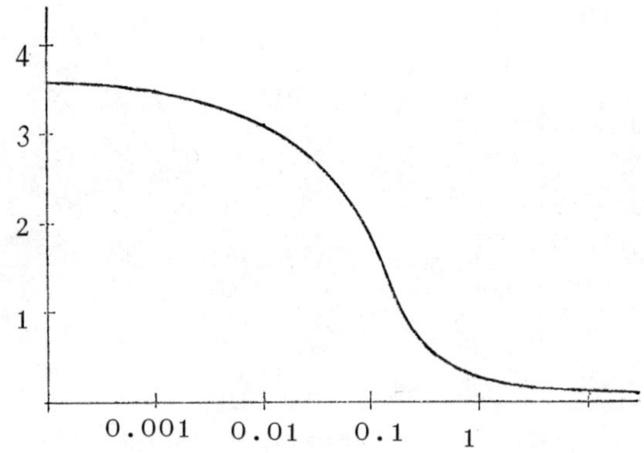

Fig.4. Control amplitudes/control parameter

relative variation of the input system allowed when a step in the reference is applied and the control parameter is shown in fig. 4.

b. Closed loop.

Let us suppose now that there is an error between the predicted output and the system one caused by

- The model parameters are not well known.
- The real system is disturbed.

In the closed loop control algorithm there are two adjustable parameters, the control parameter and the disturbance parameter.

This control algorithm has been succesfully applied to different order systems without stability problems also when a 100% error in the model system parameter is considered.

The disturbance parameters taken in these experiences have been the optimal ones computed in the identification step for the same noise conditions.

CONCLUSION

A new algorithm has been introduced The simplicity, efficiency and computations features make it suitable to be implemented in microcomputer based dedicated control systems.

The process representation, as well as the disturbances handling is very appropiate to be applied in industrial process control. The algorithm, whose theoretical aspects have been developed elsewhere, has been implemented in some experimental processes.

REFERENCES

1. Foss, A.S. (1973). Critique of chemical process control theory. IEEE trans. Auto.Control. vol. 18, n. 6, 646-652.
2. Richalet, J. and others (1976). Algorithmic control of industrial processes. IV IFAC Symp. on Identification and System parameter estimation. Tbilisi.
3. Menendez, A. (1975). Identification en temps reel des reponses impulsionnelles ... Automatisme Tome XX, N 3, 64-70.
4. Lebourgeois, F. Menéndez, A. (1973). Un nouvel algorithme de commande dynamique optimale. Automatisme Tome XVIII, N 4.
5. Witty, R.W. (1977). Dimensional flowcharting. Software-pratice and experience. Vol. 7. 553-584.
6. Unbehaven, H. Schmid, C.H.R. and Böttiger, F. (1976). Comparison and application of DDC algorithms for a heat exchanger. Automatica. Vol. 12, 393-402.

A COMPARATIVE STUDY OF SELF-TUNING REGULATORS

G. Favier and P. Guillermin

Laboratoire de Signaux et Systèmes, Equipe de Recherche Associée au C.N.R.S. n°835
41 Boulevard Napoléon III, 06041 Nice Cedex, France

Abstract. The stochastic adaptive control problem for a single-input single output system is considered. The use of a general control scheme allows us to study the servo and regulator problems at the same time. In the case where the model parameters are assumed known we present several control algorithms in a unifying framework. These algorithms are then studied using the self-tuning approach. The theoretical results are illustrated via simulated examples, and an evaluation of the different methods is given by means of performance indices, i.e. response-time, output variance and control variance.

Keywords: Adaptive control; stochastic control; parameter estimation, self-adjusting systems; discrete time systems.

1. INTRODUCTION

Most of control methods are based on the knowledge of the process and its environment. However this a priori knowledge is often not available because of the complexity of the systems under study. This leads to an adaptive control problem. One way to solve this problem is to apply an identification method to obtain a model for the process and its environment by use of observations. The control is then computed from the resulting model. An alternative way is to tune the control model parameters. These two approaches are usually implemented by the utilization of Model Reference Adaptive Systems (MRAS) and self-tuning Regulators (S.T.R.). In this paper we focus our attention to the self-tuning approach. More precisely we offer a unified presentation of a number of algorithms which have been proposed by the technical literature since the appearance of the Åström and Wittenmark (1973) paper.

The paper is organized as follows:

- In the next section we present a general control scheme to solve the servo problem, the regulator problem and the combined servo and regulator problem .
- Section 3 is concerned with the control problem for systems with known parameters.
- In section 4 we give a sufficient condition under which any stochastic pole placement problem is equivalent to a control problem where the open-loop system can be represented by use of a least-squares model.
- In section 5 the case of systems with unknown parameters is discussed and we recall the procedures for using these algorithms in a self-tuning approach.
- In section 6 the theoretical results are illustrated via simulated examples, and an evaluation of the different methods is given in terms of performance indices, i.e. response-time, output variance and control variance.
- Finally concluding remarks are given in the last section.

2. PRESENTATION OF A GENERAL CONTROL SCHEME

Before describing the different control problems for systems with known parameters we give a general control scheme. This will allow us to present, in a unifying framework, algorithms to solve the servo problem (to follow a given reference signal), the regulator problem (to eliminate a disturbance) and the combined servo and regulator problem.

We assume that the process to be controlled can be modelled by the following difference equation (CARMA model)

$$A(E)\,y(t) = E^{k+1}\,B(E)\,u(t) + C(E)\,e(t) \quad (1)$$

where E is the unit backward-shift operator
k+1 is the system time delay ($k \geq 0$)
y and u are the scalar output and control signals respectively
e is the disturbance assumed to be a stationary zero-mean white noise process with variance σ^2. Consider the control scheme in Fig. 1

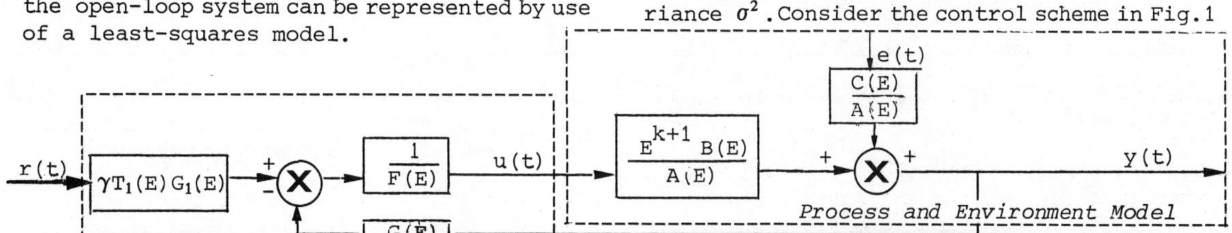

Figure 1. Control scheme

where $r(t)$ is the reference signal and γ is a gain to insure a zero steady-state error. This control system is composed of two parts

- a feedforward term: $\gamma \frac{T_1(E) G_1(E)}{F(E)} r(t)$

- a feedback term: $\frac{G(E)}{F(E)} y(t)$

Therefore we have

$$u(t) = \frac{1}{F(E)} \{\gamma T_1(E)\, G_1(E)\, r(t) - G(E)\, y(t)\} \quad (2)$$

and the closed-loop system output can be written

$$y(t) = \frac{\gamma E^{k+1}\, B(E)\, T_1(E)\, G_1(E)}{A(E)F(E) + E^{k+1} B(E) G(E)}\, r(t) + \frac{C(E)\ F(E)}{A(E)F(E) + E^{k+1} B(E) G(E)}\, e(t) \quad (3)$$

We assume that $A(E)$, $B(E)$ and $C(E)$ are polynomials in the unit backward shift operator given by

$$\left.\begin{aligned} A(E) &= 1 + a_1 E + \ldots + a_{n_a} E^{n_a} \\ B(E) &= b_o + b_1 E + \ldots + b_{n_b} E^{n_b} \\ C(E) &= 1 + c_1 E + \ldots + c_{n_c} E^{n_c} \end{aligned}\right\} \quad (4)$$

Moreover we assume that the polynomials $A(E)$ and $B(E)$ are coprime and that the polynomial $C(E)$ has all its zeros strictly outside the unit circle. No a priori hypothesis is made about the roots of $A(E)$ and $B(E)$.
The polynomial $B(E)$ will be factorized as

$$B(E) = B^+(E)\ B^-(E) \quad (5)$$

where all the zeros of $B^+(E)$ and $B^-(E)$ are respectively outside and inside the unit circle. In the sequel we write $K(n_a, n_b, n_c)$ to represent the class of models defined by equations (1) & (4) and $\mathcal{M}(A,B,C)$ for a particular element of this class.

3. THE DIFFERENT CONTROL PROBLEMS FOR SYSTEMS WITH KNOWN PARAMETERS

Tables I, II and III present, in a unifying framework, algorithms to solve one of the three control problems:

- Stochastic regulator problem ($r(t) = o$)
- Servo problem ($e(t) = o$)
- Combined servo and regulator problem

In table I we present four control strategies to solve the stochastic regulator problem:

- Control based on a stochastic pole placement design (Wellstead, Prager, Zanker-1979).
- Extended minimum variance control (Clarke, Hastings-James, 1971).
- Minimum variance control (Åström, 1970).
- Suboptimal minimum variance control (Åström, 1970).

Table II summarizes equations of different control methods solving the servo problem:

- Control based on a pole placement design (Åström, Westerberg, Wittenmark - 1978).
- Dead-beat control.

In table III we unify, in relation to the control system described in Figure 1, three different results initially proposed by Clarke and Gawthrop (1975) to solve the combined servo and regulator problem.

3.1. The stochastic regulator problem

Four main strategies are utilized.

The stochastic pole placement design strategy is directly deduced from the classical control theory. The idea is to make the poles of the transfer function between $e(t)$ and the system output equal to the roots of an a priori given polynomial $A^M(E)$.
The closed-loop system output is then given by the following ARMA process.

$$y(t) = \frac{F(E)}{A^M(E)} e(t) \quad (6)$$

where $F(E)$ is the solution of the polynomial equation

$$A(E)\ F(E) + E^{k+1} B(E) G(E) = A^M(E) C(E) \quad (7)$$

This control law can be applied to any system, and its advantages will be shown in simulated examples.

The minimum variance control law is obtained by minimizing

$$\mathcal{C}_1 = \mathcal{E}\,[y^2(t+k+1)] \quad (8)$$

with respect to the input variable $u(t)$. (Åström, 1970). The closed-loop system output is an MA process of order k

$$y(t) = F_1(E)\, e(t) = (1 + f_1 E + \ldots + f_k E^k)\, e(t) \quad (9)$$

As we have $F(E) = B(E) F_1(E)$ the system zeros are compensated by the controller poles. Therefore the application of this result is limited to minimum phase systems. We can notice that this regulator is a special case of the previous one. Indeed, in case of minimum-phase systems, if we choose to make the poles equal to the roots of $B(E)$ then the regulator based on the stochastic pole placement design is identical to the minimum variance regulator. However the control signal variance may be excessive and consequently it is interesting to include a weighting of the control variance in the cost to minimize.

The extended minimum variance control strategy is obtained by minimizing

$$\mathcal{C} = \mathcal{E}\,[y^2(t+k+1) + \lambda\, u^2(t)] \quad (10)$$

that is by minimizing a weighted sum of output and control variances. In this case, the closed-loop system output is

$$y(t) = \frac{b_o F(E)}{\lambda A(E) + b_o B(E)} e(t) \quad (11)$$

and the closed-loop system poles are equal to the roots of the following equation

$$\lambda\, A(E) + b_o B(E) = o \quad (12)$$

As the regulator doesn't cancel the system zeros, this result can be applied to a greater class of systems :

- stable, non minimum phase systems will be closed-loop stable by choosing a large enough λ ;
- unstable, minimum phase systems will be closed-loop stable by choosing a small enough λ ;
- stable, minimum phase systems will be closed-loop stable for any value of λ.

Taking into account the expression

$$F(E) = B(E)F_1(E) + \frac{\lambda}{b_o} C(E) \quad (13)$$

and using the polynomial equation satisfied by C(E)

$$C(E) = A(E)F_1(E) + E^{k+1} G_1(E) \quad (14)$$

We get the output variance

$$\mathcal{E}[y^2(t)] = \\ = \mathcal{E}\left\{\left\{\left[F_1(E) + E^{k+1} \frac{\lambda G_1(E)}{\lambda A(E) + b_o B(E)}\right] e(t)\right\}^2\right\} \quad (15)$$

or

$$\mathcal{E}[y^2(t)] = \\ = \sigma^2[1+f_1^2+\ldots+f_k^2] + \mathcal{E}\left\{\left[\frac{\lambda G_1(E)}{\lambda A(E) + b_o B(E)} e(t)\right]^2\right\} \quad (16)$$

Moreover, we can verify that, for a given λ and under certain hypotheses which we shall explicit, the resulting control law is identical to the control law obtained via the pole placement approach by choosing

$$A^M(E) = \frac{\lambda}{b_o} A(E) + B(E) \quad (17)$$

Indeed by taking into account equations (7), (13) and (14), the degrees of the F and G polynomials are equal for each of the two methods to

$$\left.\begin{aligned} n_f &= n_b + k \\ n_g &= n_a - 1 \end{aligned}\right\} \quad (18)$$

if

$$\left.\begin{aligned} n_c &\leqslant n_a + k \\ n_c &\leqslant n_b + k \\ n_{A^M} &\leqslant n_a + n_b + k - n_c \end{aligned}\right\} \quad (19)$$

By replacing (13) into (7) and using (14), we get (17) with the supplementary condition $n_{A^M} = \sup(n_a, n_b)$.

Consequently we have the following result: the extended minimum variance control law corresponds to the particular stochastic pole placement (17) when

$$n_c \leqslant \mathrm{Min}(n_a, n_b) + k \quad (20)$$

The suboptimal minimum variance control has been introduced by Åström (1970) to solve the regulator problem for non minimum phase systems. This solution has been associated by Jeanneau and De Larminat (1975) with the solution obtained in correspondence with the requirement :

- finite control variance
- output conditional expectation equal to zero in a minimum time.

In this case, equation (14) is replaced by

$$C(E) = A(E)F_2(E) + E^{k+1}B^-(E)\,G_2(E) \quad (21)$$

By subtracting (14) from (21) we get

$$A(E)[F_2(E) - F_1(E)] = E^{k+1}[G_1(E) - B^-(E)G_2(E)] \quad (22)$$

E^{k+1} and A(E) being coprime polynomials, we define

$$E^{k+1}F'(E) = F_2(E) - F_1(E) \quad (23)$$

with

$$F'(E) = f'_{k+1} + f'_{k+2}E + \ldots + f'_{k+n_b^-} E^{n_b^- -1} \quad (24)$$

The closed-loop system output being $F_2(E)e(t)$, its variance is equal to the output variance of the system controlled by use of the minimum variance control strategy added with the quantity $\sigma^2[1 + f'^2_{k+1} + \ldots + f'^2_{k+n_b^-}]$.

Finally, the optimal solution to the minimum variance control problem for non minimum phase systems was given by Peterka (1972). Åström and Wittenmark (1974) applied this solution in the self-tuning approach.
We conclude this section by observing that the regulator based on a stochastic pole placement seems to be the most flexible and the most general. Indeed, it can be noticed that the three particular pole placements :

$$A^M(E) = \{B^+(E)\ ;\ B(E)\ ;\ \lambda A(E) + b_o B(E) \text{ with given } \lambda\}$$

correspond to the suboptimal minimum variance, minimum variance and extended minimum variance controllers respectively.

3.2. The servo problem

In the case of the control based on closed-loop system poles and zeros placement, this control strategy is determined by specifying the desired closed-loop behaviour in terms of poles and zeros assignment. This is equivalent to determine the polynomials F(E), $G_1(E)$, $T_1(E)$ and G(E) in such a way that the transfer function between reference signal r(t) and system output y(t) is equal to

$$\gamma \frac{E^{k+1} B^M(E)}{A^M(E)} \quad \text{with}$$

$$\left.\begin{aligned} A^M(E) &= 1 + a_1^M E + \ldots + a^M_{n_{A^M}} E^{n_{A^M}} \\ B^M(E) &= b_o^M + b_1^M E + \ldots + b^M_{n_{B^M}} E^{n_{B^M}} \end{aligned}\right\} \quad (25)$$

The implementation of this method usually requires the use of polynomial factorization algorithms and the solution of polynomial equations. The ensuring computations are considerably simplified in the following two special cases (Åström, Westerberg, Wittenmark-1978 and Åström, Wittenmark-1979) :

- Pole zero placement control with cancellation of all process zeros. In this particular case (cf. Table II) polynomial B(E) appears in the denominator of the regulator transfer function. For stability reasons, this regulator can be applied only to minimum phase systems.
- Pole placement control without cancellation of process zeros.

Methods	Polynomial equations to solve	Closed-loop System output	Output Variance	Control Variance
Pole Assignment Control	$\begin{cases} A(E)F(E)+E^{k+1}B(E)G(E)= \\ = A^M(E)C(E) \end{cases}$	$y(t)=\frac{F(E)}{A^M(E)}e(t)$	$\mathcal{E}\left\{\left[\frac{F(E)}{A^M(E)}e(t)\right]^2\right\}$	$\mathcal{E}\left\{\left[\frac{G(E)}{A^M(E)}e(t)\right]^2\right\}$
Extended minimum variance control Min $\mathcal{E}\{y^2(t+k+1)+\lambda u^2(t)\}$	$\begin{cases} A(E)F_1(E)+E^{k+1}G_1(E)=C(F) \\ F(E)=B(E)F_1(E)+\frac{\lambda}{b_o}C(E) \\ G(E)=G_1(E) \end{cases}$	$y(t)=\frac{b_oF(E)}{\lambda A(E)+b_oB(E)}e(t)$	$\mathcal{E}\{[F_1(E)e(t)]^2\}+\mathcal{E}\left\{\left[\frac{\lambda G_1(E)}{\lambda A(E)+b_oB(E)}e(t)\right]^2\right\}$ $=\sigma^2\{1+f_1^2+\ldots+f_k^2\}+\mathcal{E}\left\{\left[\frac{\lambda G_1(E)}{\lambda A(E)+b_oB(E)}e(t)\right]^2\right\}$	$\mathcal{E}\left\{\left[\frac{b_oG_1(E)}{\lambda A(E)+b_oB(E)}e(t)\right]^2\right\}$
Minimum Variance Control, Min $\mathcal{E}\{y^2(t+k+1)\}$	$\begin{cases} A(E)F_1(E)+E^{k+1}G_1(E)=C(E) \\ F(E)=B(E)F_1(E) \\ G(E)=G_1(E) \end{cases}$	$y(t)=F_1(E)e(t)$	$\mathcal{E}\{[F_1(E)e(t)]^2\}=\sigma^2\{1+f_1^2+\ldots+f_k^2\}$	$\mathcal{E}\left\{\left[\frac{G_1(E)}{B(E)}e(t)\right]^2\right\}$
Suboptimal minimum Variance Control	$\begin{cases} A(E)F_2(E)+E^{k+1}B^-(E)G_2(E) =C(E) \\ F(E)=B^+(E)F_2(E) \\ G(E)=G_2(E) \end{cases}$	$y(t)=F_2(E)e(t)$	$\mathcal{E}\{[F_2(E)e(t)]^2\}=$ $=\sigma^2\{1+f_1^2+\ldots+f_k^2+f'^2_{k+1}+f'^2_{k+n_{b^-}}\}$	$\mathcal{E}\left\{\left[\frac{G_2(E)}{B^+(E)}e(t)\right]^2\right\}$

Table I : Stochastic Regulator problem $\left\{r(t)\equiv 0 \; ; \; u(t)=-\frac{G(E)}{F(E)}y(t)\right\}$

Methods	Polynomial equations to solve	Closed-loop System output
Pole-zero placement Control	$\begin{cases} A(E)F_3(E)+E^{k+1}B^-(E)G_3(E)=A^M(E)T_1(E) \\ F(E)=B^+(E)F_3(E);G_1(E)=B^M(E) \end{cases}$	$y(t)=\frac{\gamma E^{k+1}B^-(E)B^M(E)}{A^M(E)}r(t)$ with $\gamma=\frac{A^M(1)}{B^-(1)B^M(E)}$
Pole zero placement Control with cancellation of all process zeros	$\begin{cases} A(E)F_3(E)+E^{k+1}G_3(E)=A^M(E)T_1(E) \\ F(E)=B(E)F_3(E);G_1(E)=B^M(E) \end{cases}$	$y(t)=\frac{\gamma E^{k+1}B^M(E)}{A^M(E)}r(t)$ with $\gamma=\frac{A^M(1)}{B^M(1)}$
Pole placement Control without Cancellation of process zeros	$\begin{cases} A(E)F_3(E)+E^{k+1}B(E)G_3(E)=A^M(E)T_1(E) \\ F(E)=F_3(E);\; G_1(E)\equiv 1 \end{cases}$	$y(t)=\frac{\gamma E^{k+1}B(E)}{A^M(E)}r(t)$ with $\gamma=\frac{A^M(1)}{B(1)}$
Dead-beat 1 Control	$\begin{cases} A(E)F_3(E)+E^{k+1}B^-(E)G_3(E)=T_1(E) \\ F(E)=B^+(E)F_3(E)\;;\;G_1(E)\equiv 1 \end{cases}$	$y(t)=\gamma E^{k+1}B^-(E)r(t)$ with $\gamma=\frac{1}{B^-(1)}$
Dead-beat 2 Control	$\begin{cases} A(E)F_3(E)+E^{k+1}B(E)G_3(E)=T_1(E) \\ F(E)=F_3(E)\;;\;G_1(E)\equiv 1 \end{cases}$	$y(t)=\gamma E^{k+1}B(E)r(t)$ with $\gamma=\frac{1}{B(1)}$
Simplified Dead-beat 2 Control	Particular solution for dead-beat 2 Control $F_3(E)=1-\gamma E^{k+1}B(E)$ when $G_3(E)=\gamma A(E);T_1(E)=A(E)$	$y(t)=\gamma E^{k+1}B(E)r(t)$ with $\gamma=\frac{1}{B(1)}$

Table II : Servo problem $u(t)=\frac{1}{F(E)}\{\gamma T_1(E)G_1(E)r(t)-G(E)y(t)\}$
where $T_1(E)$ is any observer polynomial ; $G(E)=G_3(E)$
$r(t)$ is a stepwise ref.signal; $B(E)$ & $B^M(E)$ are coprime

Criterium to minimize	Polynomial eq. to solve	Closed-loop system output
$I_1 = \mathcal{E}\{[A^M(E)y(t+k+1) - B^M(E)r(t)]^2 + \lambda u^2(t)\}$	$A(E)F_4(E)+E^{k+1}G_4(E) = T_1(E)A^M(E)$; $T_1(E)=C(E)$; $G_1(E)=B^M(E)$; $F(E)=B(E)F_4(E)+\frac{\lambda}{b_o}C(E)$	$y(t)=\frac{\gamma E^{k+1}b_oB(E)B^M(E)}{\lambda A(E)+b_oA^M(E)B(E)}r(t)+\frac{b_oF(E)}{\lambda A(E)+b_oA^M(E)B(E)}e(t)$ with $\gamma=\frac{A^M(1)+\frac{\lambda}{b_o}\frac{A(1)}{B(1)}}{B^M(1)}$
$I_2 = \mathcal{E}\{[A^M(E)y(t+k+1) - B^M(E)r(t)]^2\}$	$A(E)F_4(E)+E^{k+1}G_4(E) = T_1(E)A^M(E)$; $T_1(E)=C(E)$; $G_1(E)=B^M(E)$; $F(E)=B(E)F_4(E)$	$y(t)=\frac{\gamma E^{k+1}B^M(E)}{A^M(E)}r(t)+\frac{F_4(E)}{A^M(E)}e(t)$ with $\gamma=\frac{A^M(1)}{B^M(1)}$
$I_3 = \mathcal{E}\{[y(t+k+1)-r(t)]^2 + \lambda u^2(t)\}$	$A(E)F_4(E)+E^{k+1}G_4(E) = T_1(E)$; $T_1(E)=C(E)$; $G_1(E)\equiv 1$; $F(E)=B(E)F_4(E)+\frac{\lambda}{b_o}C(E)$	$y(t)=\frac{\gamma E^{k+1}b_oB(E)}{\lambda A(E)+b_oB(E)}r(t)+\frac{b_oF(E)}{\lambda A(E)+b_oB(E)}e(t)$ with $\gamma=1+\frac{\lambda}{b_o}\frac{A(1)}{B(1)}$

Table III : Combined Servo and Regulator Problem

$$u(t)=\frac{1}{F(E)}\{\gamma T_1(E)\,G_1(E)\,r(t)-G(E)\,y(t)\}$$

where γ is a gain, $G(E) = G_4(E)$

The dead-beat control strategy corresponds to the special pole placement : $A^M(E) = 1$. Two variants are possible (dead-beat 1 and dead-beat 2 controls). They are characterized by the choice of a closed-loop system step time response either equal to $(n_b - + k + 2)T$ or equal to $(n_b + k + 2)T$.

In case of stable systems $(A(E) = A^+(E))$ we can choose $\{G_3(E) = \gamma A(E), T_1(E) = A(E)\}$ and subsequently we have the following particular solution for the polynomial $F_3(E)$

$$F_3(E) = 1 - \gamma E^{k+1}B(E) \qquad (26)$$

The control law is then given by

$$u(t) = \frac{\gamma A(E)}{1-\gamma E^{k+1}B(E)}\,[r(t) - y(t)] \qquad (27)$$

with

$$\gamma = \frac{1}{B(1)} \qquad (28)$$

This result was introduced by Kalman (1954) and used by Kurz, Isermann and Schumann (1980). We can notice that, due to its special structure (no polynomial equation to solve), the parameters of this controller are obtained by means of very simple computations.

3.3. The combined servo and regulator problem

In relation to this problem we consider the following cost function (Clarke, Gawthrop-1975)

$$\mathscr{C} = \mathcal{E}\{[A^M(E)\,y(t+k+1) - B^M(E)\,r(t)]^2 + [S(E)u(t)]^2\} \qquad (29)$$

This cost function takes into account simultaneously fluctuations of reference signal and of the system's input and output. Regulators presented in Table III are special cases of the general cost function (29).

By a simple comparison of polynomial equations (see Tables II and III), we can verify that by setting $T_1(E) = C(E)$, the regulator corresponding to the pole zero placement control with cancellation of all process zeros is identical to the optimal regulator minimizing criterium I_2. Moreover the output of the system controlled by this regulator is

$$y(t) = \frac{\gamma E^{k+1}B^M(E)}{A^M(E)}\,r(t) + \frac{F_4(E)}{A^M(E)}\,e(t) \qquad (30)$$

where the polynomial $F_4(E)$ can be written as

$$F_4(E) = F_1(E)\,A^M(E) \qquad (31)$$

with $F_1(E)$ solution to equation (14).

This regulator allows us to assign the poles and zeros of the closed-loop system while minimizing the fluctuations of the output with respect to the output $y^M(t)$ of a reference model defined by

$$y^M(t) = \gamma\,\frac{E^{k+1}B^M(E)}{A^M(E)}\,r(t) \qquad (32)$$

Indeed, the error $\epsilon(t) = y(t) - y^M(t)$ is minimum with respect to criterium $\mathscr{C}_1$. Consequently, due to the degree of freedom left by polynomial $T_1(E)$ in the structure of the control system, the solution of the servo problem (pole and zero placement) allows to minimize simultaneously the output variance by choosing $T_1(E) = C(E)$.

4. A THEOREM ABOUT EQUIVALENT CONTROL PROBLEM

In this section we use a result given by Wouters (1977) to derive a theorem about equivalent Control problem in the stochastic pole placement design.

Let us denote by
$\mathcal{P}(\mathcal{M}(A,B,C) \in K(n_a,n_b,n_c), A^M(E))$
the control problem based on the stochastic pole placement to the roots of a polynomial $A^M(E)$.

Definition: We shall say that

$\mathcal{P}(\mathcal{M}^1(A,B,C) \in K(n_a,n_b,n_c), A^M(E))$ is equivalent to $\mathcal{P}(\mathcal{M}^2(A,B,C) \in K(n_a,n_b,n_c), A^M(E))$ if both problems yield the same unique solution $(F(E),G(E))$.

Now, let us consider the class $K_{LS}(n_a,n_b) \subset K(n_a,n_b,n_c)$ defined by

$K_{LS}(n_a,n_b) = \{\mathcal{M}(A,B,C) \in K(n_a,n_b,n_c)/C(E) \equiv 1\}$;
a particular element of this class is written down $\mathcal{M}_{LS}(A,B)$; we can state the following theorem.

Theorem: Any control problem
$\mathcal{P}(\mathcal{M}(A,B,C) \in K(n_a,n_b,n_c), A^M(E))$

has an unique equivalent control problem
$\mathcal{P}(\mathcal{M}_{LS}(\mathcal{A},\mathcal{B}) \in K_{LS}(n_a,n_b+k), A^M(E))$ if

- $n_{A^M} \leqslant n_a + n_b + k - nc$
- the leading coefficient β_o of polynomial $\mathcal{B}(E)$ is a priori set
- rank $\mathcal{H}_1 = n_a + n_b + k$

with

$$\mathcal{H}_1 = \begin{bmatrix} 1 & 0 & \cdots & 0 & 0 & \cdots & \cdots & 0 \\ f_1 & \ddots & & 0 & g_o & \ddots & & \vdots \\ \vdots & & \ddots & 1 & \vdots & \ddots & & 0 \\ \vdots & & & f_1 & \vdots & & & g_o \\ f_{n_b+k} & & & \vdots & g_{n_a-1} & & & \vdots \\ 0 & \ddots & & \vdots & 0 & \ddots & & \vdots \\ \vdots & & & \vdots & \vdots & & & \vdots \\ 0 & \cdots & 0 & f_{n_b+k} & 0 & \cdots & 0 & g_{n_a-1} \end{bmatrix} \quad (33)$$

$\dim(\mathcal{H}_1) = (n_a + n_b + k \;,\; n_a + n_b + k)$

where $(f_i)_{i=1,\dots,n_b+k}$ and $(g_i)_{i=o,\dots,n_a-1}$ are the coefficients of the polynomials F(E) and G(E) which satisfy the following equation

$$A(E)F(E) + E^{k+1}B(E)G(E) = A^M(E)C(E) \quad (34)$$

Proof: The solution to the control problem $\mathcal{P}(\mathcal{M}(A,B,C) \in K(n_a,n_b,n_c), A^M(E))$ gives the following regulator

$$u(t) = -\frac{G(E)}{F(E)}\, y(t) \quad (35)$$

where F(E) and G(E), solution to equation (34), can be written as

$$\left.\begin{aligned} F(E) &= 1 + f_1 E + \dots + f_{n_b+k} E^{n_b+k} \\ G(E) &= g_o + g_1 E + \dots + g_{n_a-1} E^{n_a-1} \end{aligned}\right\} \quad (36)$$

if $n_{A^M} \leqslant n_a + n_b + k - n_c$ **(37)**

Because of the symmetric part played by the polynomials $A(E)$, $E^{k+1}B(E)$ and $C(E)$ with respect to the polynomials $F(E)$, $G(E)$ and $A^M(E)$ in equation (34), the existence and the unicity of an equivalent control problem are linked to the existence and the unicity of polynomials $\mathcal{A}(E)$ and $\mathcal{B}(E)$ which are solution of the following polynomial equation

$$\mathcal{A}(E)F(E) + E\,\mathcal{B}(E)G(E) = A^M(E) \quad (38)$$

In order to insure the existence of a solution to equation (38) the structure of polynomials $A^M(E)$, $F(E)$ and $G(E)$ forces us to take $\mathcal{A}(E)$ and $\mathcal{B}(E)$ such that

$$\left.\begin{aligned} \mathcal{A}(E) &= 1 + \alpha_1 E + \dots + \alpha_{n_a} E^{n_a} \\ \mathcal{B}(E) &= \beta_o + \beta_1 E + \dots + \beta_{n_b+k} E^{n_b+k} \end{aligned}\right\} \quad (39)$$

By expliciting equation (38) we get

$$\underbrace{\begin{bmatrix} 1 & 0 & \cdots & 0 & g_o & 0 & \cdots & 0 \\ f_1 & \ddots & & 0 & \vdots & \ddots & & 0 \\ \vdots & \ddots & & 1 & \vdots & & \ddots & g_o \\ \vdots & & \ddots & f_1 & \vdots & & & \vdots \\ f_{n_b+k} & & & \vdots & g_{n_a-1} & & & \vdots \\ 0 & \ddots & & \vdots & 0 & \ddots & & \vdots \\ \vdots & & & \vdots & \vdots & & & \vdots \\ 0 & \cdots & 0 & f_{n_b+k} & 0 & \cdots & 0 & g_{n_a-1} \end{bmatrix}}_{\mathcal{H}} \begin{bmatrix} \alpha_1 \\ \vdots \\ \alpha_{n_a} \\ \beta_o \\ \vdots \\ \vdots \\ \beta_{n_b+k} \end{bmatrix} = \begin{bmatrix} a_1^M - f_1 \\ \vdots \\ \vdots \\ a_\ell^M - f_\ell \\ \\ \\ \\ \end{bmatrix} \quad (40)$$

with

$$\left.\begin{aligned} &\dim \mathcal{H} = (n_a+n_b+k, n_a+n_b+k+1) \\ &a_i^M = o \quad \forall\, i > n_{A^M} \\ &f_i = o \quad \forall\, i > n_b+k \\ &\ell = \sup(n_{A^M}, n_b+k) \end{aligned}\right\} \quad (41)$$

The existence and the unicity of an equivalent control problem can be expressed in terms of the rank of the matrix $\mathcal{H}$. Indeed, it is sufficient that the matrix $\mathcal{H}$ has a rank equal to (n_a+n_b+k) for ensuring the existence and the unicity of polynomials $\mathcal{A}(E)$ and $\mathcal{B}(E)$ (with a coefficient of $\mathcal{B}(E)$ fixed for example) which are solution of the equivalent control problem. Setting the coefficient β_o with an arbitrary value is equivalent to say that matrix $\mathcal{H}_1$ has its rank equal to (n_a+n_b+k).

Remark: the above theorem states that, in order to solve the regulator problem using the stochastic pole placement method, it is sufficient to represent the system by use of a least squares model $(C(E) = 1)$ corresponding to the equivalent control problem. This remark is very interesting in the self-tuning approach; it will allow us to simplify the procedure of computation of a controller based on a stochastic pole placement design.

5. SELF-TUNING REGULATORS

5.1 Principle

In this section we give a brief description of S.T.R.'s which have been initially introduced by Peterka (1970), and Åström and Wittenmark (1973) in the case of a minimum variance control. The following basic hypothesis is necessary: the process under study can be modelled by an element $\mathcal{M}(A,B,C)$ of the class $K(n_a,n_b,n_c)$, and the model structure is assumed known but with unknown parameters.

Self-tuning regulators are obtained by use of the certainty equivalence principle (Bar-Shalom and Tse, 1974). This principle is based on a separation between identification and control. When applying such a principle it is only necessary to provide the controller with information about the estimated model parameters. In contrast to cautious controllers (Wittenmark, 1975), the resulting controllers do not take into account the uncertainties of the parameter estimates. A block diagram of S.T.R. is shown in Figure 2.

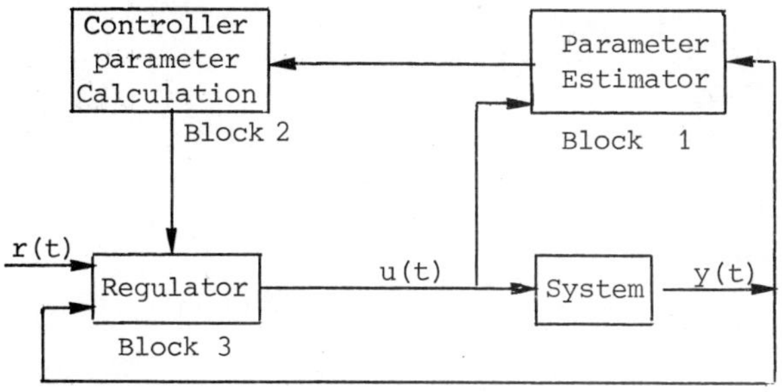

Figure 2 Block diagram of a S.T.R.

Then the procedure to obtain a S.T.R. can be summarized as follows:

First step: Estimation of process parameters (Block 1).
In this paper two parameter estimation algorithms are considered:
the recursive least squares (RLS), and the recursive maximum likelihood (RML) methods. The application of one or the other depends on the particular disturbances.

Second step: Calculation of controller parameters by using the estimated process parameters, and by solving, in most cases, a polynomial equation.

Third step: Computation of the control law.
Such algorithms are called explicit or indirect algorithms (Åström, 1980) because the controller parameters are updated indirectly via an explicit identification of the system. In certain cases and under certain conditions which we shall give in the sequel, it is possible to implement algorithms where the second step is avoided. Such algorithms are called implicit or direct algorithms because the controller parameters are updated directly via a modified model of the process. These algorithms are interesting from a point of view of computation time and of the size of the implementing computer programs.

5.2. Recursive parameter estimation algorithms

The recursive estimation methods for estimating parameters of models belonging to classes $K_{LS}(n_a,n_b)$ and $K(n_a, n_b, n_c)$ are respectively
- Recursive least squares (RLS)
- Recursive maximum likelihood (RML1)

These two methods can be unified by use of the following equations (Söderström, Ljung & Gustavsson, 1974)

$$\left.\begin{aligned} \hat\theta(t+1) &= \hat\theta(t) + K(t+1)\,\epsilon(t+1) \\ K(t+1) &= \frac{P(t)\,\varphi(t+1)}{\nu(t+1) + \varphi^T(t+1)\,P(t)\,\varphi(t+1)} \\ P(t+1) &= \left\{P(t) - \frac{P(t)\varphi(t+1)\varphi^T(t+1)\,P(t)}{\nu(t+1) + \varphi^T(t+1)\,P(t)\varphi(t+1)}\right\}/\nu(t+1) \\ \epsilon(t+1) &= y(t+1) - \varphi^T(t+1)\hat\theta(t) \end{aligned}\right\} \quad (42)$$

The forgetting factor $\nu(t+1)$ is chosen equal to

$$\left.\begin{aligned} & \nu(t+1) = 1 \\ \text{or}\quad & \nu(t+1) = \nu_o\,\nu(t) + (1-\nu_o) \\ \text{with}\quad & 0.95 \leqslant \nu(0) \leqslant 1\,,\ 0.95 \leqslant \nu_o \leqslant 1 \end{aligned}\right\} \quad (43)$$

The other variables of (42) are defined in Table VII.

Methods	$\hat\theta$	$\varphi(t+1)$
RLS	$[\hat a_1 \ldots \hat a_{n_a}\ \hat b_o \ldots \hat b_{n_b}]^T$	$[-y(t) \ldots -y(t+1-n_a)$ $u(t-k) \ldots u(t-k-n_b)]$
RML 1	$[\hat a_1 \ldots \hat a_{n_a}\ \hat b_o \ldots \hat b_{n_b}$ $\hat c_1 \ldots \hat c_{n_c}]^T$	$[-y(t) \ldots -y(t+1-n_a)\,u(t-k) \ldots$ $\ldots u(t-k-n_b)\ \epsilon(t) \ldots$ $\ldots \epsilon(t+1-n_c)]$

Table VII. Variables for RLS and RML1 methods

5.3 Presentation of the different estimation models

Owing to its structure the explicit procedure can be used to apply all the controllers, presented in section 3, in the self-tuning approach. For a given problem, the choice between an explicit and an implicit procedure depends on whether or not there exists a system model in which the controller parameters appear explicitly. Such a choice also depends on the amount of computation needed to realize the controller.
In this section we present for each of the regulators that is a solution of one of the three previous control problems, the estimation model corresponding to the procedure which seems best adapted from a point of view of computation time, computation volume and ease of realization.

- *The stochastic regulator problem*
  ★ Explicit procedure

The S.T.R. based on a stochastic pole placement design can be obtained by estimating the process parameters

$$A(E)y(t)=E^{k+1}B(E)u(t)+C(E)e(t) \quad (44)$$

The adapted estimation algorithm is then the RML1 method unless $C(E)=1$.

However, the control signal being generated by a feedback, the model identified may not be the true system (Söderström,Gustavsson & Ljung,1975). This would be the case for example when the true system is not parameter identifiable eventhough the corresponding control law may converge asymptotically to the desired law. As it was shown in the previous section there exists a unique model $\mathscr{M}_{LS}(\mathscr{A},\mathscr{B}) \in K_{LS}(n_a,n_b+k)$

which is the solution of the equivalent control problem; it is then sufficient to estimate the parameters of the following model by use of RLS method

$$\left.\begin{array}{l} \mathscr{A}(E)y(t) = E\mathscr{B}(E)u(t)+\xi(t) \\ \text{with } \mathscr{A}(E) = 1+\alpha_1 E+\ldots+\alpha_{n_a}E^{n_a} \\ \mathscr{B}(E)=\beta_0+\beta_1 E+\ldots+\beta_{n_b+k}E^{n_b+k} \\ \beta_0 \text{ fixed} \end{array}\right\} \quad (45)$$

Wellstead,Prager and Zanker(1979) have shown that under certain conditions given in the theorem about the equivalent control problem (section 4) and if the parameter estimates of model **(45)** converge to steady values,then the S.T.R. using these estimates(see Table IV) will converge asymptotically to the regulator of the stochastic pole placement problem associated with system $\mathscr{M}(A,B,C)$.

★ Implicit procedure:

- Self-tuning minimum variance regulator: The (k+1)-step-ahead prediction error of the output of a system which belongs to $K_{LS}(n_a,n_b)$ is given by

$$y(t)-\hat{y}(t/t-k-1)= F_1(E)e(t) \quad (46)$$

with $$\hat{y}(t/t-k-1)= G_1(E)y(t-k-1)+ B(E)F_1(E)u(t-k-1) \quad (47)$$

In this case we can deduce the following implicit estimation model

$$y(t)= G_1(E)\, y(t-k-1)+B(E)F_1(E)\, u(t-k-1)+F_1(E)e(t) \quad (48)$$

Or more explicitly

$$y(t)-b_0u(t-k-1)=-\alpha_1 y(t-k-1)-\ldots-\alpha_{m+1}y(t-k-m-1)+ b_0\{\beta_1 u(t-k-2)+\ldots+\beta_\ell u(t-k-\ell-1\}+\xi_1(t) \quad (49)$$

where
$$\left.\begin{array}{l} m=\sup\{n_c-k-1\ ,\ n_a-1\} \text{ (see Appendix)} \\ \ell = n_b+k \\ b_0 \text{ is assumed known to avoid identifiability difficulties (Niederlinski, 1979).} \end{array}\right\} \quad (50)$$

In the case of a system modelled by $\mathscr{M}(A,B,C)$, the equation of the (k+1)-step-ahead prediction of the output can be written as

$$C(E)\hat{y}(t/t-k-1)= G_1(E)y(t-k-1)+B(E)F_1(E)u(t-k-1) \quad (51)$$

and equation **(48)** becomes

$$y(t)=G_1(E)y(t-k-1) + B(E)F_1(E)u(t-k-1) + F_1(E)\, e(t)+[1-C(E)]\ \hat{y}(t/t-k-1) \quad (52)$$

However, Åström and Wittenmark(1973),and Keviczky and Hetthessy(1974) have shown by using two different approaches that if the estimated parameters of model **(49)** converge, then the corresponding regulator will converge asymptotically to the optimal regulator associated with model $\mathscr{M}(A,B,C)$.This can be intuitively explained by noticing that when applying RLS method to model **(52)** with the optimal control law $u(t)=-\frac{G_1(E)}{B(E)F_1(E)}y(t)$ the term $[1-C(E)]\ \hat{y}(t/t-k-1)$ disappears. Consequently, the estimation model to be used for the self-tuning minimum variance regulator will be always the model defined by equation **(49)**.

- Self tuning extended minimum variance regulator:

Define the generalized output (Clarke and Gawthrop,1975)

$$\Phi(t) = y(t) + \frac{\lambda}{b_0}u(t-k-1) \quad (53)$$

For a system belonging to $K_{LS}(A,B)$,the (k+1)-step -ahead prediction error of this generalized output is given by

$$\Phi(t)-\hat{\Phi}(t/t-k-1) = F_1(E)\, e(t) \quad (54)$$

with
$$\hat{\Phi}(t/t-k-1)=\left\{\frac{\lambda}{b_0}+ B(E)F_1(E)\right\}u(t-k-1) + G_1(E)\, y(t-k-1) \quad (55)$$

Then the implicit estimation model can be written
$$\Phi(t)=\left\{\frac{\lambda}{b_0}+B(E)F_1(E)\right\}u(t-k-1)+G_1(E)y(t-k-1) + F_1(E)\, e(t) \quad (56)$$

In the case of a system $\mathscr{M}(A,B,C)$ equations **(55)** and **(56)** become

$$C(E)\,\hat{\Phi}(t/t-k-1) = =\left\{\frac{\lambda}{b_0}C(E)+B(E)F_1(E)\right\}u(t-k-1)+G_1(E)y(t-k-1) \quad (57)$$

$$\Phi(t)= =\left\{\frac{\lambda}{b_0}C(E)+B(E)F_1(E)\right\}u(t-k-1)+G_1(E)y(t-k-1)+ F_1(E)\, e(t)+\{1-C(E)\}\hat{\Phi}(t/t-k-1) \quad (58)$$

By using the same reasoning as for the self-tuning minimum variance regulator,the following implicit estimation model can be used

$$\Phi(t)-\left(\frac{\lambda}{b_0}+b_0\right)u(t-k-1)=-\alpha_1 y(t-k-1)\ldots-\alpha_{m+1}y(t-k-m-1) +\left(b_0+\frac{\lambda}{b_0}\right)\{\beta_1 u(t-k-2)+\ldots+\beta_\ell\, u(t-k-\ell-1)\}+\xi_2(t) \quad (59)$$

with

$$\left.\begin{array}{l} m = \sup(n_a-1\ ,\ n_c-k-1) \\ \ell = \sup(n_c\ ,\ n_b+k) \\ \left(b_0+\frac{\lambda}{b_0}\right) \text{ assumed known} \end{array}\right\} \quad (60)$$

• *The servo problem:*

★ Explicit procedure

This procedure seems to be adapted more particularly to the self-tuning regulators based on

-Pole - zero placement control
-Pole placement without cancellation of process zeros.

Indeed, an implicit algorithm relative to the S.T.R. based on pole placement without cancellation of process zeros needs a step of factorization of the two estimated polynomials. This computation may be sometimes ill-conditioned. Moreover in this case the implicit approach does not contain fewer steps than the explicit approach.

★ Implicit procedure

— Pole-zero placement control with cancellation of all process zeros.
From the polynomial equation given in Table II we can deduce the following implicit estimation model

$$A^M(E)T_1(E)\,y(t)=E^{k+1}[B(E)F_3(E)\,u(t)+G_3(E)y(t)] \quad (61)$$

or more explicitly

$$y^F(t)=-\alpha_1\,y(t-k-1)\ldots-\alpha_{m+1}y(t-k-m-1)+b_0u(t-k-1)+\beta_1u(t-k-2)+\ldots+\beta_\ell u(t-k-\ell-1) \quad (62)$$

with

$$\begin{cases} m=n_a-1 \\ \ell=n_b+\sup\{n_{A^M}+n_{t_1}-n_a,k\}\end{cases} \text{ or } \begin{cases} m=\sup\{n_{A^M}+n_{t_1}-k-1,n_a-1\} \\ \ell=n_b+k\end{cases} \quad (63)$$

and $y^F(t)$ represents the filtered output signal

$$y^F(t)=A^M(E)T_1(E)y(t)=y(t)+\mu_1\,y(t-1)+\ldots+\mu_p y(t-p) \quad (64)$$

with

$$p=n_{A^M}+n_{t_1} \quad (65)$$

— Pole-zero placement control
By use of the polynomial equation associated with the pole-zero placement control (see Table II), we can derive the following implicit estimation model

$$A^M(E)T_1(E)y(t)=E^{k+1}B^-(E)\{B^+(E)F_3(E)u(t)+G_3(E)y(t)\} \quad (66)$$

Notice that this model is bilinear in the parameters and a recursive estimation procedure is given by Åström (1979 a).

— The dead-beat 1 and 2 controls being special cases of pole placement and pole placement without cancellation of process zeros respectively, the same estimation procedures can be used. In the case of simplified dead-beat 2 strategy, the control law is directly computed from the model parameters so that an explicit estimation procedure must be used.

• *The combined servo and regulator problem*
The regulators presented in table III can be realized in a self-tuning way by use of an implicit procedure. A supplementary problem then appears for determining the implicit model: the adjustment of the overall system gain. This gain is a constant which depends on the coefficients of A(E) and B(E) polynomials. However, if an algorithm with implicit identification is used these polynomials are not estimated. Therefore the reference gain must be expressed in terms of polynomials F(E) and G(E) which characterize the regulator. In the case of criterium I_3 we have

$$\frac{\gamma b_0B(E)}{\lambda A(E)+b_0B(E)}=\frac{\gamma[F(E)-\frac{\lambda}{b_0}C(E)]}{F(E)-E^{k+1}\frac{\lambda}{b_0}G(E)} \quad (67)$$

γ is computed in such a way that the transfer function from the reference value to the output has a unity steady-state gain. Hence

$$\gamma=\frac{F(1)-\frac{\lambda}{b_0}G(1)}{F(1)-\frac{\lambda}{b_0}C(1)} \quad (68)$$

Let us introduce the new generalized output

$$\Phi(t)\triangleq y(t)+\frac{\lambda}{b_0}u(t-k-1)-\gamma r(t-k-1) \quad (69)$$

In the case of a system belonging to the class $K(n_a,n_b,n_c)$ this generalized output can be written as

$$\Phi(t)=[\frac{\lambda}{b_0}C(E)+B(E)F_4(E)]u(t-k-1)+G_4(E)y(t-k-1)-\gamma C(E)r(t-k-1)+F_4(E)e(t)+[1-C(E)]\hat{\Phi}(t/t-k-1) \quad (70)$$

Defining $N(E)\triangleq\tilde{N}(E)+\gamma\triangleq\gamma C(E)$ (71)

we can deduce the following implicit estimation model

$$\Phi(t)+\gamma r(t-k-1)=F(E)u(t-k-1)+G(E)y(t-k-1)-\tilde{N}(E)r(t-k-1)+\xi_4(t) \quad (72)$$

Taking into account equations (68),(71) and (72), the estimation model can be rewritten more explicitly as

$$\Phi(t)+\gamma_{(t-1)}r(t-k-1)=-\alpha_1y(t-k-1)\ldots-\alpha_{m+1}y(t-k-m-1)+[b_0+\frac{\lambda}{b_0}]\{u(t-k-1)+\ldots+\beta_\ell u(t-k-\ell-1)\}-\mu_1r(t-k-2)\ldots-\mu_pr(t-k-p-1)+\xi_4(t) \quad (73)$$

with

$$\left.\begin{aligned} m&=\sup(n_c-k-1,n_a-1)\\ \ell&=\sup(n_c,n_b+k)\\ p&=n_c\end{aligned}\right\} \quad (74)$$

and the reference gain is computed as follows

$$\left.\begin{aligned} C_{(t)}(1)&=\frac{N_{(t)}(1)}{\gamma_{(t-1)}}\\ \gamma_{(t)}&=\frac{F_{(t)}(1)-\frac{\lambda}{b_0}G_{(t)}(1)}{F_{(t)}(1)-\frac{\lambda}{b_0}C_{(t)}(1)}\end{aligned}\right\} \quad (75)$$

For more details see Allidina and Hughes (1979) who have treated this problem in the more general case where the cost function to minimize includes a weighting of control, output and reference. The corresponding S.T.R. is called "self-tuning controller steady-state error." However an important problem remains: which

value must be given to λ for ensuring the closed-loop system stability? An a priori choice of this parameter can not be suitable in all the cases and more precisely in the case of unstable or non minimum phase systems. One solution would be to determine a confidence interval by making several successive trials. But this method is not very attractive.
Allidina and Hughes (1980) have developed a very interesting idea which consists in determining adaptively the polynomials $A^M(E)$ and $S'(E) = \frac{s_o S(E)}{b_o}$ (which are respectively the weighting polynomials of output and control in the cost function minimized by Clarke and Gawthrop (1975)). These polynomials are determined in such a way that the closed-loop system poles equal to the roots of an a priori polynomial T(E). That corresponds to extend the possibilities of application of the self-tuning controller of Clarke and Gawthrop (1975) to those of the S.T.R. based on a stochastic pole placement.
This regulator, called "generalized self-tuning controller with pole assignment", combines the robustness of a stochastic pole placement regulator with the possibility of following a reference signal.

The procedure and the realizability conditions of this algorithm are summarized in Table VI; see Guillermin (1980) for details.

6. SIMULATION RESULTS

The simulated examples discussed in the following illustrate the theoretical results presented in the previous sections. In the case of the stochastic regulator problem, and the combined servo and regulator problem, the system to be controlled is described by model **(1)** with

$A(E) = 1-1.6E + o.8\,E^2$, $k = 1$, $\sigma^2 = 1$
$B(E) = 1+ o.5E$, $C(E) = 1+ o.2E-o.48\,E^2$

For the servo problem, we have considered the following system

$A(E) = 1 - o.946E + o.741\,E^2$, $k = o$
$B(E) = o.53 + o.265E$, $C(E) = o$

In each simulation, both the time delay and the polynomial orders are assumed known. The reference signal to be followed is a square wave with 100 observations during each step change. Figure 3 shows the open-loop system output in response to this reference signal.

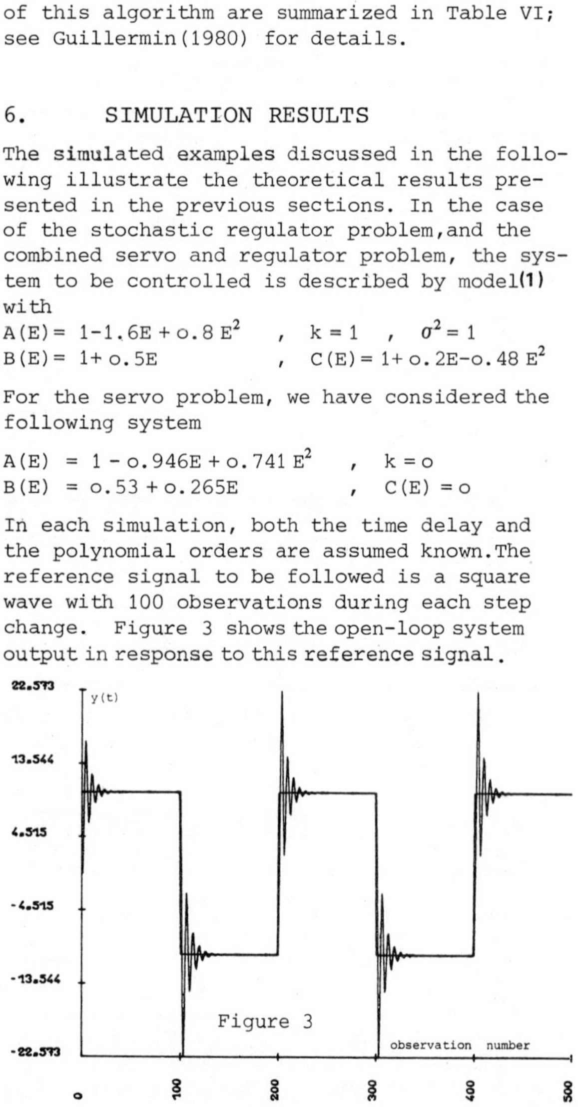

Figure 3

• In Table VIII we present numerical results for the stochastic regulator problem. Theoretical and experimental variances of output and control signals are given for the following strategies:
— pole placement, with seven different values of $A^M(E)$
— extended minimum variance, with five different values of λ, the special case $\lambda = o$ corresponding to the minimum variance strategy.

Figure 4 shows steady-state output and control signals from observation 900 to 1000:
- Figure 4.1: pole-placement to $A^M(E) = 1-o.4E$
- Figure 4.2: extended minimum variance with $\lambda = 2$
- Figure 4.3: minimum variance.

• In Table IX the combined servo and regulator problem is considered and numerical results are given in terms of experimental variance of error and control. The S.T.R. developed by Clarke and Gawthrop (1975) and the "self-tuning controller steady-state error" (Allidina & Hughes, 1979) have been used with five different values of λ .

Figure 5 represents steady-state output and control (from observation 500 to 1000) obtained by use of this last refulator for $\lambda = o$ (Figure 5.1) and $\lambda = 2$ (Figure 5.2).

• In table X, the settling time, overshoot and maximum value of control signal are given for the servo problem when using the following strategies:
— pole placement with cancellation of all process zeros. Different values of $A^M(E)$ and $T_1(E)$ are considered and the two possible solutions with respect to the degrees of polynomials $F_3(E)$ and $G_3(E)$ are taken into account.

Figures 6.1 and 6.2 show the output and control signals respectively when

$A^M(E) = 1-1.286E + o.549\,E^2$, $T_1(E) = 1$
and
$A^M(E) = 1-o.7E$, $T_1(E) = 1$

— pole placement without cancellation of process zeros, see Figure 6.3.

— simplified dead-beat 2 with two initial values of γ , see Figure 6.4 for $\gamma = o.4$.

— dead-beat 1, see Figure 6.5

Some comments are now made upon the simulation results:
• From Tables VIII & IX we can verify that the experimental results are very close to those obtained theoretically when the system parameters are known. We also verify that the pole placement design to $A^M(E) = B(E)$ corresponds to the minimum variance control from the point of view of the experimental output variance, but the control variance is less when using the pole placement strategy. Moreover the pole placement and the extended minimum variance strategies allow to decrease the control variance in great ratio but without degrading the regulation performance.
In the case of the combined servo and regulator problem, the advantage supplied by the use of an adaptive gain appears very clearly for large values of λ ($\lambda \geqslant 1$). The criteria J_2 and J_4 are related to the steady-state and therefore to the pure regulation problem, while the

criteria J_1 and J_3 take into account the transient period corresponding to a change of the reference signal value.
Finally, the compromise between the energy cost (control variance) and the quality of the regulation is pointed out.

- By comparing Figure 3 with Figures 6 we can conclude about the interest of the pole placement adaptive control for solving the servo-problem. By analyzing the results in Table X, we can notice that the influence of the observer polynomial $T_1(E)$ appears only in the self-tuning period (initial parameters estimation period) and this influence is much less important when the degree of $T_1(E)$ is lower. Moreover the symmetric solutions in $F_3(E)$ and $G_3(E)$ give the same system behaviour.
In the case of pole placement control without cancellation of process zeros, Figure 6.3 shows that the difference with respect to the pole placement controls, with cancellation of all process zeros, appears only in the self-tuning period.
Finally, when using dead-beat control, the control signal is of an impulsive and rapidly varying nature and there is a large overshoot in the self-tuning period.

7. SOME PRACTICAL ASPECTS

Some practical aspects on the different algorithms considered in this paper are now presented.

- *Stochastic Regulator Problem*

- The main drawback of the pole placement design is in the use of an explicit procedure which requires solution of a polynomial equation. But this strategy presents the definite advantage that it can be applied to any system.
- In the minimum variance strategy the parameter b_o needs to be known. Åström & Wittenmark (1973), have shown that the choice of β_o (value chosen for b_o) is crucial for the convergence of the minimum variance algorithm. Wittenmark (1973) has shown that, for a first order system, the ratio $\frac{\beta_o}{b_o}$ must be within the interval $o.5 < \frac{\beta_o}{b_o} < \infty$ for ensuring the convergence of this algorithm. However, one way to avoid the problem linked with the knowledge of b_o is to limit the control signal.
- In the extended minimum variance strategy, the parameter $(\frac{\lambda}{b_o} + b_o)$ needs to be known. So an a priori value must be chosen for this paramater and the corresponding value of λ will not give generally the wanted control weighting. Another resulting problem is the calculation of the generalized output signal which requires knowledge of the parameter $\frac{\lambda}{b_o}$.
As for the minimum variance strategy one way to ensure the convergence is to limit the control signal. Advantages of the extended minimum variance strategy is to decrease the control amplitude and to offer the possibility to regulate unstable or non minimum phase systems by successive trials.

- *Servo problem*

In this case the drawback of the different pole placement methods results from the choice of an explicit procedure which requires solution of a polynomial equation, except for the simplified dead-beat 2 control.

- *Combined servo and regulator problem*

As we have shown in the previous section, the advantage of the "self-tuning controller steady-state error" relative to the S.T.R. developed by Clarke and Gawthrop (1975) is due to the elimination of the steady-state error for reference signal step change, by use of an adaptive gain.
The existence of an extra signal (reference signal) allows us here to estimate the parameter $(\frac{\lambda}{b_o} + b_o)$; but the calculation of the generalized output signal needs the parameter $\frac{\lambda}{b_o}$. In the case where this parameter is unknown we propose the following procedure:

For a given λ, estimate $\hat{x}(t) = \widehat{(\frac{\lambda}{b_o} + b_o)}(t)$

From this estimate value, we have

$$\hat{b}_o(t) = \frac{\hat{x}(t) \pm \sqrt{\hat{x}^2(t) - 4\lambda}}{2}$$

and

$$\widehat{(\frac{\lambda}{b_o})}(t) = \begin{cases} \dfrac{2\lambda}{\hat{x}(t) + \sqrt{\hat{x}^2(t) - 4\lambda}} \; ; & i = 1 \\ \dfrac{2\lambda}{\hat{x}(t) - \sqrt{\hat{x}^2(t) - 4\lambda}} \; ; & i = 2 \end{cases}$$

if $\hat{x}^2(t) - 4\lambda > o$

When $(\hat{x}^2(t) - 4\lambda)$ becomes negative, we choose $\widehat{(\frac{\lambda}{b_o})}(t)$ such that

$$\widehat{(\frac{\lambda}{b_o})}(t) = \begin{cases} \dfrac{2\lambda_1}{\hat{x}(t)} = \dfrac{\hat{x}(t)}{2}, \text{ with } \hat{x}^2(t) - 4\lambda_1 = o \; ; & j = 1 \\ \dfrac{2\hat{x}(t)}{3} \; ; & j = 2 \end{cases}$$

For illustrating this procedure, Table XI gives numerical results corresponding to the simulated example presented in section 6 when the parameter $(\frac{\lambda}{b_o} + b_o)$ is unknown.

λ	i, j	steady-state value of $\widehat{(\lambda/b_o+b_o)}$	$E[y^2(t)]$	$E[u^2(t)]$
o.1	i=2, j=1	4.83	20.7	0.85
o.1	i=1, j=1	1.1	4.11	5.67
o.5	i=1, j=1	1.51	4.92	3.74
2	i=2, j=2	2.56	8.6	1.80

Table XI. Combined servo and regulator problem when the parameter $(\frac{\lambda}{b_o} + b_o)$ is unknown.

We can notice that the true value of b_o is such that

$$b_o = \frac{x \pm \sqrt{x^2 - 4\lambda}}{2} = \frac{\frac{\lambda}{b_o} + b_o \pm \left|\frac{\lambda}{b_o} - b_o\right|}{2}$$

and consequently the solution to be used is

$$i = 1 \quad \text{if} \quad \frac{\lambda}{b_o} - b_o < o$$

$$i = 2 \quad \text{if} \quad \frac{\lambda}{b_o} - b_o > o$$

This result is verified in Table XI.

8. CONCLUSION

Various Self-Tuning Regulators have been presented in a unifying framework. A comparison was made via simulated examples and an evaluation of the different algorithms was given by means of performance indices. In all the simulations which were done, the S.T.R.'s converged fast (after about 15 observations in the noise-free case and 100 observations in the case of stochastic disturbances) so that self-tuning control appears very attractive for industrial applications (see Åström & coworkers (1977) and Parks & coworkers (1980)).
However, as it was pointed out in the two previous sections, and in spite of its name, the self-tuning control is not completely automatic. Certain parameters of the S.T.R. need to be tuned by the user. Wellstead and Zanker (1978) gave a list of factors under the control of the user, for the identification part as well as for the control part. To simplify the tuning problem a new family of S.T.R.'s, called "simple self-tuners", has been developed; see Åström (1979 b), Wittenmark (1979), and Wittenmark & Åström (1980). This solution consists in estimating a low order model by use of least squares method and it is characterized by few tuning parameters. The corresponding regulator can be viewed as a generalized PID-controller, and this solution seems very attractive for simple servo applications.

• APPENDIX •

A comment on the solution of the identity

$$A(E)\ X(E) + B(E)\,Y(E) = C(E) \qquad (1)$$

with respect to the polynomials X(E) and Y(E).

Denote by n_a, n_x, n_b, n_y, and n_c the degrees of polynomials A(E), X(E), B(E), Y(E) and C(E) respectively.
If $n_c < n_a + n_b$, the minimal degrees of polynomials X(E) and Y(E) which satisfy the polynomial equation (1) are given by

$$\left.\begin{aligned} n_x &= n_b - 1 \\ n_y &= n_a - 1 \end{aligned}\right\} \qquad (2)$$

Otherwise, equation (1) admits the two following symmetric solutions

$$\left.\begin{aligned} n_x &= n_b - 1 \\ n_y &= \sup\{n_a - 1, n_c - n_b\} \end{aligned}\right\} \qquad (3)$$

or

$$\left.\begin{aligned} n_x &= \sup\{n_b - 1, n_c - n_a\} \\ n_y &= n_a - 1 \end{aligned}\right\} \qquad (4)$$

REFERENCES

Allidina, A.Y., & Hughes, F.M. (1979). Self-tuning controller steady-state error. Electronics letters, vol 15, n°12, 346-347.

Allidina, A.Y., & Hughes, F.M. (1980). Generalized self-tuning Controller with pole assignment. Proc. IEE, vol 127, n°1, 13-18.

Åström, K.J. (1970). Introduction to stochastic Control theory. (Academic Press).

Åström, K.J. (1979 a). New implicit adaptive pole-zero placement algorithms for nonminimum-phase systems. Dept. of Automatic Control, Lund Institute of Technology, Sweden, CODEN: LUTFD2/(TFRT-7172)/ 1-021/(1979).

Åström, K.J. (1979 b). Simple self-tuners I. Dept of Automatic Control, Lund Institute of Technology, Sweden, CODEN: LUTFD2/(TFRT-7184)/ 1-063/(1979).

Åström, K.J. (1980). Design principles for self-tuning regulators. International Symposium on adaptive systems, BOCHUM.

Åström, K.J., Borisson, U., Ljung, L. and Wittenmark, B. (1977). Theory and applications of self-tuning regulators. Automatica, vol 13, 457-476.

Åström, K.J., Westerberg, B., and Wittenmark, B. (1978). Self-tuning Controllers based on pole-placement design. Dept. of Automatic Control, Lund Institute of Technology, Sweden, CODEN: LUTFD2/(TFRT-3148)/1-052/(1978).

Åström, K.J., and Wittenmark, B. (1973). On self-tuning regulators. Automatica, vol 9, 185-199.

Åström, K.J. and Wittenmark, B. (1974). Analysis of a self-tuning regulator for nonminimum phase systems. IFAC symposium on stochastic Control, Budapest.

Åström, K.J. and Wittenmark, B. (1979). Self-tuning Controllers based on pole-zero placement. Dept of Automatic Control, Lund Institute of Technology, Sweden, CODEN: LUTFD2/(TFRT -7180)/1-43/(1979).

Bar-Shalom, Y. and Tse, E. (1974). Dual effect, certainty equivalence, and separation in stochastic control. IEEE Trans. autom. control, vol 19, n°5, 494-500.

Clarke, D.W. and Gawthrop, P.J. (1975). Self-tuning controller. Proc. IEE, vol 122, n°9, 929-934.

Clarke, D.W., and Hastings-James, R. (1971). Design of digital Controllers for randomly disturbed systems. Proc. IEE, vol 118, n°10, 1503-1506.

Guillermin, P. (1980). Thèse de 3ème cycle (à paraître).

Hetthessy, J. and Keviczky, L. (1974). Some innovations to the minimum variance control. IFAC symposium on stochastic Control, Budapest.

Jeanneau, J.L., and de Larminat, Ph. (1975). Une méthode de régulation adaptative pour les systèmes à phase non minimale. 3° congreso nacional informatica y automatica.

Kalman,R.E.(1954). Discussion of Bergen,A.R. and Ragazzini,J.R. Trans.AIEE.

Kurz,H., Isermann,R.,and Schumann,R.(1980) Experimental comparison and application of various parameter-adaptive control algorithms. Automatica,vol.16,117-133.

Niederlinski,A.(1979).The global error approach to the convergence and accuracy of complex recursive identification algorithms. Fifth IFAC symposium Identification,and system parameter estimation,Darmstadt.

Parks,P.C., Schaufelberger,W., Schmid ,Chr., and Unbehauen,H.(1980). Applications of adaptive control systems. International symposium on adaptive systems,BOCHUM.

Peterka,V.(1970). Adaptive digital regulation of noisy systems. Proceedings of the IFAC symposium on identification and process parameter estimation,Prague.

Peterka,V.(1972). On steady state minimum variance control strategy. Kybernetika,vol.8, n°3,219-231.

Söderström,T.,Gustavsson, I. ,and Ljung,L. (1975). Identifiability Conditions for linear systems operating in closed-loop. Int.J.Control,vol.21,n°2,243-255.

Söderström,T., Ljung,L.,and Gustavsson,I. (1974). A comparative study of recursive identification methods. Dept.of Automatic Control, Lund Institute of Technology,Sweden, report 7427.

Wellstead,P.E., Prager,D.,and Zanker,P. (1979). Pole assignment self-tuning regulator. Proc.IEE,vol.126,n°8,781-787.

Wellstead,P.E., and Zanker,P.(1978). The techniques of self-tuning.Report,Control systems Centre, university of Manchester Institute of Science and Technology.

Wittenmark,B.(1973). A self-tuning regulator. Dept.of Automatic Control,Lund Institute of Technology,report 7311.

Wittenmark,B.(1975). Stochastic adaptive Control methods: a survey. Int.J.Control, vol 21,n°5,705-730.

Wittenmark,B.(1979). Self-tuning PID Controllers based on pole placement.Dept.of Automatic Control,Lund Institute of Technology, Sweden,CODEN:LUTFD2/(TFRT-7179)/1-037/(1979).

Wittenmark,B.,and Åström,K.J.(1980).Simple self-tuning Controllers. International symposium on adaptive systems,BOCHUM.

Wouters,W.R.E.(1977).Adaptive pole placement for linear stochastic systems with unknown parameters.Proc.IEEE Conference on decision and Control,New Orleans,USA.

Methods	Criterium to minimize	Procedure	Conditions for applications: relative to the system	Conditions for applications: existence of a LS equivalent model	Models for estimation	Polynomial equations	Control law
Pole assignment Control	NO	Explicit	NO	$\mathrm{rank}(\mathscr{H}_1) = n_a + n_b + k$	$\mathscr{A}(E)\, y(t) = E\mathscr{B}(E)\, u(t) + \xi(t)$ with $\mathscr{A}(E) = 1 + \alpha_1 E + \ldots + \alpha_{n_a} E^{n_a}$ $\mathscr{B}(E) = \beta_o + \beta_1 E + \ldots + \beta_{n_b+k} E^{n_b+k}$ β_o fixed	$\mathscr{A}(E)F(E) + E\mathscr{B}(E)G(E) = A^M(E)$ with $F(E) = 1 + f_1 E + \ldots + f_{n_b+k} E^{n_b+k}$ $G(E) = g_o + g_1 E + \ldots + g_{n_a-1} E^{n_a-1}$ if $n_{A^M} < n_a + n_b + k + 1 - n_c$	$u(t) = -g_o y(t) - g_1 y(t-1) - \ldots$ $\ldots - g_{n_a-1} y(t-n_a+1) - f_1 u(t-1) - \ldots$ $\ldots - f_{n_b+k}\, u(t-n_b-k)$
Extended minimum Variance control	$\mathscr{E}\{y^2(t+k+1) + \lambda u^2(t)\}$	Implicit	unstable, minimum phase systems: λ small enough stable, non minimum phase systems: λ large enough (★)	NO	$\Phi(t) - (\frac{\lambda}{b_o} + b_o)u(t-k-1) = -\alpha_1\, y(t-k-1) - \ldots$ $\ldots - \alpha_{m+1} y(t-k-m-1) + (\frac{\lambda}{b_o} + b_o)\{\beta_1\, u(t-k-2) + \ldots$ $\ldots + \beta_\ell\, u(t-k-\ell-1)\} + \xi_2(t)$ with $m = \sup\{n_c-k-1, n_a-1\}$ $\ell = \sup\{n_c, n_b+k\}$ $\Phi(t) = y(t) + \frac{\lambda}{b_o} u(t-k-1)$ $(\frac{\lambda}{b_o} + b_o)$ fixed	NO	$u(t) = \frac{1}{b_o + \frac{\lambda}{b_o}} \{\alpha_1 y(t) + \ldots +$ $\alpha_{m+1}\, y(t-m)\} -$ $-\beta_1\, u(t-1) - \ldots - \beta_\ell u(t-\ell)$
Minimum Variance Control	$\mathscr{E}\{y^2(t+k+1)\}$	Implicit	Minimum phase systems	NO	$y(t) - b_o u(t-k-1) = -\alpha_1\, y(t-k-1) - \ldots$ $-\alpha_{m+1}\, y(t-k-m-1) + b_o\{\beta_1\, u(t-k-2) + \ldots$ $\ldots + \beta_\ell\, u(t-k-\ell-1)\} + \xi_1(t)$ with $m = \sup\{n_c-k-1, n_a-1\}$ $\ell = n_b + k$ b_o fixed	NO	$u(t) = \frac{1}{b_o}\{\alpha_1 y(t) + \alpha_2 y(t-1)$ $+ \ldots + \alpha_{m+1} y(t-m)\} - \beta_1\, u(t-1)$ $- \ldots - \beta_\ell\, u(t-\ell)$

(★) stable, minimum phase systems: any λ

TABLE IV : SELF-TUNING CONTROLLERS FOR STOCHASTIC REGULATOR PROBLEM

Methods	Procedure	Conditions for application	Models of estimation	Polynomial equations	Control law
Pole-zero placement Control	Explicit OR Implicit	NO	$A(E)\,y(t) = E^{k+1}B(E)\,u(t)$ OR $A^M(E)T_1(E)y(t)=E^{k+1}B^-(E)\{B^+(E)F_3(E)u(t)+G_3(E)y(t)\}$	$A(E)F_3(E)+E^{k+1}B^-(E)G_3(E) = A^M(E)T_1(E)$ with $\begin{cases} n_{g_3} = n_a - 1 \\ n_{f_3} = \sup\{n_{A^M}+n_{t_1} - n_a, n_{b^-}+k\}\end{cases}$ or $\begin{cases} n_{g_3} = \sup\{n_{A^M}+n_{t_1} - k-1-n_{b^-}, n_a-1\} \\ n_{f_3} = n_{b^-}+k\end{cases}$ OR NO	$u(t) = \frac{1}{B^+(E)F_3(t)}\{\gamma T_1(E)B^M(E)\,r(t) - G_3(E)\,y(t)\}$ with $\gamma = \frac{A^M(1)}{B^-(1)\,B^M(1)}$
Pole-zero placement Control with Cancellation of all process zeros	Implicit	Minimum phase systems	$y^F(t)=-\alpha_1 y(t-k-1) - \dots - \alpha_{m+1}\,y(t-k-m-1) + b_o u(t-k-1)+\beta_1 u(t-k-2)+\dots+\beta_\ell u(t-k-\ell-1)$ with $\begin{cases} m=n_a-1 \\ \ell=n_b+\sup\{n_{A^M}+n_{t_1}-n_a,k\}\end{cases}$ or $\begin{cases} m=\sup\{n_{A^M}+n_{t_1}-k-1,n_a-1\} \\ \ell=n_b+k\end{cases}$; $y^F(t)=A^M(E)T_1(E)y(t)=y(t)+\mu_1 y(t-1)+\dots+\mu_p y(t-p)$ where $p=n_{A^M}+n_{t_1}$	NO	$u(t)=\frac{1}{b_o}\{\gamma\{\delta_o r(t)+\delta_1 r(t-1)+\dots+\delta_{n_\delta}r(t-n_\delta)\}+\alpha_1 y(t)+\dots+\alpha_{m+1}y(t-m) - \beta_1 u(t-1)-\dots-\beta_\ell\,u(t-\ell)\}$ with $\begin{cases}\gamma=\frac{A^M(1)}{B^M(1)} \\ A^M(E)T_1(E)=\delta_o+\delta_1 E+\dots+\delta_{n_\delta}E^{n_\delta} \\ \text{where } n_\delta=n_{a^M}+n_{t_1}\end{cases}$
Pole placement control without cancellation of process zeros	Explicit	NO	$A(E)\,y(t) = E^{k+1}B(E)\,u(t)$	$A(E)F_3(E)+E^{k+1}B(E)G_3(E)=A^M(E)T_1(E)$ with the same degrees for $F_3(E)$ and $G_3(E)$ as in the previous case by changing n_{b^-} for n_b	$u(t) = \frac{1}{F_3(E)}\{\gamma T_1(E)\,r(t) - G_3(E)\,y(t)\}$ with $\gamma=\frac{A^M(1)}{B(1)}$
Dead-beat 1 Control	Identical to the case: " Pole-zero placement , Control " when $A^M(E) = B^M(E) \equiv 1$				
Dead-beat 2 Control	Identical to the case: " Pole-placement, without cancellation of process zeros " when $A^M(E) \equiv 1$				
Simplified Dead-beat 2 Control	Explicit	Stable systems	$A(E)\,y(t) = E^{k+1}B(E)\,u(t)$	NO	$u(t)=\frac{\gamma A(E)}{1-E^{k+1}\gamma B(E)}\{r(t)-y(t)\}$ with $\gamma=\frac{1}{B(1)}$

Table V: Self-tuning Controllers for servo problems { $T_1(E)$ is any observer polynomial }

CSAD - E*

Methods	Criterium to minimize	Procedure	Conditions for application	Model of estimation	Polynomial equation	Control law
Generalized self-tuning Controller with pole assignment	$\mathcal{E}[\{A^M(E)y(t+k+1) - B^M(E)r(t) + S'(E)u(t)\}^2]$ where $A^M(E)$ and $S'(E)$ are updated in order to extend the self-tuning controller to achieve pole assignment $B^M(E)$ is updated in order to eliminate the steady-state error	Implicit	$k \geqslant n_c - 1$ $n_t \leqslant n_a + n_b - 1$	$\Phi(t) + b^M_{o,(t-1)}\, r(t-k-1) = M(E)y(t-k-1) + L(E)u(t-k-1) + \tilde{N}(E)\, r(t-k-1) + \xi_5(t)$ with $\Phi(t) = A^M_{(t-1)}(E)y(t) + S'_{(t-1)}(E)u(t-k-1) - B^M_{(t-1)}(E)r(t-k-1)$, $n_M = n_a - 1$, $n_L = n_b + k$, $n_{\tilde{N}} = n_c + n_{B^M}$ the polynomial $B^M(E)$ is updated as: $B^M_{(t)}(E) = \frac{B^M_{(t)}(1)}{B^M_{(t-1)}(1)} B^M_{(t-1)}(E)$ where $B^M_{(t)}(1) = \frac{L_{(t)}(1)A^M_{(t-1)}(1) - S'_{(t-1)}(1)M_{(t)}(1)}{L_{(t)}(1) - C_{(t)}(1)S'_{(t-1)}(1)}$, $C_{(t)}(1) = \frac{b^M_{o,(t-1)} - \tilde{N}_{(t)}(1)}{B^M_{(t-1)}(1)}$	$A^M(E)L_{(t)}(E) - E^{k+1}S'(E)M_{(t)}(E) = \ell_o\, T(E)F_4(E)$ with $n_{A^M} = n_a - 1$, $n_{S'} = n_b - 1$, $n_{F_4} = k$; the leading coefficient of $A^M(E)$, $T(E)$ and $F_4(E)$ fixed to unity This polynomial identity must be solved with respect to $A^M(E)$, $S'(E)$ and $F_4(E)$ $\Rightarrow$ we obtain $A^M_{(t)}(E)$ and $S'_{(t)}(E)$	$u(t) = \frac{1}{L_{(t)}(E)}\{-M_{(t)}(E)y(t) - N_{(t)}(E)r(t)\}$ where $N_t(E) = \tilde{N}_t(E) - b^M_{o,(t-1)}$
Self-tuning Controller steady-state error	$\mathcal{E}[\{y(t+k+1) - r(t)\}^2 + \lambda u^2(t)]$	Implicit	unstable, minimum phase systems: λ small enough stable, non minimum phase systems: λ large enough stable, minimum phase systems: any λ	$\Phi(t) + \gamma_{(t-1)} r(t-k-1) = -\alpha_1 y(t-k-1) - \ldots - \alpha_{m+1} y(t-k-m-1) + (\frac{\lambda}{b_o} + b_o)\{u(t-k-1) + \beta_1 u(t-k-2) + \ldots + \beta_\ell u(t-k-\ell-1)\} - \mu_1 r(t-k-2) - \ldots - \mu_p r(t-k-p-1) + \xi_4(t)$ with $m = \sup\{n_c - k - 1, n_a - 1\}$, $\ell = \sup\{n_c, n_b + k\}$, $p = n_c$, $\Phi(t) = y(t) + \frac{\lambda}{b_o} u(t-k-1) - \gamma_{(t-1)} r(t-k-1)$ The reference gain is updated as: $\gamma_{(t)} = \frac{(b_o + \frac{\lambda}{b_o}) \sum_{i=o}^{\ell} \beta_i + \frac{\lambda}{b_o} \sum_{i=1}^{m} \alpha_i}{(b_o + \frac{\lambda}{b_o}) \sum_{i=1}^{\ell} \beta_i - \frac{\lambda}{b_o}\left(1 + \frac{\sum_{i=1}^{p} \mu_i}{\gamma_{(t-1)}}\right)}$ where $\beta_o = 1$	NO	$u(t) = \frac{1}{b_o + \frac{\lambda}{b_o}}\{\alpha_1 y(t) + \ldots + \alpha_{m+1} y(t-m) + \gamma_{(t-1)} r(t) + \mu_1 r(t-1) + \ldots + \mu_p r(t-p)\} - \beta_1 u(t-1) - \ldots - \beta_\ell u(t-\ell)$

Table VI: Self-tuning Controllers for Combined servo and regulator problems

performance / methods	theorical variance of output signal	experimental variance of output signal J_5	theorical variance of control signal	experimental variance of control signal J_6
Pole placement design to $A^M(E)$				
$A^M(E)=1-0.7E$	8.38	7.99	0.11	0.10
$A^M(E)=1-0.5E$	6.20	6.07	0.27	0.22
$A^M(E)=1-0.4E$	5.62	5.57	0.44	0.34
$A^M(E)=1$	4.55	4.54	1.89	1.15
$A^M(E)=1+0.4E$	4.26	4.09	6.66	3.62
$A^M(E)=1+0.5E$	4.24	3.97	9.25	5.88
$A^M(E)=1+0.7E$	4.36	4.25	20.10	7.83
Extended minimum variance: λ				
$\lambda=2$	6.07	6.05	0.46	0.43
$\lambda=1$	5.23	5.06	0.99	0.86
$\lambda=0.5$	4.72	4.49	1.92	1.59
$\lambda=0.1$	4.29	4.00	6.77	4.28
$\lambda=0$ minimum variance	4.24	3.92	9.25	7.18

Table VIII.

$$J_5 = \frac{1}{500} \sum_{i=501}^{1000} y_i^2$$

$$J_6 = \frac{1}{500} \sum_{i=501}^{1000} u_i^2$$

Performance / Values of λ	experimental variance of error signal ϵ		experimental variance of control signal		experimental variance of error signal ϵ with an adaptive gain γ		experimental variance of control signal with an adaptive gain γ	
	J_1	J_2	J_3	J_4	J_1	J_2	J_3	J_4
$\lambda=0$	4.03	3.87	17.12	6.94	No adaptive gain ($\gamma=1$)			
$\lambda=0.1$	4.25	3.98	11.72	5.48	4.29	4.03	12.13	5.72
$\lambda=0.5$	5.50	4.71	5.52	3.27	5.43	4.58	6.22	3.64
$\lambda=1$	7.29	6.06	3.39	2.31	6.54	5.10	4.21	2.81
$\lambda=2$	11.50	9.63	1.94	1.54	8.47	5.90	2.92	2.27

Table IX.

$$\epsilon_i = y(i) - r(i-k-1)$$

$$\begin{cases} J_1 = \frac{1}{500} \sum_{i=501}^{1000} \{\epsilon_i^2\} \\ J_2 = \frac{1}{400} \left\{ \sum_{i=521}^{600} + \sum_{i=621}^{700} + \ldots + \sum_{i=921}^{1000} \right\} \{\epsilon_i\}^2 \\ J_3 = \frac{1}{500} \sum_{i=501}^{1000} u_i^2 \\ J_4 = \frac{1}{400} \left\{ \sum_{i=521}^{600} + \ldots + \sum_{i=921}^{1000} \right\} \{u_i^2\} \end{cases}$$

performance / methods	Settling time: transient	Settling time: steady state	Overshoot: transient	Overshoot: steady state	Maximum value of control signal
Pole placement control with cancellation of all process zeros : $A^M(E), T_1(E)$					
$A^M(E)=1-1.286E + 0.549E^2$, $T_1(E)\equiv 1$	11T	8T	9.85%	16.45%	12.26
$A^M(E)=1-1.286E + 0.549E^2$, $T_1(E)=1-0.3E$, $n_{f_3}=1$, $n_{g_3}=2$	11T	8T	13.67%	16.45%	12.26
$T_1(E)=1-0.3E$, $n_{f_3}=2$, $n_{g_3}=1$	12T	8T	12.41%	16.45%	12.26
$T_1(E)=(1-0,6E)^2$, $n_{f_3}=3$, $n_{g_3}=1$	12T	8T	2%	16.45%	12.26
$A^M(E)=1-0,7E$, $T_1(E)\equiv 1$	13T	7T	0%	0%	10
Pole placement control without cancellation of process zeros					
$A^M(E)=1-1.286E + 0.549E^2$, $T_1(E)\equiv 1$	15T	8T	62.86%	16.37%	12.19
Simplified Dead-Beat 2 control					
initial value of $\gamma=1$	46T	3T	605%	0%	12
initial value of $\gamma=0.4$	43T	3T	307%	0%	12
Dead-Beat 1 control					
initial value of $\gamma=1$	8T	2T	142%	0%	34,52

Table X.

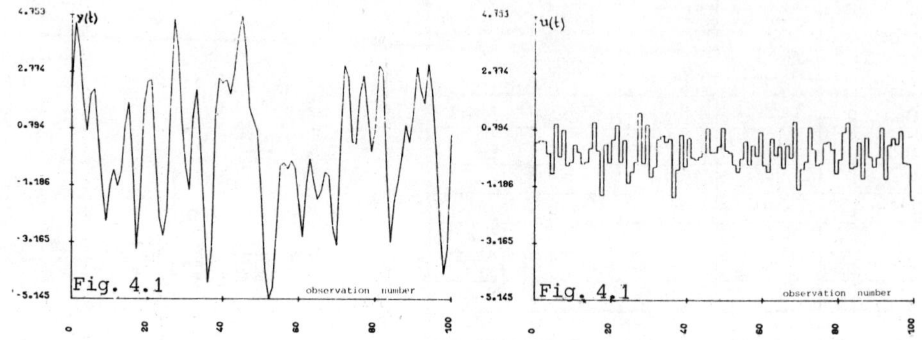

Fig. 4.1

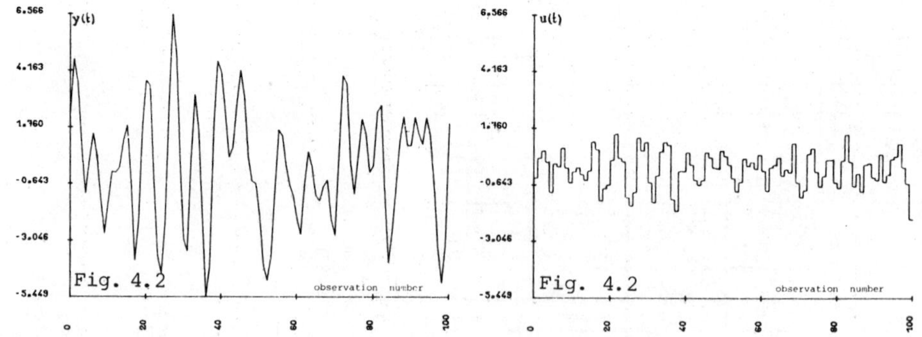

Fig. 4.2

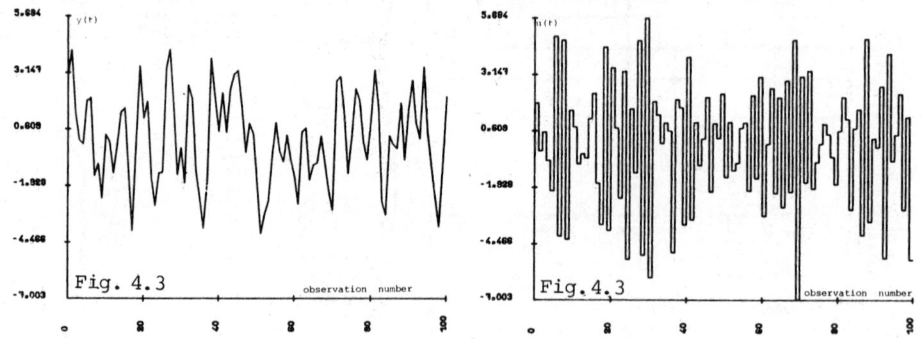

Fig. 4.3

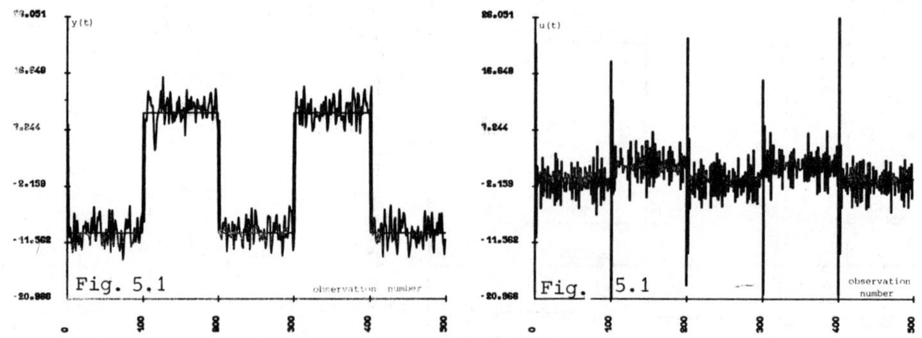

Fig. 5.1

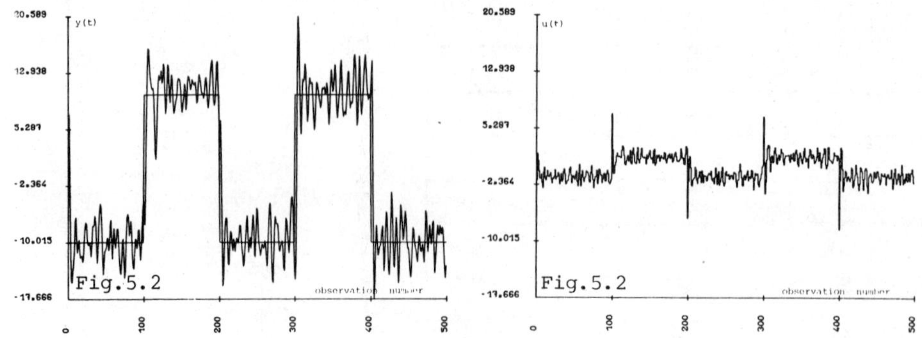

Fig. 5.2

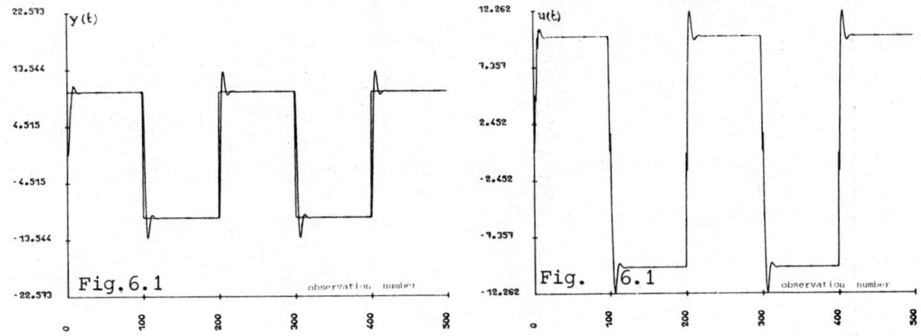

Fig.6.1

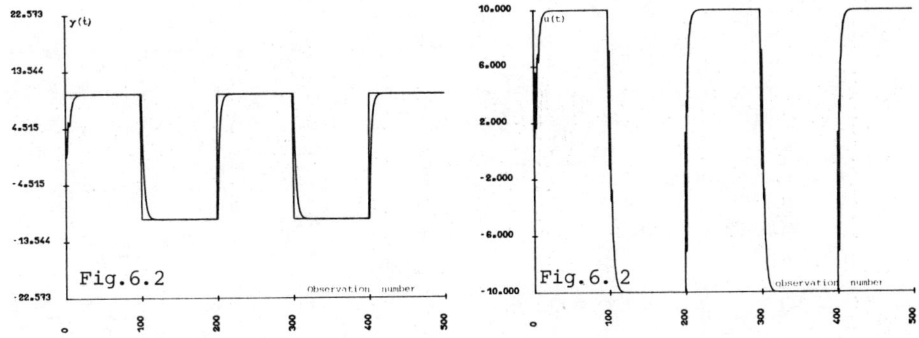

Fig.6.2

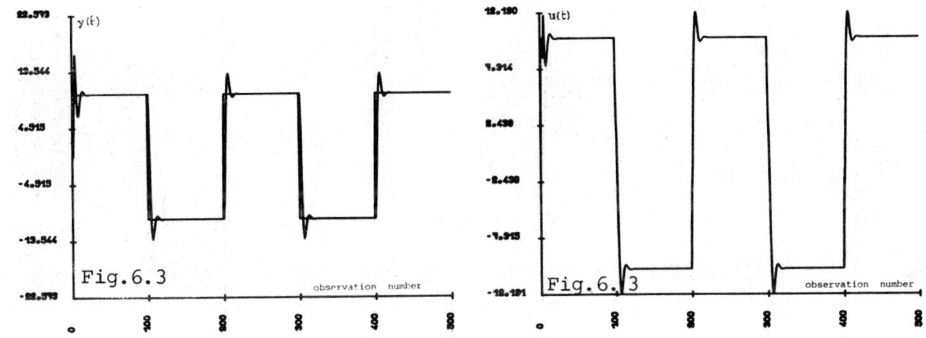

Fig.6.3

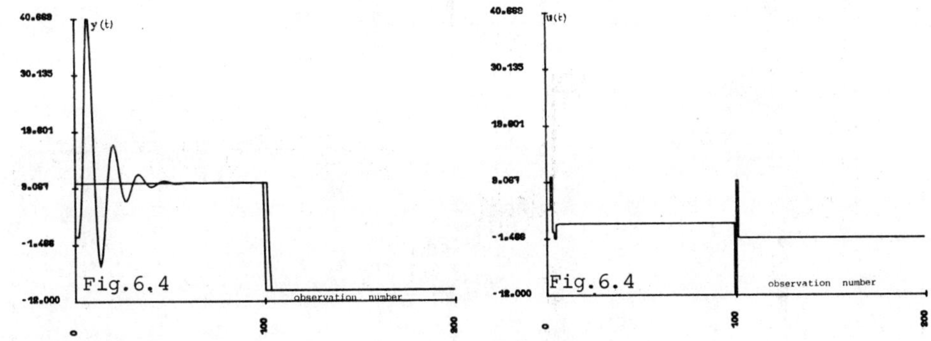

Fig.6.4

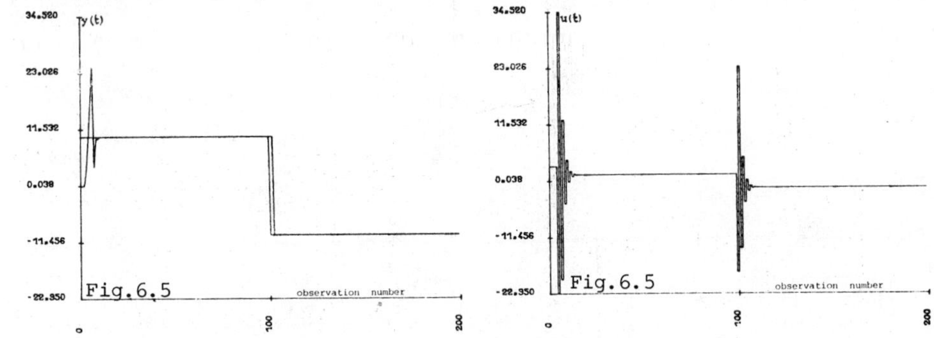

Fig.6.5

Rabat, Morocco, 1980

INERTIA AND MULTICRITERIALITY AS IMPORTANT FEATURES OF DEVELOPING SYSTEMS

V. C. Matejić

Operations Research Department, Institute Mihailo Pupin, Belgrade, Yugoslavia

Abstract. The problem of development has two main sub-problems: (1) sellection of development concept and synthesis of behavioural variables of systems which take part in development, and (2) solving various technical problems of development. In this paper two very important characteristics of developing systems are analysed with the intention to use systems approach in identification and solving planning and management problems generated by those characteristics. First characteristic to be analysed is the inertia of developing systems. It is shown how this characteristic should be incorporated in the synthesis of management behaviour. Second characteristic is the multicriterial feature of modern development problems and its implication on approach to decision making process.

Keywords: Decision theory; developing countries; inertial systems; multivariable systems; optimization, system analysis; system theory.

INERTIA, TRANSITION COST AND THEIR IMPACT ON MANAGING DEVELOPMENT

Introduction

Development can be considered as a transition process, transfering production output, standard of living, level of education, structure of industrial activities etc. from one (s_i) to other (s_j) state. The main characteristic of every real system is the resistance towards change of systems state. This feature has very important influence on "cost" and dynamic of the development.

The theory of economic and organizational systems has devoted very marginal attention to the problem of systems resistences towards changes. However, the fact that these systems function in very changing environment means that they are permanently in transient regimes, and thus, permanently loosing some quantity of input to overcome the resistances. This is the only consistent explanation of differences in efficiency between equally equipped factories in developing and developed countries. Prior to the exposition of suggestions how to control and manage the transient phenomena in development, a part of theoretical results and empirical evidence is to be presented.

Modelling of transient phenomena

Research and particilarly mathematical modelling of transient phenomena in economic and organizational systems are subject to numerous constraints. Only a very narrow class of transient phenomena in these systems provide data for the study and description in mathematical terms. However, systems approach helps us to generalize some of the results and to synthesise the appropriate control rules and mechanisms.

Let's denote production volume with x, and cost with y. Traditional theory tells us that total production cost in period T is

$$Y_s = \int_0^T f(x)t))dt \qquad (1)$$

where f(x) is usualy approximated with

$$f(x) = a_1 x + a_o \qquad (2)$$

For the case of production process,at the firm level, it is relatively easier to identify the mathematical model of production cost considered as function of production output and its changes. With the help of four basic postulates which consider the production firm as: inert, stable, able to learn and "proportional" system we have established the cost model

$$\frac{dy}{dt} + \frac{y}{T_1} = (a_1+b)\frac{dx}{dt} + \frac{a_1 x}{T_1}, \text{ for } \frac{dx}{dy} > 0 \qquad (3a)$$

$$\frac{dy}{dt} + \frac{y}{T_2} = (a_1-c)\frac{dx}{dt} + \frac{a_1 x}{T_2}, \text{ for } \frac{dx}{dt} < 0 \qquad (3b)$$

where T_1 and T_2 are time constants, b and c are cost measures of initial resistance of the production system to the increase and decrease of production output, respectively. Model (3a) and (3b) degenerates into model (2), taking dx/dt = 0. For further clarification we can compare the behaviour of production cost calculated with models (2) and (3) for linear increase of production output during period T. The comparisson results are presented illustratively in Fig. 1. It is abvious that transient cost are bigger the bigger T_1, T_2, b and c are.

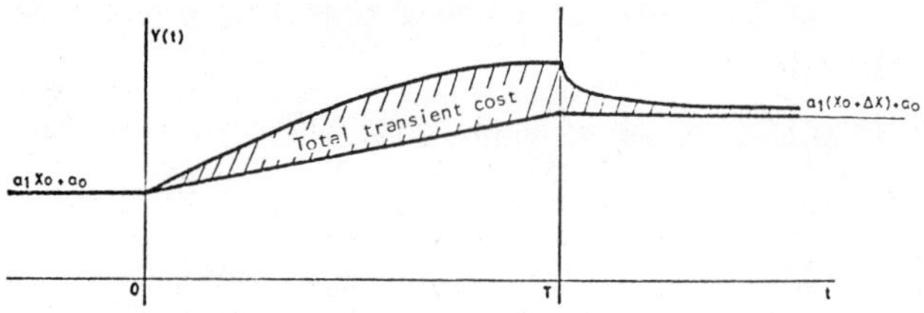

Fig. 1. An illustration of stationary and transient cost production component for linear production increase

Empirical results and conclusions

Model (3) enables very interesting theoretical conclusions to be reached but we are not dealing with them in this paper. The empirical study done in one bigger firm has shown that total transient cost i.e. losses are approximately 30% of total cost. That means that the main effort of production systems planning and managing should be devoted to the reduction of transient losses through the elimination of main factors which generate transient processes.

The most important result of theoretical analyses and its empirical verification is that total cost consists of two main components: (1) stationary, Y_s, which is dependent on production level and efficiency of production process at the level and (2) transient, Y_t, which is generated with changes in production level and dependent of amount of change and resistence characteristics of a system. "Cost" of change can become so great to influence efficiency of development to a great extent.

Second empirical study has dealt with traditional problem concerning the influence of government regulations on economic activities. There is a claim that the number of regulations is negativelly correlated with economic effectiveness. However, it is even more interesting to determine the influence of a change in the number of regulations on economic effectiveness. For this purpose we used statistics on regulations change and rates of growth of output in industry for the period of 12 years. Study of the data has shown an increase in the negative value of linear correlation coefficient between the number of regulations and rates of growth. This means that industrial organizations have losses due to adaptation to the new state of environment caused by change in government regulations. For the analysed set of data it has been found that the time constant for this process of adaptation is approximately 2 years. Thus, industrial organizations are very inert systems.

Having shown that resistances towards changes greatly affect the achievement of development goals, with acceptable use of resources, we state the following rules which are to be followed for dealing with causes of inertial features of development systems:

1. Minimize changes, i.e. don't accept them if their neccessity is not fully justified;
2. Prepare system for acceptance of changes, the better the system is prepared the smaller the losses;
3. Counteract the environment in order to minimize the disturbances which come from it;
4. Sellect the personnel, particularly the managers on the basis of their behaviour towards changes; personnel resistent to changes should be used for stable conditions and operations, whils dynamic and adaptive personnel should be used for changing environment, and
5. Maximize the prediction reliability and prefer long range plans and policies to short ones.

MULTICRITERIALITY OF DEVELOPMENT PROBLEMS AND ONE FEASIBLE AID TO DECISION MAKING

Introduction

The main purpose of models and modelling is to rationalise the search for feasible and eventually optimal solution of a given problem. Modelling of a complex development problem can be done on different levels of scientific aspiration and thus on different ways. Models which are not accepted by decision makers should be considered as infeasible. Therefore, the main aim of model buider is to develop a model which is feasible (accepted for the purpose it was intended) and which can eventually lead to acceptance of the best of all acceptable solutions. For complex decision making problems there is a discrepancy between on optimal solution extracted from the model and one which is selected by decision makers with utilization of a model. Development problems in contemporary cultural and political environment are complex decision problems. Our aim is to recognize features of such decision making process and to identify the best feasible ways of scientific aid to decision makers.

Elements and Phases of Development Decision Process

Development decisions are usually of broader social interest and can be described with following four elements.

First element is discrete set of feasible, mutually exclusive development *alternatives*, X.

Second element is a set of development *criteria*, F. Besides the social there are economis, technical, commercial and ecological criteria. Some of the criteria are quantitative in their nature, some are not. In an average situation, there are 5 to 6 criteria f_i which have substantional influence on alternative selection.

Third element is a set of *decision makers*, D. Decision makers represent interests, preferences and value systems which differ from person to person and from organization to organization. Each decision maker is characterized by: (a) preference structure concerning expected effects of development alternatives, (b) attitudes towards risk and (c) time preference structure. Value of criteria function f_i for decision maker d_k for alternative x_j is $f_i^k(x_j)$. Changes in preference structure can appear even during the decision process. Decision makers are *cooperative* in a sence that they mant to sacrifice something in order to reach the solution acceptable to all of them.

Forth element constitutes *"ideal" alternatives*. Each decision maker creates (on the basis of its own knowledge, information on decision making problem and on preference and value system) solution for the problem which is the best one for him. These "ideal" alternatives (which are not ideal in a strict sence) can become feasible with the relaxation of constraints on resources. Decision makers usually create them having in mind something which is already done somewhere. The change in preference structure which comes up during the decision making process, the change in consideration of development resource constraints, and information on "ideal" alternatives of other decision makers can change "ideal" alternative of particular decision maker. Therefore, the set A of "ideal" alternatives is, to some extent, a fuzzy set.

In all cases when decision makers are supplied with sufficient information on development alternatives, the decision making process has four distinct phases.

First phase. Decision makers (DM) analize development alternatives. Usually they have a feeling that somebody, having defined alternatives, has done the greates part of their job. The less they understand the way in which the preparation of decision process is done, the more they are ready to neglect the usefulness of prepared alternatives. At the end of this phase, decision makers are prepared to start with alternative evaluation.

Second phase. In this phase each decision maker d_k creates his "ideal" alternative a_k, and starts the search on development alternatives set X for alternative x_g^k which is, from his point of view, the best one. This best alternative for decision maker d_k is one which is closest to his "ideal" alternative a_k. For a simple case of additive criteria funciton we can take that x_g^k minimizes Euclidian distance, i.e.

$$\left\{\sum_{i=1}^{m}\left[f_i^k(x_g^k) - f_i^k(a_k)\right]^2\right\}^{1/2} = \min_{x_j} \qquad (4)$$

$$x_g^k \in X$$

The best alternative for decision maker d_k, denoted x_g^k, differs from the best solution x_g^p for $p \neq k$. Therefore the conflict situation arises between the decision makers.

Third phase. During this phase, decision makers try to relax constraints on resources in order to enlarge a set of feasible development alternatives (new set ΔX). The conflict between alternative x_g^k and x_g^p is partially resolved by supply of additional resources. During this phase, decision makers start to make trade-offs intending to come close each other, under assumption of their cooperativity.

Fourth phase, is the final phase during which the efforts are directed towards identification of development alternative which minimizes the level of conflicts. This best alternative x^* has a property to minimize the distance from all alternatives x_g^k, for k = 1,...,N. In special case that can be stated as

$$\left\{\sum_{k=1}^{N}\sum_{i=1}^{m}\left[f_i^k(x^*) - f_i^k(x_g^k)\right]^2\right\}^{1/2} = \min_{x_j} \qquad (5)$$

$$x^* \in X', \quad X' = \{X, \Delta X\}$$

In above described decision making process, three "decisions" are made before decision process is over. In the beginning, each decision maker builds its own "ideal" solution, taking into consideration the wishes and long term prospects (first "decision"). Then, decision makers confront "ideal" solution with feasible ones and each decision maker selects the best feasible, closest to his "ideal" (second "decision"). Conflicting differences between these best solutions can be lowered to some extent if some of resources or legal constraints are relaxed. Whenever the decision maker can relax constraints, he does so (third "decision"). Development decision process ends with selection of investment alternative x^*.

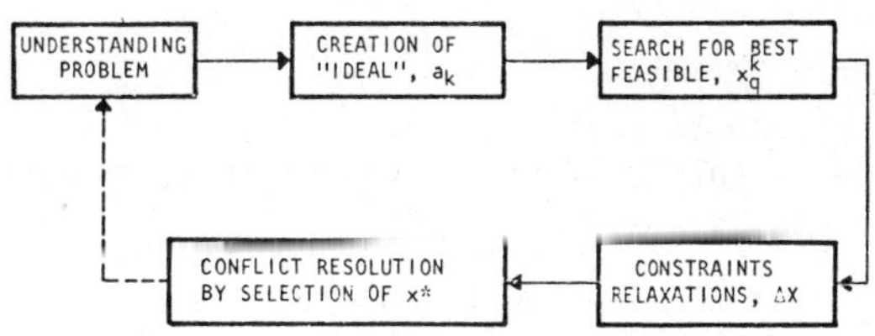

Fig. 2. Stages of development decision process

Estimation of the necessary information

The judgement on the appropriateness of complex modelling of described decision process should be based on (a) an estimation of the necessary information, (b) identification of dimension and complexity of the problem, and (c) willingness of decision makers to use the model and/or its results.

We shall first establish the number of necessary informations groups for strict modelling of a tipical development decision process. Each development alternative is described within average 20 groups of data (data on market, technology, investment costs, revenue, labour, schedules, etc.). For 5 alternatives we need 20 x 5 = 100 groups of data. For 5 alternatives and 5 evaluation criteria we need 5x5 = 25 groups of data. The decision makers evaluate alternatives and model builder should have information on their time preferences (at least 20, if time preference functions are stable), on attitudes towards risk (at least 20x5 = 100) and on index of alternative evaluation with accepted criteria (20x5 = 100). Finally we need 20x5 = 100 groups of data to present the value of criteria for each "ideal" alternative. In total, it is necessary to have

$$I_o = 100+25+20+100+100+100 = 445 \text{ groups of data}$$

It is extremely difficult to get the data on (a) decision makers criteria evaluation, (b) attitudes towards risk and (c) "ideal" alternatives. Thus we lack

$$I_\alpha = 100+100+100 = 300 \text{ groups of data}$$

The index of complexity of the problem can be defined as a product of the number of groups of necessary and frequently changeable or insufficiently reliable data, i.e.

$$C = I_o \times I_\alpha \qquad (6)$$

Complexity index for typical real development decision problem is

$$C_{or} = 445 \times 300 = 133.500$$

In order to apply mathematically based approach to this kind of a problem, an approximation of the problem is done by the simplification of original problem. It is usually taken to deal with two decision criteria (one economic and other social, both quantitative); decision making body is considered as one person with one or two risk function, one time preference function and one "ideal" alternative. The number of groups of necessary information for this problem is

$$I_o = 20x3 + 2x5 + 1 + 1x2 + 1x2 + 1x2 = 117$$

out of these insufficiently reliable are

$$I = 1x2 + 1x2 + 1x2 = 6$$

Index of complexity of the model (the problem which model builder has after model is accepted) is

$$C_m = 117 \times 6 = 702.$$

Comparing indexes of complexits we find that by simplification of the original problem, we can get a model which is (only $C_m/C_o = 0,53\%$) reliable representation of the original decision making problem. Thus, simple and manageable, mathematically based model, cannot represent a useful picture of the complex development decision process. Such models are, in fact, inappropriate because decision makers are not willing to accept them for development alternatives selection. Very reliable mathematical model of the real development problem is nearly impossible to construct.

Synthesis of a Feasible Aid to Decision Makers

The decision makers are not willing to deal with complicated models of real development problems. Simple mathematical models do not represent real problems with sufficient reliability and ecision makers then have feeling they understand problem much better than model builders. Determination of development is a process which necessarily includes creative behaviour of decision makers prior to the evaluation of proposed development alternatives. Decision makers are reluctant towards the problems where they are supposed to keep strictly within a frame of alternatives prepared by somebody else. Furthermore, they are not willing to reveal explicitly to model builders their system value, behaviour under risk and time preferences. When faced with real problems decision makers are very definite in alternative evaluation, without revealing the criteria which have accepted for the evaluation.

Bearing in mind all this, the aid to decision makers should not be directed towards identification of best development alternative simply because this alternative is best for those who prepare decision proposal. Alternatively, aid should be organized so that decision makers find by themselves the best alternative.

The Aid to Decision Makers, at least for development problems, which we have found as feasible, acceptable and sufficiently effective preserves but rationalizes described decision making process. This aid has three main tasks and prescribed procedure for each of them.

Firs task. Minimization of chanses for misunderstanding the decision making problem.

Procedure. Preparation and clear and useful presentation to decision makers of all necesary social, economic, technical, commercial and other information and justification on the development problem and its alternative solutions. Of particular importance is that norm, standards and other "basic" data are explained and justified in full. For nonquantifiable criteria an instruction for evaluation should be done.

Second task. Minimization of conflict between solution x_g^k, for k = 1,2,...,N.

Procedure. Conflict between solutions x_g^k is a consequence of (a) differences between decision makers in behaviour, attitude and value systems and (b) differences between "ideal" alternatives.

On Fig. 3a is given the set of six development alternatives evaluated by decision maker 1, with criteria f_1 and f_2. On Fig. 3b is given the same but for decision maker number 2. "Ideal" alternatives are a_1 and a_2. Best solution for decision maker 1 and "ideal" a_1 is $x_g^1 = x_6$; for decision maker 2 and "ideal" a_2 best solution is $x_g^2 = x_1$. Development alternative x_6 and x_1 are far away each from other, thus there is high level of conflict. From model builder standpoint the "ideal" alternative is a_o. The closest of all feasible alternatives to it is x_5, as shows Fig. 3c. However, decision makers are not billing to change their preference structure by acceptance of model builder evaluation and judgments. Therefore, the only way to lessen conflict is to try to influence the selection of "ideal" alternative. We have found that decision makers are willing to accept information on: (a) other similar development problems and their solution (what been done in other countries and organizations, which experience decision makers appriciate), (b) long term prospects and (c) on "final" state of the system to which their decision should lead. Accepting such information their in fact accept unique "ideal" solution, a*, and now we have decision situation given on Fig. 3a and 3b but for a_1 = a*. When evaluated with "ideals" a_1 and a_2 the best solutions for two decision makers, x_6 and x_1, have been far away each from other and decision makers in such situation first try to relax constraints in order to form new alternative, say x_7, which is candidate to be accepted as final solution of the decision making problem. However, this is not always possible; even more this is impossible in case when preparation of development alternatives is done on throughful way. Therefore, the creation of unique "ideal" alternative is more effective way for minimization of conflict between solution x_g^k. On Figures 3a and 3b we see that (in this illustrative example) after application of a*, decision makers have to choose between x_5 and x_6, which are not far away each from other. Thus the level of conflict is not high and final solution can be easily done.

Third task. Minimization of eventual inconsistency in development alternatives evaluation.

Procedure. Clear, short and full description of criteria behaviour for each development alternative and prepared justification for any of claims used in this description. For quantitative criteria clear and strict sensitivity analysis should be done in other to prevent possible speculations in alternatives evaluation durign decision making process. Model builder should try to supply decision makers with information and judgments before their publicly express their evaluation of alternatives. Decision makers stubornly keep with a merits they have given to alternatives, once they are announced.

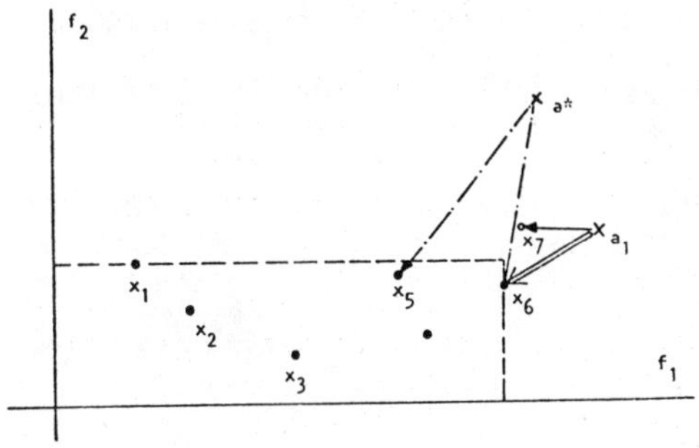

Figure 3a.

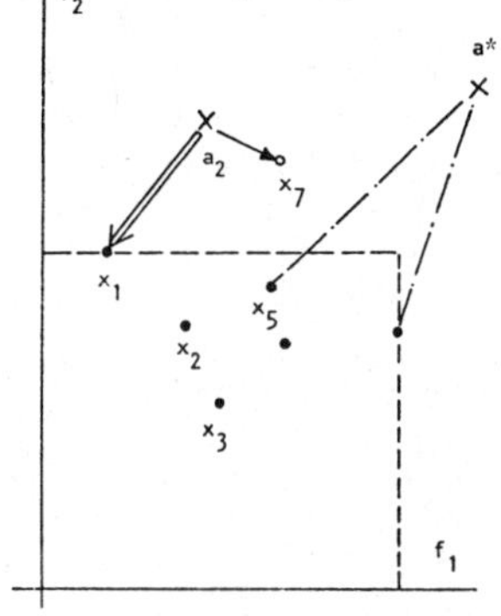

Figure 3b.

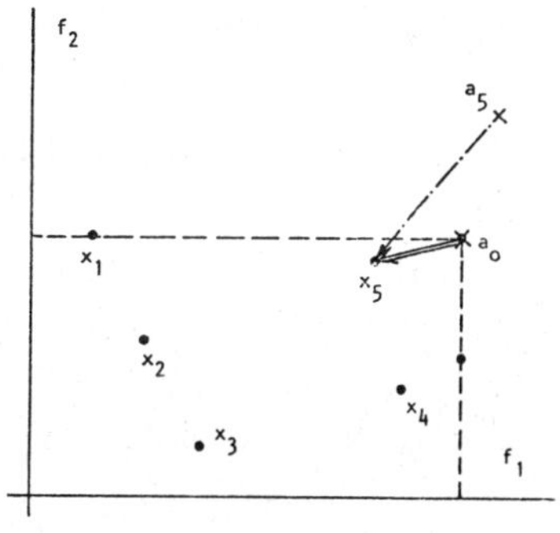

Figure 3c.

The above described Aid to Decision Makers has been applied for some time in one industrial corporation. In comparison to previous situation it is established substantial improvement of investment i.e. development decision making in terms of quality of decisions and decision making process. Mathematical formalization of the decision making process happen to be imposible.

REFERENCES

Holt C., Simon H. (1954). OPTIMAL DECISION RULES FOR PRODUCTION AND INVENTORY CONTROL, Case Institute of Technology, 1954.

Holt C., Modigliani F., Muth J., Simon H. (1960). PLANNING PRODUCTION, INVENTORIES ND WORK FORCE, Prentice Hall 1960.

Forrester J. (1961). INDUSTRIAL DYNAMICS, MIT Press, 1961.

Zelenyl M. (1976). MULTIPLE CRITERIA DECISION MAKING, Springer - Verlag, New York.

Roy B. (1977). FRAMEWORK FOR A PRESCRIPTIVE THEORY OF "DECISION-AID" (in MULTIPLE CRITERIA DECISION MAKING).

Love C. G., Mathias R. A. (1977). CHOOSING AN APPROPRIATE TYPE OF MODEL FOR USE IN STRATETIC PLANNING, XXIII International Meeting of the Institute of Management Science, Athens.

DEVELOPMENT OF SYSTEM MODELS FOR HEALTH CARE EFFICIENCY

T. Prasad, M. A. Ismail and V. H. Quintana

Faculty of Engineering, University of Waterloo, Waterloo, Ontario, Canada

Abstract. Different system approaches for modeling bisystems are outlined, emphasizing the fact that development of sophisticated techniques of interdisciplinary character is essential for effective health care studies. A stochastic modeling methodology and associated algorithms of parameter identification and state estimation are explained and discussed in this context. A newly developed computer-assisted diagnotistic strategy, which uses pattern recognition in conjunction with stochastic modeling, is reported. Applicability of the proposed approach is demonstrated for diagnosis of liver diseases using real patient data.

Keywords. Biomedical; computer applications; medical information processing; medical systems; modeling; models.

INTRODUCTION

System-theoretic formulations, mathematical models, information processing techniques, pattern recognition algorithms, image analysis and processing approaches, and artificial intelligence techniques, can effectively be utilized for improved efficiency in health care problems. Numerous individual and team researches, involvement of national and international organizations, etc. have contributed significantly in this direction during the last few years. Exposition and diffusion of fundamentals from these areas would yield methodologies and integrated concepts for more comprehensive understanding of physiological systems, as well as biomedical problems. The deliberations and interaction between the system scientists on the one hand and the physicians on the other hand are bound to generate excellent potential for development of new strategies, and enhancement of existing services in health care systems.

Mathematical modeling techniques are finding increasing acceptance in physiological research in spite of varying degrees of skepticism still existing amongst many biologists and physicians. The modeling of biomedical systems, or the living systems in general, has been, and will remain, a major challenge for researchers, in systems engineering, interested in studying such complicated processes. The purpose of this paper is to develop and explore both deterministic as well as stochastic techniques for modeling biomedical systems in general, with special emphasis on application of different methodologies, the ultimate goal being utilization of these techniques for efficiency in health care systems.

Information explosion confronting and influencing health care systems is well known. This has resulted in a tendency for physicians to seek the assistance of other experts when facing a complex problem outside of their own specialization. This is generally available either from other medical specialists or from specialists of multidisciplinary character such as computer scientists, engineers and mathematicians. Even experts in a specialized field may reach very complicated decisions concerning the management of a medical case (Shortliffe, 1979; Yu, 1979). Any physician has, in many cases, a variety of available tests for arriving at a correct diagnosis and deciding on an optimal therapeutic measure. Those tests vary in their cost, and constitute a wide range. Due to the above mentioned facts, diagnosis and subsequent therapy decided upon have shown to vary widely in their accuracy (Rosenblatt, 1973; Shortliffe, 1979). Moreover, there is lack of a structured way by which medical students can extract proper information for making decisions (Kassirer and Gorry, 1978).

There are numerous motivations and reasons for attempts to understand and computerize the diagnostic process or, in general, the clinical decision making process (Wardle and Wardle, 1978). These computerized techniques are used not only on the diagnostic level, but also on the therapeutic level, and go beyond that to prognosis. A lot of advantage can be gained by the introduction of computers in such studies. First of all, accuracy of medical diagnosis can be improved through the use of some systematic and sophisticated mathematical techniques, that can get as much information as possible out of a minimal number of tests, mostly of the noninvasive type. This leads to the balance of expenses of time, inconvenience and funds. The resulting definitive actions will improve, in general, the cost and efficiency of tests and therapy. Moreover, the reliability of clinical decisions can be improved by avoiding unwarranted influences of similar but not identical cases and by making the criteria for decisions explicit. Over and above, computerized techniques can improve our understanding of the contents of medical knowledge through the identification of inconsistencies and inadequacies, which can improve the understanding of clinical decision making.

Mathematical modes of physiological and biomedical systems can be utilized as a tool for computer-assisted diagnosis. Perhaps pharmacokinetics was one of the main avenues of interest to mathematical modelers, and a lot of programs have been proposed for optimal theraphy doses (Gorry and co-workers, 1978). The principal power of mathematical models as an aid to diagnosis and prognosis is their ability to seize mathematically sound relationships in a compact and efficient form. These relationships can approximate the physiology of the system adequately and can also locate important connections between different variables involved. In addition, higher reliability of clinical decision making can also be improved, and a better understanding of the biomedical system under study can be acquired. The most significant feature of these techniques is that they can extract information not seen by the clinician in the original system itself and which cannot be measured. This type of information is very important in the classification procedure, particularly between different abnormal groups. An outline of certain complexities encountered in modeling of biomedical systems and subsystems is presented first, followed by the development of different mathematical modeling techniques. Utilization of such models for diagnostic and prognostic studies has been illustrated next by an example.

COMPLEXITIES IN MODELING BIOSYSTEMS

Although identification techniques have been extensively applied to a variety of complex engineering problems, only a limited number of systematic techniques have been applied to biomedical problems (Bekey and Yamashiro, 1976). In recent years several books have been devoted specifically to the subject of identification (Eykhoff, 1974; Sage and Melsa, 1971; Mendel, 1973; Kagiwada, 1974; Graupe, 1976; Goodwin and Payne, 1977; Beck and Arnold, 1977). This is in addition to conferences held in the past decade also devoted to the subject.

A few reasons why some of the systematic results applied to biosystems have faced lack of acceptance are noteworthy. First of all, physiology being a descriptive experimental science, results obtained by computer, rather than measured directly by the experimenter, are considered to be doubtful until they are validated in the laboratory. Then comes the well known phenomenon that the problems of biosystems exhibit great variability; i.e., there is a variation from trial to trial in experiments and also from one subject to another. The problem of isolating inputs and outputs in biosystems is also there because of the mechanisms involved. One also may face the problem of inaccessibility of some of the state variables for the subsystem under study. Or in some cases, where one may be able to reach appropriate state variables, the measurement time required to obtain consistent estimates may be so long that stationarity can no longer be assumed. Most of the biomedical processes also display a wide range of intrinsic nonlinearities. There is also the serious problem associated with obtaining enough data of good quality from the experiment; e.g., one may find himself with a system with twelve unknowns and only five data points, presumably corrupted with noise. It could also be embarrassing to the researcher if the results are not unique, especially if he is dealing with a highly nonlinear system. On the other hand, cost of computer time consumed in biomedical problems is also expensive in contrast to engineering applications. Adaptivity of responses in biosystems could also be a serious problem facing the researcher.

However, it should be noted that the use of mathematical models in biomedical studies is increasing. Attempts continue for search of realistic models which should be capable of representing the system under consideration with improved precision. This brings the need for the general nonlinear stochastic methodology. The mathematical modeling process of a biomedical system can be best illustrated by Fig. 1 shown below.

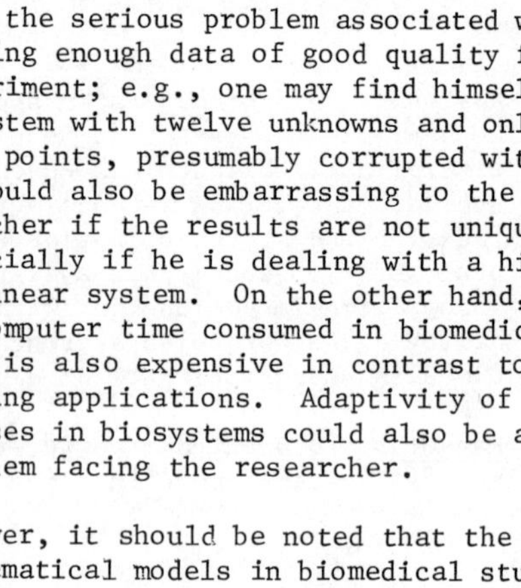

Fig. 1. Modeling in biomedical systems: schematic

DEVELOPMENT OF MATHEMATICAL MODELS

The modeling of a biomedical system utilizing observations available from measurements can be formulated as a parameter estimation problem. The model structure can only be assumed from an a-priori knowledge of the physiology of the system under investigation, as well as the available experimental data. In this case, the identification problem can be called a "grey box" problem (Bekey, 1970). This approach may be used to describe both deterministic as well as stochastic multi-compartmental models: intercompartmental relations and/or kinetic parameters may specifically be described by a set of ordinary differential equations, a set of stochastic differential equations or a set of stochastic integral equations. On the other hand, the transfer function approach can also be considered as a grey box problem if the orders of the transfer functions incorporated are obtained utilizing the data observed. If the order of the system is not known, then the problem may be considered as a "black box" problem in which orders must be assumed or otherwise obtained by making use of the data available. It is evident that the greater the amount of a-priori information is available to us, the greater the flexibility we have in both modeling the system and interpreting the results obtained.

Classical compartmental models have attracted many researchers from different disciplines due to its simplicity and mathematical tractability (Jacquez, 1972; Rubinow, 1975; Brown, 1980). The set of differential equations describing a general nonlinear deterministic multicompartmental model takes the form (Sandberg, 1978)

$$\dot{x}_i = f_{io} + \sum_{\substack{j=1 \\ j\neq i}}^{n} (f_{ij} - f_{ji}) - f_{oi}\,, \tag{1}$$

$$x_i(o) = x_{oi}, \quad i=1,2,\ldots,n$$

where n is the number of compartments, x_i is the amount of material in compartment i, x_{oi} is the initial value of x_i, and f_{rs} is the mass flow rate to compartment r from compartment s, subscript o denoting the external environment. If the kinetics are assumed to be linear, equations (1) take the form

$$\dot{x} = f_{io} + \sum_{\substack{j=1 \\ j\neq i}}^{n} (k_{ij}\, x_j - k_{ji}\, x_i) - k_{oi}\, x_i, \tag{2}$$

$$x_i(o) = x_{oi}, \quad i=1,2,\ldots,n$$

where the coefficients k_{ij}, i=0,1,...,n, j=1,2,...,n,j≠i, are referred to as the rate constants or the kinetic parameters. Solution schemes for all the studies, which employ the deterministic compartmental models, can almost be summarized by a basic question: how can the values of the kinetic parameters k_{ij} be estimated? In most of the cases, adaptive techniques have been employed for this purpose. A schematic, illustrating the procedure for estimation of these parameters using such adaptive techniques, is shown in Fig. 2.

Another technique which can be utilized is the transfer function approach. Usually in this case, one is forced to deal with a multivariable system. As an example, consider a single-input two-outpt discrete system which can be described by

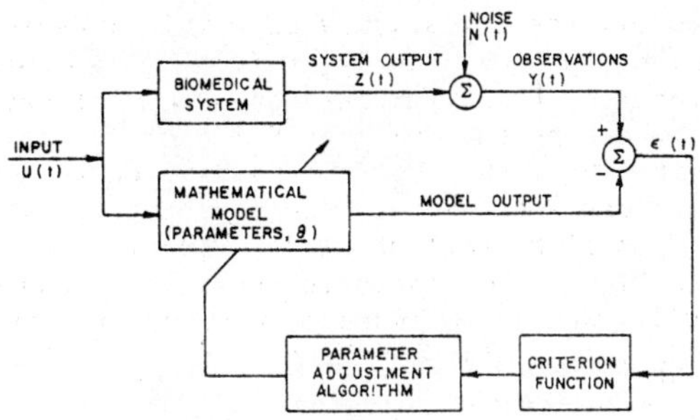

Fig. 2. Adaptive parameter estimation

$$\begin{bmatrix} y_1(z) \\ y_2(z) \end{bmatrix} = \begin{bmatrix} G_1(z) \\ G_2(z) \end{bmatrix} u(z) = G(z)\; u(z) \tag{3}$$

where G(z) is defined as the transfer matrix, u(z) is the input, and $y_1(z)$ and $y_2(z)$ are the two observed outputs. The transfer matrix G(z) consists of two transfer functions which can be identified from input-output records (Sinha and Pille, 1971). This can be achieved by assuming that each transfer function of order n has, in general, the form

$$G_i(z) = \frac{a_{oi} + a_{1i} z^{-1} + a_{2i} z^{-2} + \ldots + a_{ni} z^{-n}}{1 + b_{1i} z^{-1} + b_{2i} z^{-2} + \ldots + b_{ni} z^{-n}} \tag{4}$$

The problem now reduces to that of estimating the parameters that appear in the transfer functions given by equation (4), i.e., that a's and the b's. Although different techniques can be applied, the least-square approach gives explicit formulas in this particular case.

Considering the input to be of the impulse type, let us try to demonstrate how least-squares can be applied to identify the parameters concerned. Considering for example $G_1(z)$, it is well known from the impulse-response analysis of linear systems that

$$G_1(z) = \frac{a_o + a_1 z^{-1} + a_2 z^{-2} + \ldots + a_n z^{-n}}{1 - b_1 z^{-1} - b_2 z^{-2} - \ldots - b_n z^{-n}} \tag{5}$$

and

$$G_1(z) = g_o + g_1 z^{-1} + g_2 z^{-2} + \ldots \tag{6}$$

where g_o, g_1, g_2,... etc., are the discrete values of the impulse response y_1. It can be shown that the following relations hold true:

$$\begin{aligned} a_o &= g_o, \\ a_1 &= g_1 - b_1\, g_o, \\ &\;\;\vdots \\ a_n &= g_n - \sum_{i=1}^{n} b_i\, g_{n-i}. \end{aligned} \tag{7}$$

and

$$\begin{bmatrix} g_n & g_{n-1} \cdots g_1 \\ g_{n+1} & g_n \cdots g_2 \\ \cdot & \cdot \quad \cdot \\ \cdot & \cdot \cdots \cdot \\ \cdot & \cdot \quad \cdot \\ g_{n+i-1} & g_{n+i-2} \quad g_i \end{bmatrix} \begin{bmatrix} b_1 \\ b_2 \\ \cdot \\ \cdot \\ \cdot \\ b_n \end{bmatrix} = \begin{bmatrix} g_{n+1} \\ g_{n+2} \\ \cdot \\ \cdot \\ \cdot \\ g_{n+i} \end{bmatrix} \tag{8}$$

Using the system of equations (8), one can get the least-squares solution for the b's. Substitution in (7) yields the values of the a's. The procedure can be repeated to obtain the coefficients in $G_2(z)$ in the same manner. The choice of the proper order in each case and validation of models suggested by this approach will be published by the authors of this paper in the near future.

More recently, stochastic multicompartmental formulations have been used to study biomedical systems and as an aid for diagnosis and prognosis (Ismail and co-workers, 1978 and 1979). While the classical compartmental analysis proved to be attractive to many researchers due to its simplicity (Jacquez, 1972; Llaurado, 1973; Rubinow, 1975; Brown, 1980), the developments reported in most cases are only approximate, adopting linear and deterministic models; whereas, it is well known that information about many real-life systems is generally available in the form of measurements which are subject to random errors. Further, the overall structure may contain systemic noise, randon noise or bad data points. The associated nonlinear characteristics of input and/or states of the system are also of intrinsic interest and should be taken into consideration.

The approach reported by Ismail and co-workers (1978 and 1979) offered a stochastic approach that can handle the problem of modeling biomedical systems, given the deterministic compartmental model as a first approximation, ending up with a stochastic multicompartmental model. Taking into account the probabilistic nature of the functions involved, the generalized system model can be described as follows. The basic structure of the system constitutes of two parts: the system dynamics and the observation components. The state variables vector $x_i(t;\omega)$ could represent the amount of a certain tracer in compartment i at any time t (or the percentage from an initial dose). $f(.,.)$ is an n-column vector of functions representing the deterministic part of the state dynamic equations, and $G(.,.)$ is an n x m matrix that characterizes the transformational features of the system. The argument ω depicts the fact that x is considered along with possible random fluctuations. u(t) is an m x 1 random vector, assumed to be multiplicative in character, which accounts for stochastic surroundings, internal disturbances, etc. The quantity $\phi(.,.)$ denotes a transformation, linear or nonlinear, of the state process. The actual observation vector y, whose dimension is p x 1, is contaminated with a p x 1 additive measurement noise vector v(t). n represents the number of compartments used in the model, while p represents the numner of observable outputs.

The mathematical formulation describing the whole system readily follows as

$$\dot{x}(t;\omega) = f(t;x) + G(x;\omega)\, u(t) \tag{9}$$

which is a particular differential form of the generalized It ô-Doob integral representation

$$x(t;\omega) = c + \int_{t_o}^{t} f(\tau, x(\tau;\omega))\, d\tau + \int_{t_o}^{t} G(\tau, x(\tau;\omega))\, dw(\tau) \tag{10}$$

$$dw(\tau) = u(\tau)\, d\tau, \tag{11}$$

$$o \le \tau \le t < \infty$$

$$c = x(t = t_o) = x_o .$$

The estimation of the state variables $x(t;\omega)$ is essentially based on the observation vector y(t) with dimension p x 1. It is well known that these observed values may not be available with precision, due to some factors such as noisy experimental conditions, the use of low precision instruments, different types of uncertainties, etc. Therefore, we assume a general observation model of the form

$$y(t) = \phi(x) + v(t) \tag{12}$$

The vector v(t) may be considered to be a white Gaussian noise (WGN) vector, with covariance matrix R. A schematic showing the general nonlinear stochastic multi-compartmental structure for modeling and simulation is given in Fig. 3. The deterministic part of the system depicted by $f(.,.)$ in equation (9) can also be used to represent the compartmental structures of the type described by equations (1) and (2).

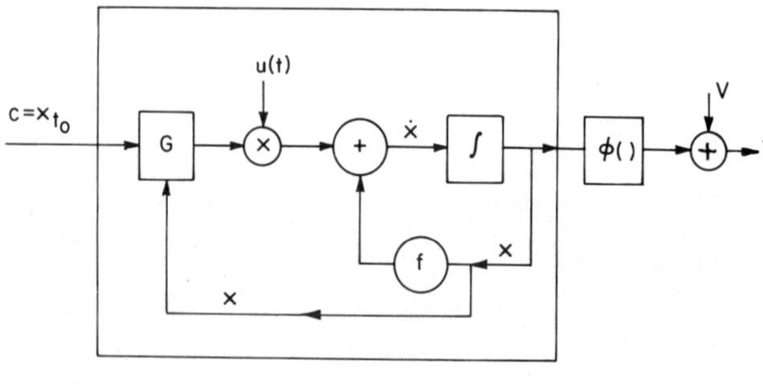

Fig. 3 A Schematic showing the stochastic multicompartmental structure

One of the schemes recommended for solving the stochastic multicompartmental models discussed above is the adaptive technique summarized as follows. The observations process y(t) can be transformed into an equivalent white noise process so that the statistical inference problem based on y(t) can be replaced by a simpler one based on the equivalent white noise process (Ibidapo-

Obe and Prasad, 1977). The method then utilizes the theories of martingales (Prasad and Sinha, 1974) and stochastic approximation yielding an adaptive algorithm (AEA) for optimal nonlinear estimation in stochastic systems. Using the well known techniques of state space representations, both the kinetic parameters and states of the model can be estimated simultaneously using this algorithm. A flowchart for the AEA is given in Fig. 4. The computational algorithm is given by

$$\dot{\hat{x}} = f(t,\hat{x}(t)) + \alpha(t)\, G(t,\hat{x}(t))\, \nu(t) \quad (13)$$

where $\nu(t)$ is the innovations process given by

$$\nu(t) = y(t) - \phi(t,\hat{x}(t)) \equiv \nu_t \quad (14)$$

and $\alpha(t)$ is a diagonal matrix with elements $(\alpha_1, \alpha_2, \ldots, \alpha_n)$, satisfying the relations

$$\int_0^\infty \alpha(t)\, dt = \infty, \int_0^\infty \alpha^2(t)\, dt < \infty \quad (15)$$

$$\alpha(t)^{i+1} = \alpha^i(t) + \gamma^i(t)\, \psi(\nu_t^i), \quad (16)$$

$$\gamma^i(t) = 1/t, \quad (17)$$

$$\psi(\nu_t^i) = \nu_t^i - E\{\nu_t^i\} \quad (18)$$

The kinetic parameters and state variables involved in the general model can be estimated using the extended Kalman filter, the invariant imbedding approach or the Bayesian estimation method.

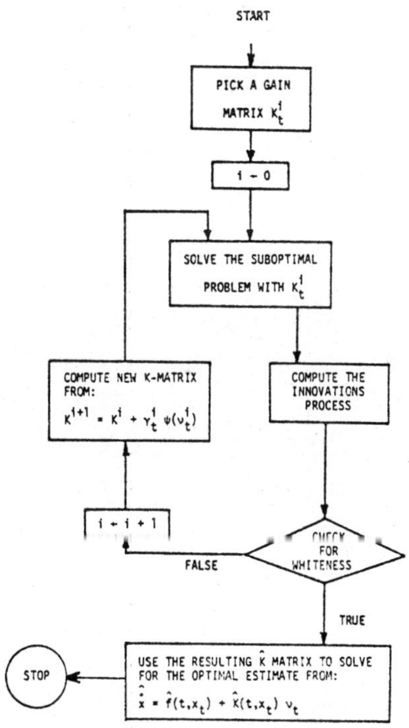

Fig. 4. A flow chart for the stochastic approximation technique

UTILIZING MATHEMATICAL MODELS FOR DIAGNOSTIC AND PROGNOSTIC STUIDES: AN APPLICATION

A blend of pattern recognition techniques (Duda and Hart, 1973) with stochastic models offers an attractive approach for both diagnostic as well as prognostic studies. The estimated quantities are utilized as features for the classification process,and some other compound features, which have clinical importance and which contribute to the decision making process can be also included.

An example of the proposed methodology has been reported recently (Ismail and co-workers 1980). This involves a stochastic model for the hepatobiliary system, where the state variables of the model have been utilized as features for the classification process. The deterministic portion of that model is shown in Fig. 5.

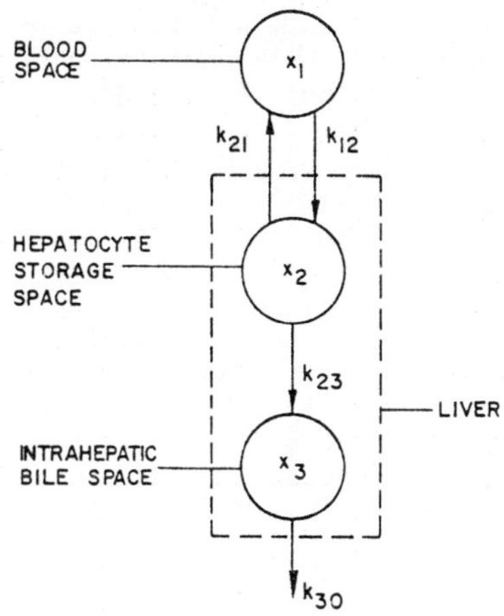

Fig. 5

The results using six state features can be best illustrated using the scatter diagram of patients shown in Fig. 6 given below. It should be noted that every patient is represented by one point in a 6-dimensional space, and on the scatter diagram also as a point in the approximate two dimensional projection in plane. Subjects numbered 1 to 7 represent normal subjects while subjects number 8 to 15 are cirrhotic patients. Patients number 16 to 21 have obstructive jaundice; all of them are of the extrahepatic type due to different reasons, except patients number 18 and 19 who have intrahepatic obstruction. Patients number 14 and 15 are cirrhotic subjects with partial obstruction.

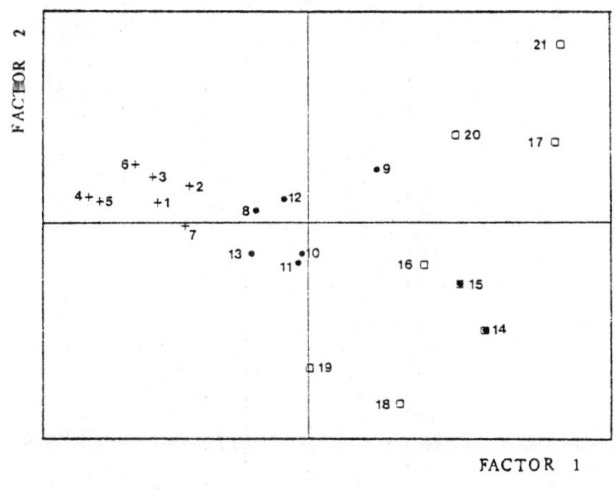

Fig. 6

The primary strength of this analysis is its attractive power in patient follow-up or prognosis. After administration of a certain drug, the physician is always interested in the evaluation of his patient state. Using the scatter diagram shown, the physician will be able to observe the state of his patient in the form of a point moving from one location to another. In other words, any moving point representing a certain patient will get closer to a certain cluster or group of patients, which can tell us exactly if the patient state has improved or not.

CONCLUSIONS

An interdisciplinary approach for studying biomedical systems for health care efficiency has been outlined in this paper. The important role that mathematical models can play in this context has been emphasized. Different techniques for modeling biosystems have been dicussed. A challenging possibility of utilizing stochastic models blended with pattern recognition techniques has been explored with the aid of an example related to the diagnosis and prognosis of liver diseases. It is envisaged that the methodologies proposed in this paper would find numerous other applications in different problems related to health care systems.

ACKNOWLEDGEMENTS

This work was supported by NSERC (Natural Sciences and Engineering Research Council) of Canada.

REFERENCES

Beck, J.V., and K.J. Arnold (1977). Parameter Estimation in Engineering and Science. John Wiley & Sons, Toronto.

Bekey, G.A. (1970). System Identification: An Introduction and a Survey. Simulation, October, pp. 151-166.

Bekey, G.A. and S.M. Yamashiro (1976). Parameter Estimation in Mathematical Models of Biological Systems. Advances in Biomedical Engineering, Vol. 6, pp. 1-43, (Brown, J.H.U. and J.F., Dickson III, eds.).

Brown, R.F. (1980). Compartmental System Analysis: State of the Art. IEEE Trans. Biomed. Eng., Vol. BME-27, No. 1, p. 1-11.

Duda, R. and P. Hart (1973). Pattern Classification and Scene Analysis. John Wiley and Sons, New York.

Eykhoff, P. (1974). Systems Identification. John Wiley, New York.

Goodwin, G.C. and R.L. Payne (1977). Dynamic System Identification: Experiment Design and Data Analysis. Academic Press, New York.

Gorry, G.A., H. Silverman, and S.G. Pauker (1978). Capturing Clinical Expertise - A Computer Program that Considers Clinical Responses to Digitalis. Amer. J. Med., Vol. 64, pp. 452-460.

Graupe, D. (1976). Identification of Systems. Van Nostrand-Reinhold, Princeton, N.J.

Ibidapo-Obe, O. and T. Prasad (1977). An Adaptive Algorithm for Optimal Nonlinear Estimation in Stochastic Systems. Int. J. Systems Sc., Vol. 8, No. 11, pp. 1209-1232.

Ismail, M.A., T.P. Prasad, and V.H. Quintana (1980). Diagnosis of the Hepatobiliary System Disease Utilizing Time Varying Features. To be published.

Ismail, M.A., T. Prasad, and V.H. Quintana (1979). An Algorithm for Studying Stochastic Compartmental Models in Biomedical Systems. Proc. 2nd International Conference on Mathematical Modeling, St. Louis, U.S.A. July 11-13.

Ismail, M.A., T. Prasad, and Quintana, V.H. (1978). A Methodology for Modeling and Simulation of Biomedical Systems and Drug Kinetics Using Nonlinear Stochastic Compartmental Analysis. Proc. Ninth Annual Pittsburgh Conference on Modeling and Simulation, Pittsburgh, U.S.A., April 27-28, pp. 373-378.

Jacquez, J.A. (1972). Compartmental Analysis in Biology and Medicine; Kinetics of Distribution of Tracer-Labled Materials. Elsevier, Amsterdam.

Kagiwada, H.H. (1974). System Identification: Methods and Applications. Addison Wesley, Reading, Massachusetts.

Kassirer, J.P. and G.A. Gory (1978). Clinical Problem Solving: A Behavioural Analysis. Anns. Int. Med., Vol. 89, pp. 245-255.

Llaurado, J.G. (1973). Compartmental Approach to Water and Electrolyte Distribution. In J.H.V. Brown and D.S. Gann (Eds.), Engineering Principles in Physiology, Vol. II, pp. 347-372, Academic Press, New York, 1973.

Mendel, J.M. (1973). Discrete Techniques of Parameter Estimation: The Equation Error Approach. Marcel Dekker, Inc. New York.

Prasad, T. and N.K. Sinha (1974). Martingales and their Application to Optimal State Estimation. Int. J. Syst. Sci., Vol. 5, No. 11, pp. 1039-1054.

Rosenblatt, M.B., P. K. Teng, and S. Kerpe (1973). Diagnostic Accuracy in Cancer as Determined by Post-Morten Examination. Prog. Clin. Cancer, Vol. 5, pp. 71-80.

Rubinow, S.I. (1975). Introduction to Mathematical Biology. John Wiley and Sons, Inc., New York.

Sage, A.P. and J.L. Melsa (1971). System Identification. Academic Press, New York.

Sandberg, I.W. (1978). On the Mathematical Foundations of Compartmental Analysis in Biology, Medicine, and Ecology. IEEE Trans. Circuits & Systems, Vol. CAS-25, No. 5, pp. 273-279.

Shortliffe, E.H., B. G. Buchanan, and Feigenbaum (1979). Knowledge Engineering for Medical Decision Making: A Review of Computer-Based Clinical Decision Aids. Proc. IEEE, Vol. 67, No. 9, pp. 1207-1224, September.

Sinha, N.K. and Pille, W. (1971). On-line Parameter Estimating Using Matrix Pseudo-inverse. Proc. IEE, Vol. 118, No. 8, August, pp. 1401-1046.

Wardle, A. and L. Wardle (1978). Computer-Aided Diagnosis: A Review of Research. Meth. Info. Med., Vol. 17, pp. 15-28.

Yu, V.L., B.G. Buchanan, E.H. Shortliffe, and co-workers (1979). An Evaluation of the Performance of a Computer-Based Consultant. Comput. Program. Biomed., Vol. 9, pp. 95-102.

MULTIVARIABLE MODELLING FOR THE CLIMATIC DESIGN COOLING LOADS IN BUILT ENVIRONMENTS

M. Barkatullah

Department of Architectural Engineering, University of Petroleum and Minerals, Dhahran, Saudi Arabia

Abstract. A collective system approach to the analysis of design cooling loads under the variable influences of climatic parameters resulted in obtaining simple multivariable linear models. It has been found that the peak cooling load bears linear correlations against four building parameters: height, length-width ratio, percentage of glazed area and roof area. Attempts to obtain linear correlations between the design cooling load and first or higher order values of the climatic parameters (e.g., mean ambient air temperature, maximum diurnal swing of ambient air temperature, mean solar radiation and maximum daily swing of solar radiation) did not succeed.

Keywords. Air conditioning; Environmental Control; Linear programming; multivariable systems; Modelling; Optimisation.

INTRODUCTION

The least square error analysis for multi-variables is widely used in econometrics for obtaining general linear model (Johnston,1963). In multiple regression analysis, the coefficient of multiple correlation is defined as follows:

Let $R_{1.234....K}$ - the coefficient of multiple correlation, the dependent variable is represented by 1 and the independent variables are represented by 2,3....k.

$\hat{\beta}$ - the least square estimators. $\hat{\beta}$ represent a colum vector and $\hat{\beta}'$ is the row vector - transposed from $\hat{\beta}$. $\hat{\beta}_1$ gives the multiple intercept.

X - the matrix of observations of the independent variables. X' is the transposed matrix X. If there are (k-1) number of independent variables and n is the number of observations, the order of X is (n,k) and $X_{i,1} = 1$.

Y - column vector for dependent variable. Order of Y is (n,1) Y' is the transpose of Y.

Y_i - the dependent variable. There are n observations.

Then, the simple correlation between Y and X_2 is given as follows:

$$r_{12} = \frac{\sum_{i=1}^{n} (y_i x_{i2})}{\sqrt{(\sum_{i=1}^{n} y_i^2)(\sum_{i=1}^{n} x_{i2}^2)}} \tag{1}$$

where $y_1 = Y_1 - \bar{Y}$ and $x_{i2} = X_{i2} - \bar{X}_2$

and the simple correlation between X_2 and X_3 is given by

$$r_{23} = \frac{\sum_{i=1}^{n} (x_{i2}\, x_{i3})}{\sqrt{(\sum_{i=1}^{n} x_{i2}^2)(\sum_{i=1}^{n} x_{i3}^2)}} \tag{2}$$

where x_{i2} is as before and $x_{i3} = X_{i3} - \bar{X}_3$.

The simple correlation between two variables, when arranged in matrix form, is given by

$$R = \begin{bmatrix} r_{11} & r_{12} & \cdots & r_{1k} \\ r_{21} & r_{22} & \cdots & r_{2k} \\ r_{k1} & r_{k}2 & \cdots & r_{kk} \end{bmatrix} \tag{3}$$

where r_{1j} (j = 2, ... k) represent the correlation between Y and X_j and r_{ii} = 1 (where i = 1,2,... k).

The least square estimators can be obtained from the simple correlation matrix as well:

$$\hat{\beta}_2 = - \frac{s_1}{s_2} \frac{R_{12}}{R_{11}} \quad (4)$$

where R_{ij} is the co-factor of r_{ij} in matrix R and s_1 and s_2 are the standard deviations

$$(\text{e.g.} \sqrt{\frac{\Sigma y_i^2}{n}} \text{ and } \sqrt{\frac{\Sigma x_{i2}^2}{n}})$$

In general,

$$\hat{\beta}_i = - \frac{s_1}{s_i} \cdot \frac{R_{1i}}{R_{11}} \quad (5)$$

The coefficient of the multiple correlation is given by:

$$R^2_{1.2,3,4,\ldots k} = 1 - \frac{R}{R_{11}} \quad (6)$$

where $R = |R|$,

and the coefficient of partial correlation between the dependent variable and any of the independent variables when the other independent variables are held constant is given as follows:

$$r_{1i.2,3,4,\ldots i-1, i+1, \ldots k} = - \frac{R_{1i}}{\sqrt{R_{i1} R_{ii}}} \quad (7)$$

Based on the above relations, computer programmes have been developed in the forms of FORTRAN Subroutine which calculates the least square estimators, multiple correlation coefficient and the partial correlation coefficients between the dependent variable and each of the independent variables when the other variables are held constant. The values of the actual and the calculated dependent variables are plotted on cartesian coordinates by the line-printer.

DEPENDENT AND INDEPENDENT VARIABLES

At an earlier study (Ullah, 1975) it was observed that the peak cooling loads varied linearly with length-width ratio, height of the building and approximately linearly with the roof area, and percentage of glazing area in buildings. These four variables are the parameters which, for a particular geographical location, can be controlled physically. For the U.K. design conditions, during the months of March to October, design solar radiation and air temperature data are available (Tables A6.8 and A6.15 to A6.2 in IHVE Guide 1970). Both the walls and the roof of the buildings considered are assumed to be made of 0.150 m thick concrete. The windows are made of single glass and no shading has been considered. The orientations of the buildings have been assumed to be the north, (i.e. the longer side of the building is perpendicular to the north). The studies have been carried out with a view to establishing (a) a linear relation for the peak cooling loads in July against the physical variables mentioned above, and (b) a general linear relation for the peak cooling load in March to July against the above mentioned physical parameters and the climatic variables such as mean outside temperature, swings of the outside temperature over 24 hours from the mean, the mean solar radiation on the horizontal, and the swing of the solar radiation on the horizontal over 24 hours from the mean, or the total outside air temperature, or the total solar radiation on the horizontal.

The peak cooling loads have been obtained by the multiple harmonic Fourier method (Ullah and Longworth, 1976) for the evaluation of heat flow through the wall and roofs and by the IHVE Guide 1970 method for the heat loads through the windows. No infiltration or other loads such as lighting, occupancy etc. have been considered.

Six cases have been studied: Table 1 shows the variables considered, Table 2 contains the coefficients of partial correlation between the peak cooling load and the dependent variables, and Table 3 shows coefficients of multiple correlation and the values of the least square regression coefficients. Case 1 is the case in which the four physical parameters of the building, e.g., the length-width ratio, the height, the window area in percentage of the wall area and the area of the roof, have been correlated to the peak cooling load in July. The value of the multiple correlation coefficient is found to be 0.925 and the regression line between the peak cooling loads calculated by the linear relation and the actual values for them is shown in Fig. 1. In the five other cases, attempts have been made to correlate the peak cooling loads in a month during March to October to the above-mentioned four physical parameters and the climatic variables. A number of combination and variation of the climatic parameters have been made, as shown in Table 1. It was observed that little change took place in the value of the coefficient of multiple correlation due to these variations in the climatic parameters. From these studies, it is observed that there is appreciable linear correlation between the mean outside air temperature and the peak cooling loads. Little correlation has been observed between the peak cooling load and the solar radiation intensities on the horizontal. The cause of this is ascribed to the fact that during the months considered, little variations were observed in the value of the solar radiation intensities on the horizontal. Total 392 cases have been considered in which there have been only 5 variations in the values of the total solar radiation intensities on the horizontal. This is because the solar radiation intensities on the horizontal are symmetrical about the month of June, and the peak occurs in June. The values of solar radiation intensities in May and July are the same, and the same is the case for April and August, March and September. The second reason may be that for buildings with the windows on the walls, the solar radiation intensity on the horizontal may not be a good independent parameter. Better correlations might have been obtained if an area-weighted mean of the solar radiation intensities over all the surfaces could have been taken, but in that case, the simplicity of the linear model would have been lost to some extent.

CONCLUSIONS

The usefulness of the multi-variable linear model may be in the mathematical optimisation of building designs (stoecker, 1971). However, the correlations between the peak design loads and the climatic variables have not been fully established here, and there is scope for greater analysis in this field. The energy consumption rate should also be analysed by multi-variable regressions. The basic procedure has been established and the fundamental relations have been programmed so that any subsequent analysis could be a simple routine procedure.

The multi-variable linear model as given here may as well be used for approximating the peak cooling loads due to the heat flow through the walls and roof and the windows (excluding infiltration)in the U.K. for the north-oriented buildings.

ABBREVIATIONS

L	Length of the building	m
W	Width of the building	m
H	Height of the building	m
P_G	Percentage of glazing	%
FL	Flat Roof area	m^2
t_a	Ambient air temperature	°C
$\bar{t}_a$	Mean ambient airtemperature over 24 hours.	°C
$t_{a_{max}}$	Peak airtemperature over the day	°C
Sol	Solar radiation on the horizontal	W/m^2
$\overline{Sol}$	Mean solar radiations of 24 hour on the horizontal	W/m^2
Sol_{max}	Peak solar radiation value over the day on the horizontal	W/m^2

REFERENCES

1. Johnston, J. (1963) Econometric Methods 2nd ed., McGraw-Hill Book Co. Inc.

2. Ullah, M.B., (1975) Ph.D. Thesis, UMIST, Manchester.

3. IHVE Guide Book A, 1970.

4. Ullah, M.B. and Longworth, A.L. (1976) 'A Simplified Multiple Harmonic Fourier Method for Calculating periodic Heat Flow through Building Fabrics' The Building Services Engineer, 44, 53-62.

5. Stoecker, W.F. (1971). Design of Thermal Systems, McGraw-Hill Book Co. Inc.

Table 1 Scheme of Variables studied and Coefficients of Multiple Correlations

$$\hat{Y}_i = \sum_{j=1}^{k} \hat{\beta}_j X_{ij} \qquad \text{Where } X_{i1} = 1.0 \text{ and } i = 1, \ldots, n \text{ (number of observations)}$$

Case	Variables									Coefficient of Multiple Correlation. $R_{1.2\,3\,4\,\ldots\,k.}$
	X_1	X_2	X_3	X_4	X_5	X_6	X_7	X_8	X_9	
1.	1.0	L/W	H	PG	FL	-	-	-	-	0.9253
2.	1.0	$\bar{t}_a-20$	$t_a-\bar{t}_{a\,max}$	Sol	$Sol-\overline{Sol}_{max}$	L/W	H	PG	FL	0.8915
3.	1.0	$\bar{t}_a-20$	$t_a-\bar{t}_{a\,max}$	Sol_{max}	$Sol-\overline{Sol}_{max}$	"	"	"	"	0.8915
4.	1.0	t_a	t_a^2	t_a^3	Sol_{max}	"	"	"	"	0.8818
5.	1.0	$\bar{t}_a$	$(t_a-\bar{t}_a)^2$	Sol_{max}	Sol^2_{max}	"	"	"	"	0.8913
6.	1.0	$t_{a\,max}$	t_a^2	Sol_{max}	Sol^2_{max}	"	"	"	"	0.8902

Table 2 Partial Correlations between the Peak Cooling Load and Different Variables

Variable	Partial Correlation $r_{1i.2\,3\,4\,\ldots\,i-1,\,i+1,\,\ldots\ldots\ldots\,k.}$					
	Case 1	Case 2	Case 3	Case 4	Case 5	Case 6
L/W	0.8055	0.7300	0.7300	0.7153	0.7298	0.7280
H	0.9274	0.8528	0.8428	0.8426	0.8527	0.8514
PG	0.7401	0.6784	0.6784	0.6626	0.6781	0.6762
FL	0.8744	0.7525	0.7525	0.7384	0.7523	0.7506
$\bar{t}_a-20$	-	0.6835	0.6835	-	-	-
$t_{a_{max}}-\bar{t}_a$	-	0.2878	0.2878	-	-	-
Sol	-	-0.2410	-	-	-	-
$Sol-Sol_{max}$		0.1830	0.2203	-	-	-
$\bar{t}_a$	-	-	-	-	0.6815	-
$t_{a_{max}}$	-	-	-	-0.0873	-	0.1021
t_a^2	-	-	-	0.1002	-	0.0137
t_a^3	-	-	-	-0.1077	-	-
$(t_{a_{max}}-t_a)^2$	-	-	-	-	0.2326	-
Sol_{max}	-	-	-0.2140	0.1696	0.3075	0.3061
Sol^2_{max}	-	-	-	-	-0.3202	0.2041

Table 3 The Values of the Multiple Linear Regression Coefficients for the Cases Studied

Case	Values of Multiple Regression Coefficients ($\hat{\beta}$)								
	1	2	3	4	5	6	7	8	9
1.	-2.927×10^{5}	2.721×10^{4}	1.203×10^{4}	3.806×10^{3}	2.247×10^{2}	-	-	-	-
2.	-4.941×10^{5}	2.106×10^{4}	4.511×10^{4}	-1.072×10^{3}	8.524×10^{2}	2.307×10^{4}	8.544×10^{3}	3.443×10^{3}	1.537×10^{2}
3.	-4.941×10^{5}	2.106×10^{4}	4.511×10^{4}	-1.072×10^{3}	1.924×10^{3}	2.307×10^{4}	8.544×10^{3}	3.443×10^{3}	1.537×10^{2}
4.	3.096×10^{5}	-1.426×10^{5}	8.660×10^{3}	-1.512×10^{2}	1.065×10^{2}	2.307×10^{4}	8.544×10^{3}	3.443×10^{3}	1.537×10^{2}
5.	-1.015×10^{6}	2.090×10^{4}	3.252×10^{3}	1.357×10^{3}	-1.110×10^{3}	2.307×10^{4}	8.554×10^{3}	3.443×10^{3}	1.537×10^{2}
6.	-1.009×10^{6}	1.788×10^{4}	6.193×10^{3}	1.290×10^{3}	-0.968×10^{3}	2.307×10^{4}	8.544×10^{3}	3.443×10^{3}	1.537×10^{2}

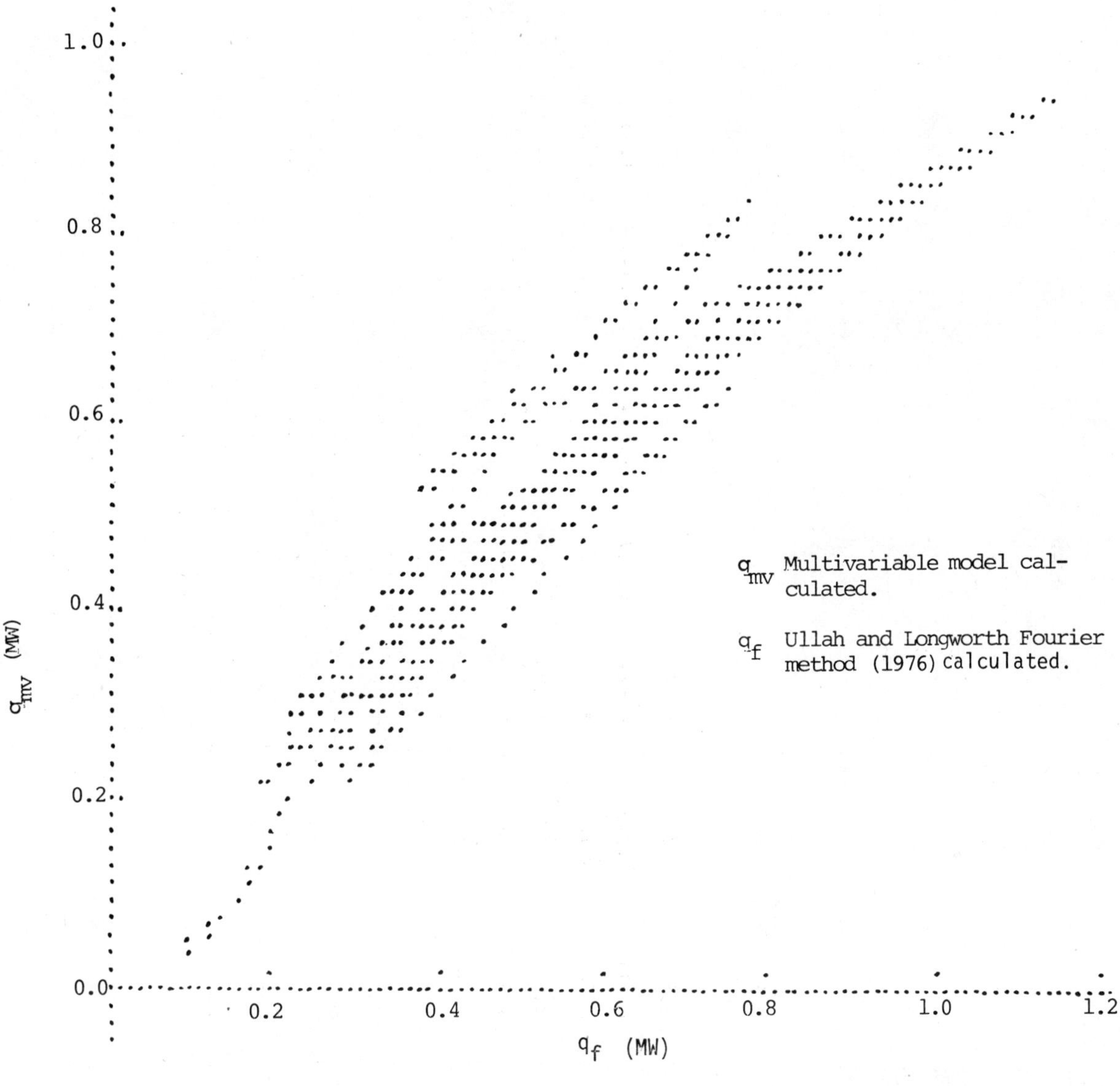

Fig. 1 The calculated values of the peak cooling load by the multivariable model vs. the peak cooling loads calculated by the multiple harmonic Fourier method (Ullah and Longworth, 1976) Independent Variables - the Building Parameters only (Case 1).

ANALYTICAL MODELLING IN SUPPORT OF DESIGN AND MONITORING OF INTEGRATED SURVEYS FOR DEVELOPMENT

W. H. de Man

ITC-Unesco Centre for Integrated Surveys, Enschede, The Netherlands

Abstract. The paper deals with the application of analytical modelling to the design and monitoring of programmes for data and information collection (surveys) for development purposes. Orientation towards policy makers is emphasised.

Keywords. Modelling; development planning; data acquisition; integrated surveys; standards; decision theory.

INTEGRATED SURVEYS AND DEVELOPMENT PLANNING

Both development planning and the actual execution of concrete development projects or implementation of development programmes ask for relevant and timely information. The complexity of investigations or surveys for development has increased considerably during the last decades. Besides physical aspects, economic and social aspects have become an essential part in these studies. Moreover, the orientation of development has gradually changed from a GDP growth strategy to ultimately, a Basic Needs Strategy.

Although there are many different definitions of the term 'integrated surveys', within the scope of this paper it comprises the following:

- all activities related to the formulation of the survey objectives, survey tasks and survey activities;
- the subsequent - often multi-disciplinary - collection, analysis and interpretation of all types of relevant information needed for the (re-) formulation of concrete action programmes (or their alternatives) for development purposes. This includes a diagnosis of the present situation, the identification of potentials, problems and potential alternative solutions;
- monitoring and subsequent evaluation of the implementation of selected concrete action programmes in the light of certain development objectives and targets. (Adapted from SCHULTZE, 1973.)

In the process of surveying, different phases or stages can be identified. These phases differ according to the different phases in development planning or to different degrees in detail required. It should be understood that the driving criterion of splitting up any survey process into phases ultimately is the need for decisionmaking. Each of the decision making phases is preceded by a distinct phase in the survey.

In the process of development planning one may recognise several phases: (SCHAAP et al., 1980)

- pre-planning phase;
- planning phase;
- implementation phase;
- monitoring and evaluation phase.

- The Pre-planning Phase. The main planning function in this phase is to arrive at a diagnosis of the situation in the light of the orientation of development planning (national and sub-national development objectives e.g. local aspirations), the identification of potentials, problems and potential alternative solutions. Some typical acitivities in this connection are the assessment of past trends, resource use and potential, and the appraisal of likely developments (prognosis) with and without alternative sets of possible interventions.

- The Planning Phase. In this phase the planning function is centered on the selection and acceptance by the government authorities of a set of government actions and interventions, the commitment of available means to achieve a certain set of objectives and targets, as well as the allocation of responsibilities and required 'resources' to the various executing agencies for the design and execution of these 'tasks' and 'activities'.

- The Implementation Phase. This phase covers the actual design and implementation of the above mentioned 'tasks'and activities' as well as the procurement of the planned (financial, manpower, etc.) public sector resources required, (quality and time!). An important planning function in this phase is an efficient - short term - adjustment to exogenous disturbances such as fluctuations in the worlds economy (commodity prices) in the yields of agriculture products etc., with a severe impact on the foreign exchange and public finance situation of the country, as well as political disturbances.

- The Monitoring and Evaluation Phase. In the situation of GDP growth oriented planning, monitoring was mainly restricted to yearly sets of national accounts and some selected key indicators. Particularly the national accounts are generally available with a delay of several years, which anyhow raises some questions concerning its function as a device to monitor actual developments versus planned developments. A change in the orientation towards multiple objectives, which e.g. in the basic needs approach are less susceptible to quantification, requires the development of new approaches and methodologies in the field of monitoring. Evaluation usually deals with a critical ex-post analysis of specific 'interventions' often of the programme and project type. Its main function is to serve as 'lessons of experience' and as a feed-back to the next planning cycle.

Often, especially with respect to the pre-planning and the (action) planning phases, several survey phases are recognised according to different degrees of details of the policy questions to be answered:

- pre-survey (fact-finding);
- reconnaissance (project formulation);
- semi-detailed (pre-feasibility);
- detailed (feasibility).

DEVELOPMENT PLANNING AS A SPECIAL CASE OF PROBLEM SOLVING

In general, the ultimate goal in (development) planning is the solution of certain problems. Therefore, development planning is closely related to the process of problem solving. Within the process of problem solving, three main phases can be identified:

- problem finding;
- solution finding;
- solution implementation.

- Problem finding refers to the phase in which the problem is identified and specified. Often, the problem may be understood as the difference between an existing situation and the desired one.

- Solution finding refers to the phase in which, from the various alternatives, the optimal solution to the previously specified problem is sought.

- Solution implementation refers to the phase in which the selected solution is implemented.

Figure 1 shows the various steps and phases within the process of problem solving and its associated information base (or system).

The phase of solution finding is governed by the clarity of difinition or firmness of the problem itself. According to Bosman, human beings meet problems at three different levels:

1. the level of the well-defined problems
2. the first-level of the ill-defined problems
3. the second level of the ill-defined problems (BOSMAN, 1973).

This classification of problems to their degree of definition is very relevant in planning. If the representation of the situation is consistent with the goal and with reality one may obtain a problem specification. In fact, the (three) different levels of structure constitute a continuum. If, on the one extreme, the situation existing in reality as well as the goals are clearly understood and (reasonably) under control, the problem is said to be well defined or well structured and consequently, the solution to the problem may be found in a programmed manner. The other extreme refers to the case in which the situation existing in reality as well as the goals are vague ('fuzzy'). In this case, the problem is said to be ill-defined or ill-structured (of second level) and consequently, only vague approximations of possible solutions can be indicated. In the case of well-structured problems, the type of information that is oriented towards solution finding is essential. In case of ill-structured problems, this type of information does not play an important role (because solution finding is only at an embryonic stage); here emphasis should be on problem finding information. Schematically, this can be visualised in Figure 2.

ANALYTICAL MODELLING AND DEVELOPMENT PLANNING

Analytical modelling techniques have been applied for many years to problem solving in various different fields. During the last decades, (theoretical) development of analytical modelling has grown considerably in such diverse fields as industrial processes, business and management systems, urban, regional and transport planning systems, economic and social planning.

Often, the major objective for the application of analytical modelling techniques has been to arrive at an optimal solution to an envisaged problem. The direct impact of a modelling exercise on the stage of solution finding or, in other words, how far the modelling results will constitute the desired and clear cut optimal solution depends on the degree of definition of the problem under consideration - whether or not one has to deal with problem solving amenable to programming. The application of analytical modelling techniques to the solution of well-defined problems has been rather well developed; their application to the other types of problem solving (to ill-defined problems) is less obvious.

In the context of development planning, analytical modelling methods can serve as an aid to policy and decision makers in analyzing the present situation, and available potentialities (problem finding) and in demonstrating consequences of some specific policy measures or interventions (problem finding and/or solution finding). However, they are never a replacement for the decision maker. Ultimately, policy and decision makers are responsible for the implementation of policies for development. Therefore, analytical modelling techniques should be used in close co-operation with the responsible parties or at least oriented towards them. In most cases this requires the application of simple methods: the way in which the results are obtained should be clear to the user. This is the only way to appreciate their relevance and to ensure their real acceptance by the decision makers.

Many analytical methods make use of sophisticated numerical procedures which can hardly be applied in situations where sufficient and reliable quantitative data are missing. It should be noted that in developing countries in particular, sufficient and reliable statistical data are often lacking. These considerations lead to the conslusion that the expectations one may have from the application of analytical modelling techniques to development (planning) problems are often limited to a global but coherent description of the main components of the (sub-) system under study.

ANALYTICAL MODELLING AND INTEGRATED SURVEYS

In cases where information programmes (survey) were related to the application of analytical modelling techniques, and vice versa, often such an excercise was set-up in two rather isolated stages: first the survey as a basis for subsequent application of analytical modelling. The underlying expectation in these cases was to arrive at an optimal or at least a set of alternative solutions to the envisaged development problem.

Recently, some work has been done in applying analytical modelling in the design and subsequent monitoring of integrated surveys. In this manner, the model provides a framework in which most relevant and quantifiable data will find their own appropriate place. Data specifications to be used in the current stage of the survey can easily be obtained with help of the model. For instance, sensitivity analysis can be used in determining which data must be collected with more precision or accuracy and which data can be estimated by order of magnitude only. The final performance of the model is not the sole and ultimate goal of the modelling exercise and as such, the meaning of 'optimum' (or 'optimal result') is less clear than in a situation in which analytical modelling has been applied in a straight foreward manner to well structured problems.

Also during the stage of data collection, the model can be very useful. In general, an analytical modelling approach involves the conversion of raw data into standard formats and tables. This both simplifies the 'dressing' of the model (formulation of details and assigning values to parameters and constants) and aids in the performance of various consistency checks, detection of errors, etc. Even at this stage, without a full-fledged model, sufficient data will often be available to make a few rough 'back-of-the-envelope' calculations in order to obtain some insight into the present situation, expressed in order of magnitude only. For instance, the area of various main crops under cultivation, a rough estimation of gaps between labour supply and labour demand. This insight may effect not only the (re-)design of the model but also the course of the current phase of the survey.

If modelling is applied to a (particular phase of the) survey, this application will effect not only data collection, but also facilitates analysis of the present situation and assessment of potentials, presentation of survey results and (if applicable) demonstration of the consequences of certain policy interventions. This has been described in detail elswhere. (DE MAN, 1978)

As said before, the driving criterion of splitting up the survey process into phases ultimately is the needs for decision making. This means that each phase in the survey process is oriented towards a specific set of policy decisions and questions and consequently, the supporting modelling has to do so. Before starting to establish the model, one should at least be aware of the type of questions to be answered by its use. What is the purpose of the excercise to be undertaken? Obviously there is no point in designing a model which may be a perfect description but is of no value for further analysis. These questions should be related - but not necessarily identical - to the set of policy questions and decisions as the particular phase of the survey is oriented towards.

The policy and decision questions to be used as guidance for designing the subsequent phase of the survey - viz. formulating survey objectives, survey tasks and survey activities - are formulated partly on the basis of the survey results in a particular phase of the survey.

Speaking in general terms, the various phases in a survey, run parallel in time with the phases in problem solving (viz. problem finding, solution finding and implementation of selected solution). Therefore, the survey will be problem finding oriented in its initial stage and gradually changing its orientation towards solution finding and solution implementation. Consequently, if modelling is applied, the orientation of the model(s) will differ along the different phases as well (viz. problem finding, etc.).

In some instances it may be possible to apply analytical modelling techniques fruitfully in all different phases of the survey. In other situation they could be applied to a few phases only. In cases where modelling could be applied to more than one phase, an efficient approach should be considered with emphasis on adaptation of the model for its use in subsequent phases of the survey instead of starting each time from the very beginning. Such an approach, for instance, facilitates linking current data collection programmes to former ones. However, each time that a decision has to be made whether or not to adapt an existing model this decision should be made very carefully. The danger exists that by adapting old models for future use, old dated, not relevant or even wrong concepts are copied uncritically.

With respect to the different phases in problem solving, models supporting a survey in the problem finding phase are in general of a broad and comprehensive type whereas models used in the solution implementation phase show a rather specific and detailed character. In the problem finding phase there may be a scope for one rather simple model, e.g. a linear programming model, providing a global but coherent descrition of the main components of the system under study. Typically here, the problem is said to be ill-defined and the survey objective is to provide a kind of a diagnosis of the present situation. In the solution implementation phase, the problem is known and thus well-defined. For reasons of efficiency and simplicity a set of independent but conceptually interrelated models is prefered over one huge, detailed and complicated model. Here, adaptation of of the model may lead to gradually splitting-up the overall model into several sub-models. Still, data formats and concepts should remain identical wherever it is possible.

CONCLUSIONS

On the strength of the observations made in the previous sections it is possible to arrive at some conclusions.

Analytical modelling techniques can make a positive contribution to design and monitoring of integrated surveys.

The modelling should be designed to help solve the specific (type of) questions which are to be asked and which are related to the policy and decision questions underlying the survey phase as a whole.

The model should be understandable to the user.

The model should permit if possible easy adaptation for future use in a subsequent phase of the survey. These considerations may lead to a plea for simple models; certainly not for over-simplification.

Applied modelling is not an isolated excercise. It should be carried out in teamwork and be user oriented. The experts in modelling should be concerned about future applicability of their models by the users.

REFERENCES

Bosman, A, 1973. Ill-structured problems and information systems. Paper delivered at the Seminar on Strategic Problem Solving: Approaches to planning in an ill-structured environment, University of Groningen, the Netherlands (mimeographed)

Daenzer, W F, 1976. Systems Engineering, Leitfaden zur methodischen Durchführung umfangreicher Planungsvorhaben. Peter Hanstein Verlag GmbH, Köin/Verlag Industrielle Organisation, Zürich

Man, W H de, 1978. Linear Programming as a Supporting Technique in Integrated Surveys. ITC Journal 1978-4, Enschede. 573-594

Schaap, A D J and W H de Man, 1980. Monograph on Information Requirements for Development Planning in Developing Countries. ITC, Enschede (mimeographed)

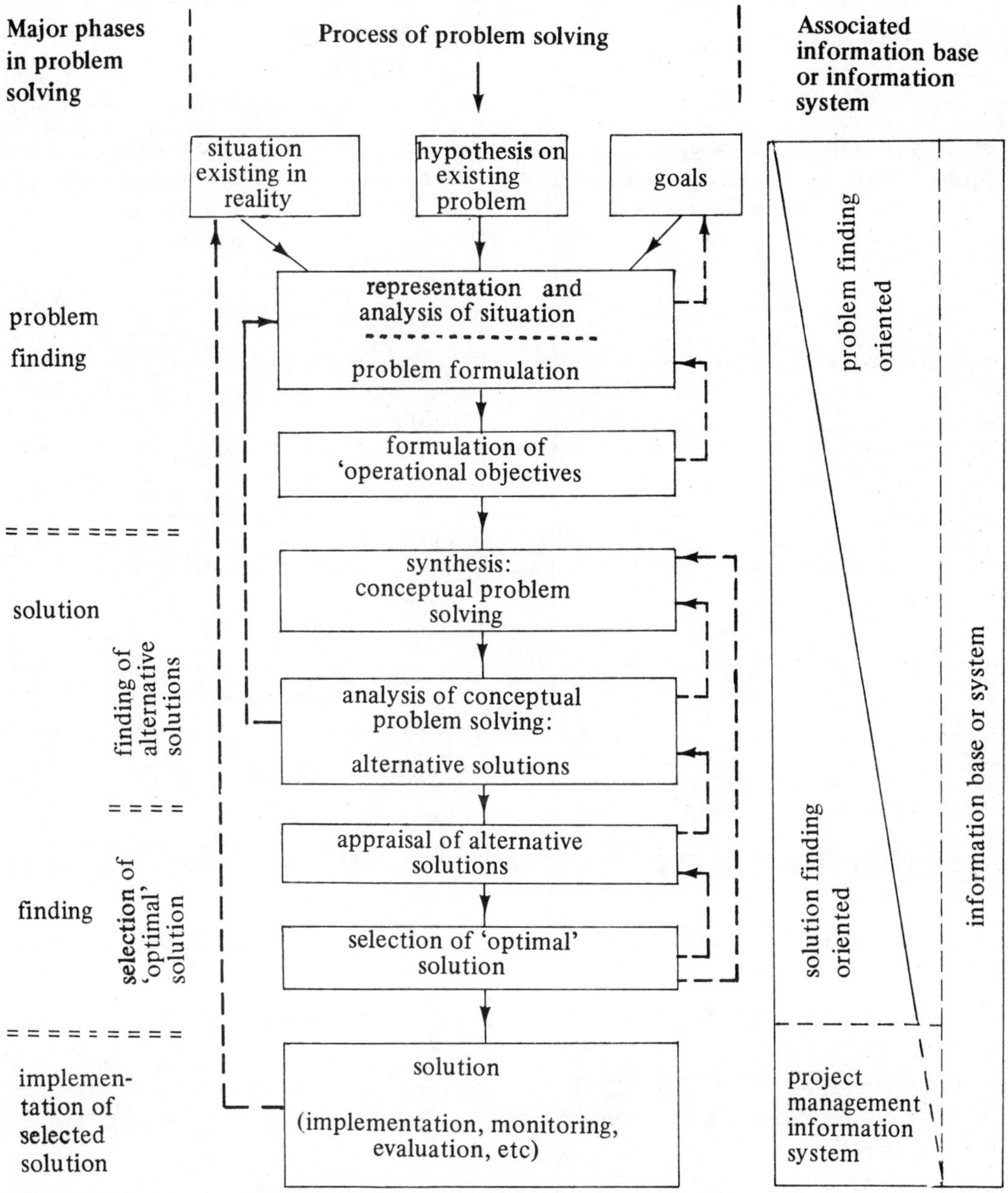

Fig. 1. Steps and phases within the process of problem solving and its associated information base or system. (Adapted from DAENZER (1978) and BOSMAN (1973)

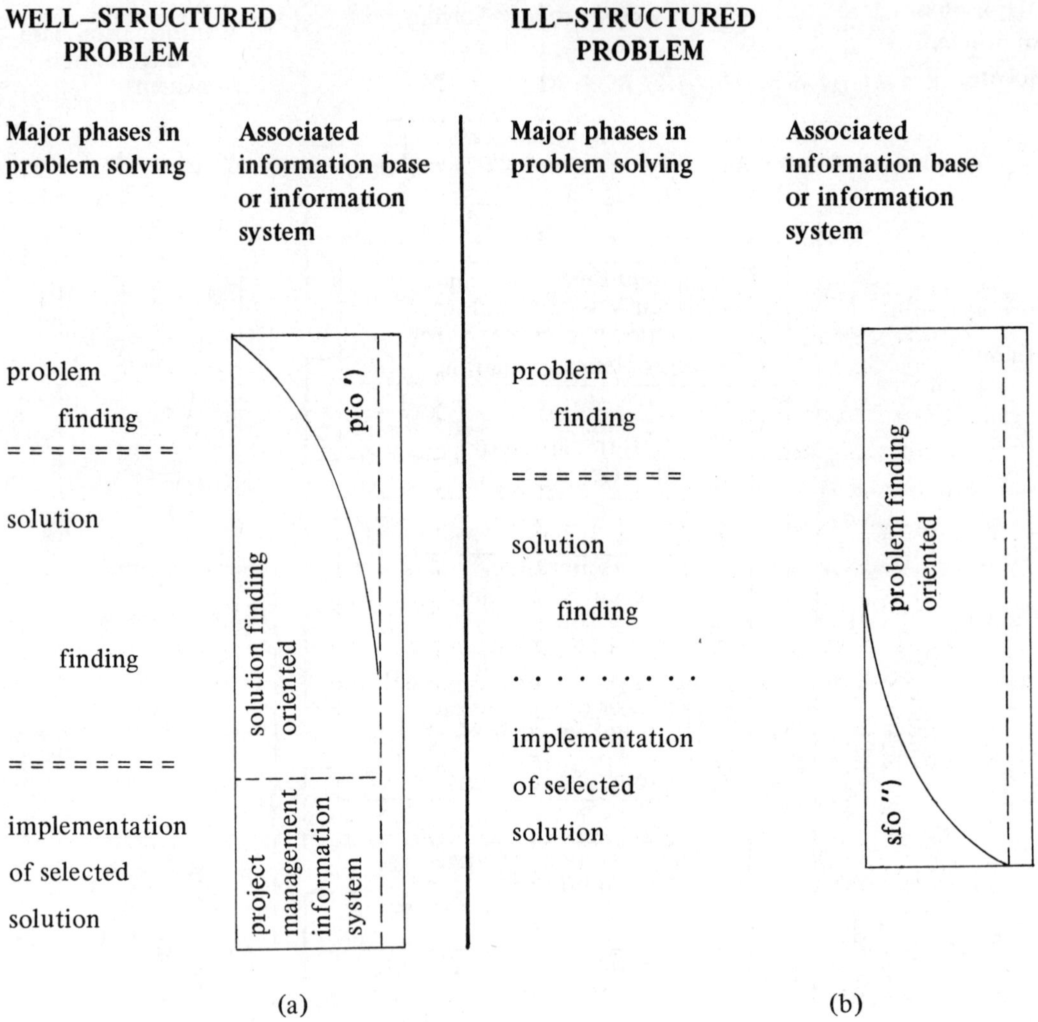

') pfo = problem finding oriented

'') sfo = solution finding oriented

Fig. 2. Phases within the process of problem solving and its associated information base or system in case of well-structed problems (a) and in case of ill-structured problems (b)

AN AUTOMATIC SYSTEM FOR GEOLOGICAL PROSPECTING DRILLING

A. G. Yarusov, G. A. Butkin, E. N. Efimov, V. A. Mazur and N. N. Novik

Institute of Technical Cybernetics, BSSR Academy of Sciences, Minsk, USSR

Abstract. A system for collection and processing of geological information in the process of oil and gas well drilling is introduced. The system is realized on the basis of control computer complex M - 6000 and provides data transfer units analysis and optimum control of drilling technology automatically or by the operator. Software includes control programs and the subsystems of:a) communication with an object under control and with the operator; b) information storage; c) optimum control; d) forecasting and preventing failures. Sampling of the drilling machine as an information source and as an object under control for hardware and software debugging and system's testing before its connection to the drilling machine are discussed.

Keywords. Computer control; process control; optimisation; modelling; real - time system; alarm systems.

INTRODACTION

Well drilling is one of the main methods for investigating rock structure and state and for mineral prospecting. The usage of computer provides an opportunity for effective extraction of information from data transfer units of drilling machine and from devices designed for the analysis of fluids and rocks appearing from the well together with the drilling liquid. Unlike other measuring methods, i.e. those based upon special devices placed into the well the method is realized in the continuous drilling process, which reduces the cost of the information obtained. Computer - aided solving the problems of drilling optimisation and preventing possible failure provides further well drilling time and cost reduction.

An automatic system for geological prospecting drilling solves the next problems.

1. Collection and analysis of processing of geological and geophysical information obtained from technological data transfer units and devices. This information is used both for optimum real - time control of drilling process and for geological prospecting enterprises data banks.

2. Optimum control of the drilling process to increase the technical and economical indexes. The accumulation of optimum technology information is of great value for subsequent well drilling in the region both with the help of computer and without it.

3. Forecasting and preventing failure of the well, drilling equipment and control system.

4. Preparing and documenting reports on technical and economical indexes of the drilling machine operation, requirement for drilling tools and materials, etc. This information is used by drilling enterprise for planning and control of the drilling works.

THE SYSTEM HARDWARE

The drilling machine is an object under control and geological information source. It is equipped with unified technological data transfer units having analog and digital outputs, local automatic control systems (ACS) and manual remote console.

The system is realized on the basis of control computer complex M - 6000 consisting of central processor M-6000, core storage (CS), direct access channel, disk storage, object - computer linking units (for input the information from data transfer units and output of controlling signals to the ACS), alpha - numerical display, punched tape input and output device and printer. Besides, the hardware includes plotter for graphic information display and the electronic simulator of drilling machine as the information source and object under control.

The system is able to control the object in two modes: 1) automatic output of controlling signal to ACS; 2) output of advises and messages to the display (operator's adviser mode).

The system is housed in a van, provided with all life conditions, permitting to work in field environment.

THE SYSTEM SOFTWARE

Software consists of control programs and set of problem program modules M functionally and informatically combined in integrated complex, designed to solve tasks faced by the system.

Control programs

The effective distribution of computing resources in multiprogramming mode is carried out by disk operating real - time system (DOS RT). Program $m_i \in M$ can be initiated in DOS RT environment:

- automatically by DOS RT dispatcher with given phase and period of running;
- by request of another program module m_j ;
- by request of initiating signal from controlled object;
- by request of the system console.

The operating system solves the tasks by queue serving for the central processor with head-of-the-line and preemptive resume disciplines. The high priority program can interrupt the low priority program run (swopping-mode) whenever preemptive discipline has been realized; the interrupted state of program is reserved in the special disk tracks (swopping-tracks). Besides, the operating system serves queues for input-output devices and loads central processor effectively in the mode of external devices interruption.

To orgonize running of a great number of interconnected programs and control a complex dynamic object, it is necessary to schedule additional resources:

- problem program m_i (the request from program m_j for usage of program m_i can appear before the m_i completes the serving of the request from the program m_k);

- object controlling parameters $\vec{U}$ (program modules can cause the "competition" mode for one of the controlling parameter u_i);

- dialog area of display (program m_i can require output of the message to the area before the operator answers the previous message received from the program m_j).

These resources are scheduled by the program "system scheduler" (SS). Any problem program m_i writes the request into the SS buffer whenever it is necessary to use any of the above resources. SS is a core-resident program, initiated periodically with a constant short period. Whenever SS is initiated it examines its buffer contents and schedules queues for resources. The period T_{ss} of starting SS must be such that at muximum intensity of arrivals the SS buffer should not be overflowed in the interval between two SS starts (scheduling without loss). Besides, the period of SS start is a part of the total system response time, determined in a maximum as the sum: $T_r = T_{in} + T_{ss} + T_s + T_{out}$, where T_{in} - time required for information input; T_{ss} - SS start period; T_s - serving time; T_{out} - time required for information output.

The following types of request can be generated by the problem program m_i :

- request for initiating program m_j ;

- request for output of the controlling signal u_i to the object;

- request for output of question or advise onto the screen of display.

Besides, the module m_i generates a message upon completing module's run (releasing m_i as a shared resource) and writes thit message in the SS buffer.

The scheduling of the requests for problem programs is realized in head-of-the-line priority discipline. The preemptive resume priority discipline is chosen for scheduling of the requests for the controlling parameters and dialog area of display. The choice of discipline is defined by time characteristic of system and by service time and memory load restrictions.

To regulate the system structure accordingly to the system tasks the set of programs M is subdevided into five problem orientated subsystems. Figure 1 shows the software organisation. Functional relation and data traffic are shown in Fig. 1. The common area in the core memory (COMMON) is designed to provide semultaneous data exchange between core-resident and disk-resident programs. The area is subdevided into the following blocks:

- program reference data block;

- the block of the main controlled process' parameters;

- the block of buffers.

While developing the system the problem of flexible organisation of technological stages completion was solved. If we assume Z to be a set of technological tasks (stages) solved by the set of problem program M, then the system must give the opportunity to form subsets $M' \subset M$, solving the necessary subsets of technological operations by the operator's instructions. Subset of tasks Z' is defined on the input of the system by a set of the text technological instructions $T' \subset T$ (T - the technological instruction vocabulary). The conversion $T' - Z' - M'$ is realized by the special program "technological dispatcher" (TD). Program TD recieves the texts T' from the operator's console accordingly with subset of tasks Z' and forms the sequence of problem program's initiating. The stage Z' can be interrupted by the programs of alarm or by the operator interference. Such a violation of the regular schedule of the technological process means transition to another subset of tasks Z'_j and must be accompanied by the reservation of the interrupted state of Z'_i in order to have an opportunity to return to subset Z'_i after the appearance of favarable conditions. The functions of switching modes are carried out by TD.

Process Optimum Control

The drilling process is the deepening of the well in rock by means of a special rotary tool (roller bit). The controlling parameter are axial load G and rotation speed of bit N. In the process of deepening the roller bit teeth and roller bit legs are subjected to wear.

Assuming drilling fluid circulation to be constant, the drilling process can be represented in the phase coordinates space (deepening - bit teeth wear - bit legs wear) and described by differencial equations system [Chefranov, 1972], which is the function of controlling parameters and drilling time t:

$$\frac{dH}{dt} = V_0(G,N)\frac{dI_t(G,N,t)}{dt}$$

$$\frac{dI_t}{dt} = EXP[-\theta(G,N)\cdot t], \quad (1)$$

$$\frac{dI_l}{dt} = \frac{G^{\alpha}\cdot N^{\beta}}{B},$$

where H, I_t, I_l is phase co-ordinate of the drilling process (deepening, bit teeth wear, bit legs wear); $V_o(G,N)$ is initial mechanical speed of deepening; $\theta = \theta_0 \left(\frac{G}{G_0}\right)^{\varepsilon} \cdot \frac{N}{N_0}$ - the teeth wear index; θ_0 - the teeth wear coefficient, controlling parameters being G_o and N_o; ε - index ($\varepsilon > 1$) defining the load influence on the teeth wear index; α, β - indexes which define the axial load and rotation speed influence on the bit legs wear. B - bit legs constant.

The following restrictions are imposed on phase coordinates I_t and I_l:

$$I_t(G,N,t) \leqslant \lambda_1, \quad (2)$$
$$I_l(G,N,t) \leqslant \lambda_2.$$

The violation of these restrictions reduces the effectiveness of process control and can cause failure. λ_1, λ_2 are limited values of bit teeth and legs wear.

The controlling parameter area is also restricted (Fig. 2) by technical abilities of the drilling equipment and requirements of accident - free drilling process.

$$G_{min} \leqslant G \leqslant G_{max};$$

$$N_{min} \leqslant N \leqslant N_{vl}; \; N_{vu} \leqslant N \leqslant N_{max}; \quad (3)$$

$$M(G,N) \leqslant M_{tol},$$

where N_{vl}, N_{vu} are lower and upper values of rotation speed range, within which intense drilling equipment vibrations occur ; $M(G,N)$, M_{tol} are rotary table torque and its tolerated value, respectively.

The optimisation criterion is the economical index of the expences (one meter deepening cost)

$$Q = \frac{C_1(T_{rp}+t)+C_2}{H(G,N,t)} , \qquad (4)$$

where C_1 is per hour cost of drilling machine's running; C_2 is the cost of the bit; T_{rp} is pulling and running the drilling string time, required to change the worn-out bit by the new one.

Mathematically the optimisation task is to determine the controlling parameters in restricted area (3) and drilling time, which correspond to the system differential relations (1), restrictions (2) and minimize the functional (4). The optimisation and drilling process controll difficulties are connected with non-observation of the process' phase co-ordinates and stochastic nature of functional (4) which is the result of stochastic bit-rock interaction. This made us develop a dialog system of optimum drilling process control, in which heuristic elements are used along with formal procedures. The optimum control is caried out using observated differencial characteristic of the deepening phase co-ordinates, i.e. mechanical deepening rate. The parameters of the big teeth wear phase co-ordinates are defined by stochastic data and are verified periodically in the process of drilling.

The operator can use two approaches for optimum control. The first one is based on searching procedures of stochastic approximation type directly in the controlling process. In the system Kiefer - Wolfowitz method modified by Fabian [Vasan, 1972] is used. In dialog mode the operator gives the initial search point in tolerable controlling parameter area, values of test and run steps, and the criterion of search stop. The traces of searching are displayed on the screen. The operator is able to output on the screen the value of criterion at any step of searching. He can interrupt search procedure and restart it with new parameters. The trace "a" of minimum criterion searching in the controlling parameter area is shown on the Figure 2.

The second approach is based on the building of regressive mechanical deepening speed model and subsequent optimisation task solving in determined form. The initial mechanical speed dependence upon controlling parameters is represented as polynomial structure:

$$V_0(G,N) = A_0 + A_1G + A_2N + A_3G^2 + A_4N^2 + A_5GN ,$$

of which $A_0 \div A_5$ are evaluted by the results of active experiment carried out by dialog approximation system. The $V_0(G,N)$ model helps to determine the controlling parameters both from constant area and from the area of time functions, which allows to increase effectiveness of control under the condition of non-stationary bit legs wear.

By using constant control function with equal entervals of constancy system (1) can be transformed

$$H(G_1,N_1,\dots,G_nN_n,\Delta t)=\sum_{i=1}^{n}V_{oi}(G_i,N_i)\cdot I_{ti}(G_i,N_i,\Delta t),$$

$$I_t(G_1,N_1,\dots,G_n,N_n,\Delta t)=\sum_{i=1}^{n}\frac{EXP[-\sum_{j=1}^{i-1}\theta_j(G_j,N_j)\Delta t]}{\theta_i(G_i,N_i)}\cdot \{1-EXP[-\theta_i(G_i,N_i)\Delta t]\}, \quad (5)$$

$$I_\ell(G_1,N_1,\dots,G_n,N_n,\Delta t)=\frac{\sum_{i=1}^{n}G_i^{\alpha}\cdot N_i^{\beta}\cdot\Delta t}{B}.$$

In this case the optimisation task is to determine constancy level values ($G_1, N_1, \dots, N_n, G_n$) of controlling parameters and intervals Δt, satisfying the (2), (3) and minimizing the optimisation criterion

$$Q(G_1,N_1,\dots,G_n,N_n,\Delta t)=\frac{C_1(T_{rp}+n\,\Delta t)+C_2}{H(G_1,N_1,\dots,G_n,N_n,\Delta t)}\rightarrow\min_{G_1,N_1,\dots,G_n,N_n,\Delta t} \quad (6)$$

The numerical method of descent procedure in the controlling parameters area was used as the method of optimisation. The number of intervals n was chosen considering the compromise between the reduction of criterion (6) value and increase in the difficulty of solving the task as n increases.

Table 1 constains the optimisation criterion values obtained in the process of solving the optimisation task (5), (6) with different n. Figure 3 shows charts of linear-constant functions of controlling parameters with $n = 5$.

The optimum control is carried out in dialog mode. The operator can set the controlling parameter determined by the first or second method and estimate the degree of model-object adequancy during the drilling process. It is assumed that the model corresponds to the real process to required degree of accuracy if the difference between the average mechanical deepening speed and its model does't exceed the given value

$$|\hat{V}_M(G,N,\bar{t})-V_o(G,N)\cdot EXP[-\theta(G,N)\cdot\bar{t}]|\leqslant\delta.$$

In case of inadequacy the operator makes an alternative decision: to carry out a new optimisation with nonzero initial values of phase co-ordinates on the basis of the corrected models, or to continue drilling with parameters chosen by expierence.

Failure Preventing

The problems of the automatic control system's reliability, accident-free work of staff and the reduction of loss due to failures are an essential part of the system's software.

The information got from data transfer units is preprocessed. It is scaled, filtered and stochastic characteristic are calculated. The system communicates with external environment through programs initiated periogically with individual period defined by frequency characteristics of signals and by responce time required by their deviations. As a result in core memory COMMON information vector $\vec{X}(t)$, describing system state at any moment of time is formed. However to identify set S of states and to start different decision making programs this information is redundant. Redundancy is eliminated by comparing co-ordinates of vector $\vec{X}(t)$ with adaptive threshold levels and forming operational state vector. The values of the threshold levels are defined in the system by calculation according to data obtained in special measuring modes or accumulated during previous drilling race.

Set S can be subdevided into two non-crossed subsets: S' - the class of correct states, S'' - the class of failures. Every element of class S'' is described by the set of features W'', defined by the vector of potentially dangerous co-ordinates $\vec{X}''(t)$. Depending upon the demage degree caused by failure, subset S'' is subdevided into levels of potential danger defining time resources and priorities in the system.

The majority of system failures develops in time and space. The evolution model of failure has three main stages: 1)complications, 2)alarm, 3)failure. The failure preventing subsystem involves all these stages. In the first stage the subsystem carries out forecasting and recognition of the system's state and prevents failure by affecting the factors which cause damages.

These factors may be of three types:

- natural (gas and oil flinging out, the drilling mud swallowing);
- technical (the breaks of computer, main and auxilary electronic equipment, data transfer units, communication lines, etc.);
- organizational (personnel errors).

In the second stage the second order protection means are involved to prevent failures. It is based on the usage of potencially dangerous coordinate vector $\vec{X}''(t)$.

There are the third order protection means which act when failure is developing and are designed to eliminate the failure, cut its consequences and warn the staff about the danger.

The failure preventing tasks in the process of deepening are in parallel with the subsystem of optimum control and restrict the controlling parameter scope. The information about complications, alarm and failure is delivered to the operator by the display. The field of the screen shows the types of complications and failures, the damages causing factors, advise to the operator and the state vector of the object under control at any given moment of the drilling process. The operator is able to input data and instructions through the console. The charts of some parameters variations and composite characteristic of the process can be also observed on the screen.

Processing, Storing and Documenting Information

The processing and analysis of initial information of the drilling process residing in COMMON allow us to form an idea about the technological modes' feature and rock characteristics. Besides, this information is a base for preparing the report about technical and economical indexes of the drilling machine. The process of the initial information vector $\vec{X}(t)$ is calculating the elements of vector $\vec{D}$ defining the main technological drilling parameters and vector $\vec{E}$ which can be used for geological analysis of rocks. Vector $\vec{D}$, for example, includes one meter deepening cost, bit teeth wear coefficient, average value of load, etc. The vector $\vec{E}$ includes such parameters as normalized mechanical speed, percentage of gas in the drilling mud, etc. The calculation of vector $\vec{D}$ and $\vec{E}$ is performed whenever $\Delta H =$ 0.1 m deepening has been made, values $\vec{D}(H), \vec{E}(H)$ being stored on disk sectors which can accomodate information about 5000 m deepening.

The software allows us in the process of drilling or after its completion to print out, display or plot data about $\vec{D}(H)$ and / or $\vec{E}(H)$ within given well depths ΔH, $0 \leqslant \Delta H \leqslant H_{max}$. The data of $\vec{D}(H)$ and $\vec{E}(H)$ are prepared for transmission by the communication line to the geological enterprice's computer to analyse the information about the group of the wells.

OBJECT SIMULATION AS A MEANS TO DEBUG AND TEST THE SYSTEM

The important stages of the system elaboration were composite debugging and testing of hardware and software. Prior to linking the complex system to the object under control (drilling machine) it is necessary to debug the software, to check the times of responce to the object state changes and operator's instructions, to estimate the tolerable levels of noise and input sygnals distortions, the resources queues, etc. This complex and time-consuming work is carried out with the help of the object model providing proper simulation of all the modes of the object operation in real environments. The realized model was an indispensable means of debugging and testing the failure preventing subsystem since the debugging of this subsystem with the drilling machine is impossible because the creation of the real failure is impermissable.

The imitation model of the drilling machine as information source and object under control could be realized by software, hardware or combined means. The main disadvantage of software simulation on the controlling computer is waste of computing resources (central processor, core memory), which leads to destortion of responce times and other dynamic characteristic. The most reasonable technique for simulation would be a software-hardware combination, where the electronic blocks generate stochastic process, realize the simple functions and contain controls; the programs of controlling computer perform calculations of complex functions.

The offered model [Yarusov, 1978] simulates signals of 25 data transfer analog and discrete units of the drilling machine under different conditions such as the type of rocks, well depth, normal and failure modes of the drilling machine operation.

In the first stage of the system's debugging the model simulated different signal of the object checked the operation of the software modules, defined the measuring track's precision, specified the scaling and filtering coefficients. The composite debugging includes the simulating in all the modes of the drilling process, evaluation of the responce times, estimation of central processor, core memory and disk filling coefficients.

Besides debugging and testing the model was succesfully used as a means for operator training and in the process of drilling for preventing checking of the system and localization of its break.

CONCLUSION

The organization of system's software and hardware, the methods of the information analysis, optimum control, forecasting and preventing failure and computing process organization supply a solution to all the tasks of an automatic system for geological prospecting drilling.

The optimum control of the drilling process is realized on the basis of a dialog approach using heuristic elements along with the formalized procedures.

The forcasting and preventing of failures are based upon a three - stage approach in accordance with the stages of failure developing.

The usage of the system resources scheduling means, methods of information and operating interaction of programs provided acceptable times of responce to input from the object and operator (TABLE 2). The filling of the main computing resources is represented by TABLE 3.

REFERENCES

Vasan, M.T. (1972). Stochastic approximation. "Mir", Moscow. pp.50-75.

Chefranov, K.A. (1972). The control of drilling. "Nedra", Moscow. pp. 7 - 47.

Yarusov, A.G., V.A. Mazur, E.N. Efimov, E.A. Bogaev. The simulation of object under control as the means for debugging of the automatic system for geological prospecting drilling. The theory and methods of mathematical simulation. "Nauka", Moscow, (1978).

TABLE 1 Optimization Criterion Value versus Number of Constancy Intervals

n	1	2	3	4	5	6
$Q(n)$	25.4	24.3	23.9	23.6	23.5	23.47

TABLE 2 The Responce Times of the System

	Average responce time (sec)	Responce time dispertion
The time of responce to signals from the drilling machine	1.12	0.594
The time of responce to signals from the operator's console	1.53	0.391

TABLE 3 Coefficients of the Resources Filling

Resource	Average value of coefficient	Muximum value of coefficient
Central processor	0,83	0,98
Direct access channel	0,54	0,65
Core memory	0,96	0,98
Disk memory	0.62	0.81

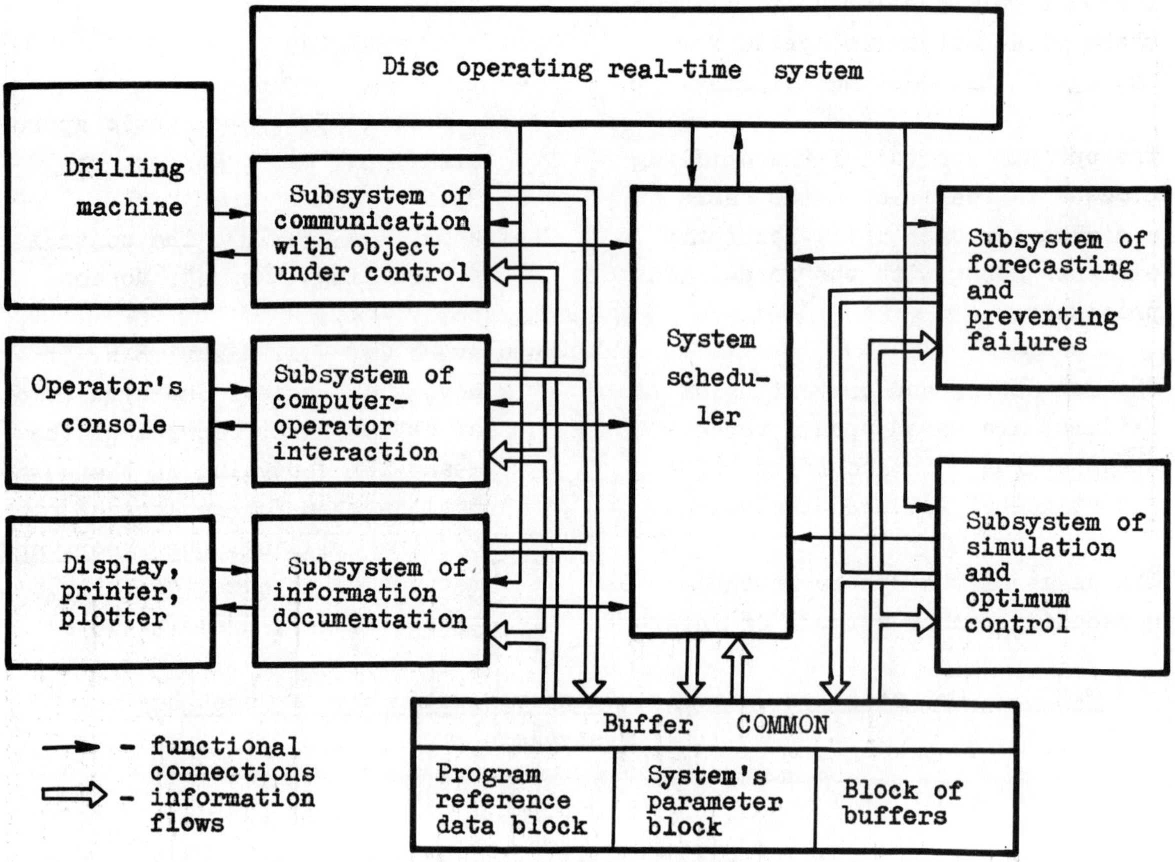

Fig. 1. Software organization.

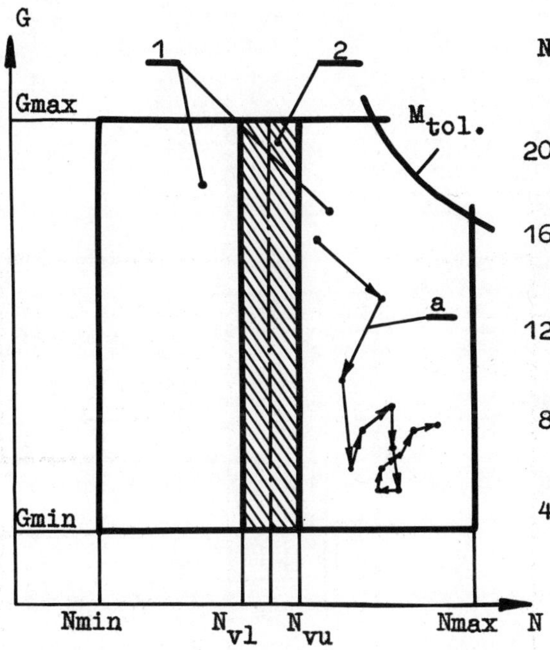

Fig. 2. Controlling parameter area.

1 - area of tolerable control;

2 - rotation speed area restricted due to vibrations.

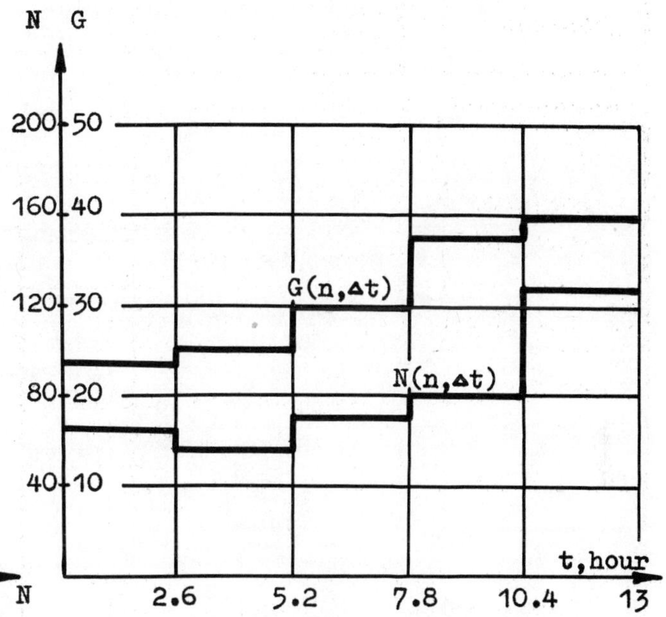

Fig. 3. Representation of controlling parameters in the form of piece-continuous time function.

AQUIFER DYNAMICS IDENTIFICATION IN NATURAL GAS STORAGES

G. L. Chierici, G. A. Gottardi and R. P. Guidorzi

University of Bologna, Bologna, Italy

Abstract - The storage of natural gas in underground reservoirs plays an economic role of increasing importance. The management of such storages requires the knowledge of accurate previsional models which are usually obtained by means of classical modelling techniques. These techniques, however, usually lead to trial-and-error procedures. This paper describes the results obtained in the construction of previsional models for water-drive natural gas reservoirs by means of identification techniques, i.e. on the basis of a direct elaboration of data regarding the reservoir past history.

Keywords - Identification; linear systems; models; multivariable systems; natural gas technology; natural resources; prediction; production control.

I. INTRODUCTION

The economic importance of storing gas in natural underground reservoirs, with the goal of its saving for a certain period of time and simplifying distribution, is well-known. The first applications of this technique can be found in Canada (1915) and in the United States (1916).

Gas storage can be carried out in depleted gas reservoirs, in aquifers, in oil reservoirs with gas cap and also in depleted oil reservoirs, in salt cavities, in worked-out mines and in caverns induced by underground nuclear explosions. The main reason underlying the development of gas storages is that this technique enables gas companies to fulfill demand for maximum delivery rates on hourly and daily bases during the coldest periods of the year and at the lowest possible cost. To cope with the high peak demand only two alternatives are possible: the first is to use natural tanks (manufactured ones would be far too expensive), usually depleted gas reservoirs, located near the zones of highest consumption. The second alternative is to connect, with large pipelines, the areas of greatest consumption with those of greatest concentration of production, in order to enable the transport of necessary peak quantities to fulfill the market demand. Obviously this solution leads to very low pipeline load factors so that the first solution is more advantageous both from a technical and an economic point of view. The presence of underground gas storages as close as possible to the areas of greatest consumption permits to the transportation system to work always at a high load factor since even in the summer, when there is less demand, the gas is transported at high rates, to be injected into the storage reservoirs. This is very important for the proper exploitation of reservoirs (and particularly of offshore fields) and also to avoid penalties in import contracts. Moreover, in case of unforeseen emergencies reducing the flow of import gas or of gas coming from reservoirs located in zones far from consumption areas, underground gas storages can play a very important strategic role. A similar situation is met when natural gas must be treated in complex and expensive plants before being transported by the pipeline system. In this case underground storage fields near the plants can avoid gas supply interruption in the event of partial or total shutdown in the plants. The technical problems involved by underground gas storage are similar to those regarding gas production. The geological factors of greatest interest are: the shape and size of the reservoir, the extension and characteristics of the aquifer, the gas-water contact and the characteristics of the hydrocarbon bearing and cap rock |1|.

This research has been supported by M.P.I. (Ministero della Pubblica Istruzione), Rome, Italy.

Underground gas storages are generally divided into three categories: conventional storages, semi-conventional storages and special storages. The first category includes storages made in depleted or nearly depleted gas reservoirs. The second category regards storages carried out in depleted oil reservoirs and aquifers while the third category comprises storages in salt cavities, worked-out mines and traps formed by nuclear explosions.

When a storage is planned in a partially or totally depleted gas field the basic factor qualifying the reservoir, determining its storage capacity and relative pressure, is the production mechanism governing the reservoir during production life. Production of a gas reservoir can result from expansion and from partial or strong water drive. Conventional underground storages, as the ones considered in this work, are usually carried out in reservoirs which have produced because of expansion and partial water drive. Storages which are made in reservoirs having a strong water drive can be compared to those carried out in aquifers.

When a mathematical model to describe the dynamical behavior of a gas reservoir is considered, it is first necessary to perform the process orientation, i.e. to define its inputs and outputs. The exchanges between a reservoir and the remaining of the world are performed through its wells, and the measurable attributes at every well are given by its gas production and its pressure. It is therefore possible to consider the cumulative gas production at the different wells as inputs and the well pressures as outputs of the system. The available data regarding the production of a given reservoir concern usually only the total production and a sort of mean pressure; a model compatible with these measures will have a single input and a single output. Moreover it can be considered that, also if the physical nature of the process is continuous, a discrete-time model is more practical because of the measurements sampling and of the possibility of its implementation on a numerical computer. The link between the reservoir cumulative gas production and its pressure can be considered as linear in absence of a water-drive mechanism. When an aquifer is present, however, the behavior is non-linear, so that a linear approximation can give good results only in a limited neighbourhood of a specific operating condition. If linear models are used to describe the reservoir dynamics, the process looks therefore as time-dependent. This problem could be overcome by using also a measure of the entered water as input of the model; it can, however, be noted that direct measures of the volume of entered water are not available since the wells are drilled where hydrocarbons, not water, can be expected. These considerations are less relevant for reservoirs used for gas storage purposes since, in this case, they are subjected to periodic injection/production schedules causing a periodic pattern also in the volume of the entered water. In fact, time-invariant linear models give accurate predictions for gas reservoirs used as storages also in the presence of strong water-drive mechanisms |2|. The traditional approach to the construction of a mathematical model for a real system is given by modelling techniques which start from the physical nature of the phenomena involved in the considered process. The model is obtained on the basis of the equations that can be written by applying the known physical laws to every part of the system. This approach leads to models which are intrinsically *realistic* since they reflect the physical nature of the process. Moreover, the parameters appearing in such models have often a well defined meaning which can be useful also to classify the process. On the other hand, modelling techniques lead to *complex* and *inaccurate* models. Complexity arises from the difficulties in the evaluation of the parts of the process which are more relevant in the determination of its input-output behavior; this leads to the introduction of non-essential relations into the model. Inaccuracy derives from the difficulties that are met in finding the correct values for the model parameters; these parameters are often obtained by means of trial and error procedures.

An entirely different approach, namely identification, is offered by system theory techniques. Identification consists in the determination, on the basis of the knowledge of input-output sequences of a given process, of a mathematical model *equivalent* to the process under test. The term equivalent means that the process and the model exibit the same input-output behavior. A model of this kind can therefore be used to predict the response of the real system to assumed inputs.

Structure of the paper

Section 2 describes some desirable features for identification procedures and the used method. The description of the performed identification experiments is also reported. In Section 3 the results obtained in the identification of a water-drive gas reservoir described in the literature and of its aquifer is reported. Section 4 describes the identification of the Italian gas storage of Brugherio, while some concluding remarks are reported in Section 5.

II. NATURAL GAS STORAGE IDENTIFICATION

Many different approaches to the identification problem can be found in the literature. It is however important to note that not all the available procedures are sufficiently general to be used with real systems. Some important features regard

a) *The class of the models*

The obvious choice for the class of the models is given by state-space representations. In fact, identification is often the first step towards the control of a process; a state-space model can be directly used since almost all modern control and optimization theory is based on the use of these representations.

b) *The class of the inputs*

The restriction of the class of inputs to particular signals as, for instance, pulse functions, step functions, pseudo-random binary sequences, white or colored noise or sinusoidal signals introduces very severe limitations to the actual applicability of an identification procedure since only seldom the previous inputs can be applied to a real process. It is on the contrary desirable to avoid limitations in the class of inputs, in order to allow the identification of a process on the basis of normal operating records. Obviously some of the input sequences that can be applied to a process are not suitable for its complete identification (e.g. constant input); apart however from the suitability of the input sequence for the identification of the process no further conditions should be introduced.

c) *Possible restrictions on the process initial state*

It is desirable to avoid restrictions regarding the process initial state (zero initial conditions are very unrealistic). It is also desirable to obtain an estimate of the initial state, not only a model of the process.

d) *The equivalence conditions between the process and the model*

Since the determination of the model equivalent to the process to be identified is performed on the basis of the process input and output, parts of the model which do not affect its output can not affect also its equivalence with the process. A model will thus be defined equivalent to a given process when it exibits the same *external behavior* i.e. when for every initial state of the process there exists a corresponding initial state of the model which allows to generate the same output sequence as the process for every possible input sequence. The considered equivalence definition leads to the exclusion of non completely observable models but not to the exclusion of non completely controllable ones (usually completely observable and controllable models are assumed). To identify a possible non controllable part of the process it is necessary that all the modes associated to the eigenvalues of this part be initially excited. This condition, which does not depend from the considered identification procedure, leads to the exclusion of some initial states of the non-controllable part of the process.

The identification procedure which has been used is the one described in |3|, |4| and |5|. This procedure satisfies all the requirements described at the previous points a-c and leads to a canonical state-space model by the following steps

1) *Structural Identification*

This step is performed on a sequence of increasing-dimension information matrices directly obtained from the noisy input-output sequences of the process to be identified. The use of the new last eigenvalue test described in |5| leads to the direct computation of performance indices (Predicted Per Cent Reconstruction Errors, PPCRE) associated to the process outputs for different structures of the model. It is thus possible to select a suitable model structure before the estimation of the model parameters.

2) *Parametric Identification*

This step |3|, |4|, is performed on the same information matrices previously constructed and leads to a consistent estimate of the coefficients of a canonical state-space model of the type

$$x(k+1) = F\,x(k) + G\,u(k) \qquad (2.1a)$$
$$y(k) = H\,x(k) + R\,u(k) \qquad (2.1b)$$

where $x(k) \in R^n$ is the state, $u(k) \in R^r$ is the input and $y(k) \in R^m$ is the output. F $(n \times n)$ is the system dynamical matrix, G $(n \times r)$ is the input distribution matrix, H $(m \times n)$ is the output distribution matrix and R $(m \times r)$ is the algebraical input-output link.

3) *Initial State Estimation*

This step computes the initial state of the estimated model (2.1) which generates the given input-output sequences |3|, |4|.

Performed Identification Experiments

The application of the previously described identification techniques to some water-drive depleted gas reservoirs used as storages |2|, has already given good results. The previsional models described in |2| have been obtained by considering as input and output the cumulative gas production and the reservoir mean pressure respectively. In this paper also the volume of the encroached water has been considered as a model output. Since a direct measure of this quantity is not available it has been computed by means of a volume balance as follows

$$W_e(k) = G_o B_{gi} - B_g(k)\left[G_o - G_p(k)\right] \qquad (2.2)$$

where $W_e(k)$ is the cumulative volume of water encroached at time k, G_o is the estimated initial gas volume at standard conditions, B_{gi} is the initial gas volume factor, $B_g(k)$ is the gas volume factor at time k and $G_p(k)$ is standard-conditions cumulative gas production at time k. Relation (2.2) can be used to compute the encroached water at time k because of the large mobility of the gas with respect to the data sampling interval.

III. GAS FIELD C IDENTIFICATION

Field C, described in |6| produced for six years prior to its conversion to storage operations. During this period 8,026,657 MCF of gas were produced accompanied by a pressure decline from 453 pounds per square inch to 267 pounds per square inch (wellhead). Recovery of pressure after short periods of high production indicated that gas-water interface was mobile. Field C is one of the several gas reservoirs found in Michigan stray sand. This sand is a series of disconnected lenses located at random, rather than a continuous layer of sandstone. For this reason, the formation has been designed as stray sand. The sand was deposited during the geologic time "Michigan" and is an integral part of the Michigan formation. Underlying the stray sand is the Marshall formation, a porous and permeable sandstone like the stray.

Although several shale lenses exist between the stray and Marshall formations, no consistent thickness of sealing shale is found between the two structures. In many places water in the Marshall is free to migrate into the stray if the pressure within the stray is reduced. The invasion of the Marshall water into the stray helps maintain the pressure in the stray. Providing that the movement of this water is controlled and is not allowed to come into a production well, thereby waterlogging the well, the water is beneficial.

The gas inventory, pressure history and computed encroached water of field C for a period of 250 months are reported in the figures 1, 2 and 3 respectively.

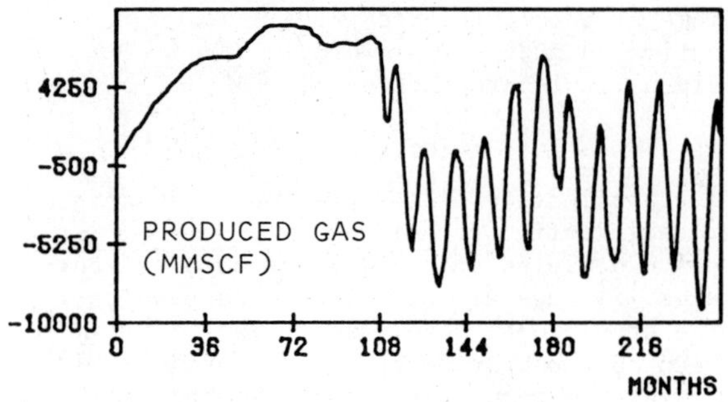

Fig. 1 - Gas Field C - Production/injection history

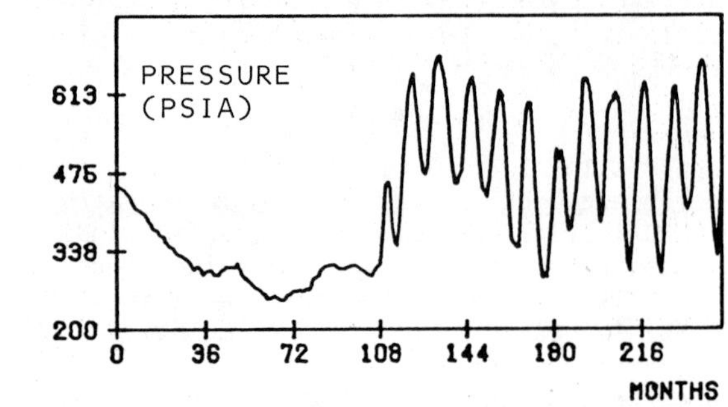

Fig. 2 - Gas Field C - Pressure history

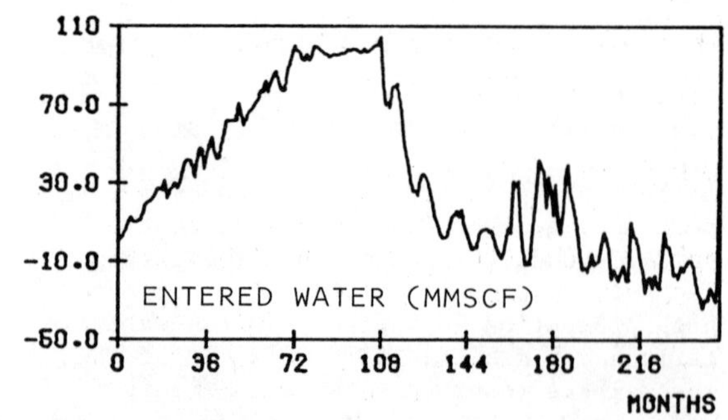

Fig. 3 - Gas Field C - Computed encroached water

The identification has been performed on the data sequences regarding the first 140 months while the remaining data have been used to check the model validity. The structural identification has been performed by means of the algorithm proposed in |5|. The sequence of computed PPCRE for the possible model structures have been

$\Pi(2,1,2) = 2.86\%$ $\Pi(3,2,3) = 2.6\%$ .....

$\Pi(2,2,2) = 8.1\%$ $\Pi(3,3,3) = 6.5\%$

These values show that the decrease in the PPCRE of the first output by increasing the order of the associated subsystem from 1 to 2 is not significant (the PPCRE is 2.86% for $\nu_1 = 1$ and 2.6% for $\nu_1 = 2$). It is then possible to assume $\nu_1 = 1$ and to compute the corresponding PPCRE associated to the second output

$\Pi(1,2,2) = 10.96\%$ $\Pi(1,3,3) = 10.74\%$ ....

These values show only a modest decrease in the PPCRE by increasing the order of the second subsystem from 1 to 2. The obtained model structure is therefore (1,1). It can be noted that the considered arguments of Π refer to the number of vectors, in the corresponding information matrices, obtained from samples first output (pressure), of the second output (encroached water) and of the input (cumulative gas production). The considered number of input vectors in the information matrices corresponds to the presence, in the model, of an algebraic input-output link. The estimated model is described by the following matrices

$$F = \begin{bmatrix} 1.00001 & -0.0285714 \\ 0.0011494 & 0.954995 \end{bmatrix} \quad G = \begin{bmatrix} 0.298388E-3 \\ 0.333136E-3 \end{bmatrix}$$

$$H = \begin{bmatrix} 1 & 0 \\ 0 & 1 \end{bmatrix} \quad R = \begin{bmatrix} -0.276871E-1 \\ 0.363359E-2 \end{bmatrix}$$

while the initial state is given by

$$x(0) = \begin{bmatrix} 453.0 \\ 0.0 \end{bmatrix} .$$

The model poles are $\lambda_1 = 0.9557$ and $\lambda_2 = 0.9993$. The PCRE (Per Cent Reconstruction Errors) given by the model on the set of 140 data used for the identification are 7.11% and 20.75%, while the PCRE on the whole set of 250 data have been 11.58% and 26.27% respectively. These errors are defined as 100 times the ratio between the error standard deviation and the considered output one.

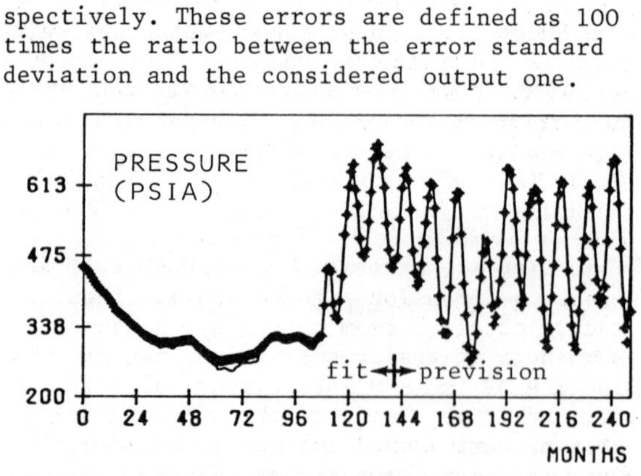

Fig. 4 - Gas Field C - Comparison between the measured pressure and the model prevision

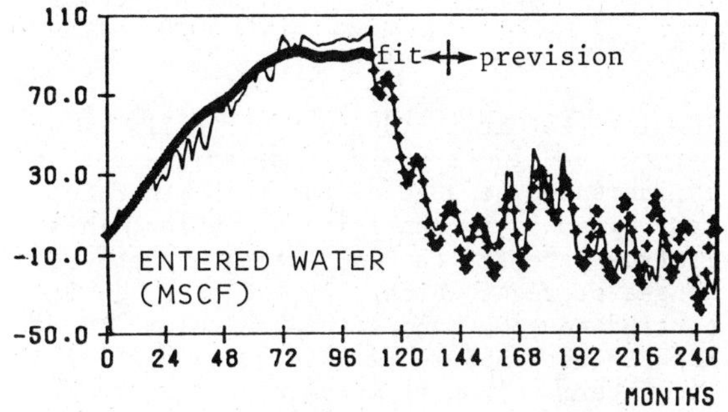

Fig. 5 - Gas Field C - Comparison between the computed encroached water and the model prevision

The actual values are higher than the predicted ones because the PPCRE are a lower bound on the PCRE. The comparison between the available data (continuous line) and the model prevision (stars) are reported in the figures 4 and 5.

IV. BRUGHERIO STORAGE IDENTIFICATION

The Brugherio field, discovered in June 1958, lies under Cinisello Balsamo, around 6 km from Milan. The hydrocarbon bearing zone (four pools) was discovered in the lower Pliocene at a depth of around 1100 meters. The main pool, which was conventionally called "A+B" and afterwards used for storage, is a stratigraphic trap determined by the deposition of coarse clastic rocks of the lower Pliocene (Sergnano gravels). The area of the reservoir is 2.5 km × 1.5 km; the main axis is located in NW-SE direction. The aquifer is on the bottom, and the original gas-water contact was at 945 meters below sea level. The maximum thickness of the hydrocarbon bearing reservoir was around 80 meters. The cap rock consists of a shaly bank, more than 260 meters thick. The production started in October 1961 and continued until June 1966, when storage operations began. In this production period around 0.7 billion cubic meters were produced equal to 35% of the original gas in place. The static pressure declined from 118.3 kg/cmq to 90 kg/cmq and the water table rose up to 935 meters. Fourteen wells have been producing, reaching a maximum daily rate of around 2 million cubic meters. Gas injection started in June 1966 utilizing only the fourteen existing wells; in 1968 the reservoir was developed by drilling three further wells. The gas in place (1.4 billion cubic meters at the beginning of storage) was increased up to around 1.6 billion cubic meters; the maximum available peak is 11 million cu m/day. The gas inventory, pressure history and computed encroached water of the Brugherio storage for a period of 160 months are reported on the figures 6, 7 and 8 respectively. The identification has been performed on the data sequences regarding the first 80 months, while the remaining data have been used to check the model validity since the good reconstruction of the data used for the model identification is not sufficient to establish the model previsional validity.

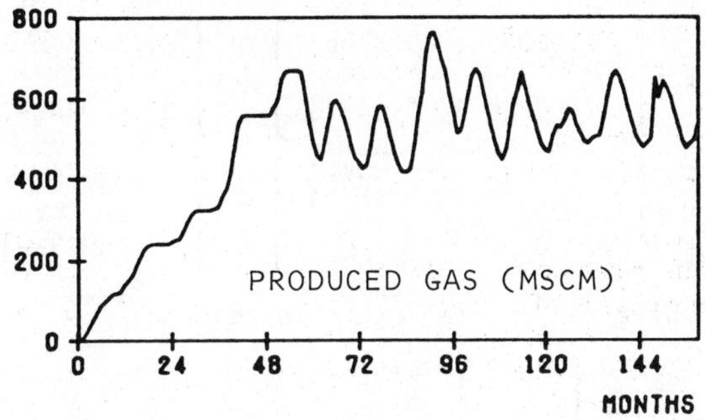

Fig. 6 - Brugherio Storage - Production/injection history

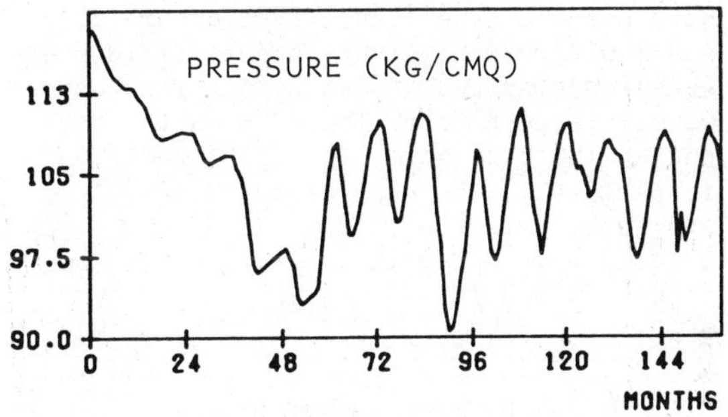

Fig. 7 - Brugherio Storage - Pressure history

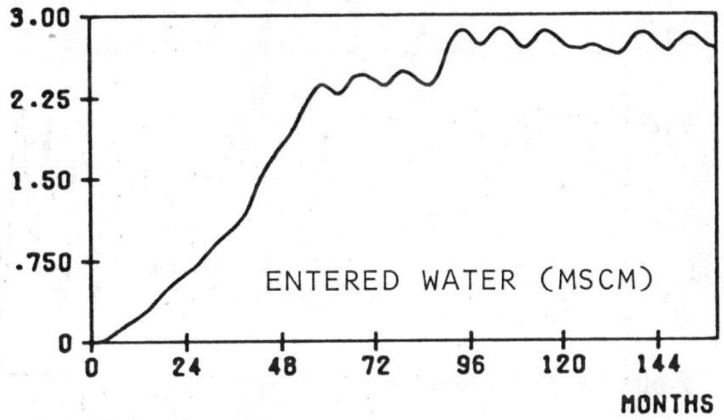

Fig. 8 - Brugherio Storage - Computed encroached water

The computed PPCRE for the possible model structures have been

$\Pi(2,1,2) = 1.03\%$ $\quad\Pi(3,2,3) = 0.77\%$ ....

$\Pi(2,2,2) = 0.63\%$ $\quad\Pi(3,3,3) = 0.40\%$

These values show that the decrease in the PPCRE of the first output by increasing the order of the associated subsystem from 1 to 2 is modest (from 1.03% to 0.77%). It is then possible to assume $\nu_1 = 1$ and to compute the corresponding PPCRE of the second output which are given by

$\Pi(1,2,2) = 0.84\%$ $\quad\Pi(1,3,3) = 0.51\%$ ....

These values show that a proper choice for the order of the second subsystem is $\nu_2 = 1$. The model structure obtained is thus (1,1). Also in this model an algebraic input-output link has been considered. The estimated model is described by the following matrices

$$F = \begin{bmatrix} 1.00048 & -0.384633 \\ -0.45928E-4 & 0.941705 \end{bmatrix} \quad G = \begin{bmatrix} 0.154530E-2 \\ 0.264936E-3 \end{bmatrix}$$

$$H = \begin{bmatrix} 1 & 0 \\ 0 & 1 \end{bmatrix} \quad R = \begin{bmatrix} -0.582268E-1 \\ 0.463578E-3 \end{bmatrix}$$

while the initial state is given by

$$x(0) = \begin{bmatrix} 118.25 \\ 0.0 \end{bmatrix} .$$

The model poles are $\lambda_1 = 0.9414$ and $\lambda_2 = 1.001$. The PCRE given by the model on the set of 80 data used for the identification are 3.45% and 3.02% for the first output (field pressure) and for the second output (encroached water) respectively. The PCRE on the whole set of 160 data have been 8.46% and 4.97% respectively. The comparison between the available data (continuous line) and the model previsions (stars) is reported in figures 9 and 10.

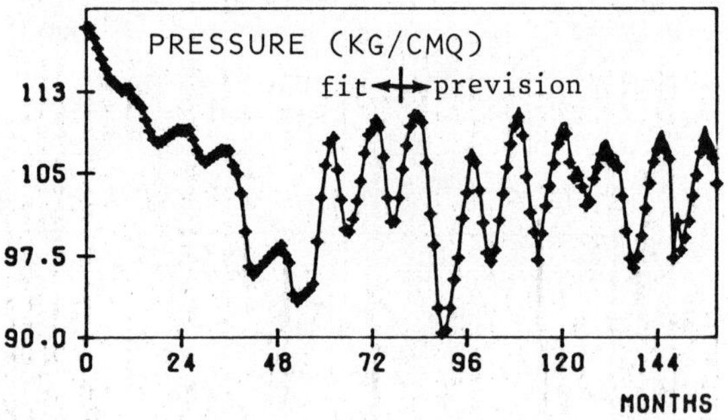

Fig. 9 - Brugherio Storage - Comparison between the measured pressure and the model prevision

Fig. 10 - Brugherio Storage - Comparison between the computed encroached water and the model prevision

V. CONCLUSIONS

The interest in the dynamic surveillance of natural and industrial processes has recently increased because of the economic role of many dynamically sensitive variables. This is true also for the monitoring and control of gas storages which play, at present, important economic and strategic roles. The results obtained in |2| had shown that, in the determination of dynamical models for natural gas storages, traditional modelling techniques could be integrated by identification procedures directly leading to accurate models starting from the past production/pressure history of the reservoir.

The results described in this paper show that linear identified models can give good long-term previsions not only on the reservoir pressure but also on the volume of encroached water.

REFERENCES

|1| G. Cau, "Underground Storage of Natural Gas in Italy", *5th Symposium for Production*, Cairo, November 1978.

|2| G. Chierici, G. Gottardi and R. Guidorzi, "Identified Models for Gas Storage Dynamics", Technical Report, Istituto di Scienze Minerarie, University of Bologna, July 1979.

|3| R. Guidorzi, "Canonical Structures in the Identification of Multivariable Systems", *Automatica*, vol. 11, n. 4, pp. 361 - 374, 1975.

|4| R. Guidorzi, "Complete Sets of Invariants and Canonical Input-Output Forms for Multivariable Systems Structural and Parametric Identification", *Proceedings of the 5th IFAC Symposium on Identification and Systems Parameters Estimation*, Darmstadt, FRG, September 1979.

|5| R.P. Guidorzi, M.P. Losito and T. Muratori, "On the Last Eigenvalue Test in the Structural Identification of Linear Multivariable Systems", *Proceedings of the V European Meeting on Cybernetics and Systems Research*, Vienna, Austria, April 1980.

|6| D.L. Katz, M.R. Tek, K.H. Coats, M.L. Katz, S.C. Jones and M.C. Miller, "*Movement of Underground Water in Contact with Natural Gas*", The American Gas Association, 1963.

MODELING OF A TUNNEL FURNACE IN THE CERAMIC INDUSTRY

D. Barreteau*, C. Laguerie*, J. P. Babary and M. Kassinopulos****

**Institut du Génie Chimique, L. A. CNRS 192, Chemin de la Loge, 31078 Toulouse, France*
***Laboratoire d'Automatique et d'Analyse des Systèmes CNRS, 7, avenue du Colonnel-Roche, 31400 Toulouse, France*

Abstract. A theoretical study was developped in order to optimise the operation of a 50m long tunnel furnace producing 30 tons of horticultural pots per day. The first step in this study consisted of modelling the thermal behaviour of the furnace by taking into account the geometrical parameters of the tunnel and of the products to be burnt and the physico-chemical properties of the different elements constituting the plant.

A mathematical model was first established for a continuous advance of the charge and then modified to account for periodic displacements in the real situation. A set of non linear hyperbolic equations was obtained and solved using the Carlson method. Thus the temperature profiles could be obtained, knowing the control distribution, which consisted of about twenty gas burners.

The model was linearised to simplify the optimisation study. A parameter identification and a numerical simulation were required. Numerical results and graphs are given.

Keywords. Tunnel Furnace - Distributed Parameter System - Modelling - Identification - Linearisation Techniques.

INTRODUCTION

When energy was cheap, only an empirical knowledge of the operation of tunnel furnaces in the ceramic industry (bricks, tiles, horticultural pots, etc) was sufficient. Today the optimal profitability of such furnaces must be determined. Two main criteria have to be taken into account.

a) minimising the quantity of broken products by optimising thermal gradients throughout the furnace.

b) minimising energy consumption by determining an optimal control law, according to the number and position of burners and thermocouples.

Such a study requires a phenomenological model of heat and mass transfer in the furnace. Some of the parameters had to be identified, using experimental data from the operating furnace.

To adapt the model for an easier optimal control study, a set of partial differential equations had to be linearised.

The work was performed in a tunnel furnace producing horticultural pots. It should be noted that one of the main problems resulted from the heterogeneity of the charge in the furnace.

PHYSICAL DESCRIPTION OF THE TUNNEL FURNACE

The ceramic industry furnace under study is a 50 meters long tunnel furnace (fig. 1). Its effective cross section is 1.64 m wide and 0.92 m high. The walls are made of refractory bricks covered by an insulating cement, and building bricks placed only on the sides. This furnace consists of four zones :

- a 6m long feeding zone (closed by a sash-door)
- a 15m long preheating zone
- a 10m long burning zone
- a 19m long cooling zone

The burning zone is equiped with three independent groups of gas burners ; the burners are distributed along the two side walls (about twenty burners).

The pots are loaded onto special carriages 3.5m long and 1.58m wide (fig. 2) and pushed through the tunnel by a hydraulic pusher (0.50 m every 30 minutes). Each carriage consists of two layers of refractory bricks separated by a space located at the level of the burners. The carriages are cooled in the cooling zone by a blow of 'secondary air' flowing in the opposite direction. This air is mixed with combustion gases in the burning zone. All these gases are drawn out at the end of the feeding zone.

The butane gas flowrates are regulated by P.I.D. controllers, from temperature measurements made with three thermocouples located in the burning zone.

The ratio of 'primary air' to butane gas for a group of burners is maintained at a constant value by means of a pressure control system.

The gas temperature profile is reconstituted by means of twelve thermocouples distributed under the ceiling of the tunnel.

A test carriage, equipped with three thermocouples, allowed us the thermal profile in the charge along the furnace to be approximately estimated.

For an optimisation study, the following parameter measurements are needed :

a) gas temperature (chiefly in the critical zones where temperature is about 830°K)

b) butane gas and primary air flowrates

c) secondary air flowrate.

TEMPERATURE MODEL

A mathematical model was formulated with the following assumptions :

a) the charge is homogeneous
b) the gas flow is one-dimensional
c) the secondary air is transparent to radiation
d) the conduction resistance is negligible with respect to the convection and radiation phenomena
e) the temperatures are defined in a section of the tunnel by their average
f) the wall surface temperature dynamics is such that its derivative with respect to the time variable is formally equal to zero.

Thermodynamic balances can be established on three levels, considering heat transfer by convection and radiation and mass transfers (pots, gas, air). For more details see KASSINOPOULOS (1980)

a) Temperature T (x,t) of the charge

$$S_T\left[\frac{\partial}{\partial t}(\rho_T C_T T) + U_T\frac{\partial}{\partial x}(\rho_T C_T T)\right] = \tag{1}$$
$$\alpha_{GT}(G-T) + \beta_{GT}(G^4-T^4) + \beta_{PT}(P^4-T^4) \triangleq C_T(T,G,P)$$

b) Temperature G(x,t) of the gas

$$S_G\left[\frac{\partial}{\partial t}(\rho_G C_G G) - \frac{\partial}{\partial x}(U_G\rho_G C_G G)\right] = \tag{2}$$
$$\alpha_{GT}(T-G) + \alpha_{GP}(P-G) + \beta_{GT}(T^4-G^4) + \beta_{GP}(P^4-G^4)$$
$$+\alpha_a(T_a-G) + \sum_{i=1}^{N} Q_i\,\delta(x-x_i)$$

c) Temperature P(x,t) of the wall surface

$$\alpha_{GP}(G-P) + \alpha_{PTO}(T_o-P) + \beta_{GP}(G^4-P^4) \tag{3}$$
$$+\beta_{PT}(T^4-P^4) \triangleq C_P(T,P,G) = 0$$

d) Mass balance

$$S_G\frac{\partial}{\partial x}(U_G\rho_G) = -\sum_{i=1}^{N}(16.5+15.5\quad)\,m_i\,\delta(x-x_i) \tag{4}$$

with $x \in\,]0, L[$ $(L = 44\ m)$

Notation :

ρ_T : density of charge
ρ_G : density of gas
C_T : specific heat of charge
C_G : specific heat of gas
U_T : charge velocity
U_G : gas velocity
S_T : vertical section of charge
S_G : vertical section of gas flow
T_a : air temperature under carriages
T_o : ambient temperature ($T_o \simeq 300°K$)
Q_i : $Q_i = 10955 \times m_i$ with
m_i : butane flowrate per burner $(i=1,2,\ldots,N)$
: excess of air

α_{GT} and α_{GP} are unit length conductances to the convective heat transfer between gas and charge, and between gas and wall.

α_a : thermal conductance between charge and air which is under carriages

α_{To}: thermal conductance between the internal wall surface and the surroundings

$\beta_{GT}, \beta_{PT}, \beta_{GP}$: radiation coefficients depending on the emissivities and the configuration of the radiative systems.

EQUIVALENT MODEL

Considering equality (4), the differential equation (2) can be written in the following form :

$$S_G\left[\frac{\partial}{\partial t}(\rho_G C_G G) - U_G\rho_G\frac{\partial}{\partial x}(C_G G)\right] =$$
$$\alpha_{GT}(T-G) + \alpha_{GP}(P-G) + \beta_{GT}(T^4-G^4)$$
$$+\beta_{GP}(P^4-G^4) + \alpha_a(T_a-G)$$
$$+\sum_{i=1}^{N}(10955 - C_G G(16.6+15.5\gamma))m_i\,\delta(x-x_i)$$
$$\triangleq C_G(T,G,P) + D_G(G) \tag{5}$$

Moreover, analytically developing the different coefficients of the model (2), we get a differential system of the type :

$$\begin{cases} A_T \frac{\partial T}{\partial t} + B_T \frac{\partial T}{\partial x} = C_T(T,G,P) & (6) \\ A_G(G) \frac{\partial G}{\partial t} + \frac{\partial G}{\partial t} + B_G \frac{\partial G}{\partial x} = C_G(T,G,P) + D_G(G) & (7) \\ C_P(T,P,G) = 0 & (8) \end{cases}$$

with :

$$A_T = S_T \rho_T C_T$$
$$B_T = U_T S_T \rho_T C_T$$
$$A_G = 114.4\ S_G(0.051\ G^{-0.78} + 0.28\ 10^{-4} G^{0.22})$$
$$B_G = -M_G(0.232 + 0.46\ 10^{-4}\ G)$$

These coefficients were obtained after calculation (Kassinopoulos (1980)).

RESOLUTION BY CARLSON'S METHOD

Carlson's discretisation method is very useful because it is unconditionally stable (at least for linear systems) and fits perfectly into the case of counterflow processes (Richtmyer, 1967).

This method consists of writing the differential system

$$\left[\frac{1}{v}\frac{\partial}{\partial t} + M \frac{\partial}{\partial x}\right] u(x,t) = S.u(x,t)$$

(M : diagonal matrix) in the form :

$$\left[\frac{\partial}{\partial t} + \hat{M}\frac{\partial}{\partial x}\right] \hat{u}(x,t) = \hat{S}.\hat{u}(x,t)$$

with $\hat{M}$ split into four parts :

$$\hat{M} \triangleq M_1 + M_2 + M_3 + M_4$$

where

M_1 contains those positive diagonal elements of $\hat{M}$ which are $> \Delta x / v\Delta t$

M_2 contains those positive diagonal elements of M which are $\leqslant \Delta x / v\ \Delta t$

M_3 contains those negative diagonal elements of $\hat{M}$ which are $\geqslant -\Delta x / v\ \Delta t$

M_4 contains those negative diagonal elements of $\hat{M}$ which are $< -\Delta x / v\ \Delta t$

Δx and Δt being the space and time discretisation step

The difference equations are :

$$\frac{\hat{u}_j^{n+1} - \hat{u}_j^n}{v\Delta t} + M_1 \frac{\hat{u}_{j+1}^{n+1} - \hat{u}_j^{n+1}}{\Delta x} + M_2 \frac{\hat{u}_j^n - \hat{u}_{j-1}^n}{\Delta x} + M_3 \frac{\hat{u}_{j+1}^n - \hat{u}_j^n}{\Delta x} + M_4 \frac{\hat{u}_j^{n+1} - u_{j-1}^{n+1}}{\Delta x} = \hat{S}\hat{u}_j^n$$

n : time index

j : space index

Although Carlson's scheme is implicit, the numerical solution can be easily performed. Boundary conditions are defined at point x = 0 ot at point x = L, according to the sign of coefficients of $\hat{M}$ (respectively positive or negative).

In the case of the tunnel furnace, equations (6) and (7) are :

$$\begin{cases} \frac{\partial T(x,t)}{\partial t} + \frac{B_T}{A_T} \frac{\partial T(x,t)}{\partial x} = \frac{C_T}{A_T} \\ \frac{\partial G(x,t)}{\partial t} + \frac{B_G}{A_G} \frac{\partial G(x,t)}{\partial x} = \frac{C_G + D_G}{A_G} \end{cases} \quad (9)$$

Or :

$$\left[\frac{\partial}{\partial t} + \begin{bmatrix} \frac{B_T}{A_T} & 0 \\ 0 & \frac{B_G}{A_G} \end{bmatrix} \frac{\partial}{\partial x}\right] \begin{bmatrix} T(x,t) \\ G(x,t) \end{bmatrix} = \begin{bmatrix} \frac{C_T}{A_T} \\ \frac{C_G + D_G}{A_G} \end{bmatrix}$$

Therefore :

$$\hat{M} \triangleq M_1 + M_2 + M_3 + M_4 = [0] + \begin{bmatrix} \frac{B_T}{A_T} & 0 \\ 0 & 0 \end{bmatrix} + [0] + \begin{bmatrix} 0 & 0 \\ 0 & \frac{B_G}{A_G} \end{bmatrix}$$

decomposition linked to the choice of the numerical values of Δx and Δt.

The global discretisation of the system leads to the following equations :

$$T_j^{n+1} = T_j^n - \frac{B_T}{A_T}(T_j^n - T_{j-1}^n)\frac{\Delta t}{\Delta x} + \frac{C_T}{A_T}\Delta t \quad (10)$$
$$j - 1, \ldots, N_L$$
$$n = 1, \ldots$$

$$G_{j-1}^{n+1} = G_j^{n+1} + \frac{A_G}{B_G}(G_j^{n-1} - G_j^n)\frac{\Delta x}{\Delta t} + \left(\frac{C_G + D_G}{B_G}\right)\Delta x \quad (11)$$
$$j = N_L, \ldots 1$$
$$n = 1, \ldots.$$

The algebric equation (8) which can be solved analytically, gives the temperature P(x,t), knowing temperatures T(x,t) and G(x,t), solutions of system (10) (11).

The boundary conditions are :

$T(0,t) = T_0$ = constant

$G(L,t) = G_L$ = ambiant temperature

We chose a space discretisation step equal to the distance between the burners ($\Delta x = 0.5$m), the time step being equal to a few seconds.

PARAMETER SENSITIVITY STUDY

The complexity of the mathematical model lies in the fact that a number of geometrical or physical parameters exist, for which the numerical value is not always well known. A preliminary study of parameter sensitivity allowed us, in fact, to retain only four of them for identification, the other parameters being determined by their mean value. These four parameters are given in decreasing order of sensitivity :

- the product $M_C.c_T$
- the secondary air flowrate
- the surface of pots per meter
- the loss coefficient under carriages

The product $M_C.c_T$ is imperfectly known since the charge consists of many different materials.

The secondary air flowrate was at first considered as a system parameter but will be chosen later as another control variable.

The measurement of the pots' surface was not easy because of the piling configuration and heterogenity of the charge throughout the furnace.

The loss coefficient under the carriages was not easily calculable for the type of carriage under consideration.

PARAMETER IDENTIFICATION

Parameter identification was made on the basis of a measurement set obtained during several carriage pushes, and by considering the mathematical model (6) to (8), in which carriage velocity is assumed constant (1m/h).

The "method model" was used, which consists of minimising the difference between the temperature measurements and the model output with respect to the unknown parameters. The minimisation algorithm was based on a discretised gradient method (Powell,1964).

In order to solve the boundary value problem leading to the steady state solution, the differential system was discretised by the Euler method and the unknown boundary conditions were determined using a classical iterative numerical method.

The numerical results of this identification are :

(a) A_T = 1270 kcal/°K
(b) Secondary air flowrate = 1450 kg/h
(c) Mean surface of pots = 28,7 m²/m
(d) Loss coefficient = 9,34 kcal/°K

MODEL LINEARISATION

The purpose of the optimisation study was the determination of a closed loop control, connected to the thermal regulation around a desired temperature profile in the charge (fig. 3).

The continuous operation of the furnace throughout the year and small temperature variations, justified a linearisation of the obtained model.

A numerical study allowed the determination of the nominal control $m_d(x)$ corresponding to the obtaining of the desired profile $T_d(x)$ in steady state, the simulation giving the desired profiles $G_d(x)$ in the gas and $P_d(x)$ on the wall.

Let :

$$\begin{cases} \alpha(x,t) = T(x,t) - T_d(x) \\ \beta(x,t) = G(x,t) - G_d(x) \\ \gamma(x,t) = P(x,t) - P_d(x) \end{cases} \qquad (12)$$

The temperature variations of charge, gas and wall, around the desired regime, and :

$$\mu_b(x,t) = m_b(x,t) - m_d(x) \qquad (13)$$

the variation of mass flowrates of butane at the burners.

According to (12) and (13), we expand the non linear terms of the model (6) to (8), retaining only terms of order less than 2. For example :

$$T^4(x,t) = (T_d + \alpha(x,t))^4 \simeq T_d^4(x) + 4\alpha(x,t)\, T_d^3(x)$$

After calculation, we got a linear model in which the algebric equation allowed us to express $\gamma(x,t)$ as a function of $\alpha(x,t)$ and $\beta(x,t)$. So, we obtained a linear differential system with two variables $\alpha(x,t)$ and $\beta(x,t)$ of the form :

$$\begin{bmatrix} \dot{\alpha}(x,t) \\ \dot{\beta}(x,t) \end{bmatrix} = \begin{bmatrix} K_{11}(x) + K_{12}\frac{\partial}{\partial x} & K_{13}(x) \\ K_{21}(x) & K_{22}(x) + K_{23}(x)\frac{\partial}{\partial x} \end{bmatrix} \begin{bmatrix} (x,t) \\ (x,t) \end{bmatrix} + \begin{bmatrix} u_1(x,t) \\ u_2(x,t) \end{bmatrix} \qquad (14)$$

with the following boundary conditions :

$\alpha(0,t) = 0 \qquad \beta(L,t) = 0$

In fact, model (14) is valid only if we consider a continuous displacement of carriages. In the real case for which carriage velocity is zero between two pushes, the former calculations lead to a model having an additional term, due to the fact that the nominal control $m_d(x)$ no longer corresponds to the steady state. This model is of the form :

$$\begin{bmatrix} \dot{\alpha}(x,t) \\ \dot{\beta}(x,t) \end{bmatrix} = \begin{bmatrix} K_{11}(x) & K_{13}(x) \\ K_{21}(x) & K_{22}(x)+K_{23}(x)\dfrac{\partial}{\partial x} \end{bmatrix} \begin{bmatrix} \alpha(x,t) \\ \beta(x,t) \end{bmatrix} + \begin{bmatrix} C_T(x) \\ 0 \end{bmatrix} + \begin{bmatrix} u_1 \\ u_2 \end{bmatrix} \qquad (15)$$

SIMULATION RESULTS

Figure 3 represents the temperature profiles in the gas and the charge compared with the desired profile in the charge for the case where the simulation is carried out for the nominal control and the carriages are moving in a continuous way. We verify that the furnace works as a heat exchanger by considering the respective position of the profiles in the left and right parts of the furnace. The distance between the desired profile and the obtained profile in the charge is relatively small. But if we wish to make this distance smaller it will be necessary to modify the number and positions of the actuator configuration.

In order to show that the linearisation of the model is valid we have represented in figure 4 the response of both the linear and non linear models at three characteristic points (one in the preheating zone, one in the burning zone and the other in the cooling zone).

The simulation results of the linear system between two consecutive advances are shown in figure 5, where the model in which the carriage velocity is zero is used. The more important variations seen on the right hand side are due to the physical differences in the wall materials. For the same type of simulation when the carriage velocity is non zero we note that the dynamic response of the model is much slower (about 3° in $\frac{1}{2}$ hour).

We conclude that both models (zero and non zero velocity of carriages) can be used, but for different optimisation problems :

- the zero velocity model for short term optimisation (between two consecutive advances)

- the non-zero velocity model for long term optimisation (time horizon of several hours).

We are now in the process of studying an optimisation problem which takes into account the short and long term optimisation and which leads to two feedback controls.

REFERENCES

Kassinopoulos M.. Modelisation et commande sous-optimale d'un four d'industrie céramique, Thèse de 3ème Cycle - Université Paul-Sabatier, Toulouse (30 juin 1980).

Powell , an efficient method of finding the minimum of a function of several variables without calculating derivatives. Comp.J. Volume 7, 1964.

Richtmyer R.D., Morton K.W. Difference methods for initial value problems, John-Wiley and Sons, 1967.

ACKNOWLEDGEMENTS

This work is supported by BERNAT SAULIERES COMPAGNIE (Castres, France).

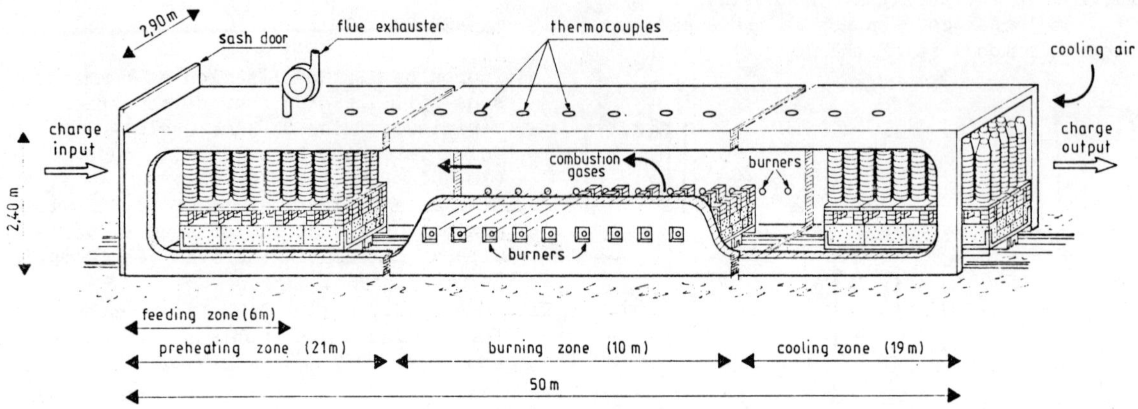

Fig. 1. Tunnel furnace for Ceramic Industry

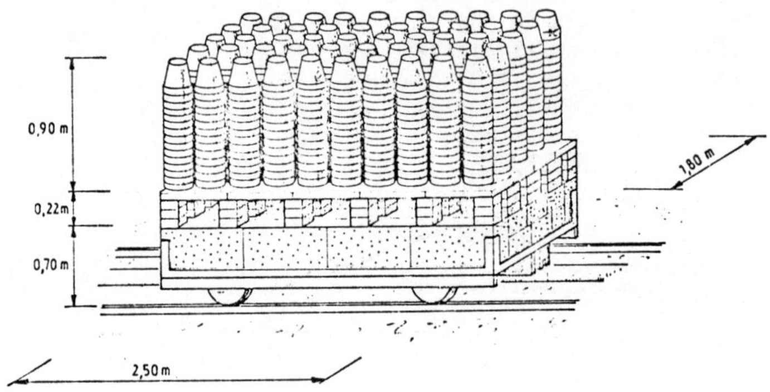

Fig. 2. Carriage

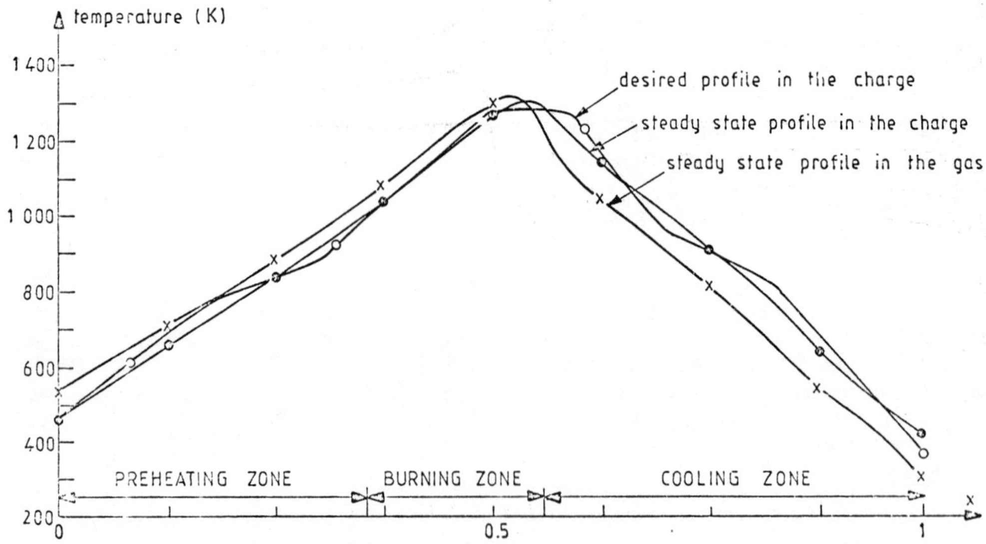

Fig. 3. Temperature profiles in steady state

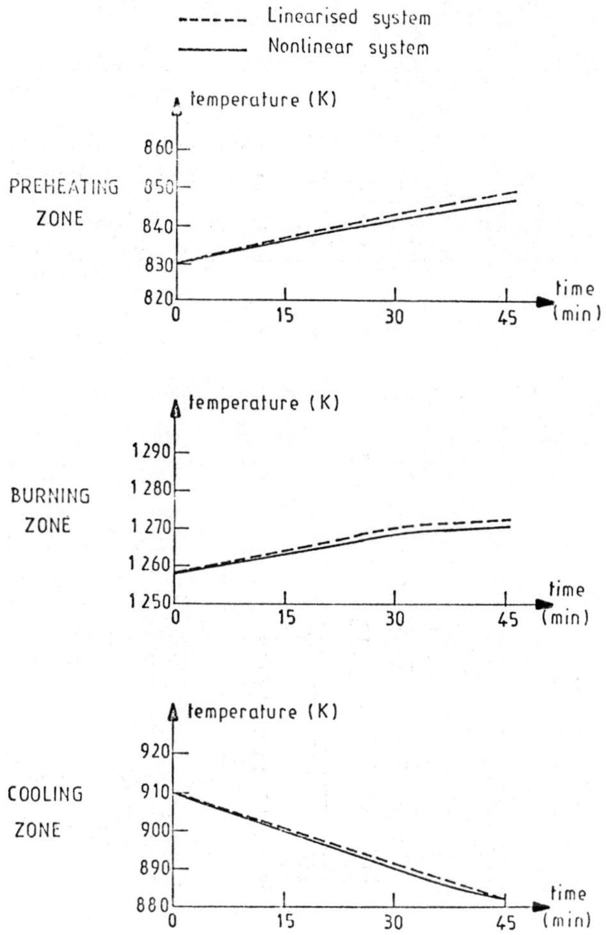

Fig. 4. Comparison of non linear and linear solutions

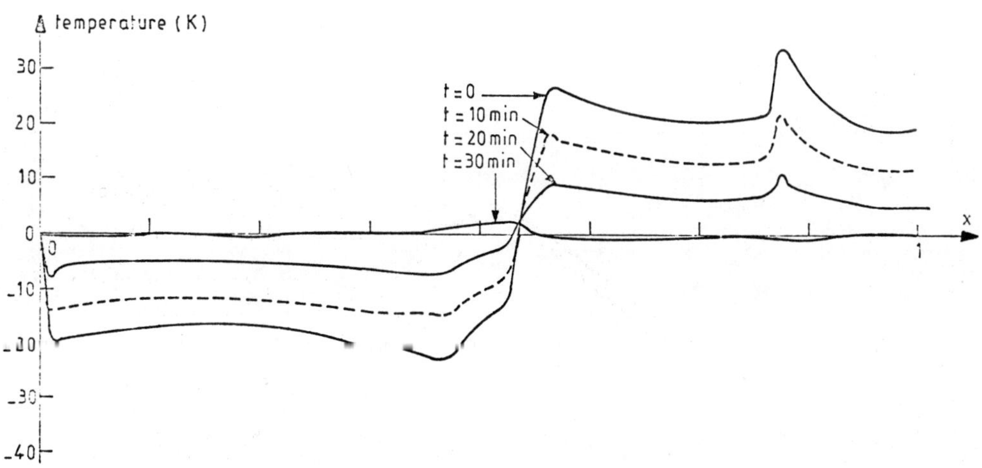

Fig. 5. Temperature variation in the charge

MODELLING AND SIMULATION OF A ROTARY CEMENT KILN - CYBERNETIC APPROACH

E. A. Dekkiche*, P. Prévôt*, A. En-Naamani* and J. P. Barbier**

**Laboratoire d'Informatique Appliquée, 502 INSA, 20 Av. Einstein, 69621 Villeurbanne, France*
***Cimenterie Lafarge de Val d'Azergues, 69380 Lozanne, France*

Abstract. An industrial process is an open system which has multiple exchanges with the environment (material, energy, information). Its running is controlled by a set of retro-actions which adapt the system to the external conditions and to the events. The modelling of a system consists in projecting it on a sub-space of representation, which corresponds to a partial apprehension of the reality. We build a methodology of cybernetic approach based on the joint utilization of two types of complementary modelling, the main purpose of which being to analyze and understand the internal phenomena of complex industrial processes, with an example of application to a rotary cement kiln using the semi dry process. The results are discussed.

Keywords. Cement industry; cybernetics; functioning modes; multivariable systems; process control; system analysis.

INTRODUCTION

"For many of us it is unusual to look on reality under two complementary aspects because we are not sufficiently trained to this way of apprehension. In our opinion, it is here one of the main reasons of the difficulty of understanding cybernetics". We think that these words of F.E. Mairlot (1974) throw a good light upon the still actual situation of the researches on the analysis and automatization of complex industrial processes. Indeed we often note that several approaches which are distinct in their principle but close with respect to their objectives keep being used separately.

Most of the time they come down to :

- an approach based on the fundamental laws of physics, adapted to the studied process, it is a theoretical approach.

- an approach by the "classical" methods of automatics, i.e. identification, records processing,... giving a main importance to the observation and the experience. The confrontation of these approaches may seem artificial insofar as a communication between the different specialists could not be established throughout the analysis.

How can we try to jointly, carry out these two types of approach ?

This paper will try to answer this question while suggesting a methodology integrating "knowledge" and "experience" on a rotary cement kiln using the semi dry process. Thus it will be possible to see the respective contributions of a mathematical modelling and of the methods of data analysis in view of the modelling and simulation of this type of process.

PROCESS DESCRIPTION

The rotary cement kiln we refer to (Fig. 1) uses the semi dry process. Its technical characteristics are : length 65 m, diameter 4 m, output capacity 1400 t/day, working with a mixture of fuel-oil and coal. The raw material consists mainly of calcium carbonate ($Ca\ CO_3$), silica ($Si\ O_2$), shale (Al_2O_3) and iron ore (Fe_2O_3). It is ground to a very fine powder, adding about 13% water, before being fed into the kiln. The product obtained is then led to a "Lepol" grate consisting in a counter-current heat exchanger composed of 2 chambers : a first chamber, called cold chamber, where the dehydration takes place, and a second chamber, the hot chamber, where the decarbonation starts, the material being brought up to about 800°C, the grate being fitted at this level with gas-burners. Introduced into the kiln, the material is subjected to higher and higher temperatures; it is then practically decarbonated at 1000°C. We assist then to reactions of formation of lime silicate and aluminate. These exothermic chemical reactions are continuing up to about 1500°C, with the formation of a molten liquid consisting of Fe_2O_3 and Al_2O_3 (Bogue, 1952; Lea, 1970). When the first liquid appears, the material starts cohering to form the "clinker". The clinker then falls into a cooler supplied with fresh air, which allows to recover calories. This

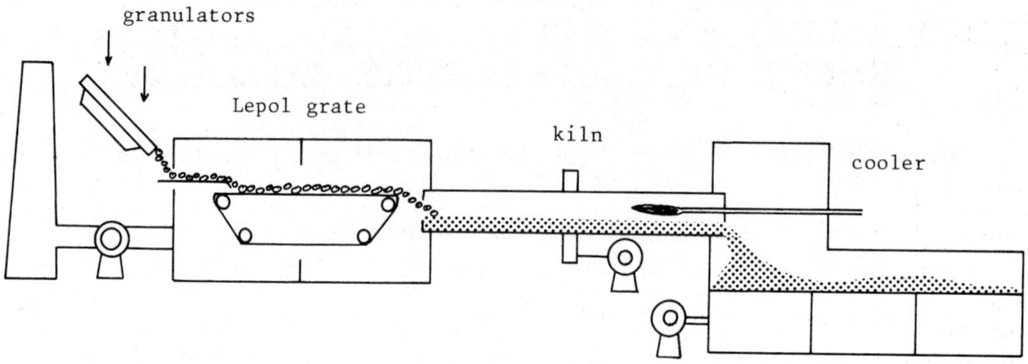

Fig. 1. Schematic representation of the process.

air is used as secondary air for the combustion.

The process is controlled by 2 computers, one ensuring the regulation loops (DDC), the other one acting as supervisor by balance programs, alarms and optimization tasks.

NECESSITY AND PURPOSES OF A CYBERNETIC APPROACH

The methodology we are suggesting is thus developped from two, momentarily distinct, approaches :

- the first one, theoretical, is based on the physical laws and takes into account the material and energy balances,

- the second one, experimental, uses a very large amount of data outcome of a data gathering which took place over a long time.

Very quickly, these approaches are going to interpenetrate and complete themselves. The complexity of the process makes it necessary to have experimental data in order to adjust the mathematical model and to give to each phenomenon its real importance. Reciprocally, the physical interpretation of the experimental results will be based on the "internal" knowledge of the running of the process. The two approaches which at the beginning seemed to be distinct, will then merge into a single approach, the only purpose of it being to go far enough in the knowledge of the process in order to better control it, i.e. to find at any time the best compromise between the required purposes (production, quality, fuel consumption, pollution, human factors, safe operation,...).

This methodology becomes a necessity due to the problems inherent in the nature and the complexity of this kind of system. Among these problems, we will mention :

- the large number of variables which are necessary to describe the state of the process.

- the multiplicity of the goals of the process control tasks which are sometimes difficult to formalize (qualitative criteria, fuzzy relationships,...).

- the nonlinear and nonstationary characters which are peculiar to some relationships between the variables.

The Figure 2. summarizes the suggested methodology. So, the apprehension of the system will be made using two complementary models which can develop and be modified to take new informations into account.

THE THEORETICAL MODELLING

Up to now, most of the mathematical models of cement kilns have been related with the wet process, process which does not need a Lepol grate and in which the water percentage in the raw material is about 35%. Results of digital and analogic simulations have been published by Min and co-workers (1961), later by Lyons and co-workers (1962). The model of Lyons has been taken up by Aamand (1969) who introduced the effect of fuel combustion. A more general dynamical model has been developped by Spang (1972) who incorporated a model of the flame.

Still in the field of rotary kilns, it is interesting to emphasize the works of Koivo and Chase (1971) who developped a model of lime kiln while fixing some parameters using a stochastic method of approximation. On the other hand, Manitius and co-workers (1974), in a study related to an aluminium oxide rotary kiln, take into account the dust conveyed by the gases. Later, their model has been used for other applications (Dumont and Bélanger, 1978; Uronen and coworkers, 1975).

Theoretical model of a cement kiln using the semi dry process

The model developped in this paper draws its inspiration from the works of Spang (1972) and has been adapted to a kiln using the semi dry process.
We recall in the Table 1 the main reactions

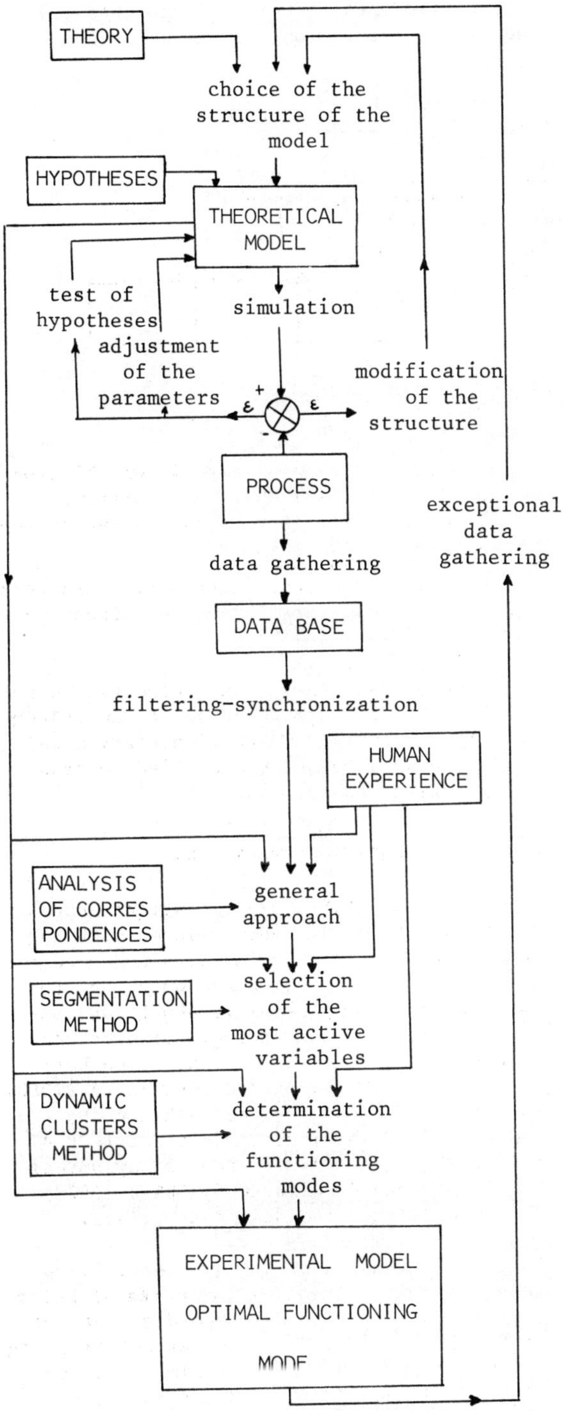

Fig. 2. Methodology of cybernetic approach of complex industrial processes.

which take place in a kiln of this type.

In the expressions of C_2S, C_3S, C_4AF and C_3A, the symbols C, S, A and F define respectively CaO, SiO_2, $A\ell_2O_3$ and Fe_2O_3.

In order to simplify the form of the equations as well as the numerical calculations, we are led to build a certain number of hypotheses :

TABLE 1 The reactions occuring in the cement kiln

Temp. (°C)	Reaction
900	completion of the decarbonation $CaCO_3 \rightarrow CaO + CO_2$
1100	start of combination of lime and silica $2\ CaO + SiO_2 \rightarrow C_2S$
1200	liquid formation $4\ CaO + A\ell_2O_3 + Fe_2O_3 \rightarrow C_4\ AF$ $3\ CaO + A\ell_2O_3 \rightarrow C_3A$
+1200	completion of cement compounds $2\ CaO + SiO_2 \rightarrow C_2S$ $C_2S + CaO \rightarrow C_3S$

- the dust conveyed by the gas is not taken into account,
- specific heats, heats of reactions and latent heats are supposed independant of temperature and position,
- the reaction rates are calculated by Arrhenius'law,
- convection and emissivity coefficients are independant of temperature and position
- the velocities of the material and the gas are constant throughout the length of the kiln.

The chemical profile

In view of simplifying the writing of the equations, we normalize all components with respect to the amount of CaO in the burning zone, i.e. kg $\overline{CaO}$. The following abreviations are thus used : $S' = $ kg SiO_2 / kg $\overline{CaO}$; $C' = $ kg CaO / kg $\overline{CaO}$; $X = $ kg C_2S / kg $\overline{CaO}$; $Y = $ kg C_3S / kg $\overline{CaO}$; $B = $ kg $CaCO_3$ / kg $\overline{CaO}$. The variation of the chemical composition of the material along the kiln is described by a system of non linear partial differential equations (see Appendix for the meaning of the used symbols)

$CaCO_3$:

$$\frac{\partial B}{\partial t} = -\frac{M_B}{M_c} k_B\ B - v_m \frac{\partial B}{\partial \ell} \qquad (1)$$

SiO_2 :

$$\frac{\partial S'}{\partial t} = -\frac{M_X}{2M_c} k_X\ S'C'^2 - v_m \frac{\partial S'}{\partial \ell} \qquad (2)$$

CaO :

$$\frac{\partial C'}{\partial t} = k_B\ B - k_X\ S'C'^2 - k_Y\ C'X - v_m \frac{\partial C'}{\partial \ell} \qquad (3)$$

C_2S :

$$\frac{\partial X}{\partial t} = \frac{M_X}{2M_c} k_X \ S'C'^2 - \frac{M_X}{M_c} k_Y \ C'X - v_m \frac{\partial X}{\partial \ell} \quad (4)$$

C_3S :

$$\frac{\partial Y}{\partial t} = \frac{M_Y}{M_c} k_Y \ C'X - v_m \frac{\partial Y}{\partial \ell} \quad (5)$$

<u>The thermic profile</u>

The temperatures of material, gas and wall are given by the following equations :

Material :

$$\frac{\partial T_m}{\partial \ell} = \frac{2}{r_1^2 \ (2\pi - \hat{p} + \sin \hat{p}) \ C_{sm} \rho_m v_m} [\alpha_1 (T_g - T_m) + \alpha_2 (T_p - T_m) + \frac{r_1^2}{2} (2\pi - \hat{p} + \sin \hat{p}) \ Q_c] - \frac{\partial T_m}{\partial t} \quad (6)$$

Gas :

$$\frac{\partial T_g}{\partial \ell} = \frac{2}{r_1^2 (\hat{p} - \sin \hat{p}) C_{sg} \rho_g v_g} [\alpha_3 (T_p - T_g) + \alpha_1 (T_m - T_g) + Q_f] \quad (7)$$

Wall :

$$\frac{\partial T_p}{\partial \ell} = \frac{1}{2\pi (r_2^2 - r_1^2) C_{sp} \rho_p} [\alpha_3 (T_g - T_p) + \alpha_2 (T_m - T_p) + \alpha_4 (T_a - T_p)] \quad (8)$$

$\hat{p}$ is defined by Fig. 3.

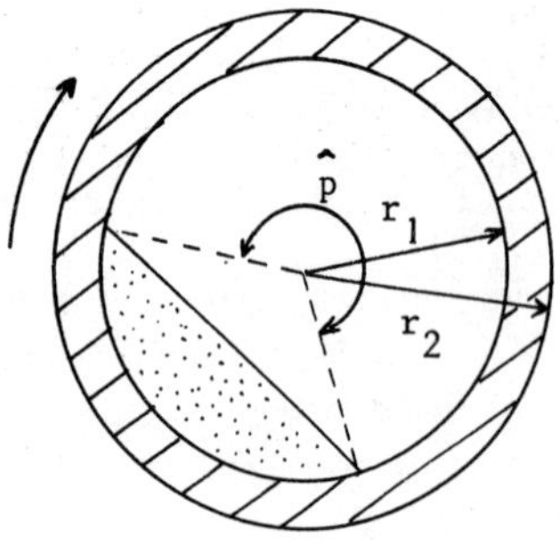

Fig. 3. Cross section of the kiln

The reaction rates are calculated according to the Arrhénius' law :

$$k_i = A_i \exp (- E_i / RT_m) \quad (9)$$

A_i : pre-exponential factor

E_i : activation energy

R : gas constant

The coefficients of heat transfer, calculated at each iteration, are given by the expressions :

$$\alpha_1 = 2r_1 \ \sin(\frac{\hat{p}}{2}) [f_2 + 1.73 \times 10^{-9} \ \varepsilon_g \varepsilon_m (T_g^2 + T_m^2)(T_g + T_m)] \quad (10)$$

$$\alpha_2 = r_1 (2\pi - \hat{p}) [f_3 + 1.73 \times 10^{-9} \ h \ \varepsilon_p \varepsilon_m (T_p^2 + T_m^2)(T_p + T_m)] \quad (11)$$

$$\alpha_3 = r_1 \ \hat{p} \ [f_1 + 1.73 \times 10^{-9} (1 - h_o) \varepsilon_g \varepsilon_p (T_g^2 + T_p^2)(T_g + T_p)] \quad (12)$$

$$\alpha_4 = 2\pi \ f_4 r_2 \quad (13)$$

$$h = 1 + \frac{2 \ h_o \ \sin (\hat{p}/2)}{2\pi - \hat{p}} \quad (14)$$

The heat generated by chemical reactions is :

$$Q_c = \frac{\rho_m}{1 + S'_i} [- \Delta H_B \ k_B \ B - \Delta H_X k_X \ C'Y - \Delta H_Y \ K_Y S'C'^2] \quad (15)$$

Q_f indicates the heat released by the flame and its expression is (Spang, 1972) :

$$Q_f = \frac{G_F(-\Delta H_F)}{\rho_g \ v_g} [\frac{M_{c1} (PM_a)^2 \psi}{\rho_F \ M_{O_2} (RTg)^2}] \ k_F \ d_o \ C_F \quad (16)$$

Since this study refers to the statical running of the process, we simplify the equations of the model, taking into account only the space variation. From the system of differential equations thus obtained, we deduct a system of difference equations with constant coefficients, noting any variation of the type $\frac{dA}{d\ell}$ under the form $\frac{A_{i+1} - A_i}{L_s}$, in which A can be B, C', X, Y or S', and L_s is the simulation step size. The thermic and chemical profiles are calculated after the conditions known at the kiln entrance. To estimate certain parameters, we used "shooting methods" which consist in adjusting the values of these parameters according to the deviation between the real values of the temperatures and of the material composition, and those obtained at the extremity of the kiln at the end of the simulation. We chose a constant simulation step size of 1 m, which is a reasonable value, taking into account the length of the kiln and the precision being searched. The thermic and chemical profiles then calculated are shown in Fig. 4.

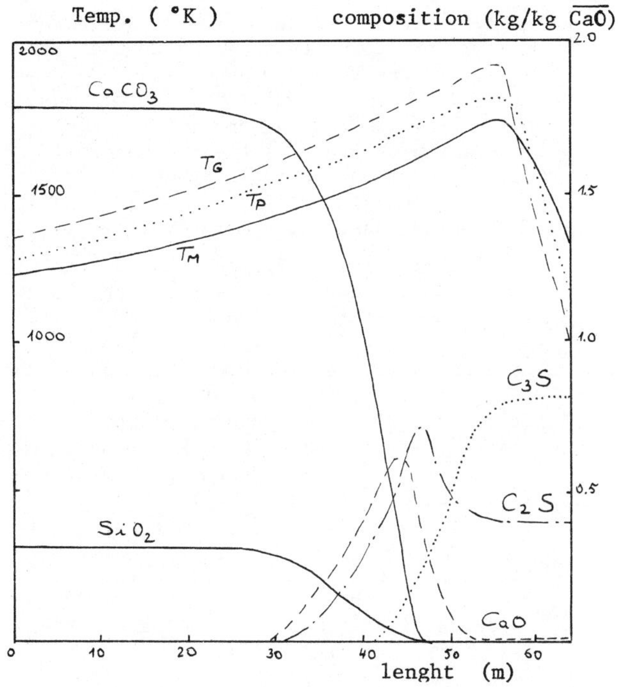

Fig. 4. Thermic and chemical profiles of the kiln.

Interpretation and justification of the results

The results obtained by the simulation of the theoretical model confirm relatively well the explanations which are usually put forward. Indeed, the decarbonation that starts in the hot chamber of the Lepol grate becomes appreciable only towards 1100°C, i.e. towards the middle of the kiln. At that time, we assist to the formation of C_2S obtained by combination of CaO released with SiO_2 in liquid phase. The formation of C_3S starts towards 1200°C while the formation of C_2S keeps going on. This remark is important because it corresponds more to the physical aspect of the clinker making while generally the two phenomena are supposed to be distinct.

The behaviour of the model, from a chemical as well as from a thermic point of view, appears reasonable.

The lack of reliable measures and the difficulties of experimentation did not always enable to confirm the numerical values of some thermic parameters.

We notice that there is an appreciable difference between these profiles and those related to cement kilns using the wet process, principally for the thermic profile.

This type of model differs particularly from the models used in engineering which are based on constant thermic variations and thus make it possible to calculate the kiln sizes.

STATICAL CHARACTERIZATION OF THE PROCESS THROUGH DATA ANALYSIS

At the beginning, it is an experimental approach. The industrial constraint of non direct intervention on the control of the process required the organization of a very long data gathering which took place over 5 months (Dekkiche and co-workers, 1980). In this study, we take into consideration 24 characteristic variables which appear in Table 2. All these variables are regularly measured, every 10 mn for the % of free CaO, every 3 mn for the others. The numerical values thus obtained constitute a data base our treatments are going to deal with.

TABLE 2. The variables

	Variables and units	
controled variable	% free CaO in clinker (%)	CAO
control variables	total water rate (m^3/h)	H2O
	feed rate (t/h)	FEE
	fuel rate (m^3/h)	FUL
	recycling fan speed (r.p.min)	RFS
	induced draft fan speed (r.p.min)	IDS
	Lepol grate speed (m/h)	LGS
	kiln rotational speed (r.p.h)	KRS
	kiln motor intensity (A)	KMI
	air rate in 1st cooler chamber (Nm^3/s)	AIR
	height of material bed (mm)	FBL
intermediate variables	% oxygen at the exhaust (%)	OXY
	exit gas temp. (°C)	EGT
	recycling temp. (°C)	RET
	cold chamber temp. (°C)	CCT
	hot chamber temp. (°C)	HCT
	secondary air temp. (°C)	SAT
	clinker temp. (°C)	CLT
	on grate pressure in hot chamber (mm H_2O)	OPH
	on grate pressure in cold chamber (mm H_2O)	OPC
	cold chamber pressure drop (mm H_2O)	CPD
	hot chamber pressure drop (mm H_2O)	HPD
	under grate differential pressure (mm H_2O)	UPD
	under grate pressure in 1st cooler chamber (mm H_2O)	UGP

Since in a first stage we only take our interest in the "average" behaviour of the process, we had to first synchronize, after filtering, the available informations in order to neutralize pure delays and time constants. The experimental study consists then in analysing the process with statistical methods and methods of data analysis, i.e. :

1. study the general behaviour of the process
2. look for the active variables
3. determine the functioning modes of the process.

Study of the general behaviour of the process

After selecting the records, filtering, synchronizing and computing the average values of the 24 variables on steady functioning zones of at least 2 hours, we obtain a set of 207 functioning points. We will use that set of points for the statical study. The first stage corresponds to a general approach of the system. Its purpose is to represent the whole of the relationships between the variables. The method which has been used here is the Analysis of Correspondences (Benzecri, 1973). It is based on a reduction of the representation space of the variables. Its application to industrial data is still quite recent (e.g. Prévôt and Vitry, 1976). The first attempts to apply this method to the kiln data gave interesting results (Dekkiche and co-workers, 1980). We restricted ourselves to its application over a one day running while representing simultaneously the experiences (samples) and the variables. The Figure 5. shows the projection of the results obtained in the plane formed by the first 2 factorial axes. On each axis is represented the % of explained variance. The experiences are numbered in the chronological order.

Let us recall here the interpretation rules of this type of representation :

- the closer two variables are to one another, the greater their mutual dependence on each other in the same sense,
- the greater the distance between two variables on both sides of the origin, the more antithetical the variable are,
- the greater the distance of the variables from the origin, the better the validity of the results.

From the results of the Fig. 5, it is possible to make the following deductions with respect to the running of the process :

1. The variables KRS, LGS, FEE and H2O, constitute a particular group, as a matter of fact, they are bound by local regulation loops, constraint one cannot rid of. Thus it is difficult to distinguish the influences on the other variables, unless taking into account the difference of the delays. In this group, the material feed rate, FEE, constitutes the main variable, it is opposed to the % of free CaO. As a matter of fact, with a constant fuel rate, if one increases FEE, the material burns less, which results in a high value of CAO, and inversely in case of a decrease of FEE.

2. The temperatures measured in the Lepol grate RET, EGT and HCT have a similar evolution, they partially characterize the degree of preparation of the raw material in the hot chamber. An important result must be here pointed out, which only appears with this method : it is the opposition between those temperatures and the variable CCT. We think this is due to the fact that, for instance, an appreciable decrease of CCT means a deficiency in calories; this situation can be coped with while decreasing the rate of the raw material which lead to an increase of CCT.

3. FUL and KMI have antithetic evolutions.

   The interpretation of this relationship involves the idea of sticking in the kiln. As a matter of fact, with a constant rate of material, an increase of the fuel rate results in a more pronounced melting of the material, a slowing down of the speed of this material, and thus a more important sticking in the internal wall of the kiln. The resulting increased inertia requires a higher power to turn the kiln, which involves an increase of the intensity KMI.

4. The evolution of the running of the kiln is clearly to be seen all along the day under consideration. Indeed the experiences gather into relatively homogeneous sub-sets. These sub-sets represent different situations which correspond to distinct functioning zones. The zone I and, particulary, the zone IV contain relatively well grouped points and therefore show a steady running of the process. This result is confirmed by the observation of the records related to the day under consideration (Fig. 6). On the other hand, the too important dispersion of the points in the zones II and III characterizes a disturbed running. The reason is that in the zone II, an increase of CAO resulted in a decrease of FEE, this decrease leading in the zone III to an action of the regulation in view of going back to a steady state.

Search for active variable

In the case of an industrial process as complex as a cement kiln, it is illusory to want to achieve well defined functioning points. Indeed the random variations of the chemical composition and of the mineralogy of the raw materials, of the nature and of the quality of the fuel, the internal state of the kiln, the surrounding conditions, ... here are all parameters which we do not measure, it is more realistic to consider the research for statical functioning modes. Before, one must determine the most active variables which are going to characterize the various modes. For this reason, we tried to explain values of CAO, a representative variable of the quality of the clinker, by means of the values of the other variables. We use here the segmentation method (Bertier and Bouroche, 1975). This method rid of the hypothesis of linearity, in opposition to the precedent method. It makes it possible to clearly show the following group of active variables according to the hierarchy CAO, KRS, FUL, FEE, KMI and SAT (Dekkiche and co-workers, 1980).

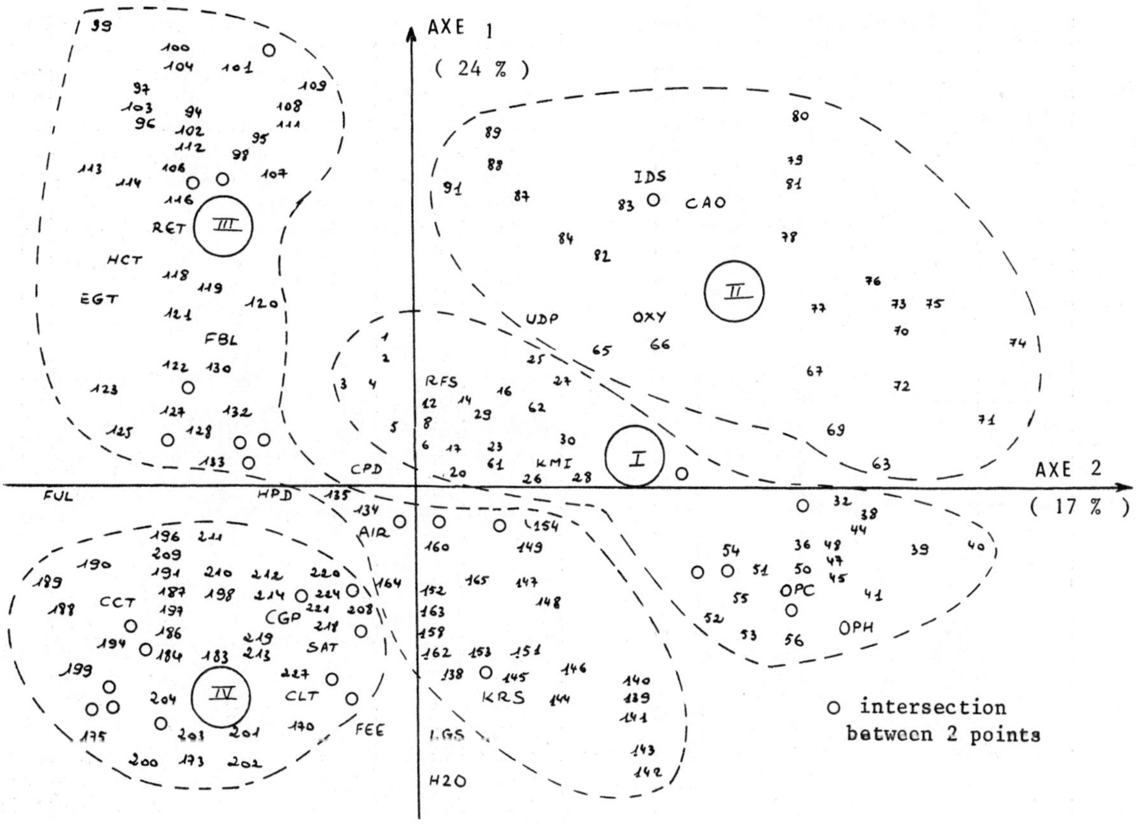

Fig. 5. Analysis of correspondences and functioning zones.

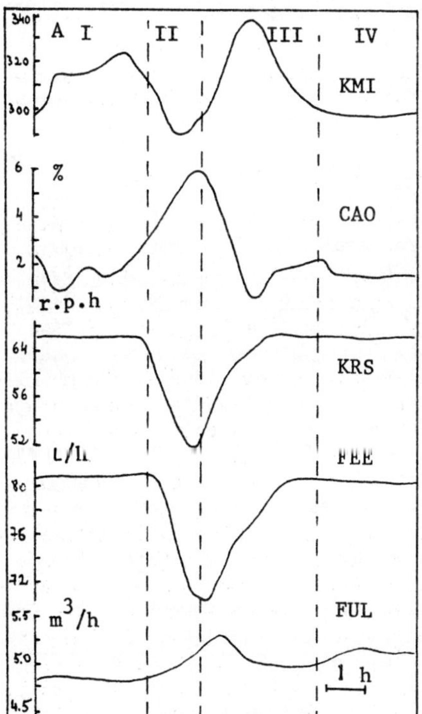

Fig. 6. Main variables and functioning zones.

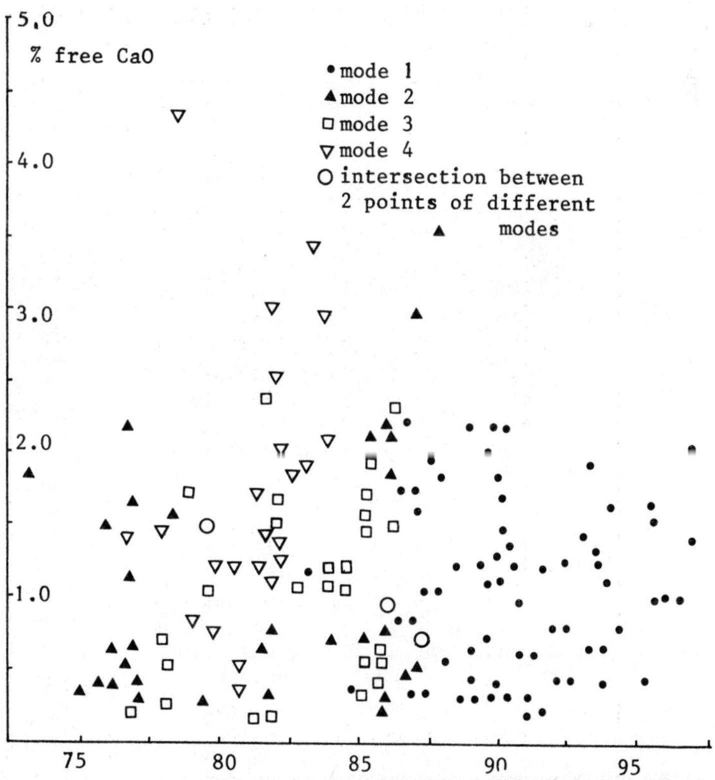

Fig. 7. The main functioning modes.

Determination of the statical functioning modes of the process

Using only that last group of variables, we were led to make a partition of the set of statical functioning points into homogeneous sub-sets while applying the Dynamic Clusters Method (Diday, 1971). Four main functioning modes have thus been found. The Table 3 shows their essential features.

TABLE 3 Characteristics of the modes.

	clinker quality	production	calorific consumption
1	good	high	high
2	good	relatively low	relatively low
3	good	middle	high
4	middle	relatively low	relatively low

In that case it has been considered that a good quality of clinker corresponds to a value of CAO between 1 and 2%, and that the production rate is proportional to the FEE.

The Figure 7 shows the variations of the value of CAO according to FEE for the 4 main functioning modes.

The analysis of Table 3 shows that the mode 1 although being characterized by an important heat consumption, is the most interesting one from a statical point of view. As a matter of fact, it enables an important hourly production and a good quality of the clinker. It is the functioning mode which can be considered as "optimal".

All the functioning modes characterized by the values of the various active variables constitute an experimental model of the process. The determination of a set of main variables will make it possible, with a view to a more accurate characterization of the model, to organize an exceptional data gathering. The contribution of this part of the analysis will be the taking into account of the phenomena which take place in the cooler, by means of the temperature of the secondary air, SAT, and this at the level of the theoretical model. In the formulation of this model, we did not take into consideration the heat supply of SAT, supply which can appreciably modify the thermic profile of the kiln in the burning zone. We are of the opinion it is an aspect which should not be neglected and which must intervene as far as the structure of the theoretical model is concerned.

CONCLUSION

After explaining the reasons which called for the use of a methodology of approach peculiar to the industrial processes, we pointed out, with an application to a cement kiln, the respective contributions of the various methods of data analysis on one hand, and of the theoretical modelling, on the other hand.

The suggested methodology, with a modular structure, partially breaks free from the industrial constraints of non direct intervention on the process, as well as from the nonlinear and nonstationary aspects which are peculiar to some involved phenomena.

Methods such as the structural analysis, the hierarchical analysis, the time series (in view of determining the short range, middle range and long range evolution) can very easily combine with this methodology according to the searched goals and to the characteristics of the process being studied (Dufour, 1979).

This methodology, like others, implies some imperfections and we do not underestimate their importance. These imperfections are essentially due to the difficulty to implement rigorous techniques leading to an accurate knowledge of the behaviour of a system as complex as a cement kiln.

REFERENCES

Aamand, C. (1969). Simulation of a cement kiln. Congrès sur la simulation des processus continus, San Francisco, 205-214.

Benzecri, J.P. (1973). L'analyse des données, Vol. II. Dunod Ed. Paris.

Bertier, P., and J.M. Bouroche (1975). L'analyse des données multidimensionnelles. Presses Universitaires de France Ed. Paris.

Bogue, R.H. (1952). La Chimie du Ciment Portland., Eyrolles Ed, Paris.

Dekkiche, E.A., A. En-Naamani, P. Prévôt, and J.P. Barbier (1980) Experimental analysis and modelling of a rotary cement kiln. Proposed to the 6th IFAC- IFIP Conference on Digital Computer Applications to Process Control, Dusseldorf.

Diday, E. (1971). Une nouvelle méthode en classification automatique et reconnaissance des formes : la méthode des nuées dynamiques. Revue de Statistique Appliquée 19, 19-33.

Dufour, J. (1979). Méthodes et méthodologie d'analyse de systèmes complexes. Applications aux procédés industriels et aux systèmes macroéconomiques. State Thesis. Lyon.

Dumont, G.A., and P.R. Bélanger (1978). Control of titanium dioxide kilns. IEEE Trans. Autom. Control., 23, 521-531

Koivo, A.J., and P.M. Chase (1971). Determination of a mathematical model for a lime kiln. Proceedings of the 3th IFAC/IFIP Conference on Digital Computer Applications to Process Control, Helsinki.

Lea, F.M. (1970). The chemisty of cement and concrete, 3rd ed. Edward Arnold. London
Lyons, J.W., H.S. Min, P.E. Parisot, and J.F. Paul (1962). Experimentation with a wet process rotary cement kiln via analog computer. I and EC Process Design and Development, 1, 25-33.
Manitius, A., E. Kurcyusz, and W. Kawecki (1974). Mathematical model of the aluminium oxide rotary kiln. I and EC Process Design and Development., 13, 132-142.
Mairlot, F.E. (1974). Propositions concernant la définition de l'objet de la cybernétique et l'extension de son contenu. Cybernetica., 17, 186-198.
Min, H.S., P.E. Parisot, J.F. Paul, and J.W. Lyons (1961). Computer simulation of a wet process cement kiln operation. Preprint n° 202-LA 61, Instrument Society of America, Fall Instrument Automation, Conf., Los Angeles, California.
Prévôt, P., and G. Vitry (1976). Analyse des correspondances et procédés complexes. Automatisme., 8, 295-300.
Spang, H.A. III (1972). A dynamical model of a cement kiln. Automatica., 8, 309-323
Uronen, P., K. Leiviskä., and H. Aurasmaa (1975). Simulation study of a lime kiln Proceedings of the International Symposium and Course Simulation - Zurich., 222-227.

APPENDIX

List of the symbols

Symbol	Notation	Value used	Unit
A_B	pre-exponential factor - $CaCO_3$	$1.64x10^{35}$	h^{-1}
A_F	pre-exponential factor - fuel	$5.6x10^{10}$	h^{-1}
A_X	pre-exponential factor - C_2S	$1.48x10^9$	h^{-1}
A_Y	pre-exponential factor - C_3S	$4.8x10^8$	h^{-1}
C_F	normalize amount of available fuel		
C_{sg}	specific heat of gas	0.28	kcal/kg°C
C_{sm}	specific heat of material	0.27	kcal/kg.°C
C_{sp}	specific heat of wall	0.27	kcal/kg.°C
d_o	% available oxygen		
E_B	activation energy - $CaCO_3$	$1.92x10^5$	kcal/kg.mole
E_F	activation energy - fuel	$1.36x10^5$	kcal/kg.mole
E_X	activation energy - C_2S	$4.61x10^4$	kcal/kg.mole
E_Y	activation energy - C_3S	$0.61x10^5$	kcal/kg.mole
f_1	coefficient of conduction-gas to wall	19.53	kcal/m^{2}h°C
f_2	coefficient of conduction-material to gas	19.53	kcal/m^{2}h°C
f_3	coefficient of conduction-wall to material	19.53	kcal/m^{2}h°C
f_4	coefficient of conduction-wall to outside air	3.42	kcal/m^{2}h°C
G_F	amount of fuel per hour	2400	kg/h
h_o	fraction of radiation not absorbed by the gas	$75.8x10^{-3}$	
k_B	reaction rate - $CaCO_3$		
k_F	reaction rate - fuel		
k_X	reaction rate - C_2S		
k_Y	reaction rate - C_3S		
M_a	molecular weight of air		g
M_B	molecular weight of $CaCO_3$	100.09	g
M_c	molecular weight of CaO	56.08	g
M_{c_1}	" " Carbone	12.01	g
M_{O_2}	" " oxygen	32.00	g
M_S	" " SiO_2	60.09	g
M_X	" " C_2S	172.30	g

Symbol	notation	value used	Unit
M_Y	molecular weight of C_3S	228.33	g
P	pressure in the air		
r_F	particle size of fuel	10^{-2}	cm
r_1	inside radius of the kiln	1.77	m
r_2	outside radius of the kiln	2.00	m
T_a	temperature outside the kiln		°K
T_g	temperature of gas		°K
T_m	temperature of material		°K
T_p	temperature of wall		°K
v_g	velocity of gas	12200	m/h
v_m	velocity of material	55	m/h
ΔH_B	heat of reaction - $CaCO_3$	708.32	cal/g
ΔH_F	" " " fuel	-7778	cal/g
ΔH_X	" " " C_2S	-211.66	cal/g
ΔH_Y	" " " C_3S	6.11	cal/g
ε_g	radiation coefficient - gas	0.273	
ε_m	" " material	0.500	
ε_p	" " wall	0.751	
ρ_F	density of fuel	920	kg/m^3
ρ_g	" " gas	0.80	kg/m^3
ρ_m	" " material	897	kg/m^3
ρ_p	" " wall	1794	kg/m^3
ψ	kg CO_2 / kg $\overline{CaO}$		

ON LINE ESTIMATION OF CLINKER QUALITY IN A ROTARY CEMENT KILN

A. En-Naamani*, P. Prévôt*, E. A. Dekkiche* and J. P. Barbier**

**Laboratoire d'Informatique Appliquée, 502 INSA - 20 Av. Einstein 69621 Villeurbanne, France*
***Cimenterie Lafarge de Val d'Azergues, 69380 Lozanne, France*

Abstract. The rotary cement kiln is a complex industrial process which is non stationary, non linear and multivariable. At the present moment it is technologically impossible to get an immediate and precise idea of the clinker quality in the burning zone; an automatic analysis sampler only measures the "free lime" after a thirty minute time lapse. It is in order to improve the control of the process by determining this variable instantaneously that we have built a free lime estimator. We describe the experimental method used to build the model. By continually adapting the model's parameters, we attempt to follow the structural evolution of the system. The principal origins of this evolution are slow integration phenomena and changes in the characteristics of the raw material.

Keywords. Cement industry; adaptative control; quality control; on line estimator.

INTRODUCTION

The studies carried out on the rotary cement kiln up to now can be placed in one the following categories :

- Theoretical approach : using the laws of thermodynamics and chemistry as a basis, Spang (1972) built a theoretical model giving the thermic and chemical profiles of the kiln. This approach leads to a better understanding of the physico-chemical phenomena and are usually used as a basis for design studies. However the model which Spang obtained is not really useable to control the process, because it would require a continual measuring of the characteristics of the raw material and of the combustible matter.

- The black box approach: for long kilns, and working from experimental data, T. Otomo et al. (1972) presented a statistical approach for the computer control of rotary kilns.

- Fuzzy logic control : I.G. Umbers et al. (1980) developed the automatic control of a cement works through the study of the behaviour of the human operator.

This shows how important the study of so complex a process is just as much for theoretical development in automatic control, as for its implementation.

In this paper we intend to start with the phenomenological study of the process and then progressively go on to its most characteristic physical aspects in order to set up a model which can estimate a clinker quality at any given moment. This experimental method is based on measuring the principal variables and on human operator's experience.

PROCESS DESCRIPTION

The operation of the cement works is schematically represented in Fig. 1.

After the raw material has been extracted from the quarry, it is broken up, homogenized, ground and stocked in silos. By adding 13 % water, the powder (in 100 μ granules) is transformed into granules of 2cm to 5cm in diameter. These granules are heated by the gases from the kiln and are thus dehydrated by passing through a double-chambered heat exchanger (Lepol grate). The decarbonation reaction at 900°C ($CaCO_3 \rightarrow CaO + CO_2$) begins in the last part of the exchanger.

The decarbonation rate (variable DEC, measured on site) thus represents the degree of heat preparation of the material as it enters the kiln itself.

The kiln is 65 meters long and 4 meters in diameter and is slightly inclined at a slope of 4 %. The rotation around its axis takes just over one minute to complete and makes the product move forward naturally. The types of reaction which take place in the kiln give rise to 2 distinct zones. The decarbonation zone is to be found in part of the Lepol grate hot chamber and takes up the first 50 meters of the kiln. The burning zone is situated, on average, along the last 15 meters of the kiln. This zone corresponds to the highest temperatures of material and gas.

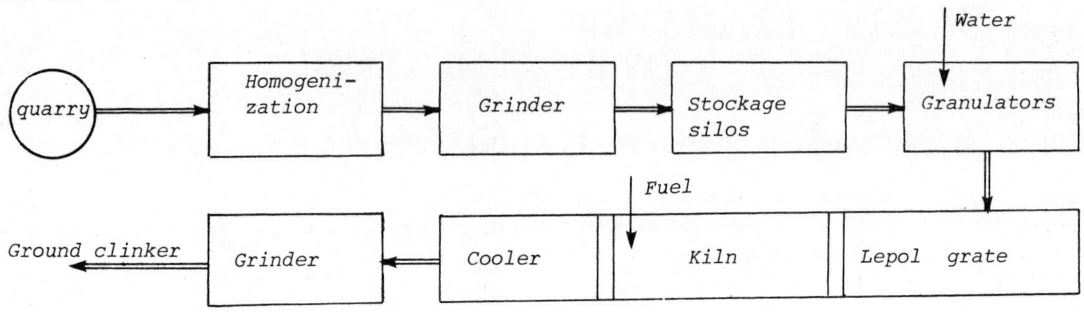

Fig. 1. Material flow

When introduced in this burning zone, the material enters a state of fusion where the CaO from the decarbonation combines with the silica (SiO_2) the iron ore (Fe_2O_3) and the shale ($A\ell_2O_3$) to form the basic constituants of the cement. This reaction, which takes place at 1450°C is globally endothermic and is referred to as clinkerization. The necessary energy comes from the fuel and the pulverized coal which are injected at material exit level by a tuyere. The quality of the clinker is determined by the percentage of free lime (variable CAO). It must be held within given limits. Unfortunately at 1450°C, there is not yet any measuring instrument capable of sampling, and making an instantaneous analysis to give the percentage of free lime. Practically, on the plant, an analyser is placed at the exit from the kiln, and the measure corresponds to the state of burning of the material about 30 minutes before, thus introducing a most regrettable delay in the running of the process. All the variables used here are measured every 3 minutes, except for variable CAO which is only known every 10 minutes.

PHENOMENOLOGICAL STUDY

Since, as a matter of principle, we do not want to perturb the process, so that production remains unaltered, we are obliged to work with data which is influenced by multiple feedbacks. When we come to study the relations between variables, we must separate the causal relations which result from the very nature of the process from those which are a result of the feedback. By the correlation and regression between the different variables of the process, we have been able to draw up a diagram showing the relations (Fig. 2) which exist between the most important variables in our study.

Control variables
- feed rate at kiln-entry (FEE)
- fuel rate (FUL)
- kiln rotation speed (KRS)

Controlled variables
- free lime content (CAO)

Intermediate variables
- decarbonation percentage at Lepol grate exit (DEC)
- kiln motor intensity (KMI)

In Fig. 2 the arrows show the direction of the relation, and the sign given to each of them, specifies the variation direction of the two variables in relation to each other. For example : an increase in FUL improves burning and leads to a decrease in CAO; an increase in KRS decreases the transit time of the material in the kiln — this leads to an increase in CAO, etc.

Thus Fig. 2 shows the existence of several different control modes which can lead to similar results for clinker quality. We have used the relations represented in the diagram to build our model.

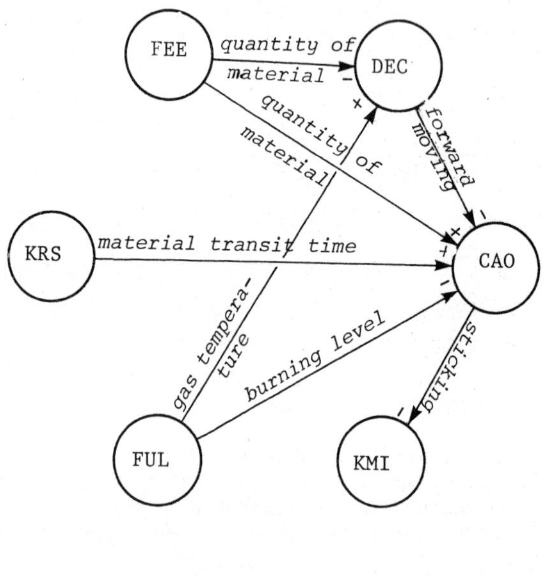

Fig. 2. Relational diagram.

Successive consideration of phenomena

Over a considerable number of days most of the linear regressions between CAO and the other variables, give KMI as being the best dependent variable. This variable represents the power which must be furnished to the drive motor in order to keep the kiln at a given rotation speed. Since the mass of the kiln itself is considered as being unvariable, KMI mainly changes as a result of the following two factors :

1. The total quantity of material in the kiln
2. The upward movement of the material along the kiln wall, principally in the burning zone, because of its semi-liquid state – sticking phenomena –(cf. Fig. 3).

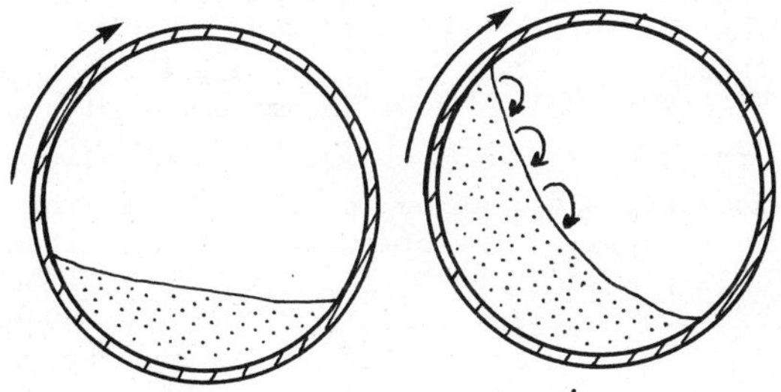

Fig. 3. Build-up position

These two phenomena are inter-dependent. Factor 1 changes as a result of change in FEE and KRS, whereas factor 2 depends upon the burning level of the material (cf. Fig.2), and consequently upon the FUL and KRS. The action of these last two is however reduced by the thermal inertia of the kiln.

Fluctuations in the composition of the raw material also act upon factor 2. Since we cannot take this composition directly into account, as it is not measured regularly, we will try to reduce its effects by adapting the model.

Phenomenon 1 : Total mass of the material. Working from variables FEE and KRS, we intend, by simulation, to estimate the distribution of the material throughout the kiln. The speed at which change occurs in the burning zone leads us to adopt a discrete step of the spatial simulation, which varies according to whether we are dealing with the decarbonation zone (2 meter lengths), or with the burning zone (0.5 meter lengths).

Phenomenon 2 : Speed at which the material moves through the kiln.

The speed at which the material moves forward is determined by 2 factors : the kiln rotation speed (KRS) and the physical state of the material. Thus, in the decarbonation zone the carbon dioxide produced causes a slide effect which is favoured by the granular nature of the material. On the other hand, in the burning zone, the viscous state of the material means that there is a slowing-down and building-up. Fig. 3 shows the position of the build-up in the two zones.

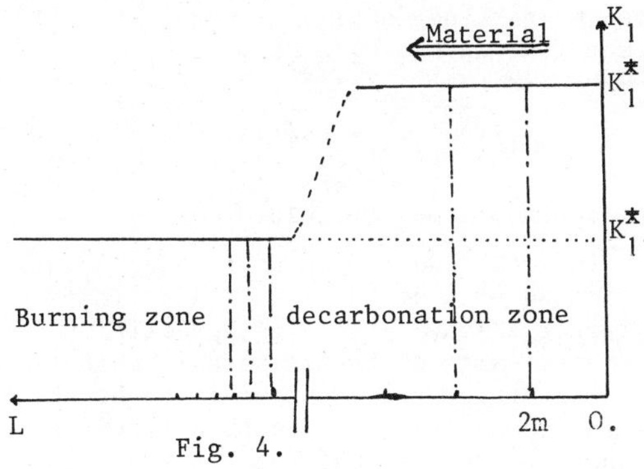

Fig. 4.

In agreement with the work carried out by Saethre (1968) who established the existence of a relationship of 2 : 1 between the speed at which the material moves in the two zones, we give the forward speed of the material accross 'i' at instant 't' by :

$$V_\ell(i,t) = K_1(i)\ KRS(t) \qquad (1)$$

where $K_1(i)$ represents the distance covered by the material in section i for one round of the kiln. (cf. Fig. 4).

The estimation of the value of the parameter K_1^* here uses the work of Peray and Waddell (1972) which express the mean forward movement speed for a sample of the material in the following form (see Appendix for the meaning of the used symbols) :

$$T = \frac{0.095\ L}{KRS\ R\ S} \qquad (2)$$

We thus deduce the value of K_1^* as 1.03 m.

If FA(i, t) represents the quantity of material in section i at t, the material balance in a section of the decarbonation zone is given by :

$$FA(i,t) = FA(i,t-\Delta t)\left[1 - \frac{V_\ell(i,t)\Delta t}{DTR1}\right] + FA(i-1,t-\Delta t)\frac{V_\ell(i-1,t)\ \Delta t}{DTR1}\ K_p \qquad (3)$$

The ignition loss, represented by K_p, corresponds to the reduction in the mass of the material caused by the release of CO_2 during decarbonation. In the cement industry this reduction is generally taken to be approximately 0.36 %. We therefore take K_p as 0.9856.

In the burning zone the mean travel speed of the material is 0.72 m/min and the length of each section is 0.5 m. So, in this zone, the material balance is as follows :

$$FA(i,t) = FA(i-1,t-\Delta t)\left[2 - \frac{V_\ell(i-1,t)\ \Delta t}{DTR2}\right] + FA(i-2,t-\Delta t)\left[\frac{V_\ell(i-2,t)\ \Delta t}{DTR2} - 1\right] \qquad (4)$$

Thus by using expressions (3) and (4), we can calculate the material profile at any given moment and deduce the total quantity of material in the kiln by :

$$V_1(t) = \sum_{i=1}^{N} FA(i,t) \qquad (5)$$

Phenomenon 3 : material sticking

Into the model we are going to incorporate the second part of the kiln load referred to as factor 2 above, which is essentially due to the sticking of the material. This second part therefore expresses the state of the burning zone directly. Saethre (1968) expresses the resisting torque of the kiln as :

$$M = \int_0^L W(x).\ R.\ \mathrm{Sin}[\beta(x)]\ dx \qquad (6)$$

where $W(x)$ = quantity of material at abscissa x
$\beta(x)$ = build-up angle at x

It is not easy to measure angle $\beta(x)$. By approximating, the author came to take the angle as 21° in the decarbonation zone and as 45° in the burning zone. Since $\beta(x)$ is in permanent evolution, and because it is important for the kiln load, we have adopted a representation which is near that used by Saethre, but only taking the sticking phenomenon into account. The term "R.sin$\beta(x)$" is then replaced by the normalized curve shown in Fig. 5.

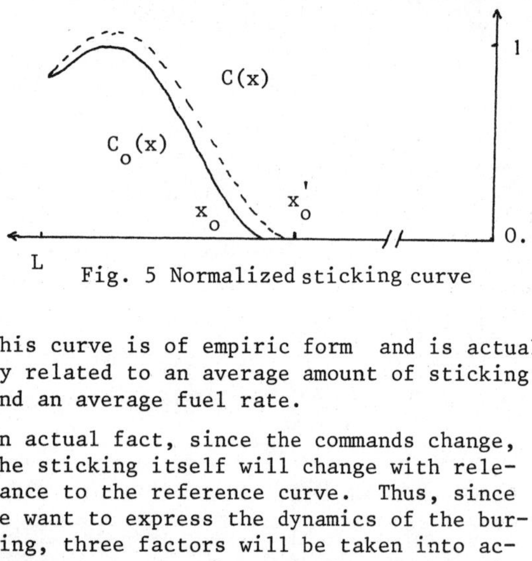

Fig. 5 Normalized sticking curve

This curve is of empiric form and is actually related to an average amount of sticking and an average fuel rate.

In actual fact, since the commands change, the sticking itself will change with relevance to the reference curve. Thus, since we want to express the dynamics of the burning, three factors will be taken into account to express the evolution of the sticking with reference to Co (x).

- Movement of the burning zone : by refering to the work of Saethre (1968) who considers the variations in KMI as being a reflexion of the movement of the start of the burning zone, we carry out a horizontal dilation of the Co(x) curve with reference to :

  $$DKMI = KMI - KMI_o.$$

  where KMI_o is the value of KMI which corresponds to Co(x). If x_o designates the start of the burning zone, the modified curve can be expressed by :

  $$C_o^{*}(x) = C_o\left(65 - \frac{(65-x)(65-x_o)}{65-x'_o}\right) \qquad (7)$$

  where $x'_o = x_o - K'\ DKMI$, designates the new abscissa of the start of the burning zone. How the coefficient of dilation K' may be determined will be seen further on.

- Variation in the heat supply : the movement of the burning zone often comes from a variation in FUL. By substituting KMI for FUL in the expression (7), we only really take account of the thermal inertia of the kiln which slows down the movement of the zone.

  However, a variation in FUL modifies the heat supply. The sticking will therefore be expressed from $C_o(x)$, not only by a horizontal deforming as result of the movement of the burning zone, but also by a vertical deforming of $C_o(x)$ which depends directly upon :

  $$DFUL = FUL - FUL_o$$

  where FUL_o corresponds to $C_o(x)$. This is expressed in (8) by the coefficient K_2.

- Speed of the movement of the material : by considering the speed of the movement of the material, the model takes what the burning conditions were into account. So we use a sticking factor which depends upon 'i' and 't' for the material during its travel time through the burning zone. Finally, these 3 aspects give a sticking factor C(i,t) defined by the expression (8).

$$C(i,t) = C^i_{i-1}(t) + C^i_{i-2}(t) \qquad (8) \quad \text{where:}$$

$$C^i_{i-1}(t) = [C(i-1,t-\Delta t) + C_o^{*}(i-1)\ K_2\ [FUL(t) - FUL(t-\Delta t)] + [C_o^{*}(i) - C_o^{*}(i-1)][1 + K_2\ [FUL(t) - FUL_o]]][2 - \frac{V_\ell(i-1,t)\ \Delta t}{DTR2}]$$

$$C^i_{i-2}(t) = [C(i-2,t-\Delta t) + C_o^{*}(i-2)\ K_2\ [FUL(t) - FUL(t-\Delta t)] + [C_o^{*}(i) - C_o^{*}(i-2)]\ [1 + K_2\ [FUL(t) - FUL_o]]][\frac{V_\ell(i-2,t)\ \Delta t}{DTR2} - 1]$$

The kiln load is then made up of two factors : $V_1(t)$, expressed by (5), which gives the quantity of material in the kiln,

$$\text{and : } V_2(t) = \sum_{i=1}^{N} FA(i,t) \star C(i,t) \qquad (9)$$

which expresses the sticking.

How to determine the coefficients K' and K_2 :

The variations of the total load made up of $V_1(t)$ and $V_2(t)$ are expressed experimentally by the variable KMI and we can then envisage the building of a linear estimator for this variable in the following form :

$$\widehat{KMI}(t)=K_3\ V_1(t)+K_4\ V_2(t) + K_5 \qquad (10)$$

Coefficients K' and K_2 then correspond to the following expressions :

$$\frac{\partial d(KMI,\widehat{KMI})}{\partial K'} = 0 \qquad (11)$$

$$\frac{\partial d(KMI,\widehat{KMI})}{\partial K_2} = 0 \qquad (12)$$

By carrying out a linear regression between $[V_1(t_i), i = 1,m]$, $[V_2(t_i), i = 1,m]$ which are independent variables and $[KMI(t_i), i=1,m]$, we obtain values for K_3, K_4 and K_5, and, consequently, for $[\widehat{KMI}(t_i), i = 1,m]$. From this, we obtain a numerical value for the following expression :

$$d(KMI, \widehat{KMI}) = \frac{1}{m} \sum_{i=1}^{m} [KMI(t_i) - \widehat{KMI}(t_i)]^2 \qquad (13)$$

where $d(KMI, \widehat{KMI})$ is the distance between KMI and $\widehat{KMI}$, and is dependent upon K' and K_2.

For different values of K' and K_2, we select those which minimize $d(KMI, \widehat{KMI})$. Here this last step was simply carried out by drawing the values of the Fisher variable in association with the successive regressions (Fig. 6).

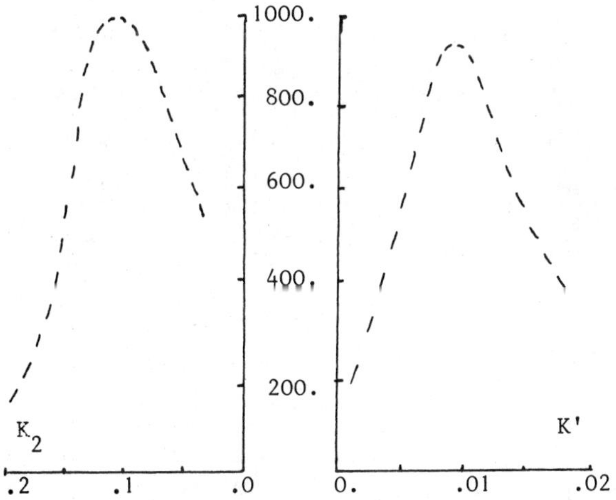

Fig. 6. Fisher variable

The Fig. 10 represents variables KMI and $\widehat{KMI}$.

FREE-LIME ESTIMATOR

Estimator quality test

At any moment, the measure given by the free lime analyser indicates what quality of the clinker at the exit of the kiln was 30 minutes before. In as far as KMI gives an indication of the material's sticking and can be known instantaneously, we could be led to use this variable to deduce CAO.

We must however make the following comments about the direct use of this variable :

- At one and the same time, it represents the quantitative aspect of the material in the kiln and the sticking aspect whilst only the later is explicative of CAO.
- It takes account of the sticking factor associated with all the material present in the burning zone, whilst what we really want is a very localized estimate for the free lime.

The same thing is not true is we use the previously established model. From expressions (3), (4) and (8), we can build a new variable $V_2^\star(t)$, which enables us to estimate the free lime:

$$V_2^\star(t) = \sum_{i=m_0}^{m_1} FA(i,t)\ C(i,t) \qquad (14)$$

In this variable, m_0 and m_1 are chosen with reference to the number of sections which we want to take in the burning zone.

Fig. 11 represents the behaviour of the simple regressive model using $V_2^\star$ given by (14). The dependent variable is the physical measure of CAO.

How to adapt the model

We are perfectly well aware that the differences between the measurements and the estimates are partially due to the following factors :

- precedent expressions do not take account of the make up of the raw material, nor do they take account of the thermal preparation of the material before it enters the burning zone.
- Furthermore we have taken the material's travel speed as constant in the 2 zones. Whereas physically speaking, the forward movement is perturbed by the momentaneous formation of plaques which stick to the kiln wall and then fall off when they became too heavy.

So the preceding evaluation of the coefficients may appear somewhat precarious, but the aim of our research was to advance experimentally. We do not want to substitute the estimate :

$$\widehat{CAO}(t) = K_6\ V_2^\star(t) + K_7 \qquad (15)$$

for CAO permanently. We simply want to

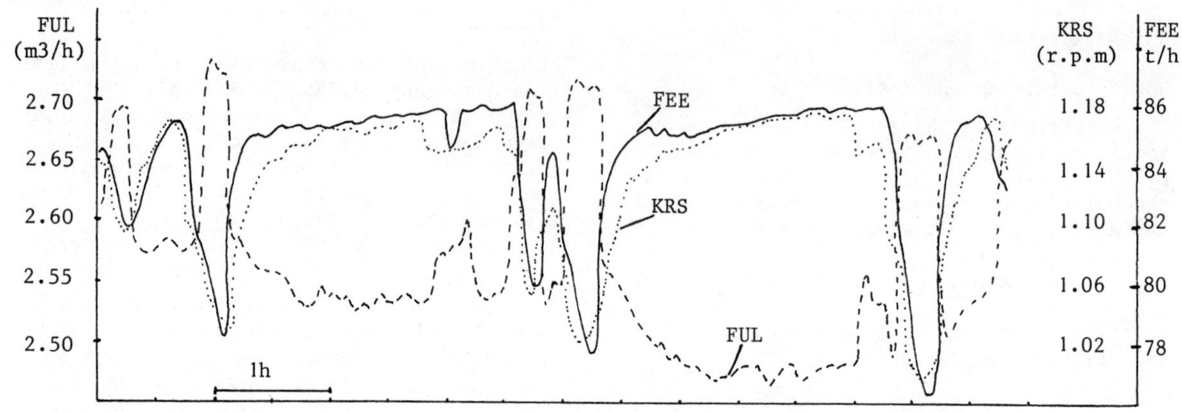

Fig. 7. Time variation of the kiln control variables

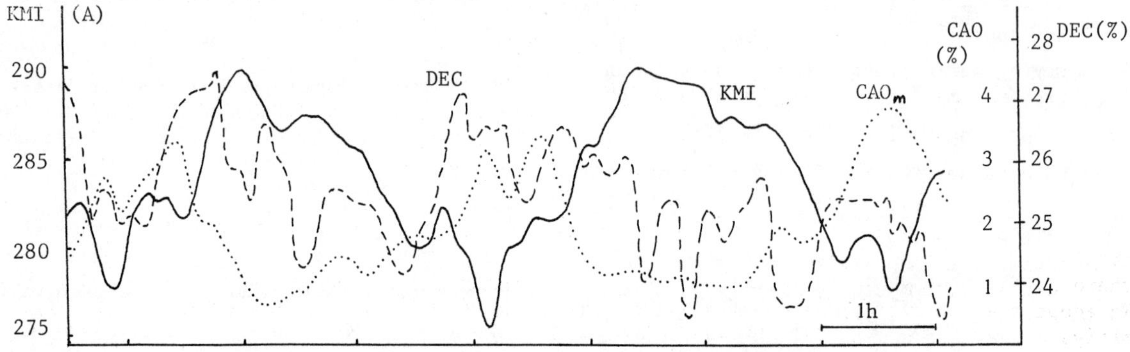

Fig. 8. Time variation of CAO KMI and DEC

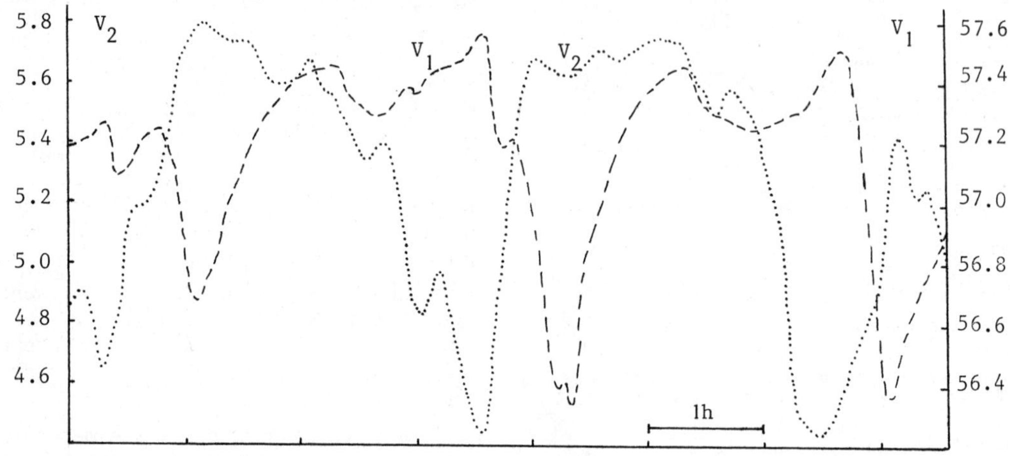

Fig. 9. Time variation of the two parts of the kiln charge.

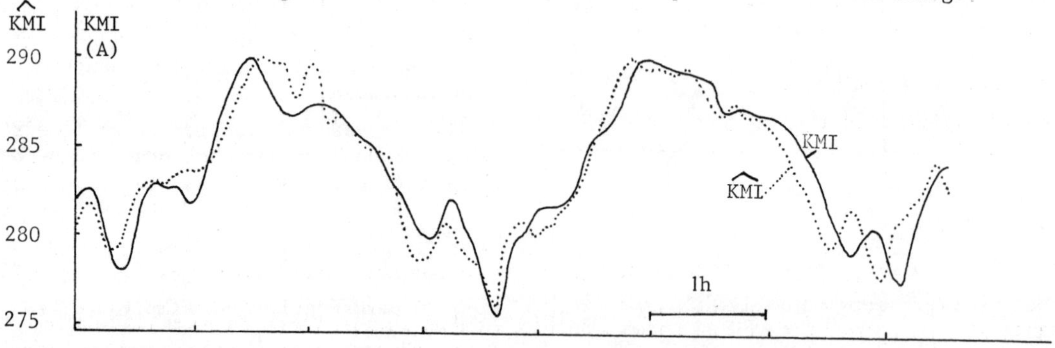

Fig. 10. Linear estimation of KMI by V_1 and V_2 (θ_H=1h)

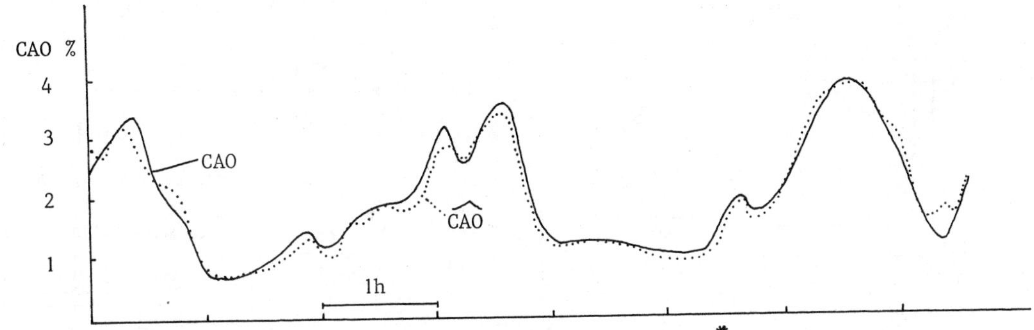

Fig. 11. Linear estimation of CAO by V_2^* (θ_H=1h)

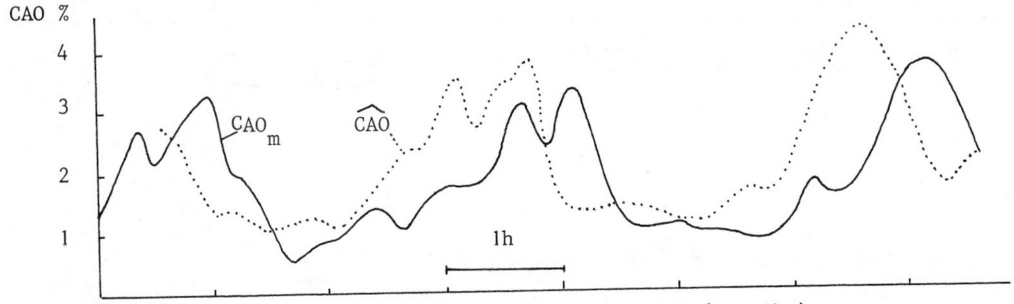

Fig. 12. On line estimation of CAO (θ_H=1h)

obtain an instantaneous estimate without there being a 30 minute delay.
We adopt the following procedure in order to mitigate the consequences of the evolution of the process :
The factors which have not been taken into account in the model lead us to recalibrate the estimator regularly. Thus, K_6 becomes a time varying coefficient. At instant 't', it is possible to calculate $K_6(t - \theta)$ with the help of a regression between CAO known from $t - \theta$ to $t - \theta - \theta_H$, and $V_2^*(t)$ calculated upon the same interval by (14).
The numerical value of $K_6(t - \theta)$ is then used to estimate CAO(t) from $V_2^*(t)$ by (15) — i.e. we adopt the value $K_6(t) = K_6(t - \theta)$ by assuming that this parameter do not vary rapidly.
The Figure 12 shows the result of an on line estimation made on a horizon of 8 hours. ($CAO_m(t)$ designates the measure of free lime known at instant 't').
The parameters $K_4(t)$ and $K_6(t)$ varying in the same sense, it may be interesting in the future to estimate $K_6(t)$ precisely by taking into account what happened between $'t - \theta'$ and 't', which is defined by $K_4(t)$.

Figure 13 represents this general methodology.

CONCLUSION

In this paper, a modelling of the running of a rotary cement kiln has been approached experimentally. We have been able to take the most significant phenomena into account by relying on the experience of the human operator and on observation. The model obtained has the advantage of being very simple and useable on line. It has been adjusted by simulation, using real running data.

The delayed knowledge of the clinker quality introduces oscillatory phenomena in the running of the process and they are due to brutal variations in the control variables. As fig. 7 shows, these variations are often misused, since the real value of the free lime is very different from the measure known at that particulary instant.
By giving an immediate CAO estimate, the model contributes to stabilizing the burning, and as a result, to economizing fuel for a more regular clinker quality.
The model's limitations come from :

- Its linear formulation. The linear regressive model (11) and the expression of $V_2(t)$ are particularly result of an a priori choice.
- The sensitivity of the model when a delay appears between V_2 and CAO (caused by irregular material moving).
- The initial assumptions about travel time and the empiric form of the reference sticking curve.
- Unknowing of the raw material characteristics which are unmeasured and very important for chemical reactions.
- The existence of other phenomena, which change the running of the kiln radically. So, the formation of rings may very slowly lead to the blocking up of the kiln. The breaking of such a ring, whether this

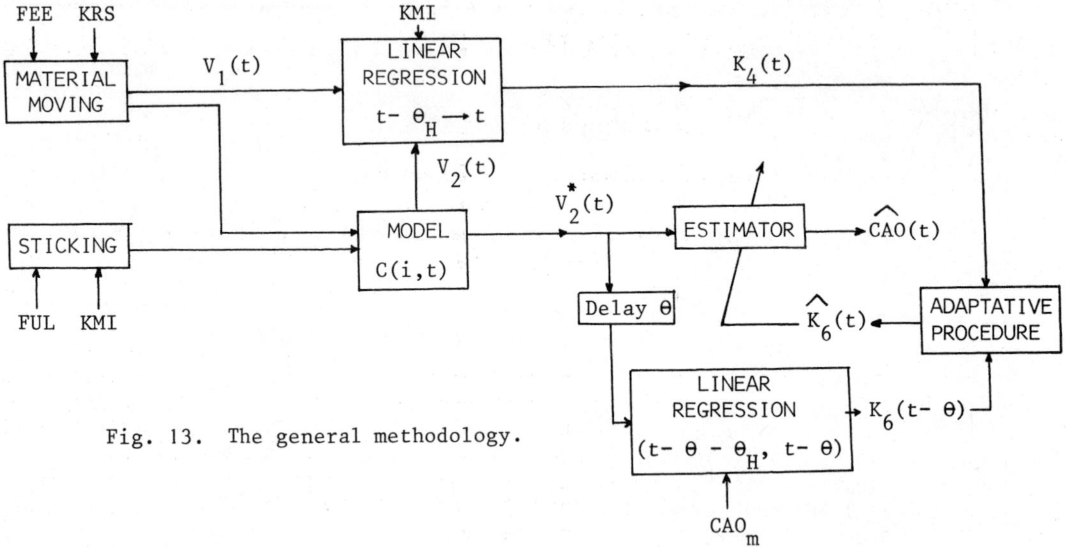

Fig. 13. The general methodology.

comes about naturally, or has to be brought about, leads to a dysfunction for which there would be no point in building a free-lime estimation model.

The model presented here is actually completed by taking account of the evolution of DEC throughout the decarbonation zone to obtain initial values for the sticking curve ($C_0(x)$). It is presently being tested over several weeks of running of the process, before being used directly in the industrial plant.

REFERENCES

Otomo, T., T. Nakagawa, and H. Akaike (1972). Statistical approach to computer control of cement rotary kilns. Automatica., 8, 35-48.

Peray, E., J. Waddell. (1972). The rotary cement kiln., chemical publishing Co., Inc., New York.

Saethre, A., (1968). On the relation between the burning zone temperature in a cement kiln and the kiln drive power. International Seminar of automatic control in lime, cement and connected industries. Brussells. 179-182.

Spang, H. A. III. (1972). A dynamic model of cement kiln. Automatica., 8, 309-323.

Umbers, I. G., and P. J. King (1980). An analysis of human decision-making in cement kiln control and the implications for automation. Int.J.Man-Machine Studies, 12, 11-23.

APPENDIX

List of the symbols

Symbol	Notation	Value	Unit
DTR1	length of a section in decarbonation zone	2	m
DTR2	length of a section in burning zone	0.5	m
K_1^*	distance covered by the material for 1 round of the kiln in the decarbonation zone	1.03	m
K_p	parameter of ignition loss	0.9856	
KRS	mean kiln rotation speed	1.16	r.p.m.
L	total length of the kiln	55	m
m	number of samples used by regression	20	
N	number of sections	55	
R	internal radius of the kiln	2	m
S	kiln slope	0.04	m/m
T	mean transit time	66	min

ON LINE IDENTIFICATION OF A FLUIDIZED BED HEAT EXCHANGER SYSTEM

T. Jeevanandam* and S. S. Lamba**

**System Engineering and Computer Services Department, Engineers India Limited*
26 Kasturba Gandhi Marg, New Delhi 110 001, India
***Centre Nationale de la Recherche Scientifique, Laboratoire des Signaux et Systemes*
Plateau du Moulon, 91190 Gif-Sur-Yvette, France

Abstract. Experimental investigations on a laboratory model fluidized bed heat exchanger system, when coupled on-line to a process computer are reported in this paper. The two input - two output process system model is identified through an on-line correlation plus least squares parameter estimation technique. The time series model, thus obtained, is utilized for evaluating a minimal discrete state-space model and a transfer matrix for the system. The validity of the identified model is carried out by comparing the transient response of the model and the system for a step change in each of the two inputs.

Keywords. Process control; identification; multivariable systems, digital computer applications.

PREAMBLE

Computer controlled industrial processes require process models for effectively carrying out the control actions. Usually the control computer is also used for identifying a simple model from input - output measurements. Excellent reviews on identification of process systems using off-line and on-line techniques have appeared in the literature (Astrom and Eykoff, 1971; Gustavsson, 1975). This paper describes the identification of a dynamic model for a laboratory model fluidized bed heat exchanger using a process computer.

Since the last few years fluidized bed operation of nuclear reactor and coal combustors is being considered as the most appropriate technology for power generation (Trupp 1968; Nack et al., 1976). In a fluidized bed nuclear reactor, a heat exchanger system is superimposed upon the reactor and supplies steam to a secondary unit to generate electrical power. Due to the problems that arise in the turn-down procedures for such fluidized bed steam raising plants, it has been suggested that reactor part of the process may be separated from the heat exchanger portion (Botterrill and Bessan, 1976). For such an arrangement, by controlling the flow of fluidized solids between the reactor and the heat exchanger, the rate of heat extraction can be regulated. An important design objective in a nuclear power generation plant is its capability to respond to load changes in the interconnected electrical network. Large and fast load changes require simultaneous coordinated control actions for both the reactor and control valves of the turbine. Other major disturbances in the system may occur as a result of the variations in the chemical properties of the raw materials used in the reactor and the time varying nature of the various heat transfer coefficients. An adaptive control strategy becomes necessary for operating the plant at its optimum conditions in the presence of such disturbances (Frogner and Rao, 1978).

In the application of fluidized bed nuclear reactor heat exchanger system, the plant is operated in a recirculatory manner with the fluidized solid particles as heat carriers between the reactor and the heat exchanger. As the solid recirculatory type model nuclear reactor is difficult to construct, a simplified laboratory model (sand-water) fluidized bed heat exchanger system has been fabricated with a facility of continuous solid moment. The heat generating effect of the reactor part of the system was simulated by electrically heating the solid particles and liquid just prior to their entry into the heat exchanger column. A tubular heat exchanger is inserted inside the fluidized column to recover heat from the bed. This laboratory model fluidized bed heat exchanger system is described in detail in the next section. The overall system can be envisaged as a multivariable system in which the controlled variables are the outlet cooling water temperature and the bed porosity and the manipulative variables are the two components of liquid flow rate and solid feed rate.

It has already been reported (Jeevanandam, 1979) that the dynamic model of the multivariable fluidized bed heat exchanger, when derived from physical principles, is too complicated and is also load dependent. Further, some of the system parameters such as, the heat transfer coefficients may have time varying nature. However, linear system models can be conveniently identified for such systems on a continual basis by employing system identification techniques based on input output data obtained around the operating point.

It has been reported in the literature (Saridis, 1974; Isermann et al. 1974) that, for single input-single output systems, correlation analysis with least squares algorithm gives favourable results compared to other on-line techniques. An extension of this algorithm has been developed and applied for the two input-two output fluidized bed heat exchanger system. The system inputs are perturbed around the optimum operating point (which was calculated earlier for optimum heat transfer coefficient) using PRBS generated from the digital processor. The time series model, thus obtained, is utilized to evaluate a transfer matrix of the system. Finally, a minimal discrete state-space representation of the process system is derived.

PROCESS SYSTEM DESCRIPTION

The laboratory model set-up includes a fluidization column in the form of 46 mm dia. and 1000 mm long glass tube fitted with a brass conical section at the bottom. The outlet of the fluidization column is connected to a moving bed section of the set up. A two-pass heat exchanger tube, made up of copper, is immersed inside the bed in order to recover heat from the system by passing cold water through it. By insulating the rest of the tube inside the column, it is ensured that the heat transfer between the column and the heat exchanger tube takes place only in the region up to 620 mm height from the bottom of the bed.

The fluidizing medium is hot water and its temperature is controlled at $50^{\circ}C \pm 10\%$. Temperature of the incoming sand particles from the moving bed section is also controlled at $50^{\circ}C \pm 15\%$. The tolerance limits simulate a situation similar to the variations in heat generation in an actual reactor part of the system. The complete flow sheet of the laboratory set up and associated accessories are shown in Fig.1. A process computer (PDP 8/e) with associated input / output hardware as shown in Fig. 2 is used for carrying out the experimental investigation. Focal, an interactive higher level language, in its modified version (HYFOC.P) is used to implement the software package developed for on-line identification.

IDENTIFICATION OF TIME SERIES MODEL FROM INPUT-OUTPUT DATA

The identification of system dynamics from input-output data using PRBS is well known and of all the techniques proposed for identifying the system with PRBS, cross correlation of the input and output signals is most often suggested. Identification of the dynamic model of an industrial type heat exchanger using three techniques is reported (Isermann and Baur, 1974). No information is available on the identification of a dynamic model for a continuous type fluidized bed heat exchanger system using real time data. The technique employed here uses input-phase separation feature which facilitates the excitation of all the process input simultaneously. The shift and add property of PRBS is employed for generating several sequences (delayed PRBSs) shifted in phase by $\nu\Delta t/m$ for an m input system from a master sequence of period $\nu\Delta t$ where ν is number of individual binary components and Δt is digit interval. Data for the PRBS used in experimental investigations for the two-input two-output heat exchanger system is given in Table 1

TABLE 1 Choice of PRBS parameters and system variables for Experimental Investigation

Master PRBS	
No. of stages	LM = 7
No. of bits	ν = 127
Digit interval	Δt = 1 sec
Delayed PRBS	
Delay introduced	$= \frac{\nu}{2} \doteq 64$ bits
Signal amplitude	$= \pm 10\%$ around normal operating point

A multivariable continuous time process with sampled input signal vector w(k) or order p and output signal vector, y(k) of order p can be represented by a linear difference equation as

$$y(k)+A_1y(k-1) + \dots A_my(k-m)=B_1w(k-1)+\dots + \dots B_m\,w(k-m) \quad (1)$$

where A_i's and B_i's are pxp matrices. The order of the polynomial is m. The transfer matrix for this representation may be derived in z domain as

$$H(z)= \left[I+A_1z^{-1}+\dots+A_m\,z^{-m}\right]^{-1}\left[B_1z^{-1}+\dots+B_mz^{-m}\right] \quad (2)$$

$$\text{where}\quad y(z) = H(z)\,w(z) \quad (3)$$

The algorithm derived in the sequel is used for identifying the elements of the matrices A_i and B_i from sampled input-output data. When a stationary discrete input signal $w_i(k)$ is used in the identification procedure, the auto-correlation function is defined by

$$\phi_{w_iw_i}(\tau)=\lim_{N\to\infty}\frac{1}{N+1}\sum_{k=0}^{N} w_i(k)w_i(k-\tau) \quad (4)$$

where τ is the shift interval. The cross-correlation function of the input $w_i(k)$ and output $y_j(k)$ is defined by

$$\phi_{w_i y_j}(\tau) = \lim_{N \to \infty} \frac{1}{N+1} \sum_{k=0}^{N} w_i(k-\tau) y_j(k) \quad (5)$$

The auto and cross-correlation functions can also be calculated recursively as follows

$$\phi_{w_i w_i}(\tau,k) = \phi_{w_i w_i}(\tau,k-1) + \frac{1}{k+1}\left\{w_i(k-\tau)w_i(k) - \phi_{w_i w_i}(\tau,k-1)\right\} \quad (6)$$

and

$$\phi_{w_i y_j}(\tau,k) = \phi_{w_i y_j}(\tau,k-1) + \frac{1}{k+1}\left\{w_i(k-\tau)y_j(k) - \phi_{w_i y_j}(\tau, k-1)\right\} \quad (7)$$

for $k = 0, 1, 2 \ldots$

with $\phi_{w_i w_i}(\tau, -1) = \phi_{w_i y_j}(\tau, -1) = 0$.

For the identification in the sequel, we assumed that the inputs are uncorrelated.

i.e. $\phi_{w_j w_i}(\tau) = 0 \quad i = 1, \ldots, p \quad i \neq j \quad (8)$

With the definitions as stated above, the identification algorithm is derived and briefly explained as follows.

Expanding (1) for output y_1 in two input-two output system and multiplying by $w_1(k-\tau)$ and taking expectation leads to

$$\begin{aligned}\phi_{w_1 y_1}(\tau) = & -a_{11}^{(1)} \phi_{w_1 y_1}(\tau-1) - \ldots \\ & - a_{11}^{(m)} \phi_{w_1 y_1}(\tau-m) - a_{12}^{(1)} \phi_{w_1 y_2}(\;-1) \\ & - \ldots - a_{12}^{(m)} \phi_{w_1 y_2}(\tau-m) \\ & + b_{11}^{(1)} \phi_{w_1 w_1}(\tau-1) + \ldots + b_{11}^{(m)} \phi_{w_1 w_1}(\tau-m) \\ & + b_{12}^{(1)} \phi_{w_1 w_2}(\tau-1) + \ldots + b_{12}^{(m)} \phi_{w_1 w_2}(\tau-m)\end{aligned} \quad (9)$$

Similarly, multiplying the expanded version of (1) for output y_1 by $w_2(k-\tau)$ and taking expectation and adding with (9) yields

$$\begin{aligned}\phi_{w_1 y_1}(\tau) + \phi_{w_2 y_1}(\tau) = & -a_{11}^{(1)}\left\{\phi_{w_1 y_1}(\tau-1) + \phi_{w_2 y_2}(\tau-1)\right\} - \ldots \\ & -a_{12}^{(1)}\left\{\phi_{w_1 y_2}(\tau-1) + \phi_{w_2 y_1}(\tau-1)\right\} - \ldots \\ & +b_{11}^{(1)}\left\{\phi_{w_1 w_1}(\tau-1) + \phi_{w_2 w_1}(\tau-1)\right\} + \ldots \\ & +b_{12}^{(1)}\left\{\phi_{w_1 w_2}(\tau-1) + \phi_{w_2 w_2}(\tau-1)\right\} + \ldots\end{aligned} \quad (10)$$

Similarly for output y_2, we can apply the same procedure. For $0 \leq \tau \leq M$, we can write a system of equations with the assumption that for $\tau > M$, $\phi_{w_i y_j}(\tau) \doteq$ constant.

After simplification, the system of equation may be written in a concise form as

$$\phi_1 = S_1 \beta_1$$
$$\text{and} \quad \phi_2 = S_2 \beta_2 \quad (11)$$

where β_1 and β_2 represent the parameter vectors containing coefficients of the matrices A_i's and B_i's. The matrices ϕ_2, S_2 and β_2 are given in the succesive page.

Since a number of approximations are involved in the above derivation an estimation in the least squares sense of the parameter vector can be obtained as

$$\beta_1 = (S_1^T S_1)^{-1} S_1 \phi_1$$
$$\beta_2 = (S_2^T S_2)^{-1} S_2 \phi_2 \quad (12)$$

The model parameters A_i and B_i for a system model TS obtained by performing the identification experiment are given in Table 2.

TABLE 2 Time Series Process Model (TS) Parameters

$$A_1 = \begin{bmatrix}-0.849 & -0.091 \\ 0.017 & -0.648\end{bmatrix} \quad A_2 = \begin{bmatrix}0.615 & 0.019 \\ 0.011 & 0.24\end{bmatrix} \quad A_3 = \begin{bmatrix}-0.094 & -0.006 \\ -0.014 & 0.017\end{bmatrix}$$

$$B_1 = \begin{bmatrix}-0.01 & 0.012 \\ -0.035 & -0.058\end{bmatrix} \quad B_2 = \begin{bmatrix}-0.006 & 0.024 \\ 0.019 & 0.014\end{bmatrix} \quad B_3 = \begin{bmatrix}0.006 & 0 \\ 0 & -0.008\end{bmatrix}$$

In order to verify the model TS obtained from the experiment, transient responses of the system output to a step change in each of the system inputs are compared with the model responses. These are shown plotted in Fig. 3.

TRANSFER MATRIX REPRESENTATION OF THE SYSTEM MODEL

Substituting the values of the matrices A_1, A_2, A_3, B_1, B_2, and B_3 of the model TS in (2). We obtain the pulse transfer matrix for the system as

$$H(z) = \left[I + A_1 z^{-1} + \ldots + A_3 z^{-3}\right]^{-1} \left[B_1 z^{-1} + \ldots + B_3 z^{-3}\right] = \frac{1}{\Delta}\begin{bmatrix}h_{11}(z) & h_{12}(z) \\ h_{21}(z) & h_{22}(z)\end{bmatrix} \quad (14)$$

where

$$h_{11}(z) = 10^{-3}\left[-8.2z^{-1} + 1.3z^{-2} + 5.6z^{-3} - 2.8z^{-4} - 0.8z^{-5} + 0.1z^{-6}\right]$$

$$\phi_2 = \begin{bmatrix} \phi_{w_1y_2}(0) + \phi_{w_2y_2}(0) \\ \vdots \\ \phi_{w_1y_2}(M) + \phi_{w_2y_2}(M) \end{bmatrix} \qquad \beta_2 = \begin{bmatrix} a_{21}^{(1)} \cdot\cdot\, a_{21}^{(m)}\; a_{22}^{(1)} \cdot\cdot\, a_{22}^{(m)}\; b_{21}^{(1)} \cdot\cdot\, b_{21}^{(m)}\; b_{22}^{(1)} \cdot\cdot\, b_{22}^{(m)} \end{bmatrix}^T$$

$$S_2 = \begin{bmatrix} -\{\phi_{w_1y_1}(-1)+\phi_{w_2y_1}(-1)\} \cdot\cdot -\{\phi_{w_1y_2}(-1)+\phi_{w_2y_2}(-1)\} \cdot\cdot \{\phi_{w_2w_1}(-1)+\phi_{w_1w_1}(-1)\} \cdot\cdot \{\phi_{w_1w_2}(-1)+\phi_{w_2w_2}(-1)\} \cdot\cdot \\ \vdots \qquad\qquad \vdots \qquad\qquad \vdots \qquad\qquad \vdots \\ -\{\phi_{w_1y_1}(M-1)+\phi_{w_2y_1}(M-1)\} \cdot\cdot -\{\phi_{w_1y_2}(M-1)+\phi_{w_2y_2}(M-1)\} \cdot\cdot \{\phi_{w_2w_1}(M-1)+\phi_{w_1w_1}(M-1)\} \cdot\cdot \{\phi_{w_1w_2}(M-1)+\phi_{w_2w_2}(M-1)\} \cdot\cdot \end{bmatrix}$$

$$h_{12}(z) = 10^{-3}\{10z^{-1}+12.8z^{-2}-11.96z^{-3}+5.2z^{-4} + -0.56z^{-5}-1.3z^{-6}\}$$

$$h_{21}(z) = 10^{-3}\{-40z^{-1}+63\,z^{-2}-44z^{-3}-3.2z^{-4} +14.5z^{-5}+6.4z^{-6}\}$$

$$h_{22}(z) = 10^{-3}\{-82z^{-1}-27z^{-2}+3.85z^{-3}+3.1z^{-4} -16z^{-5}+6.3z^{-6}\}$$

and

$$\Delta = 1+0.8z^{-1}-0.38z^{-2}-0.776z^{-3}+0.1489\,z^{-4} +0.112z^{-5}-0.0455z^{-6}$$

From the transfer matrix representation, the poles of the system are placed at −0.05, 0.7, −0.8 ± j0.6 and 0.3 ± j0.2. The transfer matrix is particularly useful in designing the system controllers in z-domain.

STATE SPACE REPRESENTATION

Regardless of how the transfer matrix is obtained, it is often desirable to have a time domain representation of the system model. As two types of realisations are possible, viz. controllable and observable, it is important to obtain a minimal realisation for the system dynamics. A minimal realisation can be obtained by combining the controllable (observable) part of an observable (controllable) realisation.

In this section a procedure outlined in the text (Wolovich, 1973) is adopted and a discrete state-space model in the following form is obtained.

$$\begin{aligned} x(k+1) &= F\,x(k) + G\,w(k) \\ y(k) &= C\,x(k) \end{aligned} \qquad (14)$$

where x is n dimensional state vector
w is p dimensional input vector
y is p dimensional output vector
and F, G, C are nxn, nxp and pxn matrices respectively.

For the time series model TS, the following minimal form of the discrete state equation is obtained.

$$F = \begin{bmatrix} 0 & 0 & 0.083 & 0 & 0 & 0.012 \\ 1 & 0 & -0.523 & 0 & 0 & -0.003 \\ 0 & 1 & 0.951 & 0 & 0 & -0.021 \\ 0 & 0 & 0.008 & 0 & 0 & -0.028 \\ 0 & 0 & -0.024 & 1 & 0 & -0.248 \\ 0 & 0 & 0.084 & 0 & 1 & 0.722 \end{bmatrix} \qquad (15)$$

$$G = \begin{bmatrix} 2.1 & 0 \\ -4.6 & 20 \\ -8.2 & 10 \\ -8.0 & -16 \\ 25.0 & 24 \\ -40.0 & -61 \end{bmatrix} \times 10^{-3} \; ; \; C = \begin{bmatrix} 0 & 0 & 1 & 0 & 0 & 0 \\ 0 & 0 & 0 & 0 & 0 & 1 \end{bmatrix}$$

A digital simulation of the discrete state space model (15) is carried out to assess the validity of the model. The responses of the actual system and of the model (15) are plotted in Fig.4. The model responses agree favourably with the system responses.

CONCLUSIONS

The identification of a dynamic model for a two input-two output fluidized bed heat exchanger is carried out. The implementation of the software package for on-line correlation and least square estimation technique utilizes a memory space of 8 K words on a PDP8 computer. The parameters of the dynamic model has been obtained after one complete sequence of the PRBS corresponding to 127 seconds.

It may be noted that the on-line identification of such heat exchanger systems can conveniently be carried out utilizing cost-effective microprocessor based systems. The package developed for this purpose may be implemented either in assembly language or a higher level language such as PASCAL or BASIC.

REFERENCES

Astrom, K.J., and P. Eykoff (1971). System identification – A survey. Automatica, 7, 123–162.

Botterrill, J.S.M., and D.J. Bessan (1976). The flow properties of fluidized solids, Fluidization Technology, vol.II, Hemisphere Publishing Company, 7-12.

Frogner, B.,and H.S. Rao (1978). Control of nuclear power plants, IEEE Trans. Autom. Control, AC-23, 405-417.

Gustavvsson, I. (1975). Survey of applications of identification in Chemical and Physical Processes, Automatica, 11,3-24.

Isermann, R., U. Baur, W. Bamberger, P.Kneppo, and H. Siebert (1974). Comparison of six on-line identification and parameter estimation methods, Automatica, 10, 81-103.

Isermann, R. and U. Baur (1977). On-line identification of a heat exchanger with a process computer - a case study,Automatica, 13, 487-496.

Jeevanandam, T. (1979), Real time computer studies on a fluidized bed heat exchanger system, Ph.D. Thesis, Dept. of Elect. Engg., Indian Institute of Technology, Delhi.

Nack, H., K.D. Kiang, K.T. Liu, K.S.Murthy, G.R. Smithson and J.H. Oxley (1976). Fluidized bed combustion review. Fluidization Technology, vol.II, Hemisphere Publishing Company, 339-379.

Saridis, G.N.(1974). Comparison of six on-line identification algorithms, Automatica, 10, 69-79.

Trupp, A.C. (1968). Dynamics of liquid fluidized beds of spheres, Institution of Chemical Engineers (London) Symposium series, 30, 182-190.

Wolovich, W.A. (1973). The determination of state space representations for linear multivariable systems. Automatica, 9, 97-106.

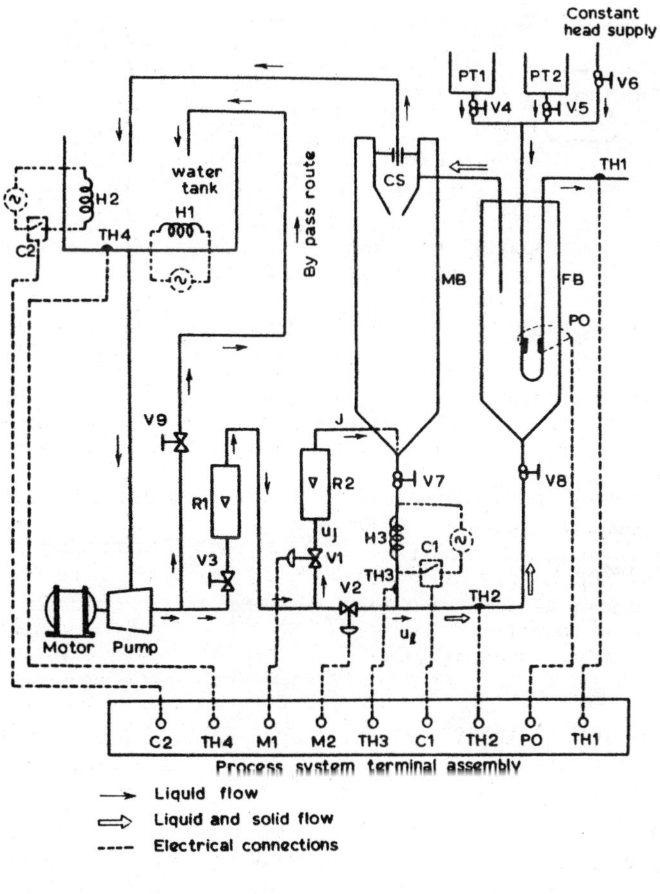

Fig.1. Flow sheet of the fluidized bed heat exchanger System

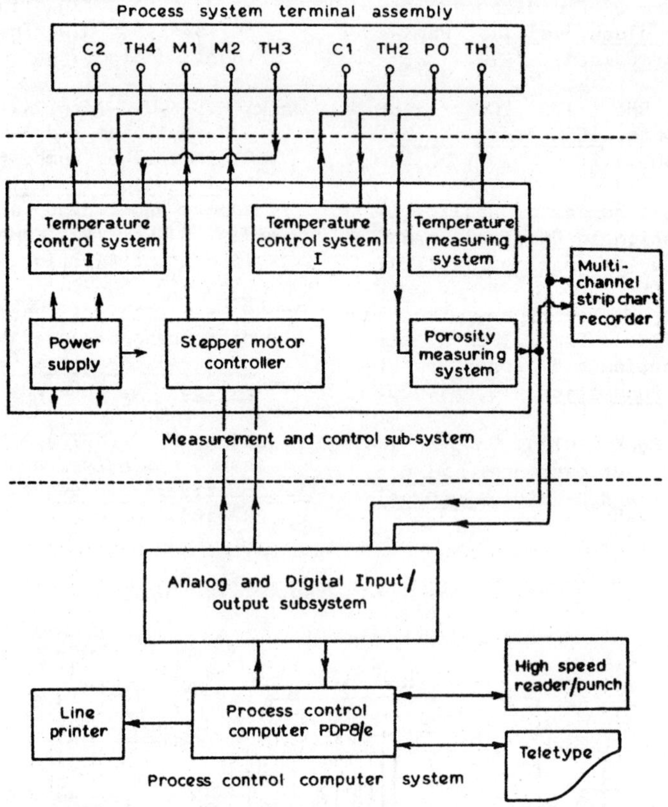

Fig. 2. Functional Flow Diagram of the Set Up

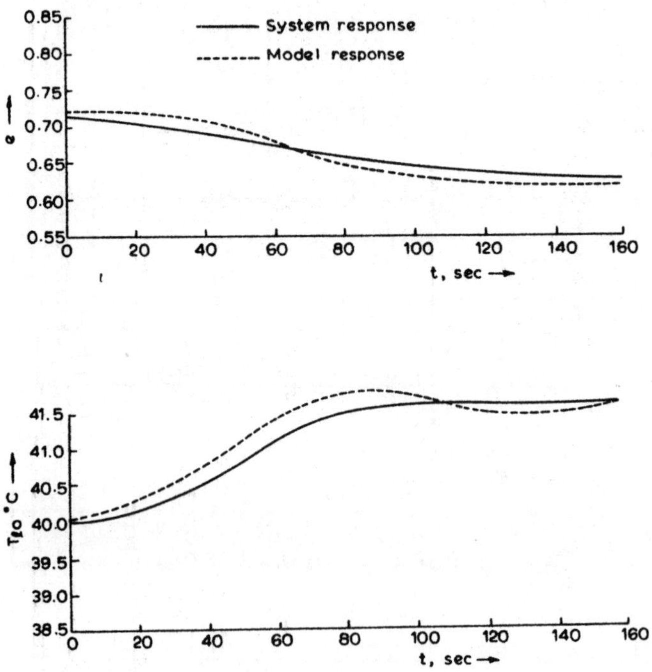

Fig. 3(a) System and time series model responses for a step change in u_ℓ from 1.72 to 1.89 cm/sec

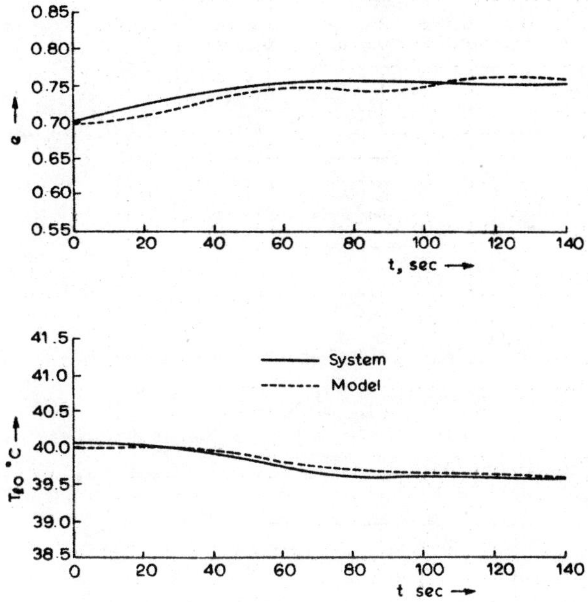

Fig.3(b). Responses of system and model TS for step change in U_j from 0.4 to 0.44 cm/sec

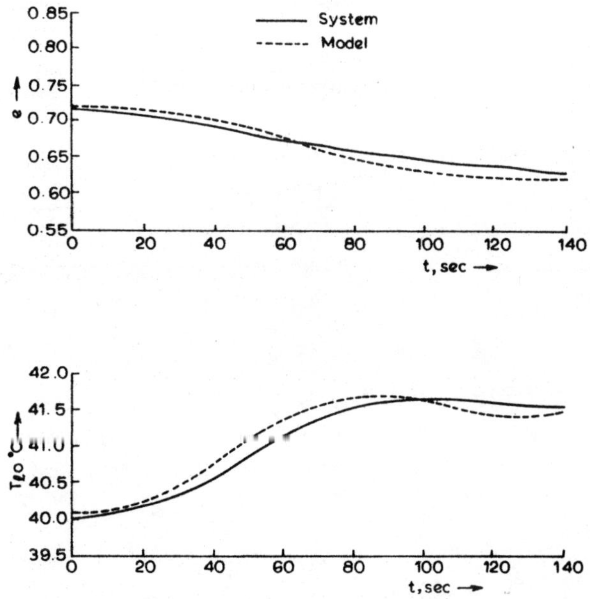

Fig. 4. Responses of system and state space model for a step change in U_ℓ from 1.72 to 1.89 cm/sec

IDENTIFICATION OF AN ETHANE-ETHYLENE DISTILLATION COLUMN

A. Andreoni*, S. Beghelli, M. Di Martino*, R. Guidorzi**, L. Pacinotti* and R. Rossi****

**Datamont S.p.A., Milan, Italy*
***University of Bologna, Bologna, Italy*

Abstract - The identification of complex industrial processes on the only basis of input-output sequences obtained during normal operating conditions plays, at present, an important role. A distillation column, for instance, can work in many different conditions (feed flow, composition, temperature etc.) also if it has been designed with reference to a standard situation. This requires the presence of a controller based on an accurate previsional model of the process. This paper describes the construction of a mathematical model for a ethane-ethylene distillation column by identification techniques.

Keywords - Chemical industry; chemical variables control; identification; linear systems; multivariable systems; petro-chemical control; state-space methods.

I. INTRODUCTION

The design of a distillation column is usually performed with reference to standard operating conditions which are, however, seldom met in practice. Possible design freedom degrees can be exploited to optimize the process yield but optimal values for reference conditions can lead to poor performances in other situations where, for instance, the feed flow, composition or temperature have varied. In these conditions the desired purity degrees of the distillate can be assured only varying other control variables such as the reflux ratio, the reboiler duty, the top and bottom spills etc. The implementation of this control requires the knowledge of a reliable previsional model. The classical methods to obtain a mathematical description of the links among the variables of interest are given by modelling techniques. These techniques are based, when distillation columns are considered, on thermal and mass balances and lead to rather complex models with a well defined physical meaning but with possible calibration and accuracy problems. A different approach, namely identification, is offered by modern system theory. Identification consists in the determination, on the basis of the knowledge of input-output sequences of a given process, of a mathematical model *equivalent* to the process under test. The term equivalent means that the process and the model exibit the same input-output behavior. A model of this kind can therefore be used to predict the response of the real system to assumed inputs. This work presents the results obtained in the identification of a ethane-ethylene distillation column. In Section II the characteristics of this process and of the available data can be found. Section III describes the considered identification procedure, the results obtained in the column identification and in the model validation. Some concluding remarks are presented in Section IV.

II. DESCRIPTION OF THE PROCESS

The considered column is a ethylene-ethane fractionator which can not be considered as classical since an intermediate spill of distillate is present at the 112-th plate. Moreover a gaseous spill is performed in the condenser to strip H_2 and CH_4 and an heat input is present at the 15-th plate. The standard operating conditions of the column are the following

Feed = 40250 Kg/h
Distillate = 29227 Kg/h
Reflux ratio = 3.549
Reboiler duty = 79850 Kg/h
Top ethane residuals = 790 ppm
Bottom ethylene residuals = 480 ppm

The variables which have been considered as inputs of the process are

1) The ratio between the reboiler duty and the feed flow (V/F)
2) The ratio betweer the reflux rate and the feed flow (R/F)
3) The ratio between distillate and the feed flow (D/F)
4) The input ethane composition (EFC)

5) The top pressure (TP)

The variables which have been considered as outputs of the process are

1) The top ethane composition (TEC)
2) The bottom ethylene composition (BEC)
3) The top-bottom differential pressure (TBDP = TP - BP)

The previous variables are measured as deviations from the nominal values expressed in per cent of the measuring device ranges. A diagram of the considered distillation column is reported in Figure 1.

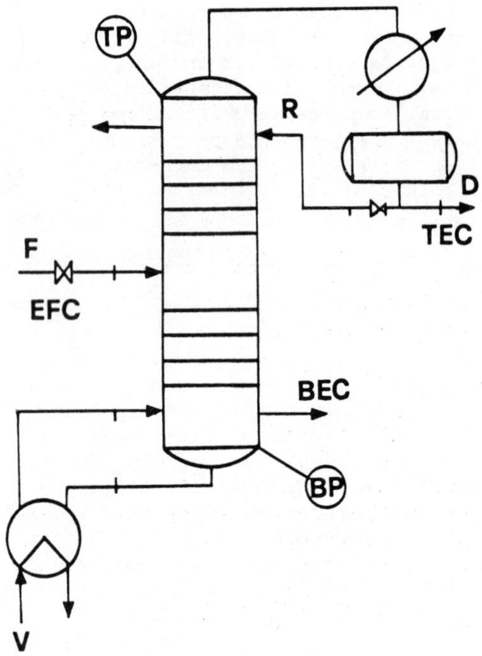

Fig. 1 - Ethane-ethylene distillation column and considered variables

The measures performed on the process have a sampling interval of 15 minutes.
The sequence which has been used for the model identification (Data Set 1) is given by 90 samples. The five input sequences and the three corresponding output sequences are represented in the Figures 2 - 6 and 7 - 9 respectively. Another sequence of 133 samples (Data Set 2) has been used for validation purposes and is represented in the Fig. 10 - 17.

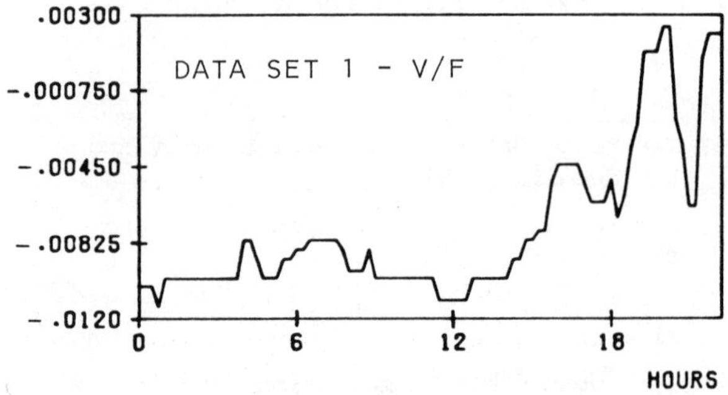

Fig. 2 - First process input

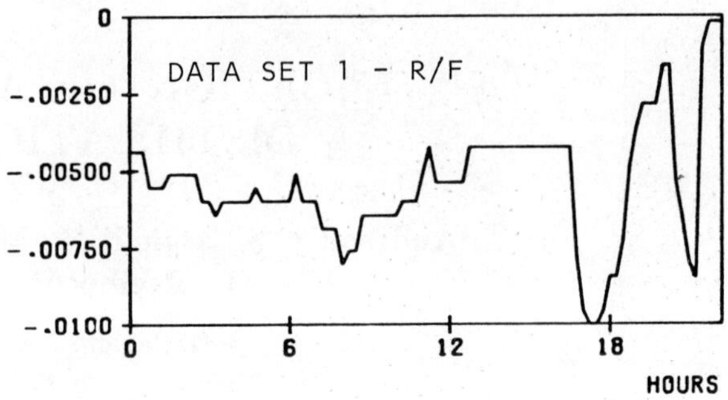

Fig. 3 - Second process input

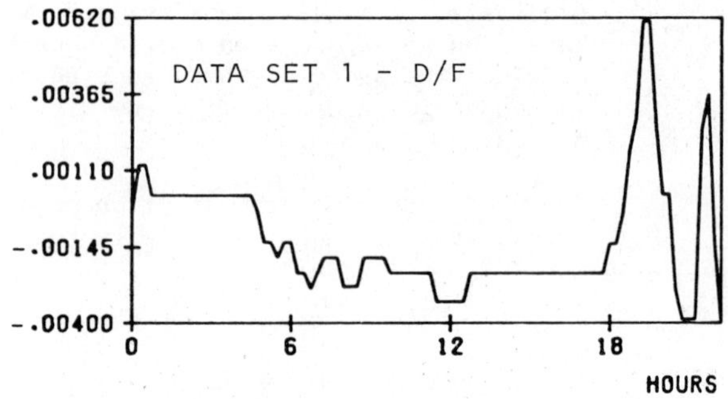

Fig. 4 - Third process input

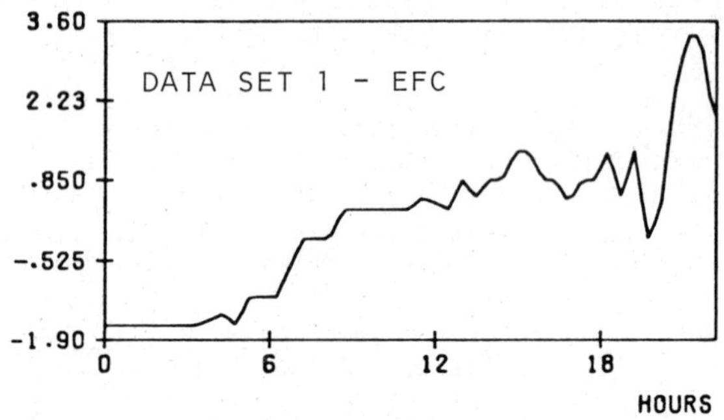

Fig. 5 - Fourth process input

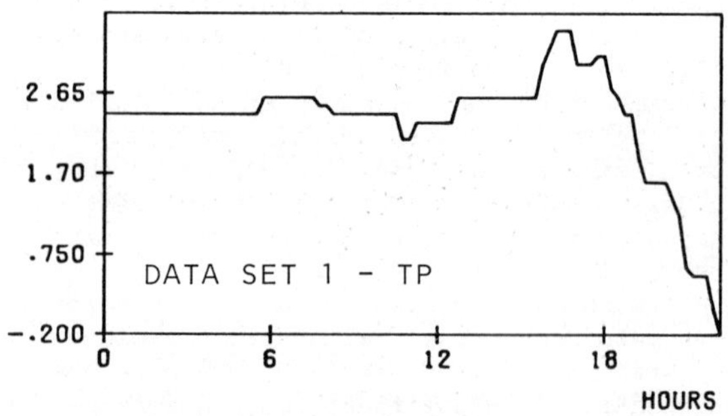

Fig. 6 - Fifth process input

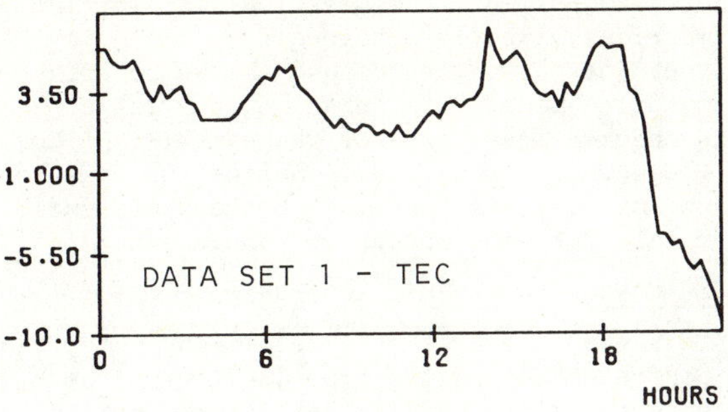

Fig. 7 - First process output

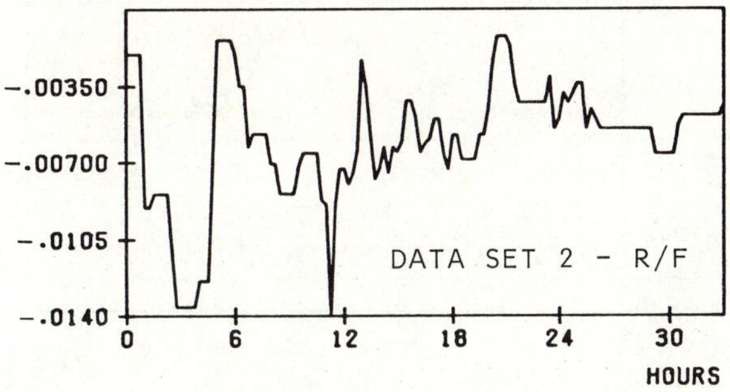

Fig. 11 - Second process input

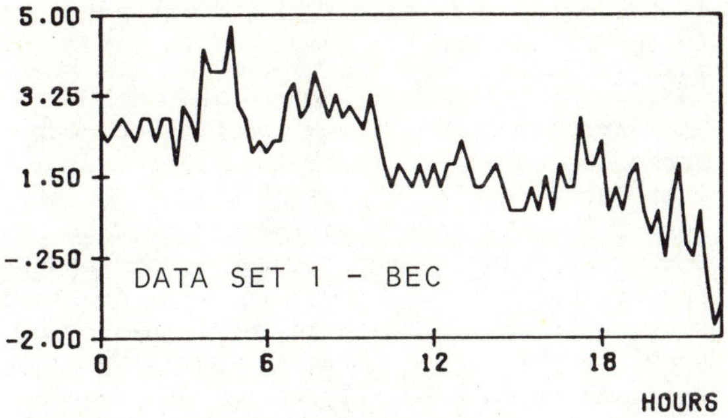

Fig. 8 - Second process output

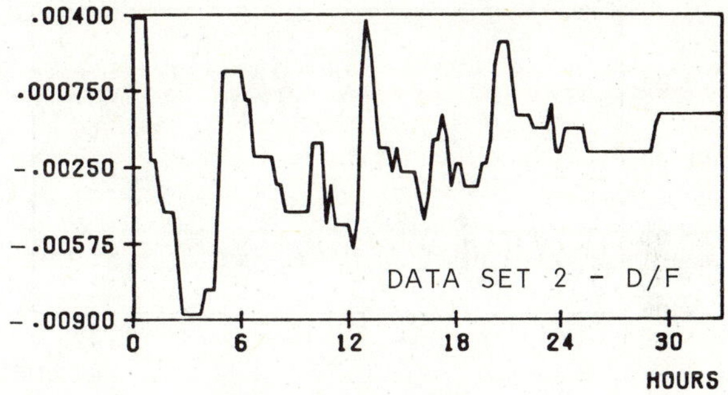

Fig. 12 - Third process input

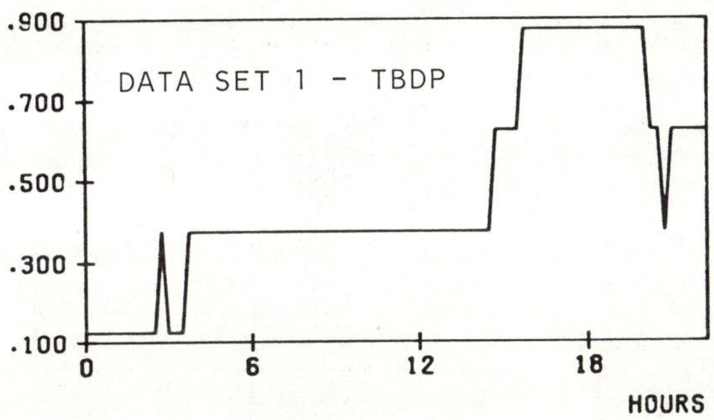

Fig. 9 - Third process output

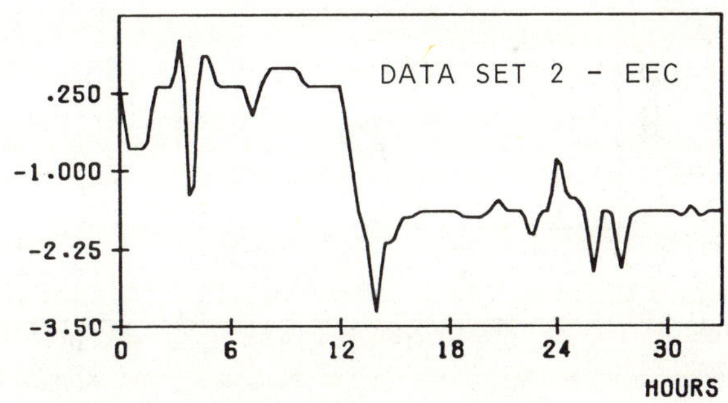

Fig. 13 - Fourth process input

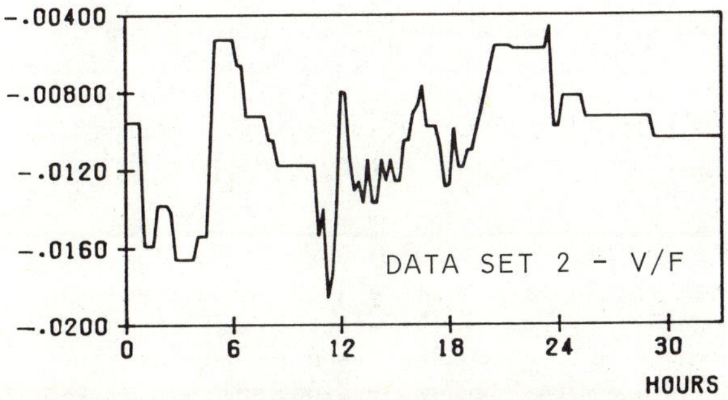

Fig. 10 - First process input

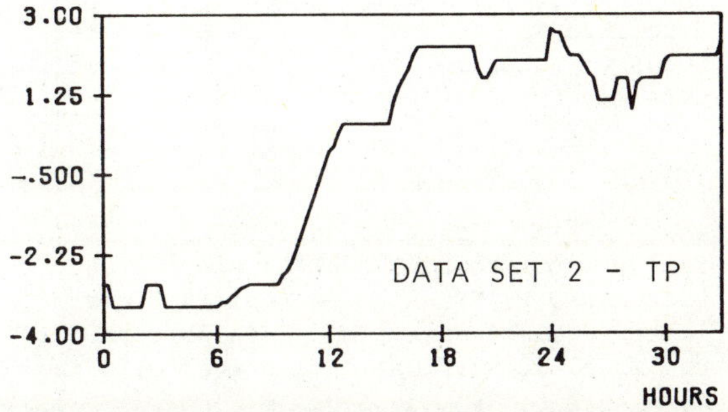

Fig. 14 - Fifth process input

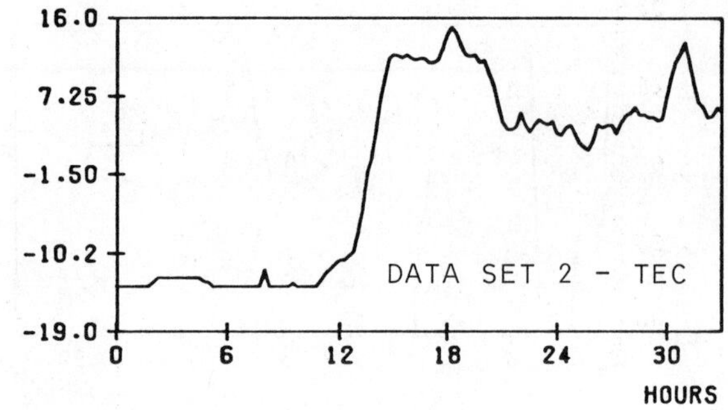

Fig. 15 - First process output

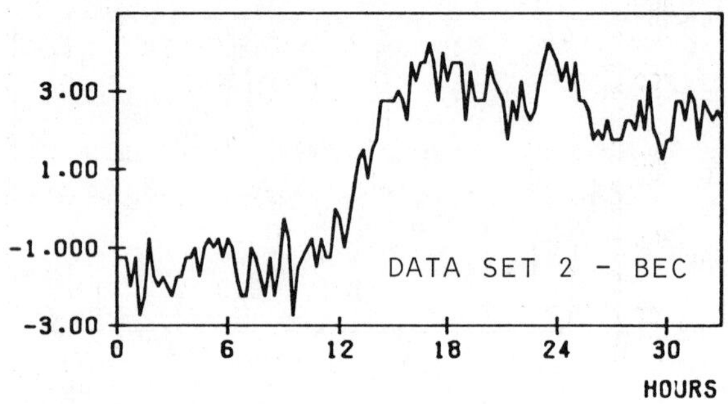

Fig. 16 - Second process output

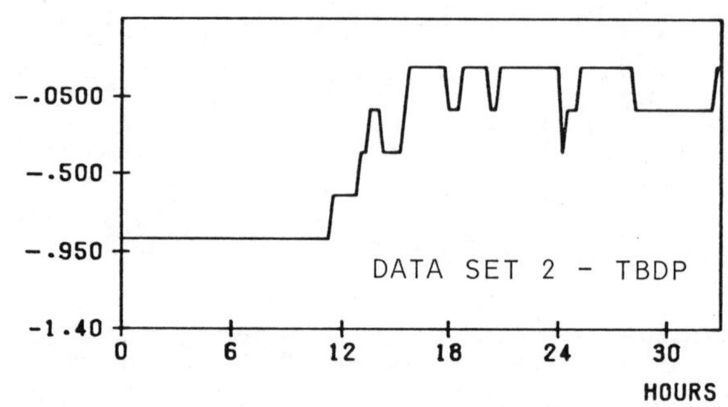

Fig. 17 - Third process output

III. PROCESS MODEL IDENTIFICATION

The used identification procedure is the one described in |1| and |2|. This procedure leads to canonical state-space models of the type

$$x(k+1) = F\,x(k) + G\,u(k) \qquad (3.1a)$$
$$y(k) = H\,x(k) \qquad (3.1b)$$

where $x(k) \in R^n$ is the state vector, $u(k) \in R^r$ is the input vector, $y(k) \in R^m$ is the output vector and F, G and H are real matrices of suitable dimensions. The steps performed in the model identification are the following

1) *Structural Identification*

This step is performed on a sequence of increasing-dimension symmetrical information matrices directly obtained from the noisy input-output sequences and allows to estimate the model structure i.e. the set of Kronecker invariants of the pair (F^T, H^T). These integers $\nu_1, \ldots, \nu_m$ define the order of the m subsystems into which the canonical model (3.1) decomposes the whole system.

2) *Parametric Identification*

This step computes the coefficients of a canonical model of the type (3.1) with the structure obtained at step 1). The obtained estimate is a maximum likelihood one if the additive noise is not correlated with the input-output sequences.

3) *Initial State Estimation*

This step allows to obtain the initial state, in the estimated model (3.1), corresponding to the input-output sequences used in the identification.

The structural identification has been performed by using the last eigenvalue test introduced in |3|. This test allows to estimate a Predicted Per Cent Reconstruction Error (PPCRE) on the model outputs corresponding to the different possible structures and thus to select the more suitable ones. The PPCRE is a lower bound on the actual Per Cent Reconstruction Errors (PCRE) of the output reconstruction with a model having the corresponding structure; both the PPCRE and the PCRE are defined as 100 times the ratio between the standard deviations of the reconstruction errors and the standard deviations of the output sequences.

The computed PPCRE for the three model outputs for increasing subsystem structures are given by

	Subsystem Order			
	1	2	3	4
Output 1	22%	21%	19%	17%
Output 2	48%	46%	42%	39%
Output 3	28%	26%	25%	23%

These values show that the performance of an order 3 model with structure (1,1,1) is not substantially improved by higher order models like, for instance, the ones with structure (2,2,2) or (3,3,3). In fact the PCRE of the three outputs with a model having a (1, 1,1) structure are given by 53%, 52% and 38% respectively while the corresponding values obtained with a (3,3,3) model are given by 45%, 51% and 45%. The reconstructions of the output sequences with a (1,1,1) model and with a (3,3,3) model (stars) are compared, in the Figures 18 - 20 and 21 - 23, with the real sequences (continuous line). When however the validation of the two considered models is performed on the second sequence of available data, the (3,3,3) model gives results quite similar to the ones observed on the first sequence used to estimate the model, with PCRE of 46%, 35% and 47% respectively while the (1,1,1) model gives extremely high reconstruction errors (183%, 121%

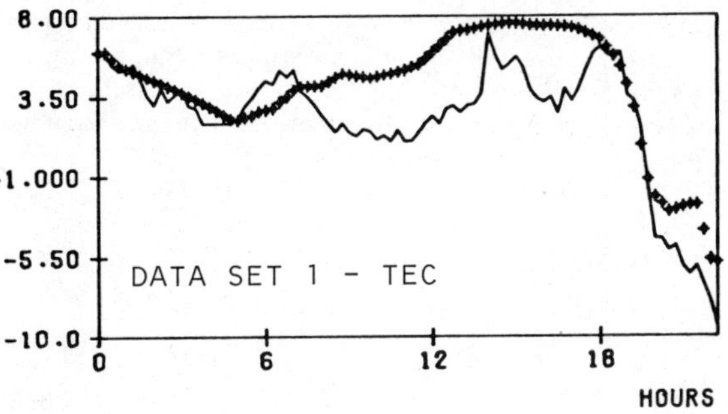

Fig. 18 - First output reconstruction with a (1,1,1) model

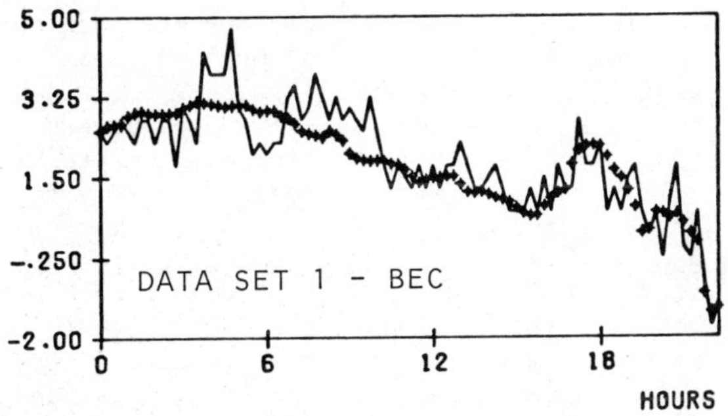

Fig. 19 - Second output reconstruction with a (1,1,1) model

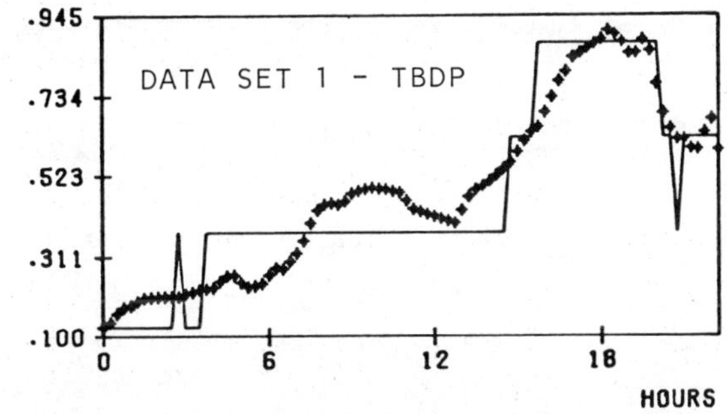

Fig. 20 - Third output reconstruction with a (1,1,1) model

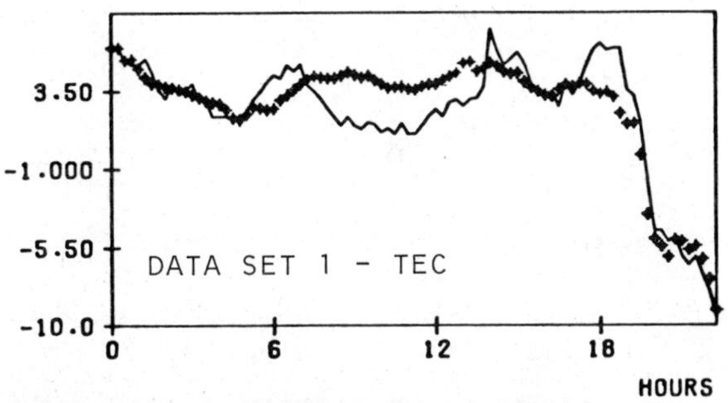

Fig. 21 - First output reconstruction with a (3,3,3) model

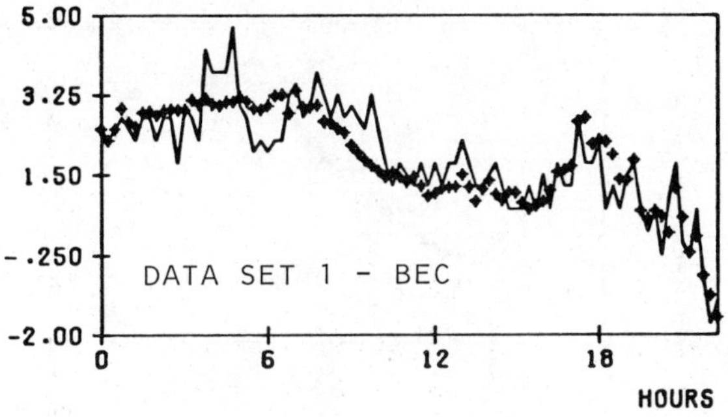

Fig. 22 - Second output reconstruction with a (3,3,3) model

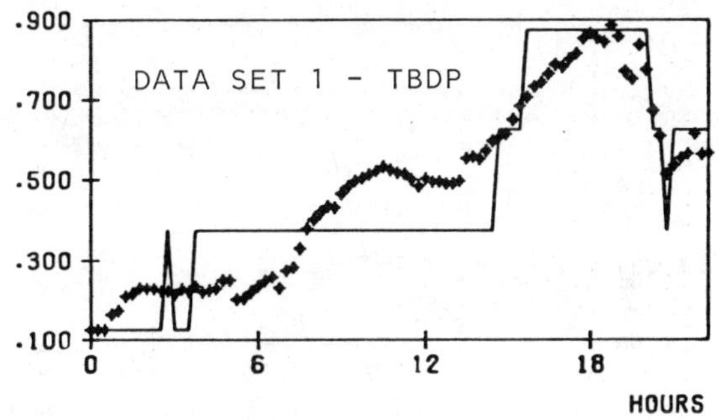

Fig. 23 - Third output reconstruction with a (3,3,3) model

and 54%). The model with structure (3,3,3) has therefore been used to describe the column dynamical behavior. The (3,3,3) model prevision (stars) is compared with the measured outputs (continuous line) in the Figures 24 - 26.

The poles of the (3,3,3) model are given by

$$\lambda_1 = 0.2781 + j\ 0.2565$$
$$\lambda_2 = 0.2781 - j\ 0.2565$$
$$\lambda_3 = -0.4159 + j\ 0.05917$$
$$\lambda_4 = -0.4159 - j\ 0.05917$$
$$\lambda_5 = -0.2448 + j\ 0.5504$$
$$\lambda_6 = -0.2448 - j\ 0.5504$$
$$\lambda_7 = 0.7331$$
$$\lambda_8 = 0.8435$$
$$\lambda_9 = 0.9364$$

The model poles are plotted in Figure 27. The previous model has proved to be reliable and has also permitted to detect improper working conditions in the distillation column. An example of this kind is given in Figure 28 where the model prevision on the first output (stars) shows an extremely large error with respect to the observed values (continuous line). The values predicted by the model are however the values which should have been obtained with a properly working column; the large previsional error has permitted, in this case, to detect the abnormal situation.

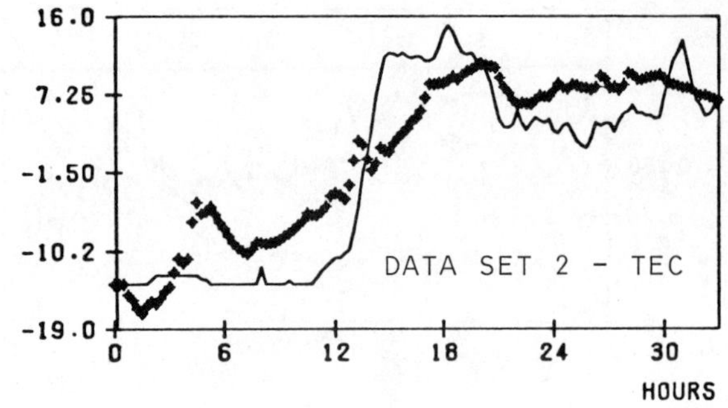

Fig. 24 - First output prevision with a (3, 3,3) model

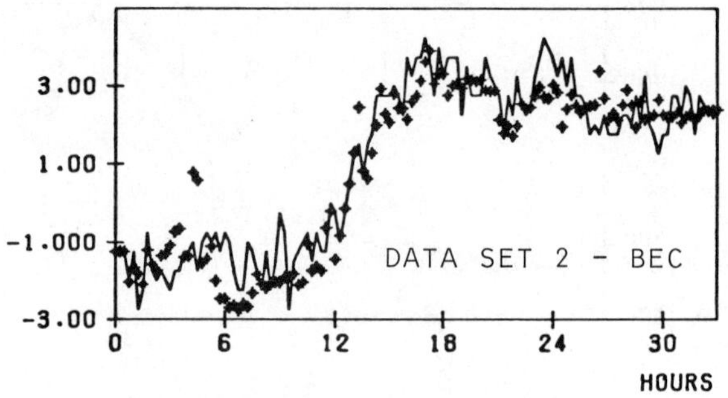

Fig. 25 - Second output prevision with a (3, 3,3) model

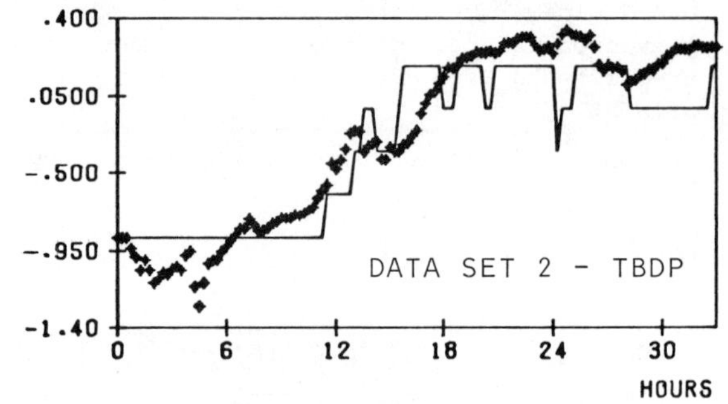

Fig. 26 - Third output prevision with a (3, 3,3) model

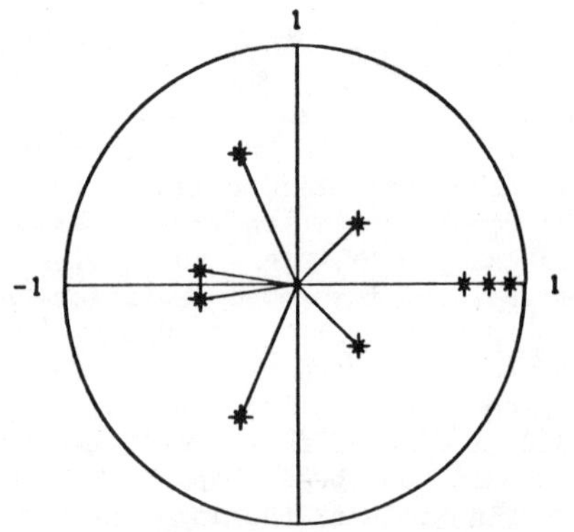

Fig. 27 - Poles of the (3,3,3) model

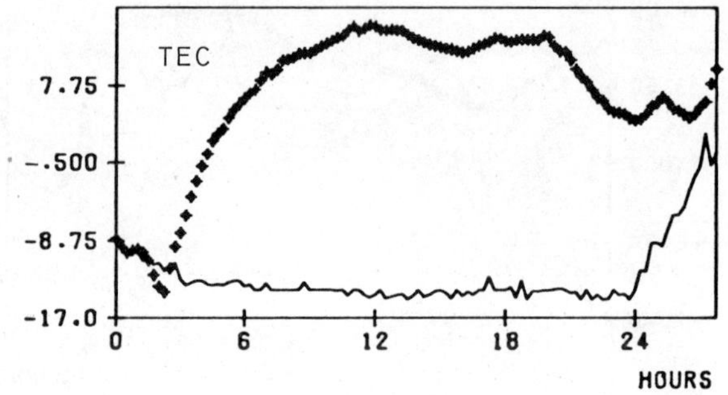

Fig. 28 - Abnormal column working conditions

IV. CONCLUDING REMARKS

The interest in the dynamic surveillance and control of many industrial processes has been rising in recent years. This fact is due both to the increased ecological consciousness and to the evolution of microcomputer technology which offers the tools necessary for the construction of efficient surveillance and control systems. The implementation of surveillance and control procedures requires, however, the knowledge of accurate previsional models for the considered processes. To this purpose classical modelling techniques are widely used today and the chemical field makes no exception. Modern identification techniques are less popular because of the possible absence of physical meaning in the internal variables of the obtained models. These techniques can however lead, if carefully applied, to simple and accurate models of complex multivariable processes. To this purpose the results described in this paper can be considered as typical. It can also be noted that the used identification procedure, because of its limited storage and computing-time requirements, can be easily implemented also on microcomputers to perform on-line identifications.

REFERENCES

|1| R.P. Guidorzi, Canonical Structures in the Identification of Multivariable Systems, *Automatica*, vol. 11, n. 4, pp. 361-374, July 1975.

|2| R.P. Guidorzi, Complete Sets of Invariants and Canonical Input-Output Forms for Multivariable Systems Structural and Parametric Identification, *Proceedings of the 5th IFAC Symposium on Identification and System Parameter Estimation*, pp. 429 - 436, Darmstadt, FRG, September 1979.

|3| R.P. Guidorzi, M.P. Losito and T. Muratori, On the Last Eigenvalue Test in the Structural Identification of Linear Multivariable Systems, *Proceedings of the V European Meeting on Cybernetics and Systems Research*, Vienna, Austria, April 1980.

DEVELOPMENT OF A NEW HEAT POWER SYSTEM

H. Chou

Department of Automation, Shanghai Institute of Technology, Shanghai, People's Republic of China

Abstract. The basic thermal cycles of all heat power systems are stutied in view of control theory, and thereby systems may be classified into three kinds accordingly. The first kind is the ideal carnot engine which serves as standard for comparison. The second kind is characterized by its work feedback process. All conventional heat power systems belong to this class. The third hind is distinguished by its heat feedback process. The present discussed new system belongs to this class. In comparison with all existing heat power systems, it will be found that the new system has higher thermal efficiency and power to weight ratio. We have first time the thermal cycle of power system under direct control, while others so far appeared are not, its thermal cycles are not controlled or at most indirectly controlled. The system is therefore adaptable to wide power utilizing fields. The concepts that underlie the new system are appicable not only to industrial systems, but the natural systems in biology and meteorology as well. Of course the latter systems are not within the scope of this paper, but are very interesting subjects worthy of study.

Keywords. Heat engine of third kind; Thermal variables; System analysis; Biology; Meteorology; Fluidic devices; Fluid control.

1. Introduction

Heat power systems are generally of open thermal systems with mechanical means to realize the thermal processes. In the past fifty or more years, the developments in the thermodynamical side and mechanical side are not balanced. Much achievements have been appeared in mechanical side than the thermodynamical side. The result is that further inpprovement of thermal efficiency and other performance of the systems becomes more and more difficulty by simple mechanical measures. Take internal combustion engines as example, mechanical features of modern engines bear little resemblance to their prototypes, but the thermal cycles remain the original forms in practice (Kolin, Ivo, 1972) We shall see in following sections that almost all conventional heat engines belong to a same thermal model, which is the source that prevents power system from any noticeable progress. We are probabily in a position now to reconsider the thermal model that have lasted more than a century.

The gist of this paper is to introduce a new heat power system. Since the new system relates with various techniques and sciences. It is impossible to deal with every phase in a short paper. We limit our attention on the basic principles and present it in an introductory form. System formation and thermofluid control are emphasized. In section 2 thermal efficiency of heat engines is generally studied in view of system control. An attempt is made in section 3 to reveal the baricade that prevents conventional engines from further improvement. Sections 3 and 4 introduce the concepts of the new system and its control. In the last section we classify power systems into three kinds and propect that heat engines of third kind are the rightful form of comming engines in which control engineering will probably play an important role.

2. Control of the thermal efficiency

Improvement of thermal efficiency defined by ratio of work output to heat input of a power system is the central problem since the time of Carnot. However, the problem has not been satisfactorily solved since then. At early nineteenth century, Sadi Carnot motivated by the desire to answer the much discussed question of how and to what extent the heat engines could be improved, studied a reversible cycle which is now called Carnot cycle, as depicted in Fig.Ia, in which AB and CD are isothermal processes, BC and DA are adiabatic expansion and compression respectively.

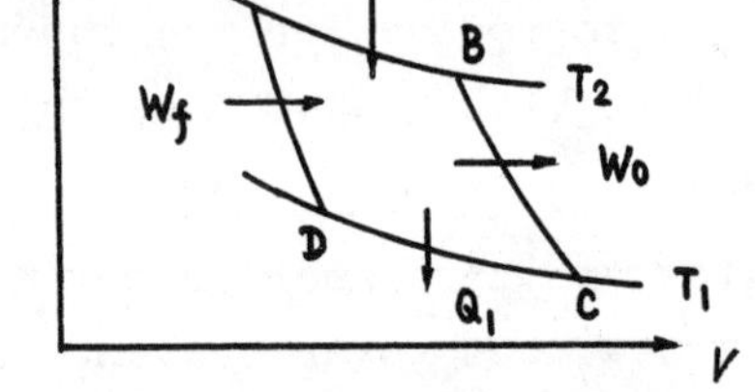

Fig. 1. Carnot Cycle

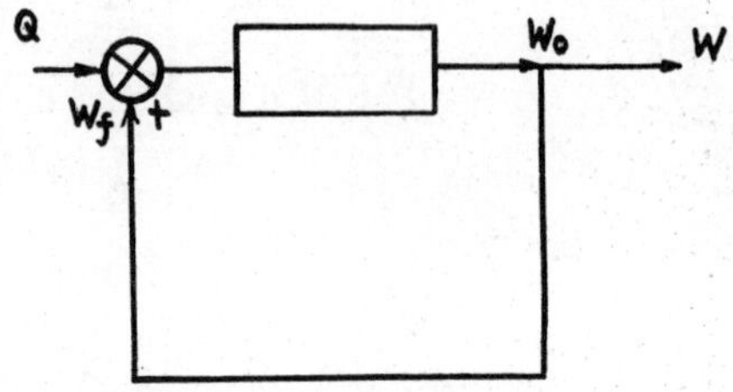

Fig. 2. Heat power systems are feedback systems

Carnot principle indicates that thermal efficiency of the cycle is

$$\frac{W}{Q} = \frac{T_2 - T_1}{T_2} \qquad (2.1)$$

where W is the difference between the work W_o delivered in BC, and work W_f feedback in DA, $W = W_o - W_f$. Q is the difference between heat Q_2 added in AB and heat Q_1 rejected in CD, $Q = Q_2 - Q_1$. T_2 and T_1 are the high and low level temperatures between which the cycle is taken place.

(2.1) indicates the maximum efficiency expectably in a reversible cycle which depends upon the high and low temperatures of the heat reservoirs only. In pratice thermal efficiencies of real engines are much lower, only about one third of the correponding Carnot efficiencies. Low in thermal efficiency means wastage of fuel used as well as superfluity of metals employed to fabricate the machines. In this limited energy resource world the significance of Carnotize the power system becomes more and more important.

In the view of control theory, power systems are invariable an energy feedback system. The corresponding control block diagram may be represented by fig.2. The overall transfer function of the feedback system is expressed by the same formula (2.1).

The difficulties of control of thermal

efficiency lie on the fact that feedback signal used in power system is quite different from that of most process control: first, it shears a great portion of energy delivered in work producing process, such as BC in fig. 2a; second, the thermal signal is restricted by both first and second thermodynamic laws, and therefore the signal dissipated rapidly; third, there must be suitable mechanical means to guide the thermo-fluid in the feedback path, which are generally labourous and occupy large portion of the mechanical parts.

Energy employed in the feedback path may have two forms to be selected: heat and work. In a reversible cycle such as Carnot proposed, there is no difference between heat and work, since they are equivalent. But in the real cycle, they are quite different. Heat and work stand in different thermal grades. For the benefit of thermal efficiency, it is preferable to select feedback signal in the form of heat. But almost all conventional engines take the form of work in its feedback path. In the next section we will try to analyse the conventional engines by a model and to see how defficiencies of engines are in alliance with the work feedback, and to confirm ourselves that heat feedback is a better choice in developing of efficient power system.

3. Defficiencies of conventional heat engines

Conventional heat engines have very low thermal efficiency and rather high weight to power ratio. They are less adaptable for different fields of applications. For instance, a jet engine for aircraft is scarcely possible to power a ship though they are all running in fluid environments. In the final analysis, most of these efficiencies may be traced to the form of feedback process used in the engines. A thermal model will help us to see it clearly.

Virtually all conventional heat engines, serving in various fields for different porposes, are of work feedback type, i.e., heat energy of working medium is transformed into external mechanical work first, then transformed into heat again and put into the system. They can be reprented by a same thermal model as shown in fig.3a. Practical engines are open systems, for simplicity, we take it as a closed one, which will not obviate the basic proporties to be studied.

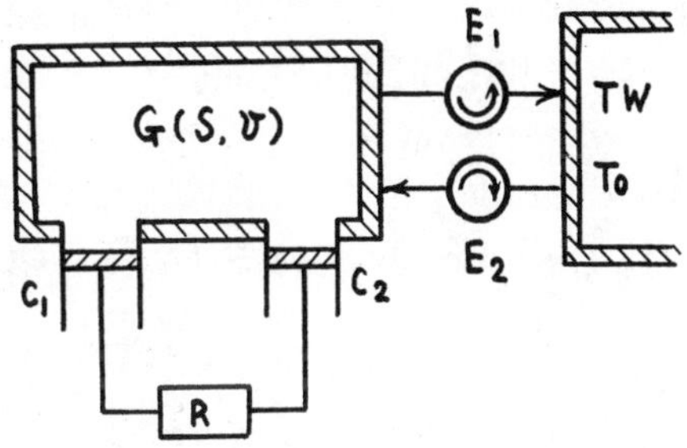

Fig.3a. Thermal model of conventional heat engines

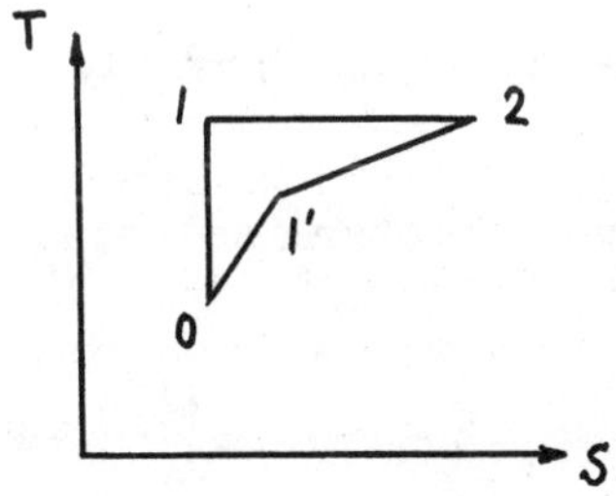

Fig. 3b. The 0 - 2 processes

In fig.3a, G is a compartment with adiabatic walls. It contains working medium of thermal state variables S and v, where S is the total entropy and v is its volume. E_1 and E_2 are reversible engines working between G and thermal well TW of temperature T_0. The thermal well is large enough so that a finite quantity of heat added to or withdrawn from it changes its temperature only by a negligible amount. C_1 and C_2 are mechanical pumps, R is a work reservoir which stores work from or supplies work to the pumps. The system has low and high temperatures indicated by states 0 and 2 in fig. 3b.

Define thermal potential of G by (Keenan, J.H., 1951)

$$G(S, v) = E + P_0 v - T_0 S \tag{3.1}$$

where E is the energy of G, v its volume, S its entropy, P_0 the pressure of atmosphere and T_0 its temperature.

We may write

$$dG = \left(\frac{\partial G}{\partial S}\right)_v dS + \left(\frac{\partial G}{\partial v}\right)_S dv \tag{3.2}$$

Since

$$\left(\frac{\partial G}{\partial S}\right)_v = \left(\frac{\partial E}{\partial S}\right)_v - T_0 = T - T_0 \tag{3.3}$$

$$\left(\frac{\partial G}{\partial v}\right)_S = \left(\frac{\partial E}{\partial v}\right)_S + P_0 = -P + P_0 \tag{3.4}$$

Substitute (3.3) and (3.4) into (3.2), we have

$$dG = (T - T_0)dS - (P - P_0)dv \tag{3.5}$$

For any two predetermined states 0 and 2, fig. 3b.

$$\Delta G = G_2 - G_1 = \int_0^2 (T - T_0)dS - \int_0^2 (P - P_0)\,dv \tag{3.6}$$

ΔG represents the maximum useful work of the system for a change of states from 2 to 0. The first term on the right side represents work delivered by engine E_1, fig.3a. The second term represents displscement work stored in R due to pumps C_1 and C_2. (The storage may be positive or negative.) Both integrations take along any reversible path between 2 and 0. For comparing purposes, it is assumed that $G_2 - G_0$ is constant.

Now consider a change of states in the system from 0 to 2, while it receives heat from TW through E_2 and work from R through pumps. Then we have

$$\Delta G = G_0 - G_2 = \int_2^0 (T - T_0)dS - \int_2^0 (P - P_0)dv \tag{3.7}$$

The first term on the right of (3.7) is the heat absorbed through E_2, and the second term is the work received through pumps, fig.3a. For reversible processes each term in the right of (3.7) equals but opposits to the corresponding term of the right side of (3.6).

In pratice C_1 and C_2 are irreversible machines. The amount of irreversible work must be compensated by the useful work delivered. Moreover, heat can not be added to the system at constant temperature. Therefore the real processes between 0 - 2 are most probabily like 0-1'-2, in which 1'-2 is the process of heat absorbing. The amount of heat absorbed in 1'-2 is less than 1-2.

Some results may be deduced from the model as follows.

a. The presence of C_1 and C_2 as well as its irreversibility imply that heat absorded by the system will be less than it would be when they are reversible machines. It is understood that less heat absorbing means less work conversed for the same system.

b. The feedback energy passes through the mechanical parts C_1 and C_2. The energy transfers in forms first from heat to mechanical work, and then mechanical to heat. The energy is under double penalties of irreversibility between these two transformations. The irreversible feedback works are compensated by the useful work delivered by the system. Therefore, the contributions of the feedback processes on the thermal efficiency will soon be cancelled when the amount of irreversible work is large.

c. C_1 and C_2 are of moving mechanical parts in real engines. More than half of the rotating mechanisms are provided for the feedback purpose, or more than half of the working time of the parts is occupied by the feedback processes.

d. The feedback processes of conventional engines are generally not controlled in processing. Efficiency and performance of the system vary with load and working conditions.

All aforementioned defficiencies are closely related with the mechanisms C_1 and C_2 in fig. 3b. Only by removal of these parts can the system be improved fundamentally. In following sections we will introduce a new system without moving part with controlled feedback.

4. A new system

We are convinced by the above analysis that efficiency and the general structures of conventional heat power systems cannot be promoted in any noticeable degree without change the working thermal system of the present state. We have developed a new heat power system in which heat feedback is used instead of work feedback as used in usual cases. Thermal processes are fluid controlled in the new system and therefore no moving mechanical parts except moving fluid. The system was first presented in Chou, H., (1977). Fig.4a is a line diagram of the system.

The whole system consists of three subsystems and a fluidic control system. The three subsystems are: the main jet subsystem, the feedback subsystem and the heat supply subsystem as shown in the diagram.

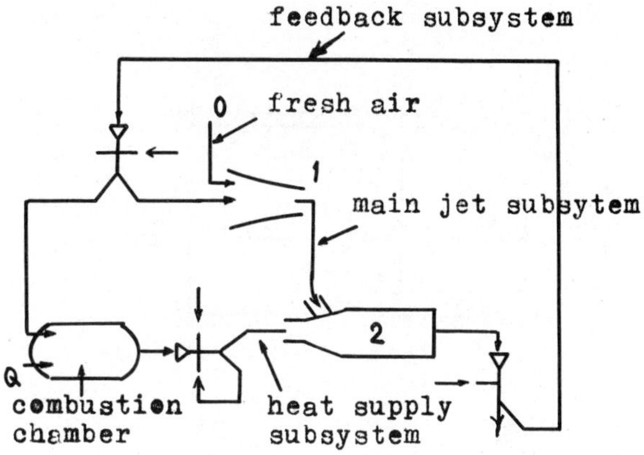

Fig. 4a. Line diagram of a new system

The working precedures of the system are as follows. A quantum of fresh air necessary with certain surplus for combustion is taken into the system by a feedback flow of the previous cycle. After these two fluids are mixed heat is injected into it from heat supply subsystem by a heat pump. The incomming and feedback fluid are both fresh air. After the mixed air is heated, the portion that is necessary for combustion is metered and sent to the combustion chamber, and the amount of heated fresh air equivalent to the previous feedback fluid is metered and sent to the following cycle. The remaining surplus air is mixed with burnt gas and sent out of system for power purpose. The cycle is thus completed.

We have analysed the system. For simplicity we represent the system by an equivalent thermal model shown in .
The heat supply subsystem is replaced by a large thermal well TW. The feedback subsystem is replaced by a large mass supply subsystem C. Both subsystem are at temperature T_o and en-

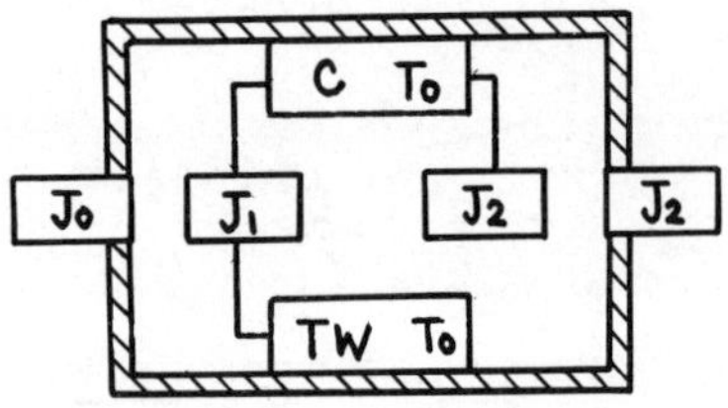

Fig. 4b. Equivalent thermal model of the new system

closed in an adiabatic wall. Main ject subsystem of fig. 4a, is replaced by J, subscript i denotes places where it locates in the compartment. J is a constant volume open system, when it enters the compartment, heat and mass are injected into it under control from TW and C by heat pump[(1)] and mass pump. Heat acquired by the mass is provided by the heat pump and is injected into the mass while in its way to J, and therefore the temperature of C is uneffected when mass is drawn from or put into it. When J leaves the compartment, same amount of mass as it received is returned to C, and the same amount of heat carrying with it is returned to TW which is large enough so that its temperature is uneffected with finite addition of withdrawal.

C,J, and TW form a closed system, no mass or heat is added to the closed system, but mass and heat may be transfered within it, The only consequence of travel of J through compartment is that it draws heat from TW while C is uneffected. The amount of heat drawn and mass circulated are under controlled.

If reversible works accomplished on J is by heat pump only, we define the collective potential V of the system by (Biot, M.A., 1977)

$$V = U + H \tag{4.1}$$

where U is the collective energy of J and C together, H is the heat acquired by the thermal well TW. We may take values of V and U relative to the initial conditions of the system, the latter are assumed zero. The V and U in (4.1) represent the increase of the values.

The total change of entropy of the isolated system is zero, therefore

$$S - \frac{H}{T_0} = 0 \tag{4.2}$$

where S is the entropy increase of J and C. From (4.1) and (4.2) we have

$$V = U - T_0 S \tag{4.3}$$

that is, the collective potential or availabity of the thermal model fig. 4b depends upon U and S defined above. The thermal state of J can be represented by two state variables and the mass involved. Mass incolved in J is composed of two parts: the original mass content of J before it enters the compartment, and the additional mass in feedback from the previous cycle. In consistent with V and U, we take the additional mass as mass content of J. Therefore the state of J is wholly determined by two thermal variable of it, say its volume v and temperature T, and the feedback mass m. While the state

(1) The name follows from Biot , M. A., (1977). We used the name 'self-exciting jet' since 1974 to denote thermofluid systems that transmit heat by shock waves. Heat power systems with self-exciting jet was developed since then, some of the results were included in Chou,H., (1977), and other publications of SIT University press.

variable of C are simply m. State of the system is determined by state variables of J entirely. We have

$$U = U(\vartheta, T, m)$$

$$S = S(\vartheta, T, m)$$

and $V(\vartheta, T, m)$

$$= U(S, T, m) - T_0 S(\vartheta, T, m)$$

V is an extentive variable of the system. One important property of V is its additive property. Temperature T of J is an intensive variable. When the independent variable T of V is increased by any cause, then any subdivision of V taking from it will have same increase of temperature.

Now if a reversible Carnot Engine is coupled with subsystem J, the engine will deliver work just equivalent to the heat absorbed from the thermal source. Therefore the thermal process carrying J through the compartment up to J_2 is internally reversible (Modell, M., 1974). It is corresponding to the 02 thermal processes of Carnot engine, shown in fig. 3b. The efficiency of subsystem J is comparable with Carnot cycle so far as 02 processes is concerned.

Our next step is to find in what way the thermal efficiency can be controlled when our system is used as power supply system, for instance use it as a power jet or to couple a turbine.

Suppose dQ is an infinitesimal heat injected into J by heat pump, then the maximum useful work available of J at state 2 with temperature T_2 is

$$\frac{T_2 - T_0}{T_2} = \theta \, dS_j \qquad (4.5)$$

where $(T_2 - T_0)/T_2$ is Carnot efficiency, $\theta = T_2 - T_0$, the temperature difference between output temperature and environmental temperature, S_j is the entropy increase of J due to heat received. Therefore, the availability or collective potential is increased with heat injection by

$$dV = \theta \, dS_j \qquad (4.6)$$

Thermal efficiency of the heat conversion system depends upon θ, which is increased with S_j,

$$\frac{\partial V}{\partial S_j} = \theta > 0 \qquad (4.7)$$

If mass dm is injected into J in addition to direct heat injection, then the increase of collective potential is (Biot, M.A., 1978)

$$dV = \Phi \, dm + \theta dS \qquad (4.8)$$

where Φ is the convective potential, which is function of state variables of the feedback mass in its way bringing from mass reservoir C to J, when heat is injected into it until it is in thermal equilibrium with J. Moreover.

$$\left(\frac{\partial V}{\partial \vartheta}\right)_{m,S} = -p$$

and $$\left(\frac{\partial V}{\partial S}\right)_{\vartheta, m} = \theta \qquad (4.9)$$

where p is the pressure and S is the total entropy expressed by

$$dS = S_f dm + dS_j \qquad (4.10)$$

where s_f is the relative specific entropy of feedback fluid, which represents the entropy increase per unit mass, and depends upon heat injected into dm by heat pump while it is bringing into J.

Similar to (4.7), we have

$$\left(\frac{\partial V}{\partial S}\right)_{\vartheta, m} = \theta > 0 \qquad (4.11)$$

that is, the temperature difference θ, and therefore thermal efficiency of our power system may be controlled not only by heat directly injected which is propertional to S_j, but by feedback mass m as well as heat carrying with m, For given S_j, that is with given output, thermal efficiency of the system may be controlled by the states of feedback fluid.

It is important to note that feedback in our new system is an internal process in the compartment, and its effect on efficiency is due to the heat carrying with the feedback mass as indicated in (4.8) We call it heat feedback process, while in conventional heat engines, feedback takes a path with heat and work interaction, that is, through a heat-work-heat processes. Mechanical parts other then the flowing fluid must be introduced in the path. We call it a work feedback.

Heat feedback system as we proposed and the work feedback system, as most conventional engines adopted, working between same intial and final states 0 and 2, have quite different in relative (to Carnot) efficiencies, structures and adaptability to control. From these point of wiew, heat feedback system has superiorities in every respect. In the following section we will see how the heat feedback system can be controlled.

5. Control of the system

After introducing the general features of the system, we shall give some ideas of control of the set. New system is adaptable to various control requirements. In fact some of the control problems are intresting. It is not proper to discuss these problems in this short paper, Emphesis is here given on the analytical relation between the parameters to be controlled and the related measurable state variables, so that control system may be planned accordingly.

We have used a fluidic system to control the heat conversion processing. It seems that for sensing and controlling of thermofluid system, like the present one, fluidics have definite superiorities, for coding, transmission, or data processing fluidics have enough capability for small units. We do not exclude the possibility of development of hybrid control systems. for large units, that is, sensing and controlling by fluidics and remaining wares by electroni pieces. For this purpose we must develop necessary transducers, which are still lacking now. Fundamental requirements of the control are to get mass and heat balances under specified power output. Fig. 5a shows the idea of how the three subsystems of fig.4a are coupled in heat transfer. Point 0,1,2 in fig.5a correspond the same points of fig.4a, in which point 0 represents the atmospheric state, and point 1 represents the intermediate state where feedback fluid mixed with the fresh air, point 2 represents the temperature output where exhaust gas mixed with residual air.

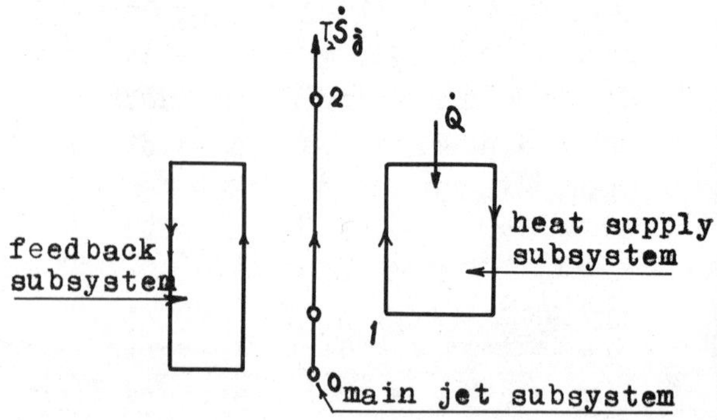

Fig.5a. Heat transfer between subsystems.

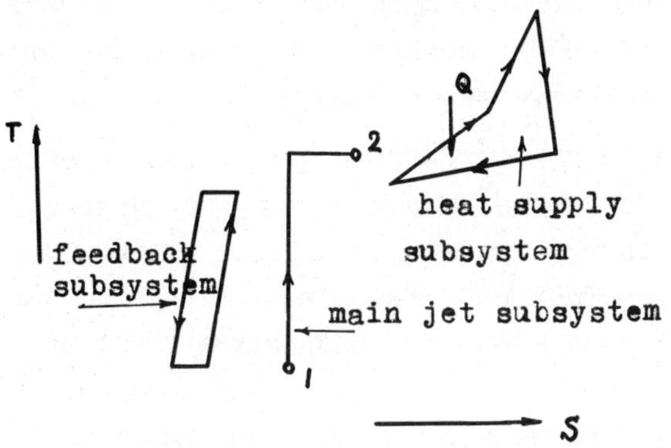

Fig.5b. Entropy temperature diagram of the subsystems (not in scale)

Fig.5b shows the temperature entropy changes of the subsystems. The area included in the feedback subsystem loop represents energy that circulated in this subsystem. The area included in heat supply subsystem loop represents heat supply Q of the whole system, which is equal to the output $T_2 S_j$. The output temperature is T_2, shown in the main jet subsystem. The main jet subsystem is internally reversible as proved before. It stores an amount of energy equal to the energy delivered by heat supply subsystem, its thermal potential may be controlled by feedback system. The power output of the system is $T_2 \dot{S}_j$ where dote denote the time derivative.

It is to be noted that in fig.4b, heat is supplied by a large thermal well, mass is supplied by a mass reservoir, both at temperature T_o. It seems different from real system which has its thermal properties as shown in fig.5b. Nevertheless, so far as the availability of the system is concerned, the effects are same.

The most important parameter to be controlled is the exit temperature T_2 of the main jet. T_2 is a function of entropy increase of the main jet and feedback mass,

$$T_2 = T_2 (m,S) \qquad (5.1)$$

where S is the sum of entropy injected S_j and entropy supplied by feedback mass s_f m,

$$S = s_f m + S_j \qquad (5.2)$$

where s_f is a function of temperature and pressure of main jet,

$$s_f = S_f (p,T) \qquad (5.3)$$

(5.1) and (5.3) depend upon the properties of working medium. We can plot T versus S taking m as parameter for (5.1) as shown in fig.5c, and plot T versus s_f taking p as parameter for (5.3) as shown in fig.5d.

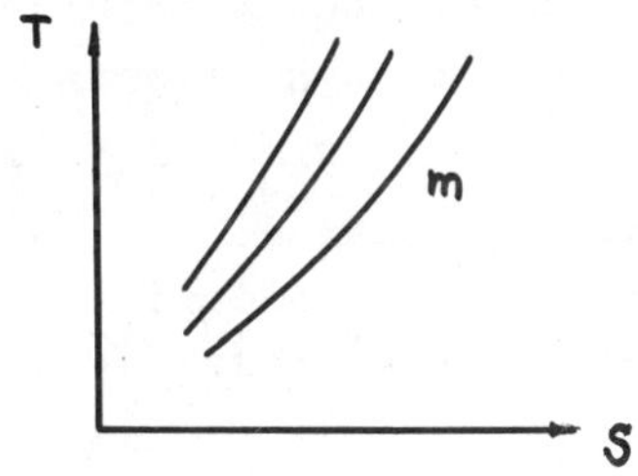

Fig.5c. Equesion (5.1)

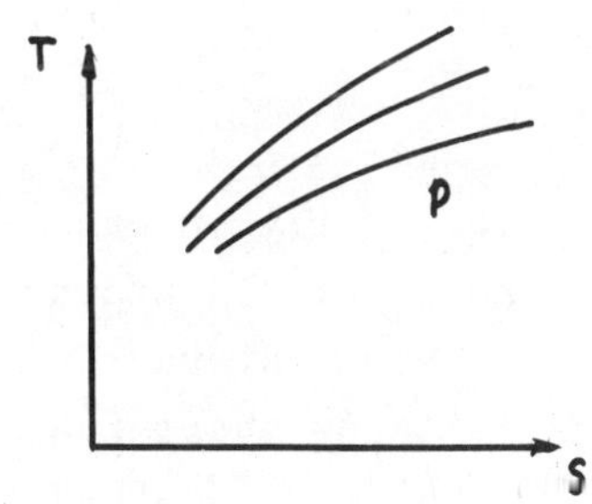

Fig.5d. Equation (5.3)

For a given work output $T_2 S_j$, S_j is preset by the heat pump. T_2 may be controlled as follows: from measured p and intended T_2, s_f can be obtained from fig.5d. Using (5.2) and fig.5c, necessary feedback mass m may be obtained by cut and try. Mass in different parts of the circuit is controlled by fluidic valves. In pratice, it is easy to plan the fluid control circuit.

6. Heat engines of third kind - a prospect

Energy feedback plays an important role in every heat power system. Heat and work are two process variables which give quite different results when used in feedback control. Classification of heat engines according to the process variables has obvious technical significances.

We call work feedback power system as heat engine of second kind, virtouly all conventional engines belong to this category. We call heat feedback system as heat engine of third kind, the new system proposed in this paper belongs to this kind. And we call the reversible Carnot engine as heat engine of first kind. Though reversible Carnot engine doesn't exist in real world, it deserves the name due to the fact that it provides a standard for comparison of the thermal efficiencies of other engines.

Heat engines of second kind have dominated the power production world more than a century. It inherits ingenious mechanisms from its ancesters and accumulated a great wealth of technological skill in manufacturing, It looks mature in appearance. Yet it is very weak internally, only to think that it has thermal efficiency of 30% - 40% of heat engine of first kind.

Heat engine of third kind will probably the next choice of heat power system, (Chou, H., 1978). It has thermal efficiency much near that of the corresponding first kind engine, less metal to be used in fabricating, and more adaptability to different fields of applications. Heat engine of third kind is the natural outgrowth of modern technology and will promote relative techniques in the course of its development.

Nature selects the best way in heat energy conversion. Principles of heat engines of third kind may be followed by many natural phenomena, both in micro-biological structures and in macro-meterorological systems. But this is only a conjecture, we must take time to prove it.

References

Biot. M.A., (1977). New fundamental concepts and results in thermodynamics with chemical applications. Chemical Physics 22, 183-198 (1977)

Biot,M.A., (1978). Variational irreversible thermodynamics of heat and mass transfer in porous solids. Quart. Appl. Math, 19-38, April 1978.

Chou,H., (1977), Self-Exciting Jet Systems, in chinese, SIT university Press.

Chou, H., (1978). Heat Engine of third kind, in chinese, SIT university Press.

Keenan,J.H., (1951). Availability and irreversibility in the thermodynamics, British J.Appl. Phys., vol.2, July. 1951.

Kolin, Ivo. (1972) The evolution of heat engines, Longman, London, 1972.

Modell, M.(1974) and Reid, R. C., Thermodynamics and its applications, p. 77, Prentice Hall, Englewood Cliffs, N.J., 1974.

DYNAMIC MODELLING AND SIMULATION OF PAPER MACHINE MULTICYLINDER DRYING SECTION

J. L. Videau*, A. Lemaitre and C. Foulard***

**Laboratoire d'Automatique de Grenoble, B.P. 46, 38042 St Martin-d'Hères*
***Centre Technique du Papier, B.P. 7110, 38020 Grenoble Cedex*

Abstract. The control of the drying section of a paper machine is very important, because a better regulation of the main variables reduces the moisture content variations of the sheet at the end of the machine, and improves the drying, which has a strong influence on the final paper quality. This study describes a mathematical dynamic model based on a theoritical analysis of the physical phenomena that occur in a multicylinder paper machine drying section. This model takes into account the main design and working parameters, the different evaporation phases, the heat and mass transfer conditions, the thermal effect of the cylinder walls and also the main ventilation variables. A computer integration of the differential equations leads to an overall model which is also the basis of a parameter estimation method. Some simulation results are presented here, on an industrial example. The responses of the main variables (sheet temperature and moisture content, air temperature and moisture content , etc...) to step changes of the different input parameters (insufflated and extracted air flows, etc) can be observed.

Keywords. Paper industry, modelling, system analysis, parameter estimation, simulation.

INTRODUCTION

The aim of the papermaker has ever been the improvement of the quality of his paper, in order to satisfy the requirements of his customers and to increase his profit.
In his contribution to this task, the control specialist introduced in the process, at first, some local regulation loops and then some more elaborated systems such as basis weight and moisture control systems. Meanwhile the best improvements have been achieved in the first phase of the paper manufacture, i.e. the wet end of the paper machine.
In this part, the physical phenomena which occur (hydraulic and pneumatic) are well known and the corresponding equations are easy to find. On the other hand, the dry end in view of the complexity which characterizes the multicylinder drying section, has never been really considered in this way, judging by the very few publications about this subject. However, the control of this part of the paper-machine is very important, because a better regulation of the main variables reduces the moisture content variations at the end of the machine and improves the drying, which has a strong influence on the final paper quality. The recent process analysis techniques, namely modelling, identification and simulation offer new prospects for such studies. The present contribution endeavours to act in this direction.

Paper drying is the last part of the water removal phase. The sheet moisture content at the entrance of the drying section is generally 60 to 70 per cent (water weight/total weight). At the end, the moisture content must be about 5 to 10 per cent. The multicylinder drying section is the most common drying process in the paper industry. It is constituted of steam heated cylinders on which the sheet is successively applied by clothings (Fig. 1). The whole drying section is generally set in a hood so that the atmosphere in which the evaporation takes place can be controlled acting on the insufflated and extracted air flows. The ventilation of the pockets, between cylinder and clothing is made by hot air blowing boxes or rolls.

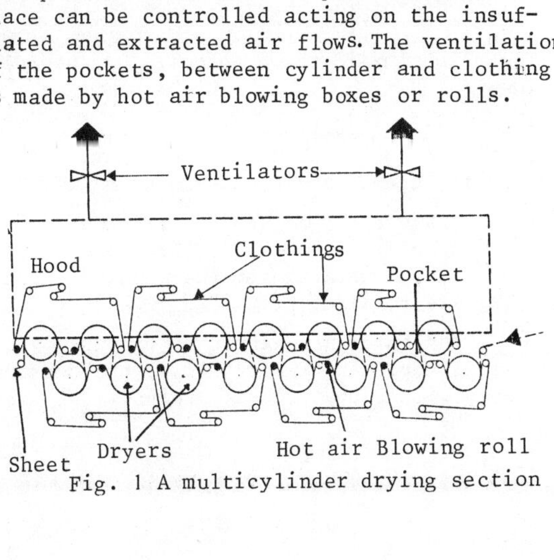

Fig. 1 A multicylinder drying section

The drying section, divided in several parts, is supplied with steam at a pressure of 3 to 10 bars. The condensates of the main section is flashed and the steam thus produced, is used by the former section.

Numerous studies have been carried out concerning the drying section in a steady state and the theoritical knowledge about the mecanisms that occur is quite large (1), (2), (3). Some authors used these theoritical results to analyse the whole process behaviour, by means of mathematical models (4), (5), but very few authors modelized the drying section to study the dynamical behaviour of the main variables of the process. Some workers proposed a numerical derivation of the FOURIER equation through the cylinder wall, to compute the dryer surface temperature (6), (7). They used these very simple models to compare different control algorithms. Others give some simulation results concerning the dynamical behaviour of the drying section (4), (8).

The aim of the work which is presented here is to elaborate a workable model of the whole drying section. This is the first step of the research, the main object of which, is to improve the dyring section control. This model takes into account the heat and mass transfert conditions, the drying characterists of the paper and also the ventilation parameters such as the insufflated and extracted air flows, moisture contents and temperatures. The initialization of this model is carried out with a steady state model which is computing the operating point. This model is also the basis of a parameter estimation method which has been developped to evaluate the heat and mass transfer coefficients on industrial equipments. Some simulation results are presented on a particular industrial example and some system responses to step changes of the main variables are examined.

1. MATHEMATICAL MODEL OF THE DRYING SECTION

The modelization is devided in three differents parts : the paper sheet, the cylinders and the ventilation system.

1.1. Model of the sheet

The main simplifying hypothesis are as follows :

- two typical portions of the paper web are distinguished : web on the cylinder, and web in the draw, as can be seen from Fig. 2
- temperature and moisture content are supposed to be uniform in the thickness of the web. Only variations in the machine direction (x) are considered
- all the internal phenomena (vapour diffusion, capillary transport...) are considered only with global relationships.

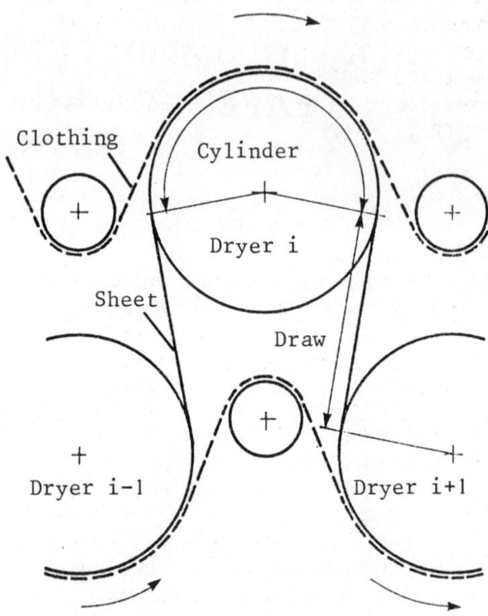

Fig. 2 : The two positions of the paper web

At the abscissa x in the machine direction, and at time t, we consider a sheet element of length dx, width 1 and moving at the speed U. It receives heat from the dryer surface (of course, only in the case of web on the cylinder) and it exchanges heat with the ambient medium : air (web in the draw) or clothing (web on the cylinder). It also loses some heat because of water evaporation.

Then, the thermal balance gives :

$$G.\{c_f + X_f(x,t).c_w\}.\{U.\frac{\theta_f(x,t)}{\partial x} + \frac{\theta_f(x,t)}{\partial t}\} = + \alpha_{cf}.\{\theta_c(x,t) - \theta_f(x,t)\} + \alpha.\{\theta(t) - \theta_f(x,t)\} - r\{\theta_f(x,t)\}.B(x,t) \qquad (A)$$

For the two typical portions of the web, we have :

- web on the cylinder

  $\alpha = \alpha_{fb}$: heat transfer coefficient between the sheet and the clothing

  $\theta(t) = \theta_b(t)$: temperature of the clothing

- web in the draw

  $\alpha = 2\alpha_{af}$: heat transfer coefficient between the sheet and the air (the heat exchange with the air takes place on the two sides of the sheet).

  $\theta(t) = \theta p(t)$: temperature of the air in the pocket

  and we have :

  $\alpha_{cf} = 0.$

The water inside the paper is removed by surface evaporation due to the difference of partial pressure of water between the sheet surface and the ambient medium. In the air, the evaporation rate B is given by the STEPHAN diffusion law :

$$B(x,t) = \frac{\beta}{R.T_f(x,t)} .P.\mathrm{Log}\ \frac{P - p}{P - p_f\{\theta_f(x,t), X_f(x,t)\}}$$

The vapour pressure at the sheet surface p_f is not only depending on the temperature θ_f and the moisture content X_f, but also on the drying rate B which is changing with the sheet moisture content (5), (9).

- When X_f is higher than a critical moisture content X_{cri} :

  $p_f = p_{sat}\{\theta_f\ (x,t)\}$ saturated vapour pressure

- When X_f is lower than X_{cri} :

$$p_f = p_{sat}\{\theta_f(x,t)\}.e^{-k_f.\{\frac{1}{X_f} - \frac{1}{X_{cri}}\}}$$

Then, the water balance for the sheet element gives :

$$G.\{U.\frac{\partial X_f(x,t)}{\partial x} + \frac{\partial X_f(x,t)}{\partial t}\} = -B(x,t) \qquad (B)$$

For the two typical portions of the web, we have :

- Web in the draw

  $\beta = 2\beta_{af}$: mass transfer coefficient between the sheet and the air

  $p = p_p$: vapour pressure in the pocket air

- Web on the cylinder

  $\beta = \beta_{fb}$: mass transfer coefficient between the sheet and the clothing

  $p = p_p$: vapour pressure in the air over the clothing

1.2. Model of the cylinders

The main hypotheses are as follows :

- the heat transfer by conduction in the cylinder wall is neglected in the cross direction y and in the machine direction x. Only the heat transfer in the direction z is taken into account
- simulation results show that the difference between the temperature of the surface in contact with the sheet and the surface in contact with the air is never higher than 0,5 °C. So, we assume that this surface temperature is constant, all around the cylinder (machine direction x).

$$\theta_c(Z,x,t) = \theta_c(Z,t)$$

For a given position in the direction x, the evolution of the temperature in the direction z is given by the FOURIER law :

$$\frac{\lambda_c}{\rho_c.c_c} . \frac{\partial^2\theta_c(z,t)}{\partial z^2} - \frac{\lambda\theta_c(z,t)}{\partial t} = 0 \qquad (C)$$

For the two portions of the cylinder, we have the following boundary conditions :

- Cylinder under the sheet :

$$\lambda_c.\frac{\partial\theta_c(z,t)}{\partial z} - \alpha_{vc}.\{\theta_c(z,t) - \theta_v(t)\} = 0$$

for $z = 0$

$$\lambda_c.\frac{\partial\theta_c(z,t)}{\partial z} + \alpha_{cf}.\{\theta_c(z,t) - \theta_f(x,t)\} = 0$$

for $z = Z$

- Cylinder in the air :

$$\lambda_c.\frac{\partial\theta_c(z,t)}{\partial z} - \alpha_{vc}.\{\theta_c(z,t) - \theta_v(t)\} = 0$$

for $z = 0$

$$\lambda_c.\frac{\partial\theta_c(z,t)}{\partial z} + \alpha_{ca}.\{\theta_c(z,t) - \theta_p(t)\} = 0$$

for $z = Z$

For $t = 0$, we have : $\theta_c(z,0) = f(z)$

The integration of the differential equation (C) is carried out by means of the classical finite difference method. The cylinder wall is divided in three parts in the thickness as can be seen in Fig. 3.

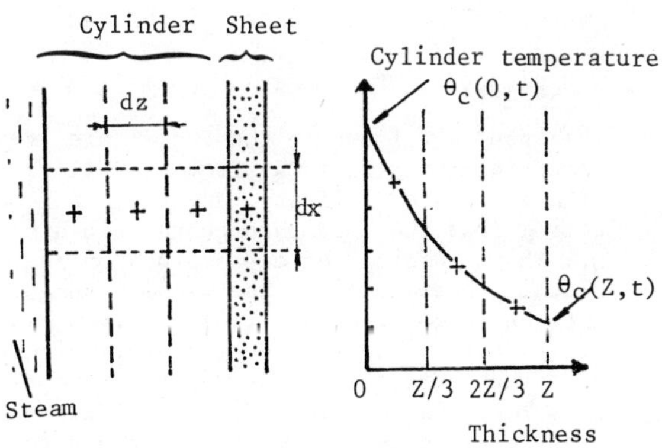

Fig. 3 : Cylinder wall division - Parabolic interpolation of the cylinder surface temperature

The cylinder surface temperature $\theta_c(z,t)$ is then computed by means of a parabolic method from the three values of the internal temperatures.

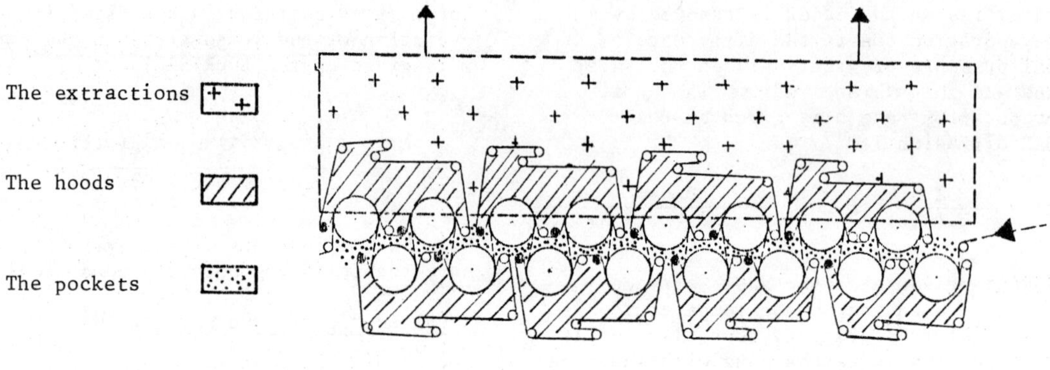

Fig. 4 : The three kinds of volume in the ventilation system

1.3. Model of the ventilation system

The main hypotheses are a follows :

- the mixing of the different air flows in a considered volume is assumed to be perfect and without delay
- we assume that the evaporation flow of vapour from the sheet on a cylinder or in a draw is going entirely in the corresponding pocket
- the mass transfer by diffusion and by convection in the air is neglected.

We consider three different kinds of volumes (Fig. 4) :

- the pockets : volumes limited by the dryer the sheet and the clothing
- the hoods : volumes limited by the upper and the lower clothings
- the extractions : volumes located over the upper clothings. The number of extractions in equal to the number of ventilators.

For each volume, we write the thermal balance, the water balance and the dry air balance.

* Pockets :

The input air flows in the pocket are, the insufflated air flow Q_s, from special ventilation systems, if any, and the air flow coming from the hood through the clothing Q_r. This "pumping" effect depends on the clothing permeability, the machine speed, and the pocket geometry (10).
It is possible to introduce the efficiency τ of the ventilation system ; only the air flow $\tau.\ Q_s$ is entering the pocket. The air also receives heat from the sheet and the cylinder, and evaporated water from the paper web.

Then, the water balance gives :

$$\rho.V_p.\frac{\partial X_p(t)}{\partial t} = S.B(t) + Q_r.X_h(t) + \tau.Q_s.X_s(t) - Q_p.X_p(t) \qquad \text{(D)}$$

The thermal balance gives, after reducing the equation, of the latent heat terms :

$$\begin{aligned}\rho.V_p.\frac{\partial\{c_a+c_v.X_p(t)\}.\theta_p(t)}{\partial t} &= \alpha_{pf}.S_1\{\bar\theta_f(t)-\theta_p(t)\}\\ &+ \alpha_{pc}.S_2.\{\theta_c(t) - \theta_p(t)\}\\ &+ S.B(t).c_v.\bar\theta_f(t)\\ &+ Q_r.\{c_a+c_v.X_h(t)\}.\theta_h(t)\\ &+ \tau.Q_s.\{c_a+c_v.X_s(t)\}.\theta_s(t)\\ &- Q_p.\{c_a+c_v.X_p(t)\}.\theta_p(t)\end{aligned} \qquad \text{(E)}$$

* Hoods :

The input air flows in the hood are the insufflated air flows not entering the corresponding pockets $\Sigma(1-\tau).Q_s$, the output air flows from these pockets ΣQ_p and the air flow coming from the outside of the machine Q_e.

The water balance gives :

$$\begin{aligned}\rho.V_h.\frac{\partial X_h(t)}{\partial t} &= \Sigma Q_p.X_p(t) + \Sigma(1-\tau).Q_s.X_s(t)\\ &+ Q_e.X_e(t) - \{Q_h+\Sigma Q_r\}.X_h(t)\end{aligned}$$

(F)

and the thermal balance :

$$\begin{aligned}\rho.V_h.\frac{\partial\{c_a+c_v.X_h(t)\}.\theta_h(t)}{\partial t} &= \alpha_{vh}.S_3\{\theta_v(t)-\theta_h(t)\}\\ &+ \Sigma Q_p.\{c_a+c_v.X_p(t)\}.\theta_p(t)\\ &+ \Sigma(1-\tau).Q_s.\{c_a+c_v.X_s(t)\}.\theta_s(t)\\ &+ Q_e.\{c_a+c_v.X_e(t)\}.\theta_e(t)\\ &-\{Q_h+\Sigma Q_r\}.\{c_a+c_v.X_h(t)\}.\theta_h(t)\\ &+\alpha_{mh}.S_4\{\theta_m(t) - \theta_h(t)\}\end{aligned} \qquad \text{(G)}$$

The last term of this equation takes into account the thermal effect of the mass M of the machine (walls and frames). The variations of the machine temperature is given by the following equation :

$$M.c_m.\frac{\partial\theta_m(t)}{\partial t} = \alpha_{mh}.S_4.\{\theta_h(t) - \theta_m(t)\} \qquad (H)$$

* Extractions :

The input air flows of each extraction are the output air flows from the corresponding hoods ΣQ_h, and the air flows from the adjacent extractions $Q_{aa_{i-1}}$ and $Q_{aa_{i+1}}$.
We also introduce a mass transfer by condensation on the hood walls (β_a : mass transfer coefficient) and a heat loss by conduction (α_a : heat transfer coefficient between the outside medium and the air in the considered extraction).
Then, the mass balance gives :

$$\begin{aligned} \rho.V_a.\frac{\partial X_a(t)}{\partial t} = {} & \Sigma\, Q_h.X_h(t) + Q_{aa_{i-1}}.X_{a_{i-1}}(t) \\ & + Q_{aa_{i+1}}.X_{ai+1}(t) \\ (I) \qquad & - \beta_a.S_5.X_a(t) \\ & - (Q_{ab} + Q_{ba} + Q_a).X_a(t) \end{aligned}$$

and the thermal balance :

$$\begin{aligned} \rho.V_a.\frac{\partial\{c_a+c_v.X_a(t)\}.\theta_a(t)}{\partial t} = {} & \Sigma\, Q_h.\{c_a+c_v.X_h(t)\}\theta_h(t) \\ & + Q_{aa_{i-1}}.\{c_a+c_v.X_{a(i-1)}(t)\}.\theta_{a_{i-1}}(t) \\ & + Q_{aa_{i+1}}.\{c_a+c_v.X_{a_{i+1}}(t)\}.\theta_{a_{i+1}}(t) \\ (J) \qquad & + \alpha_a.S_5.\{\theta_e(t) - \theta_a(t)\} \\ & + \{r(0) - c_w.\theta_a(t)\}.S_5.\beta_a.X_a(t) \\ & - \{Q_{ab}+Q_{ba}+Q_a\}.\{c_a+c_v.X_a(t)\}.\theta_a(t) \end{aligned}$$

The Fig. 5 shows the relations between the different ventilation variables.

This model has been programmed on a mini-computer SOLAR 16/65. The integration of the differential equations from (A) to (J) carried out by means of the classical finite difference method is successively done for each cylinder from the beginning to the end of the drying section.

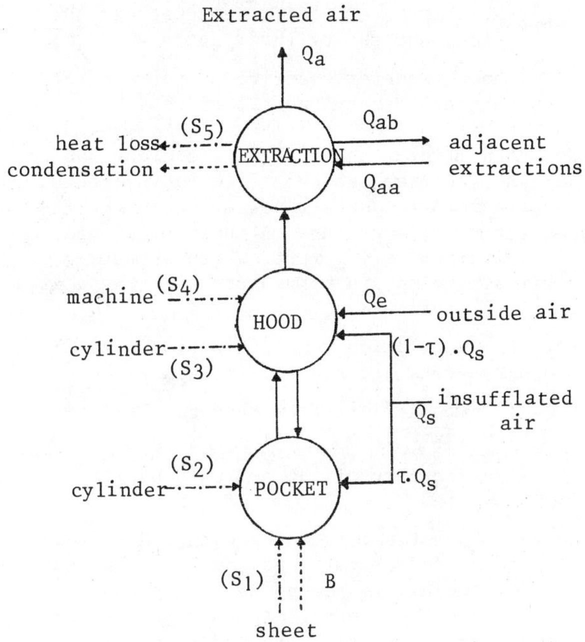

Fig. 5 : Schematic representation of the ventilation system

2. INITIALIZATION OF THE MODEL

At the beginning of the simulation, it is necessary to compute the operating point, by means of a steady state model. The numerical values of many of the parameters are unknown and their estimation is necessary, to use it.

2.1. Parameters estimation

In the sheet model (equations (A) and (B)), three different parameters are needed :

α_{cf} : heat transfer coefficient between the sheet and the cylinder

β_{fh} : mass transfer coefficient on the cylinder

β_{af} : mass transfer coefficient in the draw between the air and the sheet (the transfer being convective in the draw, it is possible in this case to assume that COLBURN analogy is valid between the heat transfert coefficient α_{at} and the mass transfer coefficient β_{af} ; then only one must be computed).

From comparison with experimental data, it is possible to estimate, cylinder by cylinder, these three coefficients. For that, we need at least three different measures on the sheet (two temperatures and one moisture content on each draw). This parameters adaptation is carried out by means of a non-linear programming method.
For the cylinder model (Equation C), two parameters are needed :

α_{vc} : heat transfer coefficient between the steam and the cylinder surface

α_{ca} : heat transfer coefficient between the air and the cylinder surface

For each dryer, the first coefficient can be computed from the heat flow coming from the steam to the sheet, measuring the cylinder surface temperature. A theoritical value of the second coefficient, which is more difficult to obtain by measurement, is taken in the litterature.

For the ventilation model (Equation D to J), two parameters are needed for each pocket :

- Q_r : "pumping" air flow through the clothing
- τ : efficiency of the pocket ventilation system

These both variables can be obtained by measuring the temperature and the moisture content of the air in the pockets, the characteristics of the insufflated air (Q_s, X_s and θ_s) and of course the heat and mass flows coming from the sheet and the cylinder.

For each extraction, two parameters are needed :

α_a : heat transfer coefficient (heat loss by conduction on the hood walls)

β_a : mass transfer coefficient (condensation on the hood walls)

These two coefficients can be determined by measuring the temperature and the moisture content of the extracted air.

2.2. Steady state model

All these parameters or coefficients can be determined for a given operating point. This identification method has been tested on a pilot paper machine and on different industrial drying sections. The influence of some working conditions such as steam pressure, basis weight and machine speed has been studied (9). These trials show that it is possible to keep constant the numerical values of the estimated parameters, for simulations around the selected operating point. The integration of the steady state equations system is carried out by means of a non linear programming method.

3. SIMULATIONS

The chosen example concerns a printing paper machine. This machine is 3,8 m wide. The drying section consists of 29 dryers, divided in three sections of 8,8 and 13 cylinders. The diameter of these dryers is 1,5 m. The section is set in a closed hood and two ventilators extract the wet air.
The paper basis weight was 0,083 kg/m² and the machine speed was 9 m/s.

3.1. Parameters estimation results

To take into account the divisions into steam sections and into clothing sections, we have divided the drying section into, of course, 29 pockets, 7 hoods and 2 extractions.
The parameters estimation has been carried out on the whole drying section, but we are presenting only the results and the simulations concerning the first extraction and the second hood composed of 5 cylinders and 5 pockets. This particular section is very representative because the evaporation rate is high and quite constant, and because the pockets are well ventilated by hot air blowing boxes, in this part of the machine.

The main measurements data, and operating conditions appear in table (1) :

Sheet moisture content :	
- entrance of the section	0,884 kg/kg
- end of the section	0,616 kg/kg
Sheet temperature (entrance)	86°C
Cylinder surface temperature	102,2°C
Steam temperature	132,7°C
Insufflated air (5 boxes)	
- flow	7,5 kg/s
- temperature	100°C
- moisture content	0,008 kg/kg
Extracted air :	
- flow	10 kg/s
- temperature	80,1°C
- moisture content	0,083 kg/kg
Pockets conditions (average)	
- temperature	94,1 °C
- moisture content	0,200 kg/kg

Table (1) : Operating conditions and measurements data for the considered section

The parameters estimation results appear in table (2)

α_{cf}	2080 W/m²°C
β_{fh}	0,019 m/s
β_{af}	0,020 m/s
α_{vc}	940 W/m².°C
Q_r(5 boxes)	4 kg/s
τ	0,2
α_a	20 W/m²°C
β_a	# 0

Table (2) : Parameters estimation results for the considered section

The value of the heat transfer coefficient α_{ca} between the cylinder and the air has been taken in the litterature (140 W/m^2°C). We can observe with these results that the "pumping" effect of the clothing is very important and that only 20 % of the insufflated air is really efficient.

3.2. Influence of the ventilation variables

Some computer simulations have been carried out to study the time responses of the system to step changes of the main ventilation variables. Step changes of the steam temperature have not been considered because the corresponding responses of the drying section are well known, in this case. We are presenting here the time evolution of the following variables :

- the extracted air characteristics (θ_a and X_a),
- the pocket air conditions (θ_p and X_p) and
- the sheet variables (θ_f and X_f).

Three different step changes have been chosen :

- decrease of the insufflated air flow (7,5 to 5 kg/s) (curves ___)
- decrease of the insufflated air temperature (100 to 90°C) (curves —·.—)
- increase of the extracted air flow (10 to 12,5 kg/s) (curves ---).

The figures (6) to (11) show the responses of the 6 variables.

We can see on these figures that a negative step change of the insufflated air flow leads to an increase of the pockets moisture content and to a decrease of the pockets temperature. Then, the water evaporation decreases and consequently the sheet moisture content and temperature increase. The evaporated water being smaller, the moisture content of the extracted air decreases and also the temperature.
A negative step change of the insufflated air temperature leads to a very small decrease of the air temperature in the pockets and in the extraction. The influence of this variable is also very small on the sheet temperature and on the evaporation rate : the sheet moisture content increases a little, but the effect on the air moisture content in the pockets and in the extraction is quite negligible.

A positive step change of the extracted air flow leads to a decrease of the air temperature in the pockets and in the extraction. Indeed, this is due to the increase of the outside air flow which is colder than the insufflated air. Consequently, the sheet is also cooled by the air in the pocket. The figures (8) and (10) show the decreases of the air moisture content in the pockets and in the extraction, the dilution being greater. Then, the evaporation rate increases owing to the decrease of the pockets moisture content, but this is partly compensated by the sheet temperature decrease. It explains the particular shape of the sheet moisture content response (Figure 6).

The analysis of all these responses also shows that the time constants are very small, especially when the step changes are applied to the insufflated air characteristics.

3.3. Influence of the heat transfer coefficient between the air and the machine

In this dynamical model, the thermal effect of the machine mass is taken into account by means of the equation (H). In this equation, we introduce the heat transfer coefficient α_{mh} between the air and the machine walls, and the contact surface S5. The figure (12) shows the influence of the product K ($K = \alpha_{mh} \cdot S5$), on a particular response of the drying section.

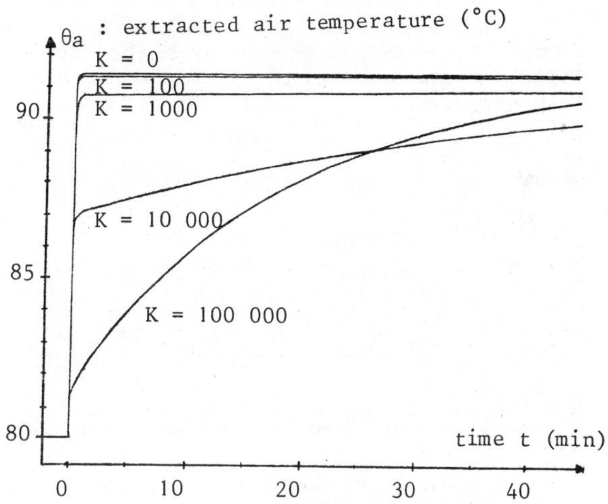

Fig. 12 : Influence of the thermal effect of the machine mass (step change of the insufflated air flow : 7,5 to 10 kg/s)

The selected example is a positive step change of the insufflated air flow (7,5 to 10 kg/s) and the considered variable is the temperature of the extracted air. K is varying from 0 to 100 000 W/°C.

For K equal to 0, we can observe the real time constant of the air temperature. When K is different from 0, a drift appears and the response time increases. Then, this coefficient K must be identified on the process because it has a very strong influence on the dynamics of the whole drying section. An average value of 3 000 W/°C has been chosen for the simulations presented here.

CONCLUSION

A dynamic mathematical model of a complete multicylinder drying section has been devellopped from a theoritical analysis of this process. Some simplifications in the description of the phenomena are introduced, which lead to an overall model easily workable.
It takes particularly into account the heat and mass transfer conditions, the drying

characteristics of the paper, the thermal effect of the cylinder wall and the main ventilation variables. The initialization of this model and the computing of the operating point is carried out by means of a steady state model. This one is the basis of a parameter estimation method which evaluate the heat and mass transfer coefficients and some other unknown parameters, on industrial equipements. Some simulation results are presented on a particular example. The time responses of the sheet characteristics and of the air conditions, to step changes of the main ventilation variables are examined. The analysis of these responses shows particularly the strong influence of the thermal inertia of the machine.

These simulation results show that the time constants of the ventilation variables are small in comparison with the cylinder responses or with the thermal inertia of the machine. But, they are not negligible and it will be necessary to take them into account for the future control of the whole drying section.

REFERENCES

(1) A.H. NISSAN and W.G. KAYE
"An analytical approach to the problem of drying of thin fibrous sheets on multicylinder machines"
TAPPI, vol. 38, n° 7 (Juillet 1955)

(2) W.D. BAINES
"Analysis of transients effect in drying of paper"
Pulp and Paper Magazine of Canada, Vol. 74, n°2 (1973).

(3) F.T. HARTLEY and R.J. RICHARDS
"Hot surface drying of paper. The development of a diffusion model"
TAPPI, vol. 57, n° 3 (Mars 1974)

(4) R.L. KNIGHT and L.A. KIRK
"Simulation of the papermachine drying section"
International Water Removal Symposium British Paper and Board Industries Federation - London (Mars 1975)

(5) A. LEMAITRE, J. VEYRE, B. LEBEAU and C. FOULARD
"Method for systematic analysis of paper machine multicylinder drying section"
4th IFAC-PRP4 Conference
Ghent - Belgium (3-5 June 1980).

(6) J. GUIGNABODET
"Digital control of a paper machine drying section"
Paper, Rubber and Plastics Industries International Congress on Automation - ANTWERP (October 1966)

(7) J.E. MILLER and E.J. FREEH
"Fourdrinier paper machine dynamics"
American Paper Industry (October 1968)

(8) O.I. LEHTIKOSKI
"A mathematical model of the dryer section"
Paperi Ja Puu, vol. 52, n° 2 (1970)

(9) A. LEMAITRE
"Modélisation et identification d'une sécherie multicylindrique de machine à papier : optimisation de sa conception et de son fonctionnement"
Thèse de Docteur-Ingénieur, Grenoble, France (Septembre 1978).

(10) M. SOININEN and J. NURMINEN
"An orientation to the pocket ventilation design in a drying section with open mesh fabrics"
Paper ja Puu, vol. 56, n° 4a (April 1974).

NOTATIONS

B : evaporation rate ($kg/m^2.s$)

c : specific heat (J/kg.°C)

G : basis weight (kg/m^2)

K : heat transfer coefficient between the air and the machine (W/°C)

M : mass of the machine (kg)

P : total surrounding pressure (P)

p : vapour partial pressure (P)

Q : air flow (kg/s)

r : latent heat of vaporization (J/kg)

R : perfect gas constant

S : surface (m^2)

t : time (s)

T : absolute temperature (°K)

U : machine speed (m/s)

V : volume (m^3)

x : machine direction

X : moisture content (kg/kg)

z : thickness direction

Z : cylinder thickness (m)

α : heat transfer coefficient ($W/m^2.°C$)

β : mass transfer coefficient (m/s)

λ : thermal conductibility (W/m.°C)

θ : temperature (°C)

τ : ventilation efficiency

ρ : specific mass (kg/m^3)

-----

a : extracted air	m : machine
b : clothing	p : pocket
c : cylinder	r : "pumping" air
e : outside air	s : insufflated air
f : sheet	v : steam
h : hood	w : water

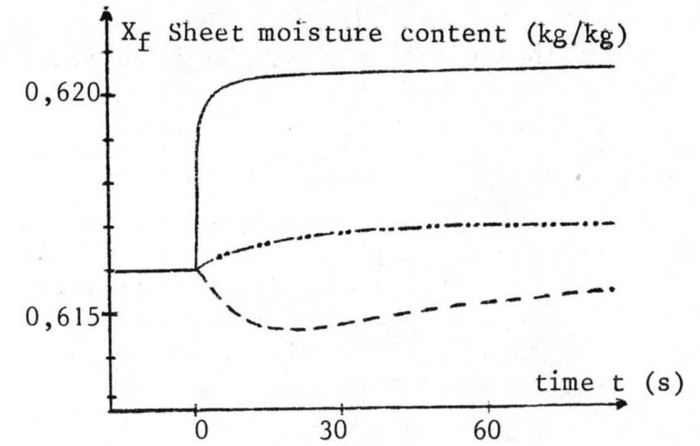

Fig. 6 : Time responses of the sheet moisture content to the three considered step changes

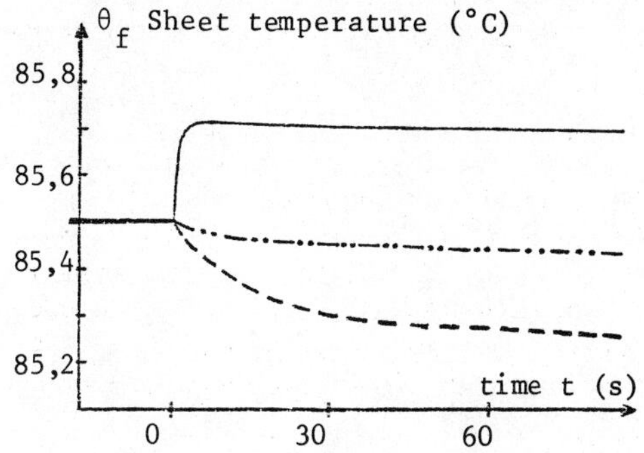

Fig. 7 : Time responses of the sheet temperature to the three considered step changes

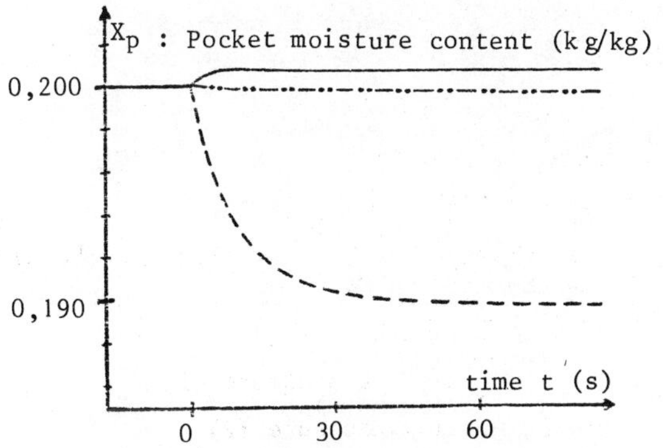

Fig. 8 : Time responses of the pocket moisture content to the three considered step changes

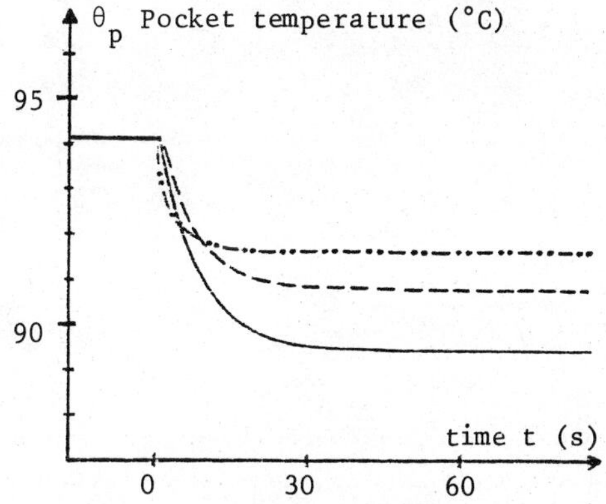

Fig. 9 : Time responses of the pocket temperature of the three considered step changes

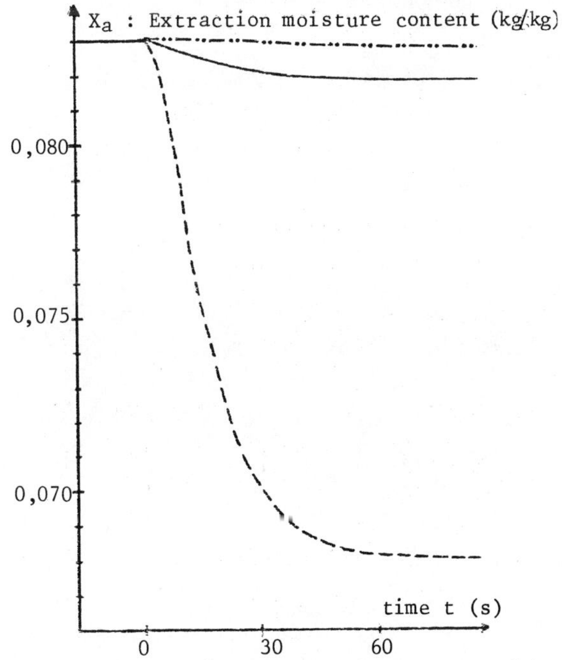

Fig. 10 : Time responses of the extraction moisture content to the three considered step changes

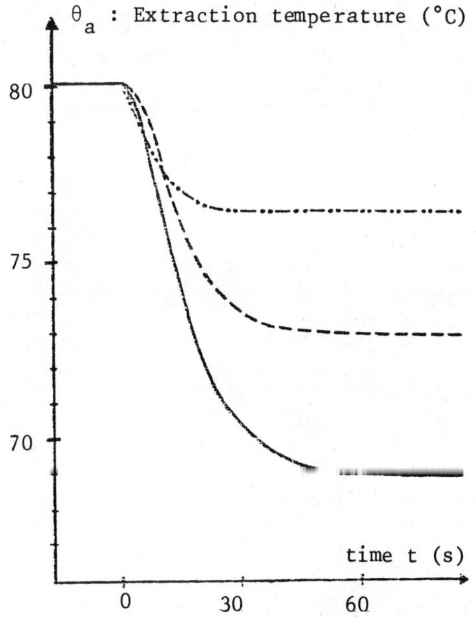

Fig. 11 : Time responses of the extraction temperature to the three considered step changes

——— : decrease of the insufflated air flow

—·— : decrease of the insufflated air temperature

----- : increase of the extracted air flow

STABILIZATION AND CONTROL OF ABSORBER TOWER CHEMICAL PROCESS

M. Darwish* and J. Fantin**

**Department of Electronics and Communication Engineering, Cairo University, Giza - Egypt*
***Laboratoire d'Automatique et d'Analyse des systèmes du CNRS 7 avenue du colonel Roche, 31400 Toulouse, France*

ABSTRACT. Stabilization and control of dynamic reactions in gasabsorber tower of chemical process are considered in this paper. Two methods are investigated and applied. In the first one a control strategies developed to insure prescribed degree of stability using pole placement techniques. The second method is a decentralized stabilization permitting to determine a set of decentralized controllers stabilizing the overall system. Results of application to a six plates absorber tower is presented and discussed.

INTRODUCTION AND SYSTEM DESCRIPTION

A gas absorber tower is an important element in several chemical processes. A typical gas absorber system consists of a number of vertically arranged plates enclosed within a chemical tower,Fig1

The chemical reactions which takes place in the tower are affected by the inlet feed compositions corresponding to a downward liquid stream and an upward vapour stream(L.LAPIDUS, N.R. AMUNDSON, 1950). These reactions may give rise to instability if the inlet feed compositions are not properly chosen, hence stabilization and control of such reactions become a very important problem(R.A. BAXLEY 1977, D.J. FRAADE 1977, DJ.B. PETROVSKI1978

The material balance for each plate gives the following set of equations Fig 2 (L.LAPIDUS,N.R,AMUNDSON 1950).

$$h_r \frac{dy_1}{dt} + h_e \frac{dx_1}{dt} = L_f(u_1 - x_1) + L_g(y_2 - y_1)$$

$$h_r \frac{dy_i}{dt} + h_e \frac{dx_i}{dt} = L_f(x_{i-1} - x_i) + L_g(y_{i+1} - y_i) \quad (1)$$

$$(i = 1, 2, \ldots, N-1)$$

$$h_N \frac{dy_N}{dt} + h_e \frac{dx_N}{dt} = L_f(x_{N-1} - x_N) + L_g(u_2 - y_N)$$

x_i, y_i are the composition of liquid and vapour leaving the ith plate.
h_r, h_e are the inlet vapour and liquid hold ups on each plate (assumed constant)
L_f, L_g are the flow rates of inlet liquid absorbent and inlet gas stream (assumed constant).
The liquid and vapour compositions are related by

$$y_i = a x_i \quad (2)$$

where a is a constant.
Hence the set of equations (1) can be rewritten as :

$$\frac{dx_1}{dt} = -d_2(1+d_1)x_1 + d_2 x_2 + d_1 d_2 u_1$$

$$\vdots$$

$$\frac{dx_i}{dt} = -d_2(1+d_1)x_i + d_1 d_2 x_{i-1} + d_2 x_{i+1}$$

$$(i = 1, 2, \ldots, N-1) \quad (3)$$

$$\vdots$$

$$\frac{dx_N}{dt} = -d_2(1+d_1)x_N d_1 d_2 x_{N-1} + \frac{d_2}{a} u_2$$

where :

$$d_1 = \frac{L_f}{aL_g}$$

$$d_2 = \frac{aL_g}{ah_r + h_e}$$

Equations (3) can be rewritten in the form :

$$\dot{X} = AX + Bu \quad (4)$$

where

$X \in R^N$ is the state vector, $X = [x_1, x_2, .., x_N]^T$

$u \in R^2$ is the control vector, $u = [u_1, u_2]^T$

$A \in R^{N \times N}$ matrix, $B \in R^{N \times 2}$ matrix

$$A = \begin{bmatrix} -d_2(1+d_1) & d_2 & 0 & & 0 \\ d_1 d_2 & -d_2(1+d_1) & d_2 & 0 & 0 \\ 0 & d_1 d_2 & -d_2(1+d_1) & d_2 & 0 \\ & & & & \\ 0 & & & d_1 d_2 & -d_2(1+d_1) \end{bmatrix}$$

$$B = \begin{bmatrix} d_1 d_2 & 0 \\ 0 & 0 \\ & \\ 0 & \frac{d_2}{a} \end{bmatrix}$$

Consider o six plates system with the following constants : $L_f = 40.8$, $L_g = 66.7$

$h_e = 0.75$, $h_r = 1.0$, $a = 0.72$

Then the matrices A and B are given by :

$$A = \begin{bmatrix} -1.173 & 0.634 & 0 & 0 & 0 & 0 \\ 0.538 & -1.173 & 0.634 & 0 & 0 & 0 \\ 0 & 0.538 & -1.173 & 0.634 & 0 & 0 \\ 0 & 0 & 0.538 & -1.173 & 0.634 & 0 \\ 0 & 0 & 0 & 0.538 & -1.173 & 0.634 \\ 0 & 0 & 0 & 0 & 0.538 & -1.173 \end{bmatrix}$$

$$B = \begin{bmatrix} 0.538 & 0 \\ 0 & 0 \\ 0 & 0 \\ 0 & 0 \\ 0 & 0 \\ 0 & 0.880 \end{bmatrix}$$

The eizen valuee of the matrix A are given by :

$$\{ 2.23 , 1.9 , 1.44 , 0.9 , 0.443 , 0.117 \}$$

Hence initially the system is instable Fig 3 shows the trajectories of the states x1 - x6. Therefore it is necessary to stabilize and control the system dynamics through proper choice of the control signaler. The objective of the present paper is to develop and apply techniques for stabilizing and controlling the transients of such chemical reactions. In this way, two methods are examined, in the first a control strategy is developed to insure prescribed degree of stability using pole placement techniques through the determination of the state weighting matrix of the objective function. The second method is a decentralized stabilization methodology permitting to determine a set of decentralized controllers each observe local outputs and controls local inputs, all the controllers however, are designed in a way to insure stabilization of the overall system.

STABILIZATION AND CONTROL THROUGH POLE PLACEMENT TECHNIQUES

Associate to the system (4) a quadratic criterion of the form :

$$J = \frac{1}{2} \int_0^\infty (X^T Q X + u^T R u) \, dt \qquad (5)$$

where A is positive semi definite matrix, R is positive definite matrix. The problem is how to determine the control U which minimize (5) subject to system dynamics given by (4). Application of optimal control techniques gives U of the form :

$$u = - R^{-1} B^T K X = G X \qquad (6)$$

where K is the solution of matrix Riccati equation

$$A^T K + K A - K B R^{-1} B^T K + Q = 0 \qquad (7)$$

and the closed loop system is of the form

$$\dot{X} = (A + B G) X \qquad (8)$$

Hence it is clear that the matrix Q affect the closed loop system certain prescribed stability margins through pole placement can be achieved by proper choice of the state weighting matrix Q. The objective is to give the system a prescribed closed loop eigen values. We adopt here the technique proposed by Solheim (D. A. SOLHEIM 1972)
In this technique iteration method is used to assign the desired eigen values one after one as follows :

(1) assuming that the matrix $\tilde{Q} = M^T Q M$ where M is the modal matrix corresponding to the initial system (4) having all elements zero except for $\tilde{q}_{jj}$ which can be determined by the relation (O.A. SOLHEIM 1972)

$$\tilde{q}_{jj} = \frac{S_j^2 - \lambda_j^2}{h_{jj}}$$

where Sj is the desired eigen value, λ_j is the actual eigen value, h_{jj} is the element of the matrix H given by

$$H = M^{-1} B R^{-1} B^T (M^{-1})^T$$

(2) using $\tilde{Q}$ we can solve the following RICCATI equation :

$$\tilde{K}\Lambda + \Lambda\tilde{K} - \tilde{K} M^{-1} B R^{-1} B^T (M^{-1}) \tilde{K} + \tilde{Q} = 0 \quad (9)$$

where Λ is the matrix of eigen values hence the control U is given by :

$$u = -(R^{-1} B^T (M^{-1})^T \tilde{K} M^{-1}) X = GX \quad (10)$$

and the closed loop system will be :

$$\dot{X} = (A + BG) X = A X \quad (11)$$

(3) restarting with the new system(11) to assign another eigen value, and repeat the above steps till all the desired eigen values are assigned.
Suppose it is required to assign the closed loop eigen values of the above example to :

$$\{-2, -4, -5, -6, -7, -8\}$$

to insure good stabilization of the system. Choosing the matrix R to be R

$$R = \begin{bmatrix} 1 & 0 \\ 0 & 1 \end{bmatrix}$$

Then apliying the above technique, the resulting matrix is given by :

$$Q = \begin{bmatrix} 136 & 0.17\times10^4 & 0.54\times10^4 & 0 & 0 & 0 \\ 0.17\times10^4 & 0.23\times10^5 & 0.8\times10^5 & 0 & 0 & 0 \\ 0.54\times10^4 & 0.8\times10^5 & 0.29\times10^6 & 0 & 0 & 0 \\ 0 & 0 & 0 & 0.17\times10^7 & 0.3\times10^6 & 0.13\times10^5 \\ 0 & 0 & 0 & 0.3\times10^6 & 0.58\times10^5 & 0.28\times10^4 \\ 0 & 0 & 0 & 0.13\times10^5 & 0.28\times10^4 & 185 \end{bmatrix}$$

The resulting trajectories for the states X_1 to X_6 and the control signals u_1 and u_2, are shown in Fig.4, from which we see that the system has good transient behaviour which satisfy the desired stabilization and control objectives.

DECENTRALIZED STABILIZATION AND CONTROL

In this appoach instead of calculating global control as given by (6) we derive a set of decentralized controllers which comperate together to control and stabilize the overall system. This technique has the advantages of overcoming the computational difficulties which may arise in case of global solution, besides its practical implementation is more easier than global control implementation.
In this technique, the original system is decomposed into a group of interconnected subsystems, where for this example of six plates absorber we decompose the system into two subsystems, hence the dynamic of system (4) can be rewritten in the form :

$$\dot{X}_i = A_i X_i + B_i u_i + H_i X \qquad i = 1,2,..$$

$$J_i = \frac{1}{2}\int_0^\infty (X_i^T Q_i X_i + u_i^T R_i u_i)\, dt \quad (12)$$

where :

$$A_1 = A_2 = \begin{bmatrix} -1.173 & 0.634 & 0 \\ 0.538 & -1.173 & 0.634 \\ 0 & 0.538 & -1.173 \end{bmatrix}$$

$$B_1 = [0.538 \quad 0 \quad 0]^T$$

$$B_2 = [0 \quad 0 \quad 0.88]^T$$

$$H_1 = \begin{bmatrix} 0 & 0 & 0 & 0 & 0 & 0 \\ 0 & 0 & 0 & 0 & 0 & 0 \\ 0 & 0 & 0 & 0.634 & 0 & 0 \end{bmatrix}$$

$$H_2 = \begin{bmatrix} 0 & 0 & 0.538 & 0 & 0 & 0 \\ 0 & 0 & 0 & 0 & 0 & 0 \\ 0 & 0 & 0 & 0 & 0 & 0 \end{bmatrix}$$

where Hi represent the interconnetion matrix between the ith subsystem and the rest of the system.
In this technique the controller for each subsystem is given by :

$$u_i = -W_i u_i , \qquad i = 1,2,\ldots \quad (13)$$

that is the controller depends only on the states of its subsystem.
We adopt here the technique proposed by Darwish et al (M. DARWISH, H. SOLIMAN, J. FANTIN, 1978, 1979)to determine Wi as follows :

$$W_i = -R_i^{-1} B_i^T K_i , \qquad i = 1,2,\ldots \quad (14)$$

Ki is the solution of matrix Riccati equation :

$$A_i^T K_i + K_i A_i - K_i B_i R_i^{-1} B_i^T K_i + 2\alpha K_i + Q_i = 0 \quad (15)$$

(2) starting with α = 0 and increasing α gradually we get Ki solution for (15) till the following condition is satisfied.

$$P = \left[2\alpha K + Q + K B R^{-1} B^T K - (KH + H^T K) \right] \quad (16)$$

is positive definite,

where $K = \begin{bmatrix} K_1 & 0 \\ 0 & K_2 \end{bmatrix}$, $Q = \begin{bmatrix} Q_1 & 0 \\ 0 & Q_2 \end{bmatrix}$ $R^{-1} = \begin{bmatrix} R_1^{-1} & 0 \\ 0 & R_2^{-1} \end{bmatrix}$

and $H = \begin{bmatrix} H_1 \\ H_2 \end{bmatrix}$

condition (16) insures the stability of the overall system.
The application of this technique to the numérical values of the present example of gas absorber gives the trajectories shown in Fig. 5 the value of which satisfies the stabilization requirement is given by α = 1.6

CONCLUSION

Stabilization and control problems of gas absorber tower used in chemical process are considered in this paper. The presented methodology insure stabilization of the system, besides, it may give prescribed stability properties for the closed loop system. These techniques reduces system transient behaviour and improve system performance.

REFERENCES

(1) L.LAPIDUS,N.R. AMUNDSON "Stagewise absorption and extraction equipment: Transient and unsteady state operation", Ind.Eng.Chem. vol. 42, 1950 pp.1071-1078.

(2) R.A. BAXLEY "the treatment of distillation systems in on line optimization",proc IFAC conference on digital computer applications to process control The Hauge, June 1977, pp. 121-127

(3) D.J. FRAADE "the application of process computer control to pilot plant operations, an automation concept development", proc. IFAC conference on digital computer applications to process control, the Hauge, june 1977 pp127-141.

(4) DJ. B. PETROVSKI "Calculation of optimum feedback gains for output constrained regulators",IntJ.numerical methods in Eng. vol 12, 1978 pp. 1873-1878.

(5) O.A. SOLHEIM "Design of optimal c ontrol systems with prescribed eigen values", int J. control 1972

(6) M. DARWISH,H. SOLIMAN, J.FANTIN stabilization techniques for large scale systems" proc 4th EHCAC Conference on automatic control system Enginéering, Cairo, Nov 1978.

(7) M. DARWISH, H. SOLIMAN,J.FANTIN "Decentralized stabilization of large scale dynamical systems", IEEE, Trans on Sept. man and Cyb, volSMC 9 N° 11, Nov 1979, pp. 717-720.

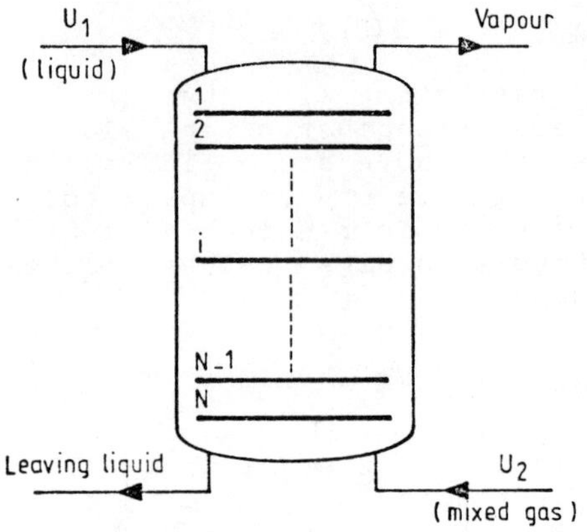

Figure 1.

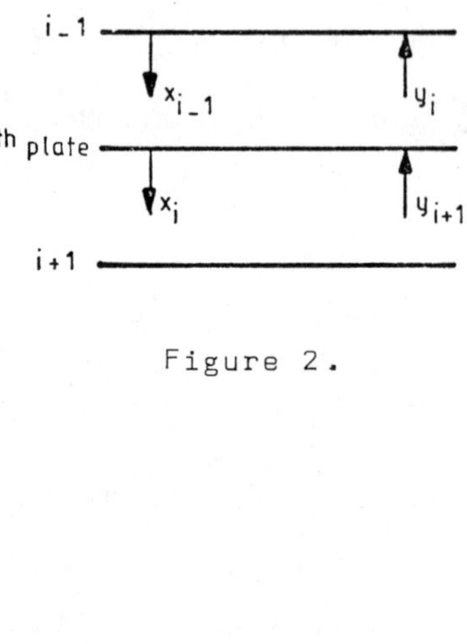

Figure 2.

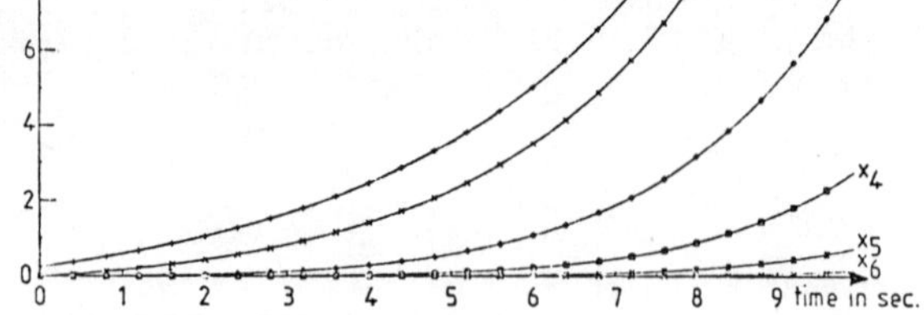

Figure 3.

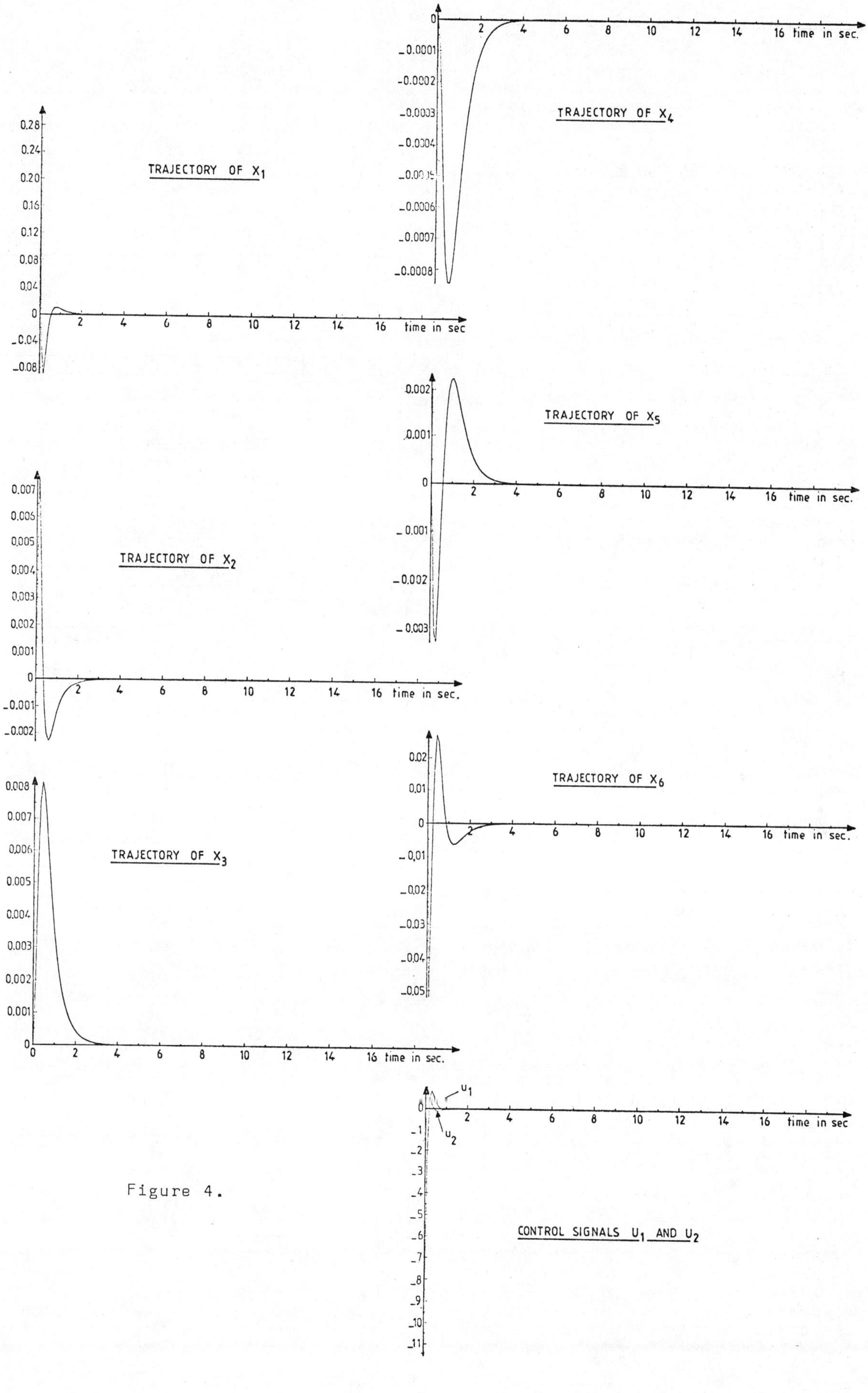
TRAJECTORY OF X_1
TRAJECTORY OF X_2
TRAJECTORY OF X_3
TRAJECTORY OF X_4
TRAJECTORY OF X_5
TRAJECTORY OF X_6
CONTROL SIGNALS U_1 AND U_2
u_1
u_2
time in sec.

Figure 4.

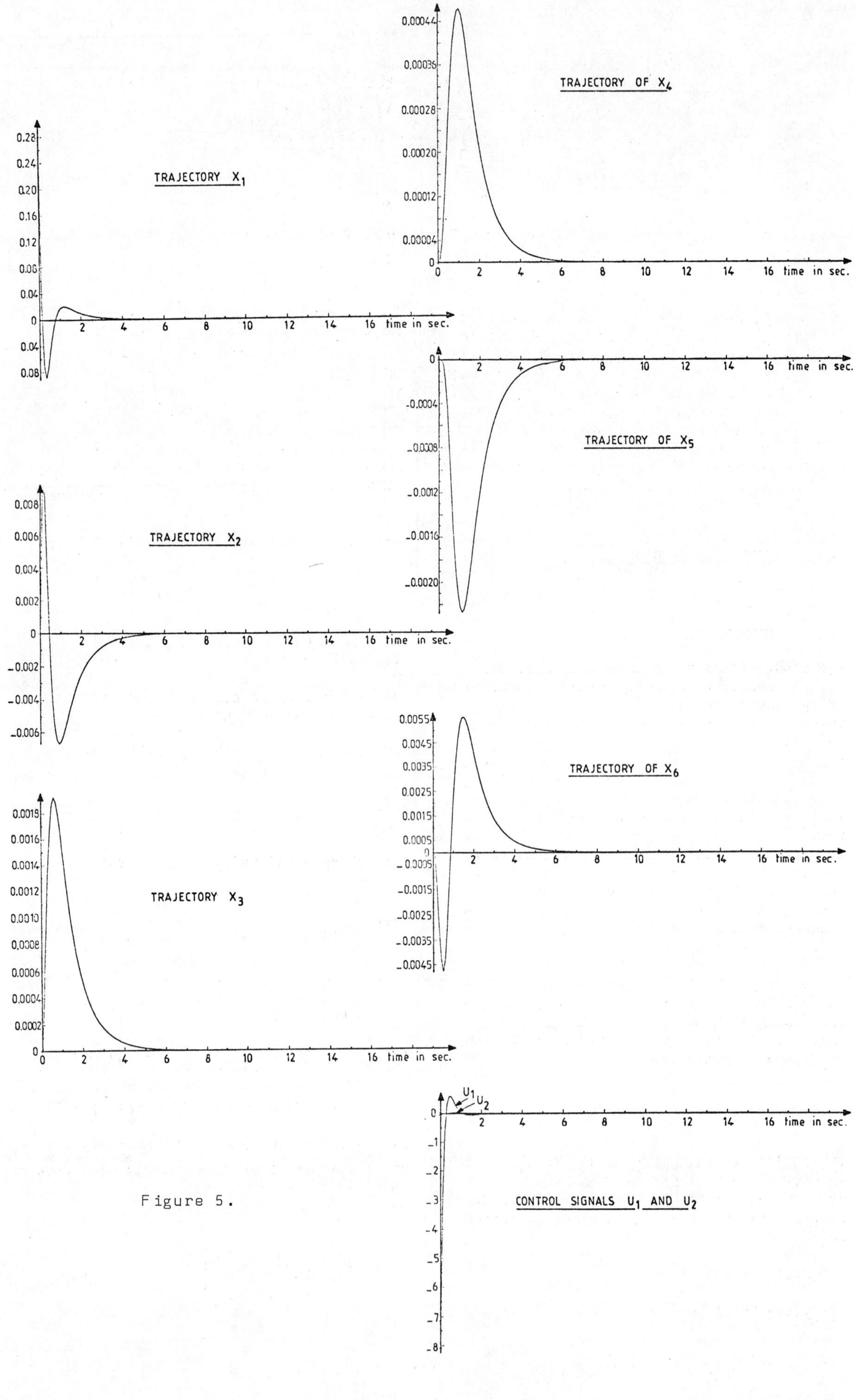
TRAJECTORY X1
TRAJECTORY X2
TRAJECTORY X3
TRAJECTORY OF X4
TRAJECTORY OF X5
TRAJECTORY OF X6
CONTROL SIGNALS U1 AND U2
U1
U2
time in sec.

Figure 5.

DESIGN OF SENSITIVITY-REDUCING OUTPUT FEEDBACK CONTROLLER FOR A DISTILLATION COLUMN

N. N. Sorial

Department of Electrical Engineering, Faculty of Engineering, Alexandria University, Alexandria, Egypt

Abstract. The design of output feedback controller for a distillation column which satisfy a sensitivity-reduction criterion is presented in this paper. The control variables of the distillation column are the heating power and the reflux flow rate. The column output is the concentration of the overhead product. The dynamic compensator which is a state observer, to estimate the unmeasured states of the column, is designed to maintain the composition of the overhead product constant not only when the system is subjected to step disturbance in the feed flow rate, but also when small variations in the parameters of the controlled plant occur. Graphs illustrating the predicted output response of the distillation column after implementing the proposed controller are also included.

Keywords. Multivariable control systems; optimal control; process control; observers; sensitivity analysis.

INTRODUCTION

The study of the control system behaviour due to nonzero tolerances in the parameters of the controlled plant is of major concern in any system design. One of control design aims is to keep the closed-loop system insensitive to small variations in the system parameters. Moreover, in some systems it is not possible to measure all the states and in others their measurement may not be economic. A method to overcome this problem is to build a dynamic observer to reconstruct the unavailable states from system outputs and inputs. Cruze and Krough (1978) developed a design technique of compensators for output feedback systems which satisfy a sensitivity-reduction criterion.

In the present paper, a distillation column is considered, and an output feedback controller using a state observer, is designed so that satisfication of sensitivity reduction criterion is guaranteed.

THE DESIGN PROCEDURE

The design procedure (Cruze and Krough, 1978) is summerized in the following steps:

step 1: Consider the linear time-invariant controllable and observable system represented in the state-space form

$$\dot{x}(t) = Ax(t) + Bu(t)$$

$$y(t) = Cx(t) \qquad (1)$$

where $x(t) \in R^n$, $y(t) \in R^1$, $u(t) \in R^m$ with rank(CB) = m_1

The plant matrices A,B and C are then transformed to plant canonical form

($T^{-1}AT$, $T^{-1}BU$ and CT) where the nonsingular matrices T, nxn, and U', mxm, are calculated from

$$CT = [I_1 \mid 0] \qquad (2)$$

$$\text{and} \quad T^{-1}BU = [B_1^{*'} \mid B_2^{*'}]' \qquad (3)$$

$$\text{where} \quad B_1^* = \begin{bmatrix} I_{m_1} & 0 \\ 0 & 0 \end{bmatrix}, \quad B_2^* = \begin{bmatrix} 0 & I_{m-m_1} \\ 0 & 0 \end{bmatrix}$$

step 2: Suppose the control u(t) is given by

$$u(t) = Hz(t) + Ky(t) \qquad (4)$$

where z(t) is the state of the kth order dynamic observer represented by

$$\dot{z}(t) = Fz(t) + Gy(t) + Du(t) \qquad (5)$$

The constant matrices of the observer are to be determined (Luenberger, 1971) placing the poles as desired .

step 3: Partition the matrix $T^{-1}AT$ as

$$T^{-1}AT = \begin{bmatrix} A_{11} \cdots A_{14} \\ \vdots \\ A_{41} \cdots A_{44} \end{bmatrix} \qquad (6)$$

and form the pair of matrices

$$\left\{ \begin{bmatrix} A_{44} & A_{42} \\ A_{24} & A_{22} \end{bmatrix}, \begin{bmatrix} A_{41} & A_{43} \\ A_{21} & A_{23} \end{bmatrix} \right\} \quad (7)$$

Determine a matrix W_1 which stabilizes the pair (7) using the pole assignment technique. Form the matrix

$$W = [W_1 \mid W_2] \quad (8)$$

where W_2 is an arbitrary matrix.

step 4: Combining the observer and plant canonical equations, the composite system is formed as

$$\dot{\bar{x}}(t)=\bar{A}\bar{x}(t) + \bar{B}u(t)$$

$$\bar{y}(t)=\bar{C}\bar{x}(t) \qquad \bar{x}(t)\in R^{n+k}, \bar{y}(t)\in R^{1+k} \quad (9)$$

where

$$\bar{A} = \begin{bmatrix} F & GCT \\ 0 & T^{-1}AT \end{bmatrix}, \quad \bar{B} = \begin{bmatrix} D \\ T^{-1}BU \end{bmatrix} \quad (10)$$

with the output feedback control (4) yields

$$u(t)= \bar{K}\bar{y}(t) \quad \text{where } \bar{K}=[H \quad K] \quad (11)$$

Convert the composite system equations(10) to the composite system canonical form

$$\hat{A} \triangleq T_s\bar{A}T_s^{-1}, \quad \hat{B} \triangleq T_s\bar{B} = \begin{bmatrix} I_m \\ 0 \end{bmatrix}$$

and

$$\hat{C} \triangleq T_o\bar{C}T_o^{-1} = [I_{k+1} \mid 0] \quad (12)$$

Using W from step 3, define the following matrices

$$\tilde{W} \triangleq \begin{bmatrix} I_m & W[I_{k+1-m} \mid 0] \\ 0 & I_{k+n-m} \end{bmatrix}$$

and

$$\tilde{A} \triangleq \tilde{W}^{-1}\hat{A}\tilde{W} = \begin{bmatrix} \tilde{A}_{11} & \tilde{A}_{12} \\ \tilde{A}_{21} & \tilde{A}_{22} \end{bmatrix}, \quad \tilde{A}_{11}\in R^{m\times m},$$

$$\tilde{A}_{22}\in R^{(n+k-m)\times(n+k-m)} \quad (13)$$

A positive semi-definite symmetric matrix Q, $(k+n-m)\times(k+n-m)$ is selected and the Lyapunov equation

$$P\tilde{A}_{22} + \tilde{A}'_{22}P + Q = 0 \quad (14)$$

is solved for P. Then, the Riccati equation given by

$$P^oA^o + A^{o\prime}P^o - P^oR^oP^o + Q^o = 0 \quad (15)$$

where $R^o \triangleq R^{-1}-\tilde{A}_{12}Q^{-1}\tilde{A}'_{12} > 0$

$Q^o \triangleq V + \tilde{A}'_{21}PQ^{-1}P\tilde{A}_{21} > 0$

and $A^o \triangleq \tilde{A}_{11}+ \tilde{A}_{12}Q^{-1}P\tilde{A}_{21}$

is solved for P^o.

The output feedback control, with sensitivity reduction criterion satisfied, is then given by

$$\bar{K} = -R^{-1}P^o[I \mid -W]T_o \quad (16)$$

DEVELOPMENT OF THE DISTILLATION COLUMN MODEL AND RESULTS

A distillation column for processing a binary mixture of methanol-water is considered (Aly and Fawzy, 1978). The control variables of the column are the heating power(H) and the reflux flow rate(R) . The column output variable is the concentration of the overhead product(x_D).

The accurate mathematical model of the column is nonlinear. Perturbing the plant variables about a specific operating condition, a linearized model representing the dynamics of the distillation column is obtained in the form

$$\dot{x} = Ax + Bu + DF$$

$$y = Cx + EF \quad \text{with } x(0) = 0 \quad (17)$$

where $x = [\, x_1 \quad x_2 \quad x_3]'$ is the state vector

$u = [\, H \quad R]'$ is the control vector

$y = x_D$ is the output variable

and F is the feed flow rate, considered as disturbance function.

Assuming that the feed flow rate is constant and the column is to operate with the steady state values

$F_{s.s.} = 156$ cc/min, $H_{s.s.} = 2.2$ kw &

$x_{D_{s.s.}} = 97\%$, then A, B and C matrices take the values

$$A= \begin{bmatrix} -0.167 & 0 & 0 \\ 1.667 & -1.667 & 0 \\ 0 & 4.034 & -0.3 \end{bmatrix} \quad B= \begin{bmatrix} 0.167 & 0 \\ 0 & 0 \\ 0 & -0.3 \end{bmatrix}$$

$$\text{and } C= [0 \quad -13.23 \quad 0.937] \quad (18)$$

The corresponding block diagram of the column is shown in Fig. 1.

The design algorithm explained above is applied to the distillation column represented by Eq. (18) . The observer is designed such that the poles are located at -0.167 and $-10.9 \pm i5.6$. The W_1 matrix is determined to stabilize the pair of matrices (7); then the W matrix of Eq. (8) is formed as

$$W = \begin{bmatrix} 0 & 1 \\ -10 & 0 \end{bmatrix}$$

The Q , R and V matrices are chosen as

$$Q=\begin{bmatrix}1&0&0&0\\0&1&0&0\\0&0&1&0\\0&0&0&1\end{bmatrix},\quad R=\begin{bmatrix}10&0\\0&10\end{bmatrix}\text{ and } V=\begin{bmatrix}1&0\\0&1\end{bmatrix}$$

Accordingly, the Lyapunov Eq. (14) is solved for P to give

$$P=\begin{bmatrix}0.1124&-0.0126&0.0078&-0.0484\\-0.0126&0.03&-0.0046&-0.0063\\0.0078&-0.0046&2.865&-0.0127\\-0.0484&-0.0063&-0.0127&0.211\end{bmatrix}$$

Then the Riccati Eq. (15) is solved for P^o resulting in

$$P^o=\begin{bmatrix}0.819&2.1\\2.1&0.861\end{bmatrix}$$

Consequently, the composite feedback control (16), with sensitivity-reduction criterion guaranteed, is given by

$$\bar{K}=[H\,|\,K]=\left[\begin{array}{ccc|c}0.0819&-0.21&-2.1&-0.164\\0.21&-0.086&-0.861&-0.42\end{array}\right]$$

The transient response of the closed-loop system is shown in Fig. 2 . To illustrate the sensitivity-reduction properties of the proposed controller, the nominal operating point is deviated by 5% and the new operating point data is used to recalculate the corresponding feedback gains. Fig. 2 shows the time response of the output variable for the perturbed case.

The nominal and perturbed cases are considered again, and the conventional Riccati design method is adopted to determine the feedback gains in both cases. The resulting output responses are shown in Fig. 3 . Comparison of the transient responses of Fig. 2 and Fig. 3 displays the sensitivity-reduction properties of the considered method.

CONCLUSIONS

In this paper , an output feedback compensator using a dynamic observer is designed for a distillation column. The resulting closed-loop system is insensitive to small variations in the parameters of the controlled plant. The use of dynamic observer provides direct means of allocating some eigenvalues of the feedback system as desired and introduces a useful interpretation of the sensitivity weighting matrix.

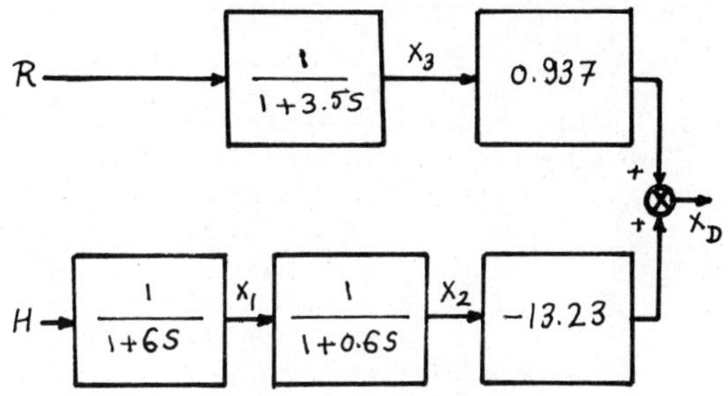

Fig. 1. Block diagram of distillation column

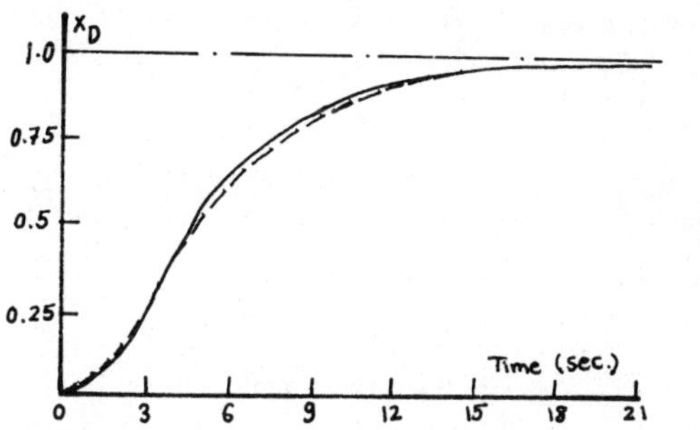

Fig. 2. Normalized output response due to 10% step change in H.
— Nominal parameters
--- perturbed parameters.

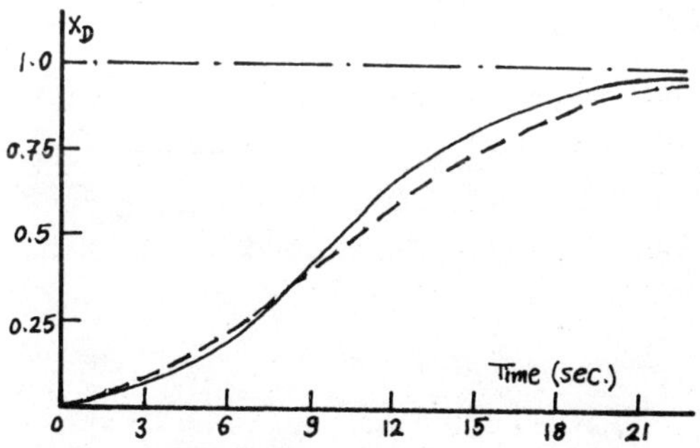

Fig. 3. Normalized output response for conventional Riccati solution due to 10% step change in H.
— Nominal parameters.
--- Perturbed parameters.

REFERENCES

Cruz, J. B. ,and B. Krogh (1978). IEEE Trans. Autom. Control, 23, 1058-1062.
Luenberger, D. G. (1971). IEEE Trans. Autom. Control, 16, 596-602.
Aly, G. M. ,and A. S. Fawzy (1978). Conference on Measurement and Control MECO'78, Athens, Vol 2, 767 pp.

OPTIMIZATION OF DISCRETE ADDITIONS OF SUBSTRATE DURING THE BATCH BIOSYNTHESIS OF ANTIBIOTICS

J. Černetič*, S. Strmčnik*, M. Pokorny** and R. Modic***

*J. Stefan Institute, P.O.B. 199-IV, YU-61000 Ljubljana, Yugoslavia
**KRKA - Chemical and Pharmaceutical Works, Novo mesto, Yugoslavia
***Department for Chemistry and Chemical Technology, University of Edvard Kardelj, Ljubljana, Yugoslavia

Abstract. A case history is described in which a practical systems engineering approach was applied to the industrial biosynthetic production of antibiotics, with the aim to develop a better operating policy for a single fermentor. The control variables to be optimized are: the sequence of discrete additions of substrate, the initial concentration of substrate and the duration of the batch (batch time). Three different criteria were considered, namely batch profit, production rate and substrate utilization efficiency. A relatively simple (formally a chemical-kinetic) model is shown, giving good agreement with the measured process outputs. For optimization, a simulation study and the Quasi-Newton method (with simple bounds on variables) were combined. From the results obtained, the following practical recommendations are given: 1. the initial concentration of substrate should be remarkably increased, 2. a falling sequence for substrate additions should be introduced, 3. for the first two criteria, higher substrate addition rate is desirable, whereas for the third criterion substrate additions later during the batch are to be canceled, 4. batch time is generally not a critical variable; eventually it can be optimized by an univariable search. In the light of these conclusions, further research of substrate inhibition in this process is highly desirable.

Keywords. Optimization; system analysis; nonlinear systems; chemical industry; biosynthetical production; antibiotics.

INTRODUCTION

The semibatch reactor to which reactants are added continuously after the initial charging, is generally not a very frequent type of reactors in the present-day chemical engineering practice. Yet less frequently such type of operation can be met in which some reactants are added in discrete amounts and many times during the operation cycle. However; just this operation procedure is used in the industrial production of certain antibiotics, with the main aim to increase product yield.

Why this process and its operation deserves the efforts and troubles of optimization? Firstly because of the high value of the product, and secondly because the quantity of the subsequently added reactant (substrate) comprises from 15 up to 40% of its initially charged quantity. As the latter is somewhat cheaper, this gives approximately 40% in value on the average. In such a situation, the question naturally arises; why not to give all substrate at once, i.e. at batch beginning? Un-fortunately, it is considered that higher initial concentrations of substrate act in some sense retarding to product formation. Therefore it should be more reasonable to give additional substrate later on in this process, when a part of the initial quantity has already been consumed.

It is not simple to check this proposition in operation and so to improve the process directly because of variations in antibiotic yield from batch to batch. This is due mainly to the inherent variability of living microorganisms and also to other uncontrollable influences (e.g. incidental contamination; variations in raw materials and operation variables).

For this reason we decided to develop an optimization strategy by means of mathematical modelling. So the main benefit expected from this study was considered to be an improvement of the existing (imported) technology which was actualy designed for an another set of technical and economic conditions. Although some improvements and adaptions have already been made, it was considered that this process has yet a potential for improvement in our specific working conditions in Yugoslavia, particularly in relation to substrate additions.

In two recent decades, a multitude of works in the area of mathematical modelling biosynthetic processes has been published. Among and besides these, there are also some publications dealing with optimization or optimal control of semibatch biosynthetic processes. So e.g. D'Ans, Gottlieb and Kokotović (1972) have developed an optimal control law for a simple Haldane - Monod model of microbial growth, with the aim to maximize the production rate of bacteria by variation of nutrient flow into a growth wessel. Their results show that in the space of substrate and microorganism-cell concentrations a large area exists where this control law recommends zero nutrient flow, i.e. essentially batch operation.

More recently, Yamane, Sada and Takamatsu (1979) have found in a theoretical consideration of the time-optimal start-up of a chemostat (essentially a continuous fermentation) that an exponentially increasing feed rate of substrate is advisable in cases of growth inhibition by substrate, but pure batch operation otherwise.

Similarly it has been reported (Wang H.Y., Cooney and Wang D.I.C., 1979) that in on-line experiments with an anticipatory computer control of bakers' yeast production in a fed-batch fermentation, where the purpose is to maximize cell yield, a growing flow rate of trients (molasses) results.

On the other side Heijnen, Roels and Stouthamer (1979) who use a somewhat similar approach in modelling the penicillin fermentation, state on the basis of their simulation study that an increasing feed rate of glucose as a function of time should be superior as compared with a constant or a decreasing feed rate, for obtaining hi final penicillin yields.

In the following discussion, we first describe briefly the process itself and how the problem of optimizing substrate additions has been formulated. Then the mathematical model, the procedure of searching the optimum, and lastly the obtained results are given. We conclude with practical recommendations concerning process improvement and with some guidelines for further work.

PROBLEM FORMULATION BY SYSTEM ANALYSIS

In this chapter we will show how, by the common systems engineering approach, we have formulated our optimization problem, starting from an analysis of the process and its operation.

The concerned industrial-scale production of antibiotic is carried out in special bioreactors, called fermentors, of several ten cubic meters of volume, and by batchwise operation. Each cycle of a batch consists of a preparatory phase and an operational phase. During the first phase, the fermentor is emptied from the previous contents, cleaned, sterilized with superheated steam, charged with reactants and water, again sterilized and cooled down to the usual operating temperature. This phase can also include some pure idle time between these operations. The following operational or production phase begins with the so-called inoculation, i.e. adding a certain amount of antibiotic-producing microorganisms. During this phase, the fermetor is aerated and its contents well mixed, whereas temperature, acidity (pH) and foam level are under steady control by conventional instrumentation. The production phase is finished when, after a specified time period, mixing and aeration is stopped by operating personnel. After that, the contents of the fermentor (fermentation broth) is transferred over to other equipment where the isolati-on of the pure antibiotic from the broth takes place.

At present, the operation of the entire plant with several fermentors is supervised by an experinced bioengineer who monitors the biosynthetical production process in each fermentor on the basis of an operation chart. In this document the time profiles of relevant component concentrations (specifically: antibiotic, carbohydrates and soluble nitrogen) are plotted in 24-hours intervals. Using this information, the engineer decides whether or not the particular batch can be considered as successful and when it is reasonable to supplement it with additional reactants, i.e. carbohydrates or soluble nitrogen compound. For sake of convenience, we call these two added components simply "substrate" and "nitrogen", respectively.

Normally, the batch is supplemented in the middle third of the operational phase. Because of practical reasons, the additions follow each other approximately every 24-hours, so the time instants for all possible additions are almost fixed in relation to batch beginning, as determined by the time of fermentor inoculation.

Initially, the goal of our study was to optimize only the sequence of substrate additions. The addition of nitrogen was also considered in the mathematical model, in or-der to be able to simulate the process as correctly as possible, but it was excluded from the subsequent optimization because nitrogen is added during the fermentation subject to an "intangible" criterion, namely when it is seen from its concentration in the fermentation broth, that the microorganism and its production ability could suffer from nitrogen deficiency.

Already during the initial problem analysis it became evident that the sequence of substrate additions cannot be optimized separately from the initial concentration of substrate and from the duration of operational phase of the batch because these variables are esentially interdependent. In the following, we will designate the latter variable briefly as "batch time".

Starting from the given state of evidence and using a system engineering approach, the problem of optimizing the discrete additions of substrate during the batch biosynthesis can be briefly formulated as follows.

Process to be modeled. The fermentor is represented as an isothermal batch chemical reactor in the form of a mixed tank in which a simplif-ied reaction sequence takes place. This reaction sequence (which is yet to be identified) should take into account a minimal number of chemical species. Preferably it should consider only the overall antibiotic production rate and eventually those subprocesses most closely related to it.

Perfect mixing in a single liquid phase is assumed, thereby neglecting the presence of small quantities of gaseous, solid and other possible liquid phases. This implies also that all internal mass and heat transfer phenomena, intrinsic biochemical reactions (like pathways and mechanisms), oxygen consumption, and the like, is to be neglected.

System state variables. The dependent variables of the system are mass and volume of reactor contents and concentrations of chemical components considered. The independent variable is the current batch operation time, measured from the time of fermentor inoculation.

System control variables. These are: a) time instants at which additions of substrate are given into the fermentor, in the following discussion termed "addition times", b) quantity, in mass units, of each substrate addition, termed subsequently with "addition mass", c) initial concentration of substrate, and d) batch time.

Criteria used for optimization. In order to facilitate the implementation of results, i.e. in adapting the production of antibiotic to different market or sales conditions, and encouraged by the ideas of Hinger and Blenke (1975), three different criteria were considered. The first and the principal one was batch profit, defined here, in monetary units per batch, as the value of antibiotic produced by a particular batch, diminished by expenses for: - the initially present and the subsequently added substrate, - the value of the nonutilized substrate at the end of the batch, and - current operation costs, including aeration, mixing, cooling, and equipment amortization. Not included herein are the labor costs needed to realize each addition because no extra manpower must be engaged for this purpose.

The second criterion used was the average production rate, in weight units of antibiotic produced in a specified (longer) time period, e.g. a month, during which the settings of the optimized control and other variables can be assumed to be approximately constant. In our computations an average duration of the nonperating or preparatory phase, as defined previously, of 24-hours per average batch has been taken.

The third criterion was the so-called "effectivity of substrate utilization", defined here in the form of a quotient, and un weight or monetary terms as the weight (or value) of antibiotic produced per unit weight (or value) of substrate invested into a particular batch. Initially also the final concentration of the antibiotic in the broth was used as an auxiliary criterion for estimating thebatch production efficiency.

MATHEMATICAL MODEL

It is assumed that the production of this antibiotic can be described by the following minimal set of overall biochemical subprocesses: growth of biomass, basic metabolism of biomass, production of antibiotic and autolysis of biomass. These subprocesses can be formally recorded by means of the usual chemical kinetic notation, i.e. by a proper "stoichiometry" and the corresponding reaction rate expressions:

$$\begin{aligned} 2\,S + 2\,N &\longrightarrow 4\,B + M \\ S &\longrightarrow M \end{aligned} \qquad (1)$$

$$S \longrightarrow M + 2A$$
$$B \longrightarrow 2S + N + M \quad (1)$$

$$r_1 = k_1 c_S c_N c_B$$
$$r_2 = k_2 c_S c_B \quad (2)$$
$$r_3 = k_3 c_S c_B$$
$$r_4 = k_4 c_B c_M^2$$

where $\underline{a}^T = (S,N,B,M,A)$ is the vector of "chemical" components (species), representing substrate, nitrogen, biomass, a complex of primary metabolytes, and antibiotic. Next: $\underline{r}^T = (r_1, r_2, r_3, r_4)$ is the vector of formal "reaction" rates, in weight units per unit of time and reactor volume: $\underline{k}^T = (k_1, k_2, k_3, k_4)$ is the vector of specific reaction rates, termed here also as "reaction rate parameters", and $\underline{c}^T = (c_S, c_N, c_B, c_M, c_A)$ is the vector of component concentrations, in dimensionless units as weight fraction.

The determination of the kinetic model structure, as defined here by stoichiometric coefficients and particular reaction rates (i.e. powers on concentrations occurring in the corresponding expressions) is based on a general method for modeling complex chemical kinetics (Černetič and Modic, 1979). The kinetic structure shown in Eqs. (1) and (2) was found previously in a slightly different form (Černetič, Modic and Pokorny, 1980).

The main proposition in this modelling procedure is that this overall kinetic structure holds for this proccess in general, and that it is possible to fit the data of each individual batch, with its individual initial and operating conditions, nly by different reaction rate parameters (vector $\underline{k}$). We have verified this assumption first for the pure batch operation (Černetič, Modic and Pokorny, 1980) and recently also for batches with aduitions of substrate and nitrogen.

Once the kinetic structure of the process is fixed, the differential equations can be derived quite directly as component mass balances for an ideal homogeneous batch reactor. Here they are shown in the convenient and short-hand vector-matrix notation:

$$d\underline{c}/dt = (V/M) \quad \underline{\underline{S}}^T \cdot \underline{r} \quad (3)$$

$$\underline{\underline{S}} = \begin{bmatrix} -2 & -2 & 4 & 1 & 0 \\ -1 & 0 & 0 & 1 & 0 \\ -1 & 0 & 0 & 1 & 2 \\ 2 & 1 & -1 & 1 & 0 \end{bmatrix} \quad (4)$$

where t is the current batch time, V and M are volume and mass of reactor contents, respectively, and $\underline{\underline{S}}$ is the matrix of stoichiometric coefficients.

During the numerical integration of these differential equations, discrete additions of principally any reactant (or component) are taken into account by means of the following algebraic equations:

$$M' = M + M_D \quad (5)$$
$$V' = V + M_D/\rho_D \quad (6)$$
$$c' = c_1 \, M/(M + M_D) \quad (7)$$
$$c_i' = (c_i M + c_{iD} M_D)/(M + M_D) \quad (8)$$

where the prime (´) stands for values of system variables after each addition and index "D" stands for added quantities. So M_D is mass or quantity of the current addition, ρ_D is its density, in kilograms per cubic meter, and c_{iD} is the concentration or weight fraction of the i-th component in the current addition. Index "1" in Eg. (7) runs over all non added components, whereas index "i" in Eg. (8) designates the component currently added, supposed that each component is added separately. The simulated time course of the main concentrations for a particular batch is given in Fig. 1.

OPTIMIZATION

Before the actual optimization with a specific algorithm and as a preparation to it, the multidimensional "surface" of the objective function was investigated, using the derived mathematical model, by variation of control variables and by observation of the selected criteria.

First, the effect of addition times on the final concentration of antibiotic was analysed. In a series of simulation runs, a fixed amount of substrate or nitrogen was added at the beginning of the batch and then at various later times. It has turned out that the highest concentration of antibiotic results when substrate or nitrogen is added right at batch beginning. The later the addition, the less it contributes to the production of antibiotic. This seems to be rather obvious, but the curve of final antibiotic concentration versus addition time is very informative. For variations in the addition time of nitrogen, it is almost linear and with a small slope. But for addition ofsubstrate, after an initial delay, it falls steeply towards low final concentrations of antibiotic. This indicates that the addition of nitrogen has not much influence on production - at least with the model derived so far. This confirms the initial decision that addition of nitrogen should not be included in this optimization.

This preliminary finding indicated that, as principally to the addition time, the "good" sequence of additions will be such that the greatest additions will be crowded at the batch beginning. Unfortunately, with regard to the actual operation conditions, any nonsterilized additions during the first quarter of batch operating time are unacceptable because the concentration of the antibiotic is yet too low to protect the batch from possible contamination during the manipulations at the opened fermentor. For that reason the further analysis was based on fixed adition times, with essentially the same sequence as in the original operation procedure.

Next in our analysis, preparations were made to optimize the initial concentration of substrate, final batch time and quantities of substrate additions at the four fixed instants. For this purpose, a complete computer program was set up for nonlinear optimization. After the testing phase, this program was first used with a "dummy" optimization subroutine, for the computation of the three selected criteria at various points in the space of the tree variables. For sake of simplicity and to keep the overall number of computations reasonably small, only an uniform addition sequence was constructed, i.e. with equal quantities added each time in a sequence.

At these runs also the effect of variations in the main economic parameters was investigated, namely variation in hourly batch operation cost (either with or without amortization) and variation in the selling price of antibiotic (either with or without premiums). In this stage it was found that an extremal point could exist, so it was decided to go further, i.e. to try simultaneous optimization of all free variables. This was done by the routine E04JAF from the NAG FORTRAN library (NAG, 1978) which uses the Quasi-Newton algorithm and simple bounds on variables (Gill and Murray, 1972, 1976).

RESULTS AND DISCUSSION

Let us introduce the discussion with the results obtained by the variation of economic parameters, in runs without substrate additions. Tabulation of batch profit versus the initial concentration of substrate and versus the final batch time, respectively, with the dummy optimization routine, reveals first, that batch profit is very sensitive (almost linearly and positively correlated) to the variation of the first mentioned variable. This sensitivity is not changed by different batch operation cost, but is markedly decreased by the lower price of the antibiotic.

The dependence of batch profit on final batch time is not a straight line but resembles either to a saturation-type curve at low operating costs, or gives a curve with a more or less expressed maximum. When the antibiotic is cheaper (price without bonuses), the maximum lies at lower batch times. Otherwise, batch profit does not change very much after a certain operation period.

The same two dependencies, for a fixed hourly operation cost (amortization included) and for an antibiotic price with bonuses, are shown in Fig. 2 and in Fig. 3. The lowest curve in both diagrams is for operation without additions (0% additions) whereas the other three curves correspond to 1, 2 and 3% additions, respectively, added at the four fixed addition times. In this notation the quantity of each particular addition is expressed as a variable weight percent in respect to the initial fermentor contents. Note from the previous chapter that all added quantities in a particular addition sequence are equal.

It is seen from Fig. 2 and Fig. 3 that batch profit grows the higher, the greater the added quantities, but also that at higher addition rates the final batch time must be longer if the full advantage of the added substrate can be taken. In Fig. 3 the dashed lines of similar shape as the solid ones are valid for an (by 2%) increased initial concentration of substrate. For this condition, the previous statement is even more accentuated.

As the next typical result, the dependence of final antibiotic concentration on substrate addition time is shown for various quantities of additions. A graph of the mentioned dependence alone would be less comparative, therefore the same information was expressed in Fig. 4 as the ratio of final antibiotic concentrations for a specified final batch time (176 hours in our case).

This ratio shows directly how much the final concentration of antibiotic is decreased when a fixed quantity of substrate is added at once later during the batch, instead of to be given at batch beginning. Curves are given for addition quantities 1, 2, 3, 4 and 5 wt % of initial fermentor contents, respectively. The addition times in these runs were chosen from the previously fixed addition sequance, i.e. at 82, 106, 130 and 154 hours of current batch time. It can be seen from Fig. 4, first that the less antibiotic is produced in a fixed batch time, the later a fixed amount of substrate is added to the batch, and second that this effect is very expressed at greater quantities of substrate additions.

Let us now discuss briefly and only qualitatively results for the other two criteria. It was found that the average production rate of a fermentor, as defined previously, is growing very fast with increased initial concentration of substrate and with higher quantities of added substrate. When depicted graphically, this dependence gives a plot very similar to that of Fig. 2. Next, a plot of production rate versus batch time shows an expressed maximum which, for higher substrate addition rates, is displaced towards higher batch times. It is also interesting to note that the position of the maximum does not change wery much at another initial concentration of substrate.

The third criterion, i.e. that of substrate utilization efficiency, gives a totaly reversed results in respect to substrate additions. Wheras the value of this criterion still increases with both, the initial concentration of substrate and with batch time, it is reamarkably decreased for batches with later additions of substrate.

With all these preliminary findings at hand, it is not surprising that in all optimization runs the initial concentration of substrate jumps to the upper bound already in 14 i-terations. On the other hand, the final batch time is left almost unchanged at initial values of a particular batch, i.e. approximately 176 hours.

As to the addition quantities, they are finally (almost all) at upper bounds. It is very interesting to follow the course of iterations during the optimization run. Generally the quantity of the first substrate addition (at the 82-th hour) grows the fastest and the last one the slowest. For a specific batch (0-584-F9), this "growth rate" can be experessed in terms of iteration numbers as follows. The particular additions reach an upper bound of 5% in:

1-st addition ... in 19 iterations
2-nd addition ... in 26 iterations
3-rd addition ... in 33 iterations
4-th addition ... in 114 iterations.

To test how this picture is changed by another economic situation, the optimization runs were repeated with the following changes in coefficients of the objective function:

- lower batch operation cosis (without amortization)
- lower value of antibiotic (without any bonuses)
- higher price of added substrate (twice the initial value).

As expected, for lower operation costs, batch time was nonsignificantly increased. For lower value of antibiotic, the changes are chown in Fig. 5. For the higher price of added substrate, the first two additions become unreasonably great (approx. 55 and 20%, respectively), whereas the third and fourth addition fall practically to zero.

CONCLUSION

In summing up the results obtained, the following recommendations for improving the operation of this process can be given, based on the present evidence and in respect to a single fermentor.

1. In any market situation the initial concentration of substrate should be increased to an upper limit which yet remains to be determined (see also below).
2. In normal conditions, a falling sequence of substrate additions should be introduced, such that as much as possible of substrate is added as early as possible during each batch. Late additions of substrate should be diminished or completely avoided.
3. Higher substrate addition rates are favourable in terms of batch profit and production rate. Conversely, if the criterion of substrate utilization efficiency becomes prevalent, any additions of substrate after the inoculation are to be canceled.
4. Batch time should generaly be increased at a higher initial concentration of substrate and also at higher substrate addition rates. When either operating without substrate additions or when higher production rate is desired, the optimal batch time can be determined by a simple univariate search procedure. Generally this control variable is not a very critical one.

It can be seen from our results that they do not coincide completely with the findings of other investigators mentioned initially. This could be partly due to the fact that our model does not account for substrate inhibition and partly due to the use of another objective function. Anyhow, these results can be taken as valid only for this special biosynthesis and for the data used, taking into account also all assumptions made during the procedure.

In this sense our results cannot be taken as ultimate, but as an incentive for further research. As it has been shown that high initial concentrations of substrate are very desirable, it would be worthwile to find out the conditions at which the production of antibiotic becomes seriously inhibited by the substrate. Done by combined laboratory and plant experimentation and coupled with computerized analysis, this would be a challenging task for a further interdisciplinary work in which the contributions of bioengineers and biochemists would be indispensable. Having gathered more data and knowl dge about this process, it is possible to repeat the entire optimization procedure with essentially the same methology.

Another interesting possibility for further research would be to analyse (and subsequently to optimize) the overall production plant including all fermentors and the corresponding equipment for olation of the antibiotic from the fermentation broth (Prokop and Votruba, 1979).

Acknowledgement. This work was in part supported by the Research Council of Slovenia. The authors are also grateful to J. Petravić, P. Kovač and F. Novosel from KRKA for their help in problem formulation.

REFERENCES

Černetič, J., and R. Modic (1979). A method for modelling of complex chemical kinetics. Preprints 12-th Simp.Comput.Appl. in Chem. Eng. (C.A.C.E. 79, Montreux, Switzerland, April 8-11, 1979), 1, 271-179.

Černetič, J., R. Modic, and M. Pokorny (1980). Formulation of a multiresponse model for the biosynthesis of antibiotics. Presented at the 13-th Europ.Symp. of the Working Party "Routine Calculations and Use of Comput. in Chem. Eng." (CHEMPLANT '80 Symp., Heviz, Hungary, September 3-5, 1980).

D'Ans, G., D. Gottlieb, and P. Kokotović (1972). Optimal control of bacterial growth. Automatica, 8, 729-736.

Gill, F.E., and W. Murray (1972). Quasi-Newton methods for unconstrained optimization. J.Inst. Math. & Appl.,9, 91-108.

Gill, P.E., and W. Murray (1976). Minimization subject to bounds on the variables. National Physical Laboratory Report, No. NAC 72.

Heijnen, J.J., J.A. Roels, and A.H. Stouthamer (1979). Application of balancing methods in modelling the penicillin fermentation. Biotechnol. Bioeng., 21, 2175--2201.

Hinger, K.-J., and H. Blenke (1975). Wahl der Zielgrosse zur wirtschaftlich optimalen Auslegungvon Chemiereaktoren (Choice of the objective function for the economically optimal design of chemical reactors). Chemie - Ing. - Techn., 47 (3), 79-120.

NAG (1979). NAG FORTRAN library Manual, Mark 7, 2, Chapter Eo4. Numerical Algorithms Group, GB.

Prokop, A., and J. Votruba (1979). Computer-aided scheduling of multiproduct batch fermentation plants. Biotechnol. Bioeng., 21, 1689--1694.

Wang. H.Y., C.L. Cooney, and D.I.C. Wang (1979). Computer control of bakers' yeast production. Biotechnol. Bioeng., 21, 975-995.

Yamane, T., E. Sada, and T. Takamatsu (1979). Start-up of chemostat, application of fed-batch culture. Biotechnol. Bioeng., 21, 111-129.

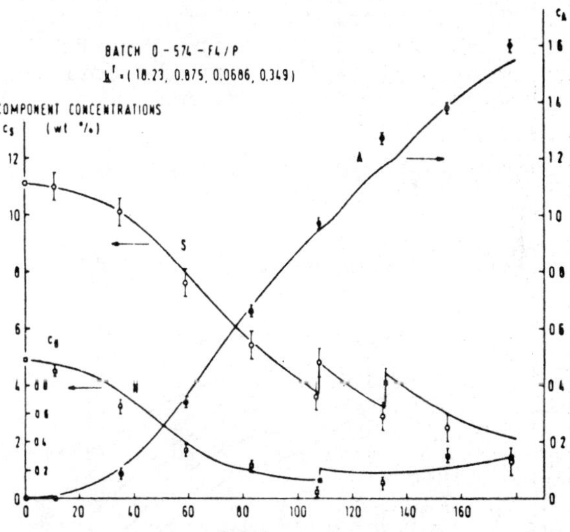

Fig.1. Comparison of model and process outputs for the following component concetrations substrate (symbols S and o), nitrogen (N and ▫) and antibiotic (A and •).

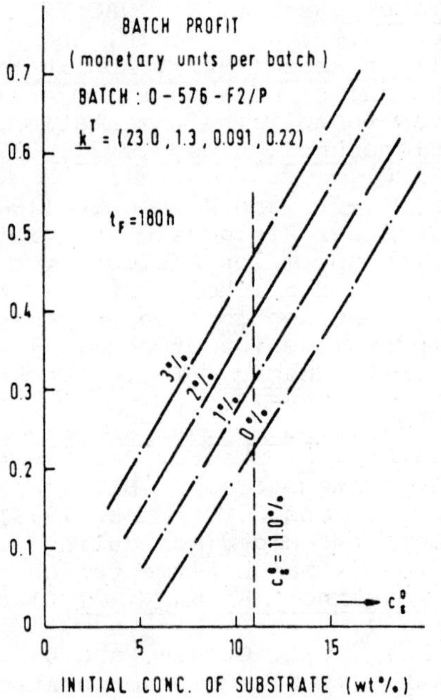

Fig.2. Batch profit versus the initial concentration of substrate (c_S^o), at various substrate addition rates: 0,1,2 and 3%, respectively (fixed addition times and uniform addition sequence). Coefficients in the objective function: (14.0, 9.25, 0.030, 14.0, 1239.0). The vertical dashed lines in Fig. 2 and Fig. 3 designate the present average value of the corresponding control variable.

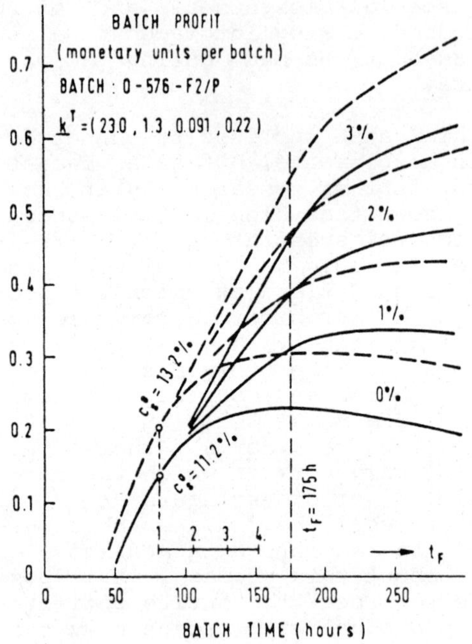

Fig.3. Batch profit versus the final batch time (t_F) (batch duration) at various substrate addition rates: 0,1,2 and 3%, respectively (fixed addition times and uniform addition sequence). Coefficients in the objective function: (14.0, 9.25, 0.030, 14.0, 1239.0). Solid curves for the initial concentration of substrate 11.2%; dashed curves for the init.conc. of substr. 13.2%.

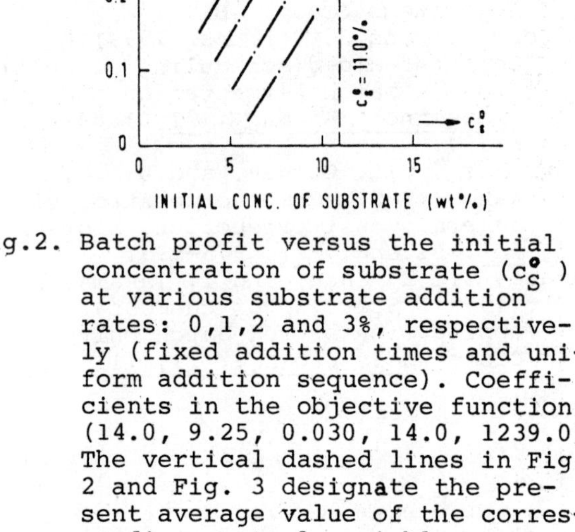

Fig.4. Effect of substrate addition time (t_D) on final concentration of antibiotic (c_{AF}) at various quantities of added substrate (1,2,3,4 and 5%) shown in relative scale as the fraction of c_{AF} which is obtained, when the same quantity of substrate is added just at batch beginning (ratio = 1) or at a specified later time during the batch (ratio less than one).

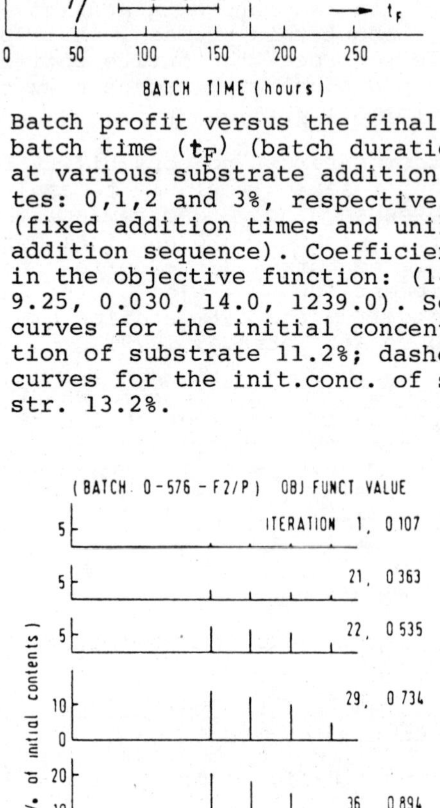

Fig.5. Changes of the substrate addition sequence during the optimization (fixed addition times, nonuniform sequence) at the lower value of antibiotic.

USE OF A MICRO-COMPUTER FOR THE CONTROL OF A PHOSPHATE DRYING FURNACE

M. Najim, K. Najim, D. El Jouhari and L. Radouane

L.E.E.S.A.; Université Mohamed V, Faculte des Sciences de Rabat, B.P. 1014, Rabat, Morocco

Abstract : Dynamic control of a phosphate drying furnace aims at regulating the moisture content of produced phosphate around a certain desired value (1%), while maintaining a constant rate of production. It is further desired to minimize the rate of fuel consumption used in phosphate drying.
A suboptimal dynamic controller is proposed using microprocessors. The controller is a multivariable P.I. (proportional and integral).

Keywords : Phosphate drying furnace, microprocessor, P.I. Control, moisture content, Fuel consumption, state prediction.

Introduction

It is well known that the linear optimal regulator in its classical formulation gives process state variables feedback /1/. The practical usefulness of this result when the assumptions about the process are reasonably satisfied has been demonstrated by many reported results. However, the steady-state error may become considerably large. This is not due to an insufficient mathematical formulation but to an insufficient formulation of the control problem for the application in question /2/. The most typical situation in process control is that the process model parameters may differ from the nominal ones...Under these conditions a simple feedback loop can not garantee a zero stationary error. This is important particulary in an industrial process constrained to follow a certain reference input vector under varying conditions.

The integral controller can be looked upon as a device which estimates the bias of the disturbances as seen from the output and takes the necessary action to bring the stationary control error to zero. This points out the need of adding integrators to yield a proportion al-integral-derivative controllers. In the literature proportional-integral multivariable feedback control has been studied from the deterministic viewpoint, either by including the time derivative of the control in the performance index /3/, or by simply extending the original system with integrators /4/.

In this paper we have adopted the later method.

Process description

Damp phosphate when extracted from the mine has a moisture ratio of 15%. Heating should reduce this ratio to about 1%. The drying process consists of the following elements (Fig.1), combustion chamber, drying tube and dust chamber.

1- Combustion chamber: the fuel is initially heated at 100°C pulverized by means of water vapour and then burned. The so-called primary air gives necessary oxygen for a complete combustion. The produced heat is carried along the tube by the air current called secondary air.

2- Drying tube: it is an horizontal tube of 25 m length with a constant rotation velocity. The internal structure is in the form of helix disposed along the tube facilates thermal exchange and moving the phosphate forward to the tube output.

3- Dust chamber: by means of a depression created at the end point of the tube the produced dust carried by the air current is collected in the dust chamber.

Process identification

Using the instrumental matrix and the generalized least mean squares methods, the furnace has been considered as a black box (fig.2) and identified from input-output observations in

the neighborhood of a nominal working point /5/.
The model is linear and discret as follows :

$$y(k)+a_1y(k-1)+a_2y(k-2) = b_{01}u_1(k)+b_{02}u_1(k-1)+b_{03}u_1(k-2) + b_{11}u_2(k)+b_{12}u_2(k-1)+b_{13}u_2(k-2) + b_{21}u_3(k)+b_{22}u_3(k-1)+b_{23}u_3(k-2) \quad (1)$$

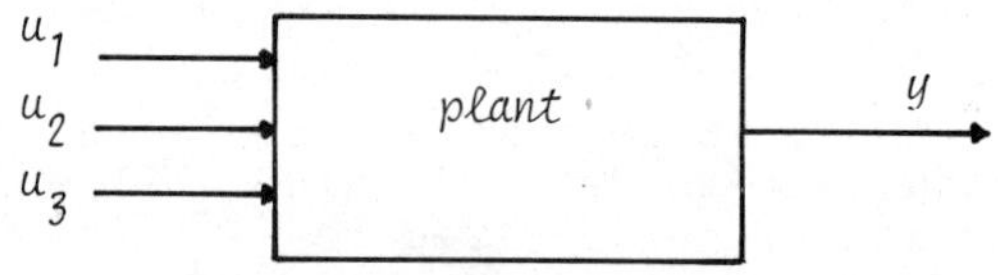

Figure .2.

$(k, k+1) = 30"$

Where $\underline{U} \begin{cases} u_1 & \text{fuel flow} \\ u_2 & \text{phosphate flow} \\ u_3 & \text{secondary air flow} \end{cases}$

The relation (1) may be written in the state space form which is convinient for the linear quadratic gaussian control problem.
It is known that for a system that is described by a number of input output couples, there exist a unique equivalent state model of minimal dimension when the necessary conditions are satisfied /6/.

After applying the observable canonical form, the process is described in the state space by :

$$\underline{X}(k+1) = \begin{bmatrix} 0 & 1 \\ 0.0061 & 0.8767 \end{bmatrix} \underline{X}(k) + \begin{bmatrix} -0.3276 & 0.0929 & 0.0005 \\ -0.1271 & 0.0912 & 0.00004 \end{bmatrix} \underline{U}(k)$$

$$y(k) = \begin{bmatrix} 1 & 0 \end{bmatrix} \underline{X}(k) + \begin{bmatrix} -0.1849 & 0.0643 & 0. \end{bmatrix} \underline{U}(k)$$

Moisture measurement

For product moisture measuremet, we were confronted with the problem of ambiant temperature. The infrared moisture sensor can not work correctly at a temperature over 35°C. This parameter reaches 70°C at the furnace output. To solve this problem, we have looked for a correlation between the phosphate moisture ratio and the dust moisture ratio for which measurement is possible . This correlation is shown in fig.3. So the output process is the dust moisture ratio which is related to furnace output moisture ratio.

Control problem formulation

Let the discret linear dynamic system be described by :

$$\underline{X}(k+1) = A\underline{X}(k) + B\underline{U}(k) + \underline{f}$$

$$\underline{Y}(k) = C\underline{X}(k) + D\underline{U}(k)$$

Where $\underline{f}$ is a constant vector (called in general the dynamic perturbation vector).
A,B,C and D are matrices of proper dimension .
The optimal control problem is to determine the control law which minimizes the objective function :

$$J = \sum_{k=0}^{N-1} \underline{U}^T(k) R \underline{U}(k) + \underline{E}^T(k) Q \underline{E}(k)$$

Where $\underline{E}(k)$ the error between the constant reference input vector $\underline{Z}$ and the actual process output, i.e $\underline{E}(k) = \underline{Z} - \underline{Y}(k)$.
O,N the interval of control
R positive definite matrix
Q definite non negative matrix
R and Q are the weightening matrices.
It is known that the optimal control law is composed of :

a) a feedback control from be dynamic disturbance vector

b) a feedback from the state variables or their estimates.

$$\hat{\underline{U}}(k) = -L(k)\underline{X}(k) + \mu(k)$$

L(k) and $\mu(k)$ are given by the system of equations :

$$L(k) = \left[R + BG(k+1)B + D^TQD\right]^{-1}\left[B^TG(k+1)A + D^TQC\right]$$

$$\mu(k) = \left[R + B^TG(k+1)B + D^TQD\right]^{-1}\left[D^TQZ - B^T(\underline{g}(k+1) + G(k+1)\underline{f})\right]$$

$$\underline{g}(k) = \Omega\left\{\underline{g}(k+1) + G(k+1)\left[B\mu(k) + \underline{f}\right]\right\} - \emptyset Z$$

$$\text{with } \Omega = A^T - C^TQD\left[R + D^TQD\right]^{-1}B^T$$

$$\Phi = C^TQ - C^TQD\left[R + D^TQD\right]^{-1}D^TQ$$

The initial conditions are G(N)=O, $\underline{g}$(N)=O, corresponding to a free final state.

Steady state solution

We suppose that the interval of control is infinite and that the system has reached the permanent state. It is demonstated /7/ that the feedback matrix L(k) and the vector $\mu(k)$ tend to stationary values when the system is controllable. The stationary solution is interesting from a practical point of view.

Proportional and integral control

The disavantage of the previous control strategy is the presence of steady state errors. This is not satisfactory, particulary in industrial

processes where the reference input vector must be reached.This problem can be solved by adding integrators. A necessary and sufficient condition that the extended system remains controllable is that the number of integrators must be less or equal to the number of outputs /8/.
Here we consider that we have an integrator for each output .
The integrators are approximated by :

$$\underline{V}(k+1)=\underline{V}(k)+T\underline{E}(k) \quad T:\text{sampling period}$$
$$=\underline{V}(k)+T(\underline{Z}-\underline{Y}(k))$$
$$=\underline{V}(k)+T(\underline{Z}-C\underline{X}(k)-D\underline{U}(k))$$

diving both sides by T, we get :

$$\underline{V}(k+1)=\underline{V}(k)+\underline{Z}-C\underline{X}(k)-D\underline{U}(k)$$

The extended system becomes :

$$\begin{bmatrix}\underline{X}(k+1)\\ \underline{V}(k+1)\end{bmatrix}=\begin{bmatrix}A & O\\ -C & I\end{bmatrix}\cdot\begin{bmatrix}\underline{X}(k)\\ \underline{V}(k)\end{bmatrix}+\begin{bmatrix}B\\ -D\end{bmatrix}\underline{U}(k) \quad \begin{bmatrix}\underline{O}\\ \underline{Z}\end{bmatrix}$$

$$\begin{bmatrix}\underline{Y}(k)\\ \underline{V}(k)\end{bmatrix}=\begin{bmatrix}C & O\\ O & I\end{bmatrix}\begin{bmatrix}\underline{X}(k)\\ \underline{V}(k)\end{bmatrix}+\begin{bmatrix}D\\ O\end{bmatrix}\underline{U}(k)$$

In a compact form :

$$\underline{X}_1(k+1)=A_1\underline{X}_1(k)+B_1\underline{U}(k)+\underline{f}_1$$
$$Y_1(k)=C_1\underline{X}_1(k)+D_1\underline{U}(k)$$

The performance index is then :

$$J=\sum_{k=o}^{N-1} \underline{u}^T(k)R\underline{u}(k)+\underline{E}^T(k)Q\underline{E}_1(k)$$

where

$$\underline{E}_2(k)\begin{bmatrix}\underline{Z}\\ \underline{O}\end{bmatrix}-\begin{bmatrix}\underline{Y}(k)\\ \underline{V}(k)\end{bmatrix}, \quad Q_1=\begin{bmatrix}Q & D\\ O & Qv\end{bmatrix}$$

Where Qv is the error integral weightening matrix . following the same reasoning as before, the optimal control is :

$$\hat{\underline{U}}(k)=-L_1(k)\underline{X}_1(k)+\mu_1(k)$$

$L_1(k)$ and $\mu_1(k)$ are given by the same equations as before.
The optimal control computed from the previous criterion does not have a unique solution in permanent state. However we can transform it into the following :

$$J_1=\sum_{k=o}^{N-1}\left(\underline{U}(k)-\underline{U}^*\right)^T R\left(\underline{U}(k)-\underline{U}^*\right)+ E_1^T(k)Q_1\underline{E}_1(k)$$

This means that some inputs are fixed to desired values.In our case, we have fixed the production to 170 tons/hour. The feedback matrix $L_1(k)$ and the vector $\mu_1(k)$ are the stationary solution of the following recurrent system:

$$L_1(k)=\left(R+B_1^TG_1(k+1)B_1+D_1^TQ_1D_1\right)$$
$$\left(B_1^TG_1(k+1)A_1+D_1^TQ_1C_1\right)$$
$$\mu_1(k)=\left(R+B_1^TG_1(k+1)B_1+D_1^TQ_1D_1\right)$$
$$\left\{D_1^TQ_1Z_1-B_1^T\left(g_1(k+1)+G_1(k+1)\underline{f}_1\right)+R\underline{u}^*\right\}$$
$$G_1(k)=L_1^T(k)RL_1(k)+\left(C_1-D_1L_1(k)\right)^T$$
$$Q_1\left(C_1-D_1L_1(k)\right)+$$
$$\left((A_1-B_1L_1(k)\right)^TG_1(k+1)\left(A_1-B_1L_1(k)\right)$$
$$g_1^T(k)=-(\mu_1-\underline{u}^*)RL_1(k)-$$
$$\left(\underline{Z}_1-D_1\mu_1(k)\right)^T Q_1\left(C_1-D_1L_1(k)\right)+$$
$$\left(B_1'\mu_1(k)+\underline{f}_1\right)G_1(k+1)\left(A_1-B_1L_1(k)\right)+$$
$$g_1^T(k+1)\left(A_1-B_1L_1(k)\right)$$

Stochastic formulation

The optimization problem is formulated as a stationary stochastic control problem.Let us consider the following system :

$$\underline{X}(k+1)=A\underline{X}(k)+B\underline{u}(k)+\underline{w}(k)$$
$$\underline{y}(k)=C\underline{X}(k)+D\underline{u}(k)+\underline{v}(k)$$

where $\underline{w}(k)$ and $\underline{v}(k)$ are zero-mean independent gaussien noises. A Kalman predictor is employed to predict the system state. The filter is governed by the equation :

$$\hat{\underline{X}}(k+1)=A\hat{\underline{X}}(k)+B\underline{u}(k)+F\left(\underline{y}(k)-C\hat{\underline{X}}(k)-D\underline{u}(k)\right)$$

It is known that the gain F depends on the noise covariances which are often unknown.However, there exist algorithms for estimating the stationary gain from observed data. To get the invariant predictor gain, we have used the Faurre algorithm /9/.This algorithm is based on the system output correlation.The complete closed-loop block diagram is shown in Fig.4.

The microprocessor

The micro-computer is based on the MOTOROLA 6800 microprocessor which is a general-purpose eight-bit machine. The microcomputer has been equiped with all the elements for simulations and coupling to the process, including:

- 18 K bytes of memory
- line printer(HP 9866A)
- Analog input-outputs
- Digital imput-outputs
- Real-time clock.

Numerical Simulations

a simulated example is presented. The system is described by :

$$\underline{X}(k+1) = \begin{bmatrix} 0 & 1 \\ 0.0061 & 0.8767 \end{bmatrix} \underline{X}(k) +$$

$$\begin{bmatrix} -0.3276 & 0.0929 & 0.0005 \\ -0.1271 & 0.0912 & 0.00004 \end{bmatrix} \underline{U}(k)$$

$$y(k) = (1 \quad 0)\underline{X}(k) + (-0.1849 \quad 0.0643 \quad 0)\underline{U}(k)$$

After adding an integrator, the system becomes :

$$\underline{X}_1(k+1) = \begin{bmatrix} 0 & 1 & 0 \\ 0.0061 & 0.8767 & 0 \\ -1 & 0 & 1 \end{bmatrix} \underline{X}(k) + \begin{bmatrix} -0.3276 & 0.0929 & 0.0005 \\ -0.1271 & 0.0912 & 0.00004 \\ 0.1849 & -0.0643 & 0 \end{bmatrix} \underline{U}(k) + \begin{bmatrix} 0 \\ 0 \\ 0.1 \end{bmatrix}$$

$$\underline{Y}_1(k) = \begin{bmatrix} 1 & 0 & 0 \\ 0 & 0 & 1 \end{bmatrix} \underline{X}_1(k) + \begin{bmatrix} -0.1849 & 0.0643 & 0 \\ 0 & 0 & 0 \end{bmatrix} \underline{U}(k)$$

The weightening matrices are shosen as follows :

$$R = \begin{bmatrix} 0.0001 & 0 & 0 \\ 0 & 0.1 & 0 \\ 0 & 0 & 1 \end{bmatrix}, \quad Q_1 = \begin{bmatrix} 200 & 0 \\ 0 & 15 \end{bmatrix}$$

with $\underline{z}_1 = \begin{bmatrix} 0.1 \\ 0 \end{bmatrix}$: the reference input vector.

The stationary solution of the feedback matrices are :

$$L_1 = \begin{bmatrix} -0.1301 & -2.236 & 0.460 \\ 12.27 & -5.20 & -1.934 \\ -0.014 & 0.0060 & 0.002 \end{bmatrix}$$

$$\mu_1 = \begin{bmatrix} 5.02 \\ 27.19 \\ 0.04 \end{bmatrix}$$

The steady state filter gain is :

$$F = \begin{bmatrix} 0.21 \\ 0.32 \end{bmatrix}$$

The curves corresponding to the following cases show the system behaviour for different initial states.

1- initial state :

$$\underline{X}(o) = \begin{bmatrix} 0 \\ 0 \\ 0 \end{bmatrix}, \underline{U}(o) = \begin{bmatrix} 0 \\ 0 \\ 0 \end{bmatrix}, \underline{Z}_1 = \begin{bmatrix} 0.1 \\ 0 \end{bmatrix}$$

Fig. 5-1 and Fig. 5-2

2- initial state :

$$\underline{X}(o) \begin{bmatrix} 0.5 \\ 0.9 \\ 0 \end{bmatrix}, \underline{U}(o) \begin{bmatrix} 2 \\ 2.8 \\ 0 \end{bmatrix} Z_1 \begin{bmatrix} 0.1 \\ 0 \end{bmatrix}$$

Fig.6-1 and Fig. 6-2

It is demonstrated that the transient time is shorter when the system is closer to the steady-state. Also, the steady-state error is zero in all cases.

Response to a variation in the phosphate flow

After the system has reached the permanent state, the phosphate flow was perturbated as shown below :

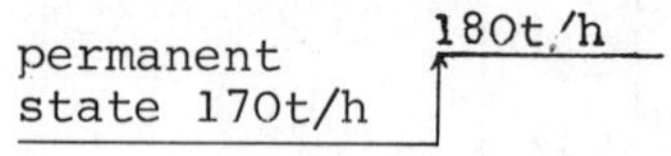

The system response is depicted in Fig.7-1 and 7-2 . We can observe that the system returns to the initial permanent state which proves that the solution is unique.

Filter behaviour

Having considered the stationary solution of the filter gain, we can note that during the transient time, the state prediction is not good. It is good however in the steady-state. The filter response is shown in Fig.8-1 and 8-2.

Real-time algorithm

The flow chart of the proposed real-time algorithm is shown in Fig.9. It is written in the assembly micro-processor language, and uses a 24 bits flotting point program for arithmetic calculations. In fig.10,11, and 12 are given the evolutions of the drying furnace inputs and output in the case when it is micro-processor controlled and in the case when it is open-loop controlled.

Experimental results

Fuel consumption

During the considered time, the economy in energy consumption is 600 liters. Hence the gain is 11%.

Fuel consumption		
open-loop controlled	microprocessor controlled	gain
5454 liters	4854 liters	11%

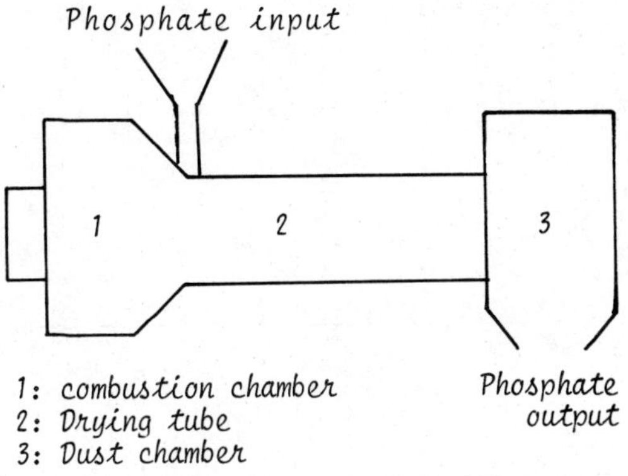

Figure 1.

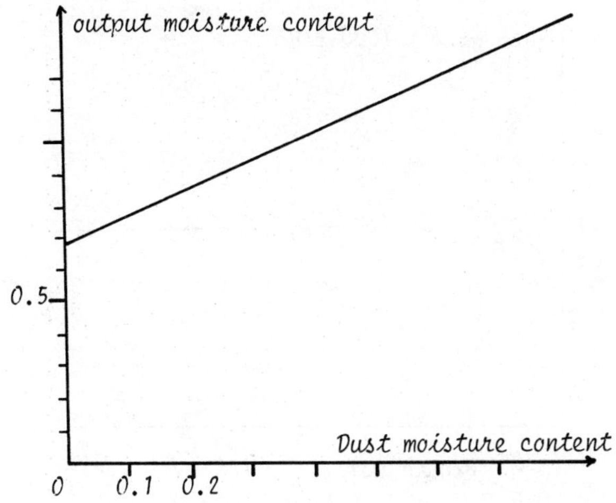

Figure 3.

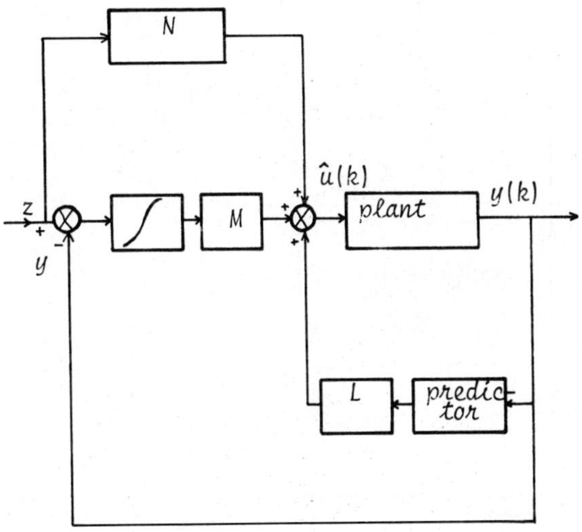

Fig.4 : Closed-loop block diagram

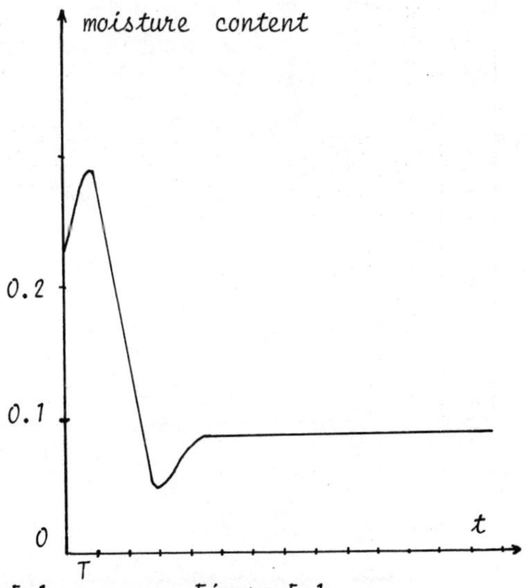

Fig.5.1 Figure 5.1.

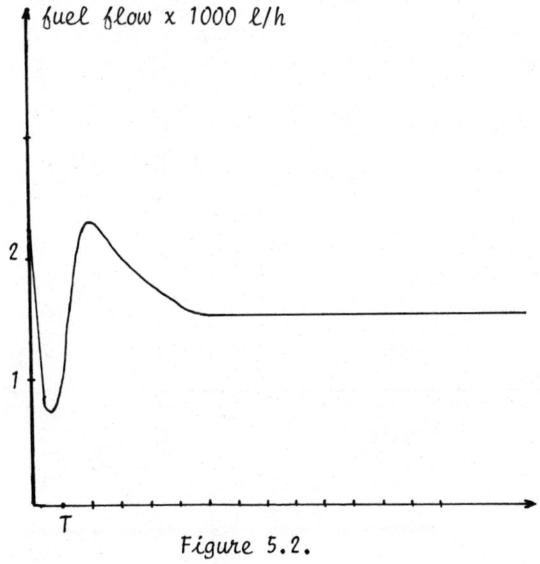

Figure 5.2.

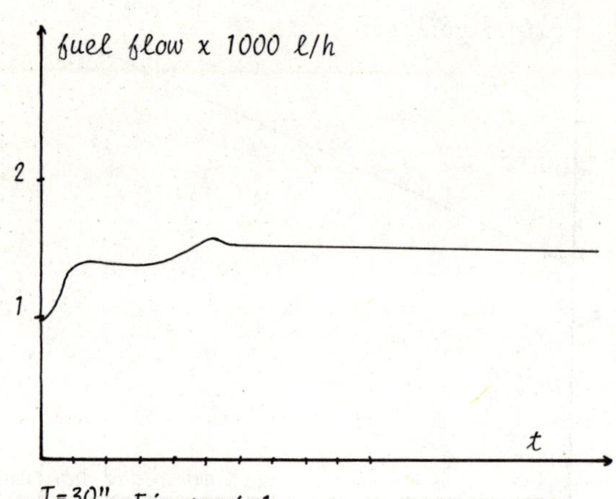

Figure 6.1.

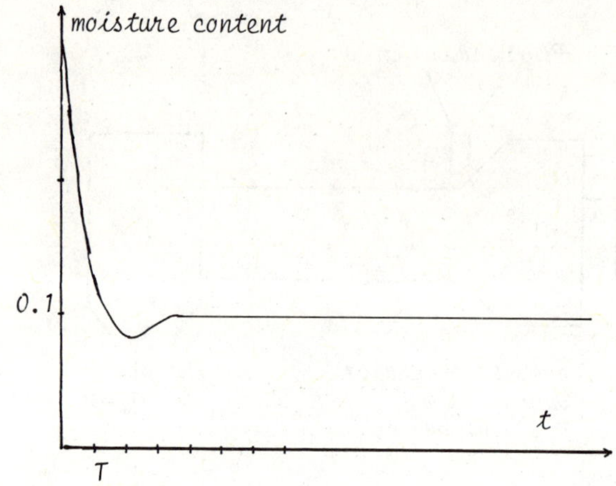

Figure 6.2.

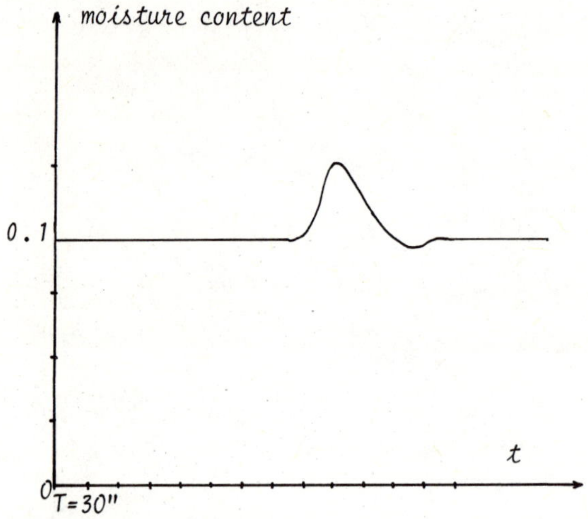

Figure 7.1.

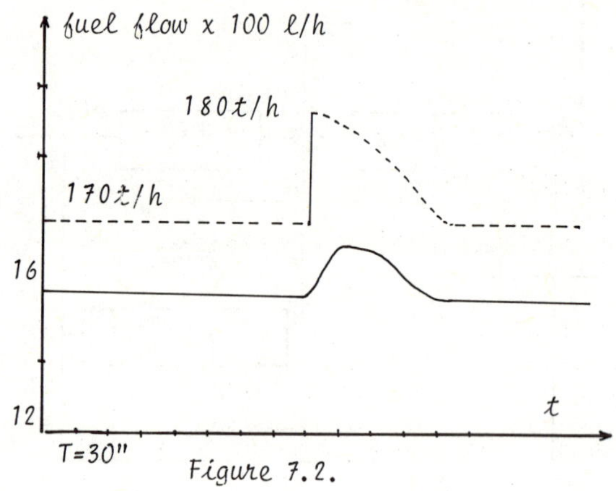

Figure 7.2.

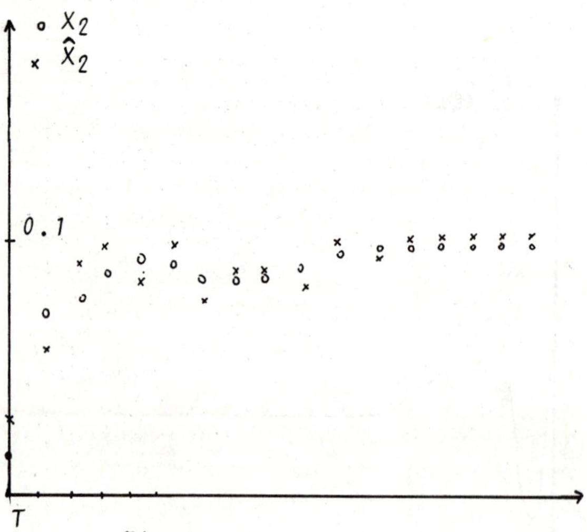

Figure 8.1

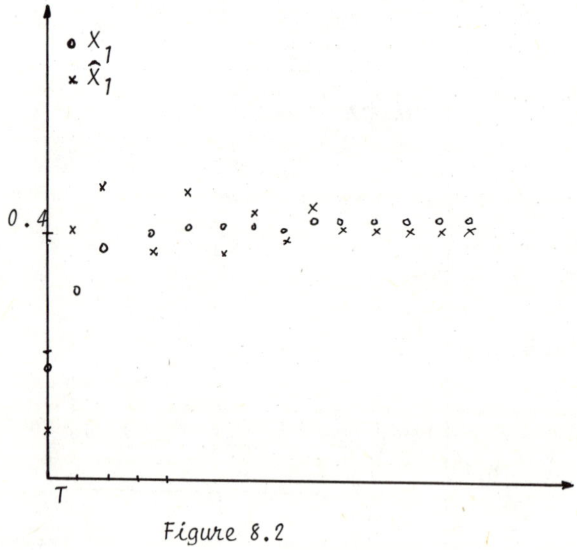

Figure 8.2

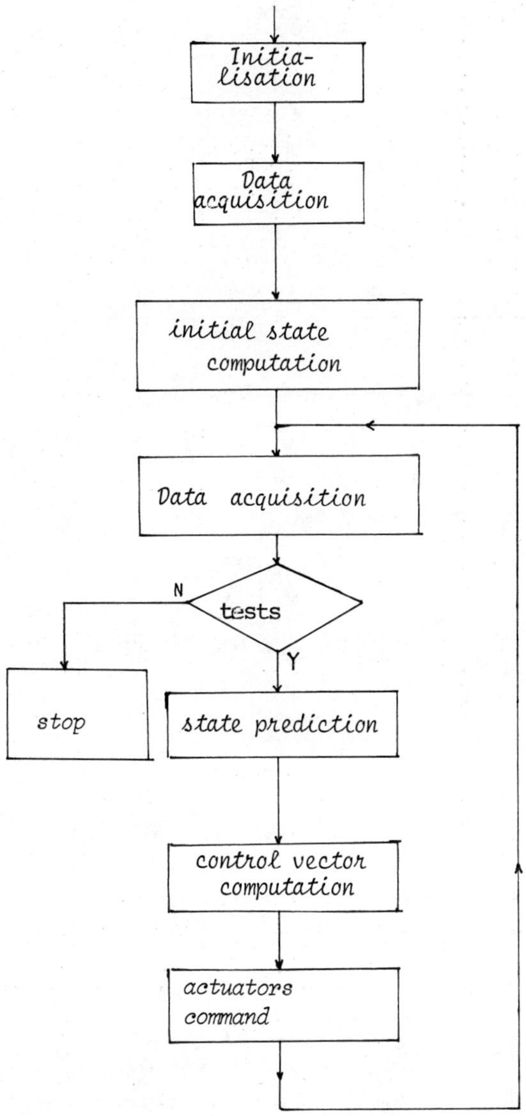
Initia-
lisation
Data
acquisition
initial state
computation
Data acquisition
N
tests
Y
stop
state prediction
control vector
computation
actuators
command

Figure 9.

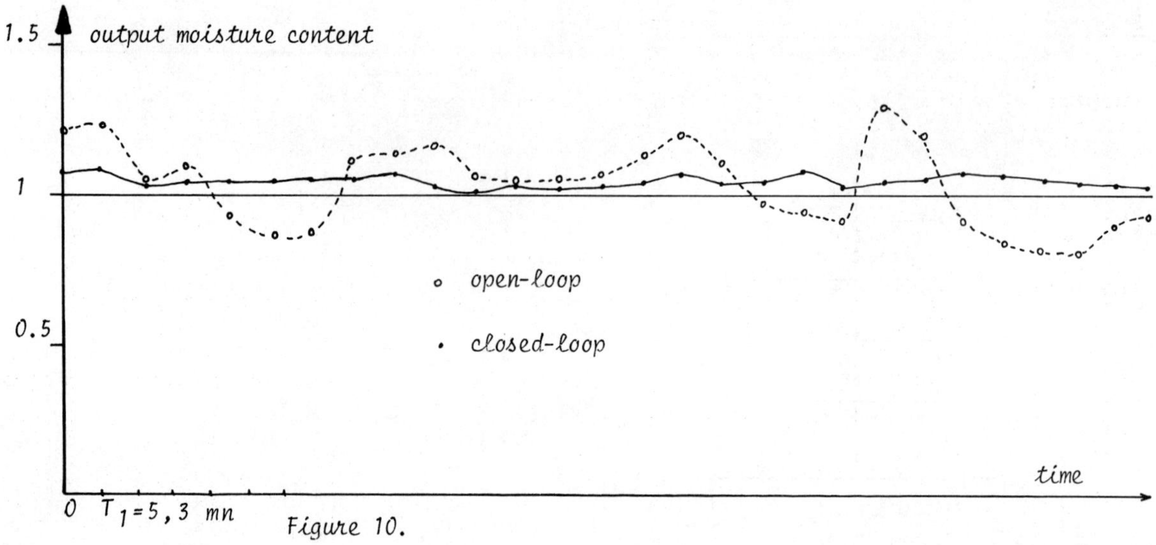

Figure 10.

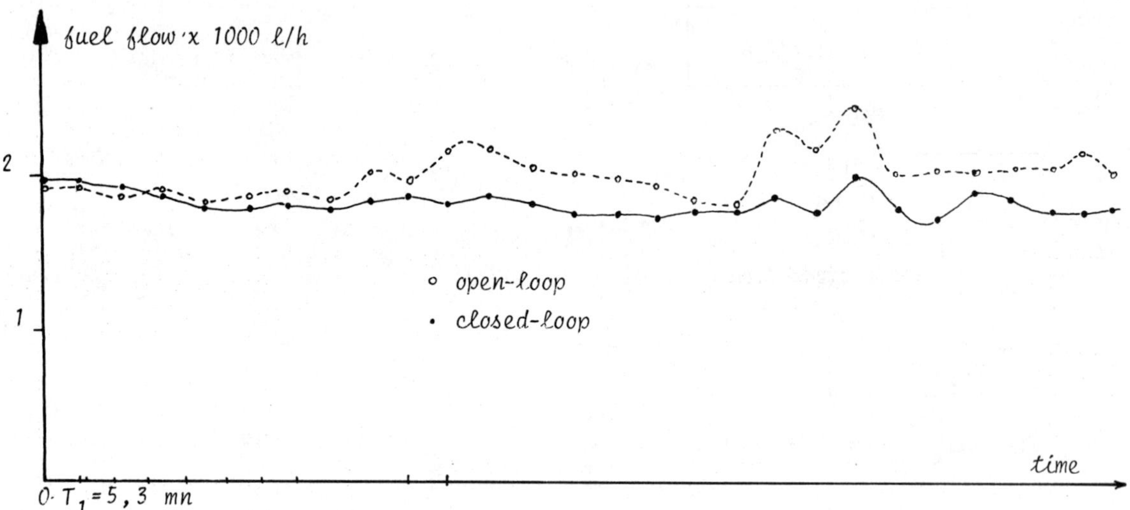

Figure 11.

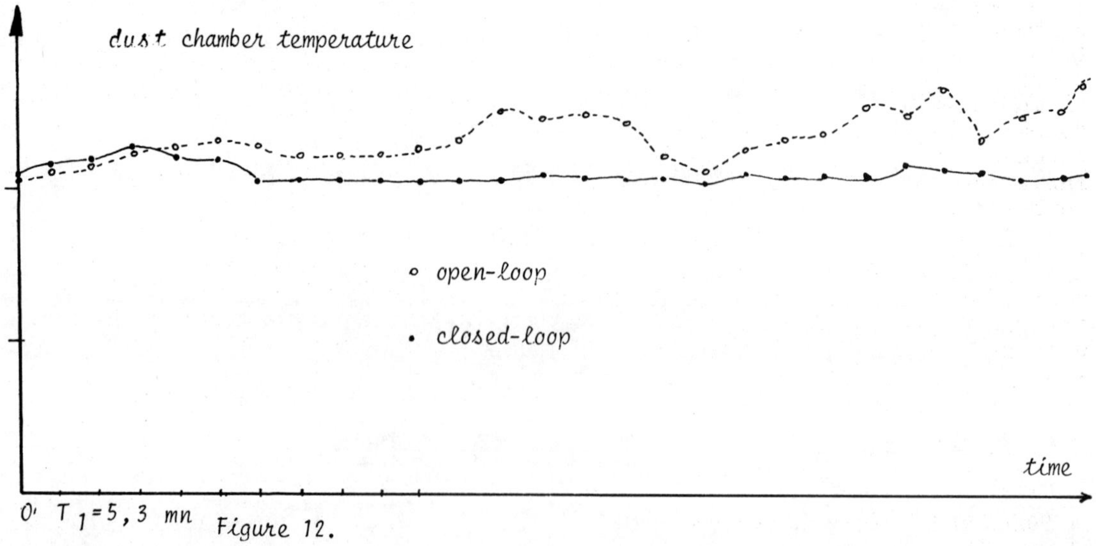

Figure 12.

Economy estimation over one year

If we consider that the drying furnace availability-time is 89%,we can estimate the gain in fuel consumption as follows :

Given that :
The mean production is 170 t/h
The mean fuel consumption is 13,7 Kg/t
The fuel cost is 93$ /t
The economy in energy would be :

economy/furnace/year
195000 $

Set-point tracking

The regulation of the output near the 1% moisture ratio is the second objective to realize.It appears from Fig. 10 that with feedback-control, the output tracks the desired ratio, where as in the case of open-loop control, the moisture ratio is oscillatory. Such oscillations are not desired for the following reasons :

1- When the moisture ratio is greater than the set-point, the production constraint is not satisfied.

2- When the moisture ratio is smaller than the set-point, the product is overdried.The loss in heat produces an increase in combustion chamber temperature which decreases the endurance of the drying furnace, particulary, the interior covering of the combustion chamber.

Due to the increase in comubstion chamber temperature, the interior covering(bricks) has been replaced once after eight years of working. The replacement has caused the following expenses :

	expenses
furnace n° 64	67100 $
furnace n° 68	86900 $

In the case when the furnace behaviour is regular, as when it is microprocessor-controlled, the replacement period would be twelve years as specified by the constructor. So we may estimate the economy in expenses as follows :

economy/furnace/year	
furnace n°64	197800 $
furnace n°68	198700 $

Conclusion

A multivariable P.I. controller has been successfully employed to control the product moisture content of a phosphate drying furnace in both simulation and experimental studies. It can be concluded that whenever, the process is operating near the nominal conditions, the parameters are relatively constant.However during conditions changes,(production constraints) the furnace may be forced away from the nominal working point due to strong non linearities. Thus an adaptation mechanism is necessary. On the other hand, the study has allowed to demonstrate the feasability of the control and the possibilities for saving energy.

REFERENCES

/1/ M. ATHANS
"The role and use of the stochastic linear-quadratic-gaussian problem in control system design"
IEEE Trans. on Automatic Control vol AC-13; n°6 December 1971

/2/ HOWARD H. ROSENBROCK and PETER D. McMORRAN,
"Good, Bad, or Optimal"?
IEEE Trans. on Automatic Control vol,AC-16 n°6,December 1971.

/3/ M. ATHANS
"On the design of P.I.D controllers using optimal linear regulator theory"
Automatica vol.7, pp. 643-647
Pergamon Press 1971.

/4/ H.SERAJ
"Design of proportional-plus-integral controllers for multivariable systems"
INT. J. Control 1979,vol.29 ,n°1

/5/ D. EL JOUHARI
"Identification d'un four industriel de séchage de phosphate"
Thèse de 3° Cycle Faculté des Sciences,Rabat,Juillet 1977.

/6/ A. BARRAUD
"Réalisation minimale et approximation optimale des systèmes dynamiques linéaires invariants"
Thèse d'Etat Université de Nante
Janvier 1975.

/7/ C. FOULARD, S. GENTIL et J.P.
J.P. SANDRAZ
"Commande et Régulation par calculateur numérique"
Editions EYROLLES , PARIS 1977.

/8/ B. PORTER
"Optimal control of multivariable linear systems incorporating integral feedback"
Electronic Letters 22nd.,April 1971, vol 7, n°8

/9/ BURIAN CAREW and PIERRE R. BELANGER
"Identification of optimum filter steady-state gain for systems with unknown noise covariances"
IEEE Trans. on Automatic Control vol, AC-18, n°6 December 1973.

OPTIMIZATION OF INDUSTRIAL TEXTILE DRYERS

A. Touzani*, M. Storoswiecki and C. Melin*****

**L.E.E.S.A. Faculte des Sciences, B.P. 1014, Rabat, Morocco*
***Centre d'Automatique, Université Lille I. B.P. 36, 59650 Villeneuve d'Ascq, France*
****Université de Compiegne, Royal - lieu, 60200 Compiegne, France*

Abstract : The study concerns convection dryers used in the textile industry for the extraction of the superficial and the inflating water from the fabric material. These dryers are characterized by an important energy consumption. By means of an original model of the heat and mass transfers between the fabric material and the drying air, we determine the optimal control of dryer which minimizes the production costs and realizes a compromise between the product specifications and the constraints on the process.

I. Introduction

In the textile industry, drying is an important operation. It consists in extracting the superficial and the inflating water from the fabric materiel Different kinds of machines are used, the most commonly used is the tenter frame dryers. Energy transfer in those machines is effected through convection and the moisture is extracted by means of a gas (a mixture of dry air and vapor) and evacuated by ventilators located in chemneys. The dryers used are thermic machines which are characterized by an important energy consumption, a great part of it is devoted to water evaporation (40% to 70% according to the fabric material being dried) /1/.

The problem then arises of reducing the energy consumption. At the same time it is compulsory to maintain a sufficient production to avoid engorgement of the drying section and respect the constraint upon the residual moisture of the fabric material. The residual moisture corresponds to the equilibrium moisture of the fabric material with the atmosphere.

From an economic point of view, two important parameters of the drying operation must be taken into account, namely :

1) The drying speed, defined as the the mass of evaporated water per unit time.

2) The drying efficiency, defined as the mass of evaporated water per unit energy put into the machine.

These parameters may be computed a posteriori using a global enthalpy balance /2/ but one presently has only particular case models /3/,/4/ which cannot be used for optimization.

The construction of a model must take into account the parameters which influence the drying effeciency as well as the presence of different drying areas /5/.

We shall present a model distributed along one direction of the dryer The parameters of this model will then be identified and used for determining the optimal controls in the steady state operation.

II. Process Modelling

The dryer studied is essentially a dynamic distributed parameter process.

- Considering the fabric material, the initial moisture at machine's input varies not only with time but also along the width of the material.
- The differences humidity and temperature of the drying air at different points of the dryer give rise to convection winds.
- Considering the dryer itself the heat ratiadet by the wall of tenter and the chains which transport the fabric material yields a superdrying of the material at the edges.
- Taking into account the above caracteristics would lead to a very complex model;which cannot be identified and used. In that sense, we construct a model which is limited to a single spatial distribution according to the length of the dryer, using the following simplifications.

1) The state of the drying air (temperature and humidity) varies only

along the longitudinal axes of the dryer.

2) The fabric material is assimilated to a sandwich formed by two dry layer materials separated by a wet layer material. There is no transversal gradient of temperature and moisture.

3) The vaporisation takes place in the wet layer. The dry layers do not hinder the vaporization transfer outside the fabric material.

4) All the transformations inside the dryer are achieved under atmospheric pressure.

2.1. Construction of model

Applying the conservation principle to a section of abscissa l and of length dl; we obtain the following equations.

2.1.1. Heat transfer in the dry layers

$$(1) \quad \frac{dte(l)}{dl} = \frac{h}{V.\rho.Cpt}\left(Ta(l)-te(l)\right) - \frac{h_1}{V.\rho.Cpt}\left(te(l)-ti(l)\right)$$

where

V: the constant speed of the fabric material (m/min).

ρ: the surface density of the dry fabric material (kg/m^2)

Cpt: the specific heat constant of the dry material (k cal/kg°c)

h: the head transfer coefficient between air and dry layers. (k cal/min. m^2.°C).

h_1: the heat transfer coefficient between dry and wet layer (K cal/min. m^2.°C).

Ta(l): the dry temperature of the air at abscissa (°C)

te(l): the superficial temperature of the fabric material at the abscissal (°C).

2.1.2. Heat transfer in the wet layer.

$$(2) \quad \frac{dt_i(l)}{dl} = \emptyset\left(x(l)\right)\left[\frac{h_1}{V.\rho}\left(te(l) - t_i(l)\right) + \lambda.\frac{dx(l)}{de}\right]$$

$$(3) \quad \emptyset\left(x(l)\right) = \frac{1}{Cpt+Cpe.x(l)}$$

where x(l): moisture of the material at abscissa (kgH_2O/kg dry material)

Cpe specific constant of the water (Kcal/Kg°C)

λ: latent heat of evaporation of the water (k cal/kg) . The moisture variation is given by :

$$(4) \quad \frac{dx(l)}{dl} = -\frac{h_2}{V.\rho}\left(X_2(l)-X(l)\right)\times \left(1-\exp\left(-\frac{x(l)-x_f}{x_c - x_f}\right)\right)$$

x_f: residual moisture

h_2: mass transfer coefficient (k gH_2O evap/m^2.min).

X(l): humidity of the drying air (kgH_2O/kg dry air).

x_2(l): humidity of the interface between air and water, depending on the temperature ti(l) and of the pressure of the drying air.

$$X_2(l) = (18/29).\frac{P_a(t_i)}{760-P_a(t_i)}$$

where $P_a(t_i)$: the vapor pressure of water at the temperature t_i(l) (mmHg).

2.1.3. State of recycled air

Let X' and T'a be the humidity and the temperature of the air after having hit the fabric material. One has:

$$(5) \quad X\ (l) = X(l) + \frac{V.\rho}{Va.\rho a}\frac{dx(l)}{dl}$$

va: speed of the air blown projected on the fabric material (m/mib)

fa: density of the dry air (Kg/m^3).

$$(6) \quad T'a(l)(Cpa+X'(l).Cpv) = \lambda.(X(l)-X'(l))+Ta(l)(Cpa+X(l).Cpv) \frac{h}{Va\rho a}(Ta(l)-te(l))+\frac{V.\rho}{Va.\rho a}(\lambda+Cpv.t_i(l)).\frac{dx(l)}{dl}.$$

Cpa, Cpo specific mass of air and vapor (k cal/kg°C).

2.2. Parametric Identification

The use of the proposed model requires, for each type of fabric material , the determination of the coefficients h, h_1,h_2. These coefficients are obtained using a parametric identification procedure. Such procedure may correspond to determining the parameters which minimize the square error between real and computed outputs of the process.

Taking into account the great difficulties encountered to obtain reliable experimental data, the identification criterion is based on the quadratic error between the real output moisture and the computed one, and the real superficial temperature profile te(l) and the computed one /5/.

Two optimization algorithms have been used : the gradient method and the alternating variable search method.

Figures (1-a)(1-b) and (1-c) give some example of the identification results.

III. Simulations

The simulation of the model allows to obtain the output variables for different profiles of the drying air temperature and humidity and for different speeds of the fabric material through the dryer.Particularly, the energy consumption can be computed for each configuration, and diagrams can be constructed given the drying efficiency as a function of the control variables.

Figures (2-a) and (2-b) give two examples of such diagrams.

A more rigourous characterization of the optimal control can be obtained by applying optimization methods to a mathematical model of the process and the energy consumption.The later is modelized by means of the profiles of humidity and temperature of the drying air along the dryer /5/ .
For the particular dryer studied,the problem is to define the values of eight control variables.
V. speed of the fabric material through the dryer, Ta, Ta_2 air temperature in the two controlled areas, X_1 to X_5 air moisture in the fives parts of the dryer.
The constraint concerning the final moisture of the fabric material is taken into account by means of a penalty method /7/ .
For an acrylic material defined by :
$\rho = 0,27\ kg/m^2$
$S = 1,5\ m^2/m$
$x(o) = 80\%$, $xf=4\%$.

table (3) gives the initial values (actual values without optimization) and the optimal values of the control variables, as well as the corresponding of energy consumption. It can be seen that a gain of 20% has been achie achieved on the drying efficiency. Figure (4) depicts the profiles of the material's moisture along the dryer.

IV. Conclusions

This study shows the possibility of an important energy economy in the drying of fabric materials . However, the determination of optimal controls is conditionned on the adequacy of the process model.

The model adaptation is presently limited to the level of the transfers between the fabric material and the drying air. Such limitation arises from the difficulty of obtaining any measure on the state of the recycled air just after having hit the material. Besides, the work done up to now does not allow to realize a specified profile for the drying air moisture .The reasons lie for one part in the difficulty of modelizing the rec cling process, and for the other part in impossibility of obtaining the measures mentioned above.

REFERENCES

/1/ A. Touzani,M.Staroswiecki J. Laneres, J. Perrin "Etude de fonctionnement des sechoires textiles"Bull.scientif.ITF vol 7, n°27, Août 1978.

/2/ A. Dascalescu "Le sechage et ses applications industrielles"editions Meridianes" Dunod, Paris 1969.

/3/ J.N. Beard "More efficient tenter frame operation through mathematical modeling chemical engineering," Vol8 n°3,Mars 1976.

/4/ J.F. Berail . M. Guien "Sechoir par convection de vapeur surchauffée"EDF departement OAP, note P42/24, nov. 76.

/5/ A. Touzani "Optimisation du rendement thermodynamique des sechoirs textiles" (rames planes) Thèse de 3°Cycle Univer. de Lille I. Mars 1979.

/6/ R.E. Treybal "Mass transfer operation" Mc Graw Hill N.Y. 1968.

/7/ K. Okamura "Some Mathematical theory of the penalty method for solving optimum control problems" J. Siam control, vol2, n°3,1965.

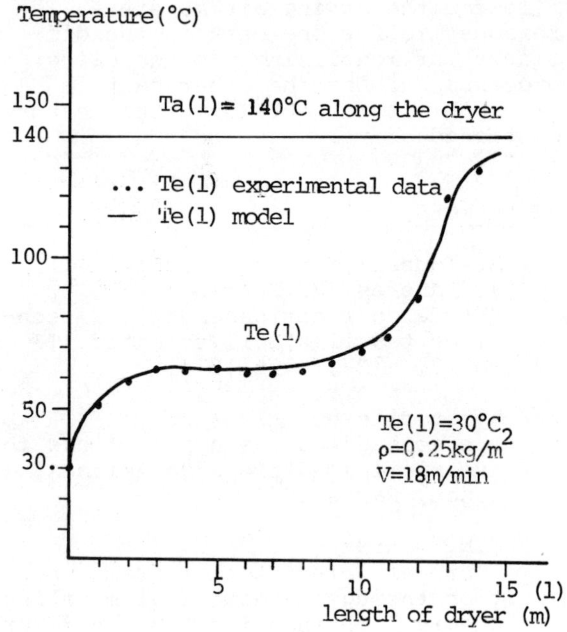

Fig. 1.a

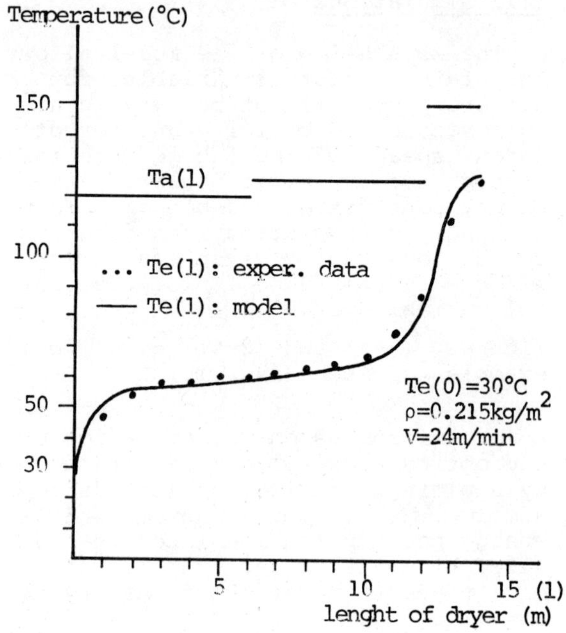

Fig. 1.b

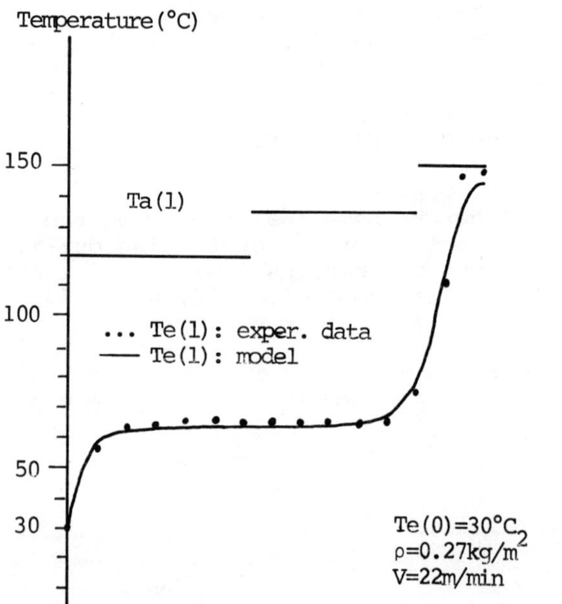

Fig. 1.c

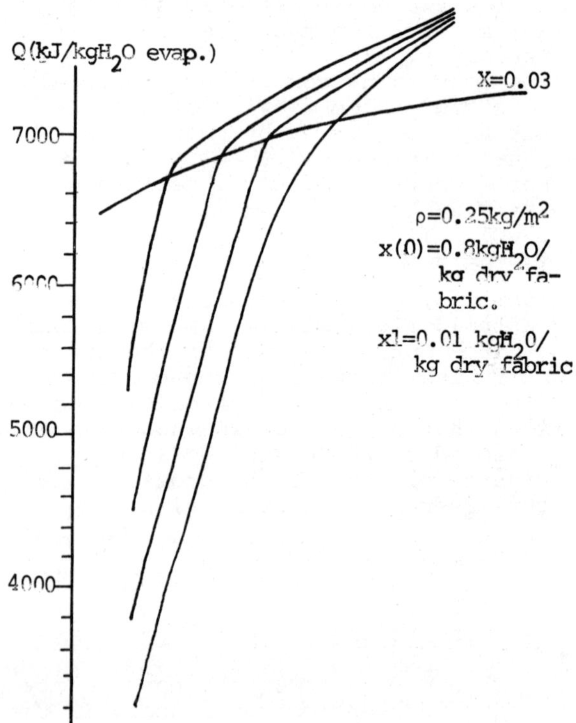

Fig. 2.a

	V	Ta1	Ta2	X1	X2	X3	X4	X5	Q
u°	18	140	140	0.03	0.03	0.03	0.03	0.03	7100
u	30	161	138	0.086	0.068	0.06	0.062	0.053	5700

Table (3)

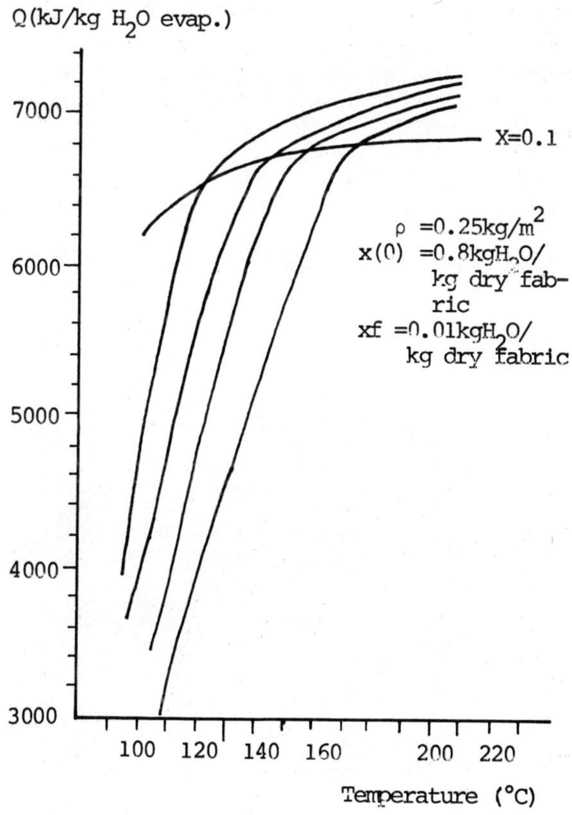

Fig. 2.b

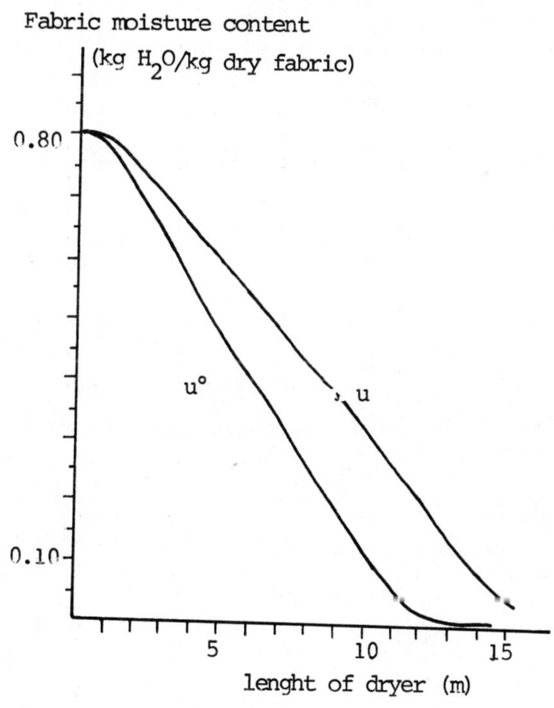

Fig. 4.

SELF-TUNING REGULATOR APPLIED TO AN INDUSTRIAL PHOSPHATE DRYING FURNACE

K. Najim and H. Youlal

L.E.S.A., Faculté des Sciences, B.P. 1014, Rabat, Morocco

Abstract. This paper presents an application of the self-tuning regulator to an industrial phosphate drying furnace. The purpose of the regulator is to keep the moisture content of the dried phosphate close to a certain desired value, and at the same time to minimize the fuel consumption. Several simulation experiments were carried out. A single-input single-output model of the furnace is assumed. A controller minimizing the variance of an auxiliary output is employed. Good control performance is achieved. The results demonstrate the feasibility of the presented algorithm.

Keywords. Adaptive control, Computer control, Microprocessors, Moisture control, Recursive parameter estimation, Self-adjusting systems, Time varying systems.

PROCESS DESCRIPTION

Damp phosphate when extracted from the mine has a moisture content of about 15 %. Drying should reduce that content to about 1 %. The furnace consists of the following elements Fig. 1.; combustion chamber, drying tube and dust chamber. These parts are detailed below.

Combustion chamber. The fuel is initially heated at 100°C, pulverized by means of water vapour then burned. The necessary oxygen for combustion is brought by the so called primary air input. The calories produced by the fuel combustion are carried by the current labelled secondary.

Drying tube. This is an horizontal tube of 25m lenght, its rotation velocity is constant, its production capacity is in the order of 170 t/hr. The cascades disposed in the form of a helix along the inside of the tube facilitates the thermal exchange and help move the phosphate forward. Unlike in cement furnaces the movement of the phosphate and the hot gas are in the same direction; moving from the combustion chamber toward the dust chamber.

Dust chamber. Consists of means of collecting the fine phosphate particles carried by the hot gas current. At its end a draught ventilator avoids saturation of the tube. The final product is received at the exit of the dust chamber by the main conveyor.

PROCESS MODEL

The phosphate drying furnace multi-input multi-output model (Najim and Jouhari , 1977), may be reduced to a single-input single-output one, by letting input phosphate to be kept constant (merely maximum production). So the fuel (control variable) and the moisture content of the dried phosphate (output) are the key parameters for a suitable single-input single-output model of the furnace. Simple representation of the so simplified process, suitable for computer control can be written as,

$$A(q^{-1})y(t) = q^{-k}B(q^{-1})u(t) + C(q^{-1})e(t) \quad (1)$$

where A,B and C are polynomials in backward shift operator q^{-1}, of order n_a, n_b and n_c respectively.
k ($\geq$ 1) is the process time delay in integer number of sample intervals.
$u(t)$, $y(t)$ and $e(t)$ are the control signal, the measured process output and the random disturbances respectively. $e(t)$ is assumed to be of zero mean and finite variance.

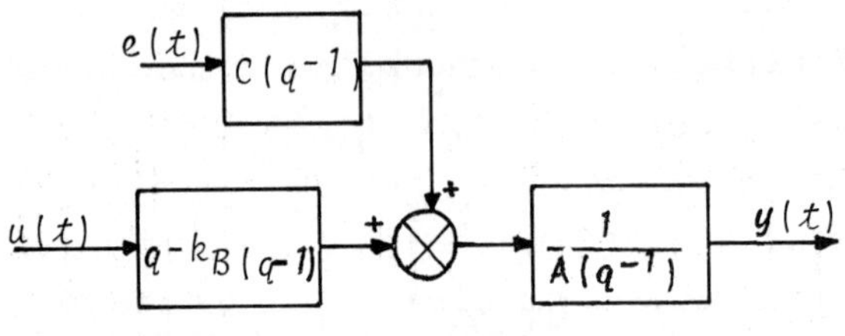

Fig. 2. Plant model

CONTROL PROBLEM

The model of the plant given in (Najim and co-workers, 1977), was identified under the steady state condition. So it can be regarded as an approximation of the actual non linear system (Najim and co-workers, 1976), obtained by linearisation arround the nominal operating point.
The nominal operating point is determined mainly by the furnace production capacity (namely by input phosphate).

The effect of the varying input conditions, the time dependent effects of environment disturbances and the variable moisture content of input phosphate stocks, dictate the necessity to design an adaptive control law. Since one principle objective of the control is to keep the moisture content of the output product close to a certain fixed value (i.e. to minimize the output's variance), and at the same time to minimize the fuel consumption, self-tuning control formulation is employed.

SELF-TUNING CONTROL

In this paragraph only a brief out line of the basic theory of the self-tuning control will be presented. Recalling that the fundamental theory of these adaptive schemes was first developed by Aström and Wittenmark (1973) and Wittenmark (1973) based on minimum variance control. A more general scheme was proposed later by Clarke and Gawthrope (1975, 1979) based on a minimization of an auxiliary output; function of the plant weighted input control, measured output and set point.

Minimum variance control.

Based on the structure of the model Eq. (1), self-tuning algorithm is derived, when the control objective is to minimize the variance of the plant output y(t) in two steps:
1- Estimate the k-step ahead prediction of y(t).
2- Set the current control u(t) to nullify the predicted output.

Self-tuning controller.

One extention of the above scheme is the algorithme based upon the minimization of a cost function given by:

$$I = E.(\Phi(t+k)^2), \quad (2)$$

where

$$\Phi(t+k) = Py(t+k) + Qu(t) - Rw(t) \tag{3}$$

is an auxiliary output.

Good closed-loop behaviour may be fixed by an appropriate choice of the weighting functions P,Q and R on system output y(t), control effort u(t) and set point w(t) respectively.

It can be shown (Clarke, Cope and Gawthrop, 1975; Clarke and Gawthrop, 1975, 1979), that the best k-step ahe ahead prediction of the weighted output can be written, by means of the definitions:

$$P = P_n/P_d \tag{4}$$

$$P_n C/P_d A = E(q^{-1}) + q^{-k}F(q^{-1})/AP_d \tag{5}$$

as,

$$(Py(t+k/t))^{\circ} = \frac{F}{CP_d}y(t) + \frac{EB}{C}u(t) \tag{6}$$

with remnant,

$$Py(t+k/t) - (Py(t+k/t))^{\circ} = Ee(t) \tag{7}$$

Hence the prediction of the auxiliary output is:

$$\Phi(t+k/t)^{\circ} = \frac{F}{CP_d}y(t) + \frac{EB}{C}u(t) + Qu(t) - Rw(t) \tag{8}$$

The control law is derived by setting this predicted auxiliary output equal to zero, to give in terms of the feedback of weighted predicted output,

$$u(t) = \frac{1}{Q}\,(Rw(t) - (Py(t+k/t))^{\circ}). \tag{9}$$

The system closed loop characteristic equation is given by

$$PB + QA = 0. \tag{10}$$

and the response is:

$$y(t) = \frac{q^{-k}BR}{PB+QA}w(t) + \frac{EB + QC}{PB + QA}e(t) \tag{11}$$

The control law Eq. (9) also minimize the cost function Eq. (2), and it can be written in recursive form using the definitions,

$$G = EB + QC \tag{12}$$

$$H = -CR \tag{13}$$

as,

$$u(t) = \frac{-1}{G}\left(\frac{F}{P_d}y(t) + HRw(t)\right), \tag{14}$$

where the controller polynomials in q^{-1}, F, G and H are continuously being update for slowly changing plant dynamics.

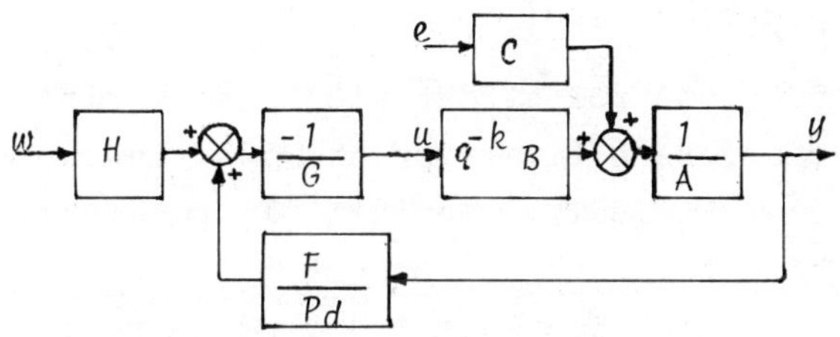

Controller Plant

Fig. 3. Block diagram of self-tuning controller.

SIMULATION RESULTS

The specific case used in the simulation was defined by:

-Plant model (Eq. 1.),

$$A(q^{-1}) = 1.+0.8767q^{-1}+0.0061q^{-2}$$

$$B(q^{-1}) = -0.1849-0.1655q^{-1}+0.1612q^{-2}$$

$$C(q^{-1}) = 1.$$

-Cost function (Eq. 2.),

$$I=E.((y(t+k)-w(t))^2+\lambda u(t)^2)$$

i.e. $P=1.$, $R=1.$ and $Q=\lambda \equiv Constant.$

-Controller polynomials,

$$F(q^{-1})=f_0+f_1q^{-1}+f_2q^{-2}$$

$$H(q^{-1})=h_0+h_1q^{-1}+h_2q^{-2}$$

$$G(q^{-1})=g_0+g_1q^{-1}+g_2q^{-2}$$

with $g_0 \neq 0.$, being fixed for algorithm convergence consideration (Clarke, Cope and Gawthrope, 1975).

One of the three updating parameters methods can be used, namely;

-Recursive least squares filter.

-Covariance square root factorisation filter (Maybeck, 1979).

-U/D Covariance factorisation filter (Maybeck, 1979).

Many simulation runs were carried out using the plant model specified above, by means of the MC6800 microprocessor. Results are summarized by the illustrative figures 4. a-b-c and 5. a-b-c, commented below.

Figs. 4. show the results of self-tuning control for a constant set point. The algorithm converges after the first 20 steps.

Fig. 4.-c, depicts the computed output, control effort and noise variances versus time, for the 500 steps run. Excluding the initial transient effects, both the output and control variances decrease and tend to reference random disturbances variance. Good behaviour of control effort can be seen on Fig. 4.-b.

The simulations carried out are beleived to represent the actual plant features. It is conceived that some offline experimentation of the algorithm may be needed to adjust the controller parameters before on line implementation. Thus avoiding the initial transient effects.

Figs. 5. Show the results of the self -tuning control, when a set point presents a rectangular profile. Notice the relatively small change in control effort after the transition of the set point signal. Fig. 5.-c, depicts the variances versus time for the same example.

Remark: All these runs correspond to zero control parameters initialisation , and a covariance matrix $P_0=\gamma I_d$ (γ= 0.1, 100.) in recursive parameter estimation algorithms.

k = 2 sampling intervals.

Similar results were obtained by both square root and U/D covariance factorisation filters, which are more accurate and less sensitive to the initial conditions.

CONCLUSION

Simulation results on a phosphate drying furnace model have shown that the self-tuning controller provides good features of both control effort and process output behaviour. It is evident that the on-site experiments would need more refinement of the presented algorithms, in order to get better performance compared with the three terms controller experiments (Najim and co-workers, 1980).

Further work on on-site implementation is in progress and more details will be reported later.

REFERENCES

Aström, K. J. and B. Wittenmark (1973). On self-tuning regulators. Automatica, 9, 185-199.

Clarke, D. W. and P. J. Gawthrope (1975). Self-tuning controller. Proc. IEE, 122, 9, 929-934.

Clarke, D. W., S.N. Cope and P. J. Gawthrope. (1975). Feasibility study of the application of microprocessors to self-tuning regulators. OUEL Report, 1137/75.

Clarke, D. W. and P. J. Gawthrope (1979). Self-tuning control. Proc. IEE, 126, 6, 633-640.

Najim, K., M. Najim, B. Koehret and T. Ouazzani. (1976). Modelization and simulation of a phosphate drying furnace. 7th.Annual Pittsburgh Conf. On Modelization and Simulation.

Najim, K. and D. Jouhari. (1977). Identification of a multivariable industrial process: A phosphate drying furnace. 2th. Midwest Symposium on Circuit and System Lubbok. Texas, USA.

Najim, M., K. Najim, D. Jouhari and L. Radouane. (1980). Use of a micro-computer for the control of a phosphate drying furnace. 3rd.IFAC Conf. On System approach for development. Rabat, Morocco.

Maybeck, P.S. (1979). In R. Bellman (Ed.), Stochastic models, Estimation and control, Vol. 1., Academic Press, New York. Chap. 7, pp. 368-406.

Wittenmark, B., (1973). A Self-Tuning Regulator. Rept. 7311, Div. Automatic Control, Lund Institute of Technology.

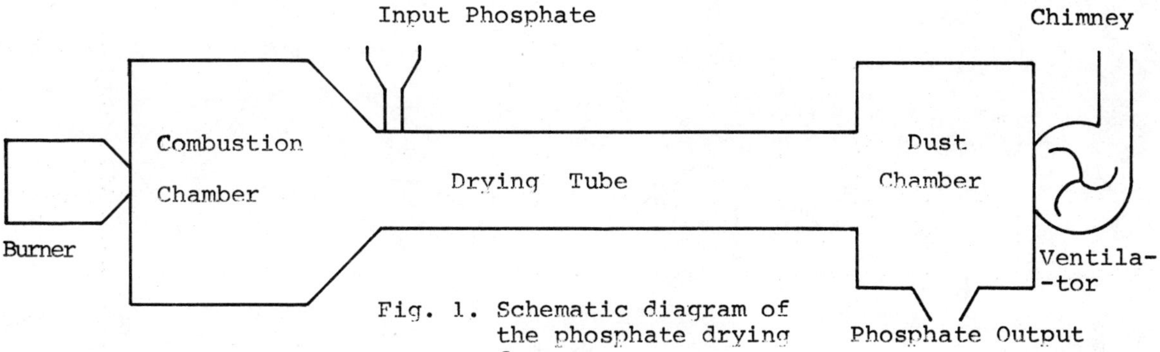

Fig. 1. Schematic diagram of the phosphate drying furnace.

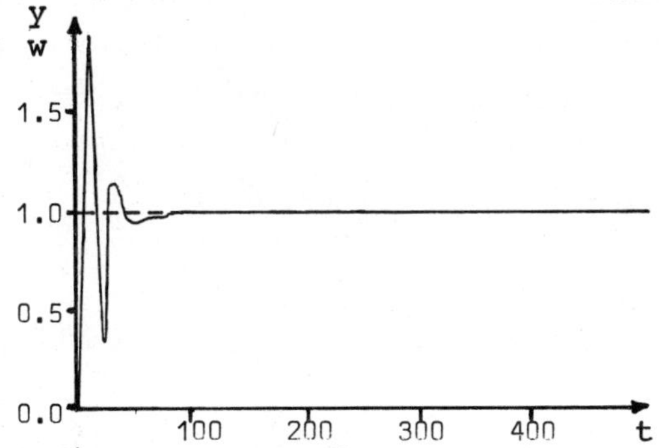

Fig. 4.-a.

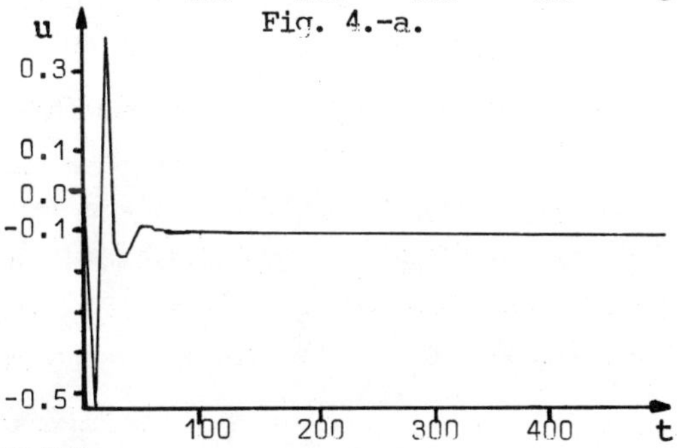

Fig. 4.-b.

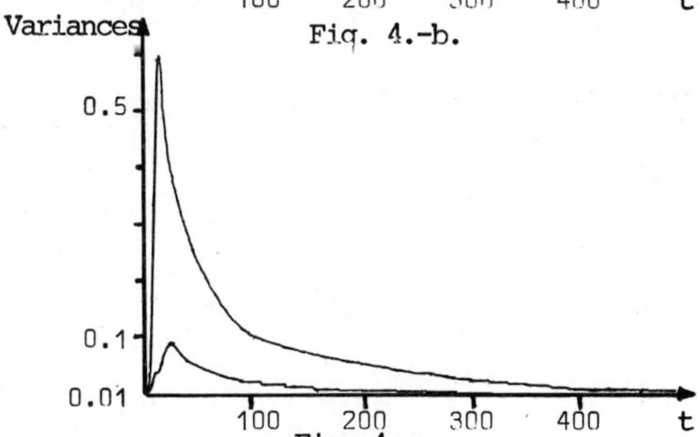

Fig. 4.-c.

Figs. 4. Self-tuning controller for constant set point signal.

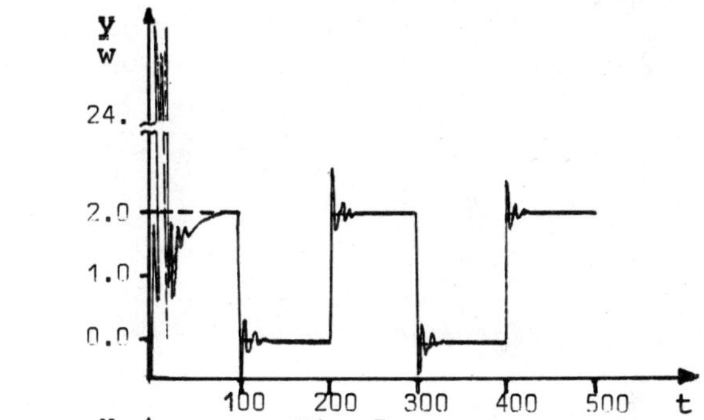

Fig. 5.-a.

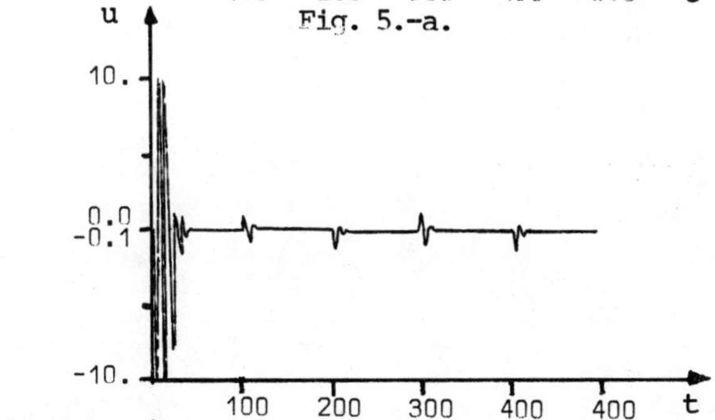

Fig. 5.-b.

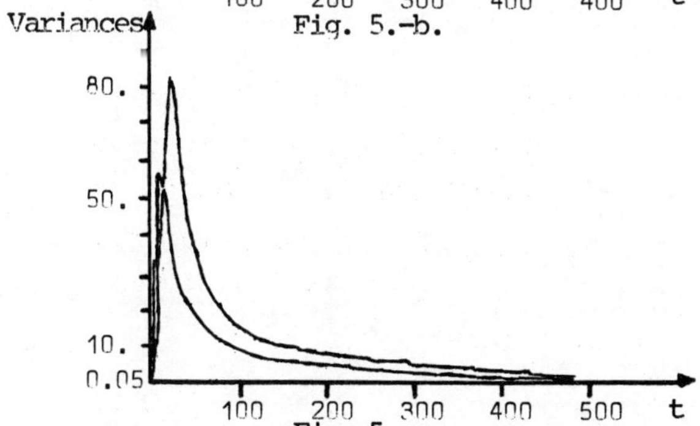

Fig. 5.-c.

Figs. 5. Self-tuning controller for a rectangular reference signal.

OPTIMIZATION PROBLEMS ASSOCIATED WITH THE THERMAL BEHAVIOUR OF AN EPITAXIAL REACTOR

J. P. Babary, A. Benhammou and B. Dahhou

Laboratoire d'Automatique et d'Analyse des Systèmes du C.N.R.S., 7 Avenue du Colonel Roche, 31400 Toulouse, France

Abstract. Among the different techniques for manufacturing electronic components such as diffusion, epitaxy (in gaseous phase, in liquid phase, by molecular jets), ionic implantation, liquid phase epitaxy is a relatively recent technique with numerous advantages from the viewpoint of component quality, but requiring on the other hand particular care and attention in relation to the optimal control of the thermodynamic behaviour of the reactor. The component manufacturing technique by liquid phase epitaxy may be decomposed into five phases :

- preheating (with maximal power) up to 750° C, necessary for the purification of the atmosphere inside the reactor ;
- raising the temperature up to about 800° C in the core of the reactor in minimum time without overshooting the desired temperature ;
- regulation of the thermal profile (as level as possible in the central zone) for a prescribed period ;
- a sequence of slow cooling periods of prescribed duration with controlled temporal and spatial temperature gradients ;
- natural cooling (reactor stopped).

Three problems of optimization associated with phases 2, 3, 4, above have been solved and implemented by microprocessor (INTEL SBC 80/20-4) or simulated.

This paper presents a development of the underlying methods as well as the results obtained on-line or by simulation on the basis of a distributed parameter model of the thermodynamic behaviour of the reactor. Problems of state observation are also considered.

Keywords: distributed parameter systems-optimal control

INTRODUCTION

Liquid phase epitaxy used in the production of electronic components is a technique which has many advantages over for example gaseous phase epitaxy. However, the constraints on the implementation of this type of epitaxy are such that the thermodynamic operation of the reactor in which it is being carried out, must be very closely controlled. This means that it is necessary to carry out a detailed automation study, leading for example to the reactor being digitally controlled. A study of this kind can only be undertaken if we have a sufficiently exact mathematical model of the thermodynamic behaviour of the reactor, which can be adapted without too much difficulty to the study of the actual control.

The aim of this paper is to show how, from a previously established mathematical model (Gouffe, 1968), we have been able to determine a sub-optimal control law of the reactor for each step of operation from heating to cooling.

LIQUID PHASE EPITAXY AND THE REACTOR (Babary, Castet, 1977)

Liquid phase epitaxy, used in the production of electronic components, is a relatively recent technique. Its rapid and constant progress allows it to rival with gaseous phase epitaxy, and to be the best technique for the fabrication of certain types of components.

Moreover, materials of type III-V are extremely useful in the fabrication of active electronic devices. Gallium Arsenide for example, has much greater electronic mobility than Silicium and Germanium. Liquid phase epitaxy has a number of advantages, with respect to gaseous phase epitaxy ; it requires simpler equipment, layers are deposited more quickly, it allows a wider choice of dopants and easier growth of III-V compounds containing aluminium, thus avoiding the presence of corrosive gases.

In order to use this technique, the operation of the reactor has to be entirely controlled, in particular so as to achieve a high level of reproducibility of phenomena as well as fairly homogeneous layers.

The reactor consists of two concentric quartz cylinders (one transparent tube and one opaque tube) heated by five identical coils distributed along them (Fig. 1). The main charge is in the middle of the central zone. It is a partitioned graphite crucible, with a sliding lid for multilayer deposits. In this crucible, a Gallium solution is saturated with a polycrystalline Ga As solution at constant temperature, generally around 800 or 900° C. Then, the equilibrium temperature having been obtained, the mixture is progressively cooled by reducing the temperature linearly by several tenths of a degree per minute. During this phase, we introduce the substrate on which a layer will be deposited on contact with the mixture.

FORMULATION OF THE OPTIMIZATION PROBLEMS

The fig. 2 shows the desired temperature evolution in the crucible.

We notice five phases, three of them needing an optimization study :

- phase 1 : preheating up to 750° C, necessary for the purification of the atmosphere inside the reactor (without optimization).
- phase 2 : heating up to 800° C, in minimum time, without overshooting the steady state temperature.
- phase 3 : regulation of a desired temperature profile in the reactor.
- phase 4 : linear cooling of the crucible (in general, linear piecewise function).
- phase 5 : natural cooling of the reactor (no optimization) :

Hence, the purpose of this paper consists of a presentation of methods and results obtained for the resolution of the three optimization problems (phases 2, 3, 4).

MATHEMATICAL MODEL

The dynamic behaviour of the reactor is modelled from a heat balance of the heat exchanges at levels of the charge, the gas, the quartz tube and the heating elements.

Basic assumptions

In order to obtain a model of reasonable complexity, it will be assumed that :

- the process has a cylindrical symmetry,
- the gas flow is laminar and one-dimensional,
- the thermal exchanges between the gas and the graphite charge on the one hand and between the gas and the quartz tube on the other, are by convection only ; the surface heat exchange coefficients do not depend on temperature in the neighbourhood of an established operating state ; we shall not therefore consider exchanges by radiation since we shall assume that the gas is transparent.
- the temperature of the gas in a section of the tube is taken to be the mean value.
- the conduction phenomenon in the gas is negligible.
- the charge has a cylindrical symmetry and is concentric with the other tubes.

Non linear model

The mathematical model which has been obtained is a non linear parabolic four equation system, with two independant variables (one space variable and the time variable). This system enables us to determine the evolution of temperature profiles in the gas, in the quartz, in the charge and at level of the heating coils, knowing the evolution of heating power profile.
The model can be written as follows (Benhammou, 1979) :

$$\frac{\partial Z}{\partial t} = A_1 \frac{\partial^2 Z}{\partial x^2} + B_1 (W^4 - Z^4) + C_1 (Z - G)$$

$$\frac{\partial G}{\partial t} = A_2 \frac{\partial G}{\partial x} + B_2 (W - G) + C_2 (Z - G)$$

$$\frac{\partial W}{\partial t} = A_3 \frac{\partial^2 W}{\partial x^2} + B_3 (T^4 - W^4) + C_3 (W^4 - Z^4) + D_3 (W - G)$$

$$\frac{\partial T}{\partial t} = A_4 \frac{\partial^2 T}{\partial x^2} + B_4 (T^4 - W^4) + D_4 (T - T_0) + E_4 P \tag{1}$$

Z = Z(x,t) : temperature of the charge
W = W(x,t) : temperature of the quartz
T = T(x,t) : temperature of the heating coils
P = P(x,t) : heating power
To = constant : ambient temperature
x : space variable such that $x \in]0,1[$

Coefficients A_i, B_i, C_i, D_i are functions of thermodynamic and geometrical parameters of the reactor. As previously mentionned, the main charge is located in the central zone of the furnace; in fact, there are also two longitudinal rods (one for maintaining the crucible and one for the thermocouple) which have a non negligible influence. Therefore the charge being non homogeneous along the tube, coefficients A_1, B_1, C_1, C_2, C_3 linked to heat exchanges at the level of the charge, will be considered as functions of x for $x \in]0,1[$.

Linear model

The model (1) is, in fact, too complex ; therefore we must consider a simplified model in order to apply classical optimization methods.

After some mathematical simplifications, reduction of the system order, linearisation and application of Galerkin's method (Dahhou, 1980), we get a model of the form :

$$\dot{a}(t) = Aa(t) + Bu(t) \qquad (2)$$

$$y(x,t) = C(x).a(t) \qquad (3)$$

a(t) being the state vector (6-dimensional) such that :

$$a(t) = \begin{bmatrix} a_1(t) \\ \hline a_2(t) \end{bmatrix} \quad \begin{matrix} (a_1(t) \text{ corresponds to the charge}) \\ (a_2(t) \text{ corresponds to the heating coils}) \end{matrix}$$

u(t) being the control vector (5-dimensional),
y(t) being the output vector such that :

$$y(x,t) = \begin{bmatrix} \Delta Z(x,t) \\ \hline \Delta T(x,t) \end{bmatrix} \qquad (4)$$

A and B are constant matrices. C(x) is a matrix whose components are the basis functions used in Galerkin's method, such that :

$$\varphi_i(x) = \sqrt{2}\sin i\pi x \quad (i = 1,2,3)$$

a(t), u(t) and y(x,t) represent the variations of state, control and output, about the linearisation profile.

FIRST OPTIMIZATION PROBLEM :
MINIMUM TIME CONTROL

The problem consists of finding a control law which transfers the charge temperature from an initial value (about 750° C) to a desired value (about 800° C) in minimum time, knowing that control variations around the nominal profile are bounded. The nominal profile of control is represented in Fig. 3 .

The maximal power on each zone being 1500 Watts, we define the extremal values for control variations :

$$\begin{aligned} u_{1\,min} &= -430\,W \leq u_1 \leq u_{1\,max} = 1070\,W \\ u_{2\,min} &= -190\,W \leq u_2 \leq u_{2\,max} = 1310\,W \\ u_{3\,min} &= -170\,W \leq u_3 \leq u_{3\,max} = 1330\,W \\ u_{4\,min} &= -180\,W \leq u_4 \leq u_{4\,max} = 1320\,W \\ u_{5\,min} &= -430\,W \leq u_5 \leq u_{5\,max} = 1070\,W \end{aligned} \qquad (5)$$

The desired temperature (800° C) corresponds to $\Delta Z = 0°C$ in the linearised system (2)-(3), at the center of the crucible ; the initial temperature (about 750° C) corresponds to $\Delta Z = -50°C$ (exactly -46° C in the studied case below).

In other words, we have to transfer $\Delta Z(x=0{,}5)$ from an initial value to 0 in minimum time, satisfying constraints (2)-(3) and (5). But :

$$\Delta Z = 0 \iff a_1 = 0$$

Therefore, we must find a control such that :

$a_1(T^*) = 0$ where T^* is the minimum time.

To solve this problem, we can apply the Maximum Principle (Athans, Falb, 1966). First of all, we verify that the system is controllable and "normal" that is to say :

$$\det\left[b_k \,\vdots\, Ab_k \,\vdots\, --- \,\vdots\, A^{n-1}b_k\right] \neq 0 \qquad \forall k = 1,2,..5 \qquad (6)$$

b_k being the kth column of B

Then we deduce the optimal control from the Hamiltonian :

$$H\left(a(t), p(t), u(t)\right) = 1 + \langle Ax(t), p(t)\rangle + \langle Bu(t), p(t)\rangle \qquad (7)$$

p(t) is the adjoint state, solution of the adjoint equation :

$$\dot{p}(t) = -\frac{\partial H}{\partial x} = -A^T p(t) \qquad (8)$$

with unknown boundary conditions.

The optimal solution is defined by :

$$u_k(t) \begin{cases} u_{k\,min} & \text{if } p^T b_k > 0 \\ u_{k\,max} & \text{if } p^T b_k < 0 \end{cases} \qquad (9)$$

Then, we have to solve a boundary value differential problem by an iterative method represented in Fig. 4.

In fact, for the physical problem, the optimal solution $\Delta Z^*(0.5,t)$ must not overshoot the desired temperature after the final time. This is the reason why it is necessary to choose an initial vector p(o) as near the optimal solution $p^*(o)$ as possible, but which avoids overshoot of the desired temperature of the charge.

A satisfactory solution is shown in Fig. 5. The optimal time is about 15 minutes.

SECOND OPTIMIZATION PROBLEM : REGULATION OF A SPATIAL PROFILE

During the phase n° 3 (Fig. 2), a desired temperature profile (about which the system equations are linearized) is to be maintained.

We want to find an optimal control law minimising the quadratic criterion :

$$J = \frac{1}{2}\int_0^\infty\int_0^1 \left[y^T(x,t)\,Q\,y(x,t) + R\,p^2(x,t)\right] dx\, dt \qquad (10)$$

With :

$$p(x,t) = \sum_{k=1}^{5} g_k(x)\, u_k(t)$$

$$g_k(x) = \begin{cases} 1 & \text{if } 0{,}2(k-1) \leq x < 0{,}2(k) \\ 0 & \text{elsewhere} \end{cases}$$

$Q \triangleq \begin{bmatrix} q_1 & 0 \\ 0 & q_2 \end{bmatrix}$ non negative definite matrix; - R positive number

It can be shown that (10) is reducible to :

$$J' = \frac{1}{2}\int_0^\infty \left(a^T(t)\,Q\,a(t) + u^T(t)\,R'u(t)\right)dt \qquad (11)$$

With :

$$Q' = \left[\begin{array}{ccc|ccc} q_1 & & & & & \\ & q_1 & & & 0 & \\ & & q_1 & & & \\ \hline & & & q_2 & & \\ & 0 & & & q_2 & \\ & & & & & q_2 \end{array}\right]; \quad R' = \begin{bmatrix} 0.2R & & & & 0 \\ & 0.2R & & & \\ & & 0.2R & & \\ & & & 0.2R & \\ 0 & & & & 0.2R \end{bmatrix}$$

The steady states obtained by the mathematical model on one hand, and the real system on the other hand, do not correspond exactly. In order to avoid errors due to their differences, we insert in the feedback loop, integrators whose inputs are these errors, then :

$$V(t) = \int_0^t \left(y(x^*,t) - y_d\right) dt \qquad (12)$$

Where y_d is the desired profile and x^* defines the location of 3 thermocouples (one in the middle of the charge ; two in the heating zones at points x = 0,3 and x = 0,7).

So, we get an augmented differential model such :

$$\dot{\xi}(t) = \tilde{A}\,\xi(t) + \tilde{B}\,u(t) \qquad (13)$$

in which : $\xi(t) \begin{bmatrix} a(t) \\ \hline V(t) \end{bmatrix}$;

$$\tilde{A} = \begin{bmatrix} A & 0 \\ C(x^*) & 0 \end{bmatrix} \qquad \tilde{B} = \begin{bmatrix} B \\ 0 \end{bmatrix}$$

where $C(x^*)$ is a constant matrix (3x6 dimensional).

We must determine an optimal control law, minimising the criterion (11) which can be written as :

$$J' = \frac{1}{2}\int_0^\infty \left[\xi^T(t)\,Q''\,\xi(t) + u^T(t)\,R'u(t)\right]dt \qquad (14)$$

with :

$$Q'' = \begin{bmatrix} Q' & 0 \\ 0 & 0 \end{bmatrix}$$

Controllability of system (13) (Power, Porter, 1970)

Rank $(CA^{-1}B) = 3$ → the system is controllable.

Control law

The minimisation of J' with respect to u(t) leads to the optimal control law:

$$u_{opt} = -R'^{-1}\tilde{B}\,K\,\xi(t) \triangleq G\,\xi(t) \qquad (15)$$

where K is the matrix solution of the Riccati equation (Boudarel, Delmas et Guichet, 1969 - Courdesses et Alengrin 1975) :

$$Q'' + \tilde{A}^T K + K\tilde{A} - K\tilde{B}\,R'^{-1}\tilde{B}^T K = 0 \qquad (16)$$

State observer (Langeland, Knudsen, 1977)

For the implementation of optimal control law (15), the knowledge of all components of the state vector $\xi(t)$ is needed. From the five measurement points in the heating zones, the three components of the sub-vector $a_2(t)$ are calculated by means of a pseudo-inverse operation. Because only one sensor is available in the charge, the three components of $a_1(t)$ are determined by means of a minimal order state observer of the form :

$$\dot{Z}(t) = F\,z(t) + S\,y_1(t) + H\,a_2(t) \qquad (17)$$

$$z(0) = z_0$$

with :

$$Z(t) = T a_1(t) + e^{Ft}\left(z_0 - T a_1(0)\right)$$

The estimated state of $a_1(t)$ is :

$$\hat{a}_1(t) = My_1(t) + Nz(t) \tag{18}$$

where :
$$\begin{cases} H = TA_{12} \\ TA_{11} - FT = SC_{11} \\ MC_{11} + NT = \mathbb{1} \end{cases} \tag{19}$$

knowing that :

$$A = \left[\begin{array}{c|c} A_{11} & A_{12} \\ \hline A_{21} & A_{22} \end{array}\right] \quad \text{and} \quad C = \left[\begin{array}{c|c} C_{11} & 0 \\ \hline 0 & C_{22} \end{array}\right]$$

The block diagram of the closed loop system (including the observer) is shown in Fig. 6.

RESULTS

In Fig. 7, the distribution of the temperature profile in the charge is represented at time t_o=o, and at time t=5mn 30s.

In Fig. 8, we have, for the same initial condition and the same control, the evolution of the temperature in the middle of the crucible.

We notice that steady state is reached without overshoot.

THIRD OPTIMIZATION PROBLEM : COOLING

The cooling of the crucible has to be linear with respect to the time, and the temperature has to be the same all along this crucible for each $t \in [t_o, t_f]$.

So, the desired profile along the reactor can be approximated as shown in Fig. 9.

$$\alpha_{d_1}(x,t) = \frac{b}{0.4}\, x t$$

with $0 \leqslant x \leqslant 0.4$ and $0 \leqslant t \leqslant t_f$

$$\alpha_{d_2}(x,t) = bt$$

for $0.4 \leqslant x \leqslant 0.6$ and $0 \leqslant t \leqslant t_f$

$$\alpha d_3(x,t) = \frac{-b}{0.4}(x-1)\,t$$

with $0.6 \leqslant x \leqslant 1$ and $0 \leqslant t \leqslant t_f$.

(20)

b being the cooling rate (in degrees/hour).

Since we transformed the distributed parameter system into a lumped parameter system, the corresponding desired state is defined as :

$$a^T_{1d} = \int_\Omega \alpha_d(x,t)\, \varphi^T dx$$

After some calculation, we get :

$$a_{1d}(t) = \begin{bmatrix} 0.6815 \\ 0 \\ -0.0468 \end{bmatrix} bt. \tag{21}$$

The performance index is somewhat complex because we must consider at the same time the temperature gradients in space and in time ; it can be expressed as follows :

$$\begin{aligned} J = &\int_{\Omega_1} (\alpha(t_f) - \alpha_d(t_f))^2 dx + q_1 \int_0^{t_f}\!\!\int_{\Omega_1} (\alpha - \alpha_d)^2 dx\,dt \\ &+ q \int_0^{t_f}\!\!\int_{\Omega-\Omega_1} (\alpha - \alpha_d)^2 dx\,dt + \lambda \int_0^{t_f}\!\!\int_{\Omega_1} \left(\frac{\partial\alpha}{\partial x}\right)^2 dx\,dt \\ &+ r \int_0^{t_f}\!\!\int_{\Omega-\Omega_1} \left(\frac{\partial\alpha}{\partial x}\right)^2 dx\,dt + r' \int_0^{t_f}\!\!\int_{\Omega} p^2(x,t)\,dx\,dt \end{aligned} \tag{22}$$

With :
$$\begin{cases} \Omega = \{x : \quad x \in\,]0,1[\,\} \\ \Omega_1 = \{x : \quad x \in [0.4, 0.6]\} \end{cases}$$

In the corresponding lumped form, J is written as :

$$\begin{aligned} J = &\frac{1}{2}\left[a^T(t_f)\, P_1 a(t_f) - 2a^T(t_f) P_1\, a_d(t_f) + a_d^T(t_f)\, P_1 a_d(t_f)\right] + \frac{1}{2}\int_0^{t_f}\left[a^T(t)\, Q\, a(t) + \right. \\ &\left. a^T(t)\, P a_d(t) + a_d^T(t)\, S a_d(t) + u^T(t)\, R u(t)\right] dt \end{aligned} \tag{23}$$

Formulating the hamiltonian and using the classical results of the tracking problem (Boudarel, Delmas et Guichet, 1969 - Athans et Falb, 1966), we obtain the optimal law :

$$u(t) = -R^{-1}B^T p(t) \tag{24}$$

where p(t) is the adjoint vector, such that a(t) and p(t) are solutions of the canonical system :

$$\begin{bmatrix} \dot{a}(t) \\ \dot{p}(t) \end{bmatrix} = \begin{bmatrix} A & -BR^{-1}B^T \\ -Q & -A^T \end{bmatrix} \begin{bmatrix} a(t) \\ p(t) \end{bmatrix} - \begin{bmatrix} 0 \\ -\frac{1}{2}Pa_d \end{bmatrix} \tag{25}$$

Moreover, we have :

$$p(t) = K(t)\, a(t) - g(t) \tag{26}$$

where K(t) is the Riccati matrix, solution of the equation (Courdesses, 1977) :

$$\dot{K}(t) = -K(t)\, A - A^T K(t) + K(t)\, B R^{-1} B^T K(t) - Q \tag{27}$$

and g(t) is the solution of the differential equation :

$$\dot{g}(t) = \left[K(t)\, B R^{-1} B^{T} - A^{T} \right] g(t) + \frac{1}{2} P a_d \quad (28)$$

The transversality conditions lead us to the final conditions :

$$\begin{cases} p(t_f) = P_1\, a(t_f) - 2 P_1\, a_d(t_f) \\ K(t_f) = P_1 \\ g(t_f) = -2 P_1\, a_d(t_f) \end{cases} \quad (29)$$

The block diagram of the controlled system is represented in Fig. 10, in which we notice that the state observer is the same as in the regulation problem (Fig. 6).

Graphical results, for a tracking problem in which b = -60 o/h and t_f = 24 minutes are shown in Fig. 11. The maximal error is about half a degree and the final error is less than 0,4°.

REFERENCES

Athans, M. and Falb, P.L. (1966). Optimal control. Mc Graw-will, 1966

Babary, J.P. and Castet, J. (1977). Modelling for sub-optimal control of an epitaxial reactor - IFAC - Le Caire - 26-29 Nov. 1977.

Benhammou, A. (1979). Identification et commande optimale d'un réacteur d'épitaxie en phase liquide - Thèse de 3è cycle, n° 2771, 1979 TOULOUSE.

Boudarel, R., Delmas, J. et Guichet, P. (1969). Commande optimale des processus - Tome 4 - Dunod Paris 1969.

Courdesses, M. (1977). Sur la commande ponctuelle des systèmes à paramètres répartis avec critère quadratique - Thèse de 3è cycle, n° 1944, 1977 TOULOUSE.

Courdesses, M. et Alengrin, G. (1975). Programme de résolution de l'équation algébrique de RICCATI - LAAS note interne n° 75 I 34 - TOULOUSE 1975.

Dahhou, B. (1980). Commande sous-optimale des séquences de fonctionnement d'un réacteur d'épitaxie en phase liquide. Thèse de 3è cycle (à paraître Juin 1980) TOULOUSE.

Gouffe, A. (1968). Transmission de la chaleur par rayonnement - Eyrolles, Gauthier Villars, 1968.

Langeland-Knudsen, J. (1977). Observers : theory and applications Servolaboratoriet. Danmarks Hojskole.

Porter, B., Power, H.M. (1970). Controllability of multivariable systems incorporating integral feedback. Electronics Letters 10th Dec. 1970, Vol. 6, n° 25.

Porter, B., Power, H.M. (1970). Mode controllability matrices of multivariable systems incorparating integral feedback. Electronics Letters. 10th Dec. 1970, Vol. 6,n° 25.

Power, H.M., Porter, B. (1970). Necessary and sufficient conditions for controllability of multivariable systems incorporating integral feedback. Electronics Letters, 29th October 1970, Vol. 6, n° 22.

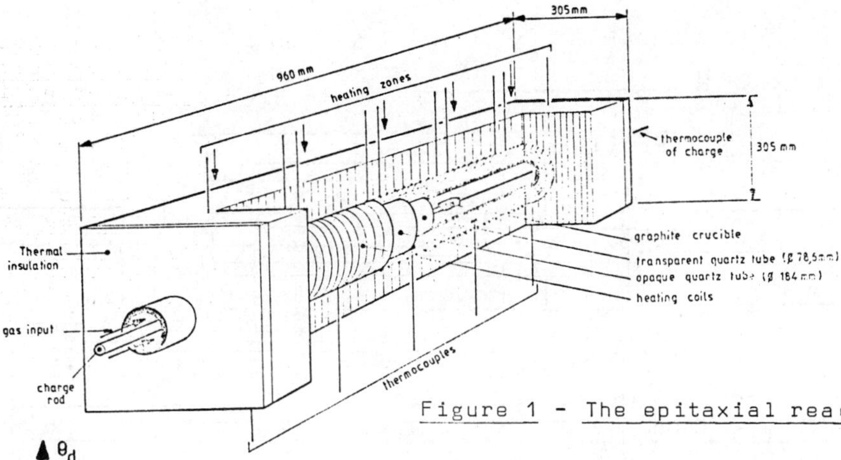

Figure 1 - The epitaxial reactor

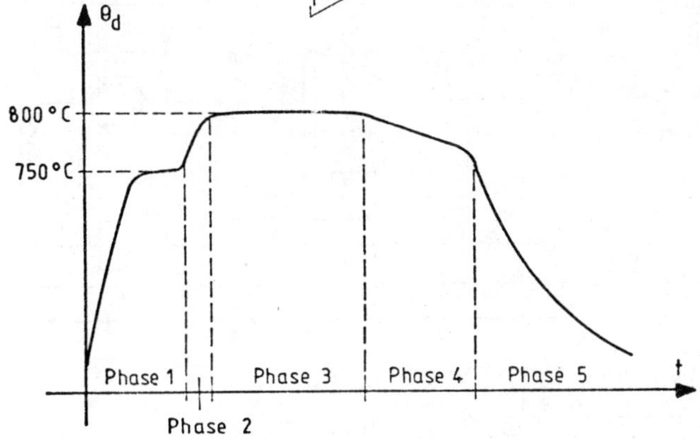

Figure 2 - Desired temperature evolution

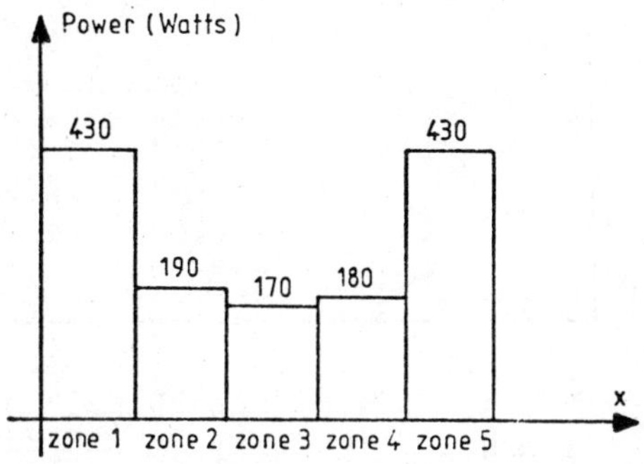

Figure 3 - Nominal control profile

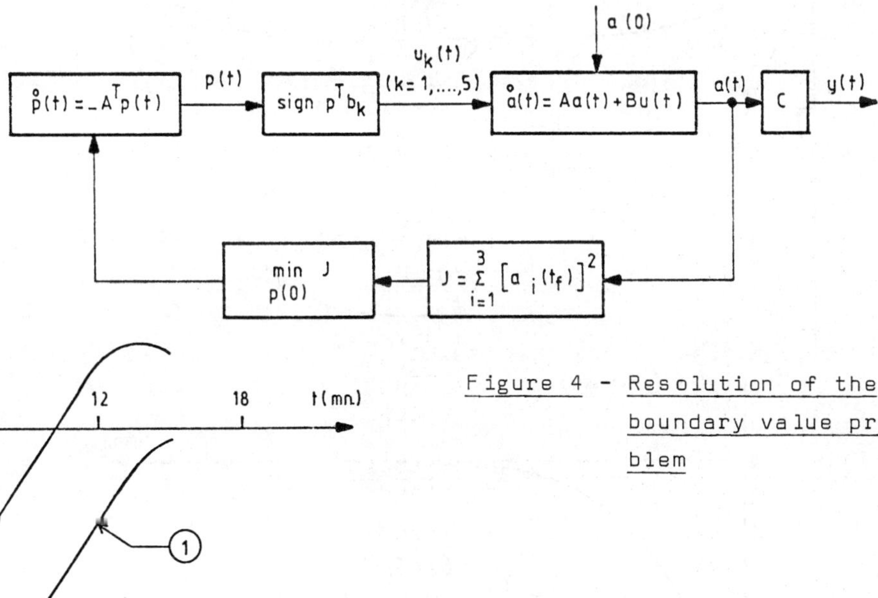

Figure 4 - Resolution of the boundary value problem

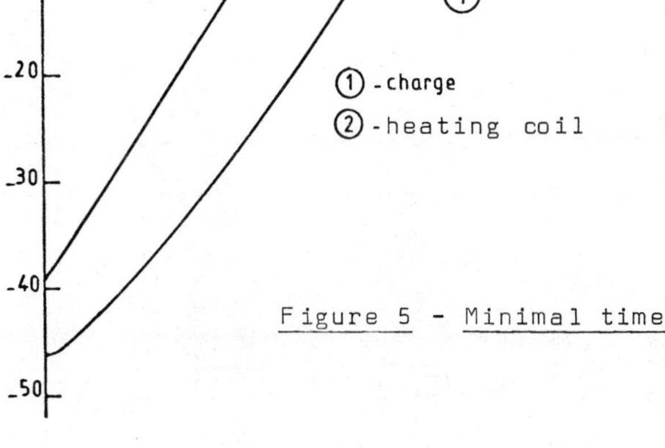

Figure 5 - Minimal time solution

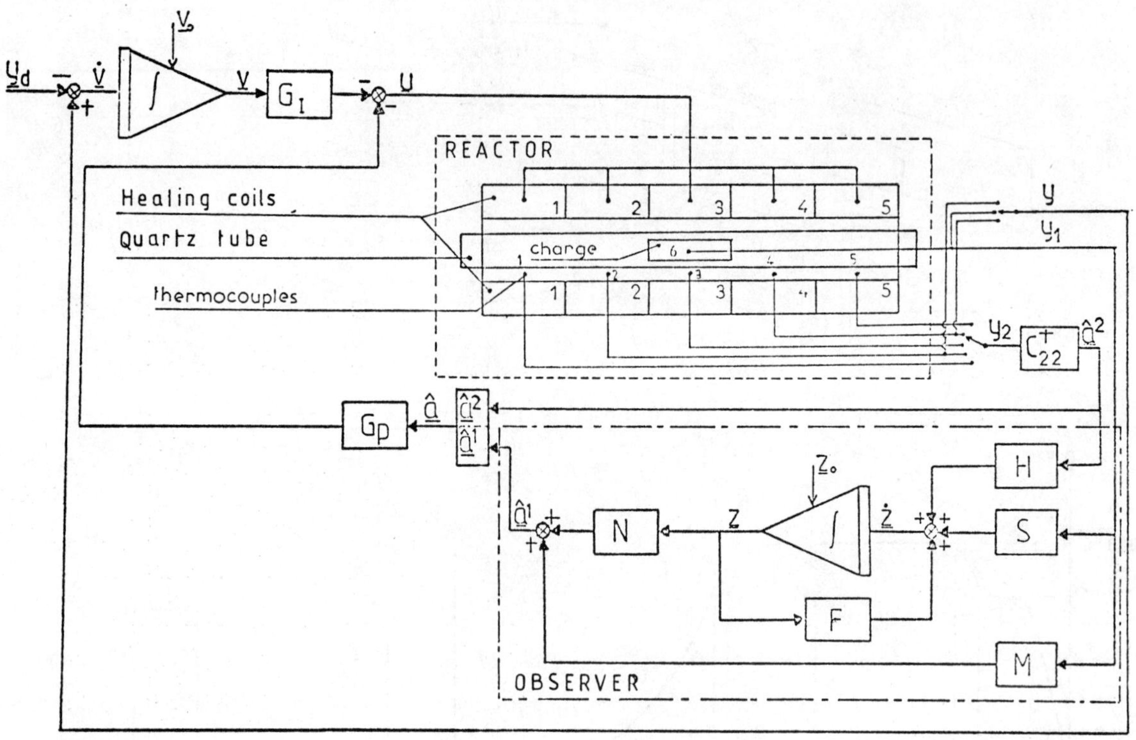

Figure 6 - Closed loop control system (regulation problem)

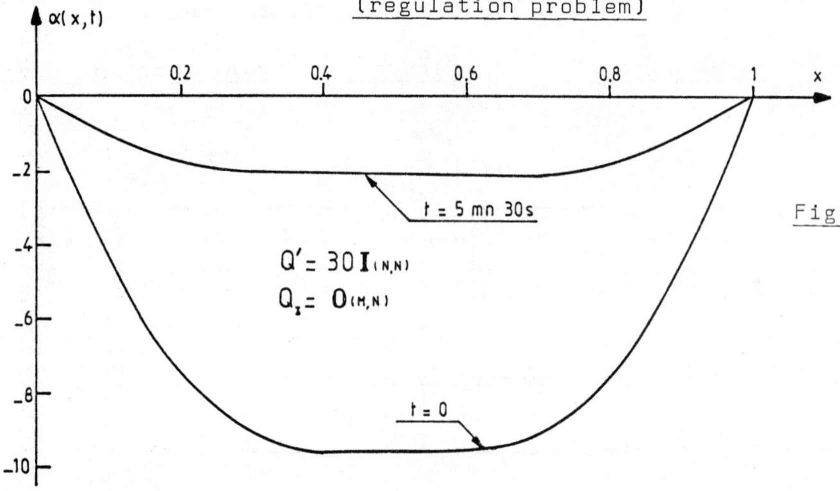

Figure 7 - Evolution of $\alpha(x,t)$

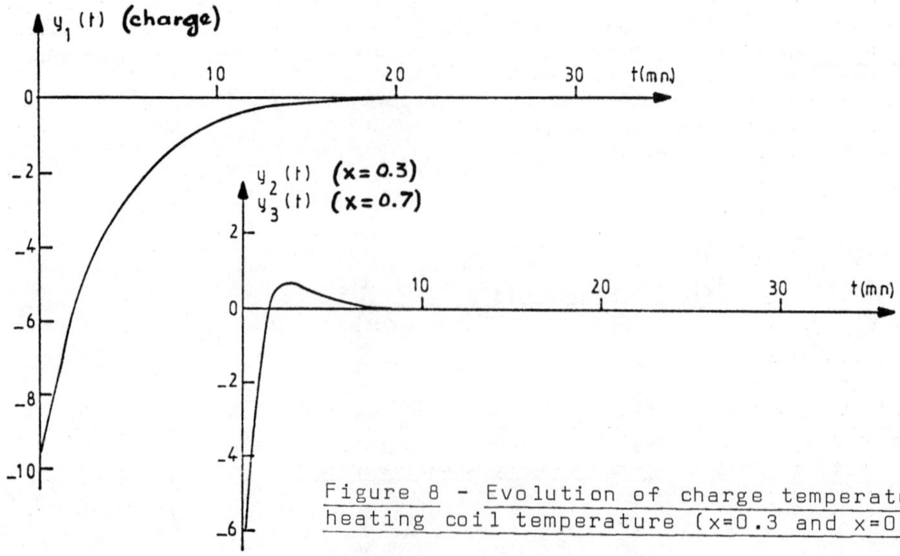

Figure 8 - Evolution of charge temperature and heating coil temperature (x=0.3 and x=0.7)

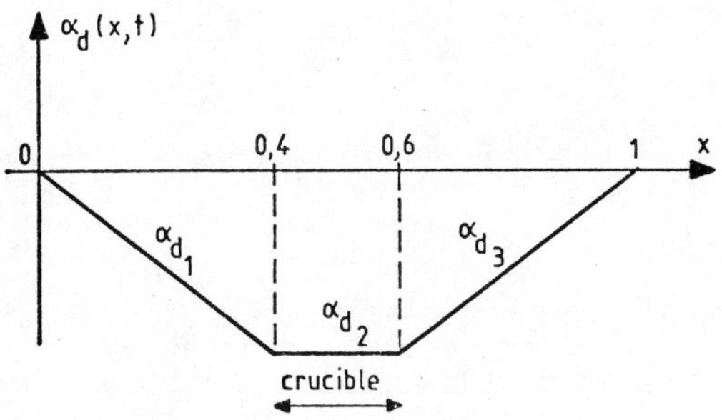

Figure 9 - Desired temperature profile

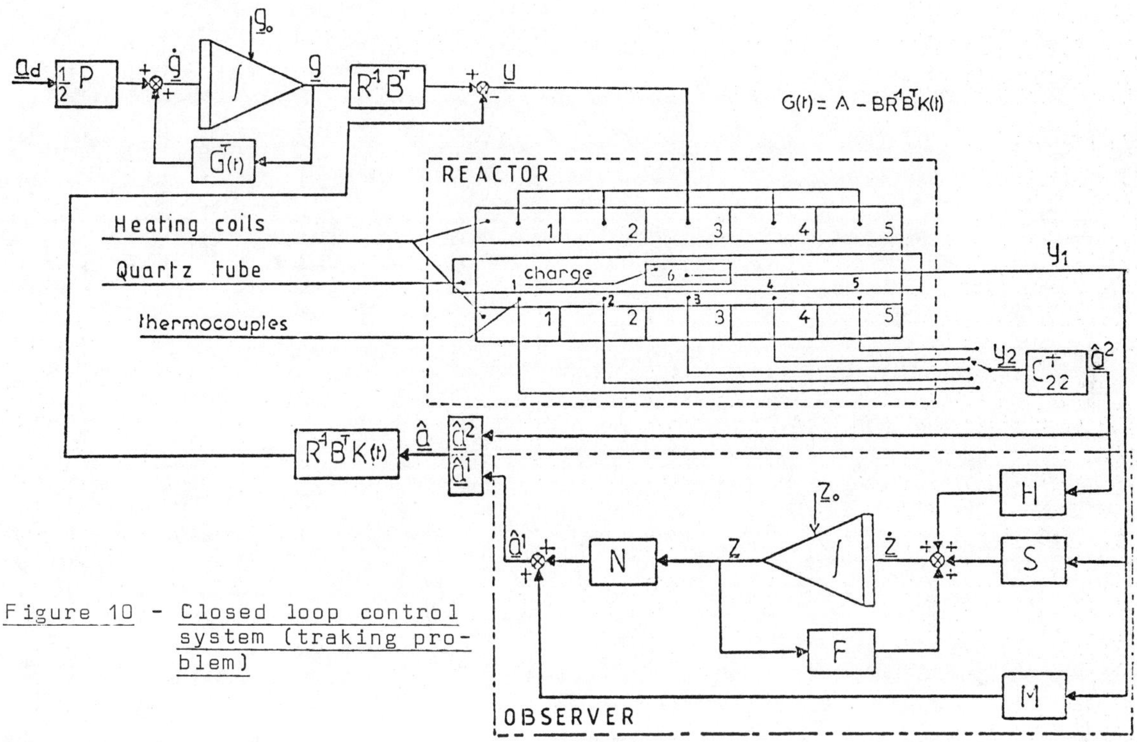

Figure 10 - Closed loop control system (traking problem)

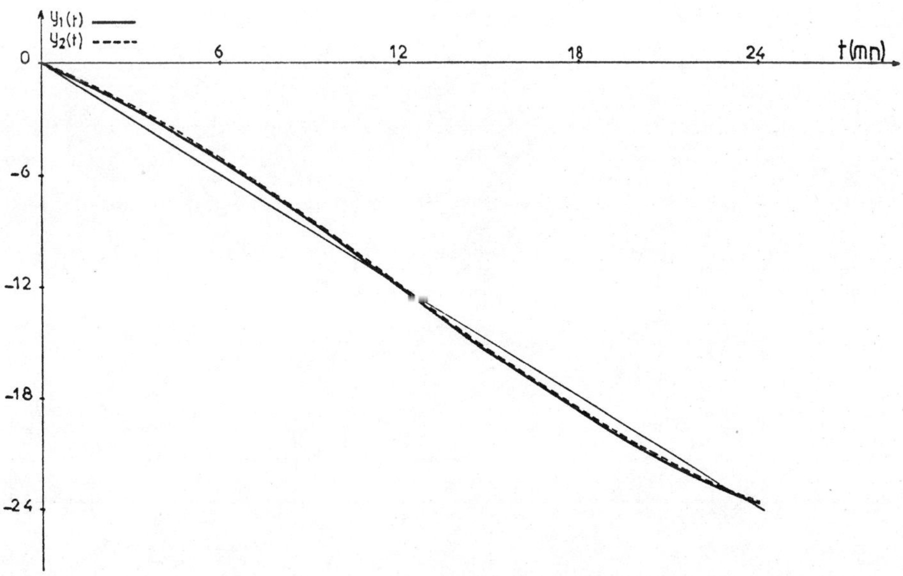

Figure 11 - Evolution of the charge temperature (x = 0.4 and x = 0.6)

MULTI-LEVEL OPTIMAL CONTROL OF A THERMAL PROCESS

G. Krzakala, B. Lang and M. Aubrun

Laboratoire d'Automatique et de Recherche Appliquée, Université de Nancy I, Institut des Sciences de l'Ingénieur de Nancy (I.S.I.N.), Parc Robert Bentz, 54500 Vandoeuvre-les-Nancy, France

Abstract. This article describes a multi-level hierarchical control method applied to a "pilot" heat process with large time constants, composed of a boiler and an exchanger rigidly coupled together, in the presence of temperature and control constraints.

The process is identified by introducing the heat flow-rate variables for which the state model, which is linear and with relatively constant parameters, allows the use of an open-loop control system. The physical coupling equation between the sub-processes, however, brings the control variable of the exchanger into operation (1)(18).

When the equations have been discretised, a three-level decomposition is used, thereby effecting coordination between the physical sub-processes, predicting the interactions between the sub-systems and the heat flow-rate local optimal control which is converted into a volumetric flow-rate control via a final and very simple level.

The rather lengthly computing time necessitates the use of a sub-optimal control for on-line operation by mini-computer.

Keywords. Hierarchical systems. Optimal control. Heat exchangers. Computer control. Temperature control.

GENERALITIES

The heat process (Fig. 1) consists of a circulating hot water boiler and a heat exchanger. The boiler is fuelled with Groningen natural gas. The exchanger is an industrial liquid-liquid type with concentric tubes operating in counter-flow. All characteristics are to be found in Table 1.

All measuring equipment (temperature and flow-rate transducers) as well as control apparatus (electrical or electro-pneumatic valves) are of the industrial type. Flow-rates are servo-controlled using an internal analog control loop.

The system is describes by the following magnitudes :

- for the gas : volumetric flow-rate Q_g
- for the primary circuit :

  volumetric water flow-rate in the boiler : Q_c

  volumetric water flow-rate in the exchanger : Q_e
- boiler outlet temperature : T_{ce}

  boiler return temperature : T_r

  exchanger outlet temperature : T_{cs}
- for the secondary circuit :

  volumetric water flow-rate : Q_f

  intake temperature : T_f

  outlet temperature : T_s

The purpose of the control device is to keep T_s at a set value T_{sc} when Q_f charge variations occur.

Q_g and Q_e flow-rates are the control variables of the boiler and the exchanger. Q_c is kept constant to restrict the interaction between boiler and exchanger.

It is obvious that the power supply cost function basically varies in accordance with the gas consumption. The optimal control problem is presented in the conventional manner.

Determining the control law u (θ) :

$$u(\theta) = \begin{bmatrix} Q_g(\theta) \\ Q_e(\theta) \end{bmatrix} \theta \in \left[t_o, \infty\right]$$

minimizing :

$$J(u(\theta)) = \int_{t_o}^{\infty} \left[(T_s(\theta) - T_{sc})^2 + \alpha\, Q_g(\theta)\right] d\theta$$

With the constraints : $Q_{g\,min} \leq Q_g \leq Q_{g\,max}$

$$Q_{e\,min} \leqslant Q_e \leqslant Q_c = \text{constant}$$

$$T_{ce} \leqslant 100°C$$

where $Q_{g\,min}$ is the minimum value of the gas flow-rate before the burners are extinguished and $Q_{e\,min}$, determined experimentally, is the limit of the water flow-rate in the exchanger below which the performance level is no longer acceptable.

PROCESS SIMULATION

The operational model of a heat exchanger is non-linear and fairly complex. A simplified approach can be obtained through linearisation about a desired steady state, but experience shows that the variations of the gains and time constants are not negligible (3).

This being the case, a conventional state simulation process, based on the flow-rates and temperatures, whereby the process is considered as consisting of two exchangers linked by a coupling equation, is only valid in the case of slight variations on either side of the operating point. But under experimental conditions, the control variation range is sometimes such that the parameters of the equipment cannot be considered as constants.

It is therefore necessary to base the model process on the power levels by introducing the heat outflow variables, obtained from the flow-rates multiplied by the temperatures, or :

$$\begin{cases} \dot{T}_{ce} = a_1 T_{ce} \cdot Q_c + b_1 T_r \cdot Q_c + c_1 Q_g \\ \dot{T}_s = a_2 T_{ce} \cdot Q_e + b_2 T_{cs} \cdot Q_e + c_2 T_f \cdot Q_f + d_2 T_s \cdot Q_f \\ \dot{T}_{cs} = a_3 T_{ce} \cdot Q_e + b_3 T_{cs} \cdot Q_e + c_3 T_f \cdot Q_f + d_3 T_s \cdot Q_f \end{cases} \qquad (1)$$

subject to the coupling equation :

$$\dot{T}_r = f(T_{ce} \cdot Q_c, T_{ce} \cdot Q_e, T_{cs} \cdot Q_e, T_r \cdot Q_c) \qquad (2)$$

A non-linear model is obtained, whose parameters vary slightly according to the operating point : in fact, the variations are due essentially to the secondary effect of the variations of the exchange coefficients.

Finally, by introducing the useful heat outflows :

$$D_f = k_g \cdot Q_g \qquad \text{kg : calorific power of gas in kcal./m}^3$$

$$D_c = (T_{ce} - T_r)\, Q_c \qquad (3)$$

$$D_e = (T_{ce} - T_{cs})\, Q_e$$

$$D_f = (T_s \cdot T_f)\, Q_f$$

The theoretical study of the energy level and the experimental results enable us to represent the process by means of the following model :

$$\dot{D}_c = A_1 D_c + B_1 D_g$$

$$\dot{D}_f = A_2 D_f + B_2 D_e \qquad (4)$$

$$\dot{D}_e = A_3 D_e + B_3 D_c$$

constituting an aggregated model of the process equipped with an open-loop control. It does not afford access to each of the system variables, and the effective control Q_e, for instance, is derived from the model control according to the graph shown in Fig. 2.

PROBLEMS LINKED WITH OPTIMAL CONTROL

With this model, the set value is :

$$D_{fc} = Q_f \left[T_{sc} - T_f\right]$$

It is derived from T_{sc} and the measurements of Q_f and T_f : it therefore varies with time. It is to be noted that a variation of this magnitude shows a set value variation or a load variation.

The initial state is known, the "model" final state is evaluated using a simple energy balance and the problem is one of quadratic optimisation with open-ended final time for which the conventional solutions are known : dynamic programming or calculus of variations for example (6).

But this problem is made more complex by control constraints, either fixed (on D_g) or linked with the state of the system (on D_e) and state constraints since temperatures are confined to set limits.

a) <u>Control constraints</u>

The constraints on the gas heat flow-rate D_g are to be deduced from those on the volumetric flow-rate Q_g.

$$D_{g_m} \leqslant D_g \leqslant D_{g_M} \quad \text{where} \quad D_{g_m} = k_g Q_{g_m}, \quad D_{g_M} = k_g Q_{g_M}$$

The constraints on heat flow-rate D_e, however, such as :

$$D_{e_m}(t) \leqslant D_e(t) \leqslant D_{e_M}(t)$$

are deduced from those on heat flow-rate Q_e by means of :

$$D_{e_m}(t) = \left[T_{ce}(t) - T_{cs}(t)\right] \cdot Q_{e_m}$$

$$D_{e_M}(t) = \left[T_{ce}(t) - T_{cs}(t)\right] Q_{e_M}$$

which are functions of inaccessible variables. Moreover, the feasible region for this control is no longer invariant.

The simulation of this region is carried out from equations (1) approximated by :

$$\left| \begin{array}{l} \dot{T}_{ce} = f_1 (D_c, D_g) \\ \dot{T}_{cs} = f_2 (D_e, D_f) \end{array} \right. \quad (5)$$

from which T_{ce} (t) and T_{cs} (t) are deducted in order to calculate the model limits of D_e (t).

In order to avoid periodical re-adjustment of this model (5), and taking into account the fact that control saturation may occur, mainly when the optimization sequence begins, the following strategy is defined :

- an optimal trajectory $\underline{u}$ (θ) is determined which is feasible in relation to the closed region D (θ) of the control space,

$$\theta \in [t_o, \infty]$$

- at time t, region D (θ) is experimentally adjusted by the measurement of T_{ce} (t) and T_{cs} (t) so that :

$$\theta \in [t, t + \Delta t]$$

and control $\underline{u}$ (θ) corresponding to this interval is in relation to the real region of the control space if Δt is small enough.

b) State constraints

All temperatures are limited, but, in practice, only temperature T_{ce} may exceed 100°C. As before, this constraint relates to an inaccessible variable.

Working along the same lines, it can be turned into a constraint on the state variable D_c, defined thus :

$$D_{c_M}(t) = Q_g \cdot [100 - T_r (t)]$$

In fact, in the decomposition-coordination method which had been used for the optimal control, the equation :

$$\dot{D}_e = A_2 D_e + B_2 D_c$$

from system (4) is the coupling equation between the boiler outlet D_c and exchanger control D_e, the corresponding dynamics being very rapid in relation to the response time of the boiler. It is therefore acceptable to approximate the state constraint D_c with constraint on D_e, thus :

$$D_{e_L}(t) = \left|\frac{B_2}{A_2}\right| \cdot D_{c_M}(t)$$

Consequently this brings us back to an optimal control problem with the following control constraints :

$$D_{g_m} \leqslant D_g(t) \leqslant D_{g_M}$$

$$D_{e_m}(t) \leqslant D_e(t) \leqslant D_{e_s}(t)$$

where D_{e_s} (t) is the lower boundary of the region (D_{e_M}, D_{e_L}).

STATING THE PROBLEM

Let us turn the optimization problem into a fixed final time problem whereby tf is assumed to be large enough to obtain a continuous operation.

With the various values defined in the annex and by applying indices 1 and 2 respectively to the boiler variables and to the exchanger variables, the problems is now as follows :

to determine $\underline{u}$ (t) $t \in [0, tf]$

minimizing $J = \int_o^{tf} (e_1^2 + e_2^2 + \alpha u_1^2)\, dt$

with :

$$\left| \begin{array}{lll} \underline{e} = \underline{X} - \underline{X}_c & & \\ \underline{X}(o) = \underline{X}_o \quad \underline{X}(tf) = \underline{X}_c & & \\ \dot{X}_j = A_j X_j + B_j u_j & j = 1,2 & \text{model equations (6)} \\ \dot{u}_2 = A_3 u_2 + B_3 X_1 & & \text{coupling equation (7)} \\ H_j (u_j) \leqslant 0 & j = 1,2 & \text{inequality constraints (8)} \end{array} \right.$$

in which the inequality constraints are expressed in the standard manner (2) by :

$$\begin{array}{l} H_1 (u_1) = (u_1 - G_2)(u_1 - G_1) \\ H_2 (u_2) = (u_2 - G_3)(u_2 - G_4) \end{array} \quad (9)$$

The choice of the optimization method takes account of the following imperatives : the control must be such that it can easily be extended to a group of several exchangers of different types operating in parallel ; a comprehensive method would fairly rapidly become unmanageable ; it must be easily applicable to pure time delay systems such as are frequently found in heat processes. Finally, an extension to the case of a non-linear model must be contemplated.

A physical breakdown of the system into two subordinate processes is therefore carried out, followed by a time discretisation of the equations of each sub-process. Thus we obtain a hierarchical control with several levels (10) corresponding to the following tasks :

0 level : local control of volumetric flow-rate.

1rst level : local control of heat flow-rate.

2nd level : coordination by prediction of the sub-systems providing an optimal local control.

3rd level : coordination between the sub-processes providing an optimal overall control.

The basic diagram of the breakdown is shown in Fig. 3.

MULTI-LEVEL BREAKDOWN (9)(10)(11)

By breaking down the finite horizon (0, tf) into N intervals Δt, the problem is to determine :

$$\min J = \sum_{k=0}^{N} \left[e_1^2(k) + e_2^2(k) + \alpha\, U_1^2(k)\right] \qquad (10)$$

with :

$$\underline{Z}(k) = \underline{T}(\underline{X}(k), \underline{U}(k)) \qquad k = 0, \dots N-1 \qquad (11)$$

$$T_j(k) = X_j(k) + \Delta t\,[\,A_j\, X_j(k) + B_j\, U_j(k)\,] \qquad k = 0, \dots N \qquad j = 1, 2 \qquad (12)$$

$$\underline{X}(k) = \underline{Z}(k-1)$$

$$\underline{Z}(N) = \underline{X}_c$$

$$\underline{X}(0) = \underline{X}_o$$

$$\underline{H}(U(k)) \leq 0 \qquad k = 0, \dots N \qquad (13)$$

and with the coupling equation :

$$U_2(k+1) = U_2(k) + \Delta t\,.\left[A_3\, U_2(k) + B_3\, Z_1(k)\right] \qquad k = 0, \dots N-1$$

with become, if $C_3 = \frac{1}{\Delta t\,.\,B_3}$

and $D_3 = \frac{1 + \Delta t.\, A_3}{\Delta t\,.\,B_3}$

$$g(k) = Z(k) - C_3\, U_2(k+1) - D_3\, U_2\,(k) = 0 \qquad k = 0, \dots N-1 \qquad (14)$$

where the vector Z(k) represents both the internal interactions between the discrete sub-systems and the interaction between the two sub-processes.

The singular nature of this coupling equation will be observed : it is a function of the control variable U_2 of the sub-system 2, and not one of the state variable.

By juxtaposing the equality constraints (11), (12), and (14) with the Lagrange parameters $\underline{\mu}(k)$, $\underline{\rho}(k)$ and $\lambda(k)$ respectively, and the inequality constraints (13) with the Kühn and Tucker parameters $\underline{\gamma}(k)$ (2), the Lagrangian L associated with criterion J is as follows :

$$L = \alpha U_1^2(o) + \sum_{k=1}^{N} \left[(X_1(k) - X_{1c})^2 + (X_2(k) - X_{2c})^2 + \alpha\, U_1^2(k)\right] \qquad (15)$$

$$+ \underline{\mu}^T(o)\,\left[\underline{T}(\underline{X}_o\,, U(o)) - \underline{Z}(o)\right]$$

$$+ \sum_{k=1}^{N=1} \underline{\mu}^T(k)\,\left[\underline{T}(\underline{X}(k)\,, U(k)) - \underline{Z}(k)\right]$$

$$+ \underline{\mu}^T(N)\,.\left[\underline{T}(\underline{X}(N)\,, \underline{U}(N)) - \underline{X}_c\right]$$

$$+ \sum_{k=1}^{N} \underline{\rho}^T(k)\,.\,\left[\underline{X}(k) - \underline{Z}(k-1)\right]$$

$$+ \sum_{k=0}^{N-1} \lambda\,(k)\,.\,\left[Z_1(k) - (C_3 U_2(k+1) - D_3 U_2(k))\right]$$

$$+ \sum_{k=0}^{N} \underline{\gamma}^T\,(k)\,.\,\underline{H}\,(\underline{U}(k))$$

DECOUPLING OF THE SUB-PROCESSES

Let us introduce L_1 and L_2 such that :

$$L_1 = \alpha U_1^2(o) + \sum_{k=1}^{N} \left[(X_1(k) - X_{1c})^2 + \alpha U_1^2(k)\right]$$

$$+ \mu_1(o)\,\left[T_1(U_1(o)) - Z_1(o)\right]$$

$$+ \sum_{k=1}^{N-1} \mu_1(k)\left[T_1(X_1(k)\,, U_1(k)) - Z_1(k)\right]$$

$$+ \mu_1(N)\left[T_1(U_1(N)) - X_{1c}\right] \qquad (16)$$

$$+ \sum_{k=1}^{N} \rho_1(k)\,\left[X_1(k) - Z_1(k-1)\right]$$

$$+ \sum_{k=0}^{N-1} \lambda(k)\, Z_1(k) + \sum_{k=0}^{N} \gamma_1(k)\,.\,H_1\left[U_1(k)\right]$$

$$L_2 = \sum_{k=1}^{N} \left[X_2(k) - X_{2c}\right]^2$$

$$+ \mu_2(o)\,.\,\left[T_2(U_2(o) - Z_2(o)\right]$$

$$+ \sum_{k=1}^{N-1} \mu_2(k)\,\left[T_2(X_2(k)\,, U_2(k)) - Z_2(k)\right]$$

$$+ \mu_2(N)\,\left[T_2(U_2(N)) - X_{2c}\right] \qquad (17)$$

$$+ \sum_{k=1}^{N} \rho_2\,(k)\,\left[X_2(k) - Z_2(k-1)\right]$$

$$- \sum_{k=0}^{N-1} \lambda_k\left[C_3 U_2(k+1) - D_3 U_2(k)\right]$$

$$+ \sum_{k=0}^{N} \gamma_2(k)\,.\,H_2(U_2(k))$$

It will be seen that for $\lambda(k) = \lambda^*(k)$ as given, the Lagrangian L is additively separable and may be broken down into two independent sub-Lagrangian L_1 and L_2 corresponding to the the two sub-processes, such that : $L = L_1 + L_2$.

The coordination task of the upper level 3 therefore consists in providing the lower level with the optimal estimation $\underline{\lambda}^*(k)$ of $\underline{\lambda}(k)$ by means of a gradient method using the duality principle of the Lagrange theorem (5) in which the optimal solution corresponds to a saddle-point of L which will accept a minimum in relation to the variables and a maximum in relation to the Lagrange parameters.

The iterative algorithm is therefore defined, for iteration i to i + 1, by :

$$\lambda^{i+1}(k) = \lambda^i(k) + \alpha^i d^i(k) \qquad k = 0, \dots N$$

where α^i is the iteration step and $d^i(k)$ the research direction, as defined by :

$$d^i(k) = g^i(k)$$

in the case of a simple gradient.

The solution is obtained when the value of $g^i(k)$ is less than or equal to the error $\varepsilon(k)$.

PREDICTION OF THE INTERACTIONS

At level 2, L_1 and L_2 are independent where $\lambda(k)$ is given. If :

$$L_1^0 = \alpha U_1^2(o) + \mu_1(o) \cdot [T_1 (U_1(o)) - Z_1(o)] + \lambda(o) Z_1(o) + \gamma_1(o) H_1 (U_1(o))$$

$$L_1^k = (X_1(k) - X_{1c})^2 + \alpha U_1^2(k) + \mu_1 (k) [T_1(X_1(k), U_2(k)) - Z_1(k)] + \rho_1 (k) [X_1(k) - Z_1(k-1)] + \lambda(k) Z_1(k) + \gamma_1(k) H_1(U_1(k)) \qquad k = 1 \dots N-1$$

$$L_1^N = (X_1(N) - X_{1c})^2 + \alpha U_1^2(N) + \mu_1 (N) [T_1(X_1(N), U_1(N)) - X_{1c}] + \rho_1 (N) [X_1(N) - Z_1(N-1)] + \gamma_1(N) \cdot H_1(U_1(N))$$

it appears that L also assumes an "additively separable" form :

$$L_1 = \sum_{k=0}^{N} L_1^k$$

where the L_1^k are independent if $Z_1 = Z^*_1$ is given.

The same applies to L_2 where $Z_2 = Z^*_2$ given, if :

$$L_2^k = [X_2(k) - X_{2c}]^2 + \mu_2(k) [T_2(X_2(k), U_2(k)) - Z_2(k)] + \rho_2(k) [X_2(k) - Z_2(k-1)] - [\lambda(k-1) C_3 - \lambda(k) D_3] \cdot U_2 (k) + \gamma_2(k) \cdot H_2 (U_2(k))$$

since $\lambda(k)$ is provided by upper level 3. The task of level 2 therefore consists in predicting the values of $\underline{Z}(k)$ by means of the following gradient algorithm :

$$\underline{Z}^{i+1}(k) = \underline{Z}^i(k) - (\beta)^T L^i_{\underline{Z}}(k)$$

where

$$\begin{cases} L_{Z_1}^i(k) = -\mu_1^i(k) - \rho_1^i(k+1) + \lambda^i(k) \\ L_{Z_2}^i(k) = -\mu_2^i(k) - \rho_2^i(K+1) \end{cases}$$

and transferring them to level 1.

We must however take into account the problem created by the presence of the non-linear constraints : the equations at level 1 admit an analytical solution -- if the $\underline{\gamma}(k)$ are known.

Level 2 must also therefore solve the inequality constraints and ensure coordination with $\underline{\gamma}(k)$. These constraints are satisfied if

$$\gamma(k^T) \cdot \underline{H} (\underline{U}(k)) = \begin{cases} \underline{\gamma}(k) \geqslant 0 \\ \text{or} \\ H(U(k) \leqslant 0 \end{cases}$$

The iterative algorithm employed must satisfy the following conditions :

$$\underline{\gamma}^{i+1}(k) \begin{cases} 0 \text{ if the constraint is saturated and } \underline{H} (U(k)) < 0 \\ \underline{\gamma}^i(k) + \delta^i \underline{H} (U(k)) \text{ in other cases where } \delta^i \text{ represents the iteration step whence eventually :} \end{cases}$$

$$\underline{\gamma}^{i+1}(k) = \text{maximum} \left\{ 0, \underline{\gamma}^i(k) + \delta^i \underline{H} (U(k)) \right\}$$

The coordination task therefore consists in estimating the vector $\begin{bmatrix} \underline{Z}(k) \\ \underline{\gamma}(k) \end{bmatrix}$ and transferring it to lower level 1.

It now remains to solve the independent sub-problems L_1^k and L_2^k $(k = 0, \dots N)$ whose solution is purely analytical and satisfies the stationary conditions of the sub-Lagrangians, thus :

$$\frac{\partial L_j^k}{\partial X_j(k)} = 0 \qquad \frac{\partial L_j^k}{\partial \rho_j(k)} = 0 \qquad k = 1 \dots N \quad j = 1, 2$$

$$\frac{\partial L_j^k}{\partial U_j(k)} = 0 \qquad \frac{\partial L_j^k}{\partial \mu_j(k)} = 0 \qquad k = 0 \dots N \quad j = 1, 2$$

The decomposition-coordination block diagram is shown in Fig. 4 where the interchanges of data between the various levels are indicated.

APPLICATIONS OF THE OPEN LOOP HIERARCHICAL CONTROL

The introduction of this control requires that tf and the period Δt be chosen beforehand. The values chosen lead to a representation of the sub-processes by a large scale system at level 1 whose solution requires a calculation time that would be too great for an on-line control. It remains therefore to determine a periodically adapted sub-optimal control, taking account of the non-stationary character of the model coefficients.

a) Choice of tf and Δt

A conventional multi-variable control using a numerical PID corrector for each sub-process in combination with a trend control shows that the overall system is stabilized after 20 minutes in response to a set level on D_f. Outlet D_f of the exchanger is in the vicinity of this set level after 3 minutes, but small oscillations may be observed around this value due to interaction between the sub-systems. This period of 20 minutes is therefore retained for tf.

The choice of sampling step Δt is a compromise between two opposing requirements : it must be small enough to ensure that an optimal open-loop control is present, but an increase in the number of sub-systems N involves a proportional increase in the number of variables to be treated, particularly at the coordinator level, thus extending the calculation time.

If we assimilate the exchanger in the conventional manner to a first-order system whose indicial response is replaced by 50 points, a value of 6 seconds is obtained, but the corresponding size N is too large. After simulation, we retain N = 60, and a step varying between 10 and 30 seconds is chosen, that is to say 10 seconds during the 4 first minutes, 20 seconds for the 4 following and 30 seconds thereafter.

b) Open-loop sub-optimal control

Iterative algorithms make use of a simple gradient method combined with an automatic step determination technique (7) which ensures convergence and speed. The stability of the coordination algorithms is obviously fundamental and can be studied in different ways (10). The initialization of the calculations is carried out based on an estimation of the trajectories of $\underline{Z}(k)$ making use of the model's linearity. The values of the parameters are those obtained during the preceding sequence.

This method implies prolonged calculation time, some 1 to 2 minutes on a Solar 16/40 minicomputer.

The use of a multi-processor and more advanced techniques, of the projected gradient type, reduce the calculation time, but a lapse of time of one minute is not compatible with on-line control.

What is actually carried out is a sub-optimal control in relation to the physical coupling between the sub-processes. In fact, the calculation of the control is completed when the gradient values of the Lagrangian operator in relation to $\lambda(k)$ and $\underline{Z}(k)$ are lower than the thresholds ε_λ and ε_Z determined experimentally as 10^{-3} and 10^{-4} : below these values, the trajectories no longer develop appreciably in view of the accuracy of the measurements. Trajectory ε_λ k), shown in Fig. 5 is determined off-line : $\varepsilon_\lambda(k)$ is the value of threshold ε_λ at the kth iteration.

The sub-problems are therefore optimal locally, but the coupling constraint between the sub-processes can only be satisfied gradually: at each iteration of k, we pass from one trajectory to another, at the same time gradually improving the criterion ; the trajectory corresponding to the preceding states being fixed definitevely. During each period Δt, the model coefficients, and therefore the set values, are adapted using a recurrent least squares method (3) whose convergence is the more rapid as the variations of the coefficients are weak and continuous.

NUMERICAL RESULTS

The on-line sub-optimal control is calculated on a Solar 16/40 mini-computer. The corresponding programme occupies 17 k and the number of processed variables at the beginning of the optimisation sequence is in the region of one thousand (1,000) of which 298 are coordination variables. This number decreases rapidly as the iterative sequence progresses.

An example of the system's response at a given variation level of load Q_f is shown in Fig. 6 to 9.

Fig. 6 shows the control trajectories $U_1(t)$ and $U_2(t)$ of the boiler and of the exchanger. Control $U_1(t)$ is saturated for one minute at the beginning of the optimisation sequence and this is repeated regularly for various indicial tests. Conversely, the exchanger control is saturated only during tests with a sizeable variation of the set value or when the initial state of the system corresponds to the limits of the nominal operating range. This result justifies the approximations made when calculating the constraints on heat flow-rate $D_e(t)$ derived from the constraints on volumetric flow-rate Q_e and temperature T_{ce}.

Fig. 7 shows the outlets of the boiler and exchanger as well as the progression of the set value D_c (t). It will be noted that the system response no longer shows the oscillations observed with the conventional control system, using PID correctors combined with open-loop trend correctors.

Fig. 8 shows the corresponding development of temperature T_s for each of the two types of control : it will be seen that with the conventional control system, a distortion in the response occurs after two minutes, owing to interaction between the sub-processes.

Finally, although there is still room for improvement here, the convergence rate proves to be sufficiently rapid, even in the case of large variations in the load or in the set value. This is an important result, as the initial trajectory exercises a considerable influence on the subsequent development of the iterative trajectories. This result may be attributed partly to the ranking-coordination method chosen (4, 9) and partly to the correct estimation of the initial trajectories of $Z_1(k)$ and $Z_2(k)$ as a result of the linearity of the model in respect of calorific flow-rates. Fig. 9 shows the development of the coefficients of this model whose variation is sufficiently small and continuous to allow for a periodically adapted open-loop control.

CONCLUSION

The algorithm proposed for the multi-level open-loop control permits on-line control of the process, given that the physical coupling of the boiler and the exchanger is only achieved gradually. However the results illustrate the validity of the method when used with rigidly coupled processes as compared with a conventional control system. The more thoroughly the model of a pilot process is developed, the more convincing the results prove.

We note a resemblance between this algorithm and that tendered by Tamura (9), with respect to discrete systems : but the problems are formulated differently and, of particular interest, there is the fact that a coupling exists between the output and control variables. Moreover, predicting the interactions has the advantage that the initial trajectories of the physical variables $Z_1(k)$ and $Z_2(k)$ are easier to estimate. Now this estimation plays an important part in the speed of convergence of the algorithm.

In order to reduce the computing time and hence the optimal conditions, this method is currently being established on a multi-processor system which brings a micro-processor into action for each physical sub-process.

Finally we should stress that the validity of the control system is dependent of the accuracy of the heat flow-rate measurements, which might create difficulties in an industrial environment. Consequently, the extension of the present project is directed towards two objectives : firstly, with respect to the simultaneous estimation of the parameters and the linear heat-flow model variables, based on the measurements ; and secondly, with respect to the application of this method to the non-linear model determined on the basis of temperature and volumetric flow-rates.

REFERENCES

(1) Aubrun, M., G. Krzakala, J. Ragot, and M. Roesch (1976). Etude expérimentale de "quatre modes de régulation d'un processus thermique à grandes constantes de temps". Automatisme.

(2) Boudarel, Delmas, and Guichet (1969). Commande optimale des processus. Dunod. Paris. t. 2.

(3) Gilles, G. (1979). Catalogue de modèles de systèmes. "Modèles de procédés" de l'A.F.C.E.T. I.R.I.A.

(4) Hassan, M. (1978). Optimisation et commande hiérarchisée des systèmes dynamiques interconnectés. Thèse d'Etat. Toulouse.

(5) Javdan, M.R. (1976). On the use of Lagrange duality in multi-level optimal control. I.F.A.C. Symposium. Udine.

(6) de Larminat, P., and Y. Thomas. (1977). Automatique des systèmes linéaires. Commande. Flammarion Sciences, Paris.

(7) Legras, J. (1980). Algorithmes et programmes d'optimisation non linéaire avec contraintes. Application au contrôle optimal. Masson. Paris.

(8) Ratefiarison, G. (1980). Applications du filtrage à l'identification et à la commande des processus. Thèse de Doctorat de Troisième Cycle. Nancy.

(9) Singh, M.G., and A. Titli (1979). Systems. Decomposition, optimisation and control. Pergamon International Library.

(10) Titli, A. (1975). Commande hiérarchisée et optimisation des processus complexes. Dunod Automatique. Paris.

(11) Titli, A. (1979). Analyse et commande des systèmes complexes. Monographie A.F.C.E.T. Division A.I. Cepadues Editions

APPENDIX

Symbol	Meaning	Order of magnitude
T_{ce}	Boiler outlet temperature	60 to 100°C
T_r	Boiler inlet temperature	35 to 50°C
T_{cs}	Outlet temperature of the exchanger primary circuit	20 to 40°C
T_s	Outlet temperature of the exchanger secondary circuit	35 to 45°C
T_f	Temperature of the circulating water	7 to 12°C
Q_g	Gas-flow in the boiler	0.1 to 2 m^3/h
Q_c	Water flow-rate in the boiler	150 l/h
Q_e	Water flow-rate in the primary circuit of the exchanger	80 to 140 l/h
Q_f	Water flow-rate in the secondary circuit of the exchanger	100 to 220 l/h
K_g	Calorific power of the gaz	8400 $kcal/m^3$
T_{sc}	Set-value of temperature T_s	40°C
D_{fc}, X_{2c}	Set-value of heat flow-rate D_f	50 to 60 kcal/min
D_{cc}, X_{1c}	Set-value of heat flow-rate D_c	60 to 70 kcal/min
$X_1(t)$	Boiler outlet heat flow-rate $D_c(t)$	50 to 75 kcal/min
$X_2(t)$	Exchanger outlet heat flow-rate $D_f(t)$	45 to 65 kcal/min
$U_1(t)$	Boiler heat flow-rate control $D_g(t)$	15 to 300 kcal/min
$U_2(t)$	Exchanger heat flow-rate control $D_e(t)$	35 to 75 kcal/min
$\underline{X}_o = \begin{pmatrix} X_{10} \\ X_{20} \end{pmatrix}$	Initial values of $D_e(t)$ and $D_f(t)$	
$\underline{e} = \begin{pmatrix} e_1 \\ e_2 \end{pmatrix}$	Deviation between set-value and measurement of $D_c(t)$ and $D_f(t)$	
G_1	Maximum value of heat flow-rate $D_g(t)$	300 kcal/min
G_2	Minimum value of heat flow-rate $D_g(t)$	15 kcal/min
G_3	Maximum value of heat flow-rate $D_e(t)$	60 to 80 kcal/min
G_4	Minimum value of heat flow-rate $D_e(t)$	25 to 40 kcal/min
α	Weighting coefficient for control $D_g(t)$	10^{-3}

Boiler	Max. power	13 000 kcal/h
	Max. gas flow-rate	2.1 m^3/h
	Nominal flow-rate	150 l/h
	Gas pressure	27 millibars
Exchanger	Water content of primary circuit	0.6 l
	Water content of secondary circuit	0.5 l
	Nominal flow-rate Q_e	110 l/h
	Nominal flow-rate Q_f	160 l/h

TABLE 1 Process characteristics

	$\dot{T}_{ce}$	$\dot{T}_s$	$\dot{T}_{cs}$
Q_g	77.7		
T_{ce}	-4.01		
T_r	3.96		
$T_{ce}Q_e$		0.197	0.09
$T_{cs}Q_e$		-0.180	-0.09
$T_s\ Q_f$		-0.205	-0.09
$T_f\ Q_f$		0.174	0.07

TABLE 2 Matrix of coefficients on the non-linear model where the variables are expressed in degrees C , m^3/h or l/h

control \ outlet	$\dot{D}_c$	$\dot{D}_e$	$\dot{D}_f$
D_g	0.021		
D_c	- 0.055	0.50	
D_e		0.57	0.46
D_f			0.48

TABLE 3 Matrix of coefficients of the linear model in heat flow-rates expressed in degrees C/kcal/min

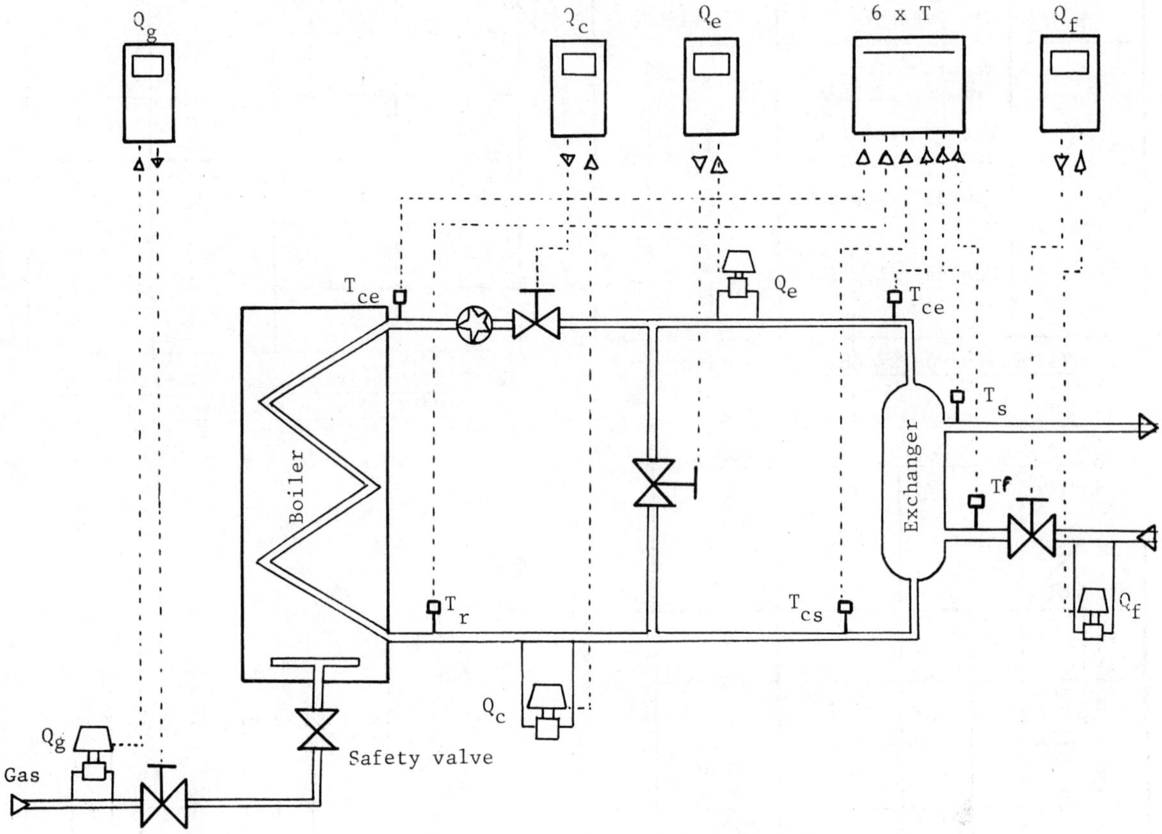

Fig. 1 Block diagram of o thermal nrocess

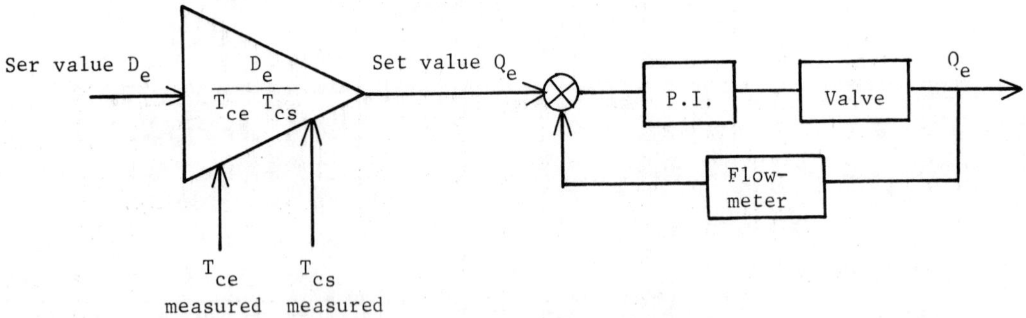

Fig. 2 Actual volumetric flow-rate control apparatus based on heat flow-rate.

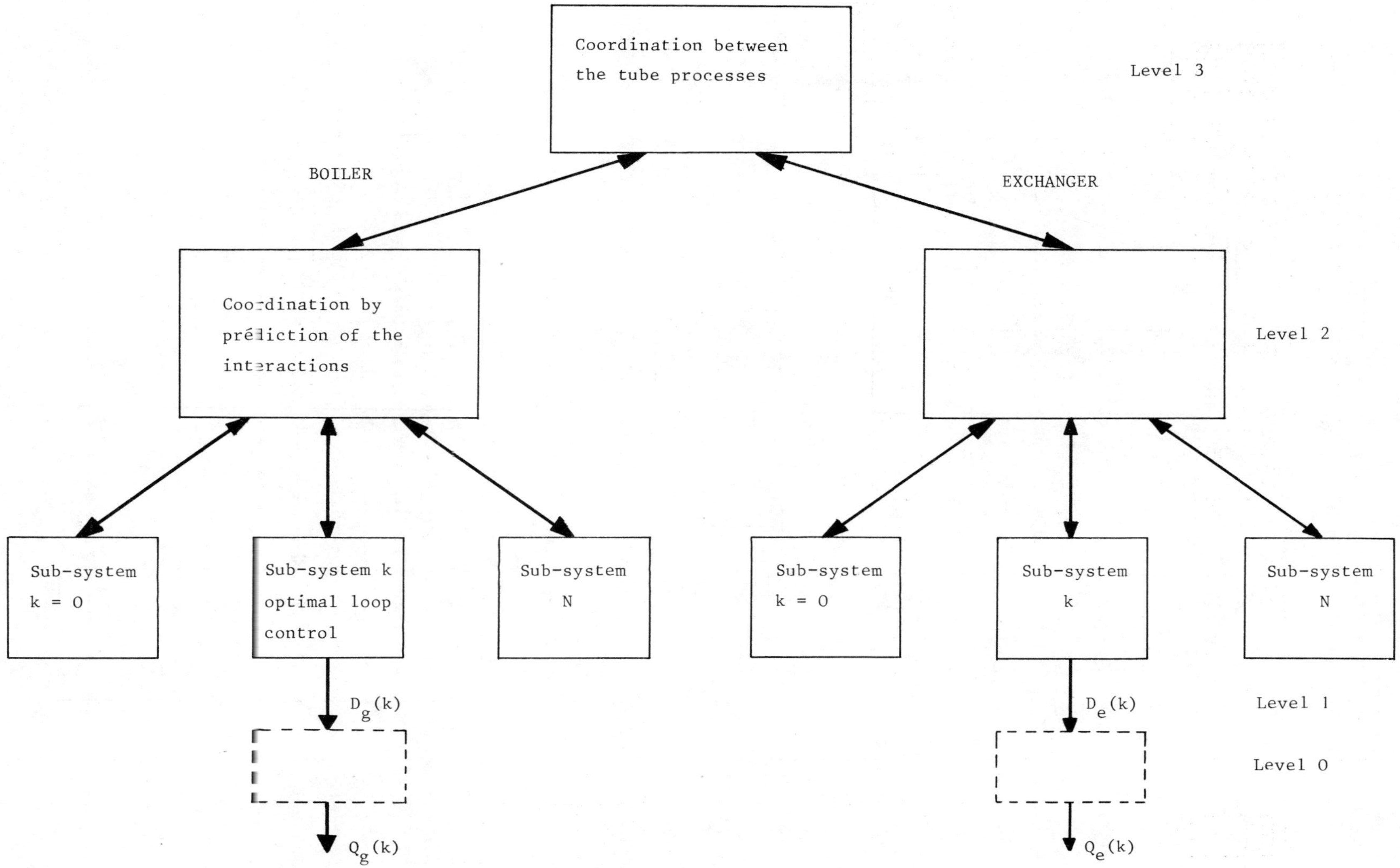

Fig 3 - Basic diagram of the multi-level decomposition - coordination

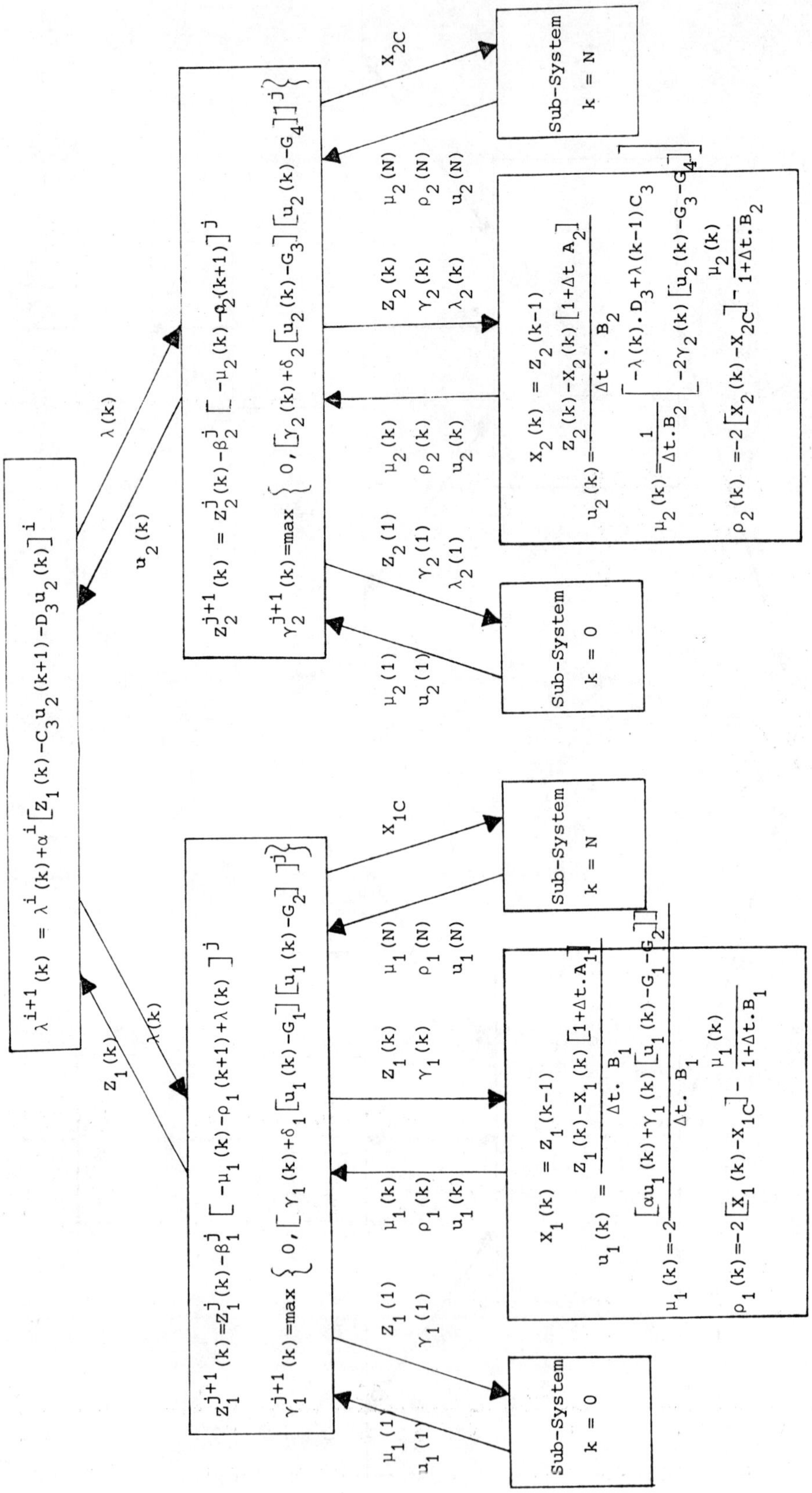

Fig. 4. Detail of the 3 level decomposition.

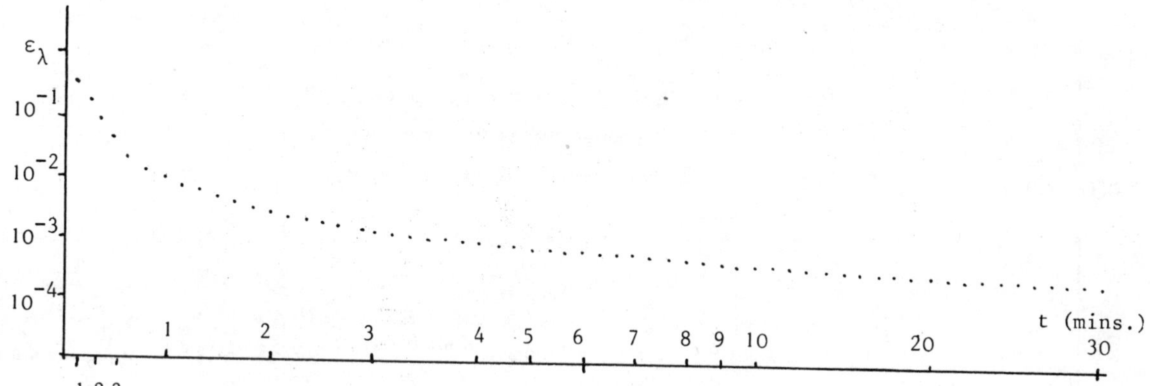

Fig. 5 Progression of threshold ε_λ overtime.

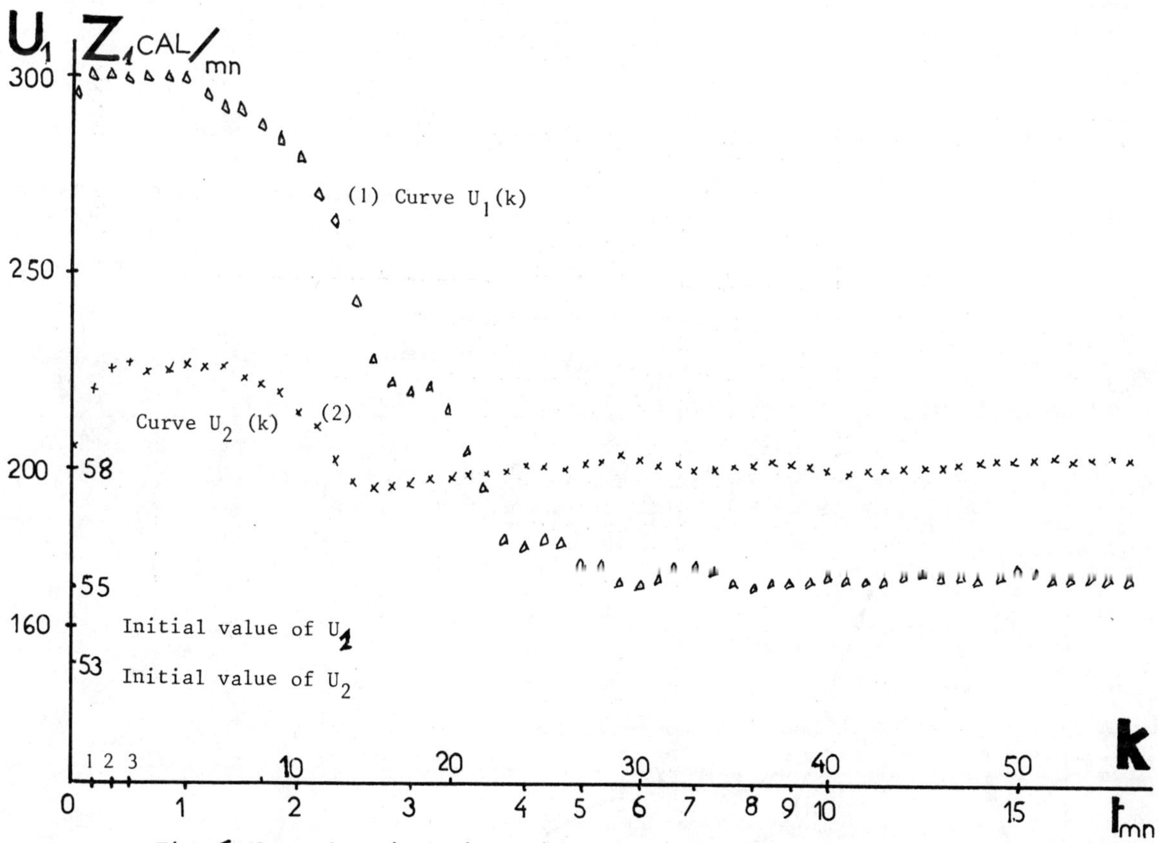

Fig. 6 Control trajectories $U_1(k)$ and $U_2(k)$ in response to a load level Q_f

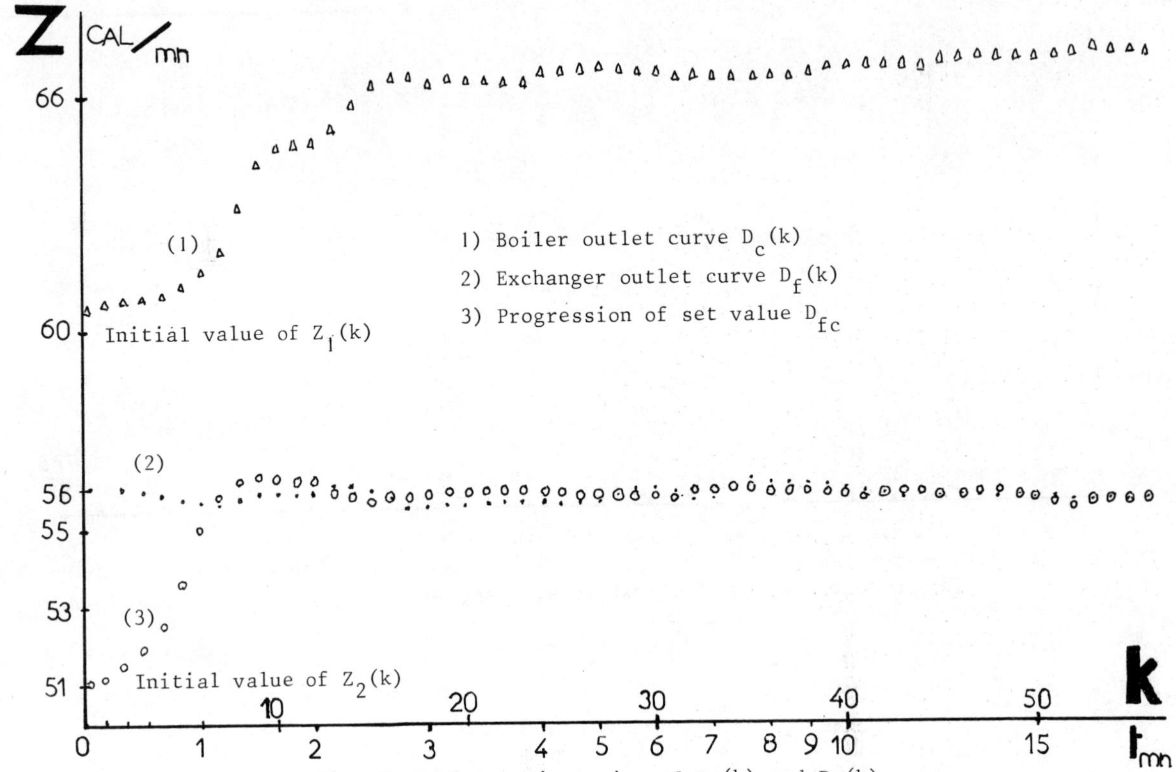

Fig. 7 Outlet trajectories of $D_f(k)$ and $D_c(k)$

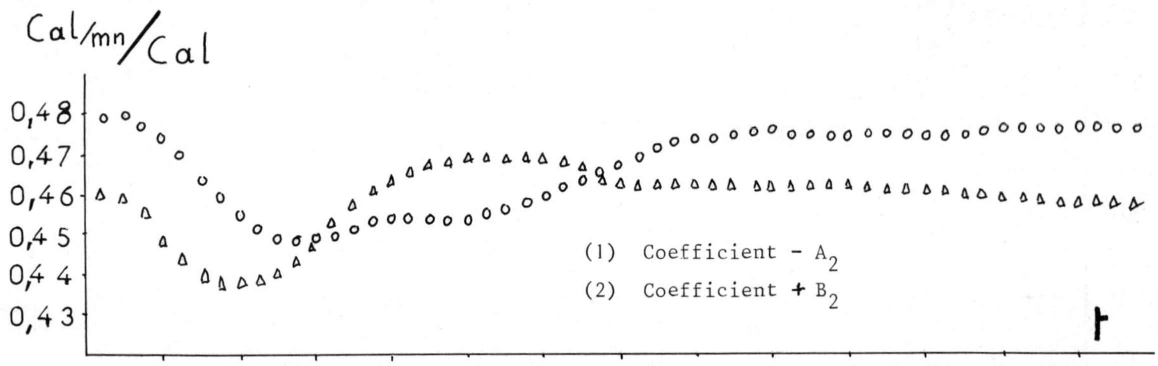

Fig. 8 Evolution of the temperature T_s. A comparison between multi-level control (1) and regulation by P.I.D. (2) (variation of the load Q(f)).

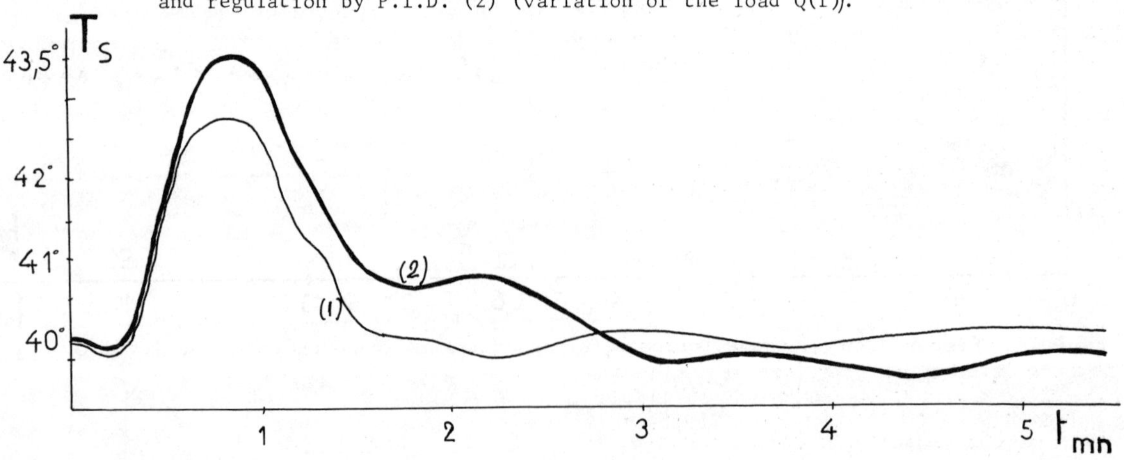

DISCRETE OPTIMISATION MODEL FOR SITING SUPER THERMAL POWER STATIONS

S. R. Gatta, T. Devnath and G. S. Adhar

Systems Engineering & Computer Services Division, Engineers India Limited, 26 Kasturba Gandhi Marg, New Delhi, India

Abstract. Long range power planning exercises have customarily treated generation and transmission planning in isolation. Location of thermal power stations are largely decided by factors such as the demand centers and movement of coal. For the presently executed super thermal power stations the practice has been to locate the generating plants close to the pit-heads to overcome the transportation bottlenecks. In light of the latest technological innovations and established practices elsewhere this paper examines the concept of pit-head generating stations.

In this paper a "systems approach" is presented for siting the super thermal power stations taking into consideration the economics of coal movement by rail/coastal shipping/slurry pipe-line transportation and Extra High Voltage transmission of electrical energy. Hitherto, studies on this subject have examined the relative economics of these modes between two given points. These results may not be applicable when the demand is distributed over a region and not concentrated at one load center.

In this paper the problem is formulated as a discrete optimization problem to take into account the spatial distribution of demand for electrical energy and also considers the existing transmission network. The cost function which is minimised includes the capital and operating costs of transportation and EHV lines, and transmission losses.

Finally, a case study is presented with representative data in the Indian context.

Keywords. Power generation; power transmission; slurry transportation of coal; digital computer applications; mathematical programming; modelling; optimisation; systems approach.

INTRODUCTION

In India at present there is a massive plan for power development and accordingly all the resources are mobilized for this purpose. The emphasis being on power generation based on coal-fired thermal power stations, the location of coal mines will have an important bearing on the choice of site for the power stations. For the presently contemplated large thermal power stations the practice has been to locate the generating stations close to the pit-heads.

Customarily, long range power systems planning is carried out in India treating generation system and transmission system planning separately. Location of coal fired thermal power stations have been largely decided by such factors as the availability of right type of coal, transportation infrastructure, consumption centers and density of load. The Central Electricity Authority (1974) (formerly a part of the Central Water and Power Commission), has recommended the location of super thermal power stations at pit-heads. CEA has found that it is cheaper to transmit electrical energy than to transport coal by rail. Likewise, the Fuel Policy Committee (1974) had recommended pit-head stations "in the overall interest of the economy and environmental consideration". It is suggested that this view needs to be reconsidered as the feasibility of coal transportation by slurry pipe line has been well established. A number of such pipe lines are in operation since 1970. This mode of transportation offers large economies of scale and becomes competitive to conventional systems where large quantities of coal movement are involved. It has been shown (Banks, 1975) in certain cases that slurry pipe line is economical compared to EHV transmission. Preliminary studies have indicated that in India slurry pipe line transportation is cheaper than power transmission by 400 KV EHV line for a

2,000 MW power station if the distance is more than about 100 KM.

The conventional approach is to compare EHV transmission of power vis-a-vis coal transportation by slurry pipe-line between two given points (Banks,1975) and work out the relative economics. Such studies ignore the existing transmission network and the generation/demand pattern, with the result the selection is non-optimal. In this paper, the authors suggest a systems approach to the solution of the location problem taking cognizance of the existing network configuration and related constraints. A discrete optimization model is developed to determine the optimal site and the associated transmission/transportation network. The cost function which is minimised includes the capital and operating costs of transportation and EHV lines, and transmission losses.

Finally, a case study is presented with representative data in the Indian context. The solution of the computer model developed above yields

(a) the most economical mode of transportation of coal to the optimum site,

(b) the transmission network augmentation plan, and

(c) the annualised minimum total system cost.

PROBLEM FORMULATION

A target year model is developed in which the total investment required to satisfy exogenously specified demand is minimised. In developing the model the following assumptions are made :

(a) Transmission losses are piecewise linearised.

(b) Spatial distribution of demand for power is exogenously specified.

(c) Reactive component of loads ignored, to reduce the complexity of the problem.

The model presupposes a set of technically feasible sites, out of which the optimal plant location is to be identified. The coal linkage for each of the feasible sites is assumed to have been established.

The input information consists of :

F_i annualised investment and fixed maintenance cost associated with location of the station at node i

C^l_{ij} annualised investment and fixed maintenance cost required for l^{th} type of line between the nodes i and j

G_i generation capacity already existing at node i (MW)

D_i load at node i (MW)

E_{ij} existing line capacity (MW) between nodes i and j

K^l_{ij} capacity of lth type of line (MW) between i and j

w_i unit cost of generation at generating port i (fuel cost only)

t^n_{ij} unit cost of transportation of coal between i^{th} node (mine) and j^{th} node (generating port) by n^{th} mode

All the costs are annualised. In the case of coal movement, the unit cost of transportation is computed based upon the quantum of coal required for one unit of electrical energy/power transferred. The type of line here refers to 220 KV (single circuit/double circuit/triple circuit/quadruple circuit), 400 KV (single circuit/double circuit/triple circuit/quadruple circuit), etc., or a transportation mode (rail/pipe line slurry transportation/coastal shipping). The objective is to arrive at the optimum site from amongst a given set of technically feasible locations, minimizing the sum of operating and investment costs.

MODEL DESCRIPTION

Definition of Variables

V_i Decision variable (zero-one) for location of super thermal station at node i
0-implies no station at node i
1-implies station to be located at node i
i-1,2,3,... (maximum value depending on the number of feasible locations)

R^l_{ij} Decision variable (zero-one) for l^{th} type of transmission line/transportation mode between nodes i and j. (type of line here refers to 220 KV S.C/D.C/T.C/Q.C; 400 KV S.C/D.C/T.C/Q.C; etc.)
0-implies no line between i and j
1-implies line to exist between i and j

P_{ij} Continuous decision variable for power flow from i to j

g_i Continuous decision variable for scheduled generation at generating port i

The transmission losses are a nonlinear function of the electrical power transmitted. This nonlinearity poses a serious problem in modelling and solution methodology. To circumvent this problem, the operating range of each transmission line is judiciously divided into two zones. In each zone, the losses are taken as a linear function of the power flowing in that line. After going through a number of trial calculations this approximation has been found to be realistic. Accordingly,

m denotes the region in which the power P_{ij} falls.

In a two-step approximation m takes the value of 1 or 2.

Objective Function

Minimise the total annualised cost

$$Z = \sum_i F_i V_i + \sum_{i,j,l} C^l_{ij} \; R^l_{ij} + \sum_i w_i \, g_i$$

System Constraints

- Power balance.

$$D_i \leq g_i - \sum_{jlm} P^l_{ijm} + \sum_{jlm} K^l_{ji\,m} \; P^l_{jim}$$

where K_{ji} = 1 - loss coefficient

- Line capacity :

$$P_{ij} \leq E_{ij} + \sum_l R^l_{ij} \; K^l_{ij}$$

- Generation capacity :

$$g_i \leq G_i \quad \text{for all } i$$

for all the existing generating stations. However, for the new super thermal power station this constraint is modified to

$$g_i = K \, V_i$$

where K is the plant factor multiplied by the plant capacity

- Unidirectional constraints :

$$\sum_l R^l_{ij} + \sum_l R^l_{ji} \leq 1$$

- As the decision is to choose only one location from among the sites specified, the constraint related to the number of plants is :

$$\sum_i V_i = 1$$

SOLUTION METHOD

The above discrete optimisation problem has been solved using a Mixed Integer Programming (MIP) package. The computer package, operational at EI computer center, uses a branch and bound technique.

CASE STUDY

The applicability and potential of the above model is illustrated by using representative network data. This network comprises of a number of contiguous State Electricity Boards of the Eastern Region of India. The network is aggregated to twenty three nodes.

The projected network of 1983-84 is taken as the 'existing' network. Fig. 1 indicates the network used in the model. It also shows the peak power demand and the available generation capacities at each node. Four alternative sites *viz.*, Farakka, Santaldih, Zeerat and Tenughat are considered for locating the thermal power stations, the installed capacity being 2,000 MW.

It is assumed that the source of coal is same for all the four alternatives and is transported by slurry pipe line. One of the locations is very close to the pit-head, while others nearer to load centres, distances ranging from 50 to 250 Km from the coal mines. The following alternatives were considered for augmenting the transmission system.

i) 220 KV S.C and D.C
ii) 400 KV S.C and D.C

Table 1 (Rikh, 1976) indicates the line cost data used in the problem. The investment costs of slurry pipe line used in this study are based on the literature available in this area.

Assumptions

The coal requirements of a 2000 MW station is estimated as 7.15 million tonnes, based on the characteristics of Rajmahal coal. A pump station is assumed to be located at every 100 Km.

The capital expenditure is assumed to be phased over three years of the project in the ratios of 20%, 30% and 50%. The annualised costs are computed using a life period of 35 years (for both pipe line and transmission systems) and a discount rate of 10%.

The other assumptions are :

- Generation cost is 5.5 paise/KWHr, uniform at all the stations
- Load factor is 70%; that is, 1 MW corresponds to 0.7 x 8760 MWHrs of energy per annum
- O & M cost of transmission line is 2% of capital cost
- Maintenance cost of pipe line is 1% of capital cost

- Scheduled generation at the proposed super thermal station is 1,500 MW

- Pipeline distances used are

Farakka - to - Rajmahal	30 km.
Rajmahal - to - Santaldih	180 km.
Farakka - to - Santaldih	180 km.
Farakka - to Hawrah (Zeerat)	285 km.

Initially, a number of new links and augmentation of existing links are considered for expanding the transmission network. The study is carried out assuming four alternative plant locations, twenty four new links and thirteen augmentations. At first, the problem is attempted treating all variables as continuous and solving by LP technique. The size of the LP model is -

158 constraints, and
261 variables (all continuous type)

The solution thus obtained helps to reduce the MIP problem to a manageable size. The reduced size of the MIP problem is

64 constraints, and
162 variables (both integer and continuous type)

DISCUSSION OF RESULTS

Figure 2 indicates the optimal solution to the plant location problem. The network shows the new links that are required and the augmentation of existing links to meet the peak power demand at each node. Table 2 compares the total investment required for all the four alternatives. Locating the power station at Santaldih yields the minimum investment condition, the optimum annualised total cost being Rs. 129.7 millions. This, however, does not include the cost of generation.

It can be seen that location of the power station based on transportation of coal alone is erroneous and the total system cost should be considered. The augmentation required depends both on the location of new station and the existing network. The optimal location of the power station is thus sensitive to the input network which is an important parameter.

LIMITATIONS

The model presented in this paper makes a substantial contribution by incorporating the effect of line losses. The losses have been piecewise linearised, which assumption is reasonably valid in a majority of the situations. This factor should, however, be checked in each specific case.

The model is run for a specific target year, which makes it static. The optimization model needs to be dynamic to take the evolutionary nature of the EHV grid, energy demand and coal availability.

This model does not incorporate security and reliability constraints. The reactive component of the electrical demand has not been considered. This is to keep the dimensions of the problem to a manageable size.

The results of this approach should be validated by simulation studies for load flow, stability and reliability of the augmented network.

CONCLUSIONS

Systems approach to the power plant location problem is presented in this paper. This method automatically takes the slack in the power grid into account to arrive at the optimum cost of network augmentation and transportation facilities. The economic trade-off between generating 2,000 MW near the coal mine vis-a-vis generation nearer the load center is determined by the model. The minimised cost is obtained after considering such factors as the capital and operating costs, line losses and project time phasing.

ACKNOWLEDGEMENT

The facilities provided by the management of Engineers India Ltd. (EI) are acknowledged. The authors wish to thank the management of EI for having given permission to publish this paper.

REFERENCES

Central Water and Power Commission, Government of India, (1974). Strategy for accelerating Power Development and Evolution of All India Grid at 400 KV by 1983-85.

Planning Commission, Government of India (1974). Report of the Fuel Policy Committee.

Banks, A. J. (1975). Pipeline Beats Transmission, Published for Ebasco Services Inc., U.S.A.

Engineers India Limited (1977). Hydraulic transportation of Solid through Pipe Line; Its development & potential in India. Presented at seminar on Transportation in Mines and Bihar Coal Fields held at Dhanbad.

Garver, Len L. (1971). Transmission Network Estimation using Linear Programming. IEEE Trans. Power Appratus and Systems.

Rikh, V.N. (1976). Economic Trends in High Voltage System Development. Journal of Institution of Engineers India, Vol 57, Part-EL 3.

TABLE 1 Transmission Line Cost Data

Line Type	Line Cost Rs. million/Km	Terminal Equipment Cost Rs. (million)
132 KV S.C	0.102	0.053
132 KV D.C	0.179	0.088
220 KV S.C	0.184	0.098
220 KV D.C	0.317	0.152
400 KV S.V	0.381	0.289
400 KV D.C	0.686	0.470

TABLE 2 Results of Optimisation Problem

Annualized Cost-Rs. (million)	Location			
	Santaldih	Farakka	Zeerat	Tenughat
Pipeline	75.16	26.74	103.97	25.28
EHV transmission	15.82	84.73	28.7	81.16
Line Losses	38.76	32.98	37.02	32.47
Generation cost (A)	1920.52	1920.52	1920.52	1920.52
Total (B)	2050.25	2064.96	2090.21	2100.41
(B) - (A)	129.73	144.45	169.69	179.89

(Generation costs have been taken uniformly same for all the four alternative cases.)

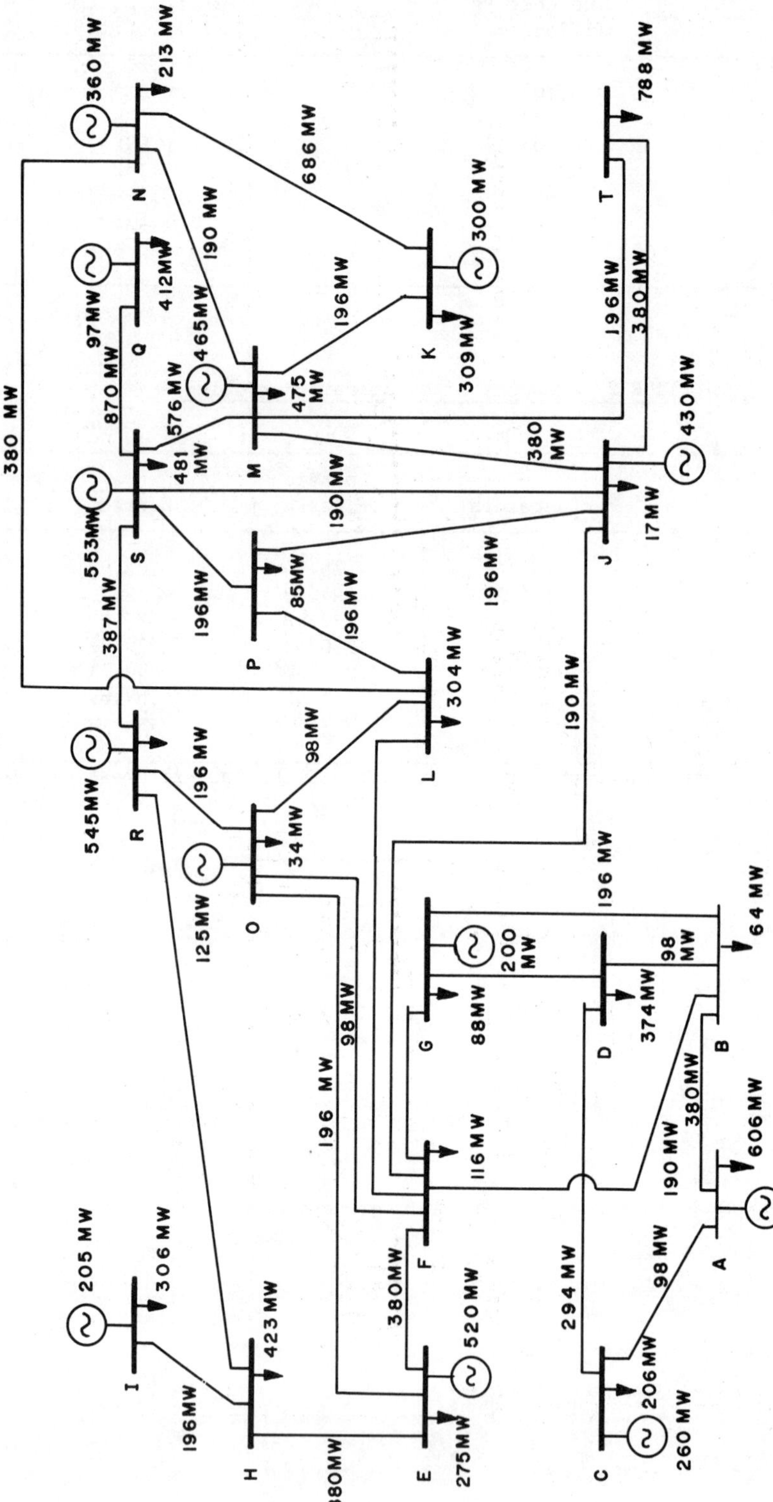

Fig. 1 Existing network configuration

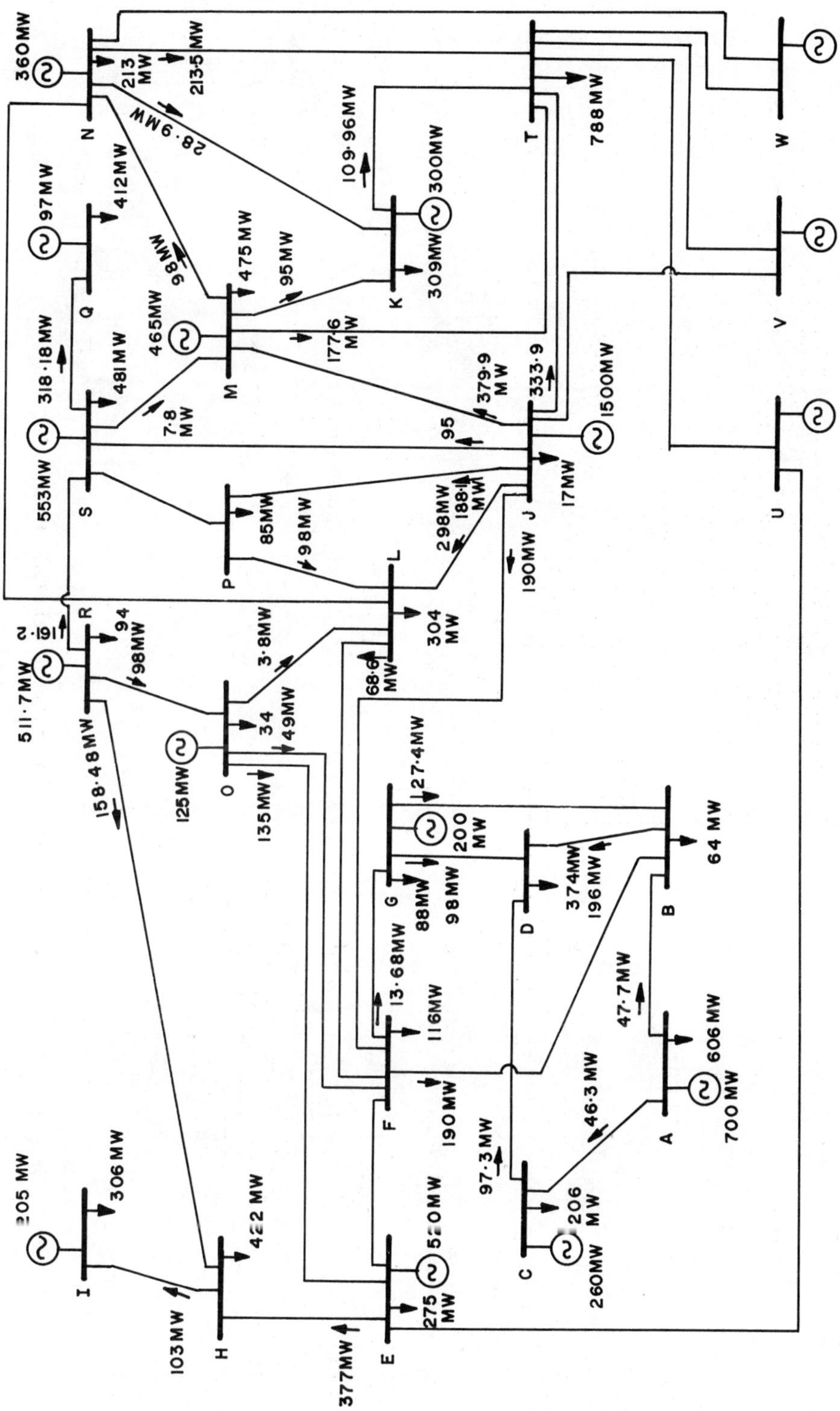

Fig. 2 Optimally augmented network configuration

OPTIMAL CONTROL OF CUTTING PROCESSES

A. S. Pogrebnyakov*, V. A. Gerasimov*, G. A. Dorrer, V. Ya. Glizer** and M. G. Dmitriev*****

**Production Centre "Krasnoyarsklesoexport", Krasnoyarsk, USSR*
***Siberian Technological Institute, Krasnoyarsk, USSR*
****Computing Centre of Siberian Branch of the Academy of Sciences, Krasnoyarsk, USSR*

Abstract. The mathematical model of a technological process used to cut several kinds of the raw materials for the manufacture of products is discussed. The optimal control of this process is formulated as a discrete optimal control problem. The model takes into account the dynamic law of the raw materials delivery and restrictions to its resources, the targets to the required assortment of products to be manufactured in the given terms, the nonlinear restrictions to the control. The criteria for the solution of the problem are maximizing of a conventional price of the product obtained from the raw material and maximizing of the output product flow uniformity. The control choice is made on the basis of a given set of the technologically available cutting schemes. The methods suggested for the solution of this problem include the base (dispatcher's) solution making followed by its improvement by the established criteria.

Keywords. Optimal control; cutting processes; optimization; iterative methods; management systems; mathematical programming; multivariable systems.

INTRODUCTION

Cutting the raw materials for the purpose of manufacturing the given product assortment is an important tecnological element of many industrial branches such as chemistry, paper industry, timber industry etc.

The cutting optimization problems have acquired special importance at the modern stage of productive forces development owing to the necessity of every kinds of the raw material economy - on the one hand and the variety of kinds of the raw materials used, the wide nomenclature of possible cutting schemes, the severe technological and target restrictions - on the other hand.

The model suggested takes into account this cutting process peculiarities and in difference to the known ones (Cantorovich, Zalgaller, 1971) includes restrictions to the simultaneous production of the several kinds of products combined with the dynamic law of the raw material delivery.

Our method of solving this model permit the optimization of cutting some kinds of raw material satisfying the technological and target restriction to the manufacture of several products.

MATHEMATICAL MODEL

The mathematical model of the cutting process can be formulated as the following discrete optimal control problem.

Find the control $u(t)$ and the corresponding phase trajectory $X(t)$, provided the optimum of criterion

$$max\, F_1 = C' \sum_{t=1}^{T} Y(u(t)) \qquad (1)$$

with the following restrictions:

a) To the material balance of products

CSAD - L

$$X(t+1)=X(t)+Y(u(t))-X^e(t) \quad (2)$$

$$X(1)=X^\circ,\ X(t)\geqslant 0,$$

$$Y(u(t))=A\,dg\{B'Q^M(t)B\}\,u(t);$$

b) to the plan of product availability

$$X(t_s+1)\geqslant X^s,\ s=1,\dots,I,\ t_I=T; \quad (3)$$

c) to the control

$$B\,u(t)=sg\{Q(t)\},\ t=1,\dots,T, \quad (4)$$

$$H'sg\{Y(u(t))\}<H'\varepsilon^N, \quad (5)$$

$$\varepsilon^{N'}sg\{Y(u(t))\}\leq d. \quad (6)$$

Model (1) - (6) is adequate enough to many real cutting processes where control $u(t)$ is selected on the basis of the given set of cutting schemes.

From the numerical realization of the control choice the model (1) - (6) has been transformed to the following form.

Find the function which provides the optimum of the criterion

$$\max F_2(u(t))=c'\sum_{t=1}^{T}Y(u(t))-\beta_o\sum_{s=1}^{I}\bar{X}^{s'}\bar{X}^{s} \quad (7)$$

with the restriction

$$u(t)\in\Delta_t, \quad (8)$$

where $\beta_o\geqslant 0$ is a coefficient of the penalty function;

$$\bar{X}^s=X^s\dot{-}\left(\sum_{t=1}^{t_s}Y(u(t))-\sum_{t=1}^{t_s}X^e(t)+X^\circ\right);$$

Δ_t is the set of controls $u(t)$ satisfying the conditions (4) - (6) at t. The capacity of this set $|\Delta_t|$ is designated by $\Lambda(t)$. Computation of $\Lambda(t)$ is made with the special recurrent procedure. For the real problem which consists of 20 kinds of products, 10 kinds of the raw materials and 100 cutting schemes have showed that $\Lambda(t)$ may be more than $4\cdot 10^6$

The model (7) - (8) takes the restrictions (2) - (3) into account by means of the penalty addition to the criterion (1) and restrictions (4) - (6) by means of the set Δ_t introduction.

The vectors $u^\lambda(t)\in\Delta_t$, $\lambda=1,\dots,\Lambda(t)$ determine correspondihg vectors $Y^\lambda(u(t))$ in N-dimensional space of variables $y_1,\dots,y_N$. The sequence $Y^\lambda(u(t))$, $\lambda=1,\dots,\Lambda(t)$ generates the local set accessibility Ω. This set shows what combinations of the different production volumes may be produced during a planning time discrete.

The function $u(t)$ definition is fulfilled in the two stages:

a) generation of a base (dispatcher's) solution which satisfies to the restrictions (8);

b) the base solution optimization based on the criterion (7).

If the set of the problem (7) - (8) solutions consists of more then one element then the best one must be selected on the basis of the criterion

$$\min F_3=$$

$$=\sum_{t=1}^{T-1}\varepsilon^{N'}\left|sg\{Y(u(t))\}-sg\{Y(u(t+1))\}\right|. \quad (9)$$

The solution of the problem (7), (8) may be multiple, for example, when the interval $(t_s+1,\ t_{s+1})$ contains at least two discretes τ_1 and τ_2 which satisfy $Q(\tau_1)=Q(\tau_2)$.

Algorithm for the problem (7) - (8) solution is based on the application of such peculiarities of the set Δ_t structure that permit to obtain the followihg optimization problems on effective solution.

Find the vector $u^\lambda(t)$ which gives the optimum of the criterion

$$\max F_4=\nu'Y(u^\lambda(t)) \quad (10)$$

with the restrictions

$$u^\lambda(t)\in\Delta_t,\ t=1,\dots,T, \quad (11)$$

where ν is N-vector determined in the process of the algorithm realization.

MAKING OF A BASE SOLUTION

The effeciency of the process of making a base solution depends sufficiently on the specifics of the given function $Q(t)$ and the number of restrictions (3) in real cutting problems. One of the typical cases is the smooth delivery of the raw materials, when $Q(t) = Q_g$ = Const. The number of restrictions (3) is determined by the number of control terms t_s, $s = 1,\dots,I$ during the plan period (for example, quarterly, monthly, decade or twenty-four hours division of the annual production plan).

A separate case is a single control term when the plan of the products availability must be fulfilled at the last discrete of the plan period:

$$X(T+1) \geqslant X^I.$$

Three methods have been worked up to provide efficient procedure of the base solution, namely;

a) SELECT - a method of a base solution making with the basis of a control $u(t)$ selection procedure (Ivakhnenko, Zaichenko, Dimitrov, 1976). This method is useful when $Q(t)$ is an arbitrary given function;
b) LINPRO - this method permits to compute the optimal intensities of the vectors $u(t)$ use when $Q(t) = Q_g$;
c) EXDEC is an heuristic method. It is useful when $Q(t) = Q_g$ and the vector of the required product availability is given only in the end of the plan period.

Let us consider briefly the main propositions on which the mentioned methods are based.

The SELECT Method

Let the G functions $u_g(\tau)$, $g = 1,\dots, G$, $\tau = 1, \dots, t-1$, be known at the discrete t. Then among the possible continuations in the interval $\tau = 1,\dots,t$ one must select G ones that give the best value of the criterion

$$\max F_5 = c' \sum_{\tau=1}^{T} Y(u(\tau)) - \sum_{s=\sigma}^{I} \beta(s,t) \bar{X}_t^{s'} \bar{X}_t^{s} \quad (12)$$

in which the penalty function depends on the time

$$\beta(s,t) = \beta_0 \frac{2^{I-s}}{(2^{I-\sigma+1}-1)(t_s - t + 1)} \quad (13)$$

$$\bar{X}_t^s = (X^s + \sum_{\tau=1}^{t} X^e(\tau)) \dot{-} (\sum_{\tau=1}^{t} Y(u(t)) + X^0), \quad (14)$$

where σ is the minimum s that $t \leq t_s$.

The selection is realized in the two stages. In the first one from each function $u_g(\tau)$, $g = 1, \dots, G$, G continuations are generated which are the best on the criterion (12). For some function $u_{g_0}(\tau)$ the criterion (12) takes the following form

$$\max F_6 = c' Y(u^\lambda(t)) - \sum_{s=\sigma}^{I} \beta(s,t) W_t^{s'} W_t^s \quad (15)$$

with restrictions

$$u^\lambda(t) \in \Delta_t, \ t = 1,\dots,T, \quad (16)$$

where $W_t^s = (\bar{X}_{t-1}^s (u_{g_0}(\tau)) + X^e(t)) \dot{-} Y(u^\lambda(t))$.

For the solution of this problem the gradient of function (15) at point $Y=0$ must be determined.

$$\nu(t, g_0) = c + 2 \sum_{s=\sigma}^{I} \beta(s,t)(\bar{X}_{t-1}^s + X^e(t)). \quad (17)$$

The selection of $u^\lambda(t) \in \Delta_t$ which gives a maximum value of the criterion F_6 is a problem (10) - (11) when ν is determined by (17).

The required G continuation of the functions $u_{g_0}(\tau)$ on the time discrete $\tau = t$ are obtained by means of the modified algorithm of the problem (10) - (11) solution.

The LINPRO Method

As $Q(t) = Q_g$ all the sets Δ_t, $t = 1,\dots,T$ are equal and determined by Δ, $|\Delta| = \Lambda$.

The optimal intensities of the vectors $u^\lambda \in \Delta$ use in the intervals $(t_{s-1}+1, t_s)$, $s = 1, \dots, I$ may be determined as solution of the following problem.

Find $z_{s\lambda}$, $s = 1,\dots, I$, $\lambda = 1,\dots,\Lambda$ which provides optimum of the criterion

$$\max F_7 = c' \sum_{s=1}^{I} \sum_{\lambda=1}^{\Lambda} z_{s\lambda} Y(u^\lambda) \quad (18)$$

with the restriction

$$\sum_{\lambda=1}^{\Lambda} \sum_{\tau=1}^{s} z_{s\lambda} Y(u^\lambda) \geqslant (X^s + \sum_{t=1}^{t_s} X^e(t)) \dot{-} X^0 \quad (19)$$

$$\sum_{\lambda=1}^{\Lambda} z_{s\lambda} = t_s - t_{s-1}, \ t_0 = 0, \ s = 1,\dots,I. \quad (20)$$

Unequalities (19) are transformed into equalities with an additional variables introduction. The restriction matrix is formed in the bloc-diagonal form with the vertical border by means of algebraic transformation. Then the dual problem is solved by Dantzig-Wolfe method (Lasdon, 1975). The solutions of the problems which correspond to the diagonal bloc of the initial matrix is made with the columns generation procedure. A column included into the basis at the current simplex-method iteration is the problem (10) - (11) solution, the vector ν being determined during the algorithm execution.

The optimal $z_{s\lambda}$ solution obtained is approximated to integer so that the conditions (20) are satisfied. Then the base solution is formed on the basis of non-zero volumes of the intensities $z_{s\lambda}$.

The EXDEC Method

If $Q(t) = Q_g$, $t = 1,\dots,T$ as before and the plan availability vector is given only in the end of the term, the making of base solution is realized in the two stages: a)the optimal intensity of the given cutting schemes achievement; b)the base solution $u(t)$, $t = 1,\dots,T$, is formed on the basis of non-zero intensities.

The intensity of i cutting scheme use is a number of the raw materials units r_i that have to be cut with i scheme.

The optimal intensities determination is the following problem.

Find M-vector R which gives

$$max\ F_8 = C'AR \tag{21}$$

with restrictions: a)to the products manufacture

$$AR \geqslant X^I + \sum_{t=1}^{T} X^e(t) - X^o, \tag{22}$$

b)to the raw materials use

$$BR \leqslant TQ_g \tag{23}$$

It may be necessary to change some elements of the vector R found for the base solution forming. This variation is fulfilled by means of the heuristic procedure whith an expert's help.

BASE SOLUTION OPTIMIZATION ON THE BASIS OF THE CRITERION (7)

The base solution of the problem (7) - (8) irrespective of the way of this ashievement must be improved on the basis of the criterion (7) with the help of the next modes.

The Coordinate Optimization Method

The criterion (7) is considered to be a function $F(u^1,\dots,u^T)$ of variables $u^t \in \Delta_t$, $t = 1,\dots,T$.

The method is realized with the following iterative procedure. An each iteration T sequent improvings of the criterion (7) are made by means of the next coordinate u^t, $t = 1,\dots,T$ variation. Let $w^1,\dots,w^T$ be the values of the variables $u^1,\dots,u^T$ at the beginning of the next iteration and $\bar{w}^1,\dots,\bar{w}^{t-1}$ be the values of the first $t-1$ changed coordinates. The value of the t coordinate w^t is obtained on the basis of the next problem solution.

Find

$$max\ F_9 = F(\bar{w}^1,\dots,\bar{w}^{t-1}, w, \bar{w}^{t+1},\dots,\bar{w}^T) \tag{24}$$

with restrictions

$$w \in \Delta_t . \tag{25}$$

The solution of the problem (24) - (25) is realized similare to the problem (15) - (16). The F increase at each iteration is determined as difference $F(\bar{w}^1,\dots,\bar{w}^T) - F(w^1,\dots,w^T)$. If $\Delta F = 0$ then the optimization process is completed. The solution obtained in this way is optimal in the criterion (7) at each coordinate $u^t, t = 1,\dots,T$.

The Variation of the Phase Trajectory Interval Method

The intervals L_p so as the vector $sg\{Y(u(t))$ is constant when $t \in L_p$, $p = 1,\dots,P$ are determined inside every fixed term t_{s-1} and t_s , $s = 1,\dots,I$.

This method provides for the variation of the trajectory $Y(u(t))$ corresponding to the base solution in the intervals L_p. These variations are based on the solution of the following linear programming problem.

Find

$$min\ F_{10} = R'\mathcal{E}^M \tag{26}$$

with restrictions

$$BR \leqslant \sum_{t \in L_p} Q(t), \tag{27}$$

$$AR = \sum_{t=1}^{T} Y(U(t)) \dot{-} \gamma^{P}, \quad (28)$$

where $\gamma^{P} = \min_{s \in [\sigma, I]} [(\sum_{t=1}^{t_s} Y(U(t)) + X^{0}) \dot{-} (X^{s} + \sum_{t=1}^{t_s} X^{e})]$.

There σ is the minimum s so that $L_P \subset [1, t_s]$.

The solution of the problem (26) - (28) gives the same degree of the restrictions (3) satisfaction as a base solution, but perhaps at the less raw material expense. The ways of the release of the raw materials use is determined by an expert who takes part in the optimization process.

In case $Q(t) = Q_g, t=1,\dots,T$ the optimization with the method effeciency may be increased by the intervals L_P prolongation by means of $U(t)$ are transposition inside the interval $[t_{s-1}+1, t_s]$.

THE SOLUTION OPTIMIZATION ON THE CRITERION (9)

Optimization on the criterion (9) of the solution obtained in the previous stage is a procedure of increasing such an important cutting process feature as the production assortiment uniformity.

The suggested optimization method may be usefull when there are multiple intervals where the function $Q(t)$ is constant. The essence of the method at $Q(t) = Q_g, t=1,\dots,T$ is as follows.

If the distance between the vectors $u(t_1)$ and $u(t_2)$ is determined by the expression

$$\delta(u(t_1), u(t_2)) =$$

$$= e^{N'} | sg\{Y(u(t_1))\} - sg\{Y(u(t_2))\},$$

then the problem of the criterion (9) minimization is the generalization of the salesman problem. All the vectors $u(t), t=1,\dots,T$ are devided into the groups which correspond to the time periods $[t_{s-1}+1, t_s]$. The sequence of the groups passing is given. The shortest route which is passed through every vector only once and retains the groups passing sequence must be obtained.

This problem may be transformed to the simple salesmans problem with the change of the distance matrix.

The above procedure completes the optimal solution of the initial problem (1) - (6) determination.

The methods described have been used many times for the solution of the real industrial cutting operation problems.

NOMENCLATURE

T - number of time intervals (discretes) in the planed period;

N - number of the kinds of products;

M - number of the given cutting schemes;

K - number of the raw material kinds;

B - $K \times M$ Boolean matrix; its element $b_{ij} = 1$ if j scheme may be applied to cutting of i raw material kind, otherwise $b_{ij} = 0$; each cutting scheme may be used only for one raw material kind;

A - $N \times M$ matrix; its element a_{ij} is expected volume of i product manufacture when one unit of raw material is cut by j scheme;

H - $L \times N$ Boolean matrix; every of its lines determines the set of products kinds the simultaneous manufacture of which is not allowed, thus sets $\{i: h_{li} = 1\}$, $l = 1,\dots,L$ are prohibited;

d - maximum number of simultaneously manufactured product kinds, $d < N$;

$Q(t)$ - given K-vector-function, its element $q_i(t)$ is the i kind of raw materials delivery for cutting at t;

$Q^{M}(t)$ - $K \times K$ diagonal matrix, $q^{M}_{ii}(t) = q_i(t)$;

C - N-vector of production prices;

$U(t)$ - Boolean M-vector, $u_j(t) = 1$ if j cutting scheme used at t time discrete, otherwise $u_j(t) = 0$;

$Y(U)$ - -vector of expected product volumes manufactured at t time discrete;

$X(t)$ - N-vector of product volumes availability in the beginning of t time discrete;

X^{s} - given N-vector of the required by the plan product volumes at times t_{s+1}, $s=1,\dots,I$; $t_1,\dots,t_I = T$ - given terms;

$X^{e}(t)$ - given N-vector of product dispatching;

$\mathcal{E}^{p}$ - p-unit vector, $\mathcal{E}^{p}_{i} = 1$, $i=1,\dots,p$;

$a \dot{-} b = \max\{0, a-b\}$;

$sg\{a\}=\begin{cases}1 & \text{if } a>0,\\ 0 & \text{otherwise;}\end{cases}$

$dg\{F\}$ - N -vector formed from the diagonal elements of $N \times N$ matrix F;

Some algebraic operations applied to the vectors are understood as the operation with its elements. Thus

$$|A-B|=(|a_1-b_1|,\ldots,|a_p-b_p|)',$$

$$min(A,B)=(min(a_1,b_1),\ldots,min(a_p,b_p))'.$$

REFERENCES

Cantorovich, L. V., V. A. Zalgaller (1971). Rational Cutting of Industrial Materials. Nauka, Novosibirsk, 299 pp.

Ivakhnenko, A. G., Yu. P. Zajchenko, and V. D. Dimitrov (1976). Decision making on the Basis of Self-Organization. Soviet Radio, M. 280 pp.

Lasdon, L. S. (1975). Optimization Theory for Large Systems. Nauka, M. 432 pp.

APPLICATION OF A MINICOMPUTER FOR DIRECT NUMERICAL CONTROL OF MULTIPLE MACHINE TOOLS

B. Parhami and K. Alvandi

Department of Mathematics and Computer Science, Sharif University of Technology, Tehran, Iran

Abstract. Numerical control is the directing of a process through the execution of a pre-coded instruction sequence. Direct numerical control refers to the real-time control of machine tools by the direct coupling of a computer to one or more machine tool controllers. In this paper, a specific direct numerical control system is presented and the corresponding implementation issues are discussed. This system uses an HP/2100S minicomputer with a 16K main memory to control up to 16 machine tools. Blocks of information are sent upon request to the machine tools, via a common bus, and the status of the entire system is maintained in tables in order to enable the control program to keep track of each part program execution. The presentation of the system's hardware and software features is followed by a discussion of the host support system which consists of programs written for a CDC Cyber 74 computer to facilitate the development and maintenance of part program libraries. Several possible extensions to the system, including the incorporation of fault tolerance features, are pointed out in conclusion.

Keywords. Computer control; Digital computer applications; Direct digital control; Industrial control; Machine tools; Mincomputers; Numerical control; On-line operation; Process control.

INTRODUCTION

Numerical control of machine tools is one of a class of applications where digital hardware (with supporting software) is used to control the sequencing of steps in a physical process. This is normally done by following a prespecified sequence of instructions and observing the feedback information derived from an instrumentation of the controlled process. Considerable flexibility and speed of operation can be achieved if the controlling hardware is directly linked to the numerically controlled (NC) machine tool, thereby obviating the need for an intermediate information transfer medium. This is commonly referred to as direct numerical control (DNC).

There are two main types of NC machine tools known as point-to-point and continuous-path machines. In point-to-point NC machines (e.g., a simple drill press), the cutter and the work-piece remain at fixed relative positions while the cutter does its work. Such a machine tool performs its operations (e.g., drilling) at specific positions and does not affect the work-piece when moving from one position to the next.

In contrast, the functioning of a continuous-path NC machine tool involves motion of the work-piece with respect to the cutter while the cutting operation is taking place. Whereas in point-to-point machines the path taken between operations is of no consequence, the entire motion must be controlled to close accuracy with respect to position and cutting speed in the case of continuous-path (contouring) NC machines. Consequently, the needed control system is more complicated and, hence, more costly.

The information processing needs of both point-to-point and continuous-path NC applications are totally within the power of even the smallest micro-processors on the market. However, if extensive support software is desired, the storage requirements may not be attainable with existing microprocessor-based systems. A possibility which is explored in this paper is the use of a mini computer to service a number of NC machines (Alvandi, 1977). This approach enables the user to develop the needed software support tools and also make use of a large number of facilities available on today's minicomputers.

An extension of this approach leads naturally to a distributed network of microprocessors, each dedicated to a single machine tool and capable of receiving and processing short sequences of instructions from a central computing facility. Such an arrangement will not only lead to improved reliability (particularly if each microprocessor is provided with some form of permanent storage; e.g., cassette tape and drive) but also will enable the minicomputer to devote most of its resources to other information processing activities within the plant.

SYSTEM STRUCTURE AND HARDWARE

The proposed DNC system consists of three main subsystems: (1) A Control Data Cyber 74 computer used as the host support system to provide facilities for developing and maintaining part program libraries; (2) A Hewlett-Packard 2100S minicomputer used to control the NC machine tools with the aid of data stored in a part program library and operator commands entered on system console; and (3) A set of NC machine tools and controllers driven through a common bus, carrying address and command information. Figure 1 shows the overall structure of the proposed DNC system.

The hardware components which comprise the proposed DNC system (excluding the host computer) are depicted in Figure 2. The HP/2100S minicomputer has 16K words of main storage (16 bits each) and communicates, via its accumulator, with the 16-bit duplex register interface card by means of input/output instructions and an interrupt mechanism.

The duplex register has three basic elements which are used to effect the communication between the computer and an external device: (1) A control flip-flop which is set by the computer as a start (strobe) signal to the device and also to enable interrupt signal from the interface card; (2) A flag flip-flop which is set by the device as a completion (service request) signal to the computer; and (3) Two 16-bit input and output buffers which can be loaded by the device and the computer, respectively, and interrogated by the other side.

The particular NC machine used in this system is a TERCO educational kit (Berg, 1973; Gustavsson, 1974) consisting of the following three components, plus a paper-tape reader whose function was replaced by direct control from the computer: (1) The register unit which is capable of storing one block of information concerning the movement of the work-piece in one or several axial directions by a specified amount relative to its current position, as well as other control functions; (2) The drive unit which, under control from the register unit, supplies the drive currents for stepper motors generating the longitudinal (X), transverse (Y), and vertical (Z) movements of the work-piece; and (3) The machine tool consisting of the machine stand and bed, a drive system for tool rotation in two directions, and a drive system for feed (work-piece) movement in three axial directions by means of DC electric stepping motors. (Refer to Figure 2).

Due to incompatibility of signal levels and logic conventions between the register unit of the NC machine and the duplex register of the HP minicomputer, a custom interface box (Figure 2) was designed to perform the needed conversions and also to act as an address decoder, enabling the control of multiple NC machines via a single bus. It generates a strobe signal for the register unit if the corresponding NC machine is selected and relays the end-of-block signal (issued by the register unit upon completion of processing tasks associated with the previous block) to the duplex register.

Figure 3 shows the information flow between the 16-bit duplex register and up to 16 machine tools with addresses 0 through 15. The computer transmits an 8-bit even-parity ASCII character to a desired machine tool by placing the character in bits 0 through 7 of the output buffer in the interface card, loading the machine tool address in bits 8 through 11 of the same buffer and setting the contorl flip-flop. The four-bit address and the device command signal (output of control flip-flop) cause the selected custom interface box to issue a strobe signal, thus loading the 8-bit character into the register unit.

After waiting for a specified period of time, the computer places another character in the output buffer and the above process is repeated. This continues until the register unit receives an end-of-block character, cancelling the corresponding request signal. The NC machine now starts to process the completed block and the computer can go on to service other machines which may need attention.

As shown in Figure 3, the logical OR of all request signals constitutes the device flag signal. In addition, each request signal is connected to a different bit of the duplex register's input lines, thus providing the computer with information as to which machine is in need of service. In case of simultaneous requests, machines with smaller logical address values are given priority. Since the data transmission time for each machine is negligible compared to the time needed for performing the actions specified in a transmitted command block, no significant delay results even for the machine with the lowest priority. Up to 16 machines can be handled with this scheme, one for each of the data input lines to the duplex register.

SYSTEM OPERATION AND SOFTWARE

The needed system hardware and software to support a DNC system with 16 NC machines, not necessarily all of the same type, has been developed and fully tested. However, the current operational hardware includes only a single TERCO machine tool and is used for educational purposes. The method of operation and the corresponding support software for the 16-machine system is described in this section. It should be noted, however, that the same approach can be used for a larger number of machines, given a wider interface register.

The implemented on-line DNC support software performs three main functions: (1) Responding to system transactions; (2) Distributing part program data to machine tool controllers on demand; and (3) storing and maintaining part program libraries. The essential features of each of these functions will be presented in the remainder of this section.

Table 1 shows an alphabetical list of the system transactions which can be entered through the console. The abbreviated (three characters) or full command code is followed by a blank, a list of parameters separated by commas, another blank and an optional descriptive comment.

When a machine tool completes the processing of the first data block, issued as a result of executing a GET command, the remainder of part program data is transmitted automatically, one block at a time, in response to request signals from the machine tool controller. The DNC system has an interrupt service routine which is activated by the request signals and gives proper service to the requesting machine tool(s). This activity is transparent to the operator working through the system console. The termination of processing for a part program causes a message to appear on the console display.

The DNC system uses a number of tables in the main memory to keep track of the current status. The most important of these is a device dynamic status table having a 48-word entry for each machine tool. As shown in Figure 4, each table entry contains information about the machine tools and the part program currently scheduled on it. Characters are transmitted to the machine tool from a 32-word buffer area in the table entry. In addition to the device dynamic status table, the DNC system uses a table for storing the opening flag, the lock switch, and the machine tool request queue and another table for keeping program constants needed at execution time.

THE HOST SUPPORT SYSTEM

The host support system consists of a batch program on a Control Data Cyber 74 computer. It can perform the following functions upon receiving appropriate user commands: (1) Editing user-written part programs prepared on punched cards; (2) Maintenance of part program libraries as permanent files; (3) Storage and retrieval of desired part programs; and (4) Generation of transferable part programs on punched cards for use with the on-line DNC system.

The host support system works with a directorized data set. Since partitioned data sets are not supported by the CDC Scope operating system, such a data set is built by the batch program using CDC direct access system files. The initialization command (+INT) is used for this purpose after the needed file on the disk storage is defined. The data set directory is created automatically and the structure will be ready to use upon execution of +INT. A set of 9 commands has been implemented for the host system.

The host system editor (called through +EDT command) accepts a character string as input and, after checking the part program definition for possible syntax errors, generates an edited part program in a format suitable for TERCO's NC machine controller. The editor's input string contains special function characters (letters P through W), coordinate directions (letters X, Y and Z, possibly followed by a minus sign if movement in the negative direction is intended), position values (up to four decimal digits each), null characters (coded as the string "NUL"), delete characters (coded as "DEL") and end-of-block characters (coded as "EOB").

The punched card output consists of two card types. The name card contains information regarding the part program name, number of records in the part program, number of data blocks and card sequence number. Data block cards contain a part program data record (64 characters) along with part program name and card sequence number. An "END" card is punched out when the task of the DNC support system is completed.

In addition to two tables (representing the editor's transition diagram and the list of valid input characters along with their corresponding state numbers) for use by the editor, the host support system works with three data files: (1) The part program library which is a direct access file; (2) The editor output result sequential file; and (3) The editor output listing sequential file on which the editor stores error-free blocks for subsequent printing.

SUMMARY AND CONCLUSION

The direct numerical control system we have described in this paper is currently operational (with one machine tool) and is being used for educational purposes at the Computer System Laboratory, Sharif University of Technology, Tehran, Iran. The system is expandable to contain up to 16 machine tools with no hardware or software modification. As the time period between successive block transfers to a machine tool is relatively long, a larger number of machine tools can easily be handled, given a longer interface register capable of storing all the request signals.

The on-line DNC system can be extended to

provide the following additional capabilities. (1) On-line reception of part programs from the host computer via a telecommunication link; (2) Collection and storage of statistical and production data either automatically or by means of new transaction codes; (3) Transmission of statistical and production data to the host computer via a telecommunication link; and (4) On-line modification of part programs by means of a patch routine.

The host system can be extended to include the following functions: (1) Part program transmission to the on-line DNC system via asynchronous or binary synchronous communication lines; (2) Receiving and processing of statistical and production information from the on-line DNC system; (3) Receiving modified part programs from the on-line DNC system for storage in the library; (4) Analysis and reduction of production data for report generation; (5) NC high-level programming language processor to simplify part description by users; (6) Part program subroutine library and linkage with part program compilers; and (7) Production planning and control.

Another possible direction for future work is the replacement of the register unit of the machine tool controller by a microprocessor or its total elimination by linking the duplex register to the machine tool's drive unit via an address decoder. With the latter approach, the minicomputer will become much more busy due to the higher service rate needed by the drive unit compared to the register unit and, therefore, may not be able to support a large number of machines.

During the early stages of implementation, numerous errors were encountered in the information received by the NC machine. Although these errors, which were mostly caused by external interference, were later eliminated by locating the source of trouble and also some re-programming and re-design to adjust the timing of pulses, they did cause us to think about the effects of errors in NC systems and possible methods of dealing with them. None of the techniques discussed here have actually been incorporated into our system.

There are at least two reasons for careful consideration of the fault tolerance aspects of NC systems: (1) In the production of highly complex parts which require several hours of machining time, a single error can render the part useless; (2) Occurrence of undetected errors may cause damage to expensive machines and/or tools due to excessive loading and heating. These points become particularly important in view of the hostile operating environment in factories (Kimble, 1977); e.g., extremes in temperature, electrical transients, and long interconnections.

Even though the simple parity check available on most NC machine controllers detects a significant portion of transmission errors, it is possible to use more effective error-detecting or error-correcting codes, particularly since the transmitted character set is quite limited. A very effective approach for dealing with transmission errors, caused by transient or permanent faults, is for each machine tool controller to re-transmit the bit-wise complement of the character it has received back to the computer so that it can be checked against the original character. Unfortunately, other types of errors (e.g., those caused by internal controller faults) are not as easy to deal with and require some form of hardware redundancy for effective detection and/or masking.

Clearly, DNC systems linked with key production facilities provide many significant advantages in terms of computerized part program library maintenance, automatic gathering and logging of production information, part program compilation, report generation, production planning and control, and considerably reduced production costs. However, it is unlikely to realize the full potential of such systems without extensive humanization of their external interfaces. It is highly desirable that DNC systems be instructed in terms familiar to the layman, possibly using natural language commands or a close approximation thereof. This is particularly important if the transfer of numerical control technology to developing countries is to be carried out successfully.

ACKNOWLEDGEMENT

This work was supported in part by Sharif University of Technology under Research Grant 22R - 004.

REFERENCES

Alvandi, K. (1977). A direct numerical control system. Master of Science Thesis, Sharif Univ. of Technology, Tehran, Iran.

Berg, G. (1973). Numerical control: theory, directions for use, exercies. TERCO Publication, Aktiebolag, Sweden.

Gustavsson, B. (1974). Numerical control textbook. TERCO Publication, Aktiebolag, Sweden.

Kimble, G. (1977). Hot time in the old factory. Digital Design, Vol. 7, No. 2 (February), pp. 36-42.

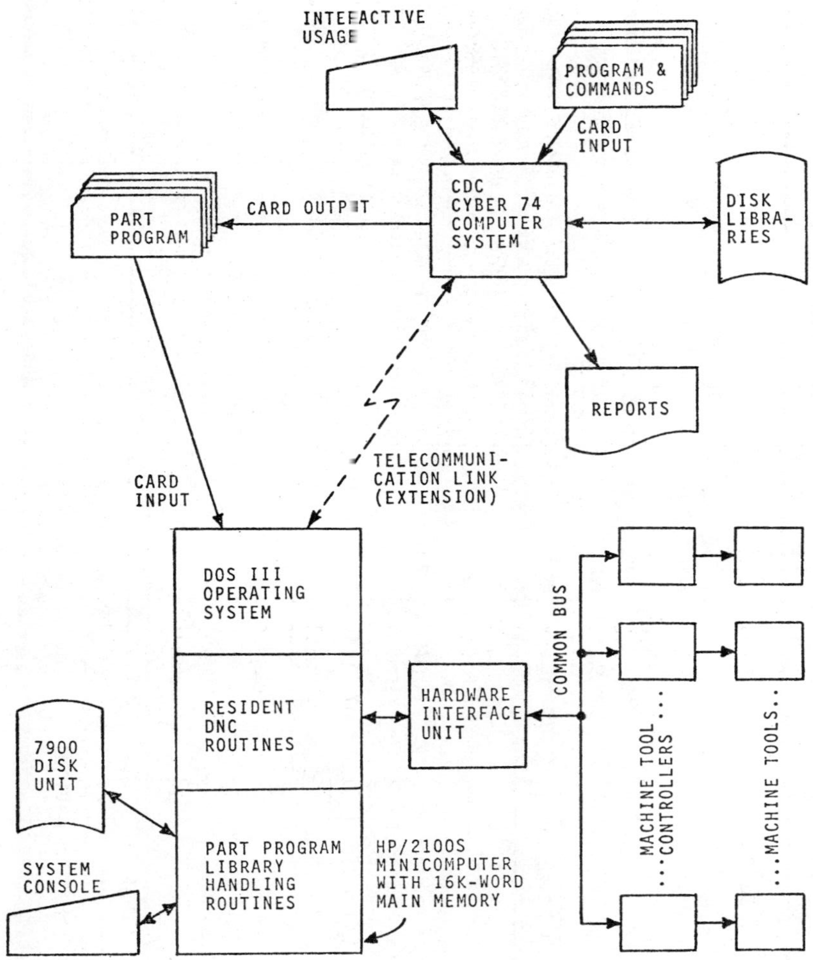

Fig. 1. Overall structure of the proposed DNC system.

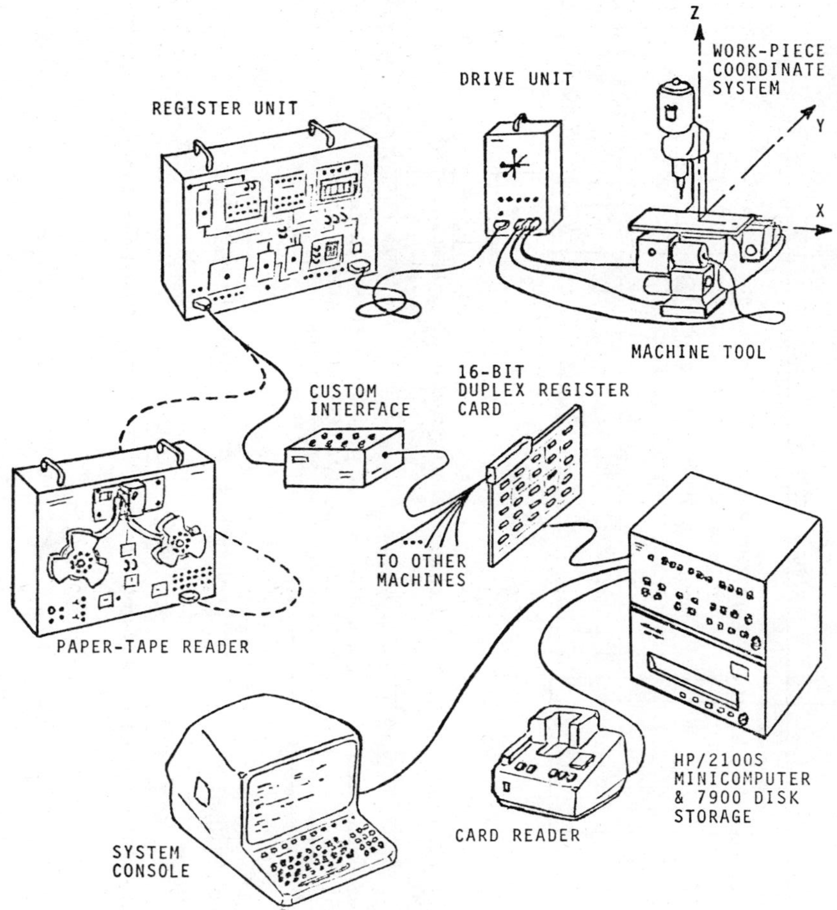

Fig. 2. Hardware components of the DNC system.

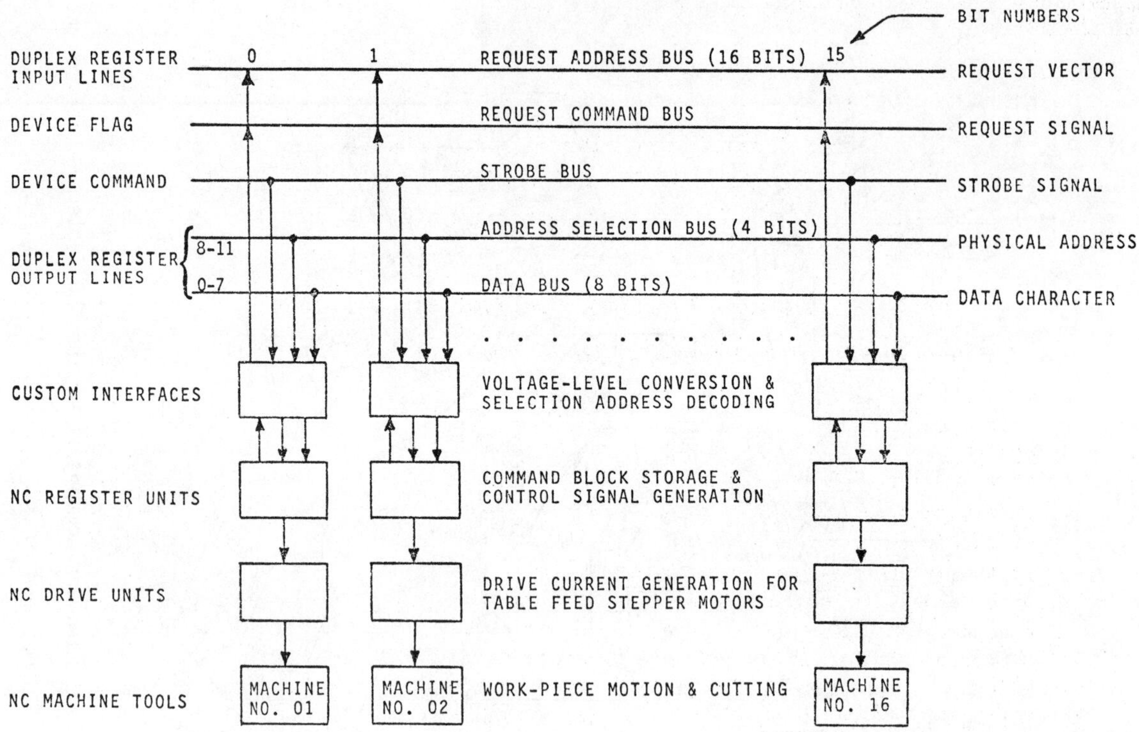

Fig. 3. Information flow schematic for the proposed DNC system.

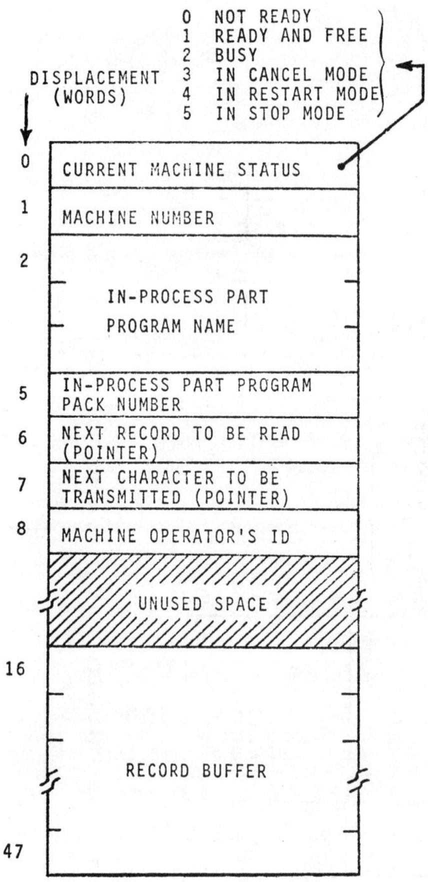

Fig. 4. Entry format in device dynamic status table.

TABLE 1 Alphabetical List of the On-Line DNC System Commands

Command Code	Parameter* List	Description of Function
ASSIGN	IIII,MM	Operator to machine tool assignment
CANCEL	MM	Terminating work on a machine tool
CHANGE	PPP,XXXXX,YYYYY	Renaming of an existing part program
CLOSE		Closing (termination) of the DNC system execution
COPY	PPP,XXXXX,QQQ	Copying a part program from one pack to another
DELETE	PPP,XXXXX	Deleting a part program's name from a pack directory
DISPLAY	MM	Displaying the status of a machine tool on the system console
DUMP	PPP,XXXXX	Listing a part program on the system console
GET	PPP,XXXXX,MM	Obtaining a part program from a pack and scheduling it on a machine tool
INITIALIZE	PPP,DD	Initializing a pack for subsequent usage
LIST	PPP	Listing a pack directory summary
LOCK	LLLL	Locking or unlocking the DNC system
OPEN		Openning (initialization) of the DNC system
PACK	PPP	Compressing the contents of a pack to reclaim the space of deleted part programs
RECEIVE	QQQ	Reading a part program from the card reader into a pack
RESTART	MM	Restarting work on a machine tool
SEARCH	PPP,XXXXX	Searching a pack directory for a specific part program name

* DD : Directory size (members)
IIII : Operator ID
LLLL : Lock switch (password)
MM : Machine number
PPP : Source pack number
QQQ : Destination pack number
XXXXX : Current part program name
YYYYY : New part program name

IMPLEMENTATION PROBLEMS IN MULTIVARIABLE ADAPTIVE CONTROL STRUCTURES

G. Bartolini, G. Casalino, F. Davoli and R. Minciardi

Istituto di Elettrotecnica, University of Genoa, Viale F. Causa 13, 16145 Genoa, Italy

Abstract. The problems arising in the implementation of two different classes of multivariable adaptive control schemes are discussed. In the first class, an estimate of the system parameters is recursively updated and the synthesis of an LQG regulator is performed on its basis. In the second class, each step of the adaptation procedure is accomplished on the basis of the identification of a "closed loop" model whose feedback matrix is the one determined in the preceding adaptation step. The main scope of the paper is to propose implementation possibilities taking into account numerical accuracy and computing time. In this sense, a special attention is given to identification and adaptation algorithms based on factorization techniques. These algorithms have proved to be particularly robust from the numerical point of view. Besides, they are shown to be well suited for possible parallel implementation.

Keywords. Adaptive control; computer control; identification; numerical methods; parallel processing.

INTRODUCTION

In the recent years, modern control theory, informatics and technology have been concurrently involved in the development and application of complex control structures. Actually, the advent of microcomputers has removed the economical constrains which have long time prevented the application of advanced control algorithms to industrial processes. For the near future, one can expect that hardware and software developments, particularly as regards multiprocessor structures, will bring the possibility of implementing complex control systems, where computing power is allocated to the various tasks to the extent that is required.

In the present paper, the implementation of some multivariable adaptive control schemes is considered, taking into account numerical stability and execution speed. More specifically, all algorithms, both for identification and control, are considered in particularly stable and robust versions (namely, those based on factorization techniques), and an effort is made in order to devise simple parallel implementations.

The adaptation schemes discussed in the following two sections belong to a rather conventional class. Namely, the estimate of the parameters of the system's prediction model is recursively updated by means of classical identification methods and an LQG-type regulator is synthesized on this basis. For such schemes, simple suggestions are given in order to accelerate the convergence of the control law.

In the fourth section, a recently introduced class of adaptive regulators is considered (Menga and Mosca, 1979) and the problems related to their implementation are discussed. In this class, each step of the adaptation procedure is performed after the identification of a "closed-loop" model of the process, which is related to the control law found in the preceding adaptation step. It is pointed out that the optimal control law has a particularly simple structure.

CONVENTIONAL APROACH TO ADAPTIVE CONTROL

In this section we shall recall some well known identification and control procedures generally used for conventional adaptive control schemes. More specifically, the techniques that will be considered are applicable to multi-input multi-output (MIMO) systems allowing the following representation

$$\bar{A}(q^{-1})y_i = \bar{B}(q^{-1})u_{i-\bar{k}} + \bar{C}(q^{-1})e_i \qquad (1)$$

where $y_i \in R^v$, $u_i \in R^p$ are the output and input vectors, respectively; $e_i \in R^v$ is the disturbance input, which is a zero-mean stationary white sequence; $\bar{A}(q^{-1})$, $\bar{B}(q^{-1})$, $\bar{C}(q^{-1})$ are polynomial matrices of suitable dimensions in the unit delay operator q^{-1}, and are represented as

$$\bar{A}(q^{-1}) = I + \bar{A}_1 q^{-1} + \ldots + \bar{A}_{\bar{n}} q^{-\bar{n}} \tag{2}$$

$$\bar{B}(q^{-1}) = \bar{B}_1 q^{-1} + \ldots + \bar{B}_{\bar{m}} q^{-\bar{m}} \tag{3}$$

$$\bar{C}(q^{-1}) = I + \bar{C}_1 q^{-1} + \ldots + \bar{C}_{\bar{r}} q^{-\bar{r}} \tag{4}$$

where $\bar{A}_i$, $i=1,\ldots,\bar{n}$, $\bar{B}_i$, $i=1,\ldots,\bar{m}$ and $\bar{C}_i$, $i=1,\ldots,\bar{r}$ are matrices with elements in R. It is assumed that det $[\bar{C}(z)]$ has zeroes outside the unit circle (as is well known, this last condition is without loss of generality in many practical situations). The orders $\bar{n}$, $\bar{m}$, $\bar{r}$, the delay $\bar{k}$, and the parameters of the matrices in (2), (3), (4) are unknown.

Then, with regard to the identification problem, consider the following general prediction model for system (1)

$$\hat{y}_i = A(q^{-1})y_i + B(q^{-1})u_{i-k} + C(q^{-1})\hat{e}_i \tag{5}$$

with

$$\hat{e}_i = y_i - \hat{y}_i$$

and

$$A(q^{-1}) = A_1 q^{-1} + \ldots + A_n q^{-n} \tag{6}$$

$$B(q^{-1}) = B_1 q^{-1} + \ldots + B_m q^{-m} \tag{7}$$

$$C(q^{-1}) = C_1 q^{-1} + \ldots + C_r q^{-r} \tag{8}$$

Define the matrix

$$\theta \triangleq \left[A_1 \ldots A_n \,\vert\, B_1 \ldots B_m \,\vert\, C_1 \ldots C_r\right] \quad ;$$

then, θ and n, m, r, k represent, respectively, estimates of the parameters, orders and fixed delay appearing in system (1).

A recursive technique to derive matrix θ from input-output measurements of the controlled process, can be obtained via the so called extended least squares (ELS) method, whose classical version has been introduced by Panuska (1969) for single-input single-output (SISO) systems; however, the method can be easily applied also for MIMO systems as in the present case (Borisson, 1979; Casalino and others, 1979). The implementation of this algorithm via matrix factorization techniques will be considered in the following section.

If $n \geq \bar{n}$, $m \geq \bar{m}$, $r \geq \bar{r}$ and if other mild assumptions about system (1) (Ljung and others, 1975) are satisfied, the sequence θ_i of estimates given by the ELS method converges to a value $\tilde{\theta}$ which causes (5) to be the optimal minimum variance predictor (see also Goodwin and others (1979) for further discussion). Note that the above mentioned assumptions are verified for almost all practical cases (Ljung, 1980; Panuska, 1980).

If $C(q^{-1})$ in model (5) is set to be a null matrix, i.e., ordinary least squares (LS) are used, the optimal minimum variance predictor cannot be theoretically achieved making use of a finite number of parameters (unless $\bar{C}(q^{-1}) = I$ in system (1)). Nevertheless, in practice, the optimal predictor can be approximated as close as desired by increasing values of n and m. This can constitute a valid alternative to ELS, since in many cases good approximations to the optimal predictor can be obtained also with relatively low values of n, m (Bartolini and others, 1980).

Observe that in conventional adaptive control schemes the adaptation procedure works in accordance with the following two successive steps (performed whenever a new estimate θ_i is obtained):
1- the prediction model is updated with the new estimate θ_i;
2- the new prediction model is regarded as the optimal minimum variance predictor and, according to an assigned control criterion, a new control law is designed.
This corresponds to enforcing the "certainty equivalence principle" and leads to the hierarchical structure depicted in Fig. 1. In the present paper we assume that the dimensionality of the adaptive control problem justifies the use of three different processors, working in pipeline mode, each one performing one of the tasks indicated in Fig. 1.

In the remaining of this section we shall consider the "regulation problem" represented by both adaptation and direct control tasks. The interest will be devoted to control laws of LQG-type only and some preliminary indications relevant to the control law implementability will be given here. The adaptation task, together with the identification task, will be considered again in greater detail in the following section.

Since the prediction model is regarded as the optimal one, system (1) can be represented as

$$y_i = \hat{y}_i + \hat{e}_i = A(q^{-1})y_i + B(q^{-1})u_{i-k} + C(q^{-1})(y_i - \hat{y}_i) + \hat{e}_i \tag{9}$$

where the prediction error sequence $\{\hat{e}_i\}$ is considered the same as the innovations sequence $\{e_i\}$ (white stationary).

A state space representation of (9) can be obtained by simply expressing its predictive part in such form, namely

$$x_{i+1} = Fx_i + Gu_{i-k} + K(y_i - Hx_i) \tag{10}$$

$$\hat{y}_{i+1} = Hx_{i+1} \tag{11}$$

$$y_{i+1} = \hat{y}_{i+1} + \hat{e}_{i+1} \tag{12}$$

where (10) can be written equivalently as

$$x_{i+1} = (F-KH)x_i + Gu_{i-k} + Ky_i \tag{13}$$

with $x_i \in R^h$, $h = v \cdot n$ and the eigenvalues of $(F-KH)$ corresponding to the inverses of the det $[I+C(z)]$. In the following, expression (13) will be regarded as the "one step state predictor" of system (1). It is worth noting that, if representation (10) is chosen to be a multicompanion observable canonical form, then matrices F, G, K, H in (10), and (F-KH) in (13), assume a very simple structure and can be obtained by directly placing the elements of $A(q^{-1})$, $B(q^{-1})$, $C(q^{-1})$ (i.e., the elements of the actual matrix of estimates) in fixed and known positions, no calculation being required (we refer to Wolovich (1973) for details about these points).

Consider now the following LQG control problem: find the input sequence $u_i(x_i, y_i)$ which minimizes the cost functional

$$J = \lim_{T\to\infty} \frac{1}{T+1} E\left[\sum_{i=t}^{t+T} (\|u_i\|_P^2 + \|y_{i+k+1}\|_V^2)\right] \tag{14}$$

where $P = P^T > 0$, $V = V^T \geq 0$.

As is well known, provided the pair (F, G) is stabilizable and the pair (F, H) is observable (this last condition is certainly met due to the particular choice of the state space representation), the solution to the stated problem turns out to be

$$u_i = - Ls_i \tag{15}$$

$$\text{where } L = (P + G^T TG)^{-1} G^T T \tag{16}$$

with T the unique positive definite solution of the associated algebraic Riccati equation, and

$$s_i = E\left[x_{i+k+1} \mid x_i, u_i = 0\right]$$

given by the following expression

$$s_i = F^k\left[(F-KH)x_i + Ky_i\right] + \sum_{j=1}^{k} F^j Gu_{i-j} \tag{17}$$

which can be derived directly from (10). Then, eqns. (13), (17) and (15) represent the overall control law which solves problem (14), insuring the closed-loop stability of the whole control system.

Some comments about the practical implementability of the above control law are now needed.

Observe first that the quantity s_i given by (17) consists of two parts: the first one is given directly by the one-step predictor (13), while the second one (denote it λ_i) can be generated by a shift register of order $p \cdot k$. The automatic retuning of the parameters appearing in (17), could be time consuming, because the computation of the matrix products F^k, $FG \ldots F^k G$ is required. A possible way to circumvent this problem consists in adopting the following form for the computation of λ_i

$$\lambda_i = \sum_{j=1}^{k} \alpha_i^j \tag{18}$$

where

$$\begin{bmatrix} \alpha_{i+1}^1 \\ \cdot \\ \cdot \\ \alpha_{i+1}^k \end{bmatrix} = \begin{bmatrix} 0 & \cdots & 0 \\ F & & \cdot \\ 0 & \ddots & \cdot \\ \cdot & \ddots & \cdot \\ 0 & \cdot\cdot F & 0 \end{bmatrix} \begin{bmatrix} \alpha_i^1 \\ \cdot \\ \cdot \\ \alpha_i^k \end{bmatrix} + F \begin{bmatrix} G \\ 0 \\ \cdot \\ \cdot \\ 0 \end{bmatrix} u_i \tag{19}$$

Then (17) requires only the computation of F^k. The use of (19) also simplifies the computational burden required to generate the control action u_i (we recall that F, H and (F-KH) are matrices of very simple structure). Moreover, observe that most of operations involved are intrinsically parallel and this fact can be exploited if particular structures (e.g., multiprocessors) are available.

Still considering computational problems, notice that even the updating of the control matrix L results to be time consuming, since it requires the solution of an algebraic Riccati equation. The amount of computation can be reduced if L is replaced by

$$L_N = (P + G^T T_N G)^{-1} G^T T_N \tag{20}$$

and T_N is obtained by N backward recursions of the Riccati equation. This corresponds to solve a finite horizon control problem where only the first control of the solution sequence is applied.

A suitable choice of N and of weighting matrices P and V can often preserve the stability properties of the infinite horizon control law.

A simple case is obtained for $N = 0$. Then

$$u_i = -(P + G^T H^T HG)^{-1} G^T H^T Hs_i \tag{21}$$

which correspond to the so called "multivariable extended minimum variance regulator", whose goal is the minimization of the cost functional

$$J = E\left[\|y_{i+k+1}\|_V^2 + \|u_i\|_P^2\right] \tag{22}$$

Moreover, letting $P = 0$ in (22) and assuming $p \geqslant v$ and rank $(HG) = v$, it turns out that the strategy

$$u_i = -G^T H^T (HGG^T H^T)^{-1} Hs_i \tag{23}$$

minimizes the cost (22) as well as the cost (14). This corresponds to the so called "multivariable minimum variance regulator", which was considered by Borisson (1979). In his work, he derives the control law by means of transfer function matrix manipulations. Besides, he shows how to identify the control law parameters directly, bypassing the adaptation step (extending the results of Aström and Wittenmark (1975) for the SISO case). Unfortunately, his results refer to a special case (namely, $p = v$ and minimum phase systems).

IMPLEMENTATION PROBLEMS OF CONVENTIONAL ADAPTIVE CONTROL SCHEMES

Considering first the parameters identification task, we recall that the use of matrix factorization techniques for least squares sequential estimation leads to algorithms (Bierman, 1977; Lawson and Hanson, 1974) which exhibit greater efficiency and numerical stability than the classical versions. The increase in numerical robustness which can be achieved by using these factorization techniques can be partially proved by theoretic considerations, and it has been experimentally demonstrated by many authors. For instance, Casalino and others (1979) have shown that "robust" least squares algorithms implemented by means of 8-bit floating-point arithmetics, are more or less equivalent, from the accuracy point of view, to classical least squares algorithms implemented by 16-bit floating-point arithmetics. Thus, factorization methods seem to be particularly suited in the cases when computing devices with a limited word length (like microcomputers) are to be used.

In this section only a particular factorization technique is considered and the implementation of the corresponding recursive identification algorithm is discussed. Refer again to the ELS method and define the following "extended" measurement vector

$$z_i^T = \left[y_{i-1}^T, \dots, y_{i-n}^T ; u_{i-k-1}^T, \dots, u_{i-k-m}^T ; \hat{e}_{i-1}^T, \dots, \hat{e}_{i-r}^T \right] \tag{24}$$

Of course, the terms $\hat{e}_i$ do not appear in z_i if $C(q^{-1})$ is chosen to be a null matrix in the prediction model (5).

If factorization methods making use of orthogonal Householder or Givens transformations are used, the estimate θ_i can be obtained by solving the linear systems

$$R_i \theta_i^T = \eta_i \tag{25}$$

where R_i is an upper triangular matrix. Matrices R_i and η_i are computed recursively as follows

$$\begin{bmatrix} R_i & \eta_i \\ 0 & \varepsilon_i \end{bmatrix} = \Psi_i \begin{bmatrix} R_j & \eta_j \\ z_i^T & y_i^T \end{bmatrix} \tag{26}$$

where Ψ_i is an orthogonal transformation which can be simply built, and R_j, η_j are the matrices corresponding to the last estimate θ_j before the time instant i.

A simple and numerically stable method (Casalino and others, 1979) can be used which allows the computation of the terms of the sequence $\{\hat{e}_i\}$, appearing in (24), directly from recursion (26), without making use of the relation

$$\hat{e}_i = y_i - \hat{y}_i$$

Thus, recursion (26) does not at all require the knowledge of the previous estimate θ_j i.e., it is not affected by the solution of systems (25) (this fact does not occur in the classical versions and represents one of the reasons of improved efficiency). Moreover, observe that if updating the regulator parameters is allowed to take place not at each recursion (26), then systems (25) can be solved only when necessary, thus saving computation time.

Of course, the correct implementation of the identification algorithm always requires that the batches of data z_i consist of successive samples of input-output measurements. However, it is possible to process the various batches asynchronously, as shown for instance in Fig. 2. It then follows that the identification algorithm is constrained to operate at the sampling frequency only as regards the acquisition of a single batch z_i, but not during its processing. Anyway, it is obvious that the convergence properties of the estimates and their quality improve when increasing the average number of processed data per sampling interval.

As to the practical implementation of the identification algorithm, it is easy to observe that the matrix operations involved in recursion (26) could be performed in some efficient way using special multiprocessor structures. Moreover, the solution of linear systems (25) can be obviously split into parallel substasks, each one corresponding to the problem of finding a column of matrix θ_i. Thus, special hardware structures (for instance, array-processor type) could be used for implementing (26) and finding the solution of (25).

Finally, coming to the task of updating the regulator parameters when a new estimate θ_i is provided by the identification program, first observe that the following decomposition can be considered, which allows a parallel implementation: i) adaptation of the predictor parameters, ii) computation of the control gain matrix L_N. It is worth noting that in most cases it is allowed to keep the adaptation rate of matrix L_N lower than that relevant to the adaptation of the predictor parameter. In fact, the computation of L_N requires the recursion of a discrete Riccati equation, while the retuning of the predictor reduces simply to substituting the elements of θ_i in fixed memory positions (actually if k>0 also the computation of F^k is required).

Since the computation of L_N seems to be a more critical task we shall consider it in greater detail.

Assuming that the weighting matrices that appear in the cost functional (14) are expressed in factorized form, namely $P=Q^TQ$, $V=Z^TZ$ the use of factorization techniques leads to the following recursions for the Riccati equation, instead of the classical one(Kailath, 1980)

$$\begin{bmatrix} S_j & W_j \\ 0 & E_j \end{bmatrix} = \Psi_j \begin{bmatrix} Q & 0 \\ Z_{j-1}G & Z_{j-1} \end{bmatrix}$$

$$\begin{bmatrix} Z_j \\ 0 \end{bmatrix} = \Lambda_j \begin{bmatrix} Z \\ E_jF \end{bmatrix} \quad ; \; Z_o = Z \tag{27}$$

where S_j and Z_j are upper triangular matrices. Here again Ψ_j and Λ_j are orthogonal transformations that can be built in several simple ways. The required matrix L_N is obtained after N iterations of (27) by solving the following linear systems

$$S_N L_N = W_N \tag{28}$$

Algorithm (27), (28) is characterized by an almost complete similarity with the identification algorithm (25), (26). Then all considerations concerning the implementation of the identification algorithm still apply here.

Even if particular hardware structures are used, the computation of L_N via N recursion of (27) and successive solution of (28) represents, in any case, one of the reasons of slackening the adaptation process.

Then, to accelerate the adaptation procedure, we propose the following very simple technique:

1 - get the i-th estimate (θ_i) at the time it is yielded by the identification process (25), (26);
2 - adapt the predictor parameters first, substitute F and G in(27) performing only few recursions;solve (28) obtaining L_i;
3 - adapt the control gain with L_i, set i=i+1 and go back to step 1.

As can be seen, this corresponds to apply a sequence of control laws, each referring to a larger control horizon as time increases. Since it is expected that θ_i will converge to a value which leads to a "good" approximation of the optimal minimum variance predictor, it is also expected that L_i will converge to a "good" approximation of the optimal control law on an infinite horizon. Simulation results have actually confirmed the validity of the above procedure.

A NEW CLASS OF ADAPTIVE REGULATORS

A new class of adaptive regulators, has been recently introduced by Menga and Mosca (1979), who developed the main characteristics of these controllers. In this section, their main results are reviewed and the differences with respect to conventional adaptive control schemes are pointed out. Besides, problems relevant to the practical implementation of this new class of regulators are discussed.

Refer again to the LQG control problem stated and discussed in the previous sections. Since it is known that the optimum state-feedback law is stationary linear, only control laws of such form can be considered. Then, assuming a stationary linear control law $u_i = Fs_i$ is operating, it is possible to define the following model

$$\begin{bmatrix} y_{t+k+T+1} \\ u_{t+T} \\ \vdots \\ y_{t+k+1} \\ u_t \end{bmatrix} = \begin{bmatrix} \Theta_T(F) & \Gamma_T(F) \end{bmatrix} \begin{bmatrix} u_t \\ s_t \end{bmatrix} + v_{t,T} \tag{29}$$

where the right hand side of the expression splits into two different parts, the first one consisting of the predictions of the input-output pairs based upon the information set $[s_t^T, u_t^T]$, the second part consisting of the prediction errors.

Menga and Mosca have shown that an alternative expression of the cost functional (14) is

$$\lim_{T\to\infty} \frac{1}{T+1} \left\| \Theta_T(F)F + \Gamma_T(F) \right\|^2_{Q_{T+1}} \tag{30}$$

where

$$Q_k = \text{Block-diag}\left\{\begin{bmatrix} V & | & O \\ - & - & - \\ O & | & P \end{bmatrix}, \dots, \begin{bmatrix} V & | & O \\ - & - & - \\ O & | & P \end{bmatrix}\right\} \quad (k \text{ times})$$

and the expression $\|K\|_Q^2$ simply means $tr(K^TQK)$. Moreover, the optimum feedback matrix, $-L$, can be obtained by solving the sequence of problems

$$\min_{F_{j+1}} \left[\lim_{T\to\infty} \frac{1}{T+1} \left\| \Theta_T(F_j)F_{j+1} + \Gamma_T(F_j) \right\|^2_{Q_{T+1}} \right] \quad (31)$$

in the sense that $-L = \lim_{j\to\infty} F_j$.

Suboptimal controllers are obtained if the following approximations are introduced: i) the horizon is finite in problems (31) (i.e., T finite), ii) the state s_i is replaced by a convenient collection of past input-output pairs, i.e.,

$$s_i^T = \left[y_i^T, u_{i-k-1}^T, y_{i-1}^T, u_{i-k-2}^T, \dots, y_{i-\delta+1}^T, u_{i-k-\delta}^T \right] \quad (32)$$

An adaptive control scheme is obtained if, for the sequence of problems (31), the matrices $\Theta_T(F_j)$ and $\Gamma_T(F_j)$ are estimated from input-output process measurements. This leads to an adaptation algorithm for which a single optimization step (i.e. the solution of a single problem (31)) requires the identification of a closed-loop model, (i.e., of matrices $\Theta_T(F_j)$ and $\Gamma_T(F_j)$).

The coefficients of each prediction model to be identified are functions of the feedback matrix F_j. This means that the identification must be performed with the system operating in closed-loop, with the feedback law $u_i = F_j s_i$. The presence of this constant state feedback law allows the reduction of the state variables set in (32). Actually, a possible reduction is

$$s_i^T = \left[y_i^T, y_{i-1}^T, \dots, y_{i-\delta+1}^T \right] \quad (33)$$

The above reduction is performed mainly to obtain a non-redundant set of state variables, in order to avoid identifiability problems in determining the coefficients of the prediction laws. Unfortunately, however, these problems are not completely avoided since the prediction model basis, namely $\left[u_t^T, s_t^T\right]$, includes u_t^T, which is a linear function of s_t. Thus, in order to identify the matrices $\Theta_T(F_j)$ and $\Gamma_T(F_j)$ correctly, an (on-line) identification procedure must be used, which operates practically open-loop at step t and closed-loop from step (t+1) onwards.

Then the following adaptive control procedure can be stated.

1 - Initialize with a feedback matrix F_o; set $j = 0$.

2 - Begin acting closed-loop for a few steps: $u_i = F_j s_i$.

3 - Initialize the estimates of $\Theta_T(F_j)$ and $\Gamma_T(F_j)$; set $i = t$.

4 - Corrupt with noise the control action for a single step: $u_t = F_j s_t + n_t$ in order to prevent a deterministic relation between u_t and s_t.

5 - Act closed-loop ($u_i = F_j s_i$) for $(T+k)$ steps, collecting the input-output pairs (u_i, y_{i+1+k}).

6 - Run the identification program to update the estimates of $\Theta_T(F_j)$ and $\Gamma_T(F_j)$ using data $\left[u_i^T, s_i^T\right]$ and the measurements (u_i^T, y_{i+1+k}) collected for $t \leq i \leq t+T$. A least squares method is used to determine the parameters of the prediction model (see Fig. 3).

7 - Test the convergence of the estimate (i.e. compare the previous and the new estimates and test if convergence is met, according to a chosen criterion). If convergence is obtained, go to the next step. Otherwise, set i=t and go back to step 4, to perform a new estimation step.

8 - Compute the matrix F_{j+1} which solves the problem

$$\min_{F_{j+1}} \left\| \Theta_T(F_j)F_{j+1} + \Gamma_T(F_j) \right\|^2_{Q_{T+1}} \quad (34)$$

9 - Test the convergence of the feedback matrix (i.e., compare F_j and F_{j+1} and test if convergence is met, according to a given criterion). If convergence is obtained, stop. Otherwise, set $j = j + 1$ and go back to step 2.

As to the implementation problems posed by the method, note that each minimization step (34) can be easily performed by means of factonization techniques. Thus, once again, the same techniques can be applied for the identification and the adaptation procedures.

Moreover, the same possibilities of parallel implementation that we mentioned in the previous sections still exist. Actually, for instance, the identification programs of the rows of matrix $\left[\Theta_T(F_j) \mid \Gamma_T(F_j)\right]$ can be run in parallel. The same observation holds for the determination of the columns of matrix F_{j+1} in (34).

It is worth noting that, for the adaptation scheme considered in this section, the controller optimization requires the identification of several closed-loop models, while the conventional approach is based on only one identification of the process model. Moreover, the above proposed identification procedure can only perform an estimation step every (T+k) sampling intervals (see Fig. 3) and not at each sampling interval as is allowed for the conventional schemes. Thus, the overall adaptation procedure here considered appears to have a rather slow convergence rate, as compared to those recalled in the previous sections.

Nevertheless, the convergence speed can be increased by reducing the control horizon T and the dimensionality of vector s_t. Actually, simulation examples have proved that the limit of the matrix sequence $\{F_j\}$ is not too sensitive to the choice of T. Moreover, only few recursions of the optimization procedure (34) are generally sufficient to give a matrix quite close to the final value. Consider, for instance, the (SISO) process

$$y_i - y_{i-1} = u_{i-1} + 0.6\ u_{i-2} + e_i + 0.4\ e_{i-1} \qquad (35)$$

and the cost functional (14) with P=0.01 and V=1. Then, Fig. 4 plots the various components of the feedback matrix F_j, for T = 7 and $\delta = 5$ in (33). In that figure, each horizontal division corresponds to 6000 sampling intervals, which were needed for a closed-loop model identification (a fixed identification time was chosen for this example).

An important feature (Menga and Mosca, 1979) of these new adaptive control schemes is the following: the adaptation procedure yields matrices F_j which correspond to nonincreasing values of the cost functional (30). Actually, this property holds only if the optimal controller is considered, without the above introduced approximations. Nevertheless, simulation results have shown that the matrix sequence $\{F_j,\ j = 0, 1, 2, \ldots\}$ indeed yields better and better performances of the control system.

Finally, note that the control law obtained by this adaptation scheme results of a very simple form, as compared to those considered in the previous sections. Clearly, this derives from the substitution of the state vector with a collection of input-output pairs. Actually, it has been noted (by means of simulations examples) that few state variables generally give prediction models whose validity is acceptable.

CONCLUSIONS

Different adaptive control schemes have been considered, from the implementation point of view. It has been shown that the efficiency of the overall adaptive controller can be improved, as regards both computing speed and numerical accuracy, by means of a suitable choice of algorithms which allow simple parallel implementation techniques. For conventional adaptive regulators, a quite attractive procedure has been presented which accelerates the convergence to the optimal control law and reduces the computational burden. Then, a recently introduced approach to adaptive control has been discussed, and the differences with the conventional schemes have been pointed out.

ACKNOWLEDGMENT

This work has been partially supported by the National Council of Research of Italy (C.N.R.) within the Research Project on Informatics (P.F.I.).

REFERENCES

Aström, K.J., and B. Wittenmark (1973). On self-tuning regulators. Automatica, 9, pp. 185-199.

Bartolini, G., G. Casalino, F. Davoli, and R. Minciardi (1980). Adaptation schemes for self-organizing controllers. Proc. 5th European Meeting on Cybernetics and Systems Research. Vienna. (To be published).

Bierman, G.J. (1977). Factorization Methods for Discrete Sequential Estimation. Academic Press, New York.

Borisson, U. (1979). Self-tuning regulators for a class of multivariable systems. Automatica, 15, pp. 209-215.

Casalino, G.,G. Menga, R. Minciardi and R. Zoppoli (1979). Microcomputer-based complex control systems. Proc. LXXX Annual Conference of the Italian Electrical Association, S. Margherita, Italy. (In Italian).

Goodwin, G.C., P.J. Ramadge, and P.E. Caines (1979). Ultimate objectives and prior knowledge in system identification. Preprints 5th IFAC Symp. Identification and System Parameter Estimation, Darmstadt, pp. 1123-1129.

Kailath, T. (1980). Linear Systems. Prentice-Hall,Englewood Cliffs, N.J.

Lawson, C.L., and R.J. Hanson (1974). Solving Least Squares Problems. Prentice-Hall, Englewood Cliffs, N.J.

Ljung, L., T. Söderström, and I. Gustavsson (1975). Counterexamples to general convergence of a commonly used identification method. IEEE Trans. Automat. Contr., AC-20, pp. 643-652.

Ljung, L. (1980). Comments on 'Non-convergence of the approximate maximum likelihood identification algorithm'. Automatica, 16, pp. 219-220.

Menga, G., and E. Mosca (1979). MUSMAR: a new class of multivariable adaptive regulators. 5th IFAC Symp.Identification and System Parameter Estimation, Darmstadt.

Panuska, V. (1969). An adaptive recursive-least-squares identification algorithm. Proc. IEEE Symp. on Adaptive Processes, Penn. State University.

Panuska, V. (1980). Non-convergence of the approximate maximum likelihood identification algorithm. Automatica, 16, pp. 215-217.

Wolovich, W. A. (1973). The determination of state-space representations for linear multivariable systems. Automatica, 9, pp. 97-106.

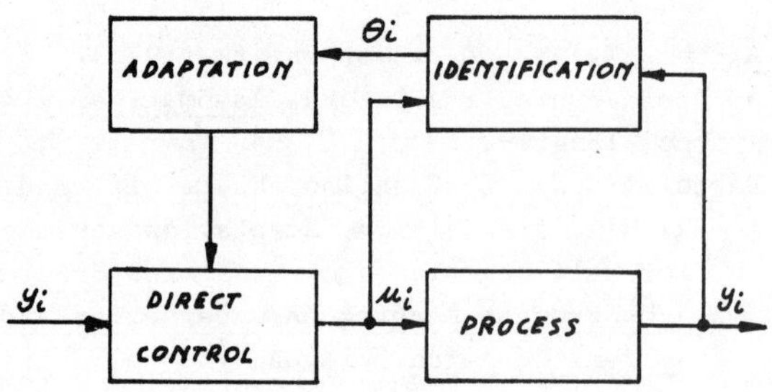

Fig. 1. Structure of the adaptive control system.

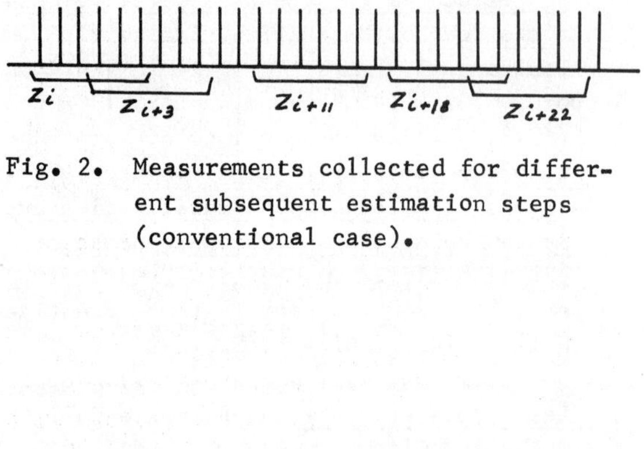

Fig. 2. Measurements collected for different subsequent estimation steps (conventional case).

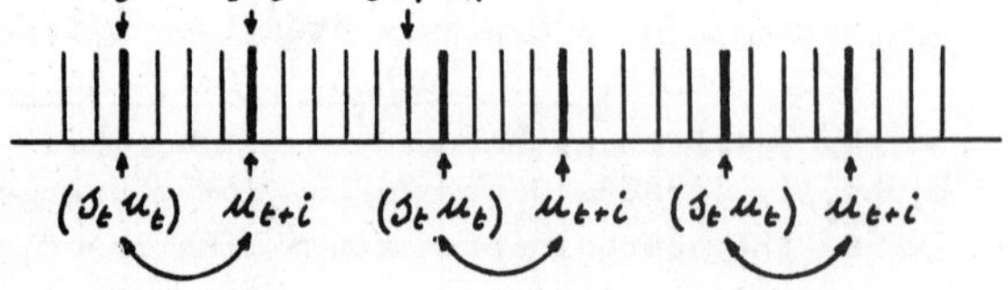

Fig. 3. Measurements collected for different subsequent estimation steps relevant to the row of matrix $\left[\Theta_T(F_j)\ \Gamma_T(F_j)\right]$ corresponding to u_{t+i}.

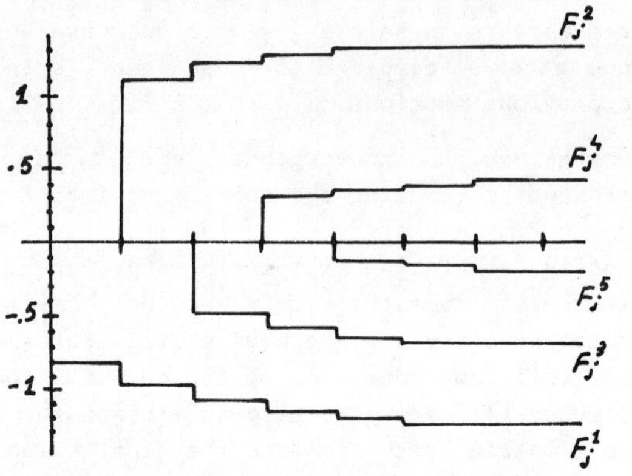

Fig. 4. Convergence of matrix sequence $\{F_j\}$ to the optimum feedback matrix.

MULTIPLICITY OF THE STEADY STATE IN FLUIDIZED BED REACTORS—VII—THE EFFECT OF TRANSPORT AND KINETIC PARAMETERS ON THE STABILITY AND CONTROLABILITY OF A COMPLEX REACTION

S. S. E. H. Elnashaie, S. S. El Shishini and M. A. El Rifai

Chemical Engineering Department, Faculty of Engineering, Cairo University, Cairo, Egypt

Abstract. The stability of a fluidized bed reactor in which a consecutive exothermic reaction takes place has been investigated earlier (Part III). The minimum controller gain which was calculated for the case of smooth heat generation functions is extended here to the case of deformed heat generation curves. The effect of kinetic parameters, interchange coefficient and mixing models on the controller parameters is discussed.

Keywords. Controlability, stability.

INTRODUCTION

In the exothermic processes the temperature level at which the reaction proceeds can be maintained by the heat of reaction alone. A steady state is established at which the heat consumption is balanced by the heat production. However, due to the complexity of the heat generation function, the solution of the heat balance equation gives rise to multiple steady states. The analysis of steady state multiplicity and dynamic behavior for the simple reaction A —— B has been the subject of many studies (Aries, 1969; Hlavacek, 1969; Uppal, 1974). Elnashaie has investigated this case in previous work (1972, 1973). In contrast less attention has been given to the study of complex reactions (Hlavacek, 1972) and especially to the selectivity and yield problems associated with such systems (Cresswell, 1970; Calderbank, 1972; Elnashaie, 1972). In a previous paper we investigated the implications of the multiplicity phenomenon on the selectivity and yield of the desired intermediate product B for the exothermic consecutive reaction A——B ——C taking place in a freely bubbling fluidized bed reactor (Elnashaie, 1977). According to Aries (1979) the solution of the heat balance equation for the case of consecutive reactions in lumped parameter systems (CSTR, fluidized bed reactors, etc...), gives rise to up to 5 steady states for certain values of the parameters.

In this paper, the problem of stabilizing an optimum unstable steady state in a fluidized bed catalitic reactor in which a complex reaction takes place, is being considered. Attention is focussed on the effect of mixing and exchange parameters on the design of the control loops for such reactors.

THE STABILITY PROBLEM FOR A COMPLEX REACTION

We have shown in an earlier paper (Elnashaie, 1977) that the problem of the determination of the steady state temperature of a fluidized bed catalytic reactor in which a consecutive exothermic reaction $A \rightarrow B \rightarrow C$ takes place can be reduced to the problem of solving a non-linear equation using the Van Heerden (1953) diagram. With regard to the stability and controlability of the optimum steady state, which coreesponds to the maximum yield of the desired intermediate product B, three different cases have been observed (Elshishini, 1979).

(1) The case of the optimum steady state temperature corresponding to a unique stable steady state.
(2) The case of the optimum steady state corresponding to a stable state in the multiple steady state region.
(3) The case of the optimum steady state corresponding to an unstable steady state.

The first case does not present a problem with regard to stability or start up. The second case, which has been discussed previously (Elnashaie, 1977), presents a problem with regard to start up but not with regard to stability. Whereas the third case presents a problem with regard to stability. In this paper we focus our attention on this third case.

The Control Scheme for the Middle Unstable Steady State

Considering the consecutive first order reaction,

$$A \rightarrow B \rightarrow C \qquad (1)$$

taking place in a catalytic fluidized bed reactor and assuming the second reaction $B \rightarrow C$

to be relatively fast

The heat balance equation for the case in which the dense phase gas is assumed to be in Plug Flow is:

$$Y-Y_f = \frac{1}{\bar{B}}(\beta_1\ EX_1\ I_A + \beta_2\ EX_2\ I_B) \qquad (2)$$

where,

$$EX_1 = \alpha_1 \exp(-\gamma_1/Y), \quad EX_2 = \alpha_2 \exp(-\gamma_2/Y) \qquad (3)$$

$$I_A = \int_0^1 x_A\, dW \ , \quad I_B = \int_0^1 x_B\, dW \qquad (4)$$

X_A, X_B are calculated as shown in Appendix 1.

When the dense phase gas is assumed to be completely mixed the heat balance equation becomes

$$Y-Y_f = \frac{EX_1}{\bar{B} + EX_1}\left(\beta_1 + \frac{\beta_2\ EX_2}{\bar{B} + EX_2}\right) \qquad (5)$$

The yield, selectivity and conversion are given by:

$$Y = (G_I\ X_B + G_C\ \bar{X}_B)/G.X_{Af} \qquad (6)$$

$$S = (G_I\ X_B + G_C\ \bar{X}_B)/(G\ X_{Af} - (G_I\ X_A + G_C\ \bar{X}_A)) \qquad (7)$$

$$CN = (G\ X_{Af} - (G_I\ X_A + G_C\ \bar{X}_A))/G\ X_{Af} \qquad (8)$$

The two phase model parameters G_I, G_C, A_I, A_C, $\bar{B}$ were calculated using the modified Partridge and Rowe model (1966) and the calculations given in Appendix 2. The values of X_A, $\bar{X}_A$, X_B, $\bar{X}_B$ are given in Appendix 1 for the case the dense phase gas being in plug flow and the case of the dense phase gas being completely mixed. The heat balance equation can be solved graphically as shown in Fig. 1. The selectivity yield and conversion are also plotted on the same figure as a function of the steady state temperature. It is clear from this figure that the desired steady state does not satisfy the slope condition and is therefore unstable and unattainable. To stabilize this desirable unstable steady state, a feed-back control system such as that shown in Fig. 2 is used (Elnashaie, in Press). The heat balance equation in this case case becomes:

$$(1+\bar{K})y - (y_f+\bar{K}y_m) = \frac{1}{\bar{B}}(\beta_1\ EX_1\ I_A + \beta_2\ EX_2\ I_B) \qquad (9)$$

Where $\bar{K}$ is the controller gain multiplied by the steady state gain of the feed heater. The heat removal line which is proportional to the L.H.S. will have a slope of $(1+\bar{K})$ and intersects the y- axis at the point $(y_f + \bar{K}\ y_m)/(1+\bar{K})$.

The heat removal line for the controlled system for a suitable value of $\bar{K}$ will give rise to a unique solution and the desired steady state will be stable provided; it statisfies the dynamic condition for stability.

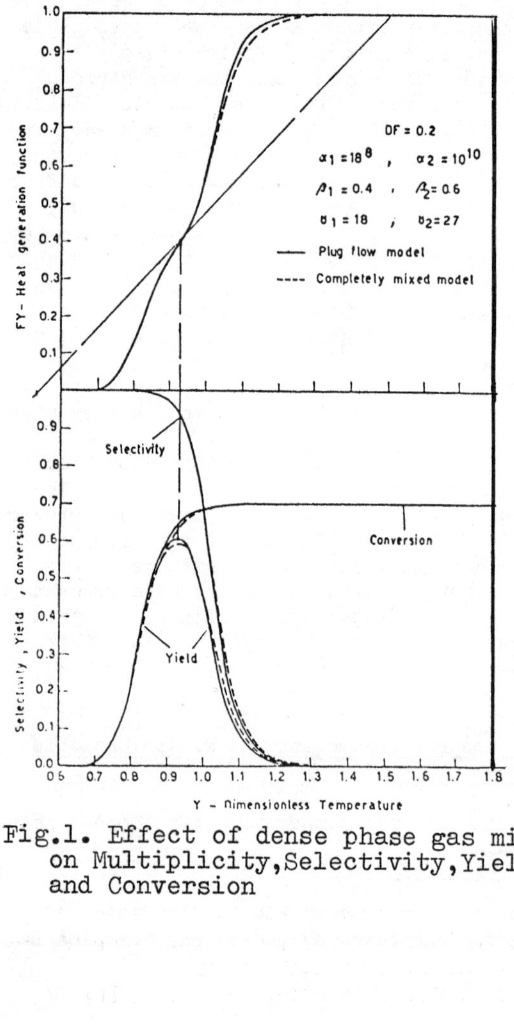

Fig.1. Effect of dense phase gas mixing on Multiplicity, Selectivity, Yield and Conversion

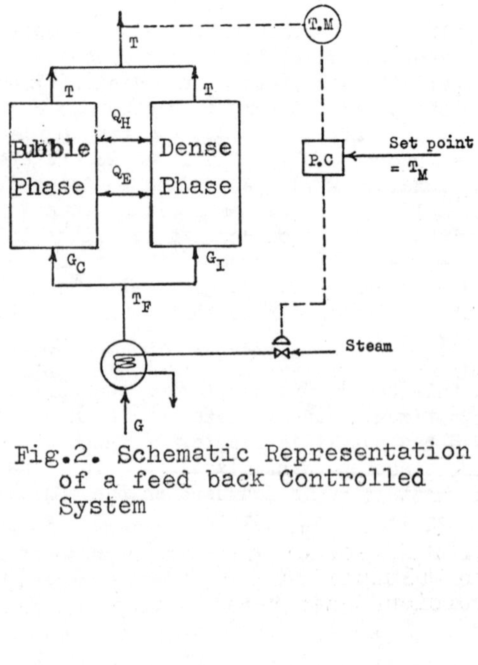

Fig.2. Schematic Representation of a feed back Controlled System

In order to obtain the minimum gain required for stability a tangent is drawn at the desired steady state Fig. 3. The minimum gain gain calculated by this method will be correct only for the case of smooth heat generation curves, which appear when the dense phase gas is assumed to be completely mixed and for values of the interchange parameter QE $\geqslant$ 2.

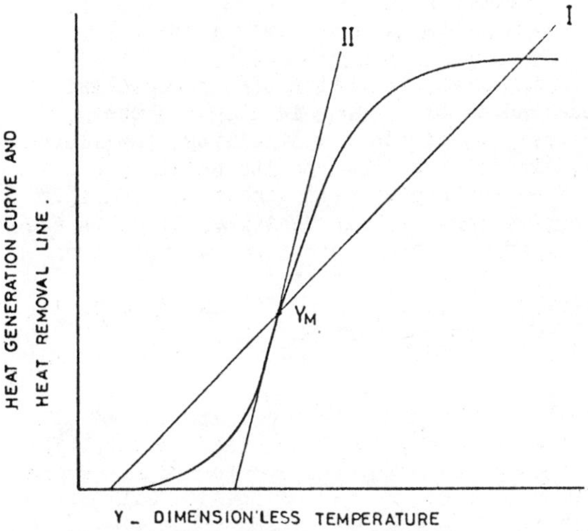

Fig. 3. Schematic representation of the minimum controller gain on a smooth heat generation curve

However, Figs. 4 and 5 show that for a value of the interchange coefficient (QE) equal to 0.02 and when the dense phase gas is assumed to be in plug flow a deformation of the intermediate branch of the heat generation function occurs. Fig. 6 represents the heat generation function for the plug flow model at QE = 0.02. It is noticed that the controller gain calculated from the slope of the tangent at the desired steady state Y_m (line 1) does not represent the minimum value over which unique stable solution occurs. Increasing the controller gain (line II), five steady states appear, then for a further increase (line III) three steady states are obtained. The minimum controller gain is obtained from the slope of line IV since any slight increase over that minimum will ensure a unique solution. Line IV is the line passing by Y_m and tangent to the heat generation function at the top part of the curve as shown in Fig. 6., i.e. it is the line passing by Y_m and having the largest slope among the group of lines passing by Y_m and tangent to the heat generation function.

Effect of Dense Phase Gas Mixing and Exchange Parameter (QE) on the Controller Gain

The effect of dense phase gas mixing on the controller gain has been studied by assuming the two limiting cases of plug flow and complete mixing. The interchange coefficient (QE) has been given the values of: 0.02, 2,6. Table I shows a comparison between the minimum controller gain obtained for the different values of QE, for the

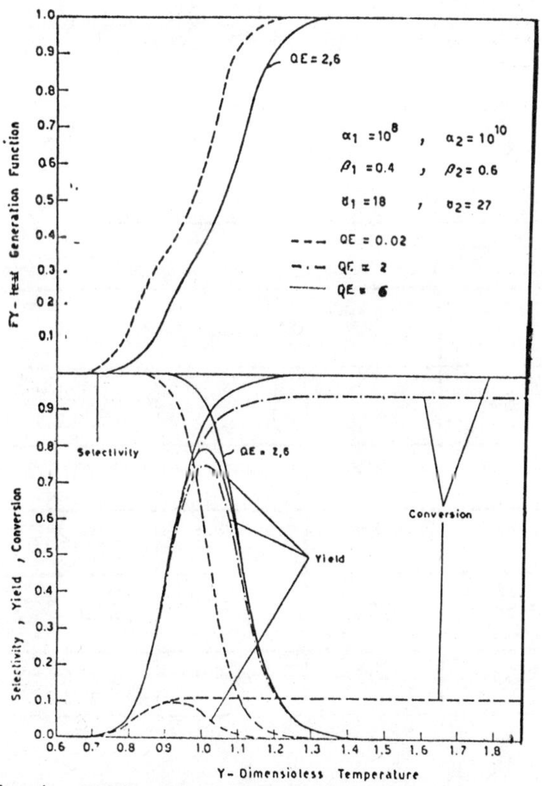

Fig.4. Effect of Interchange Coefficient on Multiplicity, Selectivity, Yield, conversion (Dense phase Completely Mixed)

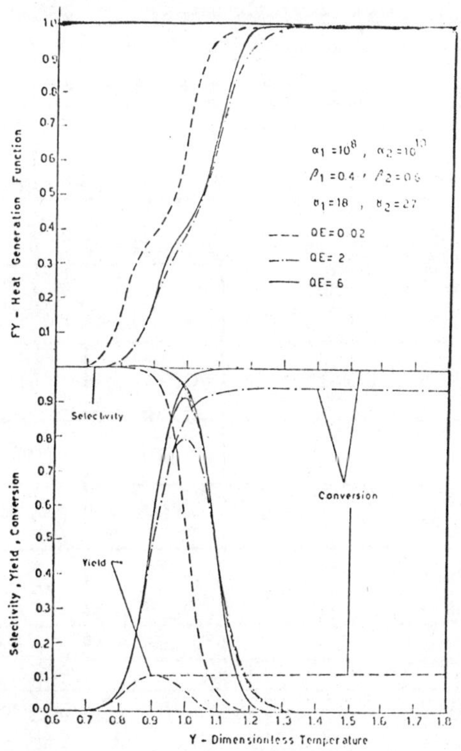

Fig.5. Effect of Interchange Coefficient on Multiplicity, Selectivity, Yield, Conversion (Dense Phase Gas in Plug Flow)

two mixing models. It is noticed that the minimum controller gain K(min.) decreases as QE increases for both mixing models and that the completely mixed model gives lower values for the minimum gain at the various QE values. It is also clear that the minimum gain obtained from the slope of the tangent has a much lower value than the true minimum slope. At QE = 0.02 for the PF model $\bar{K}_{tg}$ = 0.316 whereas $\bar{K}_{min}$ = 2.57

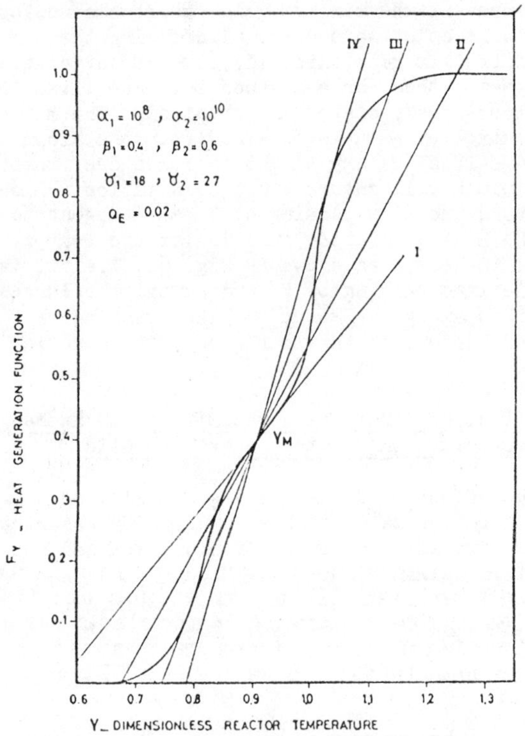

Fig. 6. Graphical calculation of the minimum controller gain for heat generation curves with deformation

CONCLUSIONS

The results presented show that the dense phase gas mixing and the exchange rate between the phases have important effect on the minimum controller gain necessary to ensure the stability of the unstable middle steady state.

For certain values of the exchange parameter, specially for the case of plug flow in the dense phase, the heat generation function is deformed causing a considerable increase in the minimum controller gain necessary for stabilizing the middle steady states.

It is important to notice that the analysis presented in this paper is based on the necessary condition for stability (the slope condition), i.e.: the results obtained are based on the steady state analysis. Further investigations is being carried to obtain the effect of mixing and exchange parameters on the minimum controller gain necessary to satisfy the sufficient condition for stability (the dynamic condition).

TABLE 1 Effect of Interchange Coefficient on Controlability of the Reactor

	PF model			CM model		
Q_E	0.02	2	6	0.02	2	6
Y_m	0.9	0.995	0.995	0.91	1.0	1.01
$(1+K)_{tg}$	1.316	1.823	1.48	1.47	1.842	1.842
$(Y_f)_{tg}$	0.505	0.603	0.58	0.54	0.64	0.64
$(1+K)_{min}$	3.57	3.33	3.33	2.857	1.842	1.842

NOTATION

A_R Cross sectional area of bed, cm^2.

A_C Area of part of bed cross-section occupied by clouds, cm^2.

A_1 Area of part of bed cross-section occupied by dense phase, cm^2.

a $= Q_E . A_C . H/G_C$.

C_A Concentration of reactant A in the dense phase, Cu.mole cm^{-3}.

C_B Concentration of product B in the dense phase, gm mole cm^{-3}.

C_C Concentration of product C in the dense phase, gm mole cm^{-3}.

$\bar{C}_A$ Concentration of reactant A in the bubble phase, gm mole cm^{-3}.

$\bar{C}_B$ Concentration of product B in the bubble phase, gm mole cm^{-3}.

$\bar{C}_C$ Concentration of product C in the bubble phase, gm mole cm^{-3}.

C_{rf} Reference concentration in gm mole cm^{-3}.

C_{Af} Concentration of component A in feed, gm mole cm^{-3}.

C_{p_f} Specific heat of gas cal/gm C.

D, D_F Gaseous diffusion coefficient cm^2 sec^{-1}.

E Activation energy cal/gm mole.

G_C Cloud gas flow rate through any horizontal section in the bed cm^3 sec^{-1}.

G_I Dense phase gas flow rate through any horizontal section the bed, cm^3 sec^{-1}.

H Bed height, cm.

K Reaction rate coefficient.

Q_E Interchange coefficient, Sec^{-1}.

Q_H Heat exchange rate between cloud and dense phase sec^{-1}.

r Rate of reaction.

T Dense phase temperature, K.

$\bar{T}$ Bubble phase temperature, K.

T_{rf} Reference temperature, K.

U_o Superficial gas velocity cm sec^{-1}.

U_C $= G_C/A_C$, cm sec^{-1}.

U_I $= G_I/A_I$, cm sec^{-1}.

U_{mf} Minimum fluidization velocity cm sec^{-1}.

U_B Bubble rising velocity, cm sec^{-1}.

X_i $= C_i/C_{rf}$ dimensionless dense phase concentration.

$\bar{X}_i$ $= \bar{C}_i/C_{rf}$ dimensionless bubble phase concentration.

Where i= A, B or C.

w = h/H dimensionless bed height.

Y $= T/T_{rf}$ dimensionless dense phase a temperature.

$\bar{Y}$ $= \bar{T}/T_{rf}$ dimensionless bubble phase temperature.

Y_F $= T_F/T_{rf}$ dimensionless feed temperature.

Greek Symbols

α $= U_B . {}_{mf}/U_{mf}$.

α_1 Dimensionless pre-exponential factor.

β $= \dfrac{(-\Delta H).\ C_{rf}}{\rho_f.\ C_{p_f}.\ T_{rf}}$ exothermicity factor.

γ $= \dfrac{E}{R.T_{rf}}$ dimensionless activation energy.

ρ_s Solids density, gm cm^{-3}.

ρ_f Gas density gm cm^{-3}.

ϵ Voidage of dense phase.

REFERENCE

Aris, R. (1969) Chem. Engng. Sci., 24, 149

Aris, R. (1979) Elementary Reactor Analysis, Prentice Hall, Englewood Cliffs, New Jersey.

Calderbanck P.H., G.I. Watt and S.N. Ellis (1972) Proc. 2nd. Int. Symp. Chem. React. Engng. Paper B2-1. Amsterdam.

Cresswell D.L. and W.R. Paterson (1970) Chem. Engng. Sci., 25, 1405.

Elnashaie S.S., M. Marek and W.H. Ray. (1972) Proc. 2nd. Int. Symp. Chem. React. Engng. Elsevier Amesterdam.

Elnashaie S.S. and J.G. Yates. (1972) Chem. Engng. Sci., 27, 1757.

Elnashaie S.S. and J.G. Yates (1973). Chem. Engng. Sci., 28, 515.

Elnashaie S.S. (1977). Chem, Engng. Sci., 32, 295.

Elnashaie S.S, S. El-Bialy. (In Press) Chem. Engng. Sci.

El shishini S.S. (1979). Analysis of a class of Fluidized Bed Catalytic Reactors. Ph.D. Thesis.

Heerden C. Van (1953) Ind. Engng. Chem., 45, 1242.

Hlavacek V., M. Kubicek and M.Marek (1969) J. Cat., 15, 17.

Hlavacek V., M. Kubicek and K. Visnak (1972) Chem. Engng. Sci., 27, 719.

Partridge B.A. and P.N. Rowe (1966). Trans. Inst. Chem. Engrs., 44, 335.

Uppal A., W.H. Ray and A.B. Poore (1974) Chem. Engng. Sci., 29, 967.

APPENDIX (1)

Dense phase and bubble phase concentrations as function of bed length.

(i) Dense phase gas Completely Mixed

$$X_A = \frac{\bar{B}}{\bar{B} + EX_1} \qquad (A1)$$

$$X_B = \frac{EX_1}{(\bar{B} + EX_1)(\bar{B} + EX_2)} \qquad (A2)$$

$$\bar{X}_A = \frac{\bar{B} + EX_1 \exp(-a)}{(\bar{B} + A_1)} \qquad (A3)$$

$$\bar{X}_B = (1 - \exp(-a))\, X_B \qquad (A4)$$

(ii) Dense phase gas in Plug Flow

$$X_A = C_1 e^{\lambda 1 w} + C_2 e^{\lambda 2 w} \qquad (A5)$$

$$\bar{X}_A = \bar{C}_1 e^{\lambda 1 w} + \bar{C}_2 e^{\lambda 2 w} \qquad (A6)$$

$$X_B = C_3 e^{\lambda 3 w} + C_4 e^{\lambda 4 w} + W_1 C_1 e^{\lambda 1 w} + W_2 C_2 e^{\lambda 2 w} \qquad (A7)$$

$$\bar{X}_B = \bar{C}_3 e^{\lambda 4 w} + \bar{C}_4 e^{\lambda 4 w} + W_1 \bar{C}_1 e^{\lambda 1 w} + W_2 \bar{C}_2 e^{\lambda 2 w} \qquad (A8)$$

where $\bar{C}_1 = \frac{a.C_1}{(\lambda_1+a)}$; $\bar{C}_2 = \frac{a.C_2}{(\lambda_2+a)}$

$\bar{C}_3 = \frac{a.C_3}{(\lambda_3+a)}$; $\bar{C}_4 = \frac{a.C_4}{(\lambda_4+a)}$

and $a = QE\, A_C\, H / G_C$

$\lambda 1$, $\lambda 2$, $\lambda 3$, $\lambda 4$ are the roots of the characteristic equations and C_1, C_2, C_3, C_4, W_1, W_2 are constants calculated by the following equations:

$$C_1 = \frac{X_{Af}\,\lambda_2\,(\lambda_1+a)}{a\,(\lambda_2-\lambda_1)} \qquad (A9)$$

$$C_2 = X_{Af} - C_1 \qquad (A10)$$

$$C_3 = \frac{(\lambda_3+a)}{(\lambda_4-\lambda_3)}\left(\frac{W_1 C_1(\lambda_1-\lambda_4)}{(\lambda_1+a)} + \frac{W_2 C_2(\lambda_2-\lambda_4)}{(\lambda_2+a)}\right) \qquad (A11)$$

$$C_4 = -(C_3+W_1C_1+W_2C_2) \qquad (A12)$$

$$W_1 = \frac{H\,EX_1\,(\lambda_1+a)}{U_I\,(\lambda_1-\lambda_4)(\lambda_1-\lambda_3)} \qquad (A13)$$

$$W_2 = \frac{H\,EX_1\,(\lambda_2+a)}{U_I\,(\lambda_2-\lambda_4)(\lambda_2-\lambda_3)} \qquad (A14)$$

The integrals I_A and I_B are given by:

$$I_A = \int_0^1 X_A\, dw = (C_1/\lambda_1)(e^{\lambda 1}-1) + (C_2/\lambda_2)(e^{\lambda 2}-1) \qquad (A15)$$

$$I_B = \int_0^1 X_B\, dw = (C_3/\lambda_3)(e^{\lambda 3}-1) + (C_4/\lambda_4)(e^{\lambda 4}-1) + (W_1C_1/\lambda_1)(e^{\lambda 1}-1) + (W_2C_2/\lambda_2)(e^{\lambda 2}-1) \qquad (A16)$$

APPENDIX (2)

Calculation of the two phase model parameters:

$$G_B = A_R\,(U_o - U_{mf}) \qquad (A17)$$

$$G_C = G_B\,(1+\epsilon/(\alpha-1)) \qquad (A18)$$

$$G_I = A_R\,U_o - G_C \qquad (A19)$$

$$A_C = \epsilon\, G_B/(\alpha-1)\, U_{mf} \qquad (A20)$$

$$A_I = (A_R - A_C) \qquad (A21)$$

$$H = H_{mf}\,A_R/(A_R - A_C) \qquad (A22)$$

where $\alpha = U_B\,\epsilon/U_{mf}$ (A23)

and $U_B = 21.2\sqrt{D_B}$ cm/sec (A24)

The data necessary for the above calculation is given below:

Bed cross sectional area, A_R = 3000 cm^2

Superfitial gas velocity, U_o = 17.5 cm/sec

Minimum fluidization velocity, U_{mf} = 3.5 cm/sec

Voidage of dense phase, ϵ = 0.5

Bubble diameter, D_B = 10 cm

Height of bed at minimum fluidization, H_{mf} = 300 cm

AN APPLICATION OF THE FILTERING METHOD IN THE IDENTIFICATION AND CONTROL STAGES OF A HEAT EXCHANGE PROCESS

G. Ratefiarison, C. Humbert and J. Ragot

Laboratoire d'Automatique et de Recherche Appliquee, B.P. 452, 54001 Nancy Cédex, France

Abstract. As a result of the various discrepancies that affect control systems, equipping a process with an adaptative control generally requires a recursive optimal filter to be designed which is capable of producing the estimators for the parameters together with those for the model variables at one and the same time.

Since we are operating in the context of both gaussian and additive discrepancies and interference, theories that have been approximately confirmed in practice, the present paper sets forth a real-time identification method for multidimensional linear systems represented in the state space :

- taking account of dynamic discrepancies, the state vector of the model is estimated at each sampling stage by applying equations of the time-discrete Kalman filter and on the basis of the estimated parameter from the previous stage.

- Using the recurrent least squares method, the parameters are modified on the basis of this estimate of the parameters and the model variables at the same sampling stage.

On the basis of this estimate the control vector is calculated quadratically, such that the output of the controlled system conforms with a known desired output within a given expectation range.

A particular application is described using the concrete example of a heat exchanger whose pulse responses are broken down on the basis of discrete Laguerre functions. We also compare the identification results with those obtained by the conventional least squares method whereby the state vector is considered to bey deterministic.

Keywords. Discrepancy, heat exchanger, identification, least squares method, model, quadratic optimal control, recursive filter.

BASIC DATA CONCERNING THE TIME - DISCRETE KALMAN FILTER .

I-1 - Formulation of the Filtering Problem.

Let us consider a discrete time system described by the following difference equation :

$$\begin{cases} x_{k+1} = \Phi(k+1,k)\,x_k + B_k\,u_k + w_k & (I-1) \\ z_k = H_k\,x_k + v_k & (I-2) \end{cases}$$

. x_k is a $(n.1)$ dimensional state vector at time k with the following notation :

$$x_k^T = [x_{1,k}\ ;\ x_{2,k}\ ;\ldots;\ x_{n,k}]$$

. $\Phi(k+1,k)$ may be assumed to be an $(n.n)$ dimensional state transition matrix allowing the transition from state x_k to state x_{k+1}

. $B_k\,u_k$ is the deterministic control of dimension $(n.1)$

. z_k is a $(m.1)$ dimensional vector of observed outputs with the following notation :

$$z_k^T = [z_{1,k}\ ;\ z_{2,k}\ ;\ldots;\ z_{m,k}]$$

. H_k is the measurement matrix of dimension $(m.n)$

. w_k is the dynamic discrepancy of dimension $(n.1)$

. v_k is the measurement discrepancy of dimension $(m.1)$

$E[.]$ denotes the mathematical expectation operator based on the following hypotheses :

The initial state x_o is considered to be normally distributed with the following known statistics :

$$E[x_o] = \mu_o$$

$$E[(x_o-\mu_o)(x_o-\mu_o)^T] = P_o$$

. w_k and v_k are white gaussian sequences with zero mean values, not correlated with each other, and with x_o :

$$E[w_k] = [0] \qquad \forall k$$

$$E[v_k] = [0] \qquad \forall k$$

$$E[w_k w_k^T] = [q_k \delta_{kj}]$$

$$E[v_k v_j^T] = [r_k \delta_{kj}]$$

$$E[w_k v_j^T] = [0] \qquad \forall_k, \forall_j$$

$$E[w_k x_o^T] = [0] \qquad \forall_k$$

$$E[v_k x_o^T] = [0] \qquad \forall_k$$

δ_{kj} is the Kronecker delta $\begin{cases}\delta_{kj} = 1 \text{ if } k = j \\ \delta_{kj} = 0 \text{ if } k \neq j\end{cases}$

We are therefore operating in the context of additive Gaussian disturbances (interference factors).

Our problem is to restore as well as possible, despite the presence of this interference, the state vector x_k, knowing the set of observations $\underline{z}_N$ with the notation :

$$\underline{z}_N^T = [z_1^T ; z_2^T ; \ldots ; z_N^T]$$

According to the value of N, the following different situations arise :

Estimated value

$N = k \rightarrow \hat{x}_k^k$ "filtering"

$N > k \rightarrow \hat{x}_k^N$ "smoothing"

$N < k \rightarrow \hat{x}_k^N$ "prediction"

I-2 - Optimal Estimator :

The criterion chosen for the estimating of x_k is the sum of the variances of the estimation error components. This sum is weighed by a symmetrical positively defined matrix Q :

$$C = E[(x_k - \hat{x}_k^N)^T Q (x_k - \hat{x}_k^N)]$$

By minimizing this criterion, the following overall result is reached :

$$\hat{x}_k^N = E[x_k / \underline{z}_N] \qquad (1\text{-}3)$$

On the basis of the Gaussian hypothesis, we may show that the optimal estimator $\hat{x}_k^N$ is a linear function of the measurements.

Let us define the covariance matrix of the estimation error :

$$P_k^N = E[(x_k - \hat{x}_k^N)(x_k - \hat{x}_k^N)^T]$$

In the case of filtering, the optimal linear estimator can be calculated in a recurrent manner following the five Kalman filter equations :

a) Prediction.

$$\hat{x}_k^{k-1} = \Phi(k,k-1)\hat{x}_{k-1}^{k-1} + B_{k-1}u_{k-1} \qquad (I\text{-}4)$$

one-step prediction

$$P_k^{k-1} = \Phi(k,k-1)P_{k-1}^{k-1}\Phi^T(k,k-1) + q_{k-1} \qquad (I\text{-}5)$$

a priori variance

b) Estimation.

$$K_k = P_k^{k-1}H_k^T[H_k P_k^{k-1} H_k^T + r_k]^{-1} \qquad (I\text{-}6)$$

filter gain

$$P_k^k = [I - K_k H_k] P_k^{k-1} \qquad (I\text{-}7)$$

a posteriori variance

$$\hat{x}_k^k = \hat{x}_k^{k-1} + K_k[z_k - H_k \hat{x}_k^{k-1}] \qquad (I\text{-}8)$$

filtering algorithm

The final equation shows that the estimation $\hat{x}_k^k$ of state x_k is given by the prediction of this state plus a linear function of the measurement production error.

The algorithm is starred using a priori estimated values : first, the values of the initial state, and then the values of the positively defined covariance matrix of the initial error.

II - Application to the Idenfication of Linear Systems

The prededing Kalman filter equations, deduced from a model in which the measurement matrix H_k, comprising the parameter of this model, in presumed known, determine the minimal variance linear estimator of the state vector.

In practive, just like state x_k, this parameter matrix cannot be determined exactly.

The state estimation can therefore be made from an estimate H_k, or, considering that the parameters are constant, from a prediction H_{k-1} of this matrix.

Having obtained an estimation of state vector x_k, and hence supposing it to be known, we examine here the estimation of the parameters, that is to say the accuracy of their prediction at each sampling stage. To this end, we shall use the recursive algorithm of multidimensional least squares.

II-1 - Criterion and overall solution :

Let us re-write the measurement equation (I-2) where the parameters are constant :

$$z_k = H x_k + v_k \qquad (II\text{-}1)$$

With the following matrix notations :

$$Z_k = [z_1, z_2, \ldots, z_k]$$
$$X_k = [x_1, x_2, \ldots, x_k]$$
$$V_k = [v_1, v_2, \ldots, v_k]$$

it follows from (III-1) that :

$$Z_k = H X_k + V_k \qquad (II-2)$$

The criterion I_k to be minimized for each stage kT is the half-sum up to this point of the squares of the deviations between the real output and the model out put. Since this real output is inobtainable, we shall minimize the half-sum of the squares of the deviations between the measured output z_i $(i = \ell, k)$, and that of the model Hx_i $(i = \ell, k)$.

Whence the following statement of this criterion :

$$I_k = Trace\ (M_k) \qquad (II-3)$$

with

$$M_k = \frac{1}{2}\ [Z_k - H X_k]\ W_k - [Z_k - H X_k]^T \qquad (II-4)$$

where matrix W_k will enable us to weigh the preceding data by attributing greater significance to the latest date acquired $(\beta > 0)$

$$W_k = \begin{bmatrix} e^{-(k-1)\beta T} & & & 0 \\ & e^{-(k-2)\beta T} & & \\ 0 & & \ddots & \\ & & & e^{-\beta T} & \\ & & & & 1 \end{bmatrix}$$

If we use $\hat{H}_k$ to denote the optimal estimation of H at point kT, in other words, after k measurements have been carried out, the estimation is given by :

$$\left[\frac{\partial I_k}{\partial H}\right]_{\hat{H}_k} = 0 \qquad (II-5)$$

which gives :

$$\boxed{\hat{H}_k = Z_k W_k X_k^T (X_k W_k X_k^T)^{-1}} \qquad (II-6)$$

II-2 - Recurrent Effects Noteds During Measurements

If $\hat{H}_k$ represents the estimation of H from the k preliminary measurements, we shall now develop a method that will enable us to calculate $\hat{H}_{k+1}$ as a function of $\hat{H}_k$ without having to calculate the overall solution on each occasion .

If we obtain an additional measurement at point $(k+1)T$ we may add it to the preceding ones, thus writing :

$$[Z_k \ \vdots\ z_{k+1}] = H [X_k \ \vdots\ x_{k+1}] + [V_k \ \vdots\ v_{k+1}]$$

or :

$$Z_{k+1} = H X_{k+1} + V_{k+1}$$

This minimization criterion is now :

$$I_{k+1} = Trace\ (M_{k+1})$$

with :

$$M_{k+1} = \frac{1}{2}\ [Z_{k+1} - H X_{k+1}]\ W_{k+1} [Z_{k+1} - H X_{k+1}]^T$$

Taking in to account the definition of W_k, we obtain :

$$M_{k+1} = \frac{1}{2} \left\{ [Z_k - H X_k \ \vdots\ z_{k+1} - H x_{k+1}] \begin{bmatrix} e^{-\beta T} W_k & \vdots & 0 \\ \cdots & & \cdots \\ 0 & \vdots & 1 \end{bmatrix} \begin{bmatrix} (Z_k - H X_k)^T \\ \cdots\cdots \\ (z_{k+1} - H x_{k+1})^T \end{bmatrix} \right\}$$

$$M_{k+1} = \frac{1}{2} \Big\{ [Z_k - H X_k]\ e^{-\beta T}\ W_k [Z_k - H X_K]^T + [z_{k+1} - H x_{k+1}] [z_{k+1} - H x_{k+1}]^T \Big\} \qquad (II-7)$$

If we consider that the estimation of H is optimal at point kT and that between points kT and $(k+1)T$, it sustains a variation ΔT, the expression is :

$$M_{k+1} = \frac{1}{2} \Big\{ [Z_k - (\hat{H}_k + \Delta H) X_k]\ e^{-\beta T} W_k [Z_k - (\hat{H}_k - \Delta H) X_k]^T + [z_{k+1} - (\hat{H}_k + \Delta H)\ x_{k+1}]\ [z_{k+1} - (\hat{H}_k + \Delta H)\ .\ x_{k+1}]^T \Big\}$$

Variation $\Delta\hat{H}_{k-1}$ is optimal and determines the optimal estimator $\hat{H}_{k+1}$ if I_{k+1}, is minimized, in other words if :

$$\left[\frac{\partial I_{k+1}}{\partial \Delta H}\right]_{\Delta\hat{H}_{k+1}} = 0 \qquad (II-8)$$

Using this relationship to make the following statement :

$$[\ x_{k+1}^T (X_k e^{-\beta T} W_k X_k^T)^{-1}\ x_{k+1} + 1]^{-1}\ x_{k+1}^T\ (X_k e^{-\beta T} W_k X_k^T)^{-1} = Ka, k+1 \qquad (II-9)$$

and :

$$(X_k W_k X_k^T)^{-1} = P_{a,k} \qquad (II-10)$$

we obtain the following results [18]:

$$K_{a,k} = [x_k^T P_{a,k-1} x_k + e^{-\beta T}]^{-1} x_k^T P_{a,k-1} \qquad (II-11)$$

$$P_{a,k} = e^{\beta T} P_{a,k-1} [I - x_k K_{a,k}] \qquad (II.12)$$

$$\hat{H}_k = \hat{H}_{k-1} + [z_k - \hat{H}_{k-1} x_k] K_{a,k} \qquad (II.13)$$

The last three formulae constitute the recursive algorithm of the multidimensional least squares, introduced by a priori estimated values of $\hat{H}_o$ and $P_{a,o}$.

Written in this order, they enable us to revise the matrices $K_{a,k-1}$, $P_{a,k-1}$ and H_{k-1} since we now know z_k and x_k. The calculation does not entail any matrix inversion since in (III-14) the quantity shown in brackets in brackets is scalar.

In practice, we cannot obtain the exact value of x_k ; so we shall modify it on the basis of estimation $\hat{x}_k^k$ obtained from the Kalman filter équations; in these equations since we do not know the matrix constant H_k, we will replace it by its prediction $\hat{H}_{k-1}$ which will therefore be corrected at each sampling stage.

We thus obtain, in the case of a systeme subjected to interference defined by equations (I-1) and (I-2), an estimation at the same sampling point of the parameters and model variables.

III - Application to the Real Time Identifica- of a Thermal Process.

III-1- Description of The Apparatus.

The process shown in figure (C-1) consists of a hot water gas boiler and of heat exchanger.

The boiler is for domestic use and is fuelled with Groningen (natural) gas whose heating capacity is 8400 kcal/m^3. It is fitted with a safety device :

- a regulating thermostat controlled by the hot water temperature ;

- a regulating thermostat controlled by the hot water temperature ;

- an electrical gas-flow isolator which comes into operation if the gas pressure is abnormally low.

The characteristics are as follows :

Nominal power	: 13200 kcal/h.
Nominal gas flow	: 1,2 m^3/h.
Water capacity	: 21 l.
Gas pressure	: 27 mb.
Water pressure	: 2 bars.

The exchanger is an industriel "liquid-liquid" type with a cluster of pipes.

The water to be heated circulates in the pipes while the heated fluid circulates in the outer casing.

The characteristics of the exchange are :

capacity of the pipe circuit : 0,6 l.
capacity of the outer circuit : 0,5 l.

Operation is defined by the following magnitudes ,

for the gas : Q_g = gas flow

for the hot water : Q_c = boiler outflow = boiler intake

Q_e = exchanger intake = exchanger outflow

T_{ce} = boiler outflow temperature = exchanger intake temperature

T_{cs} = exchanger outflow temperature

T_r = boiler intake temperature

for the cold water :

Q_f = water flow

T_f = exchanger intake temperature

T_s = exchanger outflow temperature

This process is connected up to a SOLAR 16-40 computer which operates the identification and the control. The whole is shown in figure (C-2).

III-2 - Simulation .

When the pulse response $f(t)$ of a system to be identified is aperiodic or sufficiently damped, it can be approximated at each point kT where T is the sampling period, using the sum :

$$f(kT) = \sum_{i=0}^{n-1} a_i \ell_i (kT)$$

limited to n terms where $\ell_i(kT)$ is the ith discrete Laguerre function whose representation in the Z plane is [9] :

$$Li(Z) = \frac{(e^{-\alpha T} - Z^{-1})^i}{(1 - e^{-\alpha T} Z^{-1})^{i+1}}$$

We shall select gas flow Q_g and hot water flow Q_e as control variables and exchanger outflow temperatures T_{cs} and T_s as measurements : the model used for the identification of the heat process consists of two series of discrete Laguerre filters, as shown in figure (C-3).

We shall use the following notations :

Deterministic inputs :

$$u_k = \begin{bmatrix} u_1(kT) \\ u_2(kT) \end{bmatrix} = \begin{bmatrix} Q_g(kT) \\ Q_e(kT) \end{bmatrix} \begin{matrix} \text{gas flow} \\ \text{exchanger hot water flow} \end{matrix}$$

Dynamic discrepancies : We reduce the dynamic discrepancies to discrepancies affecting the two input flow rates :

$$d_k = \begin{bmatrix} d_1(kT) \\ d_2(kT) \end{bmatrix}$$ discrepancy on Q_g, discrepancy on Q_e

Measured outputs :

$$z_k = \begin{bmatrix} z^1(kT) \\ z^2(kT) \end{bmatrix} = \begin{bmatrix} T_s(kT) \\ T_{cs}(kT) \end{bmatrix}$$ measured output temperatures

Measurement discrepancies :

$$v_k = \begin{bmatrix} v^1(kT) \\ v^2(kT) \end{bmatrix}$$ discrepancies in the measurement of T_s, discrepancies in the measurement of T_{cs}

Real outputs :

$$z_{r,k} = \begin{bmatrix} z_r^1(kT) \\ z_r^2(kT) \end{bmatrix}$$ real output temperatures

Model outputs :

$$z_{m,k} = \begin{bmatrix} z_m^1(kT) \\ z_m^2(kT) \end{bmatrix}$$ temperature T_s of the model, temperature T_{cs} of the lodel

Components of the state vector of the model :

$x_{1,i}(kT)$ output of the ith Laguerre filter in relation to the control $u_1(kT) = Q_g(kT)$

$x_{2,i}(kT)$ output of the ith Laguerre filter in relation to the control $u_2(kT) = Q_e(kT)$

Parameters of the model :

$a^1_{1,i}$: Parameter associated with $x_{1,i}$ and z^1_m

$a^2_{1,i}$: Parameter associated with $x_{1,i}$ and z^2_m

$a^1_{2,i}$: Parameter associated with $x_{2,i}$ and z^1_m

$a^2_{2,i}$: Parameter associated with $x_{2,i}$ and z^2_m

Here again, since we cannot know the real output $z_{r,k}$, we shall minimize the deviation between the measured output z_k and the output of the of model $z_{m,k}$ in order to identify the parameters.

Using the recursions obtained from the Laguerre filters, and takings in to account the discrepancy d_k affecting the system, we shall write :

$$x_k^T = [x_{1,0}(kT);\ldots;x_{1,n-1}(kT);x_{2,0}(kT);\ldots;x_{2,n-1}(kT)]$$

with the matrix notation :

$$x_k = \Phi x_{k-1} + B\,u_{k-1} + w_{k-1}$$

and :

$$\Phi = \begin{bmatrix} \begin{matrix} c_1 & 0 & \cdots & 0 \\ b_1c_1 & c_1 & & \vdots \\ \vdots & & \ddots & 0 \\ b_1c_1^{n-1} & \cdots & b_1c_1 & c_1 \end{matrix} & \begin{matrix} 0 & \cdots & 0 \\ \vdots & & \vdots \\ 0 & \cdots & 0 \end{matrix} \\ \begin{matrix} 0 & \cdots & 0 \\ \vdots & & \vdots \\ 0 & \cdots & 0 \end{matrix} & \begin{matrix} c_2 & 0 & \cdots & 0 \\ b_2c_2 & c_2 & & \vdots \\ \vdots & & \ddots & 0 \\ b_2c_2^{n-1} & \cdots & b_2c_2 & c_2 \end{matrix} \end{bmatrix}$$

$$B = \begin{bmatrix} 1 & 0 \\ c_1 & \vdots \\ \vdots & \vdots \\ c_1^{n-1} & 0 \\ 0 & 1 \\ \vdots & c_2 \\ \vdots & \vdots \\ 0 & c_2^{n-1} \end{bmatrix}, \quad w_k = B\,d_k$$

where :

$$b_1 = e^{-\alpha_1 T} - e^{\alpha_1 T}, \quad c_1 = e^{-\alpha_1 T}$$

$$b_2 = e^{-\alpha_2 T} - e^{\alpha_2 T}, \quad c_2 = e^{-\alpha_2 T}$$

By approximating output $z_{r,k}$ by means of the Laguerre model output, and tacking in to account the measurement discrepancy, the observed output may be written thus :

$$z_k = H\,x_k + v_k$$

with :

$$z_{m,k} = H\,x_k$$

where :

$$H = \begin{bmatrix} a^1_{1,0};a^1_{1,1};\ldots;a^1_{1,n-1};a^1_{2,0};a^1_{2,1};\ldots;a^1_{2,n-1} \\ a^2_{1,0};a^2_{1,1};\ldots;a^2_{1,n-1};a^2_{2,0};a^2_{2,1};\ldots;a^2_{2,n-1} \end{bmatrix}$$

III-3 - Identification

We have used two identification methods at the same time, in order to compare the results against a real time example :

- MC represents the conventional least squares algorithm, using the deterministic state :

$$\hat{x}_k = \Phi\,\hat{x}_{k-1} + B\,u_{k-1}$$

- MCF represents the least squares algorithm, taking in to account the dynamic discrepancy and, consequently, using the state estimate $\hat{x}_k^k$.

The second method therefore identifies state x_k at the same time as matrix H.

The identification programme carried out at each signal from the real time clock - whenever the signal is a multiple of the sampling period - is the following, in which matrices K_a, P_a, and H derived from the estimated state

(MCF) are respectively written

$K^F_{a,k}$, $P^F_{a,k}$, and $\hat{H}^F_k$ at point kT

Identification was carried out with :

- a sampling period : $T = 10\ s$
- outputs with 30% amplitude stages, which switch randomly after each series of 15 sampling periods
- the following poles of the Laguerre filters
$$\alpha_1 = \alpha_2 = 0,03$$
- an oversight coefficient : $\beta = 0,005$
- the following initializing values :

$$\hat{x}_o^T = [1,0,0,0,1,0,0,0]$$

$$\hat{H}_o = \begin{bmatrix} 1,0,0,0,1,0,0,0 \\ 1,0,0,0,1,0,0,0 \end{bmatrix}$$

$$P_o = P_{a,o} = 100\ I$$

Several tests enabled us to estimate that the standard deviation of measurement discrepancies is 1,5 %.

The standard deviation for dynamid discrepancies is 3 %, taking in to account the structural difference between model and system.

Measurements T_s and T_{cs} are established in relation to a steady state :

$$T_{sr}(t) = T_s(t) - T_{so},\quad T_{csr}(t) = T_{cs}(t) - T_{cso}$$

The results are indicated in figures (C-4) and (C-5) showing :

- measured output : T_s and T_{cs}
- model output from MCF: T_{sMCF} and T_{csMCF}
- model output from MC : T_{sMC} and T_{csMC}

IV- Application to Heat Process Control.

In this section, we shall apply the previous identification algorithm MCF to the real time control of water temperature T_s at the exchanger output.

IV-1 - Formulation of the Quadratic Optimal Control Problem.

Let us reconsider the model of a discrete linear system :

$$\begin{cases} x_{k+1} = \Phi x_k + B u_k + w_k & (IV-1) \\ z_k = H_k x_k + v_k & (IV-2) \end{cases}$$

We shall now examine the situation in which matrices Φ and B are constants.

The quadratic optimal control problem consists in determining the control sequence $\{u_i\}$ $(i = 0,N-1)$ which minimizes the performance criterion :

$$J(\{u_i\}) = \frac{1}{2}(z_N - y_N)^T(z_N - y_N) + \frac{1}{2}\sum_{i=0}^{N-1}[(z_i - y_i)^T(z_i - y_i) + u_i^T F u_i] \qquad (IV-3)$$

where y_i $(i = 0, N)$ are a known desired output for a given range N, in which 0 corresponds to the present moment.

Weighting matrix F, which is symmetrical and positively defined, is introduced so as not to harm the system by excessive controls.

In such conditions, retaining the same hypothèse specificed in paragraph I-1 with regard to discrepancies w_k and v_k together with the initial state x_o, the following result may be obtained [15] :

The optimal control sequence $\{u_i\}$ $(i = 0,N-1)$ which minimizes $J(\{u_i\})$ taking in to account (IV-1) and IV-2), is given by the law obtained from the deterministic problem where $v_k \equiv w_k \equiv 0$, in which x_k is replaced by the Kalman estimate $\hat{x}^k_k$

We must therefore obtain control sequence $\{u_i\}$ which minimizes $J(\{u_i\})$ with :

$$\begin{cases} x_{k+1} = \Phi x_k + B u_k & (IV-4) \\ z_k = H_k x_k & (IV-5) \end{cases}$$

In then remains for us to replace x_k by the estimate $\hat{x}^k_k$ obtained using the Kalman filter.

IV-2 - Determination of the Control Sequence

Taking in to account relationship (IV-5), the criterion (IV-3) to be minimized is written :

$$J(\{u_i\}) = \frac{1}{2}(H_N x_N - y_N)^T(H_N x_N - y_N) + \frac{1}{2}\sum_{i=0}^{N-1}[(H_i x_i - y_i)^T(H_i x_i - y_i) + u_i^T F u_i] \qquad (IV-6)$$

By introducing the constraints (IV-4), the hamiltonian operator is as follows :

$$\mathcal{H} = \frac{1}{2}(H_N x_N - y_N)^T(H_N x_N - y_N) + \sum_{i=0}^{N-1}\{\frac{1}{2}[(H_i x_i - y_i)^T(H_i x_i - y_i) + u_i^T F u_i] + \lambda_{i+1}^T(\Phi x_i + B u_i - x_{i+1})\} \qquad (IV-7)$$

where λ_{i+1} $(i = 0,N-1)$ are $[n.1]$ vectors of the Lagrange multpipliers.

By minimizing H , we obtain the following optimal control sequence [18] :

$$\boxed{u_i = -[F + B^T L_{i+1} B]^{-1} B^T [L_{i+1}\Phi \hat{x}^i_i + M_{i+1}]} \qquad (IV-8)$$

where the matrices L_{i+1} and M_{i+1} are solutions to equations :

$$\boxed{L_i = H_i^T H_i + \Phi^T L_{i+1}[I - B(B^T L_{i+1} B + F)^{-1} B^T L_{i+1}]\Phi} \qquad (IV-9)$$

$$M_i = -\Phi^T L_{i+1}[I-B(B^T L_{i+1} B+F)^{-1} B^T L_{i+1}] BF^{-1} B^T M_{i+1} + \Phi^T M_{i+1} - H_i^T y_i \quad (IV-10)$$

initialised by :

$$L_N = H_N^T H_N \quad (IV-11)$$

$$M_N = - H_N^T y_N \quad (IV-12)$$

IV-3 - Heat exchanger control.

Still using the discrete Laguerre model, we schall now apply the algorithm consisting of équations (IV-8), (IV-9), (IV-10), (IV-11), and (IV-12) to control temperature T_s.

Identification is carried out using the MCF algorithm in which T_s will be the only output taken in to account.

The notations from paragraph III-2 will be retained, but we shall abandon the second component T_{cs} (kT) of measurement z_k as well the second row of the observation matrix H and the second component of the measurement discrepancy v_k .

The control programme with identification is as follows, page .

The results of a test in which se established a periodic desired outup are shown in figure (D-1) wich shows this output and the observed temperature T_s simultaneously.

CONCLUSION

By adopting a situation with additive and gaussian discrepancies and interference, wich are hypothèses that are been verdified approximately in practice, we have endeavoured to show in this paper how to estimate simultaneously, using a combination of linear filtering and recursive least squares, the parameters and variables of a model of a multidimentional linear system.

The method proposed for the real time identification of processes subject to interference did in fact operate in the concrete cas of a thermal exchanger.

Although the convergence of the method has not been proven, and more rigorous theorical research must accompagny this study, the application of the algorithm in controlling the exhanger has shown us how the proposed solution enables the real time regulation of systems subjected to discrepancies to be carried out.

REFERENCES

[16] Athans M., Falb Pl. (1966). Optimal Control. Mc Graw-Hill Book Company, p. 750-814)

[1] Åström K.J. (1970). Introduction to Stochastic control theory. Academic Press, p. 210-237.

[2] Clarke D.W. (1967). Generalized-Least squares estimation of the parameters of a dynamic model. IFAC Symp. Identification in automatic control systems, Prague, p.3-17.

[13] Gentil S. (1972). Etude comparative de diverses méthodes statistiques d'identification des systemes dynamiques. Thèse de 3ème cycle, Grenoble.

[14] Graupe D., Krieger R.E. (1976). Identification of systems. Publicing Company Huntington, New-York, p.253-255.

[3] Guenod M.A., Fatio J.L. (1973).Cours d' introduction aux méthodes d'identification. Albin Michel.

[17] Jazwinski A.H.(1970).Stochastic Processes and Filtering Theory. Academic Press.

[4] Kalman R.E. (1960). A new approach to linear filtering and prediction problems Trans. A.S.M.E., Jnl Basic Engineering, vol. 82 D, p. 35-45.

[5] Kalman R.E., Bertram J.E. (1960). Control system analysis and design via the second method of Lyapounov. Trans.A.S.M.E., Jnl Basic Engineering, p. 371-400.

[6] Kalmann R.E., Bucy R.S. (1961). New results in linear filtering and prediction Theory. Trans.A.S.M.E., Jnl Basic Engineering, vol. 83 D, p.95-108.

[7] Kalman R.E. (1961). Contributions to the Theory of optimal control. Bol.Soc. Mat. Mex. p. 102-119.

[8] Kaufmann A. (1965). Cours de calcul des probabilités. Albin Michel.

[15] de Larminat P., Thomas Y. (1977). Automatique des systèmes linéaires. Tome III Flammarion Sciences, p. 127-205.

[9] Marmonier R. (1977). Applications des fonctions de Laguerre discrètes à l'identification et à la commande des processus linéaires. Thèse de 3ème cycle, Nancy.

[18] Ratefiarison G. (1980). Applications du filtrage à l'identification et à la commande des processus. Thèse de 3ème cycle, Nancy.

[10] Sage A.P. (1968). Optimum systems control W.L.Everitt, Editor: Series in Networks Robert W. Newcomb,Editor,Prentice-Hall Inc., p. 276-277.

[11] Sorenson H.W.(1968). Kalman filtering techniques. Advances in Control systems, edited by Leondes, vol. 3, Academic Press, p. 219-289.

[12] Sorenson H.W. (1968). Controllability and Observbility of linear, stochastic time discrete control systems. Advances in Control systems, edited by Leondes, vol. 5, Academic Press, p. 95-158.

Identification Programme

Data acquisition

$(u_{k-1} \quad et \quad z_k)$

↓

Determining state

$$\hat{x}_k = \Phi\, \hat{x}_{k-1} + B\, u_{k-1}$$

↓

MC least squares algorithm

$$K_{a,k} = [\, \hat{x}_k^T P_{a,k-1} \hat{x}_k + e^{-\beta T}]^{-1} \hat{x}_k^T P_{a,k-1}$$

$$P_{a,k} = e^{\beta T} P_{a,k-1} [\, I - \hat{x}_k K_{a,k}]$$

$$\hat{H}_k = \hat{H}_{k-1} + [z_k - \hat{H}_{k-1} x_k]\, K_{a,k}$$

↓

State estimation

$$\hat{x}_k^{k-1} = \Phi\, \hat{x}_{k-1}^{k-1} + B\, u_{k-1}$$

$$P_k^{k-1} = \Phi\, P_{k-1}^{k-1} \Phi^T + q$$

$$K_k = P_k^{k-1} \hat{H}_{k-1}^{FT} [\, \hat{H}_{k-1}^F P_k^{k-1} \hat{H}_{k-1}^{FT} + r]^{-1}$$

$$P_k^k = [\, I - K_k \hat{H}_{k-1}^F] P_k^{k-1}$$

$$\hat{x}_k^k = \hat{x}_k^{k-1} + K_k \,[\, z_k - \hat{H}_{k-1}^F \hat{x}_k^{k-1}]$$

↓

MCF least squares algorithm

$$K_{a,k}^F = [\, \hat{x}_k^{kT} P_{a,k-1}^F \hat{x}_k^k + e^{-\beta T}]^{-1} \hat{x}_k^{kT} P_{a,k-1}^F$$

$$P_{a,k}^F = e^{\beta T} P_{a,k-1}^F [I - \hat{x}_k^k K_{a,k}^F]$$

$$\hat{H}_k^F = \hat{H}_{k-1}^F + [\, z_k - \hat{H}_{k-1}^F \hat{x}_k^k]\, K_{a,k}^F$$

↓

Result output

↓

Waiting loop

Control Programme with Identification

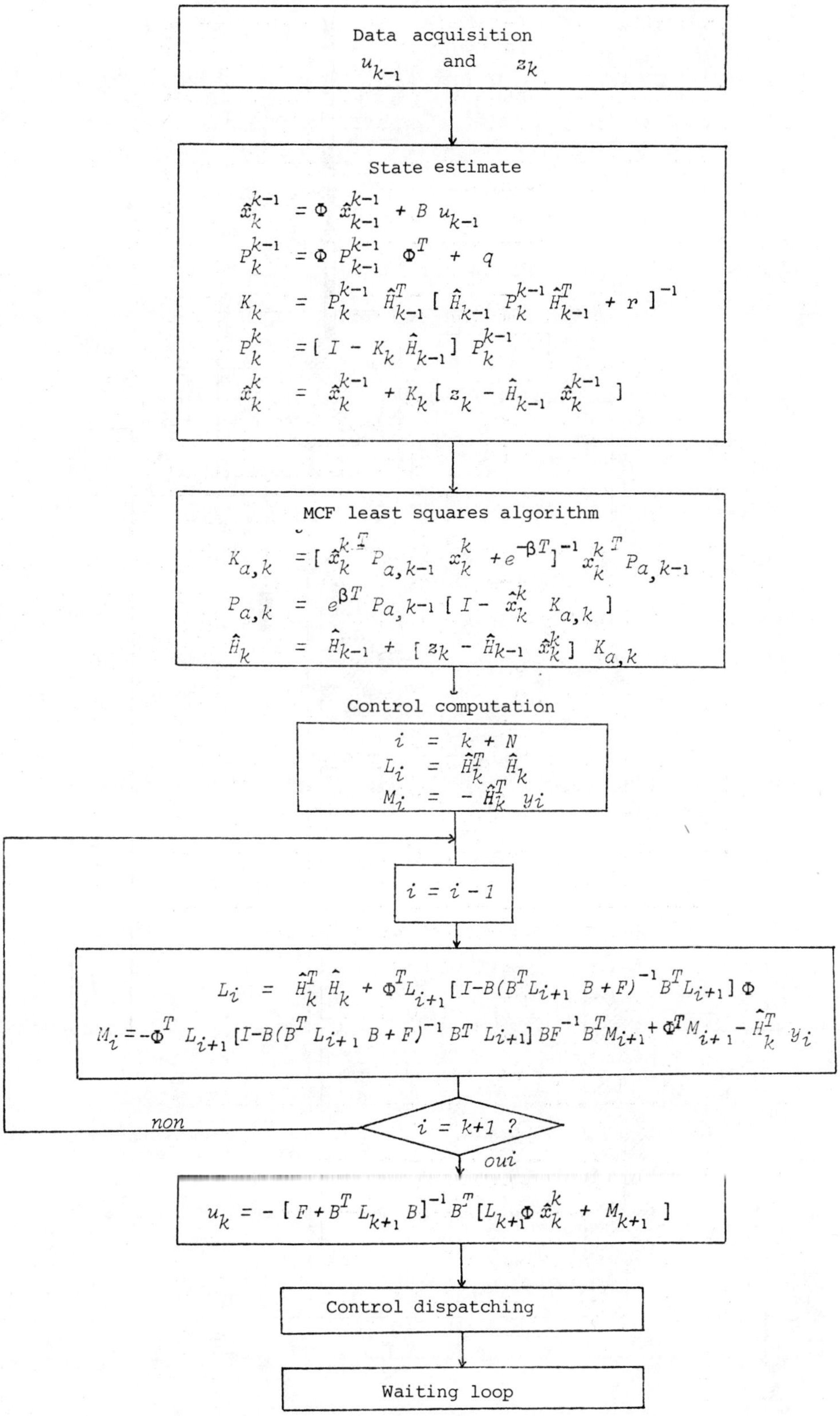

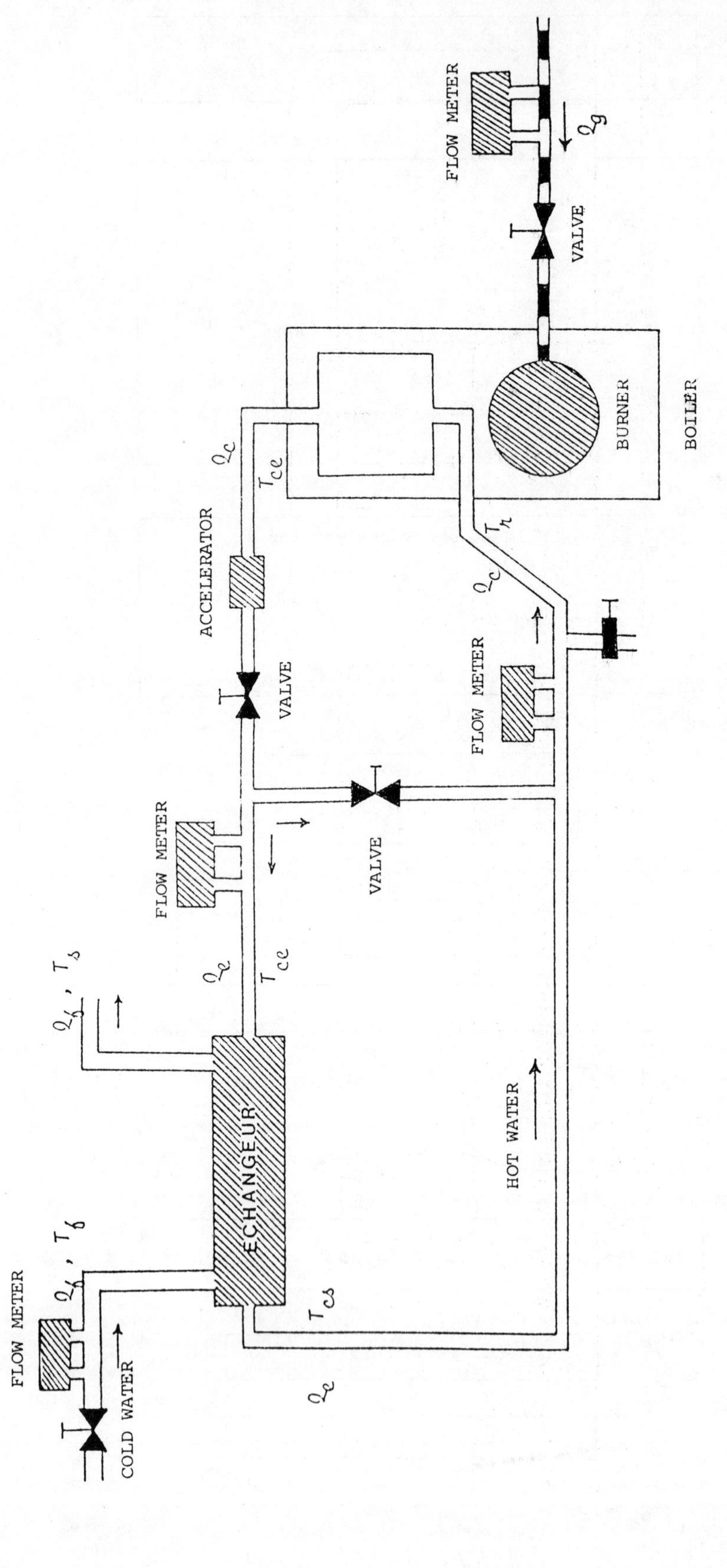

BLOCK DIAGRAM OF THE THERMAL PROCESS

Fig. C-1

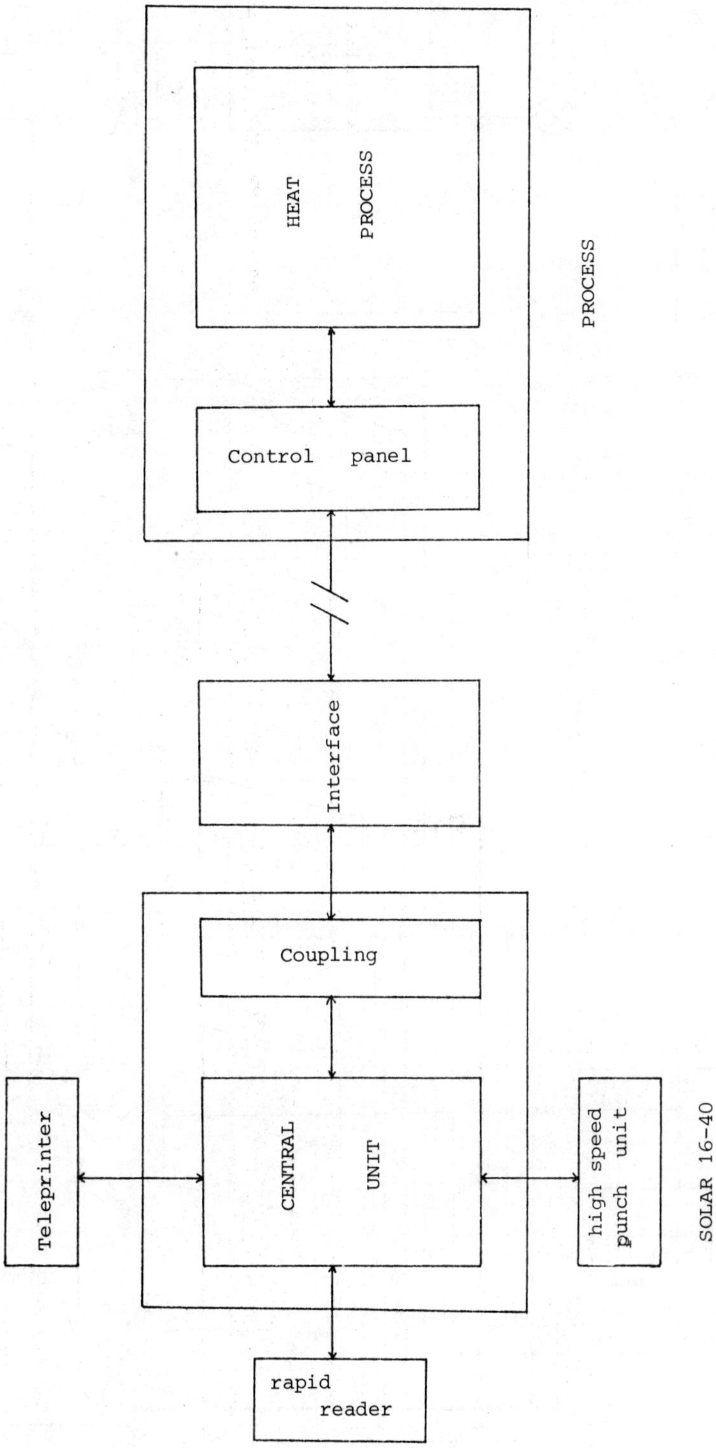
HEAT
PROCESS
PROCESS
Control panel
Interface
Coupling
Teleprinter
CENTRAL
UNIT
high speed
punch unit
SOLAR 16-40
rapid
reader

Fig. - C-2

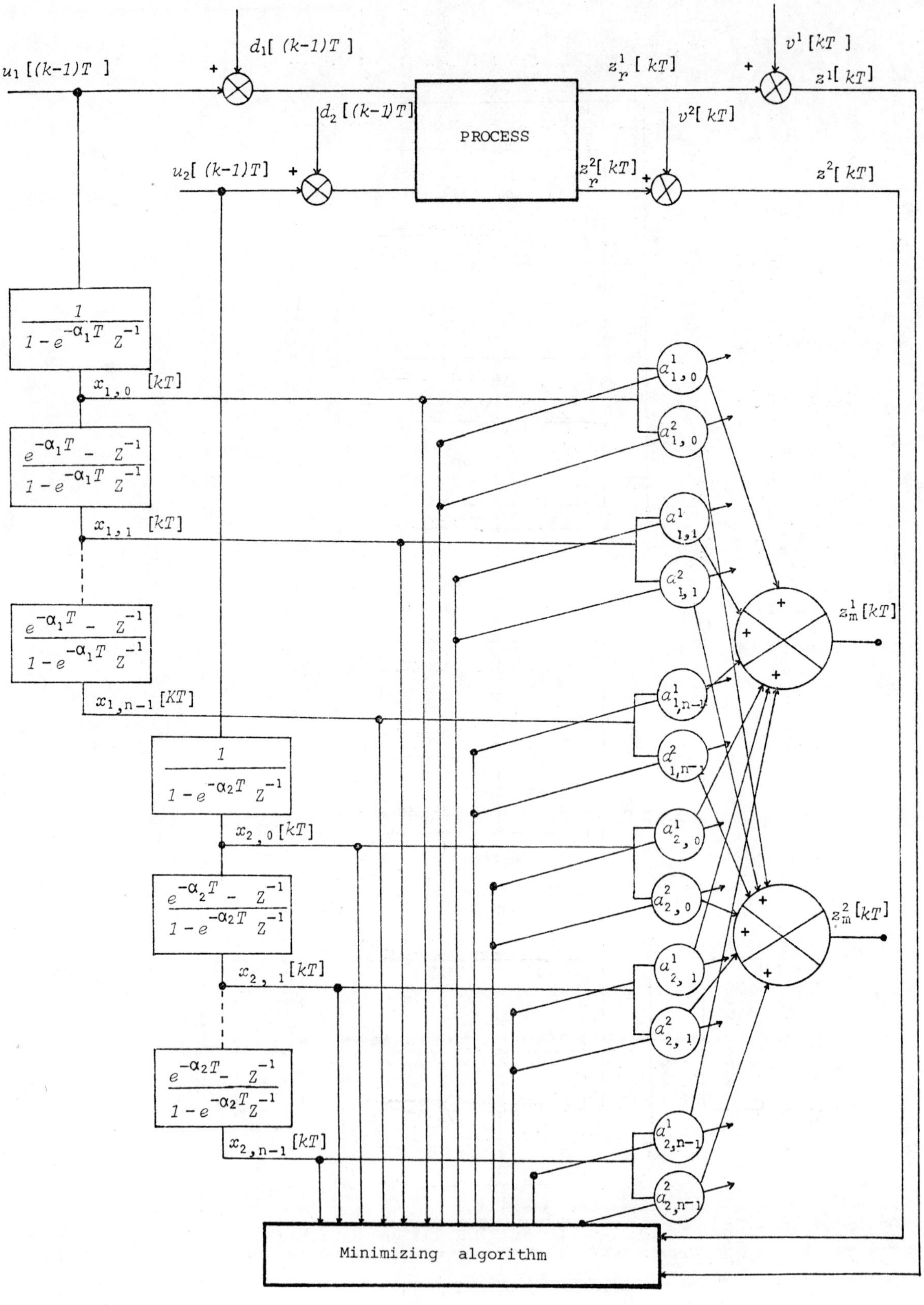
$u_1[(k-1)T]$
$d_1[(k-1)T]$
$d_2[(k-1)T]$
$u_2[(k-1)T]$
PROCESS
$z_r^1[kT]$
$z_r^2[kT]$
$v^1[kT]$
$v^2[kT]$
$z^1[kT]$
$z^2[kT]$
$\frac{1}{1-e^{-\alpha_1 T} Z^{-1}}$
$x_{1,0}[kT]$
$\frac{e^{-\alpha_1 T} - Z^{-1}}{1-e^{-\alpha_1 T} Z^{-1}}$
$x_{1,1}[kT]$
$x_{1,n-1}[KT]$
$\frac{1}{1-e^{-\alpha_2 T} Z^{-1}}$
$x_{2,0}[kT]$
$\frac{e^{-\alpha_2 T} - Z^{-1}}{1-e^{-\alpha_2 T} Z^{-1}}$
$x_{2,1}[kT]$
$x_{2,n-1}[kT]$
$a^1_{1,0}$
$a^2_{1,0}$
$a^1_{1,1}$
$a^2_{1,1}$
$a^1_{1,n-1}$
$a^2_{1,n-1}$
$a^1_{2,0}$
$a^2_{2,0}$
$a^1_{2,1}$
$a^2_{2,1}$
$a^1_{2,n-1}$
$a^2_{2,n-1}$
$z_m^1[kT]$
$z_m^2[kT]$
Minimizing algorithm

Fig. - C-3 .

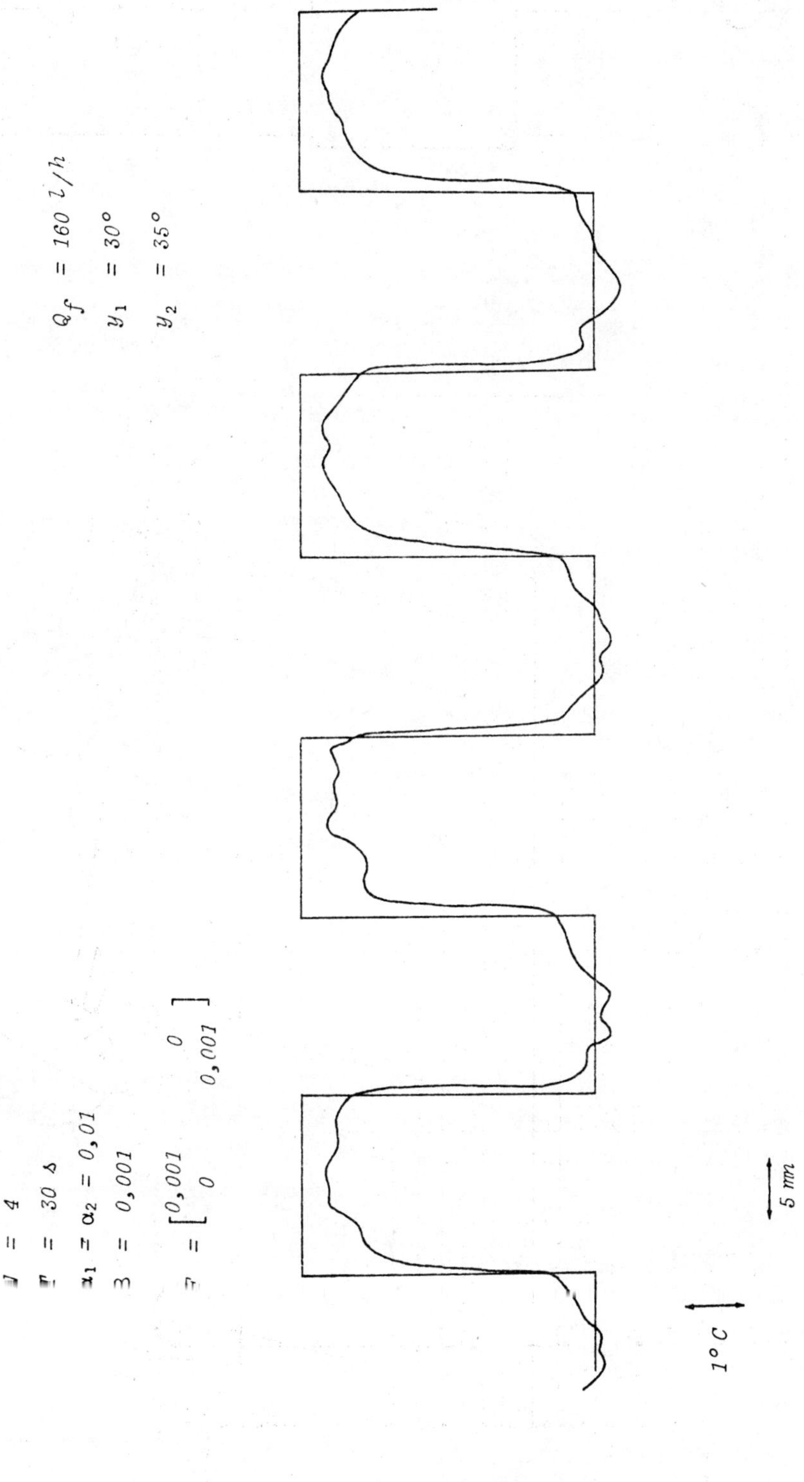

Q_f = 160 l/h
y_1 = 30°
y_2 = 35°
= 4
= 30 s
α_2 = 0,01
= 0,001
$\begin{bmatrix} 0,001 & 0 \\ 0 & 0,001 \end{bmatrix}$
1° C
5 mm

Fig. - D-1

SCIENTIFIC MANPOWER PLANNING: A METHODOLOGY FOR EGYPT*

M. S. Mahmoud* and M. I. Younis**

**Electronics and Communications Department, Faculty of Engineering, Cairo University, Giza, Egypt.*
***Systems and Information Sciences Laboratory, National Research Center, Dokki, Giza, Egypt*

Abstract. The basic problem to which this paper is addressed is that of developing a system methodology for scientific manpower planning in Egypt. We integrate the main influencing components into one unified system for the purpose of policy assessment and dynamic evolution. A linear, deterministic model is constructed based on the scientific manpower supply-utilization disequilibrium conditions. Preliminary results over the period 1976-1985 have indicated that the propsed methodology is encouraging and works satisfactorly.

Keywords. Developing Countries; Education; Manpower Planning; Modelling; Simulation; System Analysis.

INTRODUCTION

Egypt in the contemporary stage of development is facing continuous challenges to resolve many deep-rooted societal dilemmas centered around the population explosion, shortage of food production, inadequacy of housing and industrial plans and deficiency in capital production. Broadly speaking, miscoordination in the planning activities is perhaps the major influencing factor behind this intricate situation, to the extent it generates multifaceted problems.
It is fair, however, to recognize that the national authorities have been undertaking, over the past decade or so, extensive and diverse efforts to lessen the impact of these vital problems. These efforts ultimately have resulted in noticeable patterns of economic growth, educational progress and social evolution. Such patterns have reached the point where policy-makers have decided to benefit from, yet there are contradictory interrelation ships which have to be taken into consideration.

In developing countries, the widelyheld concept of "investment in human capital", is being adopted with much consciousness and interests. The principal rationale, in simple terms, is that the more investment allocations to general education, training programs and science and technology, the more qualified manpower and skilled labor a country will have and consequently more effective contribution to economic development and social returns. Indeed, this basic notion of recongizing the tangible benefits of human resources utilization has commanded the attention of numerious researchers and organizations (Correa , 1975; Hansen, 1970;OECD, 1971, 1973, 1975;, NSF, 1975; Spaey, 1971). Several approaches have been developed to evaluate, forecast and quantify the utilization of human resources components.

Our interest here is to provide a tractable methodology capable of dealing with the different aspects of scientific manpower pr probkem in developing countries. The focus for the paper, though pertains to Egypt, the conceptualizations are equally valid for other developing countries. The content of the remaining portion of the paper is organized as follow. We start by elaborating on an integrated system approach particularly developed to include the potential stream of producing scientists and technologists in Egypt. The system consists of eight interacted submodels, focusing on the generation of scientific manpower, its contribution to the labor force and economy and the capital investment expenditures. We then move to construct a dynamic model based on supply-utilization disequilibrium conditions. Preliminary results of model implementation have indicated satisfactory outcome. Indeed it needs further refinements for potential use.

It should be remarked that we have ignored temporarily the elaboration on the financial, whether fixed or current capital expenditure submodel. Also, the important problems of brain-drain and emigration, secondment or permanent, are bypassed in this work. The inclusion of these issues in addition to more detailed characterization and pertinent orientation is our immediate task.

* This work has been supported by the project (Scientific Manpower Needs for Egypt till the year 2000), sponsered by The Academy of Scientific Research and Technology, Cairo, EGYPT.

AN INTEGRATED SYSTEM APPROACH

Attaining national targets of a developing country is entirely dependent on proper procurement and utilization of scientific manpower (Spaey, 1971). Planning for researchers, scientists and technologists has been conceived as indespensible ingredient in national development policy programmes. Scientific manpower planning is essentially concerned with the employment of available human resources along with the financial and physical resources to best attain the national goals.

An integrated modelling approach is advanced in this paper to evaluate the dynamic trends of scientific manpower and to provide major guidelines for planning purposes. Basically the integrated system model is broken down into eight submodels, each is concerned with a particular aspect and is assigned a given task. They are enumerated below:
1. Student Population Submodel
2. University Structure Submodel
3. Graduation Pattern Submodel
4. Post-Graduate Profile Submodel
5. Labor Force Submodel
6. Productive Sector Submodel
7. Economic Allocation Submodel
8. Investment and Expenditure Submodel

The possible linkages between submodels are portrayed in Fig.1. Observe that this integrated system includes the main generating source, student population, of scientific personnel. By that it therefore traces the development of a researcher and possesses different policies to steer the evolution of size and nature of scientific manpower. Before going further, it seems appropriate to stress that the scientific manpower can be categorized in different ways; like

1. By field of science
   There is a coding system adopted in this regards (OECD, 1975). A one-digit code comprises the disciplines.
   Natural Sciences
   Engineering and Technology
   Medical Sciences
   Agriculture
   Social Science
   Humanities
2. By sector of performance
   Majorly consists of

   Productive Sector
   Higher Education Sector
   General Service Sector

From now on, we will use the categorization of scientific manpower by field of science. Our definition of a "Scientific Person" is similar to OECD (1975) as any person who has qualification to participate in research and development related works. In what follows, we provide a short account on some of the submodels from a statistical recording viewpoint.

Student Population Submodel

This submodel takes the admission policy as an input and produces the distribution map of student population in actual values and relative to national population. Also, the overall and annual rates are computed.

University Structure Submodel

The main function of this submodel is to appraise the potential role of teaching staff, labs, facilities, courses,... etc in generating pertinent graduates.

Up to now, there are 13 universities comprising around 131 faculties and enrolling more than 560,000 students. The number of teaching staff is around 9,000 the supporting staff (demonstrators and instructors) reachs about 15,000.

Graduation Pattern Submodel

The distribution of graduates over science disciplines and their respective rates are delivered by this submodel. At this point, the graduate/student ratios are derived and assessed. Then, by proper allocation, the supporting staff and labor force to economic sectors are evaluated.

Post-Graduate Profile Submodel

Over the fields of science, the number and rates of the persons enrolled in post-graduate work and those who went abroad (foreign missions) are computed.
A comment is in order. The problems of emigration and brain-drain should be analyzed and dealt with, in order to evaluate the losses from the proposed system.

The remaining submodels are left for the time being. It should be remarked that a mathematical model has been built based on the above consideration and using an adequate data information stored on a computer package. We present this model below.

A DYNAMIC MODEL

Our rationale in addressing a dynamic model for the scientific manpower projection stems largely from the close resemblence with scarce economic resources in developing countries. On one hand, the tangible benefits from their proper utilization are clearly manifested (Mohei El-Din and Omar, 1978; Murthy and Rangamath, 1977). On the other hand, the social cost involved and the losses to the economy as a whole as a result of their emigration or miscoordinated allocations may be extensive, especially when compared to consequences of the emigration of other occupations. This is due to the relatively large amounts of educational expenditure involved and the long gestation period needed to create a single researcher or scientific person.

In planning for scientific manpower, there are several issues come immediately into the scene. Chief among these is the questions of 'how many' and 'what kind' of manpower are required. This is a particularly cumbersome task to undertake due to the heterogenous characteristics of the process of science and technology (Joshi, 1977). To mention some of the obstacles, it is sufficient to recall the role and consequences of training programs, indoor or abroad, imports of technical know-how and brain-drain of scientists and technologists.

One way to get around these difficulties is to work in aggregate terms and on specific issues. Hereafter, we concentrate on the scientific manpower supply, production and distribution, by field of science with the premise that:

1. The demand for scientific manpower is known beforehand either from the national planning targets or by visualizing a desired profile from the international spectrum (OECD, 1975).
2. Other categorizations of scientific manpower or further details could be obtained by pertinent allocations and transformations. For example, if the required task is for universities and research institutes, then after obtaining the model results one can allocate the scientific manpower in accordance with their proportions.

In order to aid our exposition, we first define the variables to be encountered in the model building.

$\underline{h} = (h_1, \ldots, h_i, \ldots, h_m)^T$

is the vector that represents the rate of availability of scientists and technologists in different disciplines (Scientific Manpower Supply)

$\underline{q} = (q_1, \ldots, q_i, \ldots, q_m)^T$

is the vector that denotes the expected utilization rate of scientists and technologists in different disciplines (Scientific Manpower Demand)

$\underline{p} = (p_1, \ldots, p_i, \ldots, p_m)^T$

is the vector that stands for the rate at which researchers are enrolled in post-graduate work, sent in missions abroad, participated in training programs, ...etc.

$\underline{Z} = (Z_1, \ldots, Z_i, \ldots, Z_m)^T$

is the vector that accounts for the rate of production of scientists and technologists in various discplines.

$D = (d_{ij})$

is a matrix whose ijth element d_{ij} represents the rate at which scientists of ith discipline are collaborated to produce scientists (or technologists) of jth discipline per unit planning period.

$W = (W_{ij})$

is a matrix where the elements w_{ij}, $i,j=1,\ldots,m$ are the stock of scientific manpower in the ith discipline associated with the increase in the level of activity needed to produce scientific manpower in the jth discipline.

$M = (m_{ij})$

is a matrix with m_{ij} representing the probability of a scientists initially enrolled in the ith discipline transfering professionally to the jth discipline.

Intitutively, there is no dynamic equilibrium between the supply and demand components. Therefore, under disequilibrium conditions the scientific manpower supply $\underline{h}$ and scientific manpower demand $\underline{q}$ are not equal at all instants of time. This implies that an error level is allowed to exist which will be used in the dynamic model setup to produce the best derive. In this respect, scientific manpower demand is considered as an external input to the dynamic system and is created according to the future requirements of scientists and technologists. To conform with reality, we will permit q to vary with time in our model.

To exerize some regulatory policies on student population and graduation patterns, the growth of the researchers enrolled in post-graduate works or similar jobs is assumed to be:

$$\frac{d}{dt} \underline{p}(t) = A\,\underline{p}(t) + B\,\underline{u}(t) \qquad (1)$$

where the matrix A represents the internal dynamics of scientists and technologists development and the matrix B is a policy influency matrix. Whereas A reflects the nature of the change in manpower production, B signifies the extent to shape the manpower production pattern. In light of the aforementioned definitions of variables, we posit the following relation:

$$\underline{Z}(t) = D\,\underline{Z}(t) + W\,\dot{\underline{p}}(t) + \underline{h}(t) \qquad (2)$$

Where the term $D\,\underline{Z}(t)$ signifies that part of the manpower production which is used in the production of scientists and technologists in the m disciplines. It simply represents an internal utilization of scientific manpower production. The term $\underline{h}(t)$ accounts for an external utilization associated with scientific manpower production.

Finally, the term $W \dot{\underline{p}}(t)$ stands for the stock of scientific manpower in the different disciplines and can be viewed as underemployment.

By taking appropriate time constants into consideration to account for professional mobility, the following behavioral relation is induced:

$$\underline{Z}(t) = M\ \underline{p}(t) \qquad (3)$$

The error level, discrepancy, between the manpower supply and demand levels is defined as:

$$\underline{e}(t) = \underline{q}(t) - \underline{h}(t) \qquad (4)$$

Characterization of the regulatory policies entirely depends on the error and demand patterns. A linear relationship is introduced as:

$$\underline{u} = K\ \underline{e}(t) + \underline{L}(t) \qquad (5)$$

where L (t) is a linear operation on q(t), yet to be determined, and K is a gain matrix whose choice is governed by the satisfaction of an optimal performance measure.

From equation (1):

$$\underline{h}(t) = [I-D]\ \underline{Z}(t) - W \dot{\underline{p}}(t) \qquad (6)$$

Substituting equations (1) and (3) into equation (6) results in:

$$\underline{h}(t) = [I-D]\ M-WA\ \underline{p}(t) - WB\ \underline{u}(t) \qquad (7)$$

Or in a compact form:

$$\underline{h}(t) = C\ \underline{p}(t) + E\ \underline{u}(t) \qquad (8)$$

where

$$C = [I-D]\ M - WA \qquad (9a)$$
$$E = -WB \qquad (9b)$$

The set of equations (1), (8), (9a), (9b) are canonical state equations (Chen, 1970) and delineate the dynamic model. In case of Egypt, the level of underemployment in negligably small and hence $W = 0$ leading to:

$$\underline{h} = C\ \underline{p}(t) \qquad (9c)$$

To reach dynamic balance in scientific manpower evolution, the discrepancy between supply and demand should be gradually reduced. One way to assess this operation is through the minimization of

$$J = \int_0^{\infty} [\underline{q}(t)-\underline{h}(t)]^t Q [\underline{q}(t)-\underline{h}(t)]dt \qquad (10)$$

At this stage, the problem of scientific manpower planning is summarized by equations (1), (9a), (gc) and (10) which is typically a linear quadratic dynamic optimization problem (Kwakernaak and Sivan, 1972). One should remark that we have deliberately used the inifinite time-horizon for mathematical tractability. The solution procedure is rather straightforward and starts by introducing a symmetric, positive definite matrix P:

$$\frac{d}{dt}\ \underline{e}^T(t)\ P\ \underline{e}(t) = -\underline{e}(t)^T\ Q\ \underline{e}(t) \qquad (11)$$

Integrating both sides of equation (11) over the time span $[o,\infty]$ and after some manipulations, we arrive at:

$$J = \underline{e}^T(0)\ P\ \underline{e}(0) \qquad (12)$$

Assuming the initial discrepancy is uniformly distributed over a unit sphere in the m-dimensional space, the expected value of J is:

$$J = \frac{1}{m}\ tr\left\{(P)\right\} \qquad (13)$$

From equation (11), we have:

$$\dot{\underline{e}}^T(t)\ P\ \underline{e}(t) + \underline{e}^T(t) P\ \dot{\underline{e}}(t) = -\underline{e}^T(t) Q\ \underline{e}(t) \qquad (14)$$

Differentiating (4), using (4) and (5) thus:

$$[I+EK]\dot{\underline{e}}(t) = \dot{\underline{q}}(t) - CA\underline{p}(t) - CBK\underline{e}(t) - CB\underline{L}(t) + E\dot{\underline{L}}(t) \qquad (15)$$

Utilizing (4),(5) and (8) into (15), we arrive at:

$$[I+EK]\ \dot{\underline{e}}(t) = CAC^{-1} - CBK + CAC^{-1}EK\ \underline{e}(t) + \dot{\underline{q}}(t) - CAC^{-1}\ \underline{q}(t) + CAC^{-1}E\ \underline{L}(t) - E\ \dot{\underline{L}}(t) - CB\underline{L}(t) \qquad (16)$$

Finally, we conclude that:

<u>Canonic State Equations</u>

$$\dot{\underline{p}}(t) = A\ \underline{p}(t) + B\ \underline{u}^*(t) \qquad (17)$$

$$\underline{h}(t) = C\ \underline{p}(t) + E\ \underline{u}^*(t) \qquad (18)$$

<u>Error Equations</u>

$$\dot{\underline{e}}(t) = G\ \underline{e}(t) \qquad (19)$$

$$G(t) = [I+EK]^{-1}\ [CAC^{-1} - CBK + CAC^{-1}EK] \qquad (20)$$

<u>Scientific Manpower Demand Policy</u>

$$\dot{\underline{L}}(t) = E^{-1}\left\{\left[CAC^{-1}E - CB\right]\underline{L}(t) + \left[\dot{\underline{q}}(t) - CAC^{-1}\underline{q}(t)\right]\right\} \qquad (21)$$

<u>Optimal Regulatory Policy</u>

$$PG + G^TP = -Q \qquad (22)$$

$$\frac{\partial}{\partial K}\left\{tr\ (p)\right\} = 0 \rightarrow K_{opt} \qquad (23)$$

$$\underline{u}^*(t) = K_{opt}\ \underline{e}(t) + \underline{L}(t) \qquad (24)$$

Once again, in case of Egypt the level of under-employment W=o then equations (20) and (21) reduce to

$$G(t) = [CAC^{-1} - CBK] \qquad (2\ )$$

$$\underline{L}(t) = [CB]^{-1}\ \dot{\underline{q}}(t) - CAC^{-1}\underline{q}(t) \qquad (2\ )$$

A block diagram representation is delineated in Fig.2 to illustrate the different components of scientific manpower dynamic model

DATA INFORMATION AND PRELIMINARY RESULTS

Due to the critical importance of the subject of scientific manpower planning, relevant data information have been collected, organized and documented. Some statistical analysis have been performed in a manner that permits building systems models for the objective of evaluation, planning and forecasting. Because our research efforts are in the initial phase of investigation, we provide only the preliminary results. The developed dynamic model has been caliberated against the available data information.
In Fig. 3, we portray the growth of scientific manpower information in physical units. The demand pattern has been estimated based on the following considerations:

1. According to the five-year plan (1978-1982), the demand on economically active labor force is around 0.82 million per annum. The average scientific manpower/labor force stock is expected to reach about 7.2%.
2. From the international statistics (OECD, 1975), the average growth of scientists and technologists per 1000 population in case of developing countries, with conditions resemble the Egyptian situation, is around 3-5.
3. The Egyptian statistics record that the population growth is 2.5% annually. In addition, there are enormous amount of investment projects that are expected to push the economy and create enough jobs in the productive sector.

Implementation of the dynamic model over the period 1976-1985 have indicated trends that are collectively in close match with the planning targets. These results are clearly demonstrated in Fig.4. A final word to mention is that what has been developed is an aggregate modeling approach that is mainly concerned with global picture. Indeed, further details could be easily performed whenever there is data information available.

CONCLUSIONS

An aggregate dynamic modeling approach is advanced for the significant problem of scientific manpower planning in developing countries. The model stresses the supply pattern while generating the demand profile in accordance with prescribed national objectives. Preliminary results have indicated the usefulness of our approach. Future extensions of the model building are presently underway.

REFERENCES

Chen, C.T. (1970). Introduction to Linear System Theory, HRW, INC., New York.

Correa, H. (1975). Analytical Models in Educational Planning and Administration David Mckay Co., New York.

Hansen, W.L. (1970) Education, Income and Human Capital, Columbia University Press, New York

Joshi, N. (1977). A mathematical Model for Projecting the Demand for Scientists and Technologists in National R & D Institutes, R&D Management, 8,1, pp.43-47

OECD (1971). Mathematical Models for the Education Sector: A Survey, Paris

OECD (1973). Occuptational and Educational Structures of the Labor Force and Levels of Economic Development, Paris

OECD (1975). Patterns of Resources Devoted to Research and Experimental Development, Paris

Kwakernaak, H. and R. Sivan (1972). Linear Optimal Control Systems, Wiley-Interscience, New York.

Mohei El-Din, A. Omar (1978). The Emigration of Universities Academic Staff, CU/MIT Technology Planning Program, Egyptian Labor Migration Project, Cairo.

Murthy, K.K. and M.V. Rangamath (1977). Systems Approach to Technological Education, Proc. of IFAC conference on Systems Approach for Development, Cairo, pp. 145-155.

NSF (1975). Projections of Science and Engineering Doctorate Supply and Utilization, Wahington, D.C.

Spaey, J. (1977). Science for Development, UNESCO, Paris

UNESCO (1977). Statistical Year Book, Paris.

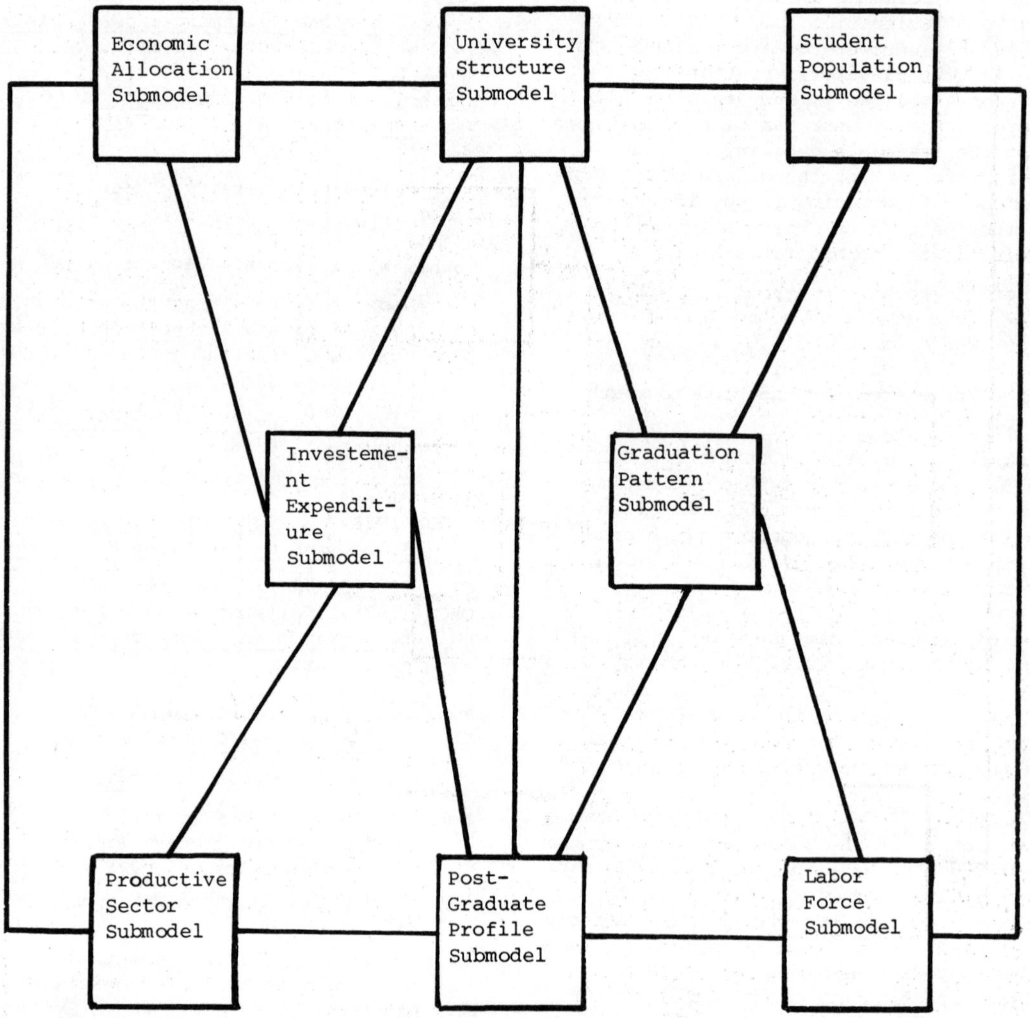

Fig. 1 A Block Diagram of The Integrated Model

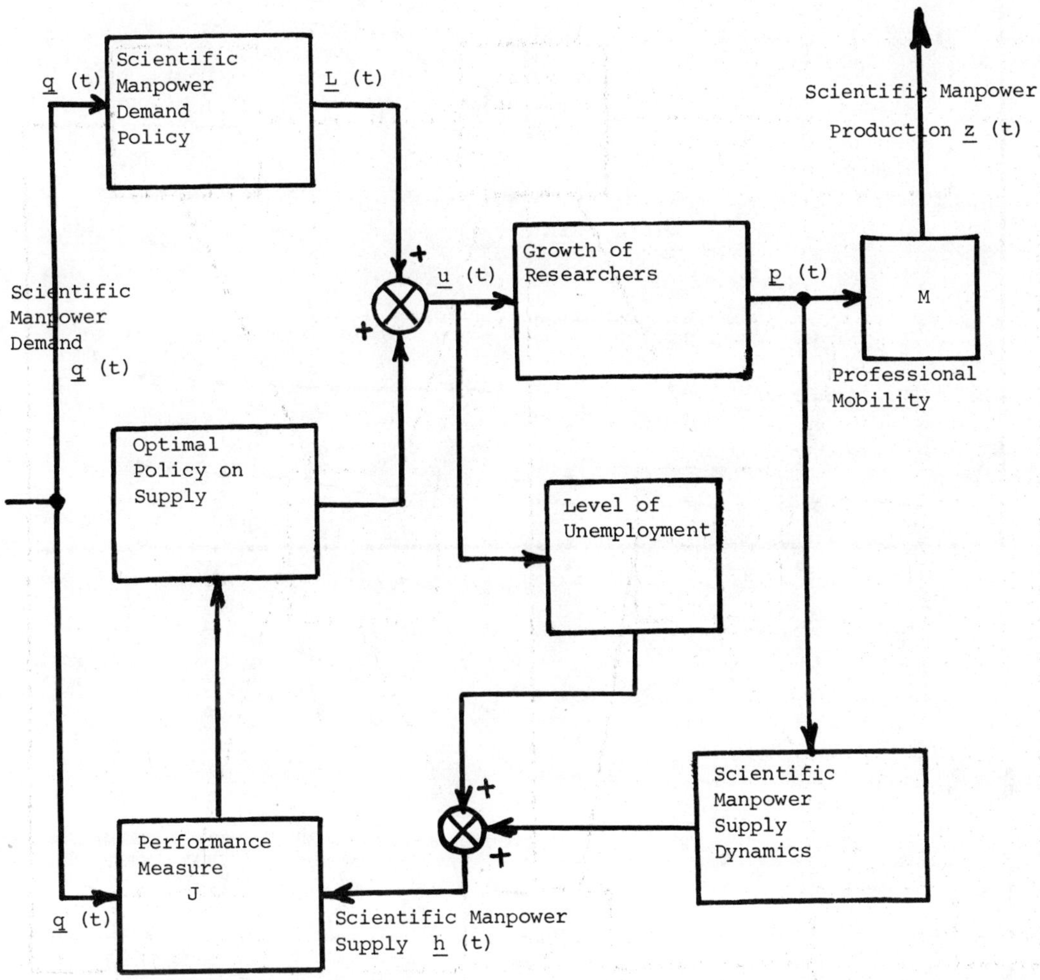

Fig. 2 The Scientific Manpower Dynamic Model

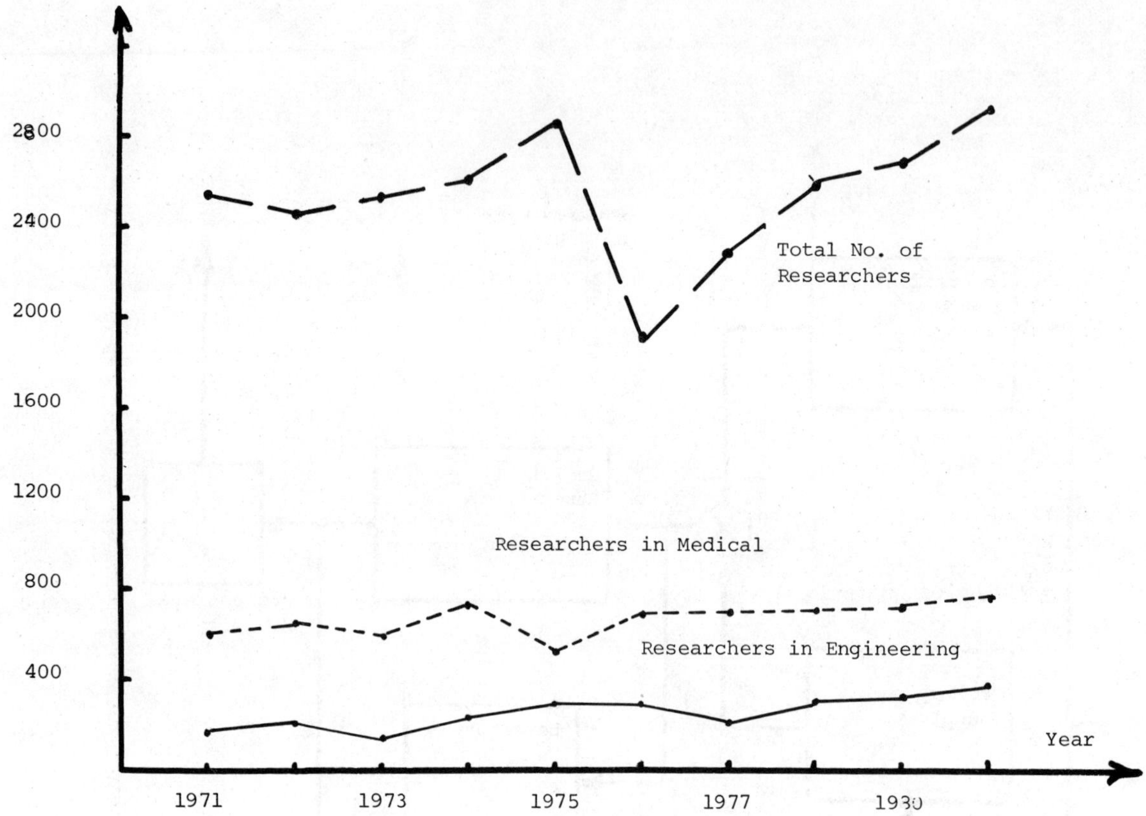

Fig. 3 Growth of Scientific Manpower

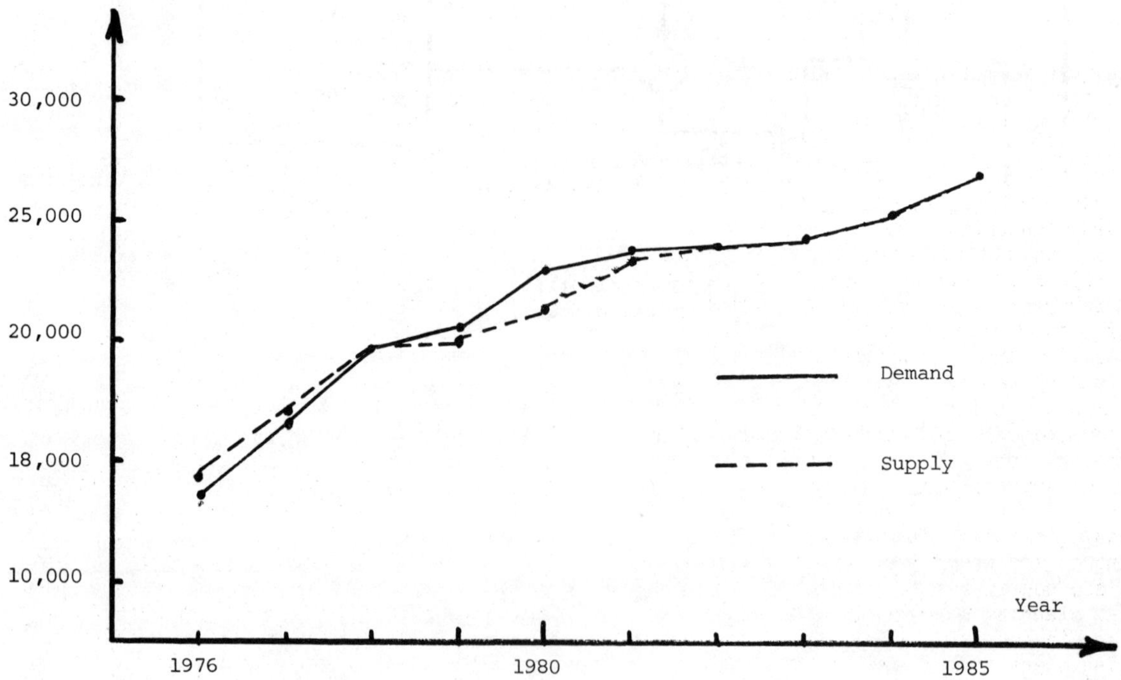

Fig. 4 Projections of Scientific Manpower

PLANNING PROCESS FOR INTRODUCING ENERGY SYSTEMS INTO THIRD WORLD COUNTRIES

J. M. Garrison

International Committee for Developing Countries, USA

Abstract. Late Developing Countries (Third World Countries) have failed to recognize the importance and crucial requirements for a system approach to development planning. Determining objectives, resulting from a thorough system assessment with possible alternatives has been the prime neglegence of Third World political regimes. Policies have not been established that reflect an understanding of realistic objectives. Development plans were not formulated by a defined methodology directing the preparation and continuous refining of alternatives for development programs. The key to successful programs to ensure development progress is the planning for effective and efficient utilization of energy systems. Development progress will stagnate without energy systems. These systems take many forms and are interdependent in their effective use. It should be noted that manpower is one of the most efficient energy system and should not be neglected in the planning process. After four years of research and collaboration, the author has derived a planning methodology that can be used by any Late Developing Country. This paper presents an overview of the planning methodology resulting from this research. The methodology represents a total systems approach to development planning. Several volumes could be written on any segment of the planning methodology; however, the objective is to define the total planning process, and identify the necessary interrelationship and intrarelationships to promote the required understanding of the importance of successful development planning.

Keywords. Planning methodology; energy systems; system analysis; planning management; system assessment; system alternatives.

INTRODUCTION

The author formulated the planning methodology after four years of collaboration and consultation with political and technical organizations in both Industrial and Late Developing Countries. During this project, the author noted that the progress of developing countries lacked the planning for a balanced, effective utilization of human and material resources. Goals for development progress were unrealistic or short-range without consideration for long-range plans. In most instances the political environment was not conducive to the understanding and balanced application of skills and knowledge available from sources outside a country's development sphere. A Late Developing Country's capability and capacity to analyze economic, social, political, and technological factors in terms of long-range strengths and weaknesses was, in most cases studied, non-existant. Emphasis on integrated planning was found to be very limited, or did not exist. Developing and implementing plans for energy utilization, re quired for all development programs, has not been recognized or its importance understood. Optimum decision making could not be substantiated with respect to a systems approach process because of the lack of a planning methodology.

The question facing the author was, "could a planning methodology be devised to relate to, and account for, all the interrelationships and intrarelationships in any Late Developing Country? The author believes that the answer to this question is affirmative.

PLANNING PROCESS AND CONSIDERATION EXPANSION (OVERVIEW)

Planning is a salient function in all countries regardless of their size or development progress. It is of prime importance to Late Developing Countries. It must be recognized that a planning process for a Late Developing Country is a continuous, never ending process. The plan is developed from a recognition and understanding of major and minor influencing factors and variables. Planning requires a total system approach, and cannot neglect any relationship regardless of how insignificant the relationship appears to be. This planning process has been develop ed by the author into a methodology. Throughout this paper, the narrative will be continuously addressing Chart 1.0, Planning

Process, and Chart 2.0, Consideration Expansion Related to the Planning Process. Chart 1.0 and Chart 2.0, together, comprise the planning methodology.

PLANNING PROCESS
(CHART 1.0)

The Planning Process begins with (A) Environmental Assessment followed by (B) Urban/Rural Sectional Assessment, (C) Requirements Dictating Energy System Objectives, and concludes with (D) Formulation of Energy Programs, Plans and Budgets. The process is continuous as illustrated by the feedback loop from (D) to (A).

Each of these headings are considered a stage of planning progression in the planning process. (A), and (B) are the formulation of the planning base. (C) and (D) are the approach and development of the plans. Each stage must be addressed and assessments made before proceeding to the next stage in the process.

In the center of Chart 1.0 are variables to be considered in each stage of planning. Seven variables are identified as : (1), (2), (3), (4), (5), (6) and (7). The interrelationships and intrarelationships of these variables must be assessed at each stage of the planning process.

At each stage of the process, consideration in the form of assessments must be given to patterns of priority allocation existing within the specific country. Actual priorities and need priorities must be assessed with respect to the influence these priorities have on each of the seven variables.

The above approach can be formulated as in the following example:

Let P = Analysis of Present Lines of Overall Progress (Stage A).
H and R = Urban and Rural Changes, Constraints and Influences.
V(1,2,3) = Variables(1),(2) and (3) to be assessed.
Pa and Pn = Pattern of priorities (actual versus need)
Then:

$$P = f\ (U,R) = f\ (V(1,2,3),\ Pa,\ Pn)$$

As the planning process continues, the results of previous assessments are not neglected nor discarded. Previous assessments are used as a base of reassessment to build on the next stage of the process. This need for reassessment becomes obvious when tracking the requirements for assessing the same variables as planning continues. Variable number (6) appears five times in the process while variable (2) appears only three times. Obviously, it was found that variable (6) is more influential in the process than variable (2).

CONSIDERATION EXPANSION RELATED TO
PLANNING PROCESS (CHART 2.0)

The seven variables to be assessed during the stages of the planning process are illustrated on Chart 2.0. The flow of the assessment is from left to right on the chart, and the headings at the top of the chart provide a guide for the assessment sequence. The assessment of the variables should always be with respect to the stage in the planning process shown on Chart 1.0. There will be a tendency, during the assessments, to look ahead in the planning stages to foresee additional relationships and influences. Caution must be exercised and discipline maintained to not look ahead or anticipate or the assessment will lead to erroneous results.

(A) ENVIRONMENTAL ASSESSMENT
(CHART 1.0)

The planning methodology begins with an Environmental Assessment of the total country starting with an evaluation of the country's historic political stability (Environmental Scale evaluation, right side of Chart 1.0). The country's political history is plotted on the environmental scale over a minimum of ten years. If the country has a history of oscillating around threshold 4 (Intense Change) for more than three years, there is a high risk of any development progress of any significance; however, the planning process should not be concluded. This risk identification will have a great influence on the remaining assessments in the planning process.

The next assessment in stage (A) is an evaluation of the Overall Economic/Political/Social Infrastructure. The objective is to identify past, present and future potential problems with respect to the country's changes and rates of change of its policies (see Policy Synthesizer, right side, Chart 1.0). Conclusions must be made with respect to policy objectives, purposes, capabilities and capacities dictated by existing military, foreign, domestic, progress, and armament policies now in effect. Two key issues must be evaluated with respect to political objectives and purposes. As an example, a military and armament policy is identified as dominant, and the capability and capacity of the country to support these policies is strained. strained. Obviously, development progress is curtailed.

Findings that the risk of introducing energy systems at this planning stage is high must not restrict the continuation of the planning process. It was found that the geographical size/population content of a country was correlated to rates of change. The smaller, more populated country, the higher probability the rate of change was inherent. This correlation may change in the future. It should be constantly considered that the planning process will identify development weaknesses and strengths, and this identification can be used for effective political persuasion of the existing regime to promote development progress with energy systems.

From the assessment of the infrastructure a base is established for an analysis of the present lines of overall progress. Note that this analysis is divided into changes, constraints and influences with respect to urban and rural sectors, and requires the first assessment of variables. The decision to divide the analysis into urban and rural sectors is extremely important. What effects one sector may effect the other sector in different degrees. Identifying this interaction will enable clear understanting of possible future cross-overs in other assessments. A good example of a cross-over is an industrial revolution drawing agricultural workers away from rural regions.

Identifying the changes, constraints and influences must inevitably consider technology trends and forecasts together with intra/inter market analysis, the next step in the process. The intra/inter markets need to be addressed with respect to each sector and to both sectors integrated within the country. The objective of this analysis is to identify markets that will be influenced by energy systems. The relationship of inter/intra markets must also consider the integration of these markets with international markets. The influences of the market structure may be unbalanced. Many developing countries have neglected this consideration opting for an autonomous market structure. The result has been development stagnation and, in some cases, a reversal of development progress. Market analysis is extremely important and has to consider the rates and influence of cost and price.

Cost is the direct cost and price includes profit. Both cost and profit will be influenced by energy systems. Cost and price are interrelated and influenced by international markets which, in turn are dictitorial of industrial progress and agricultural progress in a developing country. Associated, also, with international market assessment is the influence of technology trends developed by international markets. Thus, the planning base adds additional interrelated planning foundations to begin the Urban/Rural Sectional Assessment.

Urban and rural sections addressed in the Environmental Assessment stage have two primary objectives. These objectives are to identify the status quo with respect to these sectors and, if possible, any rate of change that may affect the status quo. It should be noted that considerations were given to only changes, constraints and influences with respect to identifying technology trends, and industrial and agricultural progress. Some weaknesses should be readily identifiable at this stage of planning; however, caution should be exercised under a strict discipline not to let these weaknesses influence future assessments at this time. The cause of a weakness and identification of a possible strength will change, and this change will become more apparent as the planning progresses. Overcoming strengths and weaknesses should not be addressed at this stage.

(B) URBAN/SECTIONAL ASSESSMENT (CHART 1.0)

The next primary stage in the process is a sectional assessment of the urban and rural areas. Socio-economic strengths and weaknesses with respect to the influence of competition should be firmly identified in both urban and rural sectors. Competition and socio-economic strengths are very much interrelated. Competition of various degrees exist in both areas and the objective is to determine where and what competition is the guiding influence to socio-economic strengths or weaknesses. Once this identification is complete, considerations can be given to how much influence each sector has on the other. It will be found that there are several influencing factors such as religion which is a great influence, in some ways, on socio-economic conditions. It should be remembered that the results of the Environmental Assessment will have some influence on the Urban/ Rural Sectional Assessment.

The assessment is now carried forth into the next step in the evaluation of urban and rural resources. These resources can take many forms such as manpower, waterpower, electric power, mining, agricultural products, and large and small industrial plants. The objective is to determine the availability of these resources versus their capabilities. If agriculture products are available, but the capability to transport the products to market is weak, then the introduction of a type of energy system, such as gasoline for trucks, may balance the availability and capability in common accord.

The next step in the process is to identify and evaluate the progress toward the country's short term goals. Short term goals are goals that are planned to be met in one or two years. In the evaluation of urban and rural sectors to meet short term goals, an assessment must be made with respect to the possibility of goals being attained in a shorter time with an energy system. As an example, the assessment may have identified a small manufacturing plant built in an urban sector because of the lack of electricity in the rural sector close to the manufacturing raw materials. Extending electricity to the rural sector may enable manufacturing development to reach short term goals faster, thus, contributing to the development of the economy. This example identifies a significant interaction with respect to urban and rural sectors.

From the assessment of progress toward short-term goals, a base is established for evaluation of progress with respect to long-range action plans. Long-range action plans represent planning for five years or more. In the urban sector the assessment should determine how much, if any, progress is industry oriented with respect to types of industry. In the rural sector the assessment should determine how much progress is agricultural oriented. An important finding from the author's research with respect to long-range actions plans in developing countries is introduced here at this time. It has been found that those countries that concentrated on long-range industry oriented action plans, and neglected long-range agricultural action plans, had a marked increase in development progress, but only over a short time period. The rate of development progress then decreased sharply.

At this stage of the planning process the planning base has been established together with indications of progress toward short-term goals and long-range progression action plans. From this base and the planning indicators, the next step is to determine the energy objectives to correct weaknesses and promote further strengthing of develop ment progress.

(C) REQUIREMENTS DICTATING ENERGY SYSTEM OBJECTIVES (CHART 1.0)

It should be noted at this stage of the planning process that all seven variables will be assessed with respect to determining energy objectives through the process of preparing strategies. As the planning base was being formulated, different variables were assessed to determine influences these variables had on the planning assessments. Determination of strategies will uncover variances in the variables influence. This is to be expected. More relevant planning assessment data is available. Now it can be understood why it is important to track variable comparisons to this stage of the planning process. The sphere of understanding variable interactions is constantly expanding.

Requirements dictating energy system objectives shall require trade-offs of availability, capability and need in the determination of strategies. The first step is to determine the energy systems direct product strategy with respect to urban and rural sectors. Direct product strategy is to determine the optimum from among several alternatives of the products required and the products produced by a given energy system. Again, manpower should be understood to be an energy system. As an example, rural electrification requires transformers, powerlines, power poles, and generating stations manned by personnel. One of the products produced could be electric heat the products produced could be electric heat for poultry farms. However, several men cutting wood and maintaining fires can produce the same amount of heat. A higher distribution of wages is then realized, available manpower is put to work, and the economy is supported. Long range and short-range trade-off assessments, obviously are important in determining product strategy.

Optimization of direct product strategy for energy systems must also consider what the author has identified as fall-out energy system product strategy. Fall-out strategy should be assessed as a by-product. As in the previous example of rural electrification, a very real fall-out is electric lights for homes and villages. Electric lights permit families to stay awake for longer hours after dark. There is, therefore, an abundance of extra time. What will those families do with this extra time? This is an influence that cannot be neglected.

Following the identification of optimum fall-out strategies, the next strategy assessment is with respect to diversification. Previous assessments with respect to technology, markets, socio-economic conditions, resources, and planning goals play an important function in determining diversification strategy. The objective is to identify and assess critical alternatives with respect to energy systems creating new products, services, and markets to join or replace the current generation of products, services, or markets, or add new markets to enable further development progress. The diversification strategy is extremely important to long-range planning programs. When should an energy system be introduced, and what advanced technology requirements should be maintained? In addition, how much outside influence (external to the country) should be maintained until competence in maintaining the eneragy system is developed?

Very closely interrelated with diversification strategy is the energy system divestment strategy. Divestment should be very carefully assessed because divestment represents change. Many developing countries have discovered that change can be a foundation of rock or sand with respect to development progress. The objective of divestment strategy development is, then, to consistently balance divestment strategies of existing systems with diversification strategies and product strategies. Divestment can easily mis-direct development progress and, in some cases, reverse progress to the detriment of society. Many industrial nations have suffered because of the lack of understanding the serious impacts of divestment on society. Serious consideration needs to be given to identfying and understanding the salient relationships between human and non-human resources. It must be understood that divestment will only be accomplished through the acceptance by a society within the country.

It is obvious that trade-offs of availability, capability and need play an important role in determining divestment strategy.

The final step in requirements dictating energy system objectives is the identification of energy priorities integrated between urban and rural sectors. The development of the previous strategies considered management, ownership, financial, legal, supply, control, labor, and technology while the assessment of availability, capability and need considered all the variables on Chart 2.0. With the determination of strategies, and the identification of energy priorities, the final stage of the planning process can now be undertaken.

(D) FORMULATION OF ENERGY PROGRAMS, PLANS AND BUDGETS (CHART 1.0)

The final plans are in the form of strategic plans, intermediate and long-range plans, and tactaical plans. The definition of these plans are on Chart 1.0. The formulation of the plans require integrated planning studies for integrating urban or rural sectors. The plans must utilize the results of the planning process. These final plans are subject to change on an annual basis, thus, the requirements for continuous planning. Maintaining the interrelationships of required change, and identifying the significance of changing assessments is the most crucial requirement in any development program. The planning studies should concentrate on cost effectiveness versus system effectiveness versus adaptability to the existing socio-economic environment.

CONCLUSION

The author could not possibly cover all the implications of the planning methodology for introducing energy systems into developing countries. What the author has accomplished is to concentrate the results of research into a methodology, and make the user of the methodology aware of previous impaired planning resulting from neglect of a system approach. The author places this methodology in the hands of those concerned. It is believed that the process outlined above will force objective planning rationale to the surface of development programs. The benefits derived resulting from four years of research to reach a specific objective. That objective is to make an awareness available to those that need ideas. Benefits leading to successful development programs can be derived from these ideas; however, there is also an immense suffering from mistakes when those making mistakes are an island unto themselves. System approach ideas are in short supply, and the author welcomes any critique with respect to the proposed methodology on contributing to the progress of any Late Developing Country.

REFERENCES

Garrison, J. M. (1974). Operations Criteria for the Internatinal Committee for Developing Countries. Dr. H. Apter, Switzerland.

CHART 1.0
ENERGY SYSTEMS FOR THIRD WORLD COUNTRIES
PLANNING PROCESS
CONSIDERATION EXPANSION REFERENCE
(See Chart 2.0)

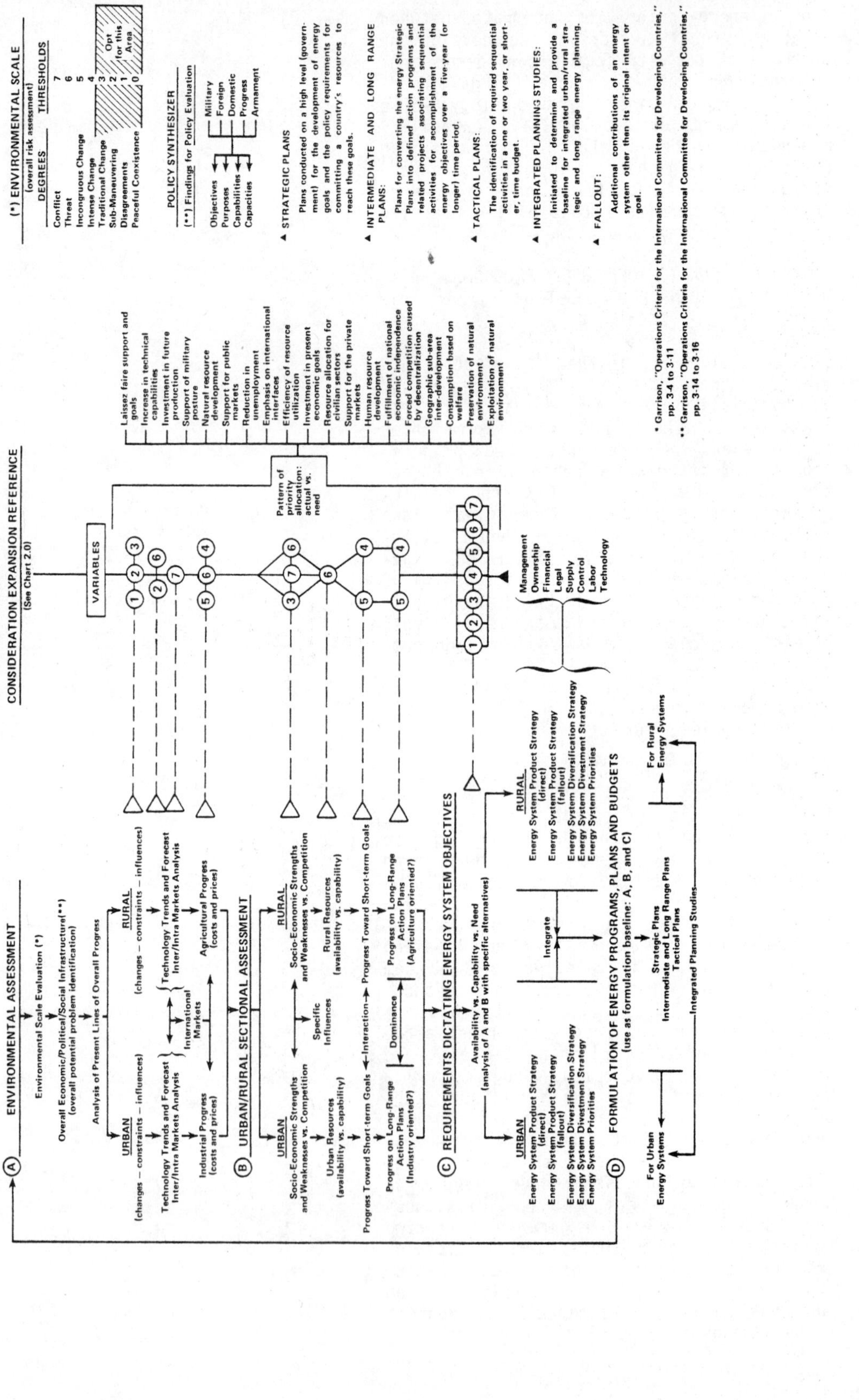

CHART 2.0
ENERGY SYSTEMS FOR THIRD WORLD COUNTRIES
CONSIDERATION EXPANSION RELATED TO PLANNING PROCESS
(See Chart 1.0)

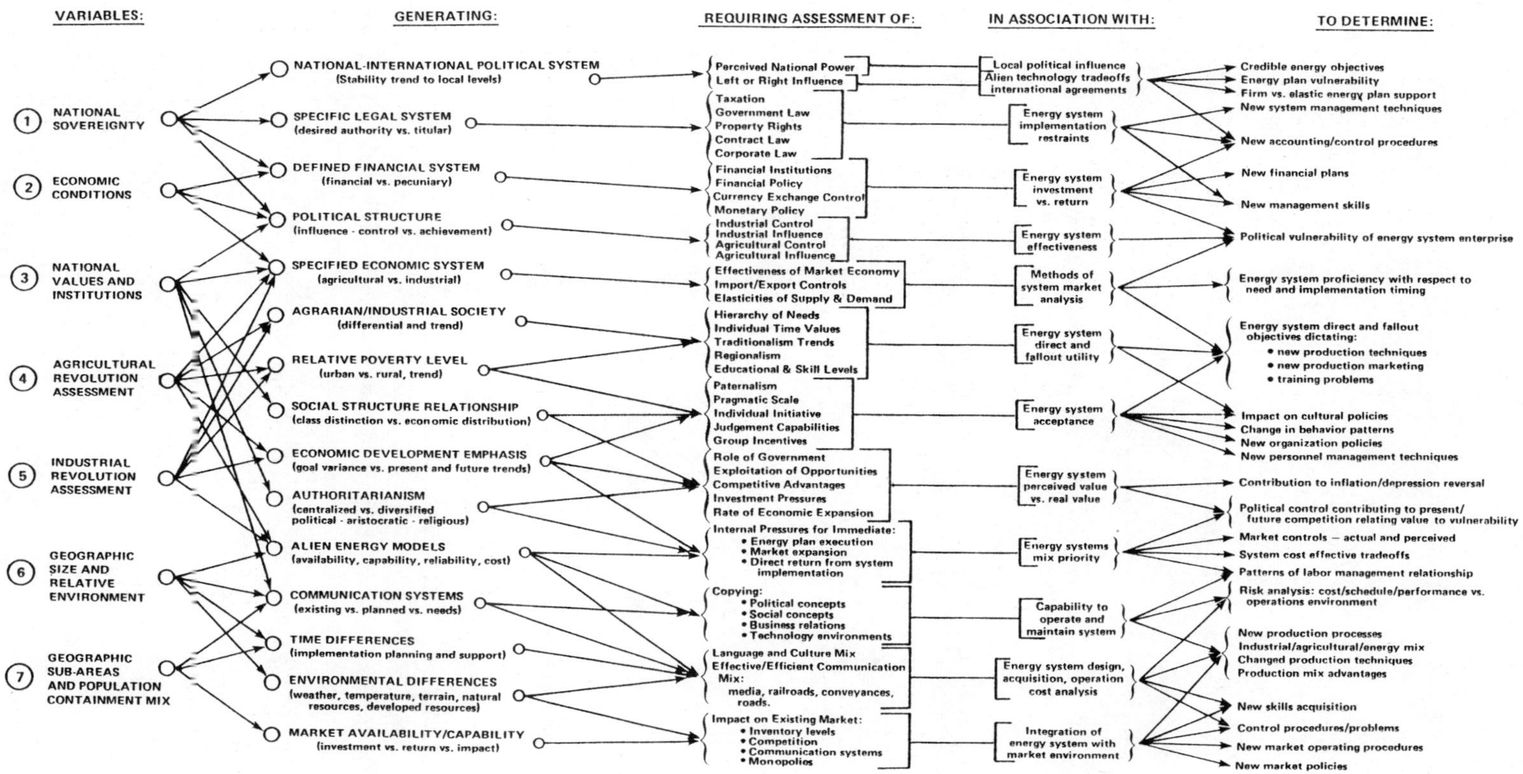

OPTIMAL TECHNOLOGY SELECTION IN RURAL ROADS CONSTRUCTION

F. Lara-Rosano

Instituto de Ingeniería, National University of Mexico, Apartado 70-418, México 20, D.F., Mexico

Abstract. In this paper an optimization model is proposed for finding the optimal labor/capital relation for rural roads construction in a given region, maximizing the socioeconomic benefits for the local population under several local and project constraints. The model has been implemented to be solved in minicomputer and has been applied with success to real cases.

Keywords. Developing countries; decision theory; economics; operations research; transportation; rural roads; appropriate technology; social and behavioral sciences; system analysis.

INTRODUCTION

The high rates of unemployment in the rural sector have inspired in some developing countries the implementation of policies prescribing labor intensive methods in public works construction, specially rural roads. However, the effects of this policy may be very different from one region to another, depending on the socio-economic situation of the region. Thus, under conditions such as overcapacity in the consumption goods industry and low opportunity cost of labor the increase of consumption due to labor payments would enhance the internal market, increasing production, lowering unemployment and estimulating economic development. (Keynes, 1936) However, if the consumption goods industry has a saturated capacity and the local pattern of income use derives most income to consumption, the increased labor payments will cause an inflationary process with all its negative effects over the economy of the low income sector that is supposed to be helped.

From all this it is argued that for every region characterized by certain socioeconomic indicators it is possible to find the optimal technology defined by the labor/capital relation such that a local objective function of socioeconomic type be maximized.

DESCOMPOSITION OF RURAL ROAD CONSTRUCTION

Construction of rural roads may be decomposed in different tasks called construction activities, for instance, excavations, embankment construction, material transport, etc. Because these construction activities are technological very different, the proposed procedure will consider each one of these activities as a separate unit.

Not every construction activity admits technology (labor-machinery) substitution. For instance, material transport is a construction activity where pure labor is out of question. Therefore we selected those activities where technological substitution was efficient under certain circunstances. The selected activities were: excavations in soft- and medium-compacted material banks, loading of material, concrete works, and embankment construction.

DIRECT COST RELATED TO LABOR AND MACHINERY

The mean direct unit costs related to labor and machinery for each construction activity were statistically determined from a sample of 71 road construction contracts covering all Mexico. These costs were deflated to 1968.

Let L be the mean deflated labor cost of a given construction activity and K be the mean deflated machinery cost of the same activity. We are interested on the sum

$$K + L \qquad (1)$$

Fortunately, K and L are not independent but related through a production function. Because the homogeneity of Cobb-Douglas function:

$$P = aL^bK^{1-b} \qquad (2)$$

P being quantity of work in activity units, this was adopted and the parameters a, b were statistically determined from the contracts sample. The results are in table 1.

TABLE 1 Parameters of the activities production functions

Activities	a	b	Correlation coeff. r	unit
Excavation (soft)	0.45	0.3322	0.95	m^3
Excavation (medium)	0.37	0.3707	0.96	m^3
Embankments	0.45	0.2798	0.94	m^3
Concrete	0.02	0.7634	0.74	m^3
Loading	0.52	0.3812	0.98	m^3

Putting K in terms of L and P we have:

$$K = \left(\frac{P}{a}\right)^{\frac{1}{1-b}} L^{-\frac{b}{1-b}} \quad (3)$$

INFLUENCE OF WEATHER CONDITIONS ON LOCAL COSTS

Calculations of unit construction prices in Mexico are based on 300 workdays. When in the region where the road would be built the number of workdays is lower because of weather conditions then the fixed local costs are higher per effective workday. Let d be the number of work days probably lost because of weather conditions then $k_{rf} = d/300$ is the proportion of expected lost workdays and $1-k_{rf}$ is the proportion of expected effective workdays. The expected fixed costs of machinery become then multiplied by a factor

$$\frac{1}{1 - k_{rf}} \quad (4)$$

INFLUENCE OF HUMAN FACTORS ON LOCAL COSTS

Consideration of human factors on local costs was made according to Ministry of Agriculture and Hydraulic Resources[1], taking skill and diligence as independent qualities and translating their qualitative appraisals into quantitative utility measures following a Likert type scale. (Table 2)

TABLE 2. Utility appraisal of skill and diligence

Qualitative appraisal	Utility
Maximum	15.00
Superior	11.25
Excelent	9.00
Very good	6.50
Good	3.50
Middle	0.00
Somewhat defficient	-3.85
Poor	-8.00
Inferior	-13.50
Infimum	-20.00

The local mean skill C_s and mean diligence C_d can be assessed consulting with experienced contractors. The correction factor for skill and diligence of local labor force k_{rh} is given, accordiing to this Ministry by

$$k_{rh} = \frac{C_s + C_d}{100} \quad (5)$$

This factor affects labor and machinery operator costs. It implies a cost reduction of 30% when both local skill and diligence are maximal and a cost increment of 40% when both are least. This calculations are based on actual observations.

INFLUENCE OF WAGES ON LOCAL COSTS

The mean wage from the contracts sample was 29.15 pesos (1968) per day. If the local wages S are different, the labor and operator costs are modified by a factor $k_{rs} = S/29.15$

INFLUENCE OF OIL PRICES ON LOCAL COSTS

The mean price of diesel oil calculated from the contracts sample was 0.42 pesos (1968) per liter. If the local price P_c is different, the consumption costs of machinery are modified by a factor $k_{rc} = P_c/0.42$

LOCAL CORRECTIONS TO MACHINERY RELATED COSTS

There are three kinds of machinery related costs: a) fixed costs k''_{MAQ}, consumption costs k''_{CON} and operator wages k''_{OP}. If these are expressed as fractions of K, then

$$k''_{MAQ} + k''_{CON} + k''_{OP} = 1 \quad (6)$$

The mean values of these fractions were statistically determined from the contracts sample. (Table 3).

TABLE 3. Fractions of machinery related costs

Activity	k''_{MAQ}	k''_{CON}	k''_{OP}
Excavation (soft)	0.69	0.17	0.14
Excavation (medium)	0.69	0.17	0.14
Embankments	0.58	0.23	0.19
Concrete	0.54	0.28	0.18
Loading	0.69	0.15	0.16

Applying the correction factor related with weather conditions $1/(1-k_{rf})$ to fixed and operator costs; the determined correction

[1]Secretaría de Recursos Hidráulicos, "Manual sobre precios unitarios de trabajos de construcción", Tomo I, México, 1963.

factor related with local oil prices to consumption costs, the total local correction factor f_1 for machinery costs would be

$$f_1 = \frac{k''_{MAQ} + k''_{OP} k_{rs} k_{rh}}{1 - k_{rf}} + k''_{CON} k_{rc} \quad (7)$$

This factor multiplied by K gives the actual local value of K.

LOCAL CORRECTIONS TO LABOR RELATED COSTS

Applying the determined correction factors related with human factors and local wages over the labor costs, the local value of L would be

$$k_{rh} k_{rs} L = f_2 L \quad (8)$$

where f_2 is the total local correction factor for labor costs.

LOCAL DIRECT ACTIVITY COST FOR THE SOCIETY

The local direct activity cost for the society will be then

$$\sigma (f_1 k + f_2 L) \quad (9)$$

where σ depends from constructor overhead and profit

INDIRECT COSTS RELATED WITH MACHINERY

Some of the indirect costs related with machinery are transportation costs for the equipment, service and maintenance workshops, spare parts storage, etc. These costs may be taken as a fraction e of K and amount therefore

$$e K \quad (10)$$

In the contracts sample the mean value of e was 0.038

The cost for the society is simply σeK

INCORPORATION OF OTHER SOCIAL BENEFITS AND COSTS

As indirect social benefit/costs we will consider: a) social benefit/cost due to employment generation, b) social benefit/cost due to increasing consumption, and c) social benefit/cost due to increasing investment.

Calling δ the daily marginal benefit to employ one worker, the social benefit due to employment generation will be

$$\delta \frac{L}{S} \quad (11)$$

δ may be estimated from the cost of subsidizing a family whose head is unemployed.

Calling β a vector whose components are the local marginal-propensities-to-consume and ε a vector whose components are the corresponding local marginal benefits of consumption and applying them to labor and operator wages, the social benefit due to increased consumption will be

$$\underline{\varepsilon} k''_{OP} K\underline{\beta} + \underline{\varepsilon} L\underline{\beta} \quad (12)$$

$\underline{\beta}$ may be estimated from a local survey and $\underline{\varepsilon}$ may be assessed from the local supply capacity: its elements may be negative (costs because saturated capacity) or positive (benefits because overcapacity).

For the mexican case we adopted vectors with following components: food, dwelling, clothing, home appliances, beverages, tabacco, transportation, education, health, other services, other goods.

Calling α_L the local marginal propensity to save corresponding to labor force and α_K the marginal propensity to reinvest in the region in terms of machinery costs K and calling γ the local marginal benefit of saving, the social benefit due to increased investment will be

$$\gamma (\alpha_L L + \alpha_K K) \quad (13)$$

γ may be determined from the local opportunity value of capital.

SOCIAL OBJECTIVE FUNCTION

Considering costs as positive terms and benefits as negative terms, the social objective function is:

$$Z = \sigma(f_1 K + f_2 L) + \sigma eK - \delta \frac{L}{S} - k''_{OP} \underline{\varepsilon\beta} K - \underline{\varepsilon\beta} L - \gamma\alpha_L L - \gamma\alpha_K K \quad (14)$$

This expression may be put in the form

$$Z = f_3 K + f_4 L \quad (15)$$

where

$$f_3 = \sigma f_1 + \sigma e - k''_{OP} \underline{\varepsilon\beta} - \gamma\alpha_K \quad (16)$$

$$f_4 = \sigma f_2 - \frac{\delta}{S} - \underline{\varepsilon\beta} - \gamma\alpha_L \quad (17)$$

Substituting for K according to (3)

$$Z = f_5 L^{-c} + f_4 L \quad (18)$$

where

$$f_5 = f_3 \left(\frac{P}{a}\right)^{\frac{1}{1-b}} \quad y \quad c = \frac{1}{1-b} \quad (19)$$

CONTRAINTS

a) Calling T the number of days available for the activity, S the daily wage and L the labor direct costs, the number of workers is L/ST and must be less or equal to the number of available local workers (OL)

$$L \le S\ T\ (OL) \quad (20)$$

b) Calling LMIN the lowest possible value of L related with the given activity, we have:

$$L \geq LMIN \tag{21}$$

c) Calling KMIN the lowest possible value of K related with the given activity and expressing it as a constraint over L we have

$$L \leq \left[\frac{(\frac{P}{a})^{\frac{1}{1-b}}}{KMIN} \right]^{1/c} \tag{22}$$

Because constrint (22) is of the same form as (20) it must be determined in each case wich one is most restrictive dropping the other one.

THE OPTIMIZATION MODEL

The optimization model takes then following form:

$$\text{Min } z = f_5 L^{-c} + f_4 L \tag{23}$$

$$L - x_1 = LMIN \tag{24}$$

$$L + x_2 = LMAX \tag{25}$$

$$x_1 \geq 0 \tag{26}$$

$$x_2 \geq 0 \tag{27}$$

where LMAX is the maximum value for L, according to constraints (20) and (22) and x_1, x_2 are slack positive variables to get (24) and (25) as equations.

OPTIMAL VALUES FOR L AND K

The type of constraints makes it possible to solve the minimization problem as if it were unconstrained, and checking if the solution satisfies constraints (24) and (25). If it don't, then the optimum will be at the extremum of feasible region, corresponding to violated constraint[2].

Equating to zero the derivative of (23), the optimal value L* of L will be

$$L^* = \left[\frac{f_5 c}{f_4} \right]^{\frac{1}{c+1}} \tag{28}$$

The condition for this point to be a minimum is f_5 be positive. This condition equivalent to f_3 positive was satisfied in the mexican case.

From (28) we can derive the optimal value K* for K:

$$K^* = (\frac{P}{a})^{\frac{1}{1-b}} \left[\frac{f_4}{f_5 c} \right] \tag{29}$$

[2] See Hadley (1970) pp 69-72.

EXAMPLE

The model will be illustrated with following example: define the optimal labor/machinery proportion according to following data:

Construction activity: Excavation in soft material bank

Quantity of work: 1 000 m^3

LMIN: 1.00 mexican peso (1968)

KMIN: 1 000.00 mexican pesos (1968)

Local workers available: 15

Marginal propensity to save: 0.22

Marginal propensity to reinvest: 0.05

Probably lost workdays because of weather: 10

Local daily wage: 20.00 mex pesos (1968)

Local diesel price: 0.40 mex pesos (1968) per liter $e = 0.04$, $\sigma = 1$

Daily marginal cost of unemployment: 5.00 mex pesos (1968) per worker

Local marginal benefit of saving: 0.50 mex pesos per peso saved (1968)

Maximal work completion time: 180 days

Local labor skill: Slightly over middle = 3

Local labor diligence: Slightly over middle = 3

Consumption data given by following table:

Concept	Marginal propensity to consume [3]	Marginal consumption cost in the region
Food	0.1190	0.05
Beverages	0.0105	0.10
Tabacco	0.0043	0.10
Dwelling	0.1292	0.10
Clothing	0.0863	0.10
Home appliances	0.0832	0.10
Transportation	0.0431	0.10
Education	0.0213	0.10
Health	0.0217	0.10
Other services	0.0466	0.10
Other goods	0.2108	0.10

We got following results:

Optimal labor direct cost: 6 308.52 mex pesos (1968)

Optimal machinery direct cost: 1 321.10 mex pesos (1968)

Total social net cost: 1 900.59 mex pesos (1968)

Optimal proportion labor/machinery (costs): 4.78

Optimal proportion of labor (costs): 82.68%

Optimal proportion of machinery (costs): 17.32%

[3] Estimations from Lara-Rosano (1973)

CONCLUSIONS

It was presented a mathematical model to define for each construction activity the quantity of labor required to optimize a social objective function. Running the model with past data it has been found that actually it was given preference to machinery over labor, although in many cases the reverse was social more convenient. The model has been computerized and can also be used for assessing different technological alternatives for rural roads construction from a social point of view.

REFERENCES

Keynes, J.M. (1936) "The general theory of employment, interest and money" Harcourt, Brace & Co., New York.

Hadley, G. (1970) "Nonlinear and Dynamic Programming". Addison-Wesley, Reading, Mass.

Lara-Rosano F. (1973) "Metodología para la evaluación de la componente extranjera del costo en la construcción de carreteras", Instituto de Ingeniería, UNAM, México.

Secretaría de Recursos Hidráulicos: (1963) "Manual sobre el Cálculo de Precios Unitarios de Trabajos de Construcción"; Tomo I; Principios y Premisas Fundamentales, México

COMPUTER-AIDED DESIGN OF THE IMPLEMENTATION OF INDUSTRIAL FACILITIES

J. Pablo Antun

Instituto de Ingeniería, U.N.A.M., Apdo. Postal 70-472, México 20, D.F., Mexico

Abstract. After having had certain preliminary discussions about industrial land development, we are presenting herewith a planning scheme for industrial states programme. The central part of it is a multicriteria algorithm for evaluation, developped on the basis of the "methode ELECTRE" proposed by B. Ray. It use in CANDE/FORTRAN IV allows a computer aided-design of the implementation of industrial facilities.

Keywords. Computer-aided system design; decision theory; economics; matrix - algebra; operations research; urban systems; multicriteria evaluation methods.

I

INTRODUCTION

Industrial Estates.

Industrial developments are those where land is prepared for industrial use. In the majority of cases, certain areas are reserved for services supportive of industrial activity - and commerce. In others there exist projects which include residencial units and areas - destined for recreative and cultural activities.

An industrial state is an area of development of land with the appropriate infrastructure for exclusively industrial use. They are provided with ground space for small common areas and for appropriate services. Sometimes there is also a provision made in the surrounding areas for the development of residencial units.

There exist different types and sizes of industrial estates, varying according to the predominating industrial structure, to demand, to the degree of technical interdependence, to the area available, to the relative location of existing centres of population.

One of the most relevant problems in planning is the location of the industrial estate. It demands a careful study to provide the most adequate location for the type of project of industrial development feasible for the adjacent centre of population and the region. - Building an industrial estate is not industrializing but more like using the existing opportunities to develop external economies. These are to be understood as technical and locational relations such as the availability of appropriate adequate land and convenient infraestructure, which will facilitate the - implementation of projects of industrial - investment.

In general, in developing countries like Mexico, the industrial developments are promoted by the public sector as yet one more strategy towards regional industrial development. - Nevertheless, this strategy is also utilized by governments of more developed countries as a way of re-distributing territory, or in - order to reactivate development in depressed or stagnant areas with serious structural - problems, such as unemployment.

The experience of Latin America is very rich. Awareness of regional differences has led to experimentation in strategies such as growth poles, in the period commonly known as that of the "internal growth" of the economies of latin american countries. Very often one of the "tools" used in regional industrialization has been of locating of industrial estates in selected areas. On occasions the concentrated investment on behalf of the public sector, combined with other socio-economic factors has produced an explosive growth[1] in these centres way beyond the limits foreseen in planning - and leading to irreversible and problematic situations.

Mexico: strategy and problems in planning.

In spite of being relatively recent, mexican experience[2] in land industrial developments is quite important. In various previous - - studies analysis has been made of the different stages of this experience and one of the studies[3] made an evaluation based on an extensive inquiry into the sectors involved.

Industrial estates strategy is contained in policies which underline two complementary objectives:

- territorial distribution
- regional industrial development.

Territorial distribution follows the regulating framework established by the National Plan for Urban Development (NPUD) lately incorporated in the regional goals of the national Plan for Industrial Development (NPID). This could be translated as follows:

- decentralization of economic activity towards the centres of greatest potential in the interior of the country
- the promotion of a "march towards the coast" in order to redistribute the population from the central plateau.
- the consolidation of intermediate centres of population with regards to productivity structure and urban services to be offered in a concentrated and diversified form.
- the preservation of the environment.

Industrial development emphasises a set of activities of prior importance: food, capital goods, equipment for transport, communication.

It becomes evident what are to be the main problems in planning from the moment when the industrial estate is an existant concentrated collection of lands with infrastructure for industrial use. These are:

- that of determining the feasible locations of industrial estates
- that of estimating the most feasible industrial structure.
- that of the type and size of industrial estate, taking into account its impact on the surrounding region
- that of optimizing the infrastructure
- that of forming a programme to introduce feasible industrial estates best designed to satisfy the political objectives of development.

This work deals with this last problem. First we shall present a sheme of the process of global planning employed in the Industrial Estates Programme for Mexico.

II

PLANNING SCHEME FOR INDUSTRIAL ESTATES PROGRAMME. (Fig. 1).

Candidate locations or candidate projects for industrial estates.

The point of departure for the analysis is the existance of a set of candidate locations for the establishment of industrial estate projects.

A candidate location for an industrial estate can come from the suggestion of the public sector or of industry. The pre-feasibility of each of the candidate projects is determined, and then a further selection is made from among these.

Pre-feasibility.

A project is considered pre-feasible when there exists a potential demand for land for industrial use. The characteristics of the existant industrial structure is taken into account.

The pre-feasibility of a project is determined by applying a filter of pre-feasibility to each candidate location. The fundamental elements of the filter are the regulating framework of the NPUD and NPID, plus an analysis of the industrial structure, existing and pre-feasible, both short and medium term, in the chosen area. That is the geographical region containing the location of the proposed industrial estate and the hinterland. For each estate the study will evaluate the location in terms of the potential industrial structure in the estate plus the contribution that this may make to the objectives of territorial distribution and support to regional industrial development.

Order of Preference of Pre-feasible Locations

The set of pre-feasible locations is put in order according to the preferences of the Secretary of Public Works, using a multicriterial algorithm for evaluation (this will be discussed later on) for this the assets of each location are reviewed. The order of preference of a pre-feasible location is a piece of information which justifies the application of resources for studies of feasibility of the locations most relevant to the policies of the industrial estates programme.

Feasibility.

A project is deemed feasible when the establishment of an industrial park is considered possible in economic terms, taking into account the market situation of land for industrial use in the locality.

The feasibility study leads to a definition of the type of industrial estate, the length of time for the establishment of the project, and the maximum amount of investment, the economic and financial feasibility of the project and it also estimates the social and environmental impact of the industrial structure.

Selection of feasible projects or first version of the Industrial Estete Programme.

Taking as a starting point the restrictions is strategy (operative and policy limitations imposed by the Secretary of Public Works) several alternative plans are defined, each of these involving a set of feasible projects.

The study of the different alternative plans

and the selection of the preferred one, as the programme of establishment of industrial developments, is realized by the designer of the alternative plans and the selector of the preferred plan. The designer of alternative plans works in agreement with the strategic restrictions. The selector of the preferred plan follows a multiple evaluation criterion. The preferred plan outlines a set of selected feasible projects.

III

MULTICRITERIAL ALGORITHM FOR EVALUATION.

The most usual method for the evaluation of projects is the costbenefit analysis. In some cases certain refinements are introduced, such as cost-effectivity analysis or shadow project analysis. Nevertheless in each case effects of the plans can no be evaluated if they can not be reduced to equivalent monetary values. In the same token it is troublesome and expensive to consider the opinions of different decision-makers involved in the final decision. These difficulties are overcome by a multicriteria analysis without dismissing the idea that the cost-benefit analysis can be utilized in a complementary fashion.

To his end a multicriteria algorithm for evaluation has been designed. It is based on "la methode ELECTRE" proposed by B. Roy[4]. As has already been shown it is applied to:

- determine the order of preference of a pre-feasible location.

and to

- selected a sub-set of feasible locations as a preferred plan for the establishment of industrial estates.

The main idea is to compare location (projects) /alternative plans in pairs; if a measure of the concordance (relative advantages) and of the discordance (relative disadvantages) is defined then there can be established the relation of dominance or preference between the compared locations/plans. Adopting a minimum level of concordance and a maximum level of discordance, the preferred location (projects) /alternative plan can be determined.

There exist also certain options in the treatment of the information on which the evaluation will be made.

Evaluation of locations (projects)/alternative plans.

The first step consists of identifying the set of locations (projects)/alternative plans to be analyzed and of evaluating them.

To determine the order of preference of the pre-feasible locations the multiple criterion for evaluation of Fig. 2 was used. Values of from 1 - 9 are assigned according to the relative ease with which the projected industrial estate fulfills the partial objectives in the multiple criterion for evaluation.

Thus we can construct a matrix

$$K = \{k_{ji}\} \qquad i = 1,\dots,I \quad j = 1,\dots,J\ (=4)$$

where k_{ji} is the value of the location i according to the partial objective j.

Specification of political preferences.

To introduce the political preferences of the decision-maker (in our case the Secretary of Public Works) a vector is specified,

$$S = \{S_j\} \qquad j = 1,\dots J$$

where s_j is the political preference of the partial objective j, with respect to the set of objectives in the multiple criterion for evaluation.

Matrix of Impact.

The comparison of locations (projects)/alternative plans on pairs is made on the basis of the impact of each one on the objectives of the industrial states programme, that is to say the matrix K. Neverthelss it is worth while to make some introductory calculations. The first is to normalize the row vectors of the matrix K at unit value.

$$R = \{r_{ji}\}\ ; \qquad r_{ji} = \frac{k_{ji}}{\sqrt{\sum_{j=1}^{j} k_{ji}^2}}$$

The second is to weight the evaluation according to the political preferences of each partial objective in the multiple criterion for evaluation,

$$P = \{p_{ji}\}\ ; \quad p_{ji} = S_j \ r_{ij}$$

This is the matrix of impact P.

Concordance/Discordance.

The comparison by pairs, allows the definition of the sets of concordance (sets of discordance) that is the set of partial objectives for which the location (project)/alternative plan i is preferred (not preferred) to the other i':

$$C_{ii'} = \{j | k_{ji} \geqslant k_{ji'} |\} \ \forall \ i \neq i'$$

$$D_{ii'} = \{j | k_{ji} < k_{ji'} |\} \ \forall \ i \neq i'$$

Defining an index of concordance (discordance) as the measure of the relative advantages (disadvantages) of i over i' we can construct the matrix of concordance (discordance):

$$C = \{c_{ii'}\}\ ; \quad c_{ii'} = \frac{\sum_{j \,\varepsilon\, C_{ii'}} s_j}{\sum_{j=1}^{J} s_j}\ ; \ i \neq i'$$

$$D = \{d_{ii'}\} \quad ; \quad d_{ii'} = \frac{\max\limits_{j\varepsilon D_{ii'}} | P_{ji}-P_{ji'}|}{\max\limits_{j\varepsilon D_{ii'}(i,i')} | P_{ji}-P_{ji'}|} ; \; i\neq i'$$

Here it is appropriate to define a differential index of concordance (discordance) as a measure of the relative net advantages (disadvantages) of i over i', we can construct the matrix of differential concordance (discordance) as follows:

$$E = \{e_{ii'}\} \quad ; \quad e_{ii'} = c_{ii'} - c_{i'i} \; ; \; i \neq i'$$

$$F = \{f_{ii'}\} \quad ; \quad f_{ii'} = d_{ii'} - d_{i'i} \; ; \; i \neq i'$$

Determination of the preferred solution.

If we define a critical index of differential concordance (discordance as e* (f*) then:

$$X = \{x_{ii'}\} \quad x_{ii'} = 1 \quad e_{ii'} \geq e^*$$
$$x_{ii'} = 0 \quad e_{ii'} < e^*$$

$$Y = \{y_{ii'}\} \quad y_{ii'} = 1 \quad f_{ii'} \leq f^*$$
$$y_{ii'} = 0 \quad f_{ii'} > f^*$$

And, introducing

$$Z = \{Z_{ii'}\} = X \cap Y$$

It can be seen that if

$$Z_{ii'} = 1 \; (i\neq i') \text{ y } Z_{i''i} = 0 \quad (i''\neq i, i''\neq i')$$

it follows the necessary condition for the location (project)/alternative plan i to possess an aggregate differential dominance, and to be the preferred solution.

Please note that the determination by proof and error of e* (f*) is an unavoidable step in order to obtain a unique solution.

Treatment of the information.

In general the decision-maker has certain difficulties in specifying the s_j and also certain "disorder" in evaluating the k_{ji}.

Optionally one can revert to the idea of entropy in information theory and incorporate it in

$$S^* = \{s_j^*\} \quad ; \quad s_j^* = \delta_j \, s_j$$

and in

$$K^* = \{k_{ji}^*\} \quad ; \quad k_{ji}^* = \delta_j \, k_{ji}$$

where

$$\delta_j = 1 - \varepsilon_j \; ; \quad i = 1,\ldots,I \; ; j = 1,\ldots,J$$

Adopting ε_j as a measure of the entropy with respect to the partial objective j at:

$$\varepsilon_j = -\frac{1}{\ell n \, I} \sum_{i=1}^{I} (\gamma_{ji} \, \ell n \, \gamma_{ji}) \; ; \; \gamma_{ji} = \frac{k_{ji}}{\sum\limits_{i=1}^{I} k_{ji}}$$

IV

APPLICATIONS TO THE INDUSTRIAL ESTATES PROGRAMME FOR MEXICO.

The Secretary of Public Works for Mexico proposed at the end of 1979 a set of 35 locations for the establishment of industrial estates. The socio-economic studies determined that only 20 were pre-feasible[5]. The order of preference of each of these was determined by using the algorithm previously described. In the following months the set of feasible locations will be applied to define their short term programming.

The employment of the multicriteria algorithm for evaluation is facilitated to a large extent by writing it in FORTRAN IV and then using it as an interactive process in terminals with CANDE (Command and Edit Language) on a Burroughs 6800 computer.

V

CONCLUSIONS.

If the present application of this proposed methodology will consolidate the favourable aspects and underline the deficiencies, there can already be seen certain advantageous characteristics:

i) it clarifies the planning process by identifying net stages.

ii) it permits the consultation of other governmental sectors involved in the problem

iii) it introduces an algorithm of plan selection based on multiple criterion

iv) it rationalizes the use of financial resources in applying them to studies whose range is defined by the importance of the inmediate decision to be taken by the decision-maker

v) it takes into account the strategic restrictions, generally of a political nature.

vi) it allows for the programming of short-term actions.

VI

REFERENCES

1. Antún, J.P. (1979) *CECREX: Un modelo de simulación de la dinámica interna de centros de crecimiento explosivo*, Primera

Convención Internacional de Informática, México.

2. SOP-NAFINSA (1975), *Ciudades Industriales*, Secretaría de Obras Públicas, México.

DGOMU (1979), *Programa de Desarrollo Urbano Industrial*", Secretaría de Obras Públicas, México.

3. Unikel, L. y otros (1976), *Ciudades Industriales: Evaluación y Perspectivas*, Centro de Estudios Económicos y Demográficos, El Colegio de México, México.

4. Roy, B. (1968) Classement et Choix en presence de points de vue multiples (La méthode ELECTRE), *Revue Francaise d' - informatique et Recherche Operationnelle*, 2eme année, No. 8, 1968, p. 57-75.

5. Antún, J. P. López Ortega, E.; Vázquez, F. (1980) *Determinación de la prefactibilidad de establecer parques industriales en 35 localidades seleccionadas*, Instituto de Ingeniería, Proyecto 0513, México.

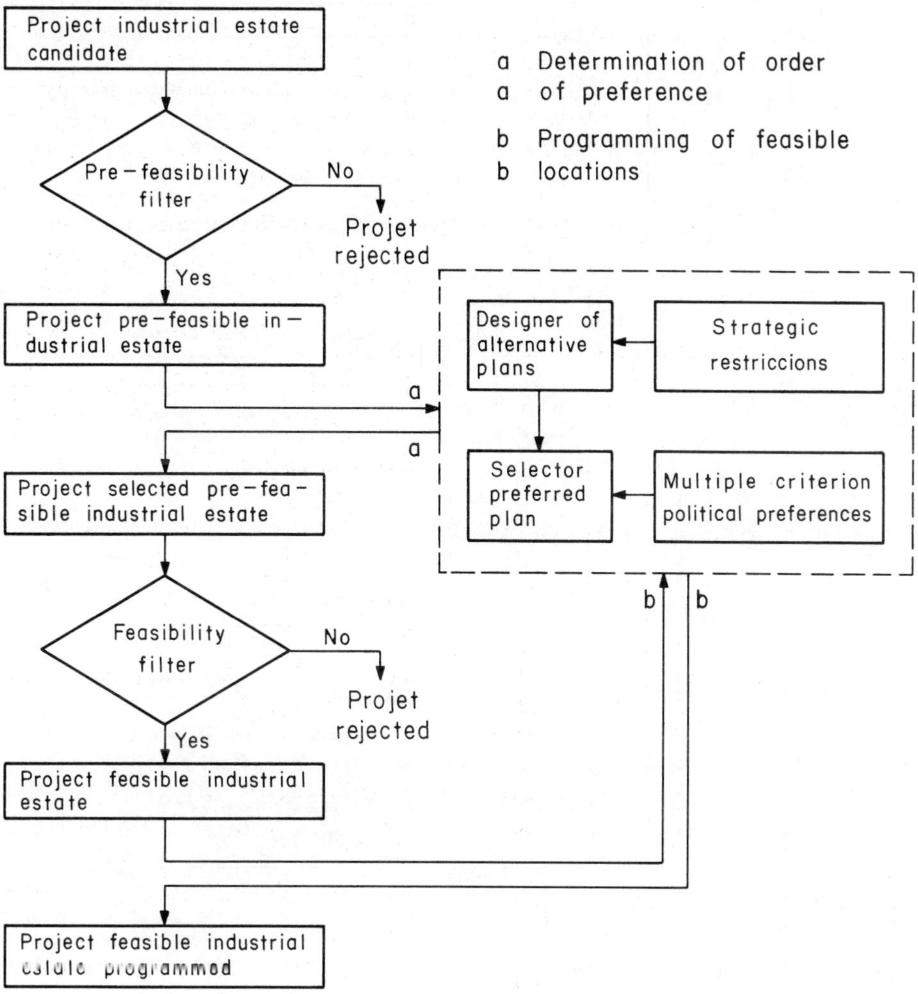

Fig 1. Planning scheme for industrial estates programme

Contribution of the industrial estate to the objectives of the National Plan of Urban Development

Relative relevance of establishing an industrial estate in the location based on the pre-feasible industrial structure of installin it with respect to:

- the spatial redistribution of economic activities over the whole territory
- the consolidation of population centres with priority

Relative relevance of establishing in industrial estate in the location based on the employment which the pre-feasible industrial structure will generate on being installed in it with respect to:

- the retention of local population and
- the generation of desired interregional migratory movement

Contribution of the industrial estate to the objectives of the National Plan of Industrial Development

Relative relevance of establishing an industrial estate in the location based on the pre-feasible industrial structure of installing itself in it with respect to

- the objectives of development of industrial activities with priority

Relative relevance of establishing an industrial estate in the location based on the economy of localization with respect to:

- momentum of development of existing industrial activity

Fig 2. Multiple criterion for evaluation

APPLICATION OF RECURSIVE ESTIMATION TO SOLAR ENERGY

H. Z. Tadros and M. M. A. Salama

Department of Electrical Engineering, Ain-Shams University, Cairo, Egypt

Abstract. The solar energy appears to be the most promising source of energy for the next few decades. The knowledge of the spectral distribution of sunlight as it passes through the atmosphere, is the key to any quantitative analysis of the solar energy conversion system. The recursive estimation technique is employed for the determination of the solar energy parameters used in the computer simulation techniques.

Keywords. Natural resources; models; computer applications; optimization; parameter estimation; least squares approximation; linearization technique.

INTRODUCTION

The ever increasing demand of more energy for industrial and public consumption is faced by the threat of the shortage and high priced energy. The solar energy appears to be the most promising source of energy for the next few decades. It is clean, unlimited and renewable source of energy. Quantitative data on available solar irradiance is an essential input parameter for the design of solar energy conversion systems. This insolation data at a particular site on ground level depends on the geometry of the site, sun earth-distance, time of the day, season of the year, cloudiness of the atmosphere, slope of the collector surface and atmospheric attenuation due to water vapour, ozone, carbondioxide and aerosols. Insolation data necessary for energy conversion can be obtained either by a direct ground measurements or a combination of ground measurements and computation from the extraterrestrial solar spectrum. Good progress has been made in the first direction (Hay, 1976) because of the continuous improvement of the experimental facilities. Calorado State University, University of Wisconsin (Trnsys, 1975) and University of Waterloo (Hollands, 1976; Orgill, 1976) have developed a computer simulation programs for solar heating system. The insolation data used in these models are based on the second direction.

Although in the last few years a remarkable steps forward have been made in utilizing the solar energy, less progress was reached in the computer simulation techniques. This is may be attributed to the fact that many of the solar energy parameters, which are commonly used in these computer models, are very difficult to be estimated accurately. Therefore the estimation techniques has to be very appropriate to the type and nature of solar energy conversion system. This paper presents a tentative mathematical model for the estimation of the attenuation coefficient of the solar irradiance.

ATTENUATION COEFFICIENT

The intensity of the direct component of the sunlight for a certain wave length λ is given by:

$$I_\lambda = I_{0\lambda} \exp(-K_{m\lambda}\, m) ,$$

where I_λ and $I_{0\lambda}$ are intensities on the ground and outside the atmosphere respectively,

$K_{m\lambda}$ the attenuation coefficient
m the airmass, $m = \operatorname{cosec} \phi$,
ϕ the solar elevation.

The attenuation coefficient depends on scattering and absorption of a variety of molecules and larger particles. The absorption and scattering mechanisms in the range of interest are:

(1) Rayleigh scattering.
(2) Mie scattering.

(3) Absorption by ozone.
(4) Absorption by water vapour.
(5) Absorption by O_2, CO_2, and CH_4, which has a little effect.

In most computer simulation models, the attenuation coefficient is evaluated by computing the loss due to absorbing and scattering by the atmosphere. The total irradiance can be determined by integrating the spectral irradiance at all wavelengths. The output of the solar conversion system computed from these models will be inaccurate due to the fact that the output of the solar cell, for example, depend on specific insolation condition and cannot be normalized by considering the total intensity.

ESTIMATION OF THE ATTENUATION COEFFICIENT

In classical estimation theory (Nahi, 1969), the observation is always assumed to contain the signal to be estimated. In practice, certain observations, or sequences of observations, may contain noise alone. The probability of occurance of such cases being available to the estimator.

The difficulty arises in estimating the attenuation coefficient is due to the lack of the statistical information about the insolation data. The knowledge about noise in the observations and the presence of a nonzero probability that any observation consists of noise only, is rarely known in the irradiance measurements. Therefore the technique to be used in estimating the attenuation coefficient will be mainly determined by the amount of the statistical information available about the noise. As an example, Table 1, gives a sample measurements of the irradiance for a certain wave length. The estimation of the attenuation coefficient can be conducted as follows:

TABLE 1 Sample of Measurements

i	m_i	I_i
1	1.015	0.9780
2	1.091	0.9769
3	1.411	0.9725
4	2.645	0.9559
5	7.168	0.9001
6	23.20	0.7607
7	78.71	0.3197
8	4.394	0.9333
9	411.4	0.92556
10	436.8	0.91916

First Approach

If in these measurements the information about the noise is completely ignored, then the best method of estimating $K_{m\lambda}$ is the averaging method.

1. Arithmetic mean value

$$\hat{K}_{m\lambda} = \frac{1}{n}\sum_{i=1}^{n} (\log I_{0\lambda} - \log I_{\lambda i})/m_i ,$$

where,
$\hat{K}_{m\lambda}$ the best attenuation coefficient for the wave length λ,
n the number of observations.

Using this method gives,

$$\hat{K}_{m\lambda} = 0.048039 .$$

2. Geometric mean value

$$\hat{K}_{m\lambda} = \left[\prod_{i=1}^{n} \log (I_{0\lambda}/I_{\lambda i}) /m_i\right]^{1/n}$$

Using this method yields,

$$\hat{K}_{m\lambda} = 0.032465 .$$

Second Approach

1. Modified arithmetic mean method.
If the noise in the measurements are completely in random distribution, the expected value of the noise is zero, and $\hat{K}_{m\lambda}$ is given by:

$$\hat{K}_{m\lambda} = \left[\sum_{i=1}^{n} (\log I_{0\lambda} - \log I_{\lambda i})\right] / (\sum_{i=1}^{n} m_i)$$

Using this method yields,

$$\hat{K}_{m\lambda} = 0.010623$$

2. Weighted least squares. If the available statistical information about noise are:

1. The components of the measurements are statistically linearly independent.
2. No correlation exists between the measurements.
3. No correlation exists between the noise.

The weighted least squares method can be used to estimate $K_{m\lambda}$. The derivation of this method is based on the Gauss-Markoff theorem (Deutsch, 1965). The above available noise information are employed in the analysis. The best value of $K_{m\lambda}$ is therefore given by:

$$\hat{K}_{m\lambda} = \sum_{i=1}^{n} (\log I_{0\lambda} - \log I_{\lambda i}) m_i / (\sum_{i=1}^{n} m_i^2)$$

Using this method yields,

$$\hat{K}_{m\lambda} = 0.009338$$

Third Approach

Optimal recursive estimation with uncertain observation. In this method all necessary statistical information about the observation and the noise are employed. These informations can be summarized as,

1. The estimate is unbaised.
2. $K_{m\lambda}$ is a linear minimum variance, unbaised estimate of K_m .
3. No correlation between the noise and the measurements.
4. The correlation matrix of the measurements is known.
5. The correlation matrix of noise is known.

The method requires the attenuation coefficient equation to be in the following linear form:

$$A = B\,K_{m\lambda} + V_i$$

where,

$$A = \log I_{\lambda i} - \log I_{0\lambda}$$

$$B = -m_i$$

V_i = noise in the ith measurement.

Therefore

$$\hat{K}_{m\lambda} = (C_K^{-1} + B^T C_V^{-1} B)^{-1} B^T C_V^{-1} A$$

With

$$C_e = (C_K^{-1} + B^T C_V^{-1} B)^{-1}$$

where

C_K the correlation matrix of the measurements,

C_V the correlation matrix of the noise,

$\hat{K}_{m\lambda}$ the optimal attenuation coefficient for the wave length λ .

In the following analysis, C_K and C_V are diagonal matrices. The diagonal element of C_K is 3.33×10^{-3} and of C_V is 1×10^{-6}. Using this method yields,

$$\hat{K}_{m\lambda} = 0.009254$$

DISCUSSION

Table 2 shows that the error in estimating $K_{m\lambda}$ is very large when very little information about noise is available. Although the optimal recursive estimation consumes a relatively large computer time, it gives very accurate results.

TABLE 2 Attenuation Coefficients

Method	$\hat{K}_{m\lambda}$	% age error
Arithmetic mean	0.048039	419.12
Geometric mean	0.032465	250.82
Modified arithmetic mean	0.010623	14.79
Weighted least squares	0.009338	0.91
Optimal recursive estimation	0.009254	reference

CONCLUSION

The attenuation coefficient of the solar irradiance is a very important input parameter to the computer simulation programs. The estimation of $K_{m\lambda}$ has to be very appropriate to the solar irradiance and has to employ very efficiently all available statistical information of the measurements. A tentative mathematical methods for estimating the attenuation coefficient is presented. The optimal recursive estimation gives best results. The choice of the estimation technique has to be selected based on the available statistical information about the observations and the noise. The informations are very essential to evaluate $K_{m\lambda}$ accurately.

REFERENCES

Deutsch, R. (1965). Estimation theory. Englewood Cliffs, N.J., Prentice Hall.

Hay, J.E. (1976). The climatology of available solar energy for Canada. Conference of the American section of ISES, Vol. I, pp. 211-225.

Hollands, K.G.T. and Orgill, J.F. (1976). Technical and economic feasibility of solar energy heating Research Institute, University of Waterloo, pp. 1-19.

Nahi, N.E. (1969). Estimation theory and application. New York, Wiley.

Orgill, J.F. and Hollands, K.G.T. (1976). Solar heating simulation and economic evaluation program. Research Institute, University of Waterloo, pp. 1-54.

Trnsys, (1975). A transient simulation program. Report 38, Solar Energy Laboratory, University of Wiscosin.

OPTIMIZATION OF THE POWER OF AN ARRAY OF PHOTOVOLTAIC SOLAR CELLS

M. Annabi*, N. B. Slama*, M. Ksouri* and J. Perard**

**ENSET Tunis, 5 av. Taha Hussein*
***INPG, Grenoble*

Abstract. The authors consider the operation of a photovoltaîc panel. Then demonstrate the necessity of matching the load and the generator internal impedance. To this end, they insert a boost chopper between the panel and the load, to maintain the output voltage constant, then show that this yields the maximum output power for a given temperature, in spite of insolation variations. Then, they show that a temperature variation lead to a variation of maximum permissible power of same sign ; therefore, it is necessary to vary reference of the output voltage if optimum power operation is to be obtained for all temperature conditions. The various elements are studies in the paper, and a convenient model of the booster is offered.

Keywords : Solar energy ; optimization ; boost chopper - modelling of static converter ; photovoltaîc cells ; temperature of junction.

INTRODUCTION

The high cost of solar photovoltaîc generators explains their association with systems, even sophisticated, that permit to place them in any circunstances in their optimal operating conditions. In particular, the caracteristics current-voltage I(V) of such a generator Fig.1 shows that for a given insolation and temperature, there exists an operating point for which the power output is maximum. Our purpose is to make the photovoltaîc panel operate always at that point.

DEVICE PRINCIPLE

The real operating point in the plane [I,V] depends upon the load and is found at the meeting point of the caracteristics of the generator and that of the load Fig.2.

It is in general far to be likely that this point coincides with Mo. Even if it were so for given conditions of insolation and temperature, any modification of one of these parameters would lead to the loss of the optimal operation. It is then necessary to insert, between the generator and the load a device that controls the impedance seen by the generator (Ben Slama, ENSET 1980).

To this end we use a DC-DC converter made of a boost chopper, the circuit of which is shown in Fig.3. The transistor, controled by its base current, operates in commutation with a variable cyclic ratio called α ($\alpha = \frac{t_{on}}{t_{off}}$), where t_{on} and t_{off} are respectively the closing and opening time of the transistor. In these conditions it is shown that in a steady state, output voltage is $V_s = \frac{V}{1-\alpha}$ and that the input impedance of the photovoltaic panel load is

$$Z_e = Z_{ch}(1-\alpha)^2$$

Where Z_{ch} is the load impedance.

The shape of the caracteristics I(V) and consequently the position of M_o depend first upon the insolation as in Fig.4. In practice, only the current I_o is affected by the variations of insolation. As a result voltage V_o coresponding to M_o is practically independant of temperature. Thus for a given junction temperature the optimal operating may be obtained by maintaining the input voltage of the boost chopper to a fixed value.

On the other hand, the junction temperature influences only the open circuit voltage, so that M_O moves as shown in Fig.5.

To take in to account the system perturbations due to the junction temperature variations, we have to modify the reference of the above mentioned voltage in relation to the temperature achieving an optimization in an opened loop. We are then led to the general block diagram shown in Fig.6 where $F_{(p)}$ represents the whole generator - converter - change, $C_{(p)}$ the elaboration apparatus of the cyclic ratio and $T_{(p)}$ that of the voltage reference from the measure θ temperature. We shall study respectively the modelling of the boost chopper, the equivalent circuit, the voltage control and the control the reference by temperature.

THE MODELLING OF THE BOOST CHOPPER

Determination of the equivalent circuit

The boost chopper is a peacewise-linear system since it gives 2 differents configurations of electric circuits in a period (T_R blocked, D conducting then the revers).Fig.7.

The modelling of such a system is in general difficult but may be effectuated by nearly the same method as the one used by Perard, Toutain and Nougaret (1979).

According to this method (used in continued conduction), the 2 configurations described by the state matrices A_1 (duration $T_{on} = \alpha T$) and A_2 (duration $t_{off} = (1 - \alpha)T$) the system is replaced by a unic circuit described by the average matrix $A_m = A_1 + (1 - \alpha)A_2$.

We show then that circuit has the same topology as the initial circuit in which the electronic switches (transistor and diode) are replaced by generators the current and the voltage of which are expressed by state variables. Fig.8 represents the obtained overage equivalent scheme that modelises the boost chopper, whateven the amplitude of the signals. On this scheme, all the ordinary circuit calculations can be done.

That is haw the transfer function $F_{(p)}$ can be determined.

Determination of $F_{(p)}$

For a cyclic fixed ratio the équivalent scheme of Fig.8 allows us to write in operational notation :

$$E = (r + r_L + Lp)I + (1-\alpha)V_s \qquad (1)$$

$$V_s = (1-\alpha)I\frac{R}{1+RCp} \qquad (2)$$

$$V = E - rI \qquad (3)$$

If the photovoltaïc panel is represented by e.m.f. E, an internal resistance r(the numerical values shaw that the internal capacity may be neglected) and if we assume that the load Z is a resistance R. Equations (1) and (2) shaw that the system is a non linear function of α. It is then nessary to linearize around an operating point and to study its response to a small variation of α.

The dynamic response of the source is different according to wether we consider the part AB or the part BC of the caracteristic curve (Fig.2). Since the optimum operating point is located on the second brench rather than in the first, we are going to study only this case.

According to equation (1), (2) and (3) and for variations of α around α_0 the transfer function $F_{(p)}$ can be written :

$$F_{(p)} = \frac{V(p)}{\alpha(p)} = \frac{-r[V_{so} + RI_0(1-\alpha_0) + V_{so}RCp]}{LRCp^2 + [L + (r+r_L)RC]p + (r+r_L) + (1-\alpha_0)^2 R}$$

Considering the following point:
$V_{so} = 17V$, $I_o = 5A$, $\alpha_o = 0{,}5$, $R = 10\,\Omega$
and with the circuit constant
$L = 0{,}7mH$, $C = 1000\,\mu F$, $r_L = 0{,}2\,\Omega$
$r = 0{,}5\,\Omega$ the transmittance becomes

$$F_{(p)} = \frac{-12\,10^3(250 + p)}{p^2 + 1{,}1\,10^3 p + 460\,10^3}.$$

The unit step response of the system for a low level of α is shown in Fig.9 and this for different values of the dynamic source resistance.

IMPUT VOLTAGE CONTROL

This control is not only to fix the tension of operating point of the panel at a given temperature but also to confer to the system satisfactory static and transient performances.

The open circuit transfer function is similar to a 1st order response $F_{(p)}=\frac{K}{1+\tau p}$ with K=0,7 and τ=1,5 ms.

We have chosen a PI corrector. The transfer function $C_{(p)}$ is given by $C_{(p)}=\frac{B}{P}(1+\frac{Ap}{B})$. The closed loop transfer function will then be $H_{(p)}=\frac{K'}{1+\tau' p}$ with K'=1, $\tau'=\frac{1}{BK}$.

with $\frac{A}{B}=\tau$; $\frac{\tau}{\tau'}=10$ we shall get $B=\frac{10}{K\tau}=9{,}5 10^3$; A=14,3; τ'=0,15ms and we shall have sensitivity $e_{o_1}=0$

transient response : $t_r 5\% = 0{,}3\tau = 0{,}45$ms.

TEMPERATURE CONTROL OF THE VOLTAGE REFERENCE

In paragraph 2 we have already mentioned the need to control the voltage by temperature (JTEA,Tunis, 1980). Let's calculate the power variation when we move from an optimum operating point Mo_2 (Fig.5) to point Mo_1 the coordinates of wich are (V_1, I_1). Without control we should go from Mo_2 to Mo_3 whose coordinates are (V_3, I_3). In the case of solar generators of the TSG MQ 40/0 (AEG telefunkun) type and for a variation of θ from 25°C to 60°C, we have

$$\frac{\Delta P}{P}=\frac{P_1-P_3}{P_1}=\frac{V_1 I_1 - V_3 I_3}{V_1\ I_1} = 56\ \%$$

or else $P_1 = 2{,}25\ P_3$. Which shows the importance of the control of voltage reference V_C by temperature.

Law of voltage variation versus temperature

The theoritical study of the PN junction (Bonnefille and Robert 1971) with widely accepted simplifying hypotheses leads to the following results :

$$V_o=\frac{KT}{q}\text{Log}(1+\frac{I_{cc}}{I_s}) \text{ with } I_s=AT^3\exp(-\frac{W_a}{KT})$$

with K : Boltzman constant
I_{cc} : Schort circuit current independant of temperature and dependant of insolation
W_a : Junction activation energy
A : Independant constant of temperature
q : Elementary charge.

The know ledge of an operating point determined by V_o, allows owing to the formula already defined, to determine A and to establish the law of the variation of V_o as a function of temperature for TSG MQ 40/0 series solar panels and an insolation E = 100 mW/cm^2
$V_o = -0{,}092\theta + 35{,}7$ Fig.10. The maximum difference between theoritical and experimental results is 2 %.

The control device

Our simple control device Fig.11 contains a non linear element sensitive to temperature (NTC), the whole set of resistances (R_1,R_2) allows the linearisation of reference V_c in relation to the temperature, R_3 and R_4 determine the requived slope.

BIBLIOGRAPHY

AEG TELEFUNKEN - Générateurs solaires TSG MQ 40/0 Fiche technique
M. ANNABI, N. BEN SLAMA, M. KSOURI, M. KACHOUKH. Identification de la station de pompage de "Hendi Zitoun" JTEA - Tunis. Mai 1980.
N. BEN SLAMA Etude d'un hacheur parallèle à transistor - adaptation aux photopiles solaires - rapport DEA E.N.S.E.T. - Tunis - 1980.
BONNEFILLE J. Robert Principe généraux des convertisseurs direct d'énergie
J. PERARD, E. TOUTAIN, M. NOUGARET Modélisation des convertisseurs d'énergie par un schéma équivalent - l'onde électrique 1979, vol 59. N°12.

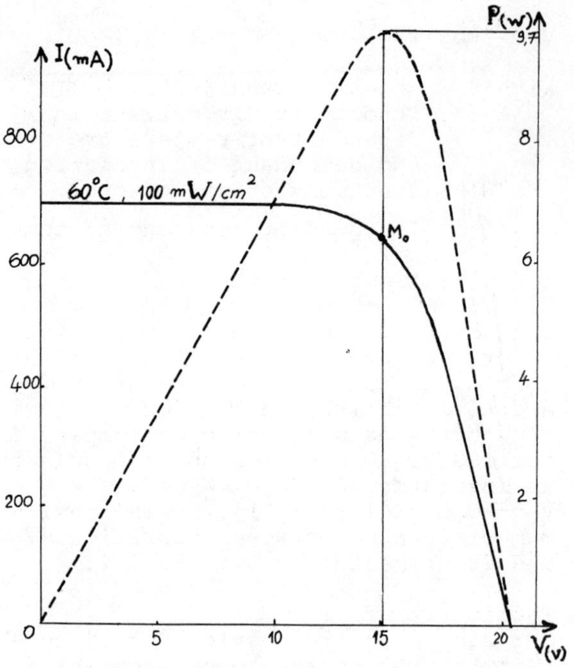

Fig.1. Current voltage curve of photovoltaïc panel (solid lines)-Variation of output power (interrupted lines).

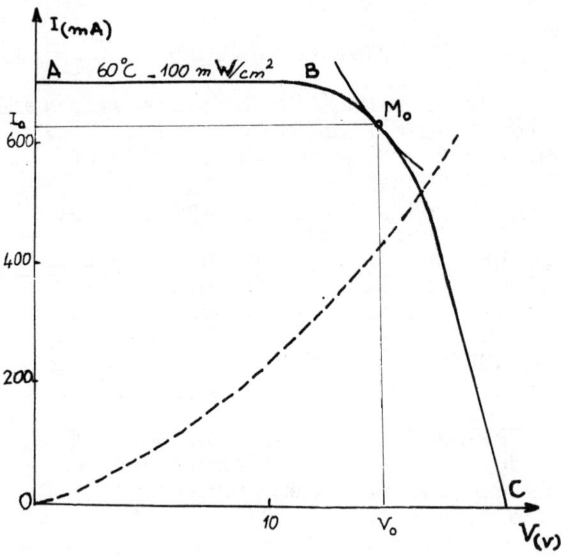

Fig.2. The actual operating point of the panel is determined by the I(V) performance curve of the load.

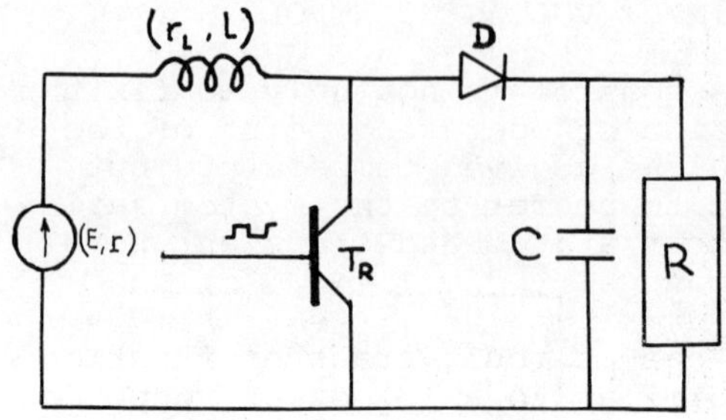

Fig.3. Principle of the transistor boost chopper.

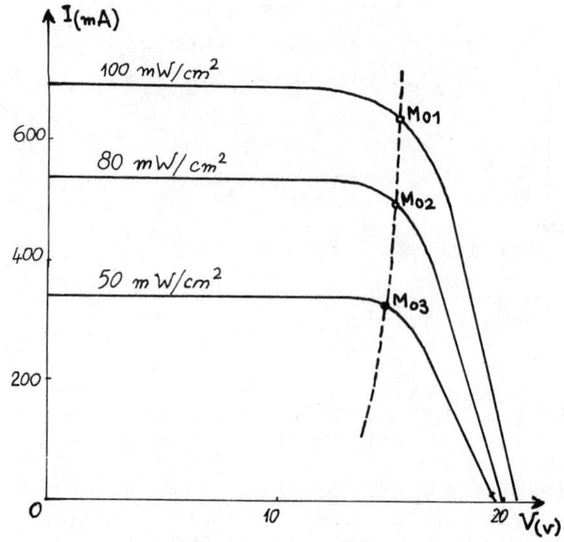

Fig.4. Caracteristic curves of a photovoltaïc panel, for several values of insolation, and a constant temperature. Optimum operating points are M_{01} M_{02} M_{03}.

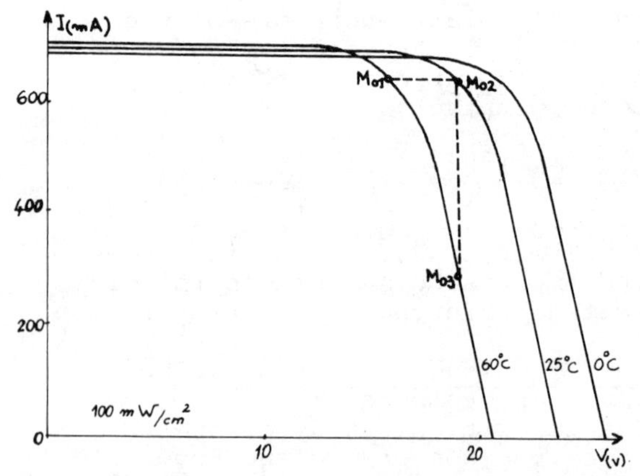

Fig.5. I(V) curve of a photovoltaïc panel for a given insolation and several temperature.

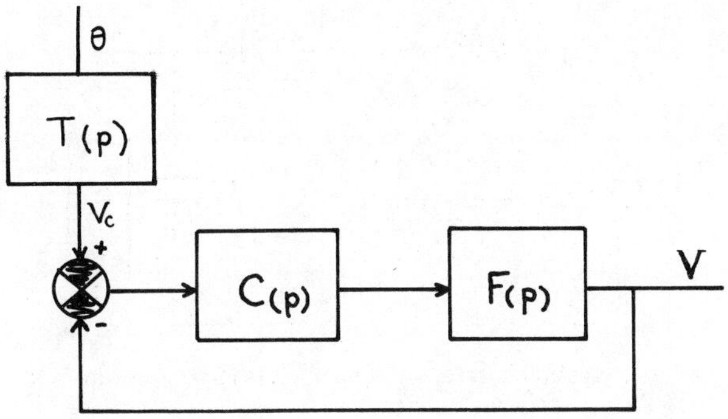

Fig.6. Control block diagram.

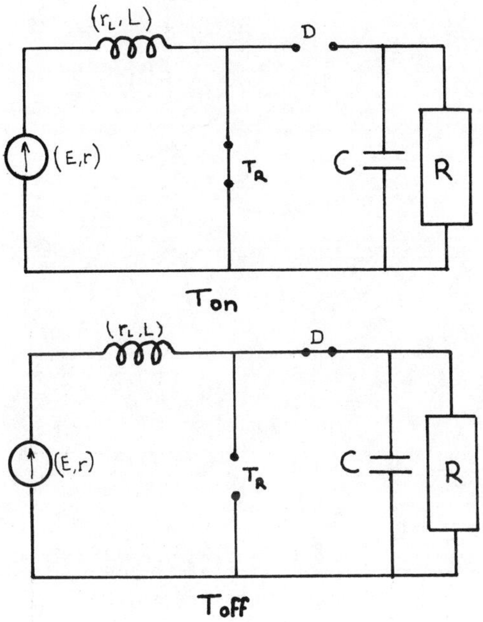

Fig.7. Two configurations correspon-ding to t_{on} and t_{off} conditions.

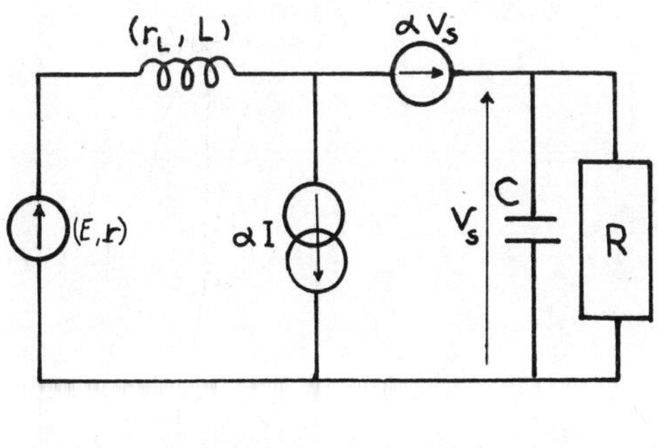

Fig.8. Average equivalent circuit.

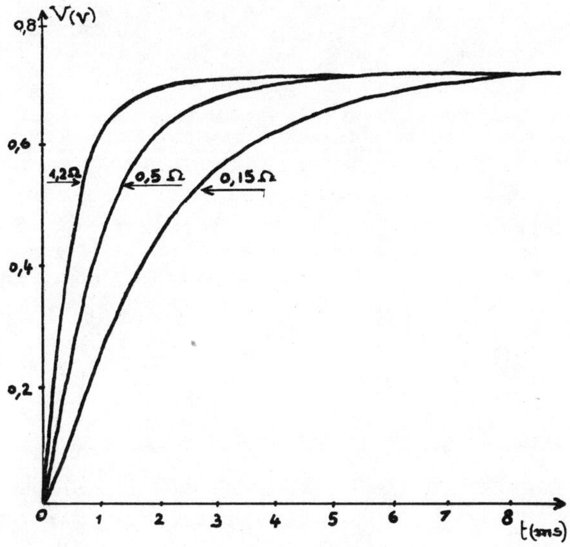

Fig.9. Opend loop step response.

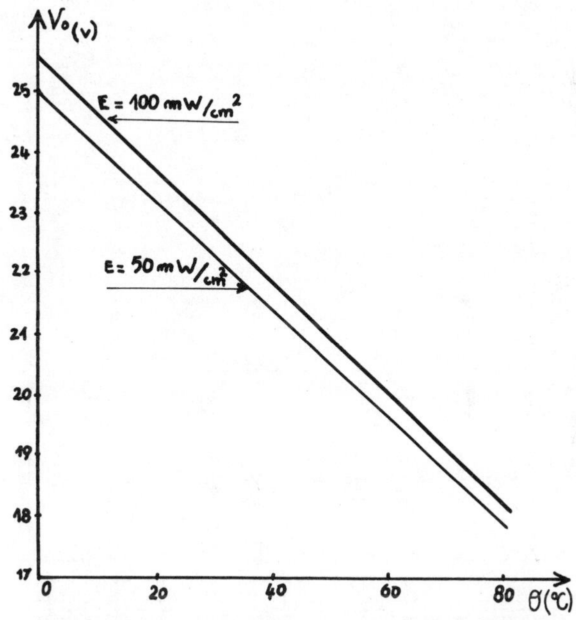

Fig.10. Vo load output voltage as a function of junction temperature.

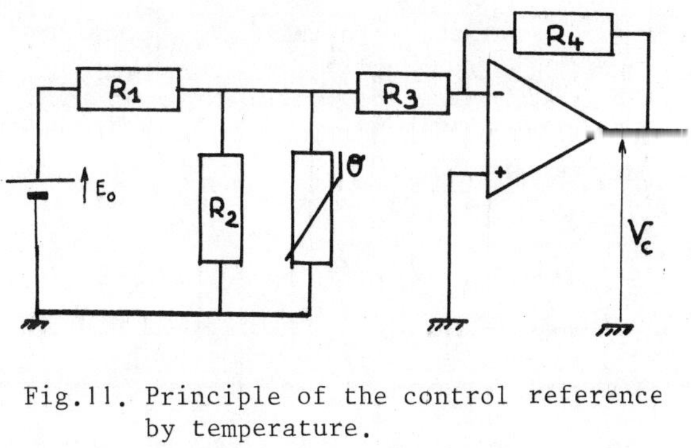

Fig.11. Principle of the control reference by temperature.

OPTIMIZING OF THE AVAILABLE SOLAR CELLS POWER BY SUNTRACKING (II)

A. A. Montasser, A. O. Attia and I. M. El-Dokany

Faculty of Electronic Engineering, Monofia University, Menouf, Egypt

Abstract: An established and well known solar cell characteristics had been used for building a specified array for obtaining the maximum power. Three different models for sun tracking had been designed, analysed and realized on small scale. This indicate satisfactory improvement in the power conversion efficiency. An intelligant mechanism for the array back rotation, and periodical cleaning had been designed for extra optimizing the gained power and fully automatically operated the system.

Keywords: Power generation; power converters; control system synthesis; self-adjusting system; target tracking; position control; optimal system.

INTRODUCTION

Suntracking technique had been used for improving the power conversion efficiency fig. 1. Three different models had been designed, analysed, and realized on small scale using high efficiency and simple fabricated Si solar cell as a building unit for the array.

The manual back rotation for the array every morning represent an additive operating difficulties, and the environmental contimination destroy the power conversion efficiency and limit the array life time. For overcoming these two major difficulties, an intelligent mechanism for back rotation and full automatic mechanism for array periodical cleaning had been designed which are the aim of this work.

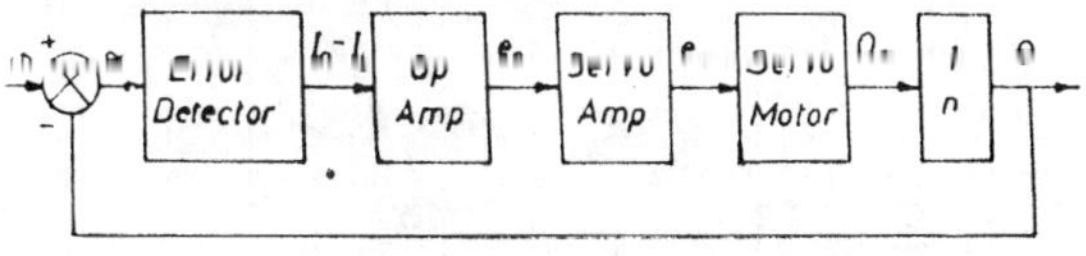

Fig. 1. Block scheme of the suntracking technique.

SUNTRACKING MODEL

The design of error detector depending upon the following statment, the intensity of sun light E at an incident angle is given by,

$$E = E_m \cos \beta \qquad (1)$$

Figure 2a, shows the case where $E_m \sin \beta$ is parallel to the array surface, which in such case is neglected; fig. 2b, shows the case where the intensity of illumination is maximum and is given by,

$$E = E_m \qquad (2)$$

Where the sun rays are normally incident to the solar cell arrays surface.

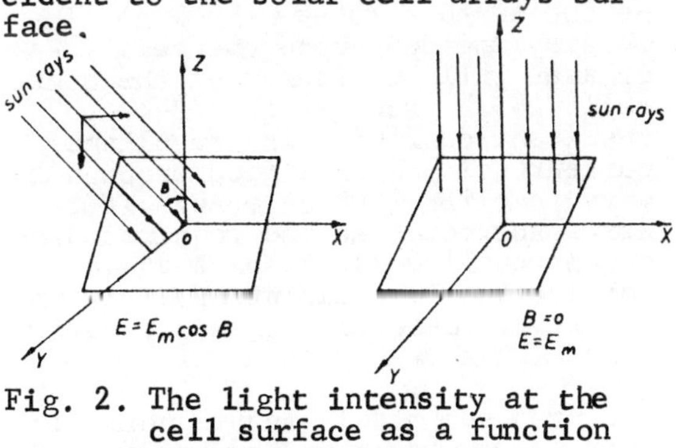

Fig. 2. The light intensity at the cell surface as a function of the incident angle.

Figures 3, 4, and 5 shows the three proposed models for the error detector units which will be now discussed.

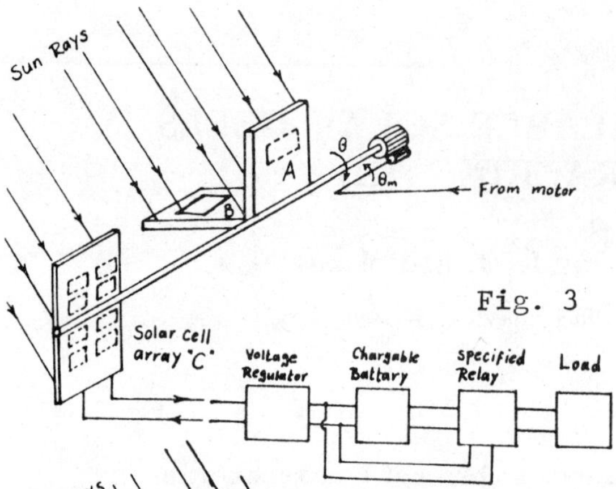

Fig. 3

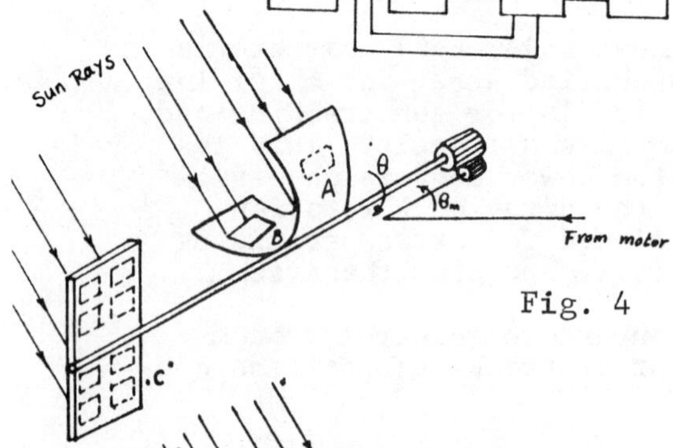

Fig. 4

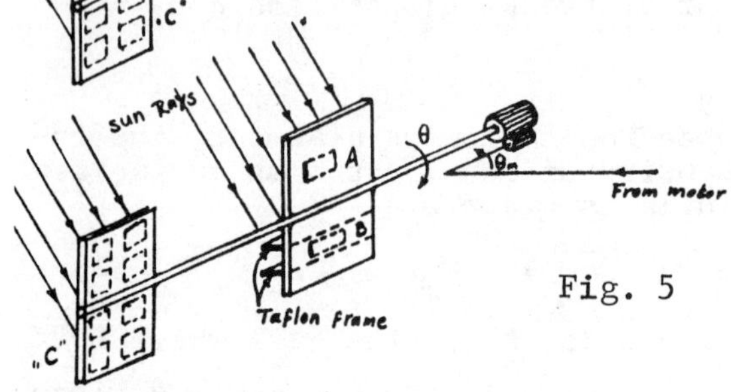

Fig. 5

Fig. 3, 4, 5. The three proposed models of the error detector units.

In the early morning when the sun rise, the cells A and B receive the same amount of light intensity which gives zero output resultant current. As the earth rotates, the light intensity incident upon one cell increases while decreases on the other cell, which produces an effective resultant current. The resultant current will be amplified through the servo amplifier which enable it to drive the motor in the required direction until cells A and B again receive the same amount of light intensity to maintain no output current.

The proposed models insure normally incident sun rays on the surface of the solar cell arrays inspite of the earth rotation. For satisfying the variation of the relative sun position with respect to the array during the different year seasons, similar mechanism can be used in a perpendicular direction. For simplicity, the latter requirement can be done manually four times a year.

OPERATIONAL AMPLIFIER MODEL

Figure 6 shows the circuit diagram of the operational amplifier,

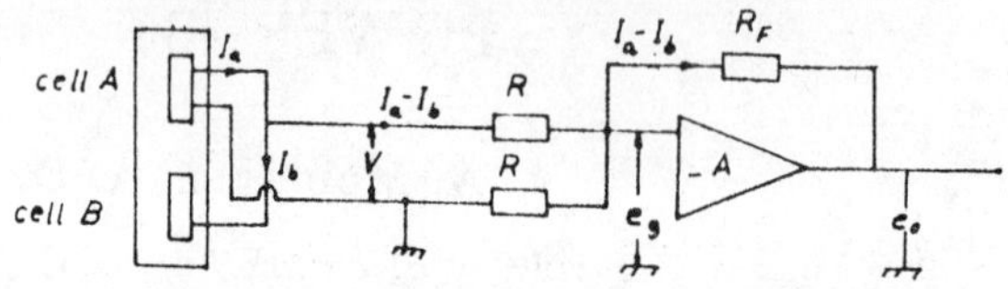

Fig. 6. The circuit diagram of operational amplifier model.

$$e_o = -A\, e_g$$

$$i_g \cong 0$$

$$I_a - I_b = \frac{V - e_g}{R} = \frac{e_g - e_o}{R_F}$$

Where e_g is much smaller than V or e_o, then;

$$I_a - I_b = \frac{V}{R} = -\frac{e_o}{R_F}$$

or;

$$e_o = -R\,(I_a - I_b) \qquad (4)$$

DC SERVO AMPLIFIER USING DARLINGTON PAIR.

Since all I.Cs OP.AMP. are limited in their power handling capablities, and the load (motor) requires a power level above 0.5 watt, then we had to use a discrete model as shown in fig. 7 as a DC servoamplifier for driving the load.

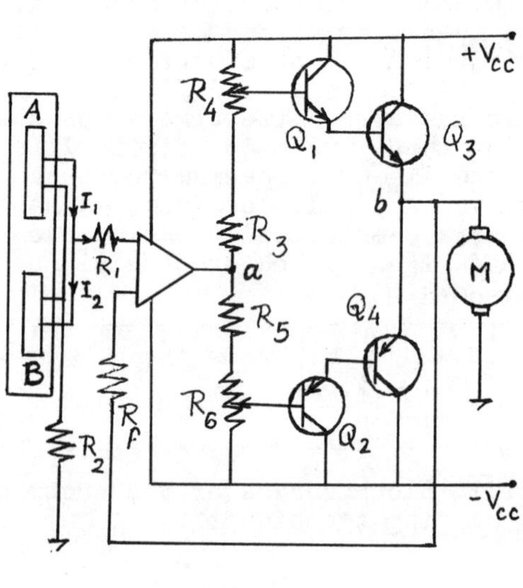

Fig. 7. The proposed DC servoamplifier model.

Transistors (Q_1, Q_3) and (Q_2, Q_4) are matched so that in absence of input current ($I_1 - I_2 = 0$), points (a) and (b) are at ground potential. Q_1 and Q_3 act as a darlington pair amplifier when a positive voltage exists at point (a), the motor rotates for example in a clockwise direction and when a negative voltage exist at point (a), the motor rotates in the opposite direction. Resistors R_F and R_I provide the feedback, path from the motor to OP. AMP. (R_3, R_4) and (R_5, R_6) act as voltage deviders to provide the base bias voltages for the transistors Q_1 and Q_2.

ARMATURE CONTROLLED DC MOTOR

The proposed model for the armature controlled DC motor is shown in fig. 8 and depends on the following parameters

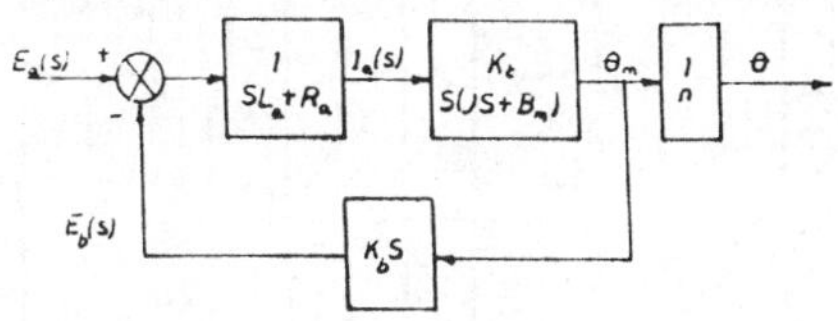

Fig. 8. Block diagram of the armature of the controlled DC motor.

- Constant current I_f is fed to the field circuit.
- The air-gap flux ϕ_m is proportional to the air-gap flux and the armature current; i.e.,

$$\theta_m = K_f I_f = \text{constant} \tag{5}$$

$$T_m(S) = K_t I_a(S) \tag{6}$$

$$E_b(S) = K_b S \theta_m(S),$$

$$= \frac{E_b(S)}{\phi_m(S)} = K_b S \tag{7}$$

Where K_f, K_t, and K_b are constants, and;

$$E_a(S) = SL_a I_a(S) + R_a I_a(S) + E_b(S) \tag{0}$$

$$\therefore \frac{I_a(S)}{E_a(S)-E_b(S)} = \frac{1}{S L_a + R_a}$$

$$T_m(S) = J S^2 \theta_m(S) + B_m \theta_m(S) \tag{9}$$

$$\therefore K_t I_a(S) = JS^2\theta_m(S) + B_m S\theta_m(S)$$

$$\frac{\theta_m(S)}{I_a(S)} = \frac{K_t}{S(J s + B)} \tag{10}$$

The gear ratio is shown in fig. 9 and is given by;

$$\frac{1}{n} = \frac{\theta}{\theta_m} = \frac{N_1}{N_2} = \frac{r_1}{r_2} \tag{11}$$

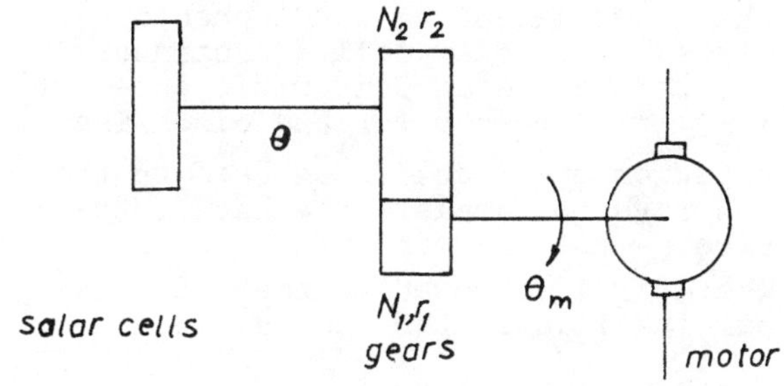

Fig. 9. The gear ratio simple representation.

BACK ROTATION MECHANISM

At sun set the solar array is directed to westwards, hence the need for a mechanism for rotate the array from the west to the east at sun rise on the next day arises. The proposed mechanism is illustrated in fig. 10 and is operated as follows,

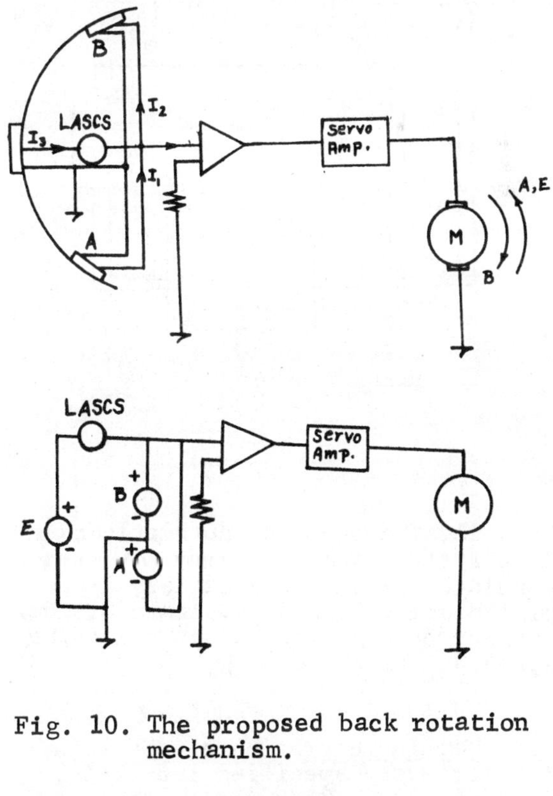

Fig. 10. The proposed back rotation mechanism.

At sun rise, the sun rays incedent on the solar cell E placed at the back of the error detector produced the initial current difference

$(I_1+I_3-I_2)$ needed for starting the suntracking system rotation from the west to the east in counter clockwise direction, as the LACS light is on activated silicon controlled switch. This direction of back rotation has been specifically choosen to avoid the entanglement of the connecting wires. The solar cell E continues to produce the current difference $(I_1+I_3-I_2)$ needed for the clockwise rotation until cell A exposed to the sun rays to complete the back rotation by the current difference I_1-I_2, where I_3 is stopped as the LASCS is off, as illustrated in fig. 10.

AUTOMATIC CLEANING MECHANISM

The effect of the environmental contomination can be minimized by using a periodical automatic cleaning system as shown in the block diagram of fig. 11. The timing circuit for the cleaning mechanism is illustrated in fig. 12.

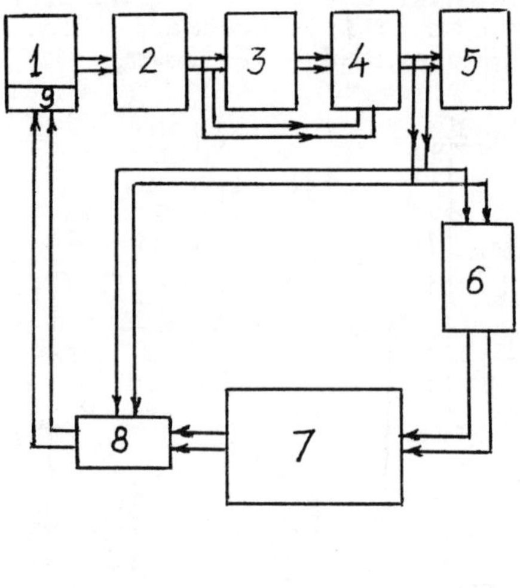

Fig. 11. The proposed automatic cleaning system, 1-solar array 2-voltage regulator 3-chargable battary 4-specified relay 5-load 6-voltage divider 7-timer 8-specified relay 9-cleaning spray and wipers.

The cleaning operation can be adjusted for every specified intervals, (i.e. 1, 3, 6, 24 hours). The conventional cleaning interval could be 24 hours. So, after the back rotation, the cleaning process started by spraying the cleaner solvant, followed by getting red of the dissolved contamination by moving the wiper blade which are automatically swing allover the array area. The sprayer and wiper blades system are illustrated in fig. 13.

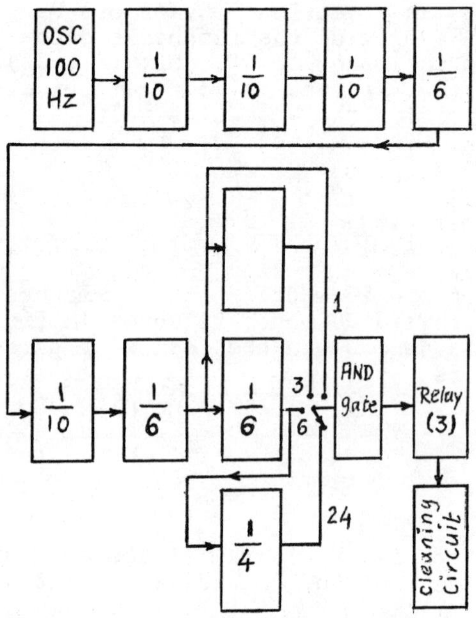

Fig. 12. Timing circuit for periodical automatic cleaning system.

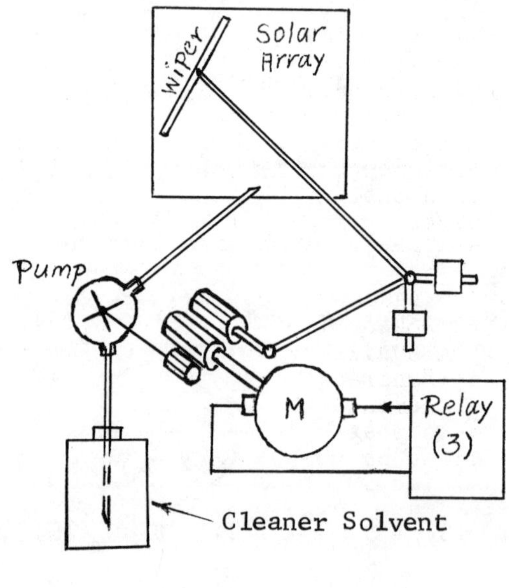

Fig. 13. The sprayer and wiper blades system.

CONCLUSION

The back rotation of the array automatically will place the array every morning towards the sun rays. The automatically cleaning mechanism will remove every deposited contamination periodically depending on the environment condition, i.e., rate of array surface contamination. This will stabilize the power conversion efficiency which had been optimized by the suntracking. These two techniques help in using the solar cells array in any place whatever its position is. The other season location of the array can be mechanically and seasonaly. We are now looking for an electromechanical system for such season location.

REFERENCES

D.M. Mosher, R.E. Boose and R.J. Soukup (1977). The advantages of sun tracking for planer silicon solar cells, Solar Energy, Vol. 19.

Wenner Luft (1962). Stationary silicon solar cell converter calculation, Report from solar energy Vol. 6, no. 1, January.

Robert Walraven (1978), Calculating the position of the sun. Solar energy, Vol. 30.

Benjamin C. Ku. (1975), Automatic control systems, 3rd (Ed.) Chap.

A.O. Attia, A.A. Montasser, and I.M. El-Dokany (1980). Protected High Efficiency and simple fabricated. Si-Solar cell, London 2nd conference for the abroad Egyptian post graduate student organized by the British Scientific Society.

A.A. Montasser, A.O. Attia, and I.M. El-Dokany (1980). Optemizing of the Available Solar Cells Power by Suntracking (1). London 2nd conference for the abroad Egyptian post graduate student organized by the British Scientific Society.

MICROCOMPUTER SYSTEMS ALLOW A REAL-TIME PARAMETER ESTIMATION FOR THE CONTROL OF A SOLAR PLANT

L. Fortuna*, A. Gallo*, G. Cammarata and M. La Cava*****

**Istituto Elettrotecnico, Università di Catania, Italy*
***Istituto di Macchine, Università di Catania, Italy*
****Dipartimento di Sistemi, Università della Calabria, Cosenza, Italy*

Abstract. This paper describes a study of the applicability of identification LS and GLS recursive algorithms to solar energy systems; a microcomputer structure is implemented. A priori physical knowledge of the considered process is utilized to determine the model structure. The recursive character of the methods is able to continuously evaluate the characteristic parameters of a linearized, time-varying, model derived from the non-linear original one. The comparison between LS and GLS algorithms is reported; the simulation of the identified models results in good agreement with the real data in spite of the introduced simplifications. The microcomputer system designed allows advantages in the cost, flexibility and adaptability in the development of effective devices for non-conventional control.

Keywords. Solar energy control; identification; models; microprocessors.

INTRODUCTION

In recent years solar energy systems have been a remarkable subject of application for the modern control theory (Dorato, 1979). Due to the variability of both the possible solar circuit configurations and the steady-state or slowly time-varying operating conditions the control problem cannot be solved by using only standard regulating units. A sort of two level control strategy seems the most suited way to satisfactorily solve the requirements deriving both from the steady-state optimization of the solar plant and from the local control problems of the single sections of the same. The wide variety both of optimization strategies and of type of disturbances in the solar input make it very proposable some form of adaptive control at the low level, which should be, moreover, effected on line due to the requirements of the respective control strategy. A basic problem in any adaptive control system is a quick and efficient parameter estimation tool.

Process computers are widely emploied in many fields of application. Large digital computers are generally used with the task of management, survey, estimation and optimum control; these functions are generally referred to high levels of a large structure.

For low level structures or for reduced dimension systems, like pilot solar plants, digital control systems have been used, mostly based on wired logic techniques. The rising development of the LSI technology has allowed for the use of control structure based on mini and microcomputers (Clark and Gawthrop, 1979). The peculiarity of adopting a microcomputer based structure for identification lies in developing a system of small dimensions which allows a real time estimation of the parameters of a process model to be carried out, through a fixed program, under the control of a microprocessor. It must be observed that, in this case, the identification function is developed directly at low level. This is different from the case of the off-line identification which is commonly worked out by high level computing devices.

The purpose of the present paper is the study of the possibility of working out specific on-line identification techniques of an experimental pilot solar energy plant using recursive algorithms implemented on a microcomputer.

The simplified model to identify is lumped parameter one and its choice is justified a priori by the physical dimensions of the plant and a posteriori by the accuracy of the obtained results. Particular attention is devoted to express some coefficients as functions of the fluid flows $\dot{M}$ and $\dot{M}'$ respectively in the collector and in the boiler because their values are the foreseeable control inputs of the system.

The proposed on-line identification algorithms have proved to be very efficient as regards accuracy and rapidity. As a consequence it makes it possible to obtain the intrinsic present parameters of the system instead of the daily mean values. This is essential for the correct tuning of the on-line local controllers which can be realized by the use of the same hardware used for identification.

In conclusion it is pointed out how the use of the microprocessor allows remarkable gain in performance at moderate costs.

PROCESS AND MODEL

The experimental set-up is given in fig. 1; it is represented of a solar flat plate collector (cupper tube and aluminium plate type), a pump for working fluid circulation and various circuit control devices, such as a heat exchanger and an electric heater. It is the same plant with given geometric parameters which we considered for off-line identification (Arcidiacono and others, 1979).

This circuit is designed both for instantaneous collector efficiency (NBS standard) and for mean daily efficiency (calorimetric method) measuring; the experimental conditions are referred to summer in Sicily. For our studies we have considered only the performance with energy accumulation in the day. The mass flow rate of the working fluid is fixed according to NBS specifications. The smallest time constant is estimated to about 300 s for the tube and plate collector and 10 hours for the boiler.

Because the solar radiation variability in the day, a sampling rate of 90 s is chosen, from 9 a.m. to 5 p.m., in according with previous time constants of circuit elements.

The structure of the analytical model is derived from physical considerations. In particular the model adopted is given by three energy balance equations respectively for the collector fluid, the absorber plate and the boiler fluid. They are derived by assuming valid the single-cell approximation and are, then, lumped parameter differential equations. Moreover, the terms due to thermal losses are suitably linearized, so that the resulting model equations are the following:

$$\begin{aligned}
.25Ln_3\pi c\rho_3 d_3^2(dT_1/dt) &= k_1 n_3 \pi d_3 L(T_2-T_1) - \\
&\quad -\gamma_1 c\dot{M}(T_1-T_3) \\
c_1\rho_1 r_1(dT_2/dt) &= -bLF_r U_L(T_3-T_0)-k_1(T_2-T_1) \\
&\quad \pi n_3 d_3 L+F_1 bLI \\
.25c\rho\pi D^2 H(dT_3/dt) &= -(\pi DH+.5\pi D^2)k_5(T_3-T_f) + \\
&\quad +\gamma_1 c\dot{M}(T_1-T_3)-\gamma_2 c\dot{M}'(T_3-T_r)
\end{aligned} \tag{1}$$

This model is, however, significant because the coefficients appearing in it are directly related to the physical behaviour of the plant. In particular it is to be considered the presence of the terms $\dot{M}$ and $\dot{M}'$ which represent the control variables.

The computation model used in the development of the identification algorithms, in compact form derived from the equations (1), is the following:

$$y_{j,k} = a_j\, y_{j,k-1} + \sum_{i=1}^{m_j} b_{ji}\, u_{ji,k-1} \qquad j = 1,2,3 \tag{2}$$

where k denotes the discrete time, j represents the equation index, $y_{j,k}$ is the process output T_j according to the corrisponding j.th equation at the kth sampling instant, m_j is the input number according to the j.th equation, that assumes the following values: $m_1 = 2$, $m_2 = 3$, $m_3 = 3$; u_{ji} are the discreted inputs as follows:

$u_{11} = T_2$, $u_{12} = T_3$, $u_{21} = T_1$, $u_{22} = T_0$, $u_{23} = I$, $u_{31} = T_1$, $u_{32} = T_r$, $u_{33} = T_3 - T_f$

a_j (with j = 1 to 3) and b_{ji} (where for j=1 i=1 to m_1, for j = 2 i = 1 to m_2, for j=3 i= 1 to m_3) represent the estimating parameters, that are strictly related with the physical coefficients. It is evident that some coefficients (a_1, a_3, b_{12}, b_{31}, b_{32}) are essential because they involve the control variables $\dot{M}$ and $\dot{M}'$.

IDENTIFICATION ALGORITHMS

In this section we develope the LS and GLS recursive algorithms devoted to the multivariable process model (2).

RLS algorithm

Defining the parameter vector of the j.th equation as:

$$\Theta_j = [a_j,\, b_{j1},\, b_{j2}\, \ldots\, b_{j\,m_j}]^T$$

where the subscript T is the transposition symbol, and the output at the (k+1).th sampling instant: $y_{j,k+1}$.

A recursive equation for the least squares $\hat{\Theta}_j$ is obtained (Sinha and Pille, 1971) as:

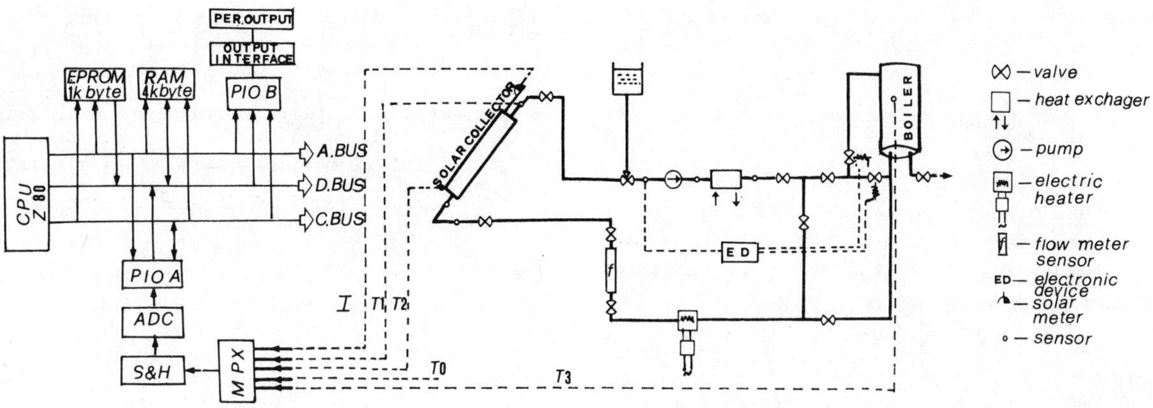

Fig. 1. Structure of the solar system connected to microprocessor hardware.

$$\hat{\Theta}_{j,k+1} = \hat{\Theta}_{j,k} + \frac{P_{j,k}\psi^T_{j,k+1}(y_{j,k+1} - \psi_{j,k+1}\hat{\Theta}_{j,k})}{\psi_{j,k+1}P_{j,k}\psi^T_{j,k+1}+1} \quad (3)$$

with:

$$P_{j,k+1} = P_{j,k} - \frac{P_{j,k}\psi^T_{j,k+1}\psi_{j,k+1}P_{j,k}}{\psi_{j,k+1}P_{j,k}\psi^T_{j,k+1}+1} \quad (4)$$

where $P_{j,k+1}$ is a $(m_j+1)(m_j+1)$ matrix and

$$\psi_{j,k+1} = [y_{j,k},\ u_{j1,k}, \ldots\ldots, u_{jm_j,k}]^T.$$

RGLS algorithm

It is well known that for noisy measurements, the LS method gives biased estimates if the residuals:

$$e_{j,k+1} = y_{j,k+1} - \psi_{j,k+1}\hat{\Theta}_{j,k} \quad (5)$$

are correlated; then an autoregressive model of the noise is developed to filter observed data to get uncorrelated residuals with zero mean (Clarke, 1967). A recursive version of this method was proposed by Hastings-James and Sage, 1969.

If δ_j is assumed as the order of noise model, the residuals are analysed by an autoregression according to a model:

$$e_{j,k} = -C_{j1}e_{k-1} - C_{j2}e_{k-2} - \ldots - C_{j\delta_j}e_{k-\delta_j} + w_k \quad (6)$$

where w_k are uncorrelated random variables and $C_j = [C_{j1}, C_{j2}, \ldots, C_{j\delta_j}]^T$ the vector of filter coefficients. Filter parameters estimation is carried out applying a LS estimation to (6) model, then inputs and outputs are filtered and a new least squares is made with the filtered inputs and outputs to get the new estimated parameters.

The RGLS consists, in effect, of two RLS estimators via-filtering: one to estimate the system parameters, the other to estimate the noise model. The RGLS estimator is expressed as follows:

$$\hat{\Theta}_{j,k+1} = \hat{\Theta}_{j,k} + \frac{\tilde{P}_{j,k}\psi^{TF}_{j,k+1}(y^F_{j,k+1} - \psi^F_{j,k+1}\hat{\Theta}_{j,k})}{\psi^F_{j,k+1}\tilde{P}_{j,k}\psi^{TF}_{j,k+1} + 1} \quad (7)$$

$$\tilde{P}_{j,k+1} = \tilde{P}_{j,k} - \frac{\tilde{P}_{j,k}\psi^{TF}_{j,k+1}\psi^F_{j,k+1}\tilde{P}_{j,k}}{\psi^F_{j,k+1}\tilde{P}_{j,k}\psi^{TF}_{j,k+1} + 1} \quad (8)$$

$$C_{j,k+1} = C_{j,k} + \frac{Q_{j,k}g^T_{j,k+1}(e_{j,k+1} - g_{j,k+1}C_{j,k})}{1 + g_{j,k+1}Q_{j,k}g^T_{j,k+1}} \quad (9)$$

$$Q_{j,k+1} = Q_{j,k} - \frac{Q_{j,k}g^T_{j,k+1}g_{j,k+1}Q_{j,k}}{1 + g_{j,k+1}Q_{j,k}g^T_{j,k+1}} \quad (10)$$

where:

$$g_{j,k+1} = [e_{j,k},\ e_{j,k-1}, \ldots\ldots, e_{j,k-\delta_j}]^T ;$$

the subscript F represents the filtering effect on the previous $\psi_{j,k+1}$ and y_{k+1}. The matrix $\tilde{P}_j$ and Q_j have respectively $(m_{j+1} \times m_{j+1})$ and $(\delta_j \times \delta_j)$ elements.

The LS algorithm, using (3) and (4), requires a number of $3(m_j+1)^2 + 4(m_j+1)$ (j=1,2,3) multiplications and a number of (m_j+1) input-output data, for each iteration. For the GLS algorithm the estimated number of multiplications is $3(m_j+1)^2+4(m_j+1)+3\delta_j^2+4\delta_j$ (j=1,2,3) for each iteration.

The recursive form of LS and GLS algorithms without any matrix inversion makes their implementation appropriated by means of a powerful software in a microcomputer system. A shifting algorithm multiplications has been chosen with 109.75 μs of computation time using a Z80 CPU with 2 MHz clock frequency.

HARDWARE IMPLEMENTATION

The hardware of the identification system consists mainly of a Z80 CPU. This microprocessor is used both for data processing and for performing the control functions of the different devices of the system.

The block diagram of the whole hardware system connected to solar plant is shown in fig. 1. With reference to its functions it may be subdived into the following sections:
a) the Z80 CPU with its I/O and bus interfaces;
b) the bus structure;
c) the data acquisition system.

As regards section (a) it may be schematically represented by the connection of the Z80 CPU with its peripherals and auxiliary memories through the bus structure. In particular the resident program is allocated into an EPROM memory of 2x1024 bytes. A working memory of 4x1024 bytes is used. Two input-output ports PIO (8 bit) are respectively dedicated to data acquisition and to the output unity. The internal structure of the Z80 CPU bus interface, an internal control logic, the Port A I/O logic, the Port B I/O and an interrupt control logic. The Ports A of the two PIO unities are used as data transfer channels and the Ports B for the status and control monitoring. The input operations are controlled by means of a polling technique.

As regards the bus structure (b) it is organized as a standard bus with 16 bits 'addressing bus' an 8 bit 'data bus' and an 8 bit 'control bus'.

As regards the data acquisition system (c) (fig. 2) it is suited to handle five channels (corresponding respectively to the temperatures T_o, T_1, T_2, T_3 and the solar radiation I.

The time constant values (very high) of the solar circuit don't require very short acquisition times of the system input quantities, with regard to the time interval which is needed for developing each cycle of identification algorithm. An AE signal enables the data acquisition system as soon as the output routine is completed. The Multiplexer, S/H and ADC circuits are accomplished by means of a mixed technology: CMOS components are used for multiplexer, control logic, dividers and delay, bipolar technology for buffers, S/H and for the eight bit analogical to digital converter ADC 800. The structure is con-

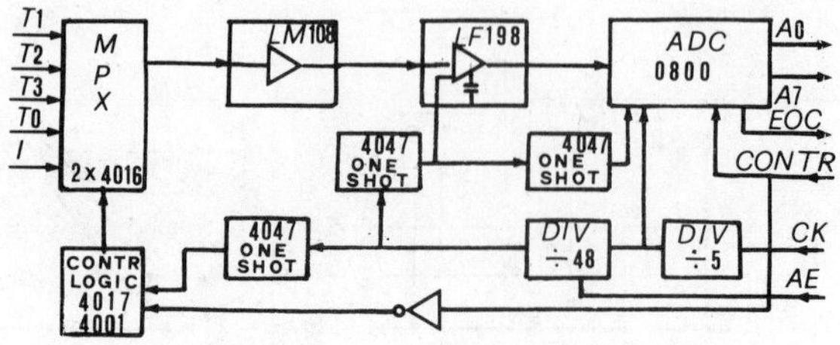

Fig. 2. Data acquisition system.

trolled by means of some suitable control signals generated by the microprocessor (Z80 CPU). Moreover the possibility of interrupt from the ADC 800 is foreseen. From the clock frequency of the microcomputer a lower frequency is obtained, which is used as converter clock. By further reducing frequency, the clock signal (4.166 kHz) for the multiplexer logic control is obtained.

RESULTS AND DISCUSSION

The trend versus time of some parameters by GLS estimated is shown in Figs. 3, 4 and 5. In table 1 the values of the parameters are given for the maximum solar radiation I_{max} and the minimum solar radiation I_{min}.

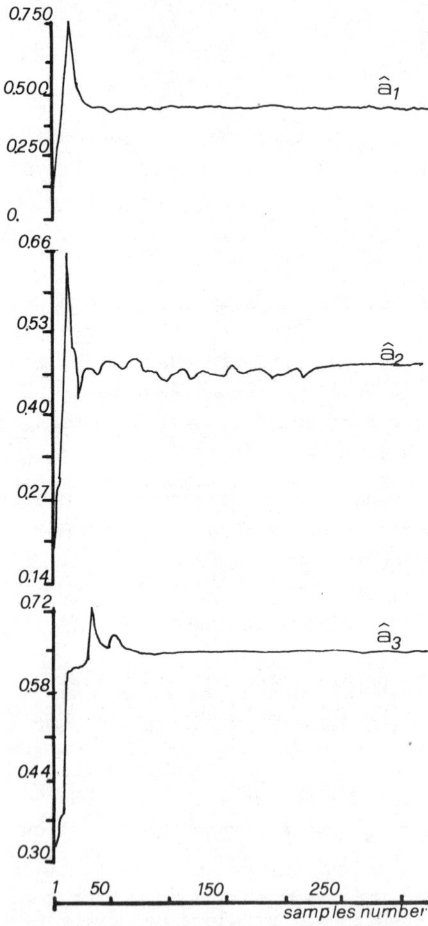

Fig. 3. Parameter estimation.

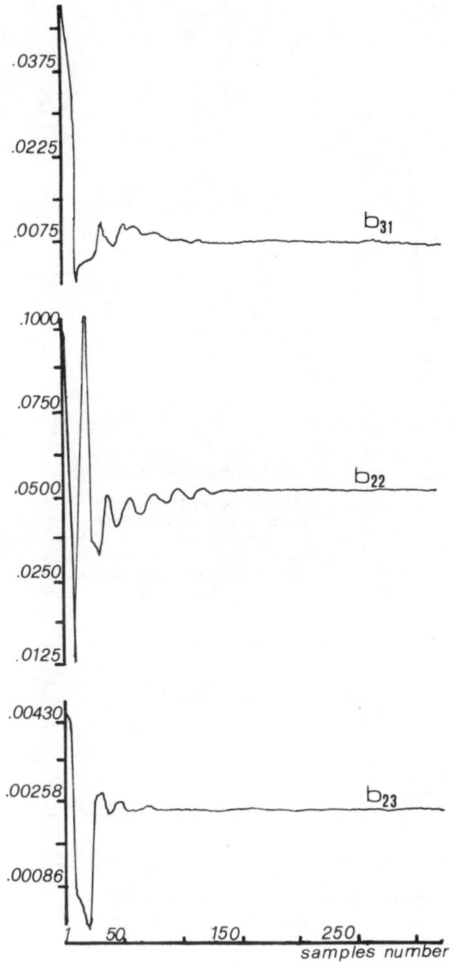

Fig. 4. Parameter estimation.

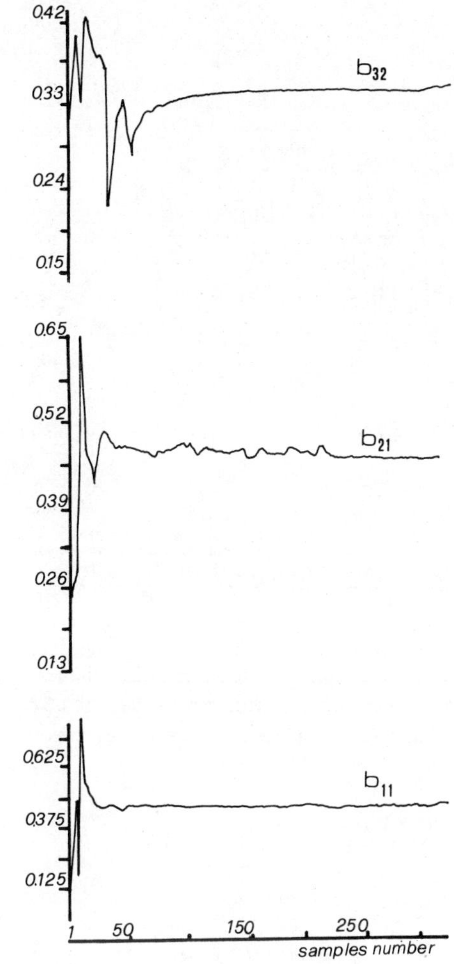

Fig. 5. Parameter estimation.

It is evident the variability of some parameters, which depend mainly from the dynamics of the solar process; in particular they evidence the behaviour of some physical coefficients: in fact the parameter b_{22} depends on the heat removal factor F_r and on the total heat loss coefficient of the collector U_L, depending also on the absorber radiativity, then the parameters b_{21} and a_2 are closely depending on the thermal transmittance k_1.

The LS and GLS estimated model simulation with experimental values is shown in Figs. 6, 7 and 8.

The process estimate by means of the LS algorithm appears sufficiently valid, it can be considered unbiased as regards the accuracy required. Best results have, however, been obtained by means of the GLS. This is demonstrated by the residual curves shown in Fig. 5a related to T_1 temperature also.

The response of the GLS estimated model mostly agrees with the experimental data. Moreover both LS and GLS on-line estimated models

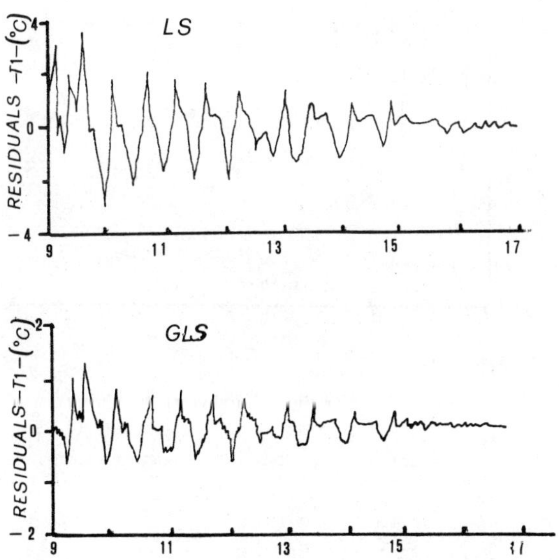

Fig.5a. Residual trend.

result different from the off-line one related to the same plant (Arcidiacono and others, 1979); but all methods have shown the adequacy and the sufficiency of the considered model structure in spite of the adopted

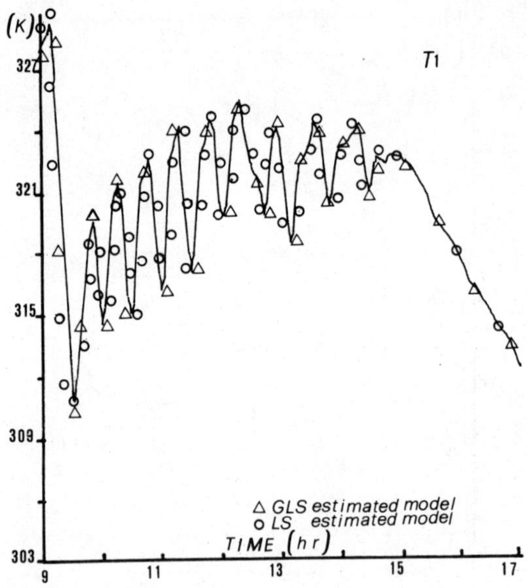

Fig. 6. Comparison between experimental (continuous) and estimated (LS and GLS) model T_1 temperature.

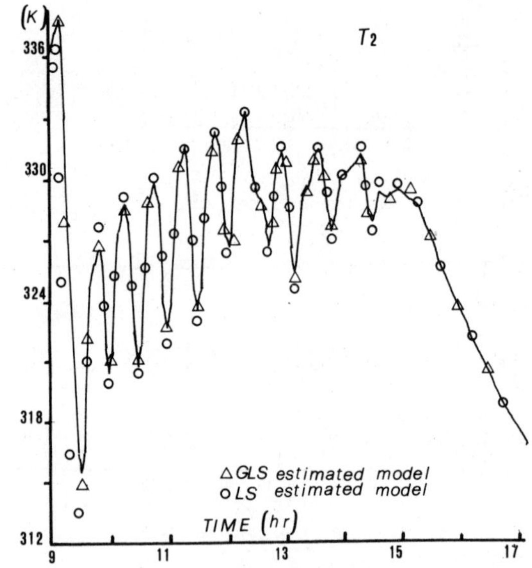

Fig. 7. Comparison between experimental (continuous) and estimated (LS and GLS) model T_2 temperature.

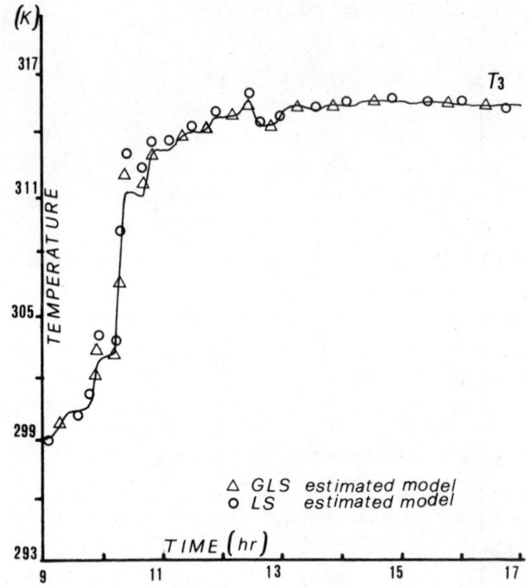

Fig. 8. Comparison between experimental (continuous) and estimated (LS and GLS) model T_3 temperature.

TABLE 1 Identified Parameters

	l_{min}	l_{max}
a_1	0.47682	0.47291
b_{11}	0.49541	0.51232
b_{12}	0.01971	0.00768
a_2	0.47162	0.47946
b_{21}	0.48112	0.47562
b_{22}	0.05226	0.05021
b_{23}	0.02700	0.00257
a_3	0.65574	0.65666
b_{31}	0.01123	0.00769
b_{32}	0.32732	0.33032
b_{33}	0.33400	0.33197

simplification hypotheses.

Generally the better estimation obtained using the GLS method is not justified by the extending amount of multiplications (respect LS method) with a corresponding increasing of computation time that makes it practically inapplicable the on-line identification method to a real process. In fact it is necessary for the on-line processing that the development of the algorithm time must be less than the sampling time: this is the strict compatibility condition between data acquisition and data processing system. On the other hand if the residuals are largely correlated,due to the noise that contaminates the observed measurements, the choise of GLS method is obliged.

The estimated model simulation has been carried out on a minicomputer HP 9825 by means of the finite difference method. The verification of the on-line identification has been carried out by implementing it on a Z80 microprocessor development system. The data relative to the observed quantities have been stored into a floppy disk; for each quantity a number of 320 samples has been considered, which have been taken at sample intervals of 90 s. The filter order δ has been appropriately chosen equal to 3. For the LS and GLS algorithms the initial parameters estimated Θ_o were set to 0 and the initial matrix P_o was chosen to be unitary. For the GLS algorithm the initial filtering vector C_o was set to 0 and the initial matrix $\tilde{P}_o$ and Q_o were assumed unitary. The trend of each identified parameter has been stored into a digital file and has been plotted by means of an HP 9862 plotter.

The microcomputer-based system is a low cost, very flexible and reliable structure for updating the model parameters of the real solar plant process. The cost of microcomputer-based identifier structure is very low in comparison with minicomputer-based ones, for no-large systems such as solar experimental pilot plant.

CONCLUSIONS

The LS and GLS recursive on-line identification algorithms have been applied to an experimental solar plant. The realization of these methods has been found to be well suited with the aid of a microprocessor structure, which have been outlined by harware formulation.

The identifier structure considered reproduces the non-linear, time-varying, plant behaviour with a linear model suitably simplified as regards the real applicability of the methods.

Both identification on-line algorithms have been analyzed versus to physical considerations. They result advantageous for the recursive character; in particular the choice of the GLS method is referred to noise/signal ratio high level.

It is important to point out as a good setting-up of algorithms and models is key-point for a network microcomputer system design dedicated to the identification and control of a solar process. The outlined controller, based on the same microprocessor structure, performs an optimal circuit energy management using the on-line identified parameters and updating the control variables at prefixed time intervals; in this way the control system is adapted to optimize system variations during the day.

The applications of microcomputer modular system, for their flexibility, expansibility and computing power, appear a useful tool to implement control systems of non-conventional process. In the further development of this work applications of the adaptive optimal control and more complex solar plant will be investigated.

NOMENCLATURE

T_a	External air temperature
T_1	Collector fluid temperature
T_2	Collector temperature
T_3	Boiler fluid temperature
T_4	Glass temperature
T_f	Room temperature
T_r	Boiler feedwater temperature
$\dot{M}$, $\dot{M}'$	Primary, secondary fluid flow-rate
c_1, c_2	Absorber plate, glass specific heat
c	Primary fluid specific heat
ρ_1	Collector absorber plate density
ρ_2	Glass density
ρ_3	Primary fluid density
F_1	Plate energy fraction
σ_1 σ_2	Absorber plate, glass radiativity
t	Time
n_3, n_4	Collector, boiler tube number
D, H	Boiler diameter and lenght
d_3	Collector tube diameter
d_4	Boiler internal tube diameter
b, L	Collector width and lenght
r_1, r_2	Absorber plate, glass thickness
V, VW	Boiler volume, wind speed
U_L	Total heat loss coefficient
F_r	Heat removal factor
k_1	Thermal transmittance between plate and primary fluid
k_5	Thermal transmittance between boiler and external air

EXPERIMENTAL DATA

L = 2,2 m	b = 1 m	V = 0.200 m^3
n_3= 8	n_4= 8	d_3= 0.008 m
d_4= 0.01 m	T_f= 293 K	T_r= 288 K
$\dot{M}$ = 0.02 Kg/s	$\dot{M}'$= 0.025 Kg/s	VW = 1.388 m/s

$\gamma_1 = \gamma_2 = 1$ for pump's ON

$\gamma_1 = \gamma_2 = 0$ for pump's OFF

REFERENCES

Arcidiacono, P., G. Cammarata, A. Faro and A. Gallo (1979). Identification of solar energy systems. Preprints 5th IFAC Symp. on Ident. and System Par. Estimation, Darmstadt.

Cammarata, G., A. Faro, A. Gallo and F. Patanè (1979). A new approach for solar energy systems design by means of the sensitivity theory. Conference on Future Energy Concepts, IEE Conference Pubblication N.171, London.

Clarke, D. W. (1967). Generalized least square estimation of the parameters of a dynamic model. Proc. 1st IFAC Symp. on Identification, Prague.

Clarke, D. W., and P. J. Gawthrop (1979). Implementation and application of microprocessor-based self-tuners. Preprints 5th IFAC Symp. on Ident. and System Par. Estimation, Darmstadt.

Dorato, P. (1979). A review of the application of modern control theory to solar energy systems. Proc. 1979 IEEE Conf. on Decision and Control, Florida.

Duffie, J. A., and W. A. Beckam (1974). Solar Energy Thermal Processes. John Wiley, N. Y.

Eykhoff, P. (1974). System Identification, John Wiley, London.

Hasting-James, R., and M. W. Sage (1969). Recursive generalized-least-squares procedure for on line identification of process parameters. Proc. IEE, 116, 2057-2062.

Mendel, J. M. (1970). Discrete Techniques of Parameter Estimation. Marcel Dekker, N. Y.

Sinha, K., and W. Pille (1971). On line parameter estimation using matrix pseudoinverse Proc. IEE, 118, 1041-1046.

Tzafestas, S. G., A. V. Spyridonos and N. G. Koumoutsos (1974). Finite-difference modelling identification and simulation of a solar water heater. Solar Energy, 16, 25-31.

OPTIMISATION OF THE USE OF WATER RESERVES IN AGRICULTURE FROM THE MODELLING WATER-SOIL-CROP SYSTEM

G. Bel*, C. Maertens and J. Puech****

**C.E.R.T., Automatic Control Department, Post Box 40 25, 31055 Toulouse, France*
***I.N.R.A., Toulouse Research Center, Post Box 12, 31320 Castanet, France*

Abstract. The aim of the study is the modelling of the water-soil-corn system from experimental data, which have been measured in full size expriments on various cultures over a period of ten years by the Agro station, in order to improve water use in agriculture. For various plots and several years, series of soil humidity at several depths (every 10 cm down to 1.6 m) have been gathered at various instants of the year (about 20 for each plot every year).The application of systematic techniques of modelling and identification to this system allowed the build up of a model simulating the time histories of the hydrous state of a soil farmed in corn. The problems encountered lead to back up the classical model error identification method by statistical data analysis methods. Good results for the specific application considered (precision better than 10%) have been obtained by alternate and iterative use of both kinds of methods. It is then possible to calculate the detailed hydrous state of the soil (every 10 cm) from easily measured data, namely rain, potential evapotranspiration and some parameters characteristic of the soil and culture. This is particularly useful for the precise forecast of the days and volumes of irrigation for a better use of the available water.

Keywords. Agriculture, data reduction and analysis, Ecology, Identification, Modelling, Moisture control, Optimisation, system analysis

INTRODUCTION

During the last few years, important progress has been made in agronomy in the areas of genetics, fertilization, crop protection and more generally in overal cultivating techniques.

Water, in subhumid regions, has terefore become one of the principal factors in determining vegetable production, especially for summer crops (various types of fodder, corn, sunflower, fruit trees...) However this water is relatively rare in agriculture and if provided by irrigation, its cost if far from negligible in relation to other crop production costs.

Its valorization is consequently a general problem which presupposes a better quantitative knowledge of the ecophysiological phenomena relating to the plant's water needs and to its manner of food intake through the roots. A better knowledge of agronomical phenomena is also necessary in order to determine the most appropriate irrigation methods.

In the purpose of meeting these ojectives, the National Institute of Agricultural Research's Agronomy Station at the Toulouse Research Center has been dedicated to the study of such hydrous problems for a period of almost ten years. Thanks to fieldwork exprimentation carried out over an area of 20 hectares, we were able to gather a considerable number of experimental moisture mesurements for soil producing various crops.

The goal of the present study is the modelling of the water-soil-crop system from this experimental data. The use of systematic modelling and identification techniques has permitted us to perfect a model that simulates the evolution in time of the hydrous state of soil yielding a corn crop. The choice of plant for this exprimental modelling was determined by its economic importance, its future development and its high sensitivity to irrigation.

So as to be easily used in sites outside the experimental one, the model must only need easily measurable data to be put into practice : daily meteorological data and certain soil and crop characteristics. It then permits one to calculate the hydric state of the different layers of soil occupied by its roots each day of the plant's life. The quantity of water necessary to satisfy the plant's needs can then be determined and, as a result, better productivity will be obtained.

DESCRIPTION OF THE EXPERIMENTAL AREA EMPLOYED AND OF THE MEASUREMENTS SELECTED

a) Experimental area

Situated near Toulouse (in south west France) the 20 hectares surface of dark brown earth, resembling "terrefors" permits a global study,

respectively calculated on the day j-1. The model permits us to determine the physiological development and hydrous state on day j.

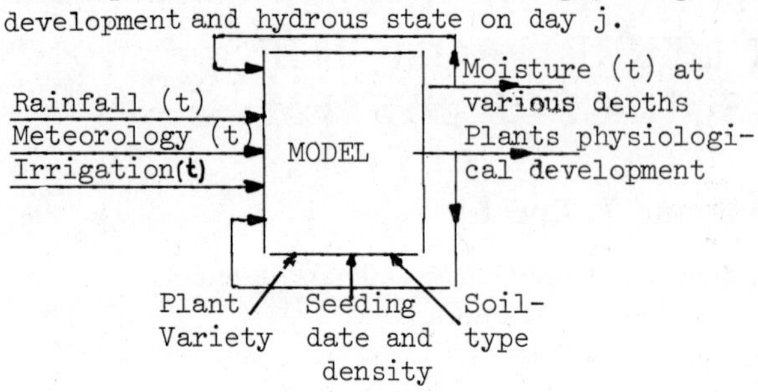

Of the various available meteorological data influencing the plant's exchanges with its exterior environment, we have chosen the potential evapo-transpiration as the model's input. Determined from evaporation levels in a (Piche) tank and from the consumption of a reference plant cover (evapo transpirometer) the PET synthesizes the effects of wind, sun and temperature.

DESCRIPTION OF THE MODEL

From acquired knowledge of the phenomena it was by induction that the INRA agronomists carried out research on the best structure to be employed for this spacio-temporel, non-linear, deterministic model. Among the various types of possible structures was chosen one that would allow the best adaptation of the parameters to measurements. This model is composed of the diverse sub models described below :

1/ The soil model

The soil is modelled as a series of cascading linear, run-off reservoirs over a threshold called field capacity FC.

The reservoirs conditions are characterized by their stock or humidity expressed in millimeters of water. The plant's roots take water directly from these reservoirs, the number of which depends on the soil in study (these are 16 in the majority of plots studies). Each reservoir corresponds to a 10 cm soil layer

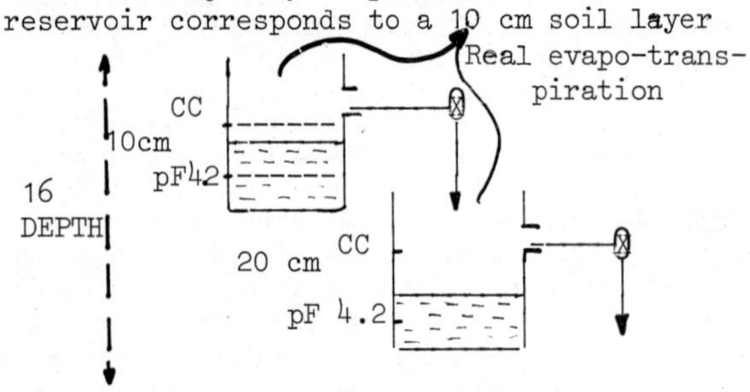

2/ Hydrous soil condition equations

The soil layers are numbered from 1 to n from the soil surface downwards.

All the hydrous variables are expressed in mm of water.

let :

$H_i(t)$: hydrous state (or moisture) of layer i (or reservoir) at time t

$DR_i(t)$: drainage of soil layer i into layer i+1 at time t

$RET_i(t)$: plants water intake of soil layer i at time t

$P(t)$: rainfall at time t

$I(t)$: irrigation at time t

$PET(t)$: potential evapo transpiration at time t

CC_i : field capacity of layer i

$PF42_i$ $PF_{4.2}$ of layer i

$x_1, x_2 \dots x_9$ parameters to be identified

Given the differential system

$$\begin{cases} \dfrac{dH_1(t)}{dt} = P(t)+I(t)-RET_1(H_1,PET(t),t)-DR_1(H_1) \\ \vdots \\ \dfrac{dH_i(t)}{dt} = DR_{i-1}(H_{i-1})-DR_i(H_i)-RET_i(H_i,PET(t),t) \\ \vdots \\ \dfrac{dH_n(t)}{dt} = DR_{n-1}(H_{n-1})-DR_n(H_n)-RET_n(H_n PET(t),t) \end{cases}$$

with

$$DR_i(H_i) = \begin{cases} \frac{1}{x_1}(H_i(t)-CC_i) & \text{if } H_i(t) > CC_i \\ 0 & \text{if } PF42_i \leq H_i(t) \leq CC_i \\ \frac{1}{x_2}(PF42_i-H_i(t)) & \text{if } H_i(t) < PF42_i \end{cases}$$

We may note that the existence of pronounced interhorizon coupling and strong non linearities complicates identification.

3/ · Plant equations

In order to calculate the plant's water consumption, one must first determine its maximum evapo-transpiration whose value is in funtion of the foliar index and the varying vegetative stages. The general law of water needs evolution for various plant covers is shown below :

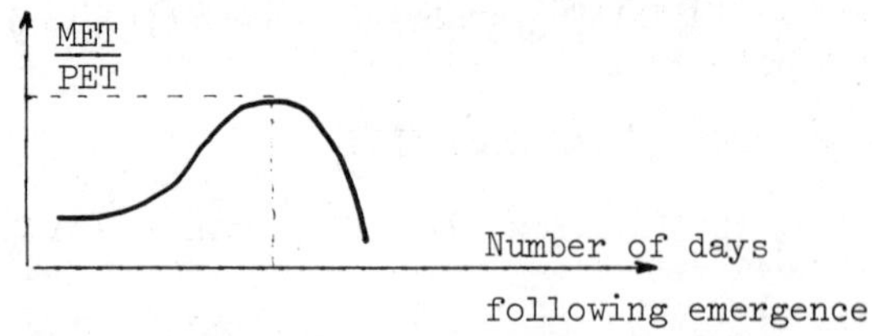

This law is characterized by an equation with two parameters (x_3 and x_4) which permit one to wary respectively the date of the maximum MET/PET relation, the width of the corresponding hump and the value of this maximum, which can exceed the value 1

$$MET = PET\ f(t, x_3, x_4)$$

The real consumption of the plant (RET) depends therefore not only on on its needs (MET) but also on the soil's hydrous condition and climatic demands. In fact /1/ a determined hydrous reserve and the soil's transfer capacity is influenced by its moisture content and the speed of its natural drying-up process. Consequently for given climatic conditions the plant's use of soil reserves over a certain threshold brings about a decrease of water consumption and has repercussions on consumption rhythms . Whatsmore, it has been noted /1/ and /10/, that the intensity of this demand has an important influence on the soil's possibilities of water transfer to the crop and eventualy cuts down the using up of the soil's water reserves during years of strong hydrous deficit. Our model provides a rough analysis: in the following series of curves /10/, the relation $\dfrac{\text{consumption}}{\text{need}} = \dfrac{RET}{MET}$

in open fields with real crop rotations, thereby comming as close as possible to large scale farming and to diverse methods of dry or irrigated cultivation. The area is divided into 196 plots of 200 m^2 each. Every year, certain plots produce the principal regional crops, some irrigated, others not. Other plots produce exprimental series of crops for interyearly assessment and irrigation policies.

b) Moisture measurements
The soil's moisture measurements (hydric profile) are taken each year, for each plot, in accordance with the periodicity adapted to each study's plan. These measurements are taken at periodic depths (every 10 cm down) by means of a neutron humidimeter.

c) Meteorology station
A meteorology station situated on the experimental area permits the gathering of classical meteorological parameters : temperature humidity , sunshine, wind , evaporation and global radiation.

d) Cultivating processes
The cultivation methods employed are similar to those of a good farming concern. The end of season yield is assessed from representative surfaces (minimum 25 m^2)

e) The soil
At the beginning of the experiment, the different types of soil of each plot, are determined by a physico-chemical analysis and by structural stability tests.

f) Available data
At the beginning of our research we had at our disposal the 1971 to 1976 hydric and meteorological informations for twelve different plots As corn is not cultivated on each of these plots every year, 21 series of hydric profiles corresponding to the various plots were available to us. It is this collection of exprimental data, exceptional in the field of agriculture, that was put into use.

DESCRIPTION OF THE SYSTEM AND METHOD EMPLOYED

The following diagram portrays one manner to describe the water-soil-crop system

Various "loops" are noticed which emphasize the plant's possibilities to adapt to climatic conditions and also the modelling difficulties.

- The plant's water needs depend on its physiological development which in turn result from the satisfying of its water needs. The plant can thus reduce its consumption and growth if the climatic conditions are harsh or if the moisture available in the crop root zone is insufficient (2nd loop).

- Root growth depends on the physiological development of the plant but also on the soil's water pressure. The easier it is to extract water from the near surface layers, the lesser will be the plant's tendency to plunge its roots deep into the earth. Consequently we can observe that, with irrigated crops the useful roots remain close to the surface while with non irrigated crops, the roots go very deep.

To make a model of this system, we used the cybernetics systems approach, or more specifically the model error identification method.

Although this method may be a classical one for use in process analysis it is much less so in the field of agronomy.

These already exist a certain number of articles relating to models of the water-soil-crop system. These are, either very detailed models connected with exprimental crops in the laboratory, or very rought models not concerned with in-depth moisture distribution, which are thus used on a site in accordance with the periodic measurements taken.

Our approach is half way between these two methods, our aim: to find a global model one that is spacio-temporal or dynamic (with timed humidity variation) of the grey box type. We will not endeavour to neatly copy reality, but rather to reproduce its effects by linking them to exterior causes.

The following diagram depicts the model's precise input and output points. At the start we know the date and density of seeding, the soil type and the plant variety. For every day j, we know the values for rainfall, irrigation and meteorology as well as the plant's development and the soil's hydrous condition as

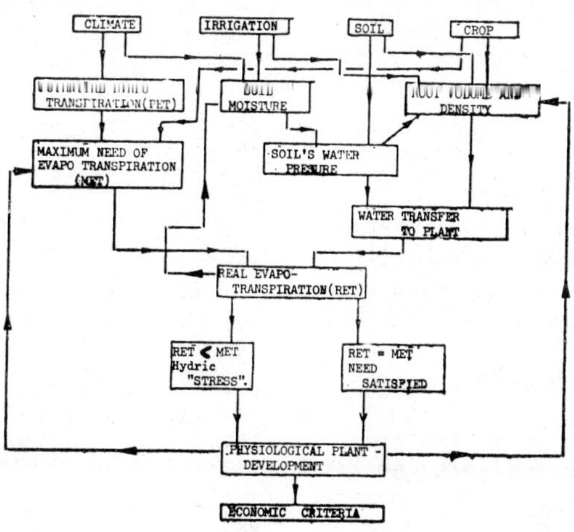

depends not only on the average soil moisture but also on the potential evapo transpiration (PET) climate

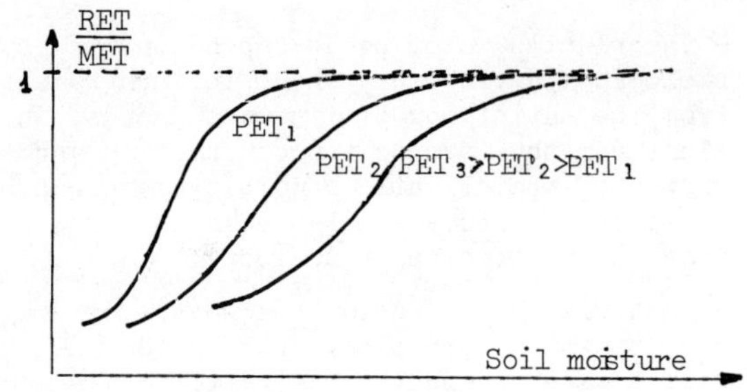

The above series of curves is depicted by an equation with parameter x_5

If
$$H_{mean}(t) = \frac{\sum_{i=1}^{n} H_i(t)}{n}$$

n = number of sections in the root area on date t - 1

$$RET = MET \quad g(H_{mean}(t), PET, x_5)$$

We must next see how this demand is distributed among the various soil layers in order to solve the differential system. The first step is to determine the plant's root depth . A measurement analysis gives us the following curve of useful root depth :

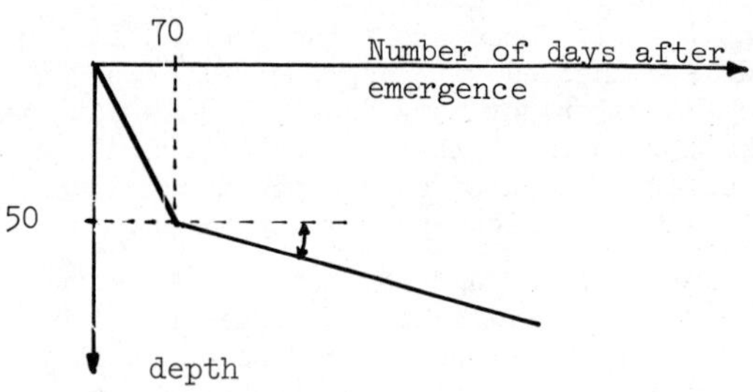

The intake distribution between layers is then inversly proportional to the depth so as to take into account the smallest degree of possibility for water transfer to the very depths. Let s (i, x_7) a function of the i^{x_7} or e^{ix_7} type. Given :

$$RET_i(t) = \frac{kRET(t)}{s(i,x_7)}$$

also
$$\sum_{i=1}^{n} \frac{k}{s(i,x_7)} = 1$$

Therefore
$$k = \frac{1}{\sum_{i=1}^{n} s(i,x_7)}$$

Thus having knowledge of the $RET_i(t)$ we have all the elements necessary to solve the differential system relative to hydrous equations of the soil and so determine the hydrous conditions immediately following.

IDENTIFICATION OF PARAMETERS

On the goal of obtaining the most realistic results for certain criteria (differences between measured moisture and calculated moisture; sum of these differences for a given depth or horizon; sum of the square of the differences at certain critical dates...) the model's parameter values were determined with the aid of Powell's Algorithm of Optimisation and by means of the following classical diagram :

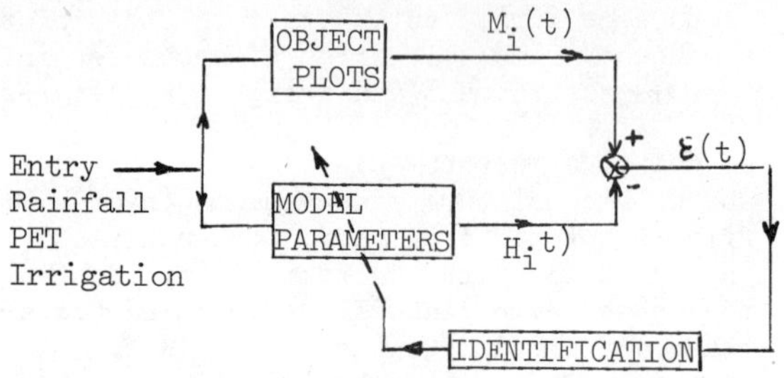

The problem dealt with various criteria, yet we decided to accept the slight deterioration of one of the criteria if this permitted the appreciable amelioration of others, notably during the critical plant growth period.

SPATIAL EVOLUTION OF PARAMETERS

After the identification phase, we had available the x_i parameter values producing the most realistic results for each plot. In order to facilitate the model's future realization on other areas, a study of the evolution of identified parameter values in relation to the various plots was carried out. The goal :

- to regroup parameters and similar plots
- to link the model's x_i parameters to easily measurable q_i physical characteristics (soil composition, planting density, corn variety..)

The method employed was that of multivariate statistical analysis : the "principal component factoral analysis method" which allowed us to underline the connections that exist between the model's parameters and certain physical characteristics. These relations enable one to obtain a model whose parameters are physical characteristics measurable once and for all on the site itself, before the cultivating season.

VERIFICATION OF MODEL'S VALIDITY

To determine the validity of the results obtained from the use of such models concerned uniquely with the soil and crop's physical characteristics we set up a simulation for each plot year. We were thus able to obtain very good results which are highly satisfactory on the agronomic point of view.

To clarify, we may add :

- average errors in layer's moisture are generally in the area of 2 mm for measurements of approximately 15 to 30 mm, and are never superior to 4 mm. This is equally the case for near surface sections where the majority of water movements take place (there are the most difficult to modelize), and for deep sections.
- Global moisture errors, accumulated on the entire hydrous profile for one day of measurements are, on average, less than 20 mm, for profiles corresponding to 500 mm of water. The maximum global moisture error value of all the various plots studied does not surpass 66 mm.

These few, but partial figures show that the information obtained is indeed sufficient in view of the fact that in a given field the soil's lack of homogeneity scarcely permits a more accurate spatial measurement evolution.

Below you will find examples of results obtained from two differents plots : plot "E2073" non irrigated of the 'Pioneer 3567' corn variety and plot "F0974" irrigated of the "INRA 400" corn variety. The curves represent on the one hand the model's calculated humidity evolution and on the other hand, the different measurements shown by small points. The individual curves correspond to various 10 cm soil layers taken at periodic depths. The peaks represent rainfall or irrigation. The figures on the left show the average error between the model and the measurement for each depth level in millimeters of water. We may note that, especially during the critical phases, the results provided by the model based on parameters calculated from physical characteristics, are very close to the original measurements.

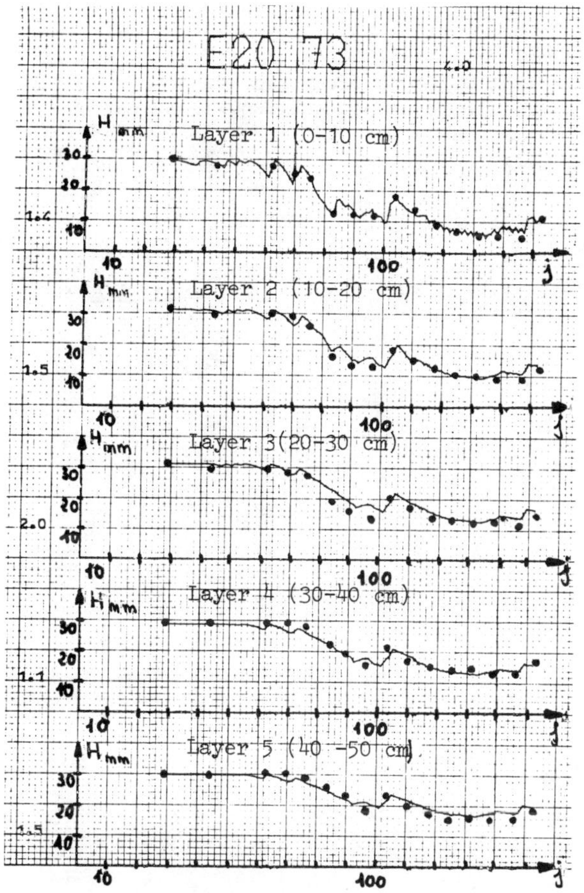

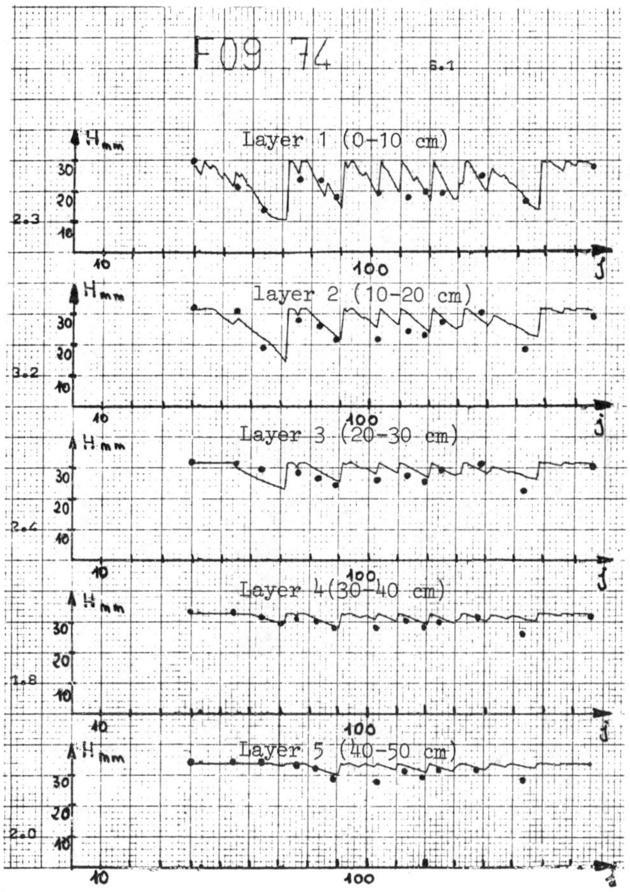

CONCLUSION

Systematic modelling methods are rather rarely applied to the field of agriculture these days. New problems encountered during the course of this ecosystems modelling study have led us to the simultaneous use of the "model method" classical in this type of study and methods relating to multivariate statistical. It is the methodology that calls for a successive and repeated use of the two above methods; that permitted us to obtain reliable results. It is therefore with confidence that we can look forward to other applications of modelling in this field.

In particular, the corn model can be carried on in two directions :

-"real time" use : model simulation replaces a moisture captor, making it possible to calculate the soil's daily hydrous condition. This could bring about:

. an improvement in functioning of planned irrigation systems

.the determining of pseudo optimal irrigation policies in open loops. With knowledge of the soil's hydrous condition and the plant's water needs (known by agronomists), one can determine irrigation dates and water quantities that will satisfy the plant's needs all of which will lead to better water management, so often rare and expensive (the profit yield of irrigated water is presently estimated at approximately 40%).

. the possibility of carrying out an automatic irrigation system that will independently start and stop water flow.

- provisional use. In this case by means of past meteorological data, one can determine the grounds for carrying out an irrigation program by calculating the level of water deficit, the time taken and the profit loss suffered. An estimate of the various necessary dimensions of irrigation facilities can likewise be reached by simulating the different irrigation policies provided by these various facilities (irrigation periodicity possible, quantity of supplied water).

With the use of other experimental data, the model's eventual applications to other types of crops can also he forseen. The advantage of such models will depend on the water sensitivity of the crops chosen. These excellent results obtained with corn, a plant highly sensitive to water, provide a good example of what can eventually be achieved.

BIBLIOGRAPHIE

/1/ Station d'Agronomie de Toulouse (INRA). Plante sol climat et irrigation. Le point de 8 années de recherches sur l'alimentation hydrique des plantes et la valorisation de l'eau - INRA, 136 pages, 1973

/2/ Maertens C., Blanchet R., Rellier J.P. Influence de la nature des sols et de leur exploitation racinaire sur l'évolution des profils hydriques - Agrochimica XVIII n°3, pp 259 265, Xème symp. Int. d'Agronomie, Bari, 1974

/3/ Maertens C, Blanchet R., Puech J. - Influence de différents régimes hydriques sur l'absorption de l'eau et des éléments minéraux par les cultures. Régimes hydriques systèmes racinaires et modalités d'alimentation en eau - Ann. Agro., 25(4)pp 575-586, 1974

/4/ Puech J., Maertens C., Efficience de l'eau consommée de quelques cultures placées dans différentes conditions écologiques - Agrochimica XVIII n°3, pp 223 230, Xème Symp. Inter. di Agrochimica "I problemi dell acqua in agricoltura" Bari, 1974

/5/ Yu O., Gintzburger G., Gounot M., Modèle de fonctionnement d'un peuplement de dactyle en phase végétale. Approche morphogénétique. Oecol. Plant, 107 139, n°2, T10, 1975

/6/ INRA Toulouse - Avertissements irrigation (1967-1975) Collaboration avec quatre départements, les chambres d'agriculture et les SUAD (Service Unité Agricole et Développement).

/7/ Jensen, Robb, Franzcy, Scheluding irrigations using climate crop sol data - Jnl Irrigation and Drainage Dsion ASCE, 70

/8/ J.C.Flinn, The simulation of crop irrigation systems, Systems Analysis in Agricultural Management, John Wiley, 1971

/9/ Hall and Dracup, Water ressources systems, Engineering MacGraw Hill, 1970

/10/ Denmead, Shaw, Availability of soil water to plants on affected by soil moisture content and meteorological conditions Agronomy Journal, vol 54, pp 385 390, 1962

/11/ Fogel, Duckstein, Kisiel, Optimum control of irrigation water application. Automatica, vol 10, 1974

/12/ Dudley, Howell, Musgrave, Optimal intra-seasonal irrigation water allocation, Water Resources Rs., 7 (770 778), 1971

/13/ Rambal, Romane, Aguilar Martin, Modélisation de la production de biomase végétale de la steppe sud tunisienne par une méthode globale d'estimation des paramètres et par filtrage non linéaire. Congrès AFCET "Analyse des systèmes" Versailles, 1977

/14/ Rawse, Stone, Gerwitz, Simulation of the water distribution in soil. Plant and Soil, 517-550, 1978

REMOTE SENSING SYSTEMS FOR WATER RESOURCE MANAGEMENT AND AGRICULTURE DEVELOPMENT

V. Cappellini

Istituto di Elettronica, Facoltà di Ingegneria, Florence University, Florence, Italy

Abstract. Different remote sensing systems are considered for earth resource evaluation. Digital processing techniques, which can be applied to remote sensing data to extract useful final results, are presented: digital filtering, enhancement, restoration, data compression, pattern recognition. Special algorithms of two-dimensional digital filtering are in more details described for image enhancement, noise reduction. A new digital technique to compare and correlate earth maps obtained from remote sensing detectors at different heights, as aboard aircrafts and satellites, is also presented. The application of the above digital processing techniques, in particular of digital filters and local space operators, to remote sensing data related to Tuscany in Italy is described. By processing aircraft photos, land classification for agriculture is shown with evaluation of oil and wine resources. By processing LANDSAT maps, water resource evaluations are finally shown.

Keywords. Agriculture; computer applications; data processing; environment control; filtering; natural resources; pattern recognition; water resources.

INTRODUCTION

Remote sensing data, as collected from the detectors aboard aircrafts or earth resource satellites, are to be processed by means of high efficiency techniques to improve the quality of the data and to obtain final useful results.

Some important digital techniques to process remote sensing data are: corrections, geographical base evaluation, scale variation, digital filtering, use of local space operators, enhancement, restoration, data compression, correlation, pattern recognition, stereoscopic representation.

In this paper, after a short description of remote sensing techniques, three data processing techniques are mainly described: two-dimensional (2-D) digital filtering, use of local space operators, data compression. It is shown how 2-D digital filters can perform image enhancement and noise reduction, local space operators can extract edges and define boundaries, data compression can reduce the increasing amount of remote sensing data.

A new digital technique to compare and correlate earth maps obtained from remote sensing detectors at different heights above the earth surface, as aboard aircrafts and satellites, is also presented.

The application of the above data processing techniques, in particular of 2-D digital filters and local space operators, to remote sensing data related to Tuscany region in Italy is finally described, showing land classification for agriculture by using aircraft photos and water resource evaluations by means of LANDSAT images.

REMOTE SENSING TECHNIQUES

Remote sensing techniques can be divided in two main classes: passive techniques, active techniques (Kazmierczak, 1977).

Passive techniques are based on the following principles: detection of the solar radiation reflected or scattered by the earth surface, detection of the thermal radiation emitted by the earth. The more important detectors of this type used aboard aircrafts or satellites are: photo camera, TV camera, infrared scanner, multispectral scanner, microwave radiometer.

Active techniques use power sources, as radar and lidar, which send suitable electromagnetic waves and detect the reflected signals. Of particular interest are the techniques using side-looking synthetic-aperture radars (SLAR).

Active or passive microwave techniques work correctly day and night despite of clouds and give complementary information with respect to optical techniques (as photo-TV cameras, multispectral scanners).

All the above remote sensing detectors give a kind of map of the observed earth region.

These maps or images contain a great amount of information elements or pixels (picture elements), each pixel corresponding to a given earth surface (typically of the order of 1 m x 1 m for aircraft detectors and of 80 m x 80 m for satellite detectors). These images are represented in digital form, each pixel gray level being a binary word, to be processed by digital computer systems: photos or analog signals can be converted to digital form through the three main steps represented by space sampling, quantization and binary coding.

Some important digital techniques used to process the above remote sensing data for obtaining useful results are: radiometric and geometric corrections to eliminate the different errors; geographical base evaluation to determine the exact position of remote sensing data; scale variations corresponding to linear or non linear gray level expansion or compression; digital filtering to smooth or enhance the data giving images of better quality; use of local space operators, in particular of 3x3 size, especially to extract edges; data compression to reduce the large amount of data contained in each map; correlations among different maps, in particular obtained by different detectors; pattern recognition to recognize some useful configurations and classify the different data; digital stereoscopic representation to obtain by two maps taken from two detectors having different position a three-dimensional (3-D) representation of the observed earth region.

In the following 2-D digital filters, local space operators, data compression and a new correlation technique are in more details presented.

2-D DIGITAL FILTERS

A 2-D digital filter can be defined by means of the following relation (Cappellini, Constantinides, Emiliani, 1978)

$$g(n_1,n_2) = \sum_{k_1=0}^{N_1-1} \sum_{k_2=0}^{N_2-1} a(k_1,k_2) f(n_1-k_1,n_2-k_2) - \sum_{m_1=0}^{M_1-1} \sum_{\substack{m_2=0 \\ m_1+m_2\neq 0}}^{M_2-1} b(m_1,m_2) g(n_1-m_1,n_2-m_2) \quad (1)$$

where: $f(n_1,n_2)$ and $g(n_1,n_2)$ are, respectively, the input and output samples; $a(k_1,k_2)$ and $b(m_1,m_2)$, defined only for positive values of k_1,k_2,m_1,m_2, are the coefficients defining the digital filter frequency response.

If in the relation (1) all the coefficients $b(m_1,m_2)$ are zero, the 2-D digital filter is called of non-recursive type and has in general a finite-impulse response (FIR); otherwise the 2-D digital filter is called of recursive type and has an infinite impulse response (IIR).

By using the z-transform, the transfer function $H(z_1,z_2)$ of the digital filter can be obtained and hence by setting $z_1=\exp(j\omega_1)$, $z_2=\exp(j\omega_2)$ (the case is considered with the space sampling interval X=1), the 2-D frequency response $H(\omega_1,\omega_2)$ can be defined.

Of particular interest for processing images as obtained by remote sensing techniques, are linear phase filters, having a symmetric impulse response, which don't introduce phase distortion in the processed image. This condition is easily verified by non-recursive FIR digital filters, which are in the following considered.

Some important design methods of 2-D non-recursive FIR digital filters are: methods using windows; frequency sampling methods; approximation methods. A relatively simple and efficient technique is represented by the window method: desired filter impulse response is truncated and its samples $h(k_1,k_2)$ are multiplied by suitable correction factors $w(k_1,k_2)$, samples of the used window function (having zero value in the region out of the truncation). Recently high-efficiency windows were introduced by Cappellini (1975, 1978): the 1-D forms of this window (2-D form can be easily obtained through rotation) are close approximations of a function giving a minimum value of the uncertainty product of modified form. In the following Table 1 a new approximation of this type is reported as a third order polynomial (in the time interval 0-1.5).

TABLE 1 Cappellini Window

$$w_3(t) = at^3 + bt^2 + ct + d$$

$0 \leq t \leq 0.75$	$0.75 < t \leq 1.5$
a = 1.783724	a = -0.041165
b = -3.604044	b = 1.502131
c = 0.076450	c = -4.591678
d = 2.243434	d = 3.651582

Efficiency comparisons were developed for the FIR digital filters using different windows: a computer program was prepared which measures the maximum ripple in band δ_1, the maximum level of fluctuations out of the band δ_2 and the width $\Delta\omega_b$ of the transition band, defined as the region in which the frequency response magnitude decreases from $1-\delta_1$ value to δ_2 value.

From the above comparisons Cappellini windows resulted to have high efficiency, better than other known windows (as Kaiser window) at lower levels of out-band attenuations with the advantage of a very simple expression.

The above 2-D digital filters can be applied to process remote sensing images to smooth the data (by means of low-pass filtering), to perform a space frequency correction or to obtain enhancement (by means of high-pass or band-pass filtering). In particular after enhancement better quality images can be in general obtained.

LOCAL SPACE OPERATORS

Local operators process in general small blocks of data in the space domain. The most important techniques in particular for edge extraction or sharpening are based on 3x3 operators.

The 3x3 operators for edge extraction are based on the evaluation of the gradient through a test on a given point and its close values. In fact if the magnitude and the direction of the gradient in a point are known and if the magnitude is greater than a given threshold, it is assumed that in the considered point there is an edge or a contour whose direction is orthogonal to the gradient direction. These techniques can be divided in two groups: to the first group belong the operators which evaluate two orthogonal components of the gradient; the second group is based on gradient detection by means of a set of templates or masks of different orientation (Bullock, 1974).

Some useful techniques of the above type are: smoothed gradient; Sobel gradient; isotropic gradient; Prewitt mask; Kirsch mask; Robinson mask; Chen Frei mask. The use of the different techniques depend on the type of examined image (gray level distribution, characteristics of the edges) and on the desired final result (edge specifications, separation of different areas). To obtain higher resolution edges, it is useful to apply two or more times subsequently the same 3x3 operator.

DATA COMPRESSION

Data compression techniques are used to reduce the amount of not useful or redundant data, to solve the problems of storage or handling which are important now and much more in the near future in the practical utilization of remote sensing maps.

Several data compression techniques have been introduced of reversible and not reversible type (reversible means here that the inverse transformation or decompression can recover all the original adata). Some important methods are: prediction, interpolation, adaptive sampling, encoding, filtering, use of transformations (Fourier, Hadamard, Walsh, Haar, Slant...). All one-dimensional data compression techniques can also be applied to image processing, if the image is processed line by line.

Two particular techniques of interest for processing remote sensing data are: 2-D digital filtering, as above described, and use of 2-D transformations. Indeed 2-D digital filtering of low-pass or band-pass type represents by itself a sort of data compression because a limited part of the space frequency spectrum is extracted, requiring a lower number of data to be maintained. Further 2-D low-pass digital filtering can be used as an useful pre-processing before the application of other compression algorithms, because the smoothed data can be more efficiently compressed.

For what regards the use of 2-D transformations, they are in general based on the fact that the transformed domain has a more compact representation requiring a lower number of significant values to represent the considered image: good compression results are therefore obtained by applying to transformed data simple procedures such as thresholding, variable length encoding, prediction-interpolation. A useful method (Cappellini, Chini, Lotti, 1976) of using 2-D Fast-Walsh-Transform (FWT) consists in dividing the transformed image data into several squares of small size (4x4, 8x8, 16x16) and in employing for each of them a minimum word length (a bit number sufficient to represent the maximum absolute value in the square plus 1 bit for the sign); an additional fixed length word is inserted before the square amplitude values in order to specify the number of bits used to represent these square values. In a modified method of this type no value is maintained for those sub-areas or squares in which the addition of the absolute values of the transformed data results below a given threshold.

The efficiency of the different data compression techniques can be evaluated by means of the following parameters: compression ratio which is the ratio of input data and final remaining data; peak error which is the maximum error between the reconstructed image after decompression and the original image and the r.m.s. error evaluated in a similar way.

A NEW CORRELATION TECHNIQUE OF MAPS

An important problem in remote sensing area is represented by the necessity of comparison and correlation among different maps of the same earth region taken by different detectors in particular set on different platforms (as aboard aircrafts or satellites).

A new technique is here presented, based on the use of suitable digital filtering procedures. Let us consider two earth maps $f_1(n_1,n_2)$ and $f_2(n_1,n_2)$, the first having higher space resolution (taken at lower altitude), the second with lower space resolution (taken at higher altitude). Let us now perform a 2-D low-pass digital filtering with a cutoff frequency $\omega_c/(2\pi) = 1/(2X_2)$, where X_2 is the space sampling interval of the lower resolution map, in such a way to obtain the filtered map $g_1(n_1,n_2)$. We can reduce or decimate the data in the map $g_1(n_1,n_2)$ up to a final space sampling interval X_2 in such a way to obtain the new lower resolution map $g_1(n_1X_2,n_2X_2)$: this data reduction or decimation corresponds to maintain only one data in the group X_2/X_1 available in the $g_1(n_1,n_2)$ map.

The above operations take out from the spectrum of the high resolution image $f_1(n_1,n_2)$ all the space frequency components greater than $\omega_c/(2\pi)$, obtaining a spectral definition equal to that of $f_2(n_1,n_2)$. In this way the two maps $f_2(n_1,n_2)$ and $g_1(n_1X_2,n_2X_2)$ are perfectly comparable and true correlations and interpretations can be developed on them.

APPLICATIONS TO AGRICULTURE AND WATER RESOURCE MAPS

In the following we give two examples of application of the above described digital techniques to the processing of remote sensing maps, with particular reference to agriculture and water resource problems.

Figure 1 shows an example of processing an aircraft photo for agriculture purposes, related to a region at South of Florence in Tuscany. Firstly an enhancement is performed on the original digitized image by means of a 2-D digital filter and after edges are extracted by using the isotropic gradient algorithm (Cappellini and others, 1980). Figure 1 (a) presents the final result of edge extraction, while Fig. 1 (b) presents the classification in forest (at left), wine and oil (at center) and other land (at right). By counting the dark points in each of the map of Fig. 1 (b), taking into account that each of them corresponds to an elementary square area, the global area of the classified regions can be easily obtained.

Figure 2 shows an example of processing a map from earth resource satellite (LANDSAT; map given by TELESPAZIO in Rome, Italy) for water resource evaluation, related to a region near Pisa in Tuscany. An enhancement is performed on the original digital image (band 7 of the multispectral scanner) by means of a 2-D digital filter and after a thresholding is applied on the gray scale levels, obtaining the final map shown in Fig. 2: the river Arno is appearing at bottom, the river Serchio at top, a water channel in the middle and the water distribution in the other area, while at left is the sea area.

CONCLUSIONS

As the above examples show, digital processing techniques using digital filters, local space operators can be very useful to process remote sensing data to perform enhancement, to extract edges and perform classifications.

Further the described techniques with data compression and special correlation procedure can give, respectively, a significant low number of data and permit useful comparisons of different resolution maps integrating the information and properties of these maps. These last techniques were extensively applied at Istituto di Elettronica in Florence to process aircraft photos and LANDSAT maps.

The above techniques can be very useful for processing remote sensing data and in particular for agriculture aspects (land classification, productivity evaluation...) and for water resource evaluations.

REFERENCES

Bullock, B. (1974). The performance of edge operators on images with texture. Hughes Aircraft Comp. Research Lab. Techn. Report.

Cappellini, V. (1975). Two-dimensional digital signal processing. I.E.E. Colloquium on Review of Digital Signal Processing, London.

Cappellini, V., A. Chini, and F. Lotti (1976). Some data compression methods for processing the images received from earth resource satellites. Proceedings 16 Convegno sullo Spazio, Rome. pp. 33-44.

Cappellini, V., A. G. Constantinides, and P. Emiliani (1978). Digital Filters and Their Applications. Academic Press, London.

Cappellini, V., M. Fondelli, F. Muntoni, and F. Vanni (1980). Alcune tecniche di elaborazione ad alta efficienza di dati di telerilevamento. Proceedings 27 Congresso per l' Elettronica, Rome. pp. 345-355.

Kazmierczak, H. (1976). Processing of image data for remote sensing application. NATO Advanced Study Institute on Digital Image Processing and Analysis, Bonas-France. pp. 63-76.

This work was developed under a Contract (SAS -79-0138) by Italian Consiglio Nazionale delle Ricerche (C.N.R.).

Acknowledgements

We thank Dr. F. Muntoni and F. Vanni for the useful cooperation in the processing of remote sensing maps at Florence Computer Center of the University.

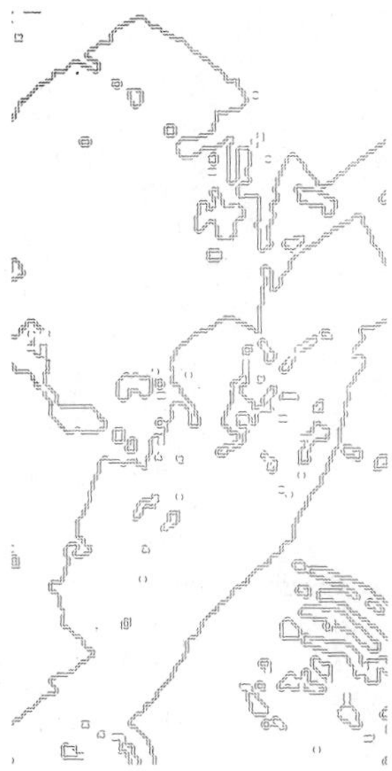

(a)

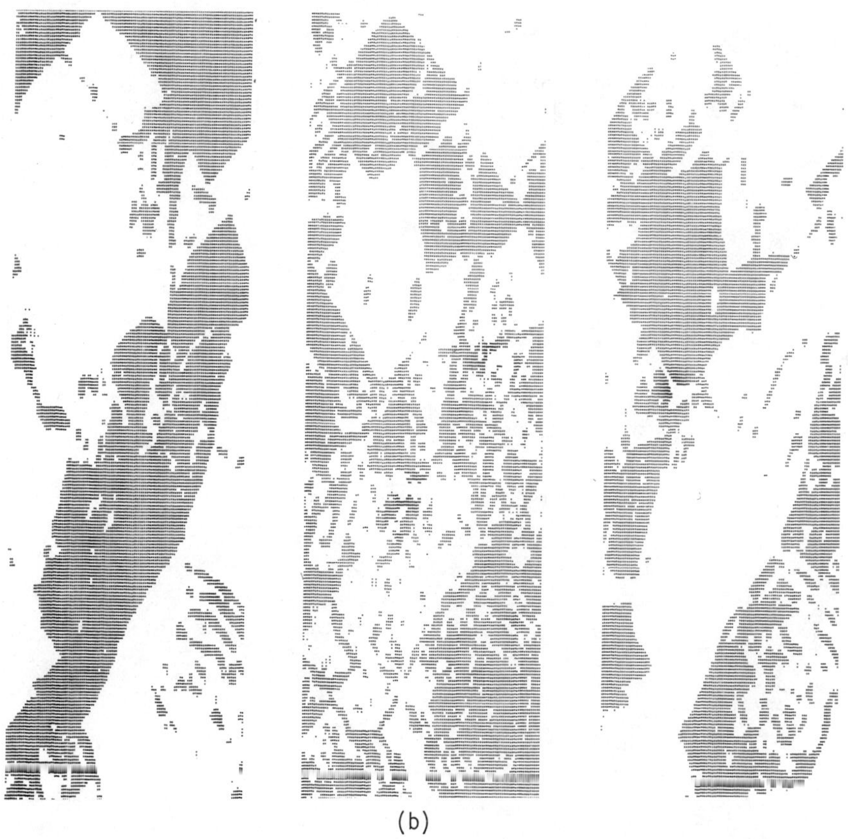

(b)

Fig. 1. Example of application of enhancement and edge extraction to an aircraft photo: (a) result of edge extraction (isotropic gradient algorithm); (b) classification in forest (at left), wine and oil (at center) and other land (at right) giving also the area of each classified region.

Fig. 2. Example of processing a map from LANDSAT satellite, by using enhancement and thresholding, for water resource evaluation: the river Arno is appearing at bottom, the river Serchio at top, a water channel in the middle and the the water distribution in the other area, while at left is the sea area.

STOCHASTIC APPROACH TO RESERVOIR DESIGN AND MANAGEMENT

M. Nouh

Department of Civil Engineering, University of Riyadh, Riyadh, Saudi Arabia

Abstract. *The linear decision rule, which has been developed recently as an effective mathematical tool in the field of reservoir design and management, is reviewed and used for solving the problem and optimal policy of reservoir operations. By adopting this tool, the optimization system can be easily transformed from a non-linear stochastic to a linear deterministic one, which could simply be solved by using the linear programming technique. The stochastic optimization system in this study was to minimize the reservoir's capacity subject to the stochastic constraints which satisfy the storage objectives. The solution of such a system gives an operational strategy to be used in order to minimize the reservoir's storage volume and fulfil its goals. It was found that this system is more flexible and efficient in comparison to some other alternative systems.*

Keywords. *Linear programming; management systems; optimization; probability; statistics, storage; system analysis; water resources.*

INTRODUCTION

Design and management of reservoirs require a high degree of accuracy in their planning because of their importance in water resources systems. Recently, due to the advances in the field of systems analysis, considerable interest has been shown in using optimisation techniques for solving the problems of water storage. Generally, two optimisation approaches are available to analyse reservoir systems, namely, deterministic and stochastic optimisation system approaches.

If inflow sequence to a reservoir is known, one can adopt the deterministic optimisation system approach, through the algorithm of linear programming, dynamic programming, or nonlinear programming, to obtain an optimal strategy which maximizes a certain objective function while simultaneously satisfying physical and operating constraints of the system. Because of the fact that inflow sequence is uncontrolled (at least partially) and, therefore, not known, this optimal strategy is meaningful and cannot be used for accurate design and long operational planning of reservoirs.

Stochastic optimisation system approach considers the fact that inflow sequence is not known in advance, and has been used by several investigators for optimum reservoirs design and management. Little (1955) developed a single-reservoir stochastic dynamic programming model. Loucks (1968) formulated a single-reservoir stochastic linear programming model. Considerable research in this area were extended by Dorfman (1962), Hassitt (1968), Schweig and Cole (1968), and Croley (1974), among others.

Recently, linear decision rules were developed to simplify the stochastic formulation and to include the risk of emptying and/or flooding reservoirs. Using these rules, a non-linear stochastic optimisation system can be transformed to a linear deterministic one. Pioneer work in this area was reported by ReVelle, C., Joeres, E., and Kirby, W.(1969), ReVelle and Kirby (1970), Eastman and ReVelle (1973), and by ReVelle and Gundelach (1975).

The objective of this study is to describe the technique of using the linear decision rules in stochastic optimisation reservoir models, and to adopt it to estimate efficient reservoir capacities and operating policies. Investigations are extended to compare this technique with the stochastic simulation method in reservoir design. Particular emphasis are placed on the reservoir of Red Deer, Alberta, Canada, and Jizan dam in Saudi Arabia. Red Deer reservoir was built, primarily, to ensure winter flows of at least 570 cfs beside the secondary objectives of recreation and flood protection. On the other hand, Jizan dam is an annual irrigation reservoir, built primarily to regulate yearly flows between flood and drought seasons.

METHODS OF ANALYSIS

Linear Decision Rules and Stochastic Optimisation System

As it was mentioned earlier, deterministic

optimisation approach can not be relied on for reservoir design and management due to the fact that future inflow sequence is not known in advance. This weak point of such a system is overcome, in the stochastic formulation, by treating inflow values as being random variables. Theory of statistics can provide methods to estimate statistical properties of any given historical inflow sequence, which could be used in a reservoir operating problem that can be viewed as a stochastic optimisation problem. In the meantime, constraints that limit a reservoir's physical and operational conditions can be viewed, in this case, as chance-constraints.

A chance-constraints (Loucks and Dorfman, 1975) insuring that some variable $\hat{x}$ is no greater than the value of a random variable X at least some fraction α of the time can be written as

$$\text{prob}\ \{\hat{x} \leqslant X\} \geqslant \alpha \quad \text{or}$$

$$\text{prob}\ \{\hat{x} \geqslant X\} \leqslant (1-\alpha) \tag{1}$$

where prob denotes probability. On the other hand, chance-constraints insuring that a variable $\hat{x}$ must be no less than the value of a random variable X at least some fraction α of the time can be written as

$$\text{prob}\ \{\hat{x} \geqslant X\} \geqslant \alpha \quad \text{or}$$

$$\text{prob}\ \{\hat{x} \leqslant X\} \leqslant (1-\alpha) \tag{2}$$

If the probability distribution of the random variable X is known, equations (1) and (2) can be transformed into deterministic form as

$$\hat{x} \leqslant x^{(\alpha)} \tag{3}$$

and $$\hat{x} \geqslant x^{(1-\alpha)} \tag{4}$$

respectively.

In general, an optimal strategy of a reservoir could be obtained if the sequences defining the reservoir's storage and release are known. Unfortunately, these sequences depend on unknown future inflow values and on the operating policy which is what is being sought (unknown). However, the problem could be solved by studying the historical inflow sequence and knowing its probability distribution function; then, deterministic equivalents to the random inflows could be obtained in a fashion similar to the transformation of equation (1) and (2) to (3) and (4), respectively. On the other hand, linear decision rules can be used to define the distributions of storage (S_t) and release (R_t) in terms of of the known distribution of inflow (I_t); then, the previously described method can be adopted to use the determined distribution of storage and release for formulating nonrandom or deterministic equivalents of chance-constraints of same. The linear decision rule used in this study is in the form of

$$R_t = S_{t1} + b_t \quad \text{for all } t \tag{5}$$

where b_t is a decision variable to be determined by using the linear programming technique. Because the mass-balance equation of a reservoir gives

$$S_t = S_{t-1} - R_t + I_t \quad \text{for all } t \tag{6}$$

storages and releases can be obtained in terms of inflows by combining the systems of equations (5) and (6) together. This yields

$$S_t = I_t - b_t \tag{7}$$

and

$$R_t = I_{t-1} + b_t - b_{t-1} \tag{8}$$

Chance-Constraints

Chance-constraints for the reservoirs under study can be formulated according to their objectives as:

1. for the flood control purpose, a free-board volume $C - S_t$ must be maintained to be at least ν_t with reliability α_1

$$\therefore\ \text{prob}\ \{C - S_t \geqslant \nu_t\} \geqslant \alpha_1 \tag{9}$$

2. recreation needs the storage S_t must be at least S_m with reliability α_2

$$\therefore\ \text{prob}\ \{S_t \geqslant S_m\} \geqslant \alpha_2 \tag{10}$$

3. irrigation activities require that release R_t must be at least q_t with reliability α_3

$$\therefore\ \text{prob}\ \{R_t \geqslant q_t\} \geqslant \alpha_3 \tag{11}$$

The minimum storage S_m, in these formulation, can be set equal zero. This leads to a reservoir's capacity C which can only ensure the desired releases. In order to obtain a reservoir's capacity that can fulfil both the releases and recreation purposes (designed reservoir's capacity), the minimum storage S_m should be added to the obtained capacity of reservoir C (ReVelle and others, 1969). In the cases under study, the system of equations (9) to (11) was simplified by setting $\nu_t = S_m = 0$, and by choosing $\alpha_1 = \alpha_2 = \alpha_3 = 0.95$.

Algorithm

Using the previous linear decision rule, chance-constraints (9-11) can be transformed into deterministic equivalents. In case of monthly flow regulated reservoir, the problem can be formulated as:

$$\min C$$

subject to

$$\left.\begin{aligned} C + b_t &\geq i_t^{0.95} \quad ; \quad t = 1,2,\ldots,12 \\ b_t &\leq i_t^{0.05} \quad ; \quad t = 1,2,\ldots,12 \\ b_t - b_{t-1} &\geq q_t - i_{t-1}^{0.05} ; \quad t = 2,3,\ldots,12 \\ b_1 - b_{12} &\geq q_1 - i_{12}^{0.05} \end{aligned}\right\} \quad (12)$$

where $i^{0.95}$ is the inverse of the distribution function of R_t evaluated at 0.95. Similar notation applies to $i^{0.05}$;

where C is the reservoir's capacity, and q_t is the release at period t.

Stochastic Simulation Method

Stochastic simulation is an experimental method. It is based on the idea of new inflow samples can be predicted from historical inflow records by using a certain stochastic model which can successfully describe the natural hydrologic time processes. While the choice and description of the appropriate model, to suit the case under study, is reported elsewhere (Nouh, 1979, 1980), the predicted samples can be used for constructing probable mass-curve diagrams. From these diagrams, the required storage capacity, corresponding to a chosen probability of failure, to satisfy a given draft rate could be obtained.

RESULTS

Jizam Dam is located in the south-west of Saudi Arabia. It was built, primarily, for irrigation purposes to ensure monthly flows of about 200 cfs. It is desired to provide an optimal strategy which fulfils the irrigation objectives with minimum amount of water storage. Data available at the dam's site are historical records of mean monthly flow. The stochastic optimisation method was adopted to solve its storage problem. The records are fitted to probability distribution functions and chi-square statistics are used for the purpose of examining their goodness of fit. It was found that log-normal probability distribution function fits, with a level of significance of 95%, the historical records of mean monthly flow. Therefore, the inverse of inflow distribution function is calculated and tabulated in Table 1.

TABLE 1 Inverse of Inflow Distribution at the Site of Jizan Dam

Month	$i_t^{0.05}$	$i_t^{0.95}$
Jan	8.97	21.06
Feb	8.78	23.43
Mar	7.76	23.39
Apr	5.14	18.14
May	3.63	14.57
June	1.46	9.88
July	1.02	7.92
Aug	0.96	5.35
Sep	2.16	6.12
Oct	2.78	12.57
Nov	3.94	13.93
Dec	6.03	17.72

Using the obtained values of $i_t^{0.05}$ and $i_t^{0.95}$, the decision variables, b_t, were calculated by solving the linear programming algorithm (system of equations number 12). These obtained decision variables (Table 2) should be used with inflow values in order to have and optimal strategy for releases which minimize the storage volume. Minimum capacity of the dam C was found, from this solution, to be 16.24 billion gallons.

TABLE 2 Decision Operating Variables of Jizan Dam in Saudi Arabia

Month	Decision Variable b_t
Jan	4.31
Feb	3.92
Mar	2.23
Apr	-0.56
May	-2.17
June	-3.16
July	-3.16
Aug	-2.31
Sep	-0.64
Oct	1.13
Nov	2.43
Dec	3.36

Dead storages needed for sediment deposition behind the dam, and for recreation are found (Nouh, 1980) to be 7.2 and 1.7 billion gallons, respectively. Therefore, the capacity needed for the dam is considered to be 25.14 billion gallons.

The reliabilities that the dam will not go dry and will not spill were calculated (Table 3) by adjusting each nonbinding constraint, in the linear programming algorithm, to be a binding one (Eastman and ReVelle, 1973).

TABLE 3 Reliabilities of Jizan Dam's Operations

Month	Dam Will Not Go Dry	Dam Will Not Spill
Jan	0.99	0.93
Feb	0.98	0.95
Mar	0.96	0.98
Apr	0.95	0.99
May	0.97	1.0
June	0.96	1.0
July	0.95	1.0
Aug	0.92	1.0
Sep	0.96	0.98
Oct	0.95	0.94
Nov	0.97	0.95
Dec	0.99	0.93

Analysis of the dam's storage capacity requirements is, also, made using the stochastic simulation method. The Integrated Autogressive Moving Average (ARIMA) models were applied to the historical average monthly flow time series for the purpose of obtaining a model which could be used in synthetic generation and forecasting of average monthly flows at the dam's site. ARIMA (1,0,1) was found to suit best such synthetic generation (Nouh, 1980) and, therefore, was used to generate 99 equiprobable inflow series, each 100-years long. The generated series were analysed for the required storages and probability of failures to satisfy a given draft rate. With a constant draft rate of 200 cfs and probability of failure of 0.05, the storage volume was found to be 13.15 billion gallons.

Same kind of analysis was used in the storage analysis of Red Deer Reservoir in Alberta, Canada. By using the stochastic optimisation and stochastic simulation techniques, the minimum storage requirement were found to be 201 and 255 acre-ft, respectively.

CONCLUSION

The stochastic optimisation method is considered by the author the best method, in principle, available to analyse the water storage problems. It has the advantage of designing a reservoir which can fulfil different goals with different reliabilities involved. It also minimizes storage requirements, in the two example considered the storage capacities needed were about 80% of same obtained by using the stochastic simulation method. Although the stochastic optimisation algorithm was simplified by using the linear decision rule, in case of two or three a multi-objectives reservoir, it is still difficult to be solved when more than three sets of chance-constraints are involved. Extended research to simplify such algorithm is needed especially when more than three chance-constraints sets are concerned.

ACKNOWLEDGEMENTS

The author wishes to express his appreciation to the Research Center of the University of Riyadh for typing the manuscript.

REFERENCES

Croley, T.E. (1974). Sequential Stochastic Optimisation for Reservoir System. Journal of Hydraulics Division, ASCE, 100 (HY 1), pp. 201-219.

Dorfman, R. (1962). Mathematical Models: The Multistructure Approach, Chapter 13. Design of Water Resource Systems. A. Maass and others, Harvard University Press, Cambridge, Mass., pp. 494-539.

Eastman, J., and ReVelle, C. (1973). Linear Decision Rule in Reservoir Management and Design 3. Direct Capacity Determination and Intraseasonal Constraints. Water Resources Research. Am. Geophysical Union., 9(1), pp. 29-42.

Hassitt, A. (1968). Solution of the Stochastic Programming Model of Reservoir Regulation. Report No. 320-3506, IBM Washington Scientific Center, Wheaton, Maryland.

Little, J.D.C. (1955). The use of Storage Water in a Hydroelectric System. Journal of the Operation Research Society of America, 3, pp. 187-197.

Loucks, D.P. (1968), Computer Models for Reservoir Regulation. Journal of the Sanitary Engineering Division, ASCE, 94 (SA4), pp. 657-669.

Loucks, D.P., and Dorfman, P.J. (1975). An Evaluation of Same Linear Decision Rules in Chance-Constrained Models for Reservoir Planning and Operation. Water Resources Research, American Geophysical Union, 11 (6), pp. 777-782.

Nouh, M. (1979). Red Deer River Basin Studies 1. Stochastic Reservoir Storage. Fourth National Hydrotechnical Conference, CSCE, Vancouver, Canada.

Nouh, M. (1979). Red Deer River Basin Studies B. Reservoir Operation. Conference on Alberta Water Resources Present and Future, Alberta Chamber of Commerce, Red Deer, Alberta, Canada.

Nouh, M. (1979). Discussion of Red Deer Dam Storage Capacity Requirements by I. Muzik. CSCE, Vancouver, B.C., Canada.

Nouh, M. (1980). Stochastic Simulation in Reservoir Design. Journal of Engineering Sciences, University of Riyadh, Saudi Arabia.

ReVelle, C., and Joeres, E., and Kirby, W. (1969). The Linear Decision Rule in Reservoir Management and Design 1. Development of the Stochastic Model. Water Resources Research, American Geophysical Union, 5(4), pp. 767-777.

ReVelle, C., and Kirby, C. (1970). Linear Decision Rule in Reservoir Management and Design 2. Performance Optimisation. Water Resources Research, American Geophysical Union, 6 (4), pp. 1033-1044.

ReVelle, C., and Gundelach, J. (1975). Linear Decision Rule in Reservoir Management and Design 4. A Rule that Minizes Output Variance. Water Resources Research, American Geophysical Union, 11 (2), pp. 197-203.

Schweig, Z., and Cole, J.A. (1968). Optimal Control of Linked Reservoirs. Water Resources Research, American Geophysical Union, 4 (3), pp. 479-497.

A SIMULATION MODEL FOR MULTIPLE REGULARIZATION OF DAMS

M. Sbihi*, F. Al Youssoufi*, R. Alami*, A. Hanafi and A. Mezzour****

**Direction de l'Hydraulique, Rabat, Morocco*
***I.ME.G, Im. Oued Ettaj, Raoud El Andalous, Rabat, Morocco*

Abstract. The purpose of this article is to present a method for modelling in order to process by computer the problem of regularization of the surface flow of any watershed. This method, having been operating for three years, allows the simulation of the rational management of the waters of a watershed. It aims at helping take decisions concerning the design of projects as well as at optimizing the exploitation of the existing hydraulic structures. This type of problem is naturally formulated as a simulation model. However, the difficulty lies in the construction of a model adaptable to any watershed, and thus capable of taking into account any configuration of multi-purpose dams, agricultural perimeters and non-regulated discharges. We have been let first to use a global approach of regularization that leads to consider a watershed as a system and to find out the characteristics of its components and its operating laws, and then to choose a type of representation that allows to take into account whatever configuration. The formalisms of Operations Research and particularly the graph theory give answer to this second aim.

Keywords. water resources, multiple reservoirs, multi-purpose dams, modelling, operation research, graph theory, system analysis.

1. INTRODUCTION

The study of the regularization of a reservoir consists in its simulated exploitation for a certain number of years in order to observe its capacity to satisfy, in an economic way, the different needs for which it is intended. These needs are the demands of agriculture, the supply of domestic and industrial water, the production of electrical power, the requirements of hygiene and the protection of the inhabited areas from floods. The simulated exploitation consists in establishing, step by step, the balance outflows inflows for each dam. This balance must take into account the losses due to the seepage and to the evaporation occuring at the surface of the reservoir.

Such a study allows us where varying the constraints (the demands to satisfy on one hand and the characteristics of the dam on the other), to bring to light the shortage in satisfying the needs fixed and consequently to revise the choices of the parameters.

Within the framework of a global study of a watershed, lies a more complex problem, that of the multiple regularization. i.e the simulation of the integrated exploitation of several dams.

This problem can be dealt within two different ways :

- a specific approach for any investigated watershed. Such an approach has been used initially for the construction of a simulation model for the Sebou watershed. Had we proceeded in this way, it would be necessary to construct as many models as watersheds to study. (That is to say ten for the whole country). The main drawbacks would be the long delays of programmation and the heavy burdens of preservation.

- a general approach : such an approach aims at elaborating a model adaptable to all watersheds. Besides the advantage of assuring a normalization of the management of different dams, it allows to avoid the shortcomings originating from a specific approach. It is the method which has been kept for the development of the Simulation Model for Multiple Regularization of Dams (SMMRD). It is characterized by a system approach to the regularization of watersheds.

2. A SYSTEM APPROACH TO THE REGULARIZATION OF WATERSHEDS

The simulation methods lead us to state the problem in the optic of system analysis then to construct a model in order to draw a solution.

The pratice of the modular analysis of systems applied to a watershed makes us consider it as a system comprising a certain number of elements (dams, agricultural perimeters, non-regulated discharges) connected to each other according to a certain order and managed in a way that can assure the optimal use of superficial water resources.

This system varies with time : evolution of needs, variation of inflows, addition or suppression of elements, change of the characteristics.

The simulation of the operating of the system is performed by a mathematical and logical model implemented in a computer program.

Some basic hypotheses relating to the system and its environment have directed the design of this model. Among them we find :

. The representation of the relation between the components by means of graphs verifying the property of convergence of the flow as well as the non-cycling of waters.

. Formalization of the different categories of the components and their characteristics.

. Choice of a period of time equal to one month.

. Formalization of some rules for the management of the reservoirs in the form of alrorithms (procedures of primary and secondary irrigation).

. Establishment of secondary hypotheses (simplified formula for the calculation of evaporation, integration of water needs and losses due to seepage).

The study resulted in a model illustrated by fig. 1.

The data provided for the model are of two different types :

. The parameters (in the optic of system analysis we can speak of variables of action). They allow us to define a water system we must study : first, by enumerating the different components and their characteristics. Secondly, by describing the interconnections. Third by fixing the needs to satisfy.

. The inputs of the system : They are series of monthly water inflows for a short or a long period. These inflows must be managed in the most economical way, taking into account the configuration of the watershed.

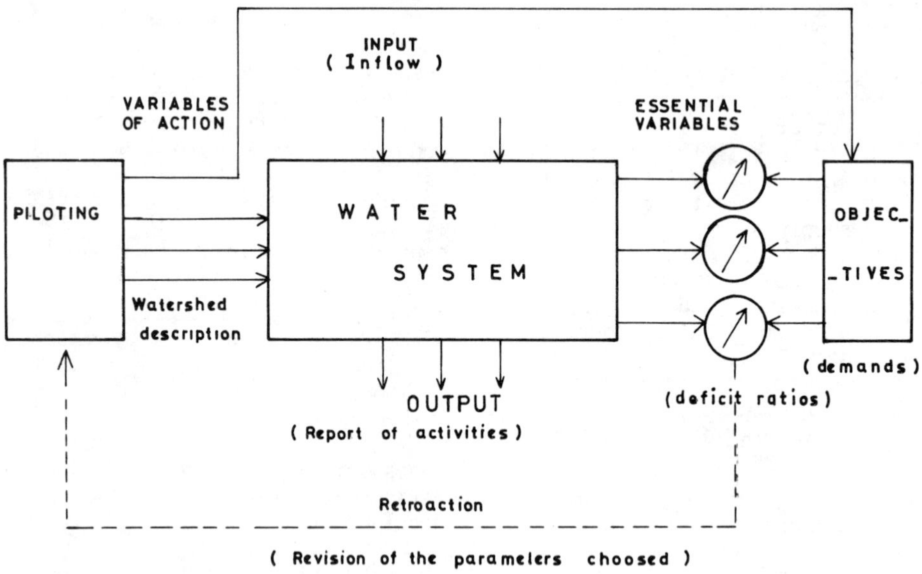

Fig. 1 : Representation of the water system of a watershed.

. <u>The outputs</u> of the model are :

- on one hand, an exhaustive report on the activity of the system during the required period.

- on the other hand, synthetic results that we can consider as the essential variables of the system. These variables are represented by the different deficit ratios that are either annual or interannual.

The analysis of these results allows the engineer in charge of the piloting of the system to adjust the variables of action.

The main characteristics of the developed model originate from the worry about the generalization and standardization, which has directed its design :

. The representation of any configuration of watersheds thanks to the application of the graph theory.

. The elaboration of standard operating rules and their implementation in specialized modules of management.

3 DESCRIPTION OF THE COMPONENTS OF THE WATER SYSTEM

The previous study leads us to bring into light three types of components : the dams, the agricultural perimeters and the non - regulated discharges. (see fig.2).

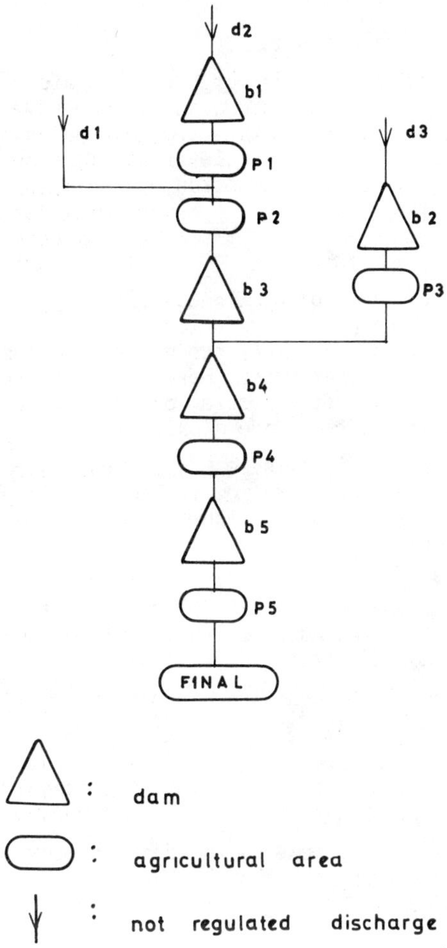

Fig. 2 : Example of configuration of a watershed

The need of standardization, generalization and simplicity makes us consider as outstanding, for each category of components, only the characteristics judged to be essential.

3.1. The dams :

The characteristics retained for a dam affect the reservoir, the site and the hydraulic structures.

The reservoir is defined by the data provided by the tables, elevation volume and elevation surface and also by its maximum volume.

Each site is characterized by some hydrological and climatological datas : the monthly value for average evaporation and the inflows that can be expected till the end of the agricultural campaign.

The technical characteristics of the hydraulic structures are the height of the intake of the bottom outlet and its maximum discharge, the height of the intake of the power-plant, the power installed and the operating curve of the power-plant.

3.2. The perimeters :

The parameters kept are the maximum coefficients for the reduction of the area to be irrigated and for endowment, the coefficient of seepage, the coefficient of pumping and finally the maximum inflow.

3.3. The non-regulated discharges; or natural inflows :

They are the entry points of the system at each step of time. Their values are the object of estimations accomplished thanks to rainfall and run off datas. These calculations are performed by other programs independant of the SMMRD.

4. DEFINITION OF THE DEMANDS

Like the choice made for the essential characteristics of the different categories of components, we have been induced to choose and also to form into groups the different demands.

4.1. For perimeters :

The agricultural needs per perimeter are concentrated into a unique annual demand modulated for twelve months. Some partial shortage can eventually be tolerated (fixation of coefficients for reduction, of endowment and surface).

4.2. For dams :

Besides the agricultural needs fixed by the demand of the perimeters situated at its downstream the dam answers the needs in domestic and industrial water. Hence it must carry out the energetic program of its plant. Moreover it must provide a salubrious flow.

In order to achieve these objectives and to respect their priority we are induced to calculate for each dam, four curves that can delimit their respective slices for each month :

- one curve which keeps a volume above the water intake intended for the supply of domestic and industrial water;

- a curve that keeps a volume for the supply of peak power;

- a curve that preserves a volume for the supply of agricultural water at the stage of primary irrigation;

- a curve that allows us to keep a hollow in the reservoir for the shock-absorbing of floods.

5. THE MANAGEMENT RULES ADOPTED

These rules are standard. All watersheds, however their configuration, will be managed according to these rules.

5.1. Definition, remarks.

We call non-regulated inflow, the inflow originating from a non-regulated discharge or the volume coming out of a perimeter or from a dam after the processing of non-regulated inflow.

The non-regulated discharges are separated into two parts :

- a part which is available representing 80% of the inflow. This volume, alone, can be used to irrigate the perimeter. The rest is directed towards the downstream element.

- A part which is not available representing 20% of the inflow. They are the quantities flowing during the peaks of the daily flow. They cannot be taken for irrigation. They are directed towards the downstream element and will not be considered available till they reach a dam that absorbs the peak flow.

5.2. The different stages for the management of watersheds.

The exploitation of a watershed is done monthly in several stages.

- The preliminary stage : a general evaluation of the situation according to the available stores and to the probable inflows before the end of the agricultural campaign taking place at that time; eventually the taking of a decision concerning the reduction of the surface or the endowment.

- The first stage : processing of non regulated inflows. The watershed is scanned from upstream to downstream. Two cases appear :

. the element is a dam : the volume stored in the reservoir in caclculated function to the inflow of the month and to the evaporated volume. The dam supplies domestic and industrial water and the salubrious flow. The plant carries out its program on both the peak power and the guaranteed power. When the inflows are important it also provides basic energy till saturation. Then, the dam releases water. The outflow is directed towards the next element i.e the dam or the perimeter downstream.

. the element is a perimeter : the utilizable part of the inflow is entirely intended to satisfy the agricultural needs. The eventual surplus and the non-available quantities are directed towards the element downstream.

- The second stage : primary irrigation. After the first stage, some demands can remain unsatisfied in some perimeters. At the stage 2, these demands are sent at the set of dams located at the immediate upstream of the perimeters we are concerned with. They are distributed among the different dams candidates in a way to obtain a balanced emptying of their reservoirs.

- The third stage: secondary irrigation. If there is a demand that remains unsatisfied in some perimeters, it is then directed towards all the dams located at their upstream. The upstream perimeters are the first to be satisfied. Every demand is distributed in a way to get a balanced release. If the volumes supplied at the stage of secondary irrigation are not sufficient, there is not resort and the residual demands will change into a deficit or shortage.

6. APPLICATION OF THE GRAPH THEORY FOR MODELLING A WATERSHED

The formalisms and results of the graph theory allow us to elaborate a model that satisfies our objectives. They help us :

. to formalize any watershed configuration as well as for the algorithm that runs from the upstream to the downstream.

. to determine the set of dams to be used for primary irrigation of agricultural perimeters.

. to define the set of dams to be used for secondary irrigation.

6.1. Formalization of a watershed configuration.

The water system of a watershed can be represented by means of a graph G = (S,L) where S is the set of nodes and L is a subset of the cartesian product S x S = S^2. The set S contains the different elements of the system.

Define : B the set of dams, P the set of agricultural perimeters and E the set of non-regulated discharges.

The hypothesis of the convergence of water allows us to introduce an extra element named FINAL which represents the sink of the water system.

We have :

$$S = E \cup B \cup P \cup \{ FINAL \}$$

$$S = \{ s_1, s_2, \dots s_i, \dots\dots, s_n \}$$

The different directed links between the vertices are represented by the elements of L. Such representation allows us to introduce any number of dams, perimeters and non regulated discharges.

Example in the case of Fig. 2, we have :

$$B = \{b_1, b_2, b_3, b_4, b_5\}$$

$$P = \{P_1, P_2, P_3, P_4, P_5\}$$

$$E = \{d_1, d_2, d_3\}$$

$$L = \{(d_1,P_1), (d_2,b_1), (d_3,b_2), (b_1,P_1), (b_2,P_2), (b_3,b_4), (b_4,P_4), (b_5,P_5), (P_1,P_2), (P_2,b_3), (P_3,b_4), (P_4,b_5), (P_5, FINAL)\}$$

According to the hypotheses stated in chapter II, the directed links must verify the following properties :

- the waters flow from upstream to downstream;
- cycling is not taken into account;
- drainage converges to a unique node. Hence we can state that the oriented graph G must be connected, with no cycles and must admit a root. These are the properties of a rooted tree, in which however all paths lead to the root.

P, B, E and F = { FINAL }are finite sets.Define disjointed index sets I_P, I_B, I_E and I_F where card (I_F) = 1. Define I_S, the index set of S. We have :

$$I_S = I_P \cup I_B \cup I_E \cup I_F$$

The configuration of a watershed will then be represented in the model by the adjacency matrix A corresponding to the graph G. We will note :

$$A = [a_{ij}] \qquad (i, j) \in I_S^2$$

$$\text{where} \begin{cases} a_{ij} = 1 \text{ if } (s_i, s_j) \in L \\ a_{ij} = 0 \text{ if } (s_i, s_j) \notin L \end{cases}$$

A is an n-order boolean square matrix where n = card (S). A indicates the paths of lenght 1 in the graph G (see figure 3).

	d1	d2	d3	FINAL	b1	b2	b3	b4	b5	P1	P2	P3	P4	P5
d1											1			
d2					1									
d3						1								
FINAL														
b1										1				
b2												1		
b3								1						
b4													1	
b5														1
P1											1			
P2							1							
P3								1						
P4									1					
P5				1										

Fig . 3 : Adjacency matrix A of the example

We note $A^{(2)}$ the result of boolean matrix product of A by itself. The existence of 1 in the matrix $A^{(2)}$ at the intersection of row i and column j means that there exists at least one 2-length path between the nodes s_i and s_j.

Similarly, the presence of 1 in the matrix $A^{(k)}$ at the intersection of row i and column j means that there exists at least one k-length path between the nodes s_i and s_j.

One of n nodes graph, with no cycle, does not admit any paths of lengh greater than $(n-1)$ directed links. Its adjacency matrix A is therefore such as :

$$(\forall k \in N) \qquad k \geq n \qquad A^{(k)} = 0$$

We define a path matric A^* such as :

$$A^* = A \cup A^{(2)} \cup A^{(3)} \cup \dots \cup A^{(m-1)}$$

where m is the smallest integer such as $A^{(m)} = 0$

The existence of 1 in A^* at the intersection of row i and column j, indicates that there exists a path which length is not greater than m.

6.2. Properties of the adjacency matrix A of a watershed

- the non-regulated discharges are the inputs of the water system; they have no antecedent in the graph G; then
$(\forall\ (i,j) \in I_s \times I_E\ )\ :\ a_{ij} = 0$

- Every dam or perimeter is connected at its upstream, with at least one component of the water system; then

$(\forall_j \in (\ I_s - I_E\ ))\ :$
$\exists\ i \in I_s\ ;\ \text{such as}\ a_{ij} = 1\ ,$

- Every component of the system, except the node named FINAL, is linked to one element and only one at its downstream, then :
$(\forall_i \in (\ I_s - I_F\ ))\ :$
$\exists\ j \in I_s\ \text{(unique)}\ \ \text{such as}\ a_{ij} = 1$

- There exists a unique root, then
$\forall_j \in I_s\ \ \text{and}\ \ i \in I_F\ :\ a_{ij} = 0$

- The graph admits no cycle then
$(\exists\ \ m \in N\ ),\ \text{such as}\ m \leq \text{card}\ (S)$
$\text{and}\ \ \ \ (k \geq m)\ \ :\ A^{(k)} = 0$

6.3. Formalization of primary irrigation

6.3.1. Definition : we call primary irrigation path each path having the form

$(s_{i_1},\ s_{t_2}\ \ \ldots.\ s_{il}\ ,\ s_{il+1},\ldots\ s_{ip})$

where

$$\begin{cases} i_1,\ i_2\ \ldots\ldots\ i_l & \in\ \ I_B \\ i_{l+1}\ldots\ldots\ldots\ i_p & \in\ \ I_P \end{cases}$$

Primary irrigation consists of providing water to an agricultural perimeter from one or several dams belonging to primary irrigation paths containing the perimeter we are concerned with. At the stage of primary irrigation for a given perimeter, the question is to identify all dams that are allowed to provide it with water at this stage.

6.3.2. Direct primary solidarity : we define a relationship named DPS in the set B U P such as

$y_1\ D\ P\ S\ y_2 \Leftrightarrow$ "there exists a primary irrigation path going through first y_1 then y_2"; (eventually $y_1 = y_2$).

The relationship DPS is a quasi-order relationship (reflexive and transitive).

6.3.3. Primary solidarity : we define a relationship named PS in the set P U B such as :

$$y_1\ PS\ y_2 : \begin{cases} y_1\ DPS\ y_2 \\ \quad \text{or} \\ \exists\ y_3 \in (B\ U\ P)\ \ \text{such as} \\ \qquad (\ y_1\ DPS\ y_3)\ \ \text{and}\ (y_2\ DPS y_3) \end{cases}$$

PS is an equivalence relationship (reflexive, symmetric, transitive). It defines in $B\ U\ P$ equivalence classes that we will call "groups (1) or subgraphs of primary solidarity".

d. Properties of the primary solidarity groups : Let C_{PS} a primary solidaroty group and let $(y_1,\ y_2,\ \ldots.\ y_m)$ the longest path in C_{PS}. y_m is called the root of the subgraph. There does not exist any path going from y_m toward another node of the subgraph. Hence:

$(\ \forall\ y\ \ \in Cps)$ *we have* $y\ DPS\ y_m$

This last property unables us to identify the primary solidarity groups in the graph $G = (S,\ L)$.

6.3.4. Identification of the primary solidarity groups :

6.3.4.1. Let M the adjacency matrix of the subgraph corresponding to the set $B\ U\ P$.

Define $I_{BP}\ \ =\ \ I_B\ \ U\ \ I_P$

We have $\ \ M\ \ =\ \ \left[m_{ij}\right]\ (i,j) \in I^2_{BP}$

where $\ m_{ij} = a_{ij}\ \ \ \ (i,j) \in I^2_{BP}$

	b1	b2	b3	b4	b5	P1	P2	P3	P4	P5
b1						1				
b2								1		
b3				1						
b4									1	
b5										1
P1							1			
P2			1							
P3				1						
P4					1					
P5										

Fig. 4 : Matrix M of the example

We will note M^{Y}_{X} the submatrix of M reduced to the sets of rows X and of columns Y. In this way, we define the matrix M' as follows

$M'^{IB}_{IP} = 0$ $\qquad M'^{IB}_{IB} = M^{IB}_{IB}$

$M'^{IP}_{IB} = M^{IP}_{IB}$ $\qquad M'^{IP}_{IP} = M^{IP}_{IP}$

Thus, the matrix M' "ignores" the directed links of the type perimeter-dam.

Let $M'^{*} = \left[m'^{*}_{ij} \right]$ the path matrix corresponding to M'. (See Fig.5 and Fig.6).

	b1	b2	b3	b4	b5	P1	P2	P3	P4	P5
b1						1				
b2								1		
b3				1						
b4									1	
b5										1
P1							1			
P2										
P3										
P4										
P5										

Fig, 5 : Matrix M' of the example

$M'^{*} =$

	b1	b2	b3	b4	b5	P1	P2	P3	P4	P5
b1						1	1			
b2								1		
b3				1					1	
b4									1	
b5										1
P1							1			
P2										
P3										
P4										
P5										

Fig. 6 : Matrix M'* of the example

we prove that :

$$m'^{*}_{ij} = 1 \Leftrightarrow s_i \text{ DPS } s_j \quad (i,j) \in I^2_{BP}$$

Proof :

We remind the reader that

$$M'^{*} = M' \cup M'(2) \qquad \cup M'^{(p)}$$

So, we have

$$m'^{*}_{ij} = 1 \Leftrightarrow \exists r \in N \text{ such as } m'^{(r)}_{ij} = 1$$

It is then sufficient to prove that :

$$m'^{(q)}_{ij} = 1 \Leftrightarrow s_i \text{ DPS } s_j$$

This property can be proved by recurrence on q i.e the length of the path :

- the property is true for 1-lenght paths of primary irrigation, by definition of M'.
- assume that the property holds for q so

$$m'^{(q)}_{ij} = 1 \Leftrightarrow s_i \text{ DPS } s_j ,$$

and let us show that it is true for $(q+1)$:

$$m'^{(q+1)}_{ij} = 1 \Leftrightarrow \bigcup_{k \in I_{BP}} (m'^{(q)}_{ik} \otimes m'_{kj}) = 1$$

$$\Leftrightarrow (\exists k \in I_{BP}) \text{ such as } (m'^{(q)}_{ik} = 1 \text{ and } m'_{kj} = 1$$

$$\Leftrightarrow (\exists k \in I_{BP}) \text{ such as } (s_i \text{ DPS } s_k) \text{ and } (s_k \text{ DPS } sj)$$

$$\Leftrightarrow s_i \text{ DPS } s_j \quad \text{(by transitivity of DPS)}$$

We will call M'^{*} <u>primary irrigation matrix</u> The existence of 1 in M' at the intersection of the row i and the column j means that there exists a primary irrigation path between the vertices s_i and s_j.

6.3.4.2. An algorithm for identification of primary solidarity groups :

We associate to M'^{*} an extra vector V which elements are the result of the arithmetical sum of the elements of the columns of M'^{*}.

We have :

$$(\forall k \in I_{BP}) : \quad V(k) = \sum_{i \in I_{BP}} m'^{*}_{ik}$$

This algorithm is an iterative process that extracts at each iteration a primary solidarity group. It can be formulated as follows (see Fig.7)

A) $I := I_B \cup I_P$

B) *Do for* $k \in I$ *: compute* $V(k)$

C) *Do while* $I \neq \emptyset$:
C.1 Search the index q such as $V(q) = Max\left\{V(k) \ / \ k \in I\right\}$

C.2 Define J such as :

$$J = \{q\} \cup \{k \in I \ / \ m'^{*}_{kq} = 1\}$$

J is the set of indexes of a primary solidarity group.

C.3. $I := I - J$

Example : in Fig. 7, for the first iteration, q corresponds to the column of p_2 indicated by $*$.

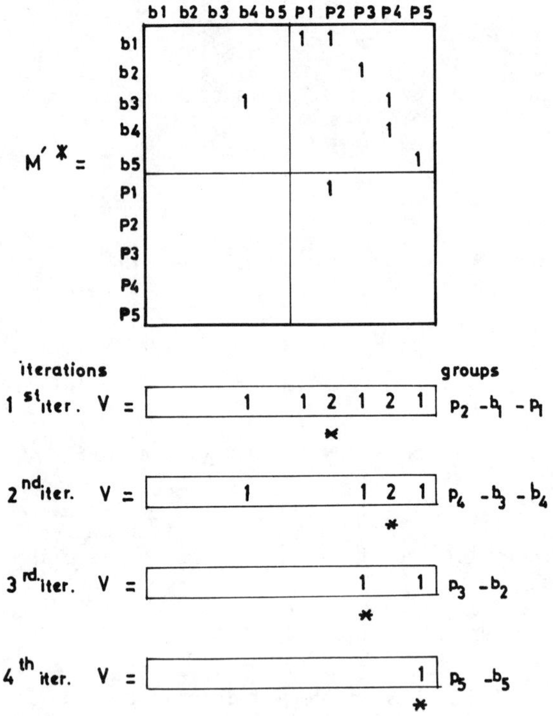

Fig. 7 : Identification of the groups of primary solidarity

At each iteration q corresponds to the index of the root of a primary solidarity group.

6.4. Formalization of the secondary irrigation

At the stage of secondary irrigation, we have to find out all dams that are located at the upstream of a given agricultural perimeter.

Let $M^{*} = [m^{*}_{ij}] \quad (i, j) \in I^2_{BP}$

the path matrix corresponding to M.

Let s_k a given perimeter : $k \in I_P$.

Let J the set of indexes of the dams located at the upstream of s_k. We have :

$$J = \{i \in I_B \ / \ m^{*}_{ik} = 1\}$$

7. EVALUATION OF THE SMMRD

Once achieved the SMMRD has been the object of many tests so as to prove its validation. These tests have consisted in comparing its output with those of almost available regularization programs.

Since that time, its utilization has imposed itself as soon as we do not deal with the simple case of the regularization of one dam with no power plant.

The SMMRD has helped a great deal the overall plan studies of water ressources. The main watersheds dealt with in these studies, are Moulouya, Tensift, Lukkos, Bou-Regreg and Sous-Massa.

The SMMRD has as well been used for regional planning for Tetouan and Taza in order to satisfy their needs in water and energy.

A three-years experience in operating the SMMRD allows us to bring into light some weaknesses of the model :

- the processing on IBM 370-125 computer is time-consuming,
- the type of configurations admitted does not cover all real cases such as cycling of water, multiple exits from the water system components,...
- the SMMRD do not take into account components such as ground-water, pumping stations...

8. CONCLUDING REMARKS

To draw a conclusion, we can say that the utilization of this model has proved the pertinency of system approach for multiple regularization of watersheds. Useful and effective in its actual version, the SMMRD paves the way for the design of new models more flexible, more operating and more general, based on a more detailed system analysis.

ACKNOWLEDGMENTS

The authors are grateful to Nourddine Dinia Director of Hydraulics Administration for his help and support and would like to thank Pr. Robert Faure for his valuable comments and discussion, and also Professor Mohamed Najim, Mohamed Jellali, and Rachid Benmokhtar, Président of I.ME.G, for their encouragement.

R E F E R E N C E S

- Ch. V. Feuvrier : "La simulation des systèmes" (DUNOD), 1971, 152 p.
- C; Berge : "graphes et hypergraphes" (DUNOD), 1970, 502 p
- Direction de l'Hydraulique - "Modèle global de simulation des retenues Sebou" (Note descriptive 1976
- R. Faure - C. Roucairol - P. Tolla : "chemins et flots, ordonnancements" (GAUTHIER - VILLARS), 1976, 234 p.
- R. Faure: "Précis de Recherche Opérationnelle" (DUNOD), 4ème édition 1979, 472 p.
- Direction de l'Hydraulique "Modèle de Régularisation Multiple des Bassins Versants" (SI 0020) 1977.
- R. Faure - J.P BOSS - A. Le Garff. "La recherche opérationnelle" (PUF - Collection que sais-je ? N° 941), 4ème édition, 1974, 128 p.
- Direction de l'Hydraulique "Dossier du programme de régularisation d'Oued Massa".
- Seminaire IFOS - Paris du 23 au 26 mai 1978 "l'Analyse de Systemes".
- Direction de l'Hydraulique "Note de synthèse du pré-plan directeur des eaux" (S P E 8/80)
- Direction de l'Hydraulique "Etude de régularisation du barrage Youssef Ben Tachfine" (S P E 44/79)
- Direction de l'Hydraulique "Etudes de régularisation des barrages" (S P E 13/78).

---

(1) this word is not used here in its mathematical understanding.

AN OPTIMAL MANAGEMENT MODEL OF A HYDRO-THERMAL POWER SYSTEM UNDER IRRIGATIONAL CONSTRAINTS: THE MOROCCO CASE

R. Missaoui*, J.-C. Bernard and A. Haurie****

**State Secretary's Office, Rabat, Morocco*
***Ecole des Hautes Etudes Commerciales de Montréal, Québec, Canada*

Abstract. A model of efficient management is proposed for a hydro-thermal power system where the water in storage is a dominant part of an agricultural system. The water reserves must meet the irrigational needs before being used as a mean of reducing the cost of satisfying the power demand. A methodology, based on the use of reference level curves, permits a closed-loop policy of operation. Simulating the system over a registered set of past hydraulicities and taking a deterministic approach to the optimization problem allows the calculation of potentially lower costs. A numerical test is performed and is based on the Morocco's power system. The general approach is very similar to the one used by researchers of Electricité de France. However the mathematical aspects of our method are extremely simple and can easily be programmed. An interactive procedure was used to "design" new reference level curves. Results are given.

Keywords. Mathematical programming; models; nonlinear programming; optimal control; power system control.

INTRODUCTION

In this article we propose a model of optimal management of the reservoirs of an hydro-thermal power system of a developping country. Each reservoir plays a double role: a primary one of satisfying the irrigational needs, then a secondary one of producing electrical power.

Our research was started from the situation which existed in Morocco during the years 1975-1976, so the data used are realistic and of public knowledge. Otherwise, the constructed model is incomplete: all of the network has not been modelized and many constraints and aspects certainly neglected.

During the recent last years, many optimal management models for power systems were proposed. One of the most recent papers on the subject and also one of the most elaborated from the mathematical point of view is by Quadrat and Delebecque [7]. In their model, the authors describe the natural inflows of each reservoir as stochastic processes of "diffusion" type. The optimal amounts of water to be processed through each powerplant are determined by solving a very large stochastic optimal control problem. A method, based on the theory of teams, using a discrete time approach and optimizing "valley by valley", permits to reach a sub-optimal solution by solving a set of optimaly controlled Markov chains.

A direct formulation of the problem in discrete time and with the use of controlled Markov chains was proposed by Pronovost [6] and Turgeon [8, 9] as a result of management models they realized for Hydro-Quebec.

An other category of models can be found in the paper of Breton and Falgarone [2]. In these models, the management policy is defined by a feed-back law which depends on a fixed set of parameters. By simulating the system over the registered set of past hydraulicities, the value of those parameters can be found by solving a deterministic optimal control problem. This approach was successfully used by the mentionned authors in their modelization of the reservoirs - networks power system of Electricité of France. The same approach also permitted to Breton, Haurie and Kalocsaï [3] the determination of the values of the optimal exchanges between interconnected networks. The positive aspects of this method are as follow:

- no stochastic control problem has to be solved
- no dynamic programming is involved
- the state variables are not discretized
- the usual limits, imposed on dimensionality by dynamic programming, are not present.

The system considered in this paper is one of irrigation needs and of hydro power producing capacities, located in the drainage bassin of a river and of its two principal affluents. Only two of its reservoirs are modulable, all the others being considered as non modulable due to the small value of their capacity. This power system is then relatively simple by comparaison with the France or Quebec systems analysed in [2], [4], [6], [9]. The reserve capacities of the reservoirs justify to limit the planning over a one-year period.

What renders the studied system particular is the double function of its reservoirs. The agricultural orientation of the country, added to its hot semi-arid climatic aspects, give to the stored water a leading importance when faced to the irrigational necessities. However, the growing cost of fossil fuels will certainly orient the utilization of the water reserves to diminish the load over the thermal powerplants. We want to know if it is possible to lower the cost of producing electricity for all of the network by using a new management policy for the reservoirs while satisfying, at all times, the irrigational demand.

Section II gives the set of conditions of the problem. In section III we describe the model utilized and the optimization techniques used. Section IV presents the results. Finally and as a conclusion, possibilities of utilizing the same model to determine the optimal policy of filling up a new reservoir to be added to the network in the near future are briefly examined.

THE PROBLEM

We wish to find an optimal policy for the management of the power system of the country over a planning period of one year. Hence we have a system operating with fixed equipments. The various technical caracteristics of each producing units are considered as constraints over the possible choices of operating policies.

We distinguish the following types of equipment:

1. hydro powerplants associated to a reservoir with a large enough capacity so as to permit a planning period of one year
2. hydro powerplants with "non-modulable" features, that is to say, associated to reservoirs of planning capacity not higher than a few days
3. thermal powerplants using fossil fuels, each particular powerplant having its own cost of production/KWH.

The drainage bassin of the river F is the central element of the hydro power system of the country[1]. Figure 1 gives a schematic view of the system and indicates where, along the different rivers, are located the power producing units and the different zones of irrigational needs. We distinguished seven powerplants, three zones of irrigational needs and two reservoirs: B_1 and B_2 .

The Hydro-Caracteristics

Table 1 gives the means and the standard deviations of the natural monthly inflows for each river of the system. Notice the small values of the inflows during the summer months when compared to those of the winter and spring periods. The variations in the monthly inflows constitute the main source of the stochasticity of the power system.

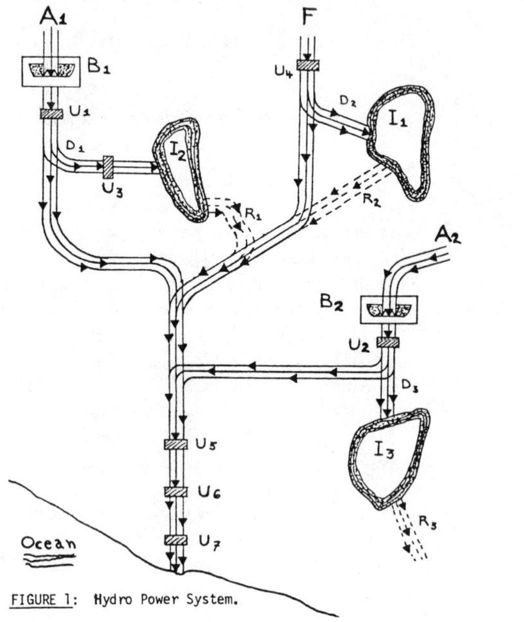

FIGURE 1: Hydro Power System.

where

A_1, A_2 : secondary rivers

B_1, B_2 : large capacity reservoirs

D_1, D_2, D_3: water inflows to irrigational zones

F : main river

I_1, I_2, I_3: zones of irrigational needs

R_1, R_2 : water outflows from irrigational zones

U_1,..., U_7: hydro powerplants.

INFLOWS / MONTH	RIVER "F"		RIVER "A_1"		RIVER "A_2"	
	Mean	S.D.	Mean	S.D.	Mean	S.D.
January	58,2	58,7	39,9	31,9	12,2	9,1
February	64,0	57,6	59,8	46,1	13,9	9,3
March	67,5	46,6	89,2	58,4	22,1	14,5
April	58,9	48,3	104,9	74,6	28,9	21,1
May	34,4	16,1	68,2	52,1	20,8	15,3
June	24,6	8,1	30,9	20,9	9,7	5,9
July	20,0	7,0	15,8	6,4	5,7	3,8
August	18,4	5,9	12,3	4,2	3,8	2,1
September	19,5	5,8	12,7	3,8	5,1	3,9
October	21,7	7,9	13,3	7,7	5,9	4,3
November	27,2	16,6	23,6	21,5	10,5	11,7
December	44,9	46,6	25,7	24,9	7,9	3,7

TABLE 1: Means and Standard Deviations of the Monthly Inflows.

[1]The data used are presented in (1).

The Demand

Generally, in most system the demand is also considered as a stochastic variable. However, given that the climate is moderate, and given that the demand comes mostly from the developped part of the country, the observed random variations of the demand are negligible in comparison with the randomness attached to the natural inflows. Hence, a first approximation is to neglect the random aspect of the demand. Figure 2 gives a representation of the demand distribution over the hourly posts of a given month.

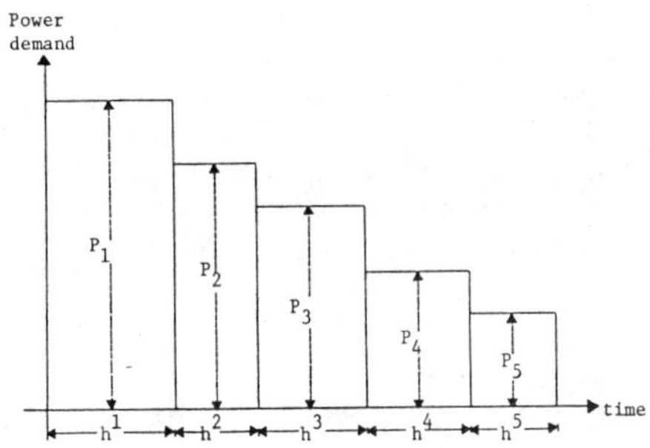

FIGURE 2: Parts of the Power Demand Attributed to the Different Hourly Posts.

Each hourly post is caracterized by the value of its associated power demand and its duration. For example, during January, the peak demand is 493,7 MW and is observed for 84 hours; the lowest demand (during the night, between 00.00 and 05.00 A.M.) is 240,7 MW and is observed for 152 hours of the month.

Caracteristics of the Thermal and Hydro Equipments

The hydro powerplants may be unable to satisfy all of the demand. Then the thermal powerplants must be utilized. There are six different types of thermal powerplants, each of these having its own maximum output power expressed in MW and its own production cost by unit of energy (MWH).

Each hydro powerplant is caracterized by:
- its nature (reservoir, "with the stream")
- its maximum output power
- its maximum rate of throughput.

For the powerplants associated with the reservoirs, we also have the available output power for a given height of fall. It is expressed by an "energy coefficient" related to the amount of water in the reservoir. The ratings of the hydro equipments are considered as constraints over the amount of water to be turbined. The energy coefficients of a reservoir will permit to calculate the quantity of energy produced for each cubic meter of water turbined.

Irrigation and Evaporation

The climate of the country is very sunny and the summer very hot. Consequently, there is an important amount of water which evaporates especially when the surfaces of the reservoirs are large. The same climatic conditions give to the irrigational needs a special importance: they must absolutely be met. The failure cost associated is considered as extremely large.

For each month of the year, the amounts of water needed to satisfy the demands of each zone of irrigational needs are known. Part of this water is returned in the system as shown in Fig. 1. The coefficients of restitution, for the two zones of irrigational needs I_1 and I_2, have values of 0,60 for the month of May through September and of 0,50 for the other months. As for the zone of irrigational needs I_3, the geographical situation is such that the coefficient of restitution can practicly be considered of value zero.

METHODOLOGY

In this section we describe a method of finding a management policy of the hydro and thermal equipments. The demand of water for irrigational needs and for power production is met so as to minimize the annual cost of operation. The principal elements of the methodology are:
- the definition of an operation policy of the reservoirs using reference level curves
- the simulation of the operations of the system for an observed subset of the hydraulicities
- the associated cost of operating the thermal powerplants after optimally allocating the hydro energy produced each month.

Once these elements are fixed in an operating mode, we consider some optimization methods which permit the determination of the best reference level curves. We wish to stress upon the fact that this methodology is very close to real life practice, the notion of reference level curve being actually used by the power company of the country mentioned.

Management of a Reservoir Using a Reference Level Curve

The management of a reservoir is ruled by the use of its associated reference level curve which indicates for a given month the amount of water we wish to keep in storage. Figure 3 shows the reference level curve use for reservoir B_1. The lowest point corresponds to a level of $104,0 \times 10^7$ m^3 (January) and the highest level is $150,34 \times 10^7$ m^3 (the maximum storage capacity) and is found during the month of June.

The management of a reservoir is done in correspondance with the following rule:
- if the actual level for a given month is higher than the reference level -- the level we seek to maintain in storage by the end of

the month -- water is turbined at a rate such that, at the end of the month, we have in storage the value indicated by the reference level. However the rate of processing the water is restricted to a maximum value equal to the maximum throughput rate of the powerplant and cannot be lower than the rate necessary to maintain downstream, due to irrigational needs.
- if the actual level is lower than the reference level, we turbine water at a rate equivalent to the downstream rate needed to meet the irrigational demands.

Figure 4 gives the reference level curve actually used to manage the storage of reservoir B_2.

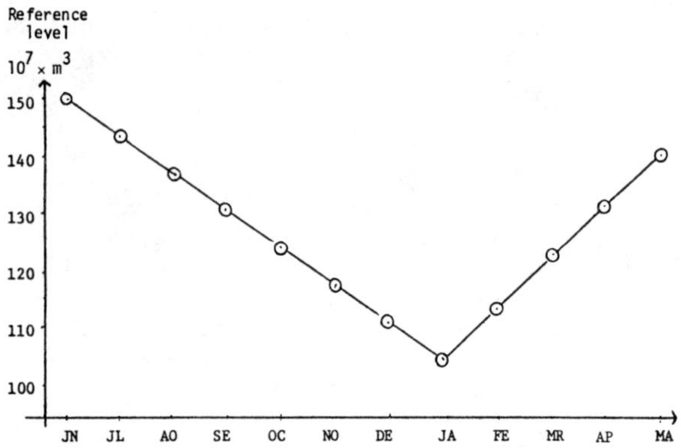

FIGURE 3: Reference Level Curve for Reservoir B_1.

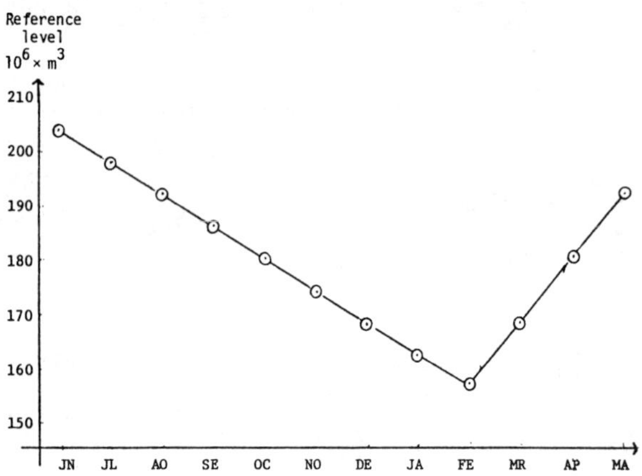

FIGURE 4: Reference Level Curve for Reservoir B_2.

The Simulation

A given set of reference level curves corresponds to a feedback law controlling the amount of water in storage in the reservoirs. The natural inflows being stochastic, each reservoir is in fact a stochastic system.

Since we possess, for 37 years, the data concerning the monthly inflows for each reservoir, it is possible to simulate, for each of these years the management of the two reservoirs under a given set of reference level curves. The principal element of this simulation is a set of equations defining the conservation of flows at each node of the network.

For each powerplant, there is also a maximum rate of turbining to be respected for each month of the planning period. For the reservoirs B_1 and B_2, there are bounds on the amount of water in storage. The higher bound is the maximum storage capacity and the lower is given by the "refill level curve".

These special curves - one for each reservoir - indicates for each month of the year, the minimum level to be maintained if we want the reservoirs to be on the average, at their maximum capacity level by the end of the planning period. This is done while considering the monthly inflows as equal to the mean of the 37 observed inflows for that particular month and taking into account only the amount of water necessary to irrigational needs.

Let us define the following quantities:

$V_1(t)$: amount of water (m^3) kept in storage at the end of month "t" in reservoir B_1. ($V_1(0)$ is given))

$V_2(t)$: amount of water (m^3) kept in storage at the end of month "t" in reservoir B_2. ($V_2(0)$ is given))

$AP_1(t)$: natural inflow for river A_1 during month "t" (m^3)

$AP_2(t)$: natural inflow for river A_2 during month "t" (m^3)

$AP_3(t)$: natural inflow for river F during month "t" (m^3)

$K(t)$: amount of water arriving at powerplant U_4 during month "t"

$EMAX_i(t)$: maximum amount of water that can be turbined at powerplant U_i during month "t", i = 1,...,7

$QD_i(t)$: amount of water spilt at powerplant U_i during month "t", i = 1,...,7

$QE_j(t)$: amount of water evaporated during month "t" at reservoir B_j, j = 1,2

$QF(t)$: amount of water lost through leaks at reservoir B_1 during month "t"

$QI_k(t)$: amount of water needed at the "k" zone of irrigational needs during month "t", k = 1,2,3

$QT_i(t)$: amount of water turbined at powerplant U_i during month "t", i = 1,...,7

We find that the equations (1-5) represent the conservation of flow at powerplants U_3 to U_7 respectively

$$QT_3(t) = QF(t) + QT_1(t) + QD_1(t) - QD_3(t) \tag{1}$$

$$QT_4(t) = K(t) - QI_2(t) \tag{2}$$

$$QT_5(t) = AP_3(t)+[RK(t)-1]\,[QI_1(t)+QI_2(t)]+QD_3(t)+QT_3(t) + QD_2(t) + QT_2(t) - QI_3(t) \quad (3)$$

$$QT_6(t) = QT_5(t) + QD_5(t) - QD_6(t) \quad (4)$$

$$QT_7(t) = QT_6(t) + QD_6(t) - QD_7(t) \quad (5)$$

At powerplants U_1 and U_2, the equations for the conservation of flow are related to the level of storage of the corresponding reservoirs and given by:

$$V_1(t) = V_1(t-1)+AP_1(t)-QE_1(t)-QF(t)-QT_1(t)-QD_1(t) \quad (6)$$

$$V_2(t) = V_2(t-1)+AP_2(t)-QE_2(t)-QT_2(t)-QD_2(t) \quad (7)$$

The values of $QT_2(t)$, $QD_2(t)$, $QT_2(t)$ and $QD_2(t)$ are obtained in the following way. We start with values for the turbined amounts of water equal to the amounts needed for irrigation and we suppose no overflow. We thus have:

$$QT_1(t) = QI_1(t) \quad (8)$$

$$QT_2(t) = QI_3(t) \quad (9)$$

$$QD_1(t) = 0.0 \quad (10)$$

$$QD_2(t) = 0.0 \quad (11)$$

We then calculate the levels we would observe if there were no demands. That is to say we find:

$$V_X \overset{\Delta}{=} V_1(t-1) + AP_1(t) - QE_1(t) - QF(t) \quad (12)$$

$$V_Y \overset{\Delta}{=} V_2(t-1) + AP_2(t) - QE_2(t) \quad (13)$$

respectively for B_1 and B_2.

These calculated values are then compared with those of the reference levels $NV_1(t)$ and $NV_2(t)$ as given by the reference level curves. For each reservoir, if the calculated value is higher than the reference level we establish the difference:

$$QX = VX - NV_1(t) \quad (14)$$

$$QY = VY - NV_2(t) \quad (15)$$

Finally, if the values of these differences are higher than the values affected to $QT_1(t)$ and $QT_2(t)$ at the beginning, we do the following modifications:

at powerplant U_1 , if $QX > QT_1(t)$

$$QT_1(t) = \text{Min}\,\{QX,\ EMAX_1(t)\} \quad (16)$$

$$QD_1(t) = \text{Max}\,\{0,\ QX - EMAX_1(t)\} \quad (17)$$

at powerplant U_2 , if $QY > QT_2(t)$

$$QT_2(t) = \text{Min}\,\{QY,\ EMAX_2(t)\} \quad (18)$$

$$QD_2(t) = \text{MAX}\,\{0,\ QY - EMAX_2(t)\} \quad (19)$$

Allocation of Hydro-Electric Energy and Calculation of the Cost of Operating the Thermal Equipment

Each type of thermal equipment is caracterized by a maximum output power and by a production cost by unit of energy. The function cost of utilization for the thermal equipment is a convex function of the power demand. Given this, it is best to minimize the maximum power demand addressed to the thermal equipment during the hourly posts of a given month, taking into account the projected total production of hydro-energy. This "Min - Max" problem is solved by an optimal allocation of the hydro-energy. The constraints acting on this allocation are the overall rates of throughput for the hydro units and hence the maximum power available from all of the hydro equipment. Remember that this power is a function of the height of fall of the water at each reservoir and consequently function of the level of storage.

For the considered hydro-system, part of the produced energy is "non modulable" since it is produced by powerplant with no associated reservoir. This non modulable part of the energy is uniformely distributed over all hourly posts of the given month. Then the "modulable" part is allocated. Figures 5 and 6 show the allocation of the hydro-energy. The values of T_1, T_2, T_3, T_4 and T_5 are the net power demands to be met by the thermal equipment during each hourly post of the month. These "net" demands are met by utilizing the different thermal equipments in the order of increasing production cost by unit of energy. If the power demand is not met, we consider it to correspond to the situation of utilizing a thermal unit with a very high marginal cost describing thus the failure cost.

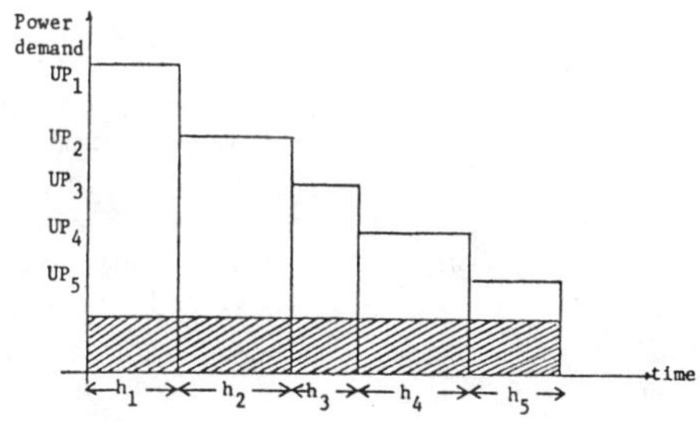

FIGURE 5: Allocation of the Non-Modulable Hydro Energy.

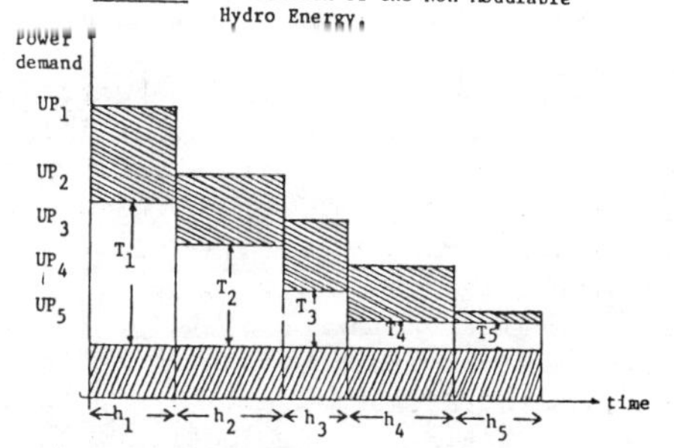

FIGURE 6: Allocation of the Modulable Hydro Energy.

So, the monthly cost has the following form:

$$CT^t = \sum_{k \in K} \sum_{i \in I} c_i \, p^t_{i,k} \, h^t_k \qquad (20)$$

where h^t_k is the length of the hourly post "k", $p^t_{i,k}$ is the power given by the thermal equipment "i" for the k-th hourly post and c_i is the value of the production cost by unit of energy for equipment "i".

The monthly cost is obtained while satisfying the power demand for each hourly post of the given month "t", that is to say satisfying

$$\sum_{i \in I} p^t_{ik} = T^t_k \; ; \; k \in K . \qquad (21)$$

In that way, given the power demand for each month and an operating policy for the reservoirs, the total annual cost of using the thermal equipment is given by:

$$J = \sum_{t=1}^{12} CT^t \qquad (22)$$

The Stochastic Optimal Control Problem and its Solution by Optimizing the Reference Level Curves

The total cost J, defined in 22, is stochastic. It depends on the amount of hydro energy available for each month of the year and this energy depends on the levels of the reservoir. These levels are dependent of the natural inflows which are stochastic. By modifying the management policy, we vary the value of the mathematical expectation of J. The optimal management policy minimizes E[J] under a set of constraints fixed by the irrigational needs, power demands and ratings of the hydro and thermal equipments.

A particular way of representing a management policy is the use of a reference level curve. If a fixed reference level curve is given for each reservoir, it is possible to simulate the management of the system for the 37 years of observed hydraulicities. Taking the power demand as that of the current year, we thus have 37 different costs of operating the system. The mean value of these costs gives a convergent estimation of the expected cost of operation as shown in [7]. We will thus seek to find the reference level curves which will minimize the mean cost for the 37 observed years of hydraulicities, given the constraints defined above.

The stochastic problem is thus replaced by a deterministic one which can be solved by classical constraint optimization techniques. The same approach was taken by Breton and Falgarone [2] and by Breton, Haurie and Kalocsai [3] who utilized a gradient projection method.

For our research, we propose an even simpler method of solving this problem, using a minimum of mathematical developments. We use Hooke's and Jeeves' direct search technique as a first approximation; then a refined solution is obtained by a steepest descent technique using numerical differentiation for the calculation of the derivatives.

The Hooke's and Jeeves' technique permits the optimization of a reference level curve represented in a piecewise manner. It is done by modifying the values of the reference levels of at most three given months. Figure 7 illustrates such a modification.

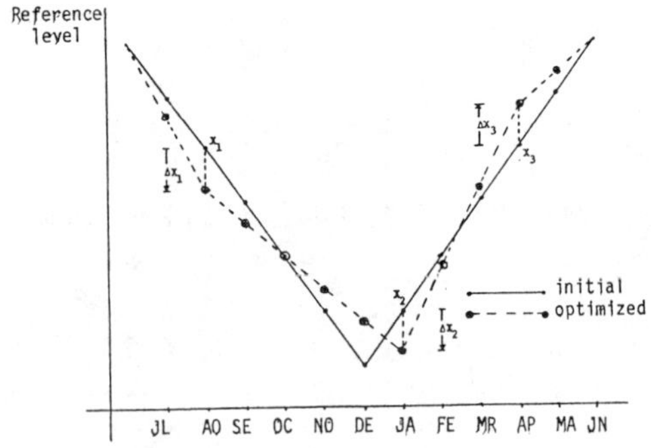

FIGURE 7: Hooke's and Jeeves' Modification of a Reference Level Curve.

This optimization technique is applied in sequence to reservoirs B_1 and B_2. Keeping constant the reference level curve associated with reservoir B_2, the best reference level curve is found for reservoir B_1. Then, this new reference level curve for B_1 is kept constant and the technique is applied to reservoir B_2. The optimization problem is, in that way, divided into two sequential optimization subproblems of lower dimensionality. For both problems, a reference level curve which does not satisfy the irrigational demand, even if only for one month of the 37 years, is considered as unfeasible and is rejected.

The steepest descent technique utilizes numerical differentiation to calculate the values of the derivatives. Every points of the reference level curves are sequentialy modified by a value of $+\delta$ and $-\delta$. Sterling's central difference formula is then applied. The negative of the gradient thus obtained determines the descent direction.

Here again, the optimization technique is successively applied to both reservoirs. The reference level curve of one of the reservoirs is kept constant while the other is optimized. To know which reservoir will be optimize first, the following rule is applied. Let $N^2(g)$ represent the squared value of the calculated gradient, then $N^2(g)$ is defined as:

$$N^2(g) \triangleq N^2[g(B_1)] + N^2[g(B_2)]$$

where $N^2[g(B_1)]$ is the part of $N^2(g)$ imputable to the reference level curve associated to reservoir B_1 and $N^2[g(B_2)]$ the part imputable to reference level curve associated to reservoir B_2. If $N^2[g(B_1)]$ has a greater value than $N^2[g(B_2)]$, then B_1 is optimised before B_2; else we begin the optimization with reservoir B_2.

Let a cycle be the application of the Hooke's and Jeeves' technique followed by the application of the steepest descent technique with respective tolerance ε_1 and ε_2. The original problem was then solved by a sequence of cycles, each time reducing the values of the respective tolerances.

RESULTS

The mathematical programming problem related to the system under study was solved on a Hewlett-Packard 3000 computer. The programs used and their descriptions will appear in a paper to be soon released.

The Initial Reference Level Curves

For each reservoir B_1 and B_2, the reference level curves indicate the amount of water we aim to maintain in storage by the end of each month of the annual planning period. The aimed levels for the month of June - beginning of the planning period - are constant and equal to the maximum storage capacities of the reservoirs. Corresponding to the actual practice, we have a feasible solution which gives the aimed levels of each reservoir and for each month. The associated cost of operation is 136 201 536.

This initial solution, as described by Table 2, corresponds with Fig. 3 and Fig. 4. The lowest levels are observed during January and February respectively for reservoirs B_1 and B_2. The levels of the other months are fixed by linear interpolation.

Month	Reservoir B_1 ($m^3 \times 10^3$)	Reservoir B_2 ($m^3 \times 10^3$)
June	1 503 400	203 700
July	1 437 680	197 925
August	[illegible]	[illegible]
September	1 306 260	186 375
October	1 240 540	180 600
November	1 174 830	174 825
December	1 109 110	169 050
January	1 043 400	163 275
February	1 135 400	157 500
March	1 227 400	168 050
April	1 319 400	180 600
May	1 411 400	192 150

TABLE 2: Reference Levels for the Initial Solution.

The Optimal Reference Level Curves

A first amelioration of the initial solution was obtained after using the Hooke's and Jeeves' method. In a few iterations, a stable feasible solution was found. The cost of operation was reduced to 132 068 960.

From this solution, a second modification was carried out by using the steepest descent technique and the obtained solution gave rise to a cost of operation of 131 681 008. Corresponding to this solution, the squared norm of the gradient was calculated and found to be 1729, too large a value to judge the obtained solution to be a local optimum.

By successively reutilizing both techniques, a solution was found with a corresponding squared norm of 228 which was judged acceptable. The associated cost was 131 264 048. Table 3 gives the aimed levels for each month thus defining the associated optimal reference level curves.

Month	Reservoir B_1 $m^3 \times 10^3$	Reservoir B_2 $m^3 \times 10^3$
June	1 503 400	203 700
July	1 436 063	193 895
August	1 322 184	203 558
September	1 205 444	192 855
October	1 089 639	183 934
November	1 013 578	174 397
December	936 822	168 286
January	907 821	164 955
February	1 018 592	156 244
March	1 180 987	165 424
April	1 343 463	187 389
May	1 503 400	180 569

TABLE 3: Reference Levels for the Optimal Solution

Figures 8 and 9 permit the comparison of the reference level curves corresponding to the initial and optimal solutions. Notice the smoothness of the optimal curve for reservoir B_1 when compared to that of reservoir B_2.

From the sequence of optimization cycles, it was found that the diminution of the operating cost is almost a total consequence of modifying the reference level curve of reservoir B_1. However, it was also found that the square of the norm of the gradient associated was lessened mostly by modifications to the reference level curve of reservoir B_2. Modifications to B_1 reference level curve acted only lightly on the value of the squared norm of the gradient. Finally, it was noticed that at the end of each cycle, the value of $N^2[g(B_1)]$ was much lower than that of $N^2[g(B_2)]$.

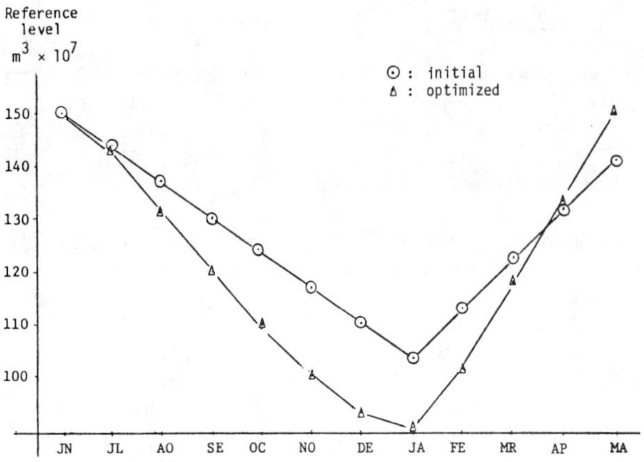

FIGURE 8: Reference Level Curves for Reservoir B_1.

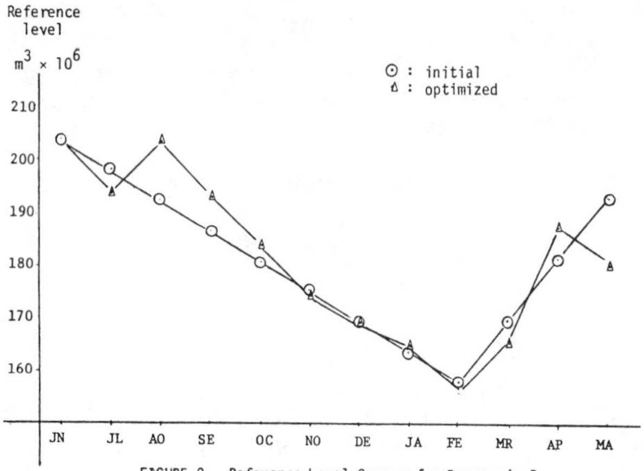

FIGURE 9: Reference Level Curves for Reservoir B_2.

CONCLUSION

The amelioration of the mean annual cost of operation is of small magnitude, more precisely 3.3%. However, this result is more significant when one bears in mind the fact that the initial solution corresponds to a sub-optimized situation based on a real life practice. The point brought forward by this research is that the reference level curve associated to B_1 can be sensibly lowered without causing, at any time, a water shortage which would prevent the irrigational demands to be met. This fact is observed even when simulating over years of very very low natural inflows, like year 1945. It thus seems that the possibility of reducing the annual cost of operation is real and can be done without increasing the risk of a water shortage from the irrigational point of view.

The model and the method of optimization used for this problem will be of particular interest for the practitionners. The simulation approach effectively permits a high level of details in the representation of the operating system and is also in complete concordance with the simulating practices generally used to evaluate the producible energy of a reservoir. An optimization technique, such as Hooke's and Jeeves' requires no derivatives and thus reduces the programming difficulties to a minimum. In addition, the steepest descent technique can be used with exact calculations of derivatives, as in [2] and [3], or with numerical differentiations, as in the present paper. Thus, an originally stochastic optimal control problem can be reduced to an extremely simple nonlinear programming problem.

The model developped could easily be used for the determination of an optimal policy of filling up a new reservoir introduced in the system. Only a few minor adaptations would be necessary. The only major modification necessitated would be the introduction of a larger planning period. A planning period of many years would certainly be necessary in order to permit the modelling of the filling-up pattern of the new reservoir. This modification of the model is likely to lengthen the computer time by a large factor. However the resulting model would certainly remain within the computable range of today's technology.

REFERENCES

[1] J.-C. BERNARD, A. HAURIE et R. MISSAOUI, "Modèle d'optimisation des courbes d'alerte pour la gestion des réservoirs en vue de l'irrigation et de la production d'électricité", Ecole des Hautes Etudes Commerciales, rapport 80-03, Montréal 1980.

[2] A. BRETON et F. FALGARONE, "Gestion de réservoirs d'une vallée hydraulique", in Col. Int. sur les Méthodes de Calcul Scientifique IRIA 73, Berlin Sringer-Verlag.

[3] A. BRETON, A. HAURIE et R. KALOCSAI, "Efficient Management of Interconnected Power Systems", Automatica, September 1978.

[4] F. DELEBECQUE et J.P. QUADRAT, "Contribution of Stochastic Control Singular Perturbation Averaging and Team Theories to an Example of Large-Scale Systems: Management of Hydropower Production", IEEE Transactions on Automatic Control, Vol AC-23, no. 2, April 1978, pp. 209 - 222.

[5] MARSCHAK et R. RADUER, Economic Theory of Teams, Yale University Press 1972.

[6] R. PRONOVOST, "Planification de la production énergétique au moyen de modèles à réservoirs multiples", in Col. Théorie des systèmes et applications à la Gestion des Services Publics, Université de Montréal et Ecole des Hautes Etudes Commerciales, Montréal 1975.

[7] J.P. QUADRAT et F. VIOT, "Méthodes de simulation en programmation dynamique, RAIRO R1, 3-22, April (1973).

[8] A. TURGEON, "Gestion optimale d'un réseau hydroélectrique", in Col. Théorie des Systèmes et applications à la Gestion des Services Publics, Université de Montréal et Ecole des H.E.C., Montréal 1975.

[9] A. TURGEON, "Optimal Operation of a Multi Reservoir Hydroelectric Power System", in Single, Titli edits, "Handbook of large scale systems engineering applications", North-Holland 1979.

IDENTIFICATION OF A POTABILIZATION PROCESS IN THE FLORENCE CITY WATER TREATMENT PLANT°

C. Manfredi and E. Mosca

SAIS, Istituto di Elettronica, Facoltà di Ingegneria, Università di Firenze Via S. Marta 3, 50139 Firenze, Italy

Abstract. This paper presents preliminary results on the determination from operating records of stochastic dynamic models that can adequately describe the clarification process of the Arno River water taking place in the Florence City (FC) Water Treatment Plant. The work was undertaken in view of ascertain the possibility of replacing the present plant manual management, based on rigid and static rules, by computerized adaptive control techniques. Besides the specific characteristics of the Arno River and the plant here considered, it is to be expected that the feasibility study here began will be of some interest wherever raw river water is potabilized for domestic use.

Keywords. Adaptive control; Chemical variables control; Identification; Stochastic systems.

1. INTRODUCTION

This paper deals with the problem of determining suitable dynamical stochastic models of a water potabilization process from input-output records. The process takes place in the Florence city water treatment plant and consists of a water clarification process whose goal is to reduce the raw water turbidity to levels acceptable for domestic use. The input to the treatment plant consists of the Arno river water. The clarification process is accomplished by adding flocculants (suitable chemicals) to raw water and sending the mixture to sedimentation tanks where the physico-chemical mechanism of flocculation occurs, yielding clarified water at the output of the treatment plant. The flocculant addition should be provided so as to guarantee a convenient chemical concentration in the sedimentation tanks. This concentration should vary according to input turbidity and other physico-chemical characteristics of the input raw water. Unfortunately, the Arno river characteristics are quickly varying and the turbidity ranges from five up to five hundred F.T.U. and no quantitative physico-chemical models of the process are available. Consequently, the chemical addition is carried out manually according to empirical rules yielding eccessive chemical feed consumption and high costs of operation. The basic goal of the study described in the paper, was to verify the possibility of identifying in real-time stochastic dynamic models from input-output records that can adequately represent the time-varying behavior of process and consequently be used in adaptive control schemes, whose implementation could provide better water quality and reduced costs of operation.

°This work was partially supported by the Florence City Administration within a SAIS-FC Water Department joint research program.

2. PLANT DESCRIPTION AND DATA COLLECTION

As shown in Fig.1, the potabilization process considered hereafter consists essentially of several physico-chemical treatments, specifically: chlorine addiction, flocculant addiction, sand filters and ozonization. Of the overall potabilization process, the study was focused on determining the dynamic behavior of the clarification process consisting of three external variables: v_t, the raw water turbidity of the Arno river feeding the potabilization plant; u_t, the flocculant (aluminum polichloride) feed-rate and y_t, the turbidity of the clarified water. The reason for concentrating the study mainly on the clarification process was motivated by the fact that the flocculant addition represents one of the most costly operations in the Plant. Therefore, the automatization of the present manual and rigid addiction of flocculants could provide high economic benefit consisting of several thousand dollars of annual savings for chemical feed only. The variables v_t and y_t were automatically recorded in continuous time, whereas u_t, the manipolable variable of the process, in all the experiments was kept at a constant value over time intervals of 30 minutes of duration. All water turbidities were expressed in F.T.U. (Formazine Turbidity Units).

3. STOCHASTIC DYNAMIC MODELS FOR THE CLARIFICATION PROCESS

The main goal was the determination of discrete-time linear stochastic dynamic models capable of describing the behavior of the clarification process. The models taken into consideration were Autoregressive Models with Exogenous Variables (ARX) (Box and Jenkins, 1970; Goodwin and Payne, 1977):

$$A(q^{-1})y_{t+k} = W(q^{-1})v_t + B(q^{-1})u_t + e_{t+k} \quad (1)$$

were:

t is the data sampling time, $t = 1,2,\ldots$;
y_t is the clarified water turbidity at time t;
v_t is the raw water turbidity;
u_t is the flocculant feed rate;
e_t is the residual;

q^{-1} is the one-step delay, e.g.,

$$q^{-1}y_t = y_{t-1}$$

$$A(q^{-1}) = 1 + a_1q^{-1} + \ldots + a_nq^{-n} \quad ;$$

$$W(q^{-1}) = w_1q^{-1} + \ldots\ldots + w_rq^{-r} \quad ;$$

$$B(q^{-1}) = b_1q^{-1} + \ldots\ldots + b_sq^{-s} \quad .$$

The time shift k is introduced in (1) to take into account the delay by which the output (clarified water) turbidity is influenced by a variation of input (raw water) turbidity and/or flocculant feed-rate. Notice that in (1) the polinomials $A(q^{-1})$, $W(q^{-1})$, $B(q^{-1})$ have different degrees, respectively n, r, s. The need of introducing in (1) polynomials with different degrees was found necessary in order to provide a better description of the experimental data. For this reason, hereafter the model (1) will be denoted as

$$\text{ARX}(n,r,s).$$

Since it is to be expected that the clarification process behaves in a higly nonlinear way, the variables y_t, v_t and u_t of (1) are "centered" variables, viz., if y_t^*, v_t^* and u_t^* are the effective recorded variables and $\bar{y}_t$, $\bar{v}_t$ and $\bar{u}_t$ their corresponding average evolutions, which will be defined in Sect.4, thus

$$y_t = y_t^* - \bar{y}_t \ , \ v_t = v_t^* - \bar{v}_t \ , \ u_t = u_t^* - \bar{u}_t .$$

In such a way (1) is regarded as modelling the variations of the process external variables from their average evolutions.

In the model (1), the variables y_{t+k}, v_t, u_t can be assumed to be known in that they are recorded by sampling the external variables of the process. On the contrary, the delay k and the model orders n, r, s, as well as the n+r+s+1 parameters a_i, w_i, b_i and

$$\sigma^2 = E[e_t^2],$$

must be determined by finding that particular model that fits the available data in the most accurate way.

The identification procedure here adopted basically consists of three different steps.

<u>Step 1 - Determination of the delay k.</u>

This is accomplished by performing the following experiment. A negative step function is superimposed at a given sampling time to the usual level of the flocculant feed-rate u_t and the corresponding evolution of the output turbidity is recorded. The delay k+1 is then set to be equal to the time delay, expressed in terms of numbers of sampling times, after which a substantial increase of turbidity is produced at the process output. The evolutions of the process variables under the experiment described above are shown in Fig.2. From this figure one obtains

$$k = 4$$

(2 hours). This number is in good agreement with what can be estimated on the grounds of physical dimensions and flow rates.

<u>Step 2 - Parameter estimation for a given ARX(n,r,s).</u>

Here it is temporarily assumed that the model orders of ARX(n,r,s) be given. Thus, the following row-vector

$$\gamma_t=[-y_{t-1}\,..\,-y_{t-n}\ v_{t-k-1}\,..\,v_{t-k-r}\ u_{t-k-1}\,..\,u_{t-k-s}], \quad (2)$$

made up of past samples of output turbidity and exogenous variables, can be constructed. In addition, let the free parameters of ARX(n,r,s) be represented by a column-vector

$$\theta = [a_1 \ldots a_n \ w_1 \ldots w_r \ b_1 \ldots b_s]' \qquad (3)$$

where the prime denotes matrix transposition. Thus, (1) can be rewritten as follows:

$$y_t = \gamma_t \theta + e_t. \qquad (4)$$

Since, as usual in ARX models, the stochastic process $\{e_t\}$ is assumed to be Gaussian, zero-mean and white, the Maximum Likelihood estimate and the Least Squares (LS) estimate of θ coincide. Therefore, the only estimate $\hat{\theta}$ of θ considered hereafter is the vector of appropriate dimensions minimizing

$$\sum_{i=1}^{T} (y_i - \gamma_i \theta)^2.$$

On the other hand, since the process under consideration is time-varying and the determination of its time variations was one of the objectives of the study, for each experiment a sequence of LS estimates $\hat{\theta}_t(N)$ of θ was found by minimizing

$$\sum_{i=t-N+1}^{t} (y_i - \gamma_i \theta)^2 \quad , \quad t=N,N+1,\ldots,T$$

where T is the number of samples of each variable recorded during a given experiment, and N is the lenght of a moving window comprising all the past data on which $\hat{\theta}_t(N)$ is based. The number N should be chosen so as to satisfy conflicting requirements of good estimate accuracy (large N) and small time variations of the model parameters (small N). By defining the following column-vector

$$Y_t(N) = [y_t \ y_{t-1} \ \ldots \ y_{t-N+1}]' \qquad (5)$$

and the following matrix

$$\Gamma_t(N) = [\gamma'_t \ \gamma'_{t-1} \ \ldots \ \gamma'_{t-N+1}]' \qquad (6)$$

the desidered estimate is given by

$$\hat{\theta}_t(N) = [\Gamma'_t(N)\Gamma_t(N)]^{-1}\Gamma'_t(N)Y_t(N). \qquad (7)$$

It is also convenient to define the corresponding residuals

$$e_t(N) = y_t - \gamma_t \hat{\theta}_t(N) \qquad (8)$$

Step 3 - Determination of model orders.
The choice of model orders is made essentially according to the AIC (Akaike Information Criterion) (Akaike, 1974). As seen at Step 2, to any ARX(n,r,s) model corresponds a sequence of estimates $\hat{\theta}_t(N)$, t=N,...,T and a related sequence of residuals $e_t(N)$. Thus, the AI-number associated to ARX(n,r,s) is given by

$$(N/2)\ln[\sigma^2(N)] + (n + r + s + 1)$$

where

$$\sigma^2(N) = (T - N + 1)^{-1} \sum_{t=N}^{T} e_t^2(N)$$

and n+r+s+1 is the number of free parameters in ARX(n,r,s), consisting of the coefficients a_i, w_i, b_i plus the variance of the process $\{e_t\}$. The AIC selects, among all possible models, the one corresponding to the minimum AI-number. However, in the presence of competitive ARX models associated to very close AI-numbers, the model exhibiting a minimum of time-variations in its characterizing parameters was selected as the most adequate one to the experimental data.

4. MODEL IDENTIFICATION FROM OPERATING RECORDS

The collection of the clarification process external variables was executed by varying the flocculant feed-rate according to the time-behavior of pseudo-random sequences, and recording simultaneously the turbidity levels of input and output water. The data sampling interval, corresponding to the unit time-step in (1), was fixed once for all at 30 minutes. This choice resulted from observing that the process variables exhibit small changes within such a time-interval. Moreover, 30 minutes seem to correspond to a reasonable time duration over which the flocculant control variable should be kept at a constant feed-rate value.

Fig.3 and Fig.4 show the typical behavior of the time-evolution of the process external variables during two experiments each of about one-day duration. Since the process was not equipped by a progammable feed-rate controller, each one-day experiment involved the presence of human operators dedicated to this specific task over successive 24 hours acting according to a preassigned program of flocculant time-variations. This was the main reason why all the performed experiments were carried out during the period July-August 1979 and thereafter interrupted in view of the introduction in the plant of more adequate sensors and programmable controllers. However,the experiments that could be carried out, yielded enough indication on the complexity of the matematical models adequately representing the clarification process and their time-dependence.

Preliminary data analysis showed that a good choice for the lenght N of the moving window is N = 35, corresponding to 17.5 hours. Accordingly, the average evolutions of the process external variables were defined as follows:

$$\begin{bmatrix} \bar{y}_t \\ \bar{v}_t \\ \bar{u}_t \end{bmatrix} = N^{-1} \sum_{i=t-N+1}^{t} \begin{bmatrix} y_i \\ v_i \\ u_i \end{bmatrix}$$

Next, AI-numbers were determined for various ARX(n,r,s) models. With reference to the experiment of Fig.3, in Fig.5 the AI-number is plotted for several values of the model orders n, r, s. From this figure, one can see that AIC selects

ARX(2,2,1)

as the most adequate model for the process under consideration. Actually, one could argue that ARX(2,1,1) and ARX(2,2,2) can still provide satisfactorily adequate models in that their corresponding AI-numbers are close enough to the minimum AI-number taken on at ARX(2,2,1). However, by comparing the time-variations of the estimated free parameter vector $\hat{\theta}_t(35)$ based on the available data, it was verified that the optimum model orders (2,2,1) yield estimates exhibiting a more constant behavior with respect to time.

The five components of the estimated free parameter vector $\hat{\theta}_t(35)$ are reported in Table 1 and Table 2, respectively for the experiments of Fig.3 and Fig.4. It is interesting to notice that going from one experiment to the other one gets a noticeable change for the values of the estimates, whereas within a single experiment the estimates tend to exhibit much more reduced variations with respect to time. In particular, a positive value for the b_1 coefficient is estimated at all steps in the experiment of Fig.3, showing that an increase of the flocculant feed-rate produces an increase of output turbidity. This behavior should not be considered as an exceptional one, in that, as is well known to the experts of the physico-chemical phenomena taking place in the clarification process, there may be conditions of excessive concentration of chemicals, under which the flocculants act in an opposite direction with respect to the desired one. On the contrary, a negative value for the b_1 coefficient is estimated at all steps except the last one in the experiment of Fig.4, showing that an increase of flocculant feed-rate yields, as desired, a decrease of output turbidity. Because of the time-varying behavior exhibited by the process during the several experiments carried out, it is to be expected that an on-line estimation of the sign of the b_1 parameter, during the normal operating conditions of the potabilization process, could provide a very relevant piece of information on which to base an effective adaptive control of the plant.

5. CONCLUSIONS AND OPEN QUESTIONS

Black-box identification techniques have been used in order to determine stochastic dynamical models for the water clarification process taking place in the Florence City Water Treatment Plant. The results obtained indicate that, according to AIC (Akaike Information Criterion), ARX(2,2,1) is the most adequate autoregressive model. Input-output data collected from the plant during July-August 1979, reveal a time-varying behavior of the process dynamics as a result, presumably, of corresponding time-variations of the physico-chemical characteristics of the raw input water and chemical concentrations in the plant. It is believed that these time-variations of the process should be tracked by suitable on-line identification algorithms, so as to be able to implement an effective chemical feed-rate automatic control, necessarily of an adaptive nature.

Further efforts should be made in order to gather larger sets of data from the existing plant, so as to test the adequacy of ARX models over different periods of the year. It would also be interesting to use suitable sensors in order to acquire other physico-chemical variables, e.g., temperature, pH, dissolved chlorine, alkalinity, hardness, etc., so as to correlate the process model time-variations with the changing characteristics of the incoming raw water of the Arno river.

REFERENCES

Akaike, H. (1974). A new look at the statistical model identification. IEEE Trans. Autom. Control, 19, 716-723.

Box, G.E.P., and G.M. Jenkins (1970). Time Series Analysis. Forecasting and Control. Holden Day, San Francisco.

Goodwin, G.C., and R.L. Payne (1977). Dynamic System Identification: Experiment Design and Data Analysis. Academic Press, New York.

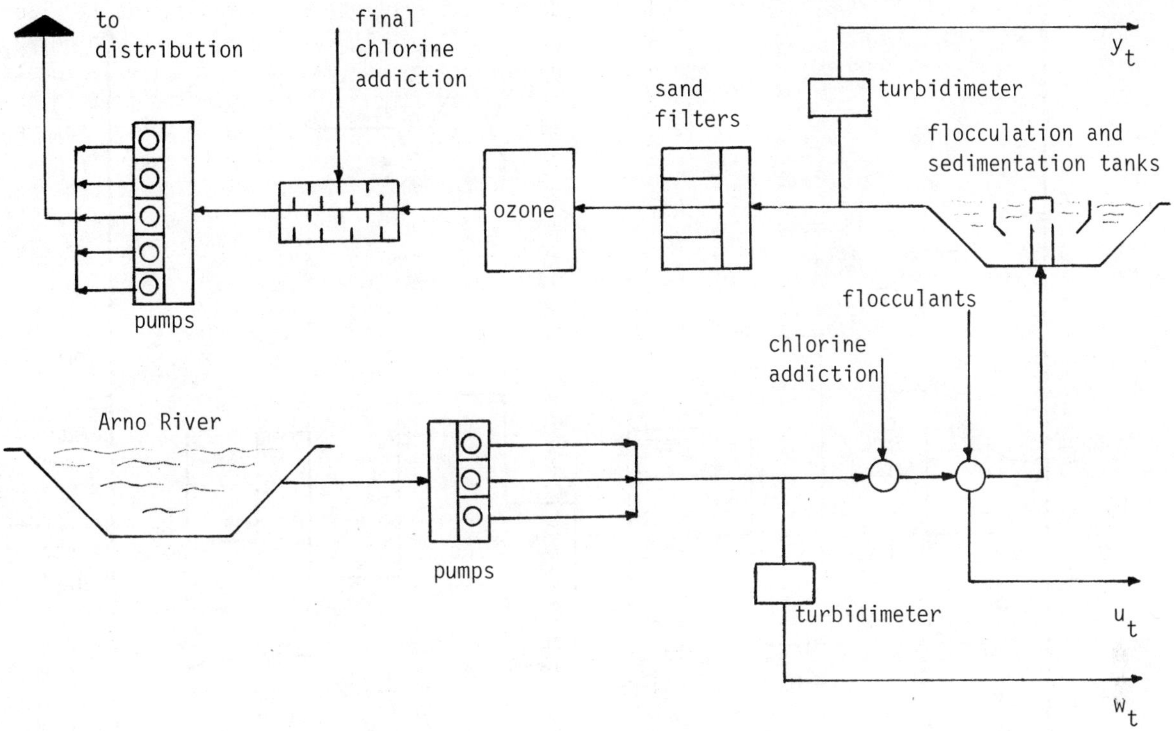

Fig. 1. Potabilization process.

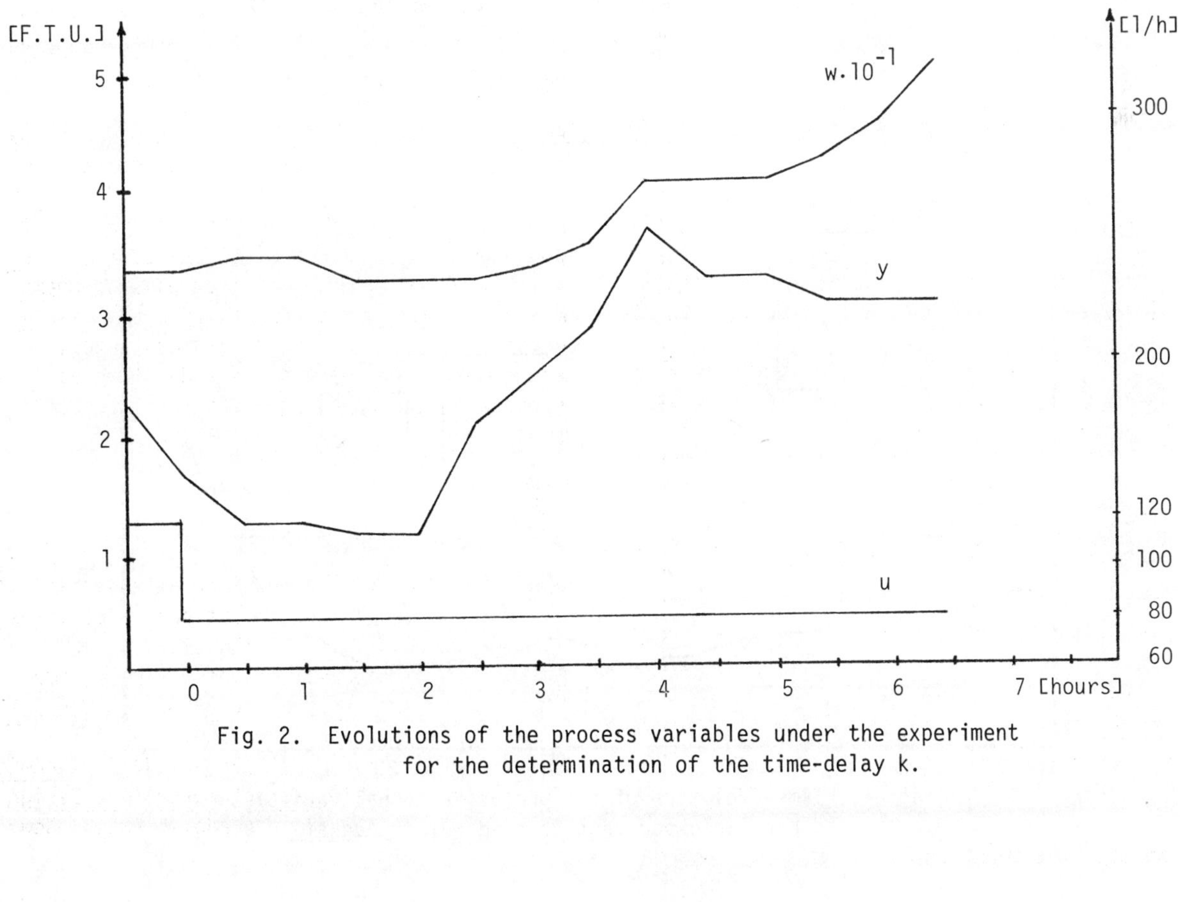

Fig. 2. Evolutions of the process variables under the experiment for the determination of the time-delay k.

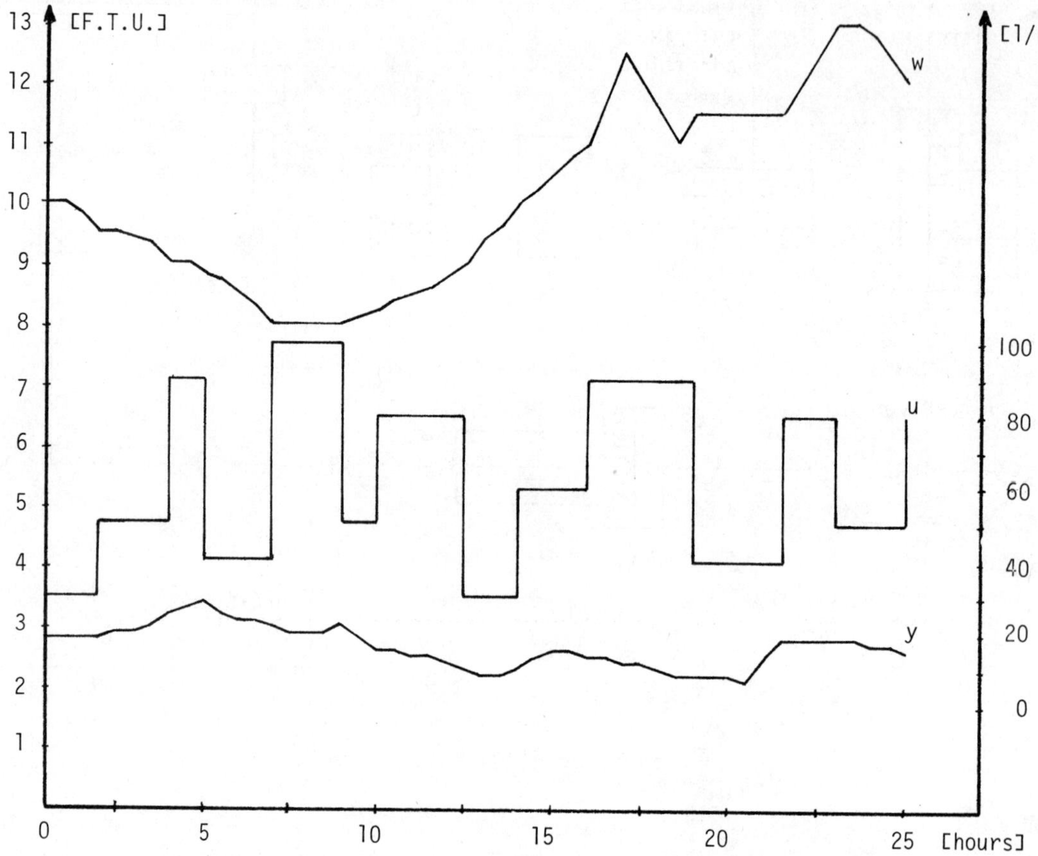

Fig. 3. Time-evolution of the process external variables during an experiment carried out in July 1979.

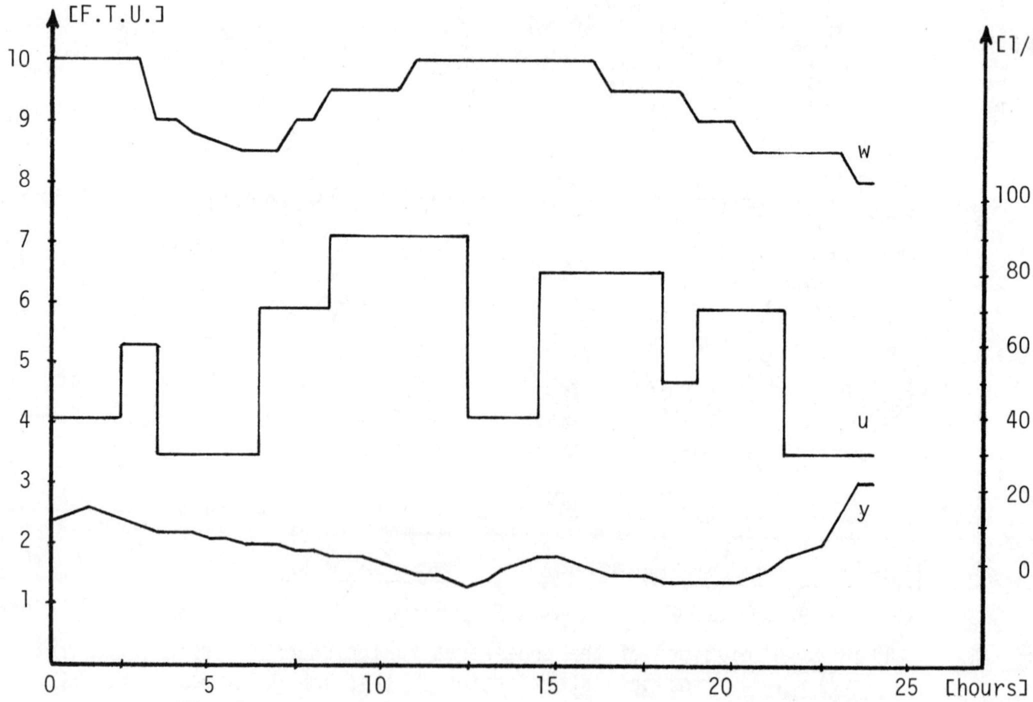

Fig. 4. Time-evolution of the process external variables during an experiment carried out in August 1979.

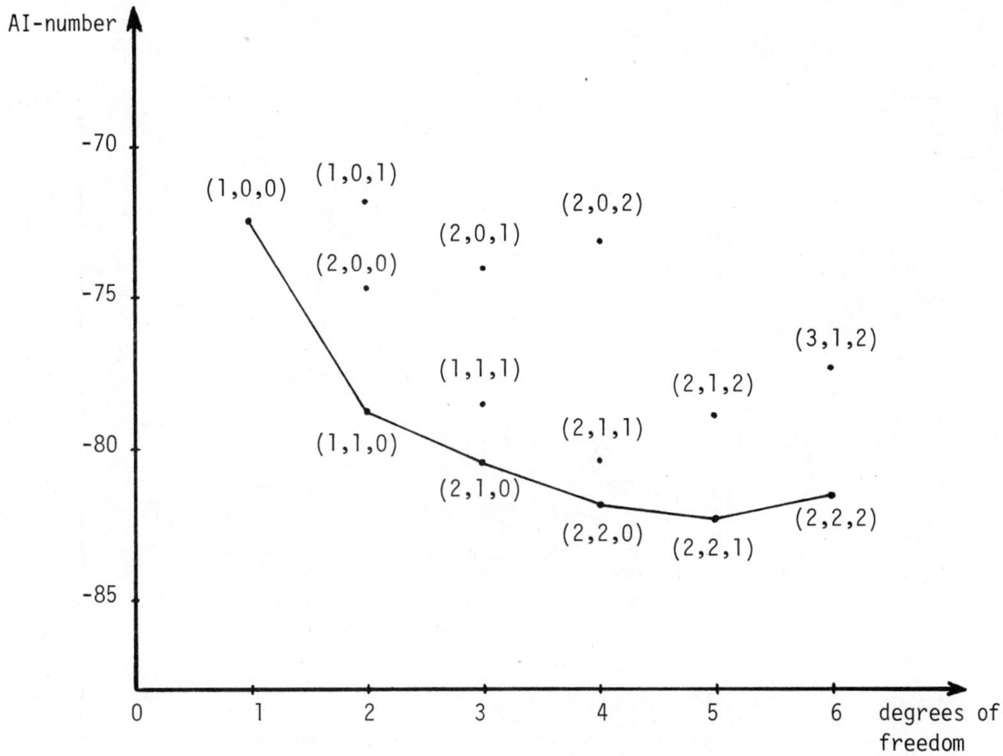

Fig. 5. Behavior of the AI-number as a function of (n,r,s) for the experimental data of Fig. 3.

TABLE 1 Estimates of the free parameters of ARX(2,2,1) based on the experimental data of Fig. 3.

a_1	a_2	w_1	w_2	$b_1 \cdot 10^3$
-.905	.341	.077	-.161	0.92
-.899	.341	.080	-.164	0.88
-.902	.339	.081	-.167	0.91
-.888	.329	.084	-.171	0.95
-.835	.304	.104	-.198	1.25
-.649	.121	.092	-.182	1.52
-.620	.153	.097	-.203	1.76
-.642	.133	.098	-.198	1.46
-.725	.138	.114	-.198	1.02
-.801	.156	.117	-.193	0.86

TABLE 2 Estimates of the free parameters of ARX(2,2,1) based on the experimental data of Fig. 4.

a_1	a_2	w_1	w_2	$b_1 \cdot 10^3$
-1.25	.339	-.081	.045	-.85
-1.29	.368	-.089	.065	-.94
-1.33	.353	-.073	.067	-1.24
-1.34	.362	-.065	.065	-1.03
-1.36	.376	-.064	.066	-.92
-1.40	.428	-.059	.066	-.65
-1.40	.459	-.074	.079	-.32
-1.43	.522	-.066	.080	.10

CASE STUDY FOR AUTOMATIC STATION FOR RESOURCES DEVELOPMENT IN IRAQ

S. Hussain and S. A. Haddad

Electronic Research and Development Centre, Electrical Engineering Department, University of Technology, Baghdad, Iraq

Abstract. The paper outlines the most important results of a case study that involved an automatic observation system implemented in Iraq by the University of Technology. The study has primary application in water resources devlopment, meteorology and agriculture. Varous parts of the system are described , together with online data processing and data transmission schemes. A critical evaluation of automatically observed data with existing (manual) data is presented. Alternative power sources have been used and their performance is outlined. The prototype has been successful. It will provide the basics for a national network of automatic observation stations.

Keywords. Automatic testing; computer applications; data processing; minicomputer; snesors; signal processing.

INTRODUCTION

This case study was undertaken by the Electronic Research and Development Centre at the University of Technology as a pilot study to preceed the installation of automatic observation station (AOS) througout IRAQ. The data acquisition and control scheme has primary application to:

1. Flood prediction and warning by monitoring continuously river guages, rain fall guages and flood roots to enable early warning to affected areas and stop floods from damaging crops.
2. Bettr allocation of water resources through understanding of real states of water resources.
3. Better meteorogical data leading to better forcasting for agriculture and aviation.
4. Better management for sahara development projects.
5. Better use of another national resource which is the voice quality telephone network.

Figure 1 shows in block form the automatic observation station while Fig. 2 shows a block diagram for the pilot system that was used in the present case study. It consists of an autoamtic observation station that has sensors for air temperature, soil temperature, rain fall, water flow, water depth, dew point,wind speed, wind direction, humidity, pressure and visibility. The automatic observation station is linked to a central station via a voice-channel quality telephone line. The central station was locally interfaced to a minicomputer. The data was also sent from the central station to a remote station with a television display. The telephone line was leased from the Public Telegraph and Telephone Establishment. Work started on the pilot scheme in January 1978 and is still being carried out. The results of the pilot scheme are encouraging. Local know-how has been developed and some valuable data has been obtained regarding alternative sources.

THE CASE STUDY

The case study included the design installation, evaluation and development of a prototype automatic observation system. The system is shown in Fig. 2 in block form. As part of system development the central station was linked to a minicomputer to facilitate additional processing capabilities on-line and off-line. Details of the various parts of the system are given below.

The Automatic Observation Station

Figure 1 gives a block diagram for the automatic observation station (Dansk, 1978). It has thirteen sensors to measure the parameters shown in Table 1. This Table also gives the ranges and sensitivities of sensors in addition to the first eleven sensors, a snow sensor is to be added to give better flood management. A dust sensor is being developed locally, since the problem of dust has its manifestations in a variety of fields from communication to aviation and from failure of electronic equipment to solar energy sources.

TABLE 1 Sensors Specifications

Sensor	Range	Sensitivity
1.Air-Temperature	-40 to 60 C	± 0.1 C
2.Soil-Temp.	-40 to 60 C	± 0.1 C
3.Relative Humidity	0 to 100%	± 1 %
4.Wind-speed	0 to 99 kn	± 1 knot
5.Wind-direction	0 to 360	± 1
6.Pressure	1050mb	±0.1 m.bar
7.Visibility	10 km	-------
8.Solar-Intensity	100 mW/cm^2	-------
9.Water - flow	250 cm/sec	± 1 cm/sec
11.Water-depth	500 cm	± 1 cm
11.Snow-Sensor	----	--------
12.Dew-point	---	--------
13.Dust-Sensor	---	--------

The sensor outputs pass through adaptors to A/D converter before being sent on a telephone line via a modem to the central station. Frequency shift keying is used. The sensor outputs to levels acceptable to the analogue multiplexer and to analogue to digital conversion. The analogue multiplexer takes the output of one sensor at a time and feeds it to the A/D converter. The analogue sensor signal is converted into twelve bit digital format. The microcomputer in the automatic observation station performs the handling of the digital signals and controls the communication process and all other operations. Scanning of the sensor outputs is achieved through a program stored in the microcomputer memory. The microcomputer also calculates the values of the dew point from the readings of air temperature and relative humidity. The calculated value of the due point is then sent to the central station along with values of other parameters. The use of microprocessing techniques gives flexibility of operation.

The automatic observation station was positioned in the same place as the manual-read sensors used by the meteoregical office. The two methods could then be directly compared. The automatic observation station housing was double shielded from the sun using stainless steel plating with louvres for ventilation. However, it was discovered that during the hot summer months, the delicate microelectronic circuits inside the AOS were affected by the heat. Therefore a third shield which consisted of a wooden hut to cover the AOS housing was necessary. It seems also necessary to make the station air-tight to prevent dust from affecting the operation of the station. Differential expansion problems in the bolts and plates of the sunshield lead to its deformation.

Alternative Power Supplies

In order to make the case study more comperhensive, it was decided in the design stage, to power the AOS from three different alternative sources with a back up battery as shown in Fig. 3. The three sources used were: (1) Mains power; (2) Wind power; (3) Solar power. The four back up 12 V batteries could keep the OAS operating for three days.

The power supply to AOS is 26 V d.c. 70W, and was derived from one of the three sources mentioned above. When mains power is used, the a.c. is rectified and regulated to 26 V d.c. and used to charge up the battery that feeds the OAS. Switching circuits made possible any desired combination of power sources. For example, wind-power and solar power used at the same time or only wind power or only solar power. Confidence gradually developed in the solar generator as a good competitor with mains power. In the initial stages of the case study we came to the conclusion that solar power is a better alternative than wind power. This is because the wind generator developed some mechanical trouble, because of the nature of wind generation.

The wind generator used was of the vertical type, i.e., its blades rotated in a vertical plane. It is thought that horizontal type blades might have been a better choice. The generator was three phase whose output was rectified and regulated at 26 V d.c. The maximum rated output of the generator is about 120 watts. The generator cuts out when the blades rotate below 140 rpm. The blades could reach speeds of up to 360 rpm (Electromat,1978).

The solar power source was a photovoltaic panel of silicon solar cells. The panel consisted of twelve solar cells and was inclined at the 45° to give optimum performance throughout the year. The automatic observation station was supplied from the solar generator from May through December 1979. Both of the wind generator and mains power were disconnected. Thus only solar power with the back up batteries were used for that period. The AOS operated satisfactorily under those conditions. The charging current was monitored automatically by using one of the spare sensor cards at the AOS. The value of the charging current was sent along with other AOS parameters to the central station as shown in Fig.4. This provided a convenient way of monitoring the charging current from the central station. What is more it enabled us to correlate the charging current which is proportional to solar generated power with solar radiation. A typical automatically obtained graph of solar generated power and solar radiation is given in Fig. 10. The highest charging current observed was 3.5 A giving a peak power from the panel of about 100 W. This corresponded to solar radiation of 100 mW/cm^2.

Data have been accumulated to enable correlation between sun intensity and solar generated power over a longer period. Some recent work (Barker, 1979) has been reported on solar generator evaluation.

During the winter months and especially when successive days were cloudy, the AOS operation was intermittent. There was not enough power to operate the AOS at night. It turned out that the back-up batteries had deterioted in less than one year of operation. In order to enable the AOS to operate continuously and to ensure the flow of valuable data from the AOS to the central station, it was decided to switch out the solar generator and switch in mains power. The AOS was operating without back up batteries and with mains power only. New batteries were ordered and solar power could be investigated further in the winter of 1980.

Data Transmission

In order to make the case study more realistic it was decided to transmit the data on ordinary voice-quality telephone line, as shown in Fig. 2. The various parts of the transmission channel are shown in Fig. 5 between the automatic observation station and the central station. The channel was a telephone line leased from the public Telephone and Telegraph Establishment. The AOS was about twenty kilometers from the central station. It can be seen from Fig. 5 that for part of that distance, there is a microwave channel.

The 12 bit microcomputer in the AOS organises the data and has it ready for transmission through the modem to the central station (Fig. 1). The microcomputer scans all the sensors in a time multiplex format. It gives commands for the analogue to digital conversion to take place. The microcomputer also calculates the dew point and then sends its value along with other parameters to the central station. The microcomputer scans the output of the wind speed and wind direction sensors every five seconds while it scans the output of other sensors every twenty seconds as shown in Table 2. This is to ensure better tracking of wind parameter changes. The output of all sensors is scanned in twenty seconds. The scanning is done in four cycles as indicated in Table 2. The duration of each cycle is five seconds. Data is transmitted via the modem in binary coded decimal format to the central station. Opto-couplers are used at the end of the line for isolation. Frequency shift keying techniques are used for the transmission of data.

There are two modes of communication between the automatic observation station and the central station. The first mode is local in which the AOS sends data continuously to the central station. Because of the extra importance of wind data, its values are transmitted every five seconds, where as values of other parameters are sent every twenty seconds as shown in Table 2.

To our knowledge this case study represents the first attempt in Iraq to transmit digital data over a voice-quality telephone line. The following tests were made on the telephone line before using it as the transmission

channel between the central station and the AOS.
1. Attenuation on the line was measured by using two pairs connected in a loop as shown in Fig. 6, over a distance of six kilometres. Maximum attenuation (20 db) occurred in a frqency of 3 kHz, over the lenght of the line.
2. A continuous sine-wave signal of 800 mV at 1 kHz was sent from the location of the AOS to the central station. The received signal was 200 mV. This gives an attenuation of 14 dB over the whole line (20 km). During the case study, the voice - channel performance was satisfactory.

The Central Station

As shown in Fig. 7, the central station consists of a modem , data receiver, data formatter, visual display unit, digital to analog converter, and chart recorders. The data from the autoamtic observation station is displayed continuously on the VDU screen, and recorded on the charts of the pen recorders. As a local development to the system, the raw data was interfaced to a minicomputer for online data processing and for storing the data on magnetic cassettes or any other form of storage. The central station was powered from the mains only. It is thought that in a realistic situation the control centre is bound to have mains power.

To make the case study a true picture of the complete national system, a remote station was installed twenty kilometres from the central station (see Fig. 7). The remote station receives the values of the parameters from the automatic observation station via the central station. At the remote station data is displayed on a visual display unit.

Interfacing the Central Station

The automatic observation station and the central station were ordered to our specifications. All the sensors were also commercial except the dust sensor which is being developed locally. The dust sensor will help in the studies of the effect of dust on perfoemance of communications channels as well as monitoring sandstorms for example. Some local interfaces have also been developed to increase the processing power of central station. The central station was linked on-line to a minicomputer via the interface shown in Fig. 8. This simple interface has as its main element a universal asynchronous receiver transmitter (UART). The interface was used to transfer digital data from central station on-line to the minicomputer at a rate of 1200 baud. Data was thus entered on-line to the minicomputer. Once the monitored data from the AOS is inside the minicomputer any desired processing could be easily performed online and the processed output could then be displayed on VDU or stored on magnetic cassette. In addition raw data could also be more efficiently stored instead of being on charts. Data could be given as print out from the minicomputer if desired.

We are currently investigating two-way communications between the minicomputer, the central station and the AOS. This will be useful when remote operation is chosen. It will also be useful for setting alarm conditions and sounding warning signals when any of the parameters reaches a predetermined danger level.

The interfacing of the minicomputer gave additional power, and made the case study approach a situation where the central station in the prototype represented the national control centre.

Additional Data Processing

The interface described above, which was designed and implemented locally gave extra on-line processing possibilities. Through this interface, data from the automatic observation station could be fed into the minicomputer and stored on magnetic cassettes or on other desired storage. Once data was inside the minicomputer a number of processing function could be easily performed. For example, programs were written to output as a printout the daily minima and maxima of some parameters like air temperature and sunshine intensity. A program was also written to obtain the total daily energy generated by the solar panel. Yet another example is the working out of daily averages, and monthly averages of some parameters. The computer was also used to correlate solar generated power with monitored sun intensity.

The manually observed data from the meteorological office notebooks were also transferred inside the computer via punch cards. A comprehensive evaluation of automatically obtained and manually obtained data was then undertaken. The main conclusions of the computer based comparison are given below.

1. The automatic and manual readings were quite close. However the automatic readings are more reliable. The manual readings through their reliance on the human observer are bound to have discrepancies. The human observer could not be rlied upon all the time especially to time the interval between readings or take the reading at a certain time. Tiredness decreases es accuracy, resolution and increases error in calculating some values.

2.The wind speed reading showed noti ceable difference between manual and automatic values. This is because of the different heights of the wind speed sensors. The automatic observation station sensor was ten metres high which is the standard height. The old (manual) sensor was six metre high. However, the difference in height is to be used to advantage to evaluate wind shear. Figure 9 gives a typical computer plot for manual and automatic wind speed variation with time.

3. The differences are small between automatic and manual readings of pressure, air temperature, soil temperature, visibility and wind direction. The automatic readings will be taken as standard measurements by the meteorological office instead of manual readings. Figure 9 illustrates some comparative typical computer plots for automatic and manual values for dew point, pressure, wind speed, and wind direction.

4. The difference in dew point plots (Fig. 9) is due to the better accuracy of calculation of the automatic method

COMMENTS AND CONCLUSIONS

1. This paper described an automatic observation system implemented in Iraq as a prototype for a national network of automatic observation stations.

2. The system has the following primary applications:
(a) Flood prediction and warning; by monitoring continuously river guages, rainfall and flood roots to enable early warning to affected areas and prevent damage to crops.
(b) Better allocation of water resources; by giving the real status of dams reservoirs.
(c) Better weather forecasting; the use of automatic station improved reliability of meteorological data, with benefits to agriculture, aviation and sahara development schemes.

3. The case study investigated the better use of another national resources; namely, the existing network of the telephone lines. The use of voice quality telephone lines for digital data transmission in Iraq has been achieved for the first time during the case study as far as is known.

4. The use of solar power and wind power has been prctically assessed, and it is thought that the solar power is more suitable in Iraq.

5. Automatic and manually obtained values have been compared using a minicomputer that has been interfaced to the system. Computer plots for the comparisons have been made available for air temperature, soil temperature wind direction, wind speed, rain fall visibility and dew point.

6. Confidence has been developed in the automatic method such that the Meteorological Office is to adopt some of the automatic readings for its official bulletins.

7. The difference in wind speed readings between manual and automatic methods has been found to be due to height difference of sensors. It is proposed to make use of this for wind shear calculations.

8. Local know-how has been developed for automatic data processing and for the use of minicomputers on-line and off-line in real-time application.

9. Some interfacing work has been done locally to give additional on-line processing capabilities.

10. Heat and dust require more rigid specifications for the integrated circuits and other electronic components that is commercially available.

11.Heat presented problems for mechanical parts as well as differential expansion problems.

12. The case study represented a successful operation beacause of the cooperation of the various end-users and beneficieries of the scheme on the one hand and the designers and experimenters on the other.

REFERENCES

Dansk, I. P. (1978). D.I.P. User's Manual of automatic observation station.

Elektromat (1978). Windpumpen Elektromat Wingenerator user's

manual.
Barker, A. S. (1979). Photovoltaic solar cell array for supplemented power generation. Solar Energy, 23, pp. 427-434.

TABLE 2 Formatting of Binary Coded Decimal Data

First Cycle	Second Cycle	Third Cycle	Fourth Cycle
Clock	Clock	Clock	Clock
Wind Direction	Wind Direction	Wind Direction	Wind Direction
Wind Speed	Wind Speed	Wind Speed	Wind Speed
Air Temperature	Dew Point	Water Flow	Snow
Soil Temperature	Precipitation	Water Level	Spare
Pressure	Sun Radiation	Visibility	Spare

ACKNOWLEDGEMENT

This work was supported by the Ministry of Communications, Ministry of Agriculture and Irrigation Reform, Ministry of Irrigation, and the Sahara Development Establishment.

The authors wish to express their gratitude for the President of The University of Technology Dr. Taha Al-Nuaimi for his continuous support and encouragement throughout the work. Thanks are also due to the Director of the Iraqi Meteorological Office Dr Assad Akrawe from the Meteorological Office, Mr. Shahab Al-Safi, Chief Engineer at the Ministry of Irrigation, Mr. Hazim Wodee, Chief Engineer at the Ministry of Communications.

The size of the project meant that many individuals devoted lots of effort to make it successful, in particular, thanks are due to to Mr. Menal Al-Kindy, Miss Inam Mahmood and Miss Wafaa Al-Shumran.

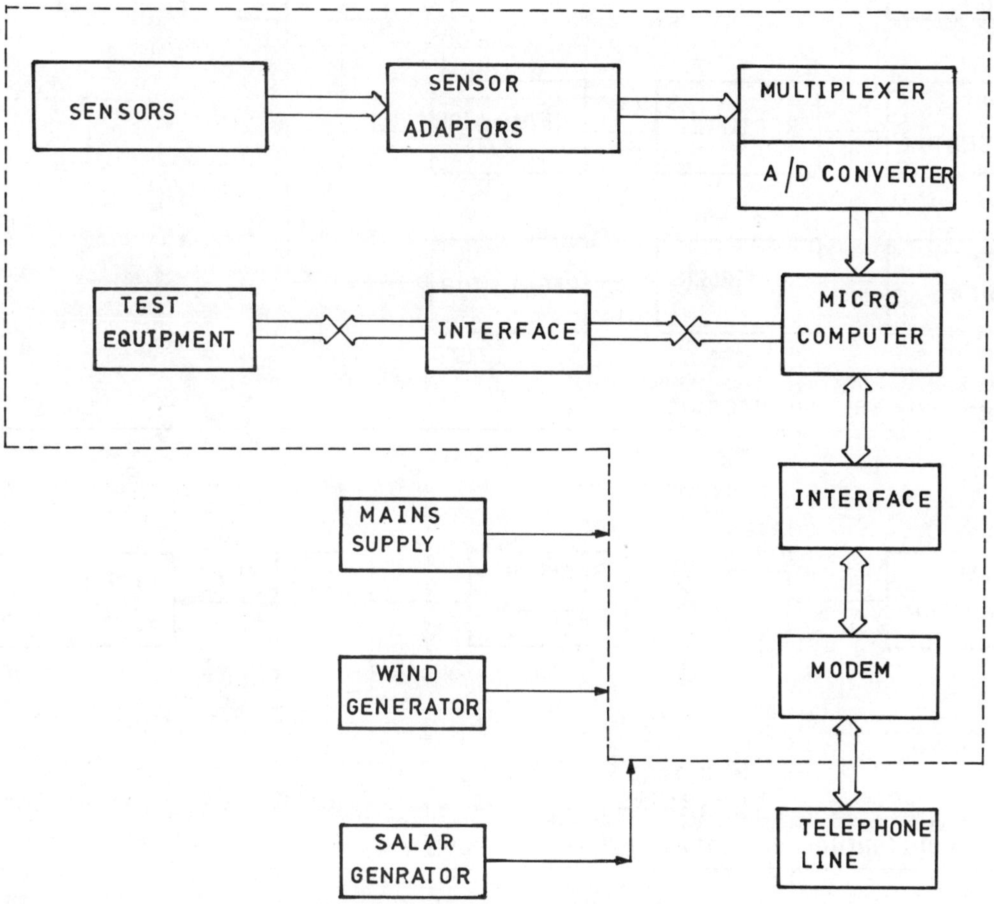

FIG. 1- SENSOR STATION.

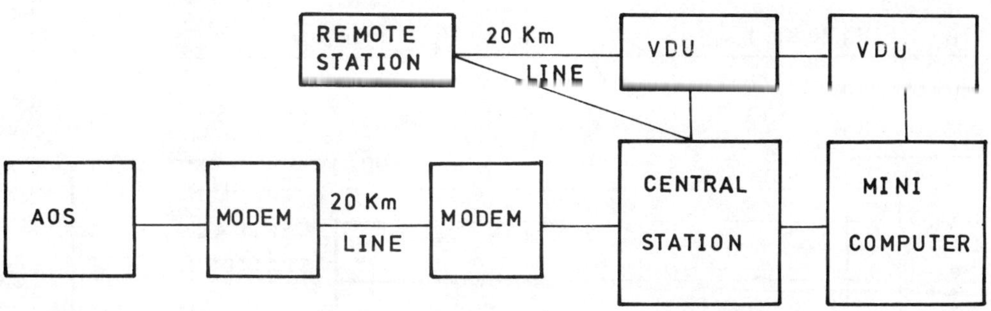

FIG. 2- PILOT SCHEME.

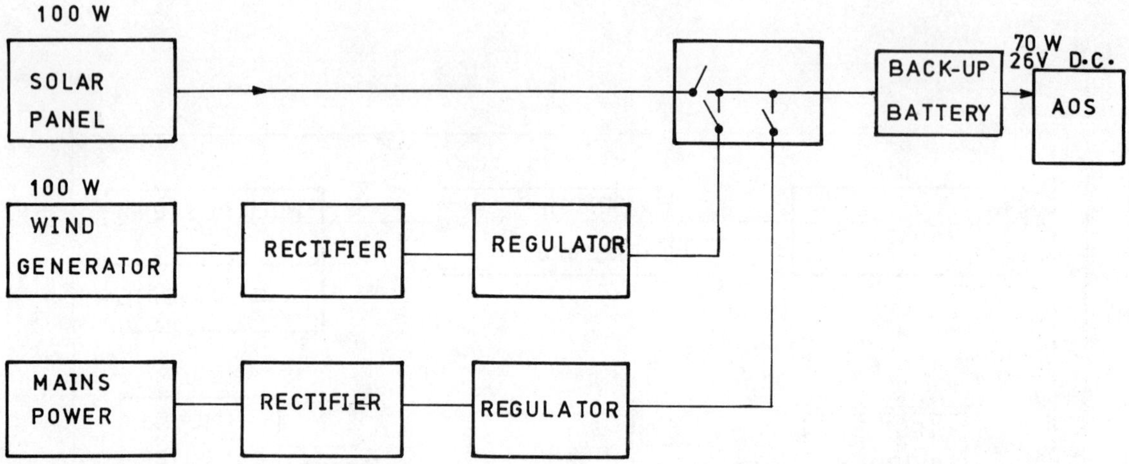

FIG. 3- POWER SOURCES.

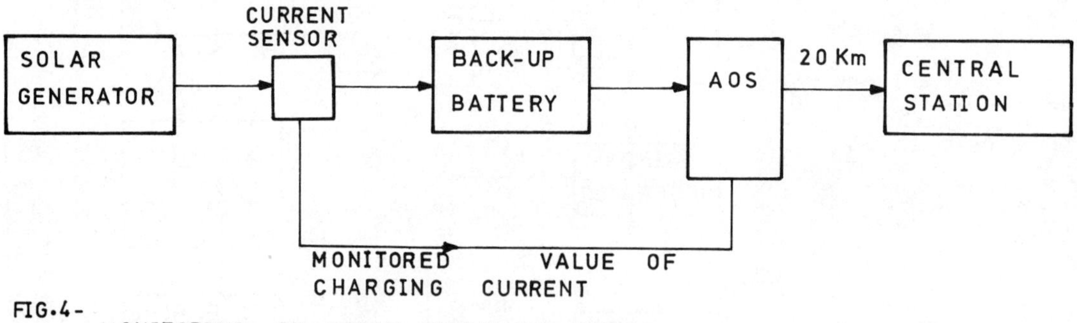

FIG. 4- MONITORING OF SOLAR GENERATOR POWER.

FIG. 5- DATA PATH (20 Km).

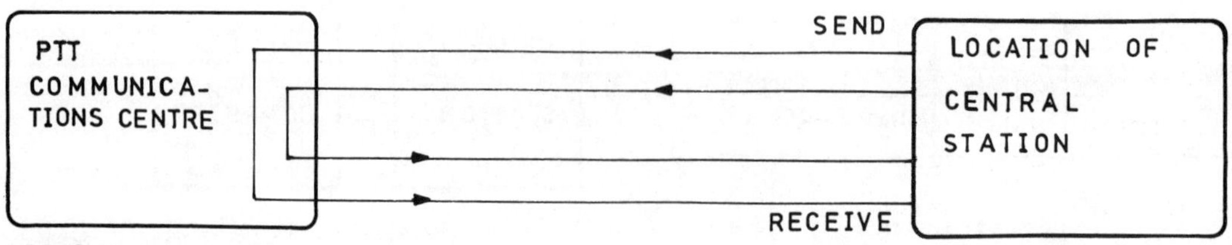

FIG. 6- LOOP TEST ON LINE (6 Km).

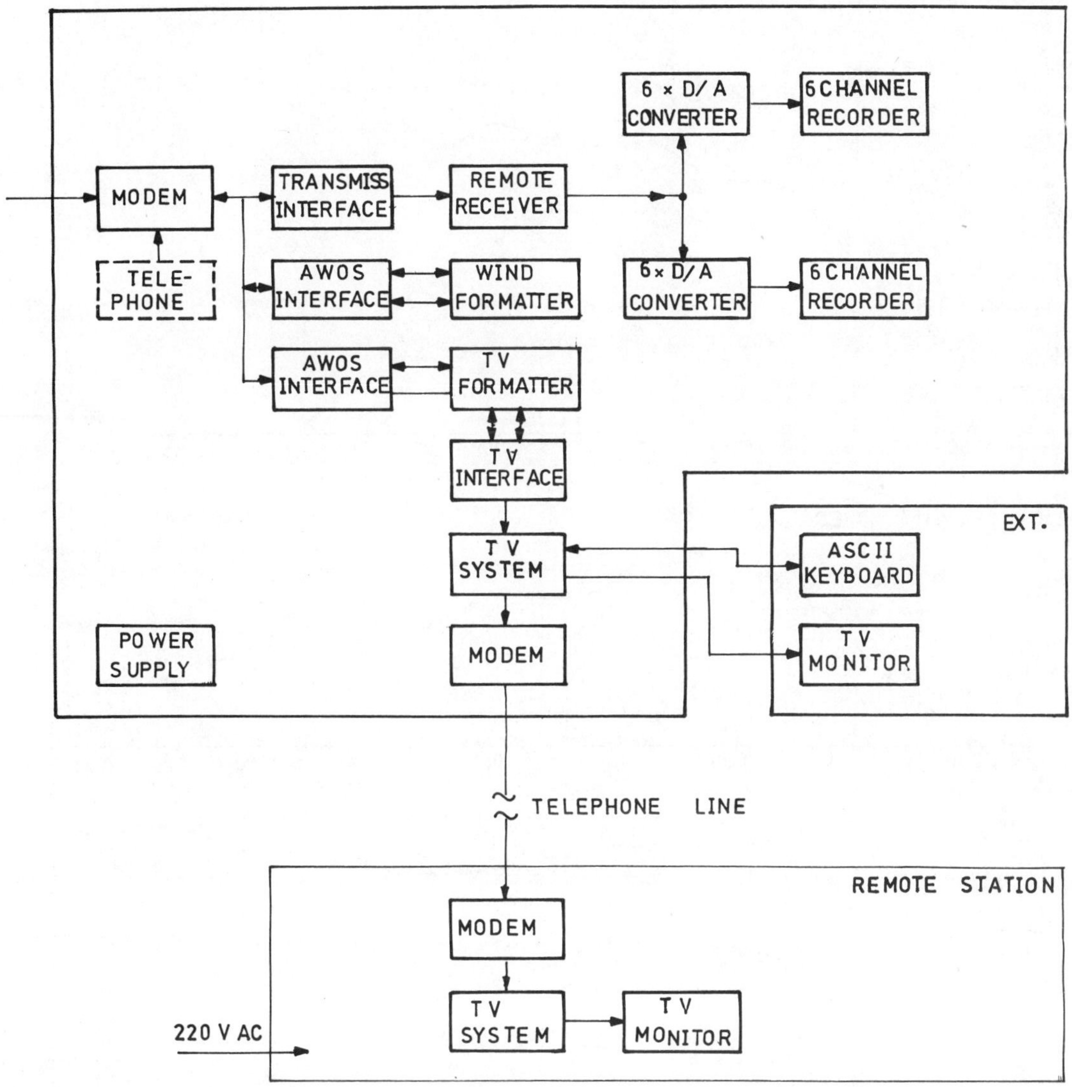

FIG. 6- CENTRAL STATION.

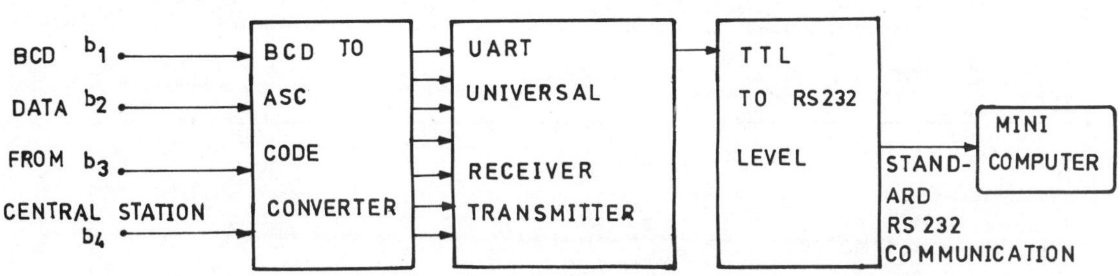

FIG. 8- INTERFACE BETWEEN MINICOMPUTER AND CENTRAL STATION.

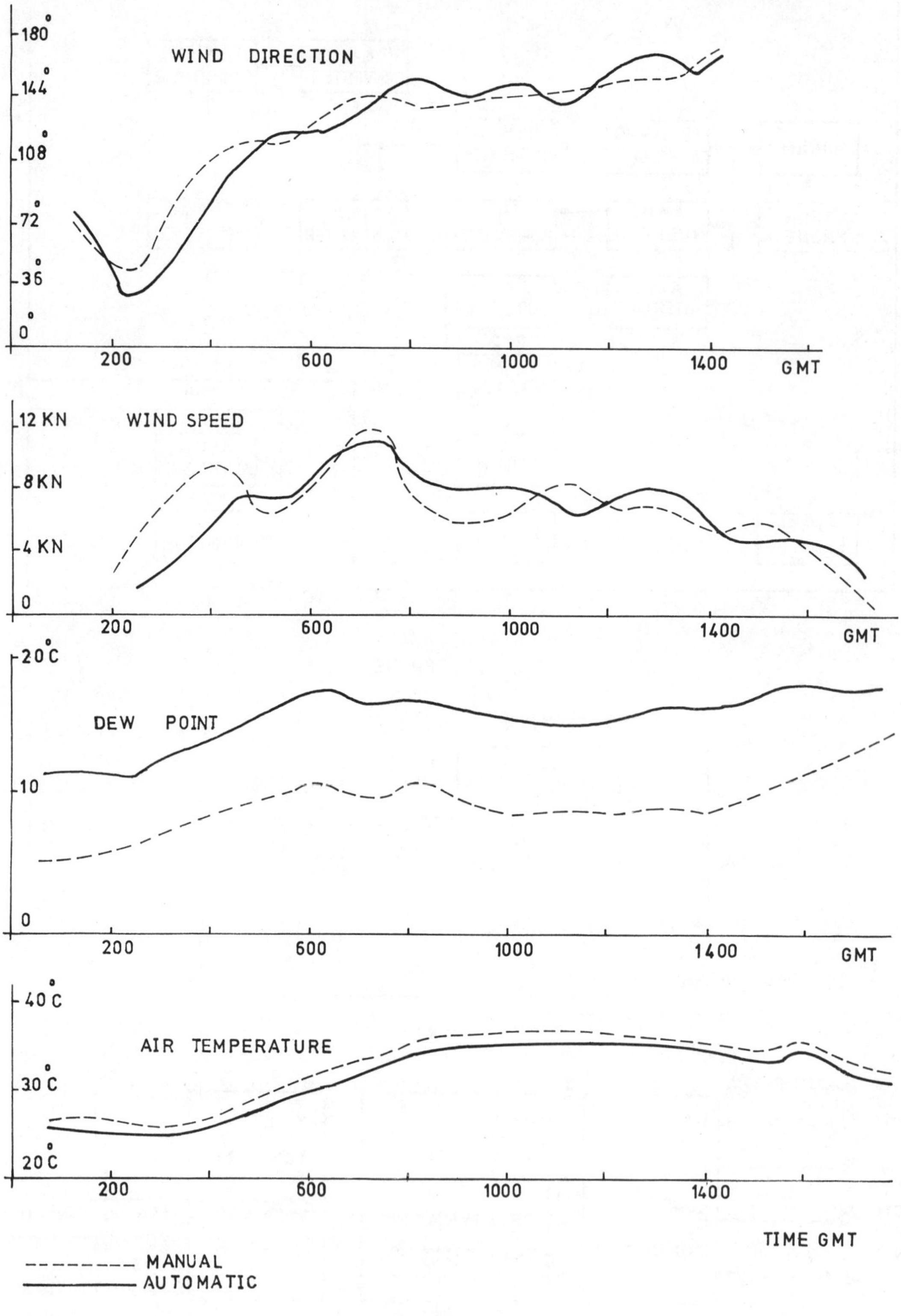

FIG. 9- COMPUTER PLOTS.

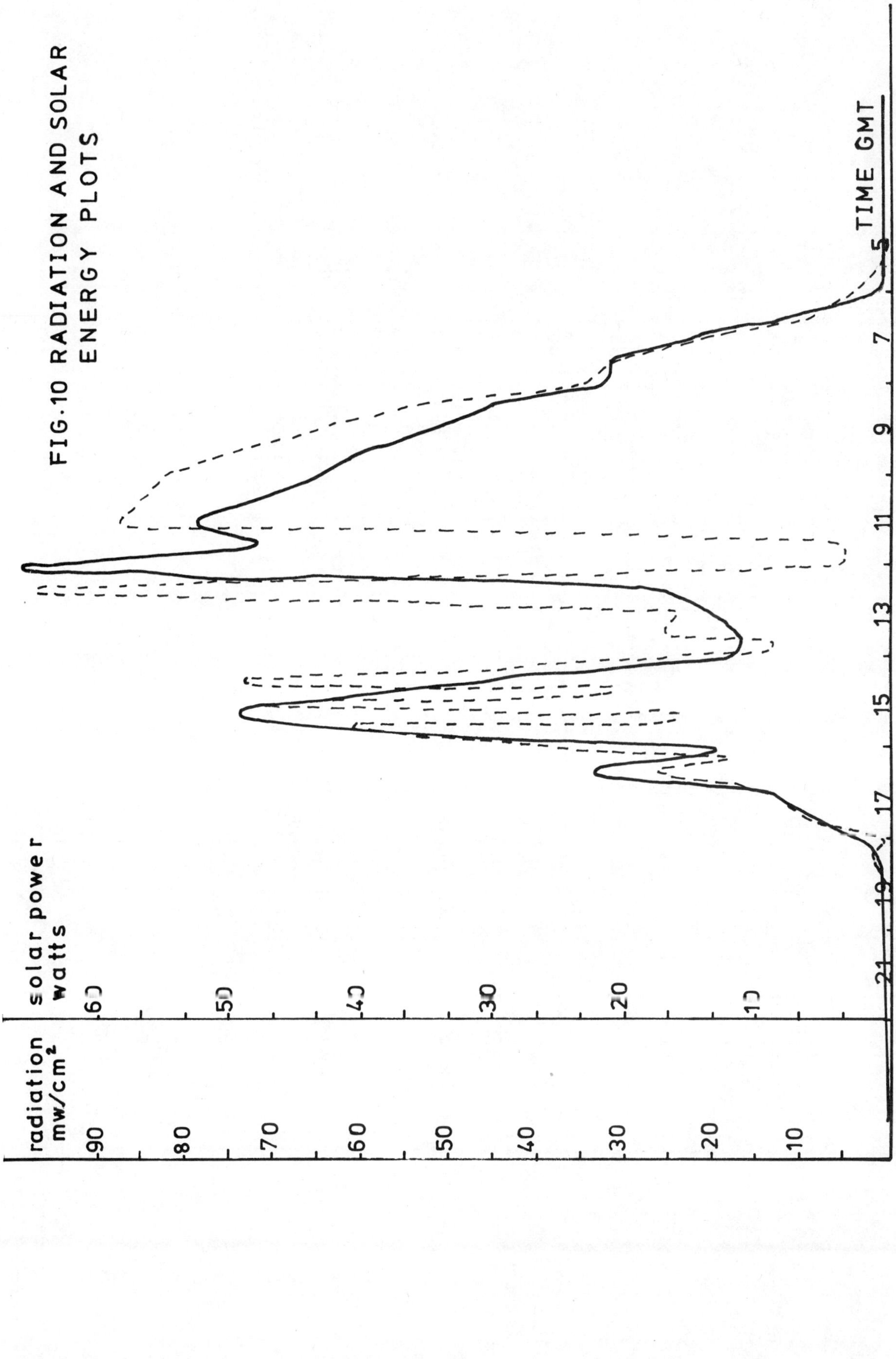

FIG.10 RADIATION AND SOLAR ENERGY PLOTS

A NEW APPROACH TO THE SIMPLIFICATION OF WATER-SUPPLY NETWORK MODELS

S. Strmčnik*, D. Matko, F. Bremšak** and J. Černetič***

**Josef Stefan Institute, Ljubljana, Yugoslavia*
***Faculty for Electrical Engineering, Ljubljana, Yugoslavia*

Abstract. The aim of the paper is to present a procedure by which a simplified linear dynamical model in the discrete or continuous form can be derived, starting from an originally complex physical model. The procedure is based on an algorithm for the identification of continuous multivariable systems from sampled data which is applied on the input-output data resulting from the simulation of the water-supply system on a digital computer. In the paper some mathematical methods are described, required for the realisation of the following procedure:

1. dynamical simulation of the water-supply system
2. recording of system outputs (reservoir levels), inputs (pump controls) and disturbances (consumer demands)
3. identification of the "unknown" system concerned, based on a discrete linear multivariable canonical model, applying the least squares method on the recorded data
4. transformation of the discrete input-output model into the statespace model
5. transformation of the discrete model into the continuous one, using the concept of the matrix logarithm function

Following the described procedure, the water-supply system of a small town near Ljubljana has been approximated by a linear time-invariant multivariable model which is very suitable for further development of optimal control algorithms. Also a comparison between the output responses of the original and simplified model is given.

Keywords. Water-supply systems, modelling, simulation, identification, simplification.

INTRODUCTION

In the last decade there is a groving trend in the water-supply technology towards the introduction of computer control because the recent supply systems are very complex and their control is very difficult. Usually such a computer system is composed of a central computer, linked by radio or phone line to all major parts of the system. These control systems are mostly designed to collect data and monitor the major variables which, together with an experienced dispatcher, results in a relatively good control. But, the mentioned hardware equipment enables also a much more efficient approach to the control where also the economical aspects of the problem can be considered. Since one of the major operational costs in water-supply is that of pumping, the introduction of optimal, minimum pumping cost control algorithms is reasonable (Coulbeck, Sterling, 1978; Fallside, Perry, 1975).

Certainly one of principal problems in the application of such control strategies is the complexity of water-supply systems. Usually they are described by large and complicated models, based on the topology of the pipeline network and on physical relations between pressures and flowrates. These models are used principally for the design and analysis (Korte, Vielhaber, 1967; Shamir, Howard, 1968). They can not be applied directly for the solving of optimal control problems because the solution procedures are very complicated and time consuming, especially if dynamic solutions are required (Coulbeck, 1979). To solve this difficulty, simplification procedures must be introduced which enables calculation of a reduced and simple mathematical model of the water-supply system.

The most commonly used method of simplification is that by physical reasoning. By experience and by considering the physical situation, certain system parts of minor importance may be approximated by simpler expressions or completely ignored (Fallside, Perry, 1975; Giadrossi and others, 1979). Unfortunatelly, this approach only seldom gives satisfactory and reliable results, especially if the considered mathematical models are very complex.

A different approach is shown by Coulbeck (1979). He uses a linearisation procedure about instantaneus operating points and applies it on the dynamical part of the original model. The simplication procedure results in a linear discrete time varying model. Unfortunatelly the time varying coefficients cannot be calculated explicitly, since they depend also on the static part of the system. Therefore a special method for their calculation has been developed.

In this paper we propose a simplification and linearization method which is based on an algo rithm for the identification of continuous multivariable systems from sampled data (Strmčnik, Bremšak, 1979). This algorithm is applied on the input-output data resulting from the dynamical simulation of the water-supply system on a digital computer.

In the first part of the paper the theoretical approach to the modelling of water-supply systems is considered and a short description of the developed computer simulation program is given.

In the second part an identification method is given which is based on a canonical model of the multivariable system in discrete space and suitable transformations from the discrete state-space to the continuous state-space.

Finally an application of the proposed procedure for simplification of a concrete water distribution system is shown.

SIMULATION OF WATER-SUPPLY SYSTEMS

The Functional Mathematical Model

Mathematical modelling of water distribution networks is a clasical tool for the design of new systems and the analysis of old ones. A network of pipes, valves, pumps, reservoirs and other elements is considered solved when the heads in all nodes and flows in all pipes are known. This is then a static solution of the problem.

The development of such a mathematical model is based on known physical laws on preservation of mass and energy, and specific relations between pressures and flows for each element in the network. The most important are, of course, the relations for pipes. The pressure drop on a pipe Δh can be expressed by means of the flow q with Eq.1

$$\Delta h = r.q|q|^{\beta} \tag{1}$$

with β lying in the interval (0,75;1). r is the hydraulic resistance of the pipe and can be calculated using many different approaches (see e.g. Korte, Vielhaber, 1967). One posibility is to use Darcy´s equation and the flow dependent friction factor according to the equation of Prandtl-Colebrook. This approach results in a flow dependent hydraulic resistance which causes some numerical difficulties. To avoid this situation the hydraulic resistance is often approximated by an average resistance. One way to do this is the introduction of Manning´s equation from which the following equation for the resistance of the pipe can be derived

$$r = \frac{10,293\ k^2 l}{d^{5,33}} \tag{2}$$

where l is pipe lenght, d the diameter of the pipe and k is the roughnes factor.

There are also other approaches, e.g. the Hazen-Williams equation (Coulbeck, 1979; Shamir, Howard, 1968).

Other elements in the distribution system are usually defined by characteristic curves or their approximations.

Considering the network topology and relations between flows and pressures for the particular elements, a set of generalised network equations can be derived in the following form.

$$\underline{f}(\underline{x},\underline{h},\underline{u},\underline{v})+\underline{q}+\underline{y} = 0 \tag{3}$$

with the symbols meaning

- $\underline{f}$ - vector function, corresponding to network topology
- $\underline{x}$ - storage node state vector, corresponding to reservoir levels
- $\underline{h}$ - non-storage node state vector, corresponding to node pressure heads
- $\underline{u}$ - pump control vector defined by pump flow control parameter
- $\underline{v}$ - valve control defined by variable resistance parameter
- $\underline{q}$ - storage flow vector, corresponding to reservoir outflows
- $\underline{y}$ - disturbance flow vector, corresponding to distributed consumer demands

The static solution of the problem for the unknown variables $\underline{h}$ and $\underline{q}$ can be obtained by Eq.3.
To get deeper insight into the system, especially when new reservoirs or pumping stations are designed, or when a control strategy is developed, a dynamic model of the water-supply system is required. It can be obtained with the introduction of dynamic relationship for reservoirs which can be expressed as

$$\dot{\underline{x}}(t) = \underline{\underline{F}}\underline{q}(t) \tag{4}$$

where $\underline{\underline{F}}$ is a diagonal matrix with elements depending on reservoir geometry. Considering Eqs. (3) and (4), a dynamic solution of the problem can be obtained.

The Simulation Program

The dynamic solution of the network involves calculation of all heads and flows for each time increment. Some integration scheme is used for the calculation of dynamic behavior of reservoir levels. The currently computed levels are then used as inputs for the static solution of the system in the next time interval.

The most commonly used methods for the computation of static solution are the Cross´ method (Korte, Vielhaber, 1968) and the Newton-Raphson method (Shamir, Howard, 1967). There exist many simulation packages on digital computers which are based on the mentioned approaches.

A similar computer simulation package was developed on the "Department for Analog-Hybrid Computation and Automatic Control" at the Faculty for Electrical Engineering in Ljubljana. It enables dynamic simulation of complex water distribution systems with different non linear elements.

By the first part of the program a system of independent nonlinear equations is derived, based on the network structure data. Then the obtained system of equations is solved used a modified Cross algorithm. Using a simple integration formula new levels in reservoirs are calculated following the Eq. (4). The same procedure is repeated until the final time is reached.
A relative large set of nonlinear elements can be considered as e. g. pumps with different characteristic curves, valves and reservoirs with different input and output flow pipes. The pressure drop in pipes is computed by the simplified Manning equation, whereas for most other elements approximations of their characteristic curves are used.

The described computer simulation package is intended principally for the analysis and design of water-supply systems. It is implemented on computers IBM 1130, PDP11/34 and CYBER 72.

IDENTIFICATION OF MULTIVARIABLE SYSTEMS

Canonical Structures

The identification method considered in this paper is based on canonical structures of multivariable systems.

For a succesful identification of a multivariable system with unknown structure a unique parameterization is of great importance (Denham, 1974). But, unlike the single-variable case, the canonical forms for multivariable systems are not unique. This problem has been considered by different authors (e.g. Denham, 1974; Luenberger, 1967; Popov, 1972). As a result, a complete set of independent invariants has been determined on which canonical forms for multivariable systems are based.

Consider a discrete observable multivariable system represented by the equations

$$\begin{aligned} \underline{\tilde{x}}(k+1) &= \underline{\underline{\tilde{F}}}\underline{\tilde{x}}(k)+\underline{\underline{\tilde{G}}}\underline{u}(k) \\ \underline{y}(k) &= \underline{\underline{\tilde{H}}}\underline{\tilde{x}}(k) \end{aligned} \tag{5}$$

with
$\underline{\tilde{x}}$ - state vektor (n)
$\underline{u}$ - input vector (r)
$\underline{y}$ - output vector (p)
$\underline{\underline{\tilde{F}}}$ - system matrix (n X n)
$\underline{\underline{\tilde{G}}}$ - input matrix (n X r)
$\underline{\underline{\tilde{H}}}$ - output matrix (p X n)

and the matrix $\underline{\underline{\tilde{H}}}$ of the form

$$\underline{\underline{\tilde{H}}} = \begin{bmatrix} \underline{\tilde{h}}_1^T \\ \underline{\tilde{h}}_2^T \\ \vdots \\ \underline{\tilde{h}}_p^T \end{bmatrix} \tag{6}$$

A canonical representation of this system can be derived, following Guidorzi (1975), by the next procedure. Consider the vector sequences

$$\begin{aligned} &\underline{\tilde{h}}_1, \underline{\underline{\tilde{F}}}^T\underline{\tilde{h}}_1, (\underline{\underline{\tilde{F}}}^T)^2\underline{\tilde{h}}_1, \dots \\ &\vdots \\ &\underline{\tilde{h}}_p, \underline{\underline{\tilde{F}}}^T\underline{\tilde{h}}_p, (\underline{\underline{\tilde{F}}}^T)^2\underline{\tilde{h}}_p, \dots \end{aligned} \tag{7}$$

and select the vectors from the sequences in the following order

$$\underline{\tilde{h}}_1, \underline{\tilde{h}}_2, \dots, \underline{\tilde{h}}_p, \underline{\underline{\tilde{F}}}^T\underline{\tilde{h}}_1, \dots \underline{\underline{\tilde{F}}}^T\underline{\tilde{h}}_p, (\underline{\underline{\tilde{F}}}^T)^2\underline{\tilde{h}}_1, \dots \tag{8}$$

where a vector $(\underline{\underline{F}}^T)^s\,\underline{h}_i$ is selected if and only if it is independent from all previously selected ones. Let n_1, n_2, ..., n_p be the numbers of vectors selected from the first, second, ..., p-th sequence in (7). Because of the complete observability of the system in Eq. (5) this set of integers completely defines the structure of the couple $(\tilde{\underline{\underline{F}}},\tilde{\underline{\underline{H}}})$ and is invariant with respect to a change of coordinates.

Introducing a new state vector

$$\underline{x} = \underline{\underline{T}}\tilde{\underline{x}} \tag{9}$$

where the transformation matrix $\underline{\underline{T}}$ is given in the following form

$$\underline{\underline{T}}^T = \left[\tilde{\underline{h}}_1, \tilde{\underline{\underline{F}}}^T\tilde{\underline{h}}_1, \ldots (\tilde{\underline{\underline{F}}}^T)^{(n_1-1)}\tilde{\underline{h}}_1 \middle| \cdot \middle| \cdot \ldots (\tilde{\underline{\underline{F}}}^T)^{(n_p-1)}\tilde{\underline{h}}_p\right] \tag{10}$$

the next system of equations is obtained

$$\begin{aligned} \underline{x}(k+1) &= \underline{\underline{F}}\underline{x}(k) + \underline{\underline{G}}\underline{u}(k) \\ \underline{y}(k) &= \underline{\underline{H}}\underline{x}(k) \end{aligned} \tag{11}$$

The matrix $\underline{\underline{F}}$ is defined in the following form

$$\underline{\underline{F}} = \underline{\underline{T}}\tilde{\underline{\underline{F}}}\underline{\underline{T}}^{-1} = \left\{\underline{\underline{F}}_{ij}\right\} \quad i,j=1,2,\ldots,p \tag{12}$$

The matrices $\underline{\underline{F}}_{ii}$ are of dimension $(n_i \text{ X } n_i)$ and the matrices $\underline{\underline{F}}_{ij}$ of dimension $(n_i \text{ X } n_j)$. The form of the submatrices can be expressed in the following way

$$\underline{\underline{F}}_{ii} \quad \begin{bmatrix} 0 & & \\ \vdots & I_{n_i-1} & \\ 0 & & \\ f_{ii,1} & \cdots & f_{ii,n_i} \end{bmatrix} \tag{13}$$

$$\underline{\underline{F}}_{ij} \quad \begin{bmatrix} 0 & \cdots\cdots\cdots & 0 \\ \vdots & & \vdots \\ 0 & & 0 \\ f_{ij,1} & \cdots f_{ij,n_{ij}} \cdots & 0 \end{bmatrix} \tag{14}$$

The matrix $\underline{\underline{G}}$ is obtained with the transformation

$$\underline{\underline{G}} = \underline{\underline{T}}\tilde{\underline{\underline{G}}} = \left[\underline{g}_1, \underline{g}_2, \ldots, \underline{g}_s, \ldots, \underline{g}_r\right] \tag{15}$$

$$\underline{g}_s = \left[g_{11,s} \cdot\cdot g_{1n_1,s} \cdots g_{p1,s} \cdot\cdot g_{pn_p,s}\right]^T$$

and the matrix $\underline{\underline{H}}$ is given in the next form

$$\underline{\underline{H}} = \tilde{\underline{\underline{H}}}\underline{\underline{T}}^{-1} = \begin{bmatrix} 1 & 0 \cdots\cdots\cdots\cdots\cdots\cdots\cdots\cdots & & & 0 \\ 0 & .0\ldots 1 \; 0 \cdots\cdots\cdots\cdots\cdots & & & 0 \\ \vdots & & & & \\ 0 & \cdots\cdots\cdots\cdots\cdots\cdots & 0 & 1 & 0..0 \end{bmatrix}$$

$$\begin{matrix} \uparrow & \uparrow & \uparrow \\ 1 & (n_1+1) & (n_1+n_2+..+n_{p-1}+1) \end{matrix} \tag{16}$$

In this way the original system has been decomposed into p interconnected subsystems, where the j-th subsystem is completely observable from the j-th output.

Generally, for the identification procedure, the input-output canonical models are more important, especially if the identification method is based on model realisation from input-output data.

Rissanen (1974) has shown that canonical forms in the state-space and in the input-output space can be equivalently treated. So the previously described state-space canonical model can be transformed into an input-output canonical model.

Consider the input-output model

$$\underline{\underline{P}}(z)\underline{y}(k) = \underline{\underline{Q}}(z)\underline{u}(k) \tag{17}$$

where $\underline{\underline{P}}(z)$ and $\underline{\underline{Q}}(z)$ are polynomial matrices in z, of dimension (p X p) and (p X r), respectively, and z^{-1} is the unitary delay operator. It can be shown (Guidorzi, 1975) that the elements of the matrices $\underline{\underline{P}}(z)$ and $\underline{\underline{Q}}(z)$ can be expressed by the elements of the triple $(\underline{\underline{F}},\underline{\underline{G}},\underline{\underline{H}})$ in the following form

$$\begin{aligned} p_{ii}(z) &= z^{n_i} - f_{ii,n_i} z^{n_i-1} - \ldots \\ &\ldots - f_{ii,2} z - f_{ii,1} \end{aligned} \tag{18}$$

$$\begin{aligned} p_{ij}(z) &= -f_{ij,n_{ij}} z^{n_{ij}-1} - \ldots \\ &\ldots - f_{ij,2} z - f_{ij,1} \end{aligned} \tag{19}$$

$$\begin{aligned} q_{ij}(z) &= \gamma_{in_i,j} z^{n_i-1} + \ldots \\ &\ldots + \gamma_{i2,j} z + \gamma_{i1,j} \end{aligned} \tag{20}$$

where f_{ij} are elements of the matrix $\underline{\underline{F}}$. The coeficients γ are related to the elements of the matrix $\underline{\underline{G}}$ according to the next equation (Strmčnik, Bremšak, 1979)

$$\gamma_{jn_j-i+1,s} = g_{j,i,s} - \sum_{k=1}^{p} \cdot \sum_{t=1}^{n_{jk}-n_j+i-1} f_{jk,n_{jk}-t+1} g_{k,n_{jk}-n_j+i-t,s} \tag{21}$$

for $s=1,2,..r$; $j=1,2,\ldots p$;
$i=1,2,\ldots,n_j$

Structural Identification

The structural identification is based

on the search of a singular input-output data matrix and is a direct consequence of the canonical structure of the system given in Eq. (11). Following the definition of Guidorzi (1975), the structural identification of a multuvariable system is the determination of the set of integers $n_1, n_2 \ldots \ldots n_p$ defining the structure of the couple ($\underline{\underline{F}}, \underline{\underline{H}}$) from the input output sequences without the intermediate construction of a parametric model.

As it has been stated before, the identification method is based on the input-output model given in Eq. (17). The s-th equation in the system (17) taking into account also Eqs. (18), (19) and (20), can be writen in the following form

$$y_s(k+n_s) = \sum_{i=1}^{p} \sum_{j=1}^{n_{si}} f_{si,j} y_i(k+j-1) + \sum_{i=1}^{r} \sum_{j=1}^{n_s} \gamma_{sj,i} u_i(k+j-1) \tag{22}$$

Consider now the matrix of input-output data

$$\underline{\underline{R}} = [\underline{y}_1(k) \underline{y}_1(k+1) \ldots | \ldots | \underline{y}_p(k) \ldots | \ldots | \ldots \underline{u}_1(k) \ldots | \ldots | \underline{u}_r(k) \ldots]$$

$$\underline{y}(k) = [y(k), y(k+1), \ldots y(k+N)]^T \tag{23}$$

$$\underline{u}(k) = [u(k), u(k+1), \ldots u(k+N)]^T$$

where N denotes the number of samples. Eq. (22) shows that the dependence relations among the vectors (23) are the same as those among the vectors (7), taking into account also the inputs. This fact allows the structural identification on the basis of vectors in Eq. (23) if the same selection procedure is used as that described in Eq. (8)

$$\underline{y}_1(k), \ldots \underline{y}_p(k), \underline{u}_1(k) \ldots \underline{u}_r(k), \underline{y}_1(k+1), \ldots \ldots \underline{y}_p(k+1) \ldots \tag{24}$$

The actual structural identification is based on the matrix $\underline{\underline{R}}^T\underline{\underline{R}}$ (with the same rank), and not on the original matrix $\underline{\underline{R}}$, since this would require a large amount of storage. Consider the mentioned matrix

$$\underline{\underline{R}}^T\underline{\underline{R}} = \underline{\underline{S}} = \underline{\underline{S}}(\delta_1, \ldots \delta_i, \ldots \delta_{p+r}) \tag{25}$$

where δ_i is the number of vectors in the i-th submatrix. Now, let us construct a sequence of matrices with increasing dimensions

$$S(2,1, \ldots 1), \quad S(2,2, \ldots, 1), \ldots \tag{26}$$

and select only the nonsingular ones.

When a singular matrix is found one of indices n_i is determined. The procedure is continued, until all p indices are determined. Now the complete structure of the system is defined.

Parametric Identification

The parameters of the multivariable system can be identified using the Eq. (22) for each subsystem separately.

Let us define a vector of parameters

$$\underline{\alpha}_s = [f_{s1,1} \ldots f_{s1,n_{s1}} \ldots f_{sp,1} \ldots f_{sp,n_{sp}} \ldots \ldots \gamma_{s1,1} \ldots \gamma_{sn_s,1} \ldots \gamma_{s1,r} \ldots \gamma_{sn_s,r}] \tag{27}$$

and rewrite the Eq. (24) in the matrix form

$$\underline{y}_s(k+n_s) = \underline{\underline{R}}_s \underline{\alpha}_s \tag{28}$$

where $\underline{\underline{R}}_s$ is the matrix of input-output data which corresponds to the s-th subsystem. The computation of the vectors of parameters $\underline{\alpha}_s$ can be interpreted also as a solution of an overdetermined system of equations. As an exact solution of such a system does not exist, a least squares approach is introduced

$$\| \underline{y}_s(k+n_s) - \underline{\underline{R}}_s \underline{\alpha}_s \| = \min \tag{29}$$

where $\| \;\|$ means the euclidean norm.

The most common procedure for the solution of Eq. (28) is based on the transformation of the system of equations into a normal form from which the following solution is obtained

$$\underline{\alpha}_s = (\underline{\underline{R}}_s^T \underline{\underline{R}})^{-1} \underline{\underline{R}}_s^T \underline{y}_s(k+n_s) \tag{30}$$

known in the control theory as the least squares estimator.

There are two problems in connection with the identification procedure mentioned. First, the normal system of equations is usually ill-conditioned (Golub, 1965). Therefore different approaches as e.g. scaling of the equations (Bauer, 1963), or the procedure with ortogonal transformations (Businger, Golub, 1965) must be used.

Second, the parameters resulting from a least squares estimator are biased if the input-output data are corrupted by noise. This problem has been treated by many authors (see e.g. Eykhoff, 1973) and as result a great number of different extensions of the least-squares estimator has been proposed.

If the identification method is used in a simplification procedure, as in the case considered, there are no problems with bias.

Transformation of an Input-Output Model into a State-Space Model

It is usually desired to transform the input-output model, resulting from an identification procedure, into a state-space model since almost all modern control theory concerns this representation. This has been considered by many authors (e.g. Guidorzi, 1975; Wolovich, 1973). In this work a procedure developed previously is used (Strmčnik, Bremšak, 1979) which is based on the Eq. (21).

In the transformation procedure the main problem lies in the evaluation of the matrix $\underline{\underline{G}}$ since the matrices $\underline{\underline{F}}$ and $\underline{\underline{H}}$ can be constructed directly on the base of known coefficients of the polynomials in the matrix $\underline{\underline{P}}(z)$, and known indices $n_1, n_2, \ldots n_p$ (Guidorzi, 1975). The elements of the matrix $\underline{\underline{G}}$ can be computed if Eq. (21) is rewritten in a slightly different form

$$g_{ji,s} = \gamma_{j,n_j-i+1,s} + \sum_{k=1}^{p} \sum_{t=1}^{n_{jk}-n_j+i-1} f_{j,k,n_{jk}-t+1}\, g_{k,n_{jk}-n_j+i-t,s} \tag{31}$$

As it can be seen, the above equation is a kind of recursive formula for the elements $g_{ji,s}$ ($s=1,2,\ldots r$; $j=1,2,\ldots p$; $i=1,2,\ldots n_j$). To assure the formula in Eq. (31) to be applicable for computing the elements of the matrix $\underline{\underline{G}}$, the indices i and j must change according to an appropriate queue algorithm, which can be derived in the following manner.

At first, the indices n_j ($j=1,2,\ldots p$) are arranged in the descending order. From these indices the vector n(l) ($l=1,2,\ldots p$) is constructed. At the same time a vector of the corresponding subsystem indices j(l) is derived. So the following triangular structure for the couple of indices (i,l) can be written.

	(i,1)	(i,2)	(i,3)
$n(1)=n_{max}$	1		
	2		
	.		
	.		
n(2)	n(1)-n(2)+1	1	
	n(1)-n(2)+2	2	
n(3)	n(1)-n(3)+1	n(2)-n(3)+1	1
.	n(1)-n(3)+2	n(2)-n(3)+2	2
.	:	:	:

(32)

Numbers in the particular columns of this structure represent the values of the index i for the given value of the index l. The correct sequence of indices (i,l) is obtained by starting at the top of the structure (32) and taking them completely from each row. Finally, the transformation $(i,l) \Rightarrow [i,j(l)]$ must be done for each couple (i,l).

The proposed method has been tested on numerous examples. It has been shown that the algorithm is very fast and accurate, (Strmčnik, 1979).

Transformation of the Disrete State-Space Model into the Continuous State-Space Model

The principal difficulty in the transformation of the discrete system into a continuous one lies in the fact that an exact reverse transformation does not exist, if the actual variations of the input between the sampling instants are not known (Smith, 1968). This is exactly the case when the identification procedure is based on sampled data. Therefore different approximations of input signals between the sampling instants must be used when performing the discrete to continuous tranformation. Various transformation methods for multivariable systems have been shown by (Strmčnik, Bremšak, 1979).

The mentioned problem vanishes if the transformation method is integrated in the simplification procedure because the input signals in this case are known. Usualy, signals of rectangular form are used for exciting the system to be simplified. In this case the transformation relations from a discrete system ($\underline{\underline{F}}, \underline{\underline{G}}, \underline{\underline{H}}, \underline{\underline{D}}d$) into a continuous system ($\underline{\underline{A}}, \underline{\underline{B}}, \underline{\underline{C}}, \underline{\underline{D}}c$) are relatively simple and are given below (Strmčnik, Bremšak, 1979).

$$\begin{aligned} \underline{\underline{A}} &= \frac{1}{T} \ln F(T) \\ \underline{\underline{B}} &= \frac{1}{T}\left[\underline{\underline{F}}(T)-\underline{\underline{I}}\right]^{-1}\left[\ln \underline{\underline{F}}(T)\right] \cdot \underline{\underline{G}}(T) \\ \underline{\underline{C}} &= \underline{\underline{H}} \\ \underline{\underline{D}}c &= \underline{\underline{D}}d \end{aligned} \tag{33}$$

In the work cited, also some algorithms for the computation of the matrix logarithm function have been treated.

The most attractive one is the series solution method

$$\ln \underline{\underline{F}}(T) = 2 \sum_{k=0}^{\infty} \frac{1}{2k+1} \underline{\underline{\emptyset}}^{2k+1} \tag{34}$$

where

$$\underline{\underline{\emptyset}} = \left[\underline{\underline{F}}(T)-\underline{\underline{I}}\right]\left[\underline{\underline{F}}(T)+\underline{\underline{I}}\right]^{-1} \tag{35}$$

The series considered in Eq. (34) converges if $\|\emptyset\| < 1$ which means that all eigenvalues of the matrix $\underline{\underline{F}}$ must lie on the right hand side of the complex plane. The negative real parts of the eigenvalues can be avoided by an appropriate choice of the sampling interval T according to the relation

$$0 < T \leqslant \pi/2\omega_k \qquad (36)$$

where ω_k is the maximal imaginary part of eigenvalues of the continuous system matrix.

To assure convergence of this series also an alternative approach can be used introduced by De Carli an La Cava (1979). If $\|\underline{\underline{\emptyset}}\| > 1$, the square root of the transition matrix $\underline{\underline{F}}(T)$ should be repeatedly computed until the norm of a $\underline{\underline{\emptyset}}$-type matrix becomes less the one. Then the real value of the continuous matrix $\underline{\underline{A}}$ is obtained simply by multiplying the logarithm of the $\emptyset$-type matrix by the final order of the root.

For the computation of the square root an iterative algorithm proposed by the same authors can be used.

The Computer Program

On the basis of the described identification and transformation mehtod, a computer program has been developed in which the following operations are implemented.

1. sampling of input and output signals of the continuous system
2. structural and parametric identification, based on the discrete input-output model
3. transformation of the input-output model to the state-space model
4. transformation of the discrete state-space model to the continuous state-space model.

We have shown previously (Strmčnik, Bremšak, 1979) that the proposed procedure is very appropriate when for the identification of the process test functions of the step form or rectangular pulses can be used. For such cases very good results have been obtained on the simulated processes. When normal operating records are used the method is less efficient.

SIMPLIFICATION OF THE WATER-SUPPLY SYSTEM

Simplification Based on an Identification Method

Simplification problems are relatively well treated in the literature. (e.g. Aplevich, 1973; Anderson, 1968; Chen, 1974). However, the greater part of the work in this field has been devoted to model order reduction. In this context we use the term simplification in a broader sense which includes both, the reduction of the number of equations and the linearisation of the model. If the simplification is understood in a such broad sence the bulk of the existing methods is not appropriate for solving this problem.

On the other hand it can be shown that the identification methods based on input-output records are very efficient also for simplification purposes. Namely, simplification can be performed by applying an identification procedure on model responses from a computer simulation whereat the original system model is excited by appropriate test functions. In this case the structural identification can be omitted and a prescribed simpler model structure is used. Alternatively the structural identification procedure is modified to a similar algorithm where the applied conditions for the singularity of the input-output data matrix must be less rigorous.

This approach is very suitable if an existing computer program for identification can be used. Note that such programs nowadays are available in almost all computer packages intended for the analysis and design of control systems.

If a computer package for the dynamic simulation of water-supply systems and a computer package for the identification of multivariable systems from input-output data are available, a general simplification algorithm can be realised by an appropriate combination of both programs. In our particular case the programs described above have been used. The resulting procedure can be summerized in the following steps.

1. dynamical simulation of the water-distribution network
2. recording of system outputs (reservoir levels), inputs (pump controls) and disturbances (consumer demands)
3. identification of the "unknown" system concerned, based on a discrete linear canonical model
4. transformation of the discrete input-output model into the state-space model
5. transformation of the discrete model into the continuous model using the concept of the matrix logarithm function.

Simulation Results

The proposed approach has been applied to the simplification of a water distribution system of a small town near Ljubljana, with 46 nodes, 61 pipelines, 5 reservoirs and 3 pumps. The water distribution system is schematically shown in Fig. 1.

The original system has been simulated using the package described previously. The system has been exited by a bynary random sequence of pump control variables. The average consumer demand is given as input data for each node. The dynamic variation of the demand is then defined with a time variable percent of average demand which is based on statistical data. This demand is interpreted as an disturbance input with known dynamic behaviour. As outputs of the system the reservoir levels have been considered. The inputs and outputs have been sampled every hour and input-output records have been generated. The described identification procedure has been then applied to this records and a simplified model of the water distribution system has been obtained. The resulting continuous simplified model is defined by the following state-space equation

$$\begin{bmatrix}\dot{x}_1\\ \dot{x}_2\\ \dot{x}_3\\ \dot{x}_4\\ \dot{x}_5\end{bmatrix} = \begin{bmatrix} -0{,}760 & 0{,}051 & -0{,}123 & -0{,}207 & 0{,}144\\ -0{,}006 & 0{,}007 & 0{,}044 & 0{,}049 & -0{,}103\\ 0{,}009 & -0{,}009 & -0{,}303 & -0{,}135 & 0{,}442\\ -0{,}014 & 0{,}038 & 0{,}004 & -0{,}066 & -0{,}026\\ -0{,}004 & 0{,}014 & 0{,}105 & 0{,}030 & -0{,}168 \end{bmatrix} \begin{bmatrix}x_1\\ x_2\\ x_3\\ x_4\\ x_5\end{bmatrix} + \tag{37}$$

$$+ \left[\begin{array}{ccc|c|c} 0{,}739 & 0{,}135 & 0{,}030 & -0{,}010 & 0{,}446\\ 0{,}706 & 0{,}133 & 0{,}115 & -0{,}020 & 0{,}406\\ -0{,}013 & -0{,}049 & 0{,}446 & -0{,}010 & 0{,}519\\ 0{,}017 & 0{,}048 & 0{,}023 & -0{,}008 & 0{,}364\\ -0{,}014 & 0{,}013 & 0{,}285 & -0{,}007 & -0{,}042 \end{array}\right] \begin{bmatrix}u_1\\ u_2\\ u_3\\ \hline c\\ \hline 1\end{bmatrix}$$

with the corresponding output matrix $\underline{\underline{C}} = \underline{\underline{I}}$.

The matrix $\underline{\underline{B}}$ is composed of three different parts. The first one which is connected with pump control signals is the real control part of the matrix. The fourth column is defining the link between consumer demands and reservoirs. The fifth column is a consequence of a special numerical approach in the identification procedure where a scaling method for the input-output data has been used (Draper, Smith, 1968). This approach results in a shift of the time responses along the ordinat axis which has no influence on the dynamic characteristics of the responses.

The obtained simplified model has been simulated using the same input functions as in the simplification procedure. Then the time responses of the original and simplified model have been compared using the following cost function

$$J_i = \left\{ \frac{\sum_{k=1}^{24} \left[y_{oi}(kT) - y_{si}(kT) \right]^2}{\sum_{k=1}^{24} y_{oi}(kT)^2} \right\}^{1/2}$$

$$i=1,2,3,4,5 \tag{38}$$

where

y_{oi} - outputs (reservoir levels) of the original model

y_{si} - outputs (reservoir levels) of the simplified model

T - sampling interval (one hour)

The results are summarized in Table 1.

TABLE 1 The Cost Function Obtained for Pump Control Sequence Used in the Simplification Procedure

i	1	2	3	4	5
J_i	0,010	0,011	0,004	0,007	0,002

The results show that a relatively good correspondence has been achieved. This has been confirmed also with the following experiment.

A new pump control sequence has been chosen, different from the sequence applied previously. The new sequence is shown in Fig. 2. where also the percent of consumer demand is given. Then a dynamic simulation with both, the original and the simplified model has been performed. Because of large computing times needed for each solution, the dynamic simulation of the original model has been computed only for each hour. The comparison of time courses of reservoir levels for reservoirs 1,2 and 4 is shown on Figs.3, 4 and 5, respectively. In the Table 2 a comparison according to Eq. (38) is shown.

TABLE 2 The Cost Function Obtained for an Arbitrary Pump Control Sequence

i	1	2	3	4	5
J_i	0,030	0,012	0,009	0,023	0,007

From the results obtained it can be seen that the simplified linear model is a good representation of the original complex model and therefore can be used for the design of control algorithms.

CONCLUSION

It has been shown that an efficient method for simplification of water-supply system models can be obtained by a combination of a simulation and an identification method. For this purpose standard computer packages for simulation of water-supply systems and identification of multivariable systems from input-output records, can be used. In our case a water-supply system simulation package is combined with a previously developed identification method based on linear multivariable canonical models. It is shown that this approach enables derivation of a simple linear state-space model given either in the discrete or in the continuous form. The results confirms the efficiency of the method.

REFERENCES

Anderson,J.H.(1967). Geometrical approach to reduction of dynamical systems. Proc.IEE,114,No.7,1014-1018.

Aplevich,J.D.(1973). Approximation of discrete linear systems. Int.J. Control,17,No.3,565-575.

Bauer,F.L.(1963). Optimally scaled matrices. Numer.Math,5,73-87.

Businger,P.,G.H.Golub(1965). Linear least squares solutions by Householder transformations. Numer.Math, 7, 269-276.

Chen,C.F.(1974). Model reduction of multivariable control systems by means of matrix continued fractions. Int.J.Control, 20, 225-238.

Coulbeck, B.,M.J.Sterling (1978). Optimised control of water distribution systems. Proc.IEE ,125, No.91, 1039-1044.

Coulbeck,B.(1979). Dynamic simulation of water distribution systems. Preprints of IMACS Congres "Simulation of systems", Sorrento, L.Dekker, G.Savastano, G.C.VanStenkiste, Eds., 545-552.

De Carli, A.,J.La Cava (1979). A direct parameter estimation procedure via the augmented state system approach. Preprints of the 5-th IFAC Symposium "Identification and system parameter estimation", Darmstadt, R.Isermann Ed., 289-296.

Denham,M.J. (1974). Canonical forms for the identification of multivariable linear systems. IEEE Trans. Control, 19, No.6, 646-656.

Draper,N.R., H.Smith (1968). Applied regression analysis. John Wiley & Sons, New York

Eykhoff,P. (1974). System identification. John Wiley & Sons, London.

Fallside,F., P.F.Perry (1975). Hierarchical optimisation of a water-supply network. Proc. IEE, 122, No.2, 202-208.

Giadrossi,G., G.Romano, D.Torriano, W.Ukowich (1979). Using simulation to define an optimal control policy for a water distribution plant. Preprints of IMACS Congres "Simulation of systems", Sorrento, L.Dekker, G.Savastano, G.C. Van Steenkiste, Eds., 553-562.

Golub,G.H.(1965). Numerical methods for solving linear least squares problems. Numer.Math,7, 206-216.

Guidorzi,R. (1975). Canonical structures in the identification of multivariable systems. Automatica, 11, 361-374.

Korte,J.W.,H.Vielhaber (1967). Ein Beitrag zur elektronischen Berechnung von Wasserversorgungsnetze I-IV. Gas-und Wasserfach, 108, 190-195; 373-376; 674-679; 781-787.

Luenberger,D.G. (1976). Canonical forms for linear multivariable systems. IEEE Trans.Autom.Control, 12, 290-293.

Popov,V.M. (1972). Invariant description of linear time-invariant controllable systems. SIAM J.Control, 10, No.2, 252-264.

Rissanen,J. (1974). Basis of invariants and canonical forms for linear dynamic systems. Automatica, 10, 175-182.

Shamir,U., C.D.A.Howard (1968). Water distribution system analysis. Proc. ASCE, Journal of the hydraulics division, 94, HY1, 219-233.

Smith,F.W. (1968). System Laplace-transform estimation from sampled data. IEEE Trans. Autom. Control, 13, No. 1, 37-44.

Strmčnik,S.,F.Bremšak (1979). Some new transformation algorithms in the identification of continuous-time multivariable systems using discrete identification methods. Preprints of the 5-th IFAC Symposium "Identification and system parameter estimation", Darmstadt, R. Isermann Ed., 397-405.

Strmčnik,S. (1979). Mathematical modelling of large scale systems by identification and simplification methods. Ph. D. thesis, University of Ljubljana.

Wolovich,W.A. (1973). The determination of state-space representations for linear multivariable systems. Automatica, 9, 97-106.

ACKNOWLEDGEMENT

The authors are greatful to the Research council of Slovenia by which this work was financially supported.

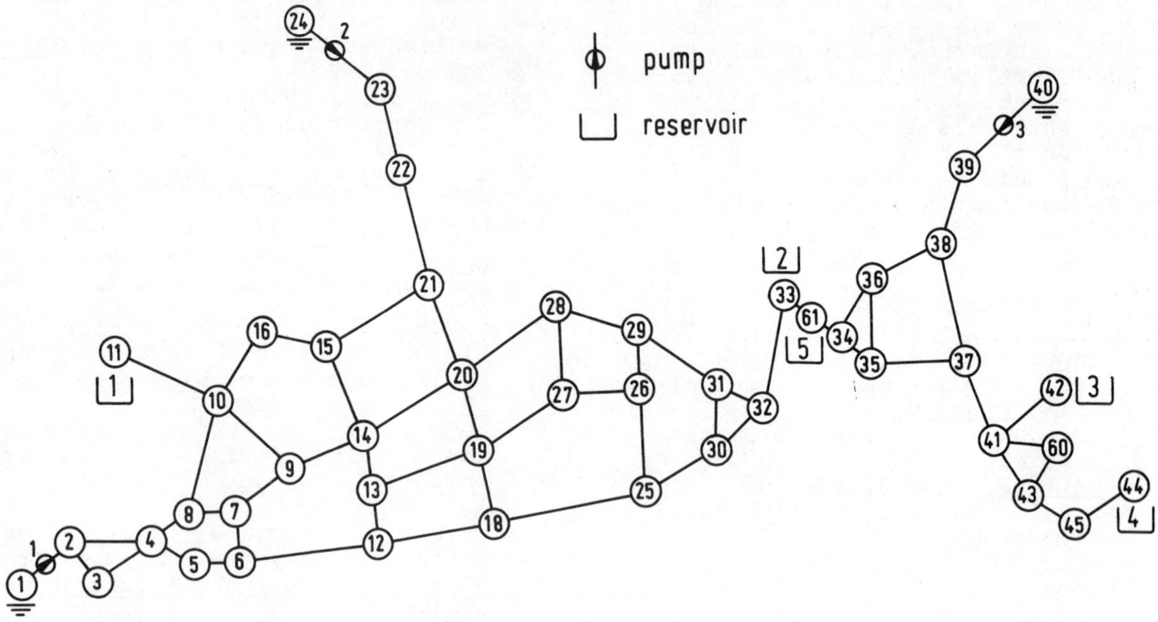

Fig. 1. Scheme of the water-supply system

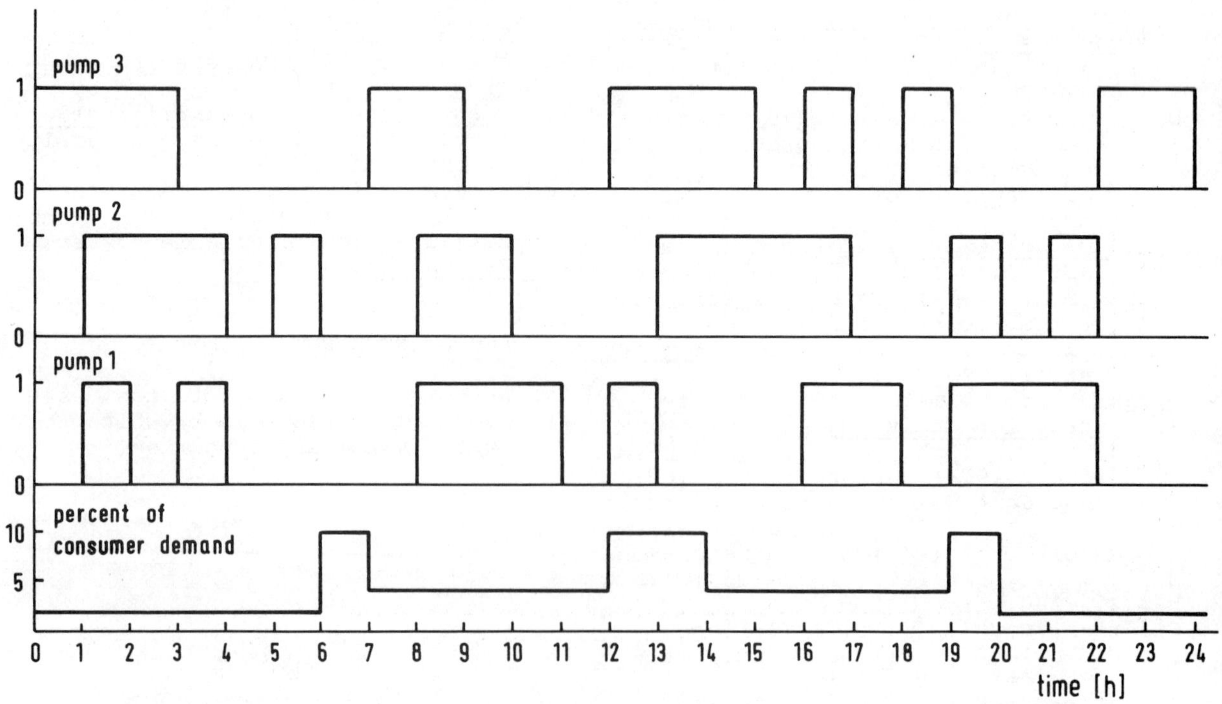

Fig. 2. The pump control and consumer demand sequence

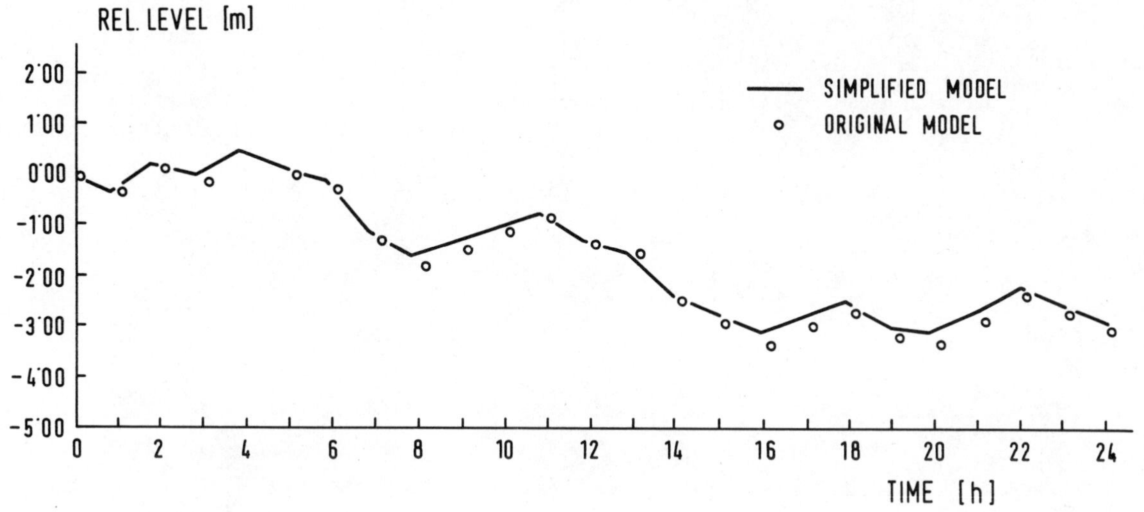

Fig. 3. Level in reservoir No.1 as a function of time

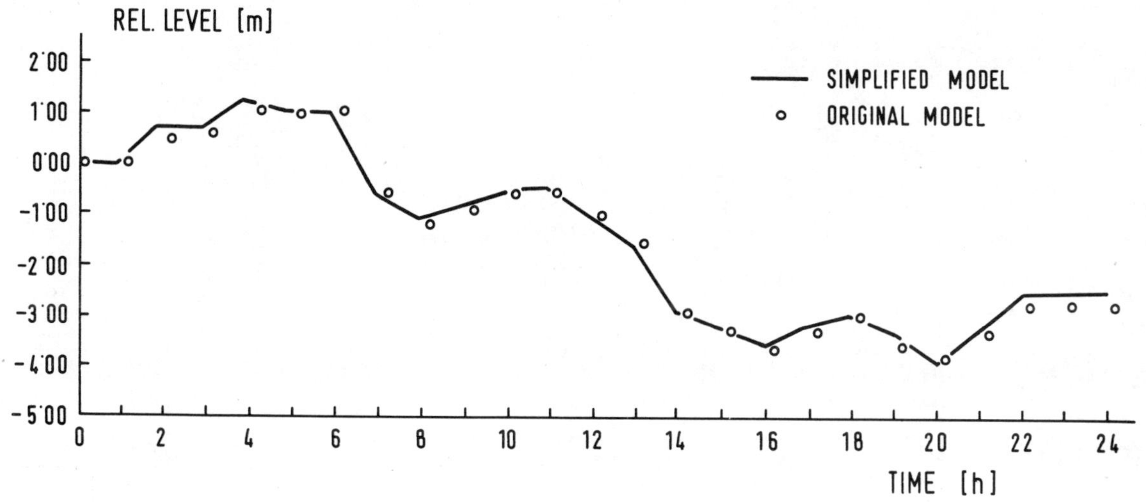

Fig. 4. Level in reservoir No.2 as a function of time

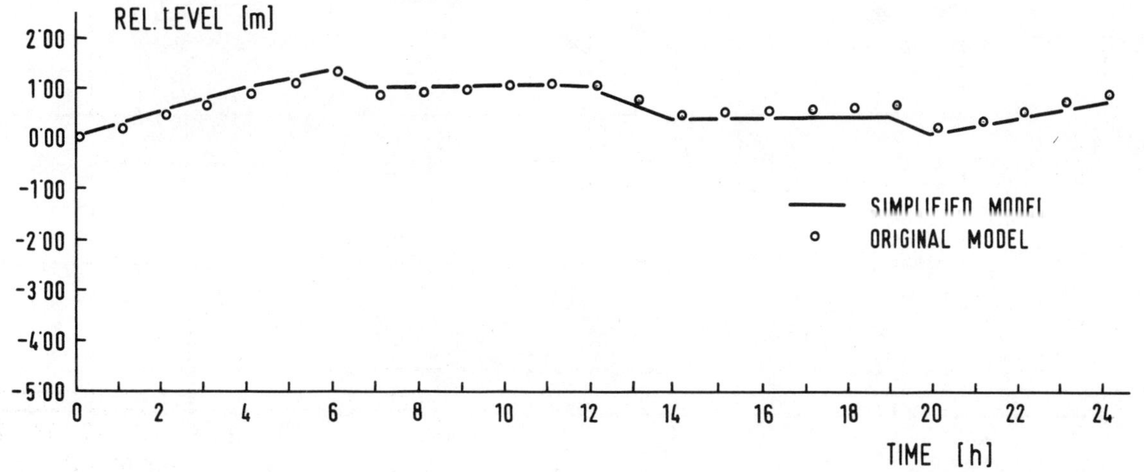

Fig. 5. Level in reservoir No. 3 as a function of time

A NEW DEVELOPED STREAM WATER QUALITY MODEL: A CASE STUDY ON THE RIVER NILE*

M. F. Hassan*, M. I. Younis and K. H. Mancy*****

**Faculty of Engineering, Electronic and Communication Department, Cairo University, Giza, Egypt*
***Systems and Information Science Laboratory, National Research Center, Dokki, Giza, Egypt*
****School of Public Health, University of Michigan, Ann Arbor, Michigan 48109, U.S.A.*

Abstract. In this paper, a new stream water quality model is developed. The model is developed to analyse accurately the water quality and at the same time is capable of generating control strategies that insures the satisfaction of water quality standards in polluted reaches. The mathematical derivation of the model is given. Then, for the purpose of validation, the model is used to analyse the water quality in the River Nile.
Finally, in order to demonstrate how the model is used for control purposes, an optimization problem is formulated. Then an algorithm based on decomposition coordination techniques is presented which can be used to give optimal open loop control stategies that maintains water quality standards with minimum cost.

Key words. Developing countries; Minimum principle; Nonlinear systems, Optimisation, Systems analysis, Water pollution.

INTRODUCTION

Research on the modelling and control of water quality in rivers and streams has been reported widely. The reason for this is that with the extension of civilization,water in streams is used for industrial purposes and effluents are discharged again in rivers. This of course disturbs the ecological balance in the system which may affect life (fish die for example). What is essentially required in a river, which is used either as a source for potable supply or for recreation, is to maintain pollution levels in polluted reaches within reasonable bounds consistent for both the community needs and for maintaining a satisfactory ecological balance as well as standards of water quality.

During the past few years, numerous mathematical models have been applied to simulate water quality conditions in rivers[4]. These models vary considerably in the degree of resolution (refinement) with which they represent the physical world. Low resolution models represent general trends for a few linked dependent variables over a limited set of boundary conditions. An example of the use of this type of models is the application of the Streeter-Phelps equation to represent average monthly dissolved oxygen (DO) concentrations in the San Antonio River [3]. High resolution models may be used to represent the responses of a large number of linked dependent variables over a much wider set of boundary conditions, for example, the application of the HEC model [1] simulating 20 constituents in the Boise River system. As the order of resolution of a model increases, so does the difficulty and cost of its application. The selection of a model for a particular situation requires value judgements of acceptable trade-offs between the practicibility and economy of model application, the amount and refinement of information to be provided by the model responses. The current trend is towards the application of the high resolution models.

In this paper, a new model is developed which can be used for both water quality analysis and control and tries to group the advantages of the already developed models and avoids their disadvantages. The rest of the paper is divided into six sections. In section 2, a new water quality analysis model is developed. It is shown that this model groups the advantages of other models, in the sense that it is sufficiently accurate. It can also be used for control purposes of polluted reaches, and avoid to some extent, the disadvantages of the other models. It is also seen that the developed model can take into consideration any number of inputs or outputs from the river system as the effect of changes in velocity and depth. The advantages and facilities given by the model as well as a comparison with the other models is given at the end of this section.

* This work was supported by THE WATER QUALITY STUDIES ON THE RIVER NILE AND LAKE NASSER PROJECT, Egyptien Academy of Scientific Research and Technology and University of Michigan, which is Sporsored by the US. Enviornmental Protection Agency.

In order to test the capability, of the developed model to simulate water quality analysis, it is applied to the first reach of the River Nile in Section 3.

The formulation of the optimization problem is developed in Section 4, and the algorithim for its solution is presented in Section 5. Finally conclusions of the findings are given in Section 6.

2- THE DEVELOPEMENT OF WATER QUALITY MODEL FOR THE RIVER NILE

The developement of such model is based on dividing the stream into reaches each one having the same hydraulic properties (e.g. same velocity, the same depth ....etc.). But in the case of the River Nile, the situation is completely different. The reason for this is, although the Nile can be divided into reaches, yet these reaches cannot be considered of the same hydraulic properties. Actually, the river is divided into reaches which are separated from one another by means of barrages.

Thus, if it is assumed that the water velocity is constant in most of the reach, one can expect a change in the velocity at a certain distance before eachbarrage due to the blocking of water infront of the barrage and hence a change in the basin area (since the discharge is assumed constant). This, of course, will affect the variations of the constituents within the stream.

In this section, a new model is developed to give a complete picture about the variation of the water quality constituents within the stream as well as taking into consideration the effect of the change of the water velocity and hence the area of the basin. Moreover, one can get any control policy from this model.

To clarify the idea of this model, let us consider a drop of water at the most upper stream part of the Nile with certain concentration of water quality constituents. Hence, under steady assumption, the model will give the variations within these concentrations by following this drop of water along the basin till the most down barrage point of the river. By this concept, the serially connected nature of the stream system is made clear and hence there is no need to use the transportation time delay. In addition to this, the model takes into account the effect of any distributed load as well as the effect of any input junction or waste load and any output canal on the water quality of the main river.

The new model is derived with the capablity of giving the variation of Algae, Ammonia nitrogen, Nitrite nitrogen, Phosphate phosphorus, Biochemical oxygen demand as well as Dissolved oxygen along the river. In general, the fundamental one dimensional transport equation which will be used in the model is given by:

$$A_x dx \frac{\partial c}{\partial t} = \frac{\partial (A_x D_L \frac{\partial c}{\partial x}) dx}{\partial x} - \frac{\partial (A_x uc) dx}{\partial x} + A_x dx \frac{dc}{dt} \pm S \qquad 2\text{-}1$$

This is a one dimensional mass balance equation representing the rate of change of the concentration of constituents with respect to absolute time as well as the distance along the basin. Using the water quality constituent reactions and interactions given in [5]. the following set of partial differential equations can be obtained:

Algae (A):

$$\frac{\partial A}{\partial t} = \frac{\partial (A_x D_L \frac{\partial A}{\partial x})}{A_x \partial x} - \frac{\partial (A_x u A)}{A_x \partial x} + \frac{S_A}{A_x dx} + (\mu - \rho - \frac{\sigma_1}{D}) A \qquad 2\text{-}2$$

Ammonia nitrogen (N_1):

$$\frac{\partial N_1}{\partial t} = \frac{\partial (A_x D_L \frac{\partial N_1}{\partial x})}{A_x \partial x} - \frac{\partial (A_x U N_1)}{A_x \partial x} + \frac{S_{N_1}}{A_x dx} + (\alpha_1 A - \beta_1 N_1 + \frac{\sigma_3}{A_x}) \qquad 2\text{-}3$$

Nitrite nitrogen (N_2):

$$\frac{\partial N_2}{\partial t} = \frac{\partial (A_x D_1 \frac{\partial N_2}{x})}{A_x \partial x} - \frac{\partial (A_x U N_2)}{A_x \partial x} + \frac{S_{N_2}}{A_x dx} + (\beta_1 N_1 - \beta_2 N_2) \qquad 2\text{-}4$$

Nitrate nitrogen (N_3) :

$$\frac{\partial N_3}{\partial t} = \frac{\partial (A_x D_L \frac{\partial N_3}{\partial x})}{A_x \partial x} - \frac{\partial (A_x U N_3)}{A_x \partial x} + \frac{S_{N3}}{A_x dx} + (\beta_2 N_2 - \alpha_1 \mu A) \qquad 2\text{-}5$$

Phosphate phosphorous (P):

$$\frac{\partial P}{\partial t} = \frac{\partial (A_x D_L \frac{\partial P}{\partial x})}{A_x \partial x} - \frac{\partial (A_x UP)}{A_x dx} + \frac{S_p}{A_x dx} + [\alpha_2 (\rho - \mu) A - \frac{\zeta_2}{A_x}] \qquad 2\text{-}6$$

Biochemical oxygen demand (L):

$$\frac{\partial L}{\partial t} = \frac{\partial (A_x D_L \frac{\partial L}{\partial x})}{A_x \partial x} - \frac{\partial (A_x UL)}{A_x \partial x} + \frac{S_L}{A_x dx} - (K_1 + K_3) L \qquad 2\text{-}7$$

Dissolved oxygen (Q):

$$\frac{\partial Q}{\partial t} = \frac{\partial (A_x D_L \frac{\partial Q}{\partial x})}{A_x \partial x} - \frac{\partial (A_x UQ)}{A_x \partial x} + \frac{S_Q}{A_x dx} [K_2 (Q^* - Q) \; (\alpha_3 \mu - \alpha \rho) A - K_1 L - \frac{K_4}{A_x} - \alpha_6 \beta_2 N_2] \qquad 2\text{-}8$$

The developed model is actually based on the following reasonable assumptions and simulations:

1- Steady state conditions i.e:

$$\frac{\partial c}{\partial t} = 0 \qquad 2\text{-}9$$

Which means that the rate of change of the constituent with respect to absolute time at any point on the river is zero.

2- The term representing the dispersion is neglected in the developement of the model i.e.:

$$\frac{\partial (A_x D_L \frac{\partial c}{\partial x})}{A_x \partial x} = 0 \qquad 2\text{-}10$$

3- At any input junction or waste load point, or at output canal the mass balance equation is satisfied as well perfect mixing at this point exists, in order to clarify this point, a junction on the river at point (a) is shown in figure (1).

Then, Q_3 and C_3 are given by the following balance equations:

$$Q_3 = Q_1 + Q_2 \qquad 2\text{-}11$$

$$C_3 = \frac{Q_1 C_1 + Q_2 C_2}{Q_1 + Q_2} \qquad 2\text{-}12$$

Assuming that the cross section area is constant, the velocity of the water is given by:

$$U_3 = \frac{Q_3}{\text{Area}} \qquad 2\text{-}13$$

Using these assumptions the differential equations (2-2) to (2-8) which represent the variation of the concentration of the constituents along the river can be written in the following simplified form:

$$\frac{d(UA)}{dx} = (\mu - \rho - \frac{\zeta_1}{D}) A + \frac{S_A}{A_x dx} \qquad 2\text{-}14$$

$$\frac{d(N_1)}{dx} = \alpha_1 \rho A - \beta_1 N_1 + \frac{\zeta_3}{A_x} + \frac{S_{N_1}}{A_x dx} \qquad 2\text{-}15$$

$$\frac{d(UN_2)}{dx} = \beta_1 N_1 - \beta_2 N_2 + \frac{S_{N_2}}{A_x dx} \qquad 2\text{-}16$$

$$\frac{d(UN_3)}{dx} = \beta_2 N_2 - \alpha_1 \mu A + \frac{S_{N_3}}{A_x dx} \qquad 2\text{-}17$$

$$\frac{d(U\rho)}{dx} = \alpha_2 (\rho - \mu) A - \frac{\zeta_2}{A_x} + \frac{S_p}{A_x dx} \qquad 2\text{-}18$$

$$\frac{d(UL)}{dx} = -(K_1 + K_3) L + \frac{S_L}{A_x dx} \qquad 2\text{-}19$$

$$\frac{d(UQ)}{dx} = -K_2 Q + (\alpha_3 \mu - \alpha_4 \rho) A - K_1 L - \alpha_5 \beta_1 N_1 - \alpha_6 \beta_2 N_2 + K_2 Q^* - \frac{K_4}{A_x} + \frac{S_Q}{A_x dx} \qquad 2\text{-}20$$

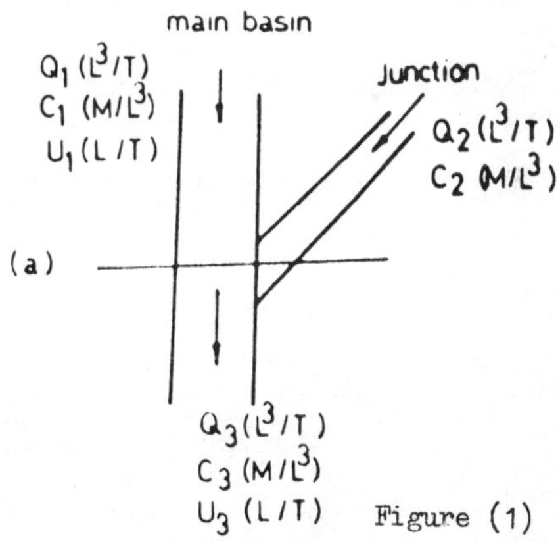

Figure (1)

As it was mentioned previously, the developed model can be take into consideration the effect of the changes in the velocity before the barrage. Thus the model will be derived for the following two different cases:

i) When the water is sufficiently far from the barrage, and it will be assumed that the stream has a constant velocity.

ii) Near the barage, and the velocity will be no longer constant and it will be assumed as a function of the distance.

In the following, a state variable water quality model [2] for these two different cases is developed.

The development of a state variable water quality model under constant stream velocity assumption:

In this case, let:

$$x = U\tau \qquad 2\text{-}21$$

Where τ is the time required for a drop of water to move a distance x

$$\therefore\ dx = U\,d\tau \qquad 2\text{-}22$$

In general, under constant velocity assumption

$$\frac{d\,(Uc)}{dx} = U\,\frac{dc}{dx} = \frac{dc}{d\tau} \qquad 2\text{-}23$$

Let $z = s/A_x\ dx = (M/TL^3)$, then,

$$\dot{A} = (\mu - \rho - \frac{\sigma 1}{D})\,A + z_A \qquad 2\text{-}24$$

$$\dot{N}_1 = \alpha_1\rho\ A - \beta_1 N_1 + \frac{\sigma 3}{A_x} + z_{N_1} \qquad 2\text{-}25$$

$$\dot{N}_2 = \beta_1 N_1 - \beta_2 N_2 + z_{N_2} \qquad 2\text{-}26$$

$$\dot{N}_3 = \beta_2 N_2 - \alpha_1 \mu A + z_{N_3} \qquad 2\text{-}27$$

$$\dot{P} = \alpha_2(\rho - \mu)\,A - \frac{\zeta_2}{A_x} + z_P \qquad 2\text{-}28$$

$$\dot{L} = -(K_1 + K_3)\,L + z_L$$

$$\dot{Q} = -K_2\,Q + (\alpha_3\mu - \alpha_4\rho)\,A - K_1 L - \alpha_5\beta_1 N_1 - \alpha_6\beta_2 N_2 + K_2 Q^* - \frac{K_4}{A_x} + z_Q \qquad 2\text{-}29$$

$$\underline{\dot{x}} = A\,\underline{x} + B\,\underline{Z} + \underline{D} \qquad 2\text{-}30$$

Where $\underline{X}^T = (A\ \ N_1\ \ N_2\ \ N_3\ \ P\ \ L\ \ Q)$

$$A = \begin{bmatrix} (\mu - \rho - \frac{\zeta_1}{D}) & 0 & 0 & 0 & 0 & 0 & 0 \\ \alpha_1\rho & -\beta_1 & 0 & 0 & 0 & 0 & 0 \\ & \beta_1 & \beta_2 & 0 & 0 & 0 & 0 \\ -\alpha_1\mu & 0 & \beta_2 & 0 & 0 & 0 & 0 \\ \alpha_2(\rho - \mu) & 0 & 0 & 0 & 0 & 0 & 0 \\ 0 & 0 & 0 & 0 & 0 & -K_1-K_3 & 0 \\ \alpha_3\mu - \alpha_4\rho & -\alpha_5\beta_1 - \alpha_6\beta_2 & 0 & & 0 & -K_1 & -K_2 \end{bmatrix}$$

$B = I_{7 \times 7}$ where I is the unit natrix

$$\underline{Z}^T = [\ Z_A\ Z_{N_1}\ Z_{N_2}\ Z_{N_3}\ Z_P\ Z_L\ Z_Q]$$

$$\underline{D}^T = [\ 0\ \ \frac{\sigma 3}{A_x}\ \ o\ \ \ o\ \ -\frac{\sigma 2}{A_x}\ \ o\ \ (KQ_2^* - \frac{K_4}{A_x})$$

One has to note that, the coefficients of the above equation are taken at standard temperature, namely, 20C. If the temperature of the basin is different from 20C, these coefficients are corrected using the following equation:

$$X_T = X_{T_s}\ \Theta^{(T - T_s)}$$

Where: X_T is the value of the coefficient at the desired temperature.

X_{Ts} is its value at the standard temp.

Θ constant

T the temperature of the basin in C

T_s = 20 C = standard temp.

Model developement under variable stream velocity:
In this case and since the discharge in the river is assumed constant, the same set of equation (2-24) to (2-29) which represent the model is still valid. However, under constant velocity assumption, a constant step length of time in the integration program is used to solve this set of equations.

On the other hand, these equations are solved for portions of the river in which velocity changes, a variable step length of time which is used depending on the stream velodity at this point and the corresponding constant interval of distance.

Discussion:

1- The developed model can be simply used for water quality analysis in a stream. This can be done by integrating the first order vector differential equation(2-30) with the given initial condition for A, N_1, N_2, N_3, P, L, Q at the most upper stream point of the river along with the coefficients of the model. When we are sufficiently close to a barrage the results of the model (2-30) will be used as initial conditions for the case in which the effect of the velocity changes is taken into consideration. Then, for the next reach, the output of the above model will be used as initial conditions to derive the original model which simulates the concentrations of the water constituents at the beginning of that reach and so on till the most downstream point of the river is reached.

2- At any input junction or waste load or any output canal, the result, of the model will be used with that of the junction to get the new discharge and concentrations at this point by using equation (2-11) and (2-12). These new concentrations will be used as new initial conditons for the model which starts to simulate the stream quality downstream of this point.

3- The model is capable of dealing with distributed loads on the river. This may be achieved by lumping the load at the starting point for each increment of integration of the differential equations. Then using equations (2-11) and (2-12) one can get the new discharge which will be used to obtain the results at the end of this integration increment. Another possibility for the treatment of this type of loads is to calculate the new discharge within this integration increment, use it to calculate the vector $\underline{Z}$ in equation (2-30) which can be considered as an input supply.

4- In general, the developed model is a non-linear one since u is a function of N_3 and P. However, in the simulation one can consider U to be constant for a specific number of simulating points along the stream or for a certain simulation interval. Then, at the end of this interval, one changes U according to the obtained concentration of N_3 and P and consider it again to be constant for the next simulation interval.

5- From the model it is easy to predict what will happen to the water quality along the stream due to any new industries that will be discharging their effluents in the river.

6- With this simualtion model, the well developed control theory may be used very easily to reach optimum policies for the treatment of the polluted sources before discharging their wastes in the river in order to maintain water quality set standards and hence the ecological balance in the river system, as will be shown in section 4.

3. DEVELOPMENT OF THE MODEL COMPUTER PROGRAM AND ITS APPLICATION ON THE RIVER NILE:

A computer program is developed for the proposed model. The program is constructed to cope with any number of input or output junctions or any number of waste load inputs. Also it can consider any distributed load along the basin or along any reach. Again, the stream is divided into reaches each one is separated from the other by means of a barrage has approximately the same hydraulic properties in most of the reach although velocity as well as area variation were admitted at the end of each reach. Also, the program may be adopted very easy to take into consideration the variation in any coefficient due to velocity changes.

The coefficients must be given to the program as input data at the standard temperature (20C°) and then changed in the program according to the temperature of each reach. The integration interval can be chosen sufficiently small to insure the stability of the algorithm used in the solution of this vector differential equation. The most important feature of the model as well as the developed program is that there is no limitation on the number of reaches chosen, the integration interval (which corresponds to the computational element in the QUAL II model) and the number of input or output junctions. In addition to this, the storage requirements of the developed program are minimal, thus it may be run on mini-computers.

These advantages make it possible to choose the increment of integration sufficiently small to have a good picture about what happens along the river. Also, adjacent junctions will not cause any problems since we do not have any limitation on the step length of integration. The only disadvantage of that as we increase the number of integration points along the river, the computation time will increase. However, this will not cause any problem, specially for control purpose, since the dynamics of the system is sufficiently slow (of the order of a day).

<u>Application on the River Nile:</u>

In order to exercise our model as well as to check its applicability, the developed program is applied on the first reach of the River Nile, the data of which is given in tables (1) and (2). The set of coefficients which gave the best results [3,7] is given in the appendix.

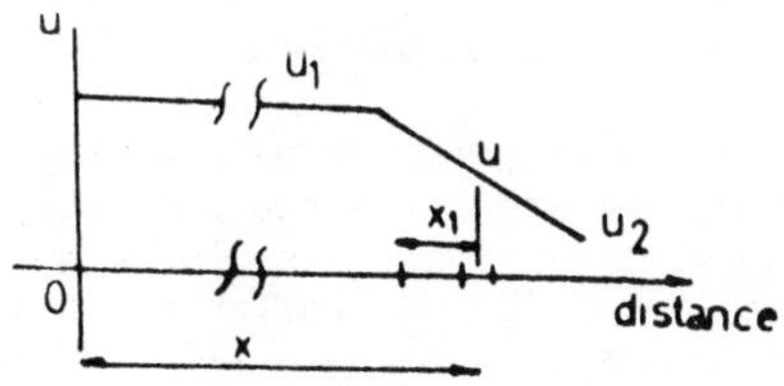

Figure (2)

To take into consideration the effect of velocity changes(stars with 48.3169 mile/day at Aswan Dam and reach a value of 16.106 miles/day at Naga Hamadi barage), it is assumed that the velocity is constant along the reach till a point which is far from Naga Hamadi barage by 40 miles. Then the velocity is allowed to vary linearly with distance as shown in figure 2. Thus we have:

$$u = u_1 + c\, x_1 \qquad 3\text{-}1$$

where

$$c = \frac{u_1 - u_2}{40} \qquad 3\text{-}2$$

u = the velocity at any point in this interval

u_1= starting velocity = 48.3169 miles/day

u_2= velocity at Naga Hamadi barage=16.1o65 miles/day

x_1= distance =x - 184 miles.

Since we considered constant interval of integration (2 miles) and in order to keep the A matrix constant (except for the element A A (1,1) which is allowed to vary each five steps) we allowed a change in the time interval of integration at each step corresponding to the new velocity, i.e.

$$\Delta t = \frac{x}{u} = \frac{2}{u} \qquad 3\text{-}3$$

Where u is as given by equation(3-1)

In addition to the above, and due to the assumption of constant discharge along the reach (except for the change in this discharge due to inputs or outputs), the cross section area of the basin is also allowed to change ($A = \frac{Q}{U}$). It is assumed that the basin has constant width, therefore, the depth is allowed to vary according to area variation.

The obtained simulation results are shown in figures (3) to (9).

4. FORMULATION OF THE OPTIMIZATION PROBLEM:

Since water in streams may be used for industrial purposes and often the resulting effluents as well as sewage are discharged again in rivers, this will naturally result in an increase in the pollution level which may disturb the ecological balance. Therefore, it seems reasobable to consider the operational control problem as one of adjusting the level of water quality constituents entering the river by way of the effluent discharge such that the in-stream water quality is maintained at some prescribed satisfactory level, to be decided by the water authority.

Suppose that we define a reach in a river system as a stretch of the river whcih receives one major controlled effluent discharge from a sewage or industrial waste treatment facility and of length equals to that required for the effect of this effluent discharge, on the water quality in the stream, to completely die out. Moreover, we assume that perfect mixing takes place in this reach, let:

ϕ_E is the flow rate of the effluent

V is the volume of the water in this reach

$U \varepsilon R^q$ [$q \le 7$] is the vector of the concentration of the water quality constituents in the effluent.

Hence equation (2-30) can be modified to the following form:

$$\dot{\underline{x}} = A\,\underline{x} + \frac{\phi_E}{V} B'\underline{u} + F\,\underline{z} + \underline{D} \qquad 4\text{-}1$$

Where B' is an 7 x 7 matrix of elements ones and zeros.

What normally happens are that effluents is pretreated to a desired level which maintain water quality requirement before their discharge in the river.

To determine the desired level of the pretreatment $\underline{u}$ *, let $\underline{x}$ * is the desired water quality standards. Thus as $\dot{\underline{x}}$ Putting $\dot{\underline{x}}$ = 0 in (4-1), we get the desired pretreatment $\underline{u}$ * from the following

$$0 = A\,\underline{x}^* + \frac{\phi E}{V} B'\,\underline{u}^* + F\underline{z} + \underline{D} \qquad 4\text{-}2$$

However, disturbances may take place which disturb the equilibrium of the system. It would, therefore, be necessary to use some form of variable treatment of the effluent such that the system returns back to its steady behaviour. As a result equation (4-1) can be modified to the following form:

$$\dot{\underline{x}} = A\,\underline{x} + \frac{\phi_E}{V} B'\,(\underline{u}^* + \Delta u) + F\underline{z} + \underline{D}$$

$$\text{or}\quad \dot{\underline{x}} = A\,\underline{x} + B\,\Delta\underline{u} + D \qquad 4\text{-}3$$

$$\text{Where } B = \frac{\phi_E}{V}\, B' \qquad 4\text{-}4$$

$$= \frac{\phi_E}{V}\, B'\, u^* + F\underline{z} + \underline{D} \qquad 4\text{-}5$$

In fact, rivers can safely absorb certain levels of effluents discharge and still keep to prescribed water quality levels. As a result we are seeking for bringing the system back to a steady state values in an optimal manner and at the same time preserving the water quality levels specified by the water authorities. A suitable objective function by which various control projectories can be compared in order to find the optimum one for this problem is :

$$J = \min_{\Delta u} \frac{1}{2} \int_{to}^{t_f} \|\underline{x} - \underline{x}^*\|_{\phi} + \|\Delta\underline{u}\|_R^2 \, dt \qquad 2\text{-}6$$

Where ϕ>o;R>o are appropriate weight, which define the importance of the various terms and $\underline{x}$* are the desired values which are chosen to be the steady state values.

5. THE DEVELOPMENT OF AN ALGORITHM FOR THE SOLUTION OF THE OPTIMIZATION PROBLEM:

The optimization problem to be solved is the following:

$$\min J = 1/2 \int_{t_o}^{t_f} \| \underline{x} - \underline{x}^* \|_\phi + \| \Delta \underline{u} \|_R^2 dt \quad 5\text{-}1$$

$$\text{S.t. } \dot{\underline{x}} = A\,\underline{x} + B\,\Delta\underline{u} + \underline{\mathcal{D}} \quad 5\text{-}2$$

The above problem is a nonlinear one since a_{11} is the matrix A is function of both x_4, x_5. However , the above problem can be decomposed into two problems as follows:

$$\text{S.t. } \dot{\underline{x}}_1 = A_{11}(\hat{x}_4, \hat{x}_5)\,\underline{x}_1 + B_{11}\,\Delta\underline{u}_1 + \underline{\mathcal{D}}_1 \quad 5\text{-}4$$

$$\dot{\underline{x}}_2 = A_{22}\,\underline{x}_2 + B_{22}\,\Delta\,\underline{u}_2 + \underline{c}\hat{x}_1 + \underline{\mathcal{D}}_2 \quad 5\text{-}5$$

$$\hat{x}_1 = x_1 \quad 5\text{-}6$$

$$\hat{x}_4 = x_4 \quad 5\text{-}7$$

$$\hat{x}_5 = x_5 \quad 5\text{-}8$$

$$\text{Where } \underline{x} = \begin{bmatrix} x_1 \\ \cdots \\ x_2 \end{bmatrix}; \quad A = \begin{bmatrix} A_{11}(x_4, x_5) & \vdots & 0 \\ \cdots & \cdots & \cdots \\ \underline{C} & \vdots & A_{22} \end{bmatrix}$$

$$B = \begin{bmatrix} B_{11} & \vdots & 0 \\ \cdots & \cdots & \cdots \\ 0 & \vdots & B_{22} \end{bmatrix}$$

Under the assumption of separable cost function, the Hamiltonian of the above problem takes the follwing form:

$$H = \frac{1}{2}\|\underline{x}_1 - \underline{x}_1^*\|^2_{\phi_{11}} + \frac{1}{2}\|\underline{x}_2 - \underline{x}_2^*\|^2_{\phi_{22}} + \frac{1}{2}\|\Delta\underline{u}_1\|^2_{R_{11}}$$

$$+ \frac{1}{2}\|\Delta\underline{u}_2\|^2_{R_{22}} + \underline{\lambda}_1^T (A_{11}\underline{x}_1 + B_{11}\Delta\underline{u}_1 + \underline{\mathcal{D}}_1)$$

$$+ \underline{\lambda}_2^T (A_{22}\underline{x}_2 + B_{22}\Delta\,\underline{u}_2 + C\hat{\underline{x}}_1 + \underline{\mathcal{D}}_2) + \underline{\pi}_1^T (\underline{x}_1 - \hat{\underline{x}}_1)$$

$$+ \underline{\pi}_2^T (\underline{x}_4 - \hat{\underline{x}}_4) + \pi_2^T (\underline{x}_5 - \hat{\underline{x}}_5) \quad 5\text{-}9$$

From the necessary conditions of optimality one gets:

$$\frac{\partial H}{\partial \pi_i} = 0 \qquad \hat{x}_i = x_i \;;\; i = 1,4,5 \quad 5\text{-}10$$

$$\frac{\partial H}{\partial \hat{x}_1} = 0 \qquad \underline{\pi}_1 = \underline{C}^T \underline{\lambda}_2 \quad 5\text{-}11$$

$$\frac{\partial H}{\partial \hat{x}_4} = 0 \qquad \pi_2 = \frac{\partial A_{11}}{\partial \hat{x}_4} x_1 \lambda_1 \quad 5\text{-}12$$

$$\frac{\partial H}{\partial \hat{x}_5} = 0 \qquad \pi_3 = \frac{\partial A_{11}}{\partial \hat{x}_5} x_1 \lambda_1 \quad 5\text{-}13$$

$$\frac{\partial H}{\partial \Delta\underline{u}_i} = 0 \qquad \Delta\underline{u}_i = -R_{ii}\, B_{ii}^T \underline{\lambda}_i \;; i = 1,2 \quad 5\text{-}14$$

$$\frac{\partial H}{\partial \lambda_1} = \dot{x}_1 \qquad \dot{\underline{x}}_1 = A_{11}\underline{x}_1 + B_{11}\,\Delta\,\underline{u}_1 + \underline{\mathcal{D}}_1;$$

$$\underline{x}_1(t_o) = \underline{x}_{10} \quad 5\text{-}15$$

$$\frac{\partial H}{\partial \lambda_2} = \dot{\underline{x}}_2 \quad \dot{\underline{x}}_2 = A_{22}\underline{x}_2 + B_{22}\Delta\underline{u}_2 + C\underline{x}_1 + \underline{\mathcal{D}}_2$$

$$\underline{x}_2(to) = x_{20} \quad 5\text{-}16$$

$$\frac{\partial H}{\partial x_1} = \dot{\lambda}_1;\; \dot{\underline{\lambda}}_1 = -\phi_{11}(\underline{x}_1 - \underline{x}_1^*) - A_{11}\underline{\lambda}_1 + \pi_1\underline{x}_1$$

$$\underline{\lambda}_1(t_f) = \underline{o} \quad 5\text{-}17$$

$$\frac{\partial H}{\partial x_2} = -\dot{\lambda}_2;\; \dot{\lambda}_2 = -\phi_{22}(x_2 - x_2^*) + A_{22}\underline{\lambda}_2 + \underline{\pi}'$$

$$\underline{\lambda}_2(t_f) = \underline{0} \quad 5\text{-}18$$

where $\underline{\pi}'^T = [\,0 \;\, 0 \;\, \pi_2 \;\; \pi_3 \;\, 0 \;\, 0\,]$

Thus we propose the following algorithm which can be represented by the schematic diagram shown in figure(10) :

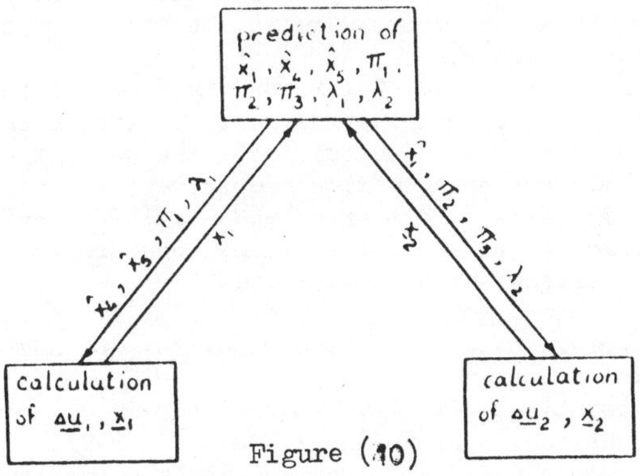

Figure (10)

1- The second level gives an initial guess to the trajectories.
and set the iteration index k=1.

2- Each subsystem calculates the control $\Delta\underline{u}_i$ and integrates the state equation forward with the given initial condition.

3- At the second level, we predict new trajectories for $\hat{x}_1, \hat{x}_4, \hat{x}_5, \pi_1, \pi_2, \pi_3, \lambda_1, \lambda_2$ using equation (5-10)-(5-13)integrating backward equation, (5-13), (5-10) to get new estimates for λ_1, λ_2 and put k=k+1. Then we calculate:

$$\sqrt{\int_{t_o}^{t_f} \| \underline{L}^{k+1} - \underline{L}^k \|^2 dt} \leq \varepsilon \;;$$

$$\underline{L}^{k^T} = [\, \hat{x}_1^k, \hat{x}_4^k, \hat{x}_5^k, \lambda_1^k, \lambda_2^{k^T}]$$

and ε is a prespecified small number. If the above inequality is not satisfied we go to step (2). Otherwise, we go to the next step.

4- Record the obtained control projectories which give the optimal response.

The convergence proof of the above algorithm is given in Hassan el al [5].

6. CONCLUSION

In this paper, a new developed model for water quality analysis is derived. The application of the developed model on a reach of the River Nile showed its capability for accurate simulation of the water quality constituents in streams. Then, an optimization problem is formulated which can be used to generate control strategies that can bring the systen back to its steady state behaviour in an optimal manner under external, disturbances. Hence, a two level algorithm is proposed to solve such an optimization problem. Research is now going on to apply this approach to the River Nile and it will be published in some elsewhere.

REFERENCES

1- Chen, C. W. and Wells (1966). Boise River water quality ecologic model for urban planning study. Tetra Tech. No. TC-368 and TC-605. Tetra Tech., Inc., Lafayette, Calif.

2- Chih-Fan chin and I. John Haas. (1968). Elements of control systems analysis classical and modern approaches. Printice Hall.

3- Computer program Documentation for the stream Quality Model QUAL II, (1978). Prepared by the Environmental Protection Agency, Systems Development Branch, Washington, D.C.

4- Grennery, W.J. Water quality relationships to flow-stream and estuaries. In stalnaker, C.B. (Ed.), Methodologies for the determination of stream resource flow requirements: An assessment. U.S. Fish and Wildlife Service, Office of Biological services, Western water allocation, Washington, D.C.

5- Hassan, M.F. and M.G. singh (1977), A two level costate prediction algorithm for non-linear systems. Automatica, Vol. 13, PP 629-634.

6- Hassan, M.F. and M. I. Younis (1979). Stream quality modeling and its application on River Nile. Water quality studies on the River Nile and Lake Nasser Project, Egyptian Academy of Science and Technology and the University of Michigan.

7- Rate, constants and kineties formulation in surface water quality modeling 1978. Prepared for U.S. Environmental Protection Agency Athens, Georgia.

8- Texas Water Development Board (1970). Simulation of water quality in stream canals. Program documentation and users manual, EPA-OWP - TEX - DOSAGI, NTIS.

APPENDIX

Coefficents used:

Ratio of chlorophyll a to algae biomass (α_o) = 75 (mgch 1-A/mgA)

Fraction of algae biomass which is N (α_1) = 0.08 (mgN/mg A)

Fraction of algae biomass which is P (α_2) = 0.015 (mgP/mgA)

Oxygen production per unit of algae growth (α_3)= 1.3 (mgO/mgA)

Oxygen uptake per unit of algae respired (α_4)= 2.3 (mgO/mgA)

Oxygen uptake per unit of NH_3 oxidation (α_5) = 3.22 (mgO/mgN)

Oxygen uptake per unit of NO_2 oxidation (α_6) = 1.12 (mgO/mgN)

Algae respiration rate (ρ) = 0.05(1/day)

Algae max. specific growth rate(μ_{max}) = 3.0 (1/day)

Nitrogen half satuation constant (K_N) = 0.4 (mg/l)

phosphorus half satuation contant(K_P) = 0.04 (mg/l)

Light half saturation constant (K_L) = 0.03 (lngly/day)

Total intensity of light (L)= 600.0(langeys)

manning coefficient (n)= 0.018($ft^{1/6}$)

rate constant for biological oxidation of $NH_3 \rightarrow NO_2$ (β_1) = 0.3 (1/day)

rate constant for biological oxidation of $NO \rightarrow NO_3$ (β_2) = 0.5 (1/day)

local settling rate for algae (σ_1) = 0.66 (ft/day)

Benthos source rate for phosphorus (σ_2) = 0.0 (mgP/day- ft)

Benthos source rate for NH_3 (σ_3) = 0.0(mgN/day- ft)

Carbonaceous BOD decay rate (K_1) = 0.23 (1/day)

Reaeration rate (K_2)= 0.5 (1/day)

Carbonaceous BOD sink rate (K_3) = 0.0(1/day)

Benthos source rate for BOD(K_4) = 0.0(mg/day- ft)

Light half extention coefficent (λ) =0.6(1/ft)

saturated oxygen (Q^*) =12.0(mg/l)

Table 1. Data avilable in spring 1979

Location	discharge ft^3/sec	velocity mile/day	Bod mg/l	Do mg/l	NH_3 mg/l	NO_2 mg/l	NO_3 mg/l	T.P mg/l	Algal biomass mg/l
D/s Aswan Dam	57222.0	48.3169	1.3	9.3	0.0	0.0	0.3	0.066	6.96
Dishns	53134.9	21.4745	2.5	10	0.0	0.012	0.13	0.6	-----
u/s Naga Hamadi	......	16.1059	1.5	10	0.0	0.0	0.704	0.116	6.6

Table 2.

Waste load identification									
Name	Distance from Aswan Dam mile	disharge $ft^3/sec.$	BOD mg/l	DO mg/l	NH_3 mg/l	NO_2 mg/l	NO_3 mg/l	T.P mg/l	Algae biamass mg/l
El sail drain	6	8.1	15.o	5	75.9	1.8	17.5	1.6	0.0
Naga Hamadi factory	103	1428.4	90	5	0.0	0.0,	0.0	4.9	0.0

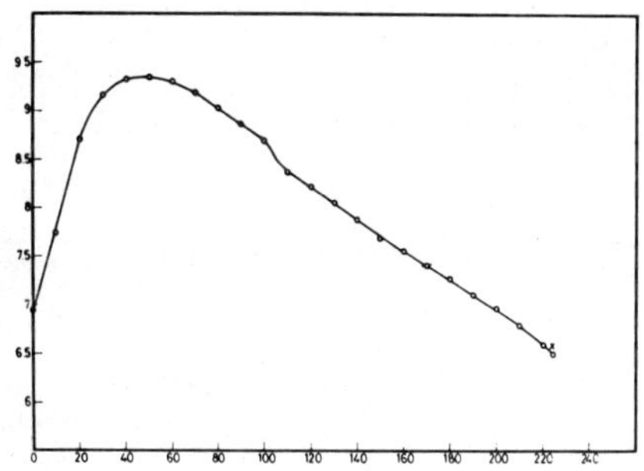
Distance in Miles
Figure (3)

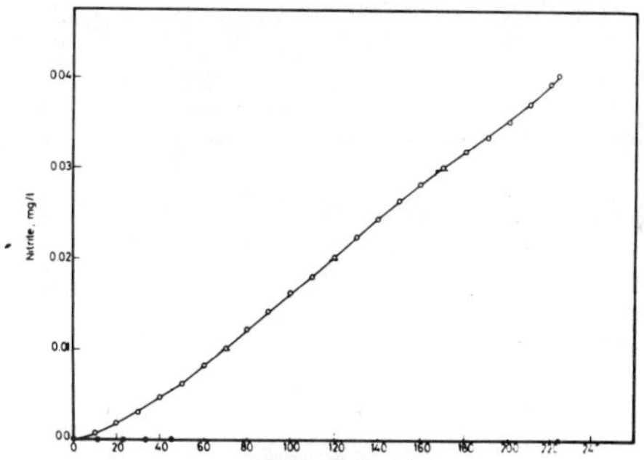
Distance in Miles
Figure (5)

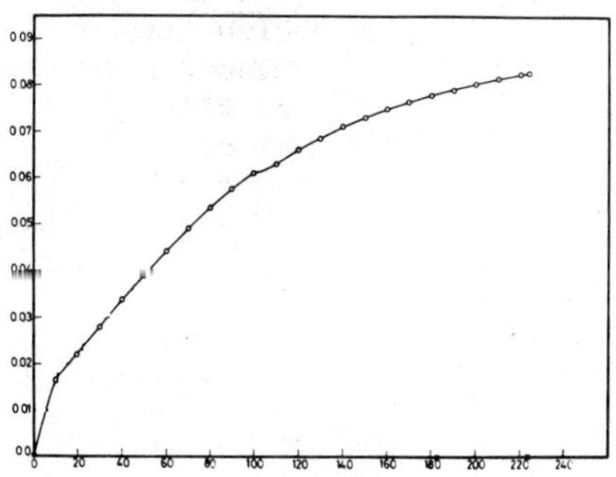
Distance in Miles
Figure (4)

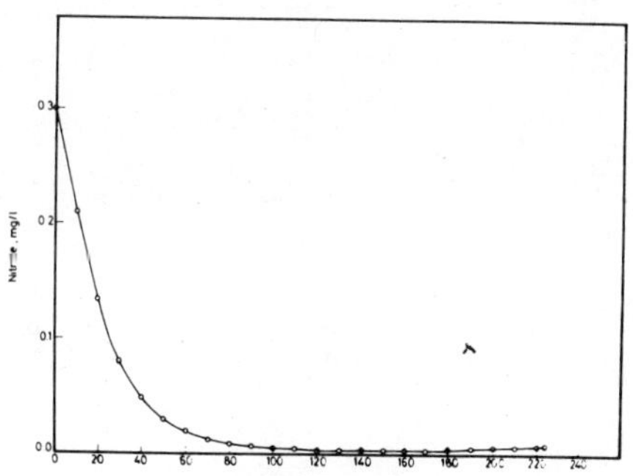
Distance in Miles
Figure (6)

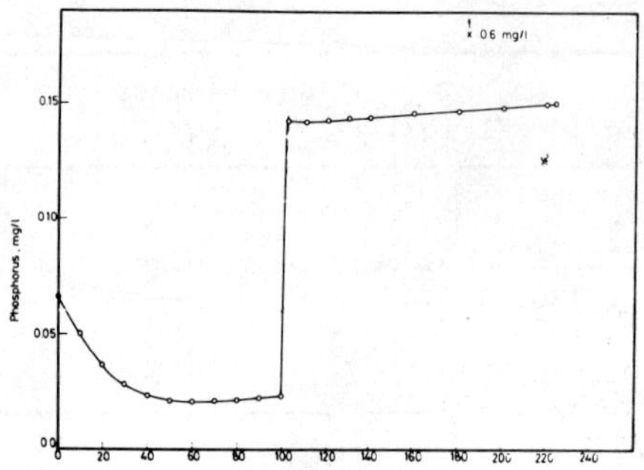

Distance in Miles
Figure (7)

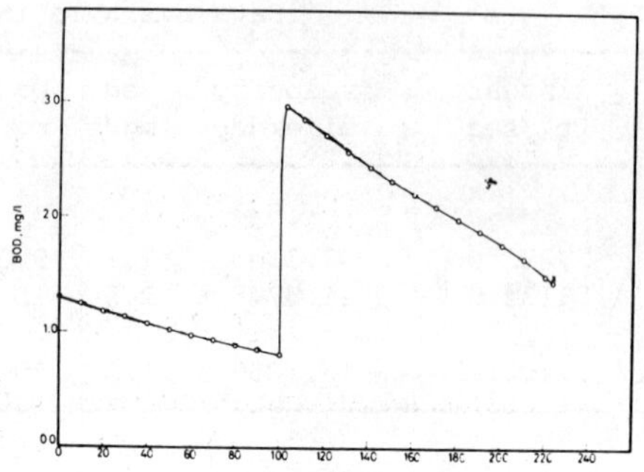

Distance in Miles
Figure (8)

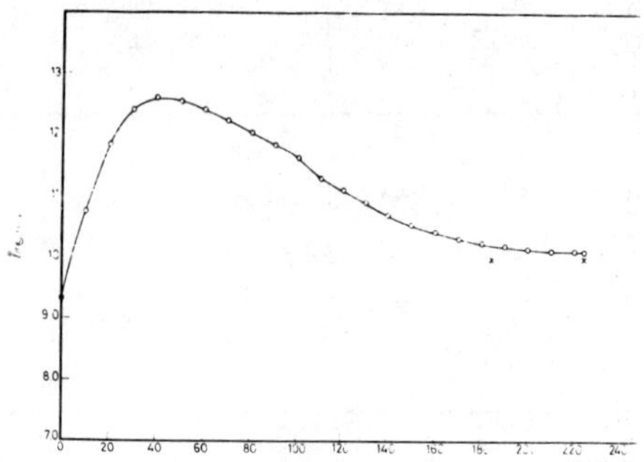

Distance in Miles
Figure (9)

THE DESALINATION BY THE MULTI STAGE FLASH PROCESS AND ITS CONTROL

H. El Hares* and M. El Hares**

**Mechanical and Industrial Department, Faculty of Engineering, Tripoli P.O. Box 13235, Libya Jamahiriya*
***L'Air Liquide, Division Internationale, Tripoli P.O. Box 12328, Libya Splaj*

Abstract. A brief description of the multistage flash type process is given followed by a review of the key parameters and other operating criteria which have to be measured , controlled and recorded .
Then the control loops used generally in such a desalting plant are described. Also the main alarms and trips are given.

Keywords. Heat exchangers ; flow control ; level control ; pressure control ; temperature control ; multistage flash process.

THE MULTI STAGE FLASH TYPE PROCESS

The Multi Stage Flash type process (MSF) is actually the most reliable and economical technique for the construction of large sea water desalination plants.
This reliability is demonstrated by the fact that today this process accounts for more than 85 % of all the sea water desalination capacity , either in operation or in construction worldwide.

The MSF technique can be rapidly described as follows :

The boiling point of a liquid varies according to the pressure in the vessel in which it is contained. Consequently , if a liquid is injected into a vessel in which the pressure is such that the appropriate boiling point is lower than the temperature of the injected liquid , it will boil and vapour will be given off.
The condensation of this water vapour on the outside of tube bundles (where the fresh sea water circulates) provides the necessary heat to warm the sea water. Such a design constitutes a flash chamber (Fig. 1).

By joining a number of flash chambers together and making suitable arrangements to ensure that the pressure in each chamber is slightly less than in the preceding chamber a great total degree of flash-off can be achieved. It is also possible to design the cooling arrangements in such a way that the coolant flows in the opposite direction to the flashing liquid and thus make it possible to raise the ultimate temperature of the coolant to within a few degrees of the boiling point in the first flash chamber.

The final coolant temperature is important as , in the MSF evaporator , the coolant of the heat recovery section , after the addition of heat in the heat input section is passed into the flash chambers to become itself the boiling liquid (Fig. 2 & 3).

The distillate , condensed in the heat exchanger of each flash chamber , drains along the distillate tray beneath the tube bundle next to an orifice which connects it with the distillate tray of the next lower pressure chamber. Here a fraction of this distillate also flashes to give further heating in this heat exchanger along with the main vapour flash from the brine.

Thus , all distillate cumulatively cascades down through all the flash chambers , progressively increasing in quantity while the brine concentration progressively increases. With the final addition of the distillate in the last flash chamber , the total is withdrawn by the distillate pump.

The purity of the distillate is assured by the fitting of wire mesh demisters in the vapour passage of each stage. These demisters trap any droplets of brine which may be carried by the vapour.

An important point is to deaerate the brine thoroughly so as to remove dissolved gases and so protect the plant against corrosion; these gases must also be removed so as not to interfere with heat transfer in the heat exchangers in each stage. The first stage is

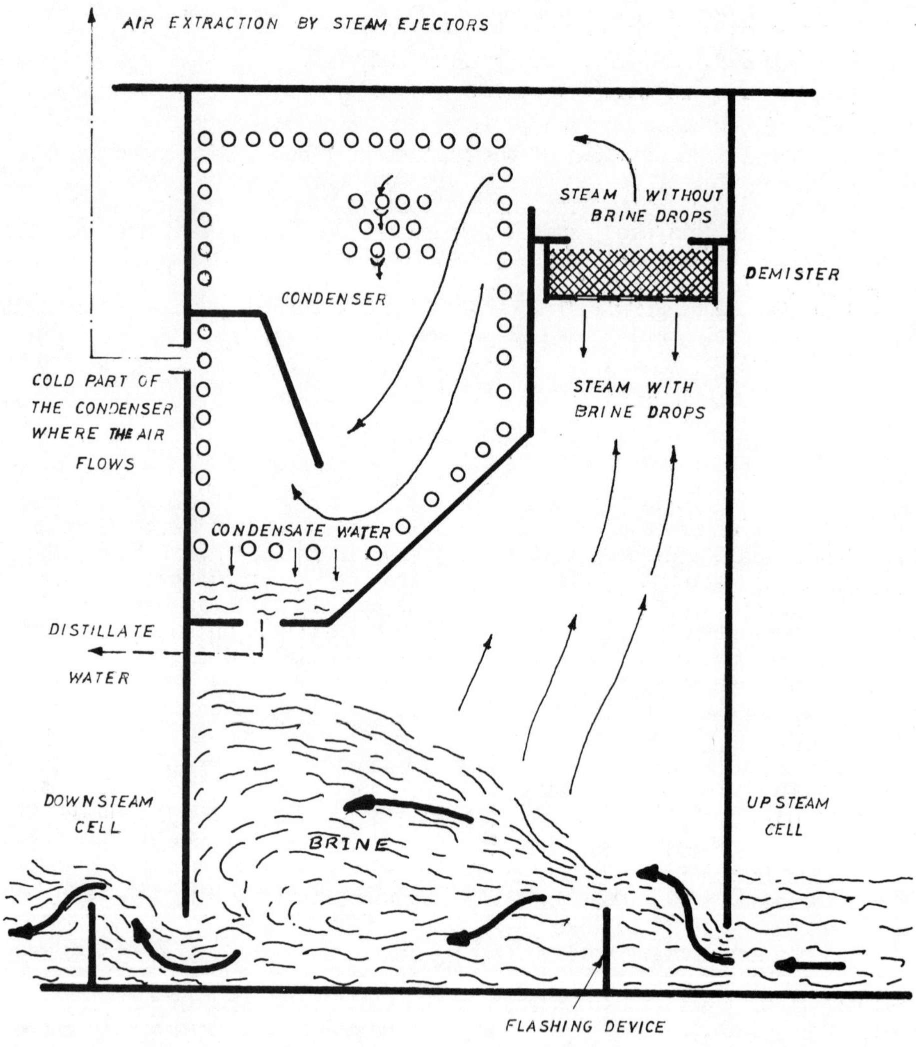

FIG.1: Cross section of the chamber of a flash evaporator

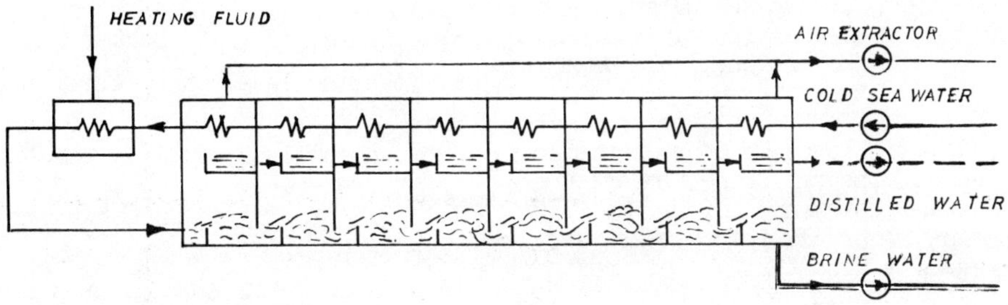

FIG. 2: Schematic of the multistage flash type process_open cycle

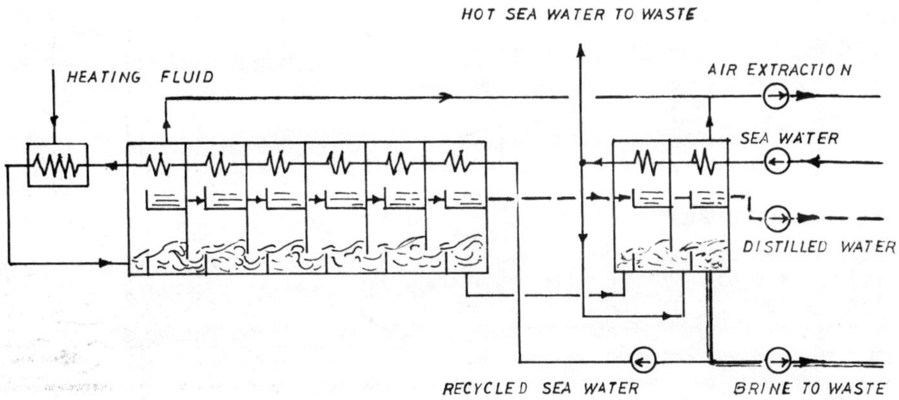

FIG.3: Schematic of multistage flash type process with brine recirculation

generally vented directly to the air ejector, the reason being that , as this is the top temperature stage , most gases will be liberated at this point. Any gases remaining are led through each succeeding stage down to the last stage and thence to an air ejector.

Figure 3 shows the so-called recirculation system where part of the concentrated sea water is not rejected but mixed up with a controlled quantity of fresh make up sea water, the mixture being recycled. This is mainly done in order to save part of the chemicals used for the sea water treatment.

We should note that a major characteristic of a MSF process is the top brine temperature , after the brine heater and before the first flash chamber. Naturally this temperature depends essentially upon the type of treatment which has been adopted for the plant to limit or prevent the formation of scale in the exchanger tubes , scale which would, if formed, rapidly deteriorate the plant performance.

The scale formation is usually governed by the following factors :

* type of chemical treatment (acid,polyphosphate,improved chemical,seeding,...)
* brine velocity inside tubes
* maximum brine concentration inside tubes
* maximum brine film temperature inside tubes

According to El-Saie and others (1980) for plants running with chemical treatment other than acid , the main factors known are as follows :

a) top brine temperature of 90°C for polyphosphate and about 110°C for other improved chemicals providing that the maximum vapour temperature is 103 °C in case of polyphosphate and 125 °C for improved chemicals to limit the brine film temperatures.

This is achieved by sensitive temperature control of the maximum brine temperature with high accuracy of the temperature controller (max. 0.1 °C accuracy) taking into account that the brine quantity is huge compared to the temperature rise in the brine heater.

b) the velocity of brine inside tubes with max. temperatures should not be less than 2 m/sec in the brine heater and 1.8 m/sec in the rest of the plant.

This can be achieved by controlling the brine recirculating quantity.

c) the maximum brine concentration inside the tubes. This is controlled by the blow-down quantity and feed make-up quantity.

This can be achieved by level control for the feed make-up quantities and salinity control for the blow-down quantities.

\- - - - - - - - - - -

MAIN PARAMETERS

The main parameters of an MSF desalting plant are (Fig. 4 & 5) :

. maximum brine temperature at the brine heater outlet

. temperature of the low pressure steam entering the brine heater

. vacuum at last stage

. level of brine heater condensates

. brine level at the last stage of the evaporator

. distillate level at the last stage of the evaporator

. steam condensate quality at the brine heater

. distillate quality at last evaporator stage

. input flow of anti-scale and anti-foam additives

. pH value of distillate output

. % of brine recirculation

Of these parameters , the following are manually controlled :

- the vacuum at the last stage of the evaporator
- input flow of anti-scale and anti-foam additives
- % brine recirculation

Also the distillate quality and steam condensate quality are automatically compared with predeterminate set values and accordingly solenoids are harnessed to operate valves in such a way that if the condensate or distillate reaches a higher conductivity than the preset value , the valve will open to send the relative liquid to waste.

KEY CONTROL LOOPS

Brine temperature at the outlet of brine heater

This temperature is controlled by means of the following (in general) :

. brine outlet temperature PID controller on which a temperature set point can be set

. steam flow PI controller which uses the output of the temperature controller in such a way that any variation in brine temperature induces a variation in the LP steam flow

Note : every flow disturbance must be corrected by the steam flow controller very quickly; every brine temperature disturbance is manipulated through the PID controller , the resultant acting as set point for the PI controller and consequently the steam flow is changed accordingly.

FIG. 4: SYMBOLS EXPLANATION

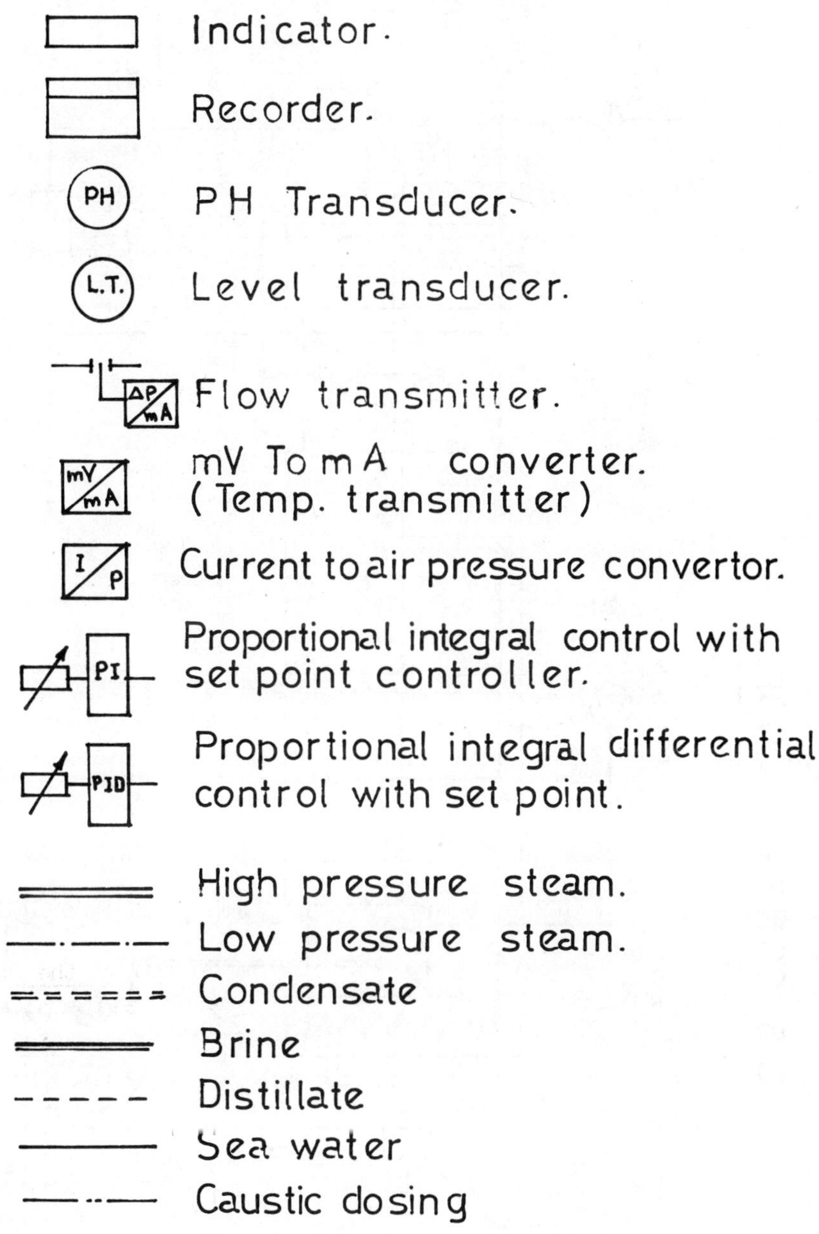

FIG. 5: CONTROL LOOPS

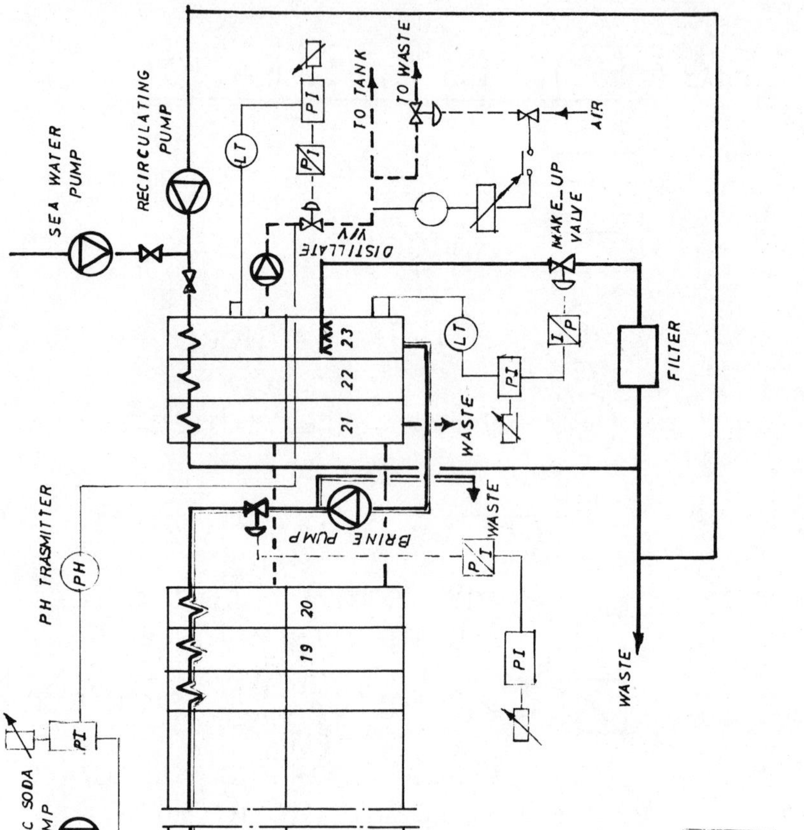

Low pressure steam temperature (desuperheater control)

A PID controller with an internal set point is generally used and takes for information the summation of two signals one of them is the temperature after the desuperheater and the other is a function of the pressure built up in the brine heater. This pressure gives a picture of the heater tubes fouling.

Condensate level control

A PI controller on which a level set point can be adjusted , is generally harnessed to the condensate valve which fits on the discharge side of the extraction condensate pump.

Any variation in the level of condensate in the brine heater induces operation of the controller so as to keep level at the set point.

Distillate level control

A PI controller on which a level set point can be adjusted , is generally harnessed to the distillate valve which fits on the discharge side of the extraction distillate pump.

Any variation in level of the distillate induces operation of the controller so as to keep level at set point.

Level of brine in the evaporator

A PI controller on which a level set point can be adjusted , is harnessed to the make up valve which fits on the sea water make up line.

Any variation in level of the brine in the evaporator induces operation of the controller so as to keep the level at set point.

Distillate pH control

situated at the discharge side of the distillate extraction pump (generally PI controller)

In case of a poor pH , soda is injected into the distillate channel of the evaporator.

Check of distillate and condensate quality

Conductivity measurement of the distillate and brine heater condensate is provided at the discharge valve of the distillate and condensate respectively and a comparator is provided with a set point. Undesirable liquids are sent to waste as explained before.

ALARMS AND TRIPS

The MSF desalination plant should have sufficient alarms, indicators and recorders to help the operators to master the situation from the control room.
Electrical valves should be operated from the control room or locally by push-buttons. The actual position of the valve stem should be adjusted to any required position by these push-buttons.

Main Trips

1) the evaporator brine heater high pressure steam supply isolating valve will close under any one of the following conditions :

- high brine heater pressure (about 1.5 bars abs.)
- high steam to brine heater temperature (about 120 °C)
- low brine pump discharge pressure (about 2 bars abs.)

2) trip of the brine heater condensate extraction pump for low condensate level (about 50 mm)

3) also trip of the desuperheater spray water pump acting on brine heater condensate for low condensate level.

4) trip of brine pump for low brine level (about 100 mm)

5) the discharge valve of the brine pump should close in case of tripping of the pump.

6) the sea water pump should stop if :

- very low level in suction pit
- low level in the lubricating water flow
- a certain time lack of lubricating water

7) if the water in the storage tanks reach a level higher than max. permissible, the distillate should go to waste.

8) the electric valve after the sea water pump should close if the pump stop.

9) if water in storage tanks reaches a predeterminate low level, a signal should stop the pumps taking water from the tanks.

10) if high ppm TDS of distillate water, it should eventually go to waste.

11) if high ppm TDS of condensate water , it should go to waste.

REFERENCES

*Ayoub,A.,F. Abushahma, and H. El Hares (1980). Control loops in the MSF desalination plant of West Tripoli. El Fateh-IFAC Workshop on Appropriate Applications of Automatic Control in the Oil Industry & Desalination, Tripoli.

*El Hares,H.(1978).Economic considerations in the choice of a desalination process in a developing country.Proceedings 6.Int.Symp.Fresh Water from the Sea,Vol.4,87-95, Las Palmas.

*El-Saie,M.H.A.,A.H. Ramadan, and S.S. Gomaa (1980).Multi Stage Flash desalination process control system. El Fateh- IFAC above Workshop.

*Murat,F., and J.F. Deschamps (1980).Control and instrumentation in the desalination plant of Homs. El Fateh- IFAC above Workshop.

EDUCATION AND ITS AIMS

J. Elizondo

Institute of Engineering, National University of Mexico, Apartado Postal 70-472, México 20, D.F., Mexico

Abstract. In planning the first step is to define the target system. For that matter it is necessary to recognize which are the elements to be affected; - those which are under control; and those which affect the first but are not under control. This paper is concerned with the definition of education and its goals as a first step for educational planning. Granted that education is more that the transfer of skills in order to furnish the productive force - required by society, education is analyzed as a transfer of culture. In this process, other social institutions besides schools play an important role. - Among them: the family, the neighborhood and the community; the churches, the mass-media and the markets; the Law, the political parties and the labor - unions; etc. The study is oriented towards general concepts, so that the results obtained may serve as a basis for the analysis and design of any national educational system.

Keywords. Education; human factors; learning system; modelling; social and behavioural sciences; teaching.

I

In pre-industrial societies, institutionalized school was a *desirable* way for the trasmission of culture, or in other words for the socialization of children, to mold their behaviour according to their family's social status. In those times, man's life was determined more by his birth than by the skills acquired. the apprenticeship of his future role in society depended more on the familiar work-a-day activities than on school. But more than a century of industrialization has deeply affected not only man's relations whith nature and artificial things but also relations between men, - specially the school's task.

By and for industrialization man has learned to incessantly modernize the production and distribution of goods, and to produce the - manpower required by this dynamical pattern. Thus, technology and education are the most important means to progress towards welfare. And consequently, more complex and institutionalized schools become a *necessity* for developed and developing societies, for it is the best known way to produce the indispensable bulk of skilled managers, technicians and workers required for the expansion of the - economy.

The increasing importance of educational systems as part of the general social transformation has a twofold source. In the first place, as outlined above, society needs competent - manpower for the convenient functioning of - its growing complexity; this is translated into the educational system as a need for - more specialized people. In the second place, and from the point of view of the individual, education is essential for his upward mobility in the social scale.

It would be naive to believe that society's and individual's wants and the educational - system's performance are all in fitted harmony. On the contrary, there are discrepancies, being two of the most important: a) matriculation and the educational system's output are greater than the offert of employment, causing in general a devaluation of social recognition to education and the consequent frustration of educated people (in some of the most developed countries a diminishing registration at university level is already observable); b) the more complex and modern society is the more specialized people it - requires and education become more narrow and alienable.

Nowadays we are, things and men, to be pro-- duced in the most economically way and almost exclusively as imputs for the economic function. We agree to appoint these processes as progress when we refer to things, but we - doubt when man is the subject. It does not - seem plausible to refuse the benefits from - technically advanced industrialization: nour ishment, shelter, health, security, comfort; but even if we act now as if humans were another instrument for materialistic production, it is no more fashionable to think of man - only as a productive force and a consumer - artifice. Costs of this kind of progress, - and of its associated educational policy, are high if one is to judge according to the - dissenchantment people feels about the attained quality of life, both in capitalist - and socialist countries. This is not new. - Thirty years ago Mannheim (1951) thought that the sources of irrationality in our societies were the manipulation of opinion through mass media and the requiered specialized knowledge which reduces the capacity to understand the complex.

Educational planning is the response to cope with these sort of problems. It is usual to speak of short and long range planning and to associate with the first decisions that aim to improve present performance of the system under planning. With long range planning are associated decisions aimed at changing the system's structure. These associations are related to the importance of changes in the environment in which the systems operate. - Thus, in the short range it is assumed that present conditions will continue to prevail and therefore, future facts (rates of change mainly) can be estimated through the principle that structural relations between environment and system will not change; the procedures - employed for decision making rest on analogies and extrapolations. On the contrary, in the long range the present conditions will - not necessarily repeat themselves and the generative mechanisms which generate economic, political, social and technological situations will change as a consequence of actual - - change's speed rate. The procedures employed to study the long range future, called prospective methods, are based on the analysis of trends by means of a set of experts' inter-- action. Every plan is normative for it determines goals and prescribes actions; but because of the fact that present circumstances will become obsolete and those that will rule the future are not yet implemented, long range planning has a greater degree of freedom to - be normative than the short range planning.

In planning it is necessary first to define - the system upon which is to be affected - - through the prescribed actions. Thus, if a national educational system is under planning - it is necessary in the first place to define what is the *integral* significance of education and its goals. This paper is concerned - with this definition in order to offer a basis for normative planning.

II

Given that education can be approached as a - social subsystem, it could be of interest to determine it through a broader study related with the social system in which education is comprised. However, it was considered in - convenient to proceed in that form for two - reasons: a) Such a task has implicit the need to select a sociological theory, but these - theories permit only partial analysis of social phenomena. Each theory reduces reality to a few variables which are considered to determine the remaining ones; in principle - this could be suitable if there was consensus among sociological theories, which is not the case; b) sociological theories are doctrinaires, and they often seem more interested in changing or preserving society than in - understanding it as theories for action they are valid, and it is doubtless that educational action (including planning) is saturated of doctrine, but here we are concerned with elucidating the essential elements of any educational system, leaving to particular plans the task of coloring it.

It is not efficient to try to analyze education without an idea, even if not accurately stated, of what its siginificance is. Having this idea we avoid the problem of finding - something that we ignore what it is. We start from the fact, recognized by historians and sociologists, that education is culture transmission. From this point of view, four elements intervene in education: an object a *subject*, *a relation* between object and subject, and a set of *criteria* which determine the other three.

III

Object.

The object of education is culture. Besides the man-as a cultural product-culture is whatever outcome of individual activity and his interactions with other men and objects. Just as individual behavior exposes the person's style, culture displays the society's style. In other words, culture is a compound of - human beings, objects and human behaviors - (as individual and social being) before himself and before objects.

Cultural objects are:

Things: every artificial object but only some natural ones. Is is said that culture is the opposite to nature, but this opinion is not absolute. The objects on an island never - tread by man are not human culture, but a flower in a vase or the tree not cultivated but wanted to be kept are part of our cul- - ture.

Ideas: any object of human thought, not only as representation of things but also as mere ideals (for instance, the concept of straight line or of unicorn).

Feelings: which are psychic objects, states of satisfaction or insatisfaction, as joy, hope, frustration and fear. Feelings are means for evaluating situations.

Values: attributes assigned to things, ideas,

feelings and behaviors, and through which objects are desired (utility), believed (truth), liked (beauty) or approved (goodness). Values are important producers of society's style. Among two objects or behaviors, equally efficient and efficacius for the satisfaction of some need, a decisor will select the one that contradicts less the society's values or, in other words, the one that is culturally better accepted. Thus, it is not enough that an object or behavior forms part of a society for it to be part of the culture. It must also be socially valued.

Events: any change in the state of things, ideas, feelings or values provided it is man made and not casual.

Man's behaviors are: a) innate, as in the case of some reflexes; b) conditioned by custom or learned, as the habit of shaking other people's hands; and c) free or less conditioned, which include creative traits. Through his behaviors, and depending on his preferences, man makes culture creating, knowing, transforming, expressing and consuming cultural objects. These behaviors are not completely free, they are neither spontaneously produced in each occasion nor are they independient of each other, forming thus a life's style and a unit integrated by culture. Therefore man makes culture culturally.

Culture and society.

Over the complex web of culture some social features are outstanding for any observer. We detach four of them as society's properties:

Behavior paradigms. Based on the professed values, society guides individuals toward behavioral ideals, opening some doors to their imagination and actuation, and closing others. So, each society develops models about what is valuable in arts, literature, science, philosophy, human interactions, quality of material and spiritual life, etc. Among other consequences of the behavior paradigms, are woman's social role, the size of the family, and the distribution of the population ages.

Forms of participation. It is a group of relations among individuals or among groups of them, which has the function to control the rest of the social interactions, distributing the valued things among the members of society, defining what is fair and what unfair, and in general, practicing the public power. Among other issues, a direct consequence of the form of participation, is the distrubution of wealth, health, education, information, justice and security.

Forms of production. It is a group of man's relations with things, including the way men organize for production purposes. It represents the degree of technical and economic development.

Foreign relations. They are dominance or independency relations held by society with respect to other social groups, in the cultural, political and economical fields. These relations define the environment of the society and partially determine its evolution.

Related to the precedent properties three forms of wisdom are distinguished:

General wisdom. Standardizing and socially needed. To this wisdom belong all the ideas, values and purposes that, in accordance to the collective feeling, do not need justification. It is different for different societies and changes with time for the same society.

The basis for the individual labor and political performance are his individual character and his personality, which are formed from the first years of life by the family and school. These institutions trasmit socially accepted behavior models, through which children learn to restrict, develop or guide their natural drives, and to integrate themselves in society. In this way man acquires conscience of what is true, beautiful, beauty, good and worth. This heritage of wisdom allows man to know what he can expect from other men, and what they expect from him. But on the other hand, this standardization could become alienable when in excess, causing social pathologies, as has been expressed by Caso, Cazeneuve, Freud, Fromm, Henry, Linton, Marcuse and many others.

Doctrinal wisdom. A group of prejudices, that supported by theories beloging to the specialized knowledge form ideologies.

The last are, in their turn, shared by people belonging to a determinate social group, as social classes, political parties or religions sects. There are different doctrinal wisdoms in the same society. In general they are antagonistic for they serve as guide of different groups' actions.

Implicitly or explicitly education trasmits doctrine and the school is, for that reason, a political institution. If the social system is totalitarian the educational criteria will be changeless; if it is liberal, it will permit some doctrinal freedom, but not to the point of setting the system in danger. Education will be tried as an instrument to prevent the generation of fundamental political change; but it will be urged to develop a new education when social reformism is in the wind. If as a consequence of a revolutionary action present social relations colapse, then education will be one of the first institutions that will be reorganized, modifying the contents and methods, but keeping its doctrine trasmission function.

Specialized knowledge. This is the case of philosophy, and of the formal and empirical sciences together with their practical applications.

Production and distribution of goods and services determines the quantity and the qualifications of the labor force. Education must provide competent and sufficient people for all classes of occupations. In last decades, governments and international organisms have promoted this educational function as one of the conditioning factors of development. Controvertible or not, this assertion has had as a consequence a considerable expansion of the demand for education, together with undesirable social effects because in countries wihout central planning it is frequently observed that the jobs' supply doesn't growth at the same rate as the jobs' demand.

IV

Subject

The subject of education is the human being. With respect to the subject we just point out that in any period of a man's life the acquisition of culture depends on: a) the inherited physical and psychical structures; b) maturity reached (maturity: capacity for acting without being teached); c) culture previously obtained.

The way in which man apprehends culture depends on: a) the cultural object appearance, or its sort of being, or the form in which is given to the observer; b) the means that man has to be aware of cultural objects. Sources of knowledge of human beings are reason, sensation and intuition. For instance, ideal objects, as mathematical or logical ones, are not known by sensation, but by reason or intuition; things which have extension are apprehended by sensation or their existance is rationally deduced or inferred; some values, beauty in particular, are perceived only by emotional intuition.

To acquire culture is to learn. To learn is the process through which a person increases his knowledge or his beliefs, and through which he modifies his behaviors. Any changes resulting from innate conditioning, maturity or provisional states caused by intoxication, emotional stress or fatigue are not included. Approaching the apprentice, he may *learn that* or *learn to*; in general, following Scheffler (1973), when the student *learns that* it implies that he *knows that* or *believes that*, and when he *learns to* then he *knows how*.

V

Culture trasmission.

The idea of the world is transmitted to the new generations mainly by four mechanisms:

Family and Social life; they mainly influence the formation of a basic personality through the injection of general wisdom. They also influence doctrinaire wisdom, for prejudices are partially determined since the first years of life. They are equally important for the attainment of superior culture because of the opportunities given by the economical and social conditions of the family and the environment.

Mass-media, which influence the basic personality formation through the reinforcement of accepted social behaviors; in particular those related with welfare. They also influence opinion, that is, the beliefs and prejudices apparently supported by facts. Mass-media are potentially capable of for the promoting or destroying aesthetic and moral values.

Teaching-learning relations in all educational institutions. Their influence comprise all wisdoms and knowledges.

Personal practice, the more efficient mechanism to obtain specialized knowledge. Scientific, technological and humanities research are included, whose influence is very important because through them culture is filtered to the other mechanisms.

Besides educational institutions, other institutions which have no explicit educational purposes participate nevertheless in culture trasmission. They belong to one of the following social spheres:

Familial. Society's basic unit grounded in kinship relations and in which the members share resources and shelter. Also included are: kinsmen or kinswomen who do not share resources and shelter; those who not being a kinsman or kinswoman share resources and shelter; other who have friendly and familiar relations.

Community. Social mit geographically differentiated in, which members reside close to each others, have frequent interactions and are conscious of their interdependence.

Colective personality. Integrated by those institutions that give character to society's style, such as the churches, mass-media information and recreation, clubs and professional associations.

Political. A set of norms, organisms, pressure groups and mechanisms for political participation, through which political power, social control and public bussines decisions making are practiced.

Production and distribution. It is a social sphere formed of units that produce goods and services, and by their markets.

Scientific and technological. Integrated by scientific and technological research institutions and by information and promotion systems which support them.

VI

Educational criteria.

To describe education as culture trasmission

is not enough to define it because in that action several factors intervene which give significance to or determine both culture and trasmission. Our descriptive definition must be completed approaching education as a set of criteria. A criterium is a rule that assigns values, used to decide what is false or true, good or bad, how efficient something is, etc. It is also employed to determine a desirable state as to guide action toward some definite state of a system; in this case, more general objectives serve as a criterium for the less general ones; as an object the criterium is confused with the function (Aristotle wrote that the function is the end and the act is the function), for instance, when it is said "to promote the critical thought of students". Approached as a criteria set, education is no more an activity as teaching is, but an expected result, that is, a desirable ex-post state to be reached by the student. To say that somebody is an educated person depends on the culture that he is expected to have acquired, and therefore on the educational criteria.

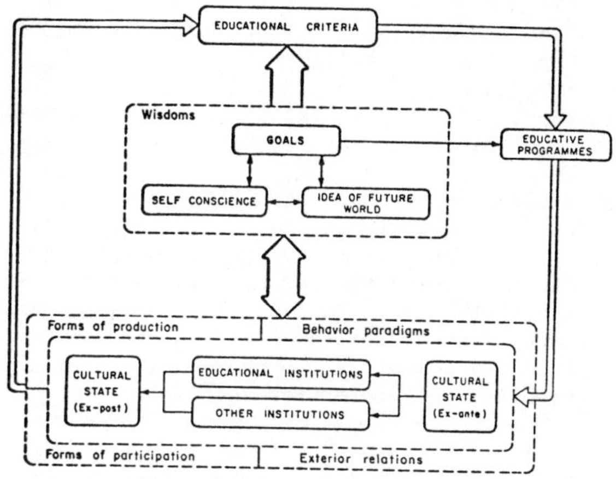

Society, professional groups, family and shools are institutions that specify what must be an educated person. The normative capacity of each group varies with their decision and action power over the educational process. In any case, criteria and power must be legal, and legality proceeds from the *social contract* that men have established at the time they meet in society; men, land and social behavior norms form the state. The course followed by society, and the educational system, is partially controled by the State, thus performing its *government* function. This task is accomplished imposing criteria or rules that according to its judgment lead toward beneficial situations. In theory the norms that state produces to govern education result from decisions which take into account the following images: a) the three wisdoms, or ideas of the world ; b) the self image, formed from the four society's properties; c) the idea of future world; and d) the goals it desires to attain (see figure).

Educational institution. With the introduction of the concept of criteria it is possible to define educational institutions as organizations that have education as their explicit purpose and that norm their activities with a set of criteria related to the ex-post cultural state of teir students and to the content and methods of culture trasmission. Beside, educational institutions requiere a set of administrative criteria for to make decisions about resources allocation.

Summing up, in order to determine education as culture trasmission two sets of criteria are necessary: ideological and pedagogical. Educational institutions require a third one, of administrative nature, for the distribution of educational services:

Idiological criteria, that define the convenient people's cultural state, and, consequently, determine educative actions. In their more general form, they take into account the basic personality acquired through the famlily and social life in order to prescribe which idea of the world must be transmitted to the commun citizen and which doctrine must be emphasized. In more particular issues, they take into account the state of production, wealth distribution, health, education, information, justice, security and other variables related to demography in order to prescribe about the social utility of a given specialized knowledge and of the educational services distribution.

Pedagogical criteria, that take into account the ideological criteria in order to prescribe to content and method (goals and means) of each educative programme. Based on these criteria the educational institutions make decisions about: a) the inclusion or not of teaching, training, indoctrination and conditioning as educative methods; and b) teacher-student relations.

Administrative criteria, through which educational institutions determine the budget which must be allocated to each educative programme.

VII

Educational aims.

We distinguish between aims and criteria. An educational aim expresses a student's cultural state property that is desired to be changed; for instance, the knowledge about material resistance. The criteria specifies the quantity and quality of the desired change and the method to be employed in order to accomplish it. Thus, the criteria norm and restrict the aims. The aims, or goals, and the means are both relative terms; means are goals that allow reaching other more general goals. We call *educational basic aims* to those educational goals that are pursued by all societies despite their behaviors' paradigms and participation and production forms. On the contrary, educational criteria are determined by these three properties of society.

We assert that the educational basic aims are the following three:

1. Socialize, through which the individual - acquires a cultural state that allows him to live with other indivuals, despite his position among them. Thus through socialization the individual develops his physical and psychical aptitudes and learns socially accepted behaviors. To that end, education promotes the properties that follow.

1.1 Rationality. which is related to the knowledge and practice of: a) the language, as a mean of reasoning and communication; b) the explanations of the world (physical and metaphysical); and c) the transactional activities with other men (personal relations and social relations) and non-human world (natural and artificial objects).

1.2 Aesthetic taste, which is influenced - - through: a) the approved values in the - social milieu; b) the experience of beauty, that is, the opportunity to have aesthetic intuitions; and c) the artistic expression, not precisely as the product of a specialized artist but in the daily occupations.

1.3 Behavior norms, which are propitiated by means of: a) the knowledge of social agreements; b) the exercise of conditioned - freedom; and c) the practice of altruism, as the acknowledgement of other's freedom and the disposition to collaborate with other men.

2. Politicize, which is the formation of the individual's social consciousness. Through this the individual acquires capacity for the analysis and evaluation of social and political situations, and is able to propose and prosecute social changes. Education oughts to promote:

2.1 Ability to be informed, or the possibility to be linked with the community's issues and problems as a basis to form one's own opinions. The ability to be informed depends on: a) the consciousness of the need to live as a political man; and b) the - selection of reliable sources of information.

2.2 Critical attitude, or a disposition to - examine and reform the established social life conditions; and which is propitiated by means of: a) the understanding of human relations; and b) the ability to judge complex social situations.

2.3 Tendency to action, which is influenced by: a) the ideals, as absolute goals; and b) the will, as determination or energy to be employed in the prosecution of one's ideals.

3. Quality, which is a improvement of knowledge and skills of the individual in order to become productive and useful for the society's and his own purposes. To that end it is necessary:

3.1 Scientific knowledge, which is the basis of any profession.

3.2 Practical apprenticeship, which is the experience obtained living the work's situation as an actor. The apprenticeship is integral if the individual has to: a) propose problems; b) transform a given situation (material or intellectual); c) propose feasible solutions; and d) make decisions.

3.3 Valuation of one's profession, or a constructive attitude towards a self development as professional and giving importance to one's social role.

Other educational institutions' aims. Besides the three basic educational aims, schools have other aims that, without being educative, are nevertheless of social character. One of the most important non-educative aims of educational systems is to reduce social tension. Education has become one of the most demanded commodities in our time. Therefore, if governments do not provide sufficient education, then social unrest could rise. Another non-educative aim of educational systems is to serve as a nursery; even if the school would not be a desirable way for culture - trasmission, modern societies would requiere an institution to take care of children and young people while parents are occupied in other tasks.

Applications and conclusions.

The outlined conceptualization of education and its aims has been useful in a current mexican educational planning process. It was mainly applied in two instances of the process: a) as an aid to define the target system and its environment, through a matrix that relates the educational aims with the social spheres which affect education; and b) as a guide to facilitate the questionaire posed to the participant people in the setting of the educational criteria.

In order to make the purpose of educational planning clearer, it is necessary to do two things; first, to define what is meant by education and its aims and to distinguish between aims and criteria; and second, to define which are the elements that influence education and which of them can be controlled by planners.

Bibliography.

Caso, A., 1975, *La persona humana y el Estado Totalitario*, UNAM, México.

Cazenueve, J., 1966, *Bonheure et Civilisation*, Gallimard, Paris

Freud, S., 1979, *El malestar en la cultura*, Alianza, Madrid.

Friedmann, J., 1973. *Retracking America*, - Anchor Press/Doubleday, New York.

Henry, J., 1967, *La cultura contra el hombre*, Siglo XXI, México.

Linton, R., 1978, *Cultura y personalidad*, - Fondo de Cultura Económica, México.

Mannheim, Karl, (1951), *Man and society in an age of reconstruction*, Harcourt, Brace and Company.

Marcuse, H., 1968, *El hombre unidimensional*, Joaquín Mortiz, México.

Sheffler, I., 1965, *Conditions of Knowledge*, Scott, Foresman and Co., Illinois.

Shofield, H., 1972, *The philosophy of education: an introduction*, Unwin Education Books, England.

"PURPOSEFUL" PHYSICS CURRICULA: A SYSTEMS MODEL

M. V. Ananthakrishnan

National Institute for Training in Industrial Engineering, Bombay, India

Abstract. Good teaching and curriculum development demand creative, careful and sensitive planning. It involves a decision on the nature of relationship between the ends that one wishes to accomplish and the means available for the purpose. It is here that curriculum development makes its presence felt.

An effective systems model will have to basically to evolve out of the inputs given by the teaching institutions, the receptivity of the students and the outputs reflected by the experiences of the end-users of the students, viz., the industries. Making use of the ends-means concept, data (collected from colleges and industries in Bombay) was used to develop a systems model. This systems model takes into account the various intrinsic and extrinsic factors that effect the college-industry interface as well as the effects of one on the other. The present paper deals with the formulation of the model, its analysis and properties vis-a-vis a typical Indian college/ industry environment - with specific reference to physics.

Keywords. Education; models; systems; physics; teaching.

INTRODUCTION

The effectiveness of any educational system or training programme in a country is evident from the economic and cultural development of that country. India is a country with an ancient cultural heritage and traditionally has an agricultural base. It is amidst this that the country has gone in for rapid industrialisation.

The present educational methodology in this country treats "pure sciences" and "applied technology" as two water-tight compartments. As a result, the interdisciplinary concept is virtually absent. Therefore, science graduates find it difficult to fit into industry without prior training (Ananthakrishnan, 1974).

Objectives lie at the very heart of the planning process whether one is planning a curriculum or a single classroom session. One has to decide after due consideration, the nature of the relationship between the "ends" that one wishes to accomplish and the "means" that are available for use (Davies, 1976), To quote Ackoff (1970): "Wisdom is the ability to see the long-run consequences of current actions, the willingness to sacrifice short-run gains for large long-run benefits, and the ability to control what is controllable and not fret over what is not".

When one speaks of education, one has necessarily to make reference to two issues viz., what are the aims and objectives of education and what are the defects in the existing system ? (Acharlu, 1975). This approach has relevance because "if everything is alright with the ends and means of education, there would be nothing left for us to do except evaluating what is being done and suggesting ways and means of improving the system in consonance with its ideals and objectives" (Acharlu, 1975).

SYSTEMS APPROACH & CURRICULUM DEVELOPMENT

Systems Approach

One of the general tenets of the general systems theory is that various properties are inherent and central to all functioning systems such as found in the mechanical, biological and social worlds. Systems are any set of components which could be looked at as working together for an overall objectives of the whole (Athey, 1977). In general, a system comprises of its components, its system boundaries and the environment and system interfaces.

Components are the primary elements which comprise a system whereas the system boundaries comprise that set of components which can be directly influenced or controlled in a design. The environment includes all those factors which have an influence on the effectiveness of the system, but which are not controllable. Interfaces are those boundaries where two systems meet and can be internal and/or external to the system itself.

Curriculum development

Curriculum is defined as all the learning experiences provided for children and youth under the direction of the schools/colleges. It, therefore, includes the programme of studies, methods, techniques and strategies of teaching.

The development of curriculum and accompanying instructional material is a continuous process in the field of education. Consequently, the steps involved in developing a new curriculum (or in revising an existing one) have been fairly established. Curriculum development can take place as a joint activity throughout the session or it can take the form of a resource unit made available as and when needed (Cook, 1966).

The constructs of any curriculum design consists of three areas, viz., the learner, the society and the organised knowledge. Whereas the learner and the society are equally important, it is the organised knowledge that calls for emphasis in any teaching situation. The organised knowledge is the actual subject-matter that is to be taught in any instructional programme (Gottesman, 1976).

AN OVERVIEW OF SOME STUDIES IN SYSTEMS APPROACH TO CURRICULUM DEVELOPMENT

The application of a systems concept to curriculum development is a well-established method by now in as far as educational planning and training methodology are concerned. de Graff (1971) came up with the Helical Physics Sequence and the FLIPS programme for teaching undergraduate physics courses at the University of Michigan (Flint). His Helical Physics sequence essentially makes use of the concept that learning is gradual and becomes more and more advanced as one goes up the helix. On the other hand, the FLIPS (Flint Introductory Physics Sequence) programme concentrates on the concept of interdependence of the various components or modules amongst themselves as well as within a module.

Ananthakrishnan and Nambi (1977) have reported of a systems approach to physical sciences teaching. According to them, integrated science teaching involves the teaching of the various components of science through the medium of group teaching whereby the teachers have to interact amongst themselves. The systems approach, as proposed by the authors, involves the teaching of the three subjects i.e., physics, chemistry and mathematics, in any one of the three ways, viz., independent coverage, interaction between two subjects or interaction amongst all the three. Figure 1 shows a schematic of this concept. From the figure it is evident that the chemistry and mathematics teachers have to co-ordinate with each other as well as with the physics teacher in order to explain the various topics and their applications. However, the physics teacher need only co-ordinate with the chemistry colleagues because the former has sufficient knowledge of mathematics to handle such requirements.

Ananthakrishnan and Garg (1978) have given a systems concept to Industrial Engineering education at the post-graduate level. According to them, the guidelines required for proper development of the sub-systems/modules and their interaction(s) are as follows:

1) All individual self-contained modules are of equal importance.
2) Every module should have a common basis for its contents.
3) Subjects that find applications in other areas must be taken as a separate module.
4) Proper interface components are needed in each module.
5) Modules should be made up of basic topics, only. Peripherals could be left as self-study assignments to the students.
6) The modules, so selected, should essentially cater to the students' specialisation requirements.
7) Every module should be flexible enough to accomodate variations in the students' background.

The authors have proposed a three-level hierarchial model viz., BASIC, Level-1 and Level-2. There are, in all, nineteen basic modules and 22 level-1 modules. The level-2 modules are the electives and depend, to a large extent, on the availability of faculty and students desiring specific areas of specialisation.

Gottesman (1976) proposes a conceptual Curriculum Network for curriculum planning. It is a systems approach model that "diagrams the influence of constructs on philosophy, goals, objectives and instructional programmes". It also shows the flow from philosophy to goals, from goals to objectives and from objectives to instructional programmes. As a finale, evaluation is seen by him as a feedback for not only making improvement during planning but also for reporting to groups on the effectiveness of the curricula.

PURPOSEFUL PHYSICS CURRICULA : SYSTEMS MODEL

The need for a systems approach

A formal education is needed by every individual in order to place him or her on a proper footing in life. In as far as science is concerned, the graduates coming out of the university have various openings to start a professional career. If we now concentrate our attention on the areas of teaching, research and industry, one should ideally find a complementary relationship between teaching/research and industry. Unfortunately, this is not the case with a majority of the institutions and industries (Ananthakrishnan, 1974).

Good teaching and curriculum development demand creative, careful and sensitive planning. In a very real sense, planning is anticipatory decision-making. It involves deciding on what to do and how to do it before any concrete action is taken. When both "ends" and "means" are known, no real problem exists. The problem, however, arises when the ends have not been determined and the means available are unknown. This leaves the teachers and curriculum development experts in a dilemma. It is here that one has to resort to the overall systems approach.

The teaching institution, the student and the end-user (industry) constitute the system with ample scope for feedback channels. An effective systems model will have to basically evolve out of the inputs (from the educational institutions), the receptivity (of the students) and the outputs (reflected by the experiences of the user-industry). Figure 2 is a schematic representation of this systems concept.

The institution-industry environment and interface

The environment existing in the institution and the industry can be said to be made up of

1) The social background of the students (or workers)
2) The educational background of the students (or workers)
3) The facilities available for practice and shop-floor experience (in the case of colleges) and pre-induction training (in the case of industries).

Figure 3 is a schematic representation of the environmental factors. In as far as the interfacing between the institution and the industry is concerned, a deeper insight is called for to study the nature of this interface. The figure is indicative of a definite interface between the two "environments". However, the size and extent of the interface is dependent on the various factors that exist in real life. It is this interface that is very vital in making or marring a good curriculum. Therefore, one needs factual information on the nature of this interface. One this interface is known (or gauged), it can be exploited to develop a "purposeful" curricula in any subject of interest.

Information systems for "purposeful" Physics curricula

It would not be out of place to once again mention the three sources of information that are essential for any curriculum development activity viz., the teaching institution, the student and the end-user industry. However, the present study will concentrate on only two of these aspects i.e., the teaching institution and the industry. This is necessarily because of the fact that one is attempting to develop a purposeful curricula - one that will help the student fit himself with very little time-lag into an industrial environment. Needless to say, such an attempt will tend to be localised to a particular place or region. But it is imperative that this model is a worthwhile starting point that could be made more broad-based in due course.

Relevant and invaluable information was requested from industries and colleges in the metropolis of Bombay. The medium of gathering information was both through questionnaires and personal discussions.

Information from colleges. The colleges were requested to give information on the following aspects:

1) Teaching methodology.
2) Laboratory work
3) Evaluation.
4) Support activities.
5) Rethinking on existing curricula.
6) Employment of science graduates.

TABLE 1 Opinions of colleges on physics curricula

Aspect	Opinions
Teaching Methodology	The teaching methodology is evolved both as individual effort as well as a group activity. There is a limited attempt to incorporate typical industrial/real-life applications in the lectures.
Laboratory work	The laboratory assignments are very rarely co-ordinated with the corresponding theory classes. Practical work is done in groups (of 2 each). Guidance is provided through instruction sheets and in-situ guidance by faculty. Laboratory assignments are usually directed towards prooving existing laws and determining well-established physical constants. About 50% (of the total time alloted to physics instructions) is spent in laboratory work.
Evaluation	The university examination questions aim at testing the memorising capacity of the students. The questions so set are a mixture of objective-and essay-types. The practical assignments (set at the university examinations) are selected from amongst those performed during the regular college classes.
Other activities	Casual visits to industries/research laboratories. Casual lectures by eminent scientists, film-shows, study tours. Scientific journals are subscribed to but referred to by students to a very limited extent. Developement of teaching aids is done in a very restricted sense.
Rethinking on existing curricula	Need for designing laboratory assignments so as to reflect the real-life/industrial applications. Suggestion for a semester system with periodic tests to discourage the students from resorting to end-of-the-term preparation as practised now. Objective-type of questions should replace the essay type - to dissuade students from "rote" learning. Moves are afoot to revise the existing curricula.
Placement of graduates	About 50% of the graduates are absorbed by industries. The balance go in for clerical, sales/marketing, research etc. The graduate do not immediately fit into an industrial environment. This is assigned to the lack of any practical bias in the college courses. Attempts are underway to remove this lacuna.

Information from industries. The enquiries directed to the industries solicited information concerning the following aspects:

1) Educational background of employees.
2) Nature of jobs handled.
3) On-the-job training needs.
4) Need of physics and its applications.
5) Avenues for improvement of academic programmes offered by the colleges.
6) Continuing education.

Analyses of information received. The opinions of the various colleges are summarised in Table 1. The aspect of "Rethinking on existing curricula" makes interesting reading. The college faculty have expressed a necessity to make the teaching more realistic and suggestions have been put forth for better teaching, effective laboratory work and purposeful evaluation.

Table 2 enlists the view-point of industries on the existing college curricula. Important observations include the need for continuing education and an industry-based project. Continuing education programmes are felt essential inorder to keep the employees abreast with the latest developments in science and technology as well as the allied basic concepts. The industry-based project is recommended as an essential prerequisite for qualifying for a degree. This suggestion surely warrants serious consideration and possible implementation.

incumbent has to be weaned out of an entirely academic environment before being fitted into the industrial work-situation and environment.

TABLE 2 Opinions of industries on science graduates/college courses.

Aspect	Opinions
Educational Background	About 30% of the employees have basic science degrees.
Nature of jobs handled	Quality control, Inspection, shop-floor supervision, clerical, sales.
On-the-job training requirements	All new incumbents (with basic science degrees) have to necessarily undergo pre-induction training in the industry. This is because of the lack of any "industrial" bias in college classes.
Need of physics and its applications	Physics, being a basic subject, finds a place in virtually every field of industrial production/activity.
Avenues for improvement of college courses	Inclusion of typical industrial applications in regular college classes and laboratory assignments. Involvement of technologists in curriculum planning, specialist lecture sessions and industrial visits. Inclusion of industry-based project work as an essential requirement for qualifying for a formal degree.
Continuing Education	Essential for keeping the employees abreast with recent developments in science and technology. The innovation of new instrumentation and processes calls for educating the users on all the basic physical principles involved.

The MVA Systems Model

The development of the systems model requires that one should be fully familiar with the various parameters that influence the system as well as other systems. Figure 4 is a more pragmatic version of the conceptual model depicted earlier in Fig. 3 and this will be used to develop the formal analogue model.

Premises. The formal model has been developed on the following premises:

1) The college provides all the knowledge to the student which is retained and recalled whenever needed. The extent of retention is a personal factor of a particular student.
2) The efficiency or rate at which a student picks up the knowledge is dependent on his inherent characteristics, social factors, pre college education and the teaching methodology.
3) Any variations in the teaching methodology or the learning characteristics will have a transient influence on the efficiency of the teaching situation.
4) The employee enters industry with a definite amount of theoretical and/or practical knowledge. This will be subsequently used and/or reinforced by the industrial training/experience.
5) Any industry has to essentially train any new recruit. During this process, the incumbent has to be weaned out of an entirely academic environment before being fitted into the industrial work-situation and environment.
6) Having entered industry, the employee is, many a time, shifted from one department to another - as a matter of company policy or with an idea to provide all-round knowledge of the organisation. This sets up temporary "readjustments" which remain as transient effects until they smoothen out.
7) The educational institution and the industry need to have a regular liaison with each other if the research activities in the former and the innovations/developments in the latter have to find a place in their dynamic policies. As a consequence, the college must necessarily update and review its syllabi and methodology. Similarly, the industry has to explore evenues to adapt the research findings to its production activities.

Analogues. The formal model for the proposed systems approach is shown in Figure 5. The analogues for the various premises are given in Table 3 with their corresponding equivalences.

The electrical analogues for the factors affecting the individual/educational institutions are represented by R', L' and C', whilst the corresponding analogues for the industrial situation are R", L" and C". The D. C. sources (in the circuit) are the

TABLE 3 Analogues for the various premises

Premise	Analogue	Equivalence
Premise 1	Capacitance	Retention of knowledge imparted in the colllege.
Premise 2	Resistance	Resistance to change depending on society, school education and teaching.
Premise 3	Inductance	Transient influence of teaching methodology etc. on efficiency of teaching.
Premise 4	Capacitance	Existing knowledge of entrant to industry - periodically updated by on-the-job experience
Premise 5	Resistance	Resistance to change from an academic environment to a purely industrial one.
Premise 6	Inductance	Temporary readjustments that are warranted by regular rotation on jobs as well as new jobs/assignments.
Premise 7	Mutual Inductance	The existence of an effective liaison between institution and industry is to be realised. Further, any change in one definitely affects the other.

analogues for the knowledge (via the teacher) and the application (from industry) in the respective environments. A D C source (and accordingly the analogue circuit) is presumed because (i) knowledge is being imparted at a steady and continuous rate and (ii) applications in industry are being imbibed by the employee in a scheduled and regularised fashion.

The mutual inductance (M) is the electrical analogue for the interaction between the institution and industry. The degree of influence depends on how closely one works with the other (the electrical equivalent of the proximity and closeness of the primary and secondary windings of a transformer). Any research/development and/or change in the gamut of education will have its influence on the industrial situation (an electrical analogy of transient currents in either of the circuits affecting the equilibrium state of the other). A similar argument can be extended for explaining the influence of industrial innovations. The other equivalences between the institution-industry cycle and the electrical analogue are mere corollaries.

A sample curriculum. The metal casting industry is one which finds a variety of applications of physics and offers employment opportunities for science graduates. Thus, the inclusion of typical examples (from this area) in an undergraduate physics curriculum would be invaluable - the actual methodology being worked out by the college and industry. As an illustration, let us consider the melting practice in the metals casting industry.

The physical phenomena involved are: heat transfer, thermal properties of solids/ liquids, magnetic induction and heating, electric arcs and their properties.

The physics inputs recommended are, therefore; heat and temperature, heat transfer, Ohm's law, electrical heating and Joule's laws, magnetic behaviour of materials, A.C. and D.C. circuits, frequency generators, temperature measurement.

CONCLUSION

The proposed systems model gives a realistic representation of the vital and oft-misunderstood institution-industry interface. What is needed is an effective control and capitalisation on the various factors that inhibit/accelerate this purposeful co-existence for the ultimate improvement of man's living conditions in this world.

A number of extensions of the proposed model are possible so as to include additional factors and/or refinements in the approach. Some of the factors could be the retention/learning rate of the student/ worker, the saturation of research/development activities, influence of political factors, etc.

ACKNOWLEDGEMENTS

The author wishes to thank the various colleges and industries who contributed their valuable advice/suggestions towards making the systems modelling more mundane. The NITIE authorities are to be thanked for providing all facilities needed for making this research see the light of the day.

REFERENCES

Acharlu, K.S. (1975). Education for Today & Tomorrow, Indian Council for Basic Education, Bombay.

Ackoff, R.L. (1970). A concept of Corporate Planning, Wiley, New York.

Ananthakrishnan, M.V. (1974). Multimedia Systems in Technical Education & Training, Procs of Intl. Conf. Frontiers in Education, pp. 275-278.

Ananthakrishnan, M.V. and K. Nambi (1977), A Systems Approach to Physical Sciences Teaching, School Science, 15(3), pp. 33-39.

Ananthakrishnan, M.V. and V.K. Garg (1978), Industrial Engineering Education: A Systems Concept, Procs. Fifth National Systems Conf. India, pp 73-76.

Athey, T.H. (1977). Fundamental Systems Concepts, J. Syst. Mgmt., 28(11), 42-45.

Cook, D.L. (1966), PERT Applications in Education, U.S. Office of Education, p. 49

Davies, I.K. (1976). Objectives in Curriculum Design, McGraw-Hill, London, pp 4-6

De Graff, D.F. (1971). Introduction to FLIPS Sequence, Univ. of Michigan, Flint.

Gottesman, A.M. (1976). Curriculum Planning: A Systems Approach, J. Syst. Mgmt. 27(9), pp. 10-12.

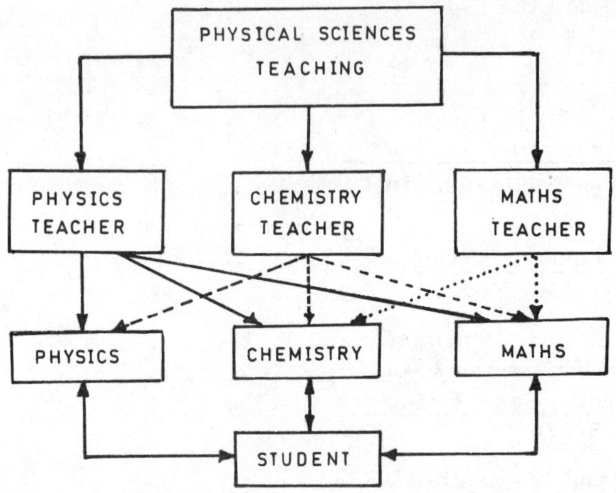

Fig. 1. A Systems Approach to Physical Sciences Teaching.

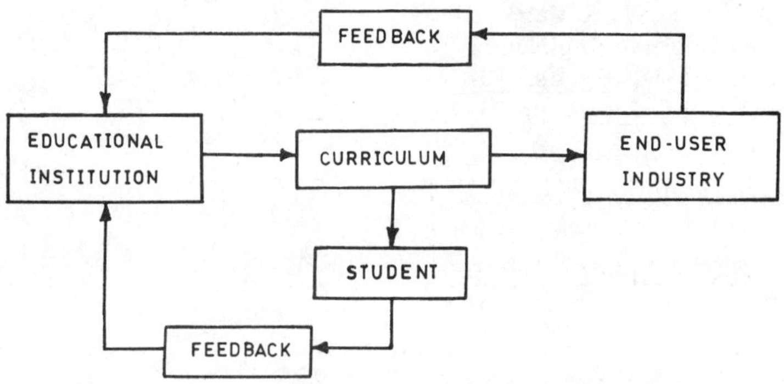

Fig. 2. The Institution-Student-Industry Feedback Model for Purposeful Curricula.

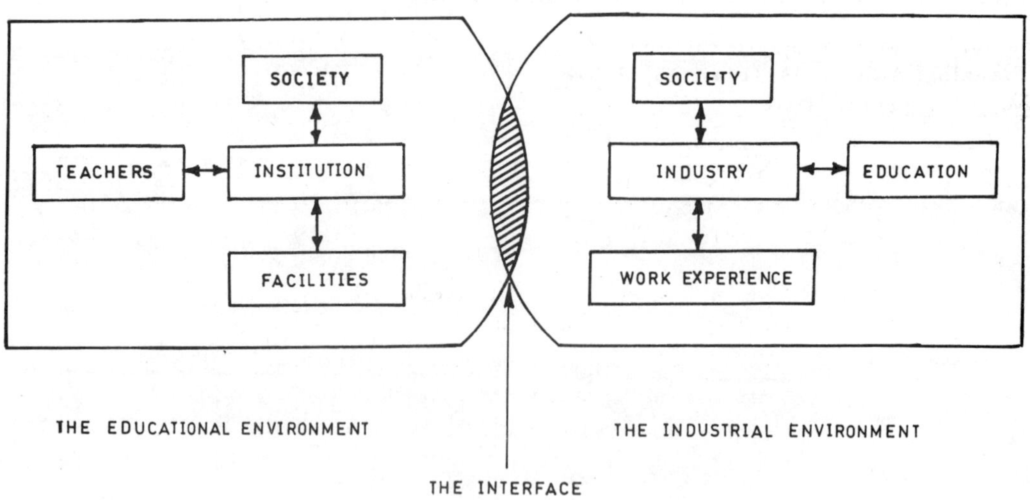

Fig. 3. The Educational and Industrial Environments and their interface - A Conceptual Model.

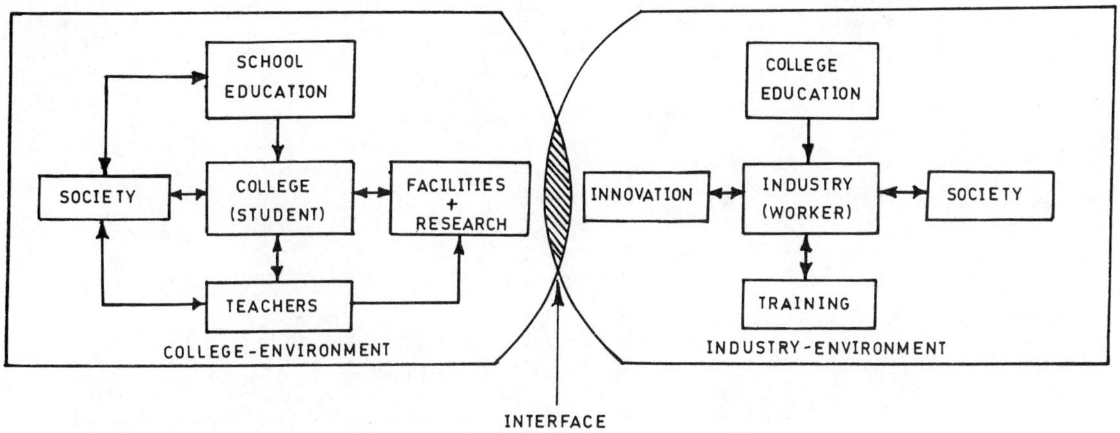

Fig. 4. A Systems Model of the Institutional-industrial Environment and Interface.

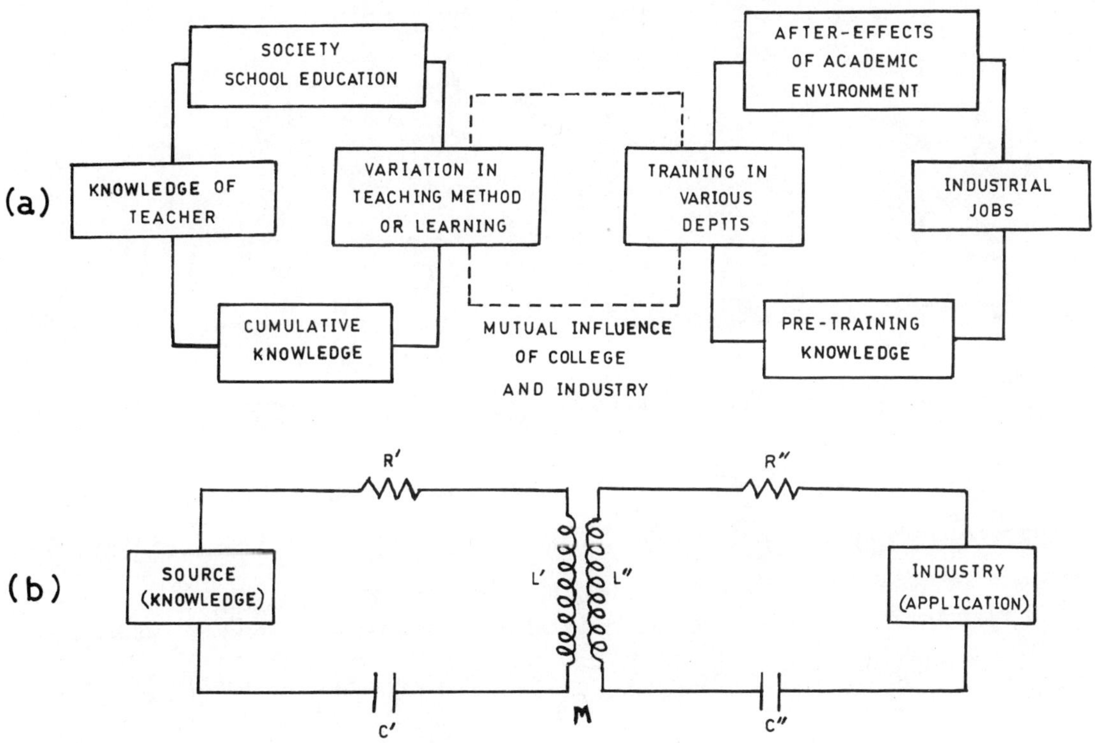

Fig. 5. (a) Formal Model of the Purposeful Approach to Curriculum Development.

(b) The Analog Model for the Purposeful Approach to Curriculum Development.

TRAINING ON ELECTRICAL ENERGY SYSTEM CONTROL: MICROSYSTEM APPROACH

E. H. T. El-Shirbeeny

Electronic Research and Development Centre, School of Electrical Engineering, University of Technology, Baghdad, Iraq

Abstract. This paper discusses the use of an experimental scaled model energy system as a research and training tool. A micropower system is presented to explore abnormal system operation and to use advanced control techniques for improving system performance. The advantages and disadvantages of different physical simulation schemes for constructing the model are given and design criteria of the micromodel are examined. Analog and digital techniques are also considered for instrumentation and control of the experimental micropower system. Special electronic devices for varying the model principal parameters are described for scaling purposes. The possibility of using simple dedicated and inexpensive microprocessors is demonstrated for different functions in the system. A broad outline of various areas of investigation and training on the micro-energy system is summarized. These include methods for improving the operational characteristics of real electric energy system using excitation and governor control.

Keywords. Power system control; voltage control; speed control; stability; minicomputers; microprocessors; electric generators; Turbines; computer applications.

INTRODUCTION

This paper describes an experimental installation directed towards the ultimate development of a physically simulated electric energy system which can be of use in research and training purposes. Methods by which a real energy system can be simulated are presented herein. A schematic diagram of an electric energy system is shown in Fig. 1. The scheme consists of three main subsystems: (1) Electric power network and machines; (2) Energy conversion and control; (3) Energy source and control.

The micro-energy model is a small scale physical system consisting of small machines and other equipments provided with adjustable features so that they can be set to form a similar model of an actual electric energy system (Hammons, 1971). The system is designed to simulate any operating condition in the power system. The items of a typical microenrgy system are mainly: microalternator, electronic exciter, microturbine, excitation controller, speed governor, and sensing devices for measuring system variables. Figure 2 shows a general block diagram of the experimental training energy model.

The paper is concerned primarily with the microenergy system and its associated auxiliary equipment for use in studies where a dynamic model power network operating in real time is required.

MICROTURBINE SYSTEM

The microturbine is a separately excited d.c. motor which has a special electonic regulator (El-Tanahi, 1978) to make the torque speed characteristics of the motor correspond to those of the turbine. Microturbine systems are usually of constant field current and with armature current so controlled that the torque can be made any desired function of both transient and steady state speed variations.

Operational amplifier circuits are used to simulate turbine-governor droop characteristics together with transfer functions of pilot valves, speed relays, high, intermediate and low pressure turbines and tie-line frequency bias control where appropriate. The voltage output of the last amplification stage of the control loop is an analogue of shaft-torque. It is converted into mechanical torque by the high response microturbine controller. A typical microturbine system is shown in Fig. 3. The analogue circuits include function generators, multipliers, limiters and operational amplifiers. Windage and friction torques are compensated by providing the microcontroller by a special bias.

MICROALTERNATOR SYSTEM

It is an electrical scale model of a large alternator. The parameters to be simulated in the model include the synchronous and transient reactances and time constants, the resistances of all circuits and the inertia constant. The microalternator is usually designed to generate 50/60 Hz three phase, 220 volts and of nominal rating up to 10 kva.

The machine's magnetic circuit is constructed from high grade magnetic material which exhibits similar magnetic properties as that in large machines. Stray iron losses are usually less than 2% under normal operating conditions. The stator is provided with fractional slot windings so that the open circuit line voltage waveform meets the standard specifications. The rotor is laminated. The field and damper windings are wound on the rotor in both direct and quadrature axis. Rotor windings are brought out together with identical auxiliary windings to slip rings on the rotor shaft. Some rotors are provided with additional quadrature-axis field windings which are used in studies on behavior of asynchronized system operation. Some designs (Dunlop 1979) propose the use of copper damper resistance. All main windings are wound to have a minimum winding resistance.

The per-unit natural and leakage reactances of the machine are the same as in large machines. The per-unit resistances are much higher than those of large alternators. The auxiliary windings enable negative resistance to be inserted in field and damper windings using 'time constant regulators'. This regulator reproduces the desired rotor time constants. The circuit of a universal regulator circuit (El-Shirbeeny, 1979) is shown in Fig. 4. Artificial time constants up to 10 seconds are available using this type of time-constant regulator.

Capability for representing a broad range of alternator excitation system can be provided by elctronic circuits utilizing one of the following designs. (1) Linear thyristor converters supplied from 50 Hz mains (Touri, 1977). (2) Three phase 1 kHz thyristor bridge which is supplied from the output of an induction alternator (Hammons, 1971). (3) Pulse width modulated convertor (Evans, 1970).

The effects of exciter saturation of any specified shape and ceiling voltage can be built using diode function generators at the exciter output. Time constants of any chosen amount can be introduced by operational amplifier-circuits at the input.

An autotransformer may be inserted between the alternator and the micronetwork transmission system. This transformer is connected to modify the machine relative rating to another in the microenergy model system. The effective rating of any alternator with a given set of parameters is proportional to the square of the autotransformer voltage ratio. This ratio is so chosen to take account of magnetic saturation of the machine's synchronous reactance. The autotransformer is connected in delta to machines of large relative rating and in star to machines of smaller rating. This type of transformer requires a rather larger range of steps than usual. Finess of tap change control is also necessary, to allow the alternator voltage level to be adjusted to a constant value for varying excitation and system voltage conditions.

The inertia constant of the microalternator is adjusted in the training model by means of detachable flywheel. This arrangement provides any desired value of the inertia constant.

LOAD MODELS

Static loads are represented by three phase inductive reactance banks. The load can be connected to any busbar in the micronetwork by busbar selector switches. Dynamic loads are provided by induction and synchronous motors in the laboratory.

Load units in the model can be chosen to simulate active or/and reactive power consumed at a load bus in a real elctric energy system.

ENERGY TRANSMISSION SYSTEM

Representation of transmission lines are implemented by series inductive and resistive sections, with shunt capacitors at each end, as shown in Fig. 5. Distributed parameters are replaced by lumped constants. Mutual coupling between phase conductors and earth return paths are to be considered. The phase and return path reactances can be simulated by iron cored reactors. Their values are selected to match inductances and resistances of actual lines. Nominal pi-representation are common in micronetworks. All line-models are terminated in three-phase-with neutral busbar selector switches.

INSTRUMENTATION AND MEASUREMENT

Measurements are required to monitor and display the system variables. Instrumentation modules should be provided on busses to measure real power, reactive power, voltage and current.

For control and recording purposes it is desirable to obtain d.c. signals which are proportional to the following quantities: voltage, current, rotor angle, speed and power. Suitable potential and current transformers connected to rectifier circuits are used for this purpose. Filters must be usually provided to eliminate ripples while retaining low frequency signal components.

For transient studies (Newton, 1976), rotor angle and its time derivative must be measured. These variables must be obtained from rectification from an a.c. tachometer fitted onto a common shaft with the main microalternator. Second order active filters are usually used in the measurement circuit without attenuation at the mechanical oscillation frequency. Digital methods of rotor angle measurements provide higher accuracy and resolution compared with analog methods.

When digital computing devices are used in the microenergy system signal conditioning is necessary. A typical analogue signal processing circuit is shown in Fig. 6.

Optical isolators can be used to reduce the noise in the data channels linking the power equipment to the digital computer. The signals are then applied to sample-hold circuits using f.e.t. buffers and finally sampled siganls are amplified to an acceptable range.

COMPUTER AIDED TRAINING

The use of digital computers for on-line control and real time processing of data is an attractive proposition in the development of a micro energy training system. This will permit considerable cost savings in terms of hardware and offer a higher degree of accuracy and flexibility not obtainable using analog circuits alone.

Figure 7 illustrates the interfacing arrangement between a minicomputer and a microalternator equipment. The analogue multiplexer is driven by a logic decoder working under program control connecting the analog sampled signals sequentially to the computer analog to digital converter. Analogue-output-scaling amplifiers and buffers are used for the control signals.

The digital input circuits allow for computer monitoring and program interrupt line-initiated conversion routines. The digital-output logic circuits are used to indicate the program execution and to control the analogue-multiplexer decoder. Teletype can serve for data setting, program preparation and results printing. Program editing is usually carried out with high-speed paper-tape reader and punch. Analogue inputs are converted by A/D converter with multiplex control unit and D/A converter for analogue output.

Microcomputers have been used recently to perform certain control actions in electric energy systems(Barrows, 1977). A minicomputer generates the main supervisory and control functions while one or more dedicated microprocessors act as data collecting and filtering units and servicing functions. Within the microprocessors, a simple digital filtering units and perform local peripheral supervision and servicing functions. Wthin the microprocessors, a simple digital filtering algorithm can be employed to remove the steady state offsets and to prescale the variables for input to the microcomputer. System security is maintained by the micrprocessor checking a "time out" situation from the minicomputer and defaulting to a simpler control algorithm in the event of loss of control information caused by noise or breakdown of the link.

TRAINING PROGRAMS

Training and research on microenergy models provide an efficient approach for verifying, enhancing and developing simulation technology through comparative testing on the physical energy model. The following demonstrates different areas of studies utiling the microenergy system.

Project 1
Short circuit problems-This project can be dedicated to investigate the dangerous stresses which may arise in the shaft under short-circuit conditions.

Project 2
Synchronisation-Synchronization training program provides real time facilities for power engineers to understand the physical characteristics of synchronization methods. Behavior of electric energy system can be studied during disturbances involving loss of synchronisation.

Project 3
Synchronous machine modelling-The microsystem model can provide a means to investigate to adequacy and accuracy of synchronous machine models which are widely used in large scale analytical tools such as the digital transient stability program.

Project 4
Transient stability tests-These tests are performed by applying a 10 cycle fault upon the transmission system. The study can take full account of automatic voltage regulators and governing systems. The connection diagram of the experimental set is shown in Fig. 8.

Project 5
Pumped storage installation-The microsystem can be used to develop experience in operating pumped storage schemes. Advanced control strategies are necessary for safe operation. A pumped storage scheme can impose a maximum hazard to security of supply during pumping conditions where the connected motors and generators swing in opposite directions following a severe system disturbance.

Project 6
Braking resistance applications-Damping methods using resistive braking during first swing following a fault can be investigated on the microsystem. Bang-bang switching of these resistors can be tried to satisfy the Pontrayagin's Maximum Principle(Sage, 1968) and suitable supplementary signals can be suggested to control the the switching process according to the state of the energy system. A typical connection diagram for the system examining this problem is given in Fig. 9.

CONCLUSION

The development of a micro energy model has been described. It is shown that the physical energy system model is a powerful tool for the experimental investigation of the dynamic behavior of a real system. It is thus possible to gain experience of different control method applications in energy system. In addition this energy training model provides a versatile experimental facility to supplement theoretical studies of novel control schemes.

It has been indicated that training and research programs can be carried out on the microenergy system for studying coordinated control of excitation and governing of the real energy system. The training on the model can give a valuable insight into the methods of applying theoretical studies to solve practical problems in an electric energy system.

REFERENCES

Burrows P. and A. Daniels (1977). Digital control of excitation of a micromachine power system model. Proc. 12th UPEC Conf., Brunel University, U.K.

El-Shirbeeny, E.H.T., and Maher, E.E., (1979). Design and analysis of a novel time constant regulator. Symposium on Control and Systems Engineering, Tehran.

El-Tanahi, S.I., and E.H.T. El-Shirbeeny (1978). Dynamic analysis of a prime-mover simulator. Proc. Int. Symp. MECO 78,Athens, pp. 124-132.

Evans, F.J., R.B. Robinson, and T. Wilson (1970). Model power systems. Electr. Engng. Trans., The Inst. of Engineers, Australia, pp. 53-59.

Hammons, T., and A. Parson (1971). Design of a microalternator for power system stability investigation. Proc. I.E.E., 118, pp.1421-1441.

Dunlop R., and A. Parkish (1979). Verification of synchronous machine modelling in stability studies. Trans. I.E.E., PAS-98, pp. 369-376.

Newton M. and B. Hogg (1976). Optimal control of a microalternator system. Trans. I.E.E.E., PAS-95, pp. 1821-33.

Malik, O.P., G.S. Hope, A.S. Farag, and D.W. Huber (1974). Digital controller for a turbine and alternator. Proc. 4th IFAC/IFIP Conference on Digital Computer Applications to Process Control, Zurich, pp. 267-278.

Touri. M. O. (1977). Regulation of the time constants of a laboratory machine. M.Sc. Report, Imperial College, London.

Sage, A.P. (1968). Optimum system control. Prentice-Hall, N.J.

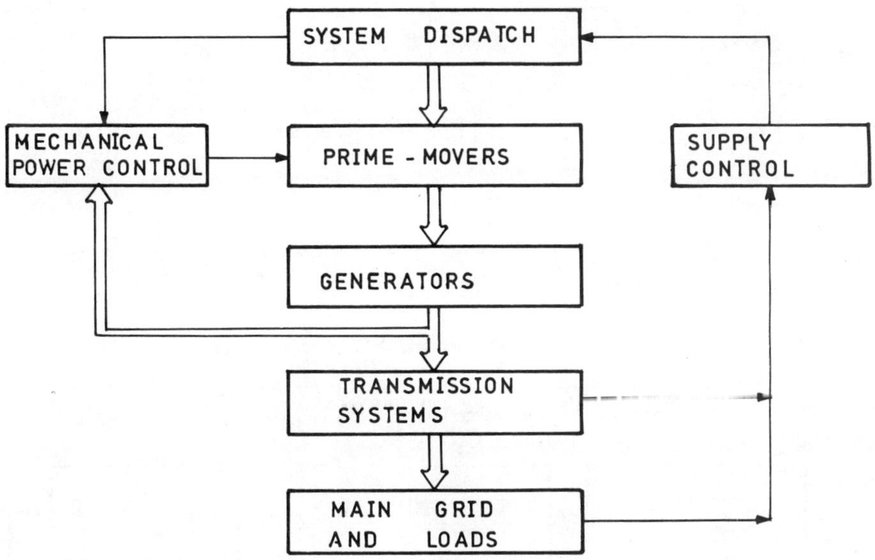

Fig. 1. Basic structure of energy system.

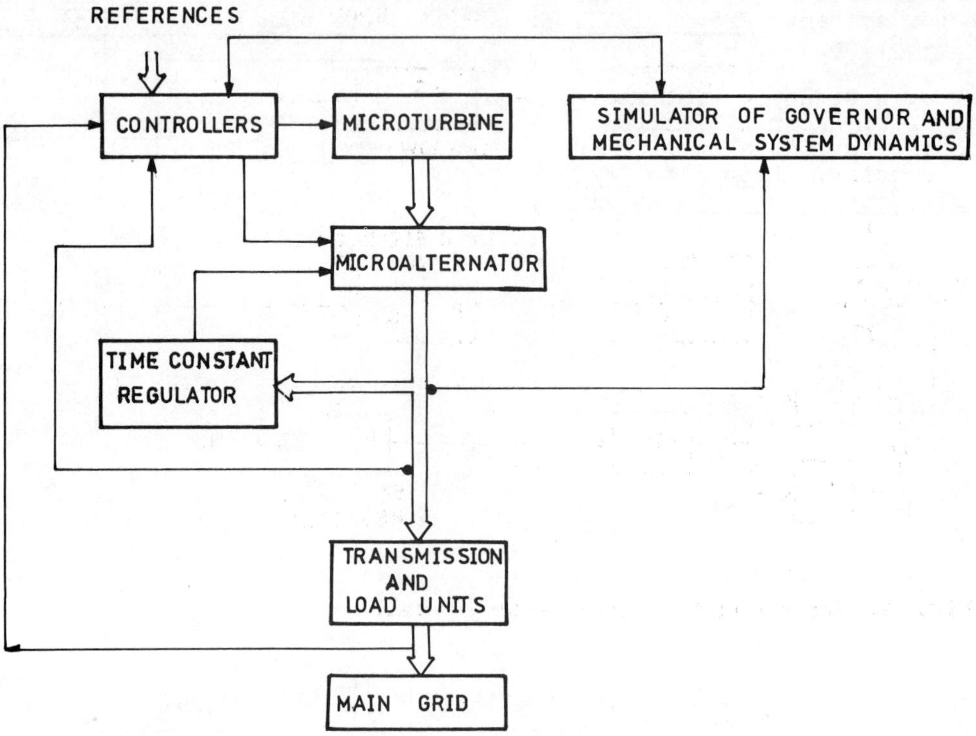

Fig. 2. Schematic layout of training microenergy system.

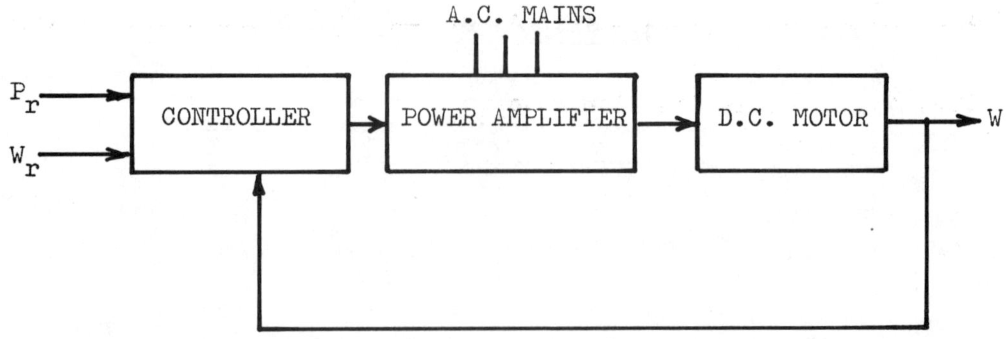

Fig. 3. Microturbine system.

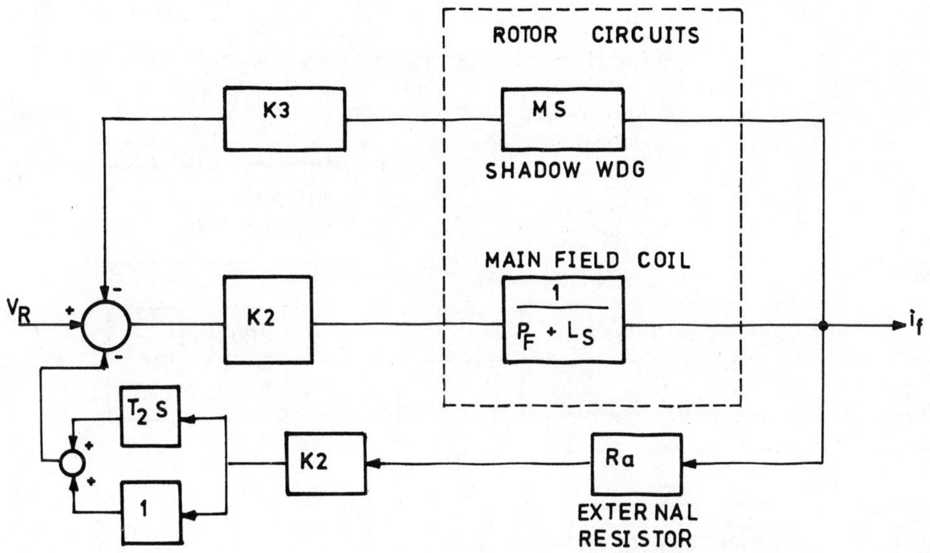

Fig. 4. Time constant regulator system.

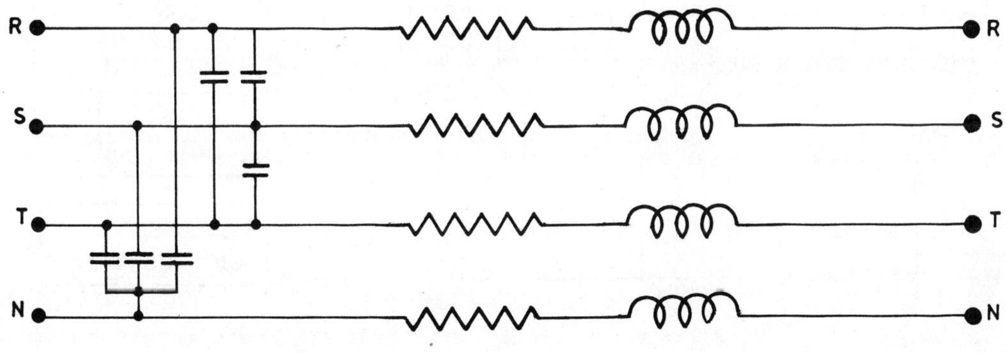

Fig. 5. Transmission line model.

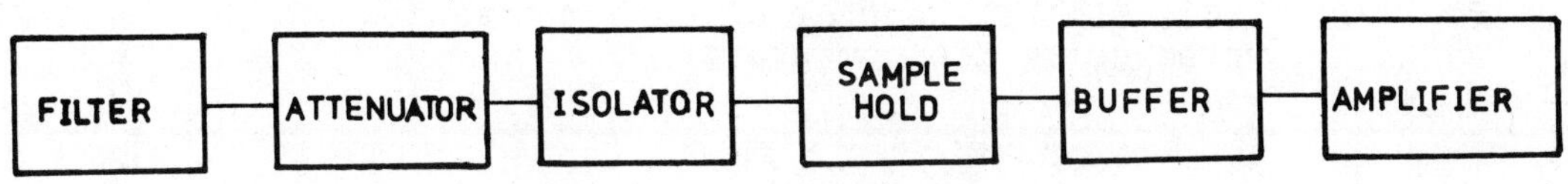

Fig. 6. Signal conditioning circuit.

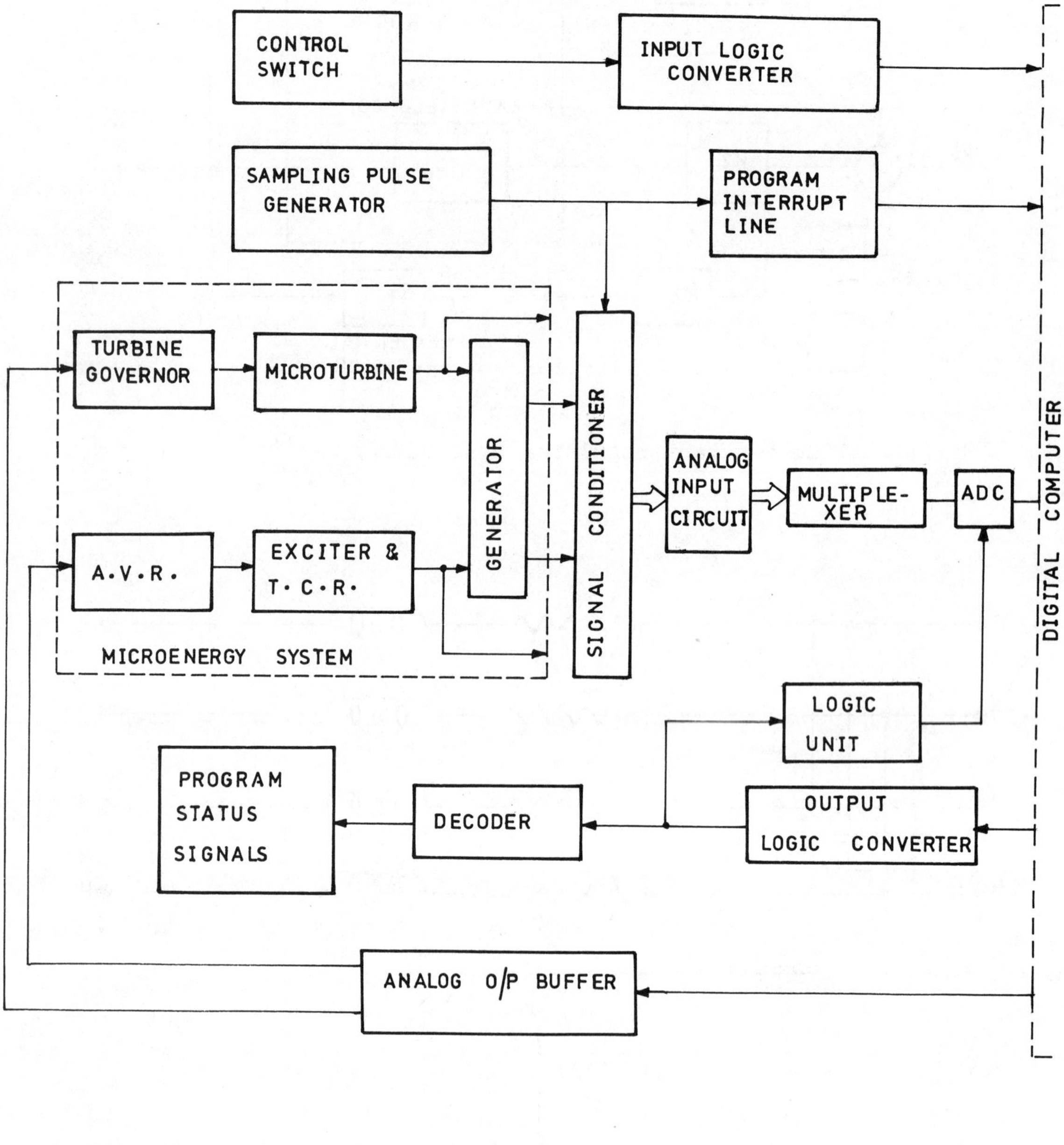

Fig. 7. Direct digital control of energy training system using minicomputer.

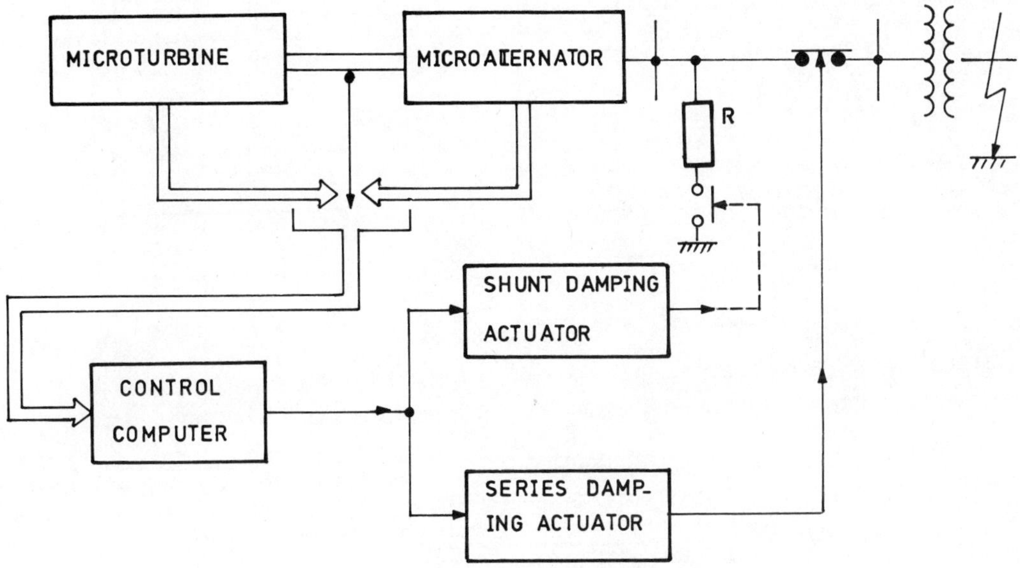

Fig. 8. Microsystem set-up for transient stability studies.

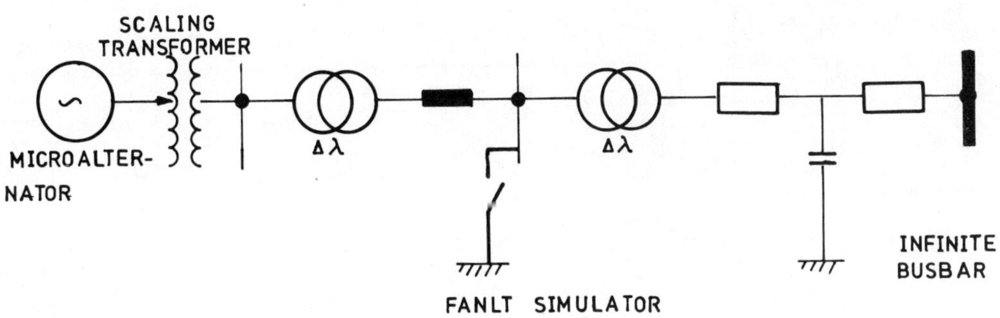

Fig. 9. Circuit arrangement of microenergy training system for Project 6.

INEXPENSIVE MICROPROCESSOR SYSTEMS, TUTORIALS AND THE FUTURE OF COMPUTER-ASSISTED INSTRUCTION

R. E. Burns*, W. I. Davisson and E. A. Kline*****

**Department of History, University of Notre Dame, Notre Dame, Indiana, U.S.A.*
***Department of Economics, University of Notre Dame, Notre Dame, Indiana, U.S.A.*
****Department of English, University of Notre Dame, Notre Dame, Indiana, U.S.A.*

Abstract. Educational computing throughout the world is rapidly coming of age. The development of hardware in the last fourteen years has been astounding, and some successful CAI materials development projects have given us a glimpse of what the educational potential of this technology can be. However, the expectations of educational theorists and technologists for the expansion of CAI into all levels of education have not been realized. The principal reason why more students are not involved with CAI is that instructional materials development has not proceeded as fast as that of hardware. For the most part, CAI applications have not exploited the enormous information delivery capacity of this technology, and quality CAI materials of any sort are in short supply.

The entry of inexpensive microprocessor systems into the educational market at once provides educators with many new instructional opportunities and threatens to compound the problem of CAI materials availability. A possible solution to this problem is a CAI application known as a tutorial developed by a group of instructional materials specialists at the University of Notre Dame. Tutorials make heavy use of the information delivery capacity and flexibility of microprocessors and are sufficiently versatile for adaptation to the teaching of any subject at any level. The marriage of inexpensive microprocessor systems with CAI tutorials seems to be a highly promising way of integrating computer technology into instructional processes and into the general understanding of our student and adult populations in the years ahead.

Examples of CAI tutorials in several subjects appear in an appendix to this paper.

Keywords. Computer-aided instruction; materials development; microprocessors; educational models; evaluations.

THE EARLY PROMISE OF CAI

Educational computing throughout the world is rapidly coming of age. The development of hardware has been astounding. Computers have become more powerful, less expensive, and smaller. Networks for the exchange of programs and data are in place, and many successful computer-assisted-instruction (CAI) materials development projects have given us a glimpse of what the general educational potential of this technology can be. The scale of CAI-related activities since 1966 has been impressive, and one can scarcely find a major college or university in the United States and Europe where some sort of computer-based instruction has not been tried.

Nevertheless, however impressive the array of hardware developed and the scale of CAI-related activities undertaken during the last fourteen years, the great expectations which educational theorists and technologists had for the expansion of CAI into all levels of education in those early years have not been fulfilled. To be sure, those expectations were unrealistic. Along with educational television, language laboratories, and teaching machines, CAI was overpromoted by enthusiasts and some equipment vendors. CAI and other electronic-based hardware were represented as being instruments of a forthcoming instructional revolution. Teacher acceptance of CAI was assumed. However, the probability of competition for valuable computer time between research interests, administrative needs, and instructional programs was not anticipated, and the difficulty of producing quality materials for CAI applications was not appreciated.

As most of us know, that instructional revolution has not occurred and the probability that it will occur in the near future is low.

The reason is fairly clear. Hardware development has proceeded much faster than that of instructional materials. Most teachers must either develop their own CAI materials or do without. The quantity of materials presently available is limited, and the range of subject matter areas is narrow. Though some strong beginnings have been made in the social sciences and humanities, most development resources and energies have gone into science and engineering CAI applications. Overall, the quality of available CAI materials is mixed and dissemination has been irregular. Until teachers are able to choose CAI materials from a broad inventory in much the same way that they now choose books and other instructional materials for their classes, the educational potential of this extraordinary technology will not be realized.

THE PRESENT STATE OF CAI MATERIALS DEVELOPMENT

The future of CAI will depend in large part upon increasing the supply of quality instructional materials. The difficulty of this task should not be underestimated. The present state of the art can perhaps best be described as childhood merging into adolescence. CAI materials development is not yet perceived as an academic discipline or as a genuine scholarly activity. There is no central body of theory or corpus of recognized knowledge. As a field, CAI materials development has few career scholars who claim it as their principal professional interest or specialty, and most of its scholarly communication outlets are limited to conference proceedings. Moreover, there is no generally recognized center in the United States or Europe where training and experience in this field can be obtained.

The only effective way of learning the skills required to write quality materials is by doing, and doing is not easy. Transforming information taken from books and articles into formats that machines will accept and that faculty and students will want to use is an extremely demanding, frustrating, and time-consuming undertaking. Probably not more than one out of ten faculty is able to do this well. Successful writers generally proceed through trial and error; that is, they search for formats and techniques whereby instruction can be improved, given new dimensions, or extended to more students.

In sum, the field of CAI materials development at present is no more and no less than a collection of projects frequently isolated from and unrelated to one another and tenuously supported by a few private foundations and government agencies. Many of these projects have progressed beyond the pioneering stage and are producing materials which have resulted in affective if not fully documented congnitive improvement. Yet if one considers the momentum building up in educational institutions throughout the world for technological answers to pedagogical problems, the future for CAI appears bright. In particular, the entry of microprocessors into the educational market is rapidly creating a demand for instructional materials that will have to be met by someone. The great instructional revolution which enthusiasts so hopefully forecast in the late 1960s may be upon us in the 1980s; and if so, microprocessors promise to be the principal mechanism of it.

THE PROMISE OF MICROPROCESSOR TECHNOLOGY

The implications for education, as indeed for communications and society as a whole, of the continuing and ever-greater miniaturization of integrated circuits are profound. Within the next ten years, because miniaturization tends to proceed exponentially, computers will become so powerful, so portable, and so inexpensive that educational technologists and teachers will be hard pressed between then and now to find ways of exploiting their enormous potential. If hardware development has greatly outstripped CAI materials production in years past, this problem will be greatly compounded in the future. For the present, however, the power and portability of available microprocessors is challenge enough to both administrators and CAI materials developers. Administrators have to determine whether or not the true costs of the present generation of microprocessors are in fact more economical for decentralized CAI applications than a centralized TSO system. Materials developers will have to try to adapt programs written for larger computers to the requirements of these smaller systems and at the same time devise new applications and modes for a range of academic disciplines in a variety of instructional situations. The educational opportunities which this new and inexpensive technology can provide appear to be limited only by the imagination and ingenuity of the people charged to develop them.

Any future must build on its past, and CAI is indeed a case in point. For the most part, the history of CAI projects is one of trying to find ways of supplementing lecture and discussion mode teaching. Computer technology has been used largely as an enrichment or extension of traditional instructional modes. Some projects have demonstrated how many tedious hours in standardized, repetitive teaching can be saved by allowing students to perform drill and practice routines on a terminal at times convenient to themselves. A few science and engineering education projects have tried to go beyond problem-solving and to stimulate creative thinking. Such projects usually provide students with data sets from actual experiments or case histories and use computer programs to recapitulate some of the reasoning which led to the solution of the problem identified in the original experiment or case history. Programs of this type frequently provide students with opportunities to access incomplete data and speculate on other solutions.

Games and simulations are also widely used as adjuncts in science, business management, and economics courses. For example, a highly successful simulation developed by one project for a large-enrollment introductory economics course enables students to confront the delicate and complex interrelationships of interest rates, taxes, employment rates, inflation, and money supply in decision-making situations. A lecturer or textbook can explain from one point of view or another how these factors interact, but a computer simulation allows students to manipulate these variables for themselves and to discover immediately how changing the value of one affects all of the others. All of these and similar CAI applications are interesting and facilitate learning, but they are not integral to the success or failure of a given course and do not exploit the enormous information delivery capacity of this technology.

THE TUTORIAL MODE

Efforts to develop applications wherein CAI shares information delivery functions with lectures and reading have been rare but encouraging and promise to make the most of microprocessor technology. Known as tutorials and pioneered by a CAI development group at the University of Notre Dame, this CAI application has the advantage of being adaptable to virtually any subject matter taught at any level. Developed primarily for humanities and social science courses, tutorials are questions presented in a multiple-choice format with programmed responses for each answer choice. Each question is preceded by a lead. Depending upon the subject matter of a given set of tutorials, this lead may provide a context for the question that follows, state a rule or principle for subsequent application, or supply special instructions or explanations. Again, depending upon the subject matter, leads vary in length from a short sentence to a full paragraph. Programmed responses to answer choices do more than indicate whether a particular choice is correct or incorrect. Each response provides an explanation of why an answer choice is right or wrong. These explanations vary in length from two or three sentences to several paragraphs. Tutorials--that is, this combination of leads, questions, and programmed responses--are organized into inventories according to subject matter. Moreover, in the last five years at Notre Dame, these inventories have grown to include tutorials for introductory courses in the history of western civilization, early American history, economics, English grammar and spelling, accounting, marketing, business management, and English as a second language. All of these tutorials, regardless of subject matter, were written to be textbook independent. Teachers can construct daily or weekly CAI lessons by selecting from these inventories those tutorials which are appropriate to their individual instructional needs. In other words, tutorials can function as supplements and adjuncts to lectures, discussions, and readings or they can be the primary source of information in a given course and the principal instructional strategy through which knowledge objectives are reached. For whatever purpose individual tutorials are written and however CAI lessons are configured, students access and run them over a data printing terminal, and upon completion of a given lesson, they have printed records of their tutorial experience for future study.

FACULTY REACTIONS TO CAI TUTORIALS

Faculty reactions to the introduction of a large number of tutorials into on-going or new courses have been mixed. Many faculty have been pleased by the experience, particularly those who are involved in teaching basic skills to groups of students who come to their subject with inadequate or unequal preparation. Other faculty have been disenchanted by tutorials and a few have absolutely refused to have anything to do with CAI at all. Our experience suggests that faculty satisfaction with CAI tutorials will be highest among teachers who are comfortable with discussion-oriented instructional modes and lowest among coverage-oriented lecturers.

For faculty who perceive their primary teaching role as that of discussion facilitator or critic, the opportunity of using CAI tutorials as mechanisms for communicating large quantities of factual information is much appreciated. For faculty who see their primary teaching role as that of lecturer and information deliverer, an experience with CAI tutorials can be a thoroughly unnerving one. Such faculty find that they are unable to deliver information more effectively than a machine and become deeply frustrated. A few become completely disenchanted with trying to do so and perceive themselves as being demeaned as persons and diminished as teachers. In most cases, frustration and disenchantment of this sort have followed from a failure by teachers to integrate the content of lectures and classroom presentations with the subject matter of whatever CAI tutorials they may be using. Those faculty most satisfied with CAI tutorials are teachers who have built their courses around them. This means that knowledge objectives are derived from tutorials and that classroom presentations are designed to expand, clarify, and provide perspectives for themes, problems, and concepts introduced and developed in the tutorials. One important and unavoidable consequence of using CAI tutorials is that all faculty involved are forced to reassess their role as teachers and to think seriously about pedagogy.

STUDENT REACTIONS TO CAI TUTORIALS

Using CAI tutorials has proven to be both successful and popular with students. Formal evaluation of this CAI application has been concerned primarily with student satisfaction, and the results have been extremely positive. Students believe that time invested in running CAI tutorials is well spent and will improve grades. One phenomenon which the developers did not anticipate is the frequency with which students running CAI tutorials deliberately select wrong answer choices in order to obtain the information and explanation contained in the programmed responses to them. Moreover, follow-up interviews with students suggest that this mode of instruction appears to facilitate internalization of abstract concepts. The quantity of information provided and the form in which it is presented in CAI tutorials appear to make abstract concepts more discussable and understandable.

CONCLUSION

The problem of integrating computer technology into instructional processes has no easy answer. However, the increasing power, diminishing size, and decreasing costs of microprocessor systems present administrators and faculty with extraordinary opportunities for innovation at all levels of education. Similarly, CAI tutorials provide instructional materials developers with an application that is sufficiently versatile to maximize exploitation of those opportunities. At the moment, a marriage between inexpensive microprocessor systems and CAI tutorials seems to be the most promising way of achieving such an integration and of introducing computer technology and applications into the general understanding of our student and adult populations in the years ahead.

APPENDIX

American and European History

The third quarter of the nineteenth century was an extremely important period in English political history. Englishmen made a series of slow, cautious advances toward political democracy. Nevertheless, the aristocracy and gentry continued to dominate politics during the period 1850 to 1870 because

1. the Reform Act of 1832 had left the urban centers overrepresented in the House of Commons. 0
2. only aristocrats and gentlemen had the financial resources to wage successful election campaigns. X
3. aristocratic candidates were more widely known than candidates from the middle class. Z
4. England had and continued to have a deferential electorate. +

0 This answer is incorrect. The opposite situation was true. Even after passage of the Reform Act of 1832, the distribution of seats in the House of Commons favored rural areas. Rural areas were overrepresented and the influence of the aristocracy and gentry was much stronger in rural areas than in urban centers. Consequently, aristocrats and gentlemen continued to be elected to the House of Commons in large numbers.

Select another answer.

X This statement is partially true, but as an answer to this question, it is incorrect. Candidates from among the aristocracy and gentry had more funds available for electioneering purposes, but the availability of campaign funds does not explain their continuing domination of English politics. Middle-class industrialists had more funds available but were unable to get nominated or elected in rural and county-town constituencies. Much more than money was involved in getting elected to Parliament. Who you were or how long your family had lived in a particular constituency counted for much more than how much money you could spend at election time.

Select another answer.

Z This statement may or may not be true depending upon the individuals involved. As a generalization and as an answer to this question,it is impossible to substantiate and is incorrect. What counted in politics was the right kind of notoriety among the right people in the appropriate constituency. Constituency electorates were by modern standards relatively small and being well known among the people who voted was more important than a more generalized national notoriety.

Select another answer.

+ Your answer is correct. In the words of Walter Bagehot, mid nineteenth-century political theorist, the English electorate was extremely deferential. By deference, Bagehot meant that English voters accepted the idea that their social betters should govern them. Voting for aristocrats and gentlemen was a habit which few mid nineteenth-century Englishmen were inclined to break.

Proceed to the next question.

The two centers of the art world in the 1920s were Paris and Berlin. Artistic rebels generally rejected the world of bourgeois reality. A wide range of artistic movements attracted many true believers and enjoyed moments of great popularity. Literature also was affected by the artistic rejection of the exterior world as it was or seemed to be. The most important expression of this trend in literature was

1. Dadaism. 0
2. The international style. X

3. Abstractionism. Z
4. Psychological realism. +

0 This answer is incorrect. Dadaism was an artistic movement rather than a literary one. Taken from the name of a child's rocking horse, it originated in 1917 as a protest against the war. It was a total rejection of contemporary society. It was anti-creative, anti-intellectual, and nihilistic. Practitioners of Dadaism insisted that the world made no sense, but that meaningful rebellion could be effected through untrammeled self-expression with the emptiest or most absurd content.
Select another answer.

X This answer is incorrect. The international style was a movement in architecture. The leading practitioners of it were Walter Gropius and Charles Le Corbusier. Their goal was an honesty of design that employed structural steel and other new construction materials in a way that was consistent with function--avoiding all purposeless details and trim. The essence of the international style was epitomized in the slogan "form follows function."
Select another answer.

Z This answer is incorrect. Abstractionism was a movement in painting and sculpture. Though the movement began in pre-war Russia, the originator of this movement in the West was Dutch painter Piet Mondrian, who developed a rigid mathematical style of lines and geometric shapes. Abstractionists concentrated on the arrangement of pure form and colors--sometimes random daubs. It all means something, if only to the painter; but that meaning cannot be grasped by the uninitiated. One must learn to understand abstract painting and sculpture.
Select another answer.

+ Your answer is correct. Psychological realism developed out of a number of masterful introspective novels written before World War I--the most important being Marcel Proust's "Remembrance of Things Past." It established the stream of consciousness technique which explicated an individual's mind and innermost personal feelings. Writers such as Virginia Woolf, Thomas Mann, William Faulkner, and James Joyce produced masterpieces using these techniques. All of these writers required intense intellectual effort by their readers. For these reasons, the appeal of psychological realism was to a relatively narrow but highly educated public.
Proceed to the next question.

The limited measures taken by President Herbert Hoover proved inadequate to deal with the Great Depression. During the 1932 Presidential campaign, the Democratic nominee Franklin Roosevelt attacked Hoover's relief record but articulated few specific proposals about how he would solve the country's economic woes. In a speech delivered to the Commonwealth Club in San Francisco on September 23, 1932, Roosevelt explained what he thought the role of Federal government in economic affairs ought to be. He argued that the Federal government had an obligation to provide employment opportunities for all people willing to work. He also insisted that regulation of business by the Federal government should come only as a last resort. Once in office, however, President Roosevelt's actions frequently went beyond the interventionist and regulatory limits set in his Commonwealth Club Speech.

The action of which of the following agencies created during Roosevelt's first administration went beyond the regulatory limits outlined in the Commonwealth Club speech?

1. Tennessee Valley Authority 0
2. Civilian Conservation Corps X
3. Federal Emergency Relief Administration Z
4. Truth-in-Securities Act +

0 This answer is incorrect. The Tennessee Valley Authority was not a regulatory agency. It was a relief effort of enormous magnitude. Proposed by FDR on April 10, 1933, it harnessed the usefulness of the entire Tennessee River through a series of dams and electric power-generating stations. The TVA was set up as a public corporation which provided economic opportunity and agricultural improvement for a multi-state region.
Select another answer.

X This answer is incorrect. The Civilian Conservation Corps (CCC) was not a regulatory agency. It was a relief program established to reduce high unemployment among American young people. The CCC provided approximately 250,000 jobs in reforestation and flood-control projects for youths between the ages of eighteen and twenty-five. Though administered by the US Army, the CCC did not provide military training; the work of this program was limited to creating employment opportunities for young people in conservation-oriented public works projects.
Select another answer.

Z This answer is incorrect. The Federal Emergency Relief Administration was not a regulatory agency. FERA was established to deal with massive unemployment. The activities of this agency illustrated the extent to which the Roosevelt administration was committed to relief efforts. Enacted into law on May 12, 1933, FERA provided $500 million in direct Federal grants for relief to cities and states. FERA had nothing to do directly with regulating American business.
Select another answer.

+ Your answer is correct. Enacted on May 27, 1933, the Federal Securities or Truth-in-Securities Act went far beyond Roosevelt's notion of government regulation of business as a last resort. This law required that corporations obtain approval of Federal Trade Commission before issuing stock. Thus,

any company with publicly held stock would become subject to at least oversight and regulation by the Federal Trade Commission.
Proceed to the next question.

Economics

The five years after 1924 were a period of prosperity for the industrial sector of the European economies. However, throughout the world, the whole decade of the 1920s was a time of such chronic agricultural depression that farmers could neither pay their debts nor purchase the quantities of manufactured goods that were being produced. The delicate and interlocking capitalist economic system was shocked by a series of financial crises in 1929 into the worst depression in modern history. The immediate cause of the Great Depression in Europe was the

1. growth of cartels. 0
2. imposition of high tariffs. X
3. failure of the lower classes to share in the prosperity of the 1920s. Z
4. collapse of the American stock market. +

0 This answer is incorrect. The growth of cartels was a product of broader technological changes known as rationalization of industrial production. This process involved the application of such techniques as assembly lines, interchangeable parts, standardization of supplies, and mergers and combinations. However, the growth of cartels was largely a German phenomenon and not characteristic of the other European economics. In any case, cartels played no special role in bringing about the financial crises of 1929.
Select another answer.

X This answer is incorrect. High tariffs were characteristic of the post-war world. International cooperation and marketing were severely hampered by the many small new nation states created out of the dissolution of the Austro-Hungarian and Russian Empires. Political independence was meaningless without economic independence, and protectionist economic policies were believed to be effective ways of assuring the latter while demonstrating the former. However, high tariffs played no special role in bringing about the financial crises of 1929.
Select another answer.

Z This answer is incorrect. A serious weakness of the European economies in the late 1920s was the failure of the lower classes to share in the prosperity of the times. Wages rose but neither as fast nor as much as prices. Working-class movements made few gains in 1920s outside of Austria. Strikes were frequent, especially in England, but the relative decline in the purchasing power of the lower classes was not perceived as a major problem by most European governments in the 1920s. In any case, the plight of workers and farmers was no worse in 1929 than it had been in 1927 or 1928. It played no special role in bringing about the financial crises of 1929.
Select another answer.

\+ Your answer is correct. The collapse of the American stock market in October 1929 sent shock waves around the world and precipitated financial crises in Europe. Wall Street's Black Thursday forced a rapid contraction of American credit and a similar shrinkage of American markets for European goods. American banks called in their loans to European companies and American importers cancelled contracts. The results were business failures and mounting unemployment. The prosperity of the late 1920s had ended and men wondered what the political and social consequences of this new and frightening situation would be.
Proceed to the next question.

One of the most fundemental elements in any national economy is population. Trends in population can greatly influence decisions and policies which govern economic activities. For example, a nation with a young growing population will employ economic policies that are different from those pursued in a nation with a more elderly population.

In contemporary America, population growth has stabilized. After a generation of rapid population increase, the United States is currently in a condition of "zero population growth." In other words, only enough infants are being born each year to replace those persons who die in that same year. A number of influences have led to this stabilization of population. Which of the following factors has contributed most to the declining birth rate in the United States during the last two decades?

1. ecology movement 0
2. new methods of contraception +
3. women's movement X
4. Malthusian theory Z

0 This answer is incorrect. It is true that some Americans have decided to remain childless or to limit the size of their families because of a fear that natural resources can no longer sustain a growing population. Energy shortages and fear of continued air and water pollution have led some to question the wisdom of commonly held beliefs that resources in the United States are infinite and that progress is inevitable. However, the influence of such thinking, while important and growing, is not the most significant cause in the decline in the birth rate in the United States.
Select another answer.

\+ Your answer is correct. In 1960, oral contraceptives first became widely available. Although the birth rate in the United States had actually begun to drop several years earlier, the introduction of the seemingly safe and effective pill led to a precipitous drop in the birth rate and to a revolution in attitudes toward sexuality, children, and women's roles in the United States. Within twenty years, the net reproduction rate for the United States has fallen by almost 50 percent. The greater availability of contraceptives has contributed most to the declining birth rate. This reality illustrates the influence that changing social values can have on economics. The declining birth rate has led to a greater number of women in the work force, to a smaller demand for products associated with children, to a greater disposable income for childless couples, and to changing pressures on the educational system. All of these results have followed from a declining birth rate.

Proceed to another question.

X This answer shows a good deal of thought, but as an answer to this question, it is incorrect. It does indicate, however, how closely sociological phenomena are related to economic change. The decline in the birth rate in the United States predates the rebirth of a conscious women's movement in the late 1960s by at least ten years. The growing interest of younger women in lifelong careers has been reinforced by an inflationary economy which seems to demand that families have two incomes. This economic reality has led many American women to conclude that their future lies in the workplace rather than in the home. In such a situation, they are choosing to have small numbers of children or no children at all. However, in so doing, they are responding more to economic necessities than to the imperatives of the women's movement. Still other women, either from choice or necessity, have decided to postpone marriage and children indefinitely. If and when these women do have children, they usually will choose to limit their families to one or two children. The women's movement would claim to have had an influence in leading women to make these choices, and it is true that the women's movement has been a major social force in the United States in the last decade. However, one is forced to conclude that the decline in the birth rate has been as much a cause as an effect of the women's movement.

Select another answer.

Z This answer is incorrect. Malthusian theory did not contribute much to the declining birth rate in the United States during the 1960s and 1970s. In 1798, Thomas R. Malthus published "Essay on the Principle of Population" in which he argued that there exists a universal tendency for population to increase geometrically unless checked by famine, disease, or war. Because, according to Malthus, the means of subsistence increase only in arithmetical progression, the number of people on earth would eventually outstrip the ability of our planet to feed the population. Malthus believed that the only way to prevent the end of human life on earth was to educate people to have fewer children and to refuse adamantly to provide assistance to the non-working poor. However, Malthus failed to consider what industrialization would do to food production and population trends. Overpopulation remains a serious problem in many parts of the world, but fear of the Malthusian nightmare cannot be considered a major cause of the declining birth rate in the United States.

Select another answer.

Business Subjects

Accounting measures in monetary terms the amount of resources controlled by an organization, the flow of resources in and out, and the claims of others against those resources. In carrying out these tasks, an accountant collects, processes, evaluates, and reports information in formats intended to facilitate economic decision-making. Conventional accounting theory and practice specifies the qualities that such reported information should have. All of the following qualities in financial reporting are desirable but in practice are not always obtainable. Which one of these qualities plays the most important role in conventional accounting?

1. precise and accurate measurement O
2. objectivity +
3. relevance X
4. clarity Z

O This answer is incorrect. Precise and accurate measurement is a desirable quality in any financial report, but obtaining it in every instance is unlikely. All of the financial affairs of business organizations are simply not amenable to precise and accurate measurement. Accounting numbers are frequently based on estimates and projections. To believe that every number in a financial report is precise and accurate is to be mistaken.

In order to understand financial reports and to interpret them wisely as a basis for decision making, one must be as aware of their limitation as of their utility. Numbers influence decisions but they do not make them. Ultimately, after numbers have been collected, processed, and analyzed, decisions in organizations have to be made by people.

Select another answer.

X This answer is incorrect. Relevance is a desirable quality in any financial report, but it is difficult to obtain. This difficulty arises from the highly subjective nature of the idea of relevance itself. Relevant information is whatever is responsive to a particular audience's information needs. Accounting theorists have only begun to resolve this problem. Apart from the generalized notions that investors and creditors need information about financial strength and operations to facilitate economic decision-making, little is known about what specific information and measurements are relevant to a user's actual or hypothesized decision-making models.

For example, in a time of continuing inflation, what dollar value should be assigned to inventory purchased three months ago--the price paid or current value? Which figure would be the most relevant one? The purchase price would be the most objective valuation, but for managers, creditors, or investors not necessarily the most relevant one.
Select another answer.

Z This answer is incorrect. Clarity is a desirable quality in any financial report but is difficult to obtain. What may be sufficiently clear for one audience may not be so for another. Accountants must presume a certain level of knowledge of financial concepts and measurement procedures by the audiences receiving their reports. While reports are prepared in formats intended to meet the information needs of managers, creditors, and investors, they are to a large extent technical documents. As technical documents, people using them must have some technical knowledge and training to do so effectively.
Select another answer.

+ Your answer is correct. Objectivity is absolutely essential in financial reporting. Financial accounting statements should be based on actual verifiable events and must be reported in an unbiased manner. In conventional accounting, the two conditions of verifiability and neutrality are the two most important components of objectivity. Financial reports which can be verified by other independent professionals and which are neutral in interpretation are objective.

Such reports will protect those audiences inside and outside of the reporting organization who lack knowledge or authority to influence the content of financial reports and will serve all audiences desiring to make their own interpretations of the information presented.
Proceed to the next question.

Outputs of accounting processes are financial reports. These reports are the means through which selected information about a business organization's economic activity is communicated to external audiences. Several different types of financial statements are regularly issued by management to external audiences. The type of statement which describes the financial condition of a business organization in terms of the equality of assets and equities or of assets and liabilities plus owner's equities is known as

1. income statement. O
2. fund statement. X
3. auditing statement. Z
4. position statement. +

X This answer is incorrect. Funds statements or statements of changes of financial position are intended to disclose major financing and investment transactions undertaken by a business organization during a specified period of time. Financing business activities has become increasingly complex. Businesses require substantial sums for operations and expansion. Such funds come from owner investment, from borrowings, from internal sources, out of earnings, or from selling non-cash assets. Funds statements show how an organization has raised capital and the major areas to which capital funds however raised have been committed. Funds statements disclose how much reliance has been placed on external as opposed to internal sources of funds, provide a basis for estimating present and future fund-generating capacities, and reveal how additions to plant and equipment have been financed. Typical funds statements provide information which serves a variety of purposes for managers, creditors, and investors.
Select another answer.

O This answer is incorrect. Income statements provide information about revenues, expenses, and net income (revenues minus expenses equals net income) during a specific period of time. Such statements indicate the revenues from sales and income from all other sources. Expenses include the cost of goods sold, operating expenses, financial expenses, and taxes payable. The botton line on an income statement will be net income and, where appropriate, earnings per share. Because income statements provide a major source of information about periodic operating activities, managers, creditors, and investors pay special attention to them. Since revenues and expenses relate to asset inflows, the income statement also provides important information about changes in asset size during a specific period.
Select another answer.

Z This answer is incorrect. Auditing statements are not issued by management. Financial reports prepared by management are presented to audiences not intimately connected with the day-to-day operations of a business organization. Such financial reports are data upon which external audiences evaluate management. Since management is largely judged on the basis of reports which management has prepared, a potential conflict of interest is avoided only by independent attestation of these reports. The function of verifying and appraising the accuracy, integrity, and authenticity of financial reports prepared by management is called auditing.

Auditing is a major accounting activity. Most published financial statements for stockholders and creditors are accompanied by an auditor's report affirming that the statements are a fair representation and have been prepared in accordance with generally accepted accounting principles.
Select another answer.

\+ Your answer is correct. Position statements report the financial position of a business organization at a particular moment in time. They are much like a photographer's snapshots. They portray in monetary terms the economic resources controlled and the claims on the interests in those resources as of a specific date. This information is always presented as a balance between assets (cash, accounts receivable, land, etc.) and liabilities (accounts payable, taxes payable, notes payable,etc.) plus owner's equity (capital stock, retained earnings, etc.). Hence, financial position statements are widely known as balance sheets.

Balance sheets are useful but they also have limitations. Balance sheets do not indicate the current value of a company's assets or of the company itself. The amount listed for any individual asset does not necessarily correspond to its market value. The balance sheets show only the unexpired cost of investments which business organizations have made. Also, because balance sheets reflect only a single moment in time, a particular balance sheet may be influenced by seasonal or exceptional circumstances. It may be accurate but not typical. Readers of balance sheets must be aware of what balance sheets show and do not show if they are to be useful in decision-making.

Proceed to the next question.

Since the end of World War II, several different approaches to the study of management theory and practice have been developed. Which one of the following groups of management theorists professes to analyze management in terms of what managers actually do?

1. operationalists +
2. empiricists O
3. behaviorists X
4. decision theorists Z

O This answer is incorrect. The empiricist approach to management theory is to look more to past managerial successes and failures than to present practices for direction. Empiricists study management case histories. They do so in order to determine why and how one team of managers is able to resolve problems that others cannot. Empiricists readily admit that history rarely ever repeats itself, but they insist that knowledge of past experiences can provide insights and guidelines for the present and the future.

Select another answer.

X This answer is incorrect. Behaviorists approach management by concentrating on understanding interpersonal relationships. Successful task performance requires that people work together efficiently. Behaviorists are especially concerned with developing motivation strategies and tactics and make heavy use of social and behavioral science research and methodologies. For behaviorists, management is primarily a matter of interpersonal dynamics, and successful managers are successful because they are able to understand these dynamics and use them to improve performance and attain both specific organizational objectives and overall goals.

Select another answer.

Z This answer is incorrect. The decision theorist approach to management problems is to identify and define all possible strategies for solving a particular management problem, and then to select from among them the strategy most appropriate in terms of time, resources, and overall organizational goals. They see management as being fundamentally a decision-making process. Consequently, decision theorists are specially concerned with such matters as information flow and reporting lines. Decision theorists have been criticized for generalizing unique business situations into universal management principles.

Select another answer.

\+ Your answer is correct. Operationalists perceive management as a process of designing and maintaining an internal environment which will facilitate organized efforts to attain group goals. Operationalists are concerned with the nature, purpose, structure, and process of function. They examine what managers actually do and then try to discover in those examinations some generalizable management principles. For operationalists, management is essentially a study of function; that is, learning how tasks are performed and then devising improvement strategies and tactics for them.

Proceed to the next question.

English

Sentence pattern 1 (S-V) is the simplest of the sentence patterns because it contains only a subject and a verb plus or minus modifiers as in: Dogs bark; The dogs barked at the cat. Which sentence below is not an example of sentence pattern 1?

1. There on the table lie the books.
2. The lake near our home often freezes in winter.
3. Did Harry leave by himself?
4. He stormed and yelled at the dog.
5. John ate the candy.

1\. Incorrect. The S is "books" and the V is "lie." The word "there" is an adverb and cannot be a subject. "Table" cannot be a direct object because it is the object of the preposition "on."

Select another answer.

2\. Incorrect. "Lake" is the subject and "freezes" is the verb. "Home" is the object of the preposition "near" and "winter" is the object of the preposition "in"; hence, neither "home" nor "winter" can be direct objects. The two prepositional phrases are modifiers.

Select another answer.

3. Incorrect. Do not be misled by word order in sentences that are questions. "Harry" is the subject and "did leave" constitutes the verb. The pronoun "himself" is the object of the preposition "by."

Select another answer.

4. Incorrect. "He" is the subject and "stormed" and "yelled" constitute a compound verb which is acceptable in pattern 1. "Dog" is the object of the preposition "at."

Select another answer.

5. Correct. The sentence contains a direct object "candy"; hence, it does not belong to pattern 1.

Proceed to the next question.

Sentence pattern 2 (S-V-DO) is a sentence containing a subject, a verb, and a direct object. The direct object is a noun or noun substitute that receives the action of the verb. Just as the subject does something as expressed by the verb, the subject does it at, on, or to the direct object. Hence, the direct object can be determined by asking S-V what or whom: The boy ate the candy.

Which of the sentences below is not of pattern 2?

1. She saw Tom.
2. What did you see on the table?
3. A display of fireworks forms part of the celebration.
4. I live in Chicago.
5. Every day he bought the Tribune and the Herald.

1. Incorrect. The direct object is "Tom" which tells whom the S saw.

Select another answer.

2. Incorrect. The direct object is "what" which tells what the S did see. Do not be misled by word order in sentences that are questions.

Select another answer.

3. Incorrect. The S is "display" and the V is "forms." The direct object "part" tells what the display forms. "Fireworks" and "celebration" are both objects of a preposition.

Select another answer.

4. Correct. This is sentence pattern 1. "Chicago" is the object of the preposition "in."

Proceed to the next question.

5. Incorrect. The compound direct object "Tribune" and "Herald" tells what the S bought.

Select another answer.

AUTOMATIC CONTROL IN MEXICO

A. Alonso-Concheiro

Department of Automation, Instituto de Ingeniería, UNAM, Apdo. Postal 70-472, México, D.F., México

Abstract. Although control system's applications in Mexico have grown in number during the last few years, automatic control is still only incipient in the country. Some of the problems encountered in the application and development of automatic control systems in Mexico are reviewed in the paper; educational and research aspects are included.

Keywords. Developing countries, appropriate technologies, education.

INTRODUCTION

Examples of automatic control system's applications have grown at an accelerated rate in Mexico during the last 5-10 years, but the achievements in this field in the country are still far from matching the possibilities of development. In industrialized countries automation is being introduced in most production processes at such a pace that some authors speak of the "Second Industrial Revolution". During the last century these countries rapidly moved towards mechanization (substitution of men and animals as the main sources of power); from the 1950's they have been rushing towards automation (substitution of man as decision maker in repetitive situations). Less developed countries are living this Second Industrial Revolution as mere spectators; their role is generally reduced to import the latest technological devices developed in industrialized countries. While in Japan, Germany and the USA one can already find factories where man is physically absent in most of the manufacture activities, in Mexico there are still many industrial sectors where the most elemental type of instrumentation is needed. And measuring the relevant variables of a given process is a first step *sine qua non* automation can not take place.

Although it may seem a paradox, the above situation might place underdeveloped countries in a somewhat privileged position. Not all the consequences of an industrial growth model based on a high level of automation need necessarily be beneficial. And while highly industrialized countries will hardly be able to turn back on the industrialization path they have chosen, Mexico and countries with a similar degree of development still have options to select the road they want to follow and there is no reason why this road should be unique or the same for all industrial and service sectors. Underdeveloped countries must try hard to be able to make use of the experience gained elsewhere, but to adopt only that which suits their particular social conditions. Thus, the development of control and instrumentation in Mexico needs not necessarily be modeled on the same way these areas have developed in highly industrialized countries.

To speak about the perspectives of instrumentation and automatic control in Mexico it would be desirable to comment first on the national objectives, ie, the goals that the majority of the members of the national community accept must be pursued. A serious discussion of these objectives (or the lack of them) is necessarily outside the scope of this paper and will not be attempted. Nevertheless, it is worth pointing out that the political system of the country is such that from one Administration to the next, goals may suffer important changes which can modify the emphasis on the technical alternatives followed. As an example, during the last presidential mandate it appeared that labour intensive technologies were looked at favorably, whereas at present it seems that government speakers lean towards capital intensive solutions to the national problems.

One of the national objectives most frequently mentioned is development. Development is often identified with industrial or economic growth, which in our opinion is incorrect. Development should be understood as the capacity of a country (or a society or an individual) to solve its problems with its own resources. Thus, indicators such as national gross product or income per capita, for example, should be related to development only if their increase is defined (explicity or implicity) as one of the national objetives and this is achieved through an adecuate use of the national resources of the country (in particular, human resources). This point of view differs from frequent statements; for example, in a document of the US National Research Council edited by HG Stever (1979, p. 3), it is mentioned that "Broadly speaking, development is concerned with two important goals: (1) eradication of extreme poverty... and (2) modernization and growth of national output both for domestic consumption and to earn income through export". This conception of the term development is esentially quantitative; the one proposed above must necessarily include qualitative criteria (Pellegrini, 1979)[1].

Let it be accepted that development is the capacity to define, pose and solve problems with one's own efforts, and let it also be accepted that development understood in this sense is the goal. Then, the processes of selection, adoption, transfer and development of technology should be modulated, no so much by direct economic benefits (although these will always certainly be wellcomed and important), but by their contribution to the enrichment of the experiences and capabilities of the national community.

The above does not imply that an absolute self-sufficiency is pretended, ignoring the accumulated international body of knowledge and experience. It only indicates the need for a search of confidence in a position where the country can reject programs of transfer of technology and "development" protecting and furthering mainly the interests of others[2].

AUTOMATIC CONTROL AND SELECTION OF TECHNOLOGIES

When referring to technologies associated with automatic control, at least two situations must be distinguished. There are processes whose operation is practically impossible without automatic control systems, be it for security reasons, speed, precision, large volumes of information handled, or complexity of the decission making algorithms. There are other processes where automation may be convenient from certain points of view but not strictly necessary. Such may be the case when production per capita can be increased by the substitution of labour intensive processes by capital intensive alternatives. Automation should be treated differently in each case. Further, in each of these two situations an evaluation of the perspectives of automation should take into account that the control and instrumentation problems and the needs of big industries differ from those encountered in small and medium industries (Regulación y Mando Automático, 1979). The capacity to adopt and to adapt technology is obviously higher in industries of great size, such as the Federal Commission of Electricity, the national oil company PEMEX, or the siderurgical complex SICARTSA, in Mexico.

This leads us to the question of where should automation be proposed as an appropriate technological solution in an underdeveloped country. The term "appropriate technology" is very fashionable nowadays and is frequently used to designate low cost, labour intensive or soft or intermediate technologies. This can create confussion and eliminate *a priori* technical solutions which, although with a high level of automation, represent the best alternative even for a poor country.

Whithin the frame of reference given in the introductory paragraphs of the paper, and given the cultural, political and economic differences between Mexico and the countries exporting control and instrumentation technology, it seems inevitable that each of these countries will interpret appropriateness differently. The need to define the term "appropriate technology" unambiguosly and to find criteria to select this technology in specific cases has been recognized and has been the subject of numerous recent contributions (Haustein, Maier and Robinson, 1979; Pellegrini, 1979; and Van Brakel, 1979). The problem is not simple. To answer if a given technology is appropriate or not, it is necessary to define, at least: (a) the ends it must fulfill; its purpose, taking into account the national goals (when defined) and those which are specific to the industrial sector under consideration; (b) the conditions under which it will be implemented; (c) the supporting technologies associated with it; (d) the available economic resources; (e) the personnel which will install, operate and give maintenance to those devices contemplated by the technology; (f) the economic, cultural and even political pressures it will have to withstand; (g) its flexibility of adaptation and its possible use in other situations. This, together with technology forecasting studies

---

[1]Defining appropriate indices to measure development understood as above is not an easy task. Cuénod (1978a) has suggested to integrate two types of criteria to measure development; one of quantitative nature, related to economic growth; and one of qualitative nature, to estimate socio-cultural achievements. As he points out, a serious and conscious effort would be required in any case to define the needs and goals of a society, both of quantitative and qualitative nature, and this should be made by the society concerned.

[2]For example, in Stever (1979, p1) the interest of the United States in helping developing countries is said to be motivated "by a desire to help build the kind of global enviroment in which the conditions for liberty and equity can grow <u>and in which US interests can best be protected</u>".

which should complete the information needed, implies that the problem to be solved should be understood with great detail and must be posed in a precise manner. With a few exceptions this is naturally but unfortunately more likely to occurr in developed countries; the decissions as to when whould and automatic control system be introduced in an underdeveloped country will generally be made on an intuitive basis. On the other hand, in Mexico as in other underdeveloped countries, those best suited to define the set of technologies which should be considered in a given situation and to select the appropriate one are seldomly those who decide. The opinion of technically qualified professionals is rarely solicited and even less frequently taken into account. In Mexico this situation is fortunately neither static nor universal, although better participation channels should be sought.

THE NATIONAL SITUATION

The availability of technology is always necesary in the developing process of any nation. But technology can be bought, transfered or developed. In the short term buying is easier than creating or copying in an intellligent manner and implies less risks and uncertanties (Reséndiz, 1979). Thus, importing technology almost always appears to be more attractive even though it generally arrests development and creates dependence. In the case of instrumentation and automatic control this is aggravated by the fact that the equipment used rapidly becomes obsolete and the fact that these fields, together with electronics to which they are closely tied today, constitute a sector which requires permanent and intensive technological innovation efforts (in these areas, the comercial life of a product is aproximately equal to the research and developing time put into it; Albarran, 1976; Magar, 1970).

A serious quantitative analysis of instrumentation and control perspectives in Mexico is difficult because reliable information is not available. The data included below can only be taken as gross estimates. Governmental sources unfortunately include instrumentation and automatic control under different headings, according to the source used, and therefore frequently differ in their figures.

Instrumentation and automatic control equipment is generally considered, in official sources, as part of the industrial electronics sector and accounts for between 25 and 30% of the total national consumption of this sector, not including computing equipment used in process control applications which would add another 1-2%. The average annual consuption growth rate in process control instrumentation during the last five years has been estimated by NAFINSA-UNIDO (Jenkner, 1978) in between 20 to 25%, reaching in 1978 according to the same source an annual figure of 80 million US dlls, with an estimated 140 million US dlls to be reached by 1982.

The market for testing equipment has grown less. The annual growth rate has probably been around 10% for this kind of equipment, reaching a net annual consumption of little over 10 million US dlls in 1978, with a projected 15 million US dlls for 1982. Of the industrial electronics sector national production, which is estimated will reach around 115 million US dlls this year, only 20% or so is in control instrumentation (communication equipment represents 70% of the sector's products). This means that around 70% of the instrumentation and control equipment annually installed in mexican industries is being imported (Jenkner, 1978).

The demand for sensing devices and transducers in the national market is high. According to the type of variable measured, the actual distribution of these devices is estimated as follows: (a) Mass, strength, volume = 25%; (b) distance, angle = 15%; (c) time, speed, acceleration = 5%; (d) temperature, heat = 10%; (e) Pressure = 10%; (f) magnetic and elasticity variables = 25%; (h) others = 10%. The demand for actuators is considerably less, pneumatic valves and electric actuators being those in greater use. Practically the total demand of the national market for electronic sensors, transducers and actuators is satisfied by imports. Of the rest, 60% is manufactured or assambled by the national industry.

Divided by industrial sectors, the main consumers of control and instrumentation equipment in Mexico are: (a) petrochemical = 25%; (b) electrical = 25%; (c) chemical = 20%; (d) metallurgical = 10%; (e) sugar and refining industries = 5%; (f) food = 3%; (g) pulp and paper = 2%. The cost of control systems in those industries where they have been implemented represents 5-10% of the total investement in plant equipment, being 60 to 80% of this cost accounted for by primary elements (sensors, transducers, actuators).

Only about 4% of the total number of computers installed in the country is being used for control purposes (2% in numerical control machines, 1% in real-time control applications, and 1% in other industrial applications); an additional 3% is being used as terminal controllers, data concentrators or front-end processors. These computers are seldomly exclusively dedicated to control functions.

A survey of 256 of the most important firms of Mexico was recently carried out (Alonso-Concheiro and co-workers, 1979) to obtain a better picture of control and instrumentation applications in the country. Of those approached only about 26% replied; of these nearly 15% declined to answer the questionnaire sent to them indicating that they did not have instrumentation and/or control systems. Some firms returned more than one questionnaire properly filled, one for each of their processes. The total number of replies was 64. The survey included only 25 questions. It was considered more important to obtain a high number of replies with general data, than detailed information of just a few firms. About 70% of the firms which replied the questionnaire have some automatic control equipment, which is generally hybrid (analog and digital). In almost all cases set points are fixed by the plant opertor based either on his experience or in laboratory analysis carried off-line. Only 7 firms reported having process control computers and only 3 of them use mathematical models as part of the control algorithm; only one has a real-time control sytem. In 75% of the industries information from the process is concentrated in one or several points, but only in 50% of them can several control commands be sent to the process from one point. Aproximately 50% of the control systems were reported to be of supervisory nature. Around 40% of the industries measure less than 50 analog and 50 digital(on-off)variables.The same applies to controlled variables. In 70% of the systems the distance between the data adquisition or control points and the control room (or pannel) is less than 100 meters.

The equipment used is almost all of foreign manufacture; 35% of the industries acquired it as an integral part of their plants, and an additional 30% bought it as a package.

It seems that the national market for control equipment is dominated by five manufacturers which account for almost 70% of the equipment installed. The figures taken from the survey are[3]: (a) Foxboro 29%; (b) Honeywell 13%; (c) Fisher & Porter 10%; (d) Taylor Instruments 9%; (e) Leeds & Northrup 8%; (f) Bayley 6%; (g) Siemmens 5%. The great majority of the industries (70%) rated the control equipment as "good". The most common equipment failures are reported to be due to environmental hazards and poor maintenance service, followed in importance by power supply failures.

According to the replies of the survey 20% of the industries have short term expansion plans for their present control systems; 39% are considering new control installations in the next five years; and a further 11% consider them in their long term plans. About 50% are considering centralized real-time computer control systems; very few of these (only 3) plan to develop them with in-house resources.

Some important sectors of industry were left out in the above mentioned survey because information about them was already available. These include: A real-time control system for the national electric power network (Alonso-Concheiro, Canales and colleagues, 1975 and 1976; Arroyo, 1978) which will be the biggest control system installed in the country and is expected to begin operation in late 1982; a real-time control system for the water supply network of Mexico City which is now under implementation (Canales and colleagues, 1979); a supervisory control system for the sewage network of Mexico City which is still in the design stages (González and colleagues, 1979); an urban traffic real-time control system for some parts of Mexico City, a partially operational project that has apparently run into some difficulties, among them financial, and which has never been fully implemented; a real-time control system for the most important oleoduct of the country (Cactus-Reynosa) which according to design specifications will have at least 16 remote terminal units under control; and the control system for the automatic long distance telephone service which tenders around 350 unattended microwave stations.

[3]By number of times mentioned in the replies received.

The number of engineers working in the country in the instrumentation and control field is difficult to estimate. There are probably about 250 engineers in the sales departments of the different companies established in Mexico, each company employing between 15 to 20 sales engineers; the average annual income of such engineers is around 1,300.00 US dlls before tax (Oropeza, 1979). There is a higher number of engineers and technicians working on instrumentation maintenance in industry; probably 1,500 of them work for the state owned oil company PEMEX; the Federal Comission of Electricity employs another 250 and the siderurgical complex SICARTSA around 80, just to mention those with the biggest instrumentation departments (it should be noted that in the three cases these figures refer to several plants and processes belonging to the same industrial complex; in the case of PEMEX, for example, the figure refers to all petroleum and refining industries of the country and a great number of the petrochemical ones).

Control and instrumentation engineers can also find jobs with the few engineering firms which implement automatic control projects. These firms help their clients to define and specify the process engineering projects, to select the manufactures which will provide the equipment, to design, develop and build support installations and civil engineering works, etc. These companies generally work with foreing licences or patents and seldomly develope their own technology.

EDUCATION

In spite of recent government efforts, higher degree education opportunities in the country are still very much those offered in Mexico City. In the last years a System of Regional Technical Schools was established in the country but its impact has been marginal up to date. The biggest University by number of students, resources, staff, etc, is the National University of Mexico, located in Mexico City, with about 13,000 students in the Engineering Faculty and a total population of over 100,000. The National Polytechnic Institute which is the second largest educational institution of the country is also located in the capital. For a general overview of engine ering education in Mexico the reader is referred to a paper by Murray-Lasso (1972). Most of his general comments are still valid today.

At present, around fifty of the institutions with first degree programs in engineering offer courses on control and instrumentation; of these only three (Autonomous University of Nuevo Leon, Mexican University of the North East, and CERETI) have programs specifically leading to a degree on control engineering (Oropeza and Grijalva, 1979). There are at least four institutions offering MSc degrees on Control; these are the National University of Mexico (UNAM), the National Polytechnic Institute (CIEA), the Institute for Technical Studies of Monterrey (ITESM), and the National Institute for Astrophysics and Electronic Optics (INAOE).

At the Faculty of Engineering of the National University of Mexico the Control Department coordinates 13 undergraduate term courses on the area. Some are compulsory for all Mechanical and Electrical Engineering students and some are optional. The subjects offered by the Control Department at undergraduate level are (numbers in parenthesis indicate the approximate number of students enrolled in each subject; students are divided in groups of fifty on average): Electromechanical Systems and Circuits I (500) and II(350); Control Engineering I (500), II (200) and III (variable); Metrology (500); Servomechanisms (variable), Industrial Control (variable); Introduction to Control and Instrumentation (100); Linear Systems Analysis (100); Control (variable); Numerical Control of Machine-tools (variable); Special Topics on Control (variable).

The contents of these and similar subjects offered by Universities and Technical Schools of the country are generally equivalent to those offered by Institutions of highly industrialized countries. In our opinion the educational problems of the country stem not so much on defficient programs or an insufficient number of them, but on the accumulation of other factors.

First, engineering courses are with a few exceptions oriented towards "information", with little attention being paid to "formative" aspects. On the long run this generates not good engineering graduates but low level specialists who, on the other hand, have not received an adequate training to become good intermediate technicians. This cannot be blamed only on educational institutions; after all they are only part of the educational system; they are inmersed in the economical, political and cultural conditions of the country, and respond to stimulus and pressures present in their environment. Instead, universities should prepare professionals with high qualifications. They should resist external pressures and influence the educational profile of their students at least as much as those who will employ the student once he graduates.

Second, Universities are literally flooded with students every term. Having a first degree is considered an important step to ascend in the social ladder, besides the obvious economic benefits that can be derived from it. The number of first degree students' applications increases every year and there are political pressures on universities to accept all those who apply, almost regardless of their school records (to try to enforce a rigorous entrance qualifications' procedure instead of the present entrance examination, which almost every student passes with little difficulty, is demagogically regarded as an antidemocratic measure). The overpopulation of students together with the universities' economic constraints create, among other problems, a serious shortage of materials and staff. To teach the 46 theoretical courses and 40 laboratory courses offered each term by the Control Department of the Faculty of Engineering of UNAM, there are 6 full-time lecturers, 3 technicians, and 50 lecturers and 30 assistants who either come to the university to impart only one or two courses per term (4 to 8 classroom hours a week) but work full-time elsewhere, or who are still working towards their first degree in engineering. Among the members of the staff (full-time or part-time) 4 hold PhD degrees, 7 MSc degrees, 53 BSc degrees, and 30 are still working towards their first degree (Oropeza and Grijalva, 1979).

In third place, it is worth emphasizing that educational programs for technicians are very important (Jamshidi and colleagues, 1979; Larsen, 1978; Apter, 1978; Cuénod, 1978b). In Mexico these programs are unexistent in most technical fields, including control and instrumentation. This is unacceptable if the field is to develop for in automation problems, perhaps more than in other technical areas, mid-level, well prepared technicians are a must. To favour the implementation of control systems of any complexity when the technical level of users and maintenance crews is low is quite difficult if not impossible.

Other problem frequently encountered in control education in developing countries is the lack of educational tools (Larsen, 1979; Jamshidi and colleagues, 1979). Although Mexico is not the exception, the situation has been improving during the last few years. Several textbooks on control have been written in Spanish specifically for courses offered by national institutions (see for example: Gerez and Murray-Lasso, 1972; Canales and Barrera, 1977; Gerez and Czitrom, 1979; Torres and Czitrom, 1979) and laboratory installations have been augmented and modernized (Oropeza and Grijalva, 1979). This is not to say that further improvements are not needed, but rather that there are positive signs that this will be done.

During the VIIth IFAC's World Congress held in Helsinki in June 1978, an interesting round table discussion took place under the tittle

"Control Technology and Developing Countries" (Apter, 1978). The speakers were almost unanimous in pointing out that the lack of adequate education and training in control topics was one of the main problems in furthering automatic control applications in developing countries. Although recognizing the importance of education and the problems encountered on the implementation of good educational programs, some of which have just been mentioned above, we differ, as we did during the round table, with the emphasis placed on this point. Institutions in charge of education are not islands isolated from the rest of society. From what was said in the introduction it should be clear that in a sense we equate development with capacity to learn. But learning has to do with more than formal education programs, and accumulated knowledge in any specific technical topic is not enough when there are other factors which deter its application to solve the national problems. Unless these factors are modified, little will be gained by more and better education and training programs in control. Information which is not used will in the end be forgotten. It is true that knowledgeable individuals in these areas are scarce in Mexico as in most underdeveloped countries; yet, those few experts available often encounter serious problems in applying what they know and in furthering their knowledge.

RESEARCH AND DEVELOPMENT

The obstacles that research and development projects find in Mexico are great but not unsurmountable (Reséndiz, 1979). In the case of instrumentation and automatic control these are probably greater than in other engineering areas. Automatic control is still a relatively new discipline, it is developing and changing rapidly, and it covers an ample spectrum of fields of knowledge, ie, those covered by its applications. Further, the research and development teams which exist in this area in Mexico are still young and have still little tradition; potential users show little understanding of control problems; and there is a strong international competition in the field. All these factors hinder the development of research projects on control in Mexico.

Before 1960 there was no institution in Mexico with research and development teams in electronics, communications, automatic control or instrumentation (Canales and Alonso-Concheiro, 1979). The little research that was done was the product of a handfull of researchers working individually. The research groups that exist today, around 12 in number, were formed almost all during the 1970's, as support groups for other research fields. Presently there are around 160 researchers and technicians in these groups. Their common research area is that of digital systems and instrumentation, in particular the application of microprocessors to data acquisition, processing and display. In many cases prototypes of local design have been built (Ramirez and Vaca, 1976; Ramirez and Alonso-Concheiro, 1979; Albarrán, Baillie and colleagues, 1979). Efforts have also been made in the analysis and synthesis of large scale systems (Alonso-Concheiro and colleagues, 1975 and 1976; González and colleagues, 1979).

Thus, more than 80% of the research and development resources have been dedicated to digital instrumentation, real-time control and data acquisition systems. Few of the activities of these groups can nevertheless be considered strictly as research. The majority of them are carried out for learning purposes or to show that some new techniques are known and can be applied with confidence to try to win the trust of those who can fund research.

CONCLUSIONS

Most industries in Mexico have yet to automate their production processes. The majority of the measuring systems installed in industry are still of the analogue type and those few control systems that do exist are small and used for local control purposes. Exceptions are: the electrical power industry, the oil and petrochemical industries, and an important siderurgical complex; there digital on-line control systems are either operating or under implementation. As long as basic industrial engineering problems remain unsolved automatic control systems will remain luxurious commodities. There are however some signs that the situation is changing. During the last few years the national market for automatic control systems has been growing at an accelerated pace and is expected to continue to do so.

Education and training programs should and could be improved. Most control courses being offered by Mexican education institutions have an adecuate technical content, but capable staff is seldomly available. Emphasis should be placed on formative aspects of education rather than exposing students to a large amount of information which in many cases will soon become obsolete.

There is no research tradition on control systems in the country. The research groups that exist today were formed during the last decade as support teams for other engineering disciplines. Today they have established themselves as independent research groups with their own objectives and needs, but are still inmature and small. Most of these groups have as their main research areas digital systems applications and large scale systems analysis and synthesis. It is thought that research will be more fruitfull if efforts are directed towards better control algorithms and their software implementation, although systems development based on mini and microcomputers will still be important.

Those who have in their hands the decisions on national technological development must be cautious before proposing a road of indiscriminate automation. In a social system such as that of Mexico, a higher level of automation will generally correspond to a bigger concentration of goods (or power) in a few hands, or at least it will favour the conditions for this to happen.

Achieving some of the national objectives, such as a better distribution of the National Product, may require the implementation of technological strategies where in some industrial sectors automation will play only a moderate role.

REFERENCES

Albarrán, J. (1976). "La Electrónica: Panorama Internacional y Transferencia de Tecnología en México", VII Biennial Congress, Colegio de Ingenieros Mecánicos y Electricistas, México.

Albarrán, J. Baillie, P, and Colleagues (1979). "TRIIE: La Unidad Terminal Remota desarrollada por el IIE", Boletín IIE, Vol. 3, No. 1, January, Mexico, pp. 27-32.

Alonso-Concheiro A., Canales R., Guillen S.T., and Colleagues (1975). "Control en Tiempo Real del Sistema Nacional de Potencia: Fase II", Internal Report, Instituto de Ingeniería, National University of Mexico, 215 p.

Alonso-Concheiro A., Canales R., and Colleagues (1976). "Control en Tiempo Real del Sistema Nacional de Potencia: Fase III", Vols. I and II, Internal Reports, Instituto de Ingeniería,National University of Mexico, 381 p.

Alonso-Concherio A., Ramírez C.V., Peñarrieta LH.E., Montaño J.L.A., Alonso J.M. (1978). "Estudio sobre aplicaciones y características de unidades terminales remotas", Internal report, Instituto de Ingeniería, National University of Mexico.

Alonso-Concheiro A., Alonso J.M., Ramírez C.V. (1979). "Resultados de una encuesta entre las principales industrias mexicanas", Internal report, Instituto de Ingeniería, National University of Mexico.

Apter H. (1978). Report of round table 3; "Control Technology and Developing Countries", Proceedings VIIth IFAC's Triennial World Congress, Helsinki, Finland, June 12-16.

Canales R., Barrera R. (1977). "Análisis de Sistemas Dinámicos y Control Automático", Limusa-Wiley, México.

Canales R., Alonso-Concheiro A. (1979). "La Investigación en Ingeniería Electrónica, de Comunicaciones, de Control Automático e Instrumentación en los Ultimos 20 Años", Meeting of the Academia de la Investigación Científica, Oaxtepec, México, September 1979.

Canales R., and Colleagues (1979). "Control en tiempo real de la red de abastecimiento de agua potable de la Ciudad de México", Internal report, Instituto de Ingeniería, National University of Mexico, February.

Cuénod M. (1978a), Personal correspondance with the author.

Cuénod M. (1978b). Contribution to Round Table 4: "Science and Technology for Development: How can IFAC Contribute?", VIIth IFAC's Triennial World Congress, Helsinki, Finland, June 12-16.

Gerez V., Murray-Lasso H.A. (1978). "Teoría de Sistemas y Circuitos", Representaciones y Servicios de Ingeniería, S.A., México.

Gerez V. Czitrom V. (1979). "Circuitos y Sistemas Electromecánicos", Representaciones y Servicios de Ingeniería, S.A., México, 350 p.

González S., Alonso-Concheiro A., Canales R., Rodríguez H. (1979). "Control en Tiempo Real de la Red de Drenaje de la Ciudad de México", Internal Report, Instituto de Ingeniería, National University of Mexico, 148 p.

Gupta M.M. (1979). "A Confluence of Feedback Loops in Social and Educational Structure (In the Context of Developing and Developed Countries)", IFAC's Symposium on Criteria for Selecting Appropriate Technologies Under Different Cultural Technical and Social Conditions, Bari, Italy, May 21-23.

Haustein H.D., Maier H., Robinson H., "Thinking About Appropriate Technologies", IFAC's Symposium Criteria for Selecting Appropriate Technologies Under Different Cultural, Technical and Social Conditions, Bari, Italy, May 21-23.

Jamshidi M., Male-Zavarei M., Vakilzadeh I. (1979). "Problems of Automatic Control Education in Developing Countries", Automatica, Vol. 15, pp. 105-112.

Jenkner K.W. (1978). "Investigation of the Electronic Capital Goods Industry in Mexico", NAFINSA-UNIDO, Preliminary report, Mexico, 353 p.

Larsen P.M. (1978). "How can Industrialized Nations Support Control Education in Developing Countries?", round table Control Technology and Developing Countries, VII IFAC's World Congress, Helsinki, Finland, June 12-16.

Larsen P.M. (1979). "Trends in Automatic Control Education", Automática, Vol. 15, No. 1, pp. 101-104.

Magar R.V., (1970). "La Electrónica", in El Perfil de México en 1980, Vol. 2, Ed. Siglo XXI, México, pp. 75-82.

Murray-Lasso M.A., (1972). "Engineering Education in Mexico", IEEE Trans. on Educ., Vol. E-15, No. 4, pp. 229-239.

Oropeza G. (1979). "Instrumentación y Control Automático en México", round table Automatic Control and Instrumentation Prospects Within the Economic Framework of Mexico", ISA Mexican Section, October 30, National University of Mexico.

Oropeza G.A., Grijalva L.P.M., (1979). "Programas de Estudio en el Area de Instrumentación y Control en la Facultad de Ingeniería, UNAM", 3rd ISA Semminar on Industrial Instrumentation and Automatic Control, México, D.F., November 26-28th.

Pellegrini U. (1979). "The Problem of Appropriate Technology", IFAC's Symposium on Criteria for Selecting Appropriate Technologies under Different Cultural, Technical and Social Conditions, Bari, Italy, May 21-23.

Ramírez C., Vaca A., (1976). "Terminal de Video de 7 Colores", Internal Report, Instituto de Ingeniería, National University of Mexico, 103 p.

Ramírez C., Alonso-Concheiro A., (1979). "Implantación de una Terminal de Video de 7 Colores a un Sistena Dual de Computadoras", Internal Report, Instituto de Ingeniería, National University of Mexico, 108 p.

"Regulación y Mando Automático",(1979). Mesa Redonda de Electricidad y Electrónica, No. 86, March 1979, pp. 30-37.

Reséndiz D., (1979). "Consideraciones Tecnológicas sobre Diseño y Producción Nacional de Maquinaria Agrícola", in Inversión y Tecnología Para Equipos Agrícolas y Agroindustriales, CONACYT, México, pp. 51-64.

Stever H.G., ed. (1978). "US Science and Technology for Development. A Contribution to the 1979 UN Conference", prepared by the National Research Council, Pub. No. 052-003-00546-B, US Government Printing Office, Washington, D. C., 212 p.

Torres F.A., Czitrom V., (1979). "Métodos Para la Solución de Problemas en Computadora Digital", Representaciones y Servicios de Ingeniería, S. A., México.

Van Brakel J., (1979). "The Doubtness of Assesing Appropriatness: Empirical Evidence", IFAC's Symposium on Criteria for Selecting Appropriate Technologies under Different Cultural, Technical and Social Conditions, Bari, Italy, May 21-23.

COMPUTER CONTROL INTEGRATED SYSTEM

J. E. Morales and J. E. Venencia

CONICET, INTEC, Control Area, C.C. 91, 3000, Santa Fe, Argentina

Abstract. This paper presents a Computer Process Control Program created for Research and Development. The project is being developed at INTEC under sponsorship of the Argentine Research Council (CONICET). The aim of it, which started in 1978, was to train young engineers and scientists in this area. The Computer Control Integrated System is intended to be the backbone of the Control Laboratory.

Keywords. Computer control; Control equipment; Educational aids; Hybrid computers; Industrial control.

INTRODUCTION

The Computer Process Control Program was established in 1977 by the Institute of Technological Development for the Chemical Industry (INTEC), to be devoted to Scientific and Technological Research on Industrial Process Control Design.

Classical steps were going from national evaluation of the state-of-the-art up to social, economical, and technical analysis. As a consequence of the previous work, the objectives and the planning of resources were defined, bearing in mind the national constraints on the matter.

It was decided to start a first four-year period in 1978 with personnel selected from universities and industries, all of them at the beginning of their professional careers. The first stage of this process was to fill the knowledge gaps of all the participants involved in the works.

Simultaneously, the so called Computer Control Integrated System was defined to be the main educational aid. The system consists of the following parts: i) the plants site, ii) the interfaces allocation; iii) the computer room, and iv) the control console. All these elements will operate in an equivalent way to any industrial installation, regarding to a generalized in-line mode, real-time system.

Based on the preliminary definitions, this communication has been structured to give an overview of the project. It starts with the statement of the problem and includes the methodological aspects of the plan.

The reader should notice that there is not a particular control problem to be solved immediately. Therefore, it is intended to provide a basic teaching framework, for the continuous creation of control scientists.

The motivation behind this is that the country has had, so far, a complete lack of this sort of facilities.

PROBLEM STATEMENT

It is important to point out that Instrumentation and Control is a common undergraduate course included into some Engineering College Syllabuses in Argentina, but there are no specific oriented postgraduate degrees -neither Master nor Doctorate- in Control Sciences.

It is obvious that, without minimal plant equipment or laboratory instrumentation, all the teaching is reduced to theoretical methodology. Long periods and large investments are required in our country, to accomplish this type of work. Furthermore, if control structure design problems are the goals of the project, and complex processes are to be handled, then all steps and theoretical updates must be included into this work.

There are some modules of an installation, e.g. digital computer, that could be a first step for a Computer Control Laboratory -that is our case- but it can not be the whole system. It is important to note that real-time processing and versatility are more important factors than high precision computation, considering our goals (Larsen, 1979).

In that way, an intermediate stage between modern control theory and industrial plant application can be reached in a relatively inexpensive and short-term project. In our case, students have a minimal Instrumentation and Control undergraduate background and they participate on the whole real world based activities.

Considering that this type of graduate education is not directly related to a particular process or control problem, the technical groups incorporated can be -perhaps must be- formed interdisciplinary and prepared integrally to give equal importance to the different parts of the real system, e.g.: process; plant; communication; instrumentation; computer; control; etc.

It is our believe that the complexity of the problems attacked, can be increased along the learning process if the personnel participates full time in the three main aspects of the project: i) analysis, ii) synthesis and evaluation, and iii) theory and practice.

Thus, we can cover the spectrum from system design to theoretical and practical aspects of modern control systems. Furthermore, regular trends can be introduced into laboratory assignments during graduate studies.

After a minimal support is obtained, it is reasonable to present and discuss more about the funds required to solve the particular industrial requirements to appear (Jamshidi and co-workers, 1979).

Based on the preliminary aspects, without going into the details of each one, there appeared some other national institutions interested in this project, with rather positive consequences. The main body of the whole problem was decomposed in such a way that all of them were induced to a team, where profits became from the coordinated tasks laying on a more complex context (Lemmen and Van Den Boom, 1979).

PROJECT DESCRIPTION

As the Computer Process Control Program was initially proposed to INTEC in 1977, a first draft of its structural chronogram and the required details for a central four-year period, was prepared in accordance with Fig. 1. It can be seen that there are five subperiods of related works, denominated with roman numbers, from I to IV, and including the following activities:

I. Preliminary feasibility studies, technically and economically justified (1); formal proposal, with some details of the objectives (2); and analysis of national industry requirements (3).

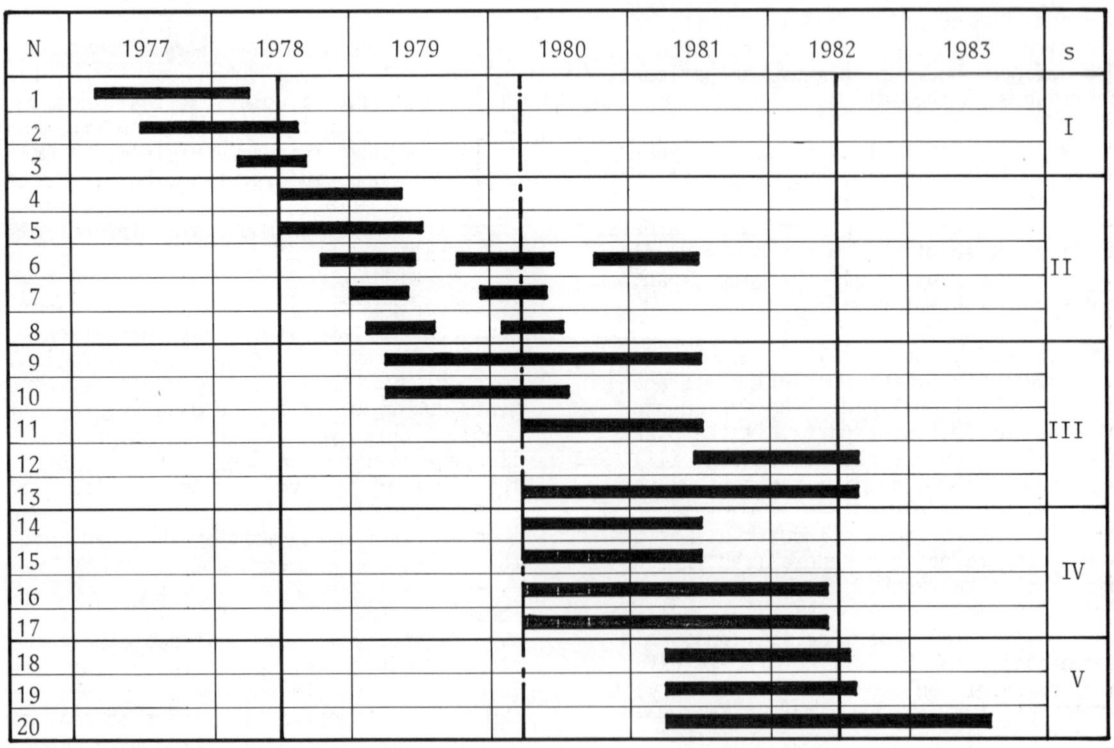

Fig. 1. Computer Process Control Program

II. Human and material resources requirements (4); national possibilities of other institutions to coparticipate, using their own resources (5); personnel incorporation, first level training, and homogenization (6); laboratory equipment and instrument forms filling solicitudes (7); and technical definitions for complete installation (8).

III. Modern control theory and practice courses, directed to the specifical human resources formation plan (9); computer control integrated system modular design and first level operational unit construction (10); command console engineering design and construction (11); parallel processor design and construction (12); and interfaces design, installation, and proofs of the integrated second level interconnections (13).

IV. Hybrid computer educational aid room design (14); electronic and control laboratories educational aids room design (15); methodological aspects of the electronic sub-systems engineering works, directed to particular applications into distributed control and its development tasks (16); and scientific computer control dedicated tasks development (17).

V. Design of an electromechanical null production pilot plant and its interconnections into the integrated system (18); installation of the pilot plant and interfaces with other pilot plants projected into INTEC academical structure (19); and final standarizations and running steps of the whole system (20).

As a consequence of INTEC projects, a new financial aid appeared in 1979, through a Contract signed by CONICET and the Interamerican Bank for Development (BID), giving a good complementary support to our possibilities of realistic R & D proposals. Global ammounts related to this source and the Control Program are indicated in Fig. 2. The main items -benefits can not be economically computed at the present time in a direct way- are mentioned, in accordance with the following description:

A. Costs involved in planning and running, directed to personnel (1); technical assistance and training plans (2); operation and other items (3).

B. Costs involved in installation and maintaing, directed to equipment (4) and building (5).

Academic personnel includes technicians; professionals and researchers. There are

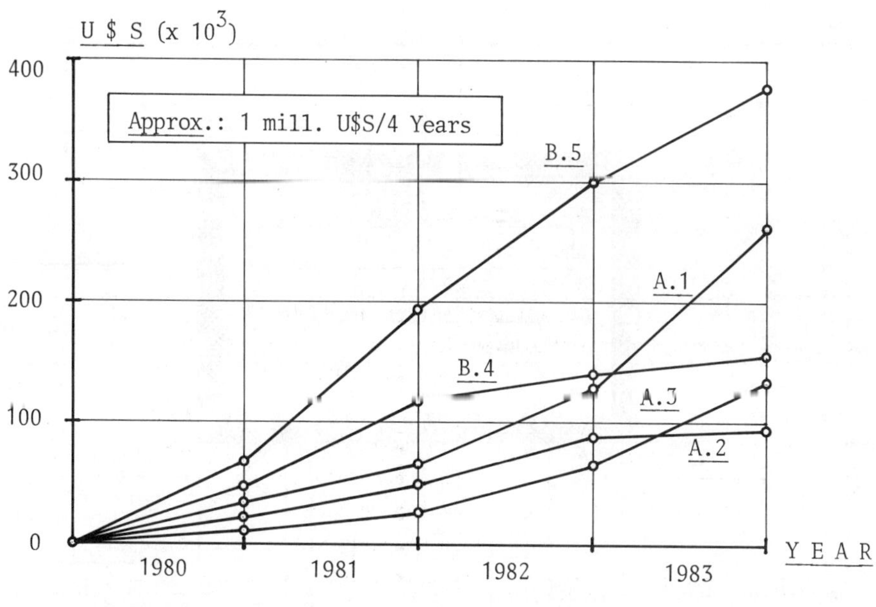

Fig. 2. Accumulative Resources BID/CONICET Plan

nine persons as of April 1st., 1980. At the end of the four-year period, there will be approximately fifteen, all of them working with this project. In regard to non-academics, INTEC is made up of 150 persons today, working by a total of 8 lines of research and specialized areas included into Chemical Engineering.

In addition, there are some other national institutions working with electronic developments to the integrated system. They are using their own laboratories and personnel, and are not included in this presentation because they are in the beginning of relative work directed to our requirements.

INTEGRATED SYSTEM

This installation will be constructed in two interrelated periods: The first one was planned to be operative in 1981 and concluded at an intermediate level; and the second, in accordance with Fig. 3, including all the parts of the integrated system. The first step, representative of the primer objective defined into the whole integrated system installation plan, is actually including:

* A Serial processor: Digital computer DEC-PDP-11/40, 128 Kw, 16 bits.
* A programmer terminal: VT-52, video input/output.
* A console terminal: LA-36, serial printer.
* A command console: VT-52, video output with independent keyboard inputs.
* An industrial control remote unit: ICR/ICS-11, one module with: 120 analog inputs; 32 digital inputs: 8 analog output; and 64 digital outputs.
* Peripherals: one LP-11 printer; two cassettes; one 800 bpi magnetic tape; two RK-05 disks; one RP-04 88 Mbytes disk-pack; one card reader; and one plotter.
* An operational unit: Totally designed and constructed by INTEC and Control Area personnel, with all communications at a distance of 30 m from the computer room. It includes a panel board to connect and switch the digital computer with all other external equipment and it is functionally represented in Fig. 4.

To obtain the complete installation in 1983, in accordance with the chronogram, the previous parts will be replaced by the follow-

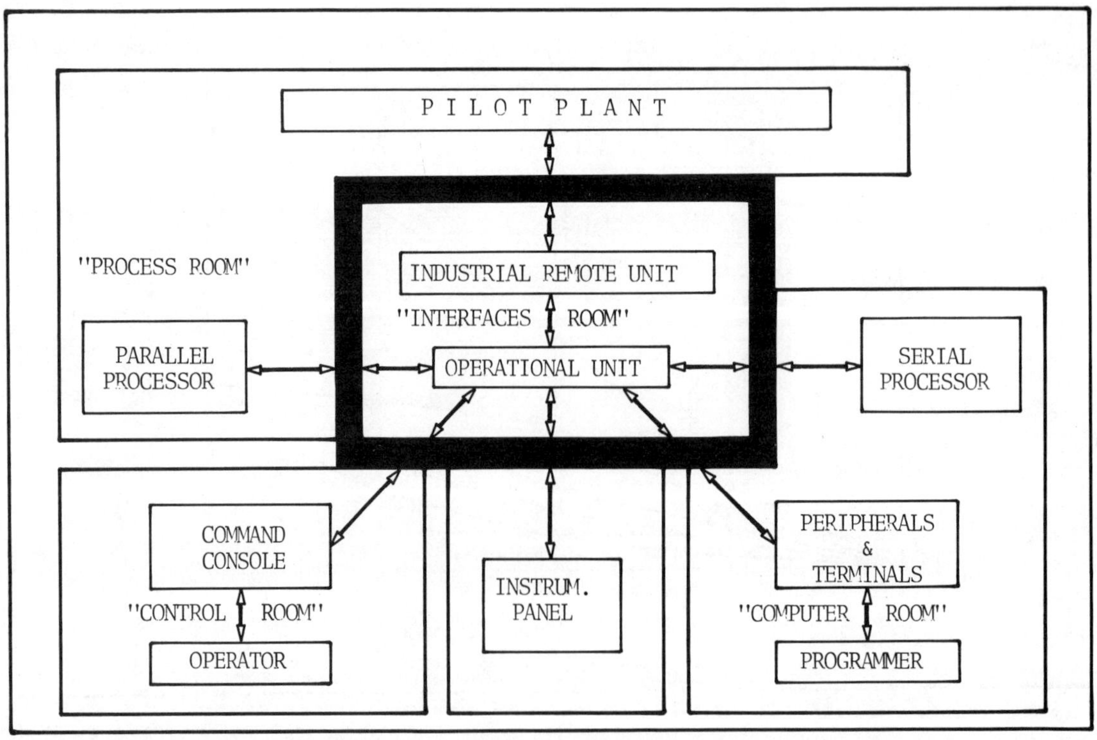

Fig. 3. Computer Control Integrated System

ing modular sub-systems:

* A command console, to be designed and constructed in conjunction with another institution (CITEFA - Force Army Technological Research Center), from which all the integrated system would be commanded in a centralized way at the control room.

* A parallel processor, to be designed and constructed in the same way that the command console, through another institution not defined totally yet (C.I.T. -Technological Research Center of the Technological National University).

* A pilot electromechanical plant of null production, to be designed into the Control Area of INTEC Institute next year, and replaced by now with two short projects on complex structure laboratory plants: one of them to be devoted to serial processes, and the other to a chemical reactor over instrumented. These sub-projects are provided by students.

System interfaces are to be designed, based on digital PDP central processor and operating system, alternatively RSX-11M or RSTS. Languages could be: FORTRAN; BASIC; or MACRO; and the Simulation Languages developed by the Area Personnel. The instrumentation panel is to be completed in accordance with sub-system level requirements.

COMPUTER CONTROL

As it can be deduced from the Integrated System description, the persons involved into the whole work pass successively through some of the intermediate stages of the control problem: digital computation; hybrid simulation; operation; and command. On the other hand, theory and practice are developed over all steps of a typical control project: evaluation; analysis; and synthesis. As a consequence, the different studies may be done on any topic of the Control Theory: data acquisition; logic; modelling; characterization; identification; simulation; stability; compensation; multivariable command; optimization; start-up and shut-down; protections; faults; etc., to express only some of all the possibilities involved in any project.

The main philosophy, resulting naturally because there is not a particular control problem to be solved -but a Control Laboratory to be implemented and a Human Resources Plan to be accomplished- is just to get knowledge in a specific area: Computer

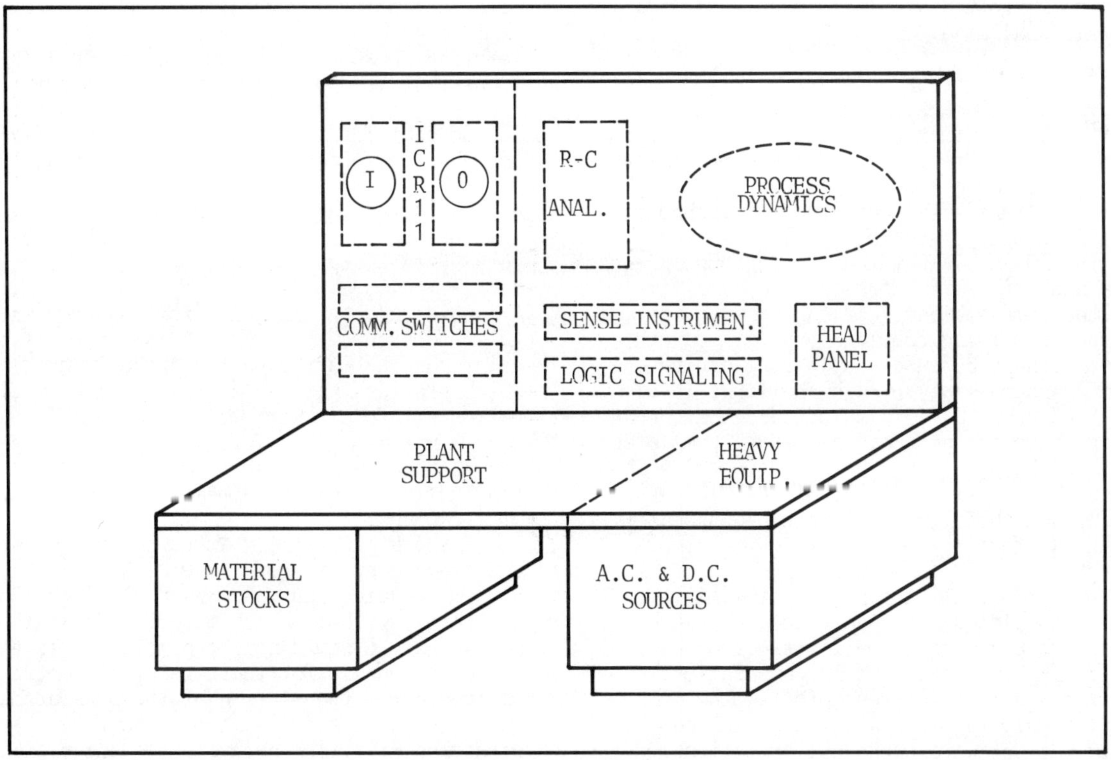

Fig. 4. Operational Unit Functions Description

Process Control. One of the reasons to do all the same time, can be encountered on personnel dedication (full-time) and a lack of industrial research groups in most of the national enterprises. There are some Automatic Control Laboratories at University structure, but the Professor-student relationship is mostly part-time and has mainly another objectives: to reach a particular Engineering Curricula non-existent in Automatic Control.

In our particular case, courses are given twice a year sequentially (e.g. Control courses: I; II; ... etc.); including from classical to modern topics; and parallel subjects (e.g. Course I -Chapter 1: Process Modelling Methods), with applications to the real world (e.g. Steady-state Mathematical Representations of: electrical; mechanical; biological; economical; etc. systems). Simultaneously, they are mathematically represented by some of the existent structures (e.g. Steady-stage and dynamic; linear and non-linear; deterministic and stochastic; continuous and discrete; distributed and lumped, etc.).

From all of them, there are cases in which a digital computer may be used to off-line simulation purposes; on others, quasi-closed loops may be defined to reach an on-line operator included system; and it can be a last one, where an in-line completed automated distributed or numerical control would be the best solution to a particular control problem. Neither one nor other can be forgotten a priori in an integral educational plan.

CONTROL LABORATORY

The Computer Process Control Program is fundamentally directed to industrial plants, and simultaneously constitued of Technological Development and Scientific Research goals, in such a way that each group -Analysis and Synthesis- works at the same time on the same subject. It can be seen that there is not any completed published work yet, perhaps because people are joining the staff and passing through the first level of studies. But, it is intended that, next year -once the academic personnel completes this background training and homogenization plan- there will be a lot of work to be made, and their manuscripts could appear in final form.

At this time, all the primary installation looks completed, the individual work or research plans would be directed to the following subjects: Hybrid computation; hierarchical structures; programming methods; simulation languages; distributed command; software packages; operational unit; command console; parallel processor; communication hardware; system software; pilot plants; optimization methods; etc. As it can be seen, in the last list, there is only a little part of the whole project. These are the particular subjects we are developing at this time, and the planned work for the near future.

CONCLUDING REMARKS

At this point, it is important to note that there are some commercial systems which are related to the Computer Control Integrated System. A sample of them, for which it is nearly impossible to have full information at present, can be trademarked as: UCS-3000; DIOGENES; MOD-III; VERITRAK-7500; TDC-2000; DCI-4000; MV-8000; SPECTRUM; BAILEY-7000; ACC; DCC; etc. It appears a reasonable query: Why not ask for just one manufacturer? -surely it will be the best way, once a particular process control problem requires a real industrial solution. By now, the whole generalized computer control installation would be our main objective, and there is no other possibility in our own training.

CONCLUSIONS

A brief outline of the Computer Control Integrated System has been given in this paper. It has been presented in order to give an overview of the system layout, and the Computer Process Control Program structure. The motivation behind this communication is to expect helpful suggestions and eventually foreign assistance. An attempt was made to bring up some of the particular applications that lead research and development work, and the necessary steps to reach them in the near future. There is not specific proposal presented here at international level. The reason for this: this project has not faced yet major difficulties, and all the intermediate steps have been taken according to the chronogram.

A considerable amount of effort was put into this endeavour. However, it is important to point out the participant's enthusiasm, which provides us with a great deal of motivation for future work.

ACKNOWLEDGEMENTS

The authors greatefully acknowledge the work and sincere efforts of Mr. W.A. Cebuhar, Mss. O.I. Jais, and the personnel of the Control Area, without which material for this communication would not be possible. To the INTEC personnel, for their relevant contribution in the translation of the present paper. To readers, we wait for their constructive criticism and valuable suggestions, for the benefits of this project.

REFERENCES

Dallimonti, R. (1973). New designs of process control consoles. Instrum. Technol., 20, 48-53.

Findeisen, W. (1975). IFAC report symposium on system engineering education in developing countries. Automatica, 11, 555-556.

Findeisen, W. (1978). On-line hierarchical control for steady-state systems. IEEE Trans. Autom. Control, Vol. AC-23.

Fisher, D.G. and D.E. Seborg (1976). Multivariable computer control -A case study. North-Holland Pub. Co.

Fossard, A.J. and C. Guéguen (1979). Multivariable system control. North-Holland Pub. Co.

Goulcher, R. and J.J.C. Long (1978). The solution of steady-state chemical engineering optimization problems using random search algorithm. Comput. & Chem., 2, 1, 33-36.

Hilburn, J.L. and P.M. Julich (1976). Microcomputers/Microprocessors: Hardware, Software, and Applications. Prentice-Hall, Inc., N.J.

Jamshidi and co-workers (1979). Problems of automatic control education in developing countries. Automatica, 15, 105-112.

Larsen, P.M. (1979). Trends in automatic control education. Automatica, 15, 101-104.

Lemmens, W.J.M. and A.J.W. Van Den Boom (1979). Interactive computer programs for education and research: a survey. Automatica, 15, 113-121.

Lowe, E.I. and A.E. Hidden (1971). Computer Control in Process Industries. Peter Peregrinus Ltd.

Mesarovic, D.M. and co-workers (1970). Theory of Hierarchical, Multilevel Systems. Academic Press, N.Y.

Mital, K.W. (1977). Optimization Methods in Operation Research and System Analysis. Wiley Eastern Ltd.

Newell, R.B. (1978). The design of operator interface for process computer systems. Aust. J. Instrum. & Control, 34, 6-9.

Rosenbrock, H.H. (1977). The future of control. Automatica, 13, 387-392.

Williams, T.J. (1967). Computers and process control. Ind. & Eng. Chem., 59, 12.

Wilson, I.D. (1979). Foundations of hierarchical control. Int. J. Control, 29, 6, 899-933.

MINICOMPUTER EVALUATED MICROCOMPUTER PROGRAMMING[1]

D. Calikoğlu

Department of Electrical Engineering, Middle East Technical University, Ankara, Turkey

Abstract. MS is a microcomputer with analog inputs and outputs, developed especially for the Discrete Time Control Systems Laboratory of the Middle East Technical University. It is intended to be used in experiments demonstrating the utilization of digital computers in automatic control systems. In order to fascillitate the teaching of programming MS, a system employing a PDP-8 minicompter is devised to evaluate the student programs automatically. A special monitor executing on the PDP-8 accepts the student programs in batch form. It loads each student program to the memory of MS directly and runs it. At the termination of each program, (or when a time-out occurs) the monitor records the job name and the degree to which the program has run correctly. When all the programs in the batch are completed, a list is generated giving the names and the corresponding coefficients of success. The coefficient of success is calculated according to a predefined rule given by the instructor for each assignment.

This program evaluation system enables the utilization of the peripherals of the minicomputer in communicating with MS, enables a relatively larger group of students to be handled with less effort, and enables MS to be used with relatively higher efficiency and effectiveness. Furthermore, the system can be used in parallel processing without hardware modifications.

Keywords. Discrete time systems; teaching; on-line operation; micropocessors; minicomputers; educational aids.

INTRODUCTION

Microcomputers, while being relatively restricted in computational speeds and capacities, are compact and inexpensive devices which offer the basic capabilities of computers. Due to this fact, they have a wide range of potential applications in many fields and particularly in automatic control. The low cost of the microcomputers also provides an oppurtunity to set up experiments in the laboratory for demonstrating the principles of the discrete time control systems. In such experiments, the students may be asked to write appropriate programs in order to realize specific digital controller functions.

For teaching microcomputer programming, one of the following possible software development approaches can be employed:

(a) Direct utilization of the microcomputer or its emulation (Maples and Fisher, 1976).

(b) Working with a simulator running on a larger computer (Mueller and Johnson, 1976).

(c) Utilization of the microcomputer via a minicomputer (Holt, Pokoski and Cordell, 1975; Lin, 1976).

The first approach would enable the students to work closely with the microcomputer. However, a comlete system would be required for each person in a simultaneously working group of students. This may mean a large investment if such a group formation is needed to handle all the students.

The second approach, besides being expensive, is inadequate for complete debugging of systems developed for real time applications, because the input-output pulses would not be available.

In cases where a minicomputer already exists, the third approach would be worth considering. The salient features in this approach can be summarized as follows:

1. The only additional equipment required besides the minicomputer system and the microcomputer is an interface between the two. This interface can usually be very

[1] This work was supported in part by the Scientific and Technical Research Council of Turkey under Projects MAG-G-456 and MAG-538

simple and inexpensive.

2. The microcomputer need not have peripheral equipment for entering programs, this can be achieved via the minicomputer.

3. One can take advantage of the minicomputer power in developing the microcomputer software.

4. If the interface allows the microcomputer to address the minicomputer's memory, then programs larger than the microcomputer's own memory size can be developed for the microcomputer.

5. The minicomputer can be employed for tracing and debugging the programs executing on the microcomputer.

6. It requires little or no modification on the minicomputer software and the interface, to replace an old microcomputer with a new one.

Despite the advantages seen in the last approach above, there is an inefficient utilization of the minicomputer system's resources which is observed in the cases reported by Holt, Pokoski, and Cordell (1975) and Lin (1976). It might be hard to justify such a utilization when it is not an alternative to the system remaining idle.

This paper presents a more efficient utilization of the minicomputer system, where it is operated under a monitor with three modes:

1. Development mode
2. Setting mode
3. Evaluation mode

In the development mode, the minicomputer accepts programs in batch form, executes them on the microcomputer and provides the outputs in the order of execution. In the evaluation mode, again the programs are accepted in batch form, however this time, instead of the outputs of the individual programs being given, the computation of each program is checked by the monitor and a coefficient of success is assigned to it according to some predefined rules. At the completion of all programs, the monitor generates a list of people who have submitted the programs, and their corresponding coefficients of success. The setting mode of the monitor enables an instructor to set the rules which are used in the evaluation mode to determine the coefficients of success.

The overall system consists of a microcomputer called MS which is developed especially for the Discrete Time Systems Laboratory of the Middle East Technical University (Çalıkoğlu, 1979), and a PDP-8 minicomputer, coupled together with a special interface.

THE MS MICROCOMPUTER

MS is based on an IM6100 microprocessor which has the same instruction set as the PDP-8/E minicomputers. This choice of the microprocessor has enabled all the software tools, readily available for the PDP-8, to be applied to the microcomputer. Particularly the assembler of the PDP-8 has served as a cross-assembler for MS with only a slight expansion of its symbol table.

MS is equipped with sixteen analog inputs which are multiplexed, and two independent analog outputs. The analog outputs are capable of generating piecewise linear waveforms. This capability can be used for driving CRT display units as well as function generation.

The memory capacity of MS is 4K words, each word being 12 bits. Any program entry to MS has to be done via the PDP-8. Although it is possible to add peripherals to MS, presently it is kept in a minimal configuration in order to keep the costs down. There is a control panel using which, the memory contents can be examined or altered one location at a time. It is also possible to set the start address of a program or to generate an external program interrupt.

To be able to operate MS at distant locations from the PDP-8, a part of the memory of MS is realized using CMOS components with battery back-up. The programs that will be used at a remote location are entered to the CMOS part of the memory in advance. Then at the remote location, the required programs are either selected by their start addresses, or program switching is performed using the external interrupt input.

THE PDP-8 MINICOMPUTER AND THE INTERFACE

The memory capacity of the PDP-8 is equal to that of MS. It has a high speed paper-tape reader for program input and a teletypewriter terminal. This computer also has analog inputs and outputs which are available through a 16-channel A/D converter and three D/A converters. The D/A converters, however, are a standard ones that supply only fixed levels. Lastly, there is a general purpose digital interface (DI) available which was implemented in an earlier project (Çalıkoğlu, 1970).

DI has proven to be a very convenient tool in various digital development projects. Here again, it is used to establish the total interface between the PDP-8 and MS. The functions performed through DI are given below.

1. The PDP-8 can,
   (a) have direct access to the memory of MS,
   (b) bring MS to an initial starting state,

(c) put MS in RUN state in case it is in the WAIT state,
(d) send a program interrupt request to MS.

2. MS can,
(a) interrupt the PDP-8 by loading a 12-bit buffer register in DI,
(b) interrupt the PDP-8 by halting.

THE MONITOR

The monitor is loaded into the PDP-8's memory and one of the modes "Development" or "Setting" is specified at the start. There is no provision for entering the evaluation mode directly. It always follows the setting mode.

Development Mode

In this mode, the monitor performs two main functions. One of these functions is to respond to the output requests coming from MS, and the other one is to keep time, so that the allowed time limit would not be exceeded by any job. In a simple stand alone microcomputer system, the common way of outputting texts is by sending characters one at a time. This operation is simulated in MS using the 12-bit DI buffer register. MS loads this register instead of a printer buffer. The characters are represented by 8-bit ASCII codes. It is possible to use the remaining 4-bits of the buffer register for control purposes or for auxiliary information. As soon as the buffer register is loaded, the PDP-8 gets an interrupt request. This request is serviced by the monitor by removing the character code from the buffer register and putting it in a printer queue. The characters in this queue are later transferred to the printer buffer of the teletypewriter at the printing speed.

Setting Mode and Evaluation Mode

When the monitor operates in the evaluation mode, the jobs executing on MS are expected to call the monitor for the inspection of their intermediate results. The points at which such calls occur are called break-points. In simple off-line exercises, the break-points may contain halt instructions. In that case the monitor puts MS back into the run state after inspection, if the job is expected to continue. In more sophisticated exercises which involve real-time control, "load the DI buffer register" instruction is used at the break-points. In this case, the monitor gets the readings using the available direct memory access fascillity, while the real-time process continues. This kind of accesses steal about 40 micro-seconds of time from MS to read a single location. In cases where the total time lost by MS due to such accesses may reach critical levels, the monitor may follow the operation of MS by sampling analog values from its analog outputs or from other points on the real-time process.

In order to enable an instructor to set the rules for determining the coefficients of success in the evaluation mode, a language called RSL (Rule Setting Language) is defined. When the monitor enters the setting mode, it begins accepting RSL statements. Following are the types of the available statements.

Checking statements. These are in the form:
n:(loc, cont,...,cont:pts)...(loc, cont,...)
where n is the break-point number. Within the paranthesis, a location and its correct contents, possibly followed by the contents of the succeeding locations and finally the points deserved for this check are included. Comments are allowed between the paranthesis. This statement can also be used to check the printings to be done by the programs, if the following paranthesis is used instead of the above ones:

(T, text:pts)

Here T is a key symbol, and pts specify the points to be given per each correct character.

Input statement. This statement is used to specify the initial values in desired locations. Only one such statement is allowed in the setting. It is in the form:
I:(loc,cont,...,cont)...(loc,cont,...,cont)
Here, the locations and contents are specified similar to the checking statements except that the contents are set instead of being checked.

Time statement. This statement declares the time limit to be enforced on each program. The form is:
TIME: units
where units is an integer and the time it specifies is, one unit of monitor time, times units.

Include statement. This statement is provided for those instructors, who may wish to include their own subroutines in order to follow-up the student programs. By using this, one can generate pseudo-random numbers as initial values and follow the computation on MS in a parallel way on the PDP-8. Also consider a case where MS is to be programmed as a digital controller in a certain discrete time control systems experiment. In order to evaluate whether the programming done has reached the desired goal (such as the optimization of a selected cost function) it is convenient to connect the control system's inputs and outputs to the analog outputs and inputs of the PDP-8, and observe its characteristics under program control.

In the course of entering the RSL statements, a previously entered statement can be corrected or altered by simply rewriting it. This, automatically removes the earlier statement from the monitor's storage area.

CONCLUSIONS

In this work, a minicomputer evaluated microcomputer programming system is developed. This system enables programs to be developed for real-time applications and also enables such programs to be evaluated automatically. Thus, the system constitutes a convenient educational aid in the field of discrete-time control systems.

Looking from a different point of view, the system enables the utilization of MS, without requiring a seperate program entry equipment for it. For some control experiments, MS could be programmed at the PDP-8 and then carried to the experimentation sight where it is to be used.

The MS-PDP-8 system is also suitable for performing parallel processing, PDP-8 as the master and MS as the slave.

The monitor on this system enables the hardware capacity to be utilized much more efficiently and effectively than similar development systems which do not have this kind of a monitor.

The hardware configuration which is presented is quite a restricted one. However, it is sufficient to put forth the main principles of the approach. For batch processing, for example, a card reader would be more suitable than a paper-tape reader.

The monitor's sophistication can be increased by bringing improvements from several directions. Expanding the editing fascillities and providing system macros for use in "Include" statements, are just a few of them. In order to expand the software, however, hardware expansions are required first.

REFERENCES

Çalıkoğlu,D. (1970). Design and Implementation Hybrid Computer Linkage System. M.Sc. Thesis, Middle East Technical University, Ankara.

Çalıkoğlu,D. (1979). Mikrohesaplayıcılı Kontrol Sistemi.Project MAG-G-456 Final Report, the Scientific and Technical Research Council of Turkey, Ankara.

Holt, O.D., J.L. Pokoski, and D.L Cordell (1975). A software development system for microcomputers. IEEE Trans. Ind. Electron. and Control Instrum., IECI-22, 279-282.

Lin, W.C. (1976). Minicomputer-driven microprocessor laboratory for teaching and system development. Proc. DISE Workshop on Microprocessors and Education.

Maples, M.D., and E.R. Fisher (1976). Microprocessor prototyping kits. Proc. IEEE, 64, 932-936.

Muller, R.A., and G.R. Johnson (1976). A generator for microprocessor assembler and simulators. Proc. IEEE, 64, 921-931.

AN APPROACH TO THE ASSESSMENT OF AUTOMATION TECHNOLOGY

H. K. Jacobsen and P. H. Jespersen

IMSOR, The Institute of Mathematical Statistics and Operations Research, Technical University of Denmark, 2800 Lyngby, Denmark

Abstract. Motives of the introduction of new technology in production are discussed and systematized, and the possibilities of controlling technological development are evaluated. It is argued, that technological development almost exclusively is directed by economic criteria, leaving little room for influence. This general discussion and systematism is specialized to the case of computer technology. Six main types of application of computers in production are described. A case study of the implementation of a computer system at the Danish National Giro is presented. The impacts of this automation on the conditions of work are discussed with the interests of the workers as a starting-point. It is concluded, that the fundamental possibility of removing restrictive work by technological change has not been turned to account at the National Giro.

Keywords. Computer applications; management systems; human factors; economy; social and behavioural sciences.

INTRODUCTION

The use of computer technology in production of goods and services is rapidly increasing.

The introduction of computer technology in production causes changes in the products, in production equipment and in the organization of work, which at the same time brings about changes in working conditions. This development has been regarded seriously by among others the trade unions (European Trade Union Institute, 1979).

Two major questions arise from this development:

- to what extent is it possible to control technological development in a market-based economy?
- towards which goals is the development of technology to be directed and by whom?

To answer these questions is a matter of technology assessment. The purpose of this paper is to give a short outline of our approach to technology assessment, based on our theoretical and practical experience, and to show the application of this approach on computer technology in production.

The organization of the paper

Knowledge of the dynamics of technological development is a necessary condition for treating the stated questions. In the following three sections we will sketch the main elements of our comprehension.

First we state that the main forces and constraints of technological development are of economic character, and we discuss the impacts of this statement on the concept of technology assessment.

Next, we discuss the nature of the economic forces and constraints, introducing the general mechanisms of technological development.

How these general mechanisms affect specific technologies is treated by the example of computer technology in the last of the sections concerning technological development. A description and systemization of the range of uses of computer technology in production is given.

The analysis described so far is restricted to a microeconomic level and cannot give an understanding of technological development in its totality. It only states the lines of direction which the individual firms have to follow in their choices of technological changes subject to various external circumstances.

However, this is sufficient for the evaluation of the impacts of technological development on the working conditions, which is the subject of the last sections of this paper.[1]

We discuss the selection of parameters describing 'conditions of work'. This is used in a case study dealing with the implementation of a computer system in the Danish National Giro. The case study is based on our experiences from a research project carried out in close collaboration with the local trade unions.

TECHNOLOGICAL DEVELOPMENT AND TECHNOLOGY ASSESSMENT

It has been our aim to develop an approach to technological assessment which involves the *dialectics* of technological development, i.e. an approach that emphasizes the interaction between the technological system and the conditions of production imposed by the organization of Society.

In this respect we differ from a very common interpretation, that technological development is *autonomous* and has its own immanent dynamics determining its direction and speed. Figure 1 is an attempt to illustrate this approach.

From the point of view of this approach, technology assessment is a question of a rational reflections of the consequences of new technology, and control of technological development a question of influencing the system parameters in accordance with the results of the assessment.

However, from our point of view, the social system surrounding the system of technological development must be taken into account. It plays a decisive part in technological development by imposing the forces and the constraints.[2]

As shown in Fig. 2 the organization of production in society is influenced by the consequences of the technological development, while the organization of production itself influences the origins of the new technology science, inventions and supply of technologies , and, most important, influences the introduction of new technology by imposing the economic conditions.

Thus, technological development is not to be set equal to the overall development of science and inventions. The central aspect in the process of development of technology is the introduction of technology in production, where its economic capability determines its dissemination. The inventors of the transistor are famous, not because they invented a fantastic device, but because the device had a fantastic dissemination through its economic capability.

We reflect this by defining technological development as changes in production processes with the purpose of increasing productivity, reducing production costs and/or changing the characteristics of the product.

This definition ties the concept of technological development to the production process and comprises not only technical changes but also changes in the use of the work force. This is important to realize, especially as we focus on the use of computer technology in production.

The dialectical understanding of technological development requires a reinterpretation of the assessment concept. Technology assessment must comprise both an evaluation of the social forces and constraints involved in technological development, an assessment of the consequences of technological development and a formulation of the direction in which the technology has to be developed.

However, economic, social and human impacts of the introduction of new technology is not unambiguous assessible. Different groups and organizations in Society may not necessarily reach consensus while assessing the consequences of the implementation of a specific technological change. Consider, for instance, the installation of a widely announced system of registration terminals at all important entrances, fotocopiers, elevators etc. in a company. This will, as it is stated in an advertisement, give management "a detailed record of who, what, where and when".[3] From a workers point of view this system will hardly be assessed in the same way.

So, in order to assess the implementation of new technology in production it is necessary to state which interests the assessment is serving.

TECHNOLOGICAL DEVELOPMENT: ECONOMICAL BACKGROUND

In this section we give a description of the economy of technological development. It is divided in two parts: One treating the external conditions that the firms have to respond to and one dealing with the ways in

[1] Still, the microeconomic approach does not suffice for the assessment of the employmental effects of new technology. Whether the potential unemployment effected by labour saving technology will become a real unemployment is a matter of overall economic conditions.

[2] This statement will not be discussed extensively. We refer to Dickson (1974); Noble (1977); Jespersen and others (1978).

[3] Advertisement for 'Identification Network' from 'Resco Electronic Systems', in Datamation, April 1979.

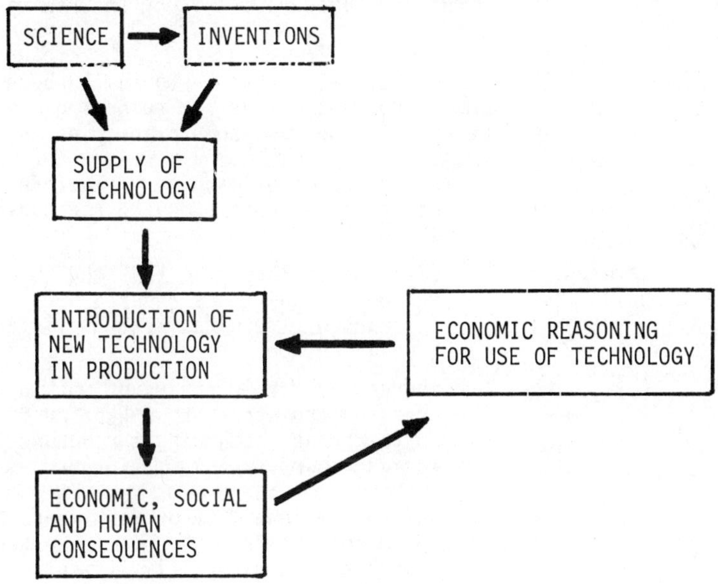

Fig. 1 Technological development as an autonomous proces

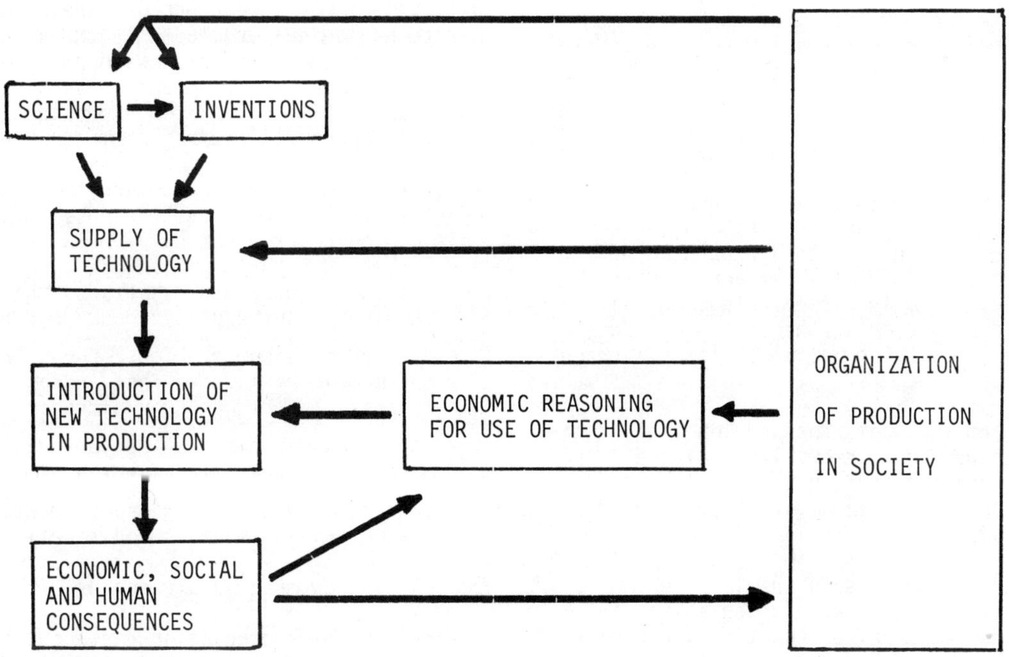

Fig. 2 Technological development as a dialectic proces

which the firms can react to the external conditions.

The global level

Production is organized in industrial branches, which abstractly can be defined as lines of business elaborating products of same or similar type, products with equal qualities of use.

Inside an industrial branch the individual undertakings will try to obtain maximum profits. So they compete for market share through advertizing, adding special features to the product, setting of the sales prize etc. The demand for profit sets a limit to these efforts. Thus a more or less stable general market price will at a given level of technological development be established through the law of supply and demand.

The market price is normally higher than the total costs of production, the cost price, leaving a gap for profits, which can be used for investments and interests on the capital outlay.

However, while the market price is general and thus more or less fixed it cannot be manipulated by the individual undertakings. However, the cost price (and thus the profit) is highly individual. So there is a constant stimulus to reduce the cost price by different methods to obtain larger profits. We will outline some of these methods later, but here we will just state, that cost price reduction is the motivation for the development of technology, and that the most important methods for cost price reduction is the increase of productivity achieved by development of the production machinery.

A firm introducing new technology to reduce cost prices will obtain an extra profit. But the new technology will sooner or later penetrate to other undertakings in the same industrial branch. As this happens a new market price will establish, and when the technology is fully disseminated, the situation is the same as in the beginning except for a lower market price and a lower average of cost prices.

Thus a firm that does not develop technologically, sooner or later will be obliged to reduce its sales prices, resulting in reduced profits or no profits at all, removing then the stimulus for production.

Consequently there is in the long run no possibility that a single undertaking can escape the 'cost price reduction imperative' and avoid technological development. Technological development is a system immanent demand in a market-based economy.

The local level

For the single firm, reduction of the production costs is the most important and indispensable means in competition. However, this reduction can be accomplished in different ways. Here we will categorize according to the objects of the cost price reduction: labour, machinery and raw materials.

Economy of labour has historically been the most important motive for technological development. We list some principles

- introduction of machinery requiring less manpower per unit produced (labour-saving machinery)
- division of labour so that the total manpower required is reduced or so that the manpower requires less skills and thus is cheaper
- increasing the intensity of work by increasing the speed of work or by filling up the breaks occurring throughout the working hours.

Increasing the intensity of work means demanding more work from the single worker and thus has some natural limits set by the physical ability of the worker and by the length of the working day. On the other side we will refer to the first and second point as development of the productive forces indicating that the amount of human work required for each product is reduced.

Economy of machinery. The machinery is permanently exposed to a physical wear and has to be renewed with certain intervals. But the machinery is also exposed to what we will call a moral wear because the continued technological development 'threatens' with making the machinery uneconomic. The numerically controlled machines e.g. have in their short time of existence required less and less operation and programming work.

Thus it is important to utilize the machinery efficiently and as fast as possible. This can be carried out by

- introduction of work shifts so that the machinery can be utilized day and night
- different measures which ensures that the machinery runs as fast and as much as possible (reduce stop times)
- using raw materials of better quality in order to increase machine speeds
- introduction of new and supplementary raw materials, such as additives.

Economy of raw materials concerns

- reduction of production waste and better exploitation of raw materials
- recirculation and reutilization of waste
- substitution with cheaper raw materials
- minimization of capital outlay for raw materials.

Figure 3 illustrates some examples of tech-

	Technological changes by technical means	Technological changes by organizational means
Economy of labour (development of the productive forces)	Introduction of machine tools Introduction of numerically controlled machines Introduction of conveyors (reduces transportation work)	Division of labour in clothing industry (less total work per product and less skill thus cheaper labour required)
Economy of labour (increasing intensity of work)	Introduction of conveyors (determines the rhythm of work	Supervision of the workers Introduction of productivity increasing wage systems (e.g. payment by the piece).
Economy of machinery	Maximization of machine speed in ore crushing by process control The use of oils to cool machine tools	Introduction of working shifts
Economy of raw materials	Recirculation of paper Use of ashes in concrete	Inventory planning

Fig. 3 Examples of technological changes

nological innovations related to the methods of economy and also related to the means (technical and organizational) by which the economy is reached.

It is evident that a given technological change usually will have more than one economical purpose, as e.g. the conveyor example in the figure. Likewise it is evident, that several of the above-mentioned methods of economy are in contrast to each other. For instance the reduction of machine stop times often will require more labour and vice versa, less labour frequently will decrease the efficiency of the machinery.

Which methods of economy to choose is a matter of the dominating costs. For instance in a labour-intensive production, the methods of economy of labour will be predominating.

AUTOMATION AND COMPUTER TECHNOLOGY

Computer technology is a multi-purpose technology. So, as the impacts of computer technology are in study the range and purposes of the applications have to be clarified.

Often the use of computers in production is confused with automation. But this is not correct, both because automation is possible without computers - Watt's governor is an example - and because computer technology in production needs not be a part of an automation (Bright, 1958).

In this paper we will not make a comprehensive discussion of the concept of automation, but only state that we consider the transition from mechanization to automation characterized by the transfer of the operation of the machine from the worker to the machinery and the only functions left for the worker to do is functions such as supply and removal of material, reparation and maintenance of machinery and supervision and control of the production process. Further automation is directed towards these working functions and the degree of automation is the extent to which these functions are taken over by machinery. (Mendner 1974)

Figure 4 illustrates some examples of the use of computer technology in production - automation as well as non-automation - and from this we derive six main types of functions (numbers refers to the figure):

1) Process automation covers the cases, where human control, correction or supervision of machinery is taken over by machinery. As such it is a development of the productive forces. Computer technology is extremely well suited for this purpose as can be seen in the numerically controlled machines, in the word processing and by certain kinds of process control.

2) Alterations in the construction of a given product can be regarded as a development of the productive forces if the result is that less work is required in the manufacturing of the product. This use of computer technology is accelerating extremely these years because of the fast reduction of

	Computer technology as technical means	Computer technology as organizational means
Economy of labour (development of the productive forces)	Alterations of the product [2] Process automation [1,5]	Production planning (reduction of manpower requirement by suitable sequencing and coordination of work)[3]
Economy of labour (Increasing intensity of work)		Collection of data with the purpose of controlling and supervising the workers [4] Production planning (minimization of the natural breaks in the working day; manpower planning) [3]
Economy of machinery	Increase production speed e.g. with process control [5] Reduce machine stop times [6]	Production planning (maximization of occupied time)[3] Reduce machine stop times [6]
Economy of raw materials	Increase the degree of utilization of raw materials with process control [5]	Production planning (Inventory planning)[3]

Fig. 4 Examples of technological developments with computers

computer size and costs. Examples of this phenomena is the production of sewing machines, where hundreds of mechanical parts are replaced by some electronic circuits.

3) Complex production planning can in practice only be made with the aid of computers. Production planning may have many rationalizing effects. Perhaps the most significant of these effects is the ability to exploit the machinery effectively by taking care that at least the bottle-neck machines are in constant use so that production is made continuous. A parallel can be drawn to the workers' side, where production planning may be used for intensifying work reducing the breaks occurring through the working hours.

However, production planning may also be used for organizational development of the productive forces, e.g. through the minimization of work by appropriate sequencing of switch-over operations.

4) Computers are used for the collection of individual data as productivity, absence etc. with the purpose of controlling or even sorting the working force. Even if it does not come to this the mere existence of the systems will force the workers to intensify their work.

"Improved shop scheduling. The system can provide a current, accurate record of the performance of each NC machine tool. This information can be most valuable to shop scheduling personnel and shop management and should help them to increase productivity. For example it is possible to determine: whether a job takes longer on one machine than on another "identical" machine (and) whether an operator on one shift is consistently out - producing other operators using the same machine." [4]

5) Process control can as mentioned above have the purpose to reduce required human work, as can be seen e.g. in the quality-control of bottle manufacturing which earlier was based on human sight.

Often, however, the primary goal of process control is the economy of machinery and raw materials. In chemical industry for instance a very common aim of introducing process control is to achieve optimal output and optimal quality of the product in relation to time and in relation to the raw materials.

6) The profitability of a given machinery depends heavily on the minimization of the total stoptime - the predictable as well as the non-predictable. This problem can be attacked in many ways, depending on the type of production. One of them is to introduce new technology to detect arising errors, for instance computers analyzing trends in number and types of error. Another is to introduce periodical repair routines (preventive maintenance). For instance Scandinavian Airlines System (SAS) is using a large computer system for controlling that repairs

[4]IBM: Computer-Aided Manufacturing, Monitor System for Numerically Controlled Machine Tools .

and exchange of parts are performed before defects are likely to occur. At the same time it schedules repair work thus intensifying the work of the skilled repair workers.

TECHNOLOGICAL DEVELOPMENT AND THE CONDITIONS OF WORK

In the case study following, and in our investigations in general, we have focused on the consequences of technological development on the conditions of work.

This choice of object of assessment, however, must be supplemented by a choice of aspects of the conditions of work which is to be studied. But it must be pointed out that although we will try to outline consequences which can be established more or less objectively, the selection of the types of consequences and the assessment of these impacts is essentially subjective.

In our investigations we have (in accordance with Kern & Schumann (1970)) chosen four parameters which we have found most relevant for the study of conditions of work in connection with the introduction of new technology:

The worker's autonomy meaning the worker's possibilities to plan and accomplish his own work, i.e. his freedom to form his job. A great autonomy will often support a positive conception of the conditions of work, whereas a restrictive work process totally determined by machines and managerial norms frequently leads to physical or psychical indisposition.

The qualifications required from the worker is another measure of the conditions of work. The development in both the qualifications narrowly connected with the machinery and in the more general qualifications as flexibility, technical intelligence and responsibility is known to have a connection with the job satisfaction.

The work load consisting of two elements, the physical and the psychical load.

Social contacts at work, both the contacts demanded by the production process, and informal contacts that are allowed and possible during the production process.

The individual workers may of course have quite different opinions concerning the same type of job, i.e. you cannot from a description of a work place in agreement with the above-mentioned parameters say anything definite about the individual worker's assessment of his work situation.

But this has not either been our aim. The guiding line in our choice of factors to describe conditions of work has rather been a conception, that technological development - and especially automation - *in principle* gives the possibility of removing restrictive work and creating work situations fulfilling demands of content and many-sidedness.

TRADE UNIONS AND COMPUTER TECHNOLOGY AT THE NATIONAL GIRO

At the National Giro, assessment of a new computer-based giro handling system has been carried out in close cooperation with the local trade unions. This was done as a part of a research project DUE[5], which was initiated by the Danish Trades Union Research Council. The primary objective of the project has been to contribute to the efforts of the trade unions to gain influence on the use of computer systems.

Analyzing and assessing various computer-based systems at three Danish workplaces have been central activities of the project.

The assessment of new technology has been based upon the interests of the local trade unions and carried out in collaboration with these.

Within the local trade unions the activities have created an understanding of the connexion between technological development and the experienced consequences of new technology on working conditions. This has been the starting point for discussions of the following questions:

- How shall future technology at the workplace be designed?
- How can the workers influence design of technology?

Technological development at the National Giro

The National Giro was one of the workplaces involved in the project. The National Giro is a public service dealing with transfer of payments. In the year 1978 205 mill. transfers were performed. Approximately 3000 persons are employed.

The National Giro is challenged by the banks which have established payment transfer systems which are competitive to the National Giro. Besides, there has been an increasing demand on the public services to be economic.

These two factors have forced the National Giro to improve customer service and at the same time reduce the costs involved in performing payment transfers. Thus the National

[5]DUE is an acronym for the Danish title of the project: Demokrati, Udvikling og EDB/ Democracy Development and EDP. One of the authors is a member of the project group. A comprehensive discussion of the project is given in:
DUE report nr. 2 (1978).
DUE report nr. 3 (1979, in English).

Giro is subject to the same demands for being economics as private enterprises.

Economy of labour

The work processes involved in performing payment transfers does mainly consist of various kinds of information handling:

Sorting various forms coming in from customers. Transport of forms at the Giro office. Registration and accounting of information written on the forms. Dispatching of information to the customers.

Until the beginning of the seventies almost all information handling processes were carried out manually. This indicates that the main source for cost reduction has been economy with labour.

Computer technology

Computer technology is the most efficient tool for information handling. Hence computer-technology has been the natural choice of management aimed at economy with labour.

In 1977 a new computer system, PATOS (Post And Telegraph Online System), was taken into use. With PATOS a number of functions formerly carried and manually have been automated. This means that the amount of labour necessary has decreased: The productive force of the human work has increased. Thus the National Giro has avoided additional employment despite increased amounts of payment transfers.

With the introduction of PATOS the entire organization of the work has been changed.

Before the work was organized almost as follows:

The incoming forms were sorted, partly manually, partly mechanical.

The sorted forms were distributed to independent groups of nine workers who carried out all information handling concerning money transfers for a limited number of customers: Further sorting, accounting of money transfers of each individual customer, writing and mailing of abstracts of the accounts and finally control of the work.

Now with PATOS the central part of the functions have been automated.

Most sorting work has now been automated. Accounting and writing of abstracts of the accounts have been totally automated.

Left is the following functions which are carried out in groups of 25 workers:

- registration and control of information written on the incoming forms. This is carried out at on-line coding machines,
- control-addition by mechanical calculaters of the numbers entered at the coding machines,
- mailing of computer written abstracts of the accounts.

Economy of machinery

Parts of the computer system has been dimensioned smaller than what is necessary if peak loads has to be worked out within normal working hours. So shift work is a calculated necessity.

Consequences

Together with groups of workers we discussed and systematized their assessment of the consequences implied by the introduction of PATOS.

The automation of the tedious work involved in sorting the incoming payment forms was positively assessed.

However, in general the workers found that the working conditions have been deteriorated, primarily because the coherence of the work functions has disappeared. It is still routine work. Before each worker carried out all information handling activities in connection with a limited number of customers.

Now the control functions have been automated and the remaining functions are incoherent and split up.

When registrating information at the coding-machines nothing else can be done. Each registration takes approximately 10 seconds.

Even if the work functions left were carried out in sequence they would not be coherent. What is left is inane routine work.

The worker's autonomy

The worker's autonomy at the National Giro has always been limited. After the introduction of PATOS the worker's autonomy have become even smaller.

Entering data at the coding-machines is an extremely well defined function leaving no room for variations. Control counting and mailing are likewise routine functions, but to a limited extent it is possible to organize these tasks within the group. And to a limited extent there is a possibility for varying the speed of work.

The qualifications required

Before the introduction of PATOS the job required flair for number handling. This is not required any more. No special skills are required. Just to be able to work steady and fast. New workers only receive a short on-the-job training.

The work load

The physical loads of the jobs are dominated by two factors. The one, that the extremely repetitive work functions are causing strain in certain muscles. The other, that the coding machines are of very bad design. They are producing a great deal of heat which is let out exactly where the workers have to place their legs. They are also quite noisy and of an uncarefully prepared ergonomic design.

The psychical loads are the dominating increased dissatisfaction in connection with technological development.

The work at the National Giro has always been repetitive with small cycle-times. The incoherence of the work is by the workers regarded as a serious psychical load.

When working at the coding machines the workers have to registrate by a personal identification. This is causing the feeling of being automatically supervised.

Social contacts at work

When working at the coding machines there is no possibility for contacts with other workers because of physical distance and noise from the terminals.

The groupsize of 25 is to the workers opinion too big to establish a solidarity within the individual groups. The solidarity within the former groups of 9 was greater in the sense that it was more common to support each other during the work.

Trade unions and technological development at the National Giro

The trade unions at the National Giro have established a working group with the purpose of evaluating the possibility of gaining influence on the technological development at the Giro.

This group has, based on assessment of recent technological development, discussed which goals technological development at the National Giro should aim at from a trade union point of view. And the group has discussed to which means the technological development should be controlled.

In discussing the goals the group has taken into account that the demands for cost reductions are inherent. Anyway the technological development should be directed so that better working conditions and increased solidarity among the workers may be established.

More specific demands have been raised for

- removal of the extreme division of work
- establishing of coherent jobs
- use of computer-terminals or coding machines of better design
- an organization of work which creates the best possibilities for solidarity among the workers.

Imposing these goals upon the development of technology is not trivial. The managers have been unwilling to share their control of the technological development. The trade unions have concluded that they have to force through their demands.

In order to improve the possibilities to gain control of the development of technology the trade unions have forwarded the following demands to the managers:

- The trade unions shall be presented for a concise survey of any initiative to introduce new technology.
- The trade unions and affected workers must be given resources to carry through assessment of any technological change during the development process.

A comprehensive discussion of the activities within the trade unions at the National Giro is given in DUE-note no.13, November 1979.

CONCLUSION

In a market based economy technological development is controlled, not by scientists, inventors or by any particular group of interest, but almost exclusively by the 'cost price reduction imperative', which is inherent in the organization of society.

The technology, which penetrates throughout the society, is solely the technology, which has shown its economic capabilities.

Managers within an enterprise are not free to choose new technology, only more or less successful in fulfilling the demands for cost price reductions. Likewise, room for variations in technology controlled by e.g. the trade unions is only left to the extent these variations do not violate the accomplishing of cost price reductions.

Cost price reductions are achieved by the use of technology for economy of labour, economy of machinery and for economy of raw materials.

Economy of labour is central. It has two aspects. On the one hand it is the mainspring for the development of the productive forces, reducing the human work necessary in producing a given product. On the other hand economy of labour implies intensification of human work.

Computer technology has several capabilities as a tool for cost price reductions. As technical means it is primarily used in product alterations e.g. substituting mechanical parts by a microprocessor and in process automation. In both cases human work involved in producing a given product is reduced, thus the productive forces are developed. As means for technical change computers are also used in optimizing the speed of machinery and process dependent use of materials.

As organizational means the use of computers are many-sided. Economy of labour may be accomplished by development of the productive forces e.g. by production planning systems stating a proper sequencing of jobs, thus reducing the amount of unnecessary work, as well as by production planning for a more intensive utilization of labour.

Economy of machinery and economy of materials may likewise be attained by complex production planning systems delivering optimal plans for utilization of machinery or minimizing cost of inventory.

The trade unions are trying to gain influence on technological development. Some of their goals are (as stated in DUE - report no. 3, 1978): Removal of restricted work. Maintaining or establishing a content of work, which demands and develops many-sided qualifications. Removal of shattering work loads. Establishing of possibilities for social contacts during the work.

Development of the productive forces by technological development should in principle open the possibilities of attaining these goals. But case studies like the one performed at the Danish National Giro show, that these possibilities are not turned to account. This is consistent with our statement that it is almost exclusively economic goals, that are setting the lines of technological development.

The possibilities of the trade unions are: either to be satisfied with minor adjustments of new technology not violating the economic imperatives or by collective action regulate or change the conditions for technological development.

FURTHER RESEARCH

In a research group at IMSOR we carry on systematization of computer technology; as well theoretical as by empirical studies in certain industrial branches (p.t. the breweries)

In particular we are treating production planning: which demands of economy are fullfilled by various methods of production planning?

We are developing further the systematism for technology assessment, in particular treating the aspects concerning 'conditions of work' and employment problems.

REFERENCES

Bright, J.R. (1958). Automation and Management. Harward University.

Dickson, D. (1974). Alternative Technology. Fontana/Collins.

DUE (1978). Demokrati, Udvikling og EDB. Report no. 2. Århus, Denmark.

DUE (1979). Systems Development and Trade Union Activities. Report no. 3. Århus, Denmark

DUE (1979). Klubarbejde og EDB på Postgiro. Note no. 13. Århus, Denmark.

European Trade Union Institute (1979). The Impacts of Microelectronics on Work and Employment in Europe in the 1980'ies. Brussels.

Jespersen, P.H. and others (1978). OR-epistemology: Some Essays. IMSOR, Lyngby, Denmark

Kern, H. and M. Schumann (1970). Industriearbeit und Arbeiterbewusstsein. Frankfurt an Main - Köln.

Mendner, J.H. (1974) Technologische Entwicklung und Arbeitsprozess. Frankfurt am Main

Noble, D. (1979). America by Design. Alfred A. Knopf, New York.

AUTOMATED AND AUTOMATORS: HUMAN AND SOCIAL COSTS OF TECHNOLOGICAL CHANGE

S. L. Hacker

Department of Sociology, Oregon State University, Corvallis, Oregon, U.S.A.

Abstract. Most studies of modernization and development ignore the differential impact of these processes on men and women. Recent resarch in developing countries indicates that women are not affected in the same way as men. This paper summarizes research on 1) women's employment patterns in corporations representing four U.S. industries and 2) the impact of automation on engineering. Mechanization and automation of factory work displaced women; simplification of craft work displaced men (telecommunications and printing); automation of clerical and service work (telecommuncations) displaced both management and non-management women. Automation and mechanization in the U.S. have complex effects paralleling those in developing countries. In general, women workers are most often disadvantaged. In engineering automation decreases the demand for highly trained engineers; increases the demand for technicians and technologists. The impact on men of such vertical differentiation in engineering is examined, as are criteria employed to "order" this process. Development does not affect all people in the same way, and may exacerbate older divisions of sex and ethnicity, divisions inappropriate in a developed society. New institutional support systems, especially for domestic activities, are suggested.

Keywords. Social and behavioral sciences; education; employment; family; telecommunications; agriculture; insurance; publishing.

INTRODUCTION

Automation and mechanization create and eliminate work. Given that a social division of labor based on sex gives rise to a sex-segregated modern work force, automation and mechanization do not affect men and women alike.

Development, in industrializing countries or the highly industrialized, is the context within which automation and mechanization usually occur. Only recently have scholars attended to the differential impact by sex.

Theories of modernization and its liberalizing effect on attitudes are generally based on male only samples. Olsen's (1980) study of Taiwanese youths includes both boys and girls, and reveals that while city vs. rural boys hold less traditional values in significant areas, the reverse is true for girls.

Development produces a differential effect on behavior as well as attitudes. Mechanization of field work opens opportunities for men, but often eliminates much of women's work, (Arizpe, 1977; Miranda, 1977), leaving for example, 40% of non-rural female workers in Latin America paid work as servants. Technological change in manufacturing often displaces women artisans. Chinchilla (1977) reports that although manufacturing may provide cheaper goods for many,it seriously affected the independent artesan industry in Guatamala, and hence women's participation in manufacturing.

Ester Boserup's classic study of Woman's Role in Economic Development (1970) documents various patterns of agriculture around the world -- female and male intensive patterns -- and ways in which some kinds of development widen the differences in power between men and women, (the right to own land, for example, became a right for men only with hte help of Missions in the Congo). Isley's study of differential mortality rates by sex in India -- with rates for women increasing relative to those of men during industrialization -- indicate the severity of the effects of these processes (1980).

These processes are complex, and not entirely consistent. Stoler, for example, (1977) shows that the poor, men and women alike, were adversely affected by agricultural mechanization in rural Java, while Andors finds such mechanization increased women's participation in Chinese agriculture (1976). Mueller describes changes in relationships between men and women in African Lesotho, wherein export mining drew men from the villages, leaving political control in the hands of women. Women however were left outside the cash economy. (1977)

In most developing countries, less than 10% of women are paid for their labor. But even highly developed countries such as Japan, Britain and the U.S., full-time, year-round working women earn from less than a half to two-thirds the wages earned by their male counterparts. Women's work is devalued economically in socialist as well as capitalist societies, as Andor (1977) indicates in her account of points given for needlework compared to agricultural labor, in China. Overall, this new body of research tends to indicate that while development may eventually bring many improvements for women, women bear a disproportionate share of the costs of social change. There are some organized efforts to alleviate the problems faced by women in developing and developed countries: the U.N. focus on International Women's Year in 1975 highlighted such programs as women's agricultural cooperatives in Sri Lanka and educational programs in Tunisia; the 1977 presidential report to the Population Council Annual Report links population and development issues to the status of women; integrated rural development is emphasized in Bangladesh; girls are allowed into agricultural educational programs in the Upper Volta; and the U.N. Protein Advisory group takes a special look at women and children.

There can be no doubt that women will eventually gain more than they lose through moderization and development. Social scientists who explore the changes in power relationships between men and women which correspond to varying forms of social organization note that women have least power in agricultural societies (Nielsen, 1978). Compared to hunting and gathering, horticultural or industrialized societies, it is in agricultural societies where the subordination of women is deepest. The studies cited above indicate that of the many ways to approach development, those which ignore or exploit the powerlessness of women should be avoided on many grounds, including those of efficiency and effectiveness of technology transfer (Boserup, 1970; Leghorn and Roodkovsky, 1977).

In this paper, I examine stratification based on sex, and, first, the role of women workers during periods of technological change in U.S. corporations representing four industries -- agribusiness, printing and publishing, insurance and especially telecommunications. Thehuman and social costs in a societal context rather than a narrowly organizational context, appear to outweigh the benefits. Second, the profession most closely related to the implementation of automated systems -- engineering -- is analyzed while in process of vertical differentiation related to automation. Again, I explore some potential costs as the nature of engineering work changes, and as sex stratification and other uncertain criteria are employed to introduce order into social and technological differention. Developing societies may want to observe these processes for insights into alternative paths toward modernization.

The best studies of the effects of automation are those which focus on individual firms, rather than those which use industry-wide data. Data and observations are presented here from a firm in each of the four industries: the telephone company, a newspaper, a poultry plant, and the home office of an insurance agency.

THE AUTOMATED: DISPLACEMENT OF WORKERS BY TECHNOLOGICAL CHANGE

Telecommuncations

The U.S. system of telephone communication is admired by developed and developing countries alike. This system provides an impressive model both for reliability of service and for change -- research, development and incorporation of technological innovation. American Telephone and Telegraph Corporation, (AT&T), the major corporation in telecommunications, integrates technologies on several levels of sophistication at once, and moves relatively smoothly through what could be disruptive changes in its occupational structure due to automation.

Even in the best of technical systems, however, with vast resources available for social problems attending automation, the costs absorbed by the labor force and the society are high. AT&T represents 2% of the employed U.S. labor force. The operating companies and Long Lines division which furnished most of the data for this paper employ close to a million people.

In the mid-20th century, the telephone company adopted technologies which reduced the need for workers at the lowest levels of its occupational structure. Systems such as Automated Message Accounting, Direct Distance Dialing, Customer Records and Billings, and the Traffic Service Position allowed a cutback of tens of thousands of operators and clerical workers. The Customer Records and Billing system, according to a Bell Labs engineer, has made for people-savings up to 50% in some areas. These systems continue to reduce employment during the seventies. Data for the Operating Companies and Long Lines for the period 1972 through 1975 show a net loss of only 8506 workers. Table 1, however, shows the differential impact by occupational level, a force loss of over 36,000 operators and lower level clerical workers. In addition, there was a decrease of 1535 lower level supervisors -- those who formerly supervised workers displaced by automation. Thus, the cost of automation was borne by management and worker, cutting across the traditional boundaries of groups usually affected adversely by the technological change. However, closer examination shows that women were far more adversely affected than men, whether management or non-management

Organizational Change in AT&T 545

TABLE 1

Occupational Distribution, All Operating Companies Plus Long Lines; 1972 and 1975

Job Category	*# Jobs 1972*	*# Jobs 1975*	*Percent Change*	*Numerical Change*
#1 3rd level mgt and above	15,780	16,610	+ 5.2	+ 830
#2 2nd level	43,168	48,297	+ 11.9	+ 5,129
#3 1st level	95,492	102,867	+ 7.7	+ 7,375
#4 Administrative positions	32,716	31,181	− 4.7	− 1,535
#5 Sales	5,813	6,541	+ 12.5	+ 728
#6 Skilled craft, outside	65,107	65,553	+ .7	+ 446
#7 Skilled craft, inside	76,542	78,047	+ 2.0	+ 1,505
#8 Gen. Servces	11,347	13,304	+ 17.2	+ 1,957
#9 Semiskilled craft, outsd.	66,104	63,549	− 3.9	− 2,555
#10 Semiskilled craft, inside	18,011	18,012	+ .0	+ 1
#11 Clerical, skilled	82,392	97,336	+ 18.1	+ 14,944
#12 Clerical, semiskilled	74,689	70,006	− 6.3	− 4,683
#13 Clerical, entry level	45,140	37,674	− 16.5	− 7,466
#14 Operators	148,622	124,431	− 16.3	− 24,191
#15 Service workers	12,365	11,374	− 8.0	− 991
Totals	793,288	784,752		

Source: EEOC Date, 1976.

employees. Table 2 shows that men gained 13,767 jobs over their 1972 level while women lost 22,273 jobs during the same time. Part of this process was due to affirmative action taken by the company to encourage men into traditionally women's work, women into traditionally men's work. ("Affirmative action" is the phrase normally used for the effort required by U.S. law and regulation to assure euqal employment opportunities for women and certain ethnic minorities. Men were more successful at breaking non-traditional barriers than were women; at the non-management level, 9,400 women moved into men's jobs while 16,300 men moved into women's jobs).

It is also apparent, however, that the Electronic Switching System and realted systems in installation, repair and maintainance will reduce the need for skilled and semi-skilled craft workers, jobs traditionally held by white men. As this work is simplified prior to automation, women and non-white men are moved into craft work. As Table 2 indicates, there are thousands fewer white male skilled craft workers than there were in 1972, and thousands more women and non-white men. There are also far fewer men (both white and non-white) in semi-skilled craft positions, many of which are now filled by women. Fibre optics transmission is another innova-

TABLE 2

AT&T Employment Shifts: Operating Companies + Long Lines
1972 vs. 1975 Profile Data

Classification		Grand Total	Total Men	White Men	Minority Men*				Total Women	White Women	Minority Women*			
					Black	SSA	AI	A/O			Black	SSA	AI	A/O
Mgt 1	1	+ 830	+ 423	+ 282	+ 77	+ 35	+ 22	+ 7	+ 407	+ 383	+ 16	+ 5	0	+ 3
Mgt 2	2	+ 5129	+ 2395	+ 1812	+ 302	+ 186	+ 50	+ 45	+ 2734	+ 2386	+ 246	+ 66	+ 14	+ 22
Mgt 3	3	+ 7375	+ 3547	+ 1982	+ 847	+ 534	+ 86	+ 98	+ 3828	+ 2229	+ 1089	+ 335	+ 60	+ 115
Super's	4	− 1535	+ 818	+ 607	+ 111	+ 82	+ 9	+ 9	2353	− 3105	+ 480	+ 202	+ 35	+ 35
Sales	5	+ 728	− 161	− 360	+ 133	+ 49	+ 10	+ 7	+ 889	+ 628	+ 207	+ 49	+ 8	− 3
Sk craft: outside	6	+ 446	− 84	− 1830	+ 832	+ 780	+ 65	+ 69	+ 530	+ 441	+ 45	+ 35	+ 8	+ 1
Sk craft: inside	7	+ 1505	− 2409	− 3486	+ 537	+ 478	+ 85	− 23	+ 3914	+ 3246	+ 467	+ 151	+ 34	+ 16
General services	8	+ 1957	+ 360	+ 193	+ 50	+ 82	+ 6	+ 29	+ 1597	+ 1149	+ 347	+ 86	+ 6	+ 9
Semisk craft: outside	9	− 2555	− 3735	− 3672	− 482	+ 282	+ 93	+ 44	+ 1180	+ 969	+ 136	+ 55	+ 12	+ 8
Semisk craft: inside	10	+ 1	− 2185	− 1143	− 722	− 269	− 11	− 40	+ 2186	+ 1575	+ 394	+ 163	+ 17	+ 37
Cler: skilled	11	+ 14944	+ 4220	+ 3164	+ 494	+ 427	+ 35	+ 100	+ 10724	+ 4514	+ 4049	+ 1580	+ 199	+ 382
Cler: semisk	12	− 4683	+ 3953	+ 2967	+ 570	+ 304	+ 8	+ 104	− 8636	− 11068	+ 1611	+ 466	+ 65	+ 290
Cler: entry	13	− 7466	+ 2536	+ 2119	+ 181	+ 188	− 9	+ 57	− 10002	− 8282	− 1623	− 162	+ 11	+ 54
Operat.	14	− 24191	+ 4867	+ 3696	+ 529	+ 539	+ 43	+ 60	− 29058	− 23058	− 6550	+ 306	+ 121	+ 123
Service workers	15	− 991	− 778	− 103	− 705	+ 22	+ 3	+ 5	− 213	− 204	− 101	+ 79	+ 2	+ 11
Grand Total		− 8506	+ 13767	+ 6228	+ 2754	+ 3719	+ 495	+ 571	− 22273	− 28197	+ 813	+ 3416	+ 592	+ 1103
% CHANGE:				+ 1.8%		+ 24.6%				− 8.4%		+ 7.5%		

Source: EEOC, 1976.

* SSA = Spanish American; AI = American Indian; A/O = American Oriental

tion which promises more reduction in traditionally male construction and repair work.

TABLE 3

Predicting Occupational Decline From Sex/Race Characteristics of Occupations (Product Moment Correlations)

Characteristics of Occupational Levels:	1972 Predicted Change by 1975: r with % change in jobs available at each occ'l level	1972 Predicted Change by 1975: r with numerical change in jobs available at each occ'l level	Actual Change by 1975: r, % change	Actual Change by 1975: r numerical change
Number of white males	04	71**	16	28
% white males	37	25	46*	35
Number of minority males	22	45*	−08	06
% minority males	41	−18	07	07
Number of white females	−59**	−18	−36	−48*
% white females	−42	−16	−38	−28
Number of minority females	−58**	−33	−51*	−69**
% minority females	−44*	−28	−64**	−55*

* sig at the .05 level.
** sig at the .01 level.

Table 3 indicates overall, although various parts of the corporation change at different rates, the best predictor of stagnation or decline in employment between 1972 and 1975 was the number and percent of non-white women employed at an occupational level. (Hacker, 1979).

The U.S. literature on technological change and organizational change rarely considers displacement or unemployment as a problem. Literature on automation and displacement tends to focus on overall employment patterns by a company, or examines the impact on management vs. workers. Such research may gloss over processes which irritate existing social divisions in a society, such as divisions based on older hierarchies of race and sex. If these divisions are heightened by automation, even inadvertently, the social problem becomes more complex than a matter of unemployment alone.

Printing and Publishing

In the newspaper firm, cold-type and other processes decreased the amount of craft work. In other areas of U.S., secretaries could be hired to type camera ready copy, formerly a job for craftsmen -- typesetters. In this case, however, a largely white male union protected craft jobs from greater access by women and non-white men, even though this work was routinized in the transition to computer or cold type process. Projected displacement of typographers was set at 35%

The experience in the newspaper corporation reflected U.S. patterns in general, where reduction in employment affected linotypists, copy cutters, proofers, make-up, mat molders, stackers, bundlers, tiers, delivery and type luggers, typists and perforators. Due to automation of other systems of work at the newspaper, employment was also reduced in mailing, circulation and billing, Employment gains came in electronic editing, systems analysis and sales -- primarily sources of male employment.

Agribusiness

In the poultry plant, technological innovation reduced the workforce -- primarily female -- by about half. White female workers gave way to Mexican-American female workers and to youths, with future plans on the part of the company to employ American-Indian labor or move to further automation. Mechanization of field harvesting brought a 15% decline in field labor in one year in a three county area of a midwestern state. Here, Mexican-American males and females were affected equally. Middle-sized farming operators faced increased cost of farming related to more sophisticated technology, (Hacker, 1977).

Insurance

Computerization in the insurance company yielded a decline in the rate of increase of clerical labor. Technologies such as magnetic cards, dictating equipment, computer storing, and video-scopes reduced the secretary/executive ratio. As in the sixties, office automation tended to open employment requiring more sophisticated technological skills, while force reduction was projected for jobs traditionally considered women's work, i.e., clerical work. Again, more of the new computer-related positions were filled predominantly by men.

COSTS OF AUTOMATION TO THE AUTOMATED

One cannot generalize about the impact of automation in terms of the effects on management vs. non-management employees, female vs. male workers, or white vs. non-white workers in these case studies. Automation takes place at different levels within one firm, and between industries -- affecting craft work at one time, clerical, unskilled or management at another. But at this point in time, females and people of color seem most adversely affected by automation in labor intensive industries. Automation in industries with a more sophisticated technological core impacts on white male craft workers and lower level (female) management, and on women workers in clerical positions.

Informal discussions with workers in urban and rural areas suggests that people react to job loss or perception of declining opportunity in various ways. White male craft workers direct anger toward groups of lower status who are "moving in". Most female workers, however, tended to individualize the experience as their "own fault", or to accept the processes of change as inevitable in the name of progress. In either case, the loss of self-esteem was severe.

Telecommunications

Most women with whom I spoke worked at the lower occupational levels in the telephone company. Some reported not being able to "keep up with" the faster pace of work preceding automation, or higher requirements for punctuality and absences. They reported low-level supervisors were increasingly "strict" about personal movement (being away from one's desk, drinking coffee, yawning or stretching on the job). They felt they were treated "like kindergartner", in the words of one, with their physical movement and their attitudes under constant surveillance and criticism. Older workers, those near retirement, were more reluctant to complain about the organization of work during automation than were younger workers. Rural and small town women, generally older, resented the choice of early retirement or "tearing up roots" to transfer to the cities as smaller telephone offices closed. White male craft workers say "a white man can't get a job at Bell these days", usually blaming legislation for equal employment for the problem. A company social worker reported resentment among colleagues. They perceived that the organization of work led to anger, frustration, drinking problems, depression, anxiety and tension among workers. Social workers were then expected to alleviate these problems through counseling and other clinical services.

Printing and Publishing

Male craft workers who managed to retain their work after reductionand simplification referred to it as "junk" workd. While these workers strongly resisted such changes, many women workers scheduled for displacement, e.g., those who took want-ads, were described by management as apathetic.

Agribusiness

Farm owners considered such processes as vertical integration, mechanization, agrochemical use and attendant increased cost of farming, as inevitable. These processes were necessary, in the words of an Iowa farm woman, "if this country is going to feed the world." The most negative result of higher technology from their point of view was the impact on famiy relations. The high cost of farming meant few if any of their children would be able to raise their own families among rural traditions. Sons and daughters migrate to urban areas for work. Farm women, in addition, felt they would be unable to continue family farming if their husbands died, due to the structure of inheritance taxes, and negative attitudes toward women in agribusiness industries (supplies, equipment, finances, etc.).

In the migrant camps, community health workers sought alternatives to field work for migrants as harvesting was mechanized -- helping migrants "settle out" in the canneries and meat packing plants in the area. In this instance, prejudice toward Mexican-Americans on the part of anglo city dwellers made such settling out difficult. One could observe semi-organized efforts among small business men and white workers to resist the intrusion of a "different" people into the already tight job market in this small midwestern town.

The human and social costs of automation extend beyond the boundaries of the corporation, and are difficult to perceive if those boundaries are accepted as a legitimate parameter for analysis. As modernization draws women into the work force for example, it weakens traditional family ties but provides few new institutional support stystems. Child-care needs and care of the elderly are prime examples. Female headed households account for an increasing proportion of families in this highly industrialized country. Where automation primarily displaces female workers, they, and the young and the old who are not in the labor force, are adversely affected.

Thus, in addition to the personal and psychological effects of displacement or simplified and repetitive work, technological change such as automation has interinstitutional ramifications, particularly in this intersection of work and family. Ideology suggests strengthening the older family patterns compatible with an agricultural society (where women perform unpaid domestic labor) while the contradictory process of development draws women into the paid labor force. Further, this research on automation in selected industries shows a pattern of interaction with existing social divisions based on race and sex. This exacerbates social tensions during a volatile period in U.S. history.

THE AUTOMATORS: VERTICAL DIFFERENTIATION IN ENGINEERING

In the process of this study of the social costs of automation, I became interested in the people who designed and implemented the technologies in question. Most of those with whom I talked were managers with an engineering background. It seemed equally important to assess the social, political and economic forces affecting those who research, develop and apply automated systems, as it was to assess those forces as they impact on workers at levels affected by displacement. I was also challenged by the disparity in world view, and in the universe of discourse -- the different "languages" we spoke. (Elsewhere I have explored the significance of mathematical abstraction in the culture of engineering, Hacker 1978).

Many had not thought of the problem of new workers, young people facing a more difficult job market. Many felt that automation would increase employment, e.g., in other sectors such as food service, but had not thought about the lower income level in those sections. Those who recognized some diffculty with displacement had not thought of the differential displacement by race and sex. Most thought of women as wives and mothers, a flexible pool of labor which could be relatively easily re-absorbed into the home. This is a special problem in developed countries such as the U.S., where three quarters of women workers either do not have husbands, or are married to men earning less than $10,000/year.

These are not unsophisticated reactions, however, since social science literature itself has until recently stemmed from a similar conceptual orientation. It may seem unreasonable to expect engineering/management to be more sensitive and aware of these problems than are academic social scientists. But the significance of decisions -- especially in an international context -- made by those who direct technological change might argue the reverse.

Also relatively poorly researched is the impact of automation on engineering itself, and the quality of the engineering/technical workforce. Specialization and stratification accompanying automation do not stop at the boundaries of this profession, but affect its organization as well (Chestnut, 1963; Rosenbrock, 1977). Automation spurs the development of new occupational groups providing technical support to engineering.

Engineering education literature and career education literature indicate complexities surrouding the development of new occupations and paraprofessions related to engineering -- demand, definitions, training, curricula, accrediting bodies and criteria. Of concern also is the status and pay appropriate for such skills.

This literature allows insight into the attempts to direct vertical differentiation within emerging technical/scientific occupations. New levels are added to a hierarchy of occupations and paraprofessions; these new groups, by definition of accrediting bodies, are to work under the supervision of a professional engineer. An intricate pattern of relationships develop between training programs, educational institutions, professional societies, and industry needs.

Vertical Differentiation in Theory

At the lowest level of new occupations are industrial technicians who may have some minimal technical/educational training in community colleges, military, trade or technical schools (public or proprietary), but most often receive training on the job, or in formal short courses offered by corporations. A local electronics firm reports this training offers one route toward upward mobility for women on the production line. Career educators (Harris and Grede, 1977) suggest no theoretical knowledge or general education is necessary at this level. They cite "industrial technician" as an excellent occupational opportunity for women.

The next level, engineering technician, usually demands a two year associate degree from a community college, entailing minimal tehoretical knowledge and general education. Harris and Grede also describe "engineering technician" as an ideal opportunity for women.

Newest, and next highest on the scale -- in theoretical and general education, status, pay and autonomy on the job -- is the engin eering technologist, a graduate of a four-year program generally offered through colleges, universities or technical institutes. The literature does not encourage women at this level.

All of these occupational groups are to work under the supervision -- direct or indirect -- of an engineer, for the most part a graduate of a BS, MS or PhD program engineering.

Accreditation occupies a significant role in maintaining orderly development of these emerging occupations. The Technician Education Yearboodk: 1963-64, traces the battle over technical program funds between vocational and higher education proponents, and over the appropriate accrediting body, with the Engineering Council on Professional Development, (ECPD, now ABET) winning out in post-secondary technical programs.

Over the years, the engineering profession has struggled with curricula and with definitions of the new middle layers of workers. In 1976, the ECPD 44th Annual Report states:

> Engineering technology is part of a continuum extending from craftsman to the engineer. Located nearest the engineer, it requires the application of scientific and engineering principles in support of engineering activities. The support is given whether or not the engineering technologist or engineering technician is working under the immediate supervision of an engineer. The "engineering technician" is applied to the graduates of the associate degree programs. Graduates of baccalaureate programs are termed "engineering technologist".

Vertical Differentiation in Practice

The above descriptions and definitions indicate how the process should work, in the abstract. Not surprisingly, the operating reality reveals some slippage. For example, when the supply of engineers dwindled in the recent past, technicians were hired in their place (mingle, 1979). Another study indicates that Bachelors of Engineering Technology (BET) are hired as engineers. Carefully as educators and accreditors distinguish between programs offering the BS in engineering and the BET degree, forty-five BET programs surveyed show that industry assigns the title "engineer" to an average of 51% of the graduates (range, 0% to 98%, Moore and Will, 1973). A representative of an Oregon electronics firm reports they will hire BET's and technicians rather than engineers, if the supply of engineering personnel continues to fall short of demand. Another firm, and faculty members of a two-year technical program, however, state the 4-year BET program produces a "misfit", someone who is not, but aspires to be an engineer. They predict the decline and ultimate demise of the BET. These notions are countered by administrators of BET programs, who report, e.g., 10 offers for every graduate in some fields, and particularly heavey demands from the electronics industry in California's Bay area.

Comparing national "degrees awarded" data in BET programs with those in engineering and associate programs, it appears as though the BET share of the technical manpower pool is far from dwindling, but in fact shows a small but steady increase over the past 10 years.

These data indicate a decline of engineering in earlier years, but a "holding pattern" since the mid-seventies. Optimism for engineering must be moderated however for two reasons: (1) the survey of engineering schools is far more thorough than that of technician/technology schools (Doigan, 1979); and (2) the latter were three times as likely to non-respond as the former in 1978 compared to 1977, years for which such information exists (Sheridan, 1979). If degrees awarded data were adjusted for these factors, the engineering share of this labor pool would likely show more marked decline.[1]

There is also concern that proper dollar distance be maintained between employees in these various levels of technical and scientific work. Calculations from data in the Engineering Manpower Bulletin yield the following premiums for additional years of education:

		Average Monthly Starting Salaries			
		1964	1977	1978	1979
BET	Assoc	x	28%	29%	40%
BS	Assoc	47%	48%	53%	55%
BS	BET	x	(10-12%)[1]	(10-12%)[1]	11%
MS	BS	22%	11%	11%	12%
PhD	MS	42%	28%	30%	28%

1 Estimates

Lower salaries are only one, but an important factor which makes technicians and technologists attractive to industry.

In an extensive Department of Engergy review of the necessary Manpower for Energy Research, Kramer Associates, inc., as (1978):

> ...do employers substitute one speciality for another (technicians for engineers, for example) when relative salaries make it attractive, or are occupational demands relatively stable regardless of relative salaries?

We must note here also, that salaries for the tow-year degree technician compared to others approaches the familiar 50% difference in income between men and women, at the same time career educators suggest technician programs are very appropriate for women.

The engineering profession has long since acknowledged the need for technicians and technologists, e.g., through standing committees in professional bodies, through publication of guidance materials for technicians and technology students. They see that

[1] Assumptions underlying this line of reasoning must be tested -- to what extent do engineers, technologists and technicians compete for similar positions? Also, the ratio of engineer to technical support personnel varies widely from field to field.

an increasingly complex organization of industry creates a need for the production of technical workers with various levels of skills below the level of engineering. Automation is cited as a prime force driving this need. According to The Technician Education Yearbook: 1978-1979:

> Automation of industrial processes and growth of new areas of work such as environmental protections and urban development, will add to the demand for technical personnel.

Engineering may be caught in the tension between these economic factors and traditional ties to management on the one hand, and values internal to the profession on the other -- values such as quality of product and fulfilling work. For example, educators in two-year programs report that many of their students are former engineering students. Some left engineering because of the difficulty; others in search of the hands-on, "tinkering" work they prefer to management or paper work. In a recent discussion with a group of engineers in industry, they expressed some resentment that technologists were getting the "fun" work, while engineers were increasingly shifted to management and paper work. They did not consider their higher pay worth the loss of interesting things to do.

Some engineers, both on and off campus, question the quality of work produced by technicians or technologists hired as engineers, ascribing large system disruptions to mistakes by those with less engineering background.

Internal strain may also result within the profession as it is charged with ordering the process which produces the potential competitors. The very selection of criteria may have a further questionable effect on the content of engineering.

Vol. 1 of the 44th Annual Report shows ECPD's increasing concern with curricula, especially the depth of coverage of concepts of calculus (1976). A three-year report reaffirmed the need for calculus for the technician and the technologist, but suggested that "approved alternatives" be allowed as a calculus substitute for the associate degree technician. Differentiation between the engineer and the technologist by educational criteria is more difficult. Some insight into the difficulty is outlined in a report from a local Dean ofEngineering concerning the State Board of Engineering Examiner's review of engineering education. (At issue is the fact that some graduates of technology programs have been able to pass the engineering examination, and claim status, therefore, as "engineers".) This Dean points out that a local four year program for the BEt does include a 3-course calculus sequence in its cirriculum, but that the last course in this sequence approaches the level of difficulty only of the first course required in the engineering calculus sequence.

Much weight appears given to what Rosenbrock (1977) argues is only one aspect of engineering -- algorithmic, mathematical, scientific techniques. While accepting that these techniques are indispensable and have great intellectual appeal he emphasizes other skills -- judgement, experience, understanding of social complexity -- are necessary for a full development of engineering. Competition affects the basic sciences as well. Restivo and Collins, 1980, trace the impact of competition and prestige on historical developments in mathematics itself.

Stronger emphasis on mathematics as a criteria for good engineering also serves, inadvertently, as a filter restricting women's access to the upper levels of the field. Special training programs however, such as those sponsored in recent years by the National Science Foundation, enable women to develop mathematical ability equal to that of men. Through these and other measures, more and less developed countries alike may be able to revertse the channeling of talent by sex, in engineering and technical occupations.

The profession is faced with conflicting expectations related to automation. Automation, implemented by engineers, is largely originated by management in its need to reduce labor costs; lesser trained workers can be hired for less pay. The same is true within engineering. Engineering is now faced with finding ways to prevent management from the inappropriate use of cheaper technical manpower. Skills of abstraction, e.g., calculus and other techniques easily quantified, become even more heavily emphasized as the important ingredient of engineering. As Rosenbrock suggests, this may have unintended effects on the nature of work remaining in the profession.

In addition to problems of definition, curricula and industry practice is the tendency for graduates of two-year technician programs to transfer into four year programs. Engineering manpower data indicate that over half of the 2-year graduates return to school following graduation.

Four year institutions are taken to task for "disgraceful recruiting" in the community college programs in the search for students. Harris and Grede (1977), insist that community college technical programs live up to their promise to provide two year graduates with employable skills, and to provide industry with qualified employees. They note a tendency for the community college programs to alter curricula in order to make transfer easier for the student. ECPD also discovered the inclusion of general education courses or theoretical courses at the two year level. In Vol. 2 of the 44th Annual Report (1976), the council viewed such alteration of two-year curricula "in conflict with ECPD accreditation criteria".

Thus a concern for the community college's adherence to vocational goals, concern for employability of students, and for greater clarity in demarcations between emerging technical occupations through educational criteria, may work against the desire of some students for cheaper education and smoother transition to four-year institutions.

At present, transfers may be even more difficult for students to obtain, as economic factors curtail engineering education. Patterns of funding education now allow expansion of educational programs at the lower levels (e.g., the community colleges and the military as a training institution), while requiring cut-backs at the state university level. These factors perhaps foretell a settling down -- fewer engineers and many more technicians. Since over half the graduates of BET programs are hired as engineers, the BET may be perceived as a competitor by engineering and a threat to the quality of engineering work. Management may perceive the BET as a cost-saver, with the 10% lower average salaries. Management may also view the BEt favorably for other reasons.

There are political issues of great significance involved in this process of vertical differentiation within and beneath the profession of engineering. For example, Harris and Grede (1977) list "union membership" as one of the likely characteristics of the industrial technician, while the engineering technician and engineering technologist, as paraprofessionals, will instead choose membership in their respective professional societies. The intermediary technologist degree may soften the distance between engineering and technicians, strengthening the likelihood of professional identification on the part of all. The loyalties and political power of the new middle and lower level management personnel entering the technical work force will attract a great deal of interest in the coming decade.

To summarize this section, management needs, expressed through automation, introduce more division of labor into engineering. This labor process may entail unexplored negative results: the division of labor may once again build on oldr divisions such as sex; it may adversely affect the quality of work; and it may reduce the satisfaction of engineers with the work itself.

Similar misgivings about the effect of extreme division of labor on quality and productivity in industry have been expressed by others as well (e.g., Sandkull,1980). As Alvin Gouldner (1976) suggests, dissatisfaction and desire for higher quality of work in engineering may foretell a fragmentation of traditional loyalties of engineers to government or industrial leadership.

CONCLUSION

Mainstream social theory explains the positive functions of differentiation and stratification which accompany processes of automation -- efficiency, coordination, interdependence. Critics have argued also for attention to the dysfunctions of highly differentiated and stratified systems. I have taken this latter appraoch. Automation in labor intensive industries studied here tends to displace women and minority workers. In more technologically sophisticated industries, management, clerical and craft workers are most affected. The overall effect, however, is a disproportionate displacement of women workers.

Social divisions of labor based on sex are most severe in agricultural societies compared with others. Agricultural societies are those which are now in the process of development. Failure to notice the differential impact on men and women may lead to even greater discord between the sexes in developing countries than, say, in the U.S., given the more rapid rate of change and differing historical patterns in the former. The interaction of work with family in particular needs research attention.

Automation also produces more vertical differentiation in engineering. This may provide work opportunities of limited creativity and autonomy "at the bottom", while engineers may be furter removed from the tinkering, or hands-on experience which draws many young people to the field. Managmeent opportunities may be attractive to some, particularly to those in the more abstract or scientific fields of engineering, but may repel others. Further, economic factors lead to the employment of lesser trained technical workers as engineers in industry, with questionable effects on the quality of work. If this is not the case, the criteria used in more developed countries to establish an orderly educational program -- e.g., the emphasis on mathematics -- are themselves called into question.

As students are transformed into a workforce, an intricate pattern of interaction emerges between industry, various educational institutions and their funding sources, accrediting societies, and professional organizations. We should continue to include in our analyses these relatively unexplored, extra-organizational and interorganizational aspects of the division of labor as automation gives new shape to the scientific and technical workforce.

REFERENCES

Andors, P. (1976). Politics of Chinese Development: The case of women, 1960-66. Signs, 2; Pt 1, 89-119.

Arizpe, L. (1977). Women in the Informal Labor Sector: The case of Mexico City. Signs, 3, Pr 1, 14-27.

Boserup, E. (1970). Woman's Role in Economic Development. St. Martin's Press, New York.

Chestnut, H. (1963). Automation -- What it is and what are the problems it poses. Automatica, 1, 241-252.

Chincilla, N.S. (1977). Industrialization, Monopoly Capitalism and Women's Work in Guatamala. Signs, 3, Pr 1, 38-56.

Doigan, P. (1979). Engineering and Technology Degrees: 1978. Engineering Education. April, 728-38.

Engineering Manpower Bulletin. (1977). No. 39. (1979). No's 44,49.

Engineers' Council for Professional Development: 44th Annual Report. (1976). 1, p 15. 2, p 3; p 13.

Gouldner, A. M. (1976). The Dialectic of Ideology and Technology: The origins, grammar and future of ideology. Seabury Press, New York.

Hacker, S. L. (1977). Farming Out the Home: Women's Role in Agribusiness. Second Wave, 5, No. 1, 38-49.

Hacker, S. L. (1978). Engineering Mental Health. Paper presented at Pacific Sociological Society meetings, Spokane, WA.

Hacker, S. L. (1979). Sex Stratification, Technology and Organizational Change: A case study of AT&T. Social Problems, 26, 539-557.

Harris, N. C., and J. F. Grede (1977). Career Education in Colleges. Jossey-Bass Publishers, London. pp. 131-165.

Isley, B. (1980). PhD dissertation, in progress, Department of Sociology, University of Oregon.

Kramer Associates (1978). Manpower for Energy Research: Design of a comprehensive manpower information system for energy research. development and demonstration. Office of Education, Business and Labor Affairs, Washington, D.C. Section III, p 18.

Leghorn, L. and M. Roodkowsky (1977). Who really starves? Women and world hunger. Friendship Press, New York.

Mingle, J. G. (1979). Engineering vs. Engineering Technology. Paper presented at West Coast Society for Automotive Engineers, Portland, Oregon.

Miranda, G. V. (1977). Women's Labor Force Participation in a Developing Society: The case of Brazil. Signs, 3, Pt 1, 261-274.

Moore, J. H., and R. K. Will (1973). Baccalaureate Programs in BET. Engineering Education. October, 34-36.

Mueller, M. (1977). Women and Men, Power and Powerlessness in Lesotho. Signs, 3, Pt 1, 154-166.

Nielsen, J. (1978). Sex and Society: Perspectives on stratification. Wadworth Publishing Co., California.

Olsen, N. (1979). The changing family attitudes of Taiwanese youths. In R. W. Wilson, A. Wilson, S. Greenblatt. Value Change in Chinese Society. Praeger, New York. pp 171-184.

Restivo, S. and R. Collins (1980). Mathematics and Civilization: A sociological theory. Paper presented at meetings of the International Society for the Comparative Study of Civilizations, Syracuse.

Rosenbrock, H. H. (1977). The Future of Control. Automatica, 13, 389-392.

Sandkull, B. (1980). Effects of Automation. University of Linkoping, Sweden, guest lecturer, Department of Sociology, Oregon State University, Corvallis, OR

Sheridan, P. (1979). Engineering and Engineering Technology Enrollments, Fall 1978. Engineering Education. October, 58-63.

Stoler, A. (1977). Class Structure and Female Autonomy in Rural Java. Signs, 3, Pt 1, 74-89.

Technicians Education Yearbook (1964). p 51 ff.

Technicians Education Yearbook. (1979). p 331.

SYSTEM APPROACH FOR THE TRANSFER OF TECHNOLOGY

M. Levi-Jakšić* and M. Jakšić**

**Faculty of Organizational Sciences, Ulica Oslobodjenja 1, 11090 Beograd, Yugoslavia*
***Economic Faculty in Belgrade, Kamenicka 6, 11000 Beograd, Yugoslavia*

Abstract. This work applies the system approach in the analysis of the technology of technology transfer. Different aspects of this important socio-economic and technological phenomenon are pointed out. The problem is viewed on the enterprise level and is considered in the circumstances of socialistic, self-management socio-economic order present in Yugoslavia today. The problem is presented by simulation model (block-diagrams and differential equations) as part of the complex methodology that should be developed in consideration of technology transfer especially in developing countries.

Keywords. System theory; technology transfer - vertical and horizontal; hierarchical systems; simulation model; differential equations; developing countries.

INTRODUCTION

The problem of the technology transfer is illustrated using the scientific methodology - systems theory, i.e. the system approach. In this way it represents a contribution towards including different socio-economic phenomena in the process of their analysis by system approach. In our opinion the appliance of system analysis and system theory on different problems of the real world is the only way to check and verify this specific methodology on the one hand, and on the other, with its powerful tools and techniques it contributes a great deal to the observing, analysing and decision making in different fields of man's activities in the real world. We stand to the point that the methodology of system approach is powerful, but some of the conclusions made today with its appliance, are partial, biased and that is where the defects can be found. The structure and dynamics of the technology transfer system at enterprise level is presented by simulation model.

The study of the process of technology transfer is vitally important as the total socio-economic development is closely related to the technology transfer (here we understand the general meaning of this process as penetration and adoption of knowledge, new techniques and technologies, new organization models and methods as well as new forms of controlling technological processes). Therefore, it is especially important to define the strategy and policy of social development and especially technological development (with detailed technology transfer policy) at all levels: on global world development level, the level of national economy, region level, and enterprise level. This complex problem is also related to the question of correct forecasting and assessment of future technology, which represents part of the methodology being used in the process of controlling the technology transfer as constituent element of technological progress.

The goal of this work would be to analyse the relevant elements of technology transfer by system approach within the dynamics of industrial organization, its functioning in circumstances of self-management, market model of economy (present in Yugoslavia today). Technology transfer is significant precondition or integral part of technological progress, which with the other production factors (labour, capital, raw materials) conditions the functioning of production, and even broader, it effects business systems which is enterprise (the organization of associated labour in our society).

Technology transfer is regarded on enterprise level owing to the significant and important role it plays in our society and owing to the fact that the mere existence of enterprise or its economic losses are effected and in close dependence of effective and efficient use of adequate technology.

DIFFERENT ASPECTS OF TECHNOLOGY TRANSFER

The technology transfer system could be viewed from different aspects,the hierarchical structure formed according to the different aspects of the same question,the technology transfer is represented on Fig.1.

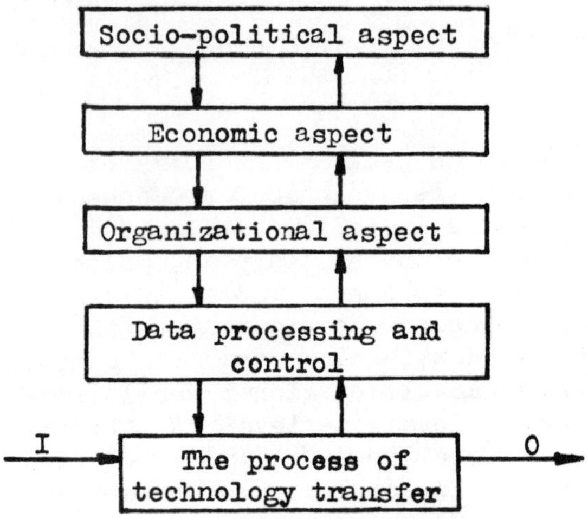

Fig.1. Aspects of technology transfer

These different aspects are significant and indispensable for a total system analysis and should be present in a complete system approach. We shall mainly deal with the data processing and problems of controlling the technology transfer on enterprise level. Nevertheless, there are two relevant questions that should be pointed out connected with the socio-political aspect which is of great importance when dealing with the technology transfer from the developed world towards the developing countries,and these questions arise related to the problem of creating optimal strategies of future development in developing countries.

The first one is how to reach the level of development of production forces and in this way reach the developed countries in the most efficient,successful way and in the shortest possible time.

This question is solved by the technology transfer from the developed countries as an inevitable process in the circumstances.

The second question is whether it is necessary and inevitable, that with the strengthening of production forces by horizontal technology transfer,to import the socio-economic relations and consequences that have already in the more developed societies shown negative and unwanted effects from the standpoint of the developing countries.

The solution and answer to this problem could be found within further progress in achieving strong basis in developing countries for controlling this process of technology transfer that is quite inevitable until a certain stage of technological progress is reached.The answer, therefore should be in encouraging the developing countries' own research and development programs which should deal with predicting,assessing,adopting and further developing technological innovations transferred from the more developed,on the one hand,and on the other, technological innovations and accomplishments should be stimulated and encouraged by R&D programs and strategies of development in developing countries.

DECOMPOSITION OF TECHNOLOGY TRANSFER SYSTEM

The technology of the technology transfer can best be illustrated at enterprise level by decomposing the technology transfer system. It should be emphasized that here we understand both the vertical (technology transfer and transfer of knowledge from fundamental science research,applied research to innovations and production) and the horizontal technology transfer (the transfer from one geographical region to the other,from developed to the developing countries).

Figure 2 illustrates the decomposed technology transfer system with its relevant subsystems according to the functions of the system.The explanations of designations on Fig. 2. is given at the end.

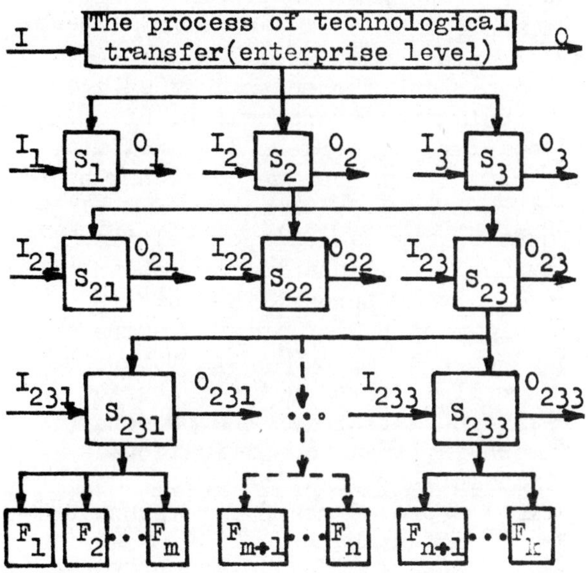

Fig.2. Technology transfer system with subsystems.

Legend:
Input: I
Output: O
State: S

Explanation:

$I_1S_1O_1$: The process of research and development aiming at achieving optimal choice of strategy of technological progress.

$I_2S_2O_2$: The process of achieving plan goals (investing in new technologies).

$I_3S_3O_3$: The process of complete adoption and further development of new technologies.

$I_{21}S_{21}O_{21}$: The process of purchasing new equipment from domestic suppliers.

$I_{22}S_{22}O_{22}$: The process of discovering new technology by enterprise R&D activities.

$I_{23}S_{23}O_{23}$: The process of obtaining new technologies by one of the modes of technology transfer (horizontal: licence buying, buying of know-how, equipment buying, and cooperation and common investments).

$I_{231}S_{231}O_{231}$: Analysing the state of enterprise concerning technological progress, technological needs, etc.

$I_{232}S_{232}O_{232}$: Forming data basis about suppliers, offered technologies, their characteristics.

$I_{233}S_{233}O_{233}$: Prediction, assessment, research of offered technologies considering the needs and state of enterprise.

$F_i \ldots F_k$ The phases of technology transfer process

F_1 Analysis of qualification structure of employed

F_2 Technological level of equipment

F_3 Financial analysis

F_4 Economic analysis

F_5 Analysis of plan achievments in enterprise

F_6 Analysis of working conditions - environment aspects as well

$F_7 \ldots F_m$ Market conditions and demand, supply of raw materials, etc.

$F_{m+1} \ldots F_n$ The processes engaged in selecting adequate technologies based on results of previous analysis (F_1-F_m). Optimal selection would be obtained according to fulfilling the needs and different requirements of enterprise, according to selecting the optimal mode of technology transfer (optimal conditions offered by partners from abroad), etc.

$F_{n+1} \ldots F_k$ The processes engaged in systematizing the information on technologies and the forming of data basis needed for appliance of methodology for technological assessment, simulation, etc.

ASSUMPTIONS AND TECHNOLOGY TRANSFER MODEL STRUCTURE

The first and always present assumption when attempt is made to model the real world is that <u>every</u> model is in a certain sense incomplete because it does not include all the relationships nor all the elements of the regarded phenomenon.

Classical market system is present in model, the free competition situation. The absence of measures and instruments of the state in limiting or stimulating the technology transfer are assumed - free market of technology transfer.

Combination of technology (capital), employed workers (labour) and technological progress are the main production factors of industrial enterprise in the model. Unlimited production factors are assumed.

Vertical technology transfer-research and development activities of enterprise- are presented and financed as part of the investment funds meant for realization of technological progress.

Horizontal transfer is understood in the model as the process of diffusion of more developed technologies into our developing self-managment society. The important question in enterprise is to define the optimal income distribution policy.

Horizontal technology transfer includes the following forms: the purchase of licences and of "know-how"; the purchase of equipment; cooperation and common investments-partnership relations with some other enterprise. Detailed explanations see in literature ad /1/.

Decisions on technology transfer in enterprise are formed on the basis of <u>demand</u> for new technologies which forms within the enterprise and depends on financial means obtained from different sources and intended for technology transfer; and on the basis of <u>supply</u> which is external and consists, from enterprise level, of complete information on offered technologies, suppliers, their requirements, etc.

The basic relationships in model are: PRODUCTION-INCOME-INVESTMENT FUNDS-TECHNOLOGY TRANSFER-PRODUCTION.

Schematically it could be presented by Fig. 3.

The model also includes the employment flow in enterprise. The departure of workers as presented here, means either their achieving the age for life pensions or leaving for another job. The model is made for Yugoslav socio-economic conditions, departure of workers does not mean losing their job due to technological progress. The changes in employment (number and structure) due to technological changes are shown exclusively through the changes of employment policy in enterprise and the opening of new posts for workers as well as through the number of workers that are in the process of improving their qualifications.

<u>Symbols in Simulation Model</u>

AMT - approved means for technological transfer
C - credits
DE - departure of employed
DTT - deciding on forms of technological transfer
E - employment
EM - employed
ENQ - employed workers in schools for new qualifications
GI - gross income
GR - graduation rate from schools
IF - investment funds
IM - import
M - man
MBE - means separated for buying the equipment from the developed
MC - machine
MLB - means separated for buying of licences from developed
MRD - means separated for enterprise research and development
MST - means separated for the technology transfer
NI - net income (gross income-material costs)
OS - obligations to the society
PA - amortization percent
PED - enterprise development plans
PF - production factors
PI - personal income
PIF - percentage of investment funds for technology transfer

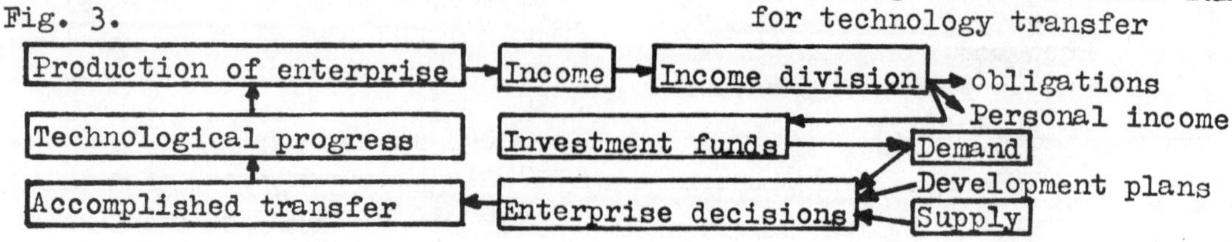

Fig. 3. Technology transfer global model

PM - production mix
PNT - purchase of new technologies
PP - sale price per product
PR - production
PSC - percent separated for cooperation and common investment
PSIF- percent separated for investment fund
PSRD- percent separated for enterprise research and development
PWC - purchase of working capital
PU - percent of usage
RC - required capacities
RCP - rate of credit payment
RD - rate of departure
RE - rate of export
REL - required employment level
RRE - required rate enlisting in schools(required qualification structure)
RS - rate of sale
RSB - rate of separation for equipment buying
RSL - rate of separation of finance for licence buying
RTT - rate for separation of finance for technological transfer
RWC - required working capital
S - sale
SDSA- self-management decisions and social agreement
SE - spending the means on equipment buying
SEB - separating the means for equipment buying
SEN - school enlistment
SFG - stock of finished goods
SG - graduating from school
SL - spending the means on licence buying
SLB - separating means for licence buying
SM - separating means for enterprise research and development
SMC - separating means for cooperation and common investments
SMRD- spending the means on research and development
SMT - separating the means for technology transfer - horizontal
STT - supply(offers) of different technologies
SWC - stock of working capital
TA - adaptation in time
TC - total capacities
TS - total sale
TCT - total costs
TU - time unit
UC - using capital
UMC - using the means to pay credits

BLOCK-DIAGRAM

Model of technology transfer system in enterprise is presented by block diagrams on Fig. 4.

SIMULATION MODEL

Block diagram consists of different flows, so the mathematical equations are grouped according to the flows. Some of the main equations of the total model have been presented here.

1. The money flow in investment funds:
   PIF.K = PIF.J +(DT)·(SMT.JK-UMC.JK)
   SMT.JK = RTT·IF.J
   UMC.JK = PIF.J·(PU + RCP)
2. The money flow in technology transfer by licence buying:
   MLB.K = MLB.J+(DT)·(SLB.JK-SL.JK)
   SLB.JK = PIF.J·RSL
   SL.JK = MLB.J·PU
3. The money flow in technology transfer by equipment buying:
   MBE.K = MBE.J+(DT)·(SEB.JK-SE.JK)
   SEB.JK = RSB·PIF.J
   SE.JK = PU·MBE.J
4. The money flow in technology transfer by cooperation and common investment:
   SMC.K = SMC.J+(DT)·(SCO.JK-SC.JK)
   SCO.JK = PSC·PIF.J
   SC.JK = SMC.J·PU
5. The money flow in technology transfer by research and development:
   MRD.K = MRD.J+(DT)·(SM.JK-SMRD.JK)
   SM.JK = PIF.J·PSRD
   SMRD.JK = MRD.J·PU

CONCLUSION

The next step within these problems would be to define the parameters in the model for concrete conditions of basic organization of associated labour functioning , and to empirically test the model on computer.

Having in mind the complexity of the problem considering the attempt for a total analysis of relationships in the technology transfer system considered on the level of basic organization of associated labour(viewed in self-management, socialistic economic system),this work represents a possible approach to these problems.

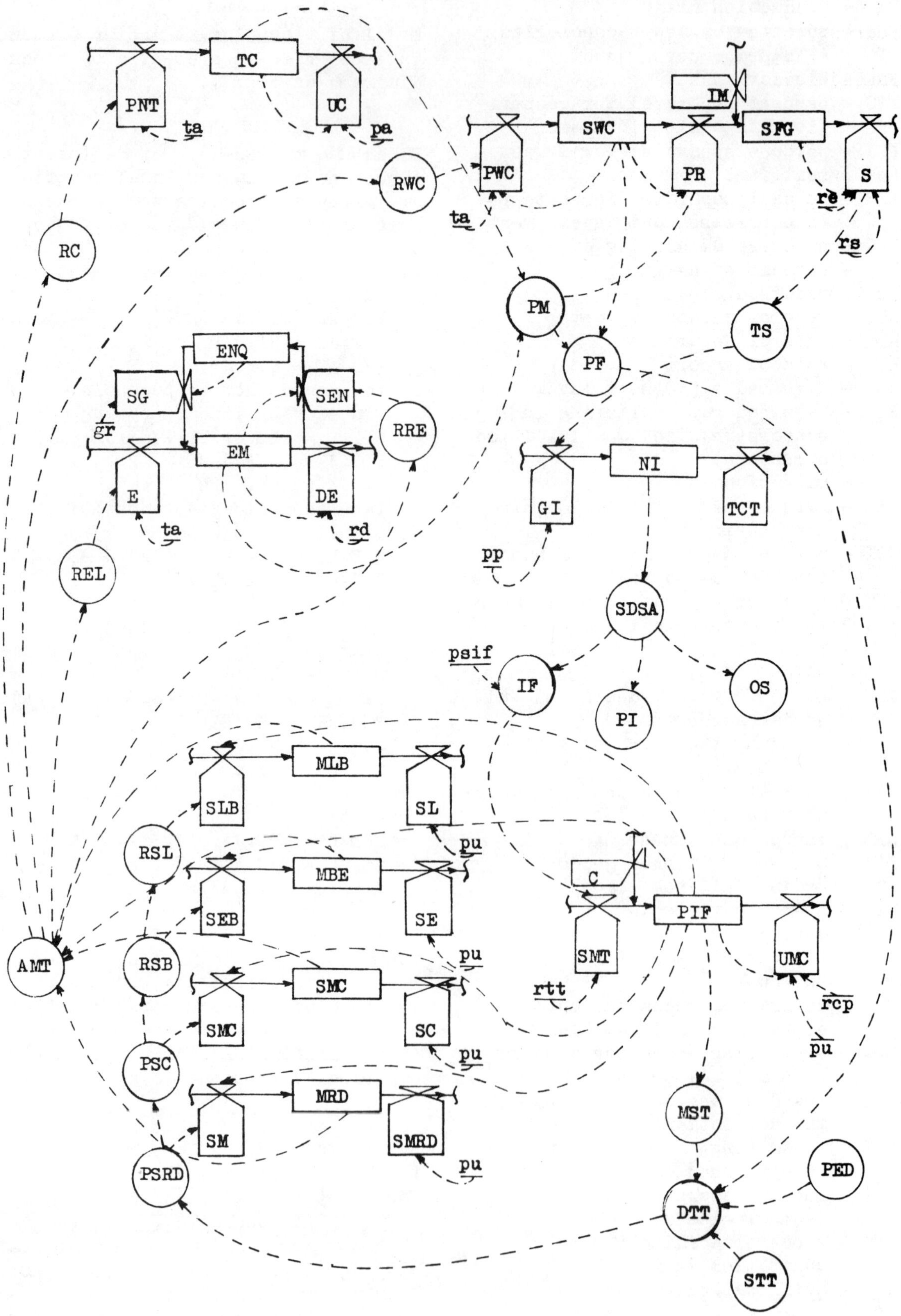

Fig.4. Block-diagram model

LITERATURE

Bodrožić, D. (1978). Tehnološki sistemi, PFV, Beograd.

Jakšić-Levi, M., Jakšić, M. Sistemsko modeliranje mehanizma ponude i tražnje, (1978). Zbornik FON, 2, 203-215.

Meadows, D. (1970). Dynamics of Commodity Production Cycles, Wright Allen Press, Cambridge.

Meadows, D. et al. (1974). Granice rasta. Stvarnost, Zagreb.

Mesarović, M., Pestel, E. Čovječanstvo na raskršću. (1976), Stvarnost, Zagreb.

OPTIMAL IMPLEMENTATION OF PROCESS CONTROL IN DEVELOPING COUNTRIES

H. El Hares* and F. Tayel**

**Department of Industrial Engineering, El Fateh University, Tripoli POB 13235, Libya*
***Planning Department, Oasis Oil Co., Tripoli POB 395, Libya*

Abstract. The lesson of the failure of many industrial projects in developing countries is leading us towards the following apparent paradox:

> the lower the degree of industrialization of a developing country, the higher the needed degree of automation of the imported equipment , provided that the foreign expertise is kept by various incentives as long as necessary.

A pattern for successful technology transfer is proposed emphasizing the technological expertise aspect.

Keywords. appropriate technology; technology transfer; automation; developing countries; continuous casting control; expertise; human factors.

TYPES OF DEVELOPING COUNTRIES

Naturally each developing country is a case in itself due to different cultural, technical and social conditions besides the great inequalities in national income.

However we can try to distinguish roughly between :

- rich (thanks mainly to natural resources) or poor developing country
- country with a history of industrialization (such as Egypt or India) and country with a history of trade activities (such as any of the Arabian Gulf states)

Now,for the next decade, the potential buyers of large industrial plants will be mainly the rich developing countries and the immense majority of them can be considered as having a history of trade activities, i.e. these countries are implementing now, and only now, their industrial structures.

In these countries the trade activities are so lucrative that there are no incentives for the natives to be involved in the industry. Consequently , it is impossible to find local skilled labour and there is a lack of this " industrial way of thinking " which will only come after at least two generations of industrialization.

Naturally the above remarks do not apply for relatively poor developing countries having a history of industrialization (it is not so difficult in Egypt, Tunisia or Morocco for instance to find cheap local relatively skilled labour).

Finally we are confronted with two main categories of developing countries :

* rich developing country with a history of trade activities
* poor developing country with a history of industrialization

COMMON PATTERN OF ACTUAL CONTROL EQUIPMENT IMPORTATION

It is a real fact that sophisticated highly automated equipment raises now and then more problems than it solves but this is due to the wrong ways of attacking the problem of technology transfer in the developing countries.
Most of these countries simply forget that it needs time (at least one generation) to develop the technical way of thinking , to ameliorate the general country infrastructure and to create all the small supporting industries.

Let us take for instance the case of Algeria where, during the last decade, the governmental agencies have tried to import the latest equipment with a very high degree of sophistication and automation.
Recently, leaks from the International Bank reports have revealed that the actual average percentage utilization of these equipments is less than 15 %.

According to the International Bank, this very low percentage is due to the great discontinuity existing between the imported sophisticated technology and the local way of living and thinking.

In our opinion, the main reason of this failure in the full exploitation of the installed capacity of these highly automated plants lies in the fact that the necessity of attraction of high level foreign expertise the necessary period of time, was not given its proper importance. Also no financial incentives were available to help the creation of qualified national counterparts.

Let us now consider as case study an exemple of the other way of thinking : in Libya for the newly created iron and steel factory of Tripoli, only semi-automatic equipment were purchased.
In fact, the consulting engineers were against the import of sophisticated highly automated equipment in order not to create too strong discontinuities in this developing country.

Unfortunately the result is disastrous; the product is always of a very bad quality and generally worthless mainly because:

a) there is no control of the level of the molten steel in the mold during the continuous casting process.
b) the control of the water-spray cooling system is done manually. This control and all other manual controls in the factory are absolutely not reliable since the great majority of the workers have practically no industrial background.

Now it is intended to introduce fully automated equipment in order to ensure a minimum level in the product quality but it will cost much more money than if it was decided first to build a completely automated production line.

For instance it is proposed to control the level of the molten steel in the mold by a series of thermocouples placed in a vertical line in the upper part of the mold. The hot junctions of the thermocouples are close to the inside surface of the copper mold.
The temperature of the copper changes abruptly at the level of the liquid steel, as shown by a significant difference in the electrical output of the thermocouples above and below the surface of the molten steel. A plot of the e.m.f. output of the thermocouples produces an electrical profile that is related to the temperature profile of the mold wall.
The abrupt change in the electrical profile corresponds to the most rapid temperature change, which is at the level of the liquid in the mold.
The signal from this device is used to continuously adjust the pinch-roll speed. If the liquid level rises the pinch rolls are accelerated, and vice versa, so that the casting is withdrawn from the mold at the rate required to hold the liquid level at a given height in the mold.

The automatic control of the water-spray slab-cooling system will use an on-line digital computer that calculates and automatically sets up all water flows throughout the spray chamber so that the desired slab-surface temperatures will be attained at different locations in the chamber.

The settings will be varied to match the type of steel to be cast and the casting speed at start.The computer subsequently will monitor the temperature at strategic locations,check the casting speed and automatically modify the flow of water to maintain the desired slab-surface temperatures.
At the end of casting, the computer follows the progress of the tail end of the slab and will shut off the water progressively to avoid overcooling of the tail end.

Naturally it is expected that expatriate technicians will be working on these proposed control systems, as long as necessary, ie until capable local counterparts will be sufficiently trained.

COMMON PATTERN OF ACTUAL TECHNOLOGICAL EXPERTISE

Unfortunately many factors specific to developing countries act towards decreasing the quality of technological expertise :

a) the unstable political,social and economical conditions of the majority of developing countries do not attract the high-level foreign expert.

b) the native decision-maker cannot generally properly judge the quality of the foreign expert ,due to a lack of experience.

c) very often, in developing countries, the supplier of the installation is chosen according to constraints which have nothing to do with quality, reputation or past-experience.
This supplier in turn will generally use low-level expatriate man-power.

d) the expertise is always desired to be minimum in time duration for two main reasons:

   1. there is generally a false idea of saving by reducing the duration of disbursing high wages.

   2. young local counterparts are always eager to reach higher levels of responsibilities.

It must also be said that the expatriate is feeling that he will be fired as soon as a "local" will begin to understand a little of what he is doing.
Consequently he has generally no interest to educate local counterpart especially if the expatriate is of low intrinsic quality and cannot easily find a similar job elsewhere.

PROPOSED PATTERN FOR SUCCESSFUL PROCESS CONTROL IMPLEMENTATION

In order to carry out the required technology transfer with the minimum cost and without disturbing the day to day life of the developing country, the following pattern is suggested provided that we take it as a whole and fulfil all its conditions :

1) importation of the latest designed equipment ,especially if it is highly automated provided that it was tested a sufficient period of time in the home country.

2) "importation" , for a period depending on the type of developing nation, of the technicians capable of running these installations and doing the appropriate maintenance.
Roughly speaking, this period in a rich developing country with a history of trade activities is at least double that required in a poor developing country with a history of industrialization.

3) absolute priority given to two things:
   - creation of high qualified centralized general workshops
   - with very good wages and other incentives, development of a qualified national counterpart for each foreign technician in the implemented industries, bearing in mind to keep these expatriate technicians (by attractive wages) as long as necessary and not to accelerate the processus of replacing them by local counterparts.

It should be noted here that the fact of keeping the foreign expertise as long as necessary will ensure that at least the implemented industries will work with a relatively good efficiency and consequently will support the overall effort of industrialization.

CONCLUSION

In this paper we have tried to emphasize both aspects of technology transfer:

men & equipment

In our opinion both directions should be given the appropriate importance, the quality of the expertise being often much more important than the quality of the equipment.

Decision-makers in developing countries should understand and apply these three facts:

a) accelerating the process of replacing the expatriate technician by local counterpart means practically always a rapidly deterioration of the production quality and the quasi-annulation of the efforts done.

b) financial incentives are necessary :
   . to attract high level foreign technician
   . to create qualified national counterpart

c) to obtain the highest efficiency, the latest sophisticated automated equipment should be adopted, provided that decision-makers in developing countries understand that technology transfer needs time and that expatriate expertise is needed for a period of time roughly proportional to the complexity of the equipment.

REFERENCES

El Hares, H. , and F. Tayel (1977a).Problems facing nuclear research in developing countries. Proceedings of the Conference on Transfer of Nuclear Technology (Persepolis/Shiraz).Atomic Energy Organization of Iran, Vol. 4.

El Hares,H. , and M. Aswed (1977b). Desalination in the development of Libya:problems of implementation and impact on environment. Desalination , 20, 249-256.

El Hares, H.(1978). Economic considerations in the choice of a desalination process in a developing country. Proceedings of the Sixth International Symposium on Fresh Water from the Sea (Las Palmas). Vol. 4, 87-95.

* * * * * * * * * * * * *

AN INTERACTIVE GRAPHIC SOFTWARE FOR COMPLEX SYSTEMS MODELING

M. Timoulali

Laboratoire d'Informatique, MIAG, Université Lyon I, 43 bd du 11-novembre-1918, 69621 Villeurbanne, France

ABSTRACT : A comparative study of different network techniques has shown the various relations among them, and has allowed to sum up facts on interactive graphics in modeling. Starting from such elements, we can conceive a software unifying the stochastic models (Q-GERT), the dynamic models (Systems dynamics), the parallel models (Petri nets), the critical paths models (PERT) and the structural models (DELTA charts, ISM).

We indicate the organization of the software with emphasis on the data structure, the algorithms for operation and control, and the languages : pictorial for the description and textual for the control of simulation and service.

We explain how it works and describe its implementation environment as well as the necessary features (computer independence, modularity, evolutivity, and personification) together with the conditions for their endowment.

We review possible fields of application of the software and illustrate, with a ferm land regrouping project, how to use some of the stated models in a given problem considered at various complexity levels.

KEYWORDS : Modeling, Simulation, Network analysers, Computer graphics, Stochastic systems.

INTRODUCTION

Network techniques are tools enabling instantaneous visualization of complex problems in the course of their analysis.

Likewise, simulation enables the study of a wide range of applications especially those which cannot be solved analytically, due to formulation problems, but which through knowledge of their working principles, may be described in the form of a network of events and activities.

Lots of descriptive models have been set up to deal with such problems but each model had its own characteristics and fiels of application. In fact, these models are not totally independent and their application range can be extended.

Moreover, interactive computer graphics provide a means of simplifying codification work, better problem formulation, good results interpretation and solution refinements.

All this has urged us to conceive a software giving the analyst a choice of different network tehcniques, enabling man-machine dialogue and providing graphic aids.

In this paper we examine certain aspects of over study in three parts. The first part is a reminder of the present state of interactive graphic modeling and give the principles underlying network reunification ; the second part is devoted to the software organization and functionning, and the last part to its fields of application with an example of graphic modeling for a farm land regrouping project.

INTERACTIVE GRAPHIC MODELING

There are two categories of network techniques :

- Discret models including : critical paths models (PERT, CPM), Parallel models (PETRI NETS, E-NETS, GRAFCET, GRAI, ORGANIPHASE, CONTROL AND DATA GRAPHS...), stochastic models (Q - GERT, GOZINTO), and structural models (DELTA CHARTS, ISM).

- Continuous models comprising dynamic models (INDUSTRIAL DYNAMICS OF FORRESTER) and thermodynamic models (BOND GRAPHS).

Relations between these tools and interconversion possibilities have been presented in (Timoulali, 1979 a). Some experimental softwares have been developed for graphic modeling : SYGNA (Marovac, 1976), "CORAL" (Sherlock, 1977), SIMILE (Delgrande and Mezei, 1979), ISMS (Hansen, Heitger and Mckell 1978) and Q-GRAF (Sabuda, 1977) (Fig. 1).

Our gool is to unify previous network techniques in a given software taking into consideration the interactive graphic modeling state and obeying four principles :

- Hierarchy of models in decreasing order generality : Q-GERT, Industrial dynamics, Petri nets, PERT.

- Models overlapping, which enables use of DELTA charts in several models, in an interpretive structural analysis environment (Warfield, 1976).

- The uniformness of the graphical representation by the multipole notions and acces dots (Hugues and Pistenon, 1973).

The bases of this work are :

- A dioîde algebraic structure permiting the generalization of many classical algorithms (Gondron, 1977).

- A process oriented simulation language (Q-GERT) or any other combined simulation language of same level, such as SLAM (Pritsker, 1979).

Other softwares QNAP (Merle, 1975) and methods (Madan and Jain, 1979) supply more possibilities in solving complex queuing systems, but are outside the range of our study, due to absence of suitable graphic symbols.

SYSTEM ORGANIZATION

Structure

The software architecture has been depicted in (Timoulali, 1979 b). We distinguish four elements : the data structure, the algorithms, the languages (Giloi, 1978) and the interface for adaptability to the environment (material, application and user) (Fig. 2).

The data structure is based on description programs (Lucas, 1977) ; it has a dynamic storage file structure (Sabuda, 1977) and utilizes five chains : four to store network graphic and a free chain for updating.

The algorithms are divided into two parts : operation and control. The first deals with description, structural analysis simulation and utility programs (optimization, subnetworks link). The second part ensure three operations (Amkreutz, 1976) :

- Evaluation, consisting essentially in the syntactique control and Petri nets analysis (Boundedness, Conservation of takens...).

- Decision making between available action and further response to an analyst request.

- Regulation in the response unavailability case to an analyst request.

The language is a network oriented one allowing model construction through specification of pictorial symbols and their manipulation. The textual language ensure simulation commands and service.

Man-machine Communication

The network is built using four different areas on the screen :

- The menu area used for displaying symbols and option actions.

- The graphic work area for creating and modifying models.

- The text area for non graphic data input.

- The message area reserved for error diagnostics.

The available input devices used are the keyset function alphanumeric keyboard and the light-pen.

The keyset is used to provide quick manual commands, the alphanumeric keyboard is used when alphanumeric data input is required. The light-pen is used as positional indicator and for identification (Lucas, 1977).

The menu is organised in three levels :
In the first one, the following commands can be selected : creation, modification, display, structural analysis, simulation, stop.
The second level lists available network techniques and the last one the symbols and the commands : selection, positioning, movement, deletion, end.

The general modeling procedure is described in the flowchart of Fig. 3. Network construction and analysis are done in an interactive mode. Structural analysis program execution or simulation must be asked for explicitely by the analyst, whereas Petri nets analysis and network conversion are executed automatically where needed.

Conversion .

Petri nets models, industrial dynamics networks and those described by DELTA charts or causal diagrams are converted into an activity and events processes coded in the simulation language input format.

Petri nets. For adequate generalisation, We consider in the first development program the class of interpreted Petri nets (labled led and timed).

In the labelled Petri nets an event $e(t_i)$ is associated to each transition t_i. only when the associated event occurs can the corresponding enabled transition be fires.

In the timed Petri nets, to each place P_i is associated a time interval during which an oncoming token cannot be available for any evolution. This time interval is a function of the place P_i and the arrival time of the token.

Petri nets are bipartite-directed graphs (bi-digraphs) that can be converted into the input format for the simulation.

DELTA charts. (Useful for results display) They are tools for describing a process by event, logic, time arrow and activity symbols. DELTA charts are equivalent to parallel flowchart having seven types of nodes : initial, operator, decision, terminal, joint and fork. The parallel flowchart and the sequentiel flowchart are Petri nets subclass (Zachariades, 1978).

Kodres,(1978) formalizes the concept of data flowgraph as a bi-digraph for analyzing parallel processing. Thanks to their identical abstract definition, we can use similar programs for Petri nets and DELTA charts conversion.

Industrial dynamics. (or system dynamics). The translation of structural models (causal diagrams) to industrial dynamics models is studied by Fitz and Hornbach, (1976).

Burns, (1977) describes an approach to the formulation of system dynamics, and the conversion of FORRESTER schematic to differencial equations. These equations would be written in the language of FORTRAN. The result is formulated into a subroutine capable of being called by a routine which performs the numerical integration and passed to the next algorithm in the execution sequence, the simulation, that can be continuous or combined.

For the simulation of system dynamics by Q-GERT analysis program, we convert them to MARKOV process (Sahin, 1979).

Implementation.

The software is in development on MITRA 125 minicomputer (128 k words), AFIGRAF CRT display and using GRIGRI Graphic language developed at GRENOBLE University (Lucas,1977) and adapted for MITRA 125 (Marc et Rueher, 1978). The system modularity enables its evolution by adding supplementary programs to the simulation language and translator (Nucleus).

The personalisation (idiosyncray) of the system is ensured by adaptation interfaces and result outputs (Translatable graphics for the structural analysis (Warfield,1979) and statistical curves for the simulation).

The computer independence or program portability is the ability to move a graphics application from one computer installation to another. The difficulties of achieving program portability are well known and are not be minimized, even if an ANSI standard language is used (Foley, 1979).

Device independence permit at particular computer installation to use a graphics application interchangeably with several different plotting, display devices, or interaction devices. Device independence is ensured by GRIGRI.

FIELDS OF APPLICATION

Many problems are studied in the literature by simulation and network techniques.

- Strategic planning and projects (Warfield, 1976).
- Production : scheduling, Inventory, Reliability, quality control (Whitehouse,1973); Ressource allocation (Pritsker, 1977) ; Computer-aided manufacturing (Runner and Leinbuhler. 1978).
- Information systems (sol. 1977).
- Data bases (Scheurman, 1977).
- Computer evaluation (Nutt, 1974).
- Societal systems (Rechenman, 1976).
- Automatic and Control (Grossetete, 1978) (Valette, 1976).

We examine the usefulness of the use of varied tools on different levels of a complex problem, in order to refine the analysis. A farm land regrouping project (Timoulali, 1978) is used to illustrate these concepts.

For a better clarity, the problem is very simplified. Figure 5 represents the DELTA charts of the project. We single out the regestering phase accomplished by the land keeping office, which is depicted using Q-GERT network (Fig. 6).

Figure 7 shows Petri nets model of land estate sale operation realized by the land keeping office.

Using this method, it is therefore possible to describe the smallest details of a system, in a global level and in the most elementary operation.

CONCLUSION

We have described a program unifying network techniques. The network can be interactively manipulated and his components can be computed.

An important advantage of the program is that different input and output modes can be chosen in the description, analysis and simulation of the real system.

In the future, we shall test the software in real applications, and in the next step to add a program for animation of the models.

REFERENCES

Amkreutz, H. A. E. (1976). Cybernetic model of the design process. Comput-Aided Des., 8,3, 187-192.

Burns, J.P. (1977). Converting Signed digraphs to FORRESTER schematics and converting FORRESTER schematics to differential equations. IEEE. Trans. Syst., Man & Cybern., SMC-7, 10, 695-707.

Delgrande, J.P., and L. Mezei (1979). An interactive system for the construction and animation of systems dynamics models. Comput. & Graphics, 4, 3, 155-160.

Fitz, R., and D. Hornbach (1976). A participative methodology for designing dynamic models through structural models. Proc. of the intern. confer. on cybernetics and society, Pittsburg, 1168-1176.

Foley, J.D., (1979). A standard computer graphics subroutine package. Int. J. comput. & Stru., 10,1/2, 141-147.

Giloi, W.K.(1978).Interactive computer graphics. Prentice-Hall.

Gondron, M.(1978). La régularisation des données. Colloque analyse structurale et modélisation. CNRS-IRIA - Lyon.

Grossetête, M. (1978). Méthodes inductives de conception des systèmes logiques complexes. Thèse d'état, Univ. PARIS VI.

Hansen, J., L.E. Heitger, and L. McKell(1978). Computer-aided modelling of decision - Support systems. J. Oper. Res. Soc., 29, 8, 789-802.

Hugues, J., and R.J. Pistenon (1973). Generation automatique des schémas industriels. Journées graphiques AFCET-IRIA.

Kodres, U.R. (1978). Analysis of real time systems by data flowgraphs. IEEE. Trans. on software eng., SE-4, 3, 169-178.

Lucas, M.(1977). Contribution à l'étude des techniques de communication graphique avec un ordinateur. Eléments du logiciel graphique de base. Thèse d'Etat, Univ. Grenoble.

Madan,S.,and J.L. Jain (1979). Cybernetic modeling of queuing complex.J.Cybern.,9, 2, 151-159.

Marovac, N.(1976). The structure of the network definition language NEDLAN. It's use in defining networks in CAD using interactive computer graphics. Comput. & graphics, 2, 1, 23-29.

Marx,G., and M.Rueher, (1978). Etude de l'infographie interactive et implémentation du logiciel graphique de base GRIGRI sur configuration MITRA 125. Mémoire DEA, Univ. LYON I.

Merle,D.(1978). Contribution à l'étude d'un analyseur de modèles à réseaux de files d'attente. Algorithmes de calcul et applications. Thèse Doc-Ing. INP.Lorraine.

Nutt,G.E.(1974). Evaluation net simulation system. Reference manual, Dep. Of comput. Science, Univ. of Colorado.

Pritsker,A.B.A.(1977). Modeling and Analysis using Q-GERT networks. Halsted press, New-York.

Pritsker,A.B.A., and C.D. Pedgen(1979). Introduction to simulation and SLAM. Halsted press, New-York.

Rechenman,F.(1976). Analyse et modélisation descendantes des systèmes socio-économiques. Thèse Doc-Ing. INP Grenoble.

Runner,J.A.,and F.F.,Leinbuhler(1978).CANSAM: A simulation analysis model for computer-Aided manufacturing systems. Proc. of summer computer confer.,Newport Beach, Caligornia.

Sabuda,J.D.(1977).A study of Q-GERT modeling and analysis using interactive computer graphics. M.S.M.E.,Purdue Univ.

Sahin,K.E.(1979). Equivalence of Markov models to a class of systems dynamics models. IEEE Trans. Syst., Man & Cybern.; SMC-9,7, 398-402.

Scheurmann,P.(1977). Concepts of a data base simulation language. Proc. ACM-SIGMOD inter. Confer. on managment of Data, 144-156.

Sherlock,J.F.(1977). Design and simulation of man-machine systems using networks models. Comput-aided design, 9,3,151-156.

Sol,H.G.(1977). SIMULA(TION) in the analysis and design of information systems. Proc. of Simulation, Acta Press, 67-71.

Timoulali,M.(1978). Conception d'un système d'aide à la décision pour le remembrement rural au Maroc. Mémoire DEA, Univ. LYON I.

Timoulali,M.(1979 a). Etude comparative des modélisations graphiques. 2ème Journées scientifiques et techniques de la production automatisée. ADEPA, Nancy.

Timoulali,M.(1979 b). Les différents modèles de réseaux. Journée représentation graphique de l'information. AFCET. Paris.

Valette,R.(1978). Sur la description, l'analyse et la validation des systèmes de commande parallèles. Thèse d'Etat, Univ. Toulouse.

Warfield,J.N.(1976). Societal systems. Planning, policy and complexity. J.WiLey, New-York.

Warfield,J.N.(1979). Some principles of Knowledge organisation.IEEE Trans. syst., Man & Cybern.,SMC-9, 6, 317-325.

Whitehouse,G.E.(1973). Systems analysis and design using network techniques.Prentice Hall, New-York.

Zachariades,M.(1977). MAS : Réalisation d'un langage d'aide à la description et à la conception des systèmes logiques. Thèse Doc-Ing., INP. Grenoble.

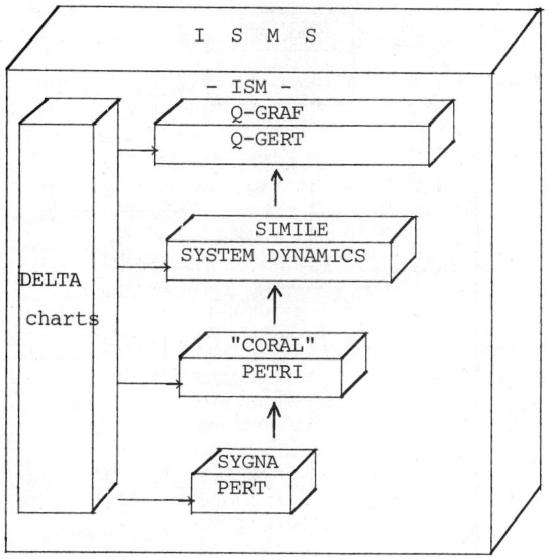

Fig.1. Network techniques relations.

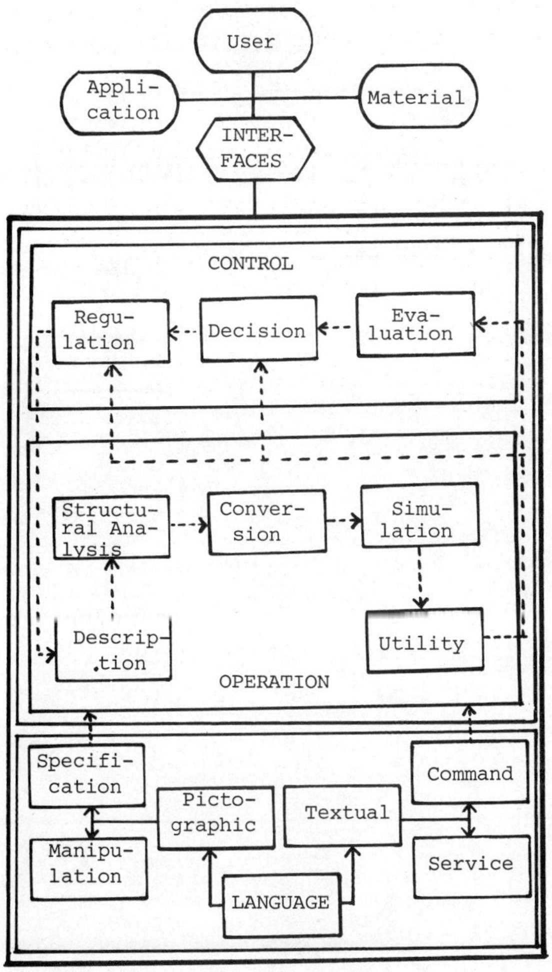

Fig.2 . Organization of the system.

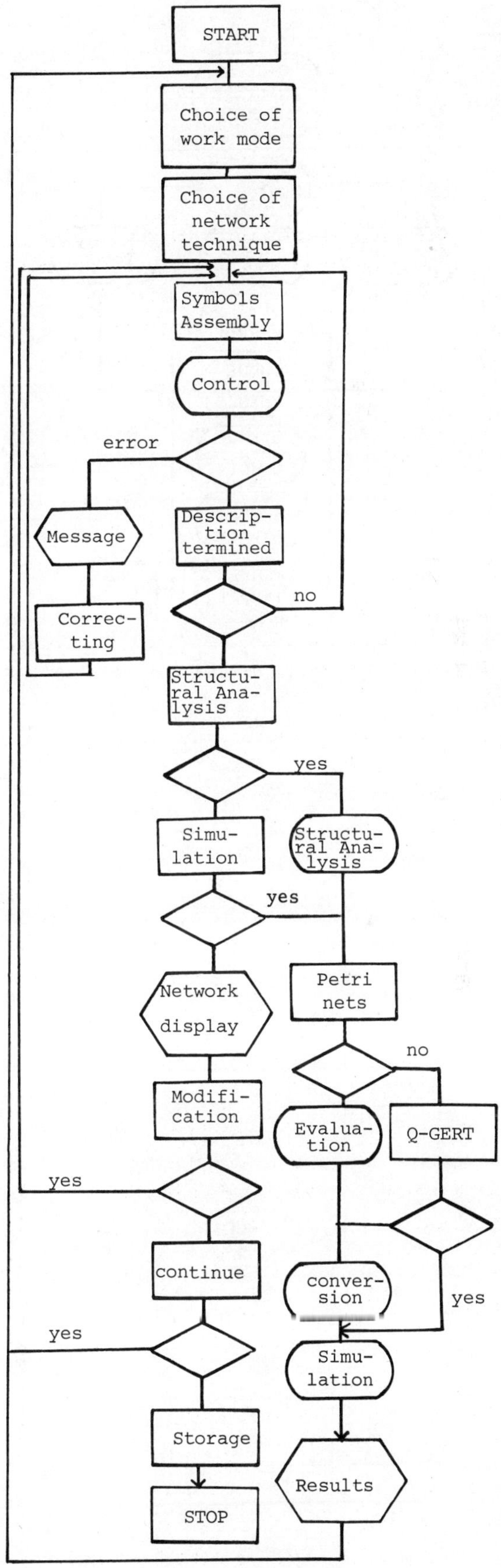

Fig.3 . Flowchart of process modeling.

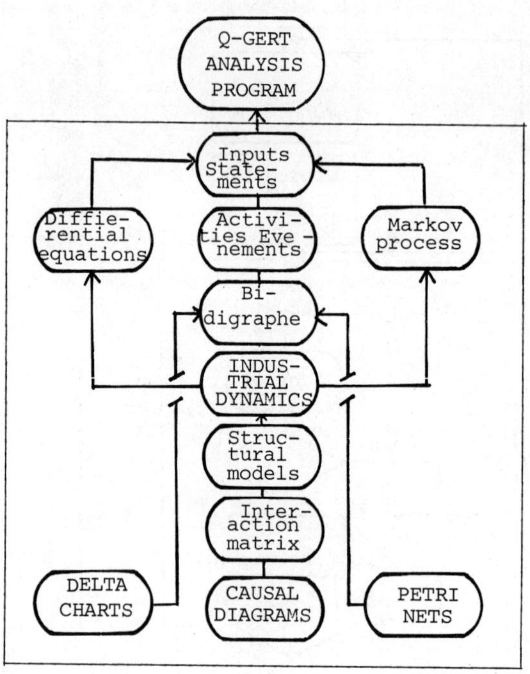

Fig.4 . Networks Translator.

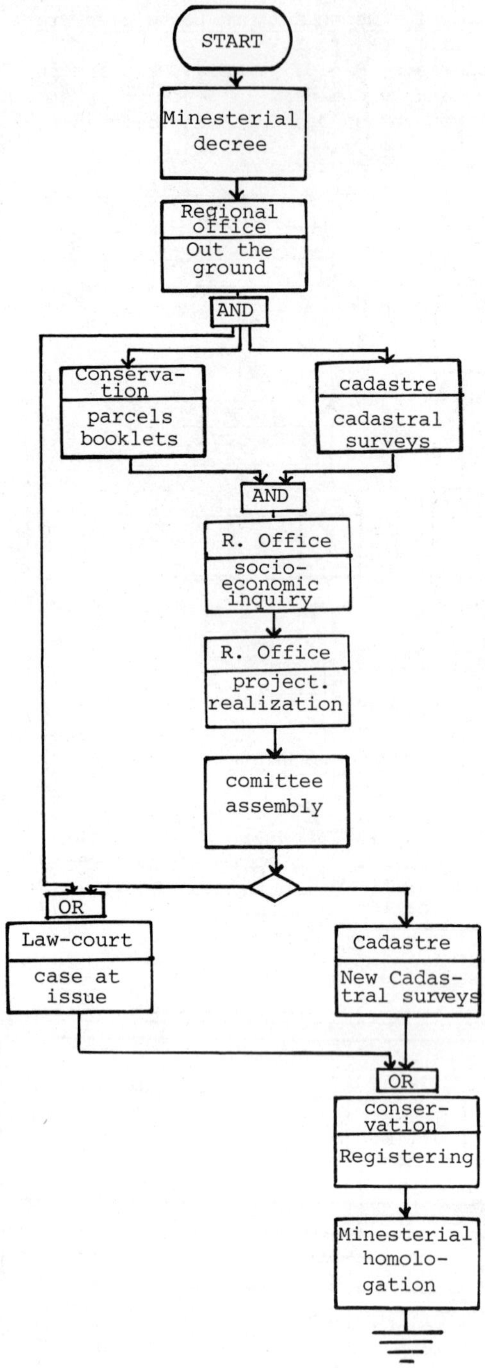

Fig.5. Schematic DELTA Charts of farm land regrouping project.

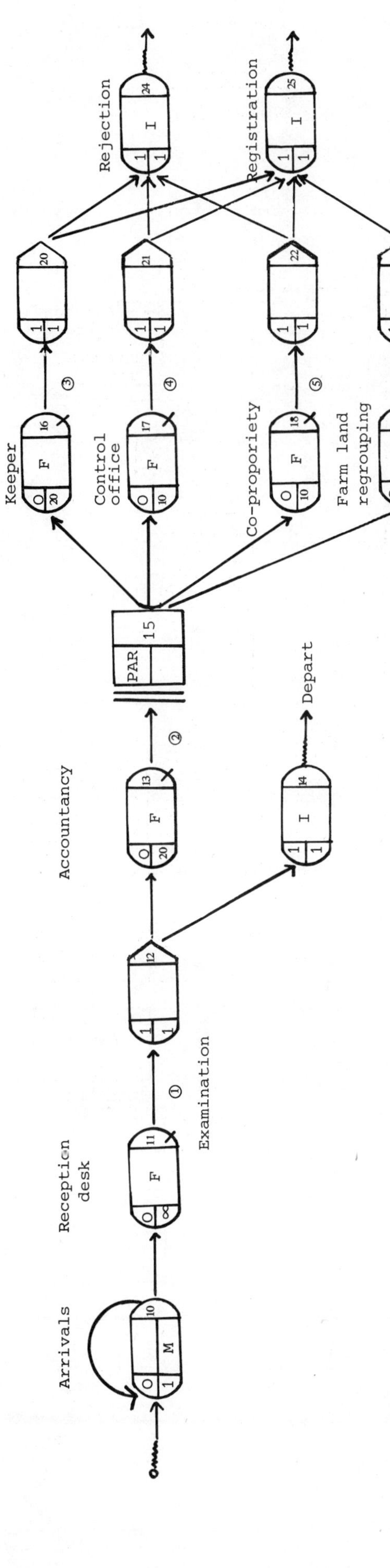

Fig.6 . Q-GERT network of Lanc Keeping System.

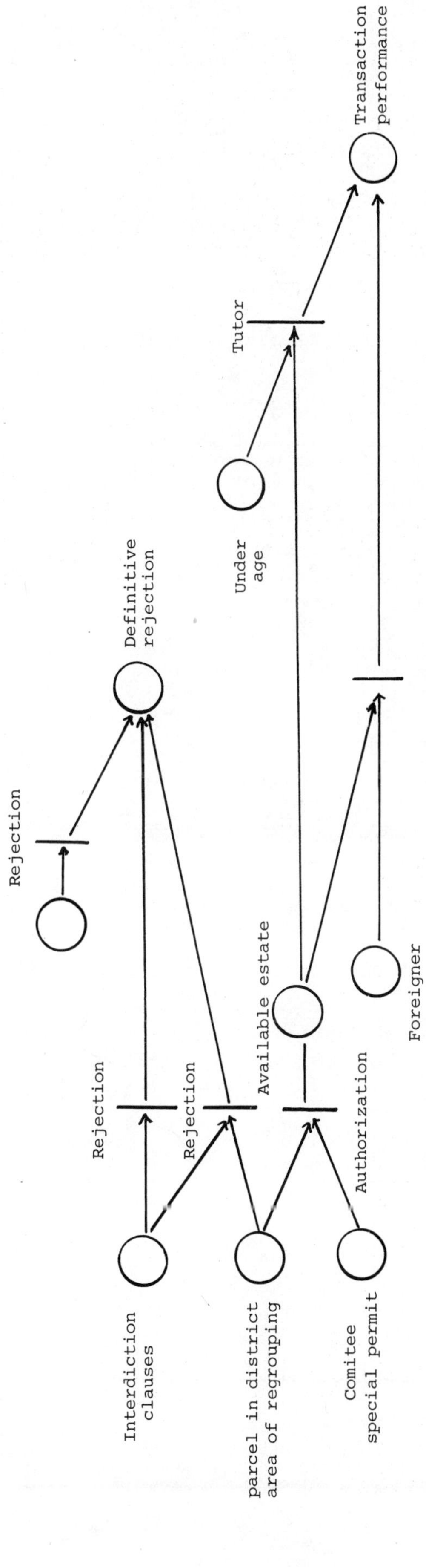

Fig.7 . PETRI net model of land estate sale operation.

THE OPERATION OF MICROCOMPUTER CONTROL SYSTEMS PERFORMING HARDWARE EMULATION

T. J. Kobylarz

Department of Electrical Engineering, University of Petroleum and Minerals, Dhahran, Saudi Arabia

Abstract. A frequent complaint by design engineers, regarding microcomputer utilization, is the difficulty they encounter during programming. Certainly higher level languages, such as FORTRAN, BASIC, and PL/M have relieved this burden. However, these languages still require an abstract association, and sometimes complicated interface design, between the microcomputer and the real time peripherals. A preferable approach is a higher level language which allows the microcomputer to be viewed as an assembly of building blocks for which a design engineer has strong familiarity; e.g. gates, flip flops, integrators, analog filters, etc. That is, to utilize a cross-compiler, which accepts inputs tantamount to a schematic, and generates a microcomputer object code which will emulate the terminal characteristics of the schematic. The operation of microcomputers which perform such hardware emulation is the subject matter of this paper.

Keywords. Microprocessors; programming languages; compilers; digital control; computer hardware.

INTRODUCTION

There is no doubt that microcomputers are experiencing a remarkable growth in useage. Their applications appear almost unbounded. Nevertheless, one obstacle has prevented an even more rapid growth. This obstacle is the difficulty in formulating and writing programs. Difficulties arise from two reasons; 1. the programming languages are incompatible with the design procedures generally used for the real time systems to which a microcomputer interfaces, 2. the design engineers have been trained and thereby more naturally reason in the mathematics concerned with the real time systems. The remedy to this problem is to eliminate the need for using an alien language and unorthodox mathematics in programming a microcomputer. This now leads to the research described in this paper. It concerns the operation of a microcomputer which has been programmed by a cross-compiler having user input data in a format akin to a schematic of a real time system. By emulating the input/output terminal properties of the system specified within the schematic, the microcomputer appears as merely additional system hardware. With such a facility, a design engineer is permitted to perform all of his work in his more natural domain of treating systems in terms of block diagrams.

The research and design in the "hardware emulation cross-compiler" is actually at its early stages. The system is presently limited to only five hardware elements, all digital, and 17 total instructions. Four of these elements perform combinational logic. The fifth is a J-K flip flop. Because the project is ongoing, the repertoire of elements will probably be larger at the time this paper is being read. The first thrust in element expansion will be digital. These are simpler to emulate and entail less microcomputer overhead memory. A second phase is being considered which expands to the emulation of continuous elements; e.g. amplifiers, integrators, filters, function generators, etc.

The underlying principal of operation is the state variable approach (Kobylarz, 1968 and McCluskey, 1965). The fundamental representation of this approach is shown in Fig. 1. Although a digital system is implied, analog systems are also representable by replacing the delay with integration and replacing the combinational logic with instantaneous function generation. This universality of Fig. 1 is quite important. It suggests that the basis of operation can be maintained, to an approximation, for emulation of continuous elements. Quantization in amplitude and in time are the main reasons for an approximation. The basic operating principle of Fig. 1 and hence, of the program is straightforward. Consequently no description is provided herein. It is however, important to point out that the present inputs $\underline{I}$ are usually from external peripherals and that the present states $\underline{Q}$ are stored within internal memory. Furthermore, the functions $\underline{f}$ and $\underline{g}$ are instantaneous (combinational) computations, where $\underline{f}$ produces the output vector $\underline{\phi}$ for external peripherals and $\underline{g}$ produces the "next" state vector $\underline{Q}'$. As will be seen later, $\underline{\phi}$ and $\underline{Q}'$ are computed by a program loop. This is termed a "computation cycle". A major significance of the computation cycle is that it represents an implicit clock signal for the emulated system; i.e. each computation cycle is tantamount to a clock cycle.

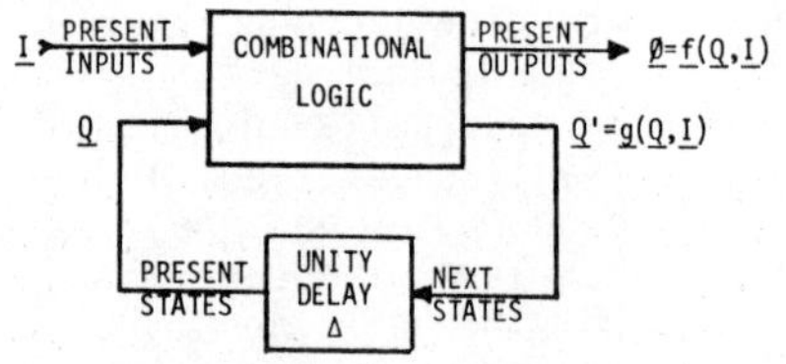

Fig. 1. Fundamental viewpoint for emulation.

The programming language for the cross-compiler is FORTRAN and therefore can be run on most general purpose computers. However, the object coding is in machine language for a National Semiconductor, 16 bit, PACE microprocessor (National Semiconductor, 1977a). The emulation programs have been run on the PACE "Low Cost Development System" (National Semiconductor, 1977b). There is no alternative but to have the cross-compiler designed for a specific microcomputer type. Nevertheless, the basic principles, described in this paper, can be universally applied to all microcomputers.

Probably the best way to provide significance to the preceding description is by means of a simple example. Therefore, the next section is concerned with an example. This is followed by sections which contain information relating to the operation of the microcomputer.

BLOCK DIAGRAM EXAMPLE

The switching elements are each specified by their function and their enumerated input and output nodes of a schematic. The general form is: LABEL, $i_1, \ldots, i_m, \phi_n$. Mneumonic labels are used, such as A for an "and" gate, V for an "or" gate, etc. Although an element type may be used a number of times within a schematic, the cross-compiler writes only one subroutine per element type within the microcomputer program. To economize on memory, no such subroutine is written when an element type is not used.

The schematic portrayal is not the only input for the cross-compiler. Other recognizable character strings exist having general form of LABEL, arguments (if any). For example, the labels ID, IH (ϕD, ϕH) can be used to prescribe the locations of peripheral inputs (outputs) to the switching system being emulated. The first argument represents an address, either in decimal (D suffix) or hexadecimal (H suffix). The remaining arguments indicate the nodes to which the input (output) bits correspond. From 1 to 14 inputs (outputs) can be assigned for each address used. The first designated node is associated with bit 1, the second with bit 2, etc. Bit 0 has a special program significance which precludes its use for peripheral input/output. Labels ID, IH (ϕD, ϕH) may be repeated, each with a unique address, whenever the quantity of inputs (outputs) exceeds 14. It is not possible to directly transfer an input to an output. This can be accomplished indirectly by means of two cascaded inverters (label N). However, a time delay of one computation cycle will transpire from input to output. At least one input and one output must be specified.

It is possible to emulate "real time" operation by means of the time delay operator D. The use of this operator will increase the microcomputer computation cycle time by an amount proportional to the argument. The PACE microprocessor, with a 2.667 MHZ crystal controlled clock, results in a proportionality factor of 60 μ-sec. Because non-negative integers of up to four digits can be designated, a single delay of about 0.6 sec. is achievable. Re-use of the D label permits an algebraic increase to longer delays.

The cross-compiler requires tnat the labels PD or PH must appear exactly once in the cross-compiler input data. The argument establishes the first address of the microcomputer program. Either a decimal (PD) or a hexadecimal (PH) number, but not both, can be specified. The PACE Low Cost Development System requires that the address not be less than $(40)_{16}$.

Two more labels must be described for the example. A Z signifies the final input data card (end of the data set) to the cross-compiler. Obviously it is to be placed at the end of the data set. Blanks may succeed the Z label. Blanks (spaces) or commas (,) should follow all labels and arguments. However, they are never to be placed within a label or argument; e.g. I,D is unacceptable, also 10 5 will be construed as 10,5 and not 105. Successive blanks, as at the end of card, merely iteratively prepare the cross-compiler for a new label or argument.

The example of an emulated hardware circuit is shown in Fig. 2. The schematic contains an oscillator, designed to have a real time period of 1 sec., and a divide by three counter (one pulse every 3 sec.). Oscillation is achieved by maintaining node 1 at a constant one value. It should be

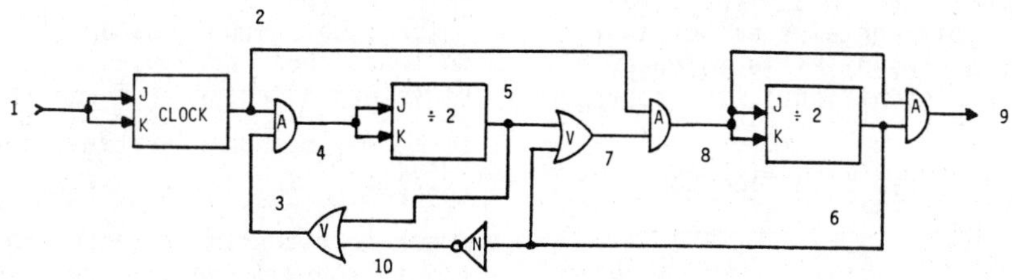

Fig. 2. Emulation example.

indicated that the J-K flip-flop corresponds to one with a "lagging edge" trigger for a computation cycle. That is to say, a state change occurs at the completion of one computation cycle, but just prior to the succeeding computation cycle.

An internal (non-peripheral) address is used for the input to node 1. The corresponding cross-compiler input data set is:

```
PH,40,IH,3FF,,1,φH,1000,2,3,5,6,8,9,D,
7990

J,1,1,2,A, 2,3,4,J,4,4,5,V,5,10,3,N,6,
10,,

V,5,6,7,A,2,7,8,J,8,8,6,A,8,6,9,Z
```

The first card specifies a microcomputer program starting address of $(40)_{16}$. The node 1 input is bit 1 of the address $(3FF)_{16}$. Six outputs are selected at the output address $(1000)_{16}$. Bit 1 corresponds to node 2 of the schematic, ..., bit 6 is used for node 9. A delay of 7990 time units or 479.4 m-sec. is indicated. This value was determined by first specifying D,0. A time measurement resulted in a half period of 20.6 m-sec. for the clock; thereby enabling the additional delay determination for a 1 HZ clock cycle. The following two cards designate the input/output nodes for the switching elements shown in Fig. 2. The last symbol (Z) terminates reading cards for the cross-compiler input data. Of course, control cards precede and succeed input data.

The microcomputer program required 246 words of memory, of which 161 are overhead The extravagance of memory and output hardware for this circuit emulation suggests that it can only be cost effective in more elaborate systems.

EMULATION STACK

The portion of the microcomputer program which is tailored to a specific switching circuit schematic is the emulation stack. The basic purpose of the emulation stack is to perform the computations suggested by Fig. 1. This stack contains the values of the present inputs ($\underline{I}$) and the present states ($\underline{Q}$). The program reads these values, either directly or indirectly, when they are necessary for logic computations. Also within the emulation stack are subroutine addresses corresponding to logic elements. As pertinent computations are performed, the microcomputer program will write the evaluations within the emulation stack. The pertinent computations are the present outputs $\underline{\phi}$, the next states $\underline{Q}'$, and values of schematic nodes which branch to more than one element input point. For example, node 8 of Fig. 2 branches to three input points and its value is written within the stack. Contrarily, node 10 will not have its value written within the emulation stack. Recomputation of nodes having multiple branching is eliminated by such a "write" instruction. After all computations of $\underline{\phi}$ and $\underline{Q}'$ have been completed, the states and the inputs are updated. Combinational logic computations are now repeated for new evaluations of $\underline{\phi}$ and $\underline{Q}'$.

The first word of the emulation stack contains a pointer, akin to a computer program counter. The second word contains a value one higher than the stack length. Remaining words correspond to instructions necessary for performing the previously described computations. These latter words are examined in sequence and, by means of relative addressing, the computer can ascertain the conclusion of the evaluations (end of stack).

Two major categories of words exist: 1. read/write and 2. subroutine function. The general bit patterns of these words are shown in Fig. 3. The sign bit (15) distinguishes between the categories. The read/write words are further subdivided into four types; two for read and two for write. Table 1 identifies the code used

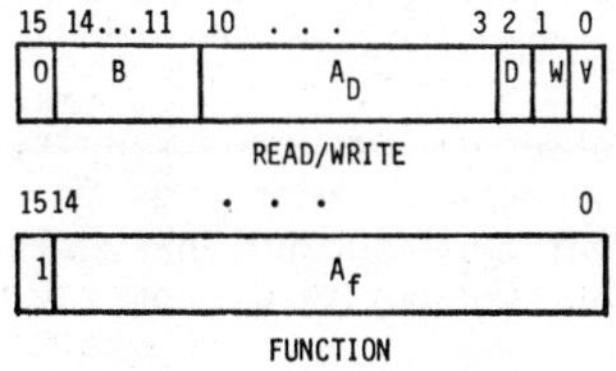

Fig. 3. Emulation stack bit patterns.

to distinguish the four types. For direct read words, the V bit contains either a present input or a present state value. Should V be for a present input, the source

TABLE 1 Read/Write Identification Code

Bit D	Bit W	Identification
1	0	Read V directly.
0	0	Read V indirectly.
0	1	Write non-state value into V.
1	1	Write state value into V.

of this value is indirectly addressed by A_D and is located in bit B. That is, an input value is originally secured from $((A_D))^*_B$. If B=0, then V corresponds to a present state. Its source is a write word, contained in the stack, at the location $((A_D))_0$. The value in $((A_D))_0$ would have been computed in a preceding cycle.

Updating V for each computation cycle is a time consuming task. Therefore only one word is reserved for each input or state value. However, such values may be used several times during the combinational logic computations. Indirect reads are employed whenever a value V must be re-used. The indirect read secures V from the emulation stack location $(B,A_D)_0$.

The pertinent computations are stored in V of the write instructions. If such a write instruction corresponds to a computation for a node having branches to more than one switching element, then an indirect read utilizes V at least once. Write instructions for outputs include an indirect address and a bit location $((A_D))_B$ for temporary storage of V. The entire word (vector) at $((A_D))$ is applied to the output peripheral address when the emulation stack computations have been completed.

If a flip-flop state is selected for an output, the microcomputer program will provide the present (not next) state. However, the value of V in a write instruction corresponds to the next state. This signifies that V should be written into the temporary storage $((A_D))_B$ just prior, rather than subsequent, to the combinational logic computations. In order to make this distinction, D=1 for a state value in V.

The other category of words, in the emulation stack, contain addresses for the emulated element type subroutines. Each unique element type, within the schematic, will have one such subroutine. The values upon which such subroutines operate are stored in two of four accumulator registers. Registers A_1 and A_2 are concatinated (A_2,A_1) to form a pushdown stack with $(A_1)_0$ at the top and $(A_2)_{15}$ at the bottom. The emulation stack operates in (reverse) Polish notation (Knuth, 1969). Hence, equivalent parantheses nesting can be 32 deep. Modifications for indirect read and write have been added to improve efficiency of operation. Values read are pushed onto the data stack and values written are pulled off of the data stack. The subroutine functions will pull off as many arguments as is needed for a computation and push the result onto the data stack. Non-commutative and non-associative operations naturally have the data stack arguments arranged in the required order.

Outputs of elements within switching networks will frequently branch (fan) out to several other element input points. A major feature of the emulation stack is that it preserves such branching and thereby reduces the amount of intermediate computations. This is accomplished by the indirect reading of prior write instructions. Such an addition to reverse Polish notation is quite simple to implement and it is believed that the procedure can be inferred from an example. Consider the computations of the output node 9 in Fig. 2. The symbols R, R', W, W' will correspond to respective instructions in Table 1. Addresses, where appropriate will be designated by node enumerations. Subroutine function calls are designated as A (for "and"), V (for "or"), and N (for negation). Observe now that nodes, 2, 5, and 6 are present states of the flip-flops and must be initially read directly into the data stack. Assuming all states have been updated, then the partial emulation stack will contain:

...A,W8,R'5,R'6,N,V,W3,R'8,R'6,A,W9,..

In the preceding, the first symbol A identifies the point at which node 8 is computed. The next instruction pulls this value from the data stack and stores it in the V bit of the emulation stack W8 instruction. Following this are instructions to indirectly read node 5 and node 6 values into the data stack. After inverting the node 6 value, an "or" computation is made to evaluate node 3. This value is then pulled from the data stack for either future computations or for the peripheral output. Rather than recompute the node 8 value for node 9, the value is now indirectly read into the data stack from the W8 address written in R'8. Subsequently, the value of node 9 is computed via the A subroutine and written in the emulation stack by means of W9.

*Parantheses are used to indicate the "contents of" in an iterative manner.

After executing the final instruction of the emulation stack, the microcomputer provides outputs to peripherals designated in the write instructions. The read instructions are later updated from peripheral inputs and the next state variables. The program now iterates the emulation stack computations.

MICROCOMPUTER PROGRAM OPERATION

An overview of the microcomputer program operation is presented in this section. The program operates in six basic steps:

1. Initialize next states,
2. First update-
   a. Transfer present states to temporary output storage,
   b. Transfer input values to emulation stack,
   c. Transfer next states to present states within emulation stack,
3. Operate emulation stack,
4. Second update- Transfer non-state computations to temporary output storage,
5. Apply outputs at peripheral addresses from temporary storage,
6. Enact computation delay, if selected by user. Return to step 2.

In step 1, the microcomputer program searches the emulation stack for all "write" instructions. When found, a "0" is placed in the V bit. This step therefore initializes all computations which are to be written within the emulation stack. As will be shortly seen, only the next state initialization will have a role in computations.

The second step is mainly used to prepare the emulation stack for its computations. In addition, it transfers the previously computed next state (present state) values to the temporary output storage. The transfer of these state values is necessary at this point, since only write instruction values can be placed into the temporary storage and next states cannot be defined outputs. Because of the initialization step, all such values will be "0" in the first pass through the program. The peripheral input values are inserted into the direct read "V" bits by means of the associated addresses and bit locations. The states are updated similarly, except that the transfer is from write state instructions to direct read instructions, within the stack, and the bit locations are "0".

After performing the computations defined by the emulation stack (step 3.), a second update takes place. The purpose of this is to complete the temporary output storage. Prior to step 4., the temporary output storage contains only those present state values which have been designated as outputs by a user. Step 4. inserts the newly computed values into this storage. When the output temporary storage has been completely updated, the values are subsequently transferred to the output peripherals selected by the user (step 5.). One peripheral at a time is addressed. This means that the new output values occur simultaneously only if they are at the same address. For multiple addresses, the time delay between new outputs for succeeding addresses is approximately 0.16 m-sec. The user should therefore maximize the number of outputs per address (14) should this delay be appreciable. The output synchronization option can substantially alter the time delay between output addresses. The program will, in this case, not provide a new output until a peripheral has read the old output (bit 15) at the address in question. The program enacts such a wait for each address individually.

The final step, prior to an iteration to step 2., is the enactment of the computation delay. This routine merely cycles through a loop of known time duration for an amount of repetitions specified by the user.

The indirect addressing entails the need of stored constants to be used as address pointers. The cross-compiler determines these constants as it composes the microcomputer program. Because of complexity in operation, only the indirect addressing for outputs to peripherals will be described. Assume that the addresses designated within the write instructions of the emulator stack are A_{D1}, A_{D2}, etc. Those instructions having non-zero bit locations ($B \neq 0$) imply that the V bit value must be provided to a peripheral. One can consider that the A_{Dj} addresses identify rows of an array, as shown in Table 2. The nomenclature "array" is used, since the bits of a word are regarded as columns. Each $(A_{Dj})=A_T+2j-1$ $(j = 1,2,\ldots)$, in the temporary address array. The contents of the temporary address array correspond to rows (addresses) of a consecutively arranged temporary output array. The contents of the temporary output array alternate with output peripheral address $A_{\phi j}$ and output vectors $\underline{\phi}'_j$, having 14 meaningful data bits. The output updating routine will place V, from the emulation stack into the bit position $((A_{Dj}))_B$ of the temporary output array. An output peripheral receives new outputs when a $\underline{\phi}'_j$ vector is transferred to the peripheral output array. Because the temporary output array is consecutively arranged in the memory, the latter transfer is accomplished by a loop. It is only necessary to identify the first address A_T

TABLE 2 Indirect Addressing Technique

Temporary Address Array		Temporary Output Array	
Address	Contents	Address	Contents
A_{D1}	$A_T + 1$	A_T	$A_{\phi 1}$
A_{D_2}	$A_T + 3$	$A_T + 1$	$\underline{\phi}'_1$
⋮		$A_T + 2$	$A_{\phi 2}$
		$A_T + 3$	$\underline{\phi}'_2$
			⋮

Peripheral Output Array	
Address	Contents
$A_{\phi 1}$	$\underline{\phi}_1$
$A_{\phi 2}$	$\underline{\phi}_2$
⋮	

and the final address $A_T + N$, where N is an odd integer. The program performs (N+1)/2 transfers of (A_T+j) into (A_T+j-1), j=1,3, ..., N.

CONCLUSION

It is believed that microcomputer hardware emulation is a potentially valueable tool for digital control system designers. Principally because it permits the incorporation of standard design procedures. Furthermore, it allows circuit elements to be regarded as having ideal properties. Therefore, the disconcerting problems arising from propagation delay, hazards, races, fan-out, cross-talk, etc. need not be considered.

The most severe limitation is computation time. It is quite apparent that applicability is restricted to slow response systems. System cost suggests that only complex systems should be emulated. However, both slow speed and large cost are governing factors for the utilization of microcomputers in most real time control systems.

REFERENCES

Knuth, D.E., (1969). The Art Of Computer Programming, Vol. 1. Addison-Wesley Publishing Co., Reading, Mass., pp. 336.

Kobylarz, T.J., (1968). Nonlinear network computer analysis. G.J. Herskowitz (Ed.), Computer-Aided Integrated Circuit Design, McGraw-Hill Book Co., N.Y. Chap. 6, pp. 197-240.

Kobylarz, T.J., (1980). A higher level language to simplify programming of microcomputer control systems. Proceedings of the IEEE Conference on Applications of Mini and Microcomputers, March, pp. 33-39.

McCluskey, E.J., Jr., (1965). Introduction To The Theory of Switching Circuits. McGraw-Hill Book Co., N.Y.

National Semiconductor Corp., (1977a). PACE Microprocessor System Design Manual. Order no. IPC-16A/928, 2900 Semiconductor Drive, Santa Clara, Calif., March.

National Semiconductor Corp., (1977b). Users Manual, PACE Microprocessor Low Cost Development System. Order no. IPC-16P/301Y, 2900 Semi-conductor Drive, Santa Clara, Calif., June.

LANGUAGE-DEPENDENT CONSIDERATIONS FOR COMPUTER APPLICATIONS IN FARSI AND ARABIC SPEAKING COUNTRIES

B. Parhami

Department of Mathematics and Computer Science, Sharif University of Technology, Tehran, Iran

Abstract. A prerequisite to economic viability and social acceptance of computer-based systems, particularly in developing countries, is their ability to accept and present information in the natural language of the user community. This paper identifies problems relating to the use of Farsi and Arabic languages in computer-based systems and outlines known approaches to their solutions. The problems discussed fall into four main categories: (1) Information representation and standardized information interchange codes; (2) Input of information through devices such as keyboards and optical readers; (3) Processing functions such as sorting, text formating and data compression; and (4) Output of information on hard-copy devices and various types of displays.

Keywords. Character recognition; Computer applications; Computer input and output; Computer peripheral equipment; Data processing; Display systems; Human factors; Printers; Standards; Text editing.

INTRODUCTION

Computers are information processing machines and much of the information we deal with in our everyday lives is generated, maintained, and used in a natural language. Furthermore, for computers to be useful, they must interact with human beings and the most convenient way of doing this, at least as far as ordinary users are concerned, is by utilizing a natural language as the medium of communication. It follows that the ability of computers to deal with information presented in a natural language is essential for their successful utilization in most environments.

Even though language considerations are not critical in activities such as programming, there are clearly instances where information storage (e.g., in the case of names and addresses), manipulation (e.g., in text processing), and presentation (e.g., for question-answering systems) can be most effectively handled in the users' native language. It is, therefore, quite natural that for a number of years, we in Iran have been dealing with language-related problems created by the rapid expansion of computer applications (Parhami and Mavaddat, 1977).

In this paper, some basic problems relating to the encoding, input, processing and output of information in Farsi and Arabic languages are identified and approaches to their solutions are outlined. Our discussion will be in terms of the Farsi language, with the understanding that almost all of these considerations apply to Arabic and Urdu as well.

ENCODING OF INFORMATION

The recording of information on computer storage media and its transmission from one place to another (be they circuits on the same board or geographically distant computer systems) usually requires some form of digital encoding. The advantages of using standard codes for these purposes are well known and have resulted in the adoption of national and international standards for various natural languages. National standards of this type must be designed according to specific guidelines, put forth by the International Standards Organization (ISO, 1973), if they are to be mutually compatible.

Following relative maturity in various application areas, the need for a standard represention of Farsi symbols was felt in Iran (Mavaddat and Parhami, 1977). In February of 1978, the Iranian Plan and Budget Organization invited a number of computer and language experts to form the "Study Committee for Standard Farsi Information Code." Even though the Committee's final report was published in August of 1979 (SCSFIC, 1979), it has not yet been adopted as a national standard.

Based on ISO's code extension guidelines, the committee set out to define a 94-symbol G1 graphic set which can either replace the standard G0 set in a seven-bit code or used in conjunction with it in an eight-bit environment (Figure 1). The following five points were taken into account in designing the Farsi G1 graphic set:

1. One-to-one correspondence between code and output symbols. This point, which proved quite important in practice, precluded the use of a reduced code set (Hyder, 1972), as originally envisaged in the Committee's preliminary proposal.

2. Inclusion of all common Farsi alphabetic and special symbols. The proposed set includes all needed numeric, alphabetic, punctuation, mathematical and commercial symbols as well as vowels which may be needed in special linguistic applications.

3. Proper alphabetic ordering of symbols for ease of sorting. Even though it has been shown that the Farsi sorting problem cannot be solved by code assignment alone (Parhami, 1978c), care has been taken to avoid complications whenever possible.

4. Maintaining the correspondence between Farsi symbols and equivalent or similar symbols in the G0 set to the extent possible. The particular set of alphabetic symbols used turned out to contain 52 members, thus exactly replacing the upper and lower case letters in G0 (Figure 1).

5. Simple recognition of numeric and alphabetic symbols and efficient packing and unpacking of numeric information. This is accomplished by assigning numeric and alphabetic sybmols to positions where G0 contains symbols of the same types.

A complete presentation of the tradeoffs involved in designing the Iranian National Standard Information Code (INSIC) and justification of the particular decomposed character set used is beyond the scope of this paper. Interested readers are referred to the Study Committee's final report for a discussion of these points as well as extensive application notes.

INPUT OF INFORMATION

The existing problems with Farsi input and some of the approaches for dealing with them have been enumerated by Parhami and Mavaddat (1977). With respect to the input of natural language information, we can visualize three distinct possibilities at the present:

1. Keyboard data entry.
2. Document data entry.
3. Voice data entry.

The third alternative is in the research and experimental development stage even for languages of technologically advanced countries. The Farsi language will undoubtedly present its own unique problems in this respect.

The second alternative, though in practical use for languages utilizing the Latin alphabet, is relatively new in the case of Farsi. The difficulties in the automatic recognition of Farsi texts are depicted in Figure 2. Parhami and Taraghi (1980) have developed algorithms for separating the symbols in connected subwords of printed Farsi texts and for recognizing the symbols from their geometric properties. Figure 3 shows the result of their algorithm applied to a three-letter subword.

The insight gained from the above-mentioned study will be helpful in the design of type fonts for various document input devices. However, the development of standards in this area does not seem urgent at present.

For most applications, the initial data entry is, and will be for the foreseeable future, through the first alternative: namely, keyboards. Even though there is no technical reason for using the present typewriter keyboard layout for computer applications, the practical factor of familiarity has prompted most manufacturers to adopt the same layout and may continue to influence keyboard designs for years to come. Nevertheless, a total reorganization may be considered desirable or even necessary in future, based on one or both of the following considerations:

1. The known inefficiency of the present typewriter layout for Farsi data entry.
2. Large scale replacement of Farsi typewriters by word-processing machines.

Figure 4 shows a proposed keyboard layout for use with the Iranian National Standard Information Code in Farsi and bilingual data entry. This layout has been proposed by the "Study Committee for Standard Farsi Information Code" (SCSFIC, 1979) based on the following considerations:

1. Maintaining the typewriter key positions to the extent possible, unless where this is in conflict with other goals described below. A total of eight changes turned out to be necessary. These changes made the keyboard layout more systematic, more efficient and easier to remember and also enabled the numerals to be placed on lower case.

2. Inclusion of all the standard symbols, and only these symbols. The only exceptions are Farsi letters which are represented by two symbols in the standard code due to the requirement for one-to-one correspondence between code and output symbols. These are entered by a single key stroke, with the needed tail symbols added automatically.

3. Placing equivalent and similar Farsi and Latin symbols in corresponding positions, to the extent possible, in order to reduce the amount of information appearing on each key and to facilitate operator training for bilingual data entry. This was accomplished quite nicely with minimal changes to conventional Farsi and Latin layouts (Figure 4).

PROCESSING OF INFORMATION

The three processing functions of sorting, text formating and data compression are briefly discussed in this section. With respect to sorting, it has been shown that the problem cannot be solved by code assignment alone, even if a reduced code set is used (Parhami, 1978c). Therefore, we either have to adjust ourselves to a new notion of sorting, which might appears somewhat unnatural, or resort to more sophisticated algorithms.

In the area of text formating, two complicating factors are observed:

1. Hyphenation is not allowed for breaking long words.
2. Words are not consistently separated by blanks.

The second problem arises since some words appear naturally disconnected when juxtaposed, enabling a human reader to determine word boundaries by paying limited attention to the context and/or meaning.

Although in the long run this problem can be rectified by proper training, the current difficulties will be with us for some time. Fortunately, however, complete decomposition is not needed in text formating applications, as long as the word boundaries that are actually recognized are not too far apart. This is usually true in practice (Nasseri, 1978). On the positive side, the possibility of extending connected Farsi letters is at times helpful for text justification.

In the area of data compression, a recent study (Tafaghodi Jami, 1980) has shown promising results. With suitable text fragment selection algorithms, a compression ratio of 45 percent was achieved. When a reduced code set enabling the encoding of more text fragments was used, the compression ratio improved to 48 percent. These results are somewhat better than those achieved for English texts.

The final aspect of processing discussed here is that of algorithm specification. There have been suggestions that we should be thinking about a high-level programming language (similar to COBOL) based on Farsi. Even though on the surface it may appear that the English-language orientation of many high-level programming languages is a deterrent to Farsi speaking programmers, there is no evidence that a Farsi-based language will make the programmer's task any easier or that our computing community will be better off in general as a result.

In response to the argument that, in the near future, interaction with computers will be done mostly by non-programmers which might be aided by the availability of a Farsi programming language, it can be stated that non-programmers are not likely to interact with computers in languages like FORTRAN and COBOL. They will probably perform simple computations using a mathematically oriented language (or simply function keys on a private, pre-programmed keyboard) or define a system by filling tables and answering multiple-choice questions.

Any new language proposed will need to be initially developed and documented for various computer systems and subsequently maintained and updated periodically. Clearly, our limited resources are better spent on developing more effective computer applications than on designing new programming languages and systems. Furthermore, for the exact reason that so many programming languages are in use today (i.e., the diversity of application areas), no single Farsi-based language can satisfy all needs. How many of these languages should we develop? The most logical answer appears to be zero!

OUTPUT OF INFORMATION

With respect to the output of natural language information, three areas need to be investigated:

1. Hard copy output of information.
2. Visual display of information.
3. Special methods (voice, braille, etc.).

We will not discuss the third area here except for saying that its importance will certainly draw some of our attention once the pressing problems in the other two areas are satisfactorily solved.

While the difficulties with Farsi codes, input and processing are of concern to computer professionals and a limited number of direct users, Farsi output is a problem of broad concern. In view of the fact that over the centuries of Islamic influence, writing of Farsi has become something of an art, extensive modification of the Farsi script for adaptation to machine printing or display is highly undesirable and has already caused some damage in the public attitude towards computers (Parhami and Mavaddat, 1977).

Most high-speed Farsi line printers produce low quality output due to the following:

1. Inter-symbol gaps make the technology of most high-speed printers unsuitable for producing the connected symbols of the Farsi script.

2. The varying widths of Farsi alphabetic symbols are either ignored for simplicity or dealt with by decomposing wider characters into two parts. The first approach reduces the readability and aesthetic quality of the resulting output while the second approach compounds the problems of connectivity and horizontal alignment.

3. Due to the large symbol set needed for printing Farsi, frequently some of the special symbols or variants of letters are eliminated. The need for mixed Farsi/Latin print in some applications has compounded this problem.

4. The similarity of many Farsi symbols, differing only in the number or places of small dots, necessitates a higher quality printout for readability. The smearing effect of most high-speed impact printers is, therefore, a serious drawback in this respect.

Farsi printer output samples in Figure 5 illustrate the above problems. Many of these difficulties can be easily overcome with character printers. Even though quite valuable in word processing applications, such devices are relatively slow and thus only of limited interest in most other environments.

For high-speed Farsi output, the new generation of non-impact printers appear to be promising, since by their method of operation (e.g., thermal, electrostatic, or xerographic processes), such devices produce sharp and highly readable output with no inherent inter-symbol gap. The only adjustment is for the need of a larger frame in the case of dot-matrix printers, since five by seven or even seven by nine matrices are inadequate for Farsi.

In the actual implementation of output devices, a great deal of flexibility can be provided. As long as the output device accepts a standard code set, it can use as many different forms for each letter as deemed necessary for a reasonable output script quality. Vowels and similar special symbols can either be overprinted on the preceding letter or form an independent symbol, with connected and separate forms, as shown in the following examples:

مـُـتـَـکـَـلـّـِـم   آ نـَـا"

مـِـشکوه"   اُ ر دَ ک

This latter approach is very easily implemented on conventional printers and results in a script with reasonable quality for applications where vowels and supplementary symbols need to be printed without cost or speed penalties.

In the visual display of Farsi information, we essentially face the same difficulties as in hard copy output. In the case of visual display units associated with intelligent terminals, the needed changes for adaptation to the special requirements of Farsi are usually not very complex. The comments made on dot-matrix printers also hold true for dot-matrix display units.

The representation of Farsi numeric and alphanumeric information on line-segment displays is also of some interest (Parhami, 1976). A prototype display unit for a single 20-character line of Farsi output is at the final stages of construction at Sharif University of Technology. The display uses 18-segment character displays as shown in Figure 6.

Even though purely numeric information can be displayed with as few as seven segments (Parhami, 1976), no design with fewer than 18 segements is known for Farsi alphabetic symbols.

In some applications, it may be desirable to present large quantities of numeric information in combined digital/analog form. This can be achieved through the use of an optically weighted numeral font. Figure 7 shows an example for decimal Farsi numerals in a seven by five matrix. The area covered by a numeral x is ax+b (i.e., linearly proportional to x) with a=3 and b=4 (b=3 if the points marked with an "o" are deleted). General guidelines for the design of such numeral fonts as well as several specific designs have been presented elsewhere (Parhami, 1978a).

CONCLUSION

Satisfactory solutions to the problems enumerated in this paper are important for the successful development of informatics technology in Iran as well as in the Arab world. In particular, immediate action on the standardization of information interchange codes is needed to assure compatibility of systems and to avoid costly replication of effort.

In the area of input devices, standard keyboard layouts need to be developed. The design of better character recognition algorithms as well as OCR type fonts is also needed, though it is less urgent.

For effective processing of Farsi information, the problems of sorting, text formating and data compression need to be further studied. The inital results in these areas are quite encouraging and point to the possibility of significant advances in the near future.

A considerable amount of effort has so far been placed on the Farsi output problem as the most immediate user concern. Hard copy devices (both impact and non-impact types) are rapidly improving in this respect. Visual display units have generally relied on large dot matrices (say with more than 150 points) to generate the Farsi symbols. This situation is not likely to improve soon and thus line-segment displays may become a viable alternative when cost is a prime factor.

As we solve the fundamental language-related problems facing us in the field of computing, we may in future attack less crucial, but perhaps more interesting, problems of generating classic Farsi scripts and analyzing the rich heritage of Farsi literature with the aid of computers. The artistic and scholarly possibilities of such undertakings are almost limitless (Parhami, 1978b).

REFERENCES

Hyder, S.S. (1972). A system for generating Urdu/Farsi/Arabic script. Information Processing 71 (Proc. of IFIP Congress). North-Holland, Amsterdam, pp. 1144-1149.

Institute of Standards and Industrial Research of Iran (1976). Character arrangement on keyboards of Persian typewriters. Document No. ISIRI 820.

International Standards Organization (1973). Code extension techniques for use with the ISO seven-bit coded character set. Document No. ISO 2022.

Mavaddat, F. and B. Parhami (1977). Informatics in Iran: problems and prospects. Proc. of the International Conf. on Computer Applications in Developing Countries, Bangkok, Thailand, pp. 121-133.

Nasseri, P. (1978). Automatic processing of Farsi texts. Master of Science Thesis, Sharif University of Technology, Tehran, Iran (in Farsi).

Parhami, B. (1976). Low-cost output displays for microcomputer applications. Proc. of the Second India Symp. on Computer Architecture and System Design, New Delhi, pp. 111-119.

Parhami, B. (1978a). Optically weighted dot matrix Farsi and Arabic numerals. Proc. of the Third Jerusalem Conf. on Information Technology. North-Holland, Amsterdam, pp. 207-210.

Parhami, B. (1978b). Impact of the Farsi language on computing in Iran. Mideast Computer, Vol. 1, No. 1 (Sep. 18), pp. 6-7.

Parhami, B. (1978c). On the use of Farsi and Arabic languages in computer-based information systems. Proc. of the Symp. on Linguistic Implications of Computer-Based Information Systems, New Delhi, India.

Parhami, B., M. J. Ashjaee, and F. Mavaddat (1977). The Farsi language and computers. Reports prepared for Plan and Budget Organization of Iran under Research Contract No. 10908201/9/120. Vol. 1, Nov. 1977. Vol. 2, June 1978 (in Farsi).

Parhami, B. and F. Mavaddat (1977). Computers and the Farsi language: survey of problem areas. Information Processing 77 (Proc. of IFIP Congress). North-Holland, Amsterdam, pp. 673-676.

Parhami, B. and M. Taraghi (1980). Automatic recognition of printed Farsi texts. Pattern Recognition (to appear).

Study Committee for Standard Farsi Information Code (1979). Final proposal for the Iranian National Standard Information Code (INSIC). Informatics Division, Plan and Budget Organization of Iran, Tehran (Farsi and English versions).

Tafaghodi Jami, A. (1980). Efficiency considerations in the storage and retrieval of Farsi texts. Master of Science Thesis, Sharif Univ. of Technology, Tehran, Iran (in Farsi).

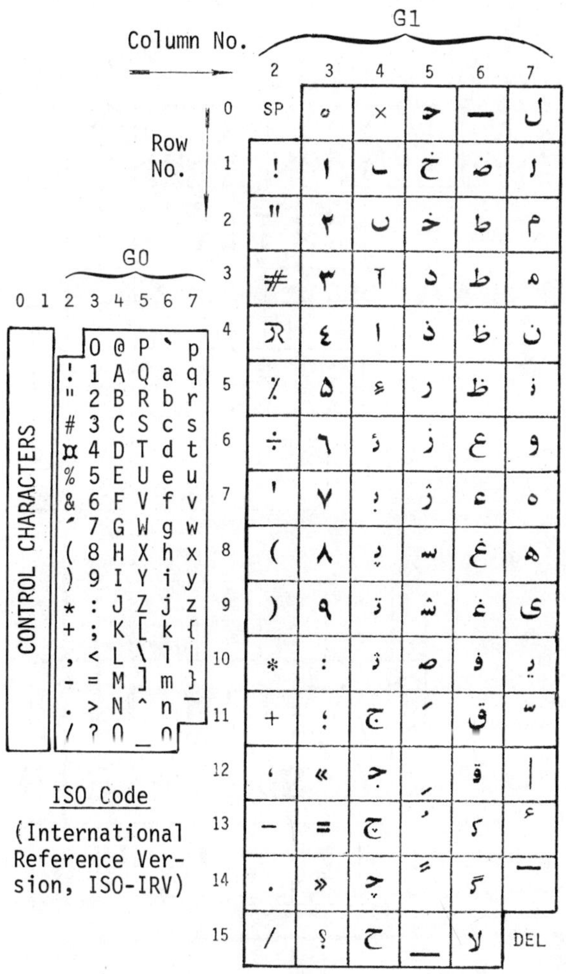

Fig. 1. The Proposed Iranian National Standard Information Code (INSIC).

Fig. 2. Difficulties in the automatic recognition of printed Farsi texts:
(a) Connectivity of symbols,
(b) Similarity of symbols,
(c) Variable-width symbols,
(d) Overlapping subwords, and
(e) Overlapping lines of texts.

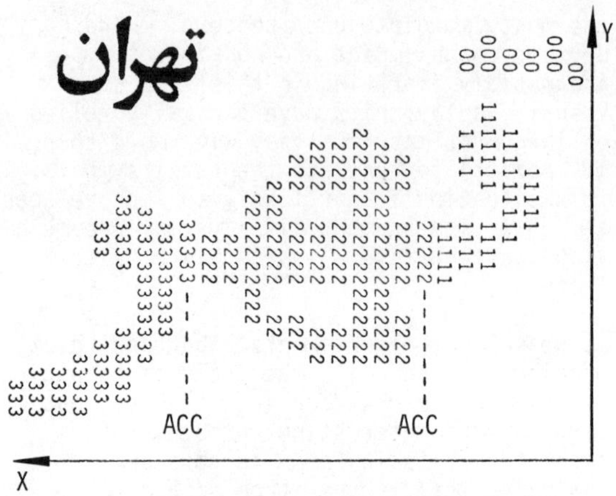

Fig. 3. Seperation of symbols within a subword in the digitized text. Actual connection columns (ACC's) are selected from a set of potential connection columns (PCC's). Dots are marked and dealt with separately.

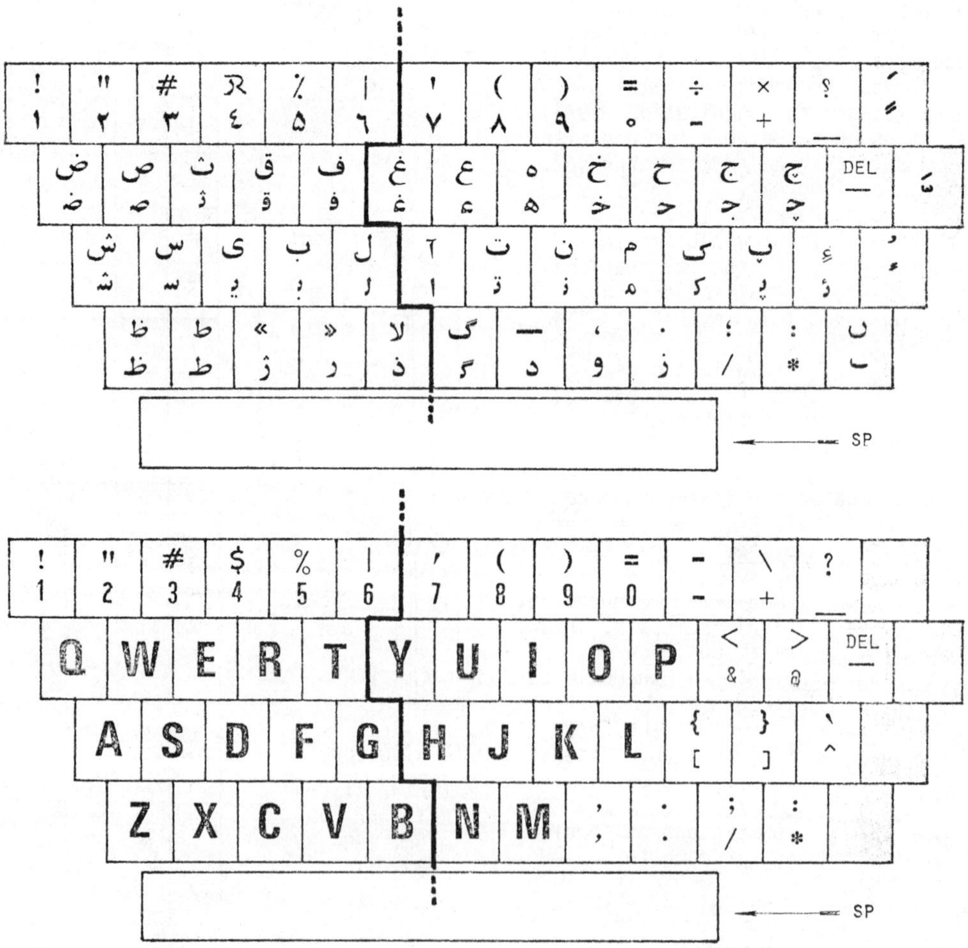

Fig. 4. The proposed layout for Farsi and bilingual (Farsi/Latin) keyboards.

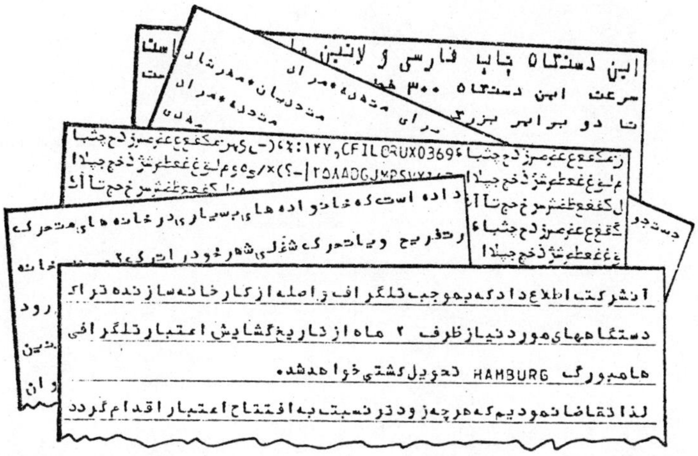

Fig. 5. Samples of hard-copy Farsi output, illustrating some of the difficulties in producing high-quality Farsi output on conventional printers.

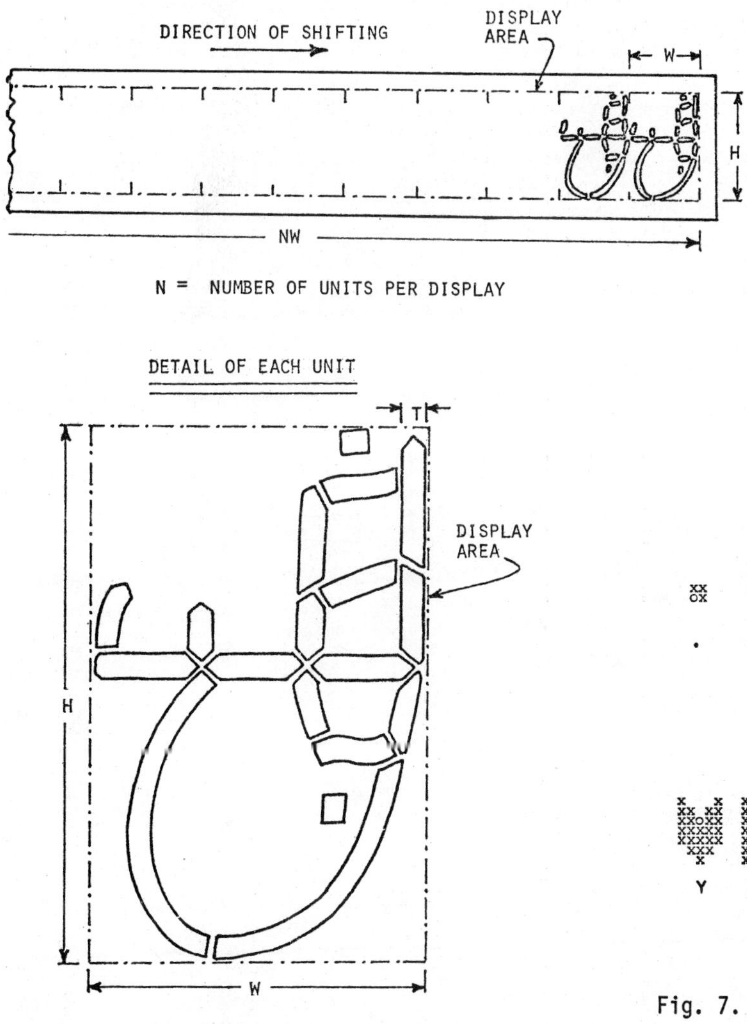

Fig. 6. The proposed line-segment display for Farsi alphanumeric information.

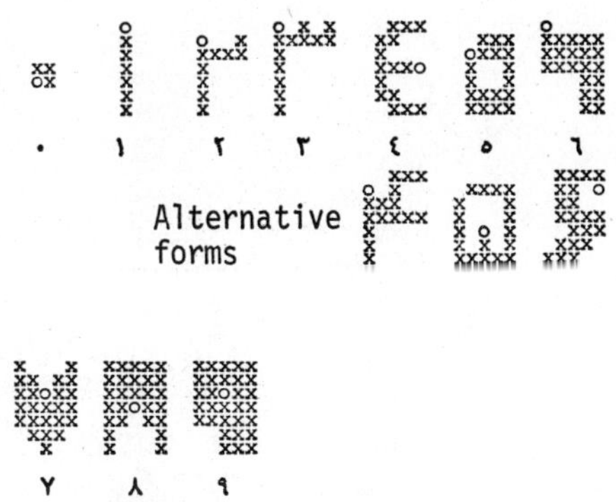

Fig. 7. Optically weighted decimal Farsi and Arabic numerals in a 7 by 5 matrix. The area covered by the numeral x is ax+b.

STATISTICAL DISTRIBUTION OF ARABIC LETTERS AIDS TO THE DESIGN OF A NEW KEYBOARD

S. H. El-Ramly* and M. A. El Hamalawy**

**Department of Electronic and Computer Engineering, Faculty of Engineering, Ain-Shams University, Cairo, Egypt*
***Systems and Computers Engineering Department, Faculty of Engineering, Al Azhar University, Cairo, Egypt*

Abstract. One of the difficulties encountered in treating arabic texts by computer is the low speed of inputing data due to the large number of keys on the keyboard(some of the arabic letters have more than one shape: it ranges from only one shape up to four distinct shapes depending on the previous and the following letters). There are many attempts to reduce the number of such keys. One method is to use an intelligence unit comprising a microprocessor, with a keyboard having only one shape for each letter, with a software program to take care of the determination of the proper letter shape to be typed. However the statistical distribution of letters enables a better arrangement of letters on the board to follow the different degrees of skillness of each finger. The present work treats all these problems, also a new alphabetic code is designed for minimum redundancy of the coded data.

Keywords. Printers; coding; statistics; artificial-intelligence; computer peripheral equipment; man-machine systems; computer interfaces optimal systems.

INTRODUCTION

Many applications of computers in the public survicing centers such as banks, insurance companies, etc... have shown the necessity to input the data in the native language of each country. The use of arabic language in communication with the computer is in growth and one of the problems encountered is the problem that some of the arabic letters have more than one shape; actually the number of different shapes can go to four of them. The large number of keys on a a keyboard increases the difficulty of feeding data by increasing the probability of error(due to confusion) and the reduction of the writing speed. This led to the need of a new keyboard having only one key for each letter, managed with an intelligence unit able to choose the proper shape of the character depending on its position in the word (El Ramly and El Hamalawy 1979). Such a problem and also others show the necessity of having the statistical distribution of the arabic letters in written texts.

STATISTICAL DISTRIBUTION OF ARABIC LETTERS

When calculating the distribution of arabic letters one must notice the following:

1. The choice of essays from different sources, different topics and of different authers.
2. Essays of both scientific and litteral nature must be considered.
3. The amount of data to be taken is determined by the study of the amount of error encountered.

Notes (1) and (2) need no comment. As for note (3), we proceed as follows: the measurement of a probability P using a limited sample size N leads to an expected root mean square error given by (Bendat, 1971 and El-Ramly, 1976):

$$e = \frac{1}{\sqrt{NP}} \qquad (1)$$

Thus the root mean square error is a function of both the sample size N and the measured probability P. The error is not the same for all the letters; the least frequent of them are the most difficult to measure.

The probability distribution obtained is tabulated in Table 1, together with the expected root mean square error encountered in the measuement. From the table it can be seen that the maximum value of error is 10% except for two letters (ظ، غ)) , actually it is less than 5% for half the letters.

CODING OF THE ARABIC LETTERS

The processing of arabic texts (storage, translation,etc...) necessitates the use of an efficient code to match the probability distribution of the arabic letters. In other words it is required that shorter codes be attached to more frequent letters and longer codes correspond to less frequent ones, (Reza,1961).

There are many methods of encoding a given probability distribution, the most efficient of them is the Huffman's code. However for encoding languages there exists alphabetical codes which preserves the alphabetical order of dictionaries in addition to the above recommendation for the length of coded messages; also it has the prefix property (i.e. no encoded letters can be obtained from each other by the addition of more symbols). The Gilbert-Moore alphapetical encoding method (Gilbert and Moore 1959) is used and the result is tabulated in Table 1 together with the result of using the Huffman's minimum redundancy code.

The average cost using Huffman's code is found to be 4.083681 digits/letter while that using the alphabetical code is found to be 4.240586 digits/ letter which is so close to that of Huffman.

THE KEYBOARD

The knowledge of the arabic letters distribution led to the layout of a new keyboard. Figure 1 is an illustration of such a keyboard where the keys are ranged in a matrix form

Table 1 :Probability Distribution of Arabic Letters and their Codes.

Letter	Probability	e %	Huffman code	Alphabetic code
space	0.188318	1.4725	00	00
أ	0.151014	1.6475	101	010
ب	0.029286	3.7300	01111	01100
ت	0.059718	2.6150	1000	01101
ث	0.005071	8.9900	11001111	011100
ج	0.013252	5.5500	010101	011101
ح	0.018324	4.7250	110001	011110
خ	0.004908	9.13	11001110	011111
د	0.027814	3.83	01110	100000
ذ	0.008835	6.80	1100110	100001
ر	0.039921	3.20	11011	10001
ز	0.005399	8.70	0110100	100100
س	0.018651	4.675	110010	100101
ش	0.007853	7.21	0110101	100110
ص	0.008835	6.80	1100001	100111
ض	0.004253	9.80	11000000	1010000
ط	0.011616	5.94	010100	1010001
ظ	0.001963	14.435	110000010	101001
ع	0.026014	3.965	01100	10101
غ	0.002617	12.50	110000011	101100
ف	0.022414	4.27	111001	101101
ق	0.014234	5.0355	011011	101110
ك	0.018815	4.66	111000	101111
ل	0.094731	2.0775	1111	1100
م	0.042866	3.0875	11101	11011
ن	0.039430	3.22	11010	110100
هـ	0.025359	4.0125	01011	110101
و	0.044829	3.020	0100	1110
ى	0.072480	2.375	1001	1111

to simplify the description . The new keypoard has the following characteristics:

1. Both Arabic and English alpha - numerals are put together on the keys . The English characters have the same position as for a traditional machine, and are written in the upper half of the key. The arabic alpha-numerals are written in the lower half of the keys.

2. Some of the keys have shift positions, which is devoted for characters written on the right-hand side of keys. The shift position for Arabic mode of operation can be different than that for English mode.

3. The keys to be pressed by the left hand have Arabic characters that have probabilities summing up to 0.516 which is so close to 0.5 which is the condition necessary for the two hands to share the work equally likely. Note that the statistical

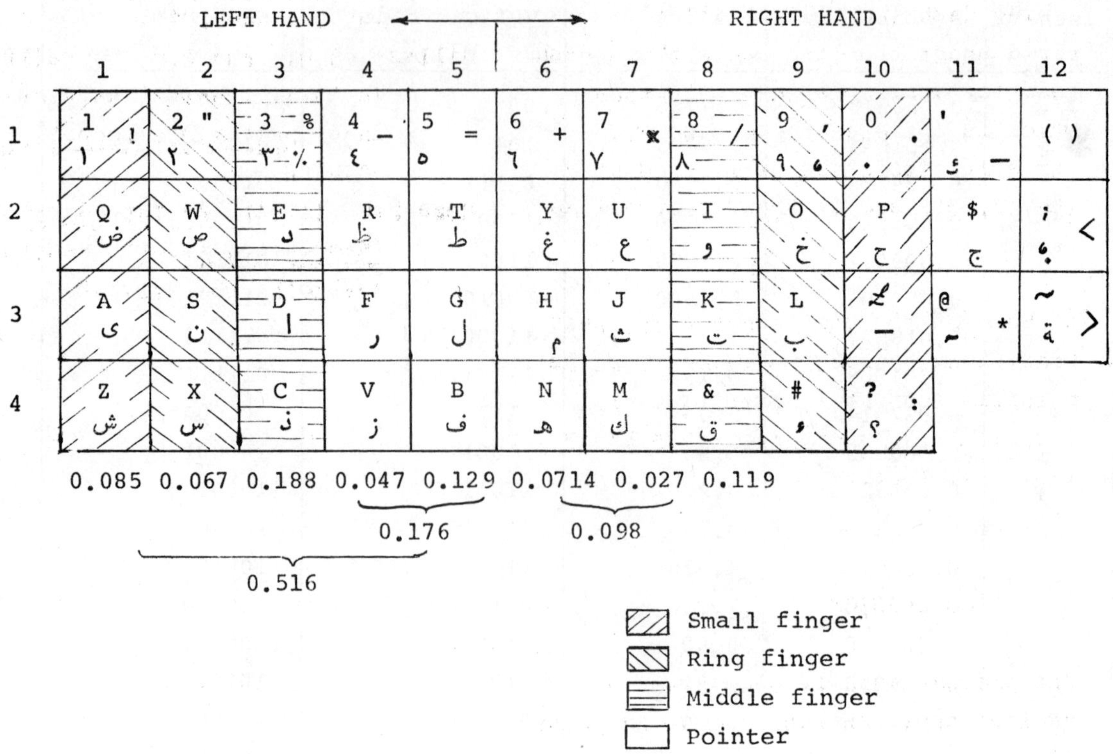

Fig.1. New bilingual keyboard

distribution given in this paper did not include neither the special signs nor the numerals (due to the very large amount of data required to calculate their probabilities). Thus the actual sum of the probabilities of the keys pressed by the left hand may approach 0.5 in a better way.

4. The pointer and the middle fingers are more skill to press the keys than both the ring and small fingers; so each of the first two takes care of characters whose probabilities sum up to 1/6 (one third of a half) while the small and ring fingers together have characters with probabilities summing up to 1/6.

5. For most of the fingers, the character with the higher probability is put in the middle raw of the alphakeys, such as to minimize up-down displacements.

6. There are 32 basic Arabic characters (the Arabic alphabet excluding لا which is composed by the machine intelligence, plus ة , ء , ئ , ~ (Madda). The compound characters ؤ , ئ, أ, لأ, ـأ, آ, لآ, and ـآ are composed by the machine intelligence.

7. Letters of the same group like (ر,ز) and (د, ذ) are assigned to the same finger for simplicity, or they may be assigned to two or three horizontally adjacent keys such as (ج, ح, خ), (ص, ض) (س, ش) (ع, غ) (ب, ت, ث) and (ط, ظ).

8. There are **three** kinds of dashes: the dash used both for English writting and as a "minus" sign in Arabic

and English modes (key(1,4)) which is typed above the line, an underlining dash (key(1,11)) which is typed under the line and a third dash used for arabic letters extention which is optional to be used when variable size characters is adopted (key(3,10)) and is typed at the line level.

9. Some of the characters kept their traditional positions whenever it is possible such as (ض، ص), (غ ، ع), (خ ، ح ، ج), ل or near to it such as (ش، س) , (ى).

CONCLUSION

The present work is essential for various applications, the probability disribution of arabic letters is measured here with a known small amount of error. The code proposed is very near to the minimum redundancy code which increases the transmission effeciency for teletype machines. Also, the suggested keyboard seems to be an optimum one for the number of keys and their positions on the board.

REFERENCES

Bendat, J.S., and A.G. Piersol (1971). Random Data Analysis and Measurement Procedures. John Wiley, New York. pp. 177-179.

El-Ramly, S.H. and P. Duhamel (1978). Prévision de l'erreur statistique dans la mesure de desités de probabilités. L'onde électrique, 58, pp. 375-382.

El-Ramly, S.H. and M.A. El-Hamalawy (1979). Low cost bilingual Microcomputer System. Proc. of 5th Conference on Automatic Control and System Engineering. Cairo, Egypt . Nov.1979.

Gilbert, E.N., and E.F. Moore (1959). Variable length binary encodings. Bell Syst. Tech.J., 38, pp. 933-968.

Reza F.M. (1961). An Introduction to Information Theory. McGraw-Hill, New York. pp. 131-138.

A MODELING TOOL FOR THE SPECIFICATIONS OF A LOGICAL AUTOMATIC SYSTEM: THE GRAFCET (FRANCE)

M. Blanchard*, P. Brard and J. P. Frachet*****

**Valoris S.A., Bordeaux, France*
***Telemecanique Nanterre, France*
****Laboratoire d'Automatique I.S.M.C.M., Saint-Ouen, France*

Abstract. After the Grafcet definition, rules of evolution and its interpretation have been given, we show, how in practice Grafcet makes the dialogue easier between automatism builder and the automatic system users, especially for introducing directions for running and securities of the system.
On the other hand, thanks to its properties and possibilities of analysis which are offered, Grafcet allows to use a methodology of " a priori " verification of specifications.
At last, Grafcet is a tool fitted to a direct synthesis, that is a direct passing from specifications to realization, whatever the techology used (programmed or hard-wired realization, programmable controllers ...). Grafcet becomes a single tool for designing a logical system, from specifications to realization, maintenance, servicing ...).
So, Grafcet is particularly adapted for C.A.D. We present main parts of such a system.

Keywords : Control system, graph theory, modelling, parallel processing, requirements, sequential machines, specifications.

INTRODUCTION

An automatic system can be broken into two parts :

- Control unit or automatism,
- Operative unit process.

Specifications of the automatism can be built up by considering 2 levels :

- Fonctionnal specifications (First level),
- Technological and operational specifications (second level).

Grafcet is a graphic tool adapted for representing the specifications of the first level or second level. It offers a methodology of verification of specifications and can also lead to a direct synthesis method for the control unit.

SPECIFICATIONS OF A LOGICAL AUTOMATISM.

Operative unit - Control unit

Generally speaking, an automatic system can be broken down into two cooperative parts : one is said to be the operative unit , the other the control unit. (Fig. 1)

The operative unit performs certain operations (transforming raw components into machined parts, vertically moving the cabin from one floor to another) when it receives orders to that effect from the control unit. Thanks to reports (position, ...) sent back by the operative unit, the control unit is kept informed of the progress of the completed operations.

In addition to this dialogue of means of orders and reports with the operative unit, the control unit exchanges information with elements outside the system, (pilot, user, supervisor,...) from whom it receives instructions and to whom it gives visual or acoustic reports.

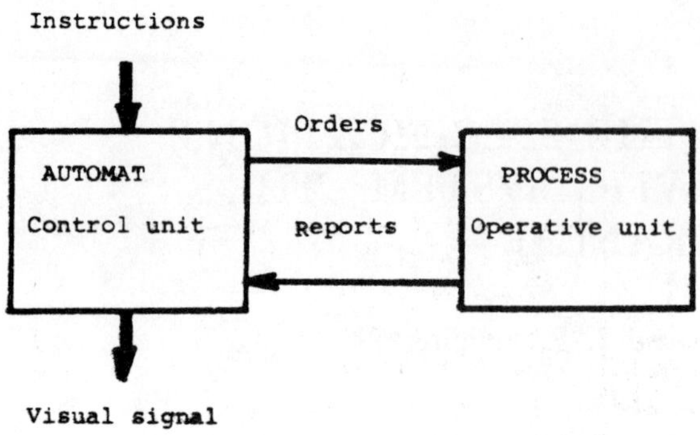

FIG 1

The operative unit is the physical process which is to be automated. The control unit is the automatic equipment which in output elaborates orders destined for the process and visual signals for the operator, according to the reports received from the process and the instructions received in input. We shall limit ourselves here to the study of logic automats for which the processed data have a binary character. The specification manual of an automatic equipment is the description of its behaviour with respect to the development of its environment, that is to say not only its inputs but also its general conditions of use.

A GRADUAL APPROACH TO THE SPECIFICATION MANUAL OF THE CONTROL UNIT

Specifications should furnish the automation engineer responsable for the design and production of the control unit, a clear, precise, unambiguous and complete description of the function and performance expected of the equipment to be designed.

To attain such a goal, it will be found useful to divide the description in two successive and complementary levels.

- The first level describes the behaviour of the control unit in relation to the operative unit : this is the role of the functional specifications which enable the designer to understand what the automatic equipment must do when confronted with the different situations it is likely to have to deal with.

The functional specifications characterise the automat's reactions to information coming from the operative unit, so that the designer may understand what will be the role of the control unit to be constructed. They therefore have to clearly and precisely define the different functions, information and controls involved in the automation of the operative unit, without prejudicing the technology to be used.

- The second level adds the technological and operational specifications to the above functional requirements i.e. the details that are absolutely essential for the running of the equipment.

The technological specifications define the manner in which the automatic equipment will have to be physically incorporated in the unit constituted by the automatic system and its environment. These are the details which have to be added as a complement to the fonctional specifications so as to enable the design of an automat that will well and truly control the operative unit.

By seriating the problems, functional on the one hand, technological on the other, such an approach allows the reader not to be submerged by a flood of details which are of more nuisance than use.

The operational specifications concern the regular running of the automatic equipment during its lifetime. These are considerations related to the equipment, once it is completed and in operation : reliability, absence of dangerous break-downs, availability, possibility of modifying the equipment in the case of alterations taking place in the operative unit, ease of maintenance, man-machine dialogue, ... Such considerations, of prime importance to the user of the process to be automated because of their repercussion on the economic level, are often underestimated in the specification manual. Though sometimes difficult to express quantitatively, they nonetheless have repercussions on the way the equipment is designed and produced.

GRAFCET ELEMENTS

The GRAFCET is a tool for describing the specifications of the control unit of an automated system, to be used both at levell and at level 2.
The working of the automatic equipment may be described by a group of:
- STEPS to which are associated certain ACTIONS.
- TRANSITIONS to which are associated certain "RECEPTIVITIES".
- DIRECTED LINKS linking the steps to the transitions and the transitions to the steps.

Steps.

A step corresponds to a situation in which the behaviour of all or part of the system with respect to its inputs and outputs in variant.

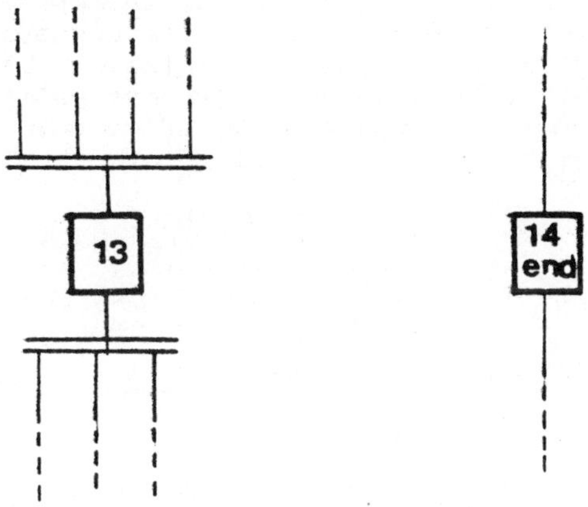

FIG. 2

The STEP is represented by a square or a rectangle identified numericaly ; a symbolic name may be added, representing the principal function of the step (e. g. WAIT, END, SYNCHRONISATION ...) (Fig. 2).

A step is either active or inactive and at a given moment the situation of the automated system is entirely defined by the total set of active steps. For each step are defined the actions to be carried out which are characteristic of this situation. Such actions may only be effective when the step is active.

These actions may be very different in nature. (Fig. 3 a)

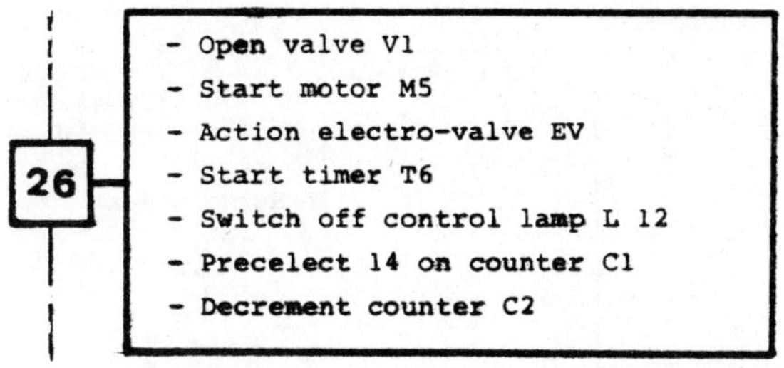

FIG. 3 a

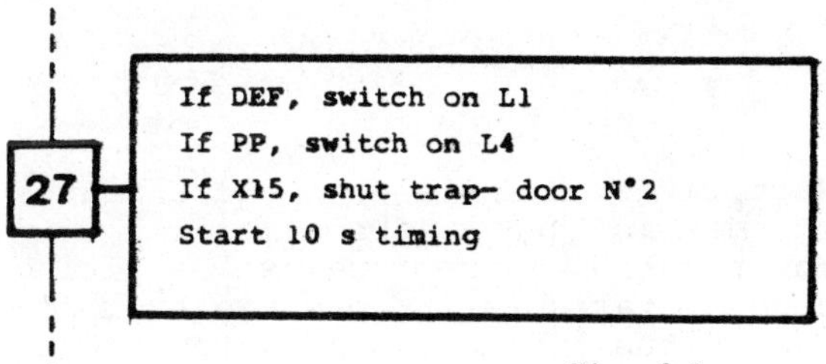

Fig. 3 b

Moreover, the carrying out of these actions may be subject to other logical variables, as a function of input variables,of internal variables orof the active or inactive state of the other steps. (Fig. 3 b)

Transitions

The TRANSITIONS indicate the possibilities of development between step.
A transition is represented by a link line (Fig. 4 a). When more links come to or from a transition, they group in two parallel lines (Fig. 4 b).
To each transition is associated a logical condition called RECEPTIVITY, which singles out among all the available information only these elements which, at a given moment, are capable of making the control unit to change.

The RECEPTIVITY, written in the form of a logical proposition, is a boolean function of external information (orders from the operator, state of pickups, counters, timers,...), of auxiliary variables or of the active or inactive state of other steps.
These receptivities may be expressed in different forms :

e.g. - upright position of moving body.
- temperature > 300° C
- Value of counter C10 > programmed value.
- Button "m" has been pressed 3 times.

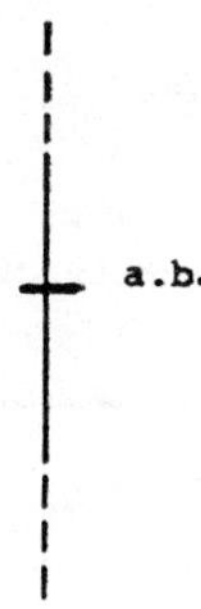

Fig. 4a

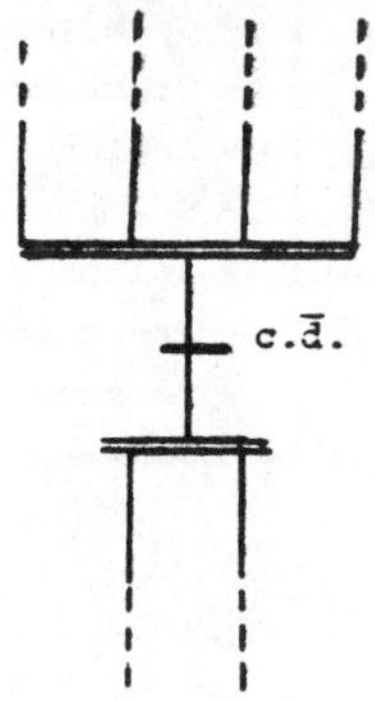

Fig. 4 b

Receptivities may also include status changes in the variables :
The notation a↑ represents the rising edge of variable a (passage from logic status "0" to logic status "1") and the notation y↓ represents the falling edge of variable y (passage from logic status "1" to logic status "0") (Fig. 5 a).

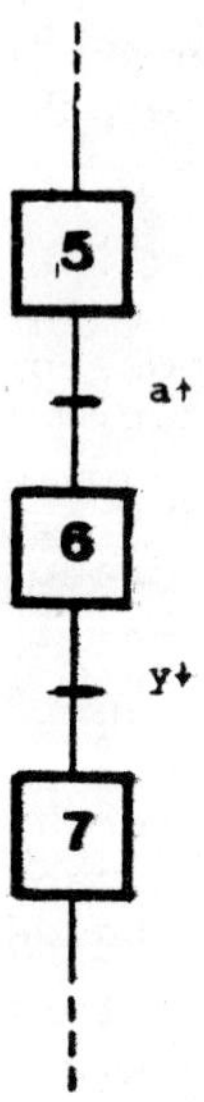

Fig. 5 a

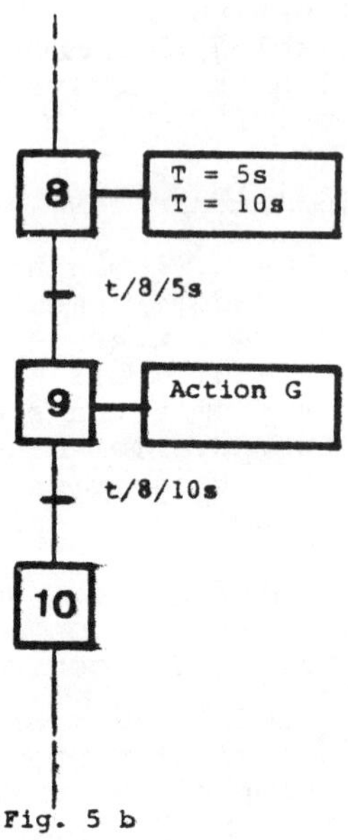

Fig. 5 b

e.g. the notationt/8/10s means 10 seconds passed since last activation of step 8.

In order to include the time factor in a receptivity, it is sufficient to indicate its origin and duration after mark t ; the origin will be the moment at which the last activation of a previous step takes place. When a step originates a time period, it can be useful to indicate the latter as an action associated to the step. (Fig. 5 b)
N.B. A receptivity which is always true is written = 1.

Directed links.

The links indicate the lines of development of the state of the Grafcet. The general direction of a description is from top to bottom and from left to right hand. Generally, input and output at or from a step are represented vertically, the arrival being above.

Rules of evolution

The active or inactive nature of each of the steps having necessarily to change, the 3 previous concepts cannot suffice to define a Grafcet.
It is necessary in addition to establish a set a rules of evolution

RULE 1
INITIALISATION specifies the active steps at the beginning of the operation. The initial steps are unconditionally activated and marked on the Grafcet by doubling the sides on the corresponding symbols. (Fig. 6)

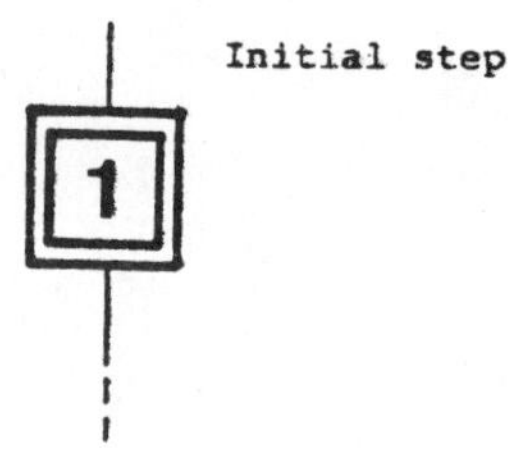

Fig. 6

RULE 2
A TRANSITION is either enabled or disabled. It is enabled when EACH immediately preceeding step is active. It cannot be cleared unless

- It is enabled
- AND its associated receptivity is true.

The TRANSITION is then necessarily cleared

RULES 3
The clearing of a TRANSITION leads to the activation all the steps immediately following and entails the disactivation of all the steps immediately preceeding.
Rule 2 and 3 are illustred on Fig 7

RULE 4
When several transitions may be cleared simultaneously, they will be cleared simultaneously.

RULE 5
If during operation the same step must be simultaneously activated and disactivated, it remains activated.

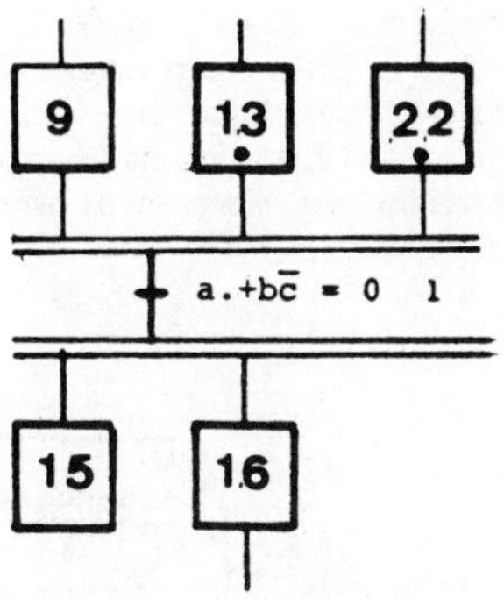

<u>Disabled transition</u> : The transition is disabled, step 9 being inactive.

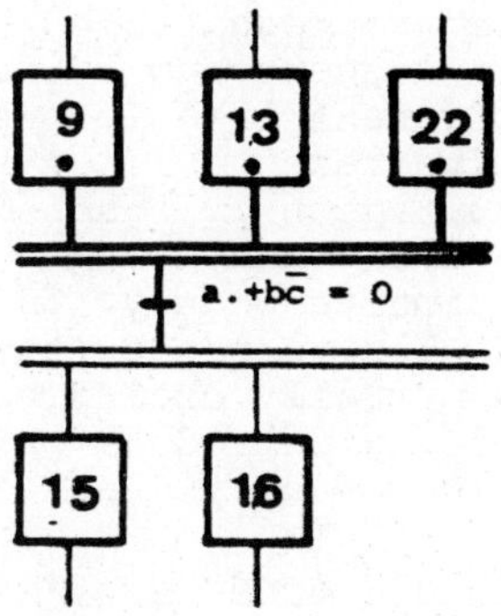

<u>Enabled transition</u> : The transition is enabled, steps 9, 13 and 22 being activy, but may not be cleared because receptivity a+bc̄ = 0.

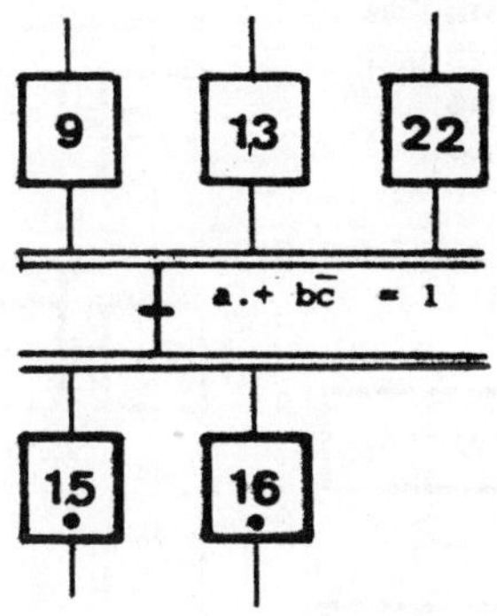

<u>Cleared transition</u> : The transition is cleared as soon as the receptivity a+bc = 1

Fig. 7

N.B. The clearing time of a transition can never be exactly zero, even though theoritically (rules 3 and 4) it may be made as short as one may wish. The same goes for the activations times of a step. Besides , Rule 5 is seldom found in practice ; these rules have been made out for internal theoritical consistency in the Grafcet.

<u>Some particular structures</u>

Switches :

Conditional selection between several possible sequences.

A Grafcet is generally made up of several sequences, that is to say of several series of steps to be performed one after the other and it is often necessary to exclusively select one of these sequences (Fig. 8).

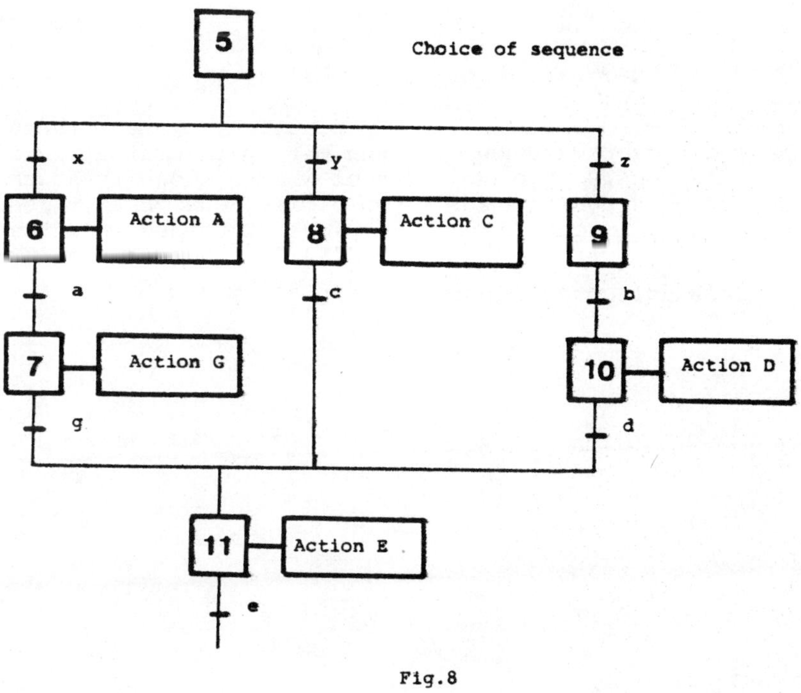

CSAD - U**

Fig.8

In the switch represented by the selection of a sequence to be carried out, the different transitions correponding with receptivities x, y, and z being simultaneously enabled by the same step 5 could according to rule 4 be cleared simultaneously. In practice one is often led to render these receptivities mutually exclusive. It is also possible to introduce priorities (For example $x = a$, $y = \bar{a}b$, $z = a\bar{b}$).

Step jumping and sequence rerunning

The conditional jump is a special switch allowing the jumping of one or more steps when the actions to be carried out are no longer of use, while rerunning a sequence allows the same sequence to be repeated one or more times as long as a fixed condition has not been obtained (Fig 9a -9 b).

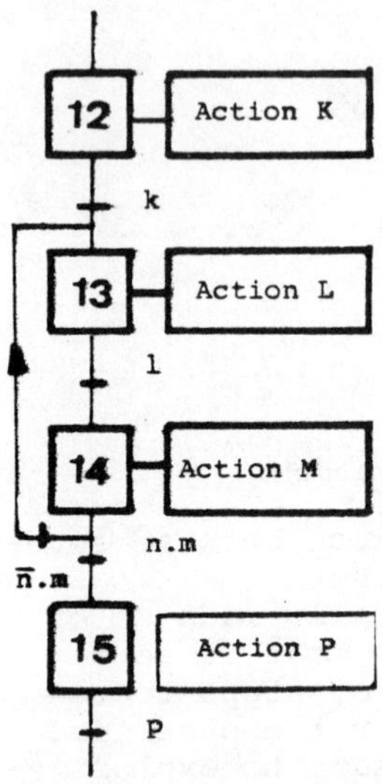

Rerun sequence 13-14 by receptivity $\bar{n}.m$ as long as receptivity n.m is not obtained.

Fig. 9.b

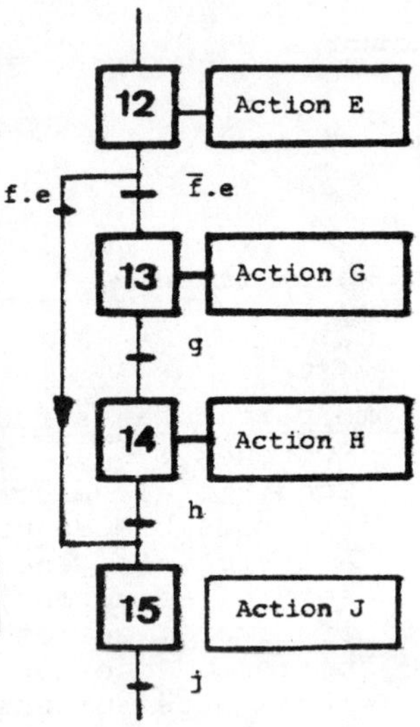

Jump from step 12 to step 15 by receptivity f.e.

Fig. 9.a

Simultaneous sequences

A Grafcet may include several sequences operating simultaneously, the development of the active steps in each branch however remaining independant.

In order to represent these simultaneous operations, a single transition and two parallel lines mark the beginning and end of the sequences, that is to say the simultaneous activation of the branches thus formed and their reciprocal waiting for a common sequence (Fig. 10).

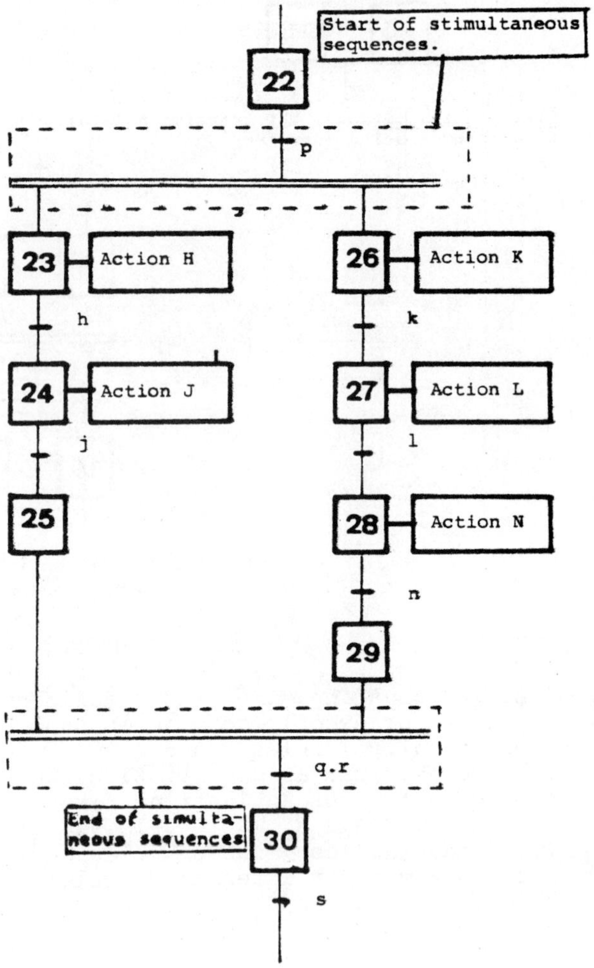

Fig. 10

In practice, these structures are often met, yet, Grafcet is a powerful tool which allows to describe complex specifications, but it does not exist typical structure which can be classified for each sort of problem.

(Fig. 11)

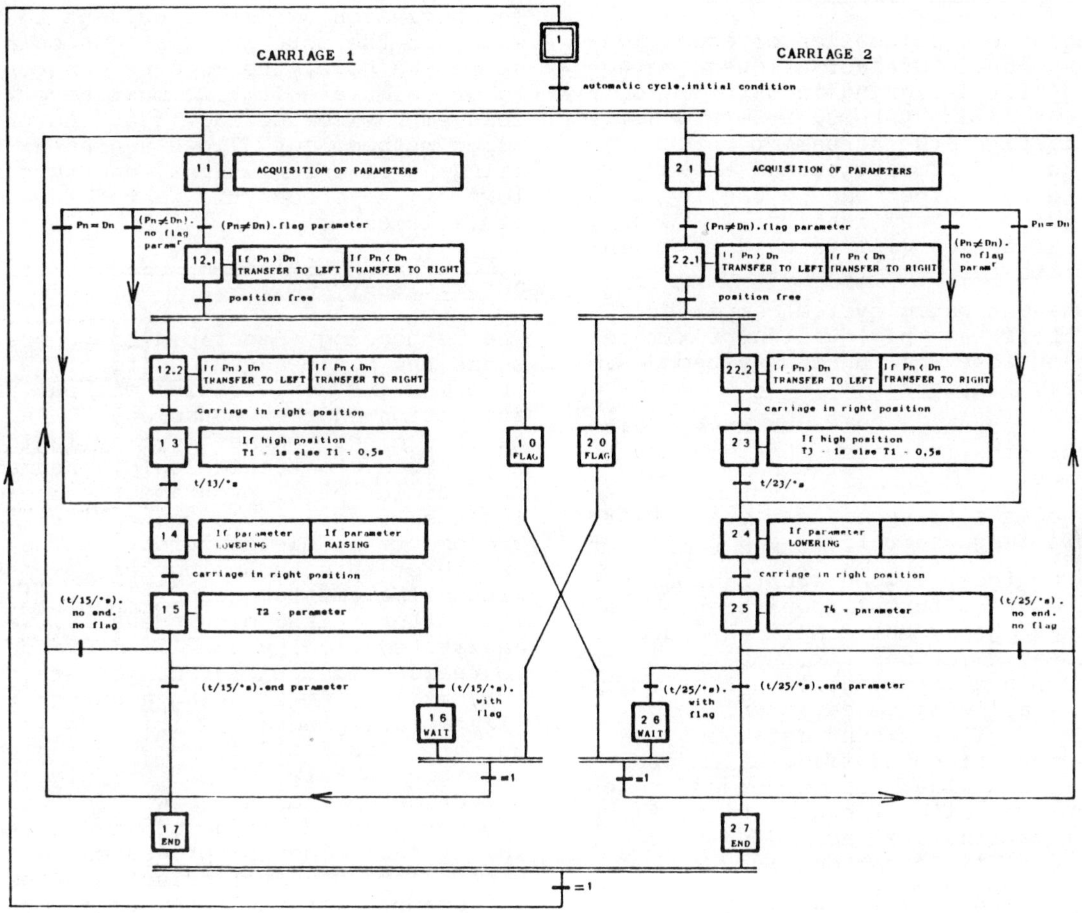

SURFACE TREATMENT

Fig. 11

Example of a GRAFCET concerning a relatively complex Automatic system.

A METHODOLOGY OF DESIGN OF THE CONTROL UNIT BASED ON GRAFCET

Verification of specifications

Beyond the modelisation of specifications manual, Grafcet allows a methodology for designing industrial logical automatic systems. Prior to all realization, the agreement between the desired properties and these described by Grafcet can be checked by a " a priori " verification. Therefore, a verification will be said to be incomplete for some reasons as :

- The automatic system components, and their assembly, will approximate more or less the running system designed by Grafcet.

- The operative unit can have a different behaviour from that Grafcet assumes.

- The initial specifications, prior to writting of Grafcet, can be incomplete.

With Grafcet, a good arbitration between great formal requirements (allowing a good, but complicated and dull verifications for industrial problems) and important liberties on the form (allowing to say every thing with a risk of misunderstanding, making the prior verification impossible). The wide, but rigourous, rules of Grafcet allow the gradation of the verification level according to the considered informations :

- The first level can be obtained by considering the intrinsic form of the graph, without considering the interpretation ("Autonomous" Grafcet). Forms probably illegal can be found on considering the graph of accessible situations : reactivation of steps, impossibility in the structure to return in previous situations, locking of evolution, instability in certain situation. In practice, at this level, a verification is complex : it gives large graphs of accessible situations and it is often difficult to find a physical interpretation of detected abnormalities.

- A second level of verification can be obtained with additionnal informations about the receptivities associated to transitions, and considering these which are similar or disjunctive. A decrease of the dimension of the really accessible situations graphs can be obtained.

- At a third level, these considerations can be extended to the disjunctions of receptivities linked to "statical" physical impossibilities on the operative unit (end of stroke in or out of a cylinder, for example). Moreover, we can consider the "dynamic" disjunctions linked to causal incompatibilities and thus obtain a true simulation of the operative system.

In conclusion, if it is evident to say that the more informations on a system we have, the more we can verify or simulate it ; we must nevertheless emphasize the fact that the proposed methodology allows a gradual bringing of informations, and thus allows us, to choose the wanted level of verification.

Grafcet and realization of industrial automatic system

The methodology described above presents the basic advantage to allow description and verification without any assumption on the technology to realize it (electromagnetic relay logic, hard-wired electronic logic, pneumatic or fluidic logic, programmable logic included programmable controller, microcomputers, minicomputers). The choice of this technology can be made with all elements needed to optimize it. After this choice, usually, we can systematically pass from Grafcet to realization. Either components of automatic sytem allowing a direct realization or programs helping us in using " classical" components can be possible. Lately, we effectively observe the emergence of two kinds of products : hard-wired modules of steps in any technology or programmable controllers code-op including steps for example for the first case, or for the second case, programs allowing to make programs for "classical" programmable controllers.

GRAFCET, BASIC TOOL FOR AN EVOLUTED SYSTEM OF C.A.D. FOR LOGICAL INDUSTRIAL AUTOMATIC EQUIPMENTS

The methodology, described above, allows us to consider a developed and modular system of C.A.D. for automatic systems. In such a system, which is now specifying by ADEPA*, Grafcet is essentially an input language and allows to make a standard internal representation of specifications. In a first time of working, one or more processors can work on the internal representation for,at first, enabling these specifications according to the indicated principles above, whereas an editing processor allows to plot or to modify graphs. In a second time of working, the operator will have to choose in a "bank", the realization module corresponding to the chosen technology to make the automatic system. The module will set up programs for programmable controllers, the lists and (or) diagrams necessary to realize the system from this standard internal representation.

CONCLUSION

Grafcet is a powerful tool, easy to use, with a lot of possibilities. Not only it is appropriated to specification of logical automats, but as we have shown it, it can become the base of designing from specifications to material realization.
In France, it begins to be widely used by important manufacturers of automatic components and numerous users.
French National Ministry of Education has officially chosen it to teach automatisms, and consequently uses it for examinations at various level.
A draft of French standardisation including the rules of Grafcet is now writting, and a discussion is under way at I. E.C. (International Electrotechnical Commitee) for a more general standardisation of Grafcet.

* ADEPA : Agence Nationale pour le Développement de la Production Automatisée 13 - 17 Rue Périer 92120 Montrouge FRANCE

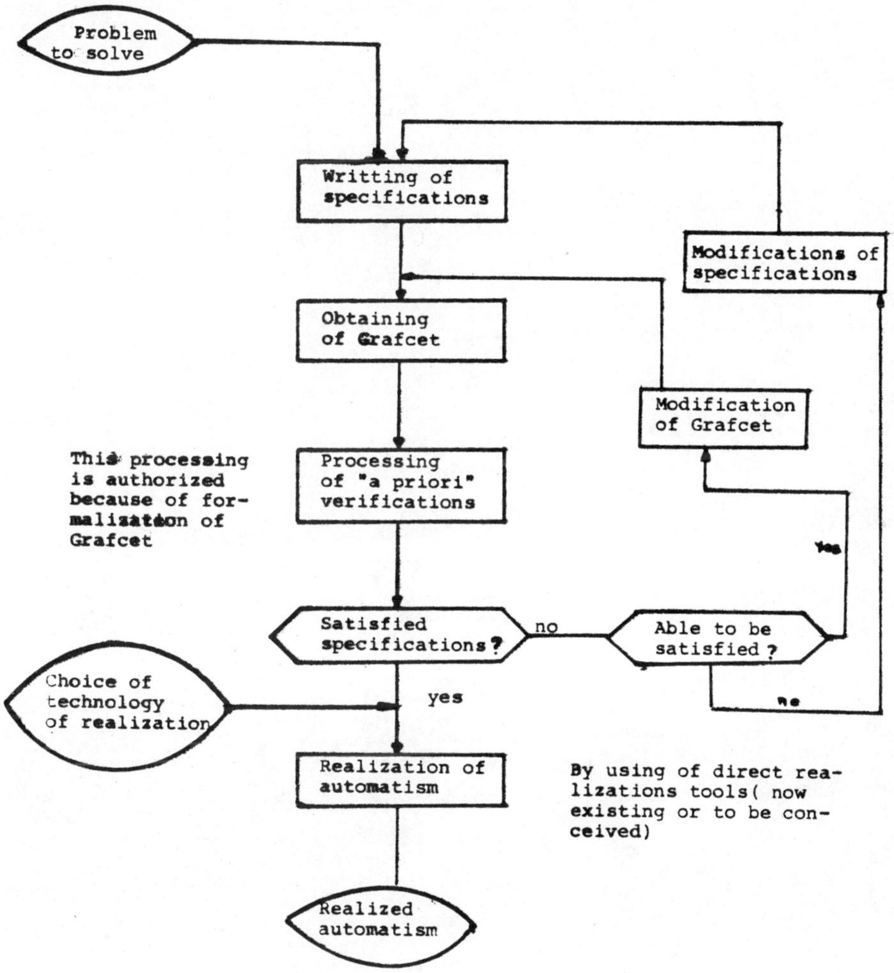

Fig. 12

LSPEST—SUBROUTINE PACKAGE FOR IDENTIFICATION AND PARAMETER ESTIMATION OF LINEAR DYNAMICAL MODELS

S. B. Maleshkov, S. I. Markov, I. G. Kalaikov and N. E. Madzarov

Department of Automatic and Remote Control, Higher Institute of Electrical and Mechanical Engineering, Sofia, Bulgaria

Abstract. In this paper a Fortran-IV subroutine package and programme system for identification and parameter estimation of linear dynamical models is disscused. Both continuous and discrete models are considered and the programmes are divided into two groups, according to the type of the model. Least squares method is the theoretical basis of the programme system. For parameter estimation of linear discrete models one of the following methods can be used: ordinary least squares method, generalized least squares, instrumental variables, least squares with correction and correlation method with least squares, which provide unbiased estimates of the parameters. The programme system provides determination of the model structure, choosing the optimal one at which the loss function is minimized. Continuous model parameter estimation is based on a modification of the generalized functions method. Users can include programmes based on their own modification of the least squares method and realize one of the following operations: data preprocessing and statistical characteristics calculations, printing and plotting all intermediate and final results, calculating impulse and step response of the optimal model, diagnostic test of the optimal model for adequacy. LSPEST consists of more than 50 modules and is organized in a way similar to the IBM scientific subroutine package and can be used on computers such as IBM 360/370 and ES, operating under DOS or OS. The realized programme system unites all the modules by using overlay structure, which provide better use of the computer and helps the user in realization of various binds of computing schemes.

Keywords. Parameter estimation; identification; optimisation; models; programme package.

INTRODUCTION

Lots of methods for parameter estimation of linear discrete models are known. The greatest part of these methods are based on the classical least squares procedure. The least squares method (LSM) requires minimal a priori information about the process to be identified and its environment. This universality in most cases is at the expense of the quality of the parameter estimates. In order to obtain unbiased estimates various modifications of LSM (Aström and Eykhoff, 1971; Isermann and coworkers, 1974; Baur, 1976; Madzarov, Maleshkov and Kalaikov, 1977; Rastrigin, Madzarov and Markov, 1978) can be applied. As a rule a digital computer is used because of the great number of computations. There are publications (Clarke, 1973; Baur, 1976) in which computer programmes for parameter estimation of linear models are given. In this paper a programme package built on a modular principle is considered, which allows the user to input the data and order the consequense of computations in a convenient manner. In many cases it is necessary to obtain a continuous model of the process. For this purpose in the package a possibility is provided for parameter estimation of such a model by the use of the modulating functions method.

METHODS FOR DETERMINATION OF THE STRUCTURE AND THE PARAMETER ESTIMATES OF LINEAR DISCRETE MODELS

The process to be identified is described by a linear difference equation

$$y_m(t)+\sum_{i=1}^{n} a_i y_m(t-i)=\sum_{i=1}^{n} b_i u(t-i-k), \quad (1)$$

where $u(t)$ is the input, $y_m(t)$ - the output of the model. The identification problem consists in estimation of the vector $\theta^T=(a_1,a_2\dots a_n, b_1,b_2\dots b_n)$ and in determination of the order n and the time delay k.
By introducing operators

$$z^{-i}u(t)=u(t-i),$$

$$A(z^{-1})=\sum_{i=1}^{n} a_i z^{-i}, \quad (2)$$

$$B(z^{-1})=\sum_{i=1}^{n} b_i z^{-i},$$

Eq. (1) can be written in the following way

$$\left[1+A(z^{-1})\right] y_m(t) = z^{-k}B(z^{-1})u(t). \quad (3)$$

The LS estimate $\hat{\theta}$ is found according to the condition for minimum of the loss function

$$J = \sum_{t=n+k+1}^{N} \varepsilon^2(t) = \varepsilon^T\varepsilon = \min, \quad (4)$$

where N is the number of observations and $\varepsilon(t)$ are residuals, defined by

$$\varepsilon(t)= \left[1+A(z^{-1})\right]\left[y(t)-y_m(t)\right] , \quad (5)$$

where $y(t)$ is the output of the process. From Eqs. (3) and (5) follows that

$$\varepsilon(t)= \left[1+A(z^{-1})\right] y(t)-z^{-k}B(z^{-1})u(t). \quad (6)$$

The residuals $\varepsilon(t)$ are correlated even if the observation noise $v(t)=y(t)-y_m(t)$ is a discrete white noise.
Getting N observations of the process input and output Eq. (6) can be written as follows:

$$\varepsilon = Y - \phi\theta, \quad (7)$$

where

$$\varepsilon^T= \left[\varepsilon(n+k+1), \dots \varepsilon(N)\right] ,$$
$$Y^T= \left[y(n+k+1), \dots y(N)\right] , \quad (8)$$
$$\phi^T= \left[\varphi^T_{n+k+1},\varphi^T_{n+k+2}, \dots \varphi^T_N\right] ,$$
$$\varphi(t)= \left[-y(t-1), \dots -y(t-n), u(t-k-1), \dots u(t-k-n)\right] .$$

The LS parameter estimate $\hat{\theta}$ is computed from

$$\hat{\theta} = (\phi^T\phi)^{-1}\phi^T Y \quad (9)$$

By substituting Eq. (7) into Eq. (9) the mathematical expectation of the estimate can be obtained

$$\hat{\theta} = \theta + (\phi^T\phi)^{-1}\phi^T\varepsilon,$$

$$E\{\hat{\theta}-\theta\} = E\{(\phi^T\phi)^{-1}\phi^T\varepsilon\} . \quad (10)$$

In more cases the estimate $\hat{\theta}$ is nonconsistent. There are well known modifications of the classical LSM, which allow to get unbiased parameter estimates, when the residuals are correlated. More of these methods assume a minimization of quadratic loss function

$$J = \varepsilon^T G \varepsilon , \quad (11)$$

where G is a weighting matrix. Then the estimate $\hat{\theta}$ is obtained by

$$\hat{\theta} = (\phi^T G \phi)^{-1}\phi^T GY . \quad (12)$$

If substitute $G=(\phi^T)^{-1}W$ into Eq.(12) it follows that

$$\hat{\theta} = (W^T\phi)^{-1}W^T Y . \quad (13)$$

If the elements of W are uncorrelated with $\varepsilon(t)$ and the matrix $E\{W^T\phi\}$ is nonsingular, the estimate $\hat{\theta}$ from Eq. (13) is unbiased. This is the well known instrumental variables method (Aström and Eykhoff, 1971; Isermann and co-workers, 1974; Rastrigin, Madzarov and Markov, 1978).
Unbiased estimates can be computed if the following loss function is minimized

$$J = \eta^T\eta \quad (14)$$

where $\eta(t)$ is a discrete white noise, obtained by a tranformation of the correlated residuals with a whitening filter

$$\left[1+C(z^{-1})\right]\varepsilon(t) = \eta(t) , \quad (15)$$

where

$$C(z^{-1}) = \sum_{i=1}^{m} c_i z^{-i}$$

$$c^T = \left[c_1,c_2\dots c_m\right] .$$

This modification is known as generalized least squares method (Clarke, 1967, 1973; Isermann and co-workers, 1974; Madzarov, Maleshkov and Kalaikov, 1977).
The bias of the estimate in LSM can be compensated by estimating the term in Eq. (10), which determines the bias (Hsia, 1976; A book of algorithms and programmes, 1980).
Unbiased parameter estimates can be obtained if the LSM is combined with correlation analysis of the process input and output. There are several modifications of this approach (Isermann and co-workers, 1974; Baur, 1976) in which a nonparametric model is used as a first step of the identification.
The order n and the time delay k can be determined by analysis of the behaviour of the loss function J. If n increases, the loss function J decreases monotonously. As a measure of the decrement of J when the number model parameters is increased from n_1 to n_2 the following test-quantity can be used

$$T = \frac{J_1 - J_2}{J_2} \cdot \frac{N - n_2}{n_2 - n_1} \qquad (16)$$

where J_1 and J_2 are the minimal values of J for models with n_1 and n_2 parameters. If the residuals are uncorrelated and $N \to \infty$, then the test quantity T has assimptotical F-distribution with degrees of freedom $N-n_2$ and n_2-n_1. For N 100 and a 5% level T must have a value not less than 3.0 in order that the decrement of J may be significant when the model order n is increased by one.

For a fixed order n the model time delay k is determined by searching the minimal value of J (starting with k=0 and stopping at the first minimum of J).

Diagnostical test of the final model can be done by cheeking the hypothesis that the residuals $\varepsilon(t)$ are uncorrelated. It is shown (Bon and Jenkins, 1970) that the quantity

$$Q = N \sum_{t=1}^{q} r_\varepsilon^2(t) , \qquad (17)$$

where $r_\varepsilon(t)$ is the normalized correlation function of $\varepsilon(t)$, is approximately χ^2distributed, if $\varepsilon(t)$ are uncorrelated. For inexact parameter estimation that frequently is a result of the correlation of $\varepsilon(t)$, the value of Q increases. Then the adequacy test for the model can be done by comparing the computed value from Eq. (17) with the value taken from χ^2-distribution table for a given level (usualy 5%).

COMPUTATIONAL ALGORITHMS FOR PARAMETER ESTIMATION OF LINEAR DISCRETE MODELS

Least Squares Method

The estimate $\hat{\theta}$ is computed from Eq. (9). The computational procedure consists of:

- formation of the matrices $\phi^T\phi$ and $\phi^T Y$,
- inversion of $\phi^T\phi$,
- computation of $(\phi^T\phi)^{-1}\phi^T Y$.

In order to reduce the required memory the matrices $\phi^T\phi$ and $\phi^T Y$ are composed by the estimates of the correlation functions:

$$\phi^T\phi = \begin{bmatrix} \hat{R}_y & -\hat{R}_{yu} \\ -\hat{R}_{yu}^T & \hat{R}_u \end{bmatrix} , \quad \phi^T Y = \begin{bmatrix} -\hat{R}_y \\ \hat{R}_{yu} \end{bmatrix} \qquad (18)$$

The standart deviation of θ is

$$\sigma_{\hat{\theta}} = \sigma_\varepsilon \sqrt{\text{diag}[(\phi^T \phi)^{-1}]} \qquad (19)$$

where σ_ε is the standard deviation of $\varepsilon(t)$.

The LSM is implemented by the subroutine LSQD.

Generalized Least Squares Method

In this method the process input and output are transformed using a whitening filter of the residuals - Eq.(15)

$$\tilde{u}(t) = [1+C(z^{-1})]u(t)$$
$$\tilde{y}(t) = [1+C(z^{-1})]y(t) \quad . \qquad (20)$$

Then from Eq. (7) follows that

$$\eta = \tilde{Y} - \tilde{\phi}\theta , \qquad (21)$$

where the matrices $\tilde{Y}$ and $\tilde{\phi}$ have the same structure as in Eq. (8), but are formed with filtered data.

The operator $C(z^{-1})$ of the whitening filter is a priori unknown. The unknown parameters c_i are usualy estimated via LSM, using expression analogous to Eq. (9) and taking into consideration the fact that

$$\varepsilon = \eta - \Psi c \quad , \qquad (22)$$

where

$$\eta^T = [\eta(n+k+1), \ldots \eta(N)]$$
$$\Psi^T = [\omega^T(n+k+1), \ldots \omega^T(N)]$$
$$\omega(t) = [\varepsilon(t-1), \varepsilon(t-2) \ldots \varepsilon(t-m)] \quad .$$

Box and Jenkins (1970) have shown, that the estimates $\hat{c}_i$ can be computed from the following recursive expressions

$$\hat{c}_l(l) = \begin{cases} \hat{r}_\varepsilon(1), & l=1; \\ \dfrac{\hat{r}_\varepsilon(l) - \sum_{i=1}^{l-1} \hat{c}_i(l-1)\hat{r}_\varepsilon(l-i)}{1 - \sum_{i=1}^{l-1} \hat{c}_i(l-1)\hat{r}_\varepsilon(i)} , & \end{cases} \qquad (23)$$

$$l=2, 3, \ldots m;$$

$$\hat{c}_i(l) = \hat{c}_i(l-1) - \hat{c}_l(l)\hat{c}_{l-i}(l-1) , \quad i=1, 2, \ldots l-1.$$

Thus the inversion of a matrix can be avoided and volume of the access memory can be reduced.

The recursive procedure (23) starts with l=1 and the order of the whitening filter is incremented in each iteration by one. Usually $m \leq n$, more frequently $m=1 \div 3$.

The whole procedure of the generalized LSM consists of the following steps:

1. Estimate the parameters using LSM (model M_o)
2. Compute the residuals ε_o for the model M_o (Eq. 6)
3. Analyze ε_o - if they are correlated estimate the parameters of the filter $\hat{c}_o$
4. Filter the process input and output to get $\tilde{u}_1$ and $\tilde{y}_1$

5. Estimate the parameters using $\tilde{u}_1$ and $\tilde{y}_1$ (model M_1)
6. Repeat the procedure from step 2. to step 5 (models M_2, ...) until the estimates become stable or any other condition for ending of the procedure is fulfiled.

The generalized LSM is implemented by the subroutine LSGD.

Instrumental Variables Method

In most of the cases the matrix W in Eq. (15) is chosen to have the same structure as matrix ϕ putting there the model output $y_m(t)$ instead of the process output $y(t)$. The procedure is iterative and it is realized in the following sequence

1. Estimate the parameters using LSM (model M_o)
2. Compute the model output $y_m(t)$
3. Form the matrices $W^T\phi$ and W^TY
4. Compute the estimates according to Eq. (13)
5. Repeat the procedure from step 2. to step 4. until the estimates become stable or any other condition for ending of the procedure is fulfiled.

The structure of the matrices $W^T\phi$ and W^TY is as follows

$$W^T\phi = \begin{bmatrix} \hat{R}_{y_m y} & -\hat{R}_{y_m u} \\ -\hat{R}_{yu} & \hat{R}_u \end{bmatrix}, \quad W^TY = \begin{bmatrix} -\hat{R}_{y_m y} \\ \hat{R}_{yu} \end{bmatrix}$$

The instrumental variables method is implemented by the subroutine IVAD.

LSM with Correction

Unbiased estimates of the parameters can be obtained if one adds up the bias to the computed estimates $\hat{\theta}$. From Eq. (10) and (22) follows that

$$\hat{\theta} = (\phi^T\phi)^{-1}\phi^T\eta - (\phi^T\phi)^{-1}\phi^T\psi\, c + \theta \tag{24}$$

and the corrected estimate $\hat{\theta}^*$ is

$$\hat{\theta}^* = \theta + E\{\hat{\theta} - \theta\} = \hat{\theta} + (\phi^T\phi)^{-1}\phi^T\psi\, \hat{c}\ . \tag{25}$$

It is assumed that the residuals $\eta(t)$ perform a discrete white noise, so that $E\{(\phi^T\phi)^{-1}\phi^T\eta\} = 0$.

The procedure includes the following steps

1. Estimate the parameters using LSM (model M_o)
2. Compute the residuals ε_o for the model M_o.
3. Compute the matrix $\phi^T\psi$
4. Compute the estimate $\hat{c}_o$
5. Compute the correction $(\phi^T\phi)^{-1}\phi^T\psi\,\hat{c}_o$
6. Compute the corrected estimate

$$\hat{\theta}^* = \hat{\theta}_o + (\phi^T\phi)^{-1}\phi^T\psi\,\hat{c}_o .$$

7. Repeat the procedure from step 2. to step 6. until the estimate becomes stable or any other condition for ending of the procedure is fulfiled.

The LSM with correction is implemented by the subroutine LSMD.

Correlation Analysis with LSM

It is known that the discrete model (1) can be replaced by an equivalent one in which the substitutions $y(t) \doteq R_{yu}(t)$ and $u(t) \doteq R_u(t)$ are made, i.e.

$$R_{yu}(t) + \sum_{i=1}^{n} a_i R_{yu}(t-i) = \sum_{i=1}^{n} b_i R_u(t-k-i) \tag{26}$$

If one considers several values of t an expression for the residuals of the model (26) can be obtained

$$e = P - S\theta . \tag{27}$$

In this case the residuals e include the errors from inexact computation of the correlation functions (in the matrices P and S). The exact structure of P and S can be seen in (Isermann and co-workers, 1974).

The procedure consists of only two steps

1. Compute the estimates of the correlation functions $\hat{R}_{yu}(t)$ and $\hat{R}_u(t)$
2. Use $\hat{R}_{yu}(t)$ and $\hat{R}_u(t)$ in the above stated LS procedure.

This method is implemented by the subroutine CORD.

In table 1 a comparative analysis of the above mentioned methods is given, which is based on two computer indexes - execution time and required access memory.

TABLE 1 Comparative Analysis

Method	Exec. time	Memory
LSQD	13%	58%
LSGD	100%	100%
IVAD	20%	65%
LSMD	60%	95%
CORD	25%	65%

PROGRAMME PACKAGE FOR PARAMETER ESTIMATION OF LINEAR DISCRETE MODELS

The parameter estimation of linear discrete models is realized by a specialized programme package LSPEST. In order to facilitate the user, a main

programme is made, by which the most frequent problems of identification can be solved. The main programme gives the possibility to be expanded including additional subroutines, for example subroutines for statistical analysis of data. Thus the user can use this package together with other ones.
The subroutines in LSPEST are made in a similar way as in the IBM Scientific Subroutine Package. The user can include every subroutine in his own main programme. In order to realize better logical and functional coordination between the subroutines and the main programme additional operators such as printing of intermediate results are introduced.
An important feature of the main programme (and of the whole package) is that the execution of the programme is controlled by directives, such as selecting of the method for parameter estimation, indication of the form of the output results, input of specific parameters and so on.
By the main programme the following problems can be solved:

1. Estimation of the structural parameters of the model (order n and time delay k)
2. Parameter estimation of the model.

These two problems are associated with one another. It is impossible to obtain the model parameters without a knowledge of the structure, and the structure is determined by comparing a sequence of models with different structure and analyzing the loss function.

The subroutines in the package LSPEST can be divided into three groups:

1. Subroutines for estimation of structural parameters
2. Subroutines for parameter estimation of the model
3. Auxiliary subroutines grouped as
   a) subroutines, related to the estimation methods
   b) subroutines, related to the organization of the main programme
   c) subroutines to generate test data
   d) some other subroutines

The interaction between the subroutines is in three levels and their execution depends on the introduced by the user directives. Figure 1 shows the flow chart of the main programme, which includes the following basic operators:

The auxiliary subroutine VHOD is executed at the beginning. It realizes several functions - input and decoding of directives, input, printing and graphical output of the values of the process input and output.

The directives must follow in a strict order. So if correct, but put not in the right order directives are used, an error is indicated. Analyzing the directives' content the subroutine gives proper values to control parameters. If an error is detected, a message is printed and the programme stops. Its execution can continue if all of the directives are correct and are put in the right order.

The next operation in the subroutine VHOD is the input of the discrete sequencse of the process input and output. This is done in the user's subroutine INDATA, where the format of the data may be arbitrary chosen and statistical analysis can be done. To print and plot these data the subroutine GRAF (A book of algorithms and programmes, 1980) is used.

The estimation of the structural parameters of the model is realized by the subroutines ORDR (for thr order of the model) and DELY (for the time delay). ORDR has higher priority level than DELY. DELY calls the auxiliary subroutine TECH, which organizes branches to one of the six subroutines, implementing the estimation methods - LSQD, LSGD, IVAD, LSMD, CORD or the user's subroutine LSEM. All of the basic operations in parameter estimation are done by these six subroutines in interaction with the auxiliary ones from the first group. Some results such as model parameters, standard deviation of the estimates, loss function and some other quantities, necessary for structure determination are computed.

The obtained model is tested for adequacy with the subroutine DIAC, which is called in every one of the six subroutines, when the best model is determined. After that an appropriate title and the final results are printed and plotted. According to the introduced directives the model output and the residuals are computed, printed and plotted. By the subroutine IMPR the step response and step impulse response can be obtained.

To determine the sequence of executed operations the user can introduce the following directives:

PRINT	- printing of all intermediate results
NOPRINT	- cancels PRINT
GRAF	- plotting and printing of the process input and output, the model output, the residuals, the step and step impulse response of the model
NOGRAF	- cancels GRAF
LSQD	- least squares method
LSGD	- generalized least squares method
IVAD	- instrumental variables method

LSMD - least squares method with correction
CORD - correlation analysis with LSM
LSEM - user's method
ALL - sequental execution of all methods
ORDER =XX. - the order of the model is given and is equal to XX.
NOORDER - the order of the model must be determined
DELAY =XX. - the time delay is given and is equal to XX.
NODELAY - the time delay must be determined
MAXDEL =XX. - maximum value of the time delay
FLTORD =XX. - the order of the whitening filter is given and is equal to XX.
CRFPTS =XX. - number of correlation function points (in CORD)
MAXITER=XX. - maximal number of iterations to correct the estimates (in LSGD, IVAD, LSMD, LSEM, ALL)

METHODS FOR PARAMETER ESTIMATION OF CONTINUOUS LINEAR MODELS

A continuous model of linear differential equation with constant coefficients is considered

$$\sum_{i=0}^{n} a_i \frac{d^i y}{dt^i} = \sum_{i=0}^{n} b_i \frac{d^i u}{dt^i}, \quad b_o=1 . \quad (28)$$

The coefficient a_o, a_1,... a_n and b_1, b_2,... b_n and the order n are to be estimated using experimental data for the process input u(t) and output y(t) at a fixed time interval $0 \div T$.

Various methods for parameter estimation of differential equations like (28) are applied. The most popular of them are the least squares method-differential approximation, the method of successive integration and the method of modulating functions (Rastrigin, Madzarov, Markov; 1978). In the first two of them the input and the output are to be differentiated. If the process is contaminated by the noise there are some difficulties of technical and of some other nature. The method of the modulating functions is free of these disadvantages. The essence of this method is the following:

A system of basic functions is given (called modulating functions) $\psi_k(t)$, k=1,2,... s, s n, which are linearly independent, differentiable and equal to zero together with their differentials of up to s-1 order on the bondories of the fixed interval $0 \div T$, i.e.

$$\psi_\kappa^{(i)}(0)= \psi_\kappa^{(i)}(T)=0, \quad i=0,1..., s-1. \quad (29)$$

For basic functions the following functions are chosen

$$\psi_\kappa(t)=(\sin\frac{\pi}{T}t)^s, \quad k=1,2..., s. \quad (30)$$

which satisfy the conditions (29). If one multiplies the two sides of Eq. (28) with $\psi_k(t)$ and integrates the result products from 0 to T, for s times successively a system of linear algebraic equations is obtained

$$\sum_{i=0}^{s} \alpha_i Y_{ik} = \sum_{i=1}^{s} \beta_i U_{ik}, \quad k=1,2..., s, (31)$$

where

$$Y_{ik} = (-1)^i \int_0^T y(t) \psi_k^{(i)}(t)dt,$$

$$U_{ik} = (-1)^i \int_0^T u(t) \psi_k^{(i)}(t)dt, \quad (32)$$

$$\alpha_i = \begin{cases} a_i & 0 \leq i \leq n, \\ 0 & n < i \leq s ; \end{cases}$$

$$\beta_i = \begin{cases} b_i & 1 \leq i \leq n, \\ 0 & n < i \leq s . \end{cases}$$

The estimates

$$\hat{a}_i = \hat{\alpha}_i \quad 0 \leq i \leq n,$$
$$\hat{b}_i = \hat{\beta}_i \quad 1 \leq i \leq n,$$

are found from the solution of Eq.(31). The order n of the model is determined from the largest index of the estimates a_i which is not equal to zero ($n=i_{max}$).

The system (31) has solution if its determinant is not equal to zero. Therefore the input u(t) must give information as much as possible in order to excite all the process frequencies.

If u(t) and y(t) are contaminated by additive noise with zero mean and finite variance, the estimates a_i and b_i are assymptotically effective and their variance decreases not slower than $1/\sqrt{T}$.

A PROGRAMME FOR PARAMETER ESTIMATION OF LINEAR CONTINUOUS MODELS

Because of the great difference in the programme realization of the methods for parameter estimation of linear continuous and discrete models, the parameters in a model of the type (28) are estimated by a separate programme XYWT. This programme carries out the following two operations:
1. For given discrete input and output sequences u(t) and y(t) to compute the integrals (32). The time interval T is divided into L equal subintervals

and the integrals (32) are estimated by the following equations

$$\hat{Y}_{ik} = \sum_{j=1}^{L} y(t) \int_{(j-1)T_o}^{jT_o} \psi_k^{(i)}(t)dt, \quad (33)$$

$$\hat{U}_{ik} = \sum_{j=1}^{L} u(t) \int_{(j-1)T_o}^{jT_o} \psi_k^{(i)}(t)dt, \quad (34)$$

$$k = 1,2\ldots, s$$
$$i = 0,1\ldots, s-1, \quad s \geq n^*;$$

where n^* is the supposed order of the model (28).
the discrete interval $T_o = T/_L$ is chosen in a way that the error from the digital integration is minimized. The error is not greater than max $|y(t)-y(t)|T$, respectively max$|u(t)-u(t)|T$, where $y(t)$ and $u(t)$ are stepwise approximations of $y(t)$ and $u(t)$ in the discrete points (zero order hold).

2. To solve the system of algebraic equations (31). If the observations u(t) and y(t) aren't contaminated by a noise, the method of Gauss elimination is used (subroutine DGELG from IBM Scientific Subroutine Package). In the case of noisy observations the least squares method is used (subroutine DMLSS).

By the input of data for the programme XYWT the following possibilities are provided:

1. The realizations of u(t) and y(t) are obtained in a normal exploatation of the process.
2. The step response of the process is obtained.
3. The step impulse response of the process is obtained.

EXAMPLES

Example 1

The programme package LSPEST has been tested on the basis of simulated data using the well known test model II (Isermann and co-workers, 1974)

$$y_m(t)-1.425y_m(t-1)+0.496y_m(t-2) = -0.102u(t-1)+0.173u(t-2) \quad (35)$$

The input signal u(t) is a coloured noise, generated by filtering of a discrete white noise $\eta_1(t)$ with zero mean and finite variance with a discrete filter

$$u(t)-1.75u(t-1)+0.83u(t-2) = \eta_1(t) \quad (36)$$

The output is corrupted by a coloured noise, generated in similar way by a discrete filter

$$v(t)-1.027v(t-1)+0.264v(t-2) = \eta_2(t-1) \quad (37)$$

The ratio of the standard deviation of the noise v(t) and the model output $y_m(t)$ is 15%.

Then the following directives are introduced

```
ALL
PRINT
NOGRAF
NOORDER
NODELAY
MAXDEL =3.
FLTORD =2.
MAXITER=5.
```

The results are shown in table 2.

Example 2

The programme package LSPEST and the programme XYWT have been tested on real data obtained from an experiment on a mazut generator in a normal operation mode. The input of the process is the mazut consumption and the output - the temperature of gasification (Velev, Popov and Ivanov, 1980). The maximal deviation of the mazut flow is $\Delta Q_{max} = 1m^3/h$ and the temperature range is 1200 ÷ 1600 °C. Pseudorandom binary sequense is used as an input signal. 128 samples of the input and output are taken at time interval of 6 s. The input and output signals are shown on Fig. 2.
The programme system LSPEST (by the use of LSQD) gives the following discrete model

$$y_m(t)-1.37y_m(t-1)+0.29y_m(t-2)+0.09y_m(t-3)=-0.37u(t-2)-0.33u(t-3)-0.37u(t-4). \quad (38)$$

The continuous linear model, which is obtained by the use of the programme XYWT has the form

$$-37.8\frac{d^2y(t)}{dt^2} - 1.86\frac{dy(t)}{dt} - 0.02y(t) = 0.01\frac{d^2u(t-8)}{dt^2} + 0.03\frac{du(t-8)}{dt} + u(t-8). \quad (39)$$

These results coincide with the models of the same process, which are obtained by other methods (Velev, Popov and Ivanov, 1980) and are used for control design.

CONCLUSION

A subroutine package and a programme system for identification and parameter estimation of linear dynamical models are exposed in this work. The programme system solves the most frequent problems of identification and the subroutines from the package can be used independent in defferent user's main programmes. The execution of the programmes is controlled by directives. The subroutine package LSPEST is tested with simulated data and after that is applied for identification of many technological processes.

REFERENCES

Astrom, K., P. Eykhoff (1971). System identification - a survey. Automatica, 7, 123-162.

Baur, U. (1976). On-line parameterschätzverfahren zur Identifikation linearer dynamischer Prozesse mit Prozeßrechnern - Entwicklung, Vergleich, Erprobung. Forschungsbericht KFK-PDV 65, Karsruhe.

A book of algorithms and programmes for analysis and control design (1980). I. Tomov 'Ed.', Technika, Sofia, Chap. 3-6.

Box, G., G. Jenkins (1970). Time series analysis, Holden-day, Cambridge. Chap. 2,3.

Clarke, D. (1967). Generalized-least squares estimation of the parameters of a dynamic model. First IFAC Symposium on identification and parameter estimation, Prague.

Clarke, D., D. Dayer, R. Hastings-James, R. Asthon, J. Emery (1973). Identification and control of a pilot-scale boiling rig. 3rd IFAC Symposium on identification and parameter estimation, Hague,PP-1.

Hsia, T. (1976). On least squares algorithms for system parameter identification. IEEE - Trans. Aut. Control, 21, 104-108.

Isermann, R., U. Baur, W. Bamberger, P. Kneppo, H. Siebert (1974). Comparison of six on-line identification and parameter estimation methods. Automatica,10,81-103.

Madzarov, N., S. Maleshkov, I. Kalaikov (1977). The least squares method in estimation of the parameters of linear discrete models. Autom. & Izchsl. Tekh. (Bulgaria), 11, №1, 7-13.

Rastrigin, L., N. Madzarov, S. Markov (1978). Parameter and state estimation of dynamical processes. Technika, Sofia.

System/360 Scientific Subroutine Package (360A-CM-03X), Version III, IBM, Tech. Publ. Dep., N.Y.

Velev,K., P. Popov, K. Ivanov (1980). Identification of a mazut generator's dynamic. Autom. & Izchsl. Tekh. (Bulgaria), 14, №1, 57-61.

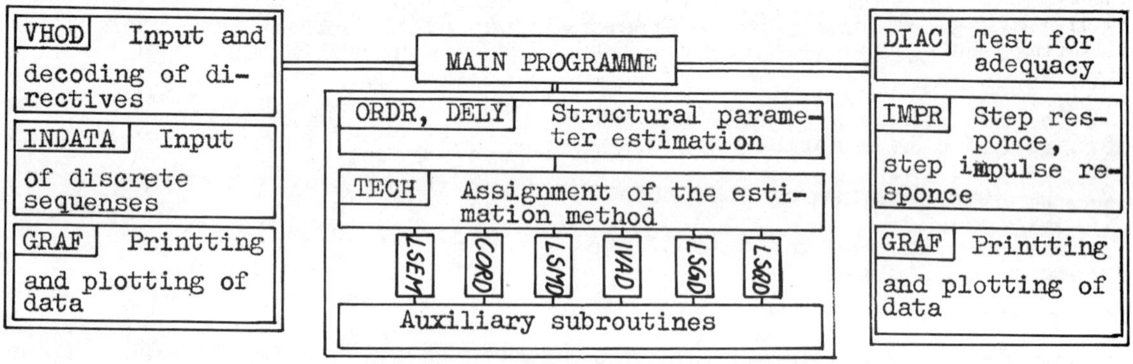

Fig. 1.Flow chart of the main programme.

TABLE 2 Parameter Estimations

Method	$\hat{a}_1$	$\hat{a}_2$	$\hat{b}_1$	$\hat{b}_2$
LSQD	-1.389	0.467	-0.134	0.207
LSGD	-1.411	0.472	-0.121	0.190
IVAD	-1.409	0.451	-0.096	0.177
LSMD	-1.417	0.489	-0.113	0.185
CORD	-1.431	0.509	-0.128	0.201

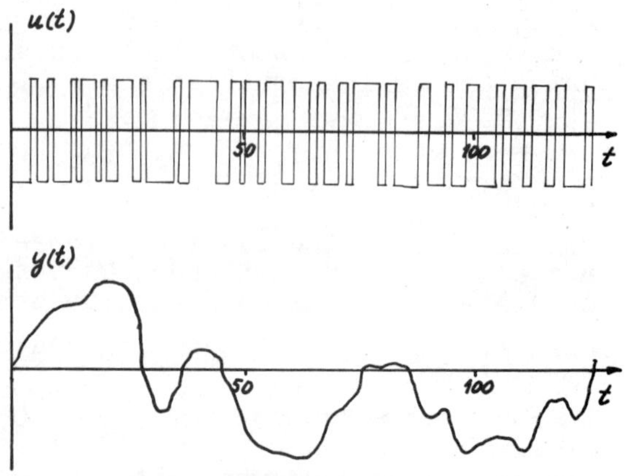

Fig. 2.Input and output signals of the mazut generator.

A MORE GENERALIZED WIENER FILTERING TECHNIQUE

H.-L. Kuo

Shanghai Institute of Technology, People's Republic of China

Abstract: Wiener filtering is a classical technique of signal estimation. It was initially applied to one-dimension continuous signals. Pratt extends it to time sampled signals. In this paper, we start from a more generalized Wiener filtering model to obtain optimum design of the system with minimum mean-square error. The analysis shows that in plotting optimum design more generalized Wiener processing produces the same minimum mean-square error performance as Wiener filtering. The more generalized Wiener filtering processing is also a system model in which the discrete signals, assumed to be composed of signal and noise components, pass through the filtering. The paper gives deduction and discussion on the optimum filtering design formula of the system.

1. INTRODUCTION

In the control system and observed data processing in communication. Serious problems arise when extracting signals from random noise and estimating of random signals. In theory, these problems have already been settled by Wiener's statistical estimation and filtering theory, which is a classical technique of signal estimation. In engineering and techology it is an important contribution. It was initially applied to one-dimension continuous signal. Pratt[2] extends it to time sampled signals. In this paper, on the basis of the above we try to perform Wiener filtering operations on a more general system.

2. THE MATHEMATICAL MODEL

Figure 1 is the block diagram of a one-dimensional more generalized Wiener filtering system. Z is the input of N-vector which is the sum of a data vector x and a noise vector W . The filter A ,B and C are in the form of (N X N) matrices. T and T^{-1} are (N X N) matrices which represent an orthogonal transform and its inverse respectively. $\hat{X}$ is the estimate of X . We will design the filter A , B and C in such a way that the expected value of the mean-square error between X and $\hat{X}$ is minimized.

3. FILTER DESIGN

In general, we can assume that the input signal have the following characteristics:
(1) The data and noise processes have zero means, i.e.

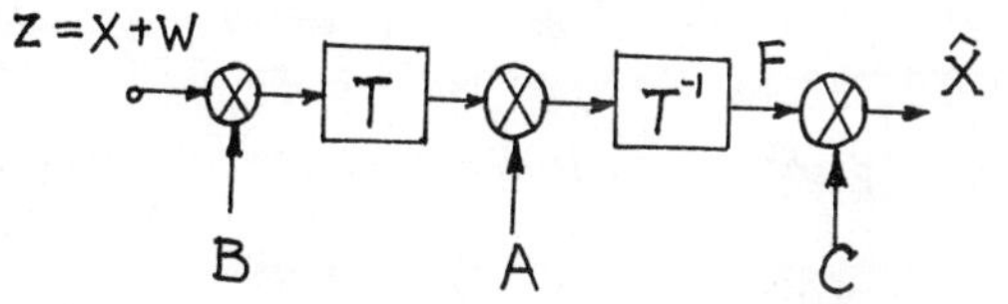

Figure 1. The more generalized Wiener filtering model

$$\overline{X} = \overline{W} = 0 \qquad \text{........(1)}$$

$$\therefore \quad \overline{Z} = 0$$

(2) The data and noise processes are uncorrelated

$$E(X W') = E(W X') = 0 \qquad \text{.......(2)}$$

Since $\hat{X}$ is the estimate of X , the error vector is

$$\mathcal{E} = \hat{X} - X \qquad \text{......(3)}$$

Thus the total mean square error in estimating X is given by

$$\mathcal{E} = E\{\|\hat{X} - X\|^2\} = E\{(\hat{X} - X)'(\hat{X} - X)\} \qquad \text{......(4)}$$

From the model in figure 1, it follows that

$$\hat{X} = C\,T^{-1} A\,T\,B\,Z$$

As the transform T is generally the orthogonal matrix, therefore

$$T^{-1} = T'$$

It follows:

$$\hat{X} = C\,T'A\,T\,B\,Z \qquad ......(5)$$

Substitution of Eq (5) in Eq (4) leads to

$$\mathcal{E} = E\{Z'B'T'A'T\,C'C\,T'A\,T\,B\,Z\} - 2E\{Z'B'T'A\,T\,C\,X\} + E\{\|X\|^2\} \qquad ......(6)$$

If $F = T'ATBZ$, Eq (6) can be expressed as

$$\mathcal{E} = E\{F'C'C\,F\} - 2E\{F'C'X\} + E\{\|X\|^2\} \qquad ......(7)$$

Now, C must be chosen in such a way that Eq (7) is minimum . Let us assume

$$C = C_0 + aV \qquad ......(8)$$

In which C_0 is a desired optimum filter. Hence it must satisfy the following equation.

$$\left.\frac{d\mathcal{E}}{da}\right|_{a=0} = 0 \qquad ......(9)$$

When we substitute Eq (8)in Eq (7) and then derive the gradient according to Eq (9).

$$\begin{aligned}\mathcal{E} &= E\{F'(C_0 + aV)'(C_0 + aV)\,F\} - 2E\{F'(C_0 + aV)'X\} + E\{\|X\|^2\} \\ &= E\{F'C_0'C_0F + aF'V'C_0F + aF'C_0'V\,F + a^2F'V'V\,F\} - 2E\{F'C_0'X + aF'V'X\} + E\{\|X\|^2\}\end{aligned}$$

$$\left.\frac{d\mathcal{E}}{da}\right|_{a=0} = E\{F'V'C_0F + F'C_0'V\,F\} - 2E\{F'V'X\} = 0 \qquad ......(10)$$

We will now use the following relations

$$F'V'X = X'V\,F$$

$$F'C_0'V\,F = F'V'C_0F$$

Eq (10) can be simplified as

$$E\{(X - C_0F)\,F'V'\} = 0 \qquad ......(11)$$

Because Eq (11) is true for any V, so it can be expressed as

$$E\{(X - C_0F)\,F'\} = 0$$

$$C_0E\{F\,F'\} = E\{X\,F'\}$$

Since $F = T'A\,T\,B\,Z$, by substituting it in the above term, we obtain

$$C_0T'A\,T\,B\,E\{Z\,Z'\}B'T'A\,T = E\{X\,Z'\}B'T'A'T \qquad ......(12)$$

consider the relation $Z = X + W$ and Eq (2), we have the following relations

$$E\{X\,Z'\} = E\{X\,X'\} = \Sigma_X$$

$$E\{W\,W'\} = \Sigma_W$$

$$E\{Z\,Z'\} = \Sigma_X + \Sigma_W \qquad ....(13)$$

Substitution of Eq (13) , leads to

$$C_0T'A\,T\,B(\Sigma_X + \Sigma_W)\,B'T'A'T = \Sigma_X B'T'A'T$$

The optimum filter now becomes

$$C_0 = \Sigma_X\{T'A\,T\,B(\Sigma_X + \Sigma_W)\}^{-1} \qquad ......(14)$$

Case 1 :

In the model of Fiqure 1, if the filter C is an orthogonal matrix, then $C'C = I$. Hence Eq (6) can be simplified as

$$\mathcal{E} = E\{Z'B'T'A'A\,T\,B\,Z\} - 2E\{Z'B'T'A'T\,C'X\} + E\{\|X\|^2\} \qquad ......(15)$$

Let $F = T\,B\,Z$ and simplify Eq (15)

$$\mathcal{E} = E\{F'A'A\,F\} - 2E\{F'A'T\,C'X\} + E\{\|X\|^2\} \qquad ......(16)$$

Then A must be chosen in such a way that the mean-square error is minimum, Assume $A = A_0 + aV$, in which A_0 is a desired optimum filter. Hence it must satisfy the following equation.

$$\left.\frac{d\mathcal{E}}{da}\right|_{a=0} = 0$$

By substitution A in Eq (16) and deriving the differential. We obtain,

$$A_0\,E\{F\,F'\} = T\,C'E\{X\,F'\}$$

Put $F = T\,B\,Z$ into the above equation

$$A_0T\,B\,E\{Z\,Z'\}B'T' = T\,C'E\{X\,Z'\}B'T'$$

the optimum filter

$$A_0 = T\,C'\Sigma_X(\Sigma_X + \Sigma_W)^{-1}\,B^{-1}\,T' \qquad ......(17)$$

Case 2 :

If the filter C and A are orthogonal matrices in the model of Figure 1. Then $C'C = I$ and $A'A = I$. The mean-square error of the system is

$$\mathcal{E} = E\{Z'B'B\,Z\} - 2E\{Z'B'T'A'T'C'X\} + E\{\|X\|^2\} \qquad ......(18)$$

The filter B must be chosen in such a way that Eq (18) is minimum . By repeating the above procedure, we obtain

$$B_0E\{Z\,Z'\} = T'A'T\,C\,E\{X\,Z'\}$$

The optimum filter becomes

$$B_0 = T'A'T\, C\, \Sigma_X (\Sigma_X + \Sigma_W)^{-1} \quad ......(19)$$

Case 3 :
In the model of Figure 1, if the filter C and A are identity matrices, i.e. C = I and A = I . The model is reduced to that originally proposed by Wiener. The mean-square error is

$$\mathcal{E} = E\{Z'B'B\, Z\} - 2E\{Z'B'X\} + E\{\|X\|^2\} \quad ......(20)$$

The filter B must be chosen in such a way that Eq (20) is minimum. By repeating the above calculation, we obtain

$$B_0 E\{Z\, Z'\} = E\{X\, Z'\}$$

The optimum filter

$$B_0 = \Sigma_X (\Sigma_X + \Sigma_W)^{-1} \quad ......(21)$$

4. MINIMUM MEAN-SQUARE ERROR

The mean-square error of the filter system can be expressed as

$$\mathcal{E} = tr\{E[(\hat{X} - X)(\hat{X} - X)']\} \quad ......(22)$$

Here "tr" denotes the trace of a matrix, i.e. the sum of the elements along the main diagonal. When we expand Eq (22), we have

$$\mathcal{E} = tr[E(\hat{X}\hat{X}') - E(\hat{X}X') - E(X\hat{X}') + E(XX')] \quad ......(23)$$

Using Eq (3), Eq (13) and Eq (14). We can deduce that

$$E\{\hat{X}\hat{X}'\} = E\{X\hat{X}'\} = \Sigma_X B'T'A'T\, C'$$

$$E\{\hat{X}X'\} = C\, T'A\, T\, B\, \Sigma_X \quad ......(24)$$

Substituting Eq (24) in Eq (23), we obtain

$$\mathcal{E} = tr\{\Sigma_X - C\, T'A\, T\, B\Sigma_X\} \quad ......(25)$$

By using the expression for the optimum filter, i.e. Eq(14) substitution of C_0 for C in Eq (25) leads to the following expression for the minimum mean-square error.

$$\mathcal{E}_{min} = tr\{\Sigma_X - C_0 T'A\, T\, B\, \Sigma_X\}$$

$$= tr\{\Sigma_X - \Sigma_X(\Sigma_X + \Sigma_W)^{-1}\Sigma_X\} \quad ....(26)$$

(i) In Case 1, the minimum mean-square error of the system is

$$\mathcal{E}'_{min} = tr\{\Sigma_X - C\, T'A_0 T\, B\Sigma_X\}$$

substituting Eq (17) and using the condition C'C = I , we obtain

$$\mathcal{E}'_{min} = tr\{\Sigma_X - C\, T'T\, C'\Sigma_X(\Sigma_X + \Sigma_W)^{-1}B^{-1}B\Sigma_X\}$$

$$= tr\{\Sigma_X - \Sigma_X(\Sigma_X + \Sigma_W)^{-1}\Sigma_X\} \quad ....(27)$$

(ii) In Case 2, the minimum mean-square error is

$$\mathcal{E}''_{min} = tr\{\Sigma_X - C\, T'A\, T\, B_0\Sigma_X\}$$

substituting Eq (19) and using the conditions C'C = I and A'A = I . We obtain

$$\mathcal{E}''_{min} = tr\{\Sigma_X - C\, T'A\, T\, T'A'T\, C'\Sigma_X(\Sigma_X + \Sigma_W)^{-1}\Sigma_X\}$$

$$= tr\{\Sigma_X - \Sigma_X(\Sigma_X + \Sigma_W)^{-1}\Sigma_X\} \quad ....(28)$$

(iii) In Case 3, the minimum mean-square error is

$$\mathcal{E}'''_{min} = tr\{\Sigma_X - B_0\Sigma_X\}$$

substituting Eq (21). We obtain

$$\mathcal{E}'''_{min} = tr\{\Sigma_X - \Sigma_X(\Sigma_X + \Sigma_W)^{-1}\Sigma_X\} \quad ....(29)$$

5. CONCLUSION

According the deduction and discussion derived from the more generalized Wiener filtering model. We can obtain some following conclusion.

(1) We observe that for the special case, when filter A and B (or C) are identity matrices, the model in Figure 1 is reduced to that originally proposed by Wiener. The expression for the desired optimum filter is given as

$$B_0(\text{or } C_0) = \Sigma_X(\Sigma_X + \Sigma_W)^{-1}$$

and the minimum mean-square error which associates with this Wiener filter is

$$\mathcal{E}_{min} = tr\{\Sigma_X - \Sigma_X(\Sigma_X + \Sigma_W)^{-1}\Sigma_X\}$$

(2) The minimum mean-square error which associates with the more generalized Wiener filtering model under the optimum design is given by Eq (26). It demonstrated that more generalized Wiener processing and some special cases provide the same minimum mean-square error performance as Wiener filtering.

(3) We can also regard the model in Figure 1 as a system, in which the signal Z , which is assumed to be composed of signal X and noise W components, passes through the filter B , transform T ,the filter A and inverse transform T^{-1} , and then be detected by Wiener filter C . The formula for the optimum filter design is given by Eq (14). The minimum mean-square error is expressed in Eq (26). It is dependent on the signal Z itself and independent of the filtering system.

(4) If filter C is an orthogonal matrix,we choose filter A as Wiener filter. The expression of optimum filter A is given by Eq(17) . Eq(27) shows the minimum mean-square error. It provides the same performance as Wiener

filtering, so the orthogonal matrix which is behind the Wiener filtering does not influences the ε_{min}.

(5) When the filter C and A are orthogonal matrices, we choose filter B as Wiener filter. The expression of optimum filter B is given by Eq (19). Its minimum mean-square error is showed in Eq (28).

REFERENCES

1. Davenport, W. B., Random signals and noise , Ner York 1968.
2. Pratt, W, K., "Generalized Wiener filtering computation technique," IEEE Trans on computers, VOL-C 21 PP 636 , 1972

it follows that the zeros and poles of $G(z)$ are respectively the eigenvalues of $(F-KH)$ and F.
From (20) we deduce that the "inverse optimal filter", whose output is the innovation process $e(k)$, has the transfer function $[1 - H(zI - F + KH)^{-1}K]$.
The characteristic parameters of the innovation model (14) constitutes the quadruplet $\mathcal{M} = (F, H, K, \sigma^2)$, with $\dim(\mathcal{M}) = n^2 + 2n + 1$.

By comparing the sets $\mathcal{P}$, $\mathcal{N}$ and $\mathcal{M}$ which are equivalent from a spectral point of view, we can conclude that the minimal number of parameters to represent the PSD $\Phi_{yy}(z)$ is equal to $(2n+1)$ with n dynamical parameters and $(n+1)$ statistical parameters. Consequently the equivalent ARMA model appears as a minimal representation for the signal $y(k)$, while the two other models (1) and (14) are not minimal in general (i.e. they use a number of parameters which is not minimal). However the transformations introduced in this section will allow finding minimal forms to these models.

• We now choose $\mathcal{L} = \mathcal{L}_2$ as the stationary solution P of the Lyapunov equation (3).
Then we have :

$$\left.\begin{aligned} Q_{\mathcal{L}_2} &= 0 \\ R_{\mathcal{L}_2} &= HPH^T + R \\ S_{\mathcal{L}_2} &= FPH^T + S \end{aligned}\right\} \quad (21)$$

Using equations (4) and defining

$$\left.\begin{aligned} \mathcal{O} &= \begin{bmatrix} H \\ HF \\ \vdots \\ HF^{n-1} \end{bmatrix} \quad \text{(observability matrix)} \\ \Gamma &\triangleq FPH^T + S = \mathcal{O}^{-1}\begin{bmatrix} \varphi_{yy}(1) \\ \vdots \\ \varphi_{yy}(n) \end{bmatrix} \end{aligned}\right\} \quad (22)$$

($\mathcal{O}^{-1}$ exists if the system is assumed completely observable).
The covariance Λ takes the form

$$\begin{bmatrix} Q_{\mathcal{L}_2} & S_{\mathcal{L}_2} \\ S_{\mathcal{L}_2}^T & R_{\mathcal{L}_2} \end{bmatrix} = \left[\begin{array}{c|c} 0 & \Gamma \\ \hline \Gamma^T & \varphi_{yy}(0) \end{array}\right] \quad (23)$$

If we replace (Q,R,S) by $(Q_{\mathcal{L}_2}, R_{\mathcal{L}_2}, S_{\mathcal{L}_2})$ in the Riccati equation (9) we get

$$\left.\begin{aligned} &\tilde{P}_{\mathcal{L}_2}(k+1) = F\tilde{P}_{\mathcal{L}_2}(k)F^T - \\ &-(F\tilde{P}_{\mathcal{L}_2}(k)H^T+\Gamma)(H\tilde{P}_{\mathcal{L}_2}H^T+\varphi_{yy}(0))^{-1}(F\tilde{P}_{\mathcal{L}_2}(k)H^T+\Gamma)^T \end{aligned}\right\} (24)$$

and by defining $\hat{P}(k) = -\tilde{P}_{\mathcal{L}_2}(k)$ (25)

equation (24) is nothing but the Riccati equation of the algorithm of Faurre (1973) to compute the Kalman gain

$$\left.\begin{aligned} &\hat{P}(k+1) = F\hat{P}(k)F^T + \\ &+(\Gamma - F\hat{P}(k)H^T)(\varphi_{yy}(0) - H\hat{P}(k)H^T)^{-1}(\Gamma - F\hat{P}(k)H^T)^T \end{aligned}\right\} (26) \; .$$

Consequently, as mentioned above this special transformation $\mathcal{L}_2 = P$ leads to the stochastic realization algorithm of Faurre, but it must be noticed that no state-space model can be associated with the triplet $(Q_{\mathcal{L}_2}, R_{\mathcal{L}_2}, S_{\mathcal{L}_2})$.

- Finally we can remark that $\mathcal{L}$ only acts on the statistical part (Q, R, S) of $\mathcal{P}$ and this fact will be used in the next section to transform the triplet (Q,R,S) into a suitable minimal form for identification.

★ Transformation $\mathcal{T}$

- That corresponds to the following transformation in the state equations (1)

$$x_{\mathcal{T}} = \mathcal{T}x \quad (27)$$

- By simple inspection we have

$$\begin{bmatrix} Q_{\mathcal{T}} & S_{\mathcal{T}} \\ S_{\mathcal{T}}^T & R_{\mathcal{T}} \end{bmatrix} = \begin{bmatrix} \mathcal{T} & 0 \\ 0 & 1 \end{bmatrix} \begin{bmatrix} Q & S \\ S^T & R \end{bmatrix} \begin{bmatrix} \mathcal{T} & 0 \\ 0 & 1 \end{bmatrix}^T \quad (28)$$

and from equation (13) we deduce

$$\left[K_{\mathcal{T}} \, , \, \sigma^2_{\mathcal{T}}\right] = \left[\mathcal{T}K \, , \, \sigma^2\right] \quad (29)$$

- Similarly to transformation $\mathcal{L}$, $\mathcal{T}$ leaves the PSD $\Phi_{yy}(z)$ unchanged.

- In the sequel, we consider a special transformation $\mathcal{T}$ such that $(F_{\mathcal{T}}, H_{\mathcal{T}}) = (F_c, H_c)$ is in the column-companion form, also called the first canonical form of Luenberger.

Let $(F_c, H_c, Q_c, R_c, S_c)$ and $(F_c, H_c, K_c, \sigma_c^2)$ be the sets $\mathcal{P}$ and $\mathcal{M}$ after application of this special transformation $\mathcal{T}$.
The structure of the pair (F_c, H_c) is given by

$$\left.\begin{aligned} F_c &= \begin{bmatrix} 0 & \cdots & 0 & -\lambda_n \\ 1 & \ddots & & \vdots \\ 0 & \ddots & \ddots & 0 \\ \vdots & & \ddots & \\ 0 & \cdots & 0 \;\; 1 & -\lambda_1 \end{bmatrix} \\ H_c &= \begin{bmatrix} 0 & \cdots & 0 & 1 \end{bmatrix} \end{aligned}\right\} \quad (30)$$

and F_c can be written as

$$F_c = J + A_c H_c \quad (31)$$

where

$$J = \begin{bmatrix} 0 & \cdots & \cdots & 0 \\ 1 & \ddots & & \vdots \\ 0 & \ddots & \ddots & \vdots \\ \vdots & & \ddots & \\ 0 & \cdots & 0 \;\; 1 & 0 \end{bmatrix} \quad (32)$$

and

$$A_c^T = \begin{bmatrix} -\lambda_n & \cdots & -\lambda_1 \end{bmatrix} \quad (33)$$

Setting $\lambda_0 = 1$, the special structure of (F_c, H_c) gives

$$H_c(zI-F_c)^{-1} = \frac{1}{\sum_{i=o}^{n} \lambda_i z^{-i}} [z^{-n} \ldots z^{-1}] \qquad (34)$$

and the transfer function of the optimal filter corresponding to the quintuplet $(F_c, H_c, Q_c, R_c, S_c)$ is given by

$$G_c(z) = \frac{\sum_{i=o}^{n} (\lambda_i + k_i) z^{-i}}{\sum_{i=o}^{n} \lambda_i z^{-i}} \qquad (35)$$

where

$$\left.\begin{aligned} K_c^T &= [k_n \ldots k_1] \\ k_o &= 1 \end{aligned}\right\} \qquad (36)$$

This leads to the following result (Favier, 1977):

The equivalent ARMA model to the quadruplet $(F_c, H_c, K_c, \sigma_c^2)$, where (F_c, H_c) is in the column-companion form, is characterized by the set of parameters $(A, B, \sigma^2) = (-A_c, -A_c + K_c, \sigma_c^2)$, or equivalently the parameters (F_c, H_c) are related to the parameters (A, B) by the simple relation :

$$\left.\begin{aligned} F_c &= J - AH_c \\ K_c &= -A + B \end{aligned}\right\}$$

★ Transformation $\mathcal{G}$

- By simple inspection we have

$$\begin{bmatrix} Q_{\mathcal{G}} & S_{\mathcal{G}} \\ S_{\mathcal{G}}^T & R_{\mathcal{G}} \end{bmatrix} = \begin{bmatrix} I & -\mathcal{G} \\ O & 1 \end{bmatrix} \begin{bmatrix} Q & S \\ S^T & R \end{bmatrix} \begin{bmatrix} I & -\mathcal{G} \\ O & 1 \end{bmatrix}^T \qquad (37)$$

and by use of equation (13) we get

$$\left[K_{\mathcal{G}}, \sigma_{\mathcal{G}}^2\right] = \left[K - \mathcal{G}, \sigma^2\right] \qquad (38)$$

- Unlike transformations $\mathcal{T}$ and $\mathcal{L}$, $\mathcal{G}$ corresponds to a modification of the PSD $\Phi_{yy}(z)$ such that the new PSD is given by

$$\Phi_{\epsilon\epsilon}(z) = \Psi(z)\, \Phi_{yy}(z)\, \Psi^T(z^{-1}) \qquad (39)$$

where

$$\Psi(z) = 1 - H(zI - F + \mathcal{G}H)^{-1}\mathcal{G} \qquad (40)$$

Indeed noticing that

$$\Psi(z)\, H(zI-F)^{-1} = H(zI-F+\mathcal{G}H)^{-1} \qquad (41)$$

we have

$$\Psi(z)\, \Phi_{yy}(z)\, \Psi^T(z^{-1}) = \left[H(zI-F+\mathcal{G}H)^{-1}\ \Psi(z)\right] \begin{bmatrix} Q & S \\ S^T & R \end{bmatrix} \begin{bmatrix} (z^{-1}I - F^T + H^T\mathcal{G}^T)^{-1} H^T \\ \Psi^T(z^{-1}) \end{bmatrix} \qquad (42)$$

and after a simple manipulation

$$\Psi(z)\, \Phi_{yy}(z)\, \Psi^T(z^{-1}) = \left[H_{\mathcal{G}}(zI-F_{\mathcal{G}})^{-1}\ 1\right] \begin{bmatrix} Q_{\mathcal{G}} & S_{\mathcal{G}} \\ S_{\mathcal{G}}^T & R_{\mathcal{G}} \end{bmatrix} \begin{bmatrix} (z^{-1}I - F_{\mathcal{G}}^T)^{-1} H_{\mathcal{G}}^T \\ 1 \end{bmatrix} \qquad (43)$$

Consequently the action of $\mathcal{G}$ on $\mathcal{P}$ corresponds to passing the output signal $y(k)$ through the filter with transfer function $\Psi(z)$.

- If $\mathcal{G} = K$, then $\Psi(z)$ corresponds to the inverse optimal filter and $\epsilon(k)$ is the innovation process $e(k)$, (see (8) and (20)).

In this case we have

$$\Psi(z)\, \Phi_{yy}(z)\, \Psi^T(z^{-1}) = \sigma^2 \qquad (44)$$

<u>Definitions</u>:

If $\mathcal{G} = \underline{K} \neq K$, $\underline{K}$ is called a suboptimal gain, $\Psi(z)$ a suboptimal inverse filter and $\epsilon(k)$ a pseudo-innovation process.

Applying successively the transformations $\mathcal{G}$ and $\mathcal{L}_1$, and using equation (38), we deduce the innovation model for this pseudo-innovation process

$$\left.\begin{aligned} x_{\mathcal{G},\mathcal{L}_1}(k+1) &= (F - \underline{K}H)\, x_{\mathcal{G},\mathcal{L}_1}(k) + (K - \underline{K})\, e(k) \\ \epsilon(k) &= H x_{\mathcal{G},\mathcal{L}_1}(k) + e(k) \end{aligned}\right\} \qquad (45)$$

where $x_{\mathcal{G},\mathcal{L}_1}(k)$ represents the difference between the optimal and the suboptimal (obtained by filtering $y(k)$ with $\underline{K}$) one-step predictors of $x(k)$.

It must be noticed that the application of $\mathcal{L}_2$ transformation acting on the quintuplet $(F_{\mathcal{G}}, H_{\mathcal{G}}, Q_{\mathcal{G}}, R_{\mathcal{G}}, S_{\mathcal{G}})$ with $\mathcal{G} = \underline{K}$, leads to stochastic realization algorithms ((Carew, Belanger, 1973) and (Bozzo, 1975)) to compute the Kalman gain K by means of a pseudo-innovation process, see Favier (1977) for a comparison of these two algorithms with that of Faurre (1973).

- An interesting case is the choice $\mathcal{G} = A_c$, when a transformation $\mathcal{T}$ was first applied to $\mathcal{P}$ in such a way that the pair $(F_{\mathcal{T}}, H_{\mathcal{T}})$ is in the column-companion form.

Then we have :

$$\left.\begin{aligned} F_{\mathcal{T},\mathcal{G}} &= F_c - A_c H_c = J \\ H_{\mathcal{T},\mathcal{G}} &= H_c \\ K_{\mathcal{T},\mathcal{G}} &= K_c - A_c = B \end{aligned}\right\} \qquad (46)$$

and

$$\Psi(z) = 1 - H_c(zI-J)^{-1} A_c = \mathcal{A}(z) \qquad (47)$$

so that

$$\Phi_{\epsilon\epsilon}(z) = \mathcal{B}(z)\, \sigma^2\, \mathcal{B}(z^{-1}) \qquad (48)$$

Consequently, in this case $\Psi(z)$ corresponds to the AR part of "equivalent ARMA model" and the signal $\epsilon(k)$ resulting from the application of $\Psi(z)$ is a Moving Average process

$$\epsilon(k) = \mathcal{B}(z)\, e(k) \qquad (49)$$

The application of this transformation $\mathcal{G} = A_c$ corresponds to filtering the output signal $y(k)$ by the AR part of the "equivalent ARMA model", i.e. $\epsilon(k)$ can be computed in the following way

$$\epsilon(k) = \sum_{i=o}^{n} a_i\, y(k-i) \qquad (50)$$

Moreover equations (46) give the following innovation model for this MA process $\epsilon(k)$

(see (Favier,Salut,1979))

$$\left.\begin{aligned} x_{\mathcal{T},\mathcal{G},\mathcal{L}_1}(k+1) &= J\, x_{\mathcal{T},\mathcal{G},\mathcal{L}_1}(k) + B\, e(k) \\ \epsilon(k) &= H_c\, x_{\mathcal{T},\mathcal{G},\mathcal{L}_1}(k) + e(k) \end{aligned}\right\} \quad (51)$$

This innovation model **(51)** is associated with the state-space model **(1)** transformed by application of $\mathcal{T}$ and $\mathcal{G}$

$$\left.\begin{aligned} x_{\mathcal{T},\mathcal{G}}(k+1) &= J\, x_{\mathcal{T},\mathcal{G}}(k) + W_{\mathcal{T},\mathcal{G}}(k) \\ \epsilon(k) &= H_c\, x_{\mathcal{T},\mathcal{G}}(k) + V_{\mathcal{T},\mathcal{G}}(k) \end{aligned}\right\} \quad (52)$$

and

$$\begin{bmatrix} Q_{\mathcal{T},\mathcal{G}} & S_{\mathcal{T},\mathcal{G}} \\ S^T_{\mathcal{T},\mathcal{G}} & R_{\mathcal{T},\mathcal{G}} \end{bmatrix} = \begin{bmatrix} I & A \\ 0 & 1 \end{bmatrix} \begin{bmatrix} Q_{\mathcal{T}} & S_{\mathcal{T}} \\ S^T_{\mathcal{T}} & R_{\mathcal{T}} \end{bmatrix} \begin{bmatrix} I & A \\ 0 & 1 \end{bmatrix}^T \quad (53)$$

III. COVARIANCES IDENTIFICATION ALGORITHMS

As it was shown in previous section the PSD $\Phi_{yy}(z)$ can be represented by means of three sets of parameters $\mathcal{P}$, $\mathcal{N}$ and $\mathcal{M}$. While the "equivalent ARMA model" is directly in canonical form, this is not the case for the two other representations. However the innovation model being characterized by only (n+1) statistical parameters (K,σ^2), a transformation $\mathcal{T}$ can be used to put the dynamical part (F,H) in canonical form like the column-companion form, as mentioned in § II.

The same transformation can be applied in the case of model **(1)** to get (F,H) in the column-companion form, but another transformation would be necessary to yield a canonical form for the triplet $(Q_{\mathcal{T}}, R_{\mathcal{T}}, S_{\mathcal{T}})$ too.
We shall use a transformation $\mathcal{L}$ to put these covariances in Toeplitz form

$$\begin{bmatrix} Q_{\mathcal{T},\mathcal{L}} & S_{\mathcal{T},\mathcal{L}} \\ S^T_{\mathcal{T},\mathcal{L}} & R_{\mathcal{T},\mathcal{L}} \end{bmatrix} = \begin{bmatrix} q_o\, q_1 & \cdots & q_n \\ q_1\, q_o & \ddots & q_{n-1} \\ \vdots & \ddots & \vdots \\ q_n & \cdots\, q_1 & q_o \end{bmatrix} \quad (54)$$

and we shall identify the triplet $(Q_{\mathcal{T},\mathcal{L}}, R_{\mathcal{T},\mathcal{L}}, S_{\mathcal{T},\mathcal{L}})$ under this special structure.
As we have shown the PSD $\Phi_{yy}(z)$ is unchanged by application of the two transformations $\mathcal{T}$ and $\mathcal{L}$, and it can be written as

$$\Phi_{yy}(z) = [H_c(zI-F_c)^{-1}\ 1] \begin{bmatrix} q_o & \cdots & q_n \\ \vdots & \ddots & \vdots \\ q_n & \cdots & q_o \end{bmatrix} \begin{bmatrix} (z^{-1}I - F_c^T)^{-1} H_c^T \\ 1 \end{bmatrix} \quad (55)$$

Taking into account equation **(34)**, we get

$$\Phi_{yy}(z) = \frac{nq_o + \sum_{\ell=1}^{n-1} (n-\ell)\, q_\ell\, (z^\ell + z^{-\ell})}{\left(\sum_{i=o}^{n} a_i z^i\right)\left(\sum_{i=o}^{n} a_i z^{-i}\right)} + \frac{\sum_{j=1}^{n} q_j z^{-j}}{\sum_{i=o}^{n} a_i z^{-i}} + \frac{\sum_{j=1}^{n} q_j z^{j}}{\sum_{i=o}^{n} a_i z^{i}} + q_o \quad (56)$$

Defining

$$\alpha_\ell = \sum_{i=o}^{n-\ell} a_{i+|\ell|}\, a_i\ , \quad \ell \in [-n, +n] \quad (57)$$

and after some manipulation, equation **(56)** can be written

$$\Phi_{yy}(z) = \frac{1}{\sum_{\ell=-n}^{+n} \alpha_\ell z^\ell} \left\{ nq_o + \left[\sum_{\ell=1}^{n-1} (n-\ell)\, q_\ell + \sum_{\ell=o}^{n-1}\left(\sum_{i=1+\ell}^{n} a_i\, q_{-\ell+i}\right) + \sum_{\ell=1}^{n}\left(\sum_{i=o}^{n-\ell} a_i q_{\ell+i}\right)\right](z^\ell + z^{-\ell}) + \sum_{\ell=-n}^{n} \alpha_\ell\, q_o\, z^\ell \right\} \quad (58)$$

By identifying this expression of $\Phi_{yy}(z)$ with the concise form **(6)** we get for $\ell \geqslant o$

$$\left.\begin{aligned} \beta_o &= (n+\alpha_o)q_o + 2\sum_{i=1}^{n} a_i\, q_i \\ \beta_\ell &= (n-\ell)q_\ell + \sum_{i=1+\ell}^{n} a_i\, q_{-\ell+i} + \sum_{i=o}^{n-\ell} a_i\, q_{\ell+i} + \alpha_\ell q_o \\ &\ell \in [1, n-1] \\ \beta_n &= a_o q_i + \alpha_n q_o \end{aligned}\right\} \quad (59)$$

Moreover by use of the "equivalent ARMA model", we can compute the β_ℓ coefficients as

$$\beta_\ell = \sum_{i=o}^{n-\ell} b_{i+|\ell|}\, b_i\ , \quad \ell \in [-n, +n] \quad (60)$$

Then a recursive estimation algorithm for the covariance matrices can be summarized in the following steps:

Algorithm 1

Step 1. Given the dynamical parameters (F,H), use a transformation $\mathcal{T}$ to put this pair into the column-companion form to find the vector $A_c = -A$.

Step 2. Process the data y(k) by use of a transformation $\mathcal{G} = A_c$ to compute the Moving Average part of the "equivalent ARMA model" as

$$\epsilon(k) = \sum_{i=o}^{n} a_i\, y(k-i) \ .$$

Step 3. Use the RML 1 algorithm, originally proposed by (Young,1968) and (Panuska,1969), to estimate the b_i parameters

$$\left.\begin{aligned}
\hat{\theta}(k+1) &= \hat{\theta}(k) + L(k+1)\,\epsilon(k+1)\\
L(k+1) &= \frac{\Sigma(k)\,\varphi^*(k+1)}{1+\varphi^{*T}(k+1)\,\Sigma(k)\,\varphi^*(k+1)}\\
\Sigma(k+1) &= \Sigma(k) - \frac{\Sigma(k)\,\varphi^*(k+1)\,\varphi^{*T}(k+1)\,\Sigma(k)}{1+\varphi^{*T}(k+1)\,\Sigma(k)\,\varphi^*(k+1)}\\
\epsilon(k+1) &= y(k+1) - \varphi^{*T}(k+1)\,\hat{\theta}(k)
\end{aligned}\right\} \quad (61)$$

where

$$\left.\begin{aligned}
\varphi^*(k) &= [\epsilon(k-1) \ldots\ldots \epsilon(k-n)]^T\\
\text{and}\quad \theta(k) &= [a_1 \ldots\ldots a_n]^T
\end{aligned}\right\} \quad (62)$$

Step 4. Compute the coefficients β_ℓ as in **(60)**.

Step 5. Solve the algebraic linear equations set **(59)** with respect to the parameters q_i. Then the triplet $(Q_{\mathcal{T},\mathcal{L}}, R_{\mathcal{T},\mathcal{L}}, S_{\mathcal{T},\mathcal{L}})$ is given by **(54)**.

Step 6. A representation (Q,R,S) corresponding to the original quintuplet(F,H,Q,R,S) is given by the inverse formulae of $\mathcal{T}$ definition, i.e.

$$\begin{bmatrix} Q & S\\ S^T & R \end{bmatrix} = \begin{bmatrix} \mathcal{T} & O\\ O & 1 \end{bmatrix}^{-1} \begin{bmatrix} Q_{\mathcal{T},\mathcal{L}} & S_{\mathcal{T},\mathcal{L}}\\ S^T_{\mathcal{T},\mathcal{L}} & R_{\mathcal{T},\mathcal{L}} \end{bmatrix} \begin{bmatrix} \mathcal{T} & O\\ O & 1 \end{bmatrix}^{-T} \quad (63)$$

Another possible way to estimate the triplet (Q,R,S) is first to apply transformation $\mathcal{T}$ to put (F,H) in the column-companion form, then to apply transformation $\mathcal{G} = -A$ to get the PSD $\Phi_{\epsilon\epsilon}(z)$ and finally to use a transformation $\mathcal{L}$ to put the triplet $(Q_{\mathcal{T},\mathcal{G}}, R_{\mathcal{T},\mathcal{G}}, S_{\mathcal{T},\mathcal{G}})$ in Toeplitz form. This procedure leads to very simple relations for computing the coefficients q_i (Favier, Salut, 1979)

$$q_i = \frac{\beta_i}{n+1-i}, \quad i \in [o,n] \quad (64)$$

This gives another recursive estimation algorithm for the triplet (Q,R,S), which can be summarized in the following steps.

Algorithm 2

Steps 1-4. Identical to steps 1-4 of Algorithm 1.

Step 5. Estimation of the triplet $(Q_{\mathcal{T},\mathcal{G},\mathcal{L}}, R_{\mathcal{T},\mathcal{G},\mathcal{L}}, S_{\mathcal{T},\mathcal{G},\mathcal{L}})$ in Toeplitz form by use of equation **(64)**.

Step 6. Computation of the corresponding triplet $(Q_\mathcal{T}, R_\mathcal{T}, S_\mathcal{T})$ as

$$\begin{bmatrix} Q_\mathcal{T} & S_\mathcal{T}\\ S^T_\mathcal{T} & R_\mathcal{T} \end{bmatrix} = \begin{bmatrix} I & A\\ O & 1 \end{bmatrix}^{-1} \begin{bmatrix} Q_{\mathcal{T},\mathcal{G},\mathcal{L}} & S_{\mathcal{T},\mathcal{G},\mathcal{L}}\\ S^T_{\mathcal{T},\mathcal{G},\mathcal{L}} & R_{\mathcal{T},\mathcal{G},\mathcal{L}} \end{bmatrix} \begin{bmatrix} I & A\\ O & 1 \end{bmatrix}^{-T} \quad (65)$$

Step 7. Estimation of the corresponding triplet (Q,R,S) of the original model

$$\begin{bmatrix} Q & S\\ S^T & R \end{bmatrix} = \begin{bmatrix} \mathcal{T} & O\\ O & 1 \end{bmatrix}^{-1} \begin{bmatrix} Q_\mathcal{T} & S_\mathcal{T}\\ S^T_\mathcal{T} & R_\mathcal{T} \end{bmatrix} \begin{bmatrix} \mathcal{T} & O\\ O & 1 \end{bmatrix}^{-T} \quad (66)$$

However by taking into account the properties of the MA process $\epsilon(k)$ we have

$$\beta_i = \varphi_{\epsilon\epsilon}(i) \quad (67)$$

Under the assumption that the ergodic hypothesis holds, $\varphi_{\epsilon\epsilon}(i) \triangleq \mathcal{E}[\epsilon(k+i)\,\epsilon(k)]$ can be estimated by means of the wellknown ergodic estimator

$$\varphi_{\epsilon\epsilon}(i) = \frac{1}{N} \sum_{k=1}^{N-i} \epsilon(k+i)\,\epsilon(k) \quad (68)$$

This remark leads to a non recursive version of the previous algorithm when replacing steps 3-4 by the computation of the coefficients β_i from equations **(67-68)**.

A block diagram of the resulting estimation algorithms is shown in figure 1.

IV. SIMULATION RESULTS

In the following some computer simulation results are described.
Two examples are first given to illustrate the proposed estimation algorithms for the particular case where the covariance matrices are in Toeplitz form.

Example 1

$$F_c = \begin{bmatrix} 0 & 0 & 0.75\\ 1 & 0 & -1\\ 0 & 1 & 0.75 \end{bmatrix} \qquad H_c = [0 \quad 0 \quad 1]$$

$$\begin{bmatrix} Q_{\mathcal{T},\mathcal{G},\mathcal{L}} & S_{\mathcal{T},\mathcal{G},\mathcal{L}}\\ S^T_{\mathcal{T},\mathcal{G},\mathcal{L}} & R_{\mathcal{T},\mathcal{G},\mathcal{L}} \end{bmatrix} = \begin{bmatrix} 1 & 0.25 & -0.1 & 0.2\\ 0.25 & 1. & 0.25 & -0.1\\ -0.1 & 0.25 & 1. & 0.25\\ 0.2 & -0.1 & 0.25 & 1. \end{bmatrix}$$

The estimated values using Algorithm 2 are represented in Figures 2 and 3, and are denoted:

q_i^3 $(i \in [0,3])$ for the non recursive version

q_i^2 $(i \in [0,3])$ for the recursive version

Example 2

In this example the same values of (F_c, H_c) as of example 1 are assumed; the triplet $(Q_{\mathcal{T},\mathcal{L}}, R_{\mathcal{T},\mathcal{L}}, S_{\mathcal{T},\mathcal{L}})$ is now chosen in Toeplitz form with the same numerical values as in example 1. The q_i parameters are estimated by use of Algorithm 1 and are denoted q_i^1 $(i \in [0,3])$. (Figure 4).

Example 3

In the third example, we start from the following system which is not in canonical form

$$F = \begin{bmatrix} 0.53 & 2.53 & 1.03\\ 0.69 & 0.69 & 1.19\\ 0.26 & -1.74 & 0.26 \end{bmatrix} \qquad H = [1 \quad 1 \quad 1]$$

$$\begin{bmatrix} Q & S\\ S^T & R \end{bmatrix} = \begin{bmatrix} 1.5 & -0.5 & 0.4 & 0.2\\ -0.5 & 1 & 0.5 & -0.3\\ 0.4 & 0.5 & 0.75 & 0.4\\ 0.2 & -0.3 & 0.4 & 2 \end{bmatrix}$$

In this case the proposed estimation algorithms are used to first estimate the triplets $(Q_{\mathcal{T},\mathcal{L}}, R_{\mathcal{T},\mathcal{L}}, S_{\mathcal{T},\mathcal{L}})$ or

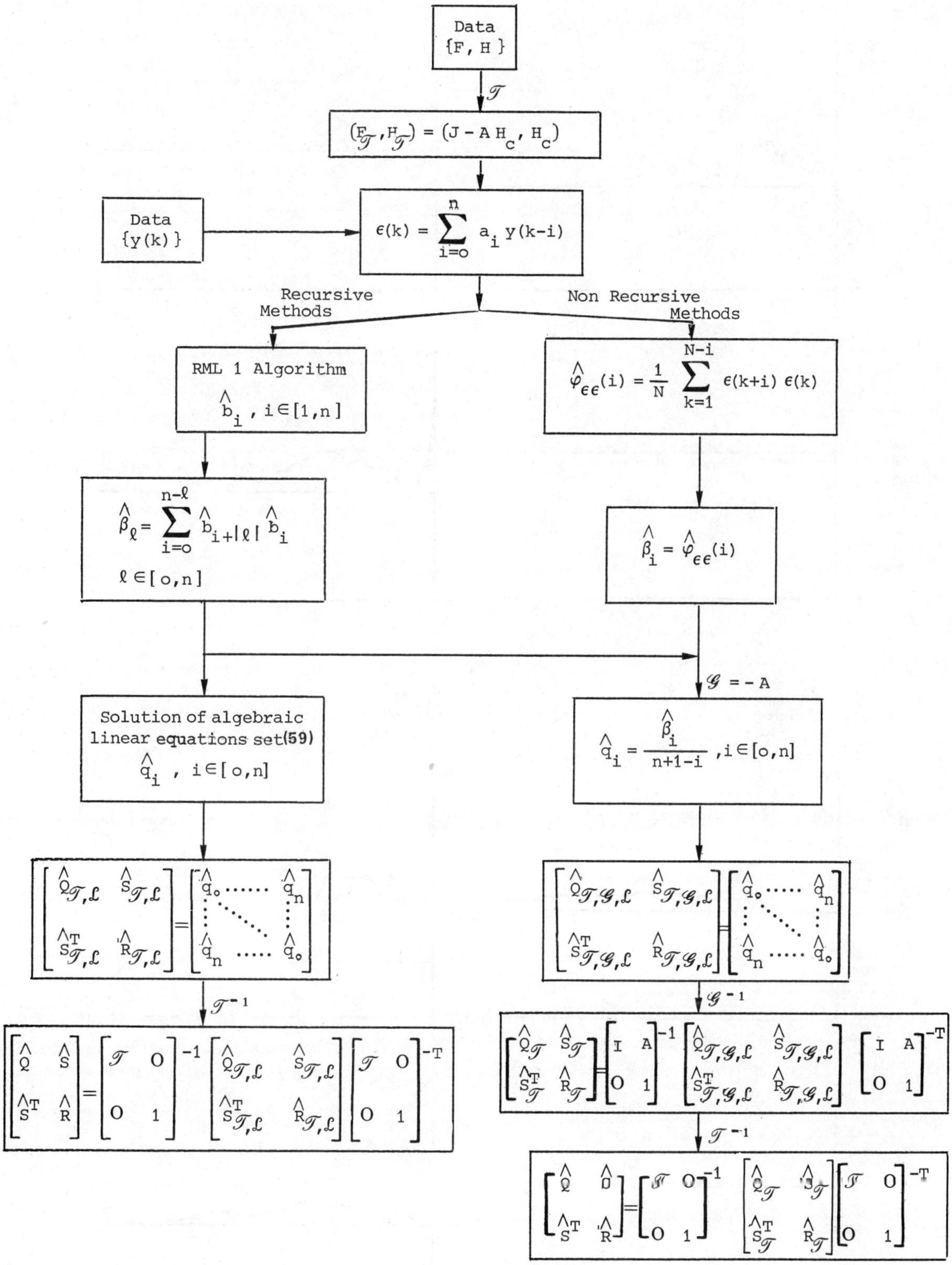

Figure 1 Covariances Estimation Algorithms

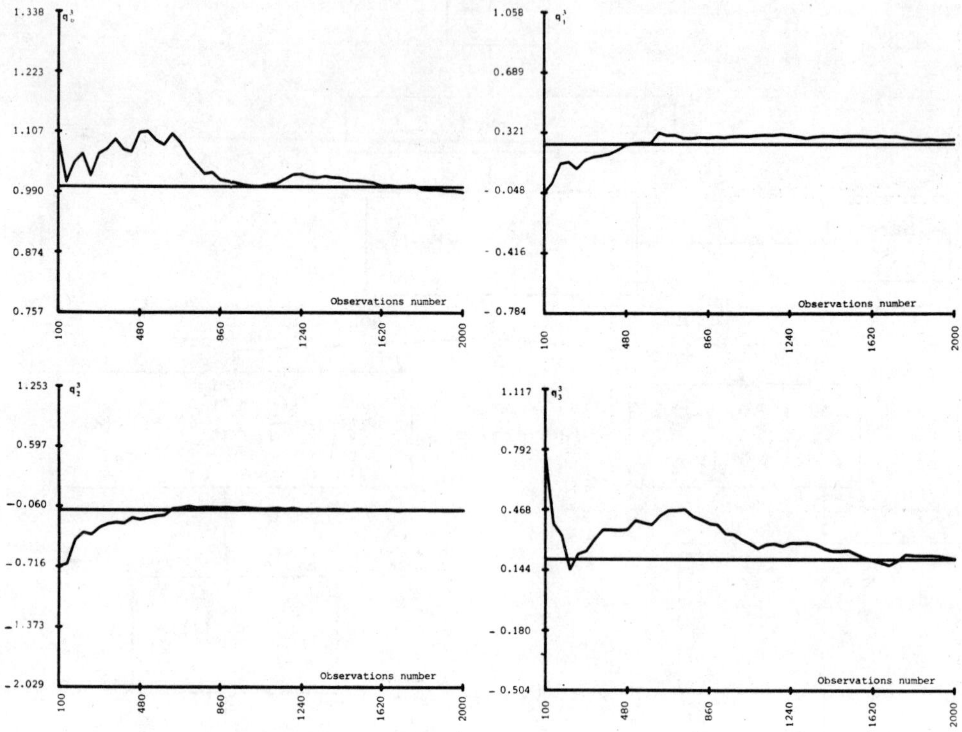

Figure 2

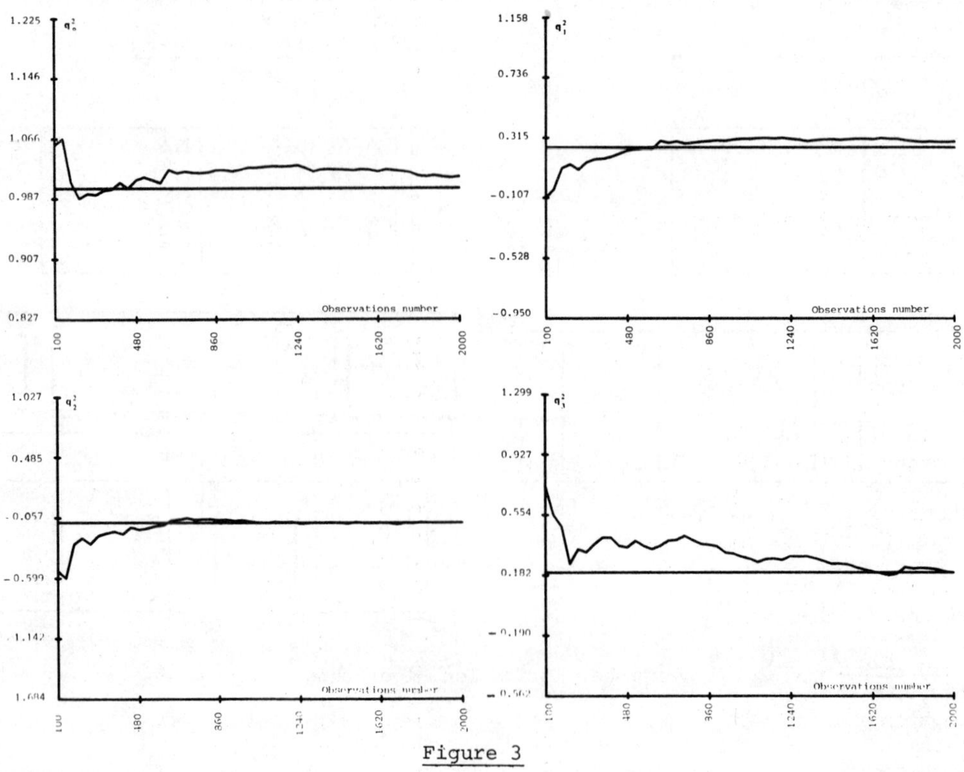

Figure 3

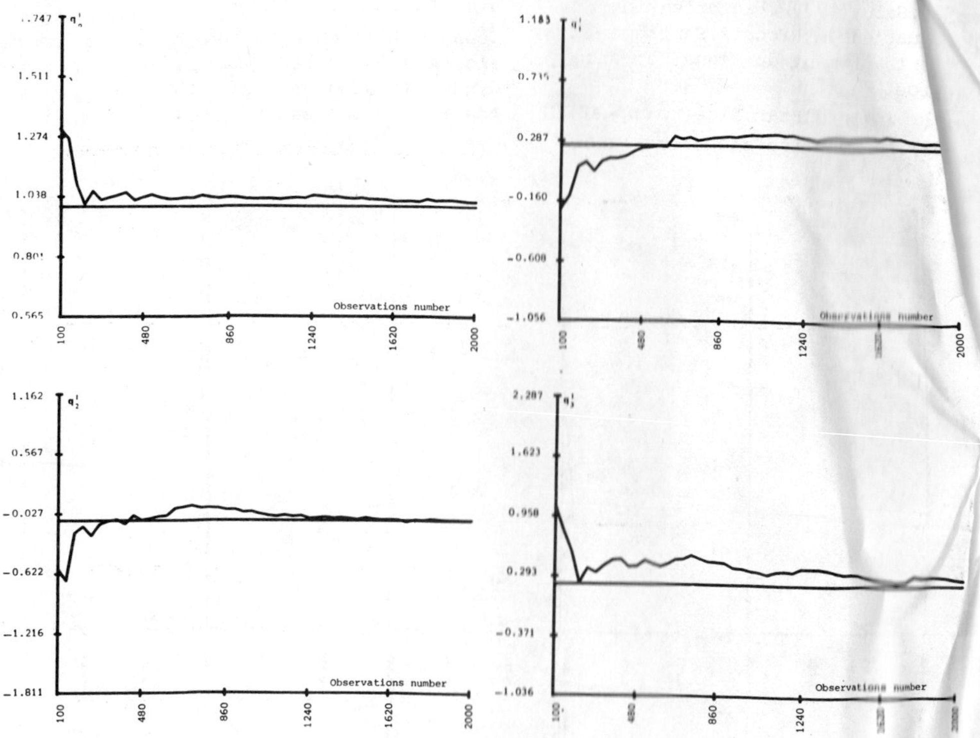

Figure 4

$(Q_{\mathcal{T},\mathcal{G},\mathcal{L}}, R_{\mathcal{T},\mathcal{G},\mathcal{L}}, S_{\mathcal{T},\mathcal{G},\mathcal{L}})$ in Toeplitz form and then to compute the corresponding triplet (Q,R,S) of the original model.The simulation results are now given in the form of Kalman gain which is computed by use of the estimated covariance matrices.

In Figure 5 the three components K_i^j of the Kalman gain obtained by each of the three estimation methods are represented with :

$i \in [1,3]$: component number

$j \in [1,3]$: {j=3 for the non recursive version of Algorithm 2, j=2 for the recursive version of Algorithm 2 and j=1 for Algorithm 1} .

V. CONCLUSION

New algorithms for estimating the noise covariances in a stochastic linear system with known dynamical parameters have been presented. These algorithms are derived by using a particular group of transformations acting on the output spectrum. The transformations are used to put the system in canonical form :the dynamical part in column-companion form,and the noise covariances in Toeplitz form. The residual sequence, corresponding to the MA part of the equivalent ARMA model, is computed by filtering the output signal through the AR part of this equivalent ARMA model. Subsequently, the parameters of Toeplitz matrices are estimated either by use of the RML 1 algorithm (recursive method) or by computing the autocorrelation function of the residual sequence (non recursive method). The main drawback of these estimation methods is related to the fact that the resulting solutions may not satisfy the positivity condition which characterizes covariance matrices. The estimated covariances may thus not correspond to real noise covariances.In this case another canonical form should be chosen. However, even if this positivity property is not satisfied,the estimated covariances can be used in the Riccati equation to compute the Kalman gain.We note that the Riccati operator can be viewed as one which effects the strong factorization of the spectral numerator for any triplet (Q,R,S).This is performed regardless the positivity condition of this triplet. Finally it may be noted that an extension of these results to the multivariable case has been developed and will be published shortly (Favier,1980).

Acknowledgements. The authors would like to express their thanks to M.Hassani for having programmed computer simulations of the proposed algorithms.

REFERENCES.

Bozzo,C.A.(1975).A discrete suboptimal adaptive estimation scheme for linear systems with unknown plant and measurement noise covariances. Proc.6th IFAC World Congress,Boston.

Candy,J.V.,T.E.Bullock and M.E.Warren (1979). Invariant system description of the stochastic realization. Automatica,vol.15,493-495.

Carew,B. and P.R.Belanger(1973). Identification of optimum filter steady-state gain for systems with unknown noise covariances. IEEE Trans.Aut.Control AC-18,n°6, 582-587.

Faurre,P.(1973).Réalisations markoviennes de processus stationnaires. Rapport Laboria n°13, IRIA,78-Rocquencourt,France.

Favier,G.(1977).Identification d'une représentation gaussienne markovienne.Algorithmes de réalisation [illegible]tique.Thèse de Docteur Ingénieur,Nice[illegible]

Favier,G. an[illegible]lut(1979).Factorisation spectrale e[illegible]ion de Riccati.Septième Colloque su[illegible]raitement du Signal et ses Application[illegible].

Favier,G.(1[illegible] Thèse d'Etat,Nice(to appear).

Panuska,V.[illegible].An adaptive recursive least squares identification algorithm.Proc.8th. IEEE Symposium on Adaptive Processes,Pennsylvania State University.

Young,P.C.(1968).The use of linear regression and related procedures for identification of dynamic processes.Proc.7th.IEEE Symposium on Adaptive Processes,U.C.L.A.

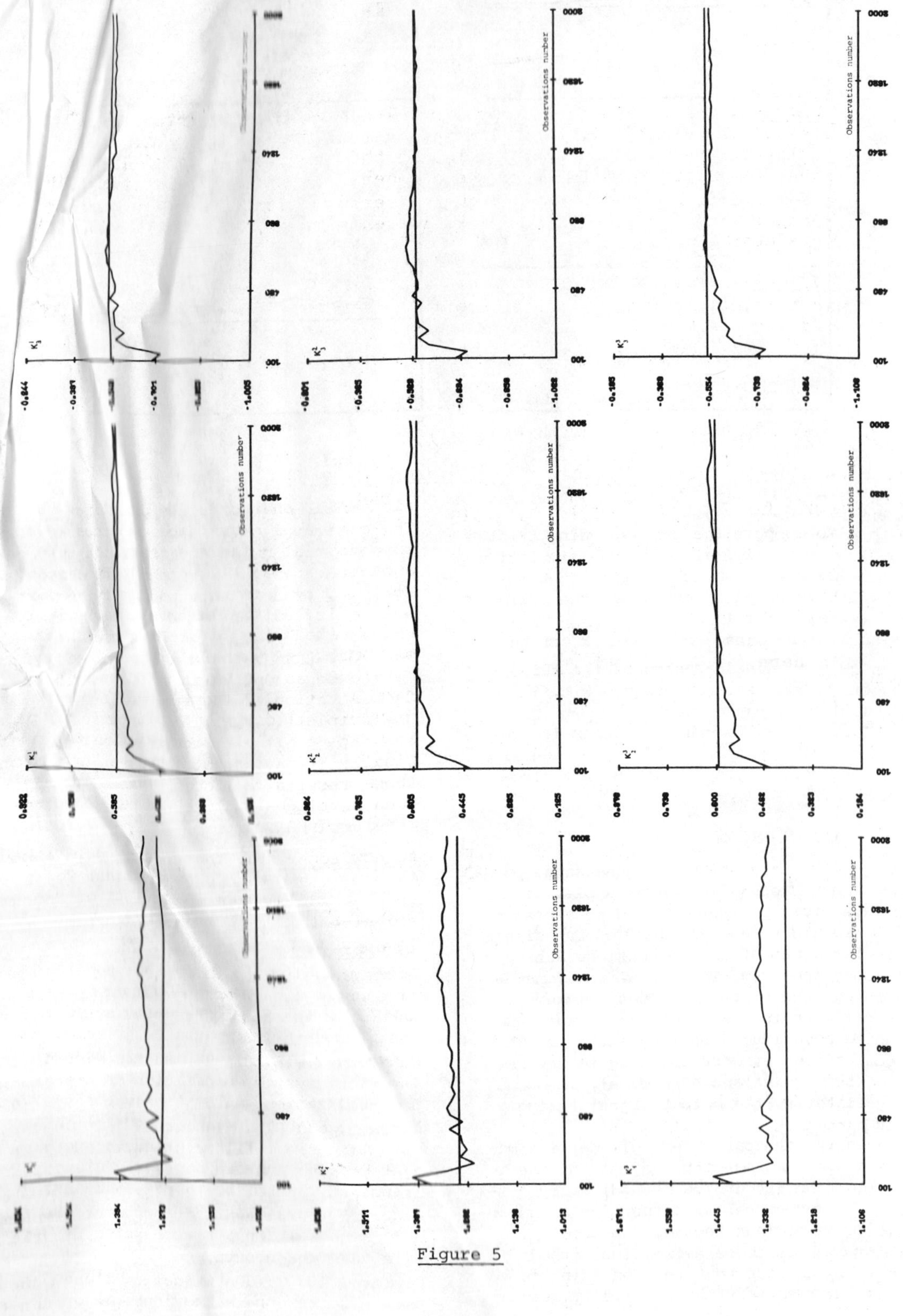

Figure 5

ON THE FREQUENCY RESPONSE OF M-D FIR DIGITAL FILTERS

N. C. Hu

Shanghai University of Technology, Shanghai, China

Abstract. The frequency response is one of the most important properties of the digital filters. The author (Hu, 1978) has obtained the basic frequency response properties of the multi-dimensional linear phase FIR digital filters and their general expressions. In this paper, a further discussion of the frequency response properties and corresponding linear phase conditions is presented, and some new results are obtained.

Keywords. Symmetric distribution; basic linear phase condition; odd-even function; frequency response mode; M-D plane symmetric property; odd-even combination.

INTRODUCTION

With the developments of the signal processing technique, the digital filters have been widely applied in many different scientific and technical regions. Linear phase FIR digital filter is one of the fundamental types of the digital filters.

Hu (1978) has obtained the basic frequency response properties of the multi-dimensional linear phase FIR digital filters, and found there are 2^{2K} different frequency response modes of the K-dimensional linear phase FIR digital filters in the K-dimensional space, in which $(K+1)^2$ modes are the basic modes, and the $2^{2K}-(K+1)^2$ other modes may be transformed from the basic modes by the substitutions of the frequency coordinates.

The general frequency response expressions of the multi-dimensional linear phase FIR digital filters are

$$H=H(e^{jw_1},\cdots,e^{jw_K}) = Q\sum_{n_1=\ddot{A}(N_1)}^{\frac{N_1-\mathring{A}(N_1)}{2}}\cdots\sum_{n_K=\ddot{A}(N_K)}^{\frac{N_K-\mathring{A}(N_K)}{2}} a(n_1,\cdots,n_K)\times$$

$$\times\prod_{i=1}^{K}\left\{\mathring{A}(R_i)\,\mathrm{Sin}\left[\left(n_i-\frac{\ddot{A}(N_i)}{2}\right)w_i\right]+\ddot{A}(R_i)\cos\left[\left(n_i-\frac{\ddot{A}(N_i)}{2}\right)w_i\right]\right\} \qquad (1)\text{a}$$

where

$$Q=\exp\left\{-j\left[\sum_{i=1}^{K}\left(\frac{N_i-1}{2}\right)w_i-\sum_{i=1}^{K}\beta_i\right]\right\} \qquad (1)\text{b}$$

is the phase factor in the K-dimensional space.

$$\left.\begin{aligned}\ddot{A}(X)&=\begin{cases}1, & \text{for even } X\\ 0, & \text{for odd } X\end{cases}\\ \mathring{A}(X)&=\begin{cases}1, & \text{for odd } X\\ 0, & \text{for even } X\end{cases}\end{aligned}\right\} \qquad (1)\text{c}$$

are the odd-even functions.

The purpose of this paper is to make a further discussion of some properties of the multi-dimensional linear phase FIR digital filters.

BASIC NOTIONS

Let $h(n_1,\cdots,n_K)$ be the K-dimensional orthogonal impulse response sequences within the region $0\leqslant n_i\leqslant N_i-1$ $(i=1,2,\cdots,K)$, the Z transform of $h(n_1,\cdots,n_K)$ may be written as

$$H(z_1,\cdots,z_K)=\sum_{n_1=0}^{N_1-1}\cdots\sum_{n_K=0}^{N_K-1}h(n_1,\cdots,n_K)\times\prod_{i=1}^{K}z_i^{-n_i} \qquad (2)$$

the relative Fourier transform is

$$H(e^{jw_1},\cdots,e^{jw_K})=\sum_{n_1=0}^{N_1-1}\cdots\sum_{n_K=0}^{N_K-1}h(n_1,\cdots,n_K)\times$$

$$\exp\left[-j\sum_{i=1}^{K} n_i w_i\right] \qquad (3)$$

this is a function of which the period is 2π in the every dimension of the K-dimensional space, i.e. we have

$$H(e^{jw_1},\cdots,e^{jw_K}) = H\left[e^{j(w_1+2\pi m_1)},\cdots,e^{j(w_K+2\pi m_K)}\right] \qquad (4)$$

here $m_1,\cdots,m_K$ are integers. Formula (3) may also be rewritten as

$$H(e^{jw_1},\cdots,e^{jw_K}) = \left|H(e^{jw_1},\cdots,e^{jw_K})\right| e^{j\theta(w_1,\cdots,w_K)} \qquad (3)a$$

where $\left|H(e^{jw_1},\cdots,e^{jw_K})\right|$ and $e^{j\theta(w_1,\cdots,w_K)}$ are the module and phase of H respectively.

To satisfy the requisition of the linear phase, it is necessary to have

$$\theta(w_1,\cdots,w_K) = -(d_1 w_1 + \cdots + d_K w_K) = -\sum_{i=1}^{K} d_i w_i \qquad (5)$$

where d_i is the group delay factor in the i-th dimension.

Now the formula (3)a can be rewritten in the following form

$$H(e^{jw_1},\cdots,e^{jw_K}) = \left|H(e^{jw_1},\cdots,e^{jw_K})\right| \times \exp(-j\sum_{i=1}^{K} d_i w_i) \qquad (3)b$$

from formula (3) and (3)b we find

$$\left|H(e^{jw_1},\cdots,e^{jw_K})\right| \cos(\sum_{i=1}^{K} d_i w_i) = \sum_{n_1=0}^{N_1-1}\cdots\sum_{n_K=0}^{N_K-1} h(n_1,\cdots,n_K)\cos(\sum_{i=1}^{K} n_i w_i) \qquad (6)a$$

and

$$\left|H(e^{jw_1},\cdots,e^{jw_K})\right| \sin(\sum_{i=1}^{K} d_i w_i) = \sum_{n_1=0}^{N_1-1}\cdots\sum_{n_K=0}^{N_K-1} h(n_1,\cdots,n_K)\sin(\sum_{i=1}^{K} n_i w_i) \qquad (6)b$$

the result is

$$\tan(\sum_{i=1}^{K} d_i w_i) = \frac{\sin(\sum_{i=1}^{K} d_i w_i)}{\cos(\sum_{i=1}^{K} d_i w_i)} = \frac{\sum_{n_1=0}^{N_1-1}\cdots\sum_{n_K=0}^{N_K-1} h(n_1,\cdots,n_K)\sin(\sum_{i=1}^{K} n_i w_i)}{\sum_{n_1=0}^{N_1-1}\cdots\sum_{n_K=0}^{N_K-1} h(n_1,\cdots,n_K)\cos(\sum_{i=1}^{K} n_i w_i)} \qquad (7)$$

There exist two possible solutions to Eq.(7). The first possibility is that

$$\tan(\sum_{i=1}^{K} d_i w_i) = 0 \qquad (8)$$

in this case we can find

$$h(n_1,\cdots,n_K)\begin{cases} =0, & n_1=\cdots=n_K=0 \\ =0, & \text{elsewhere} \end{cases} \qquad (9)$$

this is the only one date $h(0,\cdots,0)$ is existed, therefore, it has no general significance.

The other possible case is when

$$\tan(\sum_{i=1}^{K} d_i w_i) = 0 \qquad (10)$$

in this case,,from Eq.(7) we can find

$$\sum_{n_1=0}^{N_1-1}\cdots\sum_{n_K=0}^{N_K-1} h(n_1,\cdots,n_K) \sin\left[\sum_{i=1}^{K}(d_i-n_i)w_i\right] = 0 \qquad (11)$$

this result has been obtained from the linear phase condition of Eq.(5). In the general case, the phase delay in every dimension must be considered, the linear phase condition may be rewritten as

$$\theta(w_1,\cdots,w_K) = \sum_{i=1}^{K}(\beta_i - d_i w_i) \qquad (5)a$$

where β_i is the phase delay factor in the i-th dimension. Hence Eq.(11) becomes

$$\sum_{n_1=0}^{N_1-1}\cdots\sum_{n_K=0}^{N_K-1} h(n_1,\cdots,n_K) \sin\left[\sum_{i=1}^{K}(d_i-n_i)w_i - \sum_{i=1}^{K}\beta_i\right] = 0 \qquad (12)$$

Similarly, formula (3) becomes

$$H(e^{jw_1},\cdots,e^{jw_K}) = \sum_{n_1=0}^{N_1-1}\cdots\sum_{n_K=0}^{N_K-1} h(n_1,\cdots,n_K)\, e^{-j\sum_{i=1}^{K}(n_i w_i - \beta_i)} \qquad (3)c$$

LINEAR PHASE CONDITIONS

In the one-dimensional case, Eq.(12) is simplified to

$$\sum_{n=o}^{N-1} h(n)\sin\left[(d-n)w-\beta\right]=0 \tag{12)a}$$

Eq.(12)a may be solved when ß is the integral times of $\pi/2$. Therefore, we can let

$$\beta=R\frac{\pi}{2} \tag{13}$$

here R is an integer. When R and N are odd or even, four basic solutions of Eq.(12)a may be obtained, they corresponding to four symmetric or anti-symmetric distributive properties of the one-dimensional impulse response sequences (Rabiner, 1975).

In the multi-dimensional case, Eq.(12) may be solved when

$$\beta=\sum_{i=1}^{K}\beta_i \tag{14}$$

is the integral times of $\frac{\pi}{2}$. If ß is the even times of $\frac{\pi}{2}$, the conditions of the solution is

$$\left.\begin{aligned} &d_i=\frac{N_i-1}{2}, \quad (i=1,2,\cdots,K)\\ &h(n_1,\cdots,n_K)=(N_1-1-n_1,\cdots,N_K-1-n_K)\end{aligned}\right\} \tag{15)a}$$

and when ß is the odd times of $\frac{\pi}{2}$, the conditions of the solution is

$$\left.\begin{aligned} &d_i=\frac{N_i-1}{2}, \quad (i=1,2,\cdots,K)\\ &h(n_1,\cdots,n_K)=-(N_1-1-n_1,\cdots,N_K-1-n_K)\end{aligned}\right\} \tag{15)b}$$

according to Eq.(15)a, the distributive properties of the impulse response sequences $h(n_1,\cdots,n_K)$ is symmetric around the center $O(d_1,\cdots,d_K)$, and for Eq.(15)b is anti-symmetric relative to this center.

Obviously, both of the conditions Eq.(15)a and Eq.(15)b only provide the symmetric or anti-symmetric distributive properties of the impulse response sequences around one point (the center of these sequences), they generally can not to provide the multi-dimensional linear phase properties, i.e. even so the conditions of Eqs.(15)a (15)b are satisfied, the distributive properties of the impulse response sequences in every dimension may still be non-symmetric, so that, the corresponding phase properties in every dimension will also be non-linear. Therefore, the conditions of Eqs.(15)a and (15)b only are the mathematical soluble coditions of Eq.(12), and can not be become the basic conditions of the multi-dimensional linear phase properties, i.e. the conditions of Eqs.(15)a and (15)b only are the necessary conditions of the multi-dimensional properties, but are not the sufficient one.

To provide the linear phase properties may be hold in any dimension, the impulse response sequences must be distributed symmetrically or anti-symmetrically in any dimension.

To find the corresponding conditions of the multi-dimensional linear phase solutions, we consider the multi-dimensional orthogonal case. In this case, Eq. (3)c may be separated for every dimension and rewritten as

$$\begin{aligned} H(e^{jw_1},\cdots,e^{jw_K}) &= \sum_{n_1=o}^{N_1-1}\cdots\sum_{n_K=o}^{N_K-1} h(n_1,\cdots,n_K)\\ &\quad \exp\left[-j\sum_{i=1}^{K}(n_i w_i-\beta_i)\right]\\ &= \prod_{i=1}^{K}\left(\sum_{n_i=o}^{N_i-1} h_i(n_i)\times \times\exp(-jn_i w_i)\right] \end{aligned} \tag{3)d}$$

accordingly, Eq.(12) becomes

$$\sum_{n_i-1}^{N_i-1} h_i(n_i)\sin\left[(d_i-n_i)w_i-\beta_i\right]=0 \qquad (i=1,2,\cdots,K) \tag{12)b}$$

hence, we can find the linear phase properties of multi-dimensional filters for every dimension by the methods similar to the one-dimensional case. So that, we can find the necessary and sufficient basic conditions of the multi-dimensional linear phase properties to be

$$\left.\begin{aligned} &\beta_i=R_i\left(\frac{\pi}{2}\right)\\ &d_i=\frac{N_i-1}{2}\\ &h(n_1,\cdots,n_K)=\\ &\qquad=\pm h(n_1,\cdots,n_{i-1},N_i-1-n_i,n_{i+1},\cdots,n_K)\end{aligned}\right\}$$

$$(i=1,2,\cdots,K) \tag{15)c}$$

here R_i are integers; and for a given i, only one symbol before the h may be taken (positive or negative), they relative to the symmetric or anti-symmetric distributive properties in this dimension respectively.

According to Eq.(15)c, the impulse response sequences $h(n_1,\cdots,n_K)$ are distributed symmetrically or anti-symmetrically in the K-dimensional space relative to all (K-1)-dimensional planes that throught the center $O(d_1,\cdots,d_K)$.

To simplify the description, we will add one little point over R_i or N_i to indicate they are odd, and two points over them when they are even, such as $\dot{R}_1$ indicates R_1 is odd, $\ddot{N}_4$ indicates N_4 is even etc. In the following we will be further discussing from the two-dimensional case.

TWO-DIMENSIONAL PROPERTY

In the two-dimensional case, Eq.(12) becomes

$$\sum_{n_1=0}^{N_1-1}\sum_{n_2=0}^{N_2-1} h(n_1,n_2)\sin[(d_1-n_1)w_1+(d_2-n_2)w_2-\beta_1-\beta_2]=0 \qquad (12)c$$

from Eq.(15)c we have

$$\beta_1=R_1(\tfrac{\pi}{2}), \qquad \beta_2=R_2(\tfrac{\pi}{2}) \qquad (16)$$

here R_1,R_2 may be odd or even respectively, thes, there are four possible different combinations of them:

1. $\ddot{R}_1,\ddot{R}_2$; 2. $\dot{R}_1,\dot{R}_2$;

3. $\dot{R}_1,\ddot{R}_2$; 4. $\ddot{R}_1,\dot{R}_2$.

In the cases 1. and 2., $\beta_1+\beta_2$ is the integral times of π, and Eq.(12) may be rewritten as

$$\sum_{n_1=0}^{N_1-1}\sum_{n_2=0}^{N_2-1} h(n_1,n_2)\sin[(d_1-n_1)w_1+(d_2-n_2)w_2]=0 \qquad (12)d$$

In the cases 3. and 4., $\beta_1+\beta_2$ is the odd times of $\pi/2$, therefore Eq.(12) becomes

$$\sum_{n_1=0}^{N_1-1}\sum_{n_2=0}^{N_2-1} h(n_1,n_2)\cos[(d_1-n_1)w_1+(d_2-n_2)w_2]=0 \qquad (12)e$$

Corresponding to above four cases, we can obtaine four basic linear phase conditions of the solutions of Eq.(12), they are

1. $d_1=\frac{N_1-1}{2}$; $d_2=\frac{N_2-1}{2}$;

$$h(n_1,n_2)=h(n_1,N_2-1-n_2), \quad h(n_1,n_2)=h(N_1-1-n_1,n_2) \qquad (17)a$$

2. $d_1=\frac{N_1-1}{2}$; $d_2=\frac{N_2-1}{2}$;

$$h(n_1,n_2)=-h(n_1,N_2-1-n_2), \quad h(n_1,n_2)=-h(N_1-1-n_1,n_2) \qquad (17)b$$

3. $d_1=\frac{N_1-1}{2}$; $d_2=\frac{N_2-1}{2}$;

$$h(n_1,n_2)=h(n_1,N_2-1-n_2), \quad h(n_1,n_2)=-h(N_1-1-n_1,n_2) \qquad (17)c$$

4. $d_1=\frac{N_1-1}{2}$; $d_2=\frac{N_2-1}{2}$;

$$h(n_1,n_2)=-h(n_1,N_2-1-n_2), \quad h(n_1,n_2)=h(N_1-1-n_1,n_2) \qquad (17)d$$

Eqs.(17)a-(17)d show, the distributive properties of the impulse response sequences are symmetric or anti-symmetric relative to two perpendicular crossed lines that all throught the center of these sequences(i.e. the central point $O(d_1,d_2)$).

Clearly, the difference between the cases 3. and 4. and correspondingly the conditions of Eqs.(17)c and (17)d are only in the interchange of w_1,w_2 and n_1,n_2 respectively, in fact, they may be considered as a same case. So that,,there are three different basic linear phase conditions in the two-dimensional case.

Similarly, for N_1 and N_2, when they are odd or even respectively, four different odd-even combinations of N_1 and N_2 may also be found, and three combinations in which are basic.

Therefore, in the two-dimensional case, there are $4^2=16$ different linear phase solutions to Eq.(12)c, and nine of which are the basic solutions. Correspondingly, we can find the same number different frequency response modes and their expressions of the two-dimensional linear phase FIR digital filters (Hu, 1978).

The following figure shows 16 different linear phase distributive modes of the impulse response sequences in the two-dimensional case. In this figure, the white regions and shaded regions are represent the positive and negative impulse response properties respectively.

From this mode figure we can see, the parameters R_1 and R_2 of the four modes M_{11}, M_{12}, M_{21}, M_{22} are all even, so that, the properties of these modes are symmetric in the both dimensions, we call them the all-symmetric modes. In the four modes M_{33}, M_{34}, M_{43}, M_{44}, R_1 and R_2 are all odd, their properties are anti-symmetric in the both dimensions, they are the all-anti-symmetric modes. Similarly, for the other eight modes, R_1 and R_2 are not all odd or even, their properties are symmetric in one dimension and anti-symmetric in other one, they are the hybrid-symmetric modes.

The mode figure also shows, in this figure, the total mode properties are symmetric around two perpendicular diagonal lines AA and BB, the four modes that in the main diagonal line AA (M_{11}, M_{22}, M_{33}, M_{44}) have the same properties in the both dimensions, they are the isotropic modes.

HIGH-DIMENSIONAL ANALYSIS

In the three-dimensional case, Eq.(12) take the form of

$$\sum_{n_1=0}^{N_1-1}\sum_{n_2=0}^{N_2-1}\sum_{n_3=0}^{N_3-1} h(n_1,n_2,n_3)\times$$

$$\times\sin\left[(d_1-n_1)w_1+(d_2-n_2)w_2+(d_3-n_3)w_3 -(\beta_1+\beta_2+\beta_3)\right]=0 \qquad (12)f$$

according to the three phase delay factors

$$\beta_1=R_1(\tfrac{\pi}{2})$$
$$\beta_2=R_2(\tfrac{\pi}{2}) \qquad (18)$$
$$\beta_3=R_3(\tfrac{\pi}{2})$$

are the odd or even times of $\pi/2$ respectively, there are 8 different linear phase conditions of Eq.(12)f may be found.

Similar to the two-dimensional case, the linear phase conditions of Eq.(12)f are

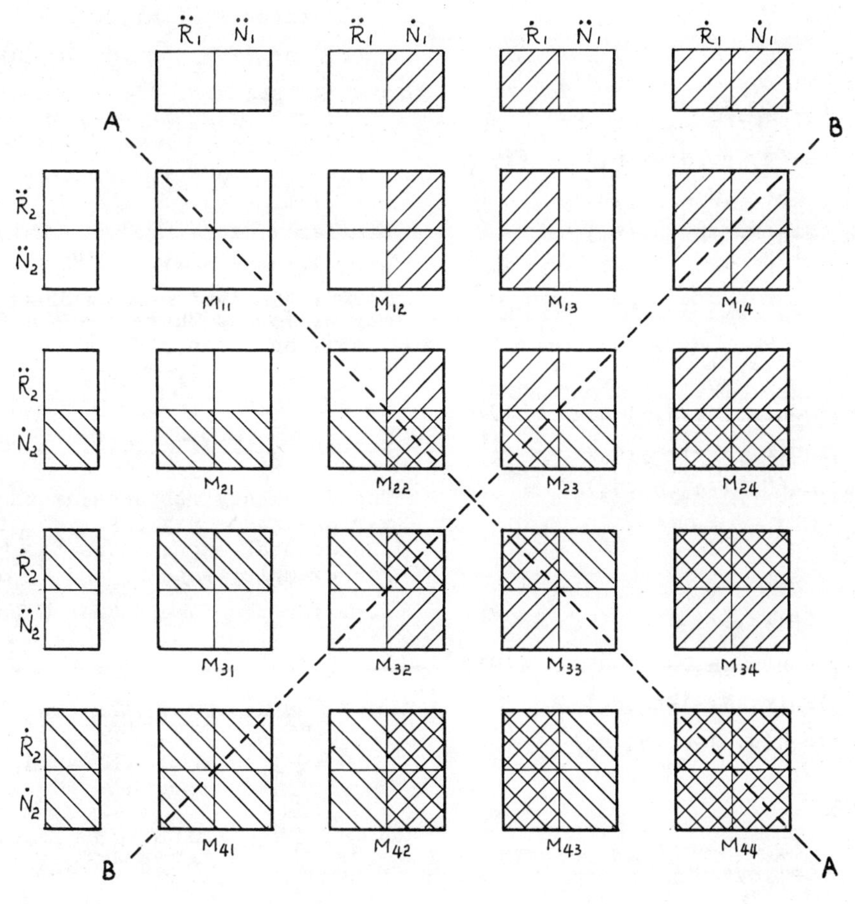

Figure. 16 different linear phase combined modes of two-dimensional FIR digital filters.

1. $\ddot{R}_1$, $\ddot{R}_2$, $\ddot{R}_3$:

$$d_i=\frac{N_i-1}{2}, \quad (i=1,2,3)$$

$$h(n_1,n_2,n_3)=h(n_1,n_2,N_3-1-n_3) \tag{19)a}$$
$$h(n_1,n_2,n_3)=h(n_1,N_2-1-n_2,n_3)$$
$$h(n_1,n_2,n_3)=h(N_1-1-n_1,n_2,n_3)$$

2. $\ddot{R}_1$, $\ddot{R}_2$. $\dot{R}_3$:

$$d_i=\frac{N_i-1}{2}, \quad (i=1,2,3)$$

$$h(n_1,n_2,n_3)=h(n_1,n_2,N_3-1-n_3) \tag{19)b}$$
$$h(n_1,n_2,n_3)=h(n_1,N_2-1-n_2,n_3)$$
$$h(n_1,n_2,n_3)=-h(N_1-1-n_1,n_2,n_3)$$

3. $\ddot{R}_1$, $\dot{R}_2$, $\ddot{R}_3$:

$$d_i=\frac{N_i-1}{2}, \quad (i=1,2,3)$$

$$h(n_1,n_2,n_3)=h(n_1,n_2,N_3-1-n_3) \tag{19)c}$$
$$h(n_1,n_2,n_3)=-h(n_1,N_2-1-n_2,n_3)$$
$$h(n_1,n_2,n_3)=h(N_1-1-n_1,n_2,n_3)$$

4. $\ddot{R}_1$, $\dot{R}_2$, $\dot{R}_3$:

$$d_i=\frac{N_i-1}{2}, \quad (i=1,2,3)$$

$$h(n_1,n_2,n_3)=h(n_1,n_2,N_3-1-n_3) \tag{19)d}$$
$$h(n_1,n_2,n_3)=-h(n_1,N_2-1-n_2,n_3)$$
$$h(n_1,n_2,n_3)=-h(N_1-1-n_1,n_2,n_3)$$

5. $\dot{R}_1$, $\ddot{R}_2$, $\ddot{R}_3$:

$$d_i=\frac{N_i-1}{2}, \quad (i=1,2,3)$$

$$h(n_1,n_2,n_3)=-h(n_1,n_2,N_3-1-n_3) \tag{19)e}$$
$$h(n_1,n_2,n_3)=h(n_1,N_2-1-n_2,n_3)$$
$$h(n_1,n_2,n_3)=h(N_1-1-n_1,n_2,n_3)$$

6. $\dot{R}_1$, $\ddot{R}_2$, $\dot{R}_3$:

$$d_i=\frac{N_i-1}{2}, \quad (i=1,2,3)$$

$$h(n_1,n_2,n_3)=-h(n_1,n_2,N_3-1-n_3) \tag{19)f}$$
$$h(n_1,n_2,n_3)=h(n_1,N_2-1-n_2,n_3)$$
$$h(n_1,n_2,n_3)=-h(N_1-1-n_1,n_2,n_3)$$

7. $\dot{R}_1$, $\dot{R}_2$, $\ddot{R}_3$:

$$d_i=\frac{N_i-1}{2}, \quad (i=1,2,3)$$

$$h(n_1,n_2,n_3)=-h(n_1,n_2,N_3-1-n_3) \tag{19)g}$$
$$h(n_1,n_2,n_3)=-h(n_1,N_2-1-n_2,n_3)$$
$$h(n_1,n_2,n_3)=h(N_1-1-n_1,n_2,n_3)$$

8. $\dot{R}_1$, $\dot{R}_2$, $\dot{R}_3$:

$$d_i=\frac{N_i-1}{2}, \quad (i=1,2,3)$$

$$h(n_1,n_2,n_3)=-h(n_1,n_2,N_3-1-n_3) \tag{19)h}$$
$$h(n_1,n_2,n_3)=-h(n_1,N_2-1-n_2,n_3)$$
$$h(n_1,n_2,n_3)=-h(N_1-1-n_1,n_2,n_3)$$

Obviously, the conditions of Eq.(19)a to Eq.(19)h provide the distributive properties of the impulse response sequences in the three-dimensional space are symmetric or anti-symmetric relative to three perpendicular crossed central planes, these central planes are perpendicular to three number axes respectively and crossed on the center $O(d_1,d_2,d_3)$ of the three-dimensional impulse response sequence.

If we neglect the order of the dimensions, according to the odd-even combinations of three parameters R_1, R_2, R_3, four basic linear phase conditions of Eq.(12) may be found, they are Eqs.(19)a, (19)b, (19)d and (19)h, and relative to four basic odd-even combinations:

(a) three R_i (i=1,2,3) all are even;
(b) one of R_i is odd and other two are even;
(c) one of R_i is even and other two are odd;
(d) three R_i all are odd.

Similarly, for three N_i (i=1,2,3), there also exist 8 different odd-even combinations, and four basic combinative types may be found. Therefore, in the three dimensional case, the total number of the different frequency response modes is $2^{2K}=2^6=64$, in which the basic mode number is $(K+1)^2=4^2=16$.

The frequency response expressions of various modes may easily be found from the general expressions Eqs.(1)a—(1)c. For example, when R_i and N_i all are odd, i.e. for the combination $\dot{N}_1$, $\dot{N}_2$, $\dot{N}_3$, $\dot{R}_1$, $\dot{R}_2$, $\dot{R}_3$, we have

$$H_o=Q\sum_{n_1=o}^{\frac{N_1-1}{2}}\sum_{n_2=o}^{\frac{N_2-1}{2}}\sum_{n_3=o}^{\frac{N_3-1}{2}}a_o(n_1,n_2,n_3)\times$$
$$\times\sin(n_1w_1)\sin(n_2w_2)\sin(n_3w_3) \tag{20)o}$$

where

$$Q=\sum_{i=1}^{K}\left[\left(\frac{N_i-1}{2}\right)w_i-\beta_i\right] \tag{21)o}$$

When R_i and N_i all are even, we have

$$\ddot{R}_1, \ddot{R}_2, \ddot{R}_3, \ddot{N}_1, \ddot{N}_2, \ddot{N}_3;$$

$$H_e = \sum_{n_1=1}^{\frac{N_1}{2}} \sum_{n_2=1}^{\frac{N_2}{2}} \sum_{n_3=1}^{\frac{N_3}{2}} a_e(n_1,n_2,n_3) \times$$

$$\times \cos\left[(n_1-\tfrac{1}{2})w_1\right] \cos\left[(n_2-\tfrac{1}{2})w_2\right]$$

$$\times \cos\left[(n_3-\tfrac{1}{2})w_3\right] \qquad (20)e$$

here Q as shown in Eq.(21)o.

In the more high (K 3) dimensional case, using the similar methods, we can find the corresponding basic linear phase conditions as shown in Eq.(13)c, the K-dimensional impulse response sequences will be distributed symmetrically or anti-symmetrically in every dimension relative to the (K-1)-dimensional central planes.

Generally, in the multi-dimensional case, there are three basic symmetric distributive types may be found:

(a) all-symmetric type, the impulse response sequences are distributed symmetrically in all dimensions;

(b) all-anti-symmetric type, the impulse response sequences are distributed anti-symmetrically in all dimensions;

(c) hybrid-symmetric type, the impulse response sequences are distributed symmetrically in some dimensions and anti-symmetrically in other dimensions.

In the one-dimensional case, there are four basic filter types of the digital filters, namely the low-pass, high-pass, band-pass and band-stop. In the multi-dimensional cases, generally, the filter can choose each one of the four basic filter types in every dimension, therefore, there are 4^K distinct combined filter modes may be formed, and in which the iso-type (i.e. the filter types are the same in all dimensions) combined filter modes will have more practical importances.

Clearly, for the multi-dimensional combined filter modes, except the all pass region and all stop region, there are K-2 different kinds of partial pass regions of where the stop orders are different.

Similar to the fan filter of two-dimensional case, in the high-dimensional cases, the multi-dimensional conical filters may be designed for some spetial purposes of multi-dimensional signal analysis or information processing.

CONCLUSION

From the above discussions we obtain the following results:

(1) The multi-dimensional linear phase frequency response properties certainly relative to the multi-dimensional symmetric, anti-symmetric or hybrid-symmetric distributions of the impulse response sequences.

(2) The basic linear phase distributive conditions of the multi-dimensional impulse response sequences are

$$h(n_1,\cdots,n_K) = \pm h(n_1,\cdots,n_{i-1},N_i-1-n_i,n_{i+1},\cdots,n_K)$$
$$(i=1,2,\cdots,K)$$

the symmetric properties are relative to all (K-1)-dimensional central p planes in the K-dimensional space.

(3) All 2^{2K} different frequency response modes of the multi-dimensional linear phase FIR digital filters may be divided in three basic types: symmetric type, anti-symmetric type and hybrid-type.

REFERENCES

Hu, N. C. (1978). Frequency response of multi-dimensional linear phase FIR digital filters. Technical References of Shanghai University of Technology. No. 7844. (1979). Proceedings of the International Conference on Information Sciences and Systems.*

Rabiner, L. R., and B. Gold (1975). Theory and Application of Digital Signal Processing. Englewood Cliffs, N. J.

Helms, H. D. (1968). Nonrecursive digital filters: design methods for achieving specifications on frequency response. IEEE Trans. on Audio. AU-16, 336-343.

Kaiser, J. F. (1966). Digital Filters. John Wiley and Sons Inc. New York.

Hu, N. C. (1978). Multi-dimensional digital windows. Technical References of Shanghai University of Technology. No. 7835. (1979). Proceedings of the International Conference on Information Sciences and Systems.*

Hu, N. C. (1979). Frequency sampling design of multi-dimensional digital filters. Technical References of Shanghai University of Technology. No. 79025. (1980). Progress in Cybernetics and Systems Research. Hemisphere Publishing Corporation, Washington, D. C..

* D. Reidel Publishing Company, Dordrecht, Holland/Boston, U. S. A..

STRUCTURE ANALYSIS OF COMPLEX FUNCTIONS: APPLICATION TO A ZINC PRODUCTION UNIT

M. Deleris, E. Chacon and A. Titli

Laboratoire d'Automatique et d'Analyse des Systèmes du Centre National de la Recherche Scientifique, 7 avenue du Colonel Roche, 31400 Toulouse, France

Résumé. L'analyse structurale des systèmes complexes à partir des données entraine, du fait d'un grand nombre de variables à considérer, une décomposition en sous-systèmes faiblement couplés. On propose une décomposition issue d'une méthode comparative des outils classiques d'analyse des données d'une part et d'une approche informationnelle d'autre part. Pour cela on fait appel aux concepts d'entropie et d'information qui permettent de définir une matrice de couplage. Cette comparaison met en valeur les différences importantes que l'on peut trouver entre la matrice des corrélations linéaires et la matrice de couplage exprimant les relations, linéaires ou non, entre les variables. Cette décomposition est validée par différentes méthodes d'analyse des données (analyse des composantes principales, segmentation) ainsi que par une méthode de partition utilisant des graphes valués, autre représentation des matrices, permettant l'estimation de relations entre sous-systèmes. L'application concerne un complexe de production de zinc dont on propose un modèle statique obtenu à partir de cette procédure.

INTRODUCTION

L'analyse des systèmes complexes passe d'abord par une analyse structurale en vue d'une décomposition en sous-systèmes faiblement couplés.

Cette décomposition est obtenue en appliquant l'algorithme de Tarjan à partir d'une évaluation des intéractions au moyen d'une approche informationnelle.

Ces sous ensembles servent de point de départ à une modélisation dont le principe proposé se base sur une méthode de comparaison entre les outils classiques d'analyse des données (corrélations) et l'analyse informationnelle (couplage). Cette méthode, qui met en valeur les linéarités d'une part, les non linéarités d'autre part, dans les relations entre les variables prises deux à deux, permet aussi la détection de points aberrants dans une série de mesures.

L'application concerne un complexe de production de zinc dont on détaille le fonctionnement dans un premier paragraphe. L'étude porte sur l'analyse d'une série de 87 relevés journaliers de 21 variables d'un de ces sous-processus récemment mis en fonctionnement : le Traitement Humide des Résidus (T.H.R.).

Le concept d'entropie est rappelé dans un deuxième paragraphe dans lequel nous définissons la matrice de couplage associée.

L'algorithme de Tarjan est ensuite développé avec son application au T.H.R. avant d'exposer l'utilisation simultanée sur ce même processus des outils statistiques et informationnels.

L'USINE DE PRODUCTION DE ZINC

- Synoptique de l'élaboration du zinc

Le processus de production de zinc électrolytique peut être décomposé en trois phases (Fig. 1) :

- grillage : le minerai de zinc (ZnS) est transformé en calcine (ZnO)
- préparation des solutions : la calcine est attaquée par de l'acide pour former une solution de sulfate de zinc ($ZnSO_4$)
- bacs d'électrolyse : le zinc métallique se dépose sur les cathodes, l'acide sulfurique régénérée par l'électrolyse est recyclée en tête de la préparation des solutions.

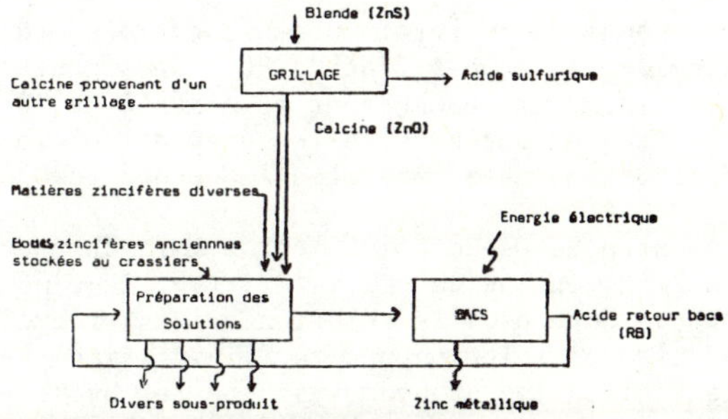

Figure 1

Evidemment, la réalité est beaucoup plus complexe : le minerai contient de nombreuses impuretés qu'il est nécessaire d'éliminer afin d'obtenir un zinc de qualité suffisante (99,995 %). C'est le second rôle de la préparation des solutions qui, par différents traitements, sépare le zinc des autres métaux (Fig. 2).

- Préparation des solutions

- traitement de matières zincifères : dissout les matières zincifères autres que la calcine par de l'acide RB

- lixiviations : dissout la calcine par de l'acide RB : la liqueur est envoyée en purification, les boues subissent un retraitement afin de récupérer le zinc soluble qui est recyclé

- purification : sépare avant l'électrolyse le zinc des autres métaux, les résidus sont envoyés au traitement des boues bleues

- traitement des boues bleues : récupère et recycle aux lixiviations le zinc emporté par les résidus de purification

- traitement humide des résidus (T.H.R.) : traitement des boues de lixiviation afin d'en séparer le zinc des autres métaux non solubles.

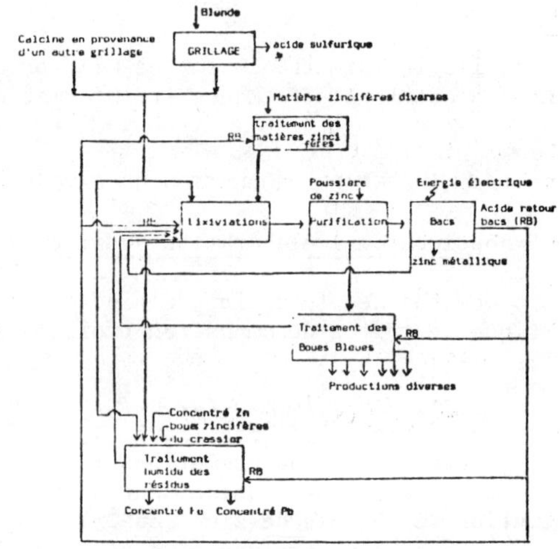

Figure 2

- Le traitement humide des résidus (T.H.R.)

Ce sous-processus, qu' est le T.H.R. (Fig. 3), n'est en fonctionnement que depuis très peu de temps.

Les boues de lixiviation subissent d'abord une première attaque à l'acide RB. Une décantation suit, dont les boues sont soumises à une deuxième attaque. Après nouvelle décantation, le résidu insoluble constitue le concentré Plomb.

Les attaques successives ont permis de récupérer le zinc contenu dans les ferrites. Ce zinc se trouve donc dans la solution qui est recyclée et mise en solution neutre (lixiviation) après neutralisation et hydrolyse. Le fer hydrolysé est précipité et entraîne avec lui certaines impuretés (As, Ge, No, K, ...).

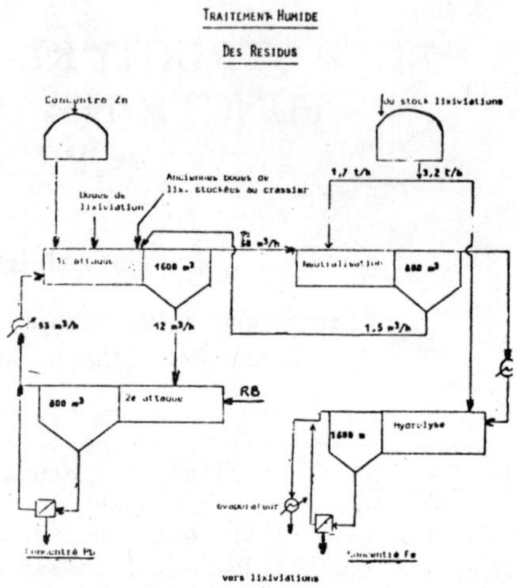

Figure 3

Cette partie hydrolyse (ou précipitation de la Goethite) est la plus compliquée du fait de la diversité des réactions s'y déroulant. De plus, c'est elle qui est la cause des plus importantes pertes zinc.

- Hydrolyse

Son objectif est la récupération du fer amené dans le circuit par le minerai de zinc.

Ce fer en solution dans la liqueur sous forme de sulfates ferreux et ferriques est précipité par apport de Calcine, en Goethite ou autres précipités parasites.

Ces réactions se déroulent successivement dans cinq bacs (ou réacteurs) en série (Fig. 4).

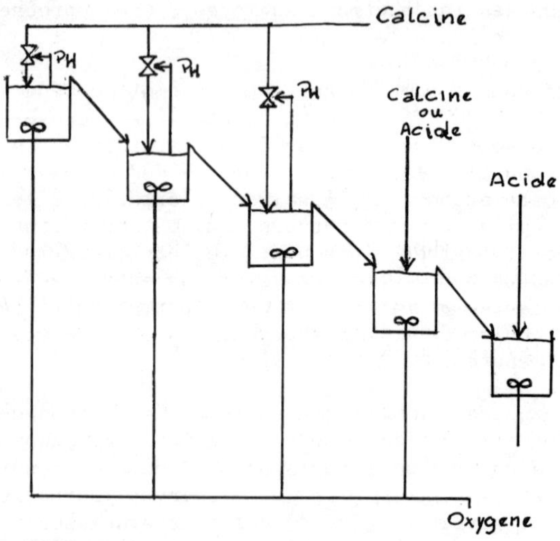

Figure 4

Le débit de Calcine est régulé dans les trois premiers réacteurs par le PH de la solution, qui doit satisfaire un point de consigne fixé de manière empirique (Fig. 5).

De même que le PH, les concentrations Fe^{++}, Fe^{+++} sont des contraintes à respecter.

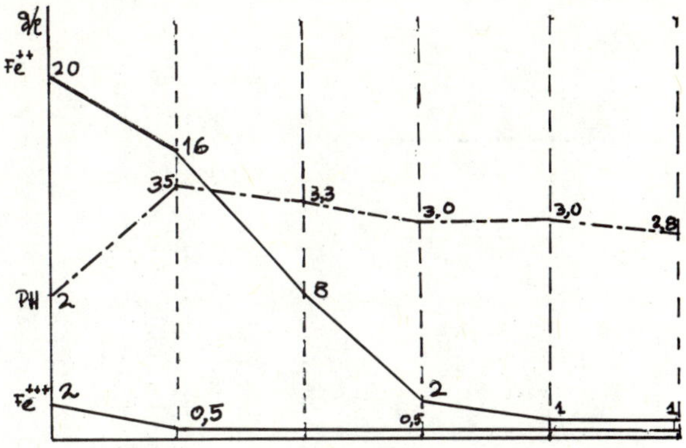

Figure 5

- Problèmes posés par le T.H.R.

Dans le T.H.R., deux transformations fondamentales se déroulent :

- mise en solution à forte acidité et haute température des ferrites de zinc contenues dans le minerai
- précipitation sous forme de Goethite du fer ainsi mis en solution.

Dans le principe adopté par l'usine de Viviez, le fer en solution sort à l'état ferreux ($FeSO_4$) de la mise en solution des ferrites. L'opération à étudier comprend :

- oxydation du fer
- précipitation de la Goethite
- neutralisation par la Calcine de l'acide libérée par la précipitation.

Ces trois réactions étant simultanées, elles peuvent s'écrire sous la forme condensée :

$$2\ FeSO_4 + H_2O + \frac{1}{2}O_2 + 2\ ZnO \rightarrow 2\ \underline{FeOOH} \downarrow + 2\ ZnSO_4$$

Or, d'une part, le fer ne précipite pas uniquement sous forme de Goethite mais aussi en Hydroxo-Jarosites et Hydroxyde Ferrique. D'autre part, le fer n'est pas le seul métal en présence et il se forme d'autres précipités tout aussi nuisibles.

Nous avons donc à faire face à un problème de cinétique chimique dans lequel un grand nombre de réactions se déroulent simultanément, dans une proportion dépendant de la teneur en métaux "parasites" du minerai.

Le précipité obtenu est évacué et filtré avant stockage afin de récupérer le zinc n'ayant pas réagi, entraîné par imprégnation plus ou moins importante suivant le précipité.

On trouve ce zinc sous forme de sulfates ($ZnSO_4$: 3 à 4%), Calcine en excès (ZnO : 3%), autres formes (2%).

En conséquence, il s'agit d'obtenir d'une part un rendement maximum de la Calcine, d'autre part une meilleure connaissance des conditions de formation du précipité fer et de sa structure.

La satisfaction de ces objectifs nécessite une modélisation du processus, modélisation qui s'avère impossible en suivant la voie d'une analyse thermodynamique et cinétique classique, ce qui nous oblige à recourir à des outils d'analyse de structure et de décomposition (approche informationnelle, outils statistiques).

Dans un premier temps, pour avoir une idée de la structure de l'ensemble T.H.R., nous travaillons sur une approche informationnelle. D'abord, nous cherchons les relations existantes entre toutes les variables prises deux à deux pour définir une matrice de couplage.

De plus, dans un ensemble de variables, celles-ci ne sont pas toutes liées ; ou bien, si elles le sont, ne présentent pas le même niveau d'intéraction. On peut alors décomposer l'ensemble en sous-ensembles faiblement couplés.

Une modélisation peut ensuite être réalisée sur chaque sous-système. Les deux étapes utilisent des concepts de théorie de l'information que nous rappelons ci-après.

RAPPELS DE THEORIE DE L'INFORMATION

Entropie

Cas discret : Soit une variable aléatoire X définie sur $x_1, \ldots, x_N$

La loi de X est $p(X = x_i) = p_i$ telle que $0 \leqslant p_i \leqslant 1$ et $\sum_{i=1}^{n} p_i = 1$

L'entropie de la variable X est définie par

$$H(X) = -\sum_{i=1}^{n} p_i \log_a p_i$$

Cas continu : On déduit facilement

$$H = -\int_a^b p(x) \log_a p(x)$$

$$p(x) \geqslant 0 \text{ et } \int_a^b p(x)dx = 1$$

$p(x)$ étant la fonction densité de probabilité.

On définit de même :

Entropie conjointe :

$$H(X,Y) = -\sum_{i=1}^{n} \sum_{j=1}^{n} p_{ij} \log_a p_{ij}$$

Entropie conditionnelle :

$$H(X/Y) = -\sum_i \sum_j p_{ij} \log \frac{p_{ij}}{p_i}$$

$$H(Y/X) = -\sum_i \sum_j p_{ij} \log \frac{p_{ij}}{p_j}$$

$$p_i(\text{resp } j) = \sum_j p_{ij} \ (\text{resp } i)$$

Nous pouvons alors définir le terme $T(X:Y) = H(X) + H(Y) - H(X,Y)$ appelée transinformation, qui représente l'information contenue dans X sur Y (ou dans Y sur X).

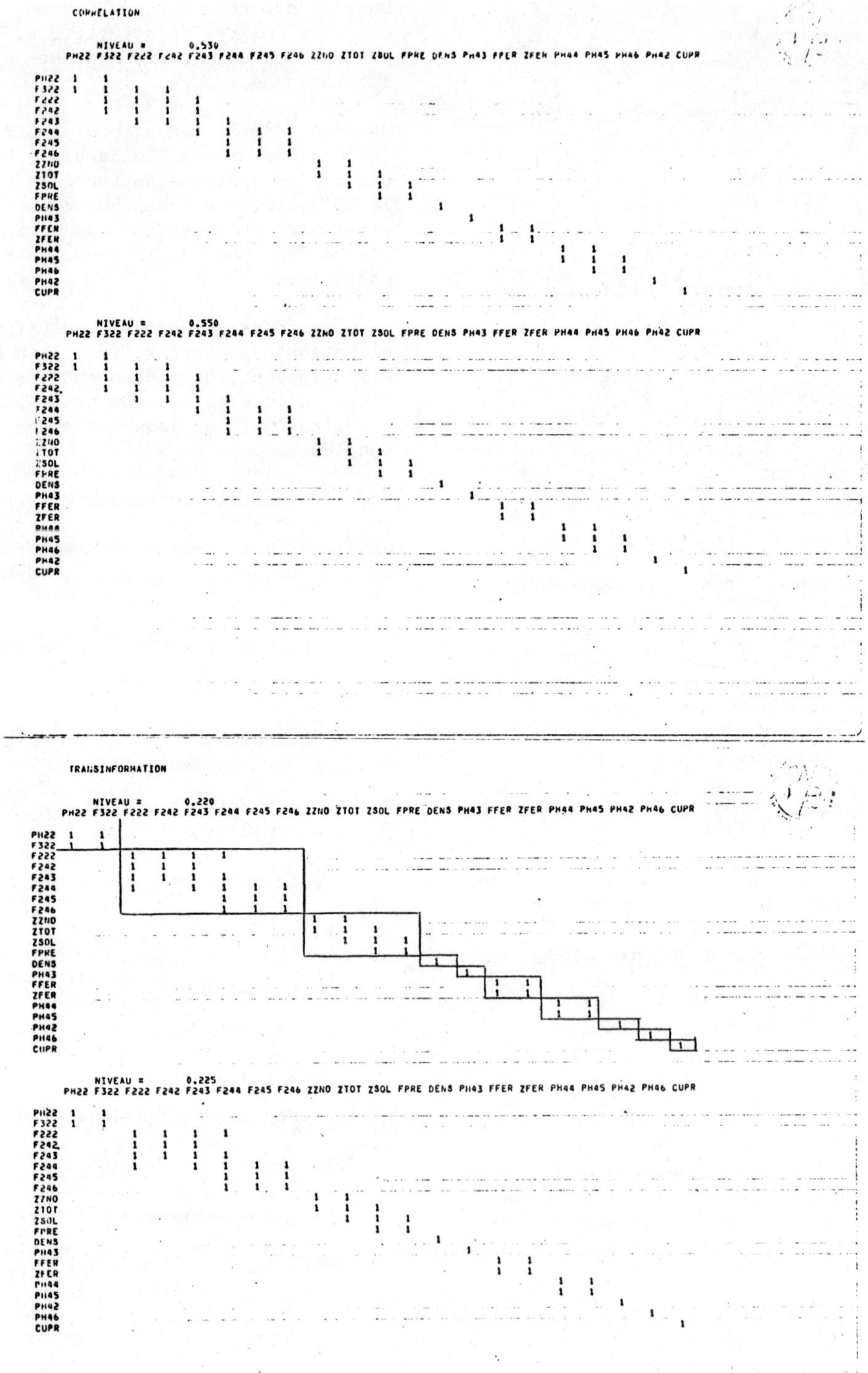
CORRELATION
NIVEAU = 0.530
PH22 F322 F222 F242 F243 F244 F245 F246 ZZNO ZTOT ZSOL FPRE DENS PH43 FFER ZFER PH44 PH45 PH46 PH42 CUPR
NIVEAU = 0.550
PH22 F322 F222 F242 F243 F244 F245 F246 ZZNO ZTOT ZSOL FPRE DENS PH43 FFER ZFER PH44 PH45 PH46 PH42 CUPR
TRANSINFORMATION
NIVEAU = 0.220
PH22 F322 F222 F242 F243 F244 F245 F246 ZZNO ZTOT ZSOL FPRE DENS PH43 FFER ZFER PH44 PH45 PH42 PH46 CUPR
NIVEAU = 0.225
PH22 F322 F222 F242 F243 F244 F245 F246 ZZNO ZTOT ZSOL FPRE DENS PH43 FFER ZFER PH44 PH45 PH42 PH46 CUPR

Figure 6

Remarques : $T(X:Y) = H(X) + H(Y)$ si X et Y sont statistiquement indépendantes |3|.
$T(X:Y) = 0$ si X et Y sont liés par une relation quelconque |3|.

Afin de pouvoir utiliser ce terme, il est préférable de le normaliser par :

$$I_{XY} = \frac{T(X:Y)}{\sqrt{H(X)H(Y)}}$$

On définit ainsi une matrice symétrique appelée matrice de couplage statique |4| de termes m_{ij}

$$0 \leqslant m_{ij} \leqslant 1$$
$$m_{ij} = 1 \quad i = j$$

Une estimation de l'entropie d'une variable peut facilement être obtenue en décomposant l'ensemble des valeurs prises par la variable considérée, en m classes. L'entropie s'écrit alors :

$$H(X) = - \sum_{i=1}^{m} \frac{n_i}{N} \log \frac{n_i}{N} \qquad \text{(a)}$$

où n_i est le nombre d'occurences de la variable X dans la classe i :

$$\Sigma n_i = N$$

$$\text{d'où } H(X) = \text{Log } N - \frac{1}{N} \Sigma n_i \log n_i \qquad |2|$$

Remarque : H(X) est maxi lorsque la distribution est uniforme :

$$H(X) = \text{Log } N$$

H(X) = 0 si tous les points sont groupés dans une même classe.

Aspects décomposition

A partir de la matrice de couplage statique, nous pouvons trouver une première décomposition en fixant un seuil minimal des valeurs des couplages et en construisant une matrice M de '0' et '1' |5||6|, qui sera la représentation d'un graphe par sa matrice d'adjacence.

En appliquant l'algorithme de Tarjan |7|, par permutations sur la matrice M, on définit une matrice diagonale par blocs.

Le principe de l'algorithme est le suivant :

1. On choisit un sommet de façon arbitraire et on recherche tous les sommets liés à celui-ci, ce qui définit un bloc.
2. Pour chaque sommet de ce bloc, on répète la procédure jusqu'à ce qu'on ne trouve plus de sommets liés.
3. On revient en 1 jusqu'à ce qu'il n'existe plus de sommet non groupé.

Les résultats obtenus par cette procédure sont donnés sur la figure 6, la définition des variables utilisées étant donnée dans le tableau 1. (en appendice)

La décomposition ainsi obtenue s'avère dans un premier temps extrêmement intéressante pour une étude qualitative du processus, car elle met en valeur certaines caractéristiques de fonctionnement souvent difficilement chiffrables, donc ne pouvant intervenir dans une modélisation.

En effet, la figure 6 nous montre les divers sous-ensembles obtenus pour un seuil donné. Nous trouvons principalement :

1. L'ensemble des concentrations Fer
{F222, F242,..., F246}
2. L'ensemble des teneurs précipité
{ZZNO, ZTOT, ZSOL, FPRE}

Cependant, certaines contraintes techniques obligent parfois à ne pas se conformer rigoureusement à la décomposition obtenue. C'est le cas dans le processus étudié où nous avons dû considérer une décomposition par réacteur, et non par facteur (Fig. 6) pour des raisons matérielles. Mais la première décomposition apporte une aide précieuse lorsqu'il est nécessaire de rajouter une ou plusieurs variables de 'pondération' sur le modèle physique.

Le principe de modélisation présenté se base sur une comparaison entre l'approche linéaire (par coeff. de corrélation) et l'approche informationnelle (par coeff. de couplage).

Modélisation

On se base sur une comparaison entre la matrice des corrélations et la matrice de couplage statique précédemment définie.

La matrice des corrélations étant constituée de termes r_{ij} tels que :

$$r_{ij} = \frac{\text{cov}(X, Y)}{\sigma_X \, \sigma_Y} \qquad 0 \leqslant r_{ij} \leqslant 1$$
$$r_{ij} = 1 \quad i = j$$

A partir d'un seuil (m) fixé a priori sur la matrice de couplage statique, nous établissons une matrice de comparaison (C), matrice diagonale dont les termes C_{ij} sont repérés par 'R', 'T', '1' ou 'blanc'.

Le seuil de corrélation étant déterminé en fonction de m comme suit, et noté R. $i < j$:

- . R représente le terme $C_{ij} = (\text{Max } r_{ij}/m_{ij} < m)$
R peut aussi être défini par $\{C_{ij}/m_{ij} < m$ et $r_{ij} \geqslant r\}$ si r est fixé a priori, alors R est un ensemble de points {R}.
- . {T} représente $\{C_{ij}/m_{ij} \geqslant m$ et $r_{ij} < \text{Max}r_{ij}\}$ (ou r)
- . {1} représente $\{C_{ij}/m_{ij} \geqslant m$ et $r_{ij} > \text{Max}r_{ij}\}$ (ou r)
- . {blanc} " $\{C_{ij}/m_{ij} < m$ et $r_{ij} < \text{Max}r_{ij}\}$ (ou r)

$i = j : \; C_{ij} = 1$

Cette matrice permet de s'affranchir du tracé des nuages de points devenant rapidement fastidieux lorsque le nombre de variables augmente.

Ainsi :

Les points notés '1' nous indiquent qu'il

existe une relation entre deux variables dont une bonne représentation peut être une fonction linéaire (Fig. 7).

Les points notés 'T' montrent qu'il existe une relation dont une représentation sera une fonction non linéaire. Cette fonction pourra généralement être formulée à partir de la connaissance physique du processus (Fig. 8).

Le(s) point(s) noté(s) 'R' indiquent qu'il existe une ou plusieurs observations aberrantes qui fixent la direction de la droite de régression (Fig. 9) |9|.

Les 'blancs' montrent qu'il n'y a pas de relation entre les variables.

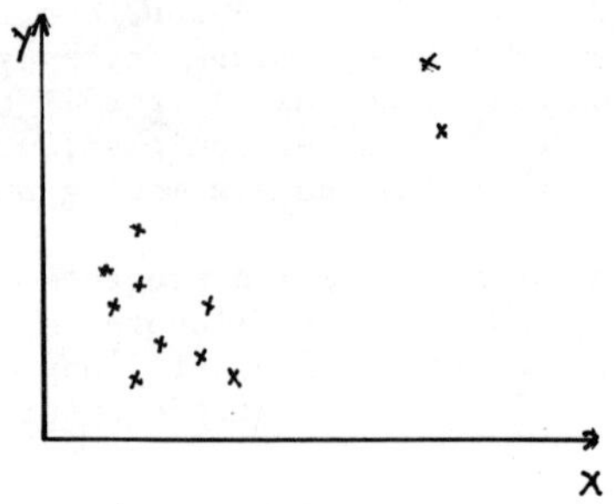

Figure 7
m faible
r fort

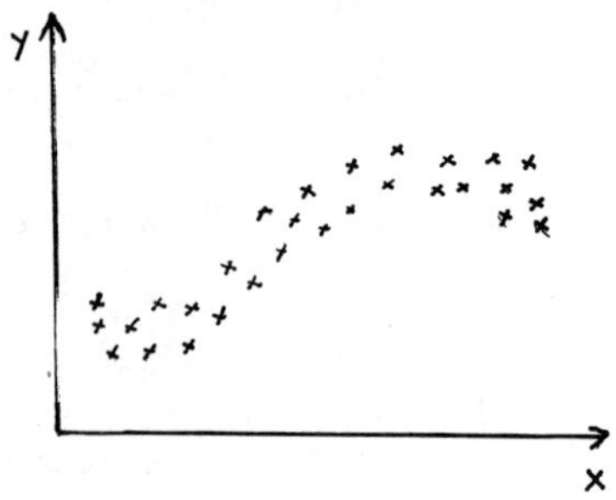

Figure 8
m fort
r faible

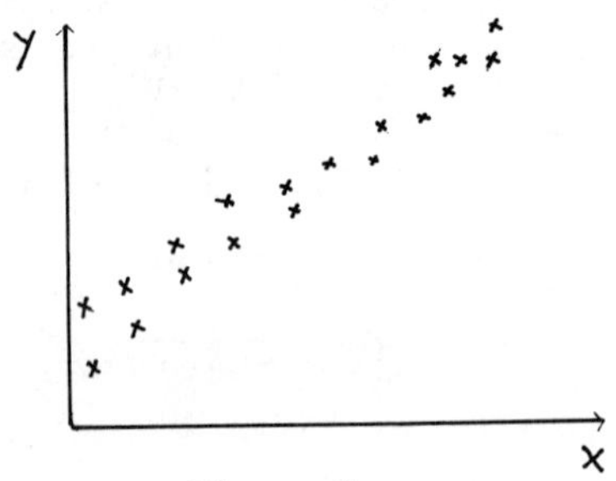

Figure 9
m fort
r fort

Un modèle peut donc être défini, sous réserve que les variables considérées soient facilement exploitables, lors d'une commande automatique.

Application

Nous avons testé cette méthode sur le processus présenté en introduction. Dans une première étude, nous disposions d'une série de 86 observations de 21 variables. La matrice C obtenue est représentée Figure 10 pour m = 0,230 et r = R = 0,628 ; figure 11, la même matrice C pour des seuils différents m = 0,225 et r = 0,46, R n'est plus un point unique.

Sur les figures 12, 13, 14, nous trouvons les nuages de points correspondant à, respectivement, R, T, 1, de la figure 10.

CONCLUSION

L'analyse et la décomposition des systèmes complexes devient une démarche nécessaire préalable à toute modélisation, à des fins de commande en particulier, ceci à cause du grand nombre de variables entrant en jeu et de la complexité des phénomènes à modéliser.

Préparant la phase modélisation, une telle démarche est très enrichissante déjà en elle-même par la vision globale qu'elle apporte et la compréhension des comportements qu'elle facilite.

L'utilisation simultanée des outils d'analyse informationnelle et d'analyse statistique permet la détection aisée des non linéarités, fréquentes dans les situations réelles et notamment les procédés chimiques, ainsi que la détection des mesures aberrantes, d'où un gain de temps sur la sélection des mesures pertinentes.

Le choix des non linéarités détectées se fait sur des considérations physiques, ce qui conduit ensuite à des problèmes d'estimation de paramètres, plus classiques en Automatique.

Les techniques avancées ont été testées sur une partie d'un complexe de production de Zinc, la phase suivante étant la concrétisation de cette modélisation par la formulation d'un problème d'optimisation dont la résolution devrait conduire à la détermination du fonctionnement de l'unité le plus économique en régime permanent.

REFERENCES

|1| Guiasu, S. Information theory with applications. Mc Graw Hill, London, 1977.

|2| Conant, R. Detecting subsystems of a complex system. IEEE Transactions on Syst., Man, Cybern., Vol. SMC.2, n° 4, pp. 550-553.

|3| Conant, R. Laws of information with govern system. IEEE Trans. on Syst., Man, Cybern. Vol. SMC.6, n° 4, 1976, pp. 240-255.

|4| Richetin, M. Analyse structurale des Systèmes Complexes en vue d'une commande

hiérarchisée. Thèse de Doctorat d'Etat, Toulouse, n° 674, 1975.

5 Dufour, J. Thèse de Doctorat d'Etat. Université Claude Bernard, Lyon, 1979.

6 Kaufmann, A. Théorie des sous-ensembles flous. Masson et Cie, Paris, 1973.

7 Duff, I.S. - Reid, J.K. An implementation of Tarjan's Algorithm for the block triangularisation of a matrix. ACM Transactions on Mathematical Software, Vol. 4, n° 2, June 1978, pp. 137-147.

8 Milgram, M. Méthodes de décomposition de Systèmes. Thèse de Doctorat de 3e cycle, Compiègne, Sept. 1975.

9 Cailliez, F. - Pages, J.P. Introduction à l'analyse des données, Paris, 1976.

Notations :

PH22 : Valeur du PH dans le décanteur 22
F322 : Concentration en Fe^{+++} dans le décanteur 22
F222 : Concentration en Fe^{++} dans le décanteur 22
F242 : Concentration en Fe^{++} dans le réacteur 42
F243 : " " 43
F244 : " " 44
F245 : " " 45
F246 : " " 46
ZZNO : % zinc sous forme ZNO dans le précipité
ZTOT : % zinc total dans le précipité
ZSOL : % zinc soluble dans le précipité
FPRE : % Fer dans le précipité

DENS : Densité
PH42 : Valeur du PH dans le réacteur 42
PH43 : " " 43
PH44 : " " 44
PH45 : " " 45
PH46 : " " 46
FFER : % Fer sous forme de Ferrites dans la Calcine
ZFER : % Zn sous forme de Ferrites dans la Calcine
CUPR : % Cuivre dans le précipité

Tableau 1

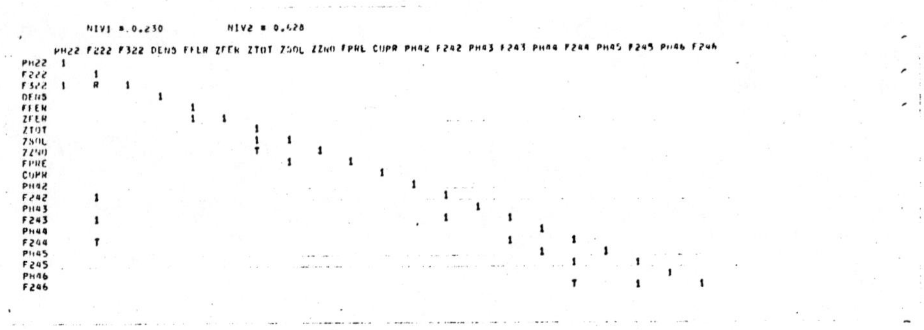

Figure 10

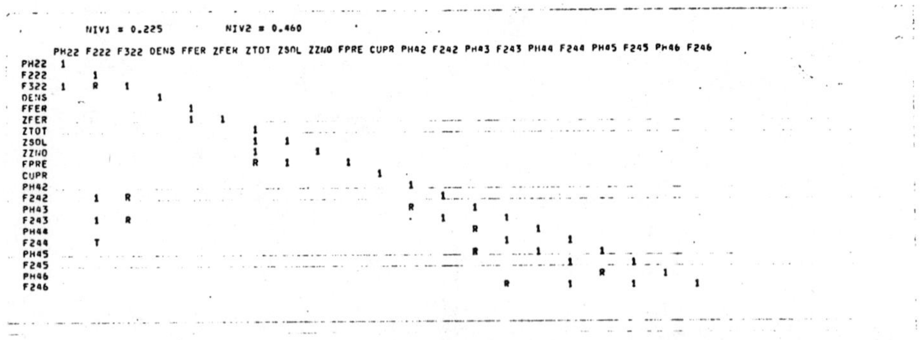

Figure 11

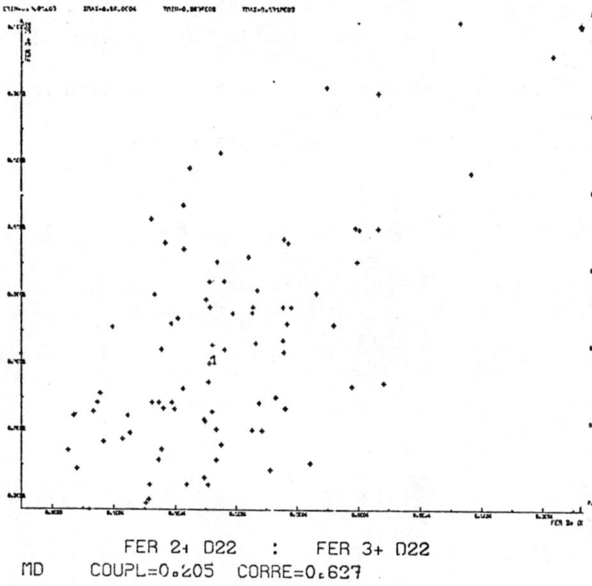

Figure 12

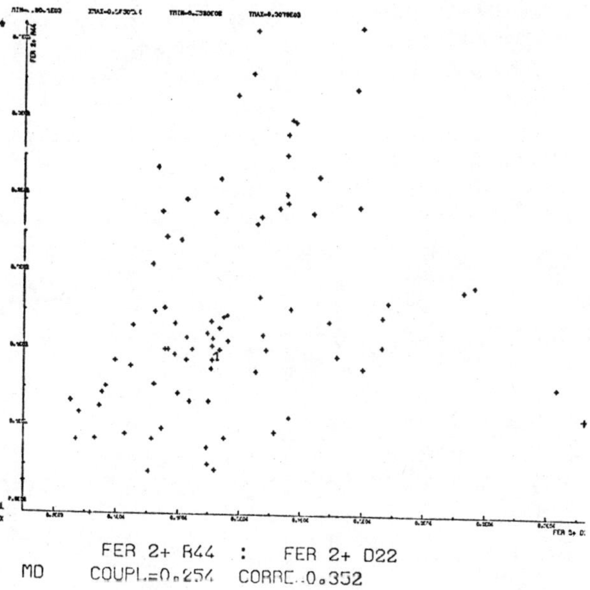

Figure 13

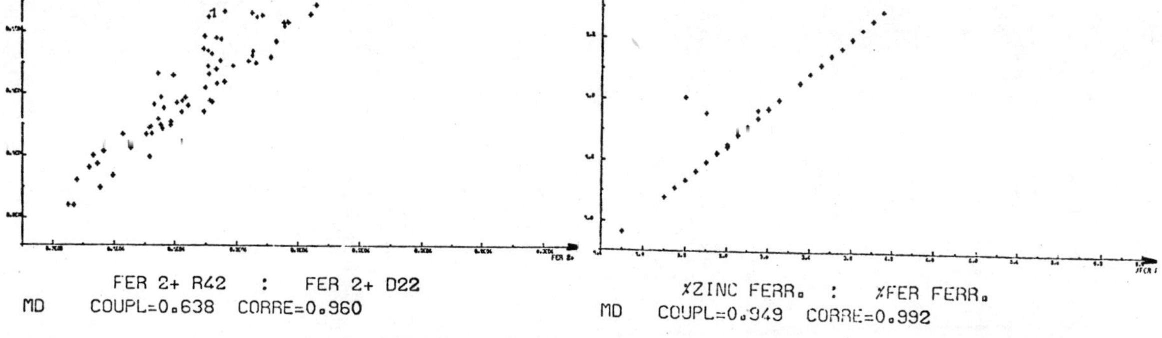

Figure 14

ROUND TABLE 1. AUTOMATIC CONTROL EDUCATION IN DEVELOPING COUNTRIES

INTRODUCTION

P.Martin Larsen *)
Electric Power Engineering Department
Techn. University, DK 2800 Lyngby Denmark

The requirements to the Automatic Control Education in Developing Countries are closely related to the level of automation in the industries of the specific country. Therefore it is of great importance to realize that there are many types of developing countries depending on their historical background. According to professor El Hares, Libya, we can distinguish between two main categories of developing countries:

- rich developing countries with a history of trade activities
- poor developing countries with a history of industrialization.

The lesson of the failures of many industrial projects in developing countries is leading towards the following apparent paradox:

- The lower the degree of industrialization of a developing country, the higher is the needed degree of automation of the imported equipment, provided that the foreign expertise is kept by various incentives as long as possible.

In order to obtain a successful implementation of automatic process control equipment the following conditions should be fulfilled as a whole:

- the latest designed automated equipment should be imported, provided it is tested a sufficient period of time in the home country
- for a period depending on the type of developing country a sufficient number of technicians capable of running the installations and doing the appropriate maintenance should be "imported".
- high absolute priority should be given to creation of a highly qualified centralized general workshop
- a qualified national counterpart for each foreign technician in the implemented industries should be educated thoroughly and attracted to stay in their jobs by means of very good wages and other incentives. The process of replacing the foreign technicians should not be accellerated.

These conditions emphasize the importance of educating very skilled technicians and engineers to supervise and maintain the imported automatic equipment.

Considering the problems of the technical universities in the developing countries the following problems should be mentioned particularly:

- in developing countries with a high population growth rate the universities suffer enormous pressure to accomodate an increasing number of students. This is partly due to the fact that technological degrees are associated with high social status.
- many control engineering courses are more informative oriented than experimentally oriented, and often designed in response to fashionable topics, instead of being closely tied to "national objectives".
- educational programs for mid-level and high-level technicians are practically non-existent in developing countries, therefore the maintenance of advanced laboratory equipment, such as process computer hardware and electronic measuring equipment is really problematic.
- control engineering textbooks written in the mother-language of the developing countries are generally very scarce. Imported books in English, French and more seldom in German and Russian are often designed for courses with objectives different from those in the developing country.

*) Chairman of the IFAC Committee on Education.

- laboratory installations and equipment are often insufficient for the large number of students. Therefore enough attension is not paid to laboratory courses, and coordination between theoretical and practical experiments is seldom present.

- the scientific and teaching staff of the engineering faculties are often overloaded by the large number of students, and the lecturers are often only parttime faculty members. It is important to attract highly qualified lecturers by high wages; they should have practical experience from industry, sufficiently pedagogic experience, and be given opportunity to do research concurrently with their teaching activities.

- the educational standard should be kept on a satisfactory high level, and maintained on this standard withstanding eventual pressure to lower it by economical or political reasons.

The problems of automatic control education in developing countries are quite differentiated. A necessary condition for solving them are extensive exchange of information between automatic control teachers in all developing countries, and eventually development of common basic control courses in geographical areas in which the countries are approximately on the same level of development. It is the hope of the IFAC Committee on Education, that Round Table Discussions also in the future can be organized in conjunction with IFAC Conferences in order to exchange points of views among teachers within the field of automatic control, and that experiences from developed countries can be properly transferred and applied in the design and evaluation of educational programs in Automatic Control in the developing countries.

EXPERIENCES FROM MEXICO

A. Alonso-Concheiro
Instituto de Ingenieria, UNAM,
Apdo Postal 70 - 472, Mexico 20, D.F. Mexico.

1. Types of developing countries

Treating all underdeveloped countries as if they were similar ("developing countries" is perhaps a polite expression but an unfortunate one) may lead to incorrect generalizations. Some of these countries have economic resources but few human resources; others have human resources but scarce economic resources.This will affect their needs for automatic control applications and thus influence education in this area. Further, "equally" underdeveloped countries may have different industrialization levels and the industrialization level is also a factor which should be taken into account in automatic control education programs.

2. Increasing number of students

Universities in underdeveloped countries with a high population growth rate generally suffer enormous pressures to accomodate large numbers of students (this is certainly the case in Mexico). This is particularly true for engineering faculties, as generally in these countries technological degrees are associated with a high social status. On the other hand, expanding industrial sectors compete on advantageous terms for them with the universities for knowledgeable individuals, thus reducing the capacity of the last to adequately pay attention to their rapidly growing student population.

3. Information versus creativity

There is a tendency to place an undue emphasis on curricula updating and administrative structural changes, overlooking in many cases other more important factors such as the quality of lecturers and the teacher-to-students ratio. With a few exceptions, courses tend to be oriented towards "information", paying little attention to "formative" aspects.

4. Relation to national objectives

Educational programs should be closely tied to "national objectives". However, many control engineering university courses are designed in response to fashionable topics being discussed in developed countries. This has obvious negative consequences.

5. Education of technicians

Good technicians are one of the resources needed for succesful control engineering applications which is less abundant in most underdeveloped countries. This is generally a problem in many other technical areas too, but it is perhaps critical in the case of automatic control. Educational programs for mid-level and high-level technicians are practically unexistent (in Mexico some efforts have been made in relatively recent times to ameliorate the situation).

6. Problems of imported installations

The design and development of control systems installed in underdeveloped countries is seldomly done by their nationals. This limits professional practice and has a negative impact on education programs (salesmen vs engineers).

7. Textbooks

Control engineering textbooks written in the mother-language of underdeveloped countries are generally scarce. For example, there are only a few books in spanish, and those available are typically not very good translations of books in english, french, and less frequently german and russian, which steamed from courses with different objectives and contents from those being offered in say Mexican institutions. These books are sometimes adopted as textbooks with unfortunate consequences.

8. Laboratory equipment

Laboratory installations are sometimes insufficient. Not enough attention is paid to laboratory courses, which tend to be seen as underrated low-quality courses. Coordination between theoretical and practical exercises is seldomly present.

9. Faculty staff

Few of the engineering lecturers are full-time faculty members. Most of them go to the university three times a week to give one-hour lectures; they rarely have the time for any kind of extra-lecture supervision of their students. Only a small minority of members of staff are engaged in research and development activities.

10.Maintenance of educational level

Social and economic conditions often drive the university students to be more interested in obtaining a degree than in learning. They thus exert pressure to lower the educational standards, thinking that they will optimize their objective function in the short term. Sometimes administration officers do not withstand this pressure and do lower the standards.

THOUGHTS FROM LIBYA

Hasan el Hares
Engineering Faculty, Triboli, P.O.Box 13235, Libya.

Automatic Control Education is offered

- in universities
- in technical intermediate schools
- as refreshment courses

Developing Countries are of two types depending on their historical background

- rich countries with a history of trade activities
- poor countries with a history of industrial activities

Why Automatic Control Education?

The aim of Automatic Control Education in Developing Countries is to educate engineers with a sound idea of the basic principles of automatic control, a person who is able to select the right components of a feedback loop and who knows when to use a PID-controller instead of a P-controller, but who probably not is able to design an electronic or a pneumatic controller.

The aim is not to educate a scientist with a Ph.D. degree or a highly qualified engineer, who probably sooner or later will emigrate to a highly developed country.

What to teach in Automatic Control Education?

The Automatic Control Education in Developing Countries should definitely contain much less theory than in developed countries. Considering the stability problem the control engineer in a developing country will probably never use the Routh-Hurwitz-Michailovsky criterion, the Cremer-Leonard criterion or the Nyquist criterion, but as he frequently will need to adjust the parameters of a controller for optimum functioning, he should be acquainted with the empirical laws such as those of Ziegler & Nichols.

It is suggested that this round table discussions should initiate the creation of a syllabus for a standard one semester undergraduate course in automatic control for the two types of developing countries mentioned above.

How to teach Automatic Control?

The Automatic Control Education in developing countries should be based on appropriate textbook such as Hale and Dorf, and the emphasis should be on selection of components and on measurement techniques. Practical examples and guided visits to factories should be included to a large extent.

CONSIDERATIONS FROM EGYPT

M.A.R. Ghonaimy
Electronics and Computer Dept., Faculty of Engineering
Ain Shams University, Cairo, Egypt

Aspects regarding Automatic Control Education in developing countries could be divided into two main categories. The first one contains aspects that are common between developed and developing countries, whereas the second contains aspects that are mainly relevant to developing countries. In the following, a brief exposition of these aspects will be given.

Common Aspects:

1. Gap between theory and practice:

If we take a glance at the recent history of Automatic Control we observe that in the 1950's a number of control techniques that have been used to solve practical and urgent problems during the second world war were refined, and Control courses started to be included in a number of curricula. During that period the link between theory and practice was quite strong.

In the 1960's a number of major theoretical contributions started to have their impact on Control. The interdisciplinary nature of Control began to be appreciated. However, the gap between theory and practice began to widen drastically. Research in Control started to concentrate on abstract mathematical techniques. Engineers in industry were unable to follow these major theoretical contributions.

In the 1970's, it was felt that theory has to be put into action. However, it was found that real world problems were quite complex. Therefore, techniques for dealing with large scale systems were tried. This necessitated looking for the developments in fields like microcomputers, computer networks, and data base management systems. However the gap still existed.

It is hoped that in the 1980's integration of all the techniques developed before will be used to tackle and solve real problems in a practically acceptable and feasible way.

2. Lack of supporting experimental work:

For a long time experiments that supported Automatic Control courses could not cope with the theoretical advances. Therefore, it was not appreciated how new control concepts affected system performance. Therefore, it is now deemed necessary to introduce more supporting experimental work, and give more emphasis on physical modeling, control components, and instrumentation systems.

3. Systems Approach:

This necessitates the emergence of a control or system engineer who concentrates on the functional behavior of systems without going very deep in the details of every component in the system. Integrated system design has to be performed first giving the desired performance characteristics of each subsystem, which has then to be designed accordingly. This approach will then guarantee that control requirements are taken into account from the onset of the design.

Aspects related to developing countries:

1. Role of control education in the national development plan

It is essential to integrate control education, whenever possible, in the national development plan. This helps in estimating the number of specialists needed, the depth and breadth of their knowledge. It will also serve as a starting point for designing the appropriate curriculum.

In order to implement the curriculum, it will be essential to have the appropriately qualified team for doing this job. In some cases there may be a nucleus of specialists available. However, it is very important to utilize part of this human energy to regenerate other nuclei of specialists. This helps to guarantee that the system will become, after a reasonable period of time, a self-propelling one.

2. Choice of applications from local environment:

This aspect will necessitate more close cooperation between local industries and the educational institutions. Also, some local publications in control applications and case studies should be supported and encouraged.

3. Integration of technical education in control

By this is meant to take care of training skilled technicians to support control engineers, together with having more emphasis on directed research and development work so that control engineers will not be isolated from the continous program in the field.

4. Directed research and development plan:

It is essential for developing countries to have a feasible and well-structured plan for research and development. This plan should concentrate only on the most needed sectors. This is actually the only way to catch up with the advanced technology and be able to benefit from the developed countries. As a matter of fact technology transfer is not a natural phenomena that is put into action automatically due to differences in the technological levels. Transfer needs a special environment to operate most efficiently.

5. Strategy for preparing personnel:

The key issue in this respect is continuing education which is actually more needed in developing countries. It is possible to organize some structures for different courses in control and other essential supporting subjects like process dynamics, instrumentation, and control components. These courses should be addressed to all levels of control personnel starting from technicians to research and development engineers.

The IFAC together with the support of some international organizations like UNESCO and UNIDO can have an active role in organizing and evaluating the performance of such courses.

THE VENEZUELAN CASE

Rafael A. Padilla
Systems Engineering Laboratory, Venezuelan Institute for Scientific Research, IVIC-Ing. II, P.O. Box 1827, Caracas 1010A, Venezuela

INTRODUCTION

In this presentation some thoughts on the setting and scope of the introduction and development of automatic control concepts in the education of engineers in developing countries are outlined. For obvious reasons, particular regard is given to the situation in Venezuela, its engineering schools, and its development situation. Both the undergraduate and graduate situations are mentioned. The impact of local research and development in the educational process in control is discussed.

Automatic control as an independent subject of study, important in its own right, has been introduced only very recently in the curricula of engineering schools at the undergraduate level in Venezuela. This fact is reflected in any survey of automatic control practice in industries, utilities, and other service companies throughout the country. Such a survey will show that automatic control as an engineering discipline is generally regarded as having reached complete maturity, and that its scope is limited to problems that may be solved with "of-the-shelf" control components and/or systems. This view will hopefully change in the near future when the most recent graduates of the engineering schools become a more representative fraction of the engineering practice community.

A brief historical sketch of the development of engineering education in Venezuela shows that it took approximately 150 years, since the begining of the XIX century when the first engineering schools was founded, until circa 1950 for the initiation of degree programs in what may called "process and device" engineering, i.e. electrical, mechanical, chemical etc. It took then another 15 years, around 1965, for the emergence of automatic control as an independent subject in engineering curricula.

The generality of the automatic control methodologies and their applicability to all problem areas in engineering warrant their inclusion in the body of basic engineering sciences, alongside with the classical subjects of physics, chemistry, mathematics. Also, they should be part of the early stages of the training of all engineering students, regardless of the field of specialization. Related to this pervasive characteristic of automatic control, and to its having grown both in width and depth, is the appearance of degree programs in systems engineering, giving an identity to Automatic Control engineering as a major problem area in engineering practice, and as an independent engineering discipline. In Venezuela, programs of this sort, in varying degrees of development, are already between 5 and 10 years old.

In connection with the teaching of automatic control concepts and methodologies within the realm of the classical engineering disciplines, and offering a simplifying view for the purpose of discussion, it may be said that the subject should be presented at three different levels. Firstly, a descriptive level where basic concepts as well as classical applications are introduced, as independently as possible of the underlying engineering discipline being taken up by the student to allow for its early scheduling in the curriculum. Secondly, a potential level describing technological applications, their reach and limitations, within the structure of the discipline, as well as the non-process type of applications. Thirdly, an advanced level, where the new automatic control methodologies are presented with their widened horizon of applications, and the potential for improving the performance in classical control problems.

The definition of the philisophy of a curriculum for automatic control without the backing of one of the classical engineering disciplines or problem areas, as is necessary in the new schools of Systems Engineering, seems to be a more difficult task. In spite of the fact of automatic control engineering having attained the width and depth necessary for its emerging as a definite problem area, its applicability is highly dependent of systems components, actuator and sensors.

Perhaps in this context the professional practice of these systems engineers will drift naturally towards the "soft" systems applications relying heavily on modeling, simulation, and "policy" decisions through the use of computers, and the programs should be designed accordingly.

The emergence, succes and effectiveness of courses of graduate studies in automatic control leading to degrees equivalent to M.Sc. and Ph.D., is highly dependent on the existence of continuous and sustained efforts in research and development in the area. In Venezuela the first degree program leading to an M.Sc. in electrical engineering with automatic control as a subfield of specialization started only approximately three years ago, within the graduate programs being sponsored by the Venezuelan Institute of Scientific Research. The effect of this efforts in education, professional practice, research and development activities will be seen in the near future.

DISCUSSIONS AND CONCLUSIONS

Dr. M. Cuenod, CH, had two proposals for improvement of the Automatic Control Education in developing countries.

1. For the universities in developing countries it is of great importance to have professors with practical experiences. The IFAC Committee on Education could promote the exchange of visiting professors by establishing a list of professors on sabbatical leave or retired engineers who would be interested in and willing to serve as visiting professors in developing countries. Furthermore a list of universities ready to invite visiting professors should be prepared and dissiminated through the IFAC channels.

2. The IFAC Committee on Education should establish a task force of 2 to 3 professors with a well prepared program to give short courses of 2 to 3 weeks duration, supported by governmental organizations like UNESCO. The aim should be to teach the teachers, assistants and graduate students in fields closely related to the needs of the relevant developing country.

P. Martin Larsen, chairman of the IFAC Committee on Education, promised to consider both proposals thoroughly. For some 7 years ago it was seriously attempted by the Education Committee to create a task force for the same purpose, but for various reasons it was never realized. However the time might be ripe now for a new attempt.

D. Aufenkamp, Chief of Informatics Section, UNESCO commented to these proposals, that his section is very much interested in the problems of education and mainly of informatics applications - especially prototype courses, regional cooperation, training centers, transfer of technology, information on informatics and application of microcomputers."I am looking forward to hearing from peoble with their comments suggestions for projects and ways in which UNESCO can help."

M. Annabi, Tunesia: In developing countries two points must be developed in connection with automatic control education: Creation of teaching aids and development of research, including exchange of professors to supervice the research projects.

A. Alonso-Concheiro answered to these comments:

1. Development of teaching aids is an important point, that was also mentioned by the members of the panel. Textbooks, sets of problems and exercises, films, laboratory equipment and experiments, etc. should be designed having in mind that translations of written aids and adoption of experimental materials designed elsewhere are not always suitable for courses in underdeveloped countries, if these have been elaborated to satisfy local conditions. A strong effort should be made by local educators to produce more and better educational tools. It is important to stress that these

should be of low-cost; lab. equipment can be developed using students of the last courses, either as part of their lab. experiments or their B.Sc. thesis.

2. Research is of particular importance for education in relation with post-graduate courses. M.Sc. and Ph.D. degrees should not be taken to be only n more years of formal education; these should be a qualitative change in the type of education involved. Post-graduate courses without research are utopia.

Benmokhtar Rachin, Morocco: Why do we only focus on high level education (graduate)? Solutions can come from the lower level education (primary and secondary schools); therefore we must link these levels with technology in developing countries.

A. Alonso-Concheiro commented: Problems in higher education are certainly a function of deficiencies on lower levels of the educational system. Technical education can and should be introduced in primary and secondary schools as suggested by Mr. Benmokhtar Rachin; this is certainly of importance in underdeveloped countries as it would increase the technological wisdom of their societies. However it is difficult for us control engineers to exert influence in this direction; we must recognize that we can hardly influence decission makers in matters closer to our field and thus to believe we can influence the contents of primary and secondary schools may be day-dreaming.

R. Padilla commented: A distinction must be kept in mind between developing countries with a western tradition (like Latin American countries) and those without it. This point of view may be more useful in eastern underdeveloped countries.

Sally Hacker, USA: Could pressure for higher level education (BS-eng.) be reduced by reducing salary differential between engineers and technicians?Hands-on laboratory work for non-traditional students (MIT-US) has been found to increase ability to deal with higher level abstractions. Wouldn't this approach versus "veeding out" students who lack such practice, be more democratic?

R. Padilla answered:

To the first question, it seems to me that the type of pressure in my country and in other underdeveloped countries towards obtaining a higher degree (BS-eng.) is more of a social nature than economic. The salary differential is not as high in my opinion as it is in the U.S.A.

To the second question the answer is, that the lack of appropriate laboratory equipment and materials is more the general rule than the exception. Although desirable, hands-on experience might not be possible in most underdeveloped countries.

P. Martin Larsen concluded the Round Table Discussion by promissing that the IFAC Committe on Education would consider the proposals from the panellist and the audience very seriously and in cooperation with the Developing Countries Committee try to initiate exchange of professors between the developing countries and eventually to create a task force for continuing education of university staff members.

On basis on the experiences from this Round Table a Round Table will be organized during the IFAC Congress in Japan, August 1981 and it is hoped, that specific results can be reported at that occation.

ROUND TABLE 2. CONTROL AND INSTRUMENTATION IN CEMENT AND CHEMICAL INDUSTRIES

M. Cuenod* and M. Zelazny**

** Co/DOLSAR Mithatpasa C19.K.7, Ankara, Turkey*
*** Ecole Mohammadia d'Ingénieur, Agdal Rabat, Morocco*

The chairman opens the session by welcoming the attendants and suggesting some discussion topics among them.

- the optimal level of automation in the cement plants
- centralized versus decentralized automation
- development of adaptive control systems in normal operation and in case of emergency

MM. Zelazny and Prevot suggest to add the problem of the measuring instruments in cement plants, especially of the gaz and solid material analyzers.

The discussion has pointed out the following items :

1. Nature and limit of the automation

According to his experience M. Cuénod wonders if it is appropriate in all cases to look for a complete automation taking into account the actual needs of the operators and the difficulty for them to take care of a sophisticated control system. In each case an optimal level of automation, according to the local conditions is to be determined.

Mr. Logdali considers as appropriate an automation of the "mechanical" operations like transport, storage, grinding operations, but for the moment an automatic operation of the kiln is not possible.
It is agreed that the actual trend is toward a decentralized automation using micro-computers and programmable controllers.

2. Automation in developing countries

The use of sophisticated control systems involves new capabilities for the operators as well as for the operation of the maintenance of the plants.
Mr. Logdali points out that the cement industry requires a relatively small amount of man power and that the level of automation has a relatively small impact on this amount.

3. Composition of the rawmeal and its impact on the kiln operation

M.Zelazny raises the problem of the homogeneity of the rawmeal. The automation of the automation of the process is facilitated if this homogeneity is stable.
It is more and more important to check this composition through appropriate measurements, all the more as it can be foreseen that less homogeneous quarries will be used in the future.
The fluctuations of the composition of the rawmeal are the main reson of the perturbances in the kiln operation and of the formation of rings.
Mr. Dekkiche mentions also the importance of other factors such as the regularity of cristallography, granulometry of the rawmeal etc..
He gives the example of 2 kilns supplied by the same quarry but through different grindings systems , which gave different results on the operation of the kiln : one was affected by the formation of rings, the other not.
In this connection the chairman points out the interest to make a systematical investigation and comparison of the origin of disturbances in different cement plants.

4. Measuring instruments

The problem of the reliability of the measuring process is discussed and the chairman points out that this reliability must be obtained more by increasing the reliability and robustness of the measuring instruments than by the redundancy of multiple measurements.
MM. Logdali and Prevot give a more generalised meaning to the concept of redundancy and state that many of these measurements are interrelated and that it is appropriate to make use of this interrelation to check the validity of the measurements issued by different mesuring devices.
Mr. En Naamani points out that taking into account the different ways of operation of the process, the redundancy of the measurement is necessary to improve its control.

Two types of measuring devices are specially discussed : the use of the measurement of the noise through an "ear" to control the grinding process and the use of a scanner to check the thermal profile of the kiln.

5. The quality of the production in the chemical industry.

To improve this quality, the actual trend is to use more efficient identification methods and more efficient mathematical models to be used on line and to implement more efficient data acquisition in view of obtaining a more precise and continuous analysis of the behaviour of the process.

Therefore,the development of multivariable control in the chemical industry is characterised by a decentralization of the decision process specially through the use of microprocessors. But the main conditions for the improvement of the control is to obtain more reliable measurements and transmission of the informations specially for continuous analysers.

6. Man-Machine Communications

The generalisation of the use of data loppers and of video screens gives the possibility to improve the communication between the process and the operators.
Answering a question of Mr. Logdali, Mr. Prevot gives the following informations about the research activity of his Laboratory concerning the kiln process :

- investigation, detection and modeling of the kiln process,

- study concerning the formation of rings : causes, detections of the start of this formation, ways to slow down or to stop this formation,

- integration of process data in data banks,

- automation of the kiln process.

In the conclusion the chairman points out the following high lights of this round table discussion :

- the actual trend of the cement and chemical industry is to implement decentralized control systems using the micro-processors technology.

- research and development are necessary to improve the automatic control of the kiln process

- developments are also necessary to improve the quality and the availability of the measuring devices, the analysers and the communication man-machines

- exchanges of experiences and information concerning the different causes of operation distrubances should be promoted

- due to the increase of fuel prices the problem of the minimization of the energy consumption becomes very important in the cement and chemical industry.

ROUND TABLE 3. CONTROL IN OIL INDUSTRY

R. Perret

Institut National Polytechnique de Grenoble, France

We shall limit these considerations to the case of oil refineries for which we have some experience. We shall study successively what are the needs for control, then what are the technological possibilities which may answer globally or partially to these needs and finally towards what kind of future objectives must be oriented this field of research.

What is called oil crisis affects oil industry according to several ways. The most important effect is naturally the increasing level of the cost of crude oil. The same increasing level of cost is true also for energy, which is used in large quantities in refineries. With such conditions any global increase of efficiency will be a benefit. About this point we shall notice that if, ten years ago, the optimal control of a refinery consisted to get the maximal possible production, nowadays it consists, for similar external conditions, to consume a minimum quantity of crude products and of energy. These need open research towards global optimization.

The irregularity and the economic optimization of supplies lead to change frequently the origin of crude products. Besides the fluctuations of the market of refined oil products, lead also to change frequently the level and the nature of the production. These needs open research towards the optimal control of dynamic states.

Working conditions in refineries are less and less well accepted by people. Continuous work day or night or during holidays, is more and more difficult to assume . These needs implies a quite sophisticated hierarchical control for stable states. Similarly a hierarchical control for transient states must be established for taking care of abnormal alarm states.

It seems that the possiblities of control which we have just mentionned could be implemented during the next decade. From the technological aspect, reliability and economy will lead probably to a large use of microcomputers.

The use in real time of distributed networks of microcomputers gives birth to a lot of problems concerning both hardware and software. Nevertheless we can already notice that in some refineries, networks connecting microprocessors and miniprocessors are used for local control, operators having at their disposal visual displays.

From the point of view of control methodology, the main ideas related to multivariable control are now well known for linear systems. The use of a hierarchy of functions will need still deeper research, specially as to automatas. Moreover the process operation covering larger and larger fields, often of a dynamic character, will imply also research about non linear system, as well about modelisation techniques as about control.

These quite sophisticated control method will have to be applied very progressively if we want to avoid prohibitive difficulties inherent to this type of progress. Highly qualified people, generally of the engineer level, intended to use such process, will have to be carefully trained.

From this last point of view we shall notice the interest that such methods present for developing countries, because it is often easier to train an engineer rather than a technician if the technical environment is not sufficient.

ROUND TABLE 4. ECONOMIC MODELLING

J. M. L. Janssen

Technische Hogeschool Twente Afdeling der Bedrijfkunde, P.O. Box 217, 7500 AE Enschede, The Netherlands

Benyakhlef

1) The curves of consumption of goods are always expressed in time series. Different algorithms have been developed to identify the more suitable mathematical models characterizing the process of growth over a long period (at least 50 years). These models are not appropriate for short term forecasts, but give the possibility to recognize the general trend of the evolution of the consumption of the relevant goods and to determine saturation phenomena much before they happen.

2) The aim of the system approach is to clarify the goals to be reached and the constraints to be taken into account for any project or development strategy. This clarification procedure is a help aggregate individual choices. In that way, system analysis can be considered as a tool to promote consensus concerning any decision related to the implementation of a project or a development policy.

Abisourour Ahmed

To establish fruitful collaboration between politician and economic researchers is a challenge not only in developing countries but also in industrialized nations.
The first step is to initiate a dialogue between designers and users of systems and of mathematical models. The aim of the present symposium is to start such a dialogue which should continue after this symposium and should point out the possiblities but also the limits of the use of the system approach applied to the definition of a development policy.

Juan Pablo Antun

1) The determination of the input and output variables of a socio-economic system is not easy . The experience of the analysis of such systems shows that these systems have generally closed loops. Their inputs are also their outputs .It is often not easy to determine which are the causes and which are the consequences of a specific economic process. But the reflections about these causes and consequences are by themselver very useful and fruitful.
2) This question is related to the problem of the validity of any scientific knowledge which must be considered as a provisional "scheme" giving suitable description of the relationship between the causes and the effects of a given phenomenon in a given environment. The only justification of the validity of this scheme is the concordance between theoretical and experimental results. In many cases it is not possible to proceed to experimental socio-economic processes. The only way available is to foressee the consequences of a development strategy using a mathematical model. But it is necessary to be aware of the limits of the accuracy of the results obtained in that way. Even by it self, this modelisation is very uself. It gives the possibility to proceed to "sensitivity analysis", and determine the impact of each parameter on the goals taken into consideration by the different development policies

J.M.L. Jansen

These observations refer to the points made so far during this round-table discussion, but they also reflect the pattern that emerged from the sessions and round-tables at the IFAC/IFORS Conference on "Dynamic Modelling and Control of National Economies", held in Warsaw, June 16-19 of this year.

If it is said that one of the purposes of modelling is to improve forecasting capabilities, one must remember that it is difficult to forecast anything, the future in particular If in this connection "scenario writing" is mentioned, it must be noted that in many organizations scenario are not used to describe a specific foreseen or planned course of action but rather a set of possible courses of external events, which may be favourable, unfavourable or in between.
For any course of events one may analyze what would happen if one would decide on a specific course of action, so that finally and hopefully one may arrive at a decision which seems good enough under any of the scenario's

considered. In this process of anlysis, economic models are sources of information about the future, in a sense.

However, economists don't always agree on the assumptions underlying the model structure, and neither is good data always available, so that quantitative results may be shaky. This, however, is no reason not to model the economy. In most cases there is not a single decision maker but rather a complex decision-making process with many actors who may have different views and goals. The process of modelling helps to clarify the issues, so that emotions may be tuned down, and so that the confrontation of these various views and goals may produce a clearer picture of the situation : "du choc des opinions jailiit la vérité".

Power systems control and automation

P. Albertos (Spain): In your talk I have heard nothing about instrumentation problems especially measurement device selection and location. Perhaps that means for you that the measurement system design techniques used nowadays are good enough?

M.Ribbens-Pavella : Instrumentation problems play of course a very important role in the field of power system control and automation even if I did not tackle it for obvious reasons of time limitations. Within the particular context of a state estimator not only the kind and quality of devices but also their location should be selected carefully. Indeed the quality of the state estimation (reliability, optimality, anomalous data analysis) highly depends upon the (local and global) redundancy and hence upon the "degree of observability" of the system.

A. Alonso-Concheiro (Mexico) : One can usually speak of an optimal solution for give problem only when one knows either the region of solutions or at least knows some properties of the solution set. I believe this is not the case in power load-flow programs(solution regions are not even convex sets).Could you please comment on what "optimal load-flow" solutions should be taken to mean ?

M.Ribbens-Pavella : Generally, one knows quite a few properties and therefore one possesses a good guess of the load-flow solution of a given power system, moreover, whenever it exists, this solution generally belongs to a convex set. Hence, apart from rare special cases, tha load-flow programs work well and converge satisfactorily . Concerning the optimal load flow solutions, there is always a trad-off between security and economy. Properly used, these optimal load-flow programs may lead to an economy of up to 30%, which of course is quite appreciable.

Artificial intelligence, automatic control and development

A.Van Cauwenberghe(Belgium) : Do you see the use of industrial robots such that new tasks will be performed by them, or rather as a substitute to perform work which is already executed now but in better conditions such that more free time could become available for mankind ?

T. Vamos : I thind the appropriate answer is mostly the second version. The development of new demands on foods and services is an other branch of progress, the two may somehow cross-fertilise each other, but the main area of robot application should be a replacement of men in unhuman works.

A. Abisourour (Morocco) : Is there any international cooperation on artificial intelligence ? How ?

T. Vamos : The main channel of cooperation is the intellectual connection of AI research through publications, meetings learning from each other.Some societies exist in this field as the AI Soc. of America, the International Pattern Recognition Conference,etc. We have also visitors, our research fellows have connections with many pioneering groups , expecially in the US, Japan,and USSR. The other channel of international cooperation is now starting by the industry standards, some big companies, which standardise the tools for application by their advanded products as IBM did in the computer field.

Y. Sevely (France) : I would like to mention to prof. Vamos that in France a national program on robotics for which research has been carried out in a number of Universities as well as CNRS Laboratoires, like the LAAS in particular where I am, under the scientific direction of Dr. Giralt. I would like also to point out that the University Paul Sabatier at Toulouse started, last year an engineering degree on Artificial intelligence, pattern recognition, and robotics, for which the program I will be glad to discuss with Prof. Vamos.

T. Vamos : I have to apologiz not mentioning the excellent French groups in this field. Prof. Simon in Paris is one of the pionners artificial intelligence, a very efficient group is working with Lacombe in Grenoble, but really the most strong groups are concen trated in Toulouse, they have a brilliant intelligent robot project, a robot equipped with all kind of sensors and with the ability of avoiding an obstacles in it's movement. Several other application projects and basic tools developed by them are well known. An important research is going on in France at very high level : AI languages, e.g.,PROLOG.

Medical information systems

A. Belmehdi (Morocco) :Comment on Adaptation and acceptance of computer system in medical information systems : Any innovation is bound to meet some resistance. One has to start to work with innovators who usually represent 5% of the target pupulation. This will save time and permit quicker generalization of the new technology. Analogy was made with extension service in Agriculture.

A. Abisourour (Morocco): How can we avoid one social problem that may arise when putting up medical subsystem namely the refusal from some medical staff to cooperate and help the system work? should the technicians in charge explain every thing to everybody or avoid explaning too much about the new system before it starts working in a regular manner ?

M.A.R.Ghonaimy : It is essential that close cooperation should always exist between information system specialist and the medical staff The only way to convince the medical staff is to implement a prototype system through which they will appreciate the importance of medical information systems.

M. Ksouri (Tunisia)

1) Introducing informatics to medical systems meets difficulty because the medical schools curricula remains archaic and unadapted to modern technological evolution.

2) I would like to know if systems like the ones you have described exist at some hospitals in developing countries.

M.A.R.Ghonaimy

1) In order to introduce medical information systems in curricula of medical schools, it is essential that the medical staff will have some hands-on experience first. That is why implementing an experimential medical information system is of utmost importance in this respect.

2) To my knowledge, no developing country has implemented its own medical information system, some countriesmay have acquired a turn-key system

Water resource allocation

A. Abisourour (Morocco)

1) The diagram of branches doesen't take into consideration exante evaluation of projects.

2) Table 1 is based on outdated information (before 1973 october war).

L. Valadares Tavares

1) No, it does : appraisal stages (c' & c"), because"exante evaluation" is called appraisal rather than evaluation.

2) All statistical tables are based on past information. However, these ones are sufficiently recent to understand the relation between water supply and demand (oil crisis was already in progress during 1973).

A. Alonso Concheiro (Mexico)

1) What is meant by "societal goal" ? How are they determined ?

2) An essential problem, particularly in developing countries, is the lack of data to

design and operate a water supply system . The importance of this fact should be stressed.

L. Valadares Tavares

1) Well defined objectives and targets to be achieved by the whole society within a given horizon.They are determined through social discussion, central & local decision.

2) Inadequacy of data is a crucial problem for developing countries. The estimation of the information value for making decisions can help to solve this problem.

PAPER CODE NUMBER I.3

S.R.Gatta (India) : What are computational aspects for implementing the algorithm in real-life situations of a complex power system

C. Alvarez : In the formulation of our algorithm we are not really engaged with computational aspects because it has been designed for off-line applications. Nevertheless, we have performed further studies in order to apply this algorithm in real time mode. In particular, in the Electric Power Systems cas, we have used a reelaborated version of this algorithm in order to adapt the measurement system configuration according to device failures happening during the state estimation process /3/. In that case, the computational efforts required for this application were not significant within the overall state estimation algorithm.

PAPER CODE NUMBER I.4

Y. Sevely (France) : I do believe you when speak about the efficiency of your algorithms. But do you think it is necessary to write as you do in your paper "state space representation is not useful in process control applications"?

J. Tornero : Of course, our intention was not to set up such a strong affirmation. The idea we want to remark is that, in a number of industrial control applications, the impulse response representation is better suited.

PAPER CODE NUMBER I.5

Y. Sevely (France): I think that Mr. Favier has presented a very good paper. Can he tell us whether his conclusions from some examples are general and can they be demonstrated ? Secondly, how fast can the parameters vary according to modern control computers (for example for one input, one output , second order system)?

G. Favier : Our experimental conclusions are general and can be deduced from the theoretical expression of the closed loop system output (tables I,III) and, for the stochastic regulator problem, by applying the theoretical variances of output and control signals numerically (table I). Only one result appears to us very difficult to explain: why the experimental variance of control signal is less when using the pole placement design to $A^M(E)=B(E)$ (corresponding to the minimum variance strategy) than when applying the minimum variance control itself ? The answer is certainly in the procedure which is used (table IV). The answer to the second question is that the process parameters must vary sufficiently slowly so that the parameters estimation has time to converge. In fact the next step in our work is to study this problem by simulating examples with time-varying parameters.

A. Van Cauwenberghe (Belgium)
1) What values of the forgetting factor did you use ? Were the results sensitive to the value of the forgetting factor ?
2) What are the computation times for the different algorithms, are they very different ?
3) It is possible to include the quality of the parameter estimates in the self-tuning algorithms. This leads to a contious self-tuning controller.See e.g.a paper on lateral dynamics of an aircraft by R. DE KEYSER, and A.Van Cauwenbergh at the MECO meeting in Athens(1979).
4) Can it be stated that for practical applications :

- one should favor direct (implicit) algorithms ?
- one should include the control energy cost, and not use pure minimum variance control algorithms.

G. Favier : 1) The forgetting factor was chosen such that

$$\nu(t+1)=\nu_0 \nu(t)+(1-\nu_0)$$

with

$$0.95 < \nu(o) < 1, 0.95, \quad 0.95 < \nu_0 < 1$$

The results did not appear very sensitive to the value of the forgetting factor for the examples which were simulatéd.
2) The computation times for the different algorithms are very different according to the fact that an implicit or an explicit procedure is used.
3) The comparison of the S.T.R presented in this paper with cautious self-tuning controllers will constitute next work for us.
4) It appears that implicit algorithms are better to use because they don't need the solution of a polynomial equation. However for certain S.T.R an implicit form is not always possible, as it is the case for certain multivariable S.T.R (see Thèse d'Etat of G. Favier 1981).
-The extended minimum variance strategy (including an energy weighting term in the cost function)allows to control unstable minimum phase systems or stable non-minimum phase systems by an appropriate choice of λ , but it is not the case by using the minimum variance strategy . Moreover , as it was pointed out from simulations, the extended minimum variance strategy allows to decrease the control variance without degrading the regulation performance.

PAPER CODE NUMBERS III.2, III.3

P. Martin Larsen(Denmark) : Has fuzzy logic control of rotary cement kilns been implemented in France? I wish to report that fuzzy logic control has been implemented in Denmark by the FLS with Company A/J with very satis factory result, and it is now included as an option to the FLS computer based supervisory and reporting system for cement kilns.

E.A. Dekkiche : The algorithm which is implemented for the control of this process integrates a lot of qualitative consideration related to the operator's know -how and ability The control of the process is not based on the fuzzy sets theory as known now but on subjective rules of human control. I must say that the results were not published till now.

A. Titli (France) : You use data analysis tools, statistical tools . Are you obliged in some steps of your approach to make assumptions on the linearity of the relationships you search ?
Did you try to use informational theory as developed, for example, by profs. Gilles and M. Dufour in Lyon (France).

E.A.Dekkiche : It is an interesting question, so I must precise that for some data analysis methods. Yes, we must make assumptions on the linearity of the relationships between variables, this is the case here for the analysis of correspondences. But the two other methods which are segmentation and dynamic clusters method, break free from this constraint.
To answer the second question, we have used the structural analysis . It is a method of system analysis based on the informational theory which allows to quantify the relationships between variables. It is the main advantage of this method. But we did not get any other interesting results than those obtained by the general methods, i.e. analysis of correspondences.

M. Cuenod (Switzerland): In the conclusion you mention that the model contributes to save fuel and to improve the clinker quality. How much ?

E.A. Dekkiche: It was not our purpose here to optimize the running of the process. From the analysis of the system from the data analysis methods, we have deduced an optimal functioning classor mode, and we have not quantified the improvement. Furthermore, we are going to implement the on-line estimator of the free lime and we have not yet measured the resulting economy of fuel and the improvement of clinker quality on the plant.

J.P. Babary (France) : Are the set points of the different variables (temperatures, kiln rotational speed,..) well known and well defined ? If it is not the case, would study by simulation or by other optimizing procedures help to better know the optimal varia bariable ?

E.A. Dekkiche : For such a complex industrial process, it is difficult to deduce from any theoretical study the optimal set points of the control variables. It is more realistic to search functioning classes. And with the dynamic clusters method, we can define the optimal functioning mode, and thus it is possible to deduce the optimal values of the main control variables.

PAPER CODE NUMBER IV.7

R. Victor Valqui Vidal (Denmark): Have you analysed the "smoothing problem" due to the change from one phase to another ?

J.P. Babary : Not yet . This is a problem we are considering at the present time. Up to now, on the practical aspect, the only consideration of the past experiences allows us a switching of algorithms without major difficulties.

F. Bargach (Morocco): The temperature read by the central thermocouple; is it the same as that of the center of the substrate ?

J.P. Babary : A method for reading temperatures at the extremes of the substrate gives an idea of the central temperature. In fact the thermocouple envisaged for the center of the furnace is indeed situated in the center of the melting pot.

PAPER CODE NUMBER IV.9

P. Martin Larsen (Denmark)

Has the results from the optimation model been realized in practice ?

S.R. Gatta : Not yet presently executed superthermal power projects in India are pit-head stations, based on Government's decision policy in mid-seventies to locate all the coal-fired stations close to the mines. However, the government is appraised of the advantages in certain situations of locating the power stations nearer load centers, particularly so in light of technological developments of transporting coal over long distances in slurry form by pipe lines.
There is nothing unique of application of these methods to power systems only. Recently we have implemented this approach for deciding the location of six gas-based fertilizer plants and the associated gas-pipe line configuration.

PAPER CODE NUMBER VIII.1

M. Cuenod (Switzerland): Which contribution do you expect from IFAC to fulfill the aims of education as defined in this paper ?

J. Elizondo : The paper deals more with basic education problems than with specialized education. The proposal aims of education as set in the paper belong to society in general and to school in particular. Not being a school IFAC can contribute only in as much as it belongs to society at large. The aims alone are nonsense because they only acquire reality through criteria and criteria have national cultural roots, thus, any international organization such as IFAC can't fulfill those aims for a particular country.
IFAC can, however contribute by discussing

the approach presented in the paper and eventually promoting its ideas so that engineering educators become aware of them. In fact, IFAC has already made a contribution by organizing this forum for the presentation of our paper

PAPER CODE NUMBER VIII.4

Alj (Morocco): Do you have empirical data of any experience with this sort of teaching and what are the results ?

R.E. Burns and E.A. Kline : We have data on student satisfaction which has revealed very high student satisfaction with this approach Upwards of 90% of those surveyed (N=800) believed that tutorial work improved grades. Follow-up interview indicate that average students benefited most from their tutorial experience. It appears that the increments of information delivered facilitates internalization of abstract concepts by average students

PAPER CODE NUMBER IX.2

A. Alonso Concheiro (Mexico) : Have you studied the relationships between the level of automation and the concentration of power?

S. Hacker : Take one concrete example : As American Telegraph and Telephone Company moves to a more sophisticated and a more centralized technology, say, the ESS, rural offices closed reducing power of rural workers. Also, decision making and the knowledge necessary appears more concentrated at the top of AT T. For example , the executive levels of the operating companies did not appear to be aware of long range planning, or the short or long range effects of this planning as it affected their own companies. They seemed puzzled by our data which showed that women were being displaced. They were sincere in their efforts to increase wemen's participation. High level systems personnel at Headquarters, then in New York , and at Bell Laboratories were not so puzzled. Internal academic literature indicates the high level decisions had been made some 15 years previously(see Social Problems ref. in bibliography).

E.R. Machol (USA) : At the end of your presentation you asserted that we might have to rethink the requirements for mathematics in engineering education. Would you please expound on that .

S. Hacker : The debate now is, or seems, more than an echo of the "shop culture"/"school culture" debate in 19th century engineering education wherein mathematics was increasingly used as a criterion for school-produced engineers. Professors hoped to democratize participation in engineering rather than have entrepreneurs hand pick successors in "shop" engineers. (See Monte Calvert's work on 19th century mechanical engineering wherein the Navy found school-produced engineers knew alot of math but "had never seen a machine"). Work in the history of mathematics informs us about the role of mathematics in preserving elite status, regardless of its usefulness to a field (see work by S. Restivo, e.g at R.P.I, Troy, NY).

L. Pun (France) : In your presentation, negative (unemployment) effects have been evaluated Have you made any evaluation of the positive effects, such as increase of efficiency, productivity ?

S. Hacker : Not sufficiently. (Given limited time here, I felt the costs (vs. benefits) were more in need of attention/discussion).

Abisourour (Morocco) : Isn't the effect of automation on employment different when we look at it from a time point of view ? (short tun vs. long-run).

S. Hacker : There are variations, fluctuations over time, but it does appear as though women bear the largest burden/cost of change in each short run period (see Isley ref. in text and bibliography, e.g., on differential mortality rates by sex in India for long range data).